UBIQUITY: TECHNOLOGIES FOR BETTER HEALTH IN AGING SOCIETIES

Studies in Health Technology and Informatics

This book series was started in 1990 to promote research conducted under the auspices of the EC programmes' Advanced Informatics in Medicine (AIM) and Biomedical and Health Research (BHR) bioengineering branch. A driving aspect of international health informatics is that telecommunication technology, rehabilitative technology, intelligent home technology and many other components are moving together and form one integrated world of information and communication media. The complete series has been accepted in Medline. Volumes from 2005 onwards are available online.

Volume 124

Recently published in this series

Vol. 123. D. Uyttendaele and P.H. Dangerfield (Eds.), Research into Spinal Deformities 5

Vol. 122. H.-A. Park, P. Murray and C. Delaney (Eds.), Consumer-Centered Computer-Supported Care for Healthy People – Proceedings of NI2006 – The 9th International Congress on Nursing Informatics

Vol. 121. L. Bos, L. Roa, K. Yogesan, B. O'Connell, A. Marsh and B. Blobel (Eds.), Medical and Care Compunetics 3

Vol. 120. V. Hernández, I. Blanquer, T. Solomonides, V. Breton and Y. Legré (Eds.), Challenges and Opportunities of HealthGrids – Proceedings of Healthgrid 2006

Vol. 119. J.D. Westwood, R.S. Haluck, H.M. Hoffman, G.T. Mogel, R. Phillips, R.A. Robb and K.G. Vosburgh (Eds.), Medicine Meets Virtual Reality 14 – Accelerating Change in Healthcare: Next Medical Toolkit

Vol. 118. R.G. Bushko (Ed.), Future of Intelligent and Extelligent Health Environment

Vol. 117. C.D. Nugent, P.J. McCullagh, E.T. McAdams and A. Lymberis (Eds.), Personalised Health Management Systems – The Integration of Innovative Sensing, Textile, Information and Communication Technologies

Vol. 116. R. Engelbrecht, A. Geissbuhler, C. Lovis and G. Mihalas (Eds.), Connecting Medical Informatics and Bio-Informatics – Proceedings of MIE2005

Vol. 115. N. Saranummi, D. Piggott, D.G. Katehakis, M. Tsiknakis and K. Bernstein (Eds.), Regional Health Economies and ICT Services

Vol. 114. L. Bos, S. Laxminarayan and A. Marsh (Eds.), Medical and Care Compunetics 2

ISSN 0926-9630

Ubiquity: Technologies for Better Health in Aging Societies

Proceedings of MIE2006

Edited by

Arie Hasman

Department of Medical Informatics, AMC-UvA, Amsterdam, the Netherlands

Reinhold Haux

Institute for Medical Informatics, Technical University of Braunschweig, Braunschweig, Germany

Johan van der Lei

Department of Medical Informatics, Erasmus MC – University Medical Center Rotterdam, the Netherlands

Etienne De Clercq

Université Catholique de Louvain, Brussels, Belgium

and

Francis H. Roger France

Université Catholique de Louvain, Brussels, Belgium

IOS Press

Amsterdam • Berlin • Oxford • Tokyo • Washington, DC

ISBN 1-58603-647-5
Library of Congress Control Number: 2006930663

Publisher
IOS Press
Nieuwe Hemweg 6B
1013 BG Amsterdam
Netherlands
fax: +31 20 687 0019
e-mail: order@iospress.nl

Distributor in the UK and Ireland
Gazelle Books Services Ltd.
White Cross Mills
Hightown
Lancaster LA1 4XS
United Kingdom
fax: +44 1524 63232
e-mail: sales@gazellebooks.co.uk

Distributor in the USA and Canada
IOS Press, Inc.
4502 Rachael Manor Drive
Fairfax, VA 22032
USA
fax: +1 703 323 3668
e-mail: iosbooks@iospress.com

PRINTED IN THE NETHERLANDS

Ubiquity: Technologies for Better Health in Aging Societies
A. Hasman et al. (Eds.)
IOS Press, 2006

Preface

Medical informatics as a discipline is not very old, yet in the fifty years of its existence some shifts in interest have occurred. The development and evaluation of information systems has been there from the beginning. This topic was driven by technology changes (new database approaches, progress in electronic communication and standardization issues, etc.). In the first decades research was also devoted to investigating basic medical and biological processes. The analysis of signals, for example EKGs, EEGs and signals from chemical analyzers and images was an important research area. Emphasis was initially on data analysis techniques but gradually shifted to interpretation of the processed data by means of statistical and pattern recognition methods. This trend continued and gradually the emphasis shifted from the provision of information to the provision of knowledge.

At present computers are ubiquitous. New technologies make computers almost invisible. Nanotechnology provides important extensions to our armamentarium. Computing based on DNA and enzymes is currently investigated in several research centers. Such computers can help us for example to determine certain pathological cells in tissues and when detected the "computers" release a drug to kill these cells. Ubiquitous computing will support our aging societies. People can stay longer at home with the help of computers that monitor them.

Some of the problems that we have to solve – apart from technical ones – in our opinion are reminiscent of the problems we had to solve in the beginning of medical informatics. But now we have the advantage of hindsight.

About three hundred manuscripts were submitted and finally 143 of them were accepted. We thank all the reviewers for their work. Without them it would have been impossible to obtain a program with a quality as the one that is presented in this Proceedings. We are also very much indebted to NWO (Netherlands Organisation for Scientific Research) that provided grants for inviting two keynote speakers.

We are sure that the program contains interesting information for a broad audience. We wish you a happy stay in Maastricht and much pleasure in reading articles from this Proceedings.

Arie Hasman
Reinhold Haux
Johan van der Lei
Etienne De Clercq
Francis H. Roger France

Contents

Preface v
Arie Hasman, Reinhold Haux, Johan van der Lei, Etienne De Clercq and Francis H. Roger France

0. Keynote Presentations

The Young Person's Guide to Biomedical Informatics 3
Jan H. van Bemmel

Reflections on Biomedical Informatics: From Cybernetics to Genomic Medicine and Nanomedicine 19
Victor Maojo and Casimir A. Kulikowski

Meeting the Challenges – the Role of Medical Informatics in an Ageing Society 25
Sabine Koch

Progress and Challenges of Ubiquitous Informatics in Health Care 32
Francis H. Roger France

1. Modern Trends in Medical Informatics

1.1 Ubiquity: Design and Applications

Facilitating Interdisciplinary Design Specification of "Smart" Homes for Aging in Place 45
George Demiris, Marjorie Skubic, Marilyn J. Rantz, Karen L. Courtney, Myra A. Aud, Harry W. Tyrer, Zhihai He and Jia Lee

Rehabilitation After Stroke Using Virtual Reality, Haptics (Force Feedback) and Telemedicine 51
Jurgen Broeren, Mark Dixon, Katharina Stibrant Sunnerhagen and Martin Rydmark

A Distributed Shared Data Space for Personal Health Systems 57
Wim Stut, Frank Wartena and Maarten van Steen

1.2 Ubiquity: Opportunities and Technologies

Ubiquitous Technologies in Health: New Challenges of Opportunity, Expectation, and Responsibility 65
Michael Rigby

Visualisation and Interaction Design Solutions to Address Specific Demands in Shared Home Care 71
Isabella Scandurra, Maria Hägglund and Sabine Koch

Safety Portal: The Safest Goes Through the Air – Ubiquitous High-Risk Reminders Bridging Out the Patient Safety in Emergency Department 77
Chang-I Chen, Cheuk-Sing Choy, Tu-Bin Chu, Yu-Sheng Lo, Wei-Kee Huang, Ching-Feng Kuan and Yu-Chuan Li

1.3 Ubiquity and Health Services

Highly Automated Documentation for Mobile Medical Services 85
Marko Hassinen, Mikko Heinonen, Maija Marttila-Kontio and Heli Tervo

Development of a System Supporting Patient Supervision and Treatment in Contemporary Home-Care: Status Report 91
Basile Spyropoulos, Maria Botsivaly, Aris Tzavaras, Konstantinos Koutsourakis, Eleutherios Koulouris, Dimitrios Liargovas and Konstantina Mertika

1.4 Clinical Bioinformatics: Modeling and Evaluation

Fuzzy Hidden Markov Models: A New Approach in Multiple Sequence Alignment 99
Chrysa Collyda, Sotiris Diplaris, Pericles A. Mitkas, Nicos Maglaveras and Costas Pappas

Modelling Epidermal Homeostasis as an Approach for Clinical Bioinformatics 105
Niels Grabe, Thora Pommerencke, Dennis Müller, Simone Huber, Karsten Neuber and Hartmut Dickhaus

Sifting Abstracts from Medline and Evaluating Their Relevance to Molecular Biology 111
Natalia Grabar, Marie-Christine Jaulent, Antoine Chambaz, Céline Lefebvre and Christian Néri

1.5 Clinical Bioinformatics: Integration

Inferring Gene Expression Networks via Static and Dynamic Data Integration 119
Fulvia Ferrazzi, Paolo Magni, Lucia Sacchi and Riccardo Bellazzi

Integrating Medical and Genomic Data: A Sucessful Example for Rare Diseases 125
Gaspar S. Dias, José Luis Oliveira, Javier Vicente and Fernando Martin-Sanchez

1.6 Health Standards

HL7 RIM: An Incoherent Standard 133
Barry Smith and Werner Ceusters

A Reference Model for Clinical Tumour Documentation 139
Udo Altmann, Frank Rüdiger Katz and Joachim Dudeck

Standards for Enabling Health Informatics Interoperability 145
Kjeld Engel, Bernd Blobel and Peter Pharow

The HL7 Reference Information Model Under Scrutiny 151
Gunther Schadow, Charles N. Mead and D. Mead Walker

Participation in the International Classification for Nursing Practice (ICNP®) Programme 157
Claudia Bartz, Amy Coenen and Woi-Hyun Hong

2. Electronic Health Records

2.1 EHR: Architectures

A User-Centred Deployment Process for ICT in Health Care Teams – Experiences from the OLD@HOME Project 167
Maria Hägglund, Isabella Scandurra and Sabine Koch

Non-Invasive Light-Weight Integration Engine for Building EHR from Autonomous Distributed Systems 173
Pere Crespo Molina, Carlos Angulo Fernández, José A. Maldonado Segura, David Moner Cano and Montserrat Robles Viejo

2.2 EHR: Research

Research Networks: Can We Use Data from GPs' Electronic Health Records? 181
Etienne De Clercq, Viviane Van Casteren, Pascale Jonckheer, Peter Burggraeve, Marie-France Lafontaine, Hans Vandenberghe, Vincent Lorant, Caroline Artoisenet and Karen Degroote

Information Technology in Clinical Research in Rheumatology Domain 187
Petr Hanzlicek, Jana Zvarova and Ctibor Dostal

2.3 EHR: Access

Method for Automatic Escalation of Access Rights to the Electronic Health Record 195
Frode Ørbeck Hansen and Rune Fensli

The French Proposal for a Health Identification Number 201
Catherine Quantin, Olivier Cohen, Benoît Riandey and François-André Allaert

Can Temporary Paper-Based Patient Records Sensibly Complete an Electronic Patient Record? 207
Petra Knaup, Jochen Pilz and Markus Thalheimer

2.4 EHR: Open Approaches

Ubiquitous Information for Ubiquitous Computing: Expressing Clinical Data Sets with *open*EHR Archetypes 215
Sebastian Garde, Evelyn Hovenga, Jasmin Buck and Petra Knaup

Towards Automatically Generating Graphical User Interfaces from *open*EHR Archetypes 221
Thilo Schuler, Sebastian Garde, Sam Heard and Thomas Beale

2.5 EHR and Clinical Management

Co-Operation, Interaction and Information-Sharing: From a Paper-Based Diabetes Pass to an Electronic Alternative 229
Heidi Buysse, Georges Van Maele and Georges De Moor

Personalized Cardiovascular Risk Management Linking SCORE and Behaviour Change to Web-Based Education 235
Selena Davis, Syed Sibte Raza Abidi and Jafna Cox

Guarantying the Quality of Chemotherapy: From the Order to the Administration 241
S. Spahni, C. Lovis, M. Ackermann, N. Mach, P. Bonnabry and A. Geissbuhler

3. Health Information Systems

3.1 Architectures of HIS

Comprehensive Management of the Access to a Component-Based Healthcare Information System 251
Christian Lovis, Stéphane Spahni, Nicolas Cassoni-Schoellhammer and Antoine Geissbuhler

Barriers Integrating Dedicated Software for Quality Management in Pain Centres 257
A. Johannes Ten Hoopen, Pieter E. Zanstra, Egbert J. Van Der Haring, Wouter S. Tuil and Marten Van Wijhe

3.2 HIS Architectures: Examples

On Feasibility and Benefits of Patient Care Summaries Based on Claims Data 265
Thorsten Brennecke, Lars Michalowski, Joachim Bergmann, Oliver J. Bott, Axel Elwert, Reinhold Haux and Ralf Schlicht

On-the-Fly Form Generation and On-Line Metadata Configuration – A Clinical Data Management Web Infrastructure in Java 271
Peter Beck, Thomas Truskaller, Ivo Rakovac, Bruno Cadonna and Thomas R. Pieber

Log Files Analysis to Assess the Use and Workload of a Dynamic Web Server Dedicated to End-Stage Renal Disease 277
Mohamed Ben Said, Loic Le Mignot, Jean Baptiste Richard, Christine Le Bihan, Laurent Toubiana, Jean-Philippe Jais and Paul Landais

3.3 HIS: Integration

Concept Representation in Health Informatics for Enabling Intelligent Architectures 285
Bernd G.M.E. Blobel

Integrating Radiology Information Systems with Healthcare Delivery Environments Using DICOM and HL7 Standards 292
Bojan Blazona and Miroslav Koncar

The Fate of Clinical Department Systems at the Dawn of Hospital-Wide Electronic Health Records in a Norwegian University Hospital 298
Eivind Vedvik and Arild Faxvaag

3.4 HIS and Electronic Health Records

Software Development for the Estimation of the Mean DRGs Related Treatment Cost 307
Maria Botsivaly, Basile Spyropoulos, Michael Marinis, Panagiotis Sarantakis, Georgios Nikoloudakis, Ilias Balabanis, Nikolaos Karagiannis, Irini Limnou, Manolis Moschidis, Aris Tzavaras and Konstantinos Koutsourakis

Apply Creative Thinking of Decision Support in Electrical Nursing Record 313
Angelica Te-Hui Hao, Chien-Yeh Hsu, Huang Li-Fang, Wen-Shan Jian, Li-Bin Wu, Ching-Chiu Kao, Mei-Show Lu and Her-Kung Chang

Benefits and Weaknesses of Health Cards Used in Health Information Systems 320
Peter Pharow and Bernd Blobel

A Socio-Technical Study of an Ubiquitous CPOE-System in Local Use 326
Helle S. Wenzer, Ulrich Böttger and Niels Boye

3.5 Evaluation of HIS and EHR Systems

HIS-Monitor: Quality of Information Processing in Hospitals 335
Elske Ammenwerth, Frauke Ehlers, Bernhard Hirsch, Gordon Gratl and Christof Machan

Development of Methods for Usability Evaluations of EHR Systems 341
Lars H. Lilholt, Signe S. Pedersen, Inge Madsen, Per H. Nielsen, Niels Boye, Stig K. Andersen and Christian Nøhr

Implementing Communication Systems in the Community Health Services. The Health Care Workers Experiences 347
Hans Tore Mogård, Eli Haugen Bunch and Anne Moen

Measurement of the Clinical Usability of a Configurable EHR 356
John Møller-Jensen, Ivan Lund Pedersen and Jesper Simonsen

Ten Years of Teledermatology 362
Nina Eminović, Nicolette F. de Keizer, Patrick J.E. Bindels and Arie Hasman

3.6 HIS: Other Topics

Integrating Anatomical Pathology to the Healthcare Enterprise 371
Christel Daniel-Le Bozec, Dominique Henin, Bettina Fabiani, Karima Bourquard, David Ouagne, Patrice Degoulet and Marie-Christine Jaulent

Medical Data GRIDs as Approach Towards Secure Cross Enterprise Document Sharing (Based on IHE XDS) 377
Florian Wozak, Elske Ammenwerth, Micheal Breu, Robert Penz, Thomas Schabetsberger, Raimund Vogl and Manfred Wurz

4. eHealth

4.1 eHealth: Applications and Approaches

A Multi-Agent Approach to the Design of an E-Health System 389
Paola Di Giacomoa and Fabrizio L. Ricci

A Site of Communication Among Enterprises for Supporting Occupational Health and Safety Management System 395
E. Velonakis, J. Mantas and I. Mavrikakis

A New Perspective in the Promotion of e-Health 404
Eva del Hoyo-Barbolla, Rita Kukafka, María Teresa Arredondo and Marta Ortega

4.2 eHealth: Architectures and Strategies

E-Health Approach to Link-Up the Actors in the Health Care System of Austria 415
Thomas Schabetsberger, Elske Ammenwerth, Ruth Breu, Alexander Hoerbst, Georg Goebel, Robert Penz, Klaus Schindelwig, Herlinde Toth, Raimund Vogl and Florian Wozak

Architecture for National e-Health Infrastructure in Lithuania 421
Egidijus Andriuskevicius, Evaldas Dobravolskas, Vytenis Punys, Vita Sinciene and Aras Valentinavicius

A National EHR Strategy Preparedness Characterisation Model and Its Application in the South-East European Region 427
Leonidas Orfanidis, Panagiotis Bamidis and Barry Eaglestone

4.3 eHealth and Economy

Simulation Based Cost-Benefit Analysis of a Telemedical System for Closed-Loop Insulin Pump Therapy of Diabetes 435
Oliver J. Bott, Ina Hoffmann, Joachim Bergmann, Patrick Kosche, Christian von Ahn, Dirk C. Mattfeld, Oliver Schnell and Dietrich P. Pretschner

E-DIMEM: An Economic Model to Estimate the Costs of e-Disease Management 441
Fabrizio L. Ricci, Luca D. Serbanati, Fabio Palazzo, Daniela Luzi and Giuseppe Stirparo

4.4 eHealth and Information Sharing

A Model for a Regional Health Information Network Sharing Clinical Information Between Professionals in Britanny 449
Marc Cuggia, Nicolas Herry, Delphine Rossille, Emmanuelle Lepage, Gilles Edan and Pierre Le Beux

Teleconsultations as a Step Towards Hospital Interoperability 455
Lacramioara Stoicu-Tivadar, Francesco Sicurello, Vasile Stoicutivadar, Victor Moga, Mariana Moga, Gianni Pellicanò and Giannluca Ronco

Health Information Exchange on a Regional Level: Dream or Reality? 461
Eline R.M. Loomans, Marja Snoeijen, Diek Scholten and Henk ten Dolle

5. Decision Support

5.1 Decision Support: Guidelines and Protocols

Enabling Protocol-Based Medical Critiquing 471
Robert-Jan Sips, Loes Braun and Nico Roos

Management of Data Quality – Development of a Computer-Mediated Guideline 477
Jürgen Stausberg, Michael Nonnemacher, Dorothea Weiland, Gisela Antony and Markus Neuhäuser

Design of a Decision Support System for Chronic Diseases Coupling Generic Therapeutic Algorithms with Guideline-Based Specific Rules 483
Vahid Ebrahiminia, Christine Riou, Brigitte Seroussi, Jacques Bouaud, Stéphane Dubois, Hector Falcoff and Alain Venot

5.2 Decision Support: Approaches

Automated Test Selection in Decision-Support Systems: A Case Study in Oncology 491
Danielle Sent and Linda C. Van der Gaag

Avoiding Literature Overload in the Medical Domain 497
Loes M.M. Braun, Floris Wiesman, H. Jaap Van den Herik and Arie Hasman

A Language Classifier That Automatically Divides Medical Documents for Experts and Health Care Consumers 503
Michael Poprat, Kornél Markó and Udo Hahn

With Good Intentions 509
Agnieszka Latoszek-Berendsen, Jan Talmon and Arie Hasman

A Method for Specification of Structured Clinical Content in Electronic Health Records 515
Knut Bernstein, Morten Bruun-Rasmussen and Søren Vingtoft

Using Treemaps to Represent Medical Data 522
Emmanuel Chazard, Philippe Puech, Marc Gregoire and Régis Beuscart

Evidence in Pharmacovigilance: Extracting Adverse Drug Reactions Articles from MEDLINE to Link Them to Case Databases 528
Nicolas Garcelon, Fleur Mougin, Cédric Bousquet and Anita Burgun

5.3 Decision Support: Other Topics

Towards the Use of Ontologies for Activity-Based Costing in Healthcare Organizations 537
Anand Kumar

From Bibliometric Analysis to Research Policy: The Use of SIGAPS in Lille University Hospital 543
Patrick Devos, Helene Lefranc, Eric Dufresne and Regis Beuscart

Dealing with an Information Overload of Health Science Data: Structured Utilisation of Libraries, Distributed Knowledge in Databases and Web Content 549
Simon Hoelzer, Ralf K. Schweiger, Joerg Rieger and Michael Meyer

A Markov Model to Describe Daily Changes in Organ Failure for Patients at the ICU 555
Linda Peelen, Niels Peek, Nicolette F. de Keizer, Evert de Jonge and Robert J. Bosman

Online Guideline Assist in Intensive Care Medicine – Is the Login-Authentication a Sufficient Trigger for Reminders? 561
Rainer Röhrig, Markus Meister, Achim Michel-Backofen, Martin Sedlmayr, Dirk Uphus, Christian Katzer and Thomas Rose

Functional Data Analysis for Gait Curves Study in Parkinson's Disease 569
Alain Duhamel, Patrick Devos, Jean Louis Bourriez, C. Preda, L. Defebvre and Regis Beuscart

A Framework for Cohesive Healthcare Coalition Formation 575
Yu-N Cheah, Yong Han Chong and Siew Lan Neoh

An Approach for Generating Fuzzy Rules from Decision Trees 581
Amir R. Razavi, Mikael Nyström, Marian S. Stachowicz, Hans Gill, Hans Åhlfeldt and Nosrat Shahsavar

Computational Representation of Alzheimer's Disease Evolution Applied to a Cooking Activity 587
Audrey Serna, Vincent Rialle and Hélène Pigot

5.4 Decision Support: Information Retrieval

Strategies for Health Information Retrieval 595
Lina F. Soualmia, Badisse Dahamna, Benoît Thirion and Stéfan J. Darmoni

A Method of Cross-Lingual Consumer Health Information Retrieval 601
Aurélie Névéol, Suzanne Pereira, Lina F. Soualmia, Benoît Thirion and Stéfan J. Darmoni

Applying an Artificial Neural Network to Predict Osteoporosis in the Elderly 609
Jainn-Shiun Chiu, Yu-Chuan Li, Fu-Chiu Yu and Yuh-Feng Wang

5.5 Evaluation of Decision Support Systems

Design and Evaluation of a Computer Reminder System to Improve Prescribing Behaviour of GPs 617
Jody D. Martens, Alied van der Aa, Bert Panis, Trudy van der Weijden, Ron A.G. Winkens and Johan L. Severens

Clinicians' Perceived Usefulness of a Support System for Patient-Centered Cancer Care 624
Cornelia M. Ruland

Evaluation of a Breast Cancer Computer Aided Diagnosis System 631
Antonis Frigas, Smaragdo Kapsimalakou, George Spyrou, Kostantinos Koufopoulos, Stamatios Vassilaros, Aikaterini Chatzimichael, John Mantas and Panos Ligomenides

5.6 Decision Support: Evaluation and Experiences

Comparison of the Impact of Cardiovascular Guidelines on a Working Population 639
Pierre Gillois, Frédérique Claudot, Gilles Chatellier, François Kohler and Marie Christine Jaulent

IT Support for Clinical Pathways – Lessons Learned 645
R. Lenz, R. Blaser, M. Beyer, O. Heger, C. Biberc, M. Bäumlein and M. Schnabel

Evidence-Based Practice in Primary Health Care 651
Anouk De Smedt, Ronald Buyl and Marc Nyssen

6. Information Management and Modeling of Healthcare

6.1 Health Information Management

Process Frame Instances for Integrating Strategic, Tactical and Operational Information Management in Hospitals 661
Lutz Issler, Gert Funkat, Stefan Smers and Alfred Winter

Web Services Based Syndromic Surveillance for Early Warning Within French Forces 666
Hervé Chaudet, Liliane Pellegrin, Jean-Baptiste Meynard, Gaëtan Texier, Olivier Tournebize, Benjamin Queyriaux and Jean-Paul Boutin

Modeling Economic Aspects of Hospital Information Systems to Give Decision Support for Strategic Information Management 672
Ansgar Kutscha, Birgit Brigl and Alfred Winter

Medical Error Management and the Role of Information Technology – A New Approach to Investigating Medical Handover in Acute Care Settings 679
Kwang Chien Yee, Ming Chao Wong and Paul Turner

Human Factors Engineering for Clinical Applications 685
Marie-Catherine Beuscart-Zephir, Peter Elkin and Sylvia Pelayo

Modeling a Health Telematics Network: Does the 3LGM² Approach Assist in Its Management and Operation? 691
Martin Staemmler

6.2 Modeling Healthcare

Agent Based Simulations in Healthcare 699
Ugur Bilge and Osman Saka

Change Readiness Research. A Qualitative Study of Variations in Participation 705
Anna Marie Høstgaard

6.3 Modeling Healthcare: Communication and Patient Flows

A Study of the Communication Notes for Two Asynchronous Collaborative Activities 713
Nathalie Bricon-Souf, Sandra Bringay, Françoise Anceaux, Saliha Hamek, Nathalie Degardin, Catherine Barry and Jean Charlet

Specific Classification of eLibrary Resources Says More About Users' Preferences 719
Judas Robinson, Simon de Lusignan, Patty Kostkova, Bruce Madge and Lesley Southgate

Cancer Patient Flows Discovery in DRG Databases 725
Nicolas Jay, Amedeo Napoli and François Kohler

7. Knowledge Representation, Ontologies, Coding, Terminology

7.1 Knowledge Representation

Non Aristotelian Categories in Medicine 735
György Surján

Referent Tracking: The Problem of Negative Findings 741
Werner Ceusters, Peter Elkin and Barry Smith

Enriching Medical Terminologies: An Approach Based on Aligned Corpora 747
Louise Deleger, Magnus Merkel and Pierre Zweigenbaum

7.2 Ontologies for Medical Disciplines

Restructuring the Foundational Model of Anatomy 755
Gergely Héja, Péter Varga, Péter Pallinger and György Surján

Four Ontological Models for Radiological Diagnostics 761
James M. Fielding and Dirk Marwede

7.3 Biomedical Ontologies

The *Derives_From* Relation in Biomedical Ontologies 769
Mathias Brochhausen

Tools for Czech Biomedical Ontologies Creation 775
Petr Kolesa and Petra Přečková

Aligning Biomedical Ontologies Using Lexical Methods and the UMLS: The Case of Disease Ontologies 781
Gwenaëlle Marquet, Jean Mosser and Anita Burgun

7.4 Concepts and Coding: Methods

Towards Automated Classification of Intensive Care Nursing Narratives 789
Marketta Hiissa, Tapio Pahikkala, Hanna Suominen, Tuija Lehtikunnas, Barbro Back, Helena Karsten, Sanna Salanterä and Tapio Salakoski

Comparison of ICHI and CCAM Basic Coding System 795
Susanne Hanser, Albrecht Zaiss and Stefan Schulz

ClaML: A Standard for the Electronic Publication of Classification Coding Schemes 801
E.J. van der Haring, S. Broënhorst, H. ten Napel, S. Weber, M. Schopen and P.E. Zanstra

7.5 Concepts and Coding: Systems

The International Classification of Primary Care (ICPC-2): An Essential Tool in the EPR of the GP 809
Marc Verbeke, Diëgo Schrans, Sven Deroose and Jan De Maeseneer

Using SNOMED CT Codes for Coding Information in Electronic Health Records for Stroke Patients 815
Judith van der Kooij, William T.F. Goossen, Anneke T.M. Goossen-Baremans, Marinka de Jong-Fintelman and Lisanne van Beek

SNOMED CT in Multidisciplinary Clinical Practice – Evaluation of Usefulness for Classification and Coding of Care-Planning Procedures 824
Anna Vikström and Gunnar Nilsson

7.6 Terminologies

Mapping of the WHO-ART Terminology on Snomed CT to Improve Grouping of Related Adverse Drug Reactions 833
Iulian Alecu, Cedric Bousquet, Fleur Mougin and Marie-Christine Jaulent

Knowledge Acquisition for Computation of Semantic Distance Between WHO-ART Terms 839
Jimison Iavindrasana, Cédric Bousquet and Marie-Christine Jaulent

Construction of a Semi-Automated ICD-10 Coding Help System to Optimize Medical and Economic Coding 845
Suzanne Pereira, Aurélie Névéol, Philippe Massari, Michel Joubert and Stefan Darmoni

Interactive Visualization and Navigation of Complex Terminology Systems, Exemplified by SNOMED CT 851
Erik Sundvall, Mikael Nyström, Håkan Petersson and Hans Åhlfeldt

Cross-Lingual Alignment of Biomedical Acronyms and Their Expansions 857
Kornél Markó, Philipp Daumke and Udo Hahn

An Ontology Driven Collaborative Development for Biomedical Terminologies: From the French CCAM to the Australian ICHI Coding System 863
Jean M. Rodrigues, Alan Rector, Pieter Zanstra, Robert Baud, Kerry Innes, Jeremy Rogers, Anne M. Rassinoux, Stefan Schulz, Beatrice Trombert Paviot, Huib ten Napel, Lucienne Clavel, Egbert van der Haring and Ceu Mateus

8. Education

8.1 Health and Biomedical Informatics Education

The Redesign of the Medical Informatics Master of Science Course at the University of Amsterdam 873
Monique W.M. Jaspers and Arie Hasman

Medical Education and Role of Medical Informatics 879
Izet Masic, Ahmed Novo, Selim Toromanovic, Almir Dzananovic, Zlatan Masic, Alisa Piralic and Sejla Deljkovic

Developing an Interactive Approach in Teaching Medical Informatics 885
Diana Lungeanu, Rochelle E. Tractenberg and George I. Mihalas

8.2 Education and Networking

The INFOBIOMED Network of Excellence: Facilitating Training and Mobility in Biomedical Informatics in Europe 893
Guillermo de la Calle, Victor Maojo and Mario Benito

Multimedia and Physiology: A New Way to Ensure the Quality of Medical Education and Medical Knowledge 899
Yvon Lessard, Pridi Siregar, Nathalie Julen, Jean-Paul Sinteff and Pierre Le Beux

ENN-ICS – Implementation and Evaluation of a Multilingual Learning Management System for Sleep Medicine in Europe 905
Brigitte Knobl, Teresa Paiva, Dieter Jungmann, Rico Böhme and Thomas Penzel

8.3 Patient Education and Consumer Informatics

What Makes an "Informed Patient"? The Impact of Contextualization on the Search for Health Information on the Internet 913
Laurence Alpay, John Verhoef, Pieter Toussaint and Bertie Zwetsloot-Schonk

The Effect of a Multimedia Health Educational Program on the Postoperative Recovery of Patients Undergoing Laparoscopic Cholecystectomy 920
Antonia Stergiopoulou, Konstantinos Birbas, Theophanis Katostaras, Mariana Diomidous and John Mantas

A New Health Strategy to Prevent Pressure Ulcer Formation in Paraplegics Using Computer and Sensory Substitution via the Tongue 926
Alexandre Moreau-Gaudry, Anne Prince, Jacques Demongeot and Yohan Payan

8.4 Health and Clinical Management

Reducing Dropouts in Outpatient Care Through an Sms-Based System 935
Andrea Battistotti, Silvana Quaglini and Enrico Cuoco

Evaluation of a Discussion Forum for Knowledge Sharing Among Emergency Practitioners: A Social Network Approach 941
Janet Curran and Syed Sibte Raza Abidi

On Neural Network Classification of Otoneurological Cases on the Basis of Recognition Results of Vestibulo-Ocular Reflex Eye Movements Signal 947
Martti Juhola, Heikki Aalto and Timo Hirvonen

9. Miscellaneous

9.1 Publication and Presentation

Publication Bias in Medical Informatics Evaluation Research: Is It an Issue or Not? 957
Christof Machan, Elske Ammenwerth and Thomas Bodner

The Multiple Faces of the e-Patient, If Not Disabled 963
M. Cristina Mazzoleni and Franco Corbella

Computing Latent Taxonomies from Patients' Spontaneous Self-Disclosure to Form Compatible Support Groups 969
Eduard Hoenkamp and Regina Overberg

9.2 Biomedical Imaging

Generation of 4D CT Image Data and Analysis of Lung Tumour Mobility During the Breathing Cycle 977
Heinz Handels, Rene Werner, Thorsten Frenzel, Dennis Säring, Wei Lu, Daniel Low and Jan Ehrhardt

Kernel Methods for Melanoma Recognition 983
Elisabetta La Torre, Tatiana Tommasi, Barbara Caputo and Giovanni E. Gigante

Pre- and Intraoparative Processing and Integration of Various Anatomical and Functional Data in Neurosurgery 989
Roland Metzner, Urs Eisenmann, Christian R. Wirtz and Hartmut Dickhaus

9.3 Professionalism

Progressing Professional Maturity in Health Informatics 997
Jean Roberts

Bridging Information Gaps Between Primary and Secondary Healthcare 1003
Habibollah Pirnejad, Arjen P. Stoop and Marc Berg

The Demographic Bias of Email as a Survey Method in a Pediatric Emergency Population 1009
Brett W. Taylor

9.4 Evaluation and Lessons Learned

Artificial Neural Network Versus Subjective Scoring in Predicting Mortality in Trauma Patients 1019
Adrian Pearl, Reuben Caspi and David Bar-Or

Clinical Pathways Development and Computer Support in the EPR: Lessons Learned 1025
Thomas Bürkle, Thomas Baur and Norbert Höss

Removal of Paper-Based Health Records from Norwegian Hospitals: Effects on Clinical Workflow 1031
Jan Tore Lium and Arild Faxvaag

Author Index 1037

0. Keynote Presentations

Ubiquity: Technologies for Better Health in Aging Societies
A. Hasman et al. (Eds.)
IOS Press, 2006

The Young Person's Guide to Biomedical Informatics[*][†]

Jan H. VAN BEMMEL[‡]

Institute of Medical Informatics, Erasmus Medical Center, Rotterdam, the Netherlands

Abstract. In a retrospective review, a parallel is drawn between the challenges by which a research department in biomedical informatics is confronted and those of a symphony orchestra. In both areas, different disciplines and different groups of instruments can be discerned. The importance of mastering one's instrument and the harmony between the team members is stressed. The conductor has to stimulate the individual players so that they can all have a successful career. Competition between orchestras and performance assessments determine survival and success. A record of refereed publications is crucial for continued existence.
Conclusions are that biomedical informatics is typically multidisciplinary, that hypotheses underlying research should be carefully formulated, that the time from research to application may easily take 20 years or more, that mutual trust and knowing each other's competences is essential for success, that a good leader gives enough room to all team members to develop their careers, and that the outcomes of assessment studies are related to the quality of publications.
Keywords: Biomedical Informatics, Refereed Publications, Quality of Research, Assessment, Research Management

1. Introduction

This will be a different lecture than I have ever given before. It will be a mixture of professional and personal experiences, of lessons learned during my path in biomedical engineering and medical informatics – let's call it for now biomedical informatics – and of experiences at the perimeter of my direct professional life[§].

To start right away with a confession: If I had the talents, I would not have studied physics and not have chosen a career in biomedical research, but it would have been music instead, the piano or the violin. So, instead, I remained just an amateur musician. However, old love never dies. Therefore, in my lecture I will make excursions to music, musicians, instruments and the orchestra as an ensemble of persons working and performing together. In particular the orchestra is a representative example of a multidisciplinary research group, in the way they harmoniously play together. There is another reason why this comparison can be made, since for both music as an art and research in biomedical informatics as a science, human creativity and perseverance are essential. But wasn't it Thomas Edison who said that creativity is 1% inspiration and

[*] The title of my lecture is inspired by Benjamin Britten's *The Young Person's Guide to the Orchestra: Variations and Fugue on a Theme of Purcell.*

[†] [In honor of Prof. van Bemmel. A special topic on Medical Informatics – Art or Science? will appear in Vol. 45, Issue 6 of Methods of Information in Medicine at the end of 2006. This paper will be part of the Special Topic. The Editors]

[‡] P.O. Box 1738, 3000 DR Rotterdam, the Netherlands. J.vanBemmel@ErasmusMC.nl.

[§] Because of the character of this lecture – a personal review of lessons learned during my path in research – this document contains primarily so-called self-references, which is in general not a good sign for any publication. I hope that the reader is lenient this time for me giving these biased citations.

99% transpiration? People who play an instrument and persons who conduct research know the truth of this expression.

During all my life I have been involved in research. The interest for it started early, around the same time that my love for music was also being aroused. It was the age that I spent much of my time as a young amateur astronomer, constructed my first telescope, and also played and enjoyed music. At school we regularly attended special concerts given by the Rotterdam Philharmonic Orchestra, where its conductor, Eduard Flipse, explained vividly how the different instruments functioned and contributed to the harmonic sound of the orchestra. Benjamin Britten's composition *The Young Person's Guide to the Orchestra: Variations and Fugue on a Theme of Purcell* is a perfect educational introduction. Although the conductor is the leader, the musical standard is given by the oboist, with his A-note (440 Hz).

In a similar way as Britten introduces the different themes and instruments, I will explain the various elements of our research. I will give some examples from the past and will sketch the circumstances and the requirements for our type of research, which is impossible without teamwork and without the fine-tuning of one's own work with that of others. As said, a research group can easily be compared with an orchestra, and I will frequently make this comparison in the following, alternating between the two areas. In both fields, all team members have to play their own party as part of the orchestra or the group, and should perfectly master the instruments they play. But that is not enough; you should carefully harmonize with the music others are playing, because only together one can create perfect music. Therefore, you have to listen attentively to each other and to pay careful attention. Before being able to play difficult compositions, lots of exercises and experiments are to be done.

Let me in the following convey to you some of my past experiences in carrying out R&D in biomedical informatics, both in playing an instrument myself and in conducting an R&D orchestra over many years. This also involved many external assessments and several contests with other orchestras, because both music and research are very competitive.

In my lecture I will discuss the following elements: (2) Instruments, (3) Exercises, (4) Difficult compositions, (5) Conducting, (6) Concerts and contests.

2. Instruments

A modern symphony orchestra consists of several groups of instruments: (1) the strings (from violin to double bass), (2) the woodwinds (e.g., piccolo and flute, oboe and clarinet, horn and bassoon), (3) the brass instruments (trumpet, trombone, tuba, French horn), (4) the percussion instruments (e.g., the drums, the triangle, the xylophone the harp and the tympani), and (5) the keyboards (e.g., the chimes, the piano, the cembalo and the organ). The groups are related by the similar ways, by which they produce sound when playing a composition. In a similar way, the members of a research group in biomedical informatics are also characterized by the different skills they contribute to research, such as people with a medical, biology, physics, mathematics, chemistry, informatics, programming, or engineering background. A research group in our field has a multidisciplinary composition.

There are also parallels with respect to the methods and the instruments used in both areas, and there are some similarities between building musical instruments, such as a violin, a flute or a grand piano, and developing instrumentation, systems and software. The same applies when playing and using the instruments. In all such cases, one need to be a good craftsman and thoroughly master one's instrument, that is, have studied at least one profession in great depth. In particular in our modern time of multidisciplinary research, each team member should be able to collaborate with people from other disciplines, but at the same time be an expert in at least one mono-discipline (see, e.g., a recent report on this matter [1]). If not, we will not be able to play our instruments properly, to conduct experiments, and to perform well in a research team.

Let me give you an example from the beginning of my own research, which I started in the field of perinatal medicine. The year I initiated this research was 1963. Computers were not yet available and we had to build the instruments ourselves from scratch.

Perinatal medicine

The goal of our research was (1) to get insight in the condition of the fetus during pregnancy and birth in order to understand the sometimes dramatic events that take place during delivery, and (2) to develop instruments to record biosignals reflecting the contractions of the uterus and the condition of the fetal heart and the circulation.

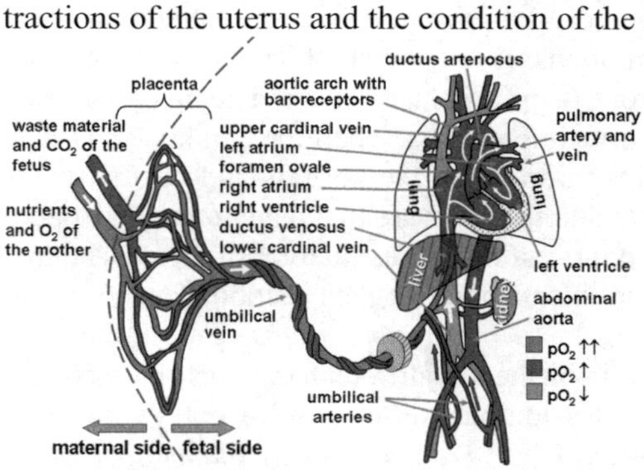

Fig. 1. Scheme of the fetal circulation (for details see the publications referred to in the text).

When a fetus is developing, it grows from a few cells only, to a newborn of about 3.5 kilograms. Of course, the same applies to all its organs, including the fetal heart, which starts beating from about 22 days after conception. After about 10 weeks it is detectable by ultrasound and as early as 12 weeks its ECG can sometimes already be recorded from the mother's abdominal wall. In the early times of our perinatal research we developed instrumentation that could detect fetal ECGs with amplitudes of a few microvolts only. This required the avoidance and cancellation of noisy disturbances from active abdominal muscles as well as from external sources, and of the maternal ECG. We even had installed a cage of Faraday within an experimental labor room in an obstetric department, to isolate all electromagnetic disturbances coming from the outside world. It was the time that we also developed the first digital fetal monitor, not yet with the chips of today, but with digital circuitry we had to construct ourselves with transistors and with hard-wired data processing.

Animal experiments

You may understand the great satisfaction we experienced, when we were able to monitor in real time the condition of the fetus during birth. The first times that we collected the perinatal data from the fetus, they were strange to all of us. The reaction of the fetus to uterine contractions sometimes appeared to be very dramatic, and it took a while and many meetings with colleagues abroad to interpret the signals. It also required more basic studies and experiments, not possible in a clinic, with women in labor. Therefore, we decided to study the physiology of the fetal circulation in depth and to conduct animal experiments with sheep and their unborn lambs, and in parallel to develop computer models of the fetal circulation, which is very different from the one after birth (see the scheme of Fig. 1). In our animal research, no sheep or lamb needed to be sacrificed. On the contrary; many lambs, once born, were brought up by the family of one of us, my PhD student and colleague Ton Veth.

Concerning the computer modeling, take note that this was the time before the first digital computers were available, and therefore we developed a model of the fetal circulation on a large hybrid – i.e., analog and digital – computer. This research, which included the development of instrumentation both for clinical and experimental data collection, for data processing and computer modeling, required a multidisciplinary team. Our team, with people from different universities, consisted of physicists like me, clinicians, electronic engineers, biochemists, anesthetists and physiologists.

Prior assumptions

What was extremely important in all these complex studies – in the animal experiment easily 10 to 15 people were involved – was to carefully formulate the hypotheses underlying our study and to ask ourselves what questions should be answered in our investigations. It is my experience, that one cannot pay enough careful attention to the formulation of hypotheses and prior assumptions. If not, we shall have to pay the price in disappointing outcomes. In short: wrong a priori's are reflected in negative outcomes. In processing the data, the same principle applies. Therefore, in all R&D it is wise to make an extensive list of all prior information we possess and on that basis design the requirements for the instruments and the software that have to be developed. The same orderliness, of course, applies to the testing and evaluation of the procedures and instruments we use. I will come back to this point in the next section.

The results of our perinatal research were published in different journals, biomedical, clinical and engineering. Medical informatics journals as such were not yet in existence, but we had several publications in the *IEEE Transactions in Biomedical Engineering* [2,3]. Different PhD dissertations were the outcome of this research, including my own [4], the first one in a row. Many publications followed, also in medical journals, e.g. [5-7], and as review chapters in handbooks [8,9]. It reflects one of the difficulties of our multidisciplinary field: we are expected to publish our results in journals belonging to at least two, sometimes more, different worlds: biomedical and scientific, the former to be discerned in the fundamental preclinical and the clinical disciplines, and the latter in physics, chemistry, or computer science, including medical informatics.

What lessons can we draw from this early experience?

- Biomedical informatics is characterized by a multidisciplinary approach.
- Each team member should be an expert in at least one mono-discipline.
- To obtain valuable results, an in-depth approach is necessary.
- Science cannot make progress without international collaboration.
- The hypotheses underlying a study should be carefully formulated.
- Wrong prior assumptions are reflected in negative outcomes.
- It is wise to make an extensive list of all available prior information.

- Results are typically published in journals belonging to different disciplines.

3. Exercises

The example I gave above was representative for both basic and applied research, for which a multidisciplinary team was indispensable. I will now come to another area, much more concise and involving data from many more patients: electrocardiography. It is a field where much basic knowledge was already available about the electrophysiological functioning of the heart.

First computers

After the first period of conducting research without the availability of even the most primitive digital computers – there were none, except for special-purpose processors, built by ourselves – the first programmable digital computer arrived at the nearby University of Utrecht. It was of a Dutch brand, called X-1, and its processing language was ALGOL-60. The only input possible was by an 8-hole paper tape. I now write the year 1965. In this case, we did not have to develop the entire hardware ourselves, but software enabled us to do at least part of the data processing. For data acquisition we still had to build the instruments ourselves. It is as if you buy your own piano, but still have to design and build your own keyboard. And, of course, in all circumstances you have to learn to play the instrument yourself. It lasted some time before the entire piano could be purchased from computer industry. Nowadays, the complete system as we had designed at that time is available on a few chips only, including well-elaborated software. But that took some decades to develop. Let me tell you part of the story.

In 1966 I attended the first conference in the field of what we now call biomedical informatics. The city was Elsinore in Denmark, where Shakespeare had located his tragedy Hamlet, with its famous sentence *To be or not to be, that's the question.* One of the reasons that I wanted to go to this conference was that I could meet there Hubert Pipberger, one of the very first who processed adult ECGs – in his case vectorcardiograms, VCGs. He had a computer, a CDC 3100, that he had installed a few years earlier in the VA hospital in Washington DC. We had already developed the instrumentation to acquire VCGs and ECGs, had started collaboration with an advanced cardiologic clinic in Utrecht, and had started the data processing on the X-1. The meeting in Elsinore with Hubert, a pioneer in the field of ECG processing, was very stimulating and in the years thereafter I had many contacts with him, during visits, workshops and conferences, and we had collaboration in several projects. In all other research projects that I have directed, I have stimulated my PhD students and colleagues to get in touch as early as possible with the best groups in the world and, if possible, to start collaboration. Again, multidisciplinary and international collaboration and cross-fertilization is of utmost value, a fact that is increasingly recognized [1]. I think it was Hubert Pipberger who taught me the short sentence that expresses it all: *Research is People.*

Electrocardiology

The first years were tough. The signals that we acquired had to be sampled, digitized and stored on paper tape, because the X-1 could only accept such type of input. This was understandable, because it was a computer for mathematicians only. Only once a day it was possible to offer two big rolls of paper tape to the operators: one with the program that I had developed, and one with one or two ECGs I wanted to process. It was a very cumbersome process, but one thing I have learned very well in those times was that every programming error was severely punished, because it would take me at least one full day before I had corrected the mistake and could offer the processing job again. But what was worse – because our ideas underlying the processing were still very innovative – was that only by experimentation, by trial and error, it was possible to assess the effects of our ideas. In other words, of the program many different versions had to be developed in order to gain more insight. Before I offered a processing job to the operators, I had already thought it over in detail and already dry-run it several times in my brain. I know that experienced musicians and conductors also do their 'dry' exercises by studying the written notes from paper, before even touching their instrument. When looking at the written music, you can even hear it in your head!

I guess that Antonio Stradivari from Cremona and Heinrich Steinweg from Seesen in Germany (called Steinway after his emigration to the USA) must also have had similar experiences when developing and building and, above all, fine tuning their instruments. This also must have taken much time, to be counted in many years. In my case, I could not wait that long and therefore got a special permission to operate the X-1 computer myself, soon to be replaced by a much more powerful X-8. Yet, although much faster, it still remained a system that could perform only one job at a time. Disk memory was nonexistent; data and programs were stored on a drum and the operating system worked with a ferrite core memory.

Luckily, we got more members in our orchestra and obtained the funding to buy our own computer for the growing amount of computing. In 1968 we purchased the first PDP-9 in our country, from Digital Equipment Corporation in Massachusetts. The cost of it was tremendous, in today's currency easily several million Euros or Dollars. It had a core memory of 16 kwords of 18 bits, two so-called DEC-tapes that could each contain 256 kwords, and no mass memory. The input was 5-hole paper tape and commands were given either by a teletype writer or by switches on the control panel. In this way, our throughput time was reduced to minutes instead of days. Besides, we were able to process data of many more patients, so that we could start a study into different cardiac diseases and collect a database. An analog-to-digital converter was still not available, so that we had to develop one ourselves. Our first results were published in Homer Warner's journal *Computers in Biomedical Research* [10-13].

The processing of ECGs is a very representative example of problems we may meet in processing biomedical data: (1) It concerns the collection of a large database of well-documented data so that the relationship between the data and disease patterns can be studied; (2) It requires the close collaboration with clinicians to get access to ECGs and patient record data and, (3) to exploit their knowledge in interpreting the data and assessing the results; (4) it is necessary to have profound knowledge of signal processing methods and pattern recognition, and to cleverly master programming. Again, without multidisciplinary collaboration all this is impossible.

Variability

When designing a computer program, it is important to take into account the same principles that I mentioned before: the careful formulation of the underlying hypotheses, and the listing of all knowledge and information you already have available. Yet, at this stage I must make an important further remark: in all biology and medicine nothing is standard; no two organisms or organs are the same, and even the same organ or organism may behave very differently the next moment in time. This phenomenon we call variability, which is quite different from noise and disturbances or from missing data. People doing research in other areas, such as physics, are not or much less confronted with such difficulties. This is the main reason why we always must collect a sufficiently large database to be able to take this variability into account. The same holds for human interpretation: never trust one observer or expert only. Instead, make use of a panel of human observers. With all this we were confronted in this line of research (see, e.g., [14]). Let me give you an example of this variability, both of the data and the observers, having a large effect on the interpretation of a database of ECGs.

Reference

From Pilate stems the expression *What is truth?* [15]. Sometimes, this also applies to the relationship between biomedical data and diseases. The problem is often, that the data may point to different disease patterns, such as different types or degrees of cardiac infarctions. In some instances, the disease may be somewhere in between normal and abnormal, e.g., a heart of a sportsman with a slightly enlarged left ventricular wall, or a patient with left ventricular hypertrophy caused by an aortic valve obstruction. All this is related to variability and our definition of what is a disease and what is normal. To assess the existence and degree of a cardiac disease, different ways may be followed to find the 'truth'. One way to obtain a proper reference for evaluation, is to rely on a panel of experts, the other is to try to find 'independent' clinical signs to diagnose the disease, e.g., by catheterization or post-mortem data. ECG interpretation systems may be based on either 'truth' (i.e., experts or clinical signs).

Evaluation

In the 1970s our system for ECG and VCG interpretation came to completeness, the 1980s were the decade of thorough evaluation, and the 1990s the decade of further diversification and industrial application. It demonstrates how long the road may take from basic research to application – if this happens at all (see, e.g. [16,17]).

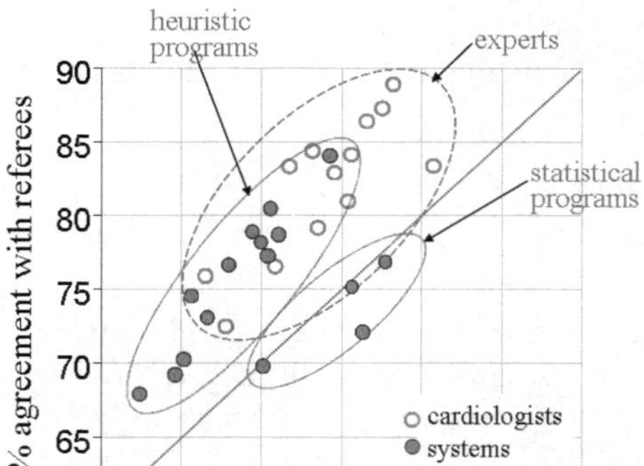

Fig.2. Total accuracy of ECG interpretation systems and cardiologists against two different reference values: ECG-independent clinical data and the group agreement of experts.

The decade of basic research was very interesting, because we were at the forefront of new knowledge in the field of electrocardiology. This pertained to both clinical and electrophysiological insights, and signal processing methods and software development. The decade of evaluation was also most exciting, because virtually all ECG interpretation systems in the world took part in a very large project, called Common Standards for Quantitative Electrocardiography (CSE), during many years subsidized by the European Union. We formed a core group of six people from different countries [18], collaborated with all existing groups in Europe, the USA and Japan, and had many fundamental meetings on the composition of the database to be used for testing, and on the results of the evaluations. The 'truth' had to remain secret (it still is!) so that no system could be tuned and optimized with the 'truth' as reference. The reference center was at that time in the hands of Jos Willems, who was independent and had no commercial interest or whatsoever in the outcomes, except scientific. Perhaps, CSE has been the largest international assessment study in our field ever done and the final results were published in top journals [19-21. The 'truth' consisted of both 'independent' clinical signs and the final weighted 'verdict' of a group of eight international experts (see Fig.2). The whole endeavor showed the complexity, duration and cost of an assessment study. It also demonstrated the importance of evaluation studies, because the outcomes of such studies are directly related to the quality of the systems we developed and to the benefit of patients and clinicians who were going to use them.

Lessons to be drawn from this experience:

- Only by many exercises one learns to play an instrument properly.
- Get in touch with the best groups in the world and, if possible, start collaboration.
- International collaboration and cross-fertilization is of utmost value.
- It is of great value to rethink a program many times before actual data processing.
- In many instances, in biomedicine a large well-documented database is essential.
- Profound knowledge of methods, such as signal processing and pattern recognition, is crucial.
- To compensate for errors in human interpretation a panel of experts is required.
- Diagnostic systems may be based either on expert knowledge or on 'independent' clinical signs.
- The time from basic research to application is very long and may easily take 20 years or more.
- The Reference in an assessment study should be kept secret.
- Assessment studies are of importance for the quality of health care and the benefit of patients.

4. Difficult compositions

Several 'lessons' to guide a young biomedical informatician may already have been derived from the two preceding cases. In the description of the third case, however, I want to convey my experience in a field that is not as fundamental as perinatal medicine or as confined as electrocardiology, but in an area that includes the whole of health care. It concerns the development of electronic health records, also called electronic or computerized patient records, in short: EHRs. We initiated our research for the development of EHRs in the beginning of the 1980s. We started in primary health care and R&D is still ongoing in clinical medicine. Let me briefly describe the different stages of our R&D and point to some important aspects that may be of wider interest.

Electronic health records

The structure of health provision in our country is, in a way, ideal for the introduction of EHRs. Each citizen has one General Practitioner (GP), who coordinates her or his health care, keeps a comprehensive patient record, and refers the patient, if necessary, to a specialist or a hospital. In the early 1980s, developments were started to introduce EHRs in primary care. Our team was deeply involved in this, and we concentrated from the very onset on the patient record itself. We had, from the beginning, close collaboration with industrial partners, because an R&D department is not able and should itself not be responsible for the implementation and maintenance of information processing systems in health care. Our goal was not only to develop an EHR system, but also to shape the infrastructure to be able to conduct research in primary care. We also had the intention to broaden later on to clinical care in hospitals.

The priorities in the first decade were on the realization of EHR systems, to start collaboration with primary care centers, and to shape a network for research [22]. In the second decade – roughly from the beginning of the 1990s onwards – we started many research projects dealing with the quality of health care, the integration of EHR systems with decision support systems [23,24], and post-marketing surveillance of drugs. This research is still continuing and expanding. Nowadays, we are no longer involved in R&D to realize an EHR system, but have built up a network in primary care for conducting research, together with many clinical partners.

In the first half of the 1990s we started a parallel development, intended to realize an EHR system for clinical use, based on the concept of structured data entry. This rather basic research also took about 10 years, before it resulted in a system that is able to be used in the entire area of clinical care, because of its conceptual approach (e.g., [25]). The system is now being implemented in several clinical departments and we have the basic software made available as Open Source on the Internet, in order to have as much as possible feedback from users worldwide [26]. We foresee that the R&D around this system will continue for some time, but in parallel clinicians are using it to start their own research projects, using the system.

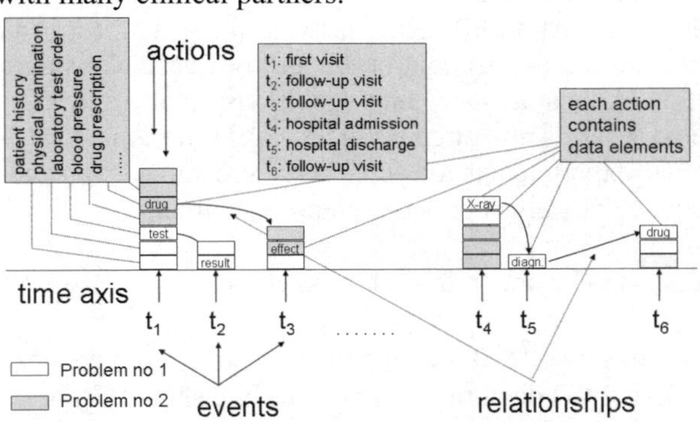

Fig. 3. Different elements in a structured EHR: events, actions, and data elements. The EHR may contain different problems, and data in the EHR may be interrelated. Time stamps on the data are essential for inferencing and decision support.

The development of a clinical EHR system that ideally comprises all patient-related data, from the patient history up to diagnostic and therapeutic results, appears to be an extremely complex enterprise (for a scheme, see Fig. 3). This is also the experience from many research groups and industries worldwide. It is an area still full of pitfalls and difficulties. Perhaps, the most complicated point is that a clinical EHR system should be implemented in an environment, where every patient history and treatment could be different, and with clinicians with different backgrounds and ideas. One has the feeling that the musicians in a clinical orchestra are all of them soloists and have sometimes only minor experience in playing together. In some instances they are even not used to obey the conductor. This is the reason why there are fundamental differences between the automation in health care and in banking, traffic control, or industry. Health care can seldom be standardized. Human beings are twice in the loop of information processing: as a subject, the clinician and as an object (or 'contra-subject'), the patient. This, of course, is related to what we have remarked in the preceding sections on variability and observer variation. Every young person who wants to follow a career in biomedical informatics should keep this in mind.

In primary care we have been very successful, and nowadays 100% of all GPs in our country use information systems containing an EHR, most of them are collaborating with their colleagues, and many of them with our team in a network. In clinical care there are still some major challenges ahead. The complexity of clinical patient care is much larger, there is much less standardization possible, and all clinical specialties are in a process of continuous change, because medical science itself is permanently renewed. EHRs in a clinical environment, therefore, should permit much more freedom to the clinician to implement her or his own ideas. Perhaps, the way a clinician uses an EHR system in a hospital environment is comparable to a *cadenza* in a piano or violin concert, or to an improvisation in a jazz concert. Here, the creativity of the musician should not be hampered by the instruments or the orchestra. This freedom, however, is in sharp contrast with standardization.

Our R&D in the field of EHRs also showed that it is important not to give up when sometimes major difficulties arise, as a result of, for instance, the conflict between freedom and standardization, or the lack of financial support or competent clinical collaboration. A conductor for motivation, inspiration and team building is then utterly indispensable. In some circumstances, R&D in biomedical informatics is comparable with atonal experimental music, where sometimes even the rhythm is absent. My advice then is: don't give up; perseverance is pivotal.

Lessons to be drawn from this experience:

- For some R&D it is important to have, from the onset, collaboration with industry.
- An R&D department is not able and should not be itself responsible for the implementation and maintenance of information processing systems in health care.
- For clinical research it may be needed to invest first in the realization of its infrastructure.
- To obtain user feedback Open Source on the Internet is a possible way to go.
- In health care one should be aware that there are no standard patients or clinicians.
- This is fundamentally different from banking, traffic control, or industry.
- Do not give up in cases of lack of finances, competent researchers, or clinical collaboration.
- A leader is indispensable for motivation, inspiration and team building.

5. Conducting

Using the three research lines as illustrations, we were already able to discuss some difficulties to which the orchestra of biomedical informatics may be confronted, the sometimes very different types of music that have to be performed, and some important lessons that could be derived on how to conduct research. The examples may have shown that the road from a first idea to success and final implementation is sometimes long and winding. While walking the road, al kinds of events in and around the team may take place for which some action may be desirable. In this section, therefore, I want to pay attention to the importance of how to maintain the spirit and the quality of the team.

Team building

The first observation is that in every environment where people work together, mutual trust and knowledge of each other's competences is of critical importance. In shaping a research group the most crucial moment is when someone is to be hired as a member of the team. Before a member of an orchestra will be nominated, she or he has to have an interview and should participate in an audition. Similarly, members of a research team will be selected on the basis of their *curriculum vitae* and an interview with the responsible leadership. Perhaps, giving a lecture may be part of the selection process. Once someone becomes a member of the team, a clear mutual agreement should be made on tasks and responsibilities, which should regularly, say once a year, be reviewed. Without trust in each other's competences and contributions, no orchestra can be composed.

Competences

It is the task of the leader of the team to find out what are the strong qualifications of all team members and, foremost, to offer them the opportunity for further enhancing their competences. The central secret of a team spirit is that everyone loves to play her or his party, in harmony with the other members of the orchestra. The conductor of the orchestra should never be afraid to recruit musicians who play some instrument better than he himself. On the contrary; a good leader is characterized by the fact that enough room is given to everyone to develop her or his career. Only then, mutual respect will grow. The only difference between senior and junior team members is their level of experience and the responsibility they have; for the rest all team members are equal. No instrument in the orchestra can be missed when a great symphony has to be performed.

Training

The advantage of the university is that the talents of young musicians and potential soloists can be discovered at an early stage. I see it as a blessing that the interest of students can be aroused very early, that they can be acquainted early in their educational track with biomedical informatics, and be offered already an apprenticeship while still being a student. Some of them may want to continue for an MSc degree and part of the latter ones for a PhD. This is exactly how several of the core staff of our department got interested in biomedical informatics. After a thorough learning period they became the section leaders of groups of musicians, e.g., of the strings or the woodwinds. Training takes time, at least 10 years or so, before being able to become a player on an international platform. The conductor should shape the conditions, whenever possible, for the further development of the careers of everyone in the orchestra, from starting musicians to experienced soloists.

Conducting

I have never seen an orchestra with two conductors (except for the separate boys' choir in the St Matthew's Passion of Johann Sebastian Bach), and certainly no ship with two captains. The same applies to a research department. Of course, projects, as part of the department's R&D, may have different principal investigators. A large research institute may have a board, but also then its chairman will be the leader, indicating the general directions. The conductor has a great responsibility, because he has to guard the quality of the orchestra and to listen attentively whether the music is harmonious and follows the professional rules. If he observes disharmony, he should not hesitate and take, if possible, immediate action. If he does not pay attention, also others in the orchestra may be influenced by the false sound and start deviating from the right track. The result is a waste of time, talent and energy.

The leader of a research group has also to pay attention to the quality of the publications, to check whether the proper methodology was applied and the correct instruments were used. Motivation and inspiration are key characteristics of a research leader. At regular times, meetings should be organized to discuss progress, and presentations should be given by all members of the team.

Fig. 4. Sailing with a tall ship (1100 m^2 of sail) from Rotterdam to the heart of London to attend Medinfo

Having respect for every individual person and paying attention to her or him is perhaps the most important task of the leader. This is the more so for the junior members of the group, such as PhD students. Always be honest, keep your promises, don't hesitate to tell when you think there is a problem, listen carefully to every single instrument. Always maintain integrity.

An important point is, to keep the team spirit going, for instance by regularly organizing social events. Celebrate the acceptance of a publication, hire a ship to sail to an international conference (Fig. 4). All these 'recommendations' seem self-evident – and indeed they are, because all major research groups in our discipline follow the same or similar rules. Nevertheless, for young musicians it is wise to make themselves aware of the importance to maintain high quality and good conduct, by the musicians and the conductor.

Funding

I remember that in the early times, when funding was not in such short supply as nowadays – there were far fewer orchestras – it was not extremely difficult to obtain financial support for one's research. This was on the one hand very advantageous, because writing proposals for grants was (at least in our country) much simpler and competition between the different orchestras was much less. This was also the situation until far in the 1980s for the funding from the European Union. On the other hand, it was in some sense also less positive, because competition between research groups and orchestras stimulates the improvement of the quality of the work that is to be performed. At present, competition and 'fishing in the same pond' is much harsher everywhere. Here is a parallel with music as well; there is a much higher demand for the top orchestras than for mediocre ones.

Funding of research comes roughly from four different resources: (1) public funding, e.g., to finance tenured positions; (2) competitive funding, e.g., from a national science foundation; (3) subsidies from non-profit research foundations, e.g., for cancer or cardiac diseases; (4) financial support from industrial companies and other private corporations. The amount of effort to be invested in writing research proposals to obtain a grant is very considerable and success is far from guaranteed. Yet, it is my experience that for solid research, sooner or later, financial resources will always become available. The most important condition for success is, however, to build an excellent publication record, that is, by publishing articles in journals with a high enough impact factor. Of course, there are many more parameters involved whether a research grant will be awarded. It depends, in general, on the outcome of regular assessments of the research of a department, to be discussed in the next section.

Lessons to be drawn from this experience:

- In research, mutual trust and knowing each other's competences is essential.
- The most crucial moment is when someone is to be appointed as a member of the team.
- A leader should not be afraid to recruit staff who are more competent than he himself.
- A good leader gives enough room to everyone to develop her or his career.
- Interest of students in biomedical informatics should be aroused early.
- Motivation and inspiration are key characteristics of a research leader.
- Having respect for every person and paying attention is the most important task of a leader.
- Competition for research funding has become severe everywhere.
- For solid research, sooner or later, financial resources will become available.
- Funding depends, in general, on the outcomes of regular research assessments.

6. Concerts and contests

The goal of the many exercises of an orchestra is, of course, the performance during a concert with live public. Only the best orchestras play in the most famous concert halls and are invited to let their music be recorded on compact disk. The very best musicians are invited to act as soloists during a concert. Reviews of a concert in newspapers may contribute to fame or failure. In a way, the same applies to research. Presentations are given for colleagues at a conference, and sometimes members of the research group are requested to give an invited lecture elsewhere. The best output, however, is a journal publication, where the article has been scrutinized by referees.

Regular assessments

The existence and survival of research institutes is increasingly dependent on regular assessments. An assessment, usually done by an independent review committee of peers, is a suitable instrument to obtain insight into the quality of research. Its outcome is directly related to the quality of publications in peer-reviewed journals. Assessments are of particular importance if the results of biomedical informatics R&D are to be applied in clinical practice.

The goals of research assessment are twofold: (1) Improvement of the quality of research; and (2) Accountability of how the resources were spent.

The rating of the quality of research is a relative matter, even when formal procedures are applied. There are no absolute standards. Only comparison with similar institutions elsewhere may offer an impression of the position of an institution on the quality scale. All evaluations contain subjective elements and, even with peer reviews, one should be cautious, because the composition of the assessment committee is essential for the outcome. This is why members of a review committee should be drawn from different centers and the committee should preferably be composed by an independent organization. Moreover, it is advisable to use a well-defined protocol with clear instructions pertaining to the intention of the assessment [27], to exclude individual opinions and to enable comparisons over time.

Personally, I have had extensive experience in participating in and undergoing external assessments. In general, four aspects are to be considered [28]:

1. The quality of the research;
2. The productivity of the scientific output;
3. The relevance of the research for academia and society;
4. The future perspective, feasibility and vitality of the research.

Although the report of a review committee usually contains judgments on the reviewed research in wording, it has also been proven useful to position the research on a five-point qualitative rating scale with scores, which are given separately for all four aspects mentioned above:

5	*Excellent*	Internationally at the forefront and with a high impact
4	*Very good*	Internationally competitive and nationally at the forefront
3	*Good*	Nationally competitive and internationally visible
2	*Satisfactory*	Solid but not exciting. Nationally visible
1	*Unsatisfactory*	Not solid, nor exciting. Not worth pursuing

In our country, we have over 25 years of experience in assessing the research in universities and semi-public research institutes. Key instruments for the assessments are well-defined protocols that are also used in an internal system for self-assessment. The experience with research assessment over the years shows that evaluation increasingly assists in improvement and accountability.

Lessons to be drawn from this experience:

- The best research output are publications that are scrutinized by referees.
- The survival of research institutes is increasingly dependent on regular assessments.
- The outcome of an assessment is greatly determined by the quality of publications.
- For the assessment, a well-defined protocol should be used

- Such protocols can also be used for self-assessment.

7. Conclusions

Most of the conclusions were already drawn at the end of each section. Therefore, no essentially different conclusions will be formulated in this concluding paragraph. The only remark I would like to make is that I am aware that the short reflection on my own research was, of course, very personal and of a confined character. I understand that other musicians and conductors will probably have their own reflections that may differ from the ones expressed above. It should also be realized that the type of research in our field is continuously changing.

For instance, since the last few years we have realized the joint venture of research between different departments of our Faculty of Medicine. Examples are the establishment of an advanced Image Processing Group between our Dept of Medical Informatics and the Dept of Radiology; the close collaboration between our Dept and the Information Processing Group in the University Hospital, Erasmus MC, on the implementation of structured EHRs; and the joint endeavor with the Dept of Epidemiology regarding post-marketing surveillance of drugs, with the help of our large primary care research database, the so-called IPCI project. As a matter of fact, a department of (bio)medical informatics would be non-existent if there would be no close collaboration and sometimes even integration with other departments in and outside the hospital. In summary, several characteristics in biomedical informatics research remain to be of central importance:

- Biomedical informatics is characterized by a multidisciplinary approach.
- Hypotheses underlying research should always be carefully formulated.
- Results are typically published in journals belonging to different disciplines.
- International collaboration and cross-fertilization is of utmost value.
- In many instances, in biomedicine a large well-documented database is essential.
- The time from basic research to application may easily take twenty years or more.
- Assessment studies are important for the quality of health care and the benefit of patients.
- In research, mutual trust and knowing each other's competences is essential.
- A good leader gives enough room to all team members to develop their careers.
- Funding depends, in general, on the outcomes of regular research assessments.
- The outcome of an assessment is greatly determined by the quality of publications.

References

[1] See, e.g., National Academy of Science. *Facilitating Interdisciplinary Research*. Washington DC: National Academy Press, 2004.

[2] Van Bemmel JH, van der Weide H. Detection procedure to represent the foetal heart rate from the electrocardiogram. IEEE Trans Biomed Engng 1966;BME-13:175-82.

[3] Van Bemmel JH. Detection of weak foetal electrocardiograms by autocorrelation and crosscorrelation of envelopes. IEEE Trans Biomed Engng 1968;BME-15:17-23.

[4] Van Bemmel JH. *Detection, Processing and Analysis of Foetal Electrocardiograms*. PhD Thesis. Nijmegen/Utrecht, 1969.

[5] Van Bemmel JH, Peeters LAM, Hengeveld SJ. Influence of the maternal ECG on the abdominal ECG

complex. Am J Obst Gynec 1968;102:556-62.

[6] De Haan J, van Bemmel JH, Stolte LAM, et al. Quantitative evaluation of foetal heart rate patterns. I, II and III. Processing methods. Europ J Obst Gynec 1971;3:95-102; 3:103-10; 4:137-46.

[7] De Haan J, van Bemmel JH, Stolte LAM, Veth AFL, Janssens J, Eskes TKAB. Trend detection in the foetal condition. Int J Obst Gynec 1972;10:202-6.

[8] Van Bemmel JH, De Haan J, Veth AFL. Instrumentation for Obstetrics, Gynecology. In: Ray Ch, ed. *Medical Engineering*. Chicago: Yearbook Med Publishers, 1974:548-88.

[9] Veth AFL, van Bemmel JH. The role of the placental vascular bed in the foetal response to cord occlusion. In: Longo LD, Reneau DD, eds. *Fetal, Newborn Cardiovascular Physiology. Vol I: Developmental Aspects*. New York: Garland Stpm Press, 1978:579-614.

[10] Van Bemmel JH, Talmon JL, Duisterhout JS, Hengeveld SJ. Template waveform recognition applied to ECG/VCG analysis. Comp Biomed Res 1973;6:430-41.

[11] Van Bemmel JH, Hengeveld SJ. Clustering algorithm for QRS, ST-T waveform typing. Comp Biomed Res 1973;6:442-56.

[12] Swenne CA, van Bemmel JH, Hengeveld SJ, Hermans M. Pattern recognition for ECG monitoring: An inter-active method for the classification of ventricular complexes. Comp Biomed Res 1973;6:150-60.

[13] Hengeveld SJ, van Bemmel JH. Computer detection of P-waves. Comp Biomed Res, 1976:9:125-32.

[14] Kors JA, van Herpen G, van Bemmel JH. Variability in ECG computer interpretation. J Electrocardiol 1992;25:264-71.

[15] Gospel of St John 18:38.

[16] Van Bemmel JH, Kors JA, van Herpen G. Methodology of the modular ECG analysis system MEANS. Meth Inf Med 1990;29:346-53.

[17] Van Bemmel JH, Zywietz Chr, Kors JA. Signal analysis for ECG interpretation. Meth Inf Med 1990:29;317-29.

[18] The so-called CSE core group consisted of Jos L. Willems (Louvain), Rosanna Degani (Padova), Pierre Arnaud/Paul Rubel (Lyon), Peter W. MacFarlane (Glasgow), Christoph Zywietz (Hannover), and Jan H. van Bemmel (Amsterdam/ Rotterdam).

[19] Willems JL, Arnaud P, van Bemmel JH et al. Establishment of a reference library for evaluating computer ECG measurement programs. Comp Biomed Res 1985;18:439-57.

[20] Willems JL, Arnaud P, van Bemmel JH et al. Assessment of the performance of electrocardiographic computer programs with the use of a reference data base. Circulation 1985;71:523-34.

[21] Willems JL, Abreu-Lima C, Arnaud P, van Bemmel JH et al. The diagnostic performance of computer programs for the interpretation of electrocardiograms. New Engl J Med 1991;325:1767-73.

[22] Van der Lei J, Duisterhout JS, Westerhof H, Van der Does E, Cromme PVM, Boon WM, Van Bemmel JH. The introduction of computer-based patient records in the Netherlands. Ann Int Med 1993;119:1036-41.

[23] Van der Lei J, Musen MA, van der Does E, Man in 't Veld A, van Bemmel JH. Review of physician decision making using data from computer-stored medical records. The Lancet 1991;338:1504-8.

[24] Van Wijk MA, van der Lei J, Mosseveld M, Bohnen AM, van Bemmel JH. Assessment of decision support for blood test ordering in primary care. A randomized trial. Ann Intern Med 2001; 134:274-81.

[25] Van Ginneken AM. Considerations for the representation of meta-data for the support of structured data entry. Meth Inform Med 2003;42:226-35.

[26] Los RK, van Ginneken AM, van der Lei J. OpenSDE: a strategy for expressive and flexible structured data entry. Int J Med Inform 2005;74:481-90.

[27] See, for instance, for a proven protocol in the Netherlands http://www.knaw.nl

[28] See, e.g., http://www.hero.ac.uk/uk/research; http://www.qanu.nl; http://www.stratresearch.se; http://www.inserm.fr/fr/inserm/evaluation; http://www.dfg.de/ranking

Ubiquity: Technologies for Better Health in Aging Societies
A. Hasman et al. (Eds.)
IOS Press, 2006

Reflections on Biomedical Informatics: From Cybernetics to Genomic Medicine and Nanomedicine

Victor MAOJO[a] and Casimir A. KULIKOWSKI[b]
[a]*Biomedical Informatics Group, Artificial Intelligence Lab, Universidad Politecnica de Madrid, Spain*
[b]*Dept of Computer Science. Rutgers Univ. New Brunswick,New Jersey, USA*

Abstract. Expanding on our previous analysis of Biomedical Informatics (BMI), the present perspective ranges from cybernetics to nanomedicine, based on its scientific, historical, philosophical, theoretical, experimental, and technological aspects as they affect systems developments, simulation and modelling, education, and the impact on healthcare. We then suggest that BMI is still searching for strong basic scientific principles around which it can crystallize. As -omic biological knowledge increasingly impacts the future of medicine, ubiquitous computing and informatics become even more essential, not only for the technological infrastructure, but as a part of the scientific enterprise itself. The Virtual Physiological Human and investigations into nanomedicine will surely produce yet more unpredictable opportunities, leading to significant changes in biomedical research and practice. As a discipline involved in making such advances possible, BMI is likely to need to re-define itself and extend its research horizons to meet the new challenges.
Keywords: Biomedical Informatics. Medical Informatics. Bioinformatics. Ontologies. Nanomedicine. Virtual Physiological Human.

1. Introduction

Several analyses of the current scientific status of Medical Informatics (MI), Bioinformatics (BI), and Biomedical Informatics (BMI), that has evolved over the last few years from the convergence between MI and BI, have been published over the last decade [1][2]. The authors of this paper have earlier contributed to this debate [3][4].

MI researchers have sometimes stated they encounter difficulties in receiving the scientific recognition that they think the discipline deserves [4]. In some cases, MI has been considered a pure engineering discipline, focusing on the technology related to health care. Could this situation be only circumstantial or might it reflect a trend that may become further accentuated? What could be the future of MI and BI? Will they merge into BMI or will they remain independent?

2. Proposal for a more systematic analysis of MI, BI, and BMI

In a series of various papers, the authors have analysed the scientific status of these three disciplines [3][4], using as a framework various ideas from the philosophy of science. We extend these ideas below, following various directions.

2.1.Historical

One of the goals of cybernetics was to study the components of what can be referred to as “human information systems”. In contrast, MI has not made this a main concern.

Instead, MI has been primarily involved in designing and developing applications for patient care, with computational methods for basic research on human physiology and pathology rarely impinging on its central agenda. BI, on the contrary, when faced with new challenges in the analysis of complex gene networks and pathways, or the influence of the environment in genetics, is connecting to the emerging subfield of "systems biology. In the process, some of the original ideas and methods from cybernetics are being re-introduced, and re-thought in the context of the latest experimental paradigms in biology. The emergence of nanobiotechnology promises to accelerate this trend, possibly catalyzing a new "nanocybernetics". Researchers such as Benenson, at Harvard University [5] propose to adapt the concept of Wiener's automata to the biomolecular world, as with an automaton built to recognize the characteristic molecular complexity of cancer cells and destroy them while leaving normal cells intact. In this sense, traditional biological research carried out "in vivo", "in vitro", and recently "in silico", can be carried out "in info".

2.2. A brief analysis from the philosophy of science

In one of our recent papers [3], we compare MI and BI based on ideas from major philosophers of science. This use is different from recent work that introduces philosophical ideas at the heart of BMI. Ontologies introduce informatics methods for creating a shared consensus about the concepts that underlie a specific domain [6]. While ontologies provide new approaches to knowledge engineering [7], they still must demonstrate that they can advance the scientific theories supporting MI. Some philosophical arguments are pertinent in considering this issue. For instance:

a) An ontology, itself, cannot be the scientific basis for triggering a paradigm shift. Kuhn suggested that a paradigm shift provoked a change in the underlying ontology [8]. An example comes from physics —a field where this kind of philosophical analysis has been carried out for well over a century. It was relativity theory that changed the meaning of the concepts "light" and "time". Without the theory, no ontology could have been developed to anticipate the new meanings of these two concepts.

b) The history of informatics and "preinformatics" disciplines illustrate some examples that should not be forgotten when attempting to redefine biomedical fields, such as physiology. Let us take an example. Recent efforts suggest the development of a broad, universal biomedical ontology, from scratch. At its core, "concepts", as used by original ontologists such as Gruber [6], would be changed to "entities". Also proposed is a new usage of the term "process", based on ideas of energy exchange and transformation. These "entities" and "processes" are then proposed to be used in redefining physiology, based on ontological assumptions. From an informatics perspective, this approach looks interesting, but from a physiological perspective they harken back to the "pre-cybernetics" period before the 1940's. One of the main achievements of cybernetics researchers was to show that physiology involves exchanges of information (with signalling in organisms) rather than simply the exchange of energy. Wiener himself considered the human body as an entire "message" [9]. The greatest drawback of returning to a strictly thermodynamical, or "energy-based" approach in physiological ontologies, as suggested above, is that it cannot represent the many information-based physiological processes that we find within organisms, such as signal messaging, feedback loops within their nervous systems and the homeostatic processes which are so pervasive at many physiological levels. .

c) Ontologists are currently taking holistic or top-down approaches in building biomedical ontologies [10]. From such a perspective, concepts (or entities) and relations (or processes) are being made explicit in computer representations. It can be argued that ontologies should be also generated using a bottom-up approach, scaling findings up from the molecular level, and both approaches, top-down and bottom-up, then combined. While building generic ontologies top-down from scratch may prove useful in describing human-designed and controlled systems and tasks such as database integration, the uncertainty and incompleteness surrounding our knowledge of continuously evolving groups of organisms in complex environments should make us very cautious in thinking of them as an a-priori valid scientific basis for deducing possible consequences about living systems.

d) More traditional ontologists such as Hartmann, or the Nobel laureate Konrad Lorenz [11] pointed out that integrative approaches in biology —and medicine— failed in their day, because they did not to consider different conceptual levels. According to Lorenz, it is not possible to deduce structure at higher levels based on characteristics of those from lower levels. This can be easily understood from an example from physics, where Einstein's theory of relativity applies and is adequate to explain phenomena at interplanetary distances, whereas quantum mechanics is needed at the atomic level. Neither can be deduced from the other and, even more fundamentally, both cannot hold true at the same time —they depend on the questions posed within their own specific experimental designs. Physicists continue to seek a new, integrated, unified theory to reconcile them. In biomedicine, terms that appear superficially alike, or even identical, frequently have very different meanings for a molecular biologist, a pathologist or an internist. Using a reductionist approach we cannot build up physiology simply from molecular biology, pathology from physiology, or public health from pathology. Similarly, the environment produces very different effects at the biochemical and the population level. Attempts to create extended ontologies that cover all levels of biomedicine, from atoms to populations, must consider such constraints and their implicit limitations for both knowledge- and software-engineering purposes.

2.3. Theoretical foundations

MI researchers have obtained outstanding results in theoretical research, related —as stated above— to topics such as natural language processing, uncertainty management, or image processing. But none of these or other advances has been strong enough to establish a scientific basis for the independent recognition of the discipline. Major challenges that are information-centered, such as understanding how consciousness works in humans, or how information is coded in the brain, have remained completely outside the mainstream of BMI. While these two challenges should indeed be primarily addressed by neuroscientists, cognitive scientists or neuro-informaticians, one might argue that a widely "integrating" field such as BMI should also consider such problems as part of its scope. They might indeed provide some of the underlying theory that the discipline needs to advance scientifically, and lead to external recognition. Insights into how brain function affects information processing would also have a deep impact in clinical medicine —especially on psychiatry, neurology, and on sensory integration and feedback for healthy subjects also.

There are areas whose research outcomes will be central to BMI. For instance, representing and modelling genotype-phenotype data and their interaction with the environment (and nutrition) is a major challenge where researchers from systems biology are focusing efforts and extending both computational as well as biological modelling. It presents an excellent opportunity for BMI researchers to participate.

2.4. Experimental research

Since the 1960s, researchers have applied a variety of techniques and tools (e.g., statistics, pattern recognition, and machine learning) to discover knowledge such as clinical prediction rules from large databases. In doing this, researchers wanted to avoid biases in the knowledge acquisition process from one or more experts. Yet, regrettably, few of these systems are actually used routinely to support medical practice. In many ways, data mining is an example of how difficult it is to encourage interactions between BMI researchers and professionals from other related areas. Biomedical statistics has developed for more than five decades, with significant impact on epidemiology and public health, but relatively little interaction with BMI. More vigorous exchanges could create a synergy, improving not only medical data analysis but also more efficient and deep knowledge integration and experimentation, facilitating the acceptance of project outcomes by medical practitioners.

2.5. Simulation

Researchers aim to predict cell behaviour by creating models that include cell features [12] as well as environmental data. Computers help to simulate the dynamic changes of cellular metabolism [13]. Virtual simulations use for this purpose data from the different "omics" projects that have proliferated over the last few years. At a higher level, BMI researchers have built disease models for over three decades. An early example is the CASNET expert system which incorporated a causal model, linking observations and pathophysiological states of the eye diseases comprising glaucoma [14]. More recently, complex simulations can reproduce pulmonary diseases using the Virtual Lung or cardiac physiology. By increasing the complexity and expanding the limits of specific simulations, researchers have proposed the development of an ambitious project, the Virtual Physiological Human (VPH) [15]. This project aims to integrate partial simulations from different systems —e.g., cardiac, lung, kidney, and others—, organs, and cells, into a whole simulation of the entire body. In the VPH, an integrated virtual human model would facilitate navigation, representation and simulation of physiological processes and, finally, carry out "in silico" experiments that can be used for testing diagnostic devices or new drugs.

Contrasting the Visible Human Project [16] and the VPH the former had as goal to build a 3-D anatomical representation of the human while the VPH aims to build a 3-D simulation of the physiological body, including metabolic pathways, biochemical reactions, electrical signaling and regulation processes, among others. While this is one of the most challenging programs that BMI researchers can address, we should keep in mind the difficulties mentioned above about jumping between the different biomedical "levels", from molecular biology to cells, from cells to tissue in systems and organs, from these to whole organisms, and from individuals to populations. Our knowledge here has significant gaps which must be filled before the VPH can be completed. Despite this, it is one of the most challenging ideas proposed during the last years, though considerably more ambitious in scale, and less clearly realizable than the Human Genome Project (HGP).

2.6. Training and education

The introduction of the technologies and new subfields such as, for instance, nanotechnology, molecular automata, or complex simulation models, will require a background that is different from that provided by the current curricula predominant in BI or MI. Since the areas of convergence are larger and quite different, new methods of modelling, analysis, and interpretation will be needed, overlapping with engineering, cognitive science, operations research, and most of the basic sciences. How good should BMI professionals be as mathematicians, physicists, or chemists? How good should they be in biological bench research? How can they possibly find the time to master the multiple disciplines and integrate the knowledge this entails? How can they compete with the specialists in increasingly narrow –omics subspecialties, and how with those in the increasingly important and related environmental and public health ones? In short, is there a good "complementarily symbiotic niche" to these other disciplines that can be designed for the biomedical informatician?

2.7. Impact in healthcare: Genomic medicine and nanomedicine, just two more "paradigm shifts"?

The success of the HGP and other *omic*s-related projects has introduced new approaches to basic medical science and healthcare, generically labelled as *Genomic Medicine*. Diseases such as cancer, a plethora of rare diseases, and many others are now being explained from different perspectives, including genetic information.

For Genomic Medicine to become real, in Europe, the European Commission has funded initiatives, such as the INFOBIOMED network of excellence and the ACGT (advanced Clinico-Genomic Trials) project, to develop methodologies for integrated analysis of biomedical information in clinical trials. In this scenario, many collaborative actions are being proposed. For instance, building biobanks that will gather clinical and genomic data; providing clinical data to bioinformaticians that will test their metabolic pathway representations; building new biomedical ontologies, or introducing omics-based data models in innovative genomic-based electronic health records. The goal is not to "deconstruct" the components of clinical practice and provide the computer tools that are needed for health professionals, but to shift clinical practice towards new approaches, deeply anchored in basic science.

In 1959, the Nobel laureate Richard Feynman proposed that there were no scientific obstacles to avoid the manipulation of atoms [17]. In his visionary speech, he also anticipated uses that this new approach could have for biological research. A good 40+ years later, a new discipline, called "nanotechnology", is slowly arising.

As nanomedicine begins to define itself, an increasing rate of discoveries will ensue, promising to deliver new diagnostic and therapeutic methods for medicine. Bionanomachines can work in the tiny space of living cells and be tailored for medical applications. Complex molecules that seek out diseased or cancerous cells, sensors for diagnosing diseased states, replacement therapy using custom constructed molecules, may be able to treat different diseases and conditions, opening a new horizon for medicine [18]. In nanotechnology, nanoparticles and nanomaterials are built as biodetection agents for deoxyribonucleic acid and proteins. In addition, nanodevices, such as nanowires and nanotubes, can be combined with nanoarrays to create automated nanodetection platforms. Molecular imaging modalities based on quantum dots and magnetic nanoparticles introduce new methods to intracranial imaging.

Paradoxically, just as the term "genomic medicine" is beginning to enjoy some currency after the completion of the HGP, a new term, "nanomedicine", comes calling at the door, claiming to be its newest, exciting frontier. Genomic medicine and nanomedicine, then have an incredible potential to transform biomedicine and healthcare but, whereas they promise to deliver methods and technologies far more ambitious, they still need the theoretical foundations that are the basis for solid scientific disciplines, suggesting a fruitful field where adventurous BMI researchers can explore with specialist colleagues in the fields needed to make concrete advances.

3.Conclusions

From a scientific perspective, it can be argued that there are no real paradigms or theories that crystallize BMI around one or a very few core principles. On the other hand, the discipline may have not promoted or encouraged sufficient debates about its underlying science. When a rare unanimity around certain issues, or "hot topics" arises among researchers, it can often lead to periods of great breakthroughs. Later, once this unanimity ceases, other issues substitute for the old ones, and a new cycle ensues.

As a final thought— while BMI has contributed to change the face of healthcare and improve the lives of people like few other scientific disciplines, it still needs the strength and self-confidence, which can be reflected in stimulating controversies characteristic of the more established sciences and their technologies.

References

[1] Kohane, I. Bioinformatics and Clinical Informatics: The Imperative to Collaborate. 2000. J Am Med Inform Assoc. 7, 5: 439-443.
[2] van Bemmel JH. Medical informatics, art or science? Methods Inf Med. 1996 Sep;35(3):157-72.
[3] Maojo, V. and Kulikowski, C. Medical Informatics and Bioinformatics: Integration, or Evolution through Scientific Crises? Methods Inf Med (in press)
[4] Maojo V, and Kulikowski CA. Bioinformatics and medical informatics: collaborations on the road to genomic medicine? J Am Med Inform Assoc 2003 Nov-Dec;10(6):515-22.
[5] Benenson, Y., Gil, B., Ben-Dor, U., Adar, R. & Shapiro, E. An autonomous molecular computer for logical control of gene expression. Nature 429, 423-429 (2004).
[6] Gruber, T.R. "A Translation Approach to Portable Ontology Specifications", Knowledge Acquisition, 1993 5(2), 199-220.
[7] Gómez-Pérez, A, Corcho, O and Fernandez, M. Ontological Engineering. Springer Verlag 2004.
[8] Kuhn, T. The Structure of Scientific Revolutions. 2nd Ed.. Chicago: Univ. of Chicago Press. 1970
[9] Wiener, N. The Human Use of Human Beings. Da Capo Press. 1988.
[10] Ceusters W, Smith B, Goldberg L. A terminological and ontological analysis of the NCI Thesaurus. Methods Inf Med. 2005;44(4):498-507.
[11] Lorenz, K. Behind the Mirror. Harvest/HBJ Book. 1978.
[12] Ishii N, Robert M, Nakayama Y, Kanai A, Tomita M. Toward large-scale modeling of the microbial cell for computer simulation. J Biotechnol. 2004 Sep 30;113(1-3):281-94. Review.
[13] Hunter PJ, Borg TK. Integration from proteins to organs: the Physiome Project.Nat Rev Mol Cell Biol. 2003 Mar;4(3):237-43. Review.
[14] Weiss,S., Kulikowski, C., Amarel, S., and Safir, A. A model-based method for computer-aided medical diagnosis, 1978. JArtif. Intell., 11, 145-172.
[15] http://cordis.europa.eu/ist/directorate_h/index.html
[16] Ackerman, M.J. The visible Human project, Proc IEEE. 1998. 86,3 :504-511.
[17] Feynman, R, There's Plenty of Room at the Bottom. Annual Meeting of the Am Phys Soc, 1959.
[18] Goodsell, D. Bionanotechnology: Lessons from Nature. Wiley, 2004.

Ubiquity: Technologies for Better Health in Aging Societies
A. Hasman et al. (Eds.)
IOS Press, 2006

Meeting the Challenges – the Role of Medical Informatics in an Ageing Society

Sabine KOCH
Centre for eHealth, Uppsala University, Sweden

Abstract. The objective of this paper is to identify trends and new technological developments that appear due to an ageing society and to relate them to current research in the field of medical informatics. A survey of the current literature reveals that recent technological advances have been made in the fields of "telecare and home-monitoring", "smart homes and robotics" and "health information systems and knowledge management". Innovative technologies such as wearable devices, bio- and environmental sensors and mobile, humanoid robots do already exist and ambient assistant living environments are being created for an ageing society. However, those technologies have to be adapted to older people's self-care processes and coping strategies, and to support new ways of healthcare delivery. Medical informatics can support this process by providing the necessary information infrastructure, contribute to standardisation, interoperability and security issues and provide modelling and simulation techniques for educational purposes. Research fields of increasing importance with regard to an ageing society are, moreover, the fields of knowledge management, ubiquitous computing and human-computer interaction.
Keywords: Medical Informatics, Ageing Society, Telecare, Smart Homes.

1. Introduction

In most developed countries, healthcare systems face a number of challenges leading to an increased demand for health and social care. These challenges are mainly related to;

1. The demographic development – a decreasing number of younger people will have to support increasing numbers of older, retired people. The old age dependency ratio in Europe is expected to double by the middle of this century[1] and the number of persons aged 80 and over (oldest-old) is expected to nearly triple, rising from 18 million in 2004 to 50 million in 2051. [1]
2. The economic development – higher standards of living will lead to greater expectations of the quality of healthcare systems. At the same time, public health services have to cope with financial resource constraints and shortage of skilled labour [2]. The cost for health and social care in Sweden is about 9.2% of GNP [3] and the corresponding figure for US is about 15% of GNP. Moreover, in Sweden, as an example, the cost of healthcare increases with

[1] This means that whereas in 2004 there was one elderly inactive person (>65 years-old) for every four persons of working age (15-64 years-old), in 2050 there would be about one inactive person for every two of working age.

about 0.8% per year due to demographic changes and another 0.8% due to medical technology development.
3. Societal factors – an increasing number of elderly people live alone, and increased mobility in society results in families/relatives distributed over large geographical areas.

In the future, effective delivery of healthcare will be more dependent on different technological solutions supporting the decentralisation of healthcare, changed life styles and increased societal demands. To prepare for this development, each European Member State is to develop a national or regional roadmap for e-Health, targeting the above mentioned challenges focusing mainly on the deployment of electronic health records, interoperability issues and reimbursement of e-health systems [4]. Besides 'enlargement of the European Union' and 'health policies and systems change', 'ageing' and 'technological development' have been identified as key drivers for change in the future healthcare delivery [5].

The goal of this article is therefore to identify trends and new technological developments that appear due to an ageing society and to relate them to research in the field of medical informatics.

2. The process of ageing

2.1. Definitions

European policies present the paradigm of *active ageing* as a coherent strategy to make ageing well possible in ageing societies. Active ageing is about adjusting people's life practices to the fact that they live longer and are more resourceful and in better health than ever before, and about seizing the opportunities offered by these improvements. In practice it means adopting healthy life styles, working longer, and remaining active after retirement. [6]

Successful ageing is a frequently discussed topic in the social, psychological and medical sciences. Biomedical models define successful ageing largely on the absence of disease and the maintenance of physical and mental functioning whereas socio-psychological models emphasise life satisfaction, social participation, functioning and personal growth [7]. Both models have been discussed mainly from the academic point of view of the respective investigators and few studies specify the meanings older people themselves attach to these models. Thus, a pilot study points out that older people perceive 'successful ageing' as coping strategies for later life [8].

2.2. Self-care models and coping strategies

Older people are not a homogenous group. In general, their needs and goals of life are not fundamentally different from those of other adults. They want to remain independent as long as possible and to keep control over their lives once outside help is needed, thereby maintaining the feeling of independence [9]. Individual needs and goals depend to a large extent on individual physical and psychological health conditions and are closely related to a person's overall attitude towards healthcare, illnesses and manner of living. This is also reflected in the way a person cares for herself and manages her life.

Backman and Hentinen describe a model with four modes of self-care characterised by different degrees of activity by home-dwelling elderly [10]. Depending on the mode of self-care, the elderly person

- continues living as an active agent (responsible self-care)
- accepts life as it comes (formally-guided self-care)
- denies getting old (independent self-care) or
- has a desire to 'give up' (abandoned self-care).

According to this study, self-care is not a separate part of elderly people's lives but is associated closely with their past life and with the future. Elderly persons' approaches to managing everyday life and the problems of ageing can be understood as various types of coping. Dunér and Nordström [11] describe active, adaptive and passive coping strategies. Those who manage actively primarily use problem-focused coping which is focused on solutions. Those who manage adaptively or passively use emotion-focused coping, in which feelings are more central than actions. Strategies depend on both internal and external resources and usually develop from active to passive towards the end of life.

2.3. Older people and information and communication technology (ICT)

A common prejudice by care professionals is that older people are unable or not willing to use new technology. 80% of Europe's home care decision makers e.g. believe that the acceptance of ICT-based services amongst older adults is very low [12]. According to SeniorWatch[2], 44 million older people (36%) have access to a computer at home and 33 million (26%) are regular computer users. However, computer involvement decreases with age: In the age range between 50 to 59 years 57% have home accesses to a computer, and 46% are regular users. At the same time, only 19% of those who are 80 years and older have home access to a PC and 6% are regular users [12]. The 2005 Eurostat ICT survey in all 25 member states revealed that 24.9% of the private individuals in age group 55-74 used the Internet over the past 12 months [13].

Moreover, older people feel unrecognised as a target group: 70% feel that in the media, ICT is only connected to younger people and 48% of the 50+ population blame manufacturers for not incorporating their needs in product characteristics. At the same time, more than half are keen on following technological developments, and half do not feel too old to familiarise with computer technology. [12]

Although we know relatively little about older people's use of technologies, research on cognitive ageing provides us with some knowledge about the challenges to confront during system development such as age-related declines in psychomotor skills, especially in dexterity and hand-eye coordination, and cognitive skills, especially processing capacity and automated response [14]. Despite these age-related declines, older adults are remarkably adaptive and can continue to perform at a high level [15].

---

[2] The SeniorWatch project consisted of an Older Population Survey (n=9661;age=50+) and a Decision Maker Survey (n=512)

3. Recent technological advances

3.1. Telecare and home-monitoring

Tele-monitoring is one way of responding to the new needs of home care in an ageing population. Being the first innovations in the field of telemedicine/telecare, vital sign parameter measurement and audio/video consultations ('virtual visits') are the available techniques most frequently described in the literature [16]. Most promising applications for tele-monitoring are chronic diseases such as cardiopulmonary disease, asthma, and heart failure or for assessing the level of activity of elderly people [16, 17].

Current research explores wearable devices for multiple parameter monitoring connected to wireless body area networks (WBAN) [18] or to be built in into smart clothes [19-21]. Developments in different scientific fields such as nanotechnology, cybernetics and artificial intelligence promise the emergence of new technologies in the future as e.g. automated consultations (person-to-machine or agent-to-machine) and self-diagnosing devices [2] moving towards chip implanted biosensors [22].

3.2. Smart homes and robotics

In smart homes ICT has already been installed for controlling a variety of functions such as opening/closing doors, switching on/off lights, smoke and gas detectors and temperature control; and for communicating with the outside world. Smart homes and telecare complement each other – smart homes prepare for the technical infrastructure that is to be filled with telecare services.

Such an infrastructure can address the prevalence of neurological and/or cognitive disorders in the elderly and enhance their ability to function independently within their residence [23]. Emergency help, prevention and detection of falls, and monitoring of physiological parameters are areas where older adults perceive that they would benefit from advanced technology. However, they also express concerns about the user-friendliness of the devices, lack of human response and the need for training tailored to older learners [24].

Current developments strive towards ambient intelligent environments where bio- and environmental sensors are combined with new methods for context-aware computing to allow for ageing in place (e.g. [25]). Research in the areas of biomedical and assistive robotics is directed towards the creation of mobile, humanoid robots to assist elderly with mobility declines [26-28].

One of the main challenges for smart homes is, however, not technological but social. It must be a sanctuary that is secure and private, and provide a harmonious space for relaxation and socialisation [29]. Moreover, the increasing introduction of medical devices into the homes of the elderly and their interaction with different kinds of ICT solutions requires close surveillance and analysis of the medical, legal and ethical responsibilities.

3.3. Health information systems and knowledge management

Increased citizen demands, decentralisation of healthcare and a shift from provider-driven to patient-centred healthcare delivery increases the demands on health information systems (HIS) and knowledge management.

Hitherto, health information systems have been to technology-centric and there is a lack of new theories and models for the design of more flexible and interoperable HIS able to support collaborative clinical work and knowledge and careflow management [30]. Elderly people, often suffering from multiple diseases, need proactive management from healthcare professionals following agreed protocols, shared care plan and personal life goals [31]. This requires an ICT infrastructure with improved support for coordination of work and cooperation, including decision support, between different healthcare professionals, especially with regard to the high number of substitutes working in the home care sector. Such a cooperative work approach should not only involve care professionals but also the patients themselves and their relatives which are today a fairly unused resource. With suitable tools and a reasonable approach, they could be engaged to a larger extent in the care process and thereby enhance its quality. Ubiquitous computing offers the possibility to provide the different users with relevant, context-dependent information and services unobtrusively [32].

However, an important prerequisite for the widespread introduction of new technologies in the healthcare sector is the acquisition of new types of skills by citizens, patients, doctors, nurses and other healthcare professionals. These skills translate to new abilities, competencies and, above all, mindsets and attitudes to new ways of working that are more responsive to the needs of citizens. [33]

4. Future scenario – Implications for medical informatics

Perhaps in 10 years most people aged over 65 will live in smart homes with the necessary assistive technology in place and with elderly-friendly Internet access guaranteeing social connectedness to the outside world. Elderly people will wear "smart" clothes and be surrounded by ambient measurement devices, surveying daily activities and monitoring their health status no matter whether they live with a chronic disease or just want to control or improve their individual fitness level. They will automatically be informed about any measures they could employ by themselves in case they deviate from their individual health profile. If necessary, the personal healthcare advisor will be alerted to take appropriate measures with their consent. The service fee will be withdrawn automatically from their healthcare bank account. Depending on the degree of severity of the present situation, the personal healthcare advisor can be a physical person, i.e. a relative, a nurse, a primary care physician or a highly qualified medical specialist, or a virtual individual with the appropriate medical skills suggesting appropriate measures to rectify the situation. The personal healthcare advisor will, of course, directly involve the older person's healthcare support team if necessary. Elderly people can rely on all-day health monitoring with individual problem handling based on approved medical and nursing practice.

This scenario – whether it becomes reality or not - reveals a number of challenges for the field of medical informatics:

- The establishment and analysis of individual health profiles requires a holistic approach to support older people's lifestyles and reliable methods for integration of information from a variety of technically and locally distributed databases including medico-technical equipment and registries on elderly care. Standards, interoperability and security still remain key challenges.

- Healthcare support teams have to be provided with the right information immediately wherever they are located, implicating that the fields of knowledge management and ubiquitous computing are crucial.
- Older people's interaction with the healthcare system requires personalised, user-friendly, context-aware interfaces that are both multi-modal and multi-lingual and adapt to the elderly person's environment and needs. The field of human-computer interaction becomes increasingly important.
- The delivery of healthcare will be re-organised, becoming more information intensive and complex. The trend towards lifestyle management, prevention, shared care and telehealth requires education of prospective users of the systems we develop. Modelling and simulation techniques can be used to support this.

The scenario also shows a strong intertwining between medical informatics and medical technology. How much and which kind of technology we want in our future homes - if we want it at all - depends on individual needs, self-care processes and coping strategies.

5. Conclusions

An ageing society requires increased accessibility of care outside traditional care settings with increased efficiency, individualisation and equity of quality-oriented healthcare with limited financial resources. This puts high demands on interoperability, usability, safety, security, availability and accessibility as well as on legal and ethical aspects. Innovative technologies do already exist and ambient assistant living environments for our ageing society are being created. However, we also have to adapt these technologies to the older peoples' self-care processes and coping strategies and to find improved ways to organise work for health professionals. Moreover, we should use ICT as an enabler for the establishment of interactive networks among the involved care professionals, patients and relatives.

ICT has the potential to bridge distances, to provide social inclusion without necessarily raising costs and to enhance quality of care, if carefully observed from a medical and ethical perspective. However, ICT should enable and not destroy independent living for the older population. Therefore, the individual person has to decide if, when and which kind of possible technology should be installed in the home.

References

[1] Lanzieri G. Eurostat Statistics in focus "Long-term population projections at national level". Newsletter 3/2006 http://www.europa.eu.int/comm/eurostat/ (last accessed May 23, 2006).

[2] Cabrera M, Burgelman J-C, Boden M, da Costa O, Rodriguez C. eHealth in 2010: Realising a Knowledge-based Approach to Healthcare in the EU. Challenges for the Ambient Care System – Report on eHealth related activities by IPTS. Technical Report EUR 21486 EN http://forera.jrc.es/documents/eur21486en.pdf (last accessed May 29, 2006).

[3] Economic portrait of the European Union, 2002. http://www.eustatistics.gov.uk/publications/publicationlist/Economic_portrait_of_the_European_Union_2002.asp (last accessed May 31, 2006)

[4] European Commission. COM(2004)356 "e-Health – making healthcare better for European citizens: An action plan for a European e-Health Area".

http://europa.eu.int/information_society/doc/qualif/health/COM_2004_0356_F_EN_ACTE.pdf (last accessed May 27, 2006).
[5] Monteagudo JL, Redondo JR. eHealth and the Elderly: a new range of products and services? IPTS Report 81- special issue on aspects of health 2004. http://www.jrc.es/home/report/english/articles/vol81/ICT6E816.htm (last accessed May 28, 2006).
[6] European Commission. Employment and Social Affairs. New Paradigm of Active Ageing. http://ec.europa.eu/employment_social/soc-prot/ageing/news/paradigm_en.htm (last accessed May 27, 2006).
[7] Bowling A, Dieppe P. What is successful ageing and who should define it? BMJ 2005; 331: 1548-51.
[8] Fisher BJ. Successful aging and life satisfaction: A pilot study for conceptual clarification. Journal of Aging Studies 1992; 6 (2): 191-202.
[9] Aronsson J. Elderly people's accounts of home care rationing: missing voices in long-term policy debates. Aging and Society 2002; 22: 399-418.
[10] Backman K, Hentinen M. Model for the self-care of home dwelling elderly. Journal of Advanced Nursing 1999; 30 (3): 564-72.
[11] Dunér A, Nordström M. Intentions and strategies among elderly people: Coping in everyday life. Journal of Aging Studies 2005; 19: 437-51.
[12] SeniorWatch Report "Older People and Information Society Technology". http://www.seniorwatch.de (last accessed May 27, 2006).
[13] Ottens M. Eurostat Statistics in focus "Use of the Internet among individuals and enterprises". Newsletter 12/2006 http://www.europa.eu.int/comm/eurostat/ (last accessed May 23, 2006).
[14] Kaufman DR, Patel VL, Hilliman C, Morin PC, Pevzner J, Weinstock RS, Goland R, Shea S, Starren J. Usability in the real world: assessing medical information technologies in patient's homes. Journal of Biomedical Informatics 2003; 36:45-60.
[15] Fisk Ad, Rogers WA. Health care of older adults: the promise of human factors research. In: Fisk Ad, Rogers WA (eds). Human factors interventions for the health care of older adults. Mahwah, NJ: Lawrence Erlbaum; 2002.
[16] Koch S. Home telehealth – Current state and future trends. Int J Med Inf 2005; digital version ahead of print.
[17] Meystre S. The Current State of Telemonitoring: A Comment on the Literature. Telemedicine and e-Health 2005; 11 (1): 63-9.
[18] Jovanov, 2005 Jovanov E, Milenkovic A, Otto C, de Groen PC. A wireless body area network of intelligent motion sensors for computer assisted physical rehabilitation. J Neuroengineering Rehabil. 2005 1; 2 (1): 6.
[19] http://www.media.mit.edu/wearables/ (last accessed May 31, 2006)
[20] http://artschool.newport.ac.uk/smartclothes/ (last accessed May 31, 2006)
[21] http://www.sensatex.com/ (last accessed May 31, 2006)
[22] http://www.kevinwarwick.org (last accessed May 31, 2006)
[23] Demiris G. Home based e-health applications. In: Demiris G (ed). E-Health: Current Status and Future Trends, IOS Press, Studies in Health Technology and Informatics, Vol 106 (2004), 15-24.
[24] Demiris, 2004 Demiris G, Rantz MJ, Aud MA, Marek KD, Tyrer HW, Skubic M, Hussam AA. Older adults' attitudes towards and perceptions of 'smart home' technologies: a pilot study. Med Inform 2004; 29 (2): 87-94.
[25] http://www-static.cc.gatech.edu/fce/ahri/index.html (last accessed May 31, 2006)
[26] www.humanoid.waseda.ac.jp (last accessed May 31, 2006)
[27] http://www.sfb588.uni-karlsruhe.de/ (last accessed May 31, 2006)
[28] http://leva.leeds.ac.uk (last accessed May 31, 2006)
[29] Punie Y. A social and technological view of Ambient Intelligence in Everyday Life: What bends the trend? EMTEL Report 2003. http://www.lse.ac.uk/collections/EMTEL/reports/punie_2003_emtel.pdf
[30] Stefanelli M. Knowledge Management in Health Care Organizations. In: Yearbook of Medical Informatics 2004. Towards Clinical Bioinformatics. IMIA and Schattauer GmbH, 2004; 144-55.
[31] Augusto JC, Black ND, McAllister HG, McCullagh PJ, Nugent CD. Pervasive Health Management: New Challenges for Health Informatics. Journal of Universal Computer Science, 2006; 12 (1): 1-5.
[32] Bott OJ, Ammenwerth E, Brigl B, Knaup P, Lang E, Pilgram R, Pfeifer B, Ruderich F, Wolff C, Haux R, Kulikowski C. The Challenge of ubiquitous Computing in Health Care: Technology, Concepts and Solutions. Methods Inf Med 2005; 44: 473-9
[33] Constantelou, A, Karounou, V. Skills and competences for the future of eHealth. ITPS Report, Special issue on aspects of eHealth, 2004. http://www.jrc.es/home/report/english/articles/vol81/ICT4E816.htm (last accessed May 30, 2005)

Ubiquity: Technologies for Better Health in Aging Societies
A. Hasman et al. (Eds.)
IOS Press, 2006

Progress and Challenges of Ubiquitous Informatics in Health Care

Francis H. ROGER FRANCE
Center for Medical Informatics, Health Services Research
School of Public Health and Cliniques Universitaires St Luc
Université Catholique de Louvain

Abstract. Ubiquitous informatics in health care can be seen as its pervasively presence everywhere, at home as well as in office and patient room, for various purposes such as patient follow up, health care professional training, aid to decision making and to public health management.
Its worldwide rapid extension could only happen in relation to major progresses such as overall availability of personal computers, diffusion through the worldwide web as well as coverage of almost all fields of medicine.
Challenges include a profound change in patient-physician relationship, a reform in health care management and financial methods, as well as the need to identify uniquely all healthcare partners, while respecting confidentiality and private life.
Keywords: Ubiquitous Informatics, Health Care, Electronic Patient Record, Distance Learning

1. The Concept of Ubiquity for Informatics

"Ubiquitous" [1] is defined in the Oxford English Dictionary as "everywhere pervasively present, as God".

Until the beginning of the 21st Century, no human being neither any object could pretend to be ubiquitous. Now, ubiquity is attributed to computer applications, in that sense that they can be obtained almost everywhere, at home as well as in office and patient room, like GPS in a car or internet at the end of a cell phone capturing also digital images.

Is such ubiquity a divine gift to humanity? Or, on the contrary, isn't it a way to emprison the information society and to reinforce the power of Big Brother ?

In furtherance to the diffusion of computerized information systems in the whole society, its extension in medicine is impressive. Informatics covers now almost all fields of health, from genomics to public health. [2][3][4][5]

After a description of progresses that led to ubiquity in health applications, we will describe some challenges and dangers to avoid.

2. What led informatics to Ubiquity ?

Computers exist only since the second world war. They were commercialized in the 1950's and applied to health care since the 1960's

The two major steps that led to ubiquity can be identified as, first, the creation of the Personal Computer (PC) or microcomputer in the 1980's in the USA and its rapid diffusion, and second, the world wide web (www) produced by the CERN[1]

Combined PCs and the www made a revolutionary change in almost all international health care information systems. They are inextrically related. Planning for the 21st Century requires a deep understanding of societal and behavioural changes in relation with this technologic environment.

3. Progresses in Health Care

3.1. Extension to all Fields of Medicine

The use of computers in medicine begun with clinical laboratories and hospital administration, soon followed by electrophysiologic signal analysis, drugs distribution and medical record summaries [6]. We still used Hollerith cards in the early 70's. We had to travel to the Computer Center. What a different world today!

3.1.1. From Medical Informatics to Health Informatics

Concepts behind words have their importance. "**Informatics**" is one of the rare French words that was accepted in the USA after the war. It is made of the conjunction of two terms:

- **Information**, ie the communication of instructive knowledge in a form that can be processed by a computer;
- and **automatic**, ie "acting by itself", or more exactly acting following sequential instructions loaded in a software and processed in a computer hardware.

As Professor François Gremy correctly observed, informatics is not a "science" but "the rational management of information by a computer". [7]

Medical informatics is still used in many countries, to describe its applications in medicine, but "**health informatics**" has been preferred by ISO, the international organization for standards, because it includes not only clinical medicine, but also biology, genetics, patient records as well as population statistics. [8]

As a matter of fact, "**health informatics"** covers now almost all fields of medicine (see figure 1), from the microscopic to the macroscopic, from bio-informatics to public health, from the cell to the society. **Telematics** and **Telemedicine** imply Informatics at distance and electronic communication for medical applications.

3.1.2. Bio-Informatics

At the **microscopic level** [9], the present knowledge of human genome could not be obtained without computers. Biochemistry as well as immunology are major fields of bio-informatics.

3.1.3. Digital Imaging [10]

Imaging benefited of spectacular innovations that resulted from interprofessional rather than interdisciplinary cross-fertilization.

---

[1] Centre Européen de la Recherche Nucléaire

Computerized Tomography or **CT Scan** or **Tomodensitometry** (using conventional Xrays) was imagined by Sir G. Hounsfield, a British electrical engineer during a discussion with physicians around a cup of tea.

Echosonography, using ultrasounds, reflecting tissues with various intensity, was introduced in the Royal Navy during the last world war, in order to detect submarines. It is also the fruit of a collaborative work between engineers and physicians, while the **RMN** (Nuclear Magnetic Resonance) was developed by the use of a technique applied by chemists to identify the composition of chemical agents.

The **PET-Scan** (Positron emission tomography) is another tool that describes organ function, like the degree of a tumoral activity in complement to anatomic description obtained by a CT Scan.

These technologies and many others in digital radiology have in common the obtention of an image of tissues that could not be seen by former radiologic or isotopic techniques, with less risk for the patient, but also a higher cost for society.

Digital imaging authorizes surgical procedures associated to visual exploration, in the so-called "**interventional radiology**". RMN is associated to stereotaxic neurosurgery, while robotics improves the surgical act, by programming quick and more refined sutures than those performed by a human being.

Most of these computerized techniques are industrialized. It is to say that the physician becomes a specialized instrumentalist of a technique marketed and diffused worldwide.

3.1.4. The EPR

The **Electronic Patient Record** is another challenge for contemporary medicine.

A medical record has been defined as "the memory of all data of a patient, both individual and collective, constantly updated". [11]

The "paper record" has become more and more difficult to manage because of the multiplicity of specialists and the large volume of results to be classified. In my institution, Cliniques Universitaires St Luc in Brussels, a 1000 beds teaching hospital of the University of Louvain Medical School, with 400,000 outpatient visits a year, we felt that we had made a first major step when we unified the folders in 1976 by making mandatory a unique identifier by patient, but we could not reach a total unique record, because of the resistance of specialists.

The unit medical record could only be obtained in 2000 when it was completely automated and became "paperless" or at least with "less paper". Such evolution could not be done easily, because laboratory computers were not fully compatible with several other systems, such as the administrative data, drug prescriptions or digital imaging.

Our EPR, called "Medical Explorer", is a nice "document retriever". However, a table of content, like a "problem list", as proposed by Larry Weed, is still missing.

3.1.5. Hospital Management

Management has been another privileged area for the development of health informatics.

It includes HIS (**Hospital Information Systems**) [4][5] that moved from **centralized** architecture in the seventies (a mammouth model that disappeared because too homogeneous) to **decentralized models** in the 1980's (a Balkanization that could not survive to lack of compatibility: it was to heterogeneous), and finally to the present

distributed architecture, since the 1990's, where each computer system can be inserted in the network as a logo piece, using the concept of client-server information system, where PCs are spread in all units, as well as wireless instruments, with access from outsiders (general practitioners, other institutions).

3.1.6. Public Health

Public health applications of informatics received attention since the beginning. They include **epidemiologic surveys, managing models, financing reform, preventive measures** and **Health Systems Research**.

Cost control has been a major objective of all governments since the seventies.

Technologic progress and new drugs cost more and more. Even if governments agree to attribute a bigger portion of their budget to health care (± 10% in Belgium in 2006) and even if patients agree to pay from their pocket a larger fraction of the bill (above 5%), it appears necessary to rationalize health care management in order to obtain the best **performance** and **quality of care.** [12]

Until 1990 hospital accounting was based on **technological productivity** (investments in personnel, buildings and techniques that could result in a number of procedures, of admissions, bed occupancy, length of stay, …).

Bob Fetter proposed to measure **case mix** [13], ie diagnoses and procedures by patient stay in order to explain resource consumption (measure of **medical productivity)**. **Hospital productivity** could be measured.

In order to reform hospital financing systems and to reduce length of stay, a **Minimum Basic Data Set** (MBDS) [14] containing all diagnoses and all procedures linked to financial data by patient stay has been proposed and accepted by the EEC. The European MBDS published in 1982 was the starting point of the hospital reform in Belgium, France, Italy, Portugal and elsewhere.

The MBDS was not designed, however, only for hospital financing, but also for quality of care development, clinical research and hospital epidemiology. Linkage to the content of the EPR could allow such developments, if terminology standards can be obtained in Europe.

3.1.7. Societal values and ubiquity applications

Two comments have to be made about Figure 1

First, if a line is going from the lower part on the left to the upper part on the right, indicating the size in scale that was just described, there is a second line going from the top on the left, to the bottom on the right, indicating that the perception of scientific value by our society [15] is mainly attributed to the results at the microscopic scale and much less to societal developments. "Impact factors" of publications and Nobel prizes are good examples of this approach.

Second, ubiquity of informatics in health care results from the multiple progresses described in so many fields. A radiologist can now consult a RMN and the electronic patient record at home in order to give his advice to a clinician in an emergency room; an orthopedist using telemedicine could instruct a general practitioner on what to do in case of fracture on someone living far away from a hospital. Even surgery could be practiced at distance, while everyone has access at home to medical literature, and students could be trained in statistics everywhere…

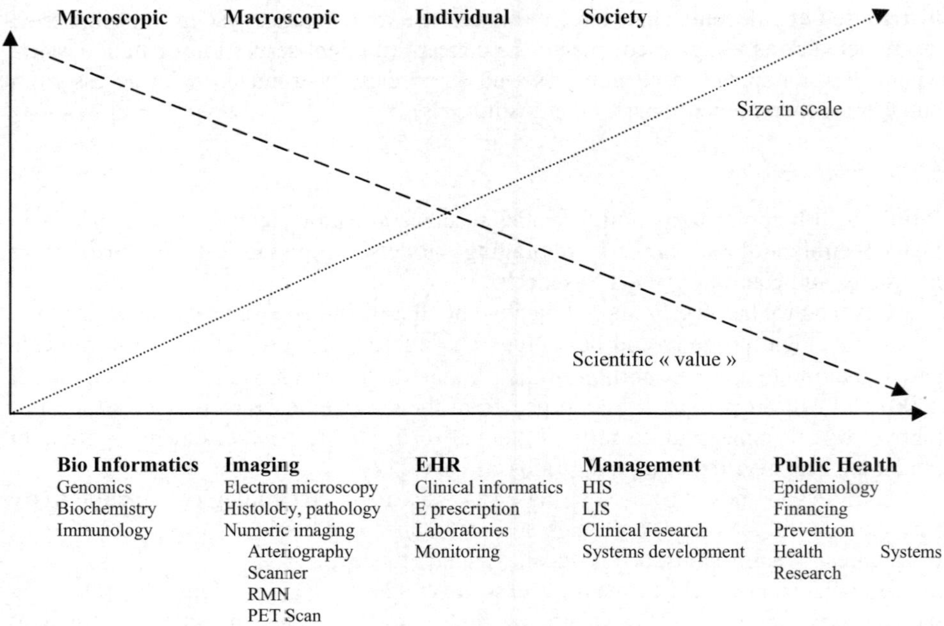

Figure 1

3.2. Geographical extension

Computers are now diffused worldwide.

Industrial countries (USA, Europe, Japan, …) have build complex fiberoptic networks and use satellites for data communication. Emerging market countries (such as "BRIC" Brazil, Russia, India, China) are following rapidly.

Africa and some developing countries in Asia and South America seem to lag behind. However, all populations need water and information, water storage and parabolic aerials, even in far away located villages.

4. Challenges of ubiquitous informatics in healthcare

The pervasively presence of health informatics everywhere has dramatically progressed but this "technological progress" introduces profound changes.

Fundamental ethical questions can be raised about patient autonomy, as well as about respect of confidentiality and private life. Laws on patient rights and on health informatics have to be adapted. Financial reforms have to take into account equity in health care and the development of quality.

4.1. A major change in patient-physician relationship

Ubiquity modifies the practice of medicine. In case of telemedicine, there is no full control locally (example: biopsy) neither at distance (example: diagnosis of pathology). Risks might be associated, to which liability issues have to be identified.

In the physician office, a screen replaces the former paper record. Hospital care can be transferred to home care, especially for chronic insufficiencies (cardiac, renal, hepatic) in old age.

Alarm systems might provide security signals to isolated persons, and infrared devices placed on mental ill patients could help health professionals to follow their movements.

Patient willingness to donate organs, or to forbid access to their record after death can be stored on a national platform, as well as their life testament, their wish not to prolong life by medical means, or their consent to euthanasia.

Patients and physicians have equal access to the literature through the web. In rare diseases, the patient might become more knowledgeable than his doctor.

The **patient empowerment** is in progress everywhere. His autonomy is increasing as well as his liability in partnership with health professionals for his care.

Most patients adapt to change, but they can be perturbated by this intrusion of machines in human relations. A few years ago, a questionnaire has been distributed in the cardiac intensive care unit of Prof. J. Col in St Luc Hospital in Brussels in order to identify patient perception of informatics for their care. Many of them were too sick to understand what the electronic devices were aimed at, but an engineer, well aware of the role of the computer, told to his physician when he entered in his room and watched vital signs on the screen: "Dear Doctor, I would rather prefer that you auscult me as before, than to watch data on the screen".

4.2. A major change in health care management

A systematic documentation on all diagnoses and procedures by patient modifies the societal approach to health care. There is more transparency of practices, better surveillance of results of care and of safety issues. Comparisons of practices allow governments to introduce incentives in order to reduce length of stay, to avoid overuse, underuse and misuse of techniques, diagnostic tests or treatments and to reduce costs of care. National platform might help to diffuse such information everywhere.

Case mix measures by APRDRGs as done in Belgium allow to finance hospital production by type of diagnoses rather than by linear administrative measures.

Several new areas of research are now being tested like **"Evidence Based Medicine"** [16] that require well established criteria to diagnose and to treat frequent diseases groups of patients.

Clinical pathways is another way to optimize the sequence of care by pathology. [17]

Equity measures and **quality of care** could also benefit from ubiquitous informatics when applied appropriately.

4.3. Conditions for changes

In order to obtain ubiquitous informatics in health care, aiming at optimal health care, patient centered, at home as well as in institutions, it appears necessary to respect fundamental ethical principles, such as:

- patient right to autonomy
- independance of clinical judgment
- protection of confidentiality and private life

- economic interest should never dominate over ethical considerations

This implies that each patient should be identified by a set of unique identifiers, to be crypted by a Trusted Third Party (TTP), where the objectives of information processing correspond to each partners role. The delivery of care should be kept separate from aggregated anonymyzed data for research, statistics and other studies, as well as from administrative data to be processed for individual payment.

Technical solutions exist, like the proposed BeHealth [18] platform in Belgium, a pilot project that intends to give a secure access to health data through a common Federal network, where a TTP distributes the personal identification numbers keys (public and private) after irreversible encryption from the Federal register numbering system by citizen, as well as verifies authentication of sources and authorizations of uses.

In conclusion, ubiquitous informatics becomes a reality in health care.

However, are our citizens, health professionals and financial bodies ready to take the necessary steps in order to meet the ethical and organizational challenges that ubiquity generates?

This meeting between medicine and informatics provokes a constant call into question.

The societal choice to accept or not unique health identifiers, that respect patient data confidentiality appears to be one of the key issues in the beginning of this century. Should we fight for a separate health identifier by patient or will we accept to leave the citizen with a unique national number for all kinds of purposes: taxation, retirement, health, justice, and others?

Another fundamental issue concerns the role of the medical profession in the progressive invasion of industrial business that gives a new face to health care. Will they be well enough public health trained in order to obtain a natural leadership in patient care or will they be infrodated to economic rules?

Finally, technologies can be the best or the worst. Will they be applied ubiquitously to help the patient to suffer less and to be healed by facilitating access to appropriate care, or will humanity be more and more prisoner of big brothers that might organize technology for their profit?

We live today a mutation of the medical practice through ubiquitous informatics. The human character of the patient-physician relation has to be preserved.

References

[1] C.T. Onions, W. Little, H.P. Fowler, J. Coulson, The shorter Oxford English Dictionary, Oxford University Press, 1972, 2516 p.

[2] E.H. Shortliffe, L.E. Peneault, G. Wiederhold, L.M. Fagan, Medical informatics, computer applications in health care and biomedicine, Springer, 2000, 856 p

[3] J. van Bemmel, Handbook of medical informatics, Bohn Staflen Van Loghum, 1996, 476 p.

[4] P. Degoulet, M. Fieschi, Introduction to clinical informatics, Springer, 1999, 245 p.

[5] R. Van de Velde, P. Degoulet, Clinical information systems, a component based approach, Springer, 2003, 294 p.

[6] F.H. Roger, Médecine et informatique, Maloine, Paris, 1979, 215 p.

[7] F. Grémy, Informatique médicale, Flammarion, Paris, 1987, 49-70.

[8] M.S. Blois, Information and medicine: the nature of medical descriptions, Univ of California Press, Berkeley

[9] R.B. Altman, Bioinformatics, in: E.H. Shortliffe et al, Medical informatics, Springer, 2000, 638-660.

[10] R.A. Greenes and J.F. Brinkley, Imaging systems, in: E.H. Shortliffe et al, Medical informatics, Springr, 2000, 485-538.
[11] F.H. Roger, Le dossier médical automatisé, in: Médecine et informatique, Maloine, Paris, 1979, 161-190.
[12] F.H. Roger, Le résumé de dossier médical, indicateur informatisé de performance et de qualité des soins, UCL, Thèse d'Agrégation, 1982.
[13] R.B. Fetter, Y. Shin, J.L. Freeman, R.F. Averill, J.D. Thompson, Casemix definition by Diagnosis Related Groups, *Med Care* **18** (2-Suppl) (1980), 1-53.
[14] F.H. Roger, The Minimum Basic Data Set for hospital statistics in the EEC, CEC Luxembourg, ref EUR 7162, 1981, 151 p.
[15] G. Fourez, Santé et médecine scientifique, *Revue des questions scientifiques* **188** (1978), 7-11.
[16] D.L. Sackett, W.S.Richardson, W.M. Rosenberg, R.B. Haynes, Evidence Based Medicine: how to practice and teach EBM, Churchill Livingstone, New York, 2000.
[17] M. Hill, The development of care management systems to achieve clinical integration, *Adv Pract Nursing Q* **4** (1998), 33-9.
[18] F.H. Roger France, E. De Clercq, M. Bangels, Purposes of health identification cards in Belgium, in: Connecting medical informatics and bio-informatics, R. Engelbrecht et al, Eds, IOS Press, 2005, 415-420.

1. Modern Trends in Medical Informatics

1.1 Ubiquity: Design and Applications

Ubiquity: Technologies for Better Health in Aging Societies
A. Hasman et al. (Eds.)
IOS Press, 2006

Facilitating interdisciplinary design specification of "smart" homes for aging in place

George DEMIRIS [a,1], Marjorie SKUBIC [a], Marilyn J. RANTZ [a], Karen L. COURTNEY [a], Myra A. AUD [a], Harry W. TYRER [a], Zhihai HE [a], and Jia LEE [a]
[a] *University of Missouri - Columbia, USA*

Abstract. "Smart homes" are defined as residences equipped with sensors and other advanced technology applications that enhance residents' independence and can be used for aging in place. The objective of this study is to determine design specifications for smart residences as defined by professional groups involved both in care delivery to senior citizens and development of devices and technologies to support aging. We assessed the importance of specific devices and sensors and their advantages and disadvantages as perceived by the interdisciplinary expert team. This work lays the ground for the implementation of smart home residencies and confirms that only an interdisciplinary design approach can address all the technical, clinical and human factors related challenges associated with home-based technologies that support aging. Our findings indicate that the use of adaptive technology that can be installed in the home environment has the potential to not only support but also empower individual senior users.
Keywords: Medical Informatics, Smart Homes, Assistive Technology, Aging, Residential Facilities.

1. Introduction

As the segment of the population over the age of 65 years keeps growing and average life expectancy increases, new models of positive ageing are being developed to allow senior citizens to adapt to degenerative changes and maintain functionality, autonomy and quality of life. Such models aim to address ways with which seniors can cope with health-related issues such as falls, sensory impairment, immobility, and medication management. The development of "smart homes" aims to meet older adults' desire to remain independent at home by proactively addressing older adults' needs. The term "smart home" refers to a residence equipped with a set of advanced electronics and automated devices designed to enhance health care delivery and remote physiological monitoring of residents, to enable early identification of possible problems or emergency situations and to maximize residents' safety and overall well-being. Smart home features usually include motion-sensing devices for automatic lighting control,

[1] Corresponding Author: George Demiris, 324 Clark Hall, University of Missouri – Columbia, Columbia, MO 65211 USA. E-mail: gdemiris@gmail.com

motorized locks, door and window openers, mobilized blinds and curtains [1], smoke and gas detectors and temperature control devices.

The study presented here is part of an initiative placed within the framework of Aging in Place, a model of long-term care for older adults [2]. In this model, seniors direct the timing and intensity of health and personal care services delivered to them in their residences, and thus, have the opportunity to 'age in place.' The specific initiative includes Tiger Place, a 34,000 square foot facility in Columbia, Missouri [3]. In preliminary work [4] assessing older adults' perceptions of the technology, seniors identified potential application areas for advanced technologies such as emergency help, prevention and detection of falls, and monitoring of physiological parameters. Overall, older adults had an overall positive attitude towards devices and sensors that can be installed in their homes in order to enhance their lives.

The objective of this study is to determine design specifications for smart residences as defined by an interdisciplinary team of experts, namely, professional groups involved both in care delivery to senior citizens and development of devices and technologies to support aging. This work lays the ground for the implementation of smart home residences and is based on the belief that only an interdisciplinary design approach can address all the technical, clinical and human factors related challenges associated with home-based technologies that support aging.

2. Material and Methods

2.1. Design

This study utilized a focus group approach which included both clinical and non-clinical domain experts. The focus group protocol was based on structured and open-ended questions defined by published literature and previous work. Both qualitative and quantitative data analysis methods were employed. Specifically, a content analysis of the sessions was performed and ratings were analyzed using SPSS.

2.2. Sample

We conducted focus groups with a convenience sample of experts in delivery of care to older adults and/or design of smart home applications. In order to include a diverse group of professionals we recruited experts from both the clinical and non-clinical sector. Experts in the clinical domain included nursing researchers with clinical gerontological nursing experience, social workers and health psychologists. Experts in the development of "smart home" devices and technologies included engineers and computer scientists. The sample was selected from an academic setting and all subjects had research experience and an extensive publication record in their domain for more than 10 years. Participants were invited to discuss the design specifications of a smart residence for the typical TigerPlace resident who is a senior in his/her eighties, retired and still independent in most basic and instrumental activities of daily living.

2.3. Data Collection

The focus group protocol included a structured questionnaire and open-ended questions. The participants were asked to rate the importance of specific devices and features for a smart residence in the context of TigerPlace and its typical residents. The devices and features included actuators for environmental control (i.e., mechanical devices such as window or door openers that are simple to use and allow users to control specific environmental attributes), heating ventilation and air conditioning, infra red sensors, iris recognition, personal data assistants, pressure pads (that send a signal to a control unit and instigate an action if triggered by the weight of a person's step), smart cards (with an embedded microprocessor for data storage) and emergency communication systems (that can be used to send an emergency alarm or allow two-way communication between resident and care provider). These items included in the protocol were identified during focus groups with senior adults in a previous study [4] and also defined by Dewsbury et al [5] as common "smart" home features. Participants were asked to provide a rating on a 5-point Likert scale for each of the items and discuss explanation for their ratings. The engineering experts were then asked to discuss the advantages and disadvantages of different types of smart home network protocols, namely bluetooth, busline based technology, X10.

2.4. Data Analysis

Ratings for the structured questions were entered into an SPSS program for analysis. Responses to open-ended questions and comments were reviewed and coded by two of the authors. The participants of every group were asked to identify the desired and undesired features of smart home technology resulting from consensus among all members of every focus group session. During this exercise focus group participants had to work as a group and identify ways to exceed their professional silos by interacting and debating with experts in other fields. These items were identified by the two coders independently and findings were compared to identify inter-rater reliability. Finally, the preferences of the experts were entered into the CUSTODIAN software system. This system was funded by the European Union and is managed by the Robert Gordon University in the UK [5]. This software suite is publicly available and provides a visualization tool that enables users to test scenarios and set-up configurations. Based on the ratings and feedback of the experts, a model apartment was created to visualize the concepts discussed during the focus group sessions and provide a blue print for the system design.

3. Results

A total of twelve subjects participated in four sessions. The sessions lasted in average 64 minutes (SD 7.3 min). Participants included three nurses, four computer engineers, two social workers and three health psychologists. Table 1 summarizes the ratings of the participants.

All participants found pressure pads to be very important. One participant stated that a pressure pad can provide means of monitoring the residents in a non-obtrusive manner and without violating their privacy. Another participant stated that it is important to have this feature on many locations throughout the residence to allow for a continuum of monitoring.

Smart cards were perceived as useful by most participants. One participant stated that a smart card can be very useful for residents who have dementia and find themselves in an unfamiliar setting assuming that the infrastructure is in place to allow for the smart card use. Two participants saw benefits in smart cards that entail parts of the medical record and would allow for residents to use these when they interact with health care providers. All participants rated emergency communication systems as very important. Infra-red sensors were also perceived as useful by participants. Two of them stated that such sensors can be utilized to detect presence of people or pets in the apartment or detect movement. Heating and air conditioning were perceived by all participants as very important; however, most commented that climate controls and thermostats are already in place.

Table 1: Participant ratings of devices & systems

Perception of importance of smart home devices and features (N=12) (1: Not Important At All; 5: Very Important)		
	Average	SD
Pressure Pads	5	0
Smart Cards	5	0
Emergency Communication system	4.91	0.3
Infra-Red sensors	4.55	0.69
Heating Ventilation and Air Conditioning	4.45	0.93
Actuators for environmental control	3.36	0.50
Personal Data Assistants (PDAs)	3.22	0.83
Iris Recognition	1.55	0.52

Actuators for environmental control were perceived as very helpful. Three participants emphasized however that a possible dysfunction of such devices can be very problematic and frustrating to users who have learned to rely on them. Two participants stated that door openers in this context appear more important than window openers. One participant emphasized that the operation of doors and windows needs to be studied carefully when considering a potential automation so as to prevent unexpected injuries. One participant stated that voice control over window or door openers would be ideal; however, voice recognition has not reached an optimal state. One participant stated that a Personal Data Assistant (PDA) could function as a reminder tool or to enable residents to control their medication. Three participants commented that the current interfaces of PDAs are not user-friendly for people with visual impairments and thus, PDAs could be difficult to operate for senior citizens. Specifically, one participant stated that the keys are too small and the finger coordination can be problematic for elderly users with functional limitations. Another participant acknowledged these problems but stated that both the interface and the operation of a PDA can be easily modified to address the user needs of senior citizens. As can be seen in Table 1, an iris recognition feature was not perceived as very important for the typical resident. One participant saw a potential implementation for the purposes of medication management if there is a need to ensure the identity of the individual handling medication. Most participants, however, saw no obvious benefit to utilizing this technology.

There was no significant difference in the scoring between professional groups, and the concordance was high (ICC Coefficient 0.85, F=1.7). Two items had less concordance between the professional groups, namely iris recognition and Personal Data Assistants, where engineers rated these in average slightly higher than nurses, social workers and health psychologists (difference in average scores of 1.2 and 0.7 respectively, $p<0.05$).

One of the respondents commented that these technologies should be integrated to support the functional independence of older adults. The sensors should trigger coordinated responses so that the sensor that recognizes the situation of a stove left on too long alerts the resident and/or the care provider. Another participant saw great benefit in a technology that can provide clinicians with information about daily and long term trends. Two participants stated that they can see increased efficiency of the use of smart home technologies if they involve family and friends. Table 2 summarizes the most important features of smart home technology as defined by the experts.

Table 2: Desired and Undesired Features of Smart Home Technology as Defined by Experts (n=12)

Desired Features:	Undesired Features:
Non-intrusive	Introduce new risks or hazards
User friendly	Place burden on the residents
Usable, accessible	Limit activities of the residents
Accurate	Increase anxiety
Reliable	"stigmatize" residents as being frail or in need of special assistance
Easy to maintain	

All engineering experts defined the network protocols as complementary rather than competing because each can play a great role in the infrastructure of a "smart" residence. Two respondents stated that the advantage of X10 is that it can control household appliances easily and in a cost-efficient way. They also emphasized that bluetooth is a wireless low power solution with limited range but with advantages in interconnecting wirelessly objects that are close together. Three respondents rated bus technology as easy and convenient and useful in minimizing electromagnetic noise. They perceived this technology as reliable. All experts stated that wire connection makes sense when there is close proximity of objects and that safety conditions need to be taken into consideration to minimize potential safety hazards (such as a resident tripping over a cord).The preferences and ratings of the experts were taken into consideration when a prototype "smart" residence was designed with the CUSTODIAN software platform. All items that received a rating higher than 3 on the five-point ordinal scale as well as other suggestions were incorporated into the design. General actuators for environmental control such as window openers, control units for doors and light were integrated in all rooms. The prototype was then reviewed and approved by the experts as a blue print that reflects their statements and views.

4. Discussion

An interdisciplinary approach is essential to design a home that is flexible and responsive to the needs and limitations of the residents. The value of interdisciplinary teams is not a new concept in gerontology. Such teams overcome the problems of the traditional health care organizational model, which reinforces functional specialties and silos of expertise. The interdisciplinary team approach promotes integrated and coordinated care for older adults in which all participants in the care-delivery process are focused on the older adult rather than their specialty. During both the design and development phases of a smart home, experts from different disciplines need to be included. Furthermore, end-users should participate in the early discussions and be able to provide feedback during the design specification phase as there may be a "disconnect" between expert beliefs and end user perceptions. The study presented here results from a previous study assessing end users perceptions of "smart" home technologies [4] and describes an effort to document experts' feedback and integrate it into the design of the smart home application. As Rogers [6] points out, we need to shift from a model of "technological determinism", namely that technology itself should be the impetus for change, to a model of the social construction of technology where technology is influenced by societal norms and needs.

5. Conclusion

The success of smart homes will depend on the level of compliance with universal design principles that are holistic and inclusive [7]. Many of the challenges that older adults face, whether functional or cognitive limitations, have been traditionally addressed by the utilization of mechanical adaptive devices, which allow the user to adequately function in their environment, but not necessarily actively participate in it. The use of adaptive and assistive technology that can be installed in the home environment has the potential to not only support but also enable and empower individual users. This study contributes to the emerging field of smart home technologies as it provides insight into the typical features of smart homes and their advantages and disadvantages as perceived by professional experts.

Acknowledgements

This work was supported in part by the National Library of Medicine Biomedical and Health Informatics Research Training Grant T15-LM07089-14 and the National Science Foundation NSF-ITR 0428420.

References

[1] Elger G, Furugren B. SmartBo-an ICT and computer-based demonstration home for disabled people. 3rd TIDE Congress: Technology for Inclusive Design and Equality Improving the Quality of Life for the European Citizen; 1998; Helsinki, Finland; 1998.

[2] Marek KD, Rantz MJ. Aging in Place: A new model for long-term care. Nurs Adm Q. 2000;24(3):1-11.

[3] Rantz MJ. Aging in place. Nurseweek, Midwest/Heartland Edition. 2003 7(4):2.

[4] Demiris G, Rantz MJ, Aud MA, Marek KD, Tyrer HW, Skubic M, et al. Older adults' attitudes towards and perceptions of "smart home" technologies: A pilot study. Med Inform Internet Med. 2004;29(2):87-94.

[5] Dewsbury G, Taylor B, Edge M. The process of designing appropriate smart homes: Including the user in the design. Equator IRC Workshop on Ubiquitous Computing in Domestic Environments; 2001; University of Nottingham, UK; 2001. p. 131-46.

[6] Rogers E. Diffusion of Innovations. 5th ed. New York: The Free Press; 2003.

[7] Edge HM, Milner J. Universal Design: A social agenda within the ecological designed built environment. Shifting Balance: Changing Roles in Policy, Research and Design, IPAS Conference; 1998; Netherlands; 1998.

Ubiquity: Technologies for Better Health in Aging Societies
A. Hasman et al. (Eds.)
IOS Press, 2006

Rehabilitation after Stroke using Virtual Reality, Haptics (force feedback) and Telemedicine

Jurgen BROEREN[a,1], Mark DIXON[b], Katharina STIBRANT SUNNERHAGEN[c], Martin RYDMARK[a]

[a] *Mednet, Institute of Biomedicine, Göteborg University, Göteborg, Sweden*
[b] *SenseGraphics AB, Stockholm, Sweden*
[c] *Institute of Neuroscience and Physiology, Göteborg University, Göteborg, Sweden*

Abstract. We have constructed a haptic immersive workbench to be placed in the patients' home for daily adjusted rehabilitation. We also propose a system for Internet based connection and communication between patients and between patients and a clinical rehabilitation center and clinical assessment/evaluation centers. The benefits of a system for rehabilitation after stroke, based on VR, Haptics and Telemedicine should be: increased quality of life, lesser isolation, feeling more secure, fewer tiring transportations, more frequent exercising, better compliance to training, lower cost for transportation. The long term recovery for a larger group of patients with motor impairments is presently under evaluation.
Key words: haptics, home care, rehabilitation, stroke, virtual reality

1. Introduction

We employ contemporary ICT (Information and Communication Technology), Virtual Reality, Haptics and Telemedicine, in our research to develop: 1) A precise quantitative assessment tool and a training device for neurological impairments, especially for stroke patients. The tool will be a low cost set-up that can be distributed to the patients' housing on a lending basis or be placed at local "activity centres". 2) Assessment and training programmes/routines for this tool. 3) Telemedical routines for daily "I see you, you see me" communication between the patient and the rehabilitation clinic, for retrieving assessment data from the patient and for tuning the patient's training exercises. Presently, we have a functional laboratory set-up for parts 1 & 2, above, and have started clinical trials with patients for part 3.

Stroke is one of our most widespread diseases and the principal cause of permanent physical impairment in the adult population (1). The number of persons that will suffer from stroke is anticipated, by 2020, to have shifted stroke from the 6th leading cause of lost disability adjusted life years to the 4th in the world (2). With rehabilitation most patients improve considerably and can increase their quality of life (1). During the post acute phase patients are living in their homes or in other non-

[1] Corresponding Author: Jurgen Broeren, Mednet - Medical Informatics & Computer Assisted Education, Institute of Biomedicine The Sahlgrenska Academy at Göteborg University, Box 420 SE 405 30 GÖTEBORG (SWEDEN).E-mail: jurgen.broeren@mednet.gu.se

hospital housing. There are well-functioning assessment and treatment procedures for rehabilitation after stroke, but they have some disadvantages. One is the fact that they seldom record movements precisely. Most of these procedures are designed for use in clinical settings, and are therefore not commonly used in the patients' homes or extra hospital situations, like local activity centres. These conditions hamper the regularity in training or force the patients to frequent travels to and from clinics. These travels are tiring for the patients and costly for the society.

A new treatment strategy that so far has not been tried to any greater extent in rehabilitation is Virtual Reality (3-7). The method implies that in a computer environment it is feasible to create a virtual three-dimensional world. Besides the visual component it is also possible to integrate "virtual" touch sensation or force feedback to the hands. This latter technique is named Haptics. Here one can actually feel the objects that are handled, this in addition to the visual perception. The user gets an impression of the three-dimensional interface as being natural. This experience is created by the co localisation of senses in the virtual set-up. The user works with tools in the hands in a realistic environment, exactly like he/she had the hands inside the computer screen. In our laboratory set-up the virtual environment consists of a haptic equipment that looks like a stylus shaped instrument attached to a lever system freely movable in all directions (figure 1). Thus, the user can see, feel and handle virtual objects like they were real objects. The user gets the feeling of being integrated in a simulated environment (7). Continuously the haptic device record all positions, movements and forces, which are stored in the computer for further processing, analysis and assessment. Built into the haptic system are motors and vibrators that can produce the sense of gravity, friction and elasticity.

Figure 1. VR and Haptics workstations.

The aim of the project is that this rehabilitative service could promote the learning of arm motor abilities at distance from the health facilities. In this way the healthcare professionals are able to continue the treatment and the initiated rehabilitation program. All components of the proposed project exist in prototypes or are developed. Presently, we cooperate with Sensegraphics (sensegraphics.se) in order to develop our hard- and software and to evaluate the possibilities to implement this system in health care organisations and in patients' housing.

2. Material and methods

2.1 Experimental System

A scenario for the use of such a telemedicine system is the following. The clinic supplies the patient with a portable telemedicine system (Figure 1) at the time of the discharge from the hospital. Back at their home the patient will be able to exercise freely at their own suitable time and in a familiar environment. At specific exercise hours the therapist can monitor and coach the patient from a distance. In this way the healthcare professionals will have an opportunity for feedback about patients' rehabilitation progress and have the opportunity to make necessary adjustments to the rehabilitation programme. The proposed system consists of three major components outlined in Figure 2; a home rehabilitation system, a larger hospital rehabilitation system and a clinical evaluation unit. The home and hospital systems use Sensegraphics virtual reality solutions that use the technology called haptics to allow users to see and feel objects that exist only in the computers memory. The rehabilitation systems will feature a library of engaging activities and "games" that are simultaneously entertaining for the patient and beneficial for rehabilitation; i.e. the games will train certain movements so that the patient can perform their daily training exercises in a fun and stimulating environment.

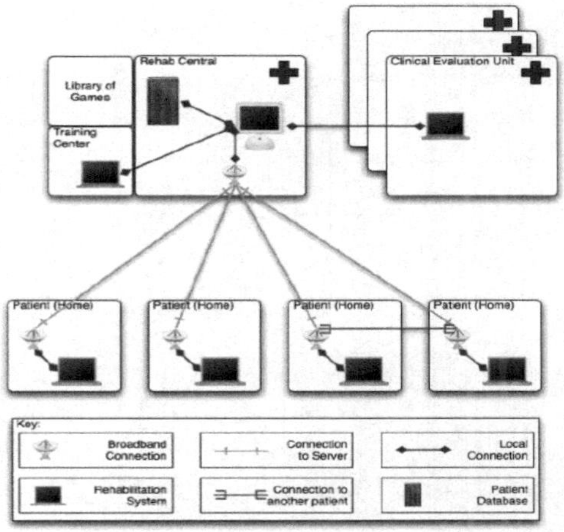

Figure 2. A home rehabilitation system, a larger hospital rehabilitation system and a database administration system.

2.2. System components and connections (Figure 3)

2.2.1 Patient Database System

The patient database system is a central database server that maintains an archive of all patient information. This database is accessible to the rehabilitation staff in order to allow them to plan a rehabilitation program for patients. The data can also be useful for research studies into various aspects of the rehabilitation process.

Data stored in the database would include;

- Results of each game, number of times run, performance for each run
- Raw hand movement data, captured at 1000Hz, for every run
- Game events time stamped to match the raw hand movement data

2.2.2 Rehabilitation Management System

The management system is a computer that acts as a front-end to the game library, patient database system and training systems. From the management system, staff can observe and graph patient's progress, prescribe games to be played by each patient and communicate (using audio and video links) with patients.

2.2.3 Patient Assessment

A Clinical Evaluation Unit will be implemented allowing rehabilitation staff to measure and monitor the patient's performance during assessment runs. Assessment runs will be performed using the standard training games run on the standard hospital based training system. All games will, by default, generate time-stamped motion data (x, y, z, yaw, pitch, roll and button press information) at 1000Hz. This data is stored together with time / date and patient information for subsequent analysis.

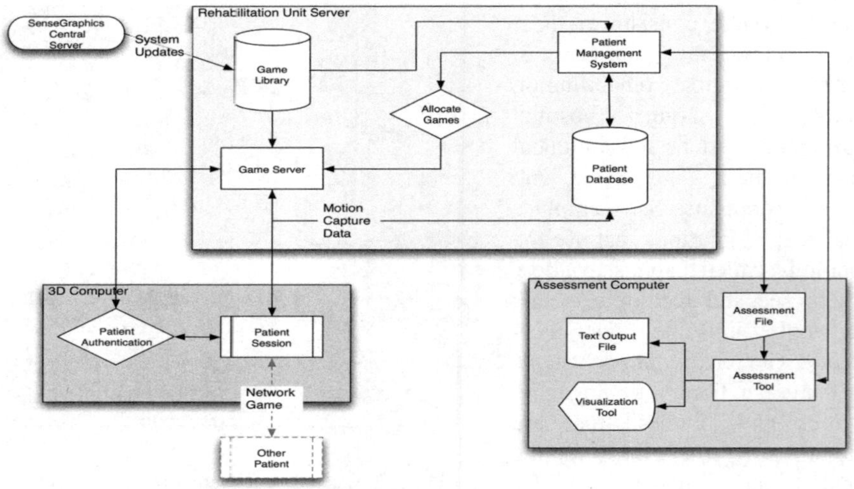

Figure 3. System components and connections.

2.3 Telemedicine platform

A client/server system will be constructed, based on two standard computers, high performance web cameras, microphones, a SQL database and a high-speed digital loop technology network. The system uses protocols within the TCP/IP suite and enables the transfer of real time system data and log files that shows the patients result, and data concerning video/audio teleconference between the patient and the clinical personnel.

3. Results

The data stored by the system can provide the rehabilitation therapist with an objective view of the patient's progress and the effect of the therapy. The report of the patient's activity includes all the relevant information and data graphs (fig 4 and 5). This version of the web data portal was intended for research use, and was designed to be easy to use and powerful enough to provide enough flexibility.

Besides raw-data graphs, the system provides the movement data that is captured in position (*x, y, z*) and orientation (*yaw, pitch, roll*) of the haptic stylus together with a time-stamp and any application event messages that may be useful in subsequent interpretation of the data. In figure 4, a movement of the stylus/hand is recorded

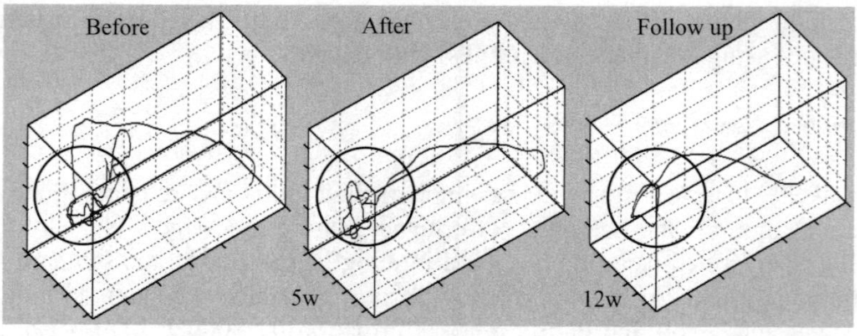

Figure 4. Raw data for the hand trajectories at baseline (A), during intervention (B), and at follow-up (C). The trajectory serves to illustrate the range of kinematic responses

illustrating the range of *kinematic responses for one movement unit* at baseline (A), during intervention (B), and at follow-up (C). The visual inspection reveals a variation in movement pattern for all patients, especially the end-point (black circle) suggests a different planning process for striking the target.

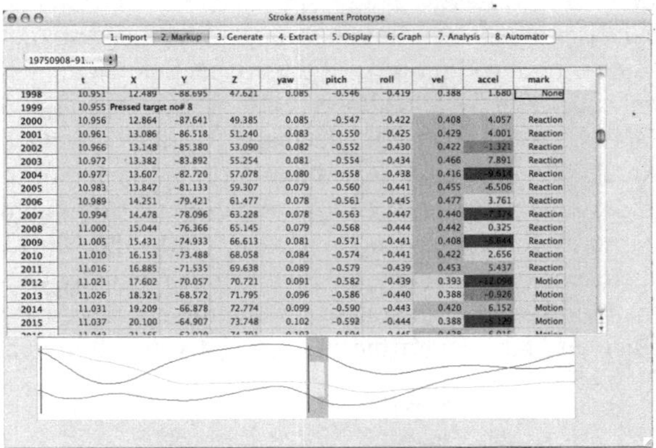

Figure 5. Movement data that is captured in position (*x, y, z*) and orientation (*yaw, pitch, roll*) of the haptic stylus together with a time-stamp and any application event messages that may be useful in subsequent interpretation of the data.

A simple utility is provided that allows for extraction of interpolated movement data at a rate slower than 1000Hz, as well as determining trajectory length, velocity and acceleration information. In figure 5 is a chart exemplifying hand movement velocity at continuous stages of rehabilitation, which present the patient's performance in each trial compiled across trials, blocks, or days. Basically, these are the graphs that show whether or not the patient is improving. Trial performances are computed after each trial is finished and stored in the database.

4. Discussion

The increasing costs of providing healthcare services to an ageing population and changing pattern of use of hospital resources, i.e. fall in the average length of stay, are shifting the focus of care from hospital to home or nearby community centres. This current trend of cost containment in health care has prompted a search for innovative methods of providing quality care. One method is the use of telemedicine. A modern telemedicine communication technique has the potential to improve the communication and prolong the contact with the patient after being discharged. Telemedicine systems minimises the barrier of distance, and makes it possible to be able to conduct evaluation and rehabilitation program at rural locations, at great distance from the clinic.

This solution is also able to monitor every tiny movement made by the patient while using the system, which can be recorded and sent back to the hospital (Figure 3). Such data is invaluable in not only monitoring the patients progress, but in creating a knowledge-base of stroke victims and even being able to validate drugs that may help stroke or even other neuromuscular victims

Keeping the application web-based brings a number of advantages, i.e. the service could be offered at the same cost without regards to long distance telecommunication facility charges. Furthermore, a web-based implementation should allow enough bandwidth when it comes to simultaneous video conferencing and precise data acquisition mode in the rehabilitation system, even when the web connection is temporarily low (buffer system). In this client/server system the server reside in the clinic and the clients in the patient's homes or at a community rehab centre. The clients, placed at a distance from the clinic, will get an input signal from the haptic device, when the patient is doing the exercises. These signals are then computed, acquired and then, within the TCP/IP protocol transmitted through a high-speed digital loop technology network to the server on the clinic side of the connection where it can be examined and evaluated in real-time. This system allows real-time monitoring of the exercises. Furthermore these data is stored in a SQL database. By this system the values of the progress in the exercises can be easy stored and look up on for further examination and evaluation. Besides sending the images from the exercise program the therapist also will have the opportunity to have an ongoing video conference through the session. With the help of this telemedicine system the therapist can actually watch the patient exercising in real time by distance at the clinic, follow the progress and be able to support and coach him/her through it with the help of the video conference service.

5. Conclusion

The benefits of a system for rehabilitation after stroke, based on VR, Haptics and Telemedicine should be: increased quality of life, lesser isolation, feeling more secure, fewer tiring transportations, more frequent exercising, better compliance to training, lower cost for transportation.

6. Acknowledgements

This study was in part supported by the Swedish Stroke Victims Association, the Hjalmar Svensson Research Foundation, Göteborg Foundation for Neurological Research, SFMI - Swedish Federation for Medical Informatics, Federation of Swedish County Councils (VG-region), The Swedish Research Council (VR K2002-27-VX-14318-01A), and VINNOVA (2004-02260).

References

[1] {HYPERLINK "http://www.riks-stroke.org/files/analyser/Rapport04.pdf "} (last accessed May 17, 2006)

[2] Kollen B. Recovery of gait after stroke. Utrecht: UMC Utrecht 2006.

[3] Broeren, J., Björkdahl, A., Pascher, R., & Rydmark, M. (2002). Virtual reality and haptics as an assessment device in the postacute phase after stroke. Cyberpsychol Behav, 5(3), 207-211.

[4] Broeren J, Georgsson M, Rydmark M and Stibrant Sunnerhagen K. Virtual Reality in Stroke Rehabilitation with the assistance of Haptics and Telemedicine. In: Sharkey P, Sik Lányi C, Standen P editors. Proceedings of The 4th International Conference on Disability, Virtual Reality and Associated Technologies; 2002 Sept18-20; Veszprém, Hungary: The University of Reading, 2002. p. 71-76

[5] Broeren, J., Rydmark, M., & Sunnerhagen, K. S. (2004). Virtual reality and haptics as a training device for movement rehabilitation after stroke: a single-case study. Arch Phys Med Rehabil, 85(8), 1247-1250.

[6] Jack, D., Boian, R., Merians, A. S., Tremaine, M., Burdea, G. C., Adamovich, S. V., et al. (2001). Virtual reality-enhanced stroke rehabilitation. IEEE Trans Neural Syst Rehabil Eng, 9(3), 308-318.

[7] {HYPERLINK "http://www.certec.lth.se/doc/touchaccess/TouchAccess.pdf "} (last accessed Jan 4, 2006).

Ubiquity: Technologies for Better Health in Aging Societies
A. Hasman et al. (Eds.)
IOS Press, 2006

A distributed shared data space for personal health systems

Wim Stut[a,1], Frank Wartena[a] and Maarten van Steen[b]
[a] *Philips Research, Eindhoven, Netherlands*
[b] *Vrije Universiteit, Amsterdam, Netherlands*

Abstract. Ubiquitous computing is a promising paradigm to support health care outside traditional care institutes. Sensor-based systems may continuously collect data on a person's health status and context, and provide immediate feedback or contact a remote physician. This paper presents a novel programming model to facilitate the development of such systems. The model, which has been inspired by tuple spaces, offers robustness for ad hoc mobile environments and explicit support for data streams
Keywords: Medical Informatics, Ubiquitous Computing, Sensor Networks, Body Area Networks, Intermittent Connectivity, Middleware.

1. Introduction

Demographic developments in society lead to an increasing demand for health care. The main reasons are that people get older and become increasingly aware of disease risks and a healthy life style. At the same time health-care budgets are under pressure. These two developments demand for new ways of delivering health care services.

One approach is to support people in having a healthy life style or managing their disease while residing outside traditional health care institutes. In this case, sensor-based systems continuously acquire data on a person's health status and context. These data can be interpreted locally by the system, which may give immediate feedback or control an actuator, or be sent to a remote care provider for further interpretation.

For example, when a patient is discharged from a hospital after heart surgery, a sensor may continuously measure the patient's heart rate. If the heart rate value exceeds a threshold, the patient's cellular phone automatically warns a remote physician. This concept may also be used to coach a person training for a marathon: while running, a small wearable computer advises the runner to adapt his or her speed based on the measured heart rate.

A system that supports an individual's health is called a personal health system. It consists of sensors, actuators, and appliances (such as cellular phones and PDAs) within the range of an individual person. A personal health system may be connected to other systems, such as hospital information systems, tele-health-care services, or fitness centres. An important requirement is that it should be able to support persons with multiple diseases, and be extendable with new hardware and software.

[1] Corresponding Author: Wim Stut, Philips Research, High Tech Campus 31, 5656 AE Eindhoven, Netherlands. E-mail: wim.stut@philips.com

As persons move around, devices or networks may become unreachable. Hence, personal health systems are in fact ad hoc mobile computing environments that show intermittent connectivity. Yet, robustness is a core requirement for these systems. Another characteristic of these systems is that they deal with data streams rather than individual data values as they continuously monitor a person's status and context.

There is currently very little support for building personal health systems. Applications are generally tailored to specific hardware and poorly integrated with other applications. We believe that lightweight middleware may solve many of the issues needed to develop open and portable applications that can operate independently of the characteristics of specific hardware devices such as sensors.

As a step in this direction, we present a novel programming model for the collection, exchange, management, and access of data in personal health systems. The paper is organised as follows. Section 2 describes the problem that we solved. Our programming model is explained in Section 3. Section 4 summarizes a prototype implementation. Finally, we discuss our work in Section 5.

2. Material and Methods

2.1. Problem description

Today, personal health systems are typically aiming at a single disease or purpose. This leads to unnecessary duplication of equipment and makes it difficult to offer integrated health support when patients have multiple diseases or goals. For example, if the patient and the runner from the previous section are the same person, with current technology he or she has to wear two heart rate sensors, and both a mobile phone and a wearable computer when running. Moreover, the sports application does not know that this person has had heart surgery, and cannot adapt its advice accordingly.

Furthermore, each personal health system has its own architecture, which makes it difficult to reuse components. Each has its own solutions for robustness and handling ad hoc mobility, or does not support these features at all. From a system development point of view, it would be much more efficient if these systems could be based on a shared architecture. Functionality that is needed by multiple applications can then be offered by a common middleware layer, which can support a range of devices.

As a contribution to solve these problems, we have developed a programming model for the collection, management, and access of data in personal health systems. The model is based on a shared data space concept, and explicitly offers robustness in ad-hoc mobile environments. The model hides sensors for applications: each application merely specifies what kind of data it needs (like heart rate) and at which frequency; the middleware finds the appropriate sensor, and warns the application if it cannot meet its request. The patient data as collected and stored in the data space can serve as basis for multiple applications, each targeted towards a specific disease.

2.2. Scenario

To further illustrate the problem domain, consider the following scenario:

Jim likes running. To improve his performance he wants to know his heart rate during his training sessions on a per-second basis; he therefore buys a heart rate sensor and a

PDA that communicate wirelessly. Unfortunately, several times a week Jim feels dizzy. His family doctor suspects a small cardio-vascular problem, and wants to know Jim's heart rate and blood pressure during the day on a per-minute basis. The heart rate sensor that Jim already has, is accurate enough, and can thus be used for this purpose too; for measuring blood pressure, another sensor is used. The collected sensor data are stored on Jim's PDA such that the doctor can inspect these data at Jim's next visit.

What makes this scenario interesting from a system's perspective?

- the sports application and the cardio-vascular application share the heart-rate sensor but have different sampling frequency requirements. When Jim is not running, the heart rate needs to be collected only once per minute. We could unnecessarily drain batteries when collecting these data every second.
- the cardio-vascular application needs sensor data only once per minute. Even if more data are available (due to the fact that Jim went running and the sports application asked for heart rate values every second), these additional data items need not be used by the cardio-vascular application.
- the sensor data must be available for future use in the order that they are produced by the sensor. In particular, when Jim, at his next visit, tells the doctor that he felt dizzy yesterday around 10 am, the doctor may connect her PC to Jim's body area network (BAN) to visualize the heart rate and blood pressure values as measured between 9 AM and 11 AM.
- the collection of data is orthogonal to the processing of data. At Jim's first visit to the doctor, the doctor sets the desired frequency to one sample per minute. The software that is used for this may run on the doctor's PC that is temporarily connected to Jim's BAN. However, the collecting of sensor data must continue when the associated application is not connected.
- the system's reaction to the removal of the heart rate sensor depends on the applications that are present. It would be no problem at all if, before his first visit to the doctor, Jim turns off his heart rate sensor after running. However, if this sensor is also used for collecting heart rate values for the cardio-vascular application, the system should warn Jim.
- details about the sensors should be hidden to the applications. The applications are merely interested in data that reflect the status of the person; whether the heart rate is measured by one or multiple sensors is irrelevant to the applications. Implementation changes in sensor technology should not affect the applications that use sensor data.

3. Programming Model

To address these issues, we have developed a programming model inspired by Linda-like shared data spaces [1]. Data spaces have shown to be a powerful concept when it comes to separating applications in time and space: the components do not need to co-exist in time for them to communicate and need not know about each other's existence. The asynchronous and connectionless programming paradigm of data spaces makes them more attractive for ad hoc mobile computing environments than the remote procedure call model (systems based on the latter model are less robust when devices or networks become unreachable).

In contrast to other approaches, our shared data space works on data streams rather than only single data items, as is normally the case. The components in our system (sensors, actuators, and applications) communicate via a distributed shared data space. These components do not need to know each other; instead they communicate via typed data elements. The components write and read data elements to and from the data space via generic operations.

Typically a data space contains data of a single person and is distributed over multiple nodes. These nodes communicate via a wired or wireless network, and may temporarily be disconnected. The data space system software is responsible for moving or replicating data elements between the nodes of the data space. This is hidden for the component developers; they can focus on application-level functionality.

Sensor data are typically processed in the same order as produced by the sensors. The processing may take place immediately, or later. To make the programming of applications easier, the model contains an explicit stream concept, where a stream is defined as a time-ordered collection of typed data elements (see Figure 1). No assumptions are made about the frequency: even the data elements produced by a sensor that measures the heart rate only once per day, can be viewed as a stream.

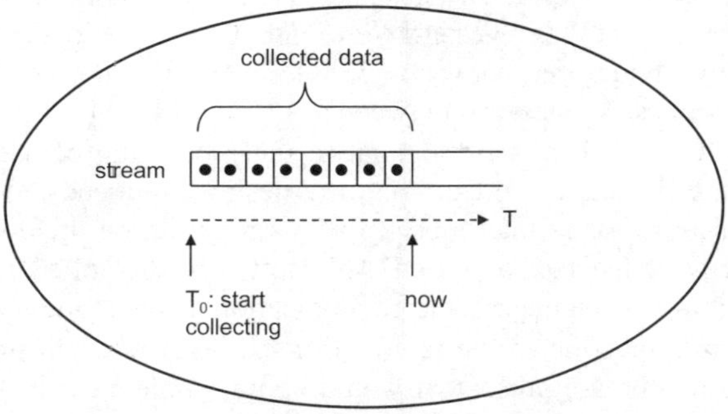

Figure 1. A stream in a distributed shared data space

Components may *create* a stream, which requires specifying the type of the data elements in the stream, such as *HeartRate*, and providing a unique name by which the stream can be identified. A stream can then be *opened* either for reading or writing elements. Since different readers may simultaneously access a stream at different positions and in different manners, opening a stream returns an application-specific *descriptor* that is subsequently used for all stream accesses by that application.

Data elements can only be appended to a stream (i.e., at the right-hand side in Figure 1), at which point they are timestamped. To read data from a stream the reader must first position itself in the stream via the *seek* operation. Data elements can then be read via the *read* operation. The effect of the read operation is that the reader gets a copy of a data element. The reader receives the data elements in the same order as they have been added to the stream.

As illustrated in the scenario, different readers of a stream may want to receive data at different frequencies. We have chosen to let the reader define its own frequency via the operation *setReadFrequency*. The side effect of the *read* operation is that the position of this reader in the stream is adapted according to its frequency. In other words, subsequent read operations return elements at the specified frequency. A read may block the caller until a sample is available.

The system may be instructed to immediately collect a data sample (which can then be read). As an alternative, an application may tell the system to start the continuous collection of data samples (at a specified frequency). We deploy call-backs to handle exceptions, such as when there is no data available.

An important observation is that the shared data space hides how it actually collects data, and from which sensors. We believe this to be an important contribution as it allows us to decouple applications from specific hardware.

4. Prototype implementation

To validate the data space concept we have built a prototype for a part of the scenario of Section 2.2 (see Figure 2). The heart rate is measured by a sensor that is connected wirelessly to a PDA via an 802.15.4 link. The PDA and laptop communicate via the network access profile of Bluetooth.

The PDA contains a sports application, which asks the data-space middleware to collect the heart rate at a frequency of 1 Hz when the user is running. The laptop (of the family doctor) contains a cardiovascular application; when Jim visits his doctor and connects his PDA to her laptop, this application asks the middleware to collect heart rate samples at a frequency of 1/60 Hz.

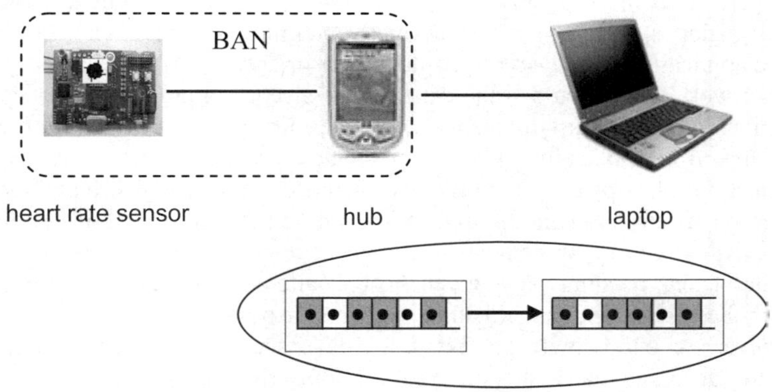

Figure 2. Overview of the prototype implementation.

Both the PDA and the laptop have data-space software. Both devices host a data-space kernel that contains stream data and that offers the data-space operations to the local applications. When the PDA and the laptop are connected, both kernels synchronize their contents by exchanging samples using an IP socket.

Once the middleware (on behalf of an application) has asked the sensor to start sampling at a certain frequency, the sensor autonomously sends the heart rate at this frequency. When a sample cannot be delivered (e.g., because the connection between the sensor and the PDA has broken), both the sensor and the PDA raise an alarm to inform the user that something is wrong. Note that this alarm is raised only after a sample could not be taken.

The prototype confirmed that the system continues to operate correctly even if a network connection breaks. The prototype did reveal that the clocks of the sensor and the PDA drift apart (about 10 ms per minute). This was an issue since the software on the sensor and the PDA need a timer to know the next sampling moment. We solved the problem by sending a clock synchronization message from the PDA to the sensor every 10 minutes to keep the drift within reasonable limits.

5. Discussion and Conclusion

This paper has presented a novel programming model to facilitate the development of personal health systems. The model, which has been inspired by tuple spaces, offers robustness for ad hoc mobile environments (data does not get lost when devices or networks become unreachable, and users are informed when the system cannot fulfil its tasks), separates applications and sensors, and explicitly supports data streams. We now briefly discuss related work for such systems.

MiLAN [2] is middleware for (medical) sensor networks. Its underlying goal is to maximize the system lifetime by reducing energy consumption. Applications explicitly specify the desired data types and quality (e.g. depending on the patient's status). MiLAN combines these requests, and determines the most feasible sensor set that satisfies the applications' requirements.

Secure UPnP [3] deals with secure access to and data transport in wireless health care systems. However, little attention is paid to data distribution with intermittent connectivity, or to the separation of applications and sensors. Fluid Computing [4] specifically addresses system robustness with intermittent connectivity, but offers no support to applications to transparently gather sensor data.

Earlier work has shown that the data space concept is a promising concept for ad-hoc mobile computing environments with intermittent connectivity. For example, LIME is aimed at applications that exhibit logical and/or physical mobility [5]. All communication takes place via transiently shared tuple spaces distributed across the mobile hosts. At any moment an application running at a host, can access the tuples located on its own host and the hosts it is connected to. The set of tuples accessible by a particular agent residing on a given host is altered transparently in response to changes in the connectivity pattern among the mobile hosts.

However, to our knowledge, existing data space models do not offer explicit support for accessing data streams and for collecting sensor data in a way that applications are shielded from low-level interfaces. We believe that by letting middleware offering this support, robust and better personal health applications can be developed. Future experiments should lead to further insight in this.

References

[1] Gelernter D. Generative communication in Linda. ACM Transactions on Programming Languages and Systems, 1985; Volume 7, Issue 1: 80–112.

[2] Heinzelman WB, Murphy AL, Carvalho HS, and Perillo MA. "Middleware to support sensor network applications". IEEE Network Magazine, 18(1):6--14, 2004.

[3] Keinänen K and Pennanen M. Secure UPnP and Networked Health Care. ERCIM News No. 63, October 2005, pp 25-26.

[4] Graf M. Fluid Computing. ERCIM News No. 54, July 2003, pp 21-22.

[5] Picco GP, Murphy AL and Roman GC. Developing Mobile Computing Applications with Lime. Proc. of the 22th International Conference on Software Engineering, ACM Press, June 2000, pp. 766-769.

1.2 Ubiquity: Opportunities and Technologies

Ubiquity: Technologies for Better Health in Aging Societies
A. Hasman et al. (Eds.)
IOS Press, 2006

Ubiquitous Technologies in Health: New Challenges of Opportunity, Expectation, and Responsibility

Michael RIGBY[1]
Centre for Health Planning and Management, Keele University, United Kingdom

Abstract. In spite of their name, 'ubiquitous' technologies are not yet ubiquitous in the true sense of the word, but rather are 'novel', being at the research, pilot, and selective use stages. In future, the proliferation in types of application, the major increase in cases and data volumes, and above all the *dependence* on ubiquitous technologies will raise practical, ethical, and liability issues. Equally significantly, it will require health service redesign, including new response services. Health informaticians need to be active in stimulating consideration of all these issues, as part of both social and professional responsibility.

1. Introduction

Ubiquitous technologies present a major new opportunity for the use of health informatics in health and healthcare, primarily by the paradigm shift of changing the monitored health environment of the patient from that of the health facility (usually hospital) to that of the patient's daily living environment (not only their home, but any location where they happen to be). It is pervasive technologies which have enabled this to happen, by making monitoring devices portable, wearable, implantable, and capable of inclusion in many domestic devices.

However, as with many innovations in the health field, the driving forces tend to be an amalgam of technological enthusiasts, and early application innovators. This is fine at the innovation stage. However, for such technologies to become truly ubiquitous, and the normal expectation for every eligible citizen, a radically different scenario needs to apply, to which little thought has yet been given. It is the responsibility of the health informatics community to stimulate consideration of the long term implications of truly ubiquitous (as opposed to currently selective) applications of these technologies.

2. Environments of Application and Client Groups of Ubiquitous Technologies

Broadly speaking, the areas of application of ubiquitous technologies fall into two distinct categories. The issues raised are quite separate.

[1] Professor Michael Rigby, Centre for Health Planning and Management, Darwin Building, Keele University, Keele, Staffordshire, ST5 5BG, United Kingdom; m.j.rigby@hpm.keele.ac.uk

2.1. Mobile Telemetric Monitoring

Telemetric monitoring by wearable, worn or implanted devices enables a specific physiological dimension such as heart function to be monitored continuously (or at regular frequent intervals) wherever the person is located. Monitoring devices worn on the body, and whose recorded data are subsequently downloaded, are already well established applications. However, the latest technologies not only enable monitoring to be done continuously and usually unobtrusively, but they can either download data, or communicate alerts of the existence of exceptional patterns as soon as they occur.

2.2. Domestic Environment

Secondly, a range of technologies is emerging which enables the monitoring of the activities, and thereby the well-being, of citizens perceived to be at risk, including whether doors are opened, lavatories flushed, and refrigerators opened. These measures are indirect and circumstantial – they indicate the apparent following of a normal pattern of daily living, but they do not prove that the 'normal' is in fact happening. For instance, a refrigerator can be opened without anything being removed. An exception is video surveillance, which uses a different technology in one of two ways: televisual contact (often known as telecare) involves the monitored person responding to a video telephone link on demand or on request, whereas remote surveillance involves involuntary monitoring of the ongoing movements of a person.

2.3. Client Groups

The client groups for these technologies also vary. The general assumption is that the elderly will be the prime beneficiaries. This is undoubtedly true, as more persons live longer, and by virtue of being maintained through chronic conditions these elderly citizens are likely to have ongoing morbidity which puts them at risk, ranging from circulatory disorders and diabetes through to confusional states. Moreover, as the elderly increase as a proportion of society so new paradigms of healthcare and health-related support for daily living will have to be developed because of the unsustainable volume and cost of the demands upon a finite health system.

However, the elderly should not be seen as the only potential beneficiaries. Many younger adults develop chronic or life-threatening conditions, and they too have the potential to benefit from specific forms of ubiquitous technology. Those with mental illness may at times be at risk, not least through failure to take essential medication, and there are arguments for applying the benefits of ubiquitous technologies here too. Some infants are already the beneficiaries of continuous monitoring, as with respiratory monitors for children considered at risk of Sudden Infant Death Syndrome.

3. The Restricted Vision of Ubiquitous Technology Advocates

Pervasive technologies, and their application into ubiquitous technologies, are as yet a young science at an early stage of development. Current conference themes tend to be about the development and proof in use of such technologies [1], or their use to devolve workloads [2]. This is understandable, and acceptable in the short term. However, the next stage, and the reason why an MIE conference theme is important, is that these technologies then need mapping forward. Once the basic technologies are proven, the discerning health policy maker (and this should include architects of health informatics policy) should next ask a series of focused questions. These include:

- What clinical conditions or circumstances render ubiquitous technology an appropriate recommendation for a care regime, and within it for an individual?
- What is the prevalence of such conditions, and what is the optimal demand?
- What is the balance between passive data recording for interpretation by a clinician expert, against algorithm-based automated interpretation of data to trigger immediate alerts?
- Where alerts are triggered, has the health care response mechanism been put in place, including adequate briefing, data access, and resourcing?
- How will the data be interlinked to the patient record, and be made available to other clinicians with a duty of care?
- How will supply be allocated and made equitable, balanced against likely demand from patients or relatives for immediate implementation?
- What will be the position regarding patients who decline such technologies, or actively circumvent them, as an intrusion into personal life?
- Are all aspects of the technical infrastructure available and able to cope, including installation and maintenance technicians?
- Have liability issues been resolved, for instance if a patient comes to harm because a device has failed or because an algorithm did not identify an adverse event?
- What are the costs, including informatics support, and how will they be met?

Until each of these questions can be answered unequivocally in the affirmative, the clear implication is that the technology cannot be made available as a routine part of the health system without serious issues of either unmitigated risk, or inequity in availability. When health informatics applications start to become part of a health system, and an assumed available part of patient care, health informaticians need to move from being merely advocates of technology and should assume roles in service design and ethics, to ensure a responsible solution.

4. Challenging Conundrums

The preceding section highlighted the practical questions that need to be asked before a technology is applied to a disease condition or client group in a specific locality as part of the fabric of regular care and support. These are essential issues in the move from novelty to healthcare tool. Several parties need to be involved, including lead clinicians, funding and policy-making bodies, and informatics staff, to agree what is both desirable and feasible.

However, the health informatics world (with the other key interest groups) also needs to address another range of issues. These are the societal and ethical issues. In general they are generic, though the specific issues may vary locally. The following are examples:

Conundrum 1: The Medical Record

User-worn monitoring devices such as ECG monitors are usually fitted and downloaded in a hospital or diagnostic environment, where the data are considered part of the medical record. Ambulatory health staff such as community nurses who are requested to visit regularly a frail elderly person to ensure their well-being are expected to keep a professional record of their calls, and for these patient records to be available to the wider care team. Failure to keep adequate records, and failure to report alerts (including not being able to make contact with the patient), are considered breaches of professional conduct with serious consequences. Therefore by definition, once a person is allocated a form of ubiquitous technology to monitor their health status or daily living at the behest of a health professional treating a condition, the ubiquitous technology (and any related software and peripheral devices) must be considered a part of medical care, and the data captured thus be considered part of the health record.

Conundrum 2: Raw or Processed and Summated Data?

In these scenarios the data volumes are potentially huge, and at raw data level some of the items (such as regular heartbeats or lavatory flushing) are trivial. On the other hand, if only summated analyses of the data are entered in the health record, detail which may later be found to be significant may be lost. These are major issues.

Conundrum 3: Circumvention, Misinterpretation, and Non-compliance

As indicated earlier, many ubiquitous technologies monitor a phenomenon that is a proxy for detecting activities of normal daily living. Others monitor physiological signs, with the assumption that the context is normal living. But older citizens, or other groups with a long-term health problem, do not necessarily live classic 'nice old person' lifestyles. They are also autonomous individuals, who may resent having every occurrence of an activity monitored. Older citizens, even those with health problems, do have illicit sexual liaisons. They may have friends round for a drinking session – indeed, they may take part in risky behaviour knowing that they have less remaining life to be compromised. They may even take part in illegal or criminal activities. Ubiquitous technologies are by definition intrusive: they may identify any of the foregoing situations. Thus while most elderly persons or disease sufferers will welcome appropriate ubiquitous devices, others will find them intrusive for good, or less good, reasons. So what are the responsibilities, how do the organizational users of ubiquitous systems react, when the following occur?

a) Detection of risky personal behaviour such as a heavy drinking party, consumption of illegal drugs, or a sexual liaison?
b) Extra-ordinary readings (in the clinical sense) trigger decision-support based automatic algorithms, caused by this type of activity. Must elderly or other persons with chronic conditions 'behave themselves' or be disqualified from health support – a situation which does not occur (except in very special situations) with regard to other health care?

c) A monitored person who is mentally competent deliberately circumvents the technology to maintain privacy, but not on a regular basis. Are they labelled as 'non-compliant' and deprived of full healthcare support, and conversely how is the health provider protected from liability?

Conundrum 4: Interested Third Parties

It is an aspect of good health care that persons take responsibility for their own health, and that families and natural communities seek to support their frail or ill members. In the traditional treatment of illness this is itself at times a difficult issue – the patient record and decisions are private, yet the next of kin are encouraged to take an active part in the support of the patient. Ubiquitous technologies bring new dimensions to this complex relationship, both positive and negative, raising issues such as:

a) A relative exerts great pressure on the health system for the installation of monitoring technology for their elderly parent living on their own. The patient argues strongly against this on grounds of privacy and inhibition of lifestyle. The relative threatens to raise questions of inadequacy of care if this equipment is not provided, whilst the elderly person threatens to jeopardize its use. What are the issues for the clinician, and the healthcare organization?
b) Next of kin request installation of monitoring equipment in a dependent person's home, and when informed it is not available from healthcare systems resources they offer to purchase this themselves. If this were a baby monitoring device for an infant this would be seen as good parenting. Is it good citizenship for family carers themselves to monitor the home of an autonomous adult who has serious health risks but whose mental capacity has not been challenged?
c) Adult children request to be able to access the monitoring records of their elderly parent, so as to adjust their support. In some parts of the United States this is now possible. Would it be legal, and ethical, within a European state, or between European states? How would a health professional respond to a request from the grown up child of the patient asking for care to be changed, and how would the emergency services respond if they received a call from the remote adult offspring requesting they make an emergency visit to the patient?
d) A private company sets up a business, offering domiciliary monitoring services using pervasive technologies. Is this ethical? Are there any controls in existence? Is this a free market, in which any relative, or indeed patient, as well as healthcare agency, can subscribe to such services?

5. Discussion

Ubiquitous technologies are currently exceptional or novel in their application – a paradox of terms and concepts. The current research related to them reflects this developmental stage of the technology, and its very early application in the applied health domain. If it is to be effective, and above all live up to its name of "ubiquitous", by definition it must be widespread and routine in those situations where it is appropriate. Thus the debate must move from focusing solely on the technologies and whether they work, through to their systematic application and their embedding into the local health system. Indeed, given the future challenges of the ageing population, and the drive to supported living of younger adults with chronic health conditions, coupled with the population's wish to reduce the risks of health problems, ubiquitous technologies must not just be bedded into the health system, but by very definition produce a paradigm shift in the design and functioning of that system [2]. New definitions of "health" and the responsibility of "health systems" are necessary, and the new technologies must show their ethical and legal responsibilities by having paradigms for ethical control and legal accountability.

Very broadly, this coincides with the maturity of health informatics as a discipline and profession. No longer are health informaticians solely technicians supporting implementation of technologies which automate paper-based systems. Health informaticians now need to be involved in the organizational development and policy development processes, indicating how health systems can be redesigned based upon technologies once these have been proved sound and reliable.

For a technology to be considered mature and validated, beta site validation and policy-supporting evidence are needed, followed by service design and linked education and training [3]. Thus health informatics needs to be engaged in the strategic thinking as to how ubiquitous technologies based on pervasive technology sciences, can be used as a building block in the next generation of health systems. If health informaticians are not going to be active in addressing the kind of issues raised in this paper, then health informatics as a discipline is not being socially or scientifically responsible. There is little evidence of this type of debate happening to date, largely because ubiquitous systems are themselves new, but now is the time for this visioning to start, followed by development of methodologies, controls, and accountabilities.

6. Conclusion

Ubiquitous health technologies are not yet living up to their name by the general definition of "ubiquitous", but need to do so if they are to achieve their purpose. The developers of pervasive and ubiquitous technologies need to interact with a policy-related group of health informaticians to identify and respond to the future issues and challenges as much as the future opportunities and benefits, and from this they need to interact with health policy makers to harness these technologies safely, ethically and efficaciously – characteristics of a mature and responsible profession.

References

[1] Wei D, Tamura T, Suk Park K. WEICONU 2005 Proceedings - The Joint International Meeting of the International Workshop on E-Health and the 2nd. International Conference on Ubiquitous Healthcare 2005; University of Aizu; Auzu-Wakamatsu, Japan, 2005.

[2] Decentralization of Healthcare via Distributed Diagnosis and Home Healthcare (D2H2): AMA, AMIA, and IEEE conference, Washington DC, April 2006 (icsl.ee.Washington.edu/d2h2/aboutd2h2.html) (accessed 14 January 2006).

[3] Rigby M. Essential Prerequisites to the Safe and Effective Roll-out of E-working in Healthcare; International Journal of Medical Informatics, 75, 138-147, 2006.

Ubiquity: Technologies for Better Health in Aging Societies
A. Hasman et al. (Eds.)
IOS Press, 2006

Visualisation and interaction design solutions to address specific demands in Shared Home Care

Isabella SCANDURRA[a, b 1], Maria HÄGGLUND[a, b] and Sabine KOCH[a, b]
[a] *Dept of Medical Sciences, Uppsala University, Uppsala, Sweden*
[b] *Centre for eHealth, Uppsala University, Uppsala, Sweden*

Abstract. When care professionals from different organisations are involved in patient care, their different views on the care process may not be meaningfully integrated. *Objective:* To use visualisation and interaction design solutions addressing the specific demands of shared care in order to support a collaborative work process. *Methods:* Participatory design, comprising interdisciplinary seminar series with real users and iterative prototyping, was applied. *Results:* A set of interaction and visualisation design solutions to address care professionals' requirements in shared home care is presented, introducing support for identifying origin of information, holistic presentation of information, user group specific visualisation, avoiding cognitive overload, coordination of work and planning, and quick overviews. The design solutions are implemented in an integrated virtual health record system supporting cooperation and coordination in shared home care for the elderly. The described requirements are, however, generalized to comprise all shared care work. *Conclusion:* The presented design considerations allow healthcare professionals in different organizations to share patient data on mobile devices. Visualization and interaction design facilitates specific work situations and assists in handling specific demands in shared care. The user interface is adapted to different user groups with similar yet distinct needs. Consequently different views supporting cooperative work and presenting shared information in holistic overviews are developed.
Keywords: Medical Informatics, Nursing Informatics, Integrated Healthcare Systems, Home Care Services, User-Computer Interface, Needs Assessment

1. Introduction

Although the need for ICT systems in home care is acknowledged, there is still a lack of support for this type of cooperative work. Clinical use is often hampered by poorly designed user interfaces [1, 2], and a large number of health information systems developed actually fail in supporting healthcare professionals in their work. In shared care the situation becomes even more complex; care professionals from different healthcare organisations participate in a care process that is often not meaningfully integrated and patient information is distributed among different information systems, impeding existence of a seamless and consistent workflow between the involved

[1] Corresponding Author: Isabella Scandurra, Dept of Medical Sciences. Medical informatics and biomedical engineering, Uppsala University, University Hospital 82:1, SE-75185 Uppsala, Sweden. Isabella.Scandurra@medsci.uu.se, http://www.medsci.uu.se/mie/project/closecare/index.htm

professionals. A patient receiving care under such conditions may not receive the appropriate care, and can be subject to unjustified interventions [3].

Shared home care requires both information at the point of care (POC) and ICT systems that are developed, not traditionally, i.e. for one care profession, but for the entire care process and all professions involved. Cooperation in shared care is successful when clinically adapted ICT systems provide care professionals with insight into other care professionals' work, support for coordination of work, POC access to documentation, and support for information sharing between different professions.

This paper focuses on describing how the specific requirements of shared home care [4] affect the demands on visualisation and interaction design of ICT systems. Illustrations of design solutions from a mobile virtual health record (VHR), developed in the Swedish action research project OLD@HOME [5], are presented.

2. Material and Methods

OLD@HOME, a 3-year action research project lasting from 2002 until 2005, resulted in a prototype system, developed for inter-organizational healthcare, which provides a seamless and consistent information and communication flow between home healthcare and primary healthcare. A VHR, integrating data from different care provider systems, is in use today and supports general practitioners (GP), district nurses (DN) and home help service personnel (HHS) with information needed at the point of need, in both web- and pocket pc-applications, the former used on a TabletPC and the latter implemented on a PDA [5].

It is argued that ICT systems in healthcare must be designed with consideration of the information requirements, cognitive capabilities and limitations of the end users, as well as considerations of daily work in process-oriented organisations [6, 7]. Therefore, a practical method, based on a participatory design [8, 9] approach, resulting in a requirement specifications that can directly be used as basis for implementation [10] was applied. In this method, real users, in this case 3 HHS, 2 DN and 2 GP, are involved in interdisciplinary working groups and assist in acquiring correct user needs in cooperative work, e.g. shared care.

In order to develop visualisation and interaction designs that actually meet the special needs in shared care, an iterative and incremental prototyping process with high end user involvement was used; in interdisciplinary working groups design solutions were iteratively improved through paper prototypes, storyboarding, design sketches and different levels of prototypes, until an adequate level of refinement was reached.

3. Results

In shared home care, specific requirements can be attributed to mobile care, ubiquitous access and cooperative work [4]. Based on existing visualization and navigations rules [1, 11-14], we propose design solutions addressed specifically to requirements for shared care, excluding problems that are common to ICT products in general.

Identifying origin of information

Care professionals working in home care need POC access not only to information from their own health record system; they also need insight in other professionals' work, irrespective of in which care provider organisation the information has its origin. When accessing shared information, it is however important that the user can easily identify who provided the information in order to interpret it correctly; *colour coding* is used for this purpose (Figures 1 and 3).

Holistic presentation of information

Some information is important for and documented by all healthcare professionals; e.g. risk factors. Access to these distributed notes not only provides a more extensive overview of the patient's health, but also acts as a tool for quality control, ensuring that contradictory information is discovered and adjusted. Risk factors are, e.g., gathered from all connected systems and accordingly colour coded (Figure 3). Each profession's documentation regarding risk factors is *presented simultaneously*, creating a *quick and holistic overview* of the patient's problems. The risk factors in the HHS' pocket pc-application are easily reached by an "[!]-icon" (Figures 1 & 2) always accessible from the menu bar, and contains identical information as the web application (Figure 3).

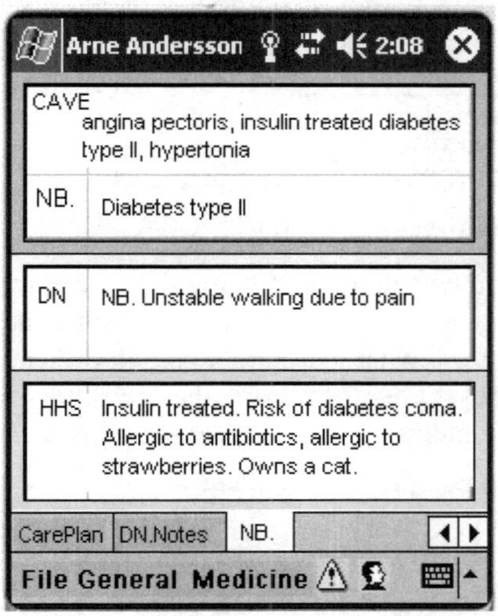

Figure 1: Risk factors on the PDA, an example of simultaneous presentation of information gathered from different feeder systems and accordingly colour coded.

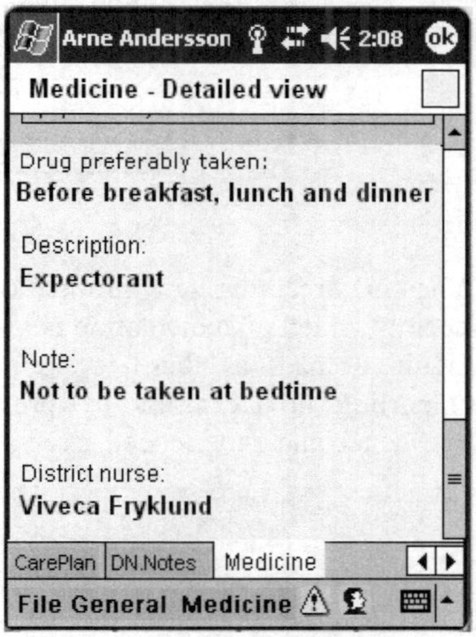

Figure 2: Avoiding cognitive information overload; only documentation needed at POC is displayed, information from feeder systems is filtered to the needs of a specific profession.

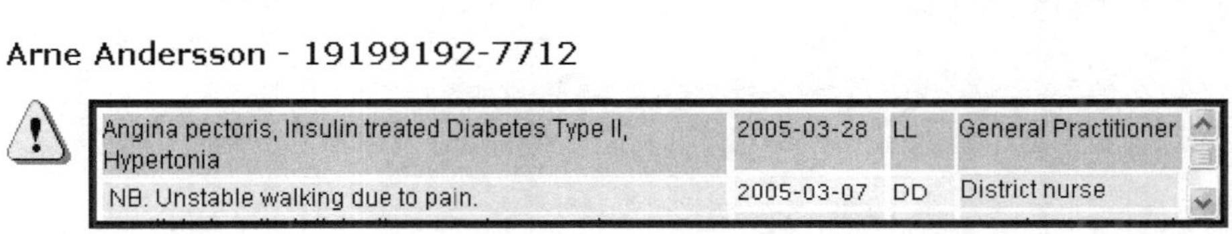

Figure 3: Risk factors on the web, placed at the top of the application; an example of simultaneous presentation of information gathered from different feeder systems and accordingly colour coded.

User group specific visualisation

Different user groups might have different requirements regarding presentation of information, and information deriving from one feeder system sometimes need to be displayed *differently for different user groups*, taking users knowledge of the domain into consideration. The prescription list e.g. is filtered and only parts that are essential to the HHS are displayed in their view; medical information such as prescribed drug, preparation, dosage and strength are gathered from the GP's system, whereas clarifying notes, adapted to the need of HHS, derive from the DN (Figure 2). The GP and DN on the other hand, need access to a more complete, traditional prescription list.

Avoiding cognitive overload

Access to integrated information gathered from several feeder systems increases the risk of cognitive information overload. However, not all information that *can* be made available is necessary to access for all healthcare professionals at POC. It is therefore important to analyse the different user groups' information needs, e.g. specific information requirements for DN, HHS and GP. Based on this analysis only documentation needed at POC is displayed and interaction is facilitated through a logical information structure, i.e. a three level architecture (Figure 4). Tabs are used as entry points; each tab indicates a specific work task and consistently contains an overview, a detailed view and, where applicable, a writing mode. This is identical for both platforms since similar situations for the users require consistent sequences of actions.

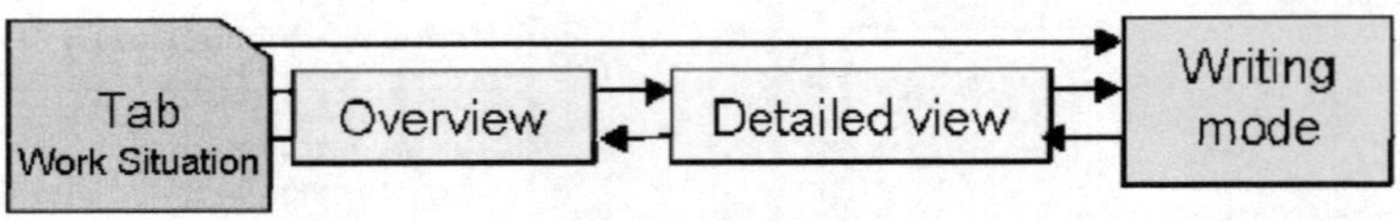

Figure 4: Three logical levels appear consistently in the VHR.

When documenting new information, e.g. writing a new note or updating the care plan, recently added documentation is available in the same view, thereby both reducing the cognitive load, as there is no need to *remember* what was written before, and supporting novice users by providing examples. This increases the personnel's confidence and experienced safety, as well as the general quality of documentation.

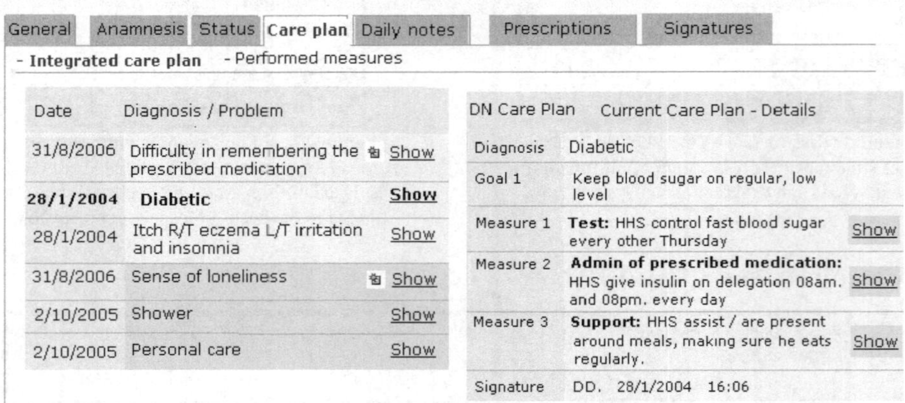

Figure 5: Integrated view of DN's and HHS' individual care plan for elderly as presented in the web application.

Coordination of work and planning

The coordination of shared care places large demands on the information processing capacity of involved healthcare providers and the efficiency of their communication. Integrated views, e.g. DN's and HHS' individual care plans for elderly (Figure 5), are developed to enhance integration of different care practices into a collaborative work process, and to keep track of jointly planned activities. A holistic view of the work process is achieved and accessible at POC. Highlighting of unread documentation (Figure 5) provides immediate notification of up-dates, and the interaction structure based on logical levels (Figure 4) clarifies the work process of care planning. The integrated care plan also works as a reminder to perform and evaluate measures, improving the care professionals work process.

Quick overview

To provide a holistic view of the patient's health is crucial, but different professionals have different needs of information content in these overviews. The VHR uses pre-defined compilations of keywords in specific views, e.g. aggregations of the most recent notes under the keywords "status" and "patient history" in the GP's and the DN's views, aggregations of the latest documentation from all care professionals and aggregations in the integrated care plan. Keywords are easily distinguishable in these aggregations by the use of bold lettering (Figure 5). Taking users knowledge of the patient into consideration is an important criterion for tailored overviews; information well known to a professional should be possible to hide, e.g. by hiding risk factors in the VHR, thereby freeing more space for tab information and focusing on information that is new to the user.

4. Discussion

Visualisation and interaction design can be used to enlighten specific requirements. For shared care the main requirements are to identify the origin of information, to get simultaneous presentations of relevant information in form of holistic overviews, and to recognize new or updated information quickly. Moreover, the information needs to be available at the point of care (POC) and accessed using mobile devices.

For POC documentation, the work process has to be supported. The presented design solutions are optimized to support cooperative work using intuitive user interfaces and easy interaction methods. There is, however, a strong dependency between the visualisation and interaction component and the underlying information structure of the ICT system. Context-dependent reduction of available information reduces the risk of information overload and supports quick overviews; thereby reducing time spent searching for information and coordinating work.

The presented design solutions are based on participatory design. Results from real target users in operational environment are, however, not yet available. Early usability and user satisfaction evaluations, performed by 14 HHS, 4 DN and 2 GP during the design phase, show promising results regarding augmented knowledge and competence, increased safety and improved interdisciplinary cooperation. Usability studies are now being conducted in laboratory settings for evaluation of the system's effectiveness and efficiency. The results thereof will later be triangulated with field studies validating the utility and further user satisfaction evaluations.

5. Conclusions

Only with deliberate attention to the user interface, can we improve the ways in which information technology contributes to the efficiency and effectiveness of healthcare providers [1].

The presented design considerations allow shared home care personnel to enhance and integrate their work processes. Applied visualization and interaction designs facilitate specific work situations in shared care, support cooperative work and present shared information in adapted views for different staff categories. Designs for identifying origin of information, quick overviews, avoiding of cognitive overload, simultaneous and/or different presentations of documentation, and coordination of work and planning are important to visualize for shared care personnel. The work process is elucidated by the navigation structure in the VHR. Early user satisfaction results exhibit that augmented knowledge and competence, gained through adapted views of required information when working in a mobile environment, result in increased safety and improved interdisciplinary cooperation. The participants emphasise *improved work situation;* less paper-work, safer documentation, more meaningful work, access to information, limiting the cognitive work load, and on *a personal level;* improved competence and feeling secure when working at a patient's home, as a result of accessing information provided from the entire team.

Acknowledgments

This work was performed within the project "OLD@HOME - Technical Support for Mobile CloseCare", sponsored by VINNOVA - Swedish Agency for Innovation Systems (P23037-1 A) and Trygghetsfonden and by its clinical and industrial partners.

References

[1] Tang PC, Patel VL. Major issues in user interface design for health professional workstations: summary and recommendations. International Journal of Bio-Medical Computing. 1994;34(1-4):139-48.

[2] Berg M. Patient care information systems and health care work: a sociotechnical approach. International Journal of Medical Informatics. 1999;55(2):87-101.

[3] Branger PJ DJ. Primary Care. In: van Bemmel J, Musen, M. A., editor. Handbook of Medical Informatics. Heidelberg: Springer-Verlag; 1997. p. 161.

[4] Scandurra I, Hägglund M, Koch S. Specific demands for developing IT systems for shared home care - a user centred approach. The Journal of Information Technology in Healthcare. 2005;3:279-85.

[5] Koch S, Hägglund M, Scandurra I, Moström D. Old@Home - Technical support for mobile closecare, Final report; November 2005. Report No.: VR 2005:14.

[6] Patel V, Kushniruk AW. Interface design for health care environments: the role of cognitive science. AMIA Symp; 1998; 1998. p. 29-37

[7] Andersson A, Hallberg N, Timpka T. A model for interpreting work and information management in process-oriented healthcare organisations. International Journal of Medical Informatics; Int J Med Info. 2003;72(1-3):47-56.

[8] Bødker S, Iversen O. Staging a Professional Participatory Design Practice - Moving PD beyond the Initial Fascination of User Involvement. In: Bertelsen, Bødker S, Kuutti, editors. NordiCHI 2002; 2002; October 19-23, Aarhus; 2002. p. 11-8.

[9] Schuler D, Namioka A, editors. Participatory Design - principles and practices. New Jersey: Lawrence Erlbaum Associates; 1993.

[10] Scandurra I, Hägglund, M., Koch, S. A practical method for efficient user needs analysis when developing ICT systems for shared care. Submitted to Methods of Information in Medicine. 2006.

[11] Shneiderman B. Designing the User Interface; Strategies for Effective Human-Computer Interaction. USA: Addison Wesley Longman; 1998.

[12] Weiss S. Handheld usability: John Wiley & Sons, Inc; 2002.

[13] Wood LE. User Interface Design: Bridging the Gap from User Requirements to Design. Boca Raton, Florida: CRC Press; 1998.

[14] Nielsen J. Usability engineering. San Francisco: Morgan Kaufmann; 1994.

Ubiquity: Technologies for Better Health in Aging Societies
A. Hasman et al. (Eds.)
IOS Press, 2006

Safety Portal: The Safest Goes Through the Air Ubiquitous high-risk reminders bridging out the patient safety in emergency department

Chang-I Chen[a,e], Cheuk-Sing Choy [b,c,e], Tu-Bin Chu[d], Yu-Sheng Lo [f], Wei-Kee Huang[e], Ching-Feng Kuan[g], Yu-Chuan Li [a,e]

[a]*Graduate Institute of Medical Informatics, Taipei Medical University*
[b]*Graduate Institute of Injury Prevention, Taipei Medical University*
[c]*Taipei Medical University Hospital, Emergency Department*
[d] *Graduate Institute of Business Administration, National Taiwan University*
[e]*Taipei Medical University, Municipal Wan Fang Hospital Emergency Department*
[f]*Taipei Medical University, Municipal Wan Fang Hospital Information Department*
[g]*Chungtai Institute of Health Sciences and Technology, Department of Health Care Administration*

Abstract. Under the safety net of hospital-based patient safety informatics (PSI) system, RFID is build to adapt emergency department safety needs and the process improvement activities. We implemented the RFID framework as the electronic patient identifier into the process improvement of ED workflow; it defined as the safety portal to introduce PSI safety features on the real-time basis. METHODS. Since 2004, we were applying RFID technology into a 200 daily visits emergency department of the regional medical center in Taipei, Taiwan. We then developed wireless web-based RFID safety portal to implement the real-time safety reminders such as the laboratory and radiology reports to the physician who can make decision promptly to the patients in the ED. RESULTS. Under the ED safety portal, the diagnosis time for physicians to make clinical decision is largely reduced nearly 40 percent due to the safety enhancement RFID system. ED physician who could actively receive patient's updated clinical data to make clinical decision via web-based informatics system reasonably decreases. CONCLUSIONS. The effectiveness of RFID system not only enhance patient identification during ED process but combine ED safety net which providing needed data for ED physician and staff who visit ED patients with valuable real-time data on time. Patient safety on ED can be clearly improved from the embracing modern technology and build up a patient centered ED environment.
Keywords: Safety Portal, High Risk Reminder (HRR), Radio Frequency Identification RFID, Emergency Department ED

Introduction

Real time critical reports may save patient's life in emergency department (ED). Over the last few years, radio frequency identification (RFID) has become the symbol of auto-identify media of various industries. In health care, there are several settings in which RFID has been useful and valuable in patient tracking and vital sign monitoring (1) (2). Adapting on-line laboratory critical value alert and real-time radiology high-risk reports as the bricks and mortar of safety guards could ED physicians to care patients more effectively. As the result, we persist to seek ways to improve safety. Via ED safety portal, ED staff execute bedside point-of-care testing for safety improvement in patients with suspected critical value of real-time reminder is an effective scheme that leads to enhanced efficiency by reducing test turnaround times and providing informative data for ED physician to make clinical decision more effectively.

Results

As most people's experiences of ED is the long waiting of the results of laboratory and radiology test, there is no sigh of major improvement on safety issue (3). This paper introduce the flow of ED safety management not only focus on the real-time reminding system which disasters or major incidents, the ED always bears the brunt of accepting and treating patients with various physical and mental problems as well (4). The ED represents the primary portal of entry into the most hospitals. Thus, emergency physicians and all ED staff play a crucial and central role in the identification their patient and perform correspond procedure correctly, subsequent ED management and correct referral to outpatient clinics follow up or observation unit as designated to its clinical results. The ability to expeditiously respond to such an event depends on the state of preparedness of the ED and its staff. The ability to adapt to the diagnosis time and effectiveness, as well as functional and structural safety management in the ED is the basic criteria of ED safety guard (5).

Discussion

The objective was to evaluate both the effectiveness and safety of High-Risks Reports (HRR) of laboratory and radiology reports. During 18 months running periods, 3,215 laboratory and 1,725 radiology alerts were issued. The alerting system has contributed

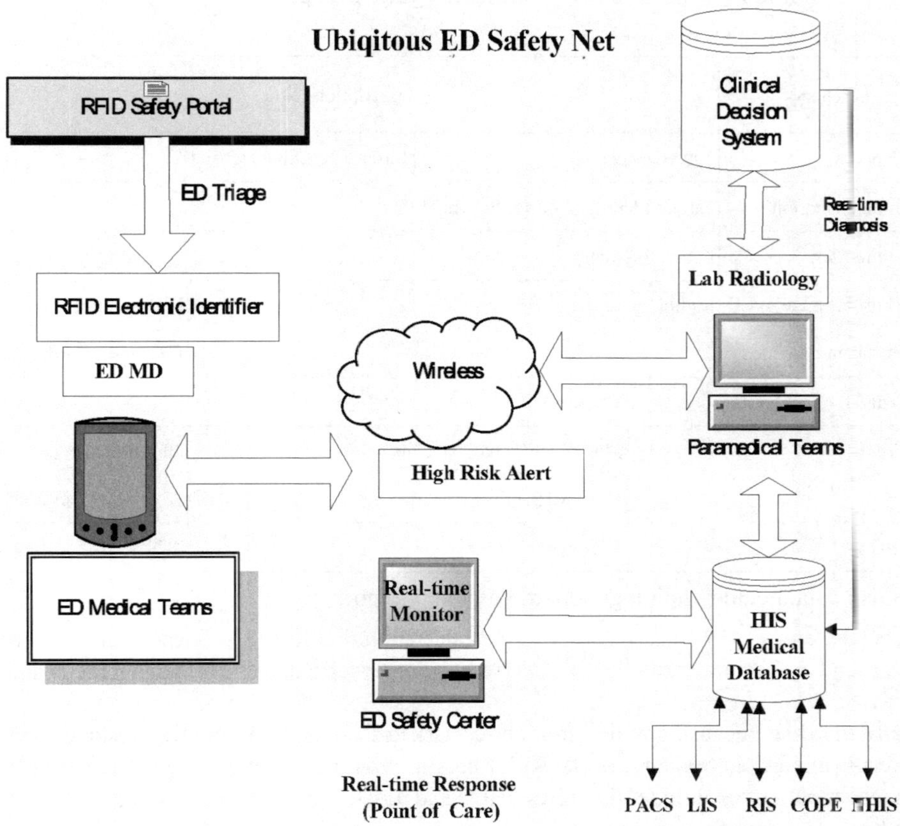

Figure. 1. Integrated Real-time High-risk Alerting System

to safety and patient care. All laboratory and radiology critical alerts are legible accomplished to prescribed physician, and potential delayed diagnosis errors have been avoided (6). Some studies have shown that point-of-care testing in the ED can reduce waiting time to making medical decisions for patients receiving appropriated treatment effectively. It needs more study to find evidences to show reductions in ED length of stay. Reducing length of stay only when decision making about patient disposition depends on the results of a specific blood test. For acute coronary syndromes, that is

Table1. RFID System in the Emergency Department

Category	Description	
ED Capacity	200 visits per day	Tertiary Teaching Hospital
ED Implementation	Patient Identification for Patient Safety	
Real-time Alert	Laboratory data	Critical values*
(short message to ED physician phone)	Radiology report	Fracture, tumor, free air, hemorrhage
RFID and mobile device		
Tag	Class II RFID passive chip 13.56 MHz	80 cents each, and 15 cents as recycled
PDA	Pocket PC	$US 450

*Critical data include: radiology report, pathology report etc.

exactly the case because any decision about patient disposition is highly dependent on the result of cardiac biomarkers (7). RFID tags in wristbands for all ED patients. As it give each patient a unique ID, physicians and nurses during busy rounds use the wristband to identify each patient, pull up an electronic medical record to verify treatment and medications. Tags allowed medical staffs on the bedside recognize right patient while reading real-time critical data from handheld PDA. HRR may transmit that patient's health information to the designated ED medical personnel who is providing point of care. The effectiveness of HRR is currently testing the system at ED, which serve 200 visits daily. RFID tags also can be used to identify medications and blood products, those items recently is scheduled for next stage of continuing improvement by current ED research team.

Today, critical single dose medications can have their packaging identified with tags. In the future, individual pills will have a tag embedded in them, allowing for even more exact checking of medications upon administration. Use of tags assists in the standard verification procedure of medication administration by helping to identify the right patient and right drug including route and dose Therefore, this system is likely to create concerns about the security of ED safety systems, due to increased data aggregation, ubiquitous access, and increasing reliance on safety methodological solutions. But we also justify the fair cost of the same technology can help building more robust, more reliable systems that may increase quality of the most ED system.

Conclusion

Safety enhancement under ED flow management is complicated and interrelated processes. The integration of RFID identification and high-risk alerts are viewed as critical in achieving effective and safe patient care. However, these systems are complex; all parts need to be aligned, and the systems must work together to produce the desired outcomes. In healthcare, automation using RFID capabilities is of growing importance because of the Institutes of Medicine study and the integrated electronic medical record.

In order to improve ED safety and successful implementation of point-of-care testing in the ED is cooperation between ED staff and the paramedical team as medical laboratory and imaging department that is responsible cooperation, which decreases many of the managerial hinders that often are linked with implementation of such point-of-care testing in the ED clinical basis (8). Under safety portal system, results of ED are more swiftly available to clinical staff than traditional laboratory-based results. What question this study addressed whether use of point-of-care testing performed by nurses could decrease ED length of observation patients. What this study adds to our knowledge In an ED that required real time reminder which may accelerate time dependent diagnosis such as acute myocardial infarction of cardiac marker results before an inpatient disposition decision was made, the use of point-of-care troponin I testing by ED nurses decreased the overall ED length of stay by approximately 2 hours. How this might change clinical practice in this ED safety practice (9), the introduction of RFID safety portal enhanced critical value into point-of-care and decreased the length of stay. Our research theme may wish to investigate whether this or other strategies that decrease laboratory turnaround time to build a safer ED care environment is the ultimate goal.

Reference

1. Alexi A. Wright, Ingrid T. Katz, Bar Coding for Patient Safety. N Engl J Med 2005; 353:329-31.
2. Drew E. Altman, Carolyn Clancy, Robert J. Blendon, Improving patient safety-Five Years after the IOM Report. N Engl J Med 2004; 351:2041-3.
3. Robert Wachter. The end of the beginning: Patient safety five years after 'To Err is Human' amid signs of progress, there is a long way to go. Quality of Care 2004; W4: 534-45.
4. Karen S. Cosby a framework for classifying factors that contributes to error in the emergency department. Annals of emergency medicine; December 2003; 815:823.
5. Georges C. Benjamin, Addressing medical errors: The key to a safer health care system. The Physician Executive; March. April 2000; 66-67.
6. Lucian L. Leape, Donald M. Berwick, Five years after To Err is Human - what have we learn? JAMA 2005; 293: 2384-90
7. Kendall J, Reeves B, Clancey M, Point of care testing: randomized controlled trial of clinical outcome. BMJ. 1998; 316:1052-1057.
8. Parvin C, Lo S, Deuser S, et al. Impact of point-of-care testing on patients' length of stay in a large emergency department. Clin Chem. 1996;42:711-717.
9. Murray RP, Leroux M, Sabga E, et al. Effect of point-of-care testing in an adult emergency department. J Emerg Med. 1999;17:811-814.

Address for correspondence

Yu-Chuan (Jack) Li, M.D., Ph.D.
President, Taiwan Assoc of Medical Informatics
Professor and Chief, Graduate Institute of Med Info
Vice Superintendent, Wan Fang Hospital
Taipei Medical University, Taipei, Taiwan
email: jack@tmu.edu.tw | http://li.tmu.edu.tw/

1.3 Ubiquity and Health Services

Ubiquity: Technologies for Better Health in Aging Societies
A. Hasman et al. (Eds.)
IOS Press, 2006

Highly Automated Documentation for Mobile Medical Services

Marko HASSINEN[a], Mikko HEINONEN[a], Maija MARTTILA-KONTIO[a] and Heli TERVO[a]
Department of Computer Science, University of Kuopio
POB 1627, FIN-70211 Kuopio, Finland
{mhassine, mjheinon, mmarttil, hitervo} @ cs.uku.fi

Abstract. Mobile Medical Services, such as Home Care and EMS (Emergency Medical Services) are to the general public probably most visible part of public health care. A great amount of expectations are placed on the quality of care given by these units. Sometimes providing this care is very intensive and all available attention has to be placed on the patient. However, documenting the treatment is very valuable for the treatment of the patient later on. In this paper we present a system that automates many tasks in documenting the treatment. Furthermore, our system is capable of producing a far more detailed documentation that has been available before. This makes reliable research of mobile medical care possible and opens new possibilities in educating paramedics and nurses.
Keywords: Documentation, Mobile Health Units, Data Acquisition, Data Security

1. Introduction

Documentation of medical care is almost as important as the care itself. Medical care without any knowledge of a patient's former medical history can be more difficult and lead to wasting resources with repeated examinations and treatments. As the medical care evolves in a more and more technical discipline, we have a whole new set of tools for monitoring and documenting both the care and physiological findings. An extensive material of measurement values combined with documentation of treatment procedures is valuable material for example for research on effectiveness of a procedure or some medicine. Another great potential with the extensive material comes with estimating the quality of treatment. With precise timing information it is possible to evaluate treatment cases in light of the decisions made by the staff. Evaluating if a procedure or administered drug really helped the patient in light of the physiological state can be very valuable in training and educating of health care providers. [1,2,3,4]

System overview is presented in Figure 1. Data is collected from measurement devices to a collector device, and the communication is managed wirelessly. The collector device is also connected to a remote application server. All data collected from devices is converted to XML form and the application server offers standard interfaces (HL7). Our aims in security are that only accredited parties can access the system and others cannot get any information from the system.

---

[a] These authors contributed equally to this work.

The rest of the paper is organized as follows: Section 2 introduces the used technology, Section 3 discusses about data processing, and security solutions of the system are presented in Section 4. Conclusions follow in Section 5.

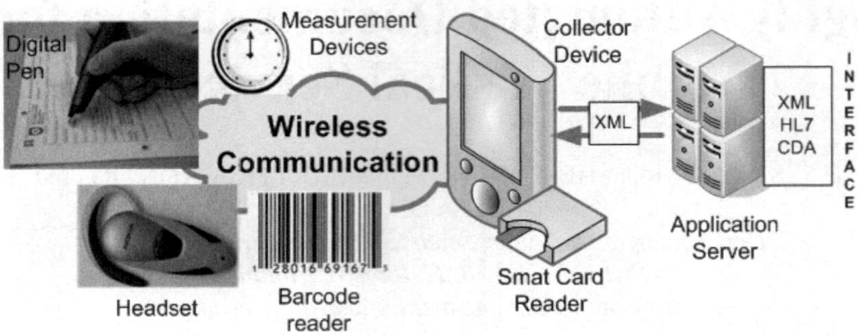

Figure 1. System overview

2. Technology

Documentation process is managed with the help of wireless data acquisition (DAQ) equipment and a collector device, see Figure 1. The collector device runs documentation management software (DMS), which gathers data from DAQ devices and also offers user interface for paramedics, nurses, medics, or patients of medical home care. Gathered data can be temporarily stored in database of the collector device or it can be transmitted wirelessly to a remote server (located in a hospital or a health care centre).

2.1. Devices

The collector device has to be light, movable, handy, and fast [5, 6]. Personal Digital Assistants (PDA) or Tablet PCs meet these demands [7]. Features like touch screen, wireless communication (Bluetooth, IrDA, WLAN, and GPRS), phone, mobility, compactness and price make both device groups competitive compared, for example, to a laptop. A PDA is a better option when only a few measurements are performed. If the collector device has to manage several incoming and outgoing data transmissions, a Tablet PC is better choice because of its better processor capacity and screen.

Digital and wireless medical measurement devices and sensors, such as, blood pressure and oxygen saturation instruments, are attached to a patient. When usability is pursued, the wireless solution ensures almost unlimited mobility of a user with the collector device. To improve the usability of our documentation system we also use a digital pen [8] and a bar code reader. With the digital pen, all written data can be saved into a database or transmitted forward. For example, when filling a required paper form of a medical treatment, written text can be transformed into an electronic form (such as a picture form) or the handwriting can even be recognized and transformed into character data with help of a capable software. A bar code reader is an excellent tool for rapid identification and documentation of, for example, given medicine ampoules or individuals with an identification card.

All medical measurement devices and other wireless data acquisition (DAQ) devices are managed by the software. With the DMS, the end user or administrator can handle data acquisition frequency and functions of each DAQ device. To get highly modifiable software for different situations, different users, and different instruments, we use LabVIEW (National Instrument) [9] for software implementation. The programming language is widely used in information technology, signal processing and handling, and data acquisition and presentation systems. Because of wide amount of predefined programming tools (for communications and data acquisition) of LabVIEW, programming errors decrease [10]. Also, the software can be rapidly modified to meet individual demands of users [11]. Mobile devices (PDA, Tablet PC) have less capacity than PC, so program code of the DMS to be as optimal and powerful as possible. LabVIEW can also be easily used for mobile development due to its PDAModule.

2.2. Communication

Distinguished from the communication between measurement and collector device, the communication between remote server and the device is bidirectional [5]. Bidirectional data transmission enables, for example, dialogical connection between medical consultant and a home care patient, or, between a paramedic and a hospital. The needed connection is handled with TCP/IP using GPRS/WLAN communication, and with Bluetooth and IrDA communications between devices. Using GPS with GPRS communication, there is a possibility to trace the location of the device and its user.

When demanding a good usability, also easy maintenance of the system has to be taken into account. Wireless communication improves use and mobilization of the system as the user does not need to try to modify or update software, or, if the device does not work properly, try to fix it. Using the bidirectional connection, most of the work can be remotely handled by an administrator.

3. Data Processing

Processing data in health care requires several matters to be considered. Not only security aspects, or constraints and commitments brought by law, but also a careful elaboration of the actual form of data. Fragmentariness of data and data systems is still a remarkable problem in using clinical information in different units of health care [4]. In many cases the data transmission means copying a paper document and sending a physical paper to a receiver, and further writing data from that paper document to some certain special and local data system.

As new measurement devices and other tools increase the amount of specific data substantially, the data handling is in a challenging situation. Standardizing data relieves data handling, processing and transfer. Health Level Seven (HL7) [12] is an international community of healthcare creating standards for the exchange, management and integration of electronic healthcare information. Standards of HL7 are based on XML that has become a universal standard form of data, in general.

3.1. Standards

XML [13] is a flexible text format designed for electronic publishing, and for the exchange of wide variety of data on the Web and elsewhere. XML techniques support interoperability of data processing; data transmission between independent services, data transformation, updating and formatting, and distribution of data in different formats. For these reasons, the use of XML is spreading all over the world among different organisations to improve their collaboration. Nowadays XML can be compared to a world wide industrial standard.

The Clinical Document Architecture (CDA) [12] is an ANSI-certified standard from HL7. CDA specifies the syntax for a clinical document, and provides a framework to define the document with full semantics. A CDA document can contain any type of clinical content. CDA defines large amount of different type of clinical documents, for example, documents for patients personal data, laboratory request and response, discharge summary, and so on. Although CDA uses XML, it also allows for a non-XML body; pdf, Word document and jpg, for instance. Several countries all over the world are already using CDA, and it is expected to become a wide-ranging global standard.

3.2. Data Flow in Mobile Medical Services

HL7 provides standards for different clinical documents, but data gathered from home care or EMS do not necessary include all information needed for that certain CDA document. On the other hand, there might not even be defined a certain CDA document yet, for example, for EMS needs, when the development of documents is still ongoing. In these cases data being in XML still brings the advantages of using standards, in general. Using XML allows the data to be processed with general XML tools and softwares, eliminating the need of using tailored solutions in the process.

In the documentation system of mobile medical services, the data is gathered from different devices and transformed into XML, see Figure 1. The data is sent to the Application Server, where it is stored. From the Application Server there are interfaces for needed CDA documents from the stored data. Using HL7 standards the data can be sent and combined easily to the third-party systems.

4. Security of the System

In any medical data system, security is of paramount importance. We have designed our system so that following requirements are fulfilled: 1) Access to the system is granted only to strongly authenticated, accredited parties. 2) Any confidential data will not be exposed to any outsiders. An authenticated accredited party must have a treatment relationship with a patient in order to access the data of that patient. 3) The system must guarantee integrity of all data within the system. 4) All documents must be signed with a digital signature that is legally valid. A user cannot deny a signature at a later time.

4.1. Treatment relationship and Role based access control

In an ideal world any person working with patient data should only be interested in those patients that she is treating. Therefore, we use the concept of Treatment relationship as an access requirement. Treatment relationships are set up using the system and all treatment relationships are recorded to the system. Once defined, a treatment relationship cannot be deleted, but it can be ended. After a treatment relationship has been ended, the treating person can access the data of that particular patient but can not change any of the data.

Defining access control can be a tremendous task in a system that has thousands of users. Errors are likely to occur if access rules are defined for each user separately. However, in an environment such as the public health care one can define different roles, which a user might have, such as a nurse or a paramedic [3, 4]. Defining access rules for each of these roles is rather straight forward as is also assigning roles to users. Thus, we have adopted the usage of roles instead of individual access rules.

The second requirement is satisfied by using the notation of treatment relationship in the access control and end-to-end encryption of all data that is transmitted outside the system. This encryption is achieved using SSL protocol with a requirement of strong mutual authentication.

4.2. Security of Bluetooth

Bluetooth communication has built-in security mechanism that provides confidentiality and integrity of data. There have been several reports about security problems in Bluetooth communication. However, in our system all the data transmitted via Bluetooth communication is measurement data and as such does not identify the patient. Even in the unlikely case of someone breaking the Bluetooth encryption, the privacy of the patient is not compromised.

4.3. FINEID and Digital Signatures

The PRC (Population Register Centre) in Finland issues electronic id cards to citizens and people permanently living in Finland. Each such card contains a private key and a corresponding public key certificate, called a Citizen Certificate. The Citizen Certificate had been issued to 96100 people by the end of year 2005 [14].

The FINEID (Finnish Electronic Identity) can be used to authenticate the holder of the card in electronic applications and it can also be used for producing legally bounding digital signatures. The FINEID infrastructure also provides a CRL (Certificate Revocation List) service [15]. We satisfy the first of our requirements by using the FINEID infrastructure to implement a two-factor authentication, hence the user is *in possession* of the smart card, and *knows* the required PIN (Personal Identification Number) to access the functions on the smart card.

Our system uses tamper resistant smart cards of the FINEID infrastructure with smart card readers on all devices which want to access any part of the system. The keys of a user are created on the smart card. It is practically impossible to extract the private key from the card. The card also contains a processor that can produce a signature using the private key on the smart card.

As with any data, especially with patient data, integrity is a very important factor. We use digital signatures for two purposes; first to be sure that we always know who has added, edited or deleted a piece of information within the system. Secondly, as the digital signature ties together both the data and the signer, it can be used to detect whether any changes has been made after signing the data. Since the holder of the FINEID card is the only one in possession of the private key, no-one else can produce a valid signature on behalf of that user [15]. This satisfies the requirement of non-repudiation stated in the requirement 4.

5. Conclusions

A lot of work has been done in the area of documentation of mobile medical services. However, adequate usability for end-users has not been achieved. The strength of our solution is full mobility enabled by wireless communication and compact devices. The system is designed with well-tried solutions, such as predefined data acquisition and handling tools, data managing with standards like XML, and strong security practices.

The main improvement to medical care that our system provides is detailed information about the treatment on the field. In terms of information this data gives solid ground to further treatment. In addition, the extensive documentation enables traceability of treatment procedures, medicines and patient's responses to aforementioned. Special benefits that follow traceability can be seen in fields of research and education. An extensively documented treatment session can bring valuable insight to training of paramedics and nurses. Furthermore, as the amount of cases grows, statistical research methods can be applied to study different phenomena, such as quality of care or effectiveness of some medicine in given circumstances. This lays ground for evidence based medicine on a field where it has been practically impossible before.

References

[1] Millan J, Park S-E, Kiefer S, Meyer J-U. TOPCARE - Implementation of a Telematic Homecare Platform in Cooperative Health Care Provider Networks. Proc. of the 2nd Joint EMBS/BMES Conf. 2002.

[2] Dwivedi A, Bali RK, James, AE, Nguib RNG. Telehealth Systems: Considering Knowledge Management and ICT Issues. Proc. of the 23rd Annual EMBS International Conf. 2001.

[3] Banaei A, Bensadon M, Vuagnat A, Chodorge M. Management of Modernizing French DRG based Information System (PMSI): How Subsidiary Principle Made It Possible. Connecting Medical Informatics and Bio-Informatics, ENMI, 2005.

[4] Hägglund M, Scandurra I, Moström D, Koch S. Integration Architecture of a Mobile Virtual Health Record for Shared Home Care. Connecting Medical Informatics and Bio-Informatics, ENMI, 2005.

[5] Hassinen M, Marttila-Kontio M, Saesmaa M, Tervo H. Secure Two-Way Transfer of Measurement Data. Proc. of 3rd International Conf. on Information Technology: New Generations. To be published on April 2006.

[6] Ping Yu, Hui Yu. Lessons Learned from the Practice of Mobile health application development. Proc. of the 28th Annual International Computer Software and Applications Conference, 2, 2004, 58-59.

[7] Myers BA, Nichols J, Wobbrock JO, Miller RC. Taking Handheld Devices to the Next Level. Computer 37(12), 2004, 40-51

[8] Despont-Gros C, Boeuf C, Geissbuhler A, Lovis C. The Digital Pen and Paper Technology: Implementation and Use in an Existing Clinical Information System. Connecting Medical Informatics and Bio-Informatics, ENMI, 2005.

[9] National Instrument's LabVIEW website http://www.ni.com/labview

[10] Myers BA. Visual Programming, Programming by Example, and Program Visualization: A Taxonomy. Proc. SIGCHI Conf. on Human factors in computing systems, Boston, Massachusetts, 1986.

[11] Hirakawa M, Ichikawa T. Advances in visual programming. Proc. 2nd IEEE Conf. on Systems Integration, Morristown, NJ, 1992, 538-543.

[12] Health Level Seven (HL7), http://www.hl7.org/.

[13] Bray T et al (eds.). Extensible Markup Language (XML) 1.0 (Third Edition), W3C Recommendation, February 2004. Available at http://www.w3.org/TR/REC-xml/.

[14] Population Register Centre: News. http://www.fineid.fi/vrk/fineid/home.nsf/pages/index_eng

[15] Population Register Centre: FINEID Specifications. http://www.fineid.fi/vrk/fineid/home.nsf/en/documents

Development of a system supporting Patient Supervision and Treatment in contemporary Home–Care: Status Report

Basile SPYROPOULOS[1], Maria BOTSIVALY, Aris TZAVARAS, Konstantinos KOUTSOURAKIS, Eleutherios KOULOURIS, Dimitrios LIARGOVAS, Konstantina MERTIKA

Medical Instrumentation Technology Department
Technological Education Institute of Athens, Athens, Greece

Abstract. The emerging amalgamation of informatics, communication technologies, and entertainment electronics in the field of Biomedical Technology combined to, first, the increase in length of the mean life expectance, and, second, the hospitalization cost avalanche, will facilitate gradually the development of a new Hi-Tec home-care environment. We have developed a home-computer based system, addressing crucial aspects of the development of contemporary home – care that comprises of: First, the employment of low-cost commercially available components, supporting home-care patient's well-being observation, including eventually vital-signs monitoring. Second, software means for the processing, the evaluation, and the targeted transmission of the acquired health-data. Third, software tools for the planning, the documentation, and the management of the corresponding home-care case. The present paper constitutes a progress report of the ongoing development efforts.
Keywords: Home–Care, Telemedicine, Nursing, Treatment Cost, Fuzzy Logic

1. Introduction

The continuous evolution, growth and integration of information, communication and electronic technologies, together with the miniaturization of Biomedical Equipment and their merging with Informatics and Medical Decision Making techniques, will eventually alter the way that health care is going to be delivered in the future. The combination of these technological innovations with the increase of mean life expectance and the hospital care cost avalanche indicates that the 21st Century Hospital will increasingly encourage home-care and especially Telemedicine supported one. The mission of this emerging Hospital will most probably be completed by a network of various associated Institutions, providing several interrelated types of preventive medicine, care for aged citizens, rehabilitation services for impaired persons and

[1] Corresponding Author: Basile Spyropoulos, basile@teiath.gr

mental health-care programs and, generally, providing care closer to home-care than to rather that of the traditional hospital-care.

Therefore, we have developed a home-computer based system, addressing crucial aspects of the development of contemporary home – care that comprises of: First, the employment of low-cost commercially available components, supporting home-care patient's well-being observation, including eventually vital-signs monitoring. Second, software means for the processing, the evaluation, and the targeted transmission of the acquired health-data. Third, software tools for the planning, the documentation, and the management of the corresponding home-care case. The present paper constitutes a progress report of the ongoing development efforts.

2. Monitoring Hardware

The hardware part of the system comprises of the following modules: First, a custom-made ECG acquisition module, equipped with a commercially available RF link (Parallax 433 MHz RF transmitter – receiver) between the custom made ECG-amplifier and the PC. Second, a finger pulse oximetry probe (disposable Nellcor) integrated into a custom-made (using an embedded Parallax BASIC Stamp controller) photo-plethysmography based Oxygen Saturation (SpO2) measuring device, enabling also the estimation of Heart Rate (HR) and Respiration Rate (RR). Third, another custom-made Carotid Sounds (CS) acquisition module, using a stethoscope combined with a miniature microphone, plugged to the PC's soundcard, for the extraction of Heart Rate (HR) and Respiration Rate (RR).

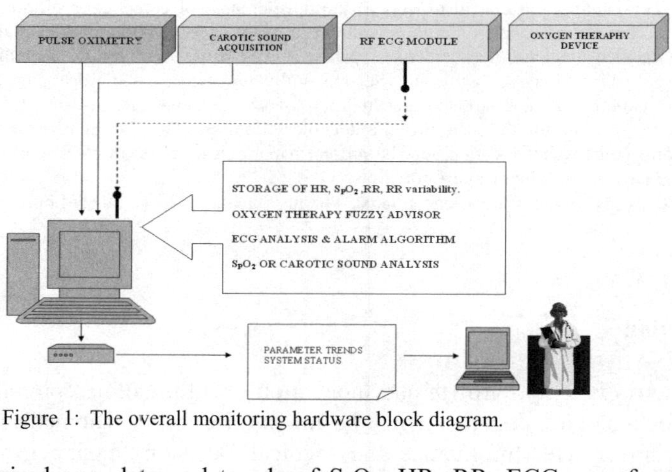

Figure 1: The overall monitoring hardware block diagram.

The acquired row data and trends of S_pO_2, HR, RR, ECG waveforms, Oxygen Therapy advice etc. are appropriately processed to produce decision supportive data, as described in the following part, while they can be easily stored in the home computer and / or transmitted to another location. The transmission can be achieved through wireless point-to-point links, modem and telephone lines on both, landlines and mobile cellular telephone lines, by a designated location through secured paths over the internet and, finally, through two satellite communication links, the Inmarsat Mini-M (connection speed 2.4 Kbits/sec), and the Inmarsat Fleet77 (connection speed 64 Kbits/sec).

It should be mentioned here that the pulse oximetry module is employed only if the patient is in need of Oxygen Therapy, as in cases of Chronic Obstructive Pulmonary Diseases (COPD). The overall monitoring hardware block diagram is presented in Figure 1, while the Carotid Sounds (CS) acquisition module and two typical acquired waveforms are presented in Figure 2.

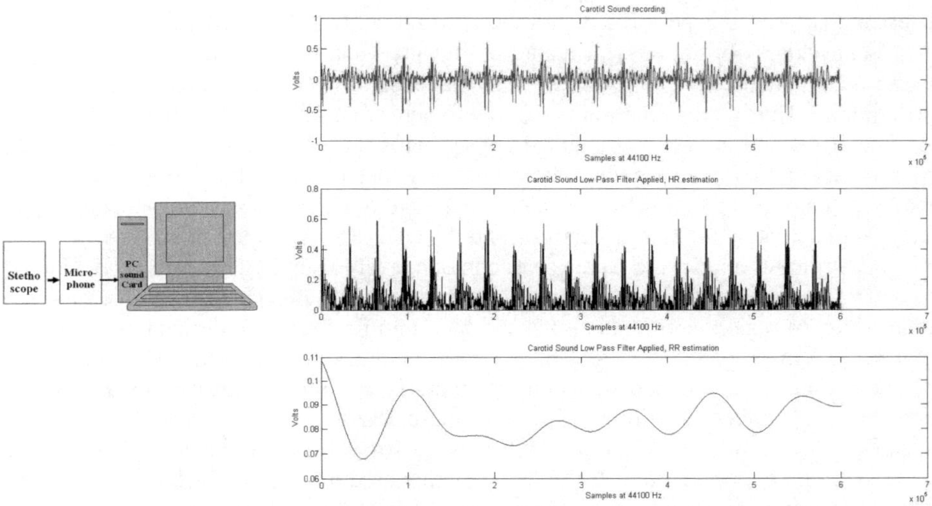

Figure 2: Details of the Carotid sound acquisition module and Heart Rate and Respiration Rate waveforms acquired through the Carotid sound acquisition module.

3. Processing, evaluation, and transmission of the acquired health-data

A Fuzzy logic algorithm was designed and implemented for producing advice on Oxygen Therapy management [1]. The system's response is best described with the graphical representation of the relationship between a pair of input parameters and the system's output as shown in Figure 3.

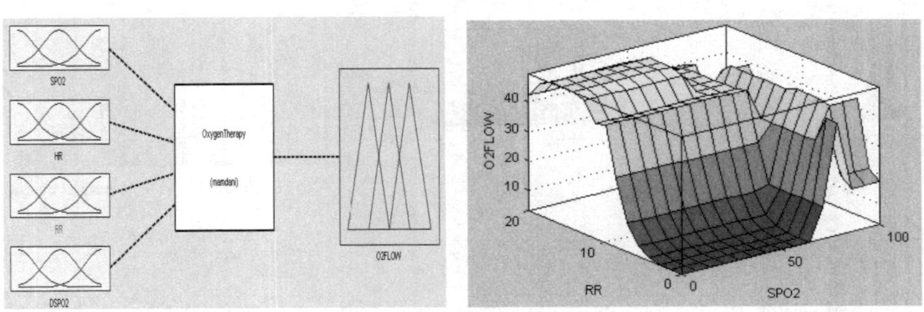

Figure 3: The fuzzy system block diagram (left) and graph surfaces of input parameters versus systems output (right).

Three of the physiology parameters acquired by the acquisition modules, the Oxygen Saturation (S_pO_2), the Heart rate (HR) and the Respiration Rate (RR) are used as inputs for the algorithm. The change in Oxygen Saturation over time (dS_pO_2), which is a reliable factor for evaluating stability, deterioration or improvement of patients' health status, is also included in the model as an indication of the effectiveness of oxygen therapy.

The inputs are translated from crisp values to linguistic variables and the system's response is dictated by a set of fuzzy rules, which were derived based on well established respiration physiology principles. The system actually produces an advice on the percentage change of Oxygen flow to the patient and the translation of its output back to crisp values is performed with the centroid defuzziation method.

Ventricular fibrillation (VF) and malignant ventricular tachycardia (VT) that causes the patient to be pulseless and unconscious are life threatening cardiac arrhythmias. Early defibrillation is the key to achieve a satisfactory survival rate from cardiac arrest and the general medical consensus is that VF and malignant VT should be shocked and all other rhythms should not. The detection of shockable rhythms was made by employing two different techniques that were previously introduced by the same authors that is the Image Analysis and the CDF-SKEW techniques [2].

According to the Image Analysis Technique, every used ECG record is considered to be an image and it is divided into 80 equal regions of interest (ROI) by dividing the image into 20 regions in the time-axis and then into four regions, defined by the limits Vmax, ½ Vmax, 0, -1/2 Vmax and –Vmax, in the amplitude axis. Discrimination between shockable and non-shockable rhythms is possible by measuring the optical density (the number of "filled" pixels) in these ROIs. Normal sinus rhythm has a specific distribution of optical densities across these regions, whereas the corresponding distributions of VF and VT rhythms are remarkably different. These 80 optical density values are used for the classification of the ECG images by the freeware commercially available neural networks program Attrasoft Decision Maker.

The CDF-SKEW technique is a combination of two Descriptive Statistics Functions. These functions allow someone to obtain a quick overview for a set of data without having to consider each observation, or datum, individually. The Cumulative Probability Distribution Function (CDF) and the coefficient of Skewness (SKEW) were used as criteria for the classification of cardiac rhythms as shockable or non-shockable.

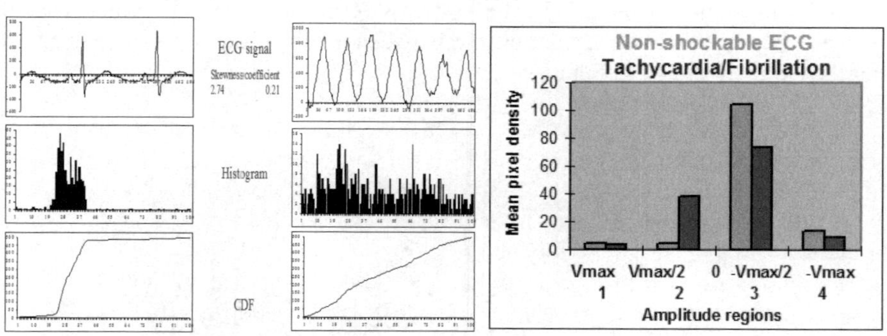

Figure 4: Classification criteria: The Cumulative Probability Distribution Function (CDF) and the Coefficient of Skewness (SKEW) (left), and the Average pixel density distribution (right)

Concerning the processing of the data acquired through the custom-made Carotid Sounds (CS) acquisition module, it should be kept in mind that the sounds and the waveforms of Heart Rate (HR) and Respiration Rate (RR) can be easily reproduced by employing one of the numerous media player freeware tools, while the extraction of direct clinical data, and FFT or otherwise transformed data, is obtained, either through custom-made software or through excellent commercially available tools such as MATLAB, DADISP etc.

4. Software tools for the planning, the documentation, and the management of home-care

The developed model allows for every Hospital Department or Medical/Nursing group, to individually assign an appropriate set of diagnostic, monitoring and treatment activities that can be actually performed in home-environment, together with a nursing – activity treatment plan to specific diagnoses codes that are coded according to Diagnosis Related Group (DRG) codification. These profiles of home-care activities are custom-made and every user, i.e. every physician responsible for discharging a patient from hospital, is actually allowed to set up his own profiles.

During the formation of these profiles the user can attach to each activity a set of nominal fees. This set consists firstly of the official Insurance Agencies reimbursement amount, which in Greece is in most ceases much lower than the actual cost of the services provided, and, secondly, by a currently valid financial rate. This later is estimated by a software tool that we have already developed and allows for a rational approximation of the effective mean cost for several elementary medical activities [3], [4]. Thus, the developed system ignites, when relevant, the approximation of the individual case-cost.

Upon the actual discharge of a patient the physician can use one of the predefined profiles, create a new one or modify an existing one in order to adapt his home-care profiles to specific instances and to emerging new demands. At this point, the appropriate patient and administrative data, as well as the relevant medical information are acquired, either by using an already installed EHR system or even manually, if no EHR is available.

After assigning the appropriate profile to the specific patient, the physician can print a number of forms, including the schedule of the home treatment, instructions for medication doses and proper administration, educational notes for the patient himself or for his relatives and diagrams of physiologic measurements, such as glucose or urine levels, blood pressure, pulse e.t.c. that the patient should monitor.

The system also provides for the production of forms that will be filled by the nursing personnel during the care visits in order to document their activities. These forms include a detailed schedule of the planned visits, which indicates the activities and / or interventions that the nursing personnel should perform during each visit and must be later filled in with the actual outcome of the visit.

The filled forms are returned to the responsible physician who evaluates them and, depending on his evaluation, can modify the care – plan of the specific patient in any suitable way. The actual flow diagram of the developed system is illustrated in the following Figure 5.

It should be mentioned here that the diagnostic and treatment activities are classified according to International Classification of Diseases Version 9 (ICD9), while the Australian Refined DRGs (AR-DRGs) have served for the case codification, since DRGs are not yet been introduced in the Greek National Health System. The Nursing Interventions taxonomy of the Clinical Care Classification (CCC) system [5], which provides a coding structure for assessing, documenting and classifying home and ambulatory care, was used for the documentation of nursing activities.

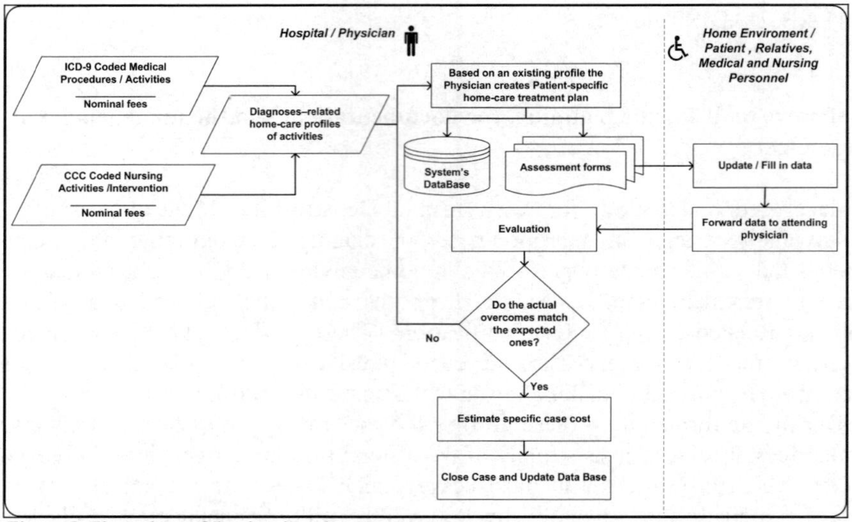

Figure 5: Flow-chart of the developed system

5. Conclusions

The described solution is set up by employing simple, low-cost, commercially available components, supporting home-care patient's vital-signs monitoring. The overall cost of the components of the prototype, beyond the PC, is less than 400 Euros. The system includes in house developed software means for the processing, the evaluation, and the targeted transmission of the acquired health-data, and software tools, for the planning, the documentation, and the management of the corresponding home-care case. The on-going testing of the system shows that it is able to contribute to an effective home – care package solution, supporting not only well organized nursing care, but further, a structured total Patient Supervision and Treatment home-care approach.

References

[1]. Tzavaras A., Spyropoulos B., Botsivali M., Gatsios K., Koufakis A. Multivariable Fuzzy Logic Advisory System. Proc. of EMBEC 2005 - 3rd European Conf. on Medical & Biological Engineering, Prague, Czech Republic, 2005.

[2]. Botsivaly M., Spyropoulos B., Marinis M., Tzavaras A., Koutsourakis K., Sarantakis P. A Software-Tool allowing for Departmental Hospital Operational Cost Estimation. Proc. of EMBEC 2005 - 3rd European Conf. on Medical & Biological Engineering, Prague, Czech Republic, 2005.

[3]. Spyropoulos B, Botsivaly M, Tzavaras A, Nikoloudakis G, Balabanis I, Karagiannis N, Limnou I Acquisition of cost-data for Surgery and Intensive medicine to be employed for drafting a DRGs hospital reimbursement system in Greece. Proc. of EMBEC 2005 - 3rd European Conf. on Medical & Biological Engineering, Prague, Czech Republic, 2005.

[4]. Botsivaly M., Koutsourakis K. Spyropoulos B. A QUASI REAL TIME BIOSIGNAL ACQUISITION PROCESSING AND TRANSMISSION SYSTEM. 4th European Symposium on Biomedical Engineering, Patras, Greece 2004

[5]. Saba VK. Home Health Care Classification of Nursing Diagnoses and Interventions. Washington, DC: Georgetown University, 1994.

1.4 Clinical Bioinformatics: Modeling and Evaluation

Ubiquity: Technologies for Better Health in Aging Societies
A. Hasman et al. (Eds.)
IOS Press, 2006

Fuzzy Hidden Markov Models: A New Approach In Multiple Sequence Alignment

Chrysa COLLYDA[a,1], Sotiris DIPLARIS[a], Pericles A. MITKAS[a], Nicos MAGLAVERAS[a] and Costas PAPPAS[a]
[a] *Aristotle University of Thessaloniki, Greece*

Abstract. This paper proposes a novel method for aligning multiple genomic or proteomic sequences using a fuzzyfied Hidden Markov Model (HMM). HMMs are known to provide compelling performance among multiple sequence alignment (MSA) algorithms, yet their stochastic nature does not help them cope with the existing dependence among the sequence elements. Fuzzy HMMs are a novel type of HMMs based on fuzzy sets and fuzzy integrals which generalizes the classical stochastic HMM, by relaxing its independence assumptions. In this paper, the fuzzy HMM model for MSA is mathematically defined. New fuzzy algorithms are described for building and training fuzzy HMMs, as well as for their use in aligning multiple sequences. Fuzzy HMMs can also increase the model capability of aligning multiple sequences mainly in terms of computation time. Modeling the multiple sequence alignment procedure with fuzzy HMMs can yield a robust and time-effective solution that can be widely used in bioinformatics in various applications, such as protein classification, phylogenetic analysis and gene prediction, among others.
Keywords: Bioinformatics, Multiple Sequence Alignment, Fuzzy Integrals, Fuzzy Measures, Hidden Markov Models, Protein Domains, Phylogenetic Analysis.

1. Introduction

Multiple sequence alignment is a powerful technique that is used by modern bioinformatics systems almost in all their applications. The biomedical methods and algorithms used in MSA have vast importance in solving a series of related biological problems. Protein function prediction exploits MSA in order to recognize patterns left behind by evolution and identify conserved regions that may be of structural or functional importance. In phylogenetic analysis, rates or patterns of change in sequences cannot be analyzed unless the sequences can be aligned [1]. The need for analyzing multiple sequences has led to the development of new methods, such as CLUSTAL-W [2], PSI-BLAST [3] and HMMER [4] that can overpower classic methods of pairwise sequence alignment [5].

The well-known and widely used statistical method of characterizing the spectral properties of the residues of a genomic or proteomic pattern is the HMM approach. Profile HMMs have proved to offer a robust solution for MSA. Their wide use in bioinformatics has led to the creation of large profile databases [6],[7] that can offer biological knowledge (alignments, phylogenetic distribution, domain organization) for

[1] Chrysa Collyda, Lab of Medical Informatics, Faculty of Medicine, Aristotle University, Thessaloniki, 54006, Greece. E-Mail: ckol@med.auth.gr

solving various problems, such as protein classification [8], building of phylogenetic trees [9], or gene function prediction.

However, an issue with the use of HMMs in MSA is its simplifying assumption of stochastic independence. This property, though, is not at all obvious when examining genomic or proteomic sequences; an underlying dependence may exist between current and previous states. Fuzzy HMMs have been introduced in speech recognition [10], in order to relax this assumption and resolve similar model definition issues.

In this paper, a novel way of profile HMM representation is defined by using fuzzy integrals and fuzzy operators in HMMs instead of probability theory. The classical HMM probabilities are replaced by fuzzy possibilities. Though, the profile HMM structure is kept intact in terms of states and observations. The Choquet integral [11] is used for the integration over the HMM states and a new fuzzy measure is used for its application. The advantage of using these fuzzy operators is that they are less constrained than classical integrals and probabilities, thus relaxing the independence assumptions that are necessary with probability functions in classical HMMs. This transformation also reduces the space of computations, thus yielding better response times. In order to perform MSA with the new model, the existing HMM algorithms that are used to train the model and score sequence alignments are also transformed for the fuzzy case.

The rest of the paper is structured as follows. Chapter 2 presents the mathematical background for HMMs and fuzzy measures. In Chapter 3, we mathematically define the new profile fuzzy HMM model for MSA and describe the transformed fuzzy HMM algorithms. In Chapter 4, we depict the potential use and impact of the new alignment method in two example bioinformatics problems, i.e. motif discovery for protein classification, and phylogenetic analysis, while conclusions and future work are highlighted in the last chapter.

2. Background

2.1. Profile Hidden Markov Models

HMMs are statistical models used for MSA that allow the comparison of one gene or protein with a group of others, therefore facilitating the production of distinct differences between itself and the others. HMMs are a generalization of the profile in terms of statistical weights, rather than scores. At each position, the profile HMM gives the probability of finding a particular residue, an insertion, or a deletion. A profile HMM is composed by a number of interconnected states, each of which is able to emit observable output symbols, i.e. residues or gaps. Each state contains symbol emission probabilities and state transition probabilities. The symbol emission probabilities b_{jk} represent the probability of emitting each possible symbol k from a state j, whereas state transition probabilities α_{ij} are the probabilities of moving from the current state i to a next one j. An observation sequence $O=O_1O_2...O_T$ can be generated by starting at an initial state and continuously changing of states, by also emitting symbols, until a specific end-state is reached at time T. The only visible outputs in this procedure are the emitted symbols, while the actual transition between states remains "hidden". Figure 1 depicts the structure of a profile HMM used for MSA, as introduced in [12]. Multiple alignments are used as a training set to build the model. One match state is assigned for each alignment column, insert states serve to insert extra symbols relative to the match states, while delete states allow for skipping positions in the training set aligned sequences. There are totally $N=3*m+3$ states in a profile HMM, where m is the number of its match states.

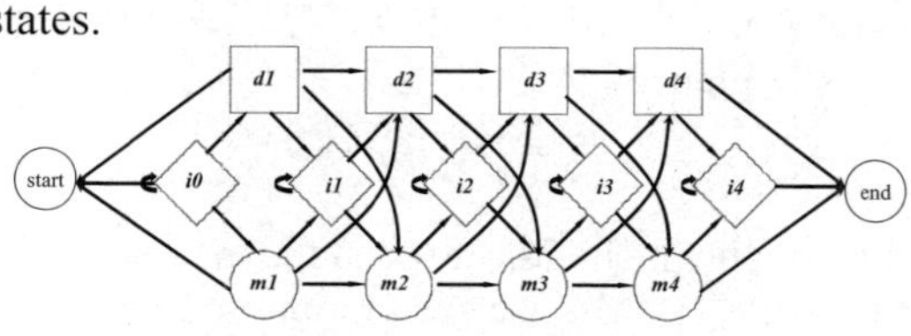

Figure 1. An example profile HMM with four match states.

We now formulate mathematically the profile HMM, according to [13] in order to better describe its fuzzyfication in the following chapter. A profile HMM $\lambda = (A, B, \pi)$ with N states and a vocabulary $V = \{\upsilon_1, \upsilon_2, ..., \upsilon_M\}$ of M output symbols can be fully defined by the matrices A, B and π, where:

- $A = \{ a_{ij} \}$, $1 \le i, j \le N$: the state transition probability distribution
- $B = \{ b_j(k) \}$, $1 \le j \le N$, $1 \le k \le M$: the observation symbol probability distribution
- $\pi = \{ \pi_i \}$, $1 \le i \le N$: the initial state distribution

The utilization of profile HMMs for MSA can be divided in three problems [13]:

- *Problem 1*: Computation of an observation probability according to the model: $P(O|\lambda)$. This is the problem of evaluation (HMM scoring), and it is usually solved using the forward-backward procedure.
- *Problem 2*: Computation of the state-sequence which fits the best to an observed sequence. This is the alignment problem. The Viterbi algorithm is usually used solve this problem and recover the hidden part of the model.
- *Problem 3*: Computation of the model parameters A, B and π to maximize the probability of one observation. This is the training problem, and the EM algorithm is usually exploited to this end.

In the next paragraph we discuss the elements of the theory of fuzzy measures and fuzzy integrals that help in the definition of the fuzzy approach to profile HMMs.

2.2. Fuzzy Measures and Fuzzy Integrals

Classical HMMs are characterized by various probability distributions inside and between the states of the model. The probability distribution, though, has a restrictive property; the property of additivity. I.e., for any subset A and B,

$$A \cap B = \varnothing \quad \Rightarrow P(A \cup B) = P(A) + P(B) \tag{1}$$

In order to relax this constraint and take the relations between subsets into consideration, a generalization in terms of fuzzy measures has been introduced [14]. For the implementation of fuzzy HMMs, a possibility measure can be defined, such that

$$P(A \cup B) = \max(P(A), P(B)) \tag{2}$$

The max operator in the above equation is the fuzzy intersection operator. For a finite fuzzy set $X = \{x_1, x_2, ... x_n\}$, the density μ^j of a fuzzy measure μ can also be defined [15] as $\mu^i = \mu(\{x_i\})$.

The fuzzy measure of Equation 2 can then be used with fuzzy integrals to compute integrations over fuzzy sets.

The fuzzy integral I of a function $h: X \rightarrow [0,1]$ over a subset $A \subset X$ with respect to the fuzzy measure μ is written as:

$$I = \int_A h(x) \circ \mu(\cdot) \tag{3}$$

where $\mu(\cdot)$ is a notation which indicates that the measure μ should be used.

For the case of profile HMMs, the integration is done over the states that are a discrete set. In such cases, the discrete Choquet integral [16] can be used:

$$I = \sum_{i=1}^{n} [h(x_i) - h(x_{i-1})] \mu_i^n \ , \tag{4}$$

with defined $x_i : 0 = h(x_0) \leq h(x_1) \leq \ldots \leq h(x_n)$, and the discrete step μ_i^j :

$$\mu_i^j = \begin{cases} \mu(\{x_i, x_{i+1}, x_j\}) & if \ \ i \leq j \\ 0 & otherwise \end{cases} \tag{5}$$

These new definition of measures and integrals are the major innovations for the new fuzzy profile HMM. The fuzzy measure replaces probabilities that are used in classical profile HMMs, and the fuzzy integral is used for the integration over the states.

3. Fuzzy Hidden Markov Models for Multiple Sequence Alignment

A fuzzy profile HMM can now be defined. Its structure, in terms of states and observations, remains the same as in the classic profile HMM, but fuzzy logic replaces probability theory, and new definitions of the model variables are required.

The fuzzy profile HMM $\bar{\lambda} = (\bar{A}, \bar{B}, \bar{\pi})$ with N states $S=\{S_1,S_2,\ldots,S_N\}$ that can be observed through a space of observations Ω with observations $O=O_1O_2\ldots O_T$ corresponding to unknown state sequences $Q=q_1,q_2,\ldots, q_T$ can be fully defined by the matrices $\bar{A}$, $\bar{B}$ and $\bar{\pi}$, where $\bar{A}$ is the fuzzy state transition matrix, $\bar{B}$ is the fuzzy observation matrix and $\bar{\pi}$ is initial state fuzzy density. Two fuzzy variables $x \in X=\{x_1,x_2,\ldots, x_N\}$ and $y \in Y=\{y_1,y_2,\ldots, y_N\}$ are used to represent the state at time t and $t+1$ [16].

In these terms, $\bar{\pi}_S(A)$ is the grade of certainty that the initial state is in A. Respectively, for $X_0 \subset X$ and $Y_0 \subset Y$, $\bar{\alpha}_Y(X_0 \mid Y_0)$ is the grade of certainty that the state at time $t+1$ is in Y_0, given that the previous state was X_0. Concerning the observation space $\Omega_0 \subset \Omega$, $\bar{b}_i(\Omega_0)$ is the grade of certainty that the current observation is in Ω_0, given a current state S_i.

The manipulation of the above defined parameters causes a different approach in resolving the three basic problems of HMMs.

Specifically, the problem of HMM evaluation can now be solved using the fuzzy forward-backward algorithm. $\bar{\alpha}_i(t)$ is the grade of certainty of $O=O_1O_2\ldots O_T$ and x_i at time t. The initialization step is $\bar{\alpha}_1(i) = \bar{\pi}_i \wedge \bar{b}_i(O_1)$, while the induction step becomes:

$$\bar{\alpha}_{t+1}(i) = \sum_{i=1}^{N} \bar{\alpha}_{ij} [\mu_i^n(t,j) - \mu_{i+1}^n(t,j)] \wedge \bar{b}_i(O_{t+1}) \tag{6}$$

where the sum is the discrete Choquet integral, the $\wedge$ operator stands for the fuzzy intersection operator, and μ_i^j is defined in Equation 5. From the above equation, it is possible to observe that the assumption of independence of the observation until time t is not necessary anymore neither is necessary the knowledge of the next state. The answer to the Problem 1 as stated in Chapter 2.2 for the forward and backward variables respectively is:

$$P(O \mid \lambda) = \sum_{i=1}^{N} \overline{\alpha}_T(i), \quad P(O \mid \lambda) = \sum_{i=1}^{N} \overline{\beta}_1(i) * \overline{b}_i(O_1) \qquad (7), (8)$$

In the fuzzy case, the grade of certainty for a sequence is used to score the model. The Choquet integral is computed over the states at each time t, where the integration step ($\mu_i^n(t,j) - \mu_{i+1}^n(t,j)$) becomes a value j at time $t+1$.

Respectively, the fuzzy Viterbi algorithm, which is used for the alignment of new sequences to the model, uses the Choquet integral and multiplication for the fuzzy intersection operator in order to define the variable δ for the fuzzy case:

$$\overline{\delta}_t(i) = \max_{q_1, q_2, \ldots, q_{t-1}} \left\{ \overline{\pi}_{q_1} b_{q_1} \prod_{\tau=2}^{t} \left[\overline{\alpha}_{q_{\tau-1} q_\tau} \rho_\tau(q_{\tau-1}, q_\tau) \right] \overline{b}_{q_\tau}(O_\tau) \right\}, \qquad (9)$$

where $\rho_t(i,j) = [\mu_i^n(t,j) - \mu_{i+1}^n(t,j)] / \overline{\alpha}_t(i)$. $\overline{\delta}_t(i)$ is the degree of certainty for a single state sequence finishing at time t in a state S_i.

Similarly, for training the fuzzy HMM model, the fuzzy version of the EM algorithm can be derived, again by using the fuzzy coefficient that multiplies the state transition coefficients and summing up using the Choquet integral.

4. Example Applications in Bioinformatics

By summarizing the diversity of information in an existing alignment of sequences, it is possible to predict whether new sequences belong to a protein family. The fuzzy profile HMMs can be used in bioinformatics to reorganize protein domains, by rescoring their profiles. This process leads to a different representation of protein data in terms of motifs, since the new scores obtained by the definition of the fuzzy HMM allow for the discovery of novel motifs that may exist in protein sequences. The protein function, as it is described by its motifs can be predicted using separate motif datasets that yield either from the classical profile HMM or the fuzzy HMM. Motif-based protein classification is then performed for each dataset respectively, thus providing the bioinformatician with enhanced quality assessment of the classification results.

Another potential impact of the proposed method is in phylogenetic analysis, a powerful tool in clinical bioinformatics that enables the identification of genetic diversity in viruses and the course of diseases progression, among others. Any effective sequence-based phylogeny inference begins by performing efficient MSA. A feature of profile HMMs, which have been used for MSA so far, is that these are finite models for the probability distribution over an infinite number of possible sequences. Profile HMMs have the great benefit on generalized profiles that they are formally built on the probability theory. Though, this theory restricts the flexibility of the models because the sum of the probability distribution over all modeled sequences must equal to one. In consequence, the probability of one sequence cannot be increased without decreasing the probability of another sequence in the profile-HMM. Fuzzy profile HMMs lack this restriction, thus enabling a better representation of the sequences common residues, and ultimately the construction of better phylogenetic trees.

5. Conclusions

In this paper we discussed a novel generalization of the profile HMM, which exploits fuzzy measures and fuzzy integrals in order to build, score and produce alignments in a more time-efficient manner. The new approach relaxes the independence restriction implied in classical profile HMMs, thus providing more biologically meaningful alignments. We have mathematically formulated the new model, by also describing the modifications of the standard HMM algorithms for the fuzzy case. Finally, we discussed the potential impact of performing MSA with this new method in two example applications in bioinformatics. Future work involves the application of the method in phylogenetic analysis, as well as the creation of a whole new series of profiles that can then be used in protein classification.

Acknowledgments

This work was supported by the EPEAEK II project: "IRAKLEITOS: Scholarships targeted in basic research. Phylogenetic analysis of biological data".

References

[1] Salemi M, Vandamme AM, editors. The Phylogenetic Handbook. Cambridge University Press; 2003.
[2] Thompson D, Higgins DG, Gibson TJ. Clustal W: improving the sensitivity of progressive multiple sequence alignment through sequence weighting, position-specific gap penalties and weight matrix choice. Nucleic Acids Res 1994; 22:4673-80.
[3] Altschul SF, Madden TL, Schäffer AA, Zhang J, Zhang Z, Miller W, Lipman DJ. Gapped BLAST and PSI-BLAST: a new generation of protein database search programs. Oxford d University Press Nucleic Acids Research 1997; 25(17):3389–3402.
[4] Eddy SR. Profile Hidden Markov Models. *Bioinformatics* 1998; 14:755-763.
[5] Pearson WR. Empirical statistical estimates for sequence similarity searches. In J. Mol. Biol. 1998; 276:71-84.
[6] Bateman A, Coin L, Durbin R, Finn RD, Hollich V, Griffiths-Jones S, Khanna A, Marshall M, Moxon S, Sonnhammer ELL, Studholme DJ, Yeats C, Eddy SR. The Pfam protein families database. Oxford University Press, Nucleic Acids research 2002; 30(1):276-280.
[7] Mulder NJ. InterPro. In: Encyclopedia of Genetics, Genomics, Proteomics and Bioinformatics, Jorde LB, Little PFR, Dunn MJ, Subramanian S, editors, John Wiley & Sons, 2435-2439; 2005.
[8] Diplaris S, Tsoumakas G, Mitkas PA, Vlahavas I. Protein classification with multiple algorithms. In Proceedings of the Tenth Panhellenic Conference on Informatics 2005; Volos, Greece.
[9] Collyda C, Maglaveras N, Pappas C. A Multiple Sequence Alignment Tool Using Hidden Markov Models For Phylogenetic Analysis. In Proceedings of the 4th European Symposium on Biomedical Engineering 2004; Patra, Greece.
[10] Tran D, Wagner M. Generalised Fuzzy Hidden Markov Models for Speech Recognition. In Proceedings of the International Conference on Fuzzy Systems 2002; Calcutta, India.
[11] Narukawa Y, Murofushi T, Sugeno T. Conditions for choquet integral representations. Proceedings of the Eighth International Fuzzy Systems Association World Congress 1999; 911–925.
[12] Krogh A, Brown M, Mian IS, Sjolander K, Haussler D. Hidden Markov models in computational biology: Applications to protein modeling. Journal of Molecular Biology 1994; 235:1501-1531.
[13] Rabiner A. A tutorial on Hidden Markov Models and selected applications in speech recognition. Proceedings of the IEEE 1989; 77(2):257-285.
[14] Wang Z, Klir G. Fuzzy measure theory. Plenum Press; 1992.
[15] Grabisch M, Murofushi T, Sugeno M, editors. Fuzzy measures and integrals : theory and applications. New York: Springer; 2000.
[16] Mohamed M, Gader P. Generalized hidden markov models. i. theoretical frameworks. IEEE Transactions on Fuzzy Systems 2000; 8(1):67–81.

Ubiquity: Technologies for Better Health in Aging Societies
A. Hasman et al. (Eds.)
IOS Press, 2006

Modelling Epidermal Homeostasis as an Approach for Clinical Bioinformatics

Niels GRABE [a,1], Thora POMMERENCKE [a], Dennis MÜLLER [a], Simone HUBER [a], Karsten NEUBER[b], Hartmut DICKHAUS [a]

[a] *Department of Medical Informatics, University Hospital Heidelberg, Germany*

[b] *University Hospital Hamburg Eppendorf, Germany*

Abstract. Modelling in systems biology currently lacks clinical applications. As a possible approach leading to clinical relevance the modelling of tissue homeostasis is proposed. As an example a model of epidermal homeostasis is presented which reproduces central morphological and kinetic characteristics of epidermal tissue. Each individual cell is modelled as an agent. The tissue arises as an emergent phenomenon from the interactions of agents. Each agent's behaviour is qualitatively modelled by a simple differentiation state-flow program. Epithelialisation under the influence of parameters concerning stem-cell location is briefly demonstrated.

1. Introduction

Systems-Biology aims at the systematic analysis and modelling of as large and complex biological networks as possible. Today, systems-biology already has become an important discipline [1, 2] although its modelling side lacks still a clear link to clinical applications. To limited intracellular systems the toolbox of non-linear dynamics has been successfully applied [3, 4]. Nevertheless, the construction of larger systems is the more difficult the larger these systems get. Therefore, here a new approach is suggested. Clinical relevance for modelling could be much better gained by switching to a higher abstraction level by modelling tissue homeostasis instead of only focusing at quantitative intracellular biochemical networks. This approach has major advantages.

Firstly, genomics and proteomics technologies work with homogenized tissue. So the spatial resolution of molecular processes in high-throughput technologies is rather neglected. In the contrary to this, in a clinical setting the morphological analysis of tissue is the prevailing criteria for defining and diagnosing diseases.

Secondly, modelling tissue homeostasis is concerned with rather slow biological processes. An average regeneration of human epidermis takes 20-25 days while signal transduction processes are executed in milli- and microseconds [5]. Slow processes have the advantage of opening the possibility of qualitative modelling. Most biological knowledge published in literature today describe biological phenomena in a qualitative way.

So, can tissue be functionally modelled today? According to the reductionist paradigm systems should be reconstructed bottom-up. Unfortunately, modelling at the tissue level on the basis of completely reconstructed intracellular networks exceeds our

[1] Corresponding Author: Niels Grabe, Department of Medical Informatics, University Hospital Heidelberg, Im Neuenheimer Feld 400, 69120 Heidelberg, niels.grabe@med.uni-heidelberg.de

technological capacity by far in the coming time. On the other hand, the analysis of tissue morphology is a corner-stone of clinical diagnosis – actually it's the gold-standard. Concluding, it is not the question if tissue functionality can be modelled. Rather it is the question of how it can be done as it will become mandatory for clinical bioinformatics and systems biology to gain a substantial understanding of many critical diseases.

Different approaches for tissue modelling are known [6, 7]. Simulation of tissue homeostasis per definitionem has to model the proliferation and differentiation activity of major parts of the tissue in order to simulate the regenerative capacity of the tissue. Until today only few works model aspects of tissue homeostasis [8-11]. Recently a first prototype has been presented for modelling both, proliferation and differentiation, while maintaining an intact tissue morphology and tissue kinetics [12].

The epidermis is an excellent example for this kind of modelling as it represents the most complex epithelium. Epithelial tissue covers all body organs like skin, liver, bladder, lung, small intestine, kidney and lines the body's cavities. Over 85% of all cancers arise from a disturbed epithelial homeostasis. Carcinomas are characterized by a loss of spatial tissue integrity. This can be generally described as an unbalance of epithelial proliferation and differentiation. Thus modelling the tissue homeostasis in epidermis may be a good field for studying systems biological modelling in a clinically relevant setting.

We here describe how qualitative modelling has been used to develop a simple agent based discrete differentiation program for epidermal keratinocytes. This discrete differentiation program is used to control the fate of individual agents in a multi-agent environment. A biomechanical model of interacting cells allows the passive spatial movement of cells by proliferative activity inside the tissue. In the following it is shown how from the interaction of all agents different morphologies of a human epidermis arise as emergent phenomena. This emergent behaviour of the agent-society can produce different epidermal homeostatic morphologies.

2. Method

The model has been implemented in the java software environment MASON [13] supporting the efficient simulation of agent societies. All agents have one common source-code implementation. The simulation environment executes the agents in a semi-parallel manner.

2.1. Biomechanical Model

The model consists of a biomechanical and a biochemical component. The biomechanical component reflects the spatial part of the simulation while the biochemical component controls each participating cell's internal program. Emergent behaviour of the simulation is enabled by simulating the tissue as a multi-agent society where each single cell is represented by an agent. The individual agents cannot move by their own force (cell migration). They can only be moved if necessary to minimize structural forces inside the tissue. As a simple biomechanical model for each cell a force linear to the distance of the cells is assumed. To allow an interactive simulation cells move heuristically in limited fractions of the forces exerted on it. If two cells overlap they try to partially restore their appropriate distance. If two cells are in a short limited distance they adhere to another. In this way both, adhesion and passive cellular movement are implemented. To avoid artificial side effects the available space for cells is the unwinded surface of a cylinder. Towards to bottom the cell's movement is restricted by a constant basal membrane. At the beginning the simulation consists only of an undulated basal membrane and a small set of stem-cells. Figure 1 shows the constants available for determining the layout of the basal membrane and the seeding of the virtual stem-cells in the simulation.

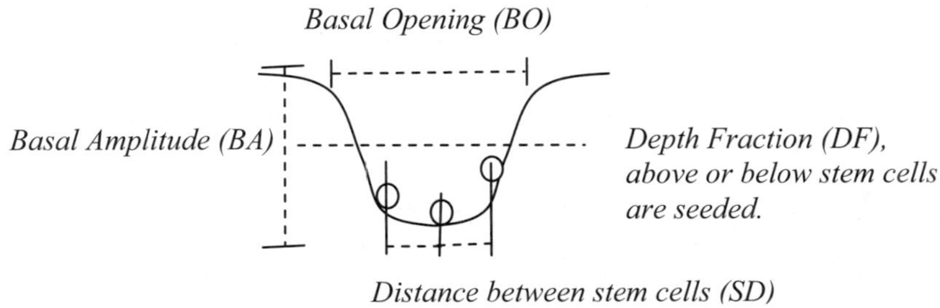

Figure 1: Constant parameters determining the basal layout of the simulations

2.2. Biochemical Model

The biochemical model of the simulation is qualitatively described by a flow of states as depicted in figure 2. Although the epidermis consists of further cell types like Langerhans-cells and melanocytes 95% of the cells are keratinocytes. Therefore, all cells included in the simulation are keratinocytes. Each agent executes the same differentiation program which is therefore common to all in-silico keratinocytes.

Stem cells are assumed to be proliferating asymmetrically which means that after completion of the cell cycle the stem cell remains a stem-cell, but has produced a differentiating daughter cell. This is consistent with a recent report [14] which shows that asymmetric proliferation is important for building a stratified epithelium. Cells which have been produced by stem cells are assumed to be transit-amplifying (TA) cells which possess a limited proliferation capacity. Proliferating TA-Cells asymmetrically produce differentiating "early spinosum cells" named after the layer stratum spinosum as described in figure 3. Under the influence of high extracellular Ca^{2+} Concentration these cells differentiate in the model into „Late Stratum Spinosum Cells".

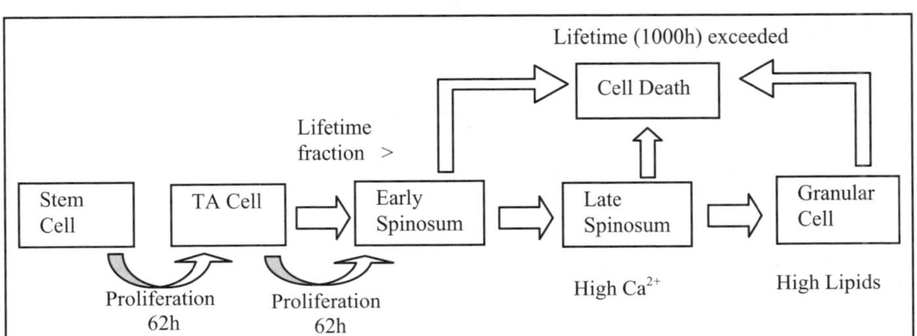

Figure 2: Qualitative differentiation model of a healthy single cell

Healthy epidermis is characterized by a Ca^{2+}-Gradient which is assumed to play a central role in regulating epidermal homeostasis. The passive transportation theory assumes that the trans-epidermal waterflow transports Ca^{2+} towards the stratum corneum under which Ca^{2+} is accumulated.

The transepidermal water flux is realised implicitly in the model. When two cells collide, the lower cell transports a fraction of its calcium environment to the upper cell. In the same way further objects like lamellar bodies are transported upwards. Lipids are produced in the model beginning from the stratum spinosum on where they are enclosed in lamellar bodies. These bodies are later secreted at the surface of the epidermis where they are secreted as a central component for building the stratum corneum. Visualising the stratum corneum requires the graphical modelling of corneocytes which has not been addressed here. Therefore in this simulation only the building of the two basic components of the stratum corneum through the underlying epidermal strata from stratum basale to stratum granulosum has been modelled.

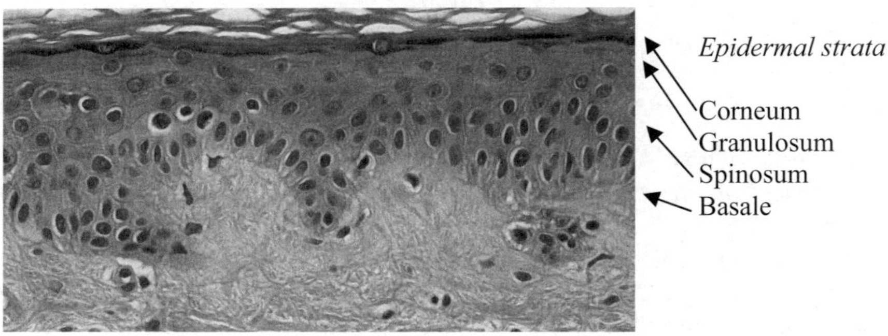

Figure 3: Section of human epidermis

The successful establishment of a stratum granulosum is set equal to successfully establishing an epidermal barrier which reduces the transepidermal waterflow. In this way a feedback of the whole system is achieved. Cells are taken out of the simulation after they reach a certain constant age independent of where there are at this moment localized in the tissue.

3. Results

Epithelialisation is shown in Figure 4. The resulting morphology represents the behaviour of the total agent society. We varied the basal parameter settings of the simulation according to table 1.

Parameter	**Scenario 1**	**Scenario 2**	**Scenario 3**	**Scenario 4**
BO	150 µm	150 µm	150 µm	150 µm
BA	40 µm	10 µm	40 µm	40 µm
DF	2%	2%	50%	50%
BD	8 µm	8 µm	8 µm	8 µm
Above/Below	Below	Below	Below	Above

Table 1: Behavioral scenarios of the simulated epidermis defined by modifications of basal membrane related parameters.

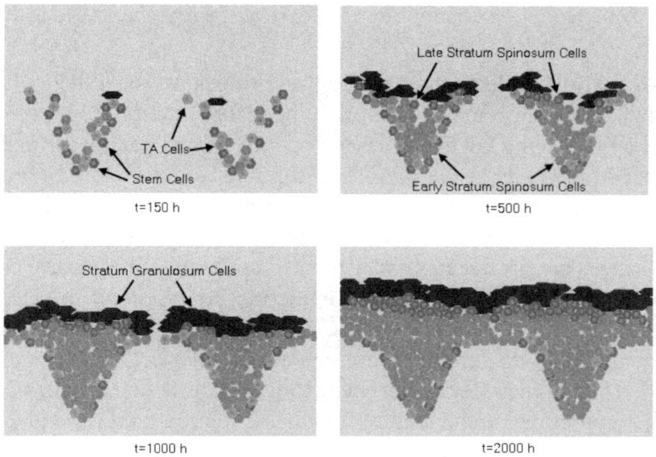

Figure 4: Intact morphology of the tissue emerges after epithelization of the model-space.

The evaluation of the four different scenarios (1, 2, 3, 4) shows four examples of morphologies obtained with the simulation (Figure 5). The standard scenario 1 shows a basal opening of 150 µm and a basal amplitude of 40 µm as shown in figure 1. The depth fraction of 2 % means in combination with the parameter below means that stem cells are seed in the lower 98% of the rete ridges. In this are stem cells are seeded with a distance of 8 µm. The other three scenarios show three different modifications of these settings. In scenario 2 the basal amplitude is reduced to create an epidermis with shallow rete ridges. In scenario 3 stem cells are only seeded in the in the upper parts of the rete ridges while in scenario 4 stem cells are exclusively seeded in the bottom of the ridges.

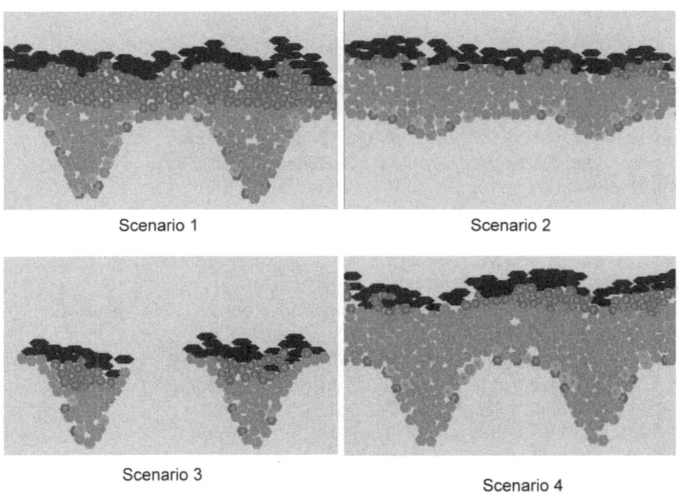

Figure 5: Different morphologies at the same timepoint t= 2100 h right after obtaining tissue homeostasis. Parameters are set according to table 1. Compared to Figure 4 a slightly lower transepidermal waterflux has been chosen to produce pronounced layers.

4. Discussion

The standard-scenario 1 reflects rather strong rete ridges as this a difficult situation for the simulation. This scenario makes clear that the differentiation of the cells may not be simply related to the aging of the cells as it could be the case with just a flat basal membrane. Instead the cells at the bottom of the rete ridges need a time-space behaviour different from the ones at the edges of the rete ridges. The well layered morphology shows that the described differentiation program can achieve this. In scenario 2 shallower rete ridges are produced similar to aged epidermis. In scenario 3 there are too few stem-cells to obtain a fully populated epidermal morphology. In scenario 4 stem-cells are located at the top of the rete ridges. The differences between scenarios 1 and 4 reflect a rather historical dispute of where epidermal stem-cells are located. It is interesting to see that the higher stem-cells seem to lead to low density populated low rete ridges, a thicker epidermis, a changed tissue differentiation pattern, and a more irregular epidermal surface.

5. Conclusion

It was shown that simulation of epidermal homeostasis with a limited and rather simple qualitative differentiation program is feasible and may yield interesting results. The approach opens the possibility to include more detailed knowledge about the behaviour of individual cells. This integration of more biomolecular knowledge in the present prototype could show a way of how the systems-biology approach could contribute to clinically relevant bioinformatics.

References

1. Hood, L., et al., *Systems biology and new technologies enable predictive and preventative medicine.* Science, 2004. **306**(5696): p. 640-3.
2. Kitano, H., *Computational systems biology.* Nature, 2002. **420**: p. 206-210.
3. Hofestadt, R. and S. Thelen, *Quantitative modeling of biochemical networks.* In Silico Biol, 1998. **1**(1): p. 39-53.
4. Sauro, H.M., et al., *Next generation simulation tools: the Systems Biology Workbench and BioSPICE integration.* Omics, 2003. **7**(4): p. 355-72.
5. Sneyd, J., J. Keizer, and M.J. Sanderson, *Mechanisms of calcium oscillations and waves: a quantitative analysis.* Faseb J, 1995. **9**(14): p. 1463-72.
6. Hunter, P. and P. Nielsen, *A strategy for integrative computational physiology.* Physiology (Bethesda), 2005. **20**: p. 316-25.
7. Noble, D., *Modeling the heart.* Physiology (Bethesda), 2004. **19**: p. 191-7.
8. Clem, C.J., M. Boysen, and J.P. Rigaut, *Towards 3-D modelling of epithelia by computer simulation.* Anal Cell Pathol, 1992. **4**(4): p. 287-301.
9. Meineke, F.A., C.S. Potten, and M. Loeffler, *Cell migration and organization in the intestinal crypt using a lattice free model.* Cell Proliferation 2001, 2001. **34**: p. 253-266.
10. Morel, D., R. Marcelpoil, and G. Brugal, *A proliferation control network model: the simulation of two-dimensional epithelial homeostasis.* Acta Biotheor, 2001. **49**(4): p. 219-34.
11. Rashbass, J., *Modelling tissues on the computer.* Trends Cell Biol, 1996. **6**(7): p. 280-1.
12. Grabe, N. and K. Neuber, *A multicellular systems biology model predicts epidermal morphology, kinetics and Ca2+ flow.* Bioinformatics, 2005. **21**(17): p. 3541-7.
13. Luke, S., et al. *MASON: A new Multi-Agent Simulation Toolkit.* in *2004 Swarmfest Workshop.* 2003.
14. Lechler, T. and E. Fuchs, *Asymmetric cell divisions promote stratification and differentiation of mammalian skin.* Nature, 2005. **437**(7056): p. 275-80

Ubiquity: Technologies for Better Health in Aging Societies
A. Hasman et al. (Eds.)
IOS Press, 2006

Sifting abstracts from Medline and evaluating their relevance to molecular biology

Natalia Grabar[a*], Marie-Christine Jaulent[a], Antoine Chambaz[b], Céline Lefebvre[c], Christian Néri[c]
[a] *Université Paris Descartes, Faculté de Médecine; Inserm, U729; SPIM, 75006 Paris, France*
[b] *Université Paris Descartes, MAP5 CNRS UMR 8145, 75270 Paris, France*
[c] *Inserm, Avenir Group, Laboratory of Genomic Biology, Centre Pail Broca, 74014 Paris, France*

Abstract. The most important knowledge in the area of biology currently consists of raw text documents. Bibliographic databases of biomedical articles can be searched, but an efficient procedure should evaluate the relevance of documents to biology. In genetics, this challenge is even trickier, because of the lack of consistency in genes' naming tradition. We aim to define a good approach for collecting relevant abstracts for biology and for studied species and genes. Our approach relies on defining best queries, detecting and filtering best sources.
Keywords: Medical Informatics, Information storage and Retrieval, Natural Language Processing, Text categorisation, Medline, Biological Sciences, Language, France

1. Introduction

Research in biology produces a huge amount of knowledge which is mostly available as raw text data. Biomedical databases and portals like Medline[1], HON[2] or CISMeF[3] provide an important effort for its capitalisation and diffusion. Querying such databases allows documents related to one's topics of interest to be reached easily and quickly. The general biomedical purpose of these databases is undoubtedly vital for progress in biomedicine. Scientists from various disciplines (psychology, cardiology, biology, etc), can find citations relevant to their questions. However, it is necessary to define the relevance of every one document to specific domains of biology and medicine. In this paper, we address the computing of document relevance to molecular biology, notably for the species and genes which are of interest in our study.

During the manual curation of the scientific literature in the context of biology, the first task is to define if the document is relevant to biology. Database curators usually define if the document may also describe experiments with given genes. The automation of this process, as it is aimed during contests on text retrieval and information extraction, follows the same schema. In the Challenge Evaluation task for the Knowledge Discovery and Data Mining (KDD) in biology [1] systems needed to return three types of information: (*a*) a ranked list of papers as for the need of their curation, (*b*) a decision on whether to curate or not each paper, and (*c*) a decision on whether the paper contains target gene products.

*Corresponding author: Natalia Grabar, U729, 15 rue de l'Ecole de Médecine, 75006 Paris, France; natalia.grabar@spim.jussieu.fr
[1]Catalog of biomedical literature: `www.ncbi.edu/entrez`
[2]Health on the Net: `www.hon.ch`
[3]Catalogue et index des sites médicaux francophones: `www.chu-rouen.fr/cismef`

During the Text REtrieval Conference (TREC) competition on biological documents, systems needed to detect Medline citations that contain a description of gene function as specified in the query [2]. The same decision, on whether documents have to be processed or not according to a given topic, has to be made during more modest tasks, *i.e.* when building relevant biological corpora for text mining. We have then to decide on whether to include or not documents into the corpora. The underlying assumption is that the quality, both precision and recall, of these corpora will define the quality of the information that may be further extracted. In the following sections, we present the context of our work (sec. 2), the source materials (sec. 3), and methods for the detection of relevant biological corpora (sec. 4). We then present and discuss the obtained results (sec. 5), and outline the perspectives of this work (sec. 6).

2. Background

Comparative biology addresses the genetic conservation and developmental diversity across species. To precisely compute genetic conservation and generate clusters of genes with a similar function, a new method, namely the Best-Balance Constraint Procedure (BBCP) has been recently developed [3]. This method combines two kinds of information: microarray experiments and comparison of protein sequences. The first one gives information about homologue genes, having similar expressions into the same organism, and the latter allows pointing out ortholog genes from different species. With two studied species, *D. melanogaster* and *C. elegans*, BBCP has generated over 700 clusters containing over 3000 genes. The biological meaning and validity of clusters involve consulting functional annotations in databases. For the studied species, we can consult *FlyBase* [4] for *D. melanogaster* and *WormBase* [5] for *C. elegans*. But these databases are often insufficient and scientific literature have to be consulted. In practice, the analysis of such literature can imply the consultation of hundreds of citations, which is extremely time-consuming. Computer work described in [3] took about 1 month for collecting, analysing, weighting the data and generating clusters. In contrast, the definition of the biological content of the resulting clusters by consulting the scientific literature took several months. For such purposes, automated methods for text mining are strongly needed.

Before the application of these methods, corpora on the topics of interest have to be built. Medline contains over 12 millions citations and represents the most used source of biomedical literature. It serves to research on text mining, information retrieval, and natural language processing. In our work, we use this database to collect scientific literature related to our topic. The main query material is composed of species and gene names. However, their use for querying Medline is error prone: gene naming lacks of discipline and consistency, and gene names are known to be ambiguous [6]. For instance, *Bazooka*, *the*, *for*, *wee*, *up* are all gene names. The purpose of this paper is to define the best approach for building relevant corpora for text mining. It relies on two points: (1) definition of queries in order to obtain the most precise results, and (2) resolving gene names ambiguity and computing the relevance of abstracts to species and genes of interest. Our approach relies as well on the detection of journals which are more relevant to the topics of interest. Working with comparative biologists, where the task of corpora building has to be performed as often as new species and genes are involved in biological phenomena and/or experiments, gives more reasons to perform research on methods to build high quality corpora.

3. Material available

Our source material consists of species names and some of their genes grouped into clusters. Current species are *D. melanogaster* and *C. elegans*. The list of species names synonyms comprises *Drosophila melanogaster*, *fruit fly*, and *Caenorhabditis elegans*, *nematode*, *worm*. A group of 3850 genes (1925 for each species) was classified into 719 clusters [3]. We took advantage of the reference databases for these two species, namely *FlyBase* and *WormBase*. 1047 gene symbols are already recorded in there, 24 for the worm and 1023 for the fly. Some of recorded genes have aliases. 12,544 is the total number of gene names with their aliases. *Gene Ontology* (GO) provides 18,315 preferred terms to describe gene functions. These terms are structured into three separate hierarchies corresponding to molecular functions, biological processes, and cellular components. Since 2003, GO is included in the UMLS [7] and about 18% of GO terms have been linked to existing Metathesaurus concepts. For these terms, UMLS provides the synonyms. We use Medline and its *eutilities*[4], tools for collecting useful abstracts. Eutilities tools send query terms to Medline and download on the local computer the resulting abstracts. It is possible to set the threshold of the number of downloaded abstracts for each query. Queries have been sent to Medline by December 2005. GO, *FlyBase*, *WormBase* and UMLS have been updated within the same period.

4. Methods for sifting and filtering of abstracts

Our query material consists of species and gene names and, for some gene names, of their aliases. The problem is that gene names, especially those of *D. melanogaster*, are known to be ambiguous [6]. Abstracts collected with queries composed with ambiguous gene names may thus be noisy. Before applying methods for gene function(s) mining we aim at defining the best method to build the corpora and being sure that our corpora are relevant to the species and genes being studied. We used a two-step process. First, we defined which queries are able to provide the most precise results. Second, we resolved gene names ambiguity and computed the relevance of abstracts to the studied species and genes. We are also interested in detecting the sources most relevant to the query content.

Our source material, species and gene names, allowed us to define four sets of Boolean queries: (1) gene symbol AND its aliases grouped together in the same query (set *GeneSynoSer*); (2) each gene symbol AND each alias as separate query (set *GeneIso*); (3) each gene symbol AND each alias AND species names (set *Gene + Species*); (4) species names AND their aliases AND words like *gene, genetics*, etc. (set *Species*). Punctuated gene names (*wee-1.3, BcDNA:GM03307, D-Fos*) are protected with quotes. The first three types of queries have been thresholded to 200 abstracts at most. For the last type of queries, we collected all the available abstracts. Once abstracts were downloaded we computed their relevance to biology area and to the query. The 'relevance score' combines three

[4] `http://ncbi.nih.gov/entrez/query/static/eutils_help.html`, to be used for numerous queries on off-peak periods, with no more than one request every 3 seconds.

parameters: (1) Centrality of abstracts for biological area. It is evaluated through the density of biological terms, like those from GO [8]. We projected GO terms and their synonyms onto abstracts and computed the useful content of each abstract. This corresponds to the percentage of GO words in a whole abstract length. The manual validation showed that the useful content rate has to be set to at least 6%; (2) Centrality of abstracts for the species: at least one mention of the species is required; (3) Centrality of abstracts for the gene being studied: at least two occurrences of gene names are required.

5. Results and discussion

In Tab. 1, we indicate, for each query, the figures on downloading abstracts from Medline and on their filtering. The first column *Query* exposes four kinds of queries. Results are indicated for both species, *D. melanogaster* and *C. elegans*: the number of downloaded abstracts (*Medline*), the number of abstracts after the filtering with the relevance rate set to 6% (*filtr*), and the reduction percentage (*%*) due to filtering. The reduction percentage is correlated with the precision of each query. The last line *Total* indicates the total number of abstracts that remain after the filtering procedure.

	D. melanogaster			***C. elegans***		
Query	**Medline**	**filtr**	**%**	**Medline**	**filtr**	**%**
GeneSynoSer	89,408	2322	2.6	10,194	1873	18.4
GenIso	138,602	4812	3.5	8513	1760	20.7
Gene + Species	14,645	4843	33.0	1626	1488	91.5
Species	21,439	4931	23.0	12,833	3103	24.2
Total		8391			3367	

Table 1: Sifting abstracts from Medline and their filtering.

The number of returned abstracts (*Medline*) is always more important for *D. melanogaster* than for *C. elegans*. We can see two reasons for this: (1) *D. melanogaster* is more studied than *C. elegans* (*Gene + Species* and *Species*); (2) gene names for *D. melanogaster* are more numerous and ambiguous that those for *C. elegans* (*GeneSynoSer* and *GeneIso*). Upon filtering (*filtr* and *%*), we noticed that the number of remaining abstracts is reduced: 2.6-33% of fly abstracts and 18.4-91.5% of worm abstracts survived the selection rate of 6%. When only gene names and aliases are used (*GeneSynoSer* and *GeneIso*), queries give the noisiest results. This is particularly true for *D. melanogaster*: precision is then 2.6 and 3.5, while for *C. elegans* we obtain up to 18.4 and 20.7 of precision. As previously observed [6] and as pointed out by our study, fly gene names are indisputably ambiguous. Our global filtering method decreased gene name ambiguity through categorising the whole abstract, thus ensuring any one abstract is related to biology.

When only species names are used, precision for both species is comparable, and it is higher compared to using gene name queries: 23% for the fly and 24.2% for the worm. These figures are difficult to explain. Only possibility is that all downloaded abstracts for these species may not refer to precise genes or may refer to genes which are not in our list. Moreover, the filtering was based on abstracts while genes and possibly their functions are mentioned in full text documents (Although the most important information of a full paper is usually concentrated in the abstract, even if the highest information coverage is located in the results sections [9]). When gene names and aliases are combined with species names (*Gene + Species*), collected abstracts show the better precision: 33.0% for *D. melanogaster* and 91.5% for *C. elegans*. In this case, species carry out the first disambiguation of gene names, like in [2]. The further filtering helped to rank abstracts according to their relevance to the biology.

Queries show different results: *Species* keywords give the best coverage, while *Gene + Species* give the best precision. But what so ever the query is, it is necessary to rate abstracts and keep only the most relevant ones. We assumed that the same strategy can be adopted for rating abstracts for other species and genes. We also assume that it will be necessary to perform further disambiguation of gene names, which will be more contextual.

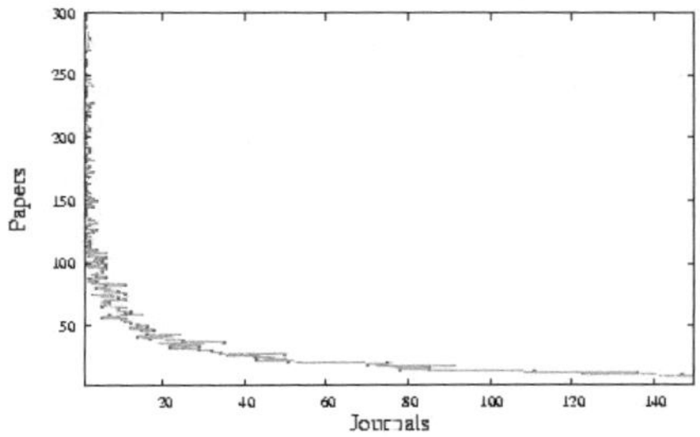

Figure 1: Abstracts downloaded (GeneIso)

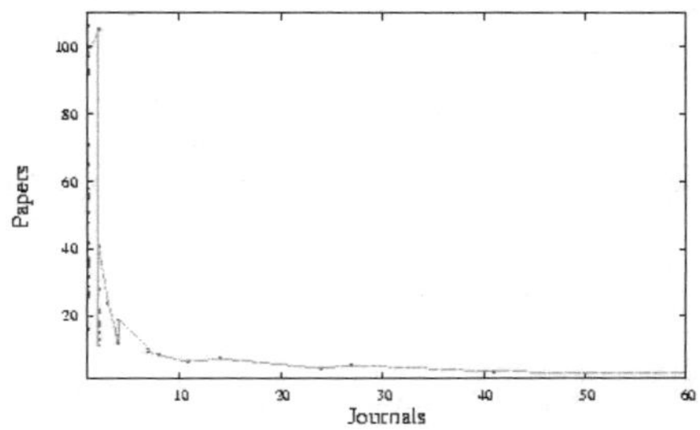

Figure 2: Abstracts after filtering

In order to define the most important sources (journals, symposia, ...) we analysed the two graphics shown above (fig. 1 and fig. 2). These figures show the distribution of sources and their papers first when downloading them with *GeneIso* queries and then after filtering. The X axis (*Journals*) indicates the number of sources and the Y axis (*Papers*) the number of papers from each of the sources. For instance, during the downloading step (fig. 1), about 20 sources provide with about 30 papers. The extremities of the curves, which are very extended, are not included in the graphics. The results are similar for all the queries: they follow the Zipf law. According to this law, the probability that an item occurs starts high and tapers off. In our experiment, few journals propose a very large number of articles while many others propose fewer articles. We noticed thus that the downloaded top ten sources are not always relevant to biology. These top ten sources include *J Biol Chem* (5289 articles), *Proc Natl Acad Sci U S A*, *Mol Cell Biol*, *Biochem Biophys Res Commun*, *Biochemistry*, *Development*, *J Virol*, *EMBO J*, *Oncogene* and *Genetics*. After filtering (fig. 2), the remaining top ten seems to be more relevant to developmental and evolutionary biology: *Development* (493 articles), *Genetics*, *Proc Natl Acad Sci U S A*, *Dev Biol*, *J Biol Chem*, *Genes Dev*, *Mech Dev*, *Mol Cell Biol*, *Cell*, *Curr Biol*. This suggests that while it is possible to enhance the biological significance of the query (searching for the best representative sources), this may result in an increase of the silence effect. Querying large bibliographic databases, appears then to be more interesting for collecting more complete corpora. Our results suggest as well that even if an article is published in a source relevant to biology, the relevance of this article to studied species and genes must be confirmed.

6. Conclusion and Perspectives

We have presented experiments on defining the best approach for collecting text corpora relevant to molecular biology and specific species and genes. When working with Medline abstracts, use of species names and their aliases gives the best coverage, while the combination of species with gene names gives the best precision. In all cases, it is necessary to rate the relevance of the abstracts and keep only the most relevant ones. We showed that it is more interesting to work with large bibliographic databases and not only taking into account the most relevant known sources. For the evaluation of biological relevance, we performed a home rating of abstracts. It would be interesting to compare our results with Medline abstracts which are already referenced in *FlyBase* or to apply our method to the abstract set of KDD contest [1]. Querying the remote Medline server is time-consuming and may overload the server. Installing a copy of Medline on local computers is advisable and should increase the flexibility of the process. The entire content of Medline is available as a set of text files[5]. Moreover, Java and Perl tools are available for loading this file into a local relational database [10]. Our future work is related to the extraction of the functional profiles of the genes by taking advantage of the GO resource.

Acknowledgements

This work was performed within the frame of the Biomathematics and Genomics working group of the University of Paris 5. This work was supported by the University of Paris 5, Inserm, and CNRS, Paris, France.

References

[1] Yeh AS, Hirschman L, and Morgan AA. Evaluation of text data mining for database curation: lessons learned from the KDD Challenge Cup. Bioinformatics 2003;19(Suppl):i331--9.

[2] Kayaalp M, Aronson AA, Humphrey SM, et al. Methods for accurate retrieval of MEDLINE citations in functional genomics. In: Proc of the TREC (Text Retrieval Conference), 2003:175--84.

[3] Lefebvre C, Aude JC, Clément E, and Néri C. Balancing protein similarity and gene co-expression reveals new links between genetic conservation and developmental diversity in invertebrates. Bioinformatics 2005;21(8):1550--8.

[4] FlyBase Consortium. The FlyBase database of the Drosophila genome projects and community literature. Nucleic Acids Research 1994;22(17):3456--8.

[5] Stein L, Sternberg P, Durbin R, Thierry-Mieg J, and Spieth J. WormBase: network access to the genome and biology of Caenorhabditis elegans. Nucleic Acids Research 2001;29(1):82--6.

[6] Hirschman L, Morgan A, and Yeh A. Rutabaga by any other name: extracting biological names. Biomed 2002;25(4):247--59.

[7] NLM. UMLS Knowledge Sources Manual. National Library of Medicine, Bethesda, Maryland, 2005. *www.nlm.nih.gov/research/umls/*.

[8] Gene Ontology Consortium. Gene Ontology: tool for the unification of biology. Nature genetics 2000;25:25--9.

[9]Schuemie M, Weeber M, Schijvenaars B, et al. Distribution of information in biomedical abstracts and full-text publications. Bioinformatics 2004;20(16):2597--604.

[10]Oliver DE, Bhalotia G, Schwartz AS, Altman RB, and Hearst MA. Tools for loading MEDLINE into a local relational database. BMC Bioinformatics 2004;5(146).

[5] `http://www.nlm.nih.gov/databases/leased.html`

1.5 Clinical Bioinformatics: Integration

Ubiquity: Technologies for Better Health in Aging Societies
A. Hasman et al. (Eds.)
IOS Press, 2006

Inferring gene expression networks via static and dynamic data integration

Fulvia FERRAZZI, Paolo MAGNI, Lucia SACCHI and Riccardo BELLAZZI[1]
Dipartimento di Informatica e Sistemistica, Università degli Studi di Pavia, Italy

Abstract. This paper presents a novel approach for the extraction of gene regulatory networks from DNA microarray data. The approach is characterized by the integration of data coming from static and dynamic experiments, exploiting also prior knowledge on the biological process under analysis. A starting network topology is built by analyzing gene expression data measured during knockout experiments. The analysis of time series expression profiles allows to derive the complete network structure and to learn a model of the gene expression dynamics: to this aim a genetic algorithm search coupled with a regression model of the gene interactions is exploited. The method has been applied to the reconstruction of a network of genes involved into the Saccharomyces Cerevisiae cell cycle. The proposed approach was able to reconstruct known relationships among genes and to provide meaningful biological results.
Keywords: systems biology, machine learning, DNA microarrays

1. Introduction

Over the last few years a noteworthy research effort has been devoted to the development of methods for the automated extraction of gene interaction networks from DNA microarray data [1]. The data can be static, i.e. snapshots of gene expression in different experimental conditions, such as in mutants, or dynamic, i.e. time series of gene expression data collected during the evolution of processes of particular interest, for example the cell cycle. The results provided by gene interaction learning algorithms are phenomenological models, which may suggest hypotheses about the topology of the underlying cellular pathways [2]. In particular, knockout experiments are considered the most effective way to reveal gene relationships. However, such experiments are typically static, since they are expensive and the number of potential knockouts is very high.

In this paper we will propose a new method for building gene interaction networks by combining Saccharomyces cerevisiae expression data coming both from static knockout experiments [3] and from wild type time series measurements [4]. We believe that the combination of these different data sources allows us to better exploit the information provided by DNA microarrays, as advocated in [5].

The method herein presented exploits a simple regression model to describe the gene expression dynamics and uses a genetic algorithm to search through the space of possible gene network structures [6]. Together with the ability to take into account both

[1] Corresponding Author: prof. Riccardo Bellazzi, Dipartimento di Informatica e Sistemistica, via Ferrata 1, 27100 Pavia, Italy. E-mail: riccardo.bellazzi@unipv.it

static and dynamic data, a distinguishing feature of the proposed approach is the inference of a set of potential interaction networks, which can be ordered on the basis of their posterior probability with respect to the data.

2. Method

In this paper we propose a method to infer gene interaction networks relative to the Saccaromyces Cerevisiae cell cycle. The basic steps of the method can be summarized as follows (Figure 1):

1. learning of an initial network topology using knockout data;
2. selection of the genes involved in cell cycle;
3. filtering of the genes on the basis of the available data about cell cycle dynamics;
4. learning of the final interaction network and of a model of gene expression dynamics through a genetic algorithm search coupled with regression models.

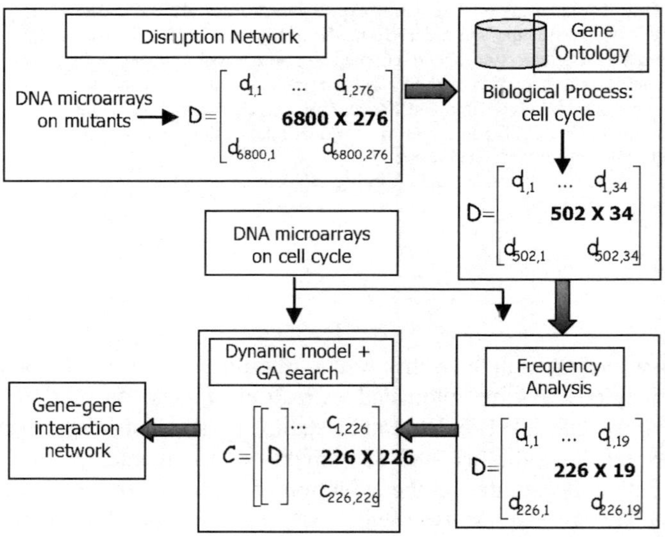

Figure 1. The proposed method

2.1. Learning the initial network topology from mutant data

This step is based on the analysis of the data made available by Hughes et al. [3]: the authors collected the data of 276 experiments in which a single gene had been knocked-out and the RNA abundance of all the other genes (about 6800) had been measured through cDNA microarrays. The goal of this study was the detection of the functional modules of each mutated gene. Starting from the knockout experiments, it is possible to derive a preliminary network of gene interactions. First of all a matrix E (6800x276) can be built, where each element e_{ij} contains the expression level for gene i measured in experiment j (that is when gene j has been mutated). The next step consists in the transformation of matrix E into a discretized matrix D, where the element $d_{ij}=1$ if

e_{ij} exceeds a gene specific threshold and $d_{ij}=0$ otherwise [3,7]. The matrix D thus obtained can be mapped into a "disruption network" [7], where an arc from gene j to gene i is drawn if $d_{ij}=1$.

2.2. Gene ontology and dynamic filtering

We decided to focus our attention on the genes belonging to the Gene Ontology biological process "cell cycle". In this way the dimension of the matrix D was reduced to 502 x 34.

In order to learn a dynamical model of the control of the genes involved in cell cycle, it is necessary to resort to dynamic data. In the case of the yeast cell cycle, the reference data are the ones coming from a well-known experiment from Spellman et al. [4], where expression values were collected in 18 different time points (one each 7 minutes, from 0 to 119 minutes). Since under the experimental conditions used by the authors the yeast cell cycle lasts 66 minutes, it is possible to observe almost two complete mitotic cycles. The knowledge on the dynamics of the cell cycle period, together with the information on the sampling time, limits the scope of the investigation to the search for relationships which can be reasonably detected in the available data. In particular, given the sampling time, we cannot detect signals with frequency components higher than (1/(2*7) min^{-1}). We therefore decided to filter out the gene profiles with energy content located in high frequencies, setting a cut-off level at 0.05 (1/20) min^{-1}. This choice is able to preserve the cell cycle frequency and its first harmonic component. In this way the matrix D has been further reduced to 226 x 19.

2.3. Learning dynamic models

Starting from the connection matrix D obtained after low-pass filtering, we implemented a novel algorithm to select the final model of the gene network interactions. Such algorithm consists of two ingredients: a) a dynamic mathematical model that describes the available data; b) a strategy to search for potential relationships in the unexplored portion of the connection links (a matrix D' 226 x 207). We have selected dynamic linear models since they are the simplest class of models which allows for periodic or damped oscillation behaviors.

The dynamics of the expression level (x) of the i-th gene is thus described using an "additive regulation model" [8] :

$$x_i(k+1) = a_{ii}x_i(k) + \sum_{j=1, i\neq j}^{n} a_{ij}c_{ij}x_j(k)$$

where the a_{ij} are the connection weights and the matrix $C=|D\ D'|^{226x226}$ is the connection matrix obtained by concatenating the known matrix D^{206x19} and the unknown matrix $D'^{206x207}$ that has to be learned from the data.

Given a certain network topology, the parameters a_{ij} can be learned from the available data through least square fitting. If also the topology is unknown, as in our case, it is possible to follow a model selection approach to compare different models, i.e. different C matrices. We used the Bayesian Information Criterion (BIC) as the score to rank the examined models and we employed a Genetic Algorithm strategy to search through the model space. Supposing that the models for each gene are independent of each other, it is possible to look for the best scoring models relative to each gene and then combine them in an overall network. In particular, the Genetic algorithm has been implemented by using the following settings: 20 “individuals” (i.e. possible models for each gene) per generation, cross-over probability = 0.85, mutation probability = 0.15, and probability of selecting the *i*-th individual which is proportional to the fitness (inverse of BIC). For each gene, we ran the algorithm 30 times, each time letting it evolve until the fitness calculated at every generation reached stability (i.e. its variation remained below a predefined threshold).

3. Results

Interesting results have been obtained in all phases of the learning process. To evaluate such results, we selected 22 genes whose role in the cell cycle is well characterized and we extracted from the inferred networks only the relationships relative to this subset of genes.

We first examined the network built exploiting the data coming from Hughes' disruption experiments. The resulting network (shown in Figure 2a) contains some links that appear to be supported by the information available in the literature: the gene *Sic1*, for example, is connected to its transcription factor *Swi5*, while *Cln1* and *Cln2* are linked to *Swi4*, a component of their transcription factor.

The network was then extended following the strategy proposed in this paper. The overall model, obtained by combining the best scoring models found for each gene, is characterized by a satisfactory goodness of fit (RMSE=0.045). Moreover several connections among the known cell cycle genes have been added (dashed arrows in Figure 2b) and a significant number of these reflects the knowledge available in the literature about gene interactions. In particular, the following interesting relationships can be observed: a) *Cdc14-Sic1*. *Cdc14* is a phosphatase required to exit from mitosis: by direct dephosphorylation of key substrates, Cdc14 promotes Sic1 accumulation, thus allowing the switch between mitosis and G1 phase of the cell cycle. This connection expresses therefore a physical link between the two genes involved. b) *Swe1-Clb5* and *Swe1-Swi4*. These links, on the other hand, suggest complex interactions between the elements involved. We can indeed interpret these connections as an indication of the co-expression of these genes, that all play an important role at the Start of the cell cycle.

We then examined the connectivity of the networks derived in the different runs of the algorithm and we observed that in each of them only a small portion of the genes is connected with more than 40 other genes. In particular, the most connected genes, conserved in all the runs, appear to be: *Swi4,* the DNA binding component of the SBF transcription factor; the B-type cyclin *Clb2*, activator of Cdc28 at the G2/M phase of the cell cycle; *Bim1*, which is the microtubule-binding protein that together with Kar9p delays the exit from mitosis when the spindle is oriented abnormally; *Rnr1* (Ribonucleotide-diphosphate reductase), which is regulated by DNA replication and DNA damage checkpoint pathways; *Dsk2,* a nuclear-enriched ubiquitin-like polyubiquitin-binding protein, required for spindle pole body (SPB) duplication and for transit through the G2/M phase of the cell cycle; *Fus3*, a mitogen-activated protein (MAP) kinase that mediates both transcription and G1 arrest in a pherormone-induced signal transduction cascade.

4. Conclusions

The approach described in this paper is an example of how different types of knowledge and data sources can be conveniently integrated in gene network learning. The current implementation of the method combines both static and dynamic gene expression data with prior information about the analyzed biological process. The algorithm was able to reconstruct known relationships among genes and to provide meaningful biological results and seems therefore suitable for further investigations and refinements. Differently from other strategies for modeling gene regulation, such as Dynamic Bayesian Networks, Boolean Networks and Differential equations, the presented algorithm offers a flexible way to explore the space of interaction networks while taking into account prior knowledge [1]. An interesting extension would be to include in the strategy also other types of genome-wide data. Several researchers have indeed recognized that expression data alone are not sufficient to build reliable models, because, given the high number of genes, it is possible to derive many models that describe the expression data equivalently well. Therefore recent approaches integrate other prior biological knowledge, such as information on the transcription factors and their binding sites and on protein interaction data [9-10].

Acknowledgments

We thank R. Amici, S. Rinaldi and S. Sotgiu for their help. This work is part of the PRIN project "Dynamic modeling of microarray gene expression profiles: clustering techniques and regulatory networks", funded by the Italian Ministry of Education.

References

[1] de Jong H. Modeling and simulation of genetic regulatory systems: a literature review. J Comput Biol 2002; 9(1): 67-103.

[2] Schlitt T, Brazma A. Modelling gene networks at different organisational levels. FEBS Lett 2005; 579(8):1859-66.

[3] Hughes TR, Marton MJ, Jones AR, Roberts CJ, Stoughton R, Armour CD, Bennett HA, Coffey E, Dai H, He YD, Kidd MJ, King AM, Meyer MR, Slade D, Lum PY, Stepaniants SB, Shoemaker DD, Gachotte D, Chakraburtty K, Simon J, Bard M, Friend SH. Functional discovery via a compendium of expression profiles. Cell 2000; 102(1):109-26.

[4] Spellman PT, Sherlock G, Zhang MQ, Iyer VR, Anders K, Eisen MB, Brown PO, Botstein D, Futcher B. Comprehensive identification of cell cycle-regulated genes of the yeast Saccharomyces cerevisiae by microarray hybridization. Mol Biol Cell 1998; 9(12): 3273-97.

[5] Bar-Joseph B. Analyzing time series gene expression data. Bioinformatics 2004; 20(16):2493-503.

[6] Mitchell TM. Machine Learning. Mac Graw Hill, 1997.

[7] Rung J, Schlitt T, Brazma A, Freivalds K, Vilo J. Building and analysing genome-wide gene disruption networks. Bioinformatics 2002;18 Suppl 2:S202-10.

[8] D'Haeseleer P, Liang S, Somogyi R. Genetic network inference: from co-expression clustering to reverse engineering. Bioinformatics 2000; 16(8):707-26.

[9] Lee TI, Rinaldi NJ, Robert F, Odom DT, Bar-Joseph Z, Gerber GK, Hannett NM, Harbison CT, Thompson CM, Simon I, Zeitlinger J, Jennings EG, Murray HL, Gordon DB, Ren B, Wyrick JJ, Tagne JB, Volkert TL, Fraenkel E, Gifford DK, Young RA. Transcriptional regulatory networks in Saccharomyces cerevisiae. Science 2002; 298(5594):799-804.

[10] Nariai N, Tamada Y, Imoto S, Miyano S. Estimating gene regulatory networks and protein-protein interactions of Saccharomyces cerevisiae from multiple genome-wide data. Bioinformatics 2005; 21 Suppl 2:ii206-ii12.

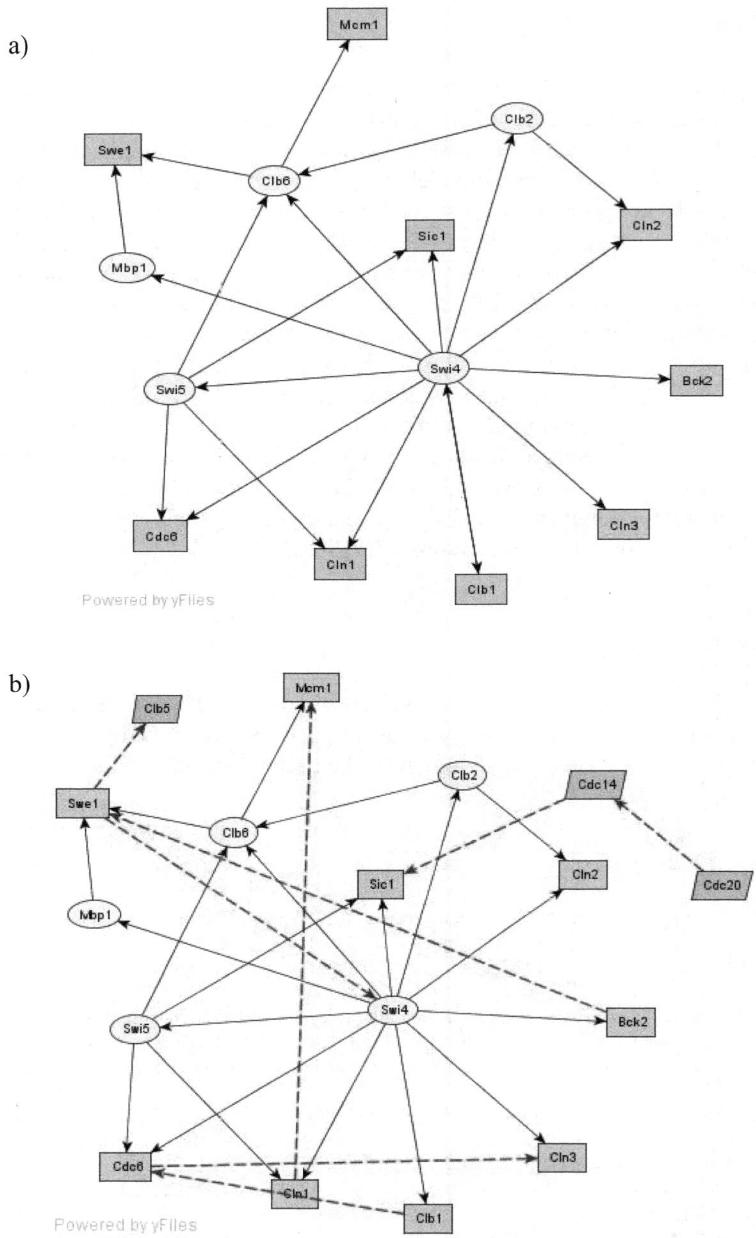

Figure 2. Results relative to a subset of well characterized cell cycle genes. a) The preliminary network inferred from the knockout data from Hughes et al. [3]. Elliptical nodes represent mutated genes; b) The final network obtained using the method proposed in this paper: the new nodes and links are represented as parallelograms and dashed arcs.

Ubiquity: Technologies for Better Health in Aging Societies
A. Hasman et al. (Eds.)
IOS Press, 2006

Integrating Medical and Genomic Data: a Sucessful Example for Rare Diseases

Gaspar S. DIAS[1], José Luis OLIVEIRA[1],
Javier VICENTE[2] and Fernando MARTIN-SANCHEZ[2]
[1] *University of Aveiro, IEETA/DET, Aveiro, Portugal*
[2] *Instituto de Salud Carlos III, Madrid, Spain*

Abstract. The recent advances on genomics and proteomics research bring up a significant grow on the information that is publicly available. However, navigating through genetic and bioinformatics databases can be a too complex and unproductive task for a primary care physician. In this paper we present diseasecard, a web portal for rare disease that provides transparently to the user a virtually integration of distributed and heterogeneous information.
Keywords: Biomedical databases, information retrieval, rare diseases.

1. Introduction

The recent advances in molecular biology and the last achievements of the Human Genome Project [1] have raised the expectations on the use of this knowledge in medicine [2]. It is also expected that the synergy between genomics and medicine will contribute to the understanding of genetic level implications to the human health [3, 4]. On this context, and due to the rapid expansion of biomedical knowledge, the reduction of computing costs and the spread of internet access, there is a huge amount of electronic data, provided by a vast amount of database implementations. These databases provide valuable knowledge for the medical practice. But, given their specificity and heterogeneity, we cannot expect the medical practitioners to include their use in routine investigations [5]. To obtain a real benefit from them, the clinician needs integrated views over the vast amount of knowledge sources, enabling a seamless querying and navigation.

In this paper we present Diseasecard (www.diseasecard.org), a public web portal system that provides a single entry point to access relevant medical and genetic information available in the Internet about rare diseases. By navigating in the entire alphabetic list, or by searching directly a disease name, a gene symbol or an OMIM code, the user can retrieve information about near to 2000 rare diseases. After the identification of the disease, the portal will present a structured report, containing the details of the pathology and providing entry points to further resources either on the clinical and genetic domains [6].

[1] José Luis Oliveira, University of Aveiro, DET/IEETA, jlo@det.ua.pt, www.ieeta.pt/~jlo

2. Material and Methods

The main concept in diseasecard is to present to the user "a card for each disease". To attain this goal the card is built upon a navigation workflow that guides the user along several well-know public databases [6]: Orphanet, Clinical Trials, OMIM, Entrez Gene, HGNC and Expasy (Figure 1). This selection of biomedical resources came from a selection made over the most significant databases registered in *The Molecular Biology Database* Collection, published by NAR (*Nucleic Acids Research*) [7]. From this core set, diseasecard's engine retrieves not only local information but also links to other resources that are relevant to build a complete pathology-to-gene card.

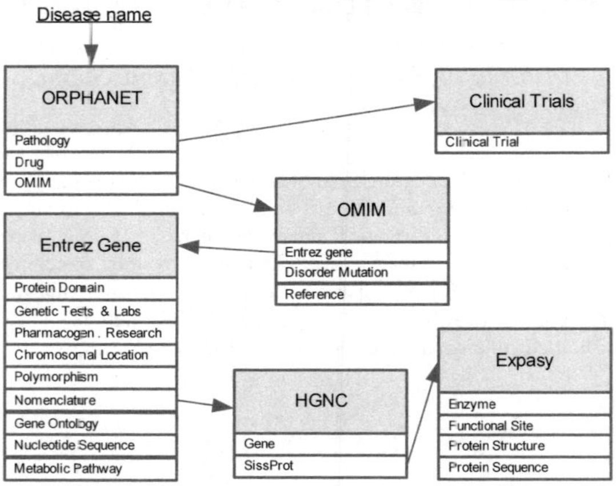

Figure 1: Diseasecard data sources network. Each box represents a data source containing respective retrievable concepts (URLs). Some of these concepts lead to other data sources and other concepts.

Since the cards' construction is based on a predefined template [8], the explored data sources are always the same so the respective queries/URLs are similar, except their query ids. Because of this feature we can map the building task into a single protocol. By this way, the hard job that is navigating through databases and retrieve information can be delegated to the system. Thus, Diseasecard application provides an engine which automatically builds card instances based on an expert-user-defined protocol descriptor. Through this protocol, an expert user can define the sources to be consulted during the querying process as well the information keywords to be retrieved for each source. This protocol is then interpreted by a parser which converts the XML syntax into process tasks. Figure 2 shows the current architecture of Diseasecard System. The XML Protocol Descriptor (XPD) engine launches a collection of web wrappers through the Internet according to the settings defined in the XPD file. After these operations are completed, the XPD engine organizes all the gathered information in a conceptual network, according to a hierarchical structure that is defined in another XML file – the Card Descriptor.

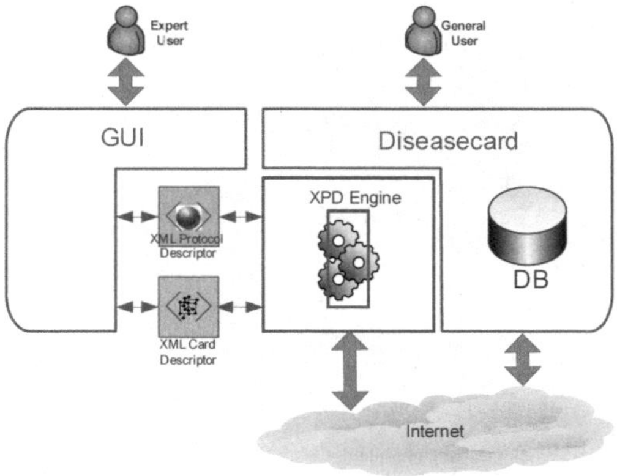

Figure 2: The Diseasecard System architecture. Diseasecard package is the core application which provides the web interfaces to the general user. XPD Engine implements and manages the retrieval tasks during information extraction. GUI is a client application which allows expert users to create, to edit and upload protocol and card descriptors through a graphical user interface. These descriptors (xml protocol descriptor and xml card descriptor) are rule files which are plugged into the XPD Engine and used to create diseasecard instances. The first descriptor defines the navigation pathways while the second one specifies the hierarchical graphical structure for each generated card.

DiseaseCard is a web application developed with Java technology (Java Server Pages and Jakarta Struts), running on Tomcat 5 web server. All system and user's information is stored and managed in a MS SQL Server relational database. To build the conceptual map of disease, the system uses the SVG model.

3. Results

Figure 3 shows a graphical representation of a small example of an XPD file. This illustration represents rules that will be processed by the XPD engine in order to explore several concepts associated to genetic diseases like disorders and mutations, drugs, polymorphisms, protein structure, gene nomenclature, protein sequences, etc. This retrieving process begins at the *morbidMap* resource, which is based on one of three start keys (disease name, gene symbol or omim code). Through this protocol, Diseasecard system can provide the answers to several questions that are relevant in the genetic diseases diagnostic, treatment and accomplishment [9], such as:

- What are the main features of the disease?
- Are there any drugs for the disease?
- Are there any gene therapies or clinical trials for the disease?
- What laboratories perform genetic tests for the disease?
- What genes cause the disease?
- On which chromosomes are these genes located?
- What mutations have been found in these genes?
- What names are used to refer to these genes?
- What are the proteins coded by these genes?
- What are the functions of the gene product?
- What is the 3D structure for these proteins?
- What are the enzymes associated to these proteins?

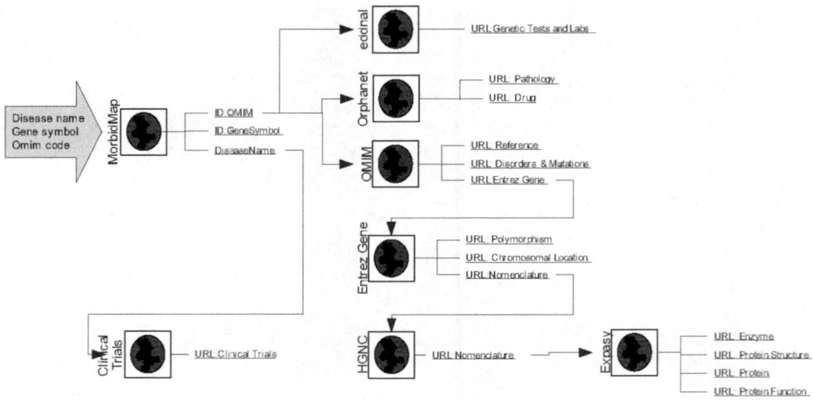

Figure 3: Partial graphical view of an XPD file. Each box represents a web source and the information that can is available in this database. The engine starts on MorbidMap resource based on a word, a gene symbol or omim code associated to a disease. The web wrappers will gather then the remaining information from one site according to the information that is collected from the previous points.

Using this set of questions we present below, as illustration, a summary of the diseasecard results for three particular rare diseases: the *Fragile X Syndrome* the *Achondroplasia* and *Fabry* disease.

Disease name: **Fragile X Syndrome**

1. What are the main features of the Fragile X Syndrome?
   - *Fragile X syndrome is the most frequent cause of inherited mental retardation. It is caused by a dynamic mutation i.e. the progressive expansion of polymeric (CGG)n trinuleotide repeats located in the non coding region at the 5' end of the FMR1 gene at Xq 27.3.[...];*
2. Are there any drugs for this disease? [Empty];
3. Are there any gene therapies or clinical trials for this disease? [Empty];
4. What laboratories perform genetic tests for this disease?
   - *University of Leipzig - Medical Faculty - Institute of Human Genetics;*
   - *Department of Clinical Genetics - Lund University Hospital;*
   - *[...];*
5. What genes cause this disease?
   - *This disease is caused by a mutation in the FMR1 gene. The official name is fragile X mental retardation 1;*
6. On which chromosome is FMR1 located?
   - *The FMR1 is located at locus Xq27;*
7. What mutations have been found in gene FMR1?
   - *5 entries have been found in the literature for this gene with omim code = 309550;*
8. What names are used to refer to this gene?
   - *FMR1 is the official symbol for the gene;*
   - *FRAXA is an alias;*
9. What are the proteins coded by this gene?
   - *1 protein has been found in SwissProt with accession number Q06787, entry name FMR1_Human and name Fragile X mental retardation 1 protein;*
10. What are the functions of the gene product?
    - *RNA-binding protein. Associated to polysomes and might be involved in the transport of mRNA from the nucleus to the cytoplasm.*
11. What is the 3D structure for this protein?
    - *The structure for this protein is available in PDB with the code 2FMR.*
12. What are the enzymes associated to these proteins? [Empty]

Disease name: **Achondroplasia**

1. What are the main features of the Achondroplasia?
   - *Achondroplasia is the most frequent form of chondrodysplasia with a prevalence of one child in every 15,000. This type of dwarfism is characterized by short limbs, hyperlordosis, short hands, and macrocephaly with high forehead and saddle nose [...];*
2. Are there any drugs for this disease?
   - *Norditropin;*

3. Are there any gene therapies or clinical trials for this disease?
   There are two complete studies sponsored by National Human Genome Research Institute (NHGRI):
   - *Study of Skeletal Disorders and Short Stature*
   - *Issues Surrounding Prenatal Genetic Testing for Achondroplasia*
4. What laboratories perform genetic tests for this disease?
   - *Università degli Studi di Verona - Laboratorio di Genetica Molecolare;*
   - *Department Center of Medical Genetics - University of Antwerp;*
   - *Centro de Análisis Genéticos (Zaragoza);*
   - *[...];*
5. What genes cause this disease?
   - *This disease is caused by a mutation in the FGFR3 gene. The official name is fibroblast growth factor receptor 3;*
6. On which chromosome is FGFR3 located?
   - *The FGFR3 is located at locus 4p16;*
7. What mutations have been found in gene FGFR3?
   - *26 entries have been found in the literature for this gene with omim code =134934;*
8. What names are used to refer to this gene?
   - *FGFR3 is the official symbol for the gene;*
   - *CEK2, JTK4 are alias and ACH is a previous symbol;*
9. What are the proteins coded by this gene?
   - *1 protein has been found in SwissProt with accession number P22607, entry name FGFR3_Human and name Fibroblast growth factor receptor 3 [Precursor];*
10. What are the functions of the gene product?
    - *Receptor for acidic and basic fibroblast growth factors. Preferentially binds FGF1.*
11. What is the 3D structure for this protein?
    - *The structure for this protein is available in PDB with the code 1RY7.*
12. What are the enzymes associated to this protein?
    - *1 enzyme has been found in Expasy with EC=2.7.1.112 called Protein-tyrosine kinase*

Disease name: **Fabry Disease**

1. What are the main features of the Fabry Disease?
   - *Fabry's disease (FD) is an X-linked inborn error of glycosphingolipid metabolism due to a deficient activity of alpha-galactosidase A, a lysosomal homodimeric enzyme. The enzymatic defect leads to the systemic accumulation of underivatized neutral glycosphingolipids in plasma and tissues [...];*
2. Are there any drugs for this disease?
   - *Fabrazyme: AGALSIDASE BETA;*
   - *Replagal: AGALSIDASE ALFA;*
3. Are there any gene therapies or clinical trials for this disease? [Empty]
4. What laboratories perform genetic tests for this disease?
   - *University Medical Center Utrecht*
   - *Unidade de Enzimologia (Porto)*
   - *Institut de Pathologie et de Génétique asbl (Loverval)*
   - *[...];*
5. What genes cause this disease?
   - *This disease is caused by a mutation in the GLA gene. The official name is galactosidase, alpha;*
6. On which chromosome is GLA located?
   - *The FGFR3 is located at locus Xq22;*
7. What mutations have been found in gene GLA?
   - *52 entries have been found in the literature for this gene with omim code =301500;*
8. What names are used to refer to this gene?
   - *GLA is the official symbol for the gene;*
   - *GALA is an alias;*
9. What are the proteins coded by this gene?
   - *1 protein has been found in SwissProt with accession number P06280, entry name AGAL_Human and name Alpha-galactosidase A [Precursor];*
10. What are the functions of the gene product?
    - *Catalytic activity: Hydrolysis of terminal, non-reducing alpha-D- galactose residues in alpha-D-galactosides, including galactose oligosaccharides, galactomannans and galactohydrolase.*
11. What is the 3D structure for this protein?
    - *The structure for this protein is available in PDB with the code 1R47 and 1R46.*
12. What are the enzymes associated to this protein?
    - *1 enzyme has been found in Expasy with EC=3.2.1.22 called Alpha-galactosidase*

The results presented above allow answering the set of questions previsously rised, for the Fragile X Syndrome, Achondroplasia and Fabry diseases. Besides these three, the diseasecard site currently links information for around 2000 rare diseases.

4. Conclusion

The Diseasecard is a web portal for rare diseases, based on an automatic information retrieval engine that provides, transparently to the user, a virtually integration of distributed and heterogeneous information – using a pathway that goes from the symptom to the gene. With this system, medical doctors can access genetic knowledge without the need to master biological databases, teachers can illustrate the network of resources that build the modern biomedical information landscape and general citizen can learn and benefit from the available navigation model.

Acknowledgements

The present work has been funded by the European Commission (FP6, IST thematic area) through the INFOBIOMED NoE (IST-507585).

References

[1] I. H. G. S. Consortium, "Finishing the euchromatic sequence of the human genome," *Nature*, vol. 431, pp. 931 - 945, 2004.

[2] A. E. Guttmacher and F. S. Collins, "Genomic medicine-a primer," *The New England Journal of Medicine*, vol. 347, pp. 1512-1520, 2002.

[3] A. Bayat, "Science, medicine, and the future: Bioinformatics," *British Medical Journal (BMJ)*, vol. 324, pp. 1018-1022, 2002.

[4] F. S. Collins and V. A. McKusick, "Implications of the Human Genome Project for Medical Science," *JAMA*, vol. 285, pp. 540-544, 2001.

[5] W. Sujansky, "Heterogeneous database integration in biomedicine " *Comput. Biomed. Res.*, vol. 34 pp. 285-298 2001

[6] G. Dias, J. L. Oliveira, J. Vicente, and F. Martin-Sanchez, "Integration of Genetic and Medical Information Through a Web Crawler System," *Biological and Medical Data Analysis (ISBMDA'2005), LNBI 3745*, pp. 78-88, 2005.

[7] M. Y. Galperin, "The Molecular Biology Database Collection: 2005 update," *Nucleic Acids Research*, vol. 33, 2005.

[8] J. L. Oliveira, G. Dias, I. Oliveira, P. Rocha, I. Hermosilla, J. Vicente, I. Spiteri, F. Martin-Sánchez, and A. S. Pereira, "DiseaseCard: A Web-Based Tool for the Collaborative Integration of Genetic and Medical Information," *Biological and Medical Data Analysis (ISBMDA'2004), Lecture Notes in Computer Science*, vol. 3337 / 2004, pp. 409, 2004.

[9] J. A. Mitchell, A. T. McCray, and O. Bodenreider, "From Phenotype to Genotype: Issues in Navigating the Available Information Resources," *Methods of Information in Medicine*, vol. 42, pp. 557-563, 2003.

1.6 Health Standards

Ubiquity: Technologies for Better Health in Aging Societies
A. Hasman et al. (Eds.)
IOS Press, 2006

HL7 RIM: An Incoherent Standard

Barry Smith[a,b,c,1] and Werner Ceusters[c]
[a] *Department of Philosophy, University at Buffalo, Buffalo NY, USA*
[b] *Institute for Formal Ontology and Medical Information Science, Saarbrücken, Germany*
[c] *Center of Excellence in Bioinformatics and Life Sciences and National Center for Biomedical Ontology, University at Buffalo, Buffalo NY, USA*

Abstract. The Health Level 7 Reference Information Model (HL7 RIM) is lauded by its authors as 'the foundation of healthcare interoperability'. Yet even after some 10 years of development work, the RIM is still subject to a variety of logical and ontological flaws, which has placed severe obstacles in the way of those who are called upon to develop implementations. We offer evidence that these obstacles are insurmountable and that the time has come to abandon an unworkable paradigm.
Keywords: HL7, RIM, standardization, ontology, realism

1. Preamble

> A message to mapmakers: highways are not painted red, rivers don't have county lines running down the middle, and you can't see contour lines on a mountain. [1]

What follows is an exegesis and critique of the HL7 Reference Information Model (RIM). We will focus primarily on the relevant portions of [2], official codex of HL7 Version 3 ("Normative Edition 2005"), in which the RIM is asserted to be 'credible, clear, comprehensive, concise, and consistent', 'universally applicable', and 'extremely stable'. These assertions not only contradict many statements to be found within HL7's own internal email forums, they are also belied by the frequent revisions to which the RIM has been and is still being subjected, and by the fact that the RIM continues to be marked by major flaws (for which further documentation is provided at [3]). Such flaws include:

- *problems of implementation*: The decision by HL7 to adopt the new RIM-based methodology was adopted already in 1996; after ten years of effort, and considerable investment in the RIM itself and in the development of associated message types, DMIMs and RMIMs, the promised benefits of interoperability remain elusive;
- *problems of usability in specialist domains*: The RIM methodology consists in defining a set of 'normative' classes (Act, Role, and so on), with which are associated a rich stock of attributes. When the RIM is applied to a new domain (for example pharmacy), one needs to select from just these pre-defined attributes, rather as if one were attempting to create manufacturing software by drawing from a store containing pre-established parts for making every conceivable manufacturable thing, from lawnmowers to hunting bows.

---

[1] Corresponding author: Barry Smith, Department of Philosophy, University at Buffalo, Buffalo NY 14260l, USA. Email: phismith@buffalo.edu; internet: http://ontology.buffalo.edu/smith.

We do not know of even one example where a methodology of this sort has been made to work successfully;

- *problems of scope*: What are called 'Acts' (roughly: intentional actions) play an overwhelmingly central role in the RIM, so that of its 'foundation classes' only Entity (defined as: 'A physical thing, group of physical things or an organization capable of participating in Acts, while in a role') is left over to comprehend those things which are not Acts. How, on this basis, can the RIM deal transparently with information about, say, disease processes, drug interactions, wounds, accidents, bodily organs, and many other phenomena central to healthcare, given that the latter are neither Entities nor Acts?
- *problems of documentation*: The RIM documentation is not only disastrously unclear, and poorly integrated with those other parts of the V3 documentation for which the RIM itself is designed to serve as backbone; it is also subject to a series of internal inconsistencies (for example in its sloppy use of terms such as 'act', 'Act', 'Acts', 'action', 'ActClass' 'Act-instance', 'Act-object') of a sort which should surely be avoided in the context of work on messaging standards.
- *problems of learnability*: Can HL7 V3 be taught, and therefore engaged with and used by a wider public, given the amateurish quality of its documentation, the massive number of special cases, the frequent amendments, and the complex hurdles that must be overcome in creating a message?
- *problems of marketing*: Are the grandiose marketing claims made on behalf of HL7 V3 as 'the data standard for biomedical informatics' justifiable, given the many still unresolved problems on the technical side?

2. Double Standards

In spite of large investments in HL7-based information systems, some of them by national governments, and in spite of the now familiar difficulties encountered in creating working implementations on the basis of such investments, there is astonishingly little secondary literature on the HL7 standard itself. One consequence of the absence of evidence-based criticism from independent outsiders is the appearance of certain symptoms of intellectual inbreeding in HL7's own literature, which is marked above all by an air of boosting self-congratulation. We believe, against this background, that the HL7 endeavor can only benefit from forthright criticism, and it is in this spirit that our remarks are offered here.

For reasons of space, we shall focus on just one set of defects, which concerns the RIM's unsure treatment of the distinction between *information about an action* on the one hand and *this action itself* on the other. The former is what is *recorded* in a message or record. The latter is what *occurs*, for example within a hospital ward or laboratory. When challenged, RIM enthusiasts will insist that the RIM is concerned exclusively with the former – with information – and of course the very title of the RIM is in keeping with this conception. Interspersed throughout its documentation, however, we find also many references to the latter, often conveyed by means of the very same expressions. This problem is of crucial importance, given the characteristic claim advanced on behalf of the RIM that it will provide a shared, rigorous semantics for HL7 V3 messages, of a type which

V2 lacked. For it shows that what the RIM in fact provides is deep confusion.

As example, consider the treatment of the RIM class LivingSubject, which is defined, confusingly, as follows:

> A subtype of Entity representing an organism or complex animal, alive or not. (3.2.5)

Examples of this class are then stated to include: 'A person, dog, microorganism or a plant of any taxonomic group.' From this we infer that a person, such as you or me, is an example of a LivingSubject. At the same time in 3.2.1.13 (which is, oddly, the only subsection of 3.2.1 in [2]) we are told that LivingSubjects, including persons, can occupy just two 'states': *normal*, defined simply as 'the "typical" state', or *nullified*, defined as: 'The state representing the termination of an Entity instance that was created in error.'

Can it really be true that we are here being invited to postulate two kinds of death for human beings: normal death, and a special kind of death-through-nullification in the case of those persons who were created in error? Or is it not much rather the case that by 'LivingSubject' the RIM means not (as is asserted at 3.2.5) 'mammals, birds, fishes, bacteria, parasites, fungi and viruses' but rather *information about* such entities? The answer, bizarrely, is that *it means both of these things*; for there are in fact, co-existing side by side within its documentation, two distinct conceptions of what terms in the RIM are supposed to designate.

What we shall call the **information model conception of the RIM** is enunciated for example in 1.1 of [2]:

> The Health Level Seven (HL7) Reference Information Model (RIM) is a static model of health and health care information as viewed within the scope of HL7 standards development activities. It is the combined consensus view of information from the perspective of the HL7 working group and the HL7 international affiliates. The RIM is the ultimate source from which all HL7 version 3.0 protocol specification standards draw their information-related content.

The RIM, according to this first conception, is intended to provide a framework for the representation of the structures of and relationships between information that is independent of any particular technology or implementation environment. It is thus designed to support the work of database schema designers, software engineers and others by creating a single environment for messaging which can be shared by all healthcare institutions.

What we shall call the **reference ontology conception of the RIM** can be inferred, first, from the many programmatic statements describing the RIM's purpose, which is to facilitate consistent sharing and usage of data across multiple local contexts. For in striving to achieve this end of consistency (and thus to rectify problems affecting implementations of HL7 V2), the RIM cannot focus merely on healthcare messages themselves, as bodies of data. Rather it must provide a common benchmark for how such bodies of data are to be formulated by their senders and interpreted by their recipients. We can conceive of only one candidate benchmark for this purpose, namely the things and processes themselves within the domain of healthcare which messages are about and which are familiar to those engaged in message formulation. That this is indeed the benchmark adopted by the RIM will become clear when we examine the many passages in its documentation in which definitions and examples are provided to elucidate the meanings of its terms.

Rather than distinguishing the two tasks, of *information model* and *reference ontology*, and addressing them in separation, the RIM seeks to tackle both simultaneously, through ambiguous use of language. Expressions drawn from the vocabulary of healthcare are used, alternately, both with familiar meanings (which enable the RIM to secure a necessary relation to corresponding activities in the healthcare domain) and with incompatible technical meanings (when the RIM is attempting to specify the relevant associated type of data). Thus for example 'stopping a medication' means both: *stopping a medication* and: *change of state in the record of a Substance Administration Act from Active to Aborted* (3.1.5).

3. 'Objects'

According to the information model conception, the RIM, and the HL7 messages defined in its terms, are about objects in information systems – hereafter called 'Objects' – which 'represent' things and processes in reality. Thus the human being named 'John Smith' is represented by an Object containing John Smith's demographic or medical data. This Object is different from John Smith himself. HL7's Glossary defines an 'Object' as:

> An instance of a class. A part of an information system containing a collection of related data (in the form of attributes) and procedures (methods) for operating on that data.

It defines an 'Instance' as: 'A case or an occurrence. For example, an instance of a class is an object.' Yet under the entry for the class 'Person' in the same Glossary we are told that 'Instances of Person include: John Smith, RN, Mary Jones, MD, etc.' For HL7, accordingly, 'John Smith, RN' is not the name of a human being; rather, it is the name of an Object which goes proxy for a human being in an information system.

Because Objects are bodies of data, and because different data about John Smith will be contained in the different information systems involved in messaging, 'John Smith, RN' will refer ambiguously to many John Smith Objects. How, then, is messaging about John Smith (the human being) possible, given that senders and recipients will associate distinct John Smith Objects with each given message?

This is, to be sure, a hard problem, and the HL7 community deserves some credit for having recognized its importance. The RIM can go part way towards alleviating it in the case of persons by insisting that information objects include corresponding unique identifiers such as social security numbers. Indeed the HL7 Glossary *defines* a 'Person' as a 'single human being who … must also be uniquely identifiable through one or more legal documents (e.g. Driver's License, Birth Certificate, etc.).' In the ideal case, at least (which would require each institution to maintain a mapping of its own locally unique patient IDs to all the locally unique patient IDs used by the institutions it is communicating with), such identifiers could serve to bind the different Objects together in such a way that they would all become properly associated with what we would normally think of as one and the same person. But what works for Persons will in almost every case not work for Objects representing things and processes of other types (for instance tumors, epidemics, adverse reactions to drugs), since unique identifiers for the latter are almost never available under present regimes for recording healthcare data [4].

4. 'Acts'

The term 'Act' refers, within those passages in the RIM which conform to the information model interpretation, not to acts (intentional actions), but rather to associated Objects. Thus at 3.1.1 'Act' is defined as meaning: 'A record of something that is being done, has been done, can be done, or is intended or requested to be done.'

Confusingly, however, we are provided, in explication of this definition, with examples not of *records* but of *intentional actions themselves* (referred to not as 'Acts' but as 'acts'):

> The kinds of acts that are common in health care are (1) a clinical observation, (2) an assessment of health condition (such as problems and diagnoses), (3) healthcare goals, (4) treatment services (such as medication, surgery, physical and psychological therapy), (5) assisting, monitoring or attending, (6) training and education services to patients and their next of kin, (7) and notary services (such as advanced directives or living will), (8) editing and maintaining documents, and many others.

At 3.1.14, similarly, the class PatientEncounter (a subclass of Act) is defined as 'An interaction between a patient and care provider(s) for the purpose of providing healthcare-related service(s).' An instance of PatientEncounter, then, is not a *record* of an action, but this action itself – as is made clear by the examples provided to elucidate this definition, which include: emergency room visit, field visit, occupational therapy. The class Procedure is defined as 'An Act whose immediate and primary outcome (post-condition) is the alteration of the physical condition of the subject' (3.1.15), with examples: chiropractic treatment, balneotherapy, acupuncture, straightening rivers, draining swamps.

The class Observation is defined as:

> An Act of recognizing and noting information about the subject, and whose immediate and primary outcome (post-condition) is new data about a subject. Observations often involve measurement or other elaborate methods of investigation, but may also be simply assertive statements. (3.1.13)

From this we can infer, dizzyingly, that we are again dealing not with *records* of actions (of observing, measuring, etc.), but rather with these actions themselves, a view supported also by the RIM's assertion, still in 3.1.13, to the effect that Observations 'are professional acts ... and as such are intentional actions', as also by its treatment of other subclasses of Act, such as DeviceTask, SubstanceAdministration, and Supply – all of which are similarly interpretable only against the background of a reference ontology conception of the RIM and thus as standing in conflict with the RIM's own definition of 'Act'. When, in contrast, we turn to other subclasses of Act, for example Account, or FinancialTransaction, then we find that the corresponding definitions revert to the information model conception.

We are told at 3.1.1 that, while an Act-instance 'represents a "statement"', 'the Act class is this attributable statement'. We can interpret this strange language as follows: each Act-instance is a statement describing what some clinician on some occasion has heard, seen, thought, or done. The Act class is the class of such coded, attributable statements. An Act-instance is, more precisely (in 3.1.1 at least), either an attributable statement or some similar attributable use of language such as an order or request, which serves as the coded accompaniment – the record – of what takes place in reality, for example when a surgical procedure is performed. There then follows a most peculiar passage, which has been criticized already in [5]:

> Act as statements or speech-acts are the only representation of real world facts or processes in the HL7 RIM. The truth about the real world is constructed through a combination (and arbitration) of such attributed

statements only, and there is no class in the RIM whose objects represent "objective state of affairs" or "real processes" independent from attributed statements. As such, *there is no distinction between an activity and its documentation.* (3.1.1, emphasis added)

The italicized passage captures what we might think of as the naked essence of the confusion at the heart of the RIM between reference ontology and information model.

5. Conclusion: Can the RIM be Saved?

Messaging standards are like telephone systems: they can function sensibly only if there is a large network of willing users. This means that to succeed such artifacts must be marked by *clear use of language* and *clear documentation*. The RIM documentation, as we have seen, is systematically ambiguous. What is needed, if HL7 V3 is to satisfy its need for a uniform information representation that is coherent, clear and implementable, are two separate, though of course related, artifacts, which might be called 'Reference Ontology of the Healthcare Domain' and 'Model of Healthcare Information', respectively. The former would include those categories, such as *thing*, *process*, *anatomical structure*, *disease*, *infection*, *procedure*, etc., needed to provide a compact and coherent high-level framework in terms of which the lower-level types captured in vocabularies like SNOMED CT could be coherently organized [6]. The latter would include those categories, such as *message, document*, *record*, *observation*, etc., needed to specify how information about the entities that instantiate the mentioned types can be combined into meaningful units and used for further processing. HL7's Clinical Document Architecture could then be related in an appropriate way to this Model of Healthcare Information, thereby avoiding the current counterintuitive stopgap, which forces a document to be an Act. And HL7's Clinical Genomics Standard Specifications could similarly be related in an appropriate way to the Reference Ontology of the Healthcare Domain, thereby avoiding the no less counterintuitive current stopgap, which identifies an individual allele as a special kind of Observation.

Acknowledgements: This paper was written under the auspices of the Wolfgang Paul Program of the Humboldt Foundation, the EU Network of Excellence SemanticMining, the Volkswagen Foundation, and the National Center for Biomedical Ontology (the latter funded by the National Institutes of Health through the NIH Roadmap for Medical Research). Thanks for helpful comments are due also to Thomas Beale, Walter Dierckx, Amnon Eden and Gunther Schadow, who however bear no responsibility for the positions here adopted.

References

[1] Kent W. Data and Reality, Bloomington IN, 2000, p. v.
[2] ANSI/HL7 V3 RIM, R1-2003: RIM Version V 02-11. Membership Normative Ballot Last Published: 11/22/2005 8:05 PM.
[3] http://ontology.buffalo.edu/HL7.
[4] Ceusters W, Smith B. Strategies for referent tracking in Electronic Health Records, J Biomed Inform, in press.
[5] Vizenor L. Actions in health care organizations: An ontological analysis. Proc MedInfo 2004, San Francisco, 1403-1407.
[6] Rosse C, Kumar A, Mejino JLV, Cook DL, Detwiler LT, Smith B. A strategy for improving and integrating biomedical ontologies. Proc AMIA Symp 2005;:639-643.

Ubiquity: Technologies for Better Health in Aging Societies
A. Hasman et al. (Eds.)
IOS Press, 2006

A Reference Model for Clinical Tumour Documentation

Udo ALTMANN[1], Frank Rüdiger KATZ, Joachim DUDECK
Institute of Medical Informatics
University of Gießen, Germany

Abstract. The definition of common system semantics is an explicit and generally accepted precondition for comparability and exchangeability of data from different systems. We have looked back on 15 years of experience with a data model that was developed in a co-operative effort by experts from various hospital cancer registries as the foundation of a new tumour documentation system (GTDS). The data model is based on the definition of a common basic data set for hospital cancer registries which is agreed by the German Association of Comprehensive Cancer Centres (ADT). This paper presents an "entity relationship" view of this model. Since data exchange among registries and with hospital or practice information systems is becoming increasingly important we describe our method to import data from such systems. We discuss the requirements that systems have to have for a most effective way of exchanging data with a hospital cancer registry. The most important feature is the possibility to associate disease phenomena and therapies with each other and with an entity that represents the tumour across encounters. The reference model we present respectively the requirements we propose for other communicating systems might also fit for other chronic diseases.
Keywords: Cancer, Tumour, Registry, Documentation, Data Model, Reference Model, Chronic disease

1. Introduction

From a technical perspective, important preconditions for the exchange of data in health care seem to be solved. For instance, Health Level Seven's (HL7) version 3 [1] (including e.g. the Reference Information Model (RIM) and the Clinical Document Architecture) provides a framework for the development and implementation of messages in health care including the exchange of documents. Yet domain specific knowledge has to be added to such frameworks to be able to develop and implement real messages. Domain specific knowledge has to describe attributes and their domains (data types etc.) as well as the coherences between the data (entity relationship model – ERD - or other models such as UML).

In this contribution we are dealing with the domain "Clinical Tumour Documentation", which is based on a standard agreed on by the German Association of Comprehensive Cancer Centres (ADT) [2, 8]. Besides the classical tasks of data collection and evaluation, it aims at a closer integration of the documentation process

[1] Corresponding Author: Udo Altmann, Institut für Medizinische Informatik, Justus-Liebig-Universität Gießen, Heinrich-Buff-Ring 44, 35392 Gießen, Germany. E-Mail: udo.altmann@informatik.med.uni-giessen.de WWW: http://www.gtds.de

into clinical routine to overcome retrospective data entry. The rationale is to provide the greatest possible benefit by the immediate use of data, thus obtaining a best possible data quality at the same time such as described by Enterline [3] and Shortliffe [4]. It is clear that such an integration needs data exchange and this can only work when all underlying systems have a common domain model of the data they exchange.

The ability for data exchange is increasingly important for cancer registries [5] since the preparedness of physicians to fill in traditional paper forms is sinking in the same extent as data become available in clinical information systems. Such, the workflow in registries will move from transcribing data to integrating and checking data from electronic resources. But not all kinds of information are available in an appropriate form in hospital information systems. In general, there is little information structured in a way that it fits directly into tumour documentation [6, 7]. Being aware that cancer reporting and quality control are obligatory some providers of hospital information systems started to integrate forms for the entry of cancer data. Such reporting systems dare not be confused with a fully operational cancer registry. Due to the fragmented cancer treatment in different institutions and the long course of diseases, with possible relapses, and their treatments, it is necessary to carry out a record linkage to produce an integrated view of cancer cases.

In 1991, a nation-wide expert group defined a requirements specification for a common registry system that was supposed to be the software basis of the evolving hospital cancer registries of the comprehensive cancer registries in the New States of Germany. The data model was developed using a CASE tool. We started to develop the system GTDS that became operational in 1993. The core of the data model has been stable since its development in 1991 and reportedly served as a template for three more registry systems. Even when the basic data set was revised in 1999 [2] only additional attributes had to be added to the existing entities and some entities were introduced for new extensions. GTDS is in use in more than 45 hospital cancer registries of various types and sizes. It supports of patient care by calculation of chemotherapy, report writing and follow-up management. [9].

The aim of this contribution is to describe the data model's most important objects in order to help designers of possible communication partners understand the specific requirements of hospital cancer registries. We do not describe attributes since those may vary from country to country and their description is rather a matter of terminology services.

2. Methods

We describe the reference model as entity relationship diagram (ERD, Figure 1). This is not the original diagram from 1991 but an abstraction that reduces complexity and aggregates similar entity types. For example, for GTDS each therapy modality was implemented as a specific entity type but all types are handled in the same way. Thus from an object oriented view they have many common attributes and methods. Medical therapy and radiotherapy both have relationships to the same entity type that stores therapy side effects. Are short term complications in surgery really something basically different than short term side effects in radiotherapy besides the way of coding? At some locations of the model we go beyond existing relationship types to make future developments towards more explicit relationships between objects (e.g. therapy and phenomena) possible.

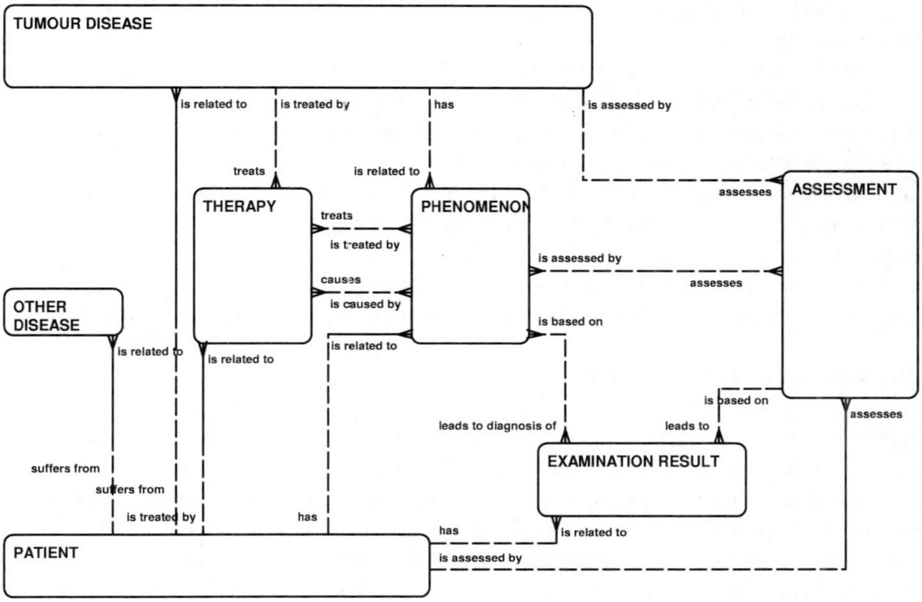

Figure 1: Entity relationship diagram of the reference model

3. Results

3.1. Description of Entity Types

Patient is the entity type that represents one single patient and encompasses all properties or information that only exist once per patient (demographic data, risk factors, history with respect to the time before the tumour). All other entity types in the model are directly related to the patient since sometimes the relationship to other entities cannot be determined, e.g. in the case of the phenomenon metastasis it may be impossible to decide to which primary tumour it is related.

Cancer registries have to distinguish all tumour diseases and relate them to the appropriate patient. The **tumour disease** is the basic concept for this distinction. The entity type encompasses all properties or information that only exists once per disease (denomination of the tumour, diagnosis date, ICD ...). It is the basis of statistics and clasps different phenomena that might appear as different problems / diagnoses in hospital information systems to a single case.

A **phenomenon** refers to each physical (primary site, metastasis, ...) or patho-physiological appearance (paraneoplastic syndrome), adverse drug reaction, and after-effect of tumour or therapy.

Assessment is a central component in which data about phenomena or the patient's state provides an overview of the course of disease. It is a subject of statistical evaluation and can be related to almost all other objects in the model. Such, with respect to the patient it may be a performance state (ECOG), or with respect to the tumour disease it may be a relapse or a partial or complete remission.

Therapy is a container for all therapy modalities. The object itself contains a denomination, some classification, begin, end, type of end (regular, etc). The implementation of therapy depends on its modality (operations, chemotherapy, etc.).

Examination results represent a wide spectrum of results from simple laboratory tests, clinical findings to imaging procedures. In implementations it may be necessary to group examinations for display or other workflow related reasons. Although statistics are usually performed on the information in assessment, it can be important to register specific examinations for quality control or reporting of a patient's course of disease.

Other diseases lists all diseases and states that are relevant for the long-term assessment of a patient's treatment but are not a phenomenon of the neoplastic process.

3.2. Description of Relationship Types

The meaning of the relationship types can be read out of the diagram in conjunction with the description of the entity types. In GTDS the relationships between therapy and phenomena are not yet realised. Reasons were to keep data entry simpler and the fact that the appropriate information actually often is not available. Nevertheless those associations exist in reality and sometimes users wished to express such facts more explicitly in the past. Thus future versions of GTDS might provide the possibility of documenting such associations.

3.3. Attribute Information

Attributes are the real bearers of information, the data items. It is clear that attribute information is important for the comparability and exchange of data. The attribute information for German hospital cancer registry is defined in the "Basisdokumentation für Tumorkranke" [2] and can be assigned to the different entity types. Since this part of the model is subject to change over time and depends on country specific requirements, we will not describe this aspect in detail.

3.4. Implementation of Communication Interface

Supposing that source and destination system have to have a structure that is compatible with the reference model with respect to the data that are to exchanged, we implemented a data structure that is parallel to the GTDS's data structure and complements it by a source identifier (for details including attribute information see http://www.med.uni-giessen.de/akkk/gtds/grafisch/doku/import.htm, in the German language). Data that have to be imported are imported into this structure. From here they are imported into the GTDS tables, partially automatically, partially under user control. User control is necessary because of the registry's character of the system which implies that data on the same subject can come from different sources.

On import, GTDS builds up a master object index of all imported objects that makes it possible to identify objects that have already been imported and to transfer relationships between data that have been imported from the same source at different times. For example, if a tumour's data was imported in 2005 and the sending system has a tumour identifier then a follow-up assessment of the tumour in 2006 from the same sending system enables an automatic record linkage on import. Up to now at least three applications of this import facility have demonstrated its feasibility:

- The import of data from a department system (radiotherapy documentation, in routine use)
- The import of data from another hospital cancer registry
- The import of data of a specialised breast cancer documentation system

4. Discussion

Merzweiler et al [10] dealt with registration in oncology and focused on design aspects of a data model for a data dictionary, i.e. the control over the contents on an attribute level. Such models complement our model and could store the definitions of the "Basisdokumentation". Blum et al described a data model with focus on the application of protocols [11]. Our reference model deals with the core aspect of registry data. In GTDS the registry is the unit that provides oncology specific functions for a hospital whereas in OCIS the registry is a separate unit. Data in OCIS are mainly patient-time related. Data in the proposed model are tumour related which is important for a registry, since more than 5 % of tumour patients suffer from more than one tumour.

HL7's version 3 [1] Reference Information Model (RIM) is an abstract static model of health and health care information as viewed within the scope of HL7 standards development activities. The Patient entity type of our model corresponds to the Person Class in the subject area Entities. All other entity types correspond to classes in the Acts subject area: Tumour Disease, Other Diseases, Phenomenon, Examination Result, and Assessment to the Observation class and Therapy to Procedure. Since Observation and Procedure are specialisations of the Act class, the ActRelationship class allows implementing any association of our model including decomposition of complex diagnostic or therapeutic interventions. The definitions of the "Basisdokumentation" which complement the model with the attributes are implementations of the HL7 concept "vocabulary domain". It is also important to note that identifiers are provided that correspond to the identifiers in our import structure. Actually our import structure can be regarded as a domain information model from which message descriptions can be derived.

The idea of continuity of care that often was claimed in publications of Dr Weed can be supported by hospital cancer registries especially in a fragmented health care system in which oncological care is carried out by a rather large number of institutions. In a sense, our model is an implementation of the problem-oriented record [12].

While the discussed models are rather frameworks, our model provides a specialisation where system designers can understand the business of tumour documentation on an ERD-level. With respect to the contents (attributes) we expect an increasing need for flexibility, e.g. to enable site-specific extensions. But such an extension, for example for breast cancer, will only need to define the contents and to declare at what position of the reference model they have to be placed. Ideally such definitions should be placed in a dictionary such as that proposed by Merzweiler et al [10] or should reference the LOINC or SNOMED-CT coding system as proposed by the HL7 standard, or, as an oncology specific tool, the NCI's Cancer Data Standards Repository (caDSR) with its Common Data Elements Browser (http://cdebrowser.nci.nih.gov/CDEBrowser/) [13].

Our model has only one entity with an oncology specific denominator: tumour disease. In fact the oncology specific aspect is mainly taken from the contents, the attributes. We suppose that other diseases with complex treatment and long-term follow-up, e.g. rheumatic diseases could also be implemented in a very similar model. Eventually the basic idea is to implement associations between data that currently are stored on a patient-time/encounter basis, e.g. to explicitly link therapies or examinations to the appropriate diagnosis.

5. Conclusion

Electronically processable domain knowledge is one of the key issues of medical informatics. We propose our model as reference for the implementation of cancer related records in systems that are possible communication partners of cancer registries. The model is more explicit than other reference models like the HL7's RIM but it is generic enough to also serve for other chronic diseases where data from different encounters or even different institutions have to be aggregated and evaluated. Although the reference model is limited to the entity relationship part of the information model, it claims important features like the association of therapies and diagnoses to the long term concept tumour disease. Such features enable exported data to be integrated into registry data in the most automated way. The contents themselves have to be defined by a common reference. For Germany this is currently the book "Basisdokumentation für Tumorkranke" but in future implementations a data dictionary such as the NCI's Cancer Data Standards Repository might provide a more flexible infrastructure.

References

[1] HL7® Version 3 Standard, © 2005 Health Level Seven (Last Published: 09/29/2005 10:16 PM)

[2] Dudeck J; Wagner G, Grundmann E, Hermanek P. Basisdokumentation für Tumorkranke. 5th revised ed., München: W. Zuckschwerdt Verlag 1999

[3] Enterline JP, Lenhard R, Blum BI. A clinical information system for oncology. Springer Verlag New York Berlin Heidelberg 1989

[4]. Shortliffe EH. An Integrated Oncology Workstation. National Cancer Institute, Maryland, Bethesda 1990

[5] Altmann U, Dudeck J, Eisinger B, Kunath H, Schott G, Kurbjuhn H, Wächter W, Tafazzoli AG, Katz FR. Development Of An Oncology Data Network In Germany. In: A. Hasman et al. (Eds.): Medical Informatics Europe '00. Amsterdam: IOS Press 2000, 959-963

[6] Gumpp V. Einführung der Elektronischen Vitalstatusabfrage für Klinische Krebsregister in Baden-Württemberg? Informatik, Biometrie und Epidemiologie in Med. u. Biol. 34/3 (2003) pp 553-554

[7] Altmann U, Katz FR. Nutzung elektronischer Datenquellen für klinische Krebsregister. German Medical Science; 2004. Doc 04gmds330

[8] Dudeck J, Wächter W, Altmann U, Katz FR. The definition of a new uniform basic data set for hospital cancer registries in Germany. MIE-Proceedings 1993, Freund Publishing House Ltd.: 489-492

[9] Altmann U, Katz FR, Tafazzoli AG, Haeberlin V, Dudeck JW. GTDS - a Tool for Tumor Registries to Support Shared Patient Care. Proc AMIA Annu Fall Symp; 1996 Oct 26-30; Washington; Philadelphia: Hanley & Belfus; 512-516

[10] Merzweiler A, Knaup P, Creutzig U, Ehlerding H, Haux R, Mludek V, Schilling FH, Weber R, Wiedemann T. Requirements and design aspects of a data model for a data dictionary in paediatric oncology. Stud Health Technol Inform. 2000;77:696-700.

[11] Blum BI, Lenhard RE Jr, McColligan EE. An integrated data model for patient care. IEEE Trans Biomed Eng 1985 Apr;32(4):277-88

[12] Weed LL. Medical Records, Medical Education And Patient Care. The Press of Case Western Reserv University, Cleveland Ohio, 1969

[13] http://cdebrowser.nci.nih.gov/CDEBrowser/ (last accessed Jan. 15, 2006).

Ubiquity: Technologies for Better Health in Aging Societies
A. Hasman et al. (Eds.)
IOS Press, 2006

Standards for enabling health informatics interoperability

Kjeld ENGEL[1], Bernd BLOBEL and Peter PHAROW
eHealth Competence Center, Regensburg, Germany

Abstract. Most of industry countries are turning their healthcare system towards integrated care paradigms for improving quality, efficiency, and safety of patients' care. Integrated care has to be supported by extended communication and co-operation between the involved healthcare establishments' information systems. The required interoperability level goes beyond technical interoperability and simple data exchange as it has been started in the early world of electronic data exchange (EDI). For realising semantic interoperability, series of standards must be specified, implemented and enforced. The paper classifies standards for health information systems needed for enabling practical semantic interoperability.
Keywords: Standards, Interoperability, eHealth, Health Information Systems

1. Introduction

To the widespread adoption of IT networks in health, providers' fragmentation represents a significant barrier. The diversity of participants in the healthcare industry and the complexity of their relationships have frustrated the voluntary adoption of industry standards. Standards are needed as convention for structure and behaviour of specific computing functions, formats, and processes. Therefore, they play an important role in computer-to-computer transmissions of electronic information. Without general adoption of industry-wide standards, the ability to speed up transactions through automation is more difficult. More encouragingly is that one of the Internet's appeals as a communications network has been its ability to reduce the need to agree on common communications protocols. It also permits the healthcare industry to draw on successful networking models in other industries. The majority of standards developed within the healthcare industry are classified into two categories: proprietary (de-facto standards) and consensus standards. In this paper consensus standards will be discussed which are developed by various committees.

2. Requirements to be met for achieving semantic interoperability

The challenge for communication and collaboration through connecting health information systems can be done at different levels of interoperability: At the lowest level, mechanical plugs including the voltage and the signals used have been

[1] Corresponding Author: Kjeld Engel, eHealth Competence Center, University of Regensburg Medical Center, Franz-Josef-Strauß-Allee 11, D-95053 Regensburg, Germany. Email: kjeld.engel@ehealth-cc.de

harmonised. We are talking of technical interoperability. At the next level, the data exchanged have been standardised providing data level interoperability. Nevertheless, different terminologies might be used. Therefore, at the next level, terminology must be agreed on. For realising a common understanding, the semantic of terms must be harmonised providing semantic interoperability. At the highest level, concepts and context of information exchanged are harmonised including the service realised based on that information. We call this highest level service oriented interoperability, i.e. the applications' function and behaviour has been adapted. Furthermore, the design process of systems meeting that level of interoperability must be comprehensively defined and standardised. In some environments such as HL7 with reference to the 1990 IEEE Standard Computer Dictionary, service-oriented interoperability has been summarised under the term of semantic interoperability.

For meeting the challenges of improving quality and efficiency of patient's care including homecare and prevention, health information systems have to provide semantic interoperability supporting seamless care. Especially in the context of long-term applications such as Electronic Health Record (EHR) systems, several crucial requirements must be realised. Thus, advanced communication and co-operation between different systems and their components in a complex and highly dynamic environment provided in a sustainable way requires:

- Openness, Scalability, Flexibility, Portability,
- Distribution at Internet level,
- Standard conformance,
- Service-oriented semantic interoperability,
- Consideration of timing aspects of data and information exchanged
- Appropriate security and privacy services.

For achieving the aforementioned characteristics, the system architecture, i.e. the system's components, their relationships and functionalities, have to meet the following paradigms established in the Generic Component Model [1]:

- Distribution, Component-orientation (flexibility, scalability),
- Model-driven and service-oriented design,
- Separation of platform-independent and platform-specific modelling → separation of logical and technological views (portability),
- Specification of reference and domain models at meta-level,
- Interoperability at service level (concepts, contexts, knowledge),
- Common terminology and ontology (semantic interoperability),
- Advanced security, safety and privacy services.

The aforementioned paradigms provide an excellent basis for classifying the standards needed to enable semantic interoperability between health information systems.

It is clear that standards form the basis for both integration and interoperability, though the standards to be used may differ for both perspectives. The development of eHealth requires well tailored standards aimed at facilitating interoperability between local, regional, and national information systems, and the availability of information systems' components. However, in a number of domains, it is extremely difficult to find on the shelves standards that can be used in part or in whole, without further developments.

It is obvious that national initiatives for introducing interoperability will not be thoroughly able to review the high number of standards as a whole let alone be conformant. This constitutes a major barrier for interoperability [1].

3. Standards for specifying, implementing and maintaining semantic interoperability for health

3.1. *Standards development organisations relevant for health informatics*

Following, only standards and publicly available specifications deployed in health information systems will be considered. The classification of specifications might be completely defined without claiming the completeness of references, however.

As relevant standards development organisations have been considered by given examples: International Organization for Standardization and International Electrotechnical Commission (ISO/IEC, ISO/TC 215 - Health informatics), European Standardization of Health Informatics (CEN/TC 251), European Telecommunications Standards Institute (ETSI), Object Management Group (OMG), and ASTM International (originally known as the American Society for Testing and Materials). Additionally, some other organisations and relevant projects have to be mentioned such as, e.g., Organization for the Advancement of Structured Information Standards (OASIS), Institute of Electrical and Electronics Engineers (IEEE), United Nations Centre for Trade Facilitation and Electronic Business (UN/CEFACT), European Committee for Electrotechnical Standardization (CENELEC), Digital Imaging and Communications in Medicine (DICOM), World Health Organization (WHO), United Nations (UN), as well as GEHR/openEHR from the openEHR Foundation, but cannot explained closer in this short overview [2-8].

3.2. *Standards Classification*

Architecture standards describe frameworks from which applications, databases and workstations can be developed in a coherent manner. The architecture of a system describes its components, their functions and relationships. Architecture standards will be used to define how a piece of hardware or software is constructed and which protocols and interfaces are required for communications. In that context, requirements are specified too. Examples of those standards are HL7 versions 2.x/3, the Common Object Request Broker Architecture (CORBA) and Model Driven Architecture (MDA) from the OMG, ASTM E1769-95: "Standard Guide for Properties of Electronic Health Records and Record Systems", ISO 18308: "Requirements for an Electronic Health Record Reference Architecture", and CEN 12967: "Service architecture (HISA)".

Modelling standards are written for formal representations of objects, concepts or processes. They are abstractions of something for the purpose of understanding it before building it. Examples are the Unified Modeling Language (UML) from OMG, ASTM E1715-01: "An object-oriented model for registration, admitting, discharge, and transfer functions in computer-based patient record systems", and CEN 15300: "CEN Report: Framework for formal modelling of healthcare security policies".

Communication standards are needed to facilitate communication between independent information systems within and between organisations, for health related purposes, and to meet specific healthcare requirements for messaging health information. Such standards are essential if healthcare services are to obtain the benefits of open systems and avoid the constraints of proprietary interfaces. Examples are CEN 13608: "Security for healthcare communication" and CEN 13606: "Electronic healthcare record communication".

Terminology and ontology standards collect terms used in a particular area and explain the meanings of these terms. Examples are the Unified Medical Language System (UMLS) maintained by the US National Library of Medicine, SNOMED which is until now provided by the College of the American Pathologists and currently moving to an international standards developing organisation, and the NHS Read Code.

Identifier schemas are necessary to clearly define and distinguish data and datasets for data exchange in healthcare as well as an efficient telematics infrastructure. Examples are LOINC: "Logical Observation Identifiers Names and Codes", ASTM E1714-00: "Standard guide for properties of a Universal Healthcare Identifier", and ISO/IEC 9834-1: "Procedures for the operation of OSI Registration Authorities: General procedures and top arcs of the ASN.1 Object Identifier tree".

Security standards deal with the controls of threats made to the integrity of a system. They describe a set of rules that specify the procedures and mechanisms required to maintain the security of a system, and the security objects and security under the domain of the policy. Examples are ASTM E2085-00a: "Standard guide on security framework for healthcare information", ETSI TS 101733: "Electronic Signature Formats", ISO 17090: "Public key infrastructure", ISO 21547: "Security requirements for archiving and backup", and CEN 12388: "Algorithm for Digital Signature Services in Health Care".

Privacy standards describe the right of individuals to control or influence what information related to them may be collected and stored and by whom and to whom that information may be disclosed. Examples are ASTM E1987-98: "Standard guide for individual rights regarding health information" and CEN 13729: "Secure user identification - Strong authentication using microprocessor cards".

In Safety standards the freedom from unacceptable risk of harm is represented. It's the expectation that systems do not, under defined conditions, enter a state that could cause human death or injury. An example is CEN 13694: "CEN Report: Safety and security related software quality standards for healthcare".

3.3. Description of an exemplary standard

In this section ISO 22600: "Privilege management and access control" will be described. It belongs to the group of security standards.

This standard defines privilege management and access control services which are required for communication and use of distributed health information across domain and security borders. It introduces principles and specifies services needed for the management of privileges and access control. The standard specifies necessary component-based concepts and should support their technical implementation. It doesn't specify the use of these concepts in specific clinical process chains.

This document was originally prepared in CEN and then, with changing the main focus to privilege management, completely revised in ISO. This standard consists of 3 parts to satisfy different user groups:

- Part 1: Overview and policy management (describes scenarios and critical parameters in cross border information exchange; it also gives examples of necessary documentation methods as basis for the Policy agreement),
- Part 2: Formal models (describes and explains, in a more detailed manner, the architectures and underlying models for the privileges and privilege management which are necessary for secure information sharing plus examples of Policy agreement templates),

- Part 3: Implementations (provides examples for the formal models of part 2).

The technical specifications of this standard should be incorporated especially in part 3. Also the adaptation ASN.1 to XML should be considered in part 3. Parts 1 and 2 were finalised in ISO/WG 4 and submitted to the ISO Secretariat for publication.

Part 3 will be pushed now. There will be two emphases of the contents: the bridge to formal models and the implementation on basis of meta-languages and tools. If possible, existing specifications should be used (e.g. OASIS). The data are usable for administration by structural and organisational roles (MAC model as pate). Though, they also based on special applications (DAC model). Both parts are role-related, both are role-based access control models (RBAC).

The new approach refers to the HL7 RIM: E (entity) - R (role) - P (participation) - A (acts). "Role" establishes the relationships between entities in the sense of structural roles, while "Participation" is the functional role which will be captured in relationship to action, and this can be delegate again in the old schema of MAC and DAC models. Working Group 1 has adopted these fundamental considerations for EHR.

4. Discussion

The demand for simplified, standardised methods to access healthcare information and services is crucial in making healthcare safe and available to all. Appropriate standards for healthcare information and systems provide the cornerstone to achieving a reasonable healthcare infrastructure.

Health informatics standards are essential to achieve goals of eHealth in Europe for many important points as, e.g., interoperability between systems and patient information exchange between healthcare organisations to improve efficiency and quality of care as well as managing eHealth services. Many health informatics standards exist, or are under adaptation, to meet many of the requirements. But there are some obstacles: their existence is not well known, they are not used enough, their interoperability is often not proven, and some of them conflict. Whereas there are many standards available from the aforementioned organisations which might meet the needs of applications, there can be no guarantees that standards will interwork unless proven in practical applications/pilots. Even where two vendors have implemented the same standard, there are similarly no guarantees that their products will interoperate without adequate interoperability tests due, e.g., to allowable options within the standard. Therefore, conformance statements have to be set up and testing and certification must be performed. So achieving interoperability is a considerable challenge.

Standards are vital to healthcare systems and the deployment of information technologies for healthcare. This has led to a number of standards development and deployment activities by numerous accredited standards developing organisations, consortia, trade associations, government agencies, and individual companies. There is a real need for formal and informal co-ordination of these efforts to leverage the synergy of the various efforts, to harmonise vocabularies, to enable interoperability, and to promote consistent testing and certification programs across and within organisations. So it is important to develop tools and prototypes to promote consistent definitions and artefact reuse, and facilitate interoperability for healthcare systems.

The task is to spread the acceptance of management of health information standards across the whole eHealth community and achieve full and appropriate integration into all information flows. To gain this requires, keeping the standards current and relevant, dissemination of the standards to the eHealth community, and the gathering of evidence to support the investment in implementation of the standards.

For instance, ETSI is revising several ASTM standard guides referenced in drafting the Health Insurance Portability and Accountability Act (HIPAA) to establish technical specifications that will help ensure compliance with HIPAA security and privacy rules. A privilege management infrastructure standard is important to address complexities of role-based access control and management of user privileges. This standard effort will establish consistent means for protecting personal health information within and across enterprises. [9]

5. Conclusion

The overall objective of standardisation is to facilitate the production, handling, and use of products or services to the best possible satisfaction of both users and suppliers. The healthcare sector's lack of general adaptation of industry standards is a technological barrier for implementing healthcare-related information technologies. There is also a lack of guidance on where and how to use standards. Without industry-wide standards, advanced information technologies' ability to speed up transactions through the elimination of human involvement is lost. In order to address the lack just mentioned and to ensure the continued promotion and advertisement of new standards as they become necessary, recommends that organisations and IT companies work with healthcare professionals to encourage more participation of individuals from the medical community, information technology providers, employee groups, employers, payers, and government officials in processes to propose and promote the voluntary adoption of industry-wide standards. It's required for every standardisation organisation to collaborate with other standards development organisations. Standardisation for eHealth should be considered as essential component of any European, national, or regional strategy for eHealth.

Acknowledgments

The authors are in debt to all of the mentioned standards organisations as well as all other partners and organisations for their support and their kind co-operation.

References

[1] Blobel B. HL7 and architecture standards of electronic health records (EHRs). In: . Ferrer-Roca (Edr.): Proceedings XIII Winter Course: Ambient Intelligence in Medicine AmI@Medicine, pp. 141-148. CATAI edition 2004, Tenerife, Canary Islands, Spain.

[2] http://www.iso.org (last accessed January 13, 2006).

[3] http://www.centc251.org (last accessed January 13, 2006).

[4] http://www.etsi.org (last accessed January 13, 2006).

[5] http://www.omg.org (last accessed January 13, 2006).

[6] http://www.astm.org (last accessed January 13, 2006).

[7] http://www.oasis-open.crg (last accessed January 13, 2006).

[8] http://www.openehr.org (last accessed January 13, 2006).

[9] CEN/ISSS e-Health Standardization Focus Group. Executive Summary. Current and future standardization issues in the e-Health domain: Achieving interoperability. 2005.

Ubiquity: Technologies for Better Health in Aging Societies
A. Hasman et al. (Eds.)
IOS Press, 2006

The HL7 Reference Information Model Under Scrutiny

Gunther SCHADOW[a,1], Charles N. MEAD[b], D. Mead WALKER[c]
[a] Regenstrief Institute, Indiana University School of Informatics., Indianapolis, USA
[b] Oracle, Redwood City, USA [c] Mead Walker Consulting, Downington, USA

Abstract. The Health Level 7 (HL7) Reference Information Model (RIM) was once incepted as an object oriented information model to harmonize the definition of HL7 messages across different application domains. On the heels of the hugely successful HL7 version 2, version 3 and the RIM has received significant attention and credit and in turn is increasingly subjected to criticism. In this paper the authors, who are among the chief designers of the RIM, respond to the major points that have been raised against the RIM in the published literature. We find that much of the criticism is based on misunderstandings and differences in point of view. We wish to advance the dialogue in the hope that when we account for those differences, effective critique may lead to real improvements of the standard.
Keywords: Medical informatics, HL7, ontology, electronic health records.

1. Introduction

At the core of the HL7 version 3 standards development methodology is the Reference Information Model (RIM), which is a static object-oriented model in UML notation. The RIM serves as the source from which all specialized HL7 version 3 information models are derived and from which all HL7 data ultimately receives its meaning. This is to establish semantic interoperability across a vast and growing number of subject domains (e.g., laboratory, clinical health record data, problem- and goal-oriented care, public health, clinical research, etc.), which are loosely but critically related. The RIM was first conceived as a *data model*, where all data elements known from HL7 version 2 and some large electronic health record data models were put on a single information roadmap. In an iterative process of harmonization, analysis, unification and extension of scope, today's RIM emerged as an abstract model, which defines the grammar of a *language* for information in healthcare. As shown in Figure 1, all data is in a form in which Entities (e.g., people places and things, nouns) are related in Roles (relators) to other Entities, and through their Participations (prepositions) interact in Acts (verbs). Through ActRelationships, networks of structurally or logically related Acts are formed, expressing composition, reason, order-fulfillment, data-derivation, etc.. The RIM outlines the logical form of data, but it has specific extension points, at which users and other organizations can contribute content dynamically. This is done through domain-specific terminologies and using a data element definition framework which is built into the RIM directly. For example, all clinical data systems will have a "master file" which defines most clinical data elements, and the HL7 RIM reflects that. [1]

[1] Corresponding Author: Gunther Schadow, Regenstrief Institute, 1050 Wishard Blvd., Indianapolis, IN 46202, USA. gschadow@regenstrief.org

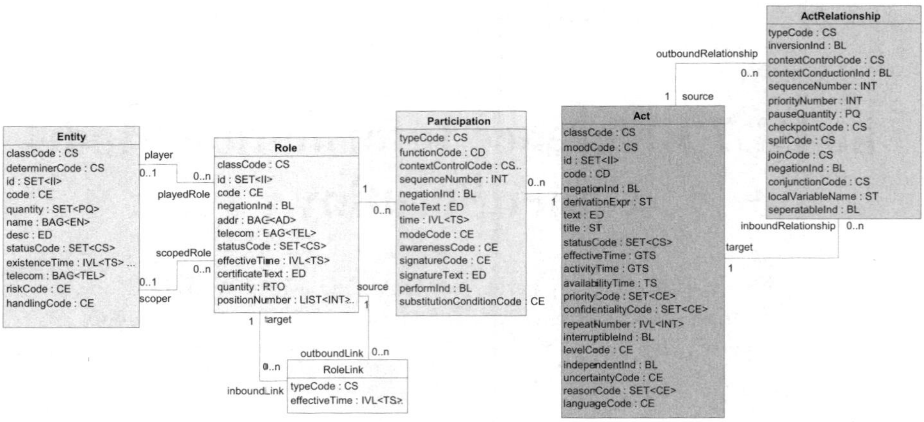

Figure 1: UML Class Diagram of the Reference Information Model "Backbone"

2. Technical Criticism

In surveying the literature (using PubMed, Google), we found that technical criticism is delivered only in passing, rarely in a fashion in which it could be dealt with to either improve the quality of HL7 or to dissipate the concerns. A paper by Fernandez and Sorgente is symptomatic; it claims that HL7 violates the Unified Modeling Language (UML), violates some ostensibly universally accepted object-oriented principles, and that it is not practically implementable. In truth all HL7 models are UML models and maintained as UML by the HL7 development tools. Domain-specific models may first be designed as unconstrained UML *domain analysis models* (DAM) and then transformed into a highly constrained normalized UML model where all classes in the constrained model must specialize one of the RIM classes and all features (attributes and associations) must derive from one of the RIM features. This guarantees that each instance-structure that complies with the specialized model also complies with the general model. This facilitates interoperability between interfaces for special domains on the one hand and general HL7 interfaces (such as for data warehouses) on the other, and thereby between two specialized interfaces where they overlap.

When mapping the special domain model to the RIM, we invoke an operation called *refinement* or sometimes "cloning", where RIM classes are shown as separate specialized classes on a new page as if they were de-novo created classes (such graph-refinement or folding operations are well established in computer science.) Some specialized classes may then not use certain attributes or associations from its original RIM class, because, while these features logically apply, the domain specialists had no business case for them in their design. Some critics perceive this as "deletion" of inherited features (which is frowned upon). However, this is not deletion but *projection*, a well-established operation in relational database theory and practically used in virtually all integrative information systems: through projection one maps the needs of a special application to the general (enterprise) database. Specialization through constraints is fully compliant with the UML notion of specialization.

HL7 has used UML models from the beginning, but has later developed another visualization technique. These "block diagrams" are easily traceable to UML, but visualize the standard RIM patterns and important constraints more concisely (e.g. vocabulary domain constraints). UML would use separate boxes with constraint annotations in the form of OCL expressions, which is much more verbose, provides less guidance, and is not suitable for the domain experts who are not computer scientists. HL7's deep experience with making model-based development work with domain specialists often creates the impression HL7 is "re-inventing the wheel" rather than using whatever is called the "industry standard" of the time. The truth, however, is that HL7 version 3 has been based on an object-oriented methodology since its inception in 1995/1996. HL7 not only actively adopted the best industry standard tools and methodologies available at the time, but also engaged the communities which developed these methodologies. When the HL7 version 3 project was launched, it had to pioneer a model-driven standard development methodology almost a decade before the first mentioning of the OMG's "Model Driven Architecture" (MDA). So, HL7 has not "re-invented" but rather originally invented many "wheels" to meet its requirements and vision at a time when the need for such wheels was not generally accepted.

That in the end the HL7 version 3 specifications may induce fear among some implementers is due to its mission of establishing semantic interoperability in loosely coupled systems. Thus, HL7 specifications address many of the practical difficulties with real world medical information processes (e.g., incomplete information, thing identification, duplicate record problems, uncertainty in data, etc.) directly, rather than deferring them to special applications. Even fundamental HL7 specifications such as the HL7 Version 3 Data Types do not remain silent about these ever challenging issues (such as uncertainty or incomplete information) where most general computing technology does not mention any of this. Implementers of special interfaces need not be concerned with the complexity that comes with HL7's full specification, as simple things usually are simple in HL7. But for more general HL7 interoperability, these issues are relevant, and need to be addressed (which would be more difficult without HL7.) Those who invest the appropriate effort in implementing general interfaces, such as the HL7 Java SIG or the Oracle Health Transaction Base, find that it can be done, that it works quite well, and they feed their improvements back to the working group.

3. Conceptual Design – Flaw or Virtue?

A range of conceptual doubts are raised, much of which have to do with the myth that HL7 is based on antiquated ideas of U.S. billing messaging, and that it therefore cannot deal effectively with ones favorite new challenge, be that "security" or the "electronic health record" (EHR). In truth HL7 has the most extensive collective experience of healthcare computing, specifically covering practical EHR; and security requirements are indeed considered very carefully. So, instead of arguing against myths, we shall focus in the rest of this paper on a core body of criticism raised by Smith et al. [2,3,4]

3.1. Incoherent Specification

Smith et al. seem to struggle with HL7's imperfect specification. Smith's ironic use of the term "exegesis" is not meant as a compliment. Smith rightfully scrutinizes the language of the HL7 definitions and discovers questions which are often overlooked in practice. Imperfections and inconsistencies in the HL7 documentation exist and need to be addressed, however, as much as one might try, inconsistency and ambiguity are deeply unavoidable in constructing a collaborative volunteer-based standard, which brings together a wide range of people from different backgrounds. As different people edit parts of the specification, inconsistencies in form and quality may emerge; as some ambiguities are clarified, other previous systematic ideas may be corrupted; and well-meant glossary entries may cause confusion. Sometimes irreconcilably opposed conceptualizations may coexist and one resorts to vague or ambiguous language in the interest of moving forward in areas where parties can not agree. Thus, standards are akin to the body of laws in pluralistic societies, which at any given point in time seems imperfect, incoherent and even contradictory. Over the long term, however, the engagement of different viewpoints in the development of the standard yields the best result because it is practically relevant and there are no alternatives to such consensus.

3.2. "Double Standard" – Reference Information Model vs. Reference Ontology

Smith's criticizes that the RIM blurs a distinction which should be made between *information model* and *reference ontology*. An information model defines what we record and communicate about the world, whereas the reference ontology would model the world itself. Smith shows how the definitions of the RIM elements switch between object-language (e.g., "person is a human being") and meta-language (person-record representing information). In this, HL7 has followed common practice in object-oriented analysis casting the result of the real world analysis into information model designs, using classes that bear intuitive names, such as "Person" or "Procedure," and yet being aware that no person will ever be stored in a computer system. Instead, HL7 documentation implicitly assumes that computer systems deal only with records of information about these real world phenomena. Since HL7 is about information management, some of its features (e.g. Entity statusCode) are about such management functions, but beyond that HL7 has revised and constantly rejected traditional data elements from specific local business needs, and only admits data elements that emerge from an analysis of what should generally be true about the entities of interest.

Why, then, does the RIM not seem to define the biomedical reality, e.g., molecular processes or disease processes, the ultimate subjects of the health information? To be sure, the RIM *does* provide structures (*form*) for modeling physical reality, but it leaves the definition of the *content* to specialty groups, either inside or outside HL7, and often in the form of terminology or ontologies. For example the HL7 drug knowledge model that is implemented with the U.S. FDA and pharmaceutical industry [8] contains classes for medicines, substances (Entities) and ingredient relationships (Roles relating the Entities). This is a general structure which supports the aggregation of a detailed and authoritative ontology of medicinal products. However, to describe the analysis of an unknown substance, or to determine the blood-level of a substance, HL7 uses laboratory measurement observation Acts instead of ingredient Roles, even though both are about the same chemical phenomenon, i.e., concentration. The Entity model is used for information that is known *a priori*, while observation acts are used for information that is only in the process of being discovered. This is why the clinical genomics model *seems* to conceptualize biomolecules as Acts, because it supports the process of identifying and discovering these molecules in particular instances.

HL7 also uses observation acts to describe that which may well be known *a priori*, but which is *not universally agreed* as to whether it is relevant or how it should be described. For example, we model the concentration of ingredients directly in the Roles but the color of medicines we model as observations. What seems like an arbitrary split between information that is supposedly known *a priori* vs. discovered or whose conceptualization is generally agreed vs. unknown, is in fact the practical thing to do. The alternative would be to leave even simple generally agreeable *a priori* information in a generic undefined structure, as it is common in "Entity-Attribute-Value" models, which are omnipotent but have no normative power, or, to create a variety of competing and evolving detail models which would require frequent modifications, such as addition or removal of special attributes. The RIM has many structural similarities with existing EHR systems' data models, featuring generic Observations structures, dynamically defined by data dictionaries, next to detailed *a priori* defined structures, which, when changed, would force expensive software updates. The modeling tradeoffs in the RIM are a reflection of the tacit knowledge and compromises of a community of practical system developers and users in health care.

A complete and integrated ontology of everything would certainly be nice to have; however, we think it is impractical and in fact dangerous to force such a model into being independently of the RIM. The moment such a model gained traction people would then expect that the RIM reflect that other model. Why should there be two models, if in the end one is to reflect the other? Instead, a single model of real world objects should suffice, but must contain well-defined features for information-management functions. These information-management functions are necessary to address the core problem of how intelligent agents (humans and heterogeneous software systems) arbitrate their perceptions of reality, how they communicate their intentions, and how they eventually effect changes to the real world. These are the core issue which HL7 addresses, and the remainder of this paper is dedicated to them.

3.3. Triple Standard – Act, Speech Act, and Documentation

We have introduced into the HL7 model the notion of *speech acts* [5], to describe the role that healthcare information plays in enabling cooperation. [6, 1] Speech acts are a generally accepted linguistic tool for understanding pragmatics, i.e., how language is used for achieving certain goals. Speech acts consist of *propositional content* and *illocutionary force* [7]. Propositional content is what the speech act *is about*; the illocutionary force is what it *accomplishes*. For instance, making assertions, demands, promises, proclamations, etc. accomplishes certain extra-linguistic effects in the behavior of others through illocutionary force. *Modality* is what the force accomplishes (e.g., assertion, demand, wish) subject to preparatory conditions, such as the relationship between the speaker and the one who is spoken to. Only if this relationship is appropriate the speech act can have its effect. For example, an order issued by an unauthorized person is invalid and will not have the intended effect. In the HL7 RIM, most of the features (attributes and participations) of the Act class (called "descriptive" features) carry the *propositional content*, and some features (called "inert" features) carry the information that substantiates the *illocutionary force*, such as the *mood-code* for modality, *author-participation* to substantiate the identity and role of the speaker and its relation to the receiver. Vizenor [2] took our speech-act analogy to further analyze cooperative healthcare actions, and he raises three major points of critique.

Firstly, Vizenor says that not all Acts in the RIM are speech-acts, e.g., an order for an injection is a speech act, but the injection itself is not a speech act, and that therefore the real act ontology and the speech-act ontology should be separately refined. We agree that the act of injecting is not a speech act, yet the medical record does not contain injections *per se*, but rather someone's talking about injection as an order or a report. The shared care record contains speech acts (even a simple assertive statement is a speech act) of which the physical act of injecting is simply reflected as propositional content. Trying to separate propositional content from its speech act is futile, because in the end, we have to represent both in a linguistic form. Propositional content is tied to speech just as one side of a coin is tied to the other.

Secondly, Vizenor says that the RIM documentations' considering Acts as "attributed statements" contradicts Acts being "speech acts", and that it is therefore wrong to tie attribution to the Act class. "Attribution" is the widely accepted practice to keep all data associated with its originator, and does not neutralize the illocutionary force of the speech acts. Instead, it simply substantiates the preparatory conditions that establish the illocutionary force in the first place. In human speech acts, the identity of the speaker is immediately given in the speech performance, the RIM only accounts for this. So, both propositional content and attribution are inseparable from speech acts.

Lastly, Vizenor identifies certain post-conditions in his analysis of speech acts, such as a promise creates an obligation of the *promiser* and a "claim" on the side of the *promisee*. Vizenor feels that records or documents represent these post-conditions, which he identifies as the "continuants" in his ontology and therefore expects to find them modeled as RIM Entities. The HL7 RIM, however, regards the post-conditions of speech acts as states inside the communicating systems that are sufficiently substantiated by the record of the speech act itself, which each system interprets given its aspect with regard to the speech act. This is why a "document" in the RIM is only a special form of speech act, in the form of a human readable text, while the functions of *documentation* and *recording* are supported by any RIM Act object.

4. Conclusion

Many formal technical criticisms as well as critique about the conceptual design of the RIM are rooted in misunderstandings, and more importantly in what seems to be a fundamental disagreement as to how a reference model for semantic and pragmatic interoperability should be created. We welcome the discussion on and off the record of published literature, as it helps to clarify the fundamental assumptions and opens up the RIM to much needed logical validation and improvement. However, we believe that if the fundamental approaches are at odds, the specific criticism will remain ineffective.

References

[1] Schadow G, Russler DC, McDonald CJ. Conceptual alignment of electronic health record data with guideline and workflow knowledge. Int J Med Inform. 2001 Dec;64(2-3):259-74.
[2] Vizenor L. Actions in health care organizations: an ontological analysis. Medinfo. 2004;11:1403-7.
[3] Smith B, Ceusters W. HL7 RIM: an incoherent standard. Medical Informatics Europe. 2006. Submitted.
[4] Ceusters W, Smith B. Strategies for referent tracking in electronic health records. Biomed Inform. 2005 Sep 27.
[5] Austin JL. How to do things with words. Oxford: Oxford University Press, 1962.
[6] Schadow G. Krankenhauskommunikation mit HL7; Analyse, Implementation und Anwendung eines Protokollstandards für medizinische Datenkommunikation. Aachen, Germany: Shaker-Verlag; 2000.
[7] Loos EE, Anderson S. Day DH, Jordan PC, Wingate JD (Eds). Glossary of linguistic terms. http://www.sil.org/linguistics/GlossaryOfLinguisticTerms (Accessed on: 1/30/2004)
[8] Schadow G, Gitterman S, Boyer S, Dolin RH eds. HL7 v3.0 structured product labeling, release 2 [standard]. Ann Arbor, MI, Health Level Seven, 2005

Ubiquity: Technologies for Better Health in Aging Societies
A. Hasman et al. (Eds.)
IOS Press, 2006

Participation in the International Classification for Nursing Practice (ICNP®) Programme

Claudia BARTZ[a1], Amy COENEN[b] Woi-Hyun HONG[c]
[a]*International Council of Nurses, Geneva, Switzerland & Clinical Associate Professor, University of Wisconsin-Milwaukee, USA*
[b]*International Council of Nurses, Geneva, Switzerland & Associate Professor, University of Wisconsin-Milwaukee, USA*
[c]*International Council of Nurses, Geneva, Switzerland & Doctoral Student, University of Wisconsin-Milwaukee, USA*

Abstract. The International Council of Nurses (ICN) is a federation of 129 national nurses associations. The International Classification for Nursing Practice (ICNP®) is a programme of the ICN. The purpose of this paper is to describe the development and maintenance processes of the ICNP® Programme that are used to increase participation. These include processes by which the ICNP® was and continues to be developed, tested, distributed and implemented worldwide, with emphasis on the current version, ICNP® Version 1.0. The ICNP® is a unified nursing language that facilitates cross-mapping among local terms and existing terminologies. ICNP® conforms to current terminology standards and criteria, for example, ISO standards and HL7. The ICNP® Alpha and Beta Versions documented the progress of concept validation and classification of nursing phenomena and interventions. The ICNP® Beta 2 Version was a combinatorial terminology organized in two multi-axial structures representing nursing phenomena and nursing actions. The ICNP® Version 1.0, launched in 2005, changed the relatively straight-forward multi-axial structure into a compositional terminology through the application of description logics using Web Ontology Language (OWL) within Protégé, an ontology development environment. ICNP® Version 1.0 is also represented in a multiaxial model (7-Axis) for nurses to compose nursing diagnosis, intervention and outcome statements. Language translations and clinical information systems applications are required to make the ICNP® Version 1.0 available to nurses at the point of healthcare delivery. ICNP® data collected in healthcare environments provide standardized terminology for nursing that allows comparison of nursing practice across health care settings, specialties and countries; facilitate data-based clinical and management decision making; and contribute to the development of guidelines and standards for best practices and optimal outcomes for patients, families and communities.

Keywords: Informatics, Nursing, Terminology, Vocabulary, Ontology

[1]Claudia Bartz, UWM, P.O.Box 413,Milwaukee, WI, USA 53201. cbartz@uwm.edu

Introduction

The International Classification for Nursing Practice (ICNP®) was developed under the auspices of the International Council of Nurses (ICN) to clearly articulate nursing practice. Florence Nightingale wrote in *Notes on Nursing* "the very elements of nursing are all but unknown".[1] More recently, Clark and Lang commented on the invisibility of nursing: "If we cannot name it, we cannot control it, finance it, research it, teach it, or put it into public policy."[2] The electronic capture of nursing diagnoses, interventions and outcomes using the ICNP® substantially increases the potential for information systems to support evidence-based practice and generate further research.

1. Material and Methods

1.1 Information Infrastructure for Evidence-based Practice

Bakken described five building blocks of an information infrastructure for evidence-based practice: [3]

- standardized terminologies and structures,
- digital sources of evidence,
- standards that facilitate health care data exchange among heterogeneous systems,
- informatics processes that support the acquisition and application of evidence to a specific clinical situation, and
- informatics competencies.

The ICNP® provides a standardized terminology for nursing which can be used to compare nursing practice across health care settings, specialties and countries. ICNP® can be cross-mapped to other nursing classification systems, thus enhancing data exchange across heterogeneous systems.

1.2 ICNP® Beta 2 Version

The Beta 2 Version of ICNP®, released in 2001, was used and analyzed extensively around the world. ICNP® project types included validation studies, computer-based information system demonstration projects, research and evaluation teams, and cross-mapping projects. An ICNP® Country Project was funded by the W.K. Kellogg Foundation starting in 1994 to expand coverage of primary care and community based concepts in ICNP®. The Beta 2 Version of ICNP® was translated into more than 25 languages. With increased dissemination of Beta 2, the ICNP® proponents recognized that the goal of a unifying nursing language system that would represent nursing practice worldwide needed new classification structures and strategies.

1.3 ICNP® Version 1.0

Following consultation with world leaders in the field of healthcare vocabularies, major recommendations for improvement of the ICNP® were to:

- provide a more formal foundation for the ICNP®, and

- apply software that would be capable of satisfying current, accepted criteria for a fully workable vocabulary. [4]

Examples of such criteria, which had not been consistently met by the ICNP® Beta and Beta 2 Versions, were to avoid:

- redundancy of terms,
- ambiguity of terms, and
- context laden identifiers (ensuring that codes do not reflect the hierarchical structure of the terminology). [4]

The ICNP® Version 1.0 reflects major reformulations aimed at making the terminology technologically more robust while still being accessible to the nurse user. ICNP® Version 1.0 was developed using Web Ontology Language (OWL) within the ontology development environment, Protégé. It is a resource that can accommodate existing vocabularies (through cross-mapping), that can be used to develop new vocabularies (as a compositional terminology), and that can identify relationships among concepts and vocabularies (as a reference terminology). [4]

2. Results

To realize vision of the ICNP®, a number of products and processes have been established to facilitate participation in the ICNP® Programme. These include:

- 7-Axis Model of ICNP®
- ICNP® Centres
- ICNP® Translations
- ICNP® Catalogues
- ICNP® Review Process

2.1 The 7-Axis Model for ICNP®

The new, simplified 7-Axis Model was derived from the ICNP® Version 1.0 and has the purpose of providing user-friendly access to ICNP® concepts and definitions. The seven axes of Version 1.0 unify the Beta 2 Version's eight axes of the nursing phenomena classification structure and eight axes of the nursing action classification structure. The seven axes are Focus, Judgment, Time, Location, Means, Action and Client. [4]

The 7-Axis Model is used by nurses to create nursing diagnosis, intervention and outcome statements for use in practice. Consistent with the ISO standard 18104, nursing diagnosis and nursing outcome statements must include a term from the Focus axis and the Judgment axis and may include additional terms as needed from the Focus, Judgment or other axes. Additionally, nursing intervention statements must include a term from the Action axis, at least one Target term, which can be from any axis except the Judgment axis, and additional terms as needed from any of the axes.

2.2 ICN-accredited ICNP® Centres for Research and Development

The ICNP® Programme is committed to worldwide dissemination of the ICNP® simultaneous with constant improvement in ICNP® content and capacity for implementation. Three ICNP® Centres for Research and Development have been accredited to date. These Centres are charged with providing resources and outcomes in terms of information, services, research and education.

2.3 ICNP® Translation

Nurses and their colleagues are participating in translation activities to increase access to the INCP®. The ICNP® Programme is developing translation guidelines and a translation browser to assist in translations to multiple languages.

2.4 ICNP® Catalogues

The purpose of ICNP® Catalogues is to support nursing documentation at the point of care by providing relevant sets of pre-coordinated nursing diagnosis, intervention, and outcome statements composed using the ICNP®. Nurses can submit their post-coordinated statements to ICN for coding. Catalogues will be organized in various ways to address nursing needs. For example, catalogues may address:

- Unit of Care: e.g. Individual, Family, Group;
- Health Problem: e.g. Specified Disease, Illness, Injury, Condition;
- Care Specialty/Setting: e.g. Medicine, Surgery, Mental Health, Woman's Health, Pediatrics;
- Nurse-sensitive phenomena: e.g. Pain, Dehydration, Incontinence, Adherence.

2.5 ICNP® Review Process

The ICNP® Review Process has been in place since 2001 to provide a transparent method to evaluate recommendations and suggestions submitted to ICN. The process is currently under review.

For further information and to volunteer as an expert reviewer see: http://www.icn.ch/icnp_review.htm

3. Discussion

The dynamic nature of ICNP® Version 1.0 supports constant interaction between and among nurses in practice, researchers, educators, health care informatics specialists, terminology and vocabulary specialists, standards development organizations, and vendors. Nurses worldwide are generating new concepts and definitions for consideration for inclusion in ICNP®. A multinational team of nurse experts serves as reviewers of new and revised terms.

The development and implementation of health care information systems and electronic health records must include a unified nursing language and the ICNP® meets that requirement.

4. Conclusion

ICN is a federation of national nurses' associations representing the millions of nurses worldwide. The ICNP® Version 1.0, as a programme of the ICN, strives to contribute to ICN's mission to represent nursing worldwide, advancing the profession and influencing health policy. There are a number of ICNP® programme initiatives to promote partnerships in research and development.

No terminology can be absolute or static. The new version of ICNP® was developed based on the several previous revisions to meet the need of nurses across the world. This unified nursing language system has great potential to continue to grow, through the participation of the international nursing and health care community.

References

[1] Nightingale F. Notes on nursing: What it is and what it is not. 1859.
[2] Clark J, Lang, N. Nursing's next advance: An international classification for nursing practice. International Nursing Review 1992; 39: 109-112.
[3] Bakken S. An informatics infrastructure is essential for evidence-based practice. JAMIA 2001; 8:199-201.
[4] ICN. International Classification for Nursing Practice Version 1.0. Geneva: ICN; 2005.

2. Electronic Health Records

2.1 EHR: Architectures

Ubiquity: Technologies for Better Health in Aging Societies
A. Hasman et al. (Eds.)
IOS Press, 2006

A user-centred deployment process for ICT in health care teams – Experiences from the OLD@HOME project

Maria HÄGGLUND[a, b ,1], Isabella SCANDURRA[a, b] and Sabine KOCH[a, b]
[a] *Dept of Medical Sciences, Uppsala University, Uppsala, Sweden*
[b] *Centre for eHealth, Uppsala University, Uppsala, Sweden*

Abstract. *Objective*: To present a user-centred method for introducing ICT in health care organisations, taking factors that influence acceptance into account. *Methods*: User centred methods are used in combination with previous research regarding factors that affect user acceptance, in order to facilitate users' acceptance of new ICT tools. *Results*: A method is presented that supports the introduction of ICT in team work. The method consists of three major steps; (1) the start-up seminar, (2) end user education and (3) continuous follow-up during the deployment phase. Important results of the start-up seminar are documentation of the users' expectations, and an agreement of ground rules that supports both the social norm factor and the users' perceived behavioural control. Education and follow-up also improve perceived behavioural control, and by involving super users perceived usefulness and ease of use can be improved through subjective norm. *Conclusion*: Key factors in the deployment process are; user participation, end user experience and education, and continuous follow-up of the process.
Keywords: Medical Informatics, Diffusion of Innovation, Technology Transfer, Computer Literacy, Technology acceptance, User Centred Methods,

1. Introduction

Not only characteristics of new ICT itself determines whether it will be adopted by the users [1], other factors also affect user acceptance. This is particularly problematic when introducing ICT in work environments where the professionals have little experience from ICT, and little personal interest in new technology. Several models, e.g. the technology acceptance model (TAM) [2], the theory of planned behaviour (TPB) [3], and innovation diffusion theory (IDT) [4], have been used to explain the motivational factors involved.

It is crucial to improve the acceptance of ICT in the health care sector since expected benefits from the investments are realized only when systems are adopted by their intended users and subsequently used. We present a deployment process that takes these factors into account and use them in practice when introducing ICT. The process is based on experiences from the action research project OLD@HOME, where a mobile virtual health record (VHR) was introduced in a home care for elderly environment [5].

[1] Corresponding Author: Maria Hägglund, Dept. of Medical Sciences; Medical informatics and biomedical engineering, Uppsala University, University Hospital 82:1, S-75185 Uppsala, Sweden. Maria.Hagglund@medsci.uu.se

2. Materials and Methods

Yi et al have developed a model of ICT acceptance, based on previous research, that measures how different factors affect users' *behavioural intention*, i.e. a persons subjective probability to perform a specific behaviour [1], e.g. accepting to use an ICT tool. The factors studied by Yi et al are presented in Table 1.

Table 1. Factors that affect users' behavioural intention (adapted from [1])

Factor	Description
Subjective (social) norm	Perception of what important others feel about adopting an ICT tool, e.g. if a superior or peer says that an innovation might be useful, the suggestion can affect the individual's perception of the usefulness of the tool.
Image	In order to create or preserve a positive image within a social group, individuals often respond to social influences, hence an individual can believe that a system is useful if it enhances their image and social status.
Perceived behavioural control	Reflects user perceptions of internal and external constraints, e.g. if they believe themselves capable of using ICT, or that factors in their environment help or prevent them using ICT.
Result demonstrability	When tangible results of the ICT are directly apparent it is easier for the users to adopt it.
Personal innovativeness in IT (PIIT)	Concerns a person's tendency to adopt new technology. Some individuals are more willing to take a risk by trying out an innovation (early adopters), whereas others are hesitant to change their practice.
Perceived usefulness	The extent to which a person believes that using the system will improve his or her work performance.
Perceived ease of use	The extent to which a person believes that using the system will be effortless

According to the study presented in [1] all of the factors listed above affect the potential user, and the authors advocate awareness of these factors when introducing ICT in a work place, in order to fully realize the expected benefits.

An important aspect in our work is *user participation*. User participation is strongly advocated in many system design approaches, e.g. in the Scandinavian approach [6], also referred to as Participatory Design [7, 8], and the User Centred Systems Design [9] approach (UCSD), all actively involving real users, acting in real settings, to improve development of ICT systems. By using experiences from this type of user centred work in combination with knowledge of influential factors described above, a deployment process which actively manages and manipulates these factors to facilitate user acceptance has been developed.

The deployment process was used when introducing a mobile VHR in home care of elderly patients in a small town in northern Sweden. The user group (n=15) was fairly homogeneous, consisting of home help service personnel (HHS), mainly assistant nurses, performing different community services, e.g. home help and delegated nursing activities. A questionnaire based user analysis was performed before initiating the deployment process, with questions relating to experience of, personal interest in, and general feelings towards ICT. Few had experience of working with ICT and the majority were not what is normally referred to as early adopters of technology. Three persons from the HHS team with a higher interest in ICT and new technology had volunteered to participate in the user centred development of the VHR, here referred to as *super users*, and played an important role in the deployment process. Evaluation of the deployment process consisted of qualitative interviews, both individual and in group sessions, with the HHS. The fears and expectations expressed at the beginning of the deployment were revisited, and the different strategies for handling these discussed.

3. Results

It is important to state the benefit of having applied a UCSD process when introducing a system. If applied, end users (i.e. super users) are involved from early on, becoming good advocates for the system, and key resources during the deployment process. If a user centred process has not been applied it is important to find interested professionals with high PIIT early in the deployment process. This requires extra educational efforts and quite a bit of luck in finding the right persons to act as super users.

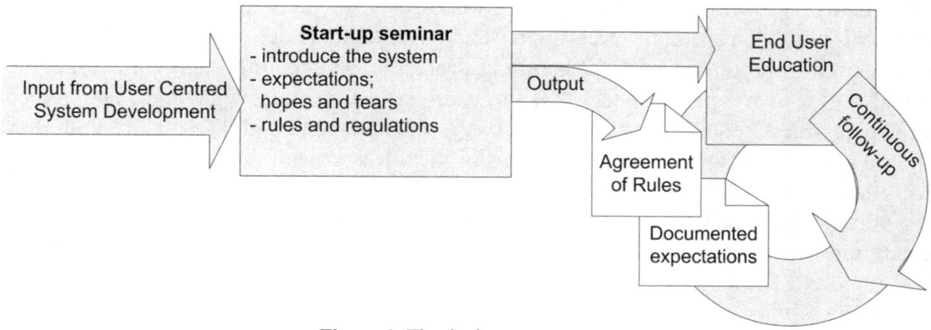

Figure 1. The deployment process

The deployment process consists of several steps (Figure 1), with the continuous follow-up continuing while the system is used in the actual work settings. The UCSD process delivers input to the deployment process both in form of super users and the benefit of having transferred information from the development team to the rest of the team during development, resulting in already informed personnel at deployment start.

3.1. The inspirational Start-up seminar

The purpose of the start-up seminar is to create a good environment for introducing the ICT and to identify possible benefits and risks. The new end users should gain a clear picture of what is expected of them, and how problems should be handled when they occur, making the personnel feel more secure and motivated. The entire work team is gathered to the start-up seminar in order to create a common ground for the personnel; to openly discuss their expectations and apprehensions, and the impact the system will have on their work.

The perceived usefulness of a system is found to be one of the most important factors influencing user acceptance, in turn affected by other factors, e.g. perceived ease of use and result demonstrability [1]. The first step of the seminar is therefore to present the ICT in an understandable way, focusing on its usefulness in the daily work. The super users are usually better at explaining how the ICT-tool is to be used in the daily work, and what impact it will have on work practices, and are therefore well suited to handle the introduction. The subjective norm is also an important factor here; acceptance from the personnel tends to be higher when ICT is introduced and advocated by their own colleagues.

The process continues by open discussions of the team's hopes and fears related to the introduction of ICT. All fears need to be taken seriously, and strategies formulated to handle them. Conflicting ideas within the team have to be ventilated without judgement. This is important in relation to the personnel's perceived behavioural control. The decision to introduce an ICT tool in health care is usually taken at a higher level in the organisation, and even if the end users have been involved, there will be conflicts. Those against the introduction are forced into a situation they fear they cannot handle and they experience a lack of control. By taking expressed objections seriously and finding ways to handle problems in advance, an increase in the perceived behavioural control can be achieved. Expected benefits and improvements need to be discussed too, so that the perceived usefulness and ease of use are not brought down, and here super users play an important role.

Finally, it is time to formulate the ground rules for the deployment; realistic and practical rules that all can agree upon. The rules contain descriptions of the team's responsibilities and strategies for handling problems. This is important with respect to social norm; everyone must feel that the team expects certain behaviour. The agreed strategies often require involvement of e.g. management and IT-support and they should commit to their respective responsibilities. It is crucial that the health care staff feel that the effort they put into the deployment of the ICT is fully supported, that they are allowed the extra time needed and that they are supported when problems occur, again improving their perceived behavioural control.

3.2. Education

Education is of essence to improve users' perceived behavioural control. Users with little experience of ICT often experience fear; fear of not being able to handle the tools and not being able to perform their work. Again the super users play a key role in explaining how the ICT tool is to be used in daily work. In concurrence with the subjective norm, acceptance from the personnel also appears to be higher if their own colleagues tell them how to use the new device. Therefore, most of the education is handled by the super users, and the development team is mostly standby to answer questions of a more technical character. The super users also continue to act as support and a link to the development team throughout the deployment process.

The education process in OLD@HOME consisted of two major parts; (1) each health care professional was given individual education by the super users, guidance through the systems functionality and walk through of manuals and support tools. Follow-up sessions were available if insecurity remained. (2) The work group then started to use the tools in their daily work, but only for practise, and with simulated data. This is to handle fears expressed by many users about making mistakes when inputting data, or accidentally deleting important information. The opportunity to use the system in real work situations but with simulated data allowed them to make mistakes, and experience that these did not have irreversible consequences.

3.3. Continuous Follow-up

Once the deployment process is launched, it is important to follow up the work done during the initial start-up seminar. By continuously revising and evaluating expectations documented at the first seminar, it is possible to adjust the deployment process on the fly, thereby making it more suitable for the specific work group at hand. Regular meetings where problems and benefits are discussed and use of techniques, e.g. diary keeping [10], to make users reflect on their work enables the deployment process to be monitored and adjusted. By ensuring that experienced problems are handled quickly, end users perceived behavioural control can also be kept high.

3.4. Qualitative evaluation of the deployment process

During the start-up seminar the HHS expressed a number of fears; *lack of support;* from management and from IT-support, *technical problems*; disturbance of daily work, *time consuming*; take too long to perform everyday tasks, the personal interaction with the patient will suffer, *lacking usability and/or lack of belief in ones own capabilities*; too difficult to learn, too difficult to handle, *not useful;* the provided functionality will not be adapted to and useful in the work situations, *personal responsibilities;* fear of making mistakes when using the ICT, fear of loosing or damaging the hardware, and finally fears related to *team work*; if members of the group do not use the tools double work loads, misunderstandings and conflicts within the team can occur.

Positive expectations were also great, and mainly related to; *improved work situation;* less paper-work, safer documentation, more meaningful work, access to information, and *personal development;* interesting to learn something new, new experiences, improved competence.

The HHS were overall positive to the deployment process. Discussing fears and concerns in a group seminar were seen as particularly important since a lot feelings were stirring under the surface and needed to be ventilated. The personnel however felt that the amount of information about the ICT and education, including time to practice, could have been increased, and come earlier, stressing the importance of these factors when introducing ICT.

Most of the fears did not become real problems during deployment of the ICT. Technical problems and a perceived lack of support however remained issues throughout the introduction. They did however feel that the positive outcomes outweighed the problems and adopted the ICT into their daily work. Personal development and improved work situations were confirmed results after the introduction, as well as an increased insight into their work processes and improved cooperation within the team.

4. Discussion

The deployment process described here is intended for use in team based care, thereby differing from the study performed by Yi et al in [1], which considers factors that affect individual professionals acceptance of new technology. Their work also focuses on the individual's voluntary adoption of a system, whereas we have worked with the mandatory situation occurring when ICT is introduced in a care provider organisation. We do however believe that similar processes occur within the individual professional, and the factors can therefore be of use in a method such as the one described here.

Another difference is the studied profession; Yi et al work with general practitioners (GP) who generally have high status, high education and high income, whereas the HHS work in a low status, low income profession requiring little education. The image factor is likely to increase in importance for the latter group, and is often used as an incitement for introducing ICT in home help service. Perceived behavioural control can also differ, where a GP has high influence and control over work, the HHS have little influence and often feel under-prioritised and overlooked.

A perhaps more important issue concerns the transfer from introducing ICT in a small organisation, as is the case here, to a larger, more complex care provider organisation, e.g. an entire hospital. By limiting the focus to the teams providing care in a large organisation, it is however possible to use this method, it then becomes a financial issue. We would in any case recommend that in order to realize the full potential of the, often very large, investment when introducing ICT in health care, it is necessary to prioritise the deployment process.

5. Conclusion

In conclusion, important success factors to be taken into consideration when introducing a computerized information system in health care are; (1) User participation; end users need to be involved throughout the process (2) Experience and education; the end users computer experience and relation to the tool they need to master are very important, education becomes increasingly important when users have little experience and low PIIT, (3) Continuous follow-up; closely monitoring the introduction over a longer period of time is crucial since problems may occur only once the system is in use in the clinical environment, and giving feedback to end users that the problems they experience are acknowledged and handled. All these factors affect the users' perceived behavioural control, which appears to be a key factor in the health care setting.

Acknowledgements

Lena Eriksson from Komanco provided invaluable support and knowledge during the start-up seminar. The work was performed in the project "OLD@HOME - Technical Support for Mobile CloseCare", sponsored by VINNOVA - Swedish Agency for Innovation Systems (P23037-1 A), Trygghetsfonden, clinical and industrial partners.

References

[1] Yi MY, Jackson JD, Park JS, Probst JC. Understanding information technology acceptance by individual professionals: Toward an integrative view. Inform Manage. 2006;43:350-63.

[2] Davis FD. Perceived usefulness, perceived ease of use, and user acceptance of information technology. MIS Quarterly. 1989;13(3):319-39.

[3] Ajzen I. The theory of planned behavior. Organ Behav Hum Dec. 1991;50(2):179-211.

[4] Rogers EM. Diffusion of Innovations. New York, NY: The Free Press; 1995.

[5] Koch S, Hägglund M, Scandurra I, Moström D. Towards a virtual health record for mobile home care of elderly citizens. In: Fieschi M, Coiera E, Li J, editors. MEDINFO 2004 - 11th World Congress on Medical Informatics; 2004; San Fransisco, USA, Sept 2004: IOS Press, Amsterdam. p. 960-3.

[6] Bødker S, Ehn P, Sjögren D, Sundblad Y. Cooperative design perspectives on 20 years with "the Scandinavian IT Design Model". In: Gulliksen J, editor. NordiCHI 2000, Design vs Design; 2000.

[7] Bødker S, Iversen O. Staging a Professional Participatory Design Practice - Moving PD beyond the Initial Fascination of User Involvement. In: Bertelsen, Bødker S, Kuutti, editors. NordiCHI 2002; 2002; October 19-23, Aarhus. p. 11-8.

[8] Schuler D, Namioka A, editors. Participatory Design - principles and practices. New Jersey: Lawrence Erlbaum Associates; 1993.

[9] Göransson B. User-Centred Systems Design: Designing usable Interactive systems in Practice: Uppsala University; 2004.

[10] Ericsson A, Tesch-Romer C, Krampe RT. The role of practice and motivation in the acquisition of expert-level performance in real life: An empirical evaluation of a theoretical framework. In: Howe MJA, editor. Encouraging the Development of Exceptional skills and Talents. Leicester: The British Psychological Society; 1990. p. 109-30.

Non-invasive light-weight integration engine for building EHR from autonomous distributed systems

Pere Crespo Molina[1], Carlos Angulo Fernández, José A. Maldonado Segura, David Moner Cano, Montserrat Robles Viejo
BET Group, Technical University of Valencia, Spain

Abstract. Pangea-LE is a message oriented light-weight integration engine, allowing concurrent access to clinical information from disperse and heterogeneous data sources. The engine extracts the information and serves it to the requester client applications in a flexible XML format. This XML response message can be formatted on demand by the appropriate XSL (Extensible Stylesheet Language) transformation in order to fit client application needs. In this article we present a real use case sample where Pangea-LE collects and generates "on the fly" a structured view of all the patient clinical information available in a healthcare organisation. This information is presented to healthcare professionals in an EHR (Electronic Health Record) viewer Web application with patient search and EHR browsing capabilities. Implantation in a real environment has been a notable success due to the non-invasive method which extremely respects the existing information systems.
Keywords: Systems integration, health records, electronic patient records, medical records, medical records linkage, hospital records, EN13606, architecture and sharing of electronic health records.

1. Introduction

Healthcare is a very data-intensive sector, producing and consuming a great amount of data. In healthcare organisations, especially hospitals, the big amount of data gets increasingly obscure due to its widely decentralised organisation which allowed different departments to meet specific or local data requirements without the coordination and/or standardisation of information systems. This led to fragmented and heterogeneous data resources, which contains health data about patients, so called islands of information, making the access and aggregation of data across systems very difficult [1]. This situation has created a large gap between the potential and actual value of the information content of EHRs.

Closing this gap by making efficient use of the health data held by these systems, could improve significantly patient care, clinical efficiency and empower research activity. Having this in mind, it results difficult to imagine a healthcare organisation without any kind of information sharing among its information systems. Even though, many hospitals already have best-of-breed departmental information systems, only a few possess an integrate workstation, allowing health professionals to access to relevant patient's healthcare information in a single and unified view. One solution is the acquisition or development of large and centralised information systems where integration gets guaranteed, but specialty services loose their freedom to select the software that fit best their requirements. On the other side, we can find some integration systems which allow each clinical service to use the software they need while getting data integrated all along the organisation. Furthermore, the less invasive the integration system is, the less modification in the integrated subsystems must be

[1] pedcremo@fis.upv.es

done. Most common integration systems must perform routing mechanisms, manage system events and queues to emulate point to point data sharing, which make them too heavy for the only purpose of getting a unified view of data; we need a lighter solution.

We present here an overview of the Pangea-LE system, emphasising its flexibility and quick organisation deployment process. Briefly, Pangea-LE is a data integration system that provides an integrated and global view over distributed health data sources. In Pangea-LE, the global view is easily customisable to different user groups whose needs may change over time. On the other hand, data source adapters are as general as possible. In an evolving environment as healthcare, the local databases may change often. The databases are designed and maintained to meet local needs, and changes are almost made independently of the integrated global view. Applications connected to Pangea-LE global data view do not need to be updated after any local data source change. These features make the Pangea-LE system a very valuable solution to achieve fast, easy and secure data sharing and integration.

2. Materials and Methods

Pangea-LE acts like a piece of glue or mediator [2] between existing health data repositories in an organisation and health professionals. It allows the definition and management of a global, integrated and structured view of all the clinical records stored for a patient in an organisation. This view is presented to the professionals through an EHR Web application. Information views are created on the fly and are role dependant so that enterprise-wide access roles can be defined for different professional profiles with specific access rights levels corresponding to different EHR views (clinicians, nurse, management, administrative, etc.)
We must highlight the purpose of this concrete use case is only to display EHR views. Information is presented in human readable way and the clinician must provide his/her particular interpretation and meaning. Nevertheless, it is worthy to notice that it is also possible to keep the original meaning of shared EHR extracts for machine interpretation purposes, guided by some EHR standards. This use case has been tested only for experimental purposes according to the European norm CEN EN13606 [3][4].
Pangea-LE can be classified as a generic middleware that integrates clinical information. It is important to notice that there are only a few basic requirements to be accomplished before Pangea-LE could be deployed into a healthcare organisation:

- ✓ Unique Patient Identification throughout all the organisation local information subsystems. In case there is none, the organisation must provide a method in order to solve patient identification conflicts.
- ✓ Organisation-wide user authentication and role assignment. This task is commonly executed by means of a directory service such a LDAP (Light-weight Directory Access Protocol) [5].

Every subsystem involved in an integration project is a potential container for a set of clinical concepts. Pangea-LE offers read-only views only. Pangea-LE architecture has four basic components:

1. Adapters: Heterogeneous data source access is reached through a set of JDBC (Java Database Connectivity) drivers for commercial and/or specific databases and through a set of configuration files to parameterise each integrated data source. The majority of data sources come from relational universe and can be accessed using one of the currently 221 different JDBC drivers available [6]. Other types of

sources, like file systems, ftp, XML [7], messaging systems or even a Web Service JDBC Wrapper can also be defined and accessed in a homogenous way.

2. EHR Extract Definitions (EED) are the primary components in Pangea-LE since they describe the clinical concepts that can be shared among the different subsystems involved in an integration project. Neither whole system nor finest atomic data but only contextualised data, containing clinical information, should be shared through Pangea-LE. This selected information is organised and specified in a clinical EED entity. Several EED's can be defined for one subsystem, each one representing a different clinical concept. Each EED can only be shared as a whole. In other words, the minimum unit of information that can be shared between two subsystems in Pangea-LE is generated by an EED entity. EED entities are defined using XML files where some fixed descriptor elements define their behaviour:

- ✓ Elements that specify a valid adapter configuration for accessing the information that the EED conforms.
- ✓ Elements specifying data to be extracted from the data source.
- ✓ Input parameters accepted by query processor to execute filters on data sources.
- ✓ Valid calls (input parameter combinations which are allowed in order to build an EHR Extract from EED).
- ✓ Elements specifying data pre-processing which allows source atomic data to be combined or transformed before XML generation.
- ✓ Elements that describe the labelling and nesting format that constitutes the resulting XML document, etc.

EED entities can be instantiated by a Pangea-LE exposed Web Service (WS) [8]. This WS accepts XML petition messages pointing to the desired EED together with the required parameters (e.g. patient identification number). When this petition arrives to Pangea-LE and it is checked as a valid call, Query Processing Module (see Figure 1) can construct the appropriate SQL statements to fill with data the EED entity. This process involves obtaining a set of data and building meantime a XML response conforming the EED labelling and nesting rules defined [9] in the EED. Finally, if the petition requires XSL transformation, this is applied and the response is sent back to the requester application.

3. XSLT Transformations: Each EED can specify one or more XSL transformation files to be applied to the response XML message. This method allows Pangea-LE to adapt the output to different application message formats or devices (PDAs, Tablet PCs, etc.). Moreover, the same clinical entity can be formatted in different styles according to specific user access roles. Transformations can be applied in the server side or alternatively in the client application if it is enabled to perform this task. A data instance of an EED entity with no transformation applied is obtained by default.

4. EHR Browsing Trees: One of the most remarkable Pangea-LE features is the ability to organise the retrieved clinical information in a very flexible data tree structure. Once information has been extracted, different views can be configured

to organise EHR in a clinical history manner (ordered by date), or according to information origin (emergencies, consults, explorations), etc. An EHR tree is also an XML document that defines how to structure, organise, aggregate, summarise and guide the extraction process of the clinical information associated to one patient. At run time, each EHR tree constitutes a health history view for a particular role and is mainly composed using attached EED entities. When one user accesses to the clinical information of a patient, the module that interprets EHR trees retrieves firstly only the minimum information needed to shape a summarised view of the patient's health history. Subsequently, user can interact with each particular tree node to get a more detailed view of the particular clinical subject described by the linked EED entity. All these possible interactions are defined and also controlled by the EHR trees. Each structural component of the EHR tree data model corresponds to nodes in the visual EHR tree control used in the EHR viewer Web application. Thus, EHR tree nodes are user interactive and EHR can be build on demand. EHR browsing trees are designed for visualization purposes only and it is possible to define a different EHR tree for each different access role.

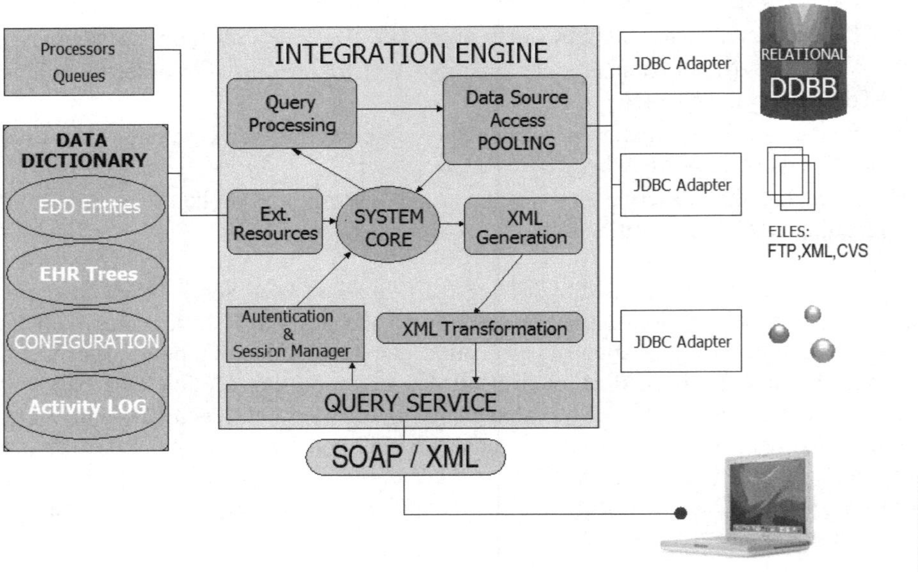

Figure 1 - Pangea-LE Architecture

Pangea-LE system core has been developed using technologies from a wide variety of free software components [10]: Eclipse as development framework, Ant for compiling and packaging, JDBC drivers for relational database source access, Struts framework and Prizetags for EHR Web viewer, Xerces, Xalan and Jdom libraries for XML parsing and transformation, Tomcat as application container, AXIS for Web Service development, JMeter for performance and functional testing, Log4j and HSQLDB for log support. The whole system has been developed in Java, so it is multiplatform. Two different clients have been Java developed; one version as a swing desktop application and the second as a Web version; both sharing the same capabilities.

3. Results

Pangea-LE deployment in a real environment such as CHGUV (General University Hospital of Valencia) has been a very worthy experience. Currently, CHGUV has 592 beds, 21 surgeries, 470 doctors and serves a population of 350.000 inhabitants.

The deployment process itself has been progressive and strictly coordinated. The most important systems have been the first to be integrated. At the moment, the majority of patient information is available electronically through Pangea-LE, including: Alerts, emergencies, inpatient episodes, outpatient consultations, laboratory results, biopsy and cytology study results, magnetic resonance reports, mammography, endoscopies and most of the discharge summaries from specialty units.

A multidisciplinary work team has been created in order to take integration project to success. The team is composed of representative specialists from clinician, management, clinical documentation and informatics departments. They are in charge of the coordination of resources, agenda and priorities during the project deployment time. Easy access, completeness and immediate retrieving of information have made EHR viewer application rapidly extends organisation-wide. One year after implantation, its number of users is 472 (approximately the half of users in the organisation), 180 of them access daily to the EHR viewer. At present, more than twenty local systems have been integrated and 36 EED have been defined. The system has served, on average, 12000 petitions per month.

Most significant use of EHR viewer occurs during patient encounters when clinicians have a quick access to past patient encounters and explorations information, but not only Pangea-LE is helping in healthcare supplying but also in clinical documentation department where some discharge report must be processed to accomplish legal requirements: Diagnostics must be codified in ICD system.

4. Discussion

The use of Pangea-LE in CHGUV is restricted to an EHR viewer, that is, to extract clinical information regarding one patient by means of the defined EED entities. Current capabilities do not meet some interoperability requirements which might be satisfied in the future either by improving existing capabilities or adding new ones:

- ✓ Asynchronous communication mechanisms are needed to support petition / subscription methods, event control and process state management. Also flexible mechanisms for message transformation, scheduling and routing between integrated systems are needed as a basis requirement in order to satisfy clinical processes natural workflow.
- ✓ Pangea-LE's infrastructure may be a helping tool in ETL (Extract, Transform and Load) processes required in clinical / economical research. Pangea-LE may ease the always laborious load stage of data warehouses / data marts.
- ✓ Due to the wide range of clinical information that it manages and its easy use, the EHR viewer has widely extended throughout the organisation leaving the door open to become the star dashboard application. It also may become a virtual medical desktop used by health professionals to access to his/her particular set of applications
- ✓ Extending Pangea-LE for inter-organisation communication is not a trivial issue since the Service Oriented Architecture must be yet defined and fine tuned. A

restricted solution is being designed based on patient identification uniqueness zones and a hierarchy of server nodes, each one as an EHR zone manager.

✓ Nowadays, the more widely extended strategy on enterprise application integration is the Enterprise Server Bus (ESB), as a logical Bus where integrated systems are plugged and applications interchange messages. Primary resolved issue with ESB is the number of point to point application connections to make them interoperable.

5. Conclusion

Pangea-LE allows a non-invasive integration mechanism, completely respectful with the already existing autonomous subsystems of a health organisation with a very specific purpose: to define secure, global and unified views, organized by flexible criteria, over all EHRs dispersed among manifold subsystems of a healthcare institution. Nevertheless and due to its design, it is prepared to evolve towards the advanced functionalities described in the discussion section of this paper. But without a doubt the strongest feature of the system is its high flexibility and scalability. Flexibility to specify multiple formats for the presentation of information, to structure the health data according to normalized formats to communicate EHRs (Hl7, openEHR, EN13606) etc, and to organize the clinical history view according to different criteria. Scalability to deploy the engine in surroundings distributed opened environments for inter-institutional interoperation.

Pangea-LE is high flexible, fully scalable and without almost any preliminary requirement and allows fast EHR deployment throughout a whole organisation. Our experience in CHGUV has shown that an integration project full specification can be done in only a few weeks, in fact, the main part of it was only a matter of a few days.

Acknowledgements

The authors would like to thank CHGUV for their always willing for help especially to Daniel Pérez, Jesús Mandingorra, Jose Vicente Dolz, Nestor Alis and Irene Abad. This work was partially funded by the Spanish Ministry of Science and Technology (MEC – TSI2004-06475-102-01)

References

[1] Bouguettaya, A., Benatallah, B., Elmagarmid, A. *Interconnecting heterogeneous information systems.* Kluwer Academic Publishers (1998).

[2] Wiederhold, G. Mediators in the Architecture of Future Information Systems. Computer 25(3), 38-49 (1992).

[3] Maldonado JA, Robles M, Crespo P. *Integration of distributed healthcare records: publishing legacy data as XML documents compliant with CEN/TC251 ENV13606.*Proceedings of the 16th IEEE Symposium on Computer Based Medical Systems. IEEE Computer Society Press, pp. 213-218 (2004).

[4] http://www.centc251.org (last visited July 27, 2005)

[5] LDAP RFC: http://www.ietf.org/rfc/rfc2251.txt (last visited December 15, 2005)

[6] JDBC driver's list. http://developers.sun.com/product/jdbc/drivers (last visited Dec. 15, 2005)

[7] http://www.w3.org/XML (last visited December 12, 2005)

[8] http://www.w3c.org/2002/ws (last visited November 29, 2005)

[9] http://www.hitsw.com/products_services/xml_platform/allora.html (last visited December 11, 2005)

[10] Apache software foundation. http://www.apache.org/ (last visited December 15, 2005)

2.2 EHR: Research

A. Hasman et al. (Eds.)
IOS Press, 2006

Research networks: can we use data from GPs' Electronic Health Records?

Etienne DE CLERCQ[a,1], Viviane VAN CASTEREN[b], Pascale JONCKHEER[c], Peter BURGGRAEVE[d], Marie-France LAFONTAINE[b], Hans VANDENBERGHE[b], Vincent LORANT[a], Caroline ARTOISENET[a], Karen DEGROOTE[b]

[a] *Université Catholique de Louvain (UCL - HSR), Brussels, Belgium*
[b] *Scientific Institute of Public Health (IPH), unit of Epidemiology, Brussels, Belgium*
[c] *Société Scientifique de Médecine Générale (SSMG), Brussels, Belgium*
[d] *Wetenschappelijke Vereniging van Vlaamse Huisartsen (WVVH), Berchem, Belgium*

Abstract: As widely discussed in the literature, there are many potential scientific usages of data extracted from the primary care Electronic Health Records (EHR), such as quality of care, epidemiological or socio-economical studies. Yet, can we use the current available data in the EHR for such purposes? In this paper, our objective is to report on the preliminary findings of the Belgian ResoPrim project (2003-2005) to answer the question. We set up a semi-anonymous network involving 26 current practices (28 volunteer GPs), 3 different EHR software systems and two Trusted Third Parties. Based on a literature overview we identified 27 research questions to be answered using 50 indicators. The study design includes retrospective (2002 – 2004) and prospective (6 weeks) data collection processes around the theme of "Hypertension and cardiovascular risk factors". For some data sets, the data extraction was a full automatic procedure, for some others, the data extraction was related to an input from the GPs allowing some comparisons between both procedures. At this stage, we performed an extended descriptive analysis of our data. Retrospectively we collected data related to 42,217 patients and 203,128 contacts. Prospectively we collected data for 9,236 patients and 15,234 contacts. Our main findings are briefly presented and discussed in this paper. The most promising fields seem to be the Health Research Information Systems assessment and the quality of care studies. It is quite too soon to reach the expected theoretical benefits for epidemiologic and socio-economic studies, yet some progresses could be made in relation with the denominator issue. Based on our preliminary findings and hypotheses, further analyses are foreseen during the second phase of the project (2006 – 2007).
Keywords: Primary Health Care, Computerized Patient Records, Data Collection

1. Introduction

In Belgium, as in other countries, there is an increasing demand of information from primary health care for various research purposes such as epidemiological studies, health care quality assessment, socio-economical studies [1-5]. This increases the number of dedicated networks such as sentinel and other morbidity networks, pharmaco-vigilance networks, networks to assess the quality of care. For these purposes, the Electronic Health Records (EHR) can be a rich source of information.

[1] Dr Etienne De Clercq, HSR – ESP – UCL, Clos Chapelle aux Champs 30.41, 1200 Brussels Belgium; Email: declercq@sesa.ucl.ac.be

Many problems however have already been documented when using data from EHR for research purposes: secondary usage of the data, great variation in completeness of the EHR content, issues related to structured or coded data entry, missing data, etc. [6-8]. Yet numerous advantages have also been described: recording on long term period, potential availability of numerous kinds of data, possibility to collect data from many GPs' practices and about many patients, etc. [9-12].

The Belgian ResoPrim project (phase I, 2003-2005) aimed at describing and analysing the potential, limits and difficulties in the implementation and use of a General Practitioners' (GP) research network regarding the daily clinical practice and the management of patient records. To preserve some of the advantages mentioned above, we wanted to allow not specifically well-trained GPs to collaborate to the network. The particular Belgian context had also to be taken into account: there are many GPs' software packages (+/- 20), patients may freely choose their GP (no list of patients), there are many home visits (+/- 40% of GPs' contacts).

The research objective developed in this paper is the assessment of the usefulness for the researchers of the currently available data in the GPs' Electronic Health Records (EHR). This was done for three main fields: Quality of care, Epidemiology and Socio-economy. We present hereafter the global method applied, most of our restricted research questions related to the three main fields and some preliminary answers.

2. Material and Methods

Based on previous experiences in Belgium and abroad [11-12], we set up a semi-anonymous network to collect data from General Practitioners' EHR. In a semi-anonymous network, the researchers can only indirectly contact the collaborating GPs through a trusted third party. Three different software producers collaborated to the network and 28 volunteer GPs (26 practices) were recruited partly at random (671 GPs contacted out of the 1700 GPs using the collaborating software systems). We also applied some technical criteria thought to be critical for the research network: "data not anymore put in a paper record", "start using software <2004", "use of coding system for diagnosis".

To reach our research objective, we firstly identified, based on our previous experience and expertise, 8 major research axes: Quality of Care, Epidemiology, Socio-Economy, Health Research Information System Assessment, denominator and sampling issues (as a basis for epidemiological, socio-economical and quality of care studies), GPs' education and GPs' benefits. Within these axes, as a result of an extended literature review and multidisciplinary working meetings, we identified more than 50 long and short terms objectives. Finally, related to these objectives, we defined 36 restricted research questions for the first phase of the ResoPrim project.

To answer all the 36 restricted research questions, 4 methods were identified: a quantitative research based on the data extraction from the EHR, a qualitative research, questionnaires sent to the GPs (sampling questionnaire, satisfaction survey) and an analysis of data collected during previous Belgian projects. In this paper we focus on the quantitative research, appropriate for 27 restricted research questions.

We implemented automatic extraction tools for the EHRs. We also set up manual procedures (questionnaire filled in by the GPs at the end of the contacts) and procedures for semi-automatic data extraction (manual validation by the GPs of extracted data before sending them). A comprehensive recording method would have implied a too heavy workload for the GPs. Moreover a theme specific method seemed more reassuring and more motivating for the GPs. Therefore we selected a theme that best suited our restricted research questions: "hypertension and cardiovascular risk factors".

To answer the 27 restricted research questions, we identified 50 indicators, such as "number of patients with a diagnosis code for hypertension / number of patients with hypertension encountered during the data collection period" (related to RO3 in table 1).

To implement the 50 indicators we defined 8 data sets to be collected in an automatic, semiautomatic or manual way. Two data sets are retrospective (2002 – 2004) ; the 6 remaining ones are prospective (data extracted at the end of each contact included in the data collection period of 6 weeks). A specific data set was dedicated to patients who refused to take part in the study. Some data sets were extracted for each contact with any other patient: demographic data and data related to diagnosis, referrals or prescribed drugs. For each of these contacts, 4 questions were asked: "location of the contact?", "educational attainments of the patient?", "civil status of the patient?" and "hypertensive patient?". Some additional data, mainly related to drugs prescribed and to cardio-vascular risk factors (CVR) were extracted for each contact with an hypertensive patient (semi-automatic and manual procedures).

Starting from these data sets, and within an iterative process we built detailed specifications for the software developers. The whole procedure to define the variables used in the quantitative research is summarized in Figure 1.

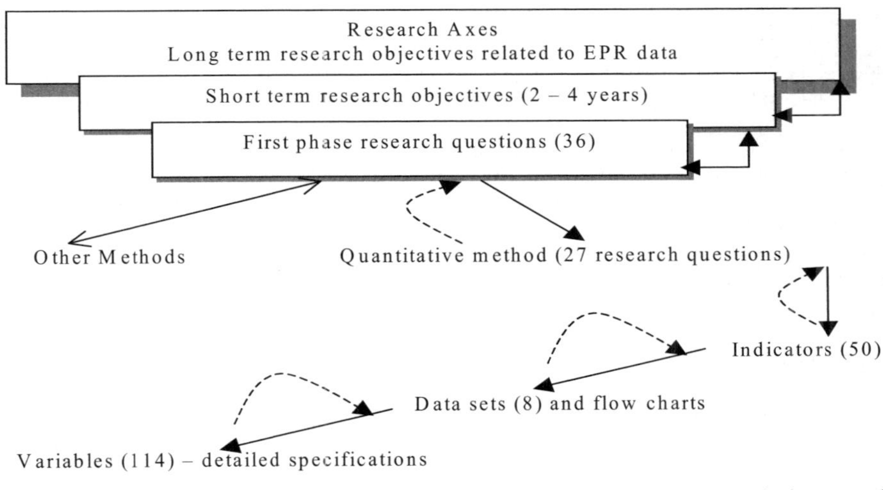

Figure 1: defining research questions and variables for the ResoPrim (phase 1) quantitative research

Quality control (4 weeks) and quality assessment procedures (using a dummy patients technique) were conducted for the extraction modules developed by each software package. The results presented hereafter are based on a first extended descriptive analysis of the collected data.

3. Results

Prospectively, we got data for 26 practices, 9,236 patients and 15,234 contacts. Retrospective data were only obtained for 2 software systems (18 Practices, 42,217 patients and 203,128 contacts). Some of our 27 restricted research questions and preliminary findings (answers) are presented in table 1. For some of them, to improve readers' understanding, additional information is provided in the text.

Table 1: restricted research questions and preliminary findings/answers (not representative sample)

Nr	*Restricted Research questions*	*Preliminary findings*
Health Research Information System Assessment		
RO2	Is it possible to extract a link between a specific drug or referral and a diagnosis?	Yes. 47% of identified referrals and 28.3% of drug prescriptions were explicitly linked to a diagnosis.
RO3	What is the sensitivity of using automatically extracted data (coded diagnoses) against question posed to GPs ("golden standard") for finding cases of hypertension?	Sensitivity: 0.56; specificity: 0.95; PPV: 92%; NPV: 87%
RO5	Has ResoPrim any impact on the coded content of the EHR?	Yes, an improvement (needs further analysis)
RO6	Does the PDA improve the number of home-contacts registered?	Yes (31.6% of all the contacts vs 2.6% in 2004)
RO7 R08	Does the PDA improve the identification of hypertensive patients or the number of documented blood pressure?	No
Denominator issue		
RO10	Is it possible to produce Yearly Contact Groups (age and sex?	Yes
RO11	Is it possible to build broader prospective denominators (Yearly Contact Groups by age, sex, diagnosis of hypertension and socio-economical status)?	Perhaps but mainly prospectively and based on semi-automatic extraction or on questionnaires.
Quality of care		
RO15	How many hypertensive patients <u>with</u> antihypertensive drugs?	Widely underestimated by automatic data extraction vs questionnaire (68.9% vs 93.7%)
RO17 R018	How many hypertensive patients with a high CVR actually have some individual risk cardiovascular factors under control and receive accurate treatment and accurate follow-up?	40% (of identified high risk patients) had their blood pressure under control (≤ 140/90 mmHg); 39% took statins; 45% had received a cholesterol check-up in the past year.
RO19	How many hypertensive patients receive accurate treatment according to some associated pathologies?	Only 34% of hypertensive patients with type 2 diabetes were taking ACE-inhibitors.
Epidemiology		
RO20	Can we disentangle prevalent and incident cases of hypertension?	No (needs further analysis; 22.2% of new cases among hypertensive patients using automatic extraction vs 2.2% using a questionnaire!)
Socio-economy		
RO22	Can we obtain patterns and determinants (patient's and GP's characteristics) of hypertensive drug prescribing and referral behaviour for hypertensive patients?	Hardly. Many data are missing. This requires further investigations.
RO24	Is the diagnosis of hypertension related to socio-demographic status?	Surprisingly no difference in the risk of hypertension by educational level. Married individual tends to have a higher prevalence of hypertension.
RO25	Are referrals and antihypertensive drug prescriptions related to socio-demographic status?	Higher educated individuals are more frequently referred to cardiologist (35% vs 30%). No significant educational inequalities regarding drug prescriptions.

(RO5) When we compared 2005 and 2004 data, coded drug prescriptions recorded in the EHR increased with 40%; prescriptions linked with a diagnosis increased from 0% to 27.5%; coded diagnoses increased with a factor 2, coded referrals increased with 18% but the proportion of referrals linked with a diagnosis remained the same.

(R06, RO7, RO8) Personal Device Assistant (PDA) and current software systems using it do not seem suitable for daily data entry into the EHR. During the prospective data collection many technical problems prevented us to highlight any significant result, except the number of home-contacts.

(RO15) As already studied elsewhere [13], the number of hypertensive patients under prescription is widely underestimated by the automatic data extraction. Yet the global GPs' prescribing pattern for hypertension seems preserved.

(RO17, RO18) To calculate the global cardiovascular risk (CVR) we used an algorithm [14] where a first assessment can be done using so-called "anamnestic" variables (age, smoking status, dyslipidemia, diabetes, personal cardiovascular event, familial cardiovascular event, BMI, hypertension). A consecutive fine-tuning is realized using biological measurements (total cholesterol, HDL, LDL, triglycerides, glycemia). For hardly 40% of patients all requested anamnestic data were gathered with the semi-automatic and manual data extractions. This fell to 15% when the complete parameter set was used (anamnestic + biological). The automatic extraction method needs further investigation.

(RO24) There is no difference in the risk of hypertension by educational level: in both educational groups, hypertension hit about 26% of the patients. This is a bit surprising given the well-known relationship between hypertension and socio-economic status and the results of the 2001 Belgian Health Interview Survey: 6.5% of the individuals with higher education had hypertension compared to 15% in the lower educated.

4. Discussion and conclusions

Our preliminary results are only valid for our sample (not representative) and need further analysis, which is planned during the second phase of the project (2006 – 2007). Yet these results already provide us with interesting findings for further work to build research hypothesis and research questions related to the usefulness of currently available data in most GPs' EHRs. In the near future, the most promising fields seem to be the Health Research Information Network assessment (RO1–RO9) and the quality of care studies (RO12–RO19). In both fields, some research hypothesis could be drawn from our results such as "Does the improvement of GPs' coding behavior last after the end of the recording period, increasing the potential for secondary usage?" "Can we assess changes in GPs' prescribing patterns?", keeping in mind that we are only dealing with documented care.

For epidemiologic studies (R020, R021), our preliminary findings tend to show that, for a ResoPrim like network, it is rather too soon to attain such goals as studying the incidence and prevalence of relevant health problems in the general population or providing policy makers with relevant information to assess the health needs and to commission services (e.g. vaccination programs or alert systems). Yet some progress could be observed regarding the denominator of epidemiologic rates (R010, R011), which could also bring benefits for socio-economical and/or quality of care studies.

In the Socio-economic field (R022–R027), much work remains to be done to get useful additional information from the EHR to enhance research addressing some important issues such as 1) the use of health care resources by the GP through prescription, referral to specialty care, request of additional examinations (particularly medical imaging and clinical biology) and hospitalization requests, 2) the employment and activity of GPs, such as the number of patients and contacts, number of activities and moonlighting, proportion of home visits and 3) last but not least equity in GP care (preventive procedures, drug prescriptions and referrals among different socio-economic and ethnic groups).

Data capture from home visit still remains a problem (R06–R08).

Acknowledgments

We thank all the GPs and the industrial partners involved in the data collection. The ResoPrim project is supported by grants of the Belgian Federal Public Planning Service Science Policy (Nr I2/AE/207 and Nr I2/2F/207).

References

[1] Fleming, D. M. and Pavlic, D. R. Information From Primary Care: Its Importance and Value. A Comparison of Information From Slovenia and England and Wales, Viewed From the 'Health 21' Perspective. Eur.J.Public Health 2002;12(4):249-53.

[2] McColl, A. and Roland, M. Clinical Governance in Primary Care: Knowledge and Information for Clinical Governance. BMJ 2000;321(7265):871-4.

[3] Macintyre, K., Lochigan, M., and Letipila, F. Understanding Referral From Primary Care Clinics in Rural Kenya: Using Health Information Systems to Prioritize Health Services. Int.J.Health Plann.Manage. 2003;18(1):23-39.

[4] Bottomley, A. Methodology for Assessing the Prevalence of Angina in Primary Care Using Practice Based Information in Northern England. J.Epidemiol.Community Health 1997;51(1):87-9.

[5] McCormick, A., Charlton, J., and Fleming, D. Assessing Health Needs in Primary Care. Morbidity Study From General Practice Provides Another Source of Information. BMJ 1995;310(6993):1534.

[6] Prins, H., Kruisinga, F. H., Buller, H. A., and Zwetsloot-Schonk, J. H. Availability and Usability of Data for Medical Practice Assessment. Int.J.Qual.Health Care 2002;14(2):127-37.

[7] Wilson, A. E., Pollock, C., Weekes, T., and Dowell, A. Can General Practice Provide Useful Information?--Evaluation of a Primary Health Care Information Project in Northern England. J.Epidemiol.Community Health 1995;49(3):227-30.

[8] Hiddema-van de Wal, A., Smith, R. J., van der Werf, G. T., and Meyboom-de Jong, B. Towards Improvement of the Accuracy and Completeness of Medication Registration With the Use of an Electronic Medical Record (EMR). Fam.Pract. 2001;18(3):288-91.

[9] Dresser MV, Feingold L, Rozenkranz SL, Coltin KL. Clinical Quality Measurement. Comparing Chart Review and Automated Methodologies. Medical Care 1997; 35(6): 539-52

[10] Okkes IM, Groen A, Oskam SK, Lamberts H. Advantages of Long Observation in Episode-oriented Electronic Patient Records in Family Practice. Method Inform Med 2001; 40: 229-35

[11] Pearson N et al. Collecting morbidity data in general practice: the Somerset morbidity project. BMJ 1996; 312: 1517-20

[12] Vlug AE, van der LJ, Mosseveld BM, van Wijk MA, van der Linden PD, Sturkenboom MC et al. Postmarketing surveillance based on electronic patient records: the IPCI project. Methods Inf Med 1999; 38(4-5): 339-344.

[13] Vandenberghe HE, Van Casteren V, Jonckheer P, Bastiaens H, Van der Heyden J, Lafontaine MF, De Clercq E. Collecting information on the quality of prescribing in primary care using semi-automatic data extraction from GPs' electronic medical records. Int J Med Inform 2005; 74(5):367-376.

[14] Boland B, De Muyldeer R, Goderis G, Degryse J, Gueuning Y, Paulus D, Jeanjean M. Cardiovascular prevention in general practice: development and validation of an algorithm. Acta-Cardiol. 2004; 59(6): 598-605

Ubiquity: Technologies for Better Health in Aging Societies
A. Hasman et al. (Eds.)
IOS Press, 2006

Information Technology in Clinical Research in Rheumatology Domain

Hanzlicek Petr[1a], Zvarova Jana[a], Dostal Ctibor[b]
[a] *EuroMISE Centre of Charles University and Academy of Sciences CR, Institute of Computer Science AS CR, Prague, Czech Republic*
[b] *Institute of Rheumatology, Prague, Czech Republic*

Abstract. The development of a functional clinical database of rheumatic diseases represents an essential step in the process of acquiring the necessary epidemiological and other information on disorders under study. In 1999-2005 the Institute of Rheumatology in cooperation with the EuroMISE Center has developed the Clinical database/National Register of selected systemic inflammatory rheumatic diseases inclusive of a bank of sera and DNA. Aims of this phase of the pilot research were gathering clinical, laboratory, genetic, pharmaco-and socio-economic data in a representative sample of patients with systemic lupus erythematosus, systemic sclerosis, polymyositis/dermatomyositis, mixed connective tissue disease; rheumatoid arthritis, juvenile chronic arthritis, ankylosing spondylitis, psoriatic arthritis and reactive arthritis. In 2002 the preset number of over 2000 registered patients had been achieved with collaboration of 34 territorial and 20 institutional rheumatologists in the whole covering the majority of the Czech Republic. Based on experiences gathered, the systems for other related studies are being developed and implemented using modern information technologies.
Keywords: medical informatics, rheumatology, electronic health record

1. Introduction

According to the World Health Organization rheumatic (muscular-skeletal) diseases take still an important place in the future worldwide development. It is presumed that for example non-inflammatory disorders, so-called osteo-arthrosis in the localization of knee joints, will affect more than 40% of people above the age of 70. Almost 25% of them will not be able to perform their basic life activities at all and more than 80% of them will be restricted in their movements in a certain level. In the case of inflammatory knee joints disorders, the incidence of rheumatoid arthritis as one of the most serious of polyarthritis should exceed 165 million of affected people worldwide [1].

Statistics from USA show that rheumatic diseases are the most frequent chronic diseases and the leading cause of disability [2]. During the years 1991-1992 arthritis or rheumatism generally made almost 18% of all causes of disability and problems of the "back pain" kind and more than 13% of other spinal problems. Troubles with heart

[1] Corresponding Author: Petr Hanzlicek, Institute of Computer Science, AS CR, Department of Medical Informatics, Pod Vodarenskou vezi 2, 182 07 Prague 8, Czech republic. e-mail: hanzlicek@euromise.cz

disorders followed – 10% and e.g. diabetes 4% [3]. The economic problem appears subsequently. Hospitalization of people with all forms of arthritis cost 3146 million dollars in the year 1995, medication cost 184 million dollars and nursery care in patients' houses consumed 12 704 dollars [4].

To provide an efficient treatment and prevention of rheumatic diseases, the effective information processing and communication is a key requirement for all involved parties. Physicians need information by means of epidemiological studies on incidence of stated health disorders to be able to establish effective prevention. Patients need to have the access to adequate information on individual diseases. And finally, the public administration, responsible for the development of special-purpose programmes, arrangement of the financial support and establishment of efficient systems [5] require the necessary information for their decision making.

These requirements can be better satisfied using the modern information and communication technologies, providing ways of fast, efficient, safe and secure information sharing and exchange. The electronic health record systems offer the possibility of formalization of structure of classical health record, which can bring a more transparent way of recording of healthcare procedures and immediate access to the health data of the patient. To achieve this, the specialists must agree on the clinical terminology, the definitions and health record structure.

2. Methods

The applied research in the field of electronic health record in the European Centre for Medical Informatics, Statistics and Epidemiology (EuroMISE Center) was inspired by several European projects and standards. During previous years, several modern approaches and ideas were proposed, analyzed and implemented as software tools. The main objective of the research in the EuroMISE Center in this field was the promotion and facilitation of conversion from free-text based electronic health records towards the electronic health records based on structured data collection combined with the intelligent systems for decision support of the physician.

The most advanced system, developed in the EuroMISE Center is the electronic health record named MUDR (MUltimedia Distributed health Record). The main architecture of the electronic health record is developed using a 3-layer architecture – database layer, the application layer and the user interface layer. The main innovation of MUDR in the field of data representation is the universal dynamically extensible and modifiable structure of healthcare concepts allowing the reorganization without change of the database structure. The architecture is based on two main structures described by the graph theory expressions – the knowledge base describing the set of concepts and relations among them, and the similar structure of the instances of concepts (values) related to their representation in the knowledge base [11], [12].

The derived system, starting from the ideas of MUDR is the MUDRLite application, simplified in the area of a system architecture and data representation and extended in the definition of user interface and customized functionality. The MUDRLite architecture is based on two layers – the relational database (MS-SQL) and the MUDRLite User Interface (MUDRLite UI) running on a MS Windows operating system. All the visual aspects and the behaviour of the MUDRLite UI are completely described in an XML configuration file by the so called MLL language (MUDRLite Language). MUDRLite operates as a MLL interpreter; it builds the user interface from set of forms and various controls placed on them and manipulates the database layer according to definitions in the configuration file. The database schema corresponds to the particular needs and varies therefore in different environments, as opposed to the fixed database schema in the MUDR data layer.

The experiences and techniques of creating the data models used during development of these systems were crucial for the implementation of database systems in the field of rheumatology.

3. Results

Considering the current situation of rheumatology and its needs in the nowadays Czech Republic, our priority task was the compilation of a clinical database which would eventually serve the purpose of the continual follow-up of the incidence and prevalence of at least the salient primary inflammatory involvement of the joints and the spine.

3.1. National Register of Systemic Inflammatory Rheumatic Diseases

The objective of the cooperation on the establishment of the clinical database was the development of a system that would smoothly and continually retrieve information from a number of territorial areas sufficient enough to be representative of the population of the whole of our country. Such a clinical database should provide us with effective sources of information:

a) on the onset, course and prognosis of selected rheumatic diseases,
b) on their anticipated, natural and also extraordinary complications,
c) on comprehensive therapeutical care including treatment for adverse reactions to the medication provided,
d) on the patients′ current social, medical and economic conditions,
e) on their quality of life in general.

The basic structure of the clinical database that came into existence under the name of National Register of Systemic Inflammatory Rheumatic Diseases 1999-2005 was created in cooperation of two institutions. The two principal constituents – Institute of Rheumatology – Department of Rheumatology, Charles University, 1st Medical Faculty, as the leading clinical, specialist and scientific research base of the field of rheumatology in the Czech Republic, and the EuroMISE Centre of the Academy of Sciences of the Czech Republic provide ample scope for the collection of clinical and laboratory data, including their storage, and help formulate clinical and scientific methods for their future processing. At this stage of pilot research we have formulated the objectives within the following pragmatic and temporal limits:

a) To collect socio-economic, clinical and laboratory (serological, genetic) data on a representative sample of a cohort of 2000 patients with the following systemic inflammatory rheumatic diseases: systemic lupus erythematodes (SLE), systemic sclerodermia (Ssc), polymyositis (PM) / dermatomyositis (DM), psoriatic arthritis (PsA), adult rheumatoid arthritis (RA), juvenile chronic arthritis (JCA), reactive arthritis (ReA), mixed connective-tissue diseases (MCTD), ankylosing spondylitis (AS) and last year also secondary amyloidosis (AA) and ANCA-associated vasculitidis.

b) To develop a sera and DNA bank for the cohort of patients, to assess the patients′ clinical condition, to specify an early diagnosis and prognosis – including the presence of infection, to analyze an autoantibody activity in selected nosological entities including the introduction of new specific tests, to analyze genetic and immunogenetic factors in selected nosological entities and finally to analyze pharmaco-epidemiological data.

c) To launch the active epidemiological investigation in two defined CR localities aimed at the incidence and prevalence in the population of the nosological entities under scrutiny as the groundwork for a permanent register of those rheumatic diseases.

The software solution of the National Register of System Inflammatory Rheumatologic Diseases was developed as a result of the analysis of requirements of specialists for each monitored disease, done by computer scientists from the EuroMISE Center. The data about patients, entering the register, are differentiated according to the disease of the patient. However, many diseases have several data items in common. Therefore, a simple common data model for examination of all monitored diseases was created. The whole database contains about 650 different monitored variables, each group defined by diagnosis contains approximately 90-100 variables.

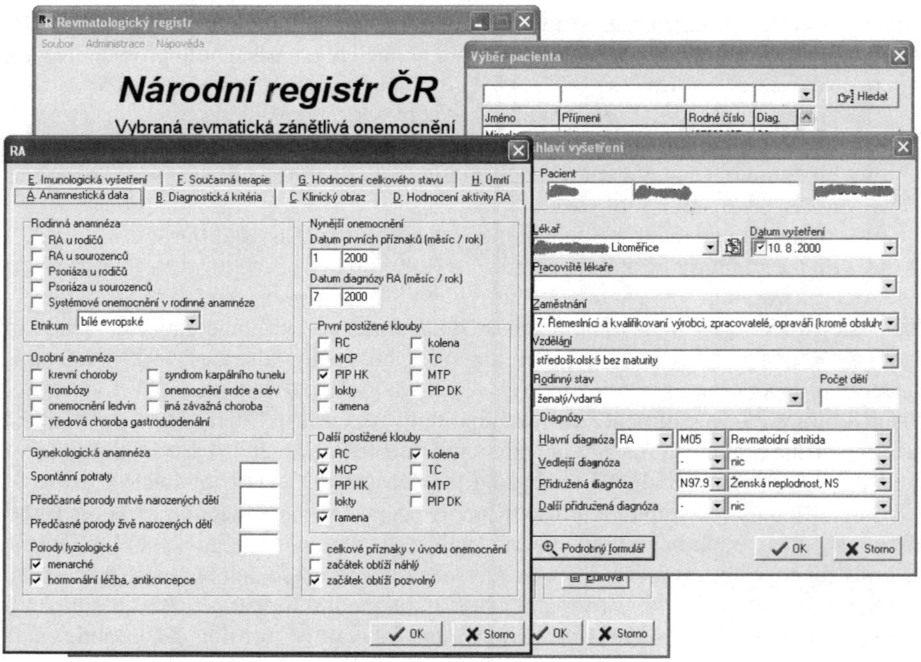

Figure 1: Electronic questionnaires of the software application of the National Register of System Inflammatory Rheumatologic Diseases

The data are initially acquired from physicians using structured paper questionnaires, which are then computerized in the Institute of Rheumatology using software, developed as part of National Register of System Inflammatory Rheumatologic

Diseases. The developed software contains the basic functionality for management of patient administrative data; it stores the history of examinations of each patient and provides simple statistical information about number of patients in each category (disease). When entering a new patient into the system, the common electronic registration questionnaire is used, and then separate forms can be used to enter specific data about the examination of a particular disease. Each form is subdivided into several parts (anamnesis, diagnosis criterions, clinical data, activity and other examinations, laboratory examination, current therapy, rating, death), the diagnostic part contains an automatic verification of diagnosis based on defined rules and entered values. The detailed statistical analysis is not a part of the developed system; the analysis is done by statistical software packages using exported data.

The system is realized using a client-server architecture, the database part uses the Microsoft SQL server 2000, the client application was created as the Win32 application, communicating over the ODBC interface with the database server. Backup and maintenance of the database is performed daily, the additional level of data safety is achieved by mirroring of a disc used for data storage. The system uses communication infrastructure of the Institute of Rheumatology LAN, secured by a firewall. The access to the non-anonymized database is limited only to selected computers within the Institute of Rheumatology with the additional level of security based on mandatory user authentication into the Windows2000 domain. The system is used exclusively by the administrative workers of the Institute of Rheumatology, the statistical analysis is done in the EuroMISE Center using anonymized data, exported from the database by an authorized operator. The data collection process is conformant to the Czech legislative act 101/2000 on personal data protection.

The preset number of over 2000 registered patients had been achieved by the end of the year 2002. In the data collection 34 territorial and 20 institutional rheumatologists from 21 territorial rheumatological offices and hospital departments had assisted, including the Rheumatological Institute in Prague. The largest share is taken by the most significant disease – rheumatoid arthritis in adults and in childhood, the latter as a separate variant, i.e. juvenile chronic arthritis.

3.2. The CYCLOFA-LUNE study

Good results of previous implementation led to the development of a new application, supporting the design of the CYCLOFA-LUNE 2002 study comparing effect of two drugs (Cyclosporine and Cyclofosfamide) on the treatment of lupus nephritis class III and IV. The study has been run for 3 years based on paper forms, the system should support its process in the following years. The system has been developed as a web application, based on so called LAMP platform (Linux, Apache, MySQL and PHP), allowing the users to access the central database in a secure and user-friendly way with minimal requirements to their equipment. Except basic functionality for entry of the results of examinations, the system has some management functions.

Each examination planned in the study has its own schedule. In addition to basic examinations, the schedule contains additional planned examinations like SLEDAI, SLICC SLE and EuroQOL and additional security check-ups of basic parameters. The execution of all planned examinations is verified for each patient, in case of examination schedule breaking for some patient, the physician responsible for the patient's data is informed. The system also provides physician a calendar showing planned and performed examinations of one or all his patients. This functionality helps the physician to plan better the needed visits and examinations. To help to supervise the continuation of the study, two groups of users are supported by the system – the ordinary users (physicians providing data about their patients) and monitors with access to the whole database, verifying quality of the data and progress of the study. The system is now being implemented in routine operation, the evaluation of its usability and effect on the progress of the study will be done at the end of the study.

4. Conclusion

The higher use of modern information technologies in clinical research in rheumatology represents an essential step for all future research activities in this field. The pilot study and testing the ability of collecting and handling clinical and laboratory data and their processing together with all other analytical approaches has started new chances for better understanding of rheumatic diseases.

Acknowledgments

The work was supported by the Research Grant of Ministry of Health of the Czech Republic No. 00000023728 and by the research institutional plan of ICS AS CR AV0Z10300504.

References

[1] A.Murray, J.Lopez (Eds). The global burden of disease: a comprehesive assessment of mortality and disability from diseases, injuries and risk factors in 1990 and projected to 2020 WHO 1996.

[2] R.C. Lawrence, C.G. Helmick, F.C.Arnnet, R.A.Deyo, D.T.Felson, E.H.Giannini, S.P.Heyse, R. Hirsch, M.C.Hochberg, GG Hunder, M.H.Liang, S.R.Pillemer, V.D.Steen, F.Wolfe Estimates of the prevalence of arthritis and selected musculoskeletal disorders in the United States, Arthritis Rheum 41 (1998) 778-799.

[3] Centers for Disease Control and Prevention, Prevalence of disabilities and associated health conditions – United States, 1991-1992. MMWR Morb Mortal Wkly Rep 43 (1994) 730-739.

[4] A.Praemer, S.Furner, D.Rice Musculoskeletal conditions, in: I.L.Rosemont (E.D.), American Academy of Orthopaedic Surgeons, 1999.

[5] R.F. , E.F.Calahan, C.G.Helmick The National Arthritis Action Plan: a public health strategy for a looming epidemic. Arthritis Care Res 12 (1999) 79-81.

[6] Česká revmatologie 10 (2002) 57-105.

[7] J.Vencovsky, L.Sedova, S.Machacek, J.Kafkova, J.Gatterova, V.Pesakova, S.Ruzickova Autoantibodies can be prognostic markers of an erosive disease in early rheumatoid arthritis. Ann Rheum Dis 62 (2003) 427-430.

[9] I.Putova, C.Dostal, R.Becvar The prevalence of anti-nucleosome antibodies detected by enzyme-linked immunosorbent assays (ELISA) in systemic lupus patients, Ann Rheum Dis 63 (2004) Suppl.1 206

[11] Hanzlicek Petr; Development of Universal Electronic Health Record in Cardiology; Health Data in the Information Society. (Ed.: Surján G., Engelbrecht R., McNair P.) - Amsterdam (Studies in Technology and Informatics), IOS Press 2002, pp. 356-360

[12] Zvarova Jana, Hanzlicek Petr, Spidlen Josef; Electronic Health Record in Cardiology: Pilot Application in the Czech Republic; Achieving a Knowledge-based Economy in Biomedicine through Informatics. - Taiwan, Taiwan Association for Medical Informatics 2002, pp. 10-13

[13] Zvarova Jana; Advanced Information Technologies in the Czech Health; E-he@lth in Common Europe. (Ed.: Zielinski K., Duplaga M.) - Krakow, Academic Computer Centre CYFRONET UST 2003, pp. 241-249

2.3 EHR: Access

Ubiquity: Technologies for Better Health in Aging Societies
A. Hasman et al. (Eds.)
IOS Press, 2006

Method for Automatic Escalation of Access Rights to the Electronic Health Record

Frode Ørbeck Hansen[a,1] and Rune Fensli[a]
[a]*Agder University College Faculty of Technology and Science, Grimstad, Norway*

Abstract. In an emergency situation, it can be vital for rescuing personnel to have access to fragmented parts of patients Electronic Health Record (EHR) shared between patients and health care services. In such situations, can Spatial Role Based Access Control combined with measurements of vital sign parameters recorded from a wireless monitoring system used by the patient and patient's physiological situation be used to facilitate for medical personnel automatic access to parts of the EHR.
Keywords: Core-EHR, Spatial Role-based Access Control, Wireless Patient Monitoring.

1. Introduction

Based on the personal data act and the requirements on privacy, the access to patient's electronic health record is strictly limited and normally dependant of the patients consent. Typically the Electronic Health Record (EHR) is not a complete database covering all needed information about the patient's illness and treatment, but is quite a fragmented part of information residing on many different systems depending upon the actual medical history. The patient will normally have regular contact with a General Practitioner (GP), where important information of the illness, treatment and medication is stored in a local EHR system. In a typical situation, the patient can be taken care of by the local community health care system which has its own isolated EHR-system. If the patient has been treated by a hospital, there will also be an important part of the EHR located in the hospital's database, and there can be several hospitals involved in the treatment, each with a different EHR-system. Today, regulations protect the patients' privacy, something which restricts automatic interaction between the different EHR-systems, and opens only for fixed message exchange [1].

New paradigms in telemedicine will, however, require new mechanisms for interaction between the caregivers [2], and it is a challenge to find new solutions for keeping updated EHR-information available whenever needed. Web-based systems are available today like iHealthRecord [3] and My Personal Health Record [4], where the patient is taking a more active part in managing his own EHR-system, and the European Commission is under the programme eEurope 2005 preparing a health insurance card to be adopted in 2008 containing medical emergency data and secure

[1] Corresponding Author: Frode Ørbeck Hansen, Agder University College, Grooseveien 36, 4876 Grimstad. E-mail: frode.hansen@hia.no

access to personal health information [5]; however those systems are quite static in the access control and require manual interactions in order to give necessary permissions. In cases of emergency, the need of quick medical assistance can be important and in life threatening situations there will be a need of quick access to the patients EHR in order to give the correct treatment. Today, this is difficult to achieve because of no interaction between the different parts of all the EHR-systems involved. That is one of the reasons for establishing a new type of EHR-system where the term "core journal" or "core electronic medical record" (C-EHR) is used to describe a “common data sets” as the minimum of information required to get an adequate overview of the patient's actual situation [6]. The journal could possibly contain demographic data, medical history, past illness history, type of allergies, medications, diagnostic data etc.

This core journal can in fact be “owned” by the patient himself, and he will have to take control of the necessary authorizations of who is obliged to have access to the various parts of the information. The patient can write his medical diary into the system and can keep tracking of medications, actual vital signs recordings, training results etc. as well as information of expected progress and personalized training or treatment programme. In cases where the patient write information in this journal, the patient's GP can read this information and write a reply if requested for the patient to read. In the future, such systems should have incorporated Spatial Role Based Access Control (SRBAC) solutions [7], where the patient is able to define necessary authorizations schemas based on his illness and the need of assistance from medical care giving personnel and the required services.

2. Patient monitoring and Access Control Issues

2.1. C-EHR and access control

In an emergency situation, it can be important for the rescuing personnel to have direct access to the patients EHR information and will in a future solution be implemented in a C-EHR to be shared between the patient and the health care services. The data in the C-EHR should be available for medical personnel (pre-approved by the patient) to access when this is necessary [8]. This information could be valuable in emergency situations, e.g. in cases where a person is treated, due to an acute change in the situation. A physician at the rescuing station that has no prior records of this patient should then be given access to the central core journal.

For patients that are remotely monitored through a wearable system like the WPR wireless ECG solution [1], it is possible that this information is automatically written into the C-EHR to be accessed by legitimate users. Also, if abnormal readings are detected and fed into the core journal, the system should detect and prioritize such information to be read by for example the patient's GP. Furthermore, if more serious conditions are detected, such as sudden cardiac arrest, the ECG readings should be written into the C-EHR associated with an alarm message. This information could be vital for the medical personnel that treat the patient in emergency situations. The personnel could then access the patient's core journal and read the ECG recordings and other medical data adequate for the treatment of the patient.

If an Automated External Defibrillator (AED) is used for treating the patient in cardiac arrest situations, the information collected by the AED during the cardiac arrest event could be written into the patient's core journal and should be available for medical personnel at the rescuing station. To assure that the ECG readings are from the patient under treatment, the WPR and AED should be able to detect each others presence[2]. The AED obtains a unique patient ID from the WPR and supplement this ID to the readings sent to the journal system such that correct ECG readings are registered to the correct patient.

2.2. Locating people and assets

To be able to locate the medical personnel suitable for treating a patient, the system needs to estimate the position to the respective persons and estimate the distance to the patient. Moreover, such an algorithm must not only take into account the distance, but also the qualifications that the medical personnel possess such that the most qualified and available person(s) can be alarmed and sent to the scene of accident. Furthermore, because in SRBAC, the system makes authorization decisions based on user location, a mediator must be able to obtain the requester's location to acquire permissions enabled for the user based the user's position (and role assignments).

The overall positioning system must be able to locate the persons both indoors and outdoors, because the monitored patients and the medical personnel may be situated in both environments at different times of day. However, using physical location information in the process of locating users and assets can be unpractical in some cases because it represents infrastructure of location detection system. This information should be represented in a universal and flexible way, such that it can be used efficiently in the access control procedure and in the decision making of locating the best qualified medical personnel and patients. For example, if a monitored patient's Wireless ECG sensor detects irregular readings from the patient and the patient is located in his house, the address of patient's residence should appear as location and not the actual ID of the positioning device (e.g. a MAC address to a Bluetooth device) that detected the patient. Also, this is necessary in cases where an AED is located near a patient (e.g. shopping malls) that experience cardiac arrest. It is therefore crucial that the location information provided to the medical personnel is accurate and explanatory such that it could be found and used as quickly as possible. Tracking of medical personnel and patients at all times will not be practical and would cause unnecessary traffic in the system. We need only to track users when users request access the medical information system and when alarms are caused by a patient's Wireless ECG sensor.

3. Framework

In this Section we present the security framework for use of handheld devices in a medical information system. The framework supports the spatial RBAC model as described by Hansen and Oleshchuk in [7]. It consists of following entities: mobile terminals, user monitor agent, authorizer, location server and role server. The

---

[2] It could be possible for the location sensing system to locate both the patient and the AED equipment, and to link the ECG readings to the patient ID. However, higher accuracy is obtained by letting the parties detect each others presence. Also, it eliminates the risk of associating the ECG readings to another patient ID if there exist another WPR equipped patient in the same location (in cases where the location system is not able to differ the position of the patients with the WPR wireless monitoring system).

framework is illustrated in Figure 1 and descriptions of the various entities are described below.

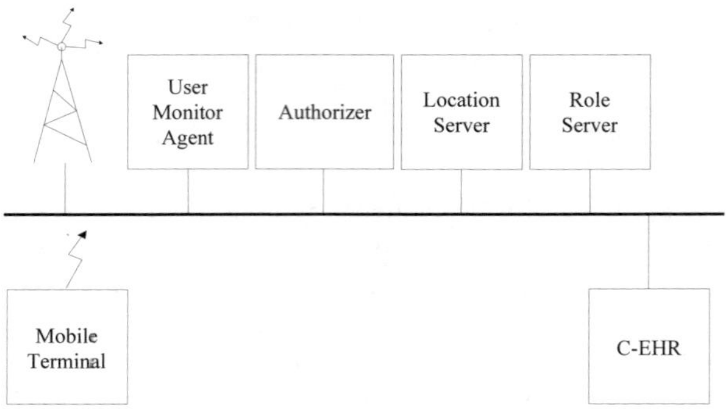

Figure 1 Authorization Framework

3.1. Logical Entities

Mobile Terminal: This represents patients, medical staff, or even mobile devices, such as a patient's mobile terminal that automatically writes ECG readings into the patient's respective C-EHR or AED's that feeds the result of a treatment directly into the C-EHR.

User Monitor Agent: UMA tracks the position of active users by collecting positioning data from location sensors.

Authorizer: The Authorizer takes care of the access requests made by mobile terminals. When a patient or medical staff requests to access some service, the Authorizer obtains an updated permission set (pairs of objects with related set of permitted operations) from the Role Server and used in the access decision process.

Location Server (LS): LS gathers physical positioning data on active users from the UMA. LS maps the physical location data into logical location domains as discussed in subsection 2.2 and communicate this information to the Role Server. LS send a location update to RS only when the user changes its position with respect to logical domain

Role server (RS): RS maintains tables containing user-role assignments, permission-role assignments, active user sessions and constraints on the SRBAC components. It maintains state information on active roles, i.e., valid permission set available for roles active in user sessions. It receives logical location information (either by request or location updates) from the LS to update role- and permission states. RS returns an updated permission set for active users to the Authorizer. More detailed information on the function of the Role Server is explained in the scenario in next Section 4.

4. Scenario

Following up patients in a Tele-home-care situation give possibilities for continuous monitoring of ECG, with an automatic arrhythmia detection alarm system which can give alarms to the emergency department at a hospital. In order to have quick and secure access to the patients EHR, we propose a method for SRBAC, and this solution is described in the scenario below which shows an escalation of the emergency and a stepwise change in the C-EHR access.

In Figure 2 this scenario shows interaction between the patient and the rescuing personnel. The Medical Information System (MIS) contains the C-EHR and the Authorization Framework, and will automatically receive heart alarm messages. In an acute situation the system will automatically detect the change in the patient's situation, and perform an escalation of the access rights for the parties involved in case of emergency event.

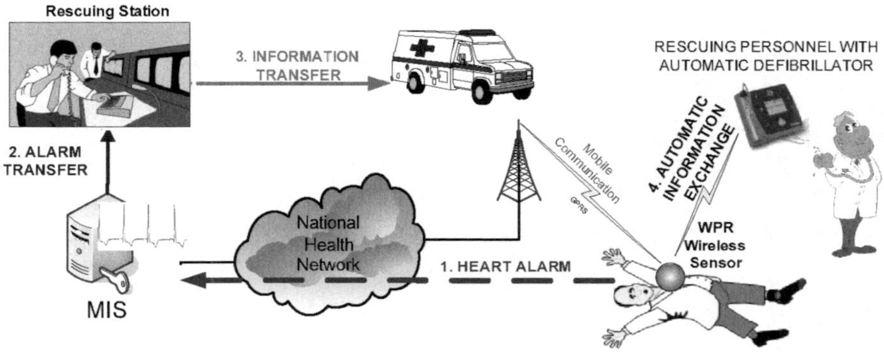

Figure 2 Automated response to a cardiac arrest event.

MIS will store the information in the patient's C-EHR, and automatically request the location from the Location Server and forward this location information together with the alarm to the Rescuing Station and the Rescuing Station alerts ambulance personnel and the nearest located doctor about the situation. The SRBAC model in the MIS will automatically give responsible personnel the appropriate permissions to update or read the current situation of the patient. For example, the Rescuing Station might need to update and give more accurate position or status of the patient due to that bystanders have called in and reported the incident.

Figure 3 shows that upon arrival of the alarm sent from the patients WPR, MIS would start to locate the nearest qualified medical personnel to treat the patient. It receives a list of qualified personnel from the Role Server and the current position of the personnel from the Location Server upon request. MIS calculates the distances and determines the person closest to the scene of accident. Based on qualification, location and availability of the located personnel, MIS makes a decision and sends an alarm message to the selected person about situation. This person has to acknowledge the alarm with a negative or a positive answer. If the person's answer is positive, this would trigger a delegation request from MIS to the Role Server. The Role Server enables roles and appropriate permissions for the person such that it is possible for him to access the patient's core journal and read the ECG recordings and other medical data adequate for the treatment of the patient.

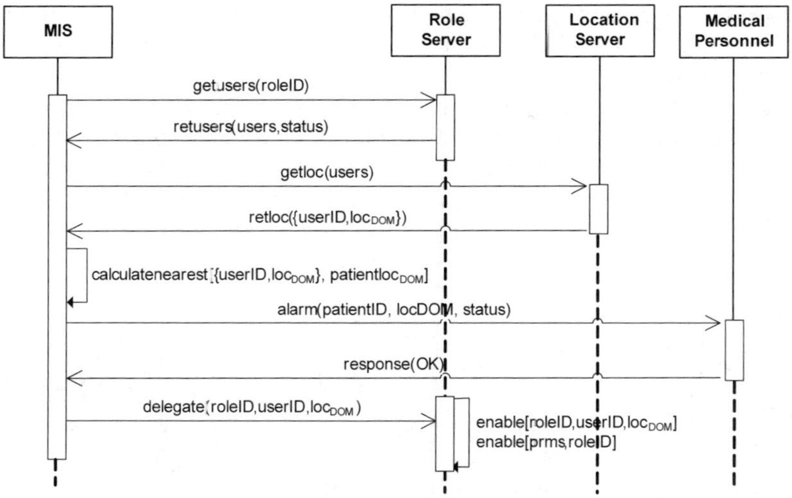

Figure 3 Locating nearest medical personnel and enable appropriate access rights.

5. Conclusions and Future Work

In an emergency situation, it can be important for the rescuing personnel to have direct access to the patients EHR information and can in a future solution be implemented in a C-EHR to be shared between the patient and the health care services. This information could be vital for the medical personnel that treat the patient in emergency situations. We have shown through a scenario how a Medical Information System (MIS) automatically, on the reception of an alarm message sent from a patient's WPR, locate the best qualified personnel close to the scene of accident, alert them, and give them the necessary access privileges to access the patient's C-EHR to be able to provide the best possible treatment for the patient.

References

[1] Fensli R, Gunnarson E. Mobile Monitoring of Vital Parameters within the Electronic Health record - Medical, Technological and Legal aspects. In: Tromsø Telemedicine and eHealth Conference - TTeC2004; 2004 21.-23. June 2004; Tromsø, Norway; 2004. p. 46.

[2] Rubel P, Fayn J, L S-C, Atoui H, Ohlsson MT, D, Adami S, et al. New paradigms in telemedicine: ambient intelligence, wearable, pervasive and personalized. Stud Health Technol Inform 2004;108:123-32.

[3] Interactive Personal Health Record; http://www.ihealthrecord.org (online 2006.01.113)

[4] My Personal Health Record; http://www.myphr.com (online 2006.01.13)

[5] European Commission – Information Society – eEurope 2005 – eHealth: http://europa.eu.int/information_society/eeurope/2005/all_about/ehealth/index_en.htm

[6] Core Data Sets for the Physician Practice Electronic Health Record (AHIMA Practice Brief). Available online at http://library.ahima.org/xpedio/groups/public/documents/ahima/pub_bok1_021607.html

[7] Hansen F, Oleshchuk V. Security Model for Wireless Environments Based on Spatial Role-Based Access Control. Conference on Information Security and Cryptology, Higher Education Press 2005; 129-138.

[8] Nystadnes T. Legemiddelopplysninger i Samtykkebasert kjernejournal, Technical report, Kompetansesenter for IT i helse- og sosialsektoren (KITH); 2005.

Ubiquity: Technologies for Better Health in Aging Societies
A. Hasman et al. (Eds.)
IOS Press, 2006

The French proposal for a health identification number

Catherine QUANTIN[a], Olivier COHEN[b] Benoît RIANDEY[c], François-André ALLAERT[d]

[a] *Service de Biostatistique et Informatique Médicale, CHU de Dijon, BP 77908, 21079 Dijon Cedex, INSERM EMI 0106*
[b] *Laboratory TIMC-IMAG UMR 5525, CNRS University of Joseph Fourier Medical School of Grenoble*
[c] *Institut National d'Etudes Démographiques(INED), 133 Bd Davout, F-75980 Paris Cedex 20*
[d] *Department of Epidemiology and Biostatistics, Mc Gill University, Montreal Canada*

Abstract: The French ministry of Health is setting up the personal medical record (PMR or DMP for Dossier Medical Personnel in French). This innovating tool is highly expected and will be extremely useful for the therapeutic follow-up as well as for epidemiological studies on which public health policies are based. Therefore the currently planned identifying process should prevent any epidemiological use of these data. Numerous scientific organisations have alerted government powers about the threat that this impairment represents, and they wish to promote some secure procedures that exist, which have already proved their efficiency at the national and international level.
Keywords: Electronic Health Records – Unique Patient Identifier - Security

1. Introduction

In a majority of industrialized countries, at the heart of many of the concerns relating to electronic processing of health information lies the problem of patient identification. The health card now being disseminated in Europe is meant to simplify the process of health insurance coverage for European nationals engaged in cross-border travel. In France the "Dossier Médical Personnel – DMP" (personal medical record) designed to promote health care coordination, proposes a patient identifier that is for the moment incompatible with the European health card. The goal of this article is, in a first section, to show the structural diversity of health information identifiers in use in some countries that can be considered relatively advanced in this area. The second section proposes a health identification number for France in the context of the European DMP. The third section proposes an alternate identification procedure that would allow France to broaden the scope of its DMP project by making it possible to contribute to public health research and policy while increasing interoperability with the European health card.

2. The international situation as seen in ten countries

A recent study by the Group for the Modernization of Hospital Information Systems (GMSIH), in the context of the "Principle and Process of Patient Identification" project, inventoried the international patient identification experience of ten countries (Germany, Australia, Canada, Denmark, Finland, Luxembourg, the United States, New

Zealand, The Netherlands and the United Kingdom) considered by the GMSIH as representative of best practices in the areas of socialized health care, electronic health care information systems or patient identification.

Only four countries (Australia [1], the United Kingdom, New Zealand and The Netherlands) have specifically health care related national patient identification numbers. In all of these countries, the national health number is used both administratively (for billing) and medically. In Australia, the number is given to each citizen at birth and to immigrants upon obtaining permanent resident status. In the United Kingdom, the National Health Service (NHS) number is used for all health related purposes [2]. In New Zealand, the New Zealand Health Information Service (NZHIS) gives every health system user (including tourists) a unique reference number managed in the NHI (National Health Index) database [3]. In The Netherlands, the unique patient health identifier (ZIN) is based on the SOFI number, a unique personal number used by taxation authorities [4]. Three countries (Denmark [5], Finland [6] and Luxembourg [7]) use a single national number in all areas including health. The Danish CPR number used in health (for administrative and medical purposes) is the identifier for the central civil registry (CRS). This number is widely used in many areas including tax collection and banking. In Germany a national patient identifier will be proposed soon [8]. Two other countries (the United States and Canada) [9-12] do not appear to be headed in the short term towards the creation of a homogeneous national patient identification system. The formats used for all of these numbers differ significantly among countries, most national patient identification numbers using 10 digits. Only the New Zealand number is made up of a combination of numbers and letters totalling seven characters. Except for Denmark, where a part of the number is derived from sex and date of birth, these identifiers do not include other patient information. In the United Kingdom a new content-free number replaced a previous number that was non content-free. However, to avoid collision problems, this number will be correlated with sex or date of birth. In New Zealand, the randomly assigned number is entirely anonymous, but is related in the NHI database to numerous personal identifiers (first and last name, date of birth, address, residency status, date of death), and the system can assign a large number of aliases for a given individual.
From another source, we know that in Belgium, the Federal Public Service for public health [13] has recognized that a patient might have several unique identifiers relating to various objectives: administrative purposes, health care, research statistics.

This brief literature review confirms the general impression of heterogeneity of patient identifiers in use among countries, making interoperability between health information systems difficult. Although harmonization of European health care systems is not yet a reality, as early as 1996 the European Parliament recommended the creation of a European health card to allow European Community nationals to benefit from adequate health care when moving between member states. In 2004, in the wake of the European Netcard project, coordinated by Mr. Noël Nader (Vitale card French Public Interest Group) the European health card was born. This card includes the individual's first and last names, date of birth and social security number in the insured's country of origin. The card is currently in paper form but is expected to shortly evolve toward a smart card format [14]. Access to the information using internet XML would be conditional upon an exchange of authorizing messages according to rules, which would at all times respect principles of information and transmission confidentiality. Thus, the goal of the creation of a European health card, could also easily serve to facilitate medical treatment by allowing access to authorized medical information. This in turn opens perspectives of Europe-wide statistical processing and epidemiological studies. The feasibility of these possibilities obviously lies in the implementation throughout Europe of a patient record that is compatible, be it only regarding patient identification, with the principles retained for the European card. The creation in France of a national patient medical system constitutes thus an original asset not yet possessed by our foreign partners in spite of their advances, most notably in New Zealand and Belgium. Only Australia has announced a project very similar to the French DMP with, contrary to France, clearly enunciated perspectives for use of information for public health research, policy and planning and for quality assurance procedures for information contained in the file.

3. French proposal for a DMP health identification number

In France, the Minister of Health has planned that by January 1, 2007 every patient will possess a personal medical record (DMP – Dossier Médical Personnel) containing information that will permit monitoring of medical acts and care. The goal of the DMP is to promote coordination, quality and continuity of care, as well as prevention. This DMP is to be housed with an approved personal health information host.

The five following basic principles of the "numéro d'identifiant santé - NIS" (health identification number) assigned to the personal medical record: unique, content-free, public, permanent, irreversible. The processes used to create and to manage identifiers must be extremely resistant to aggression (pirating of secrets related to NIS computing, destruction or pollution of related materials, software or files). If lost, the same number can be recreated. This working group proposes that at the request of the patient, an independent body known as a "trusted authority" generates a national "numéro d'identifiant santé – NIS" (health identification number). The role of the "trusted authority", which may not be a host, is to guarantee that the identifier has no duplicate. To preserve the anonymity of the identifier, the trusted authority creates the NIS by the following process: The patient (beneficiary) chooses an approved host and advises the DMP office using an NIS request form and the host using a membership form. By contacting the National Health Insurance Repertory (RNIAM), the DMP office confirms the requester's affiliation and that it is a first-time request. The trusted authority creates the NIS using the request form number that it transmits along with patient attributes to the host. The host sends the patient his NIS and access information and informs the trusted authority that the patient's NIS is operational. The patient will give the AQS (health quality address), made up of the NIS and the host identifier, to the health care professionals that he consults.

Use of the request form number and not the NIR (the national identification number) to generate the NIS is designed to guarantee on the absence of a table of correspondence directly linking NIS and NIR. The host would thus possess only those identifiers that could not be used to reconstitute the NIR.

4. Limitations of the current French proposal for a health identification number

This system is designed to ensure the uniqueness of each patient's NIS based on the bijective correspondence between the NIS and the NIS request form number. It remains to be ensured, however, that only one request form number will exist for the patient who might have made several NIS requests during his lifetime. The guarantee of uniqueness also supposes the establishment of a request form number-NIR registration table. This solution is prejudicial to the interest of patients: if 10 years from now, patient associations should decide that it is beneficial to patients to pool certain information, by request or with the written consent of the patient, it will unfortunately no longer be possible to link information from various sources on a given patient

As the president of the Academy of Science notes in the recent Academy's report relating to epidemiology, by preventing the multi-source validation necessary for epidemiological information, this system is prejudicial to health policy and prevention. Moreover, this solution is contrary to the will of the legislature as expressed on August 6, 2004 law in conformity with the position of the Council of the European Parliament in October 24 1995,: "*subsequent data processing for purposes of statistical study or scientific or historic research is deemed compatible with the initial aims of data collection*". The legal foundations of the prohibition of epidemiological processing of data contained in the DMP are as questionable as the optimality of the adopted solution. Like our british colleagues who complain about an overzealous interpretation of UL laws, stifling epidemiological research [15,16], we think that this may cause particular problems for epidemiological research.

5. An alternate proposal for an identifier (avoiding these limitations)

It would be perfectly possible to preserve the confidentiality due the patient by putting in place irreversible anonymizing procedures using a one-way hash algorithm. This process was approved by the CNIL[1] in 1996. The resulting code is but always the same for a given individual, thus allowing data linkage for a given patient. The one-way hash algorithm used is the Standard Hash Algorithm (SHA). However, while SHA-1 had a design level of protection of 280 recently discovered vulnerabilities have lowered the protection to 263 [17]. Based on this news, new applications should only use SHA-1 if necessary and should implement SHA256 if possible. A Double hashing performed with two different keys, can be applied to avoid dictionary attacks. Applications in medicine are numerous in view of the existence of regional databases (cancer, perinatality, genetic disorders…) [18-20] or national one (Medical Insurance Information System, and French [21] or Swiss hospital data [22].

Hashing of the NIR by the trusted authority would be sufficient to assure the anonymous passageway function as in the Belgian example. Although when using the hash function the risk of collision is negligible (10^{-48}), principally as a remedial measure against problems presented by NISs awaiting certification, other patient identifiers, in addition to the NIS, could be separately hashed as the date of birth of the patient. Because of the extreme sensitivity of the hashing function (to avoid collisions) results are completely different if information in one of the three fields is incorrect. Another solution would consist of including, along with the NIR, the patient's first and

[1] Commission nationale de l'informatique et des libertés (National Commission on Information Technology and Freedoms).

last names, date of birth and, where available, even parental information drawn from the family component [14]. This component was proposed recently to optimise inter-operability of the European health card identifier, validated by the CNIL (notice no. 04-006) and patented in September 2004. This is why we described the concept of an individual identifier with a family component [14] and discussed its inclusion in the European Health Card in addition to the social security number (or the NIR in France) attributed by the patient's home country. This concept draws upon the patient's first and last name and date of birth plus those of the father and mother. Indeed, familial medical information is becoming progressively more important in medical genetics and other areas such as multifactoral diseases (cancers, diabetes, arterial hypertension…) as well as in the development of new medications through pharmacogenetics that aim to treat the patient rather than the illness.

In addition, application of encryption and anonymization will result in use of this data remaining under the control of the patient. Moreover, asymmetric encoding of the anonymizing number, provided that a public patient key is used, would give control over data to the patient. As sole possessor of the private decoding key, only the patient (or the trusted authority chosen by the patient) would have the ability to retrace the number that would link that patient's medical information. This system would guarantee that links of patient DMP data useful for epidemiological studies are subject to the express consent of the patient, in compliance with the law.

6. Conclusion

French epidemiologists are thus faced with a double French and European challenge. They fear that short term political decisions will hinder the flourishing of an ambitious and prolific public health project as witnessed by the error of the designers of the "Dossier Médical Personnel" (Personal Medical Record) who believe it is necessary to prevent epidemiological processing of this data. These are the stakes in the current technical battle over the "numéro d'identification santé - NIS"(health identification number). The proposal presented by the signers of this article draws upon ten years of productive and reliable experience at the CHU de Dijon. We hope that it will be heard so that the immense potential for epidemiological progress contained in the DMP is not reduced out of hand to cinders. On the contrary, the DMP should constitute the French link in the pooling of European epidemiological resources undertaken by the Strasbourg parliament with the institution of the European Health Card.

Références:

[1] Health Connect - a health information network for all Australians. Health Connect Project Plan 2003/05. URL : http://www.healthconnect.gov.au/pdf/hcpp0305r.pdf

[2] http://www.nursingworld.org/ojin/search/ojsrchrs.cfm

[3]http://hcro.enigma.co.nz/website/index.cfm?fuseaction=articledispaly&FeatureID=040304

[4] Konijn J, Brouwer W. Inventarisatie partijen in de zorg. Versie 09, 11-12-2002. URL : http://www.nictiz.nl/kr_nictiz/uploaddb/downl_object.asp?atoom=2162&VolgNr=1

[5]http://www.telemed.no/getfile.php/259490.357.cccccetef/Ref_Patient_+data_on_demandK%F8benhavn-0-02-2006.doc

[6] www.ehto.org/ht_projects/trends/finland.html

[7] http://www.seebeyond.com/industries/organizations.asp
[8] Blobel B, Pharow P. Patient data to go. Germany's national smart card project. J Ahima 2005;76(2):48-51.
[9] Department of Health and Human Services : Unique Health Identifier for Individuals, A white paper – 1998).
[10]http://www.ircm.qc.ca/bioethique/english/telehealth/documents/cartesante_sicotte_doc.pd
[11]http://www.gdaus.com.au/pdf/Industry%20and%20Government/Health%20Cards.pdf
[12] Inforoute Santé du Canada. Inforoute fera la promotion de l'interopérabilité entre les systèmes d'information sur la santé grâce à une initiative Logiciel libre. URL : http://www.infoway-inforoute.ca/fr/News-Events/InTheNews_long.aspx?uid=147
[13] Roger France FH, De Clercq E, Bangels M. Purposes of Health Identification Cards in Belgium In: Engelbrecht R, Geissbuhler A, Clovis C, Mihalas G, eds. Connecting Medical Informatics and Bio-Informatics. Proceedings of MIE 2005. Amsterdam: IOS Press, 2005.
[14] Quantin C, Allaert FA, Gouyon B, Cohen O. Proposal for the creation of a European healthcare identifier. MIE 2005, 28-31 août 2005, Genève.
[15] Walley T. Using personal health information in medical research. Bristish Medical Journal, 2006;332 ; 130-131.
[16] Iversen A, Liddell K, Fear N, Hotopf M, Wessely S. Consent, confidentiality and the Data Protection Act. British Medical Journal 2006;332:165-9.
[17] Wang X, Cryptanalysis of SHA-1 Hash Function, Cryptographic Hash Workshop National Institute of Standards and Technology, October 31,2005. http://www.csrc.nist.gov/pki/HashWorkshop/2005/Oct31_Presentations/Wang_SHA1-New-Result.pdf accessed November 30, 2005.
[18] Quantin C., Bouzelat H., Allaert F.A., Benhamiche A.M., Faivre J., Dusserre L. How to ensure data security of an epidemiological follow-up : quality assessment of an anonymous record linkage procedure, International Journal of Medical Informatics. 1998; 49:117-122.
[19] Cohen O, Mermet MA, Demongeot J. HC Forum® : a web site based on an international human cytogenetic database. Nucleic Acids Res 2001;29:305-307.
[20] Quantin C, Binquet C, Allaert FA, Cornet B, Pattisina R, Le Teuff G, Ferdynus C, Gouyon JB. Decision analysis for the assessment of a record linkage procedure : application to a perinatal network. Methods of Information in Medicine, 2005;44:72-79.
[21] Trouessin G, Allaert FA, 1997, FOIN : a nominative information occultation function. MIE 97 ;43 :196-200.
[22] Borst F, Allaert FA, Quantin C. The swiss solution for anonymously chaining patient files. MEDINFO. 2001;10(Pt 1):1239-41.

Ubiquity: Technologies for Better Health in Aging Societies
A. Hasman et al. (Eds.)
IOS Press, 2006

Can Temporary Paper-Based Patient Records Sensibly Complete an Electronic Patient Record?

Petra KNAUP[a,1], Jochen PILZ[b] and Markus THALHEIMER[c]

[a] *University of Heidelberg, Department of Medical Informatics, Heidelberg, Germany*
[b] *Center for Information Technology and Biomedical Engineering, Heidelberg University Hospital, Heidelberg, Germany*
[c] *Department of Internal Medicine, Heidelberg University Hospital, Germany*

Abstract.
Objectives: In the department of internal medicine of Heidelberg University Hospital (HUH), medical patient records are archived electronically. Since there are still paper-based external documents, a temporary patient record is maintained for these documents during the patient's stay. Afterwards, the paper-based documents are scanned, indexed and integrated in the electronic patient record (EPR). To ensure process quality we evaluated quality and availability of scanned documents in the EPR. *Methods*: Observation study, structured interviews, systematic quantitative before-after comparison. *Results*: The workflow takes place according to the guidelines of HUH. Nevertheless, there are variations in the different wards which may influence the quality. Of 343 scanned documents about 90% showed no loss of information. Most of the documents with loss of information were ECG-curves. Four documents (1,2%) could not be found in the EPR. All documents were assigned to the correct patient and episode of care. The mean time from patient discharge to availability of scanned documents in the EPR system was 36 days. *Conclusions*: Due to external paper-based documents, a complete EPR is currently not possible. A temporary paper-based patient record in addition to the EPR is not an optimum procedure but feasible. The quality of the scanned, indexed and integrated documents in the EPR is high and the availability is sufficient.
Keywords: Medical Informatics, evaluation studies, computerized patient record systems, digital archiving, process quality

1. Introduction

Emerging technologies have led to a variety of realisations of patient records in clinical practice and research. A patient record can be defined as the collection of data and documents at a health care institution which are generated for the care process ([1]). Progress in information and communication technologies (ICT) has not only led to approaches for electronic patient records (i.e. patient records on electronic medium, EPR) but also to virtual patient records ([2]), shared patient records ([3]) and may once result in the lifelong, institution-independent health record ([4]). In any case, the major

[1] Corresponding Author: Dr. Petra Knaup, University of Heidelberg, Department of Medical Informatics, Im Neuenheimer Feld 400, D-69120 Heidelberg, Germany, petra.knaup@med.uni-heidelberg.de

aim of a patient record is to support patient care by providing relevant information at the point of care. Thus, it can influence the quality of care.

Despite emerging technologies and innovative approaches, we are still challenged by migrating from paper-based records to EPR in health care institutions. Waegemann [5] introduced 1996 five levels from an automated medical record (which is paper-based with about 50% of the information being generated by computer systems and printed) to an institution-independent electronic health record - containing e.g. also wellness data or patient notes (cf. also [6]). Since there still remain documents which are not available electronically there are generally two concepts for archiving:

- paper-based archiving: Computer-based information is printed and added to the paper-based record for archiving. This is still the most common way in Germany to fulfil legal requirements: The readability of archived data must be guaranteed for decades, also considering the value of evidence ([7]).
- digital archiving: Paper-based documents are scanned and integrated into the EPR-system for archiving ([8]) and electronic documents are collected digitally from the application systems by communication interfaces.

Heidelberg University Hospital (HUH) is well on the way to set up a complete hospital-wide EPR to enhance readability, availability and data quality. Today, records are stored electronically as well as conventionally (steps 2 and 3 of [5]). In the department of internal medicine a pilot project without a paper-based archive is running. However, there are still documents, which are not available electronically, like those from external colleagues. Therefore, a paper-based temporary patient record with all documents that are not available in the EPR is maintained for the time of the patient's stay in hospital. After discharge all paper-based documents are transmitted to a scan service partner, assigned to the electronic patient record and indexed according to document type. Finally, all documents are archived electronically in the EPR.

Objective of the paper: The workflow of running a temporary patient record in addition to an EPR is complex and error-prone. Completeness and consistent assignment of data and documents are important factors for outcome quality of patient care. To be able to judge the quality of the migration process we evaluated

- the workflow of the current handling of the temporary record
- the completeness of scanned documents in the EPR
- the consistency of the assignment of scanned documents in the EPR
- the availability of the scanned documents in the EPR.

2. Material and Methods

Currently, a temporary record is started for each inpatient by administrative staff at the time of patient admission. After patient discharge the temporary record remains in the ward until the discharge letter is written. It is transferred to a central archive of Heidelberg University Hospital where all temporary records are registered and collected for the next weekly transport to the scan service partner. The scan service partner indexes the scanned documents according to document type and assigns them to the EPR of the patient. The week later a DVD with all scanned documents is delivered to the Center of Information Technology and Biomedical Engineering, where the scanned documents are integrated into the EPR system. At this point in time the scanned documents are available electronically for patient care.

To analyze the current workflow of handling the temporary patient record we visited all 14 wards of the department of internal medicine and interviewed all staff that is working with the temporary record. We conducted structured interviews with the users, the staff in the central archive, and the staff of the scan service partner.

To evaluate the quality of the scanned documents in the EPR we drew a random sample in the central archive three weeks in a row, which were ready for transport. We copied all documents and noted if they were of a non-standard format. After availability in the EPR-system, we checked for each document if

- it was available in the EPR system
- it was assigned to the right patient
- it was assigned to the right document type
- the quality was comparable with the source document.

For each copied record we recorded the date of patient discharge, availability in the archive, transport to the scan service partner, and availability in the EPR system.

3. Results

3.1. Qualitative Analysis of the Workflow

According to our observations the temporary record is generally handled as follows:

Instantiation: After instantiation, the temporary record is available on the carriage for ward rounds. It is possible to print documents from the EPR and file them. In three wards instantiation takes place one to three days after patient admission by the ward's staff and not immediately (cf. Figure 1).

New Documents: New documents to be filed in the temporary record are labeled with a barcodes according to their type. There is one barcode for external documents and 114 more specific barcodes for internal ones. Example document types are lab result, ECG-result, chemotherapy plan. The documents are collected in a preliminary folder for ward rounds. In eight wards the documents are filed into the temporary patient record during the nightshift, in six wards the next morning from secretary staff.

New Documents after Discharge: If new documents occur after discharge of the patient and the temporary record has already been transferred, an additional temporary record is instantiated and immediately sent to the archive. But there are considerable variations in the workflow: documents after discharge are only transferred into the physician's offices, are sent separately to the archive, collected in general folder, which is submitted to the archive some time, or not transferred at all.

Variations of the workflow: Other variations in the workflow were: Two wards are labeling the documents immediately before transmitting them to the archive (cf. Figure 1). Two wards do not label documents which are printed from the electronic patient record (this could be necessary if they are modified manually). Up to four wards mention that they only label documents which they produced themselves and no external ones and that they only use up to 20 barcodes (out of 114).

Problems and Suggestions from the users' point of view: Generally spoken the temporary patient record is regarded feasible from the users' point of view. Users report on some problems with the categories or layout of the folder. The temporary record seems not to be feasible for the progress notes of the intensive care unit. Two wards constitute that the time from transferring the temporary records to the archive until availability of the documents in the EPR-system is too long. All wards regard the task of assigning barcodes to documents as complicated.

All wards, all archive staff, and the scan service report that it scarcely occurs that temporary records are necessary which are already transmitted to the scan service partner (at most once a month).

Some users suggest among others the following improvements: ward-dependent structure of the temporary record, barcode-printer for single barcodes, scanner on each ward for self-scanning and availability of documentary staff.

All together the users prefer the temporary record in addition to the EPR to having a complete paper-based record.

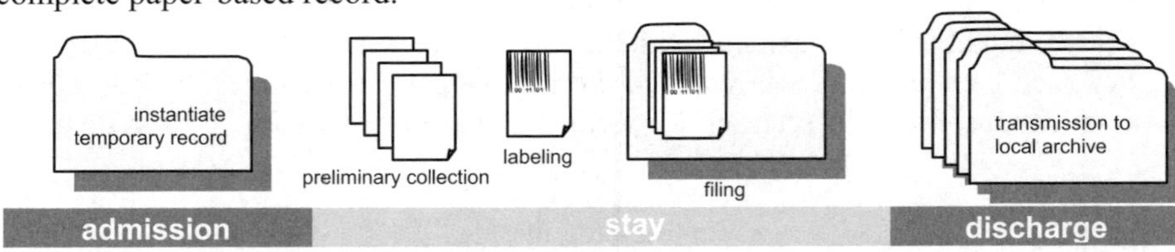

Figure 1: Filing of documents in temporary records. The temporary record is instantiated at the time of admission. Documents are collected during patient stay, labeled and filed in the temporary record. After discharge of the patient the temporary records are transmitted to a local archive for further processing. Observed variations in the process were: Instantiation after patient admission, labeling after discharge, and new documents after discharge.

3.2. Quantitative Analysis of the Scanned Documents in the Electronic Patient Record

Loss of documents: We analyzed 71 temporary records which contained 343 documents. All records were available in the EPR system (100%). Of 343 documents 28 (8,2%) could not immediately be found in the EPR. Twelve documents (3,5%) were archived within other documents. Sixteen (4,7%) could not be found at all, 12 of them being labeled as documents which have been printed from the EPR (cf. Table 1). There were no documents being assigned to the wrong patient or the wrong episode of care.

Loss of information: Of the 343 documents, 307 documents (89,5%) were scanned without loss of information, 32 documents (9,3%) with minor loss of information, 4 (1,2%) with major loss of information and no documents lost all information (cf. Table 1). Minor loss of information means that the quality of presentation is worse than the presentation in the source data but that the necessary data for patient care is still readable. Three of the four documents with major loss of information were ECG-curves as well as 9 of the 32 (28,1%) with minor loss of information. Six of the documents with minor loss of information (18,8%) contained information on anamnesis. The documents with minor loss of information contained handwriting in 59,4% of the cases and figures in 50%.

Correctness of Barcodes: Of the 343 documents 102 (29,7%) were not labeled with a barcode (cf. Figure 1). The most frequent categories that should have assigned were anamnesis (15,7%), administrative (11,8%), external results (11,8%), and referral (6,9%). Of the remaining 241 documents with barcode, 28 (11,6%) were labeled with a wrong barcode. No typical categories could be identified.

Table 1: Quality of scanned documents in the EPR.

Criteria	# docs	available	sufficient readability	correct barcode
abs.	343	327	323	213
rel.	100%	95,3%	98,8%	62,1%

Availability: We analyzed 117 temporary records. On average these records remained in the ward after patient discharge for 23,8 days (median 14 days, standard deviation 39,7), in the archive for 5,5 days (std. dev. 5,8), and at the scan service partner for 6,7 days (std. dev. 0.5). These numbers are summarized in Table 2.

Table 2: Average number of days the temporary records remains after patient discharge.

#Days	Mean	Range	Standard Dev.
Wards	23,8	1-283	39,7
Archiv	5,5	0-36	5,8
Scan Service	6,7	6-8	0,5

4. Discussion

The workflow of handling the temporary record is mostly in adherence with the guidelines from the Center for Information Technology and Biomedical Engineering. Since these guidelines are rather general there are variations in the process in the different wards. This can influence the quality of the temporary record if it is instantiated too late and of the EPR if the documents are not carefully labeled. Labeling is the most critical task in the workflow, because it is a prerequisite to efficiently retrieved documents in the EPR-system. The completeness of records and documents is rather high. Only 4 out 343 documents (1,2%) could not be found in the EPR.

The most critical factors for loss of information by scanning where the document type ECG-curve, handwriting and figures. The coloured ECG was initially scanned black and white with low resolution, after using a higher resolution the scan was well readable. Handwriting is mostly found in documents printed from the EPR and added by handwriting. Users should be motivated to use the EPR more consequently and avoid printing documents. The number of 30% missing barcodes is rather high. Currently this is compensated by the scan service, but should in future completely be done by clinical staff to ensure high quality classification from a clinical viewpoint.
After discharge the temporary record remains most of the time in the ward for writing the discharge letter. Additionally, the record is kept until it is unlikely that their content is needed for further clinical or research purposes. The average time in the archive and the scan service is based on the weekly transferral and can hardly be reduced.

Limitations of the study: The qualitative study gave the chance to identify limits of the current process and provided ideas for more detailed evaluation studies. We were able to observe all wards and a variety of staff of different professions. Thus, the results can be regarded as comprehensive.

Since loss of information and documents occurred rather seldom, the main limitations of the quantitative study is the sample size. For more information on typical types of concerned documents a higher sample would be necessary. Nevertheless, it was possible to identify possible problems and their magnitude. Another limitation was that we compared the documents in the EPR not with the original document but with a copy. We were aware of this limitations in advance, so tried to compensate this with precise descriptions on each copy if necessary.

Perspective: To further increase the quality of the workflow and the patient records it is first of all necessary to make the staff more sensitive how to efficiently handle the temporary record and how important correct labelling is. To achieve this it is a challenge to define document types as detailed as necessary but as few as possible. A possible chance to scanning documents locally in the wards has to be considered carefully with respect to feasibility, costs and quality.

5. Conclusion

Since there are still lots of external paper-based documents a complete EPR is currently not possible. A temporary paper-based patient record in addition to the EPR is not an optimum procedure but is feasible. The disadvantages are the parallel handling of two record systems and the effort and motivation for labeling documents according to their document types. The quality of the scanned, indexed and integrated documents in the EPR is high and the availability is sufficient.

Acknowledgement

The study was planned and conducted by medical informatics students in the third year of the Heidelberg/Heilbronn curriculum ([9]) under supervision of the authors. We gratefully acknowledge their thorough work. We thank the staff in the wards of the department of internal medicine, in the central archive, and the scan service partner for their helpful statements as well as Birgit Zeeh and Cornelia Homsy (HUH).

References

[1] F. Leiner, W. Gaus, R. Haux, and P. Knaup-Gregori, Medical Data Management - A Practical Guide. Health Informatics, ed. K.J. Hannah and M.J. Ball, 2003, Springer, New York, 2003.

[2] J.H. van Bemmel, A.M. van Ginneken, B. Stam, and E. van Mulligen, Virtual electronic patient records for shared care, Medinfo, 9 Pt 1 (1998) suppl 37-41.

[3] M. van der Haak, V. Mludek, A.C. Wolff, H. Bülzebruck, D. Oetzel, D. Zierhut, P. Drings, M. Wannenmacher, and R. Haux, Networking in Shared Care - First Steps towards a Shared Electronic Patient Record for Cancer Patients, Methods of Information in Medicine, 41 (2002) 419-425.

[4] A. Shabo, P. Vortman, and B. Robson, Who's Afraid of Lifetime Electronic Medical Records, in: (Eds.) Proceedings of 1st Annual Conference on Mobile and Wireless Healthcare Applications 11 - 14 November 2001, Café Royal, London., 2001, pp.

[5] C.P. Waegemann, The five levels of electronic health records, MD Comput, 13 (1996) 199-203.

[6] A. Hasman, Care for records for care, International Journal of Bio-Medical Computing, 42 (1996) 1-7.

[7] A. Häber, C. Dujat, and P. Schmücker, Leitfaden für das rechnerunterstützte Dokumentenmanagement und die digitale Archivierung von Patientenunterlagen im Gesundheitswesen, 2005, GIT Verlag, Darmstadt, 2005.

[8] C. Dujat, R. Haux, P. Schmücker, and A. Winter, Digital Optival Archiving of Medical Records in Hospital Information Systems - A Practical Approach towards the Electronic Patient Record?, Methods of Information in Medicine, 34 (1995) 489-497.

[9] F.J. Leven, P. Knaup, D. Schmidt, and T. Wetter, Medical informatics at Heidelberg/Heilbronn: status-evaluation-new challenges in a specialised curriculum for medical informatics after thirty years of evolution, Int J Med Inform, 73 (2004) 117-25.

2.4 EHR: Open Approaches

Ubiquity: Technologies for Better Health in Aging Societies
A. Hasman et al. (Eds.)
IOS Press, 2006

Ubiquitous Information for Ubiquitous Computing: Expressing Clinical Data Sets with *open*EHR Archetypes

Sebastian GARDE[a,1], Evelyn HOVENGA[a], Jasmin BUCK[b] and Petra KNAUP[b]
[a] *Central Queensland University, Rockhampton, Australia*
[b] *University of Heidelberg, Dept. of Medical Informatics, Germany*

Abstract. Ubiquitous computing requires ubiquitous access to information and knowledge. With the release of *open*EHR Version 1.0 there is a common model available to solve some of the problems related to accessing information and knowledge by improving semantic interoperability between clinical systems. Considerable work has been undertaken by various bodies to standardise Clinical Data Sets. Notwithstanding their value, several problems remain unsolved with Clinical Data Sets without the use of a common model underpinning them. This paper outlines these problems like incompatible basic data types and overlapping and incompatible definitions of clinical content. A solution to this based on *open*EHR archetypes is motivated and an approach to transform existing Clinical Data Sets into archetypes is presented. To avoid significant overlaps and unnecessary effort during archetype development, archetype development needs to be coordinated nationwide and beyond and also across the various health professions in a formalized process.
Keywords: Electronic Health Records, Semantic Interoperability, *open*EHR, Archetypes, Clinical Data Sets, Computerized Medical Record Systems

1. Introduction

In our quest for ubiquitous computing, the Health Informatics community currently investigates the use of unobtrusive, active, non-invasive technologies, including wearable devices to continuously monitor and respond to changes in the health of a patient (e.g. [1]). Ubiquitous computing, i.e. the integration of computation into the environment, rather than having computers which are distinct objects, is an exciting research field that offers potential e.g. for higher quality and more user-friendly and comprehensive data capture. On the other hand, we believe that ubiquitous computing requires first of all ubiquitous availability of information and knowledge.

A considerable amount of work has already been undertaken by the Health Informatics community along the path to ubiquitous use of patient information and health knowledge: National and international standards developments, connectivity using HL7, document exchange using HL7 CDA [2], comprehensive specifications for Electronic Health Records like the recent release of *open*EHR Version 1.0 (http://www.openEHR.org), the development of specialist Clinical Data Sets (CDSs)

[1] Corresponding Author: Dr. Sebastian Garde, Health Informatics Research Group, Faculty of Business and Informatics, Central Queensland University, Rockhampton QLD 4702, Australia. s.garde@cqu.edu.au http://healthinformatics.cqu.edu.au

on a national and sometimes international level. Creating CDSs is a tedious task and the bigger the scope or the applicable region is, the bigger is the effort [3]. To name but a few, in Germany Merzweiler and colleagues defined a Basic Data Set for Pediatric Oncology [4], Flynn and colleagues recently defined the European data standard for clinical cardiology practice [5], and the *international* Nursing Minimum Data Set (i-NMDS, http://www.inmds.org) is currently under development.

With the release of *open*EHR Version 1.0 (http://www.*open*EHR.org) there is a common information and knowledge model for Electronic Health Records available that can solve some of these still prevailing problems of insufficiently ubiquitous information and knowledge in the health sector by improving semantic interoperability and providing a common basis for decision support, data mining, etc.

The aim of this paper is to

- outline typical problems that arise when developing CDSs,
- analyse whether *open*EHR archetypes can be used to overcome some of the problems identified when developing new CDSs,
- present an approach to develop archetypes based on existing CDSs in light of typical problems that arise when transforming them into archetypes.

2. Material and Methods

2.1. Clinical Data Sets

The results in this paper stem from the analysis of various existing CDSs and typical development processes for CDSs with a special focus on German, European and Australian data sets. Further, we analysed the *open*EHR and archetypes approach with regard to their suitability for the development of CDSs. For this we investigated existing iterature for criteria for 'good' CDSs. While we found very little published literature dealing with this exact problem, some not yet published works are currently circulated, for example, the principles (e.g. system-independency, use of existing standards) for good data development in [6]. Also, desiderata for terminologies (e.g. ICD) have been postulated by Cimino [7]. These are not applicable as such to CDSs or archetypes but still helped us to identify some common problems of CDSs. In addition, we applied the process as described in this paper for the development of archetypes based on existing CDSs to develop archetypes based on the German Basic Data Set for Pediatric Oncology [4].

2.2. The openEHR and Archetypes Approach

The *open*EHR approach (http://www.*open*EHR.org) is defined in comprehensive specifications originally based on the results of the European Union's GEHR-Project, used and refined in a series of further European and Australian projects. Its design principles are described by Beale and colleagues in [8]. The main characteristic of *open*EHR is a two level modelling approach for EHRs. The first level of this approach is the reference model which is reduced to a relatively small set of classes to support the medico-legal requirements and record management functions. The second level involves the *open*EHR archetype methodology representing all the clinical knowledge that rather than being hard-coded in databases and applications is defined in archetypes. Each archetype represents one clinical concept by constraining instances of the *open*EHR reference model. By using two levels – a reference model and archetypes - *open*EHR separates technical concerns from clinical data collection. Archetypes themselves are terminology-neutral, but can link to external terminologies like SNOMED CT.

3. Results

3.1. Problems with CDSs not based on a common Methodology

Notwithstanding the importance of the work on CDSs, there are several problems that remain unsolved even if all clinical data were formally expressed in standardized CDSs but not following a common methodology: During our work on CDSs and implementing them in application systems with structured data entry we observed the following problems:

- **Problem 1:** CDSs use different kinds of basic data types.
- **Problem 2:** CDSs use different kinds of presentation formats.
- **Problem 3:** CDSs do not necessarily follow basic design principles (e.g properly modelled relationships between concepts, optionality or repeatability of concepts).
- **Problem 4:** CDSs often do not define when and how often which data item has to be captured. Neither do CDSs consistently define data that is important for the interpretation of a measurement, For example was a patient resting or exercising before a blood pressure measurement.
- **Problem 5:** CDSs often do not define integrity constraints that will allow data validation at runtime.
- **Problem 6:** Domain knowledge is replicated inconsistently in the various data sets, clinical concepts are overlapping.
- **Problem 7:** Depending on the tools used to model CDSs, no support may be provided to develop or maintain CDSs in different languages. This is becoming increasingly relevant for example in the EU.
- **Problem 8:** Tools to support or make use of CDSs are reinvented all the time, and specialist applications which are not integrated in the master health information system are developed and have to be maintained.

3.2. The Potential of openEHR to overcome the Problems with CDSs

The *open*EHR approach and archetypes address the problems associated with CDSs (Section 3.1) in the following ways:

- **Re Problem 1:** The *open*EHR approach offers a common reference model. It uses, among others, the ISO 11404 standard for basic data types. Thus with *open*EHR, CDS developers automatically adhere to existing standards.
- **Re Problem 2:** Existing tools like the Archetype Editor or the Archetype Workbench present archetypes in various ways that can be easily understood by clinicians and technicians (for example prototype graphical user interfaces, HTML rendition of the archetype). In Australia, this is used for example by major

Australian institutions like the National E-Health Transaction Authority (NEHTA) who successfully develop national specifications based on archetypes.

- **Re Problem 3:** Archetypes offer a predefined structure for expressing clinical knowledge, thus providing CDS developers with the necessary structure to build their data sets in a uniform way adhering to basic design principles.
- **Re Problem 4:** Archetypes have inbuilt support for defining time-series when recordings should take place (*History*) and relevant archetypes have *State* Data, i.e. all information that is relevant for the interpretation of a measurement can be captured in a consistent way.
- **Re Problem 5:** Archetypes are also used to define integrity constraints in a uniform way that can be used to validate captured data. In an archetype-enabled clinical system this will eventually be done automatically by adding the archetype to the system. It has to be said that not all possible types of integrity constraints are currently supported by *open*EHR archetypes alone.
- **Re Problem 6:** Processes to ensure that domain knowledge is and remains consistent are being put in place by the *open*EHR foundation to avoid the inconsistent re-definition of clinical knowledge, and to avoid the definition of clinical concepts that are overlapping. To support these processes (by Domain Knowledge Governance as discussed in [3]), the Archetype Finder – developed by the first author – can be used. The Archetype Finder has relevant meta-data for archetypes based on a Protégé-OWL ontology to enable a high precision and high recall of relevant archetypes (http://dualitysystems.com.au/archetypefinder).
- **Re Problem 7:** Since any translation occurs within one archetype only, translating an archetype is relatively easy and meaning and structure can be preserved across languages. This is comparable to the multi-language support of GALEN's Classification Workbench [9] for terminologies.
- **Re Problem 8:** Existing tools like the Archetype Editor and the Archetype Finder offer considerable support during the development of archetype-based CDSs. As agreed models of clinical or other domain specific concepts, archetypes are clinically meaningful entities with a common structure and have the same meaning no matter where they appear. Thus, archetypes can be shared by multiple health systems and authorities and information can be exchanged between different systems with increased ease and semantic meaning. This may eventually lead to clinical systems based on these archetypes that are more easily maintainable as domain knowledge needs no longer be hard-coded into the system.

3.3. An overview of the procedure to transform CDSs into archetypes

In the previous section we presented theoretical considerations why archetypes are of possible advantage when expressing CDSs. Based on this, we now present an overview of the procedure needed to transform CDSs into archetypes in order to explore the feasibility of the approach.

Initially all appropriate high-level archetype concepts present in the data set, for example 'blood pressure measurement' or 'diagnosis' should be defined. It is possible to use an Archetype Editor at this stage already, but not necessary. Also the use of existing demographics archetypes will be decided upon. These should be identical or specialized from existing *open*EHR demographics archetypes. After all high-level archetype concepts are defined – and before actually designing the archetypes from scratch using an Archetype *Editor*, it is advisable to check for existing reference archetypes. The Archetype *Finder* avoids that already defined archetypes are reinvented thus saving effort and fostering semantic interoperability. If an archetype is only available in a foreign language it can be translated. If no fitting archetype is discovered, an appropriate structure for the new archetype has to be selected: This includes the decision if a clinical concept is an *Observation*, *Evaluation, Instruction* or *Action.* Using an Archetype Editor, the archetype is then defined (or specialised respectively) by adding data items of the appropriate type, binding data items to internal codes or terminologies like SNOMED. For certain archetypes, *History* events are defined to express the points of time when the data is supposed to be captured. *State* (essential for the interpretation of a measurement) and *Protocol* information (e.g. what kind of device was used) can be added to the archetype. Broad input from clinician should be sought during this phase (however if the data set is already well defined less input is necessary). Often CDSs do not define the exact points of time when data documentation is required. If the data set does not provide all the necessary information, this information has to be gathered or agreed on by the responsible domain experts. Finally, section and composition archetypes are defined for ordering the clinical contexts into a fitting structure.

Once an archetype is defined, it should be submitted to the Clinical Review Board of *open*EHR as for semantically interoperable systems, archetype development should be coordinated to avoid incompatible archetypes for the same concept by systematic Domain Knowledge Governance. Thus it can be ensured that no overlapping concepts or otherwise incompatible archetypes are used for different health disciplines, etc.

4. Discussion

We believe that the presented results support our hypothesis that expressing CDSs as *open*EHR archetypes is (a) feasible and (b) advantageous. To overcome some of the shortcomings of current CDSs (like conflicting basic data types) a uniform model like the *open*EHR model is helpful. To overcome other shortcomings like overlapping, inconsistent concepts, systematic Domain Knowledge Governance is required in addition. As each archetype forms a clearly defined semantic unit that expresses one clinical concept, archetypes enable knowledge to be governed within clearly defined boundaries. Some of the described problems of CDS development can be overcome by other means, e.g. by the development of a Guide to Data Development [6].

Although we cannot prove it as yet, we believe that *open*EHR archetypes are more expressive than any kind of conventional structure for CDSs. Because of this and the described advantages of archetypes, we believe that *open*EHR archetypes are an adequate means for developing and structuring *new* CDSs. As shown in Section 3.3 it is also possible to transform *existing* CDSs into archetypes. Generally speaking, the better the quality of the CDS adheres to good data development guidelines, the less effort is required to transform it into a collection of archetypes. Archetypes do not solve the problem that domain experts will need to agree on the appropriate representation for a clinical concept. However, as archetypes separate informatics concerns and clinical content discussion, in our experience from archetype workshops recently conducted for clinicians in Melbourne, Sydney and Brisbane, archetypes help clinicians to focus on the discussion of *clinical* content. Thus, by using archetypes to develop new CDSs, a more streamlined development process can be adopted. Still some training for clinicians to understand the basic concepts of archetypes is necessary.

More generally, the *open*EHR two-level modelling and archetype methodology cannot overcome all of the barriers to ubiquitous computing. For example, easier, higher quality, more convenient and faster data capture at the point of care (or in fact anywhere) as envisioned by ubiquitous computing (e.g. wearable devices, and combination of technologies such as voice or handwriting recognition) are beyond the scope of this approach. The *open*EHR approach can however provide the common basis for ubiquitous *presence* of meaningful and computer-processable knowledge and information and thus contribute to the usability of clinical systems, improve data quality, and improve semantic interoperability.

5. Conclusion

In the forthcoming era of ubiquitous computing, we must not neglect the fact that ubiquitous computing involves ubiquitous access to information and knowledge. We conclude that a good common information model and an archetype-based approach can be of vital importance in realizing this. For this, existing CDSs should in our opinion be transferred to archetypes and new data sets directly defined using archetypes. This does not only enable semantic interoperability, but helps to develop archetypes that are of better quality than the original CDS.

Acknowledgements

The authors would like to thank Sam Heard and Thomas Beale for valuable discussion about *open*EHR and archetypes in general and this project in particular.

References

[1] Scheffler M and Hirt E. Wearable devices for telemedicine applications. J Telemed Telecare 2005; 11 Suppl 1:11-4.
[2] Dolin RH, Alschuler L, Boyer S, Beebe C, Behlen FM, Biron PV, and Shabo Shvo A. HL7 Clinical Document Architecture, Release 2. J Am Med Inform Assoc 2006; 13(1):30-9.
[3] Garde S, Heard S, and Hovenga E. Archetypes in Electronic Health Records: Making the case and showing the path for domain knowledge governance. In: Grain H and Wise M, editors., HIC 2005: 13th Australian Health Informatics Conference, Melbourne, 31.07.2005–02.08.2005, Brunswick East, Vic: Health Informatics Society of Australia, 2005.
[4] Merzweiler A, Ehlerding H, Creutzig U, Graf N, Hero B, Kaatsch P, Zimmermann M, Weber R, and Knaup P. Standardizing terminology in pediatric oncology--the basic data set (in German). Klin Padiatr 2002; 214(4):212-7.
[5] Flynn MR, Barrett C, Cosio FG, Gitt AK, Wallentin L, Kearney P, Lonergan M, Shelley E, and Simoons ML. The Cardiology Audit and Registration Data Standards (CARDS), European data standards for clinical cardiology practice. Eur Heart J 2005; 26(3):308-13.
[6] Standards Australia IT-014-02. A Guide to Data Development (Pre-public comment draft). 2006;
[7] Cimino JJ. Desiderata for controlled medical vocabularies in the twenty-first century. Methods Inf Med 1998; 37(4-5):394-403.
[8] Beale T, Goodchild A, and Heard S. EHR Design Principles. *open*EHR Foundation; 2001. http://titanium.dstc.edu.au/papers/ehr_design_principles.pdf (last accessed 16.01.2005).
[9] Trombert-Paviot B, Rodrigues JM, Rogers JE, Baud R, van der Haring E, Rassinoux AM, Abrial V, Clavel L, and Idir H. GALEN: a third generation terminology tool to support a multipurpose national coding system for surgical procedures. Int J Med Inform 2000; 58-59:71-85.

Ubiquity: Technologies for Better Health in Aging Societies
A. Hasman et al. (Eds.)
IOS Press, 2006

Towards Automatically Generating Graphical User Interfaces from *open*EHR Archetypes

Thilo Schuler[a,b,1], Sebastian Garde[b], Sam Heard[b], and Thomas Beale[c]
[a] *University of Freiburg, Department of Medical Informatics, Freiburg, Germany*
[b] *Health Informatics Research Group, Central Queensland University, Rockhampton, Australia*
[c] *Centre for Health Informatics and Multiprofessional Education, University College London, United Kingdom*

Abstract. One of the main challenges in the field of Electronic Health Records (EHRs) is semantic interoperability. To utilise the full potential of interoperable EHR systems they have to be accepted by their users, the health care providers. Good Graphical User Interfaces (GUIs) that support customisation and data validation play a decisive role for user acceptance and data quality. This study investigates the use of *open*EHR archetypes to automatically generate coherent, customizable, data-validating GUIs. Using the Mozilla XML User Interface Language (XUL) a series of prototypes has been developed. The results show that the automatic generation of GUIs from *open*EHR archetypes is feasible in principle. Although XUL revealed some problems, the advantages of XML-based GUI languages are evident.
Keywords: Computerized Medical Record, User-Computer Interface, Archetype, openEHR, Electronic Health Record.

1 Introduction

"Semantic interoperability" is the goal when it comes to Electronic Health Records (EHRs) and the perceived means by which medical informatics can improve today's shared healthcare environment. Standards development organisations, open source communities, research groups, and early implementers in different countries (usually funded by their governments) are putting in considerable effort to make this vision a reality.

To achieve semantically interoperable and future-proof EHR systems, the *open*EHR initiative (http://www.openehr.org/) introduced a two-level methodology ([1]). Formally expressed, distinct, reusable domain-level content models, called archetypes, are used to constrain data instances that conform to a generic reference model ([2]). The basic constraining notion of this approach has been adopted or considered by most major standardisation efforts in the EHR realm. The revised CEN EN 13606 (EHRcom) will incorporate this archetype methodology directly, while the HL7 is working on a templating methodology for this purpose - "HL7 CDA Templates and

[1] Corresponding Author: thilo.schuler@gmail.com

DICOM Structured Reporting Templates are conceptually very similar to archetypes" ([4]). However, due to features of the HL7 version 3 RIM, these templates are not able to support the clear separation between information and domain models expressed in *open*EHR's two-level methodology.

*Open*EHR archetypes are formally defined using the Archetype Definition Language (ADL). An example for an ADL modelled medical concept would be a blood sugar level archetype. The modelling process is supported for example by the Ocean Informatics Archetype Editor available under the Mozilla Tri-license (http://oceaninformatics.biz/). Using this editor, clinicians without informatics skills can define archetypes themselves and therefore guarantee flexibility and coverage of clinical documentation needs ([6]). According to the *open*EHR methodology, several archetypes are grouped to form a particular data structure e.g. to document a diabetes monitoring examination. For a user to fill this data structure it has to be presented as a Graphical User Interface (GUI) for data entry.

The GUI is arguably the most important part of an EHR system with which a health professional interacts and is therefore of utmost importance for its success ([12], [5], [7], [8]). In this paper, we present a series of prototypes that aim to automatically generate a GUI from *open*EHR archetypes expressed in ADL syntax.

2 Material and Method

Microsoft's C# and Mozilla XML User Interface Language (XUL; [13]) including adjacent technologies such as JavaScript, Document Object Model (DOM), and Cascading Stylesheets (CSS) were selected as prototyping technologies. Web-based applications have the advantage that they can be run with minimal requirements on the user side and are easily distributable to many users. Their downside is a restrictive user interface due to the intrinsic limitations of HTML and the traditional web paradigm with complete page refreshes. Compared to desktop applications, web applications are less dynamic (e.g. real-time input validation or context-sensitive interface alterations are generally not supported) and do not offer rich controls such as tabbed panels or tree components. The declarative language XUL tries to combine the strengths of both approaches. XUL uses XML markup to define the GUI in a declarative way. A number of such languages exist ([14]). XUL is one of the oldest and used in all Mozilla software projects, such as the Mozilla Firefox browser, to create the GUI.

An XML based format appears to be of advantage for generating a GUI description as there are many tools for XML editing. The reasons we selected Mozilla XUL were the comparatively long usage as part of the Mozilla projects and available documentation. Other advantages include the built-in "skinning" ability via CSS and the multi-language feature.

The proposed XUL interface communicates with a middleware implemented as a web service. The main reasons for choosing C# to implement the middleware and the GUI generating software were the good support of web service-based communication and availability of an advanced development environment.

3 Results

At the beginning of our research we produced a model of how the archetype-based generation of GUIs could function. The generation mechanism has been designed to answer the question: “To what extent can GUIs be automatically generated from *open*EHR archetypes?” Real world implementations would require many additional components like security features (encryption, user authentication, etc.). The following diagram describes the system from a high level view:

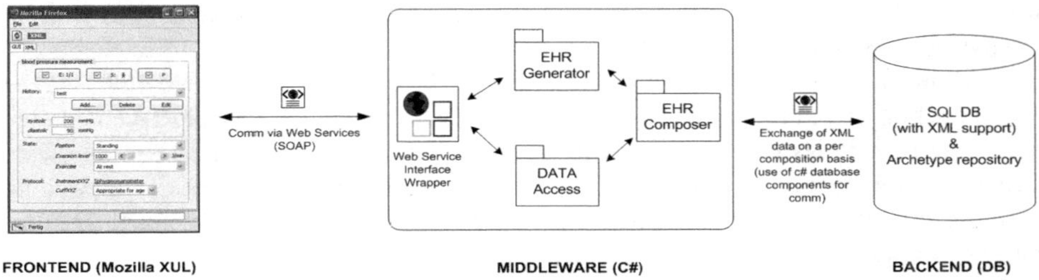

Figure 1: Archetype based GUI generator system

Using this approach as a starting point we implemented a series of prototypes to analyse different aspects of this concept. The diagram shows a backend, which has been included to get a complete view of a possible system, but the backend is not subject of our research.

3.1 Prototype 1: Handcoded GUI of blood pressure archetype

The intention of this prototype was to test the possibilities of XUL. It helped to gain experience in using the framework to create rich GUIs and to implement communication with the middleware via web service messages. The prototype is one example of what the generator could produce and therefore it helped to assess the applicability of the system design in general.

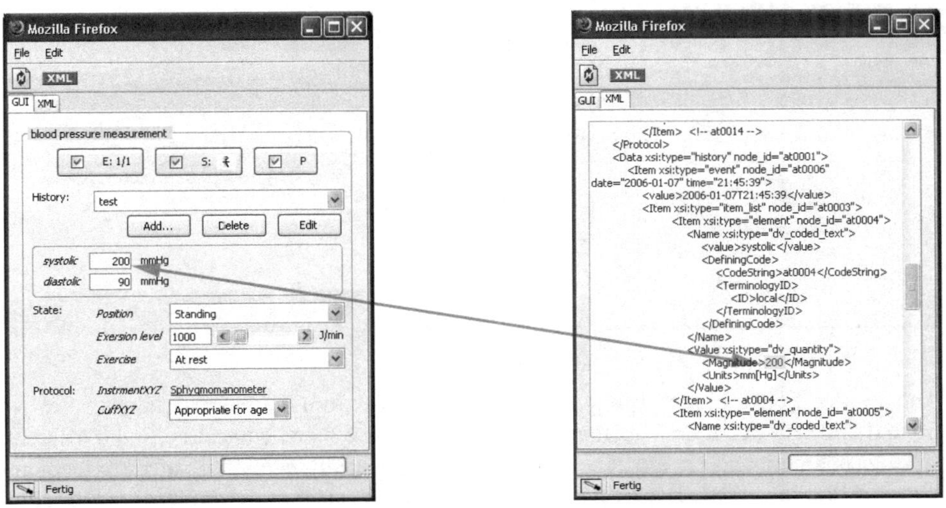

Figure 2: XUL blood pressure GUI and resulting XML data produced by middleware

Figure 2 shows a GUI of a blood pressure archetype. The archetype includes a range of information: the systolic and diastolic values; assignment of these values to a series of events (history) with recording of maximum, minimum and other aggregate values; information about the state of the patient such as position or level of exertion at the time of measurement; and protocol information about the measurement itself such as the site, or size of cuff, if a sphygmomanometer is used. In the GUI the event selector, as well as the state and protocol input fields, can be hidden while their main information is still visible as an icon. The XML tab has only been added to monitor the effects in the middleware. It shows the XML instance that the Composer produces to store in the database (persistence).

Technically the main challenge was to establish communication from XUL, via the SOAPCall object, to the middleware methods exposed as a web service. To enable the recording of a series of blood pressure measurement events, new input controls were dynamically added via the DOM. Depending on the selected event, the corresponding controls are displayed, while the others are hidden. This design decision has the advantage of swapping between events without constantly reloading content.

3.2 *Prototype 2: Custom control to validate archetype defined value range*

Good GUIs should provide as much input validation as possible. Warnings and error messages should appear as promptly as possible when syntactically or semantically inconsistent data is input. Input constraints like value ranges and cardinalities are defined in the archetype. To represent these constraints, a generic set of customized controls (one for each *open*EHR data type) has to be developed. By assembling these controls, a whole archetype can be visually represented. Later, different controls for the same data type could be developed to reflect different preferences (customisation). Specialized controls that stand for a whole archetype are imaginable as well.

This second prototype aims to show how to build a masked numerical input control that only accepts values within a certain range.

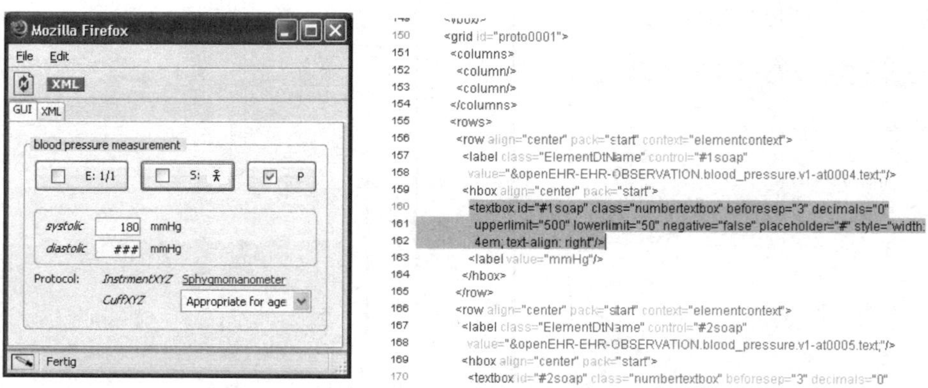

Figure 3: Custom input control and its integration into XUL

The masked fields (Figure 3) require 3 digits and accept only numerical values between 50 and 500. The XUL code example shows how a customized control is used. Different settings can be made via attributes and can therefore easily be manipulated by the generator. Mozilla's XML Binding Language (XBL) was used to define this custom control.

3.3 Prototype 3: GUI Generator

The goal of the third prototype was to build an archetype-based generator that creates GUIs similar to those seen in the previous prototypes. For this prototype a standalone C# application was built, but it would be possible to integrate the generator in a web service as shown in Figure 1.

To generate the markup, the selected archetype is passed to the kernel component. The kernel then forms an in-memory tree representation of the archetype and the GUI is generated by recursively walking through this tree. Depending on the underlying reference model class at each tree node, a suitable generator class handles the kernel information and produces XUL output.

4 Discussion

Our prototypes show that *open*EHR archetypes contain sufficient information to automatically derive functional GUIs, similar to those GUIs in existing medical applications that are hardcoded and implemented in traditional programming environments. Thus rich GUIs, including validation rules according to the specific domain, are feasible. Our research also proves that using an XML-based GUI description language is a viable approach. The common XML format made creation and handling of the generated output straightforward, due to well-understood and supported mechanisms, such as the DOM or XPath.

We also encountered several problems, including the immaturity of the Mozilla XUL language and the lack of documentation of the early *open*EHR kernel component. Although XUL is used in major software projects within the Mozilla foundation, it is still in its first version and some inconsistencies and implementation bugs (e.g. problem with references to cloned objects) still prevail. In a "once-only" programmed application it is possible to circumnavigate these problems, but to dynamically generate GUIs a more stable development framework is necessary. Another practical limitation was the lack of modern development environments for XUL. Still, we believe XML-based GUI languages are the future. The fact that the presentation layer of Microsoft's new Windows version is expected to use a XML format called XAML is one strong indicator for this ([15], [16]).

There are other approaches that aim to provide generic solutions for collecting structured data in health, for example the DOSPO project from the University of Heidelberg ([9],[11]) and OpenSDE application from the Erasmus MC in Rotterdam ([10]). As part of these systems, GUIs can be generated from predefined domain models. In distinction to our archetype-based approach, DOSPO's and OpenSDE's modelling mechanisms would need to be mapped to an EHR standard in order to achieve semantic interoperability or to provide other, non-generic, means of communication.

Although *open*EHR archetypes are not built to generate GUIs, they suit this purpose well. ADL defines knowledge concepts by nesting instances of types from a small, stable, generic reference model. This hierarchical structure can be further constrained by determining the cardinality relations between the instances, as well as by setting allowable values for primitive type instances at the leaf nodes ([3]). Since the number of reference model types is small, customized controls can be developed for each of them. Assembling them as described in the results section leads to a generic

GUI representation for every archetype. Obviously the GUI can only enforce the constraints which have been defined in the archetype.

Even before the advent of an EHR system with a fully "on the fly" generated GUI, the customisation function for GUIs becomes important. Only if GUIs can be adapted to personal and organisational workflows they will be usable, accepted and efficient. The described maximum level of customisation requires a presentation layer design that is completely independent from other layers of the system and that uses a ubiquitous and adaptable format such as a XML based GUI language.

5 Conclusion

Our research proves that the automatic generation of GUIs from *open*EHR archetypes is feasible in principle, based on three prototypes. Although the XML-based GUI languages like XUL are still in their infancy, they show a lot of potential despite the difficulties encountered. Further studies should be conducted to investigate automatic GUI generation for several archetypes based on an *open*EHR template.

References

[1] Beale T. Archetypes: Constraint-based Domain Models for Future-proof Information Systems. In OOPSLA 2002 workshop on behavioural semantics. http://www.deepthought.com.au/it/archetypes/archetypes_new.pdf (last accessed Jan. 3, 2006).

[2] Beale T, Heard S. Archetype Definitions and Principles. http://svn.openehr.org/specification/TRUNK/publishing/architecture/am/archetype_principles.pdf (last accessed Jan. 3, 2006).

[3] Beale T, Heard S. Archetype Definition Language (ADL), Revision 1.3. openEHR Specification, the openEHR foundation. http://svn.openehr.org/specification/TRUNK/publishing/architecture/am/adl.pdf (last accessed Jan. 3, 2006).

[4] Eichelberg M, Aden T, Riesmeier J, Dogac A Laleci G. A Survey and Analysis of Electronic Healthcare Record Standards. ACM Computing Surveys, Vol. V, No. N 20YY pp. 1-47.

[5] Gainer A, Pancheri K, Zhang J. Improving the human computer interface design for a physician order entry system. AMIA Annu Symp Proc. 2003;:847.

[6] Hovenga E, Garde S, Heard S. Nursing constraint models for electronic health records: a vision for domain knowledge governance. Int J Med Inform. 2005 Dec;74(11-12):886-98.

[7] Johnson CM, Johnson TR, Zhang J. A user-centered framework for redesigning health care interfaces. J Biomed Inform 38 (2005) 75-87.

[8] Kim MI, Johnson KB. Patient entry of information: evaluation of user interfaces. J Med Internet Res 6 (2004) e13.

[9] Knaup P, Garde S, Merzweiler A, Graf N, Schilling F, Weber R, Haux R. Towards shared patient records: An architecture for using routine data for nationwide research. Int J Med Inform. 2005 Aug 19.

[10] Los RK, van Ginneken AM, van der Lei J. OpenSDE: a strategy for expressive and flexible structured data entry. Int J Med Inform. 2005 Jul;74(6):481-90.

[11] Merzweiler A, Weber R, Garde S, Haux R, Knaup-Gregori P. TERMTrial--terminology-based documentation systems for cooperative clinical trials. Comput Methods Programs Biomed. 2005 Apr;78(1):11-24.

[12] Rose AF, Schnipper JL, Park ER, Poon EG, Li Q, Middleton B. Using qualitative studies to improve the usability of an EMR. J Biomed Inform. 2005 Feb;38(1):51-60.

[13] Wusteman J. About XML: from Ghostbusters to libraries - the power of XUL. Library Hi Tech, vol.23, no.1, 2005, pp.118-29. Publisher: Emerald, UK.

[14] http://en.wikipedia.org/wiki/List_of_user_interface_markup_languages (last accessed Jan. 3, 2006).

[15] http://msdn.microsoft.com/library/default.asp?url=/library/en-us/dnintlong/html/longhornch03.asp (last accessed Jan. 3, 2006).

[16] http://www.ondotnet.com/pub/a/dotnet/2004/01/19/longhorn.html (last accessed Jan. 3, 2006).

2.5 EHR and Clinical Management

Ubiquity: Technologies for Better Health in Aging Societies
A. Hasman et al. (Eds.)
IOS Press, 2006

Co-operation, interaction and information-sharing: from a paper-based Diabetes Pass to an electronic alternative

Heidi BUYSSE[a,1], Georges VAN MAELE[b] and Georges DE MOOR[a]
[a] *IBBT-Dept. of Medical Informatics and Statistics., Ghent University, Ghent, Belgium*
[b] *Dept. of Medical Informatics and Statistics., Ghent University, Ghent, Belgium*

Abstract. Diabetes is one of the most challenging problems in the 21st century, whereby research showed that reducing the concentration of blood glucose is thought to prevent or reduce the long-term complications. To do so, a multidisciplinary approach is favourable. In Belgium, a revalidation programme for the diabetic patients concerning self-regulation was introduced followed by the introduction of the Diabetes Pass in March 2003 whereby some goals were stipulated. In IBBT-COPLINTHO, a still ongoing project, an eHomeCare platform is implemented whereby the patient is the central actor. The analysis of the current paper-based Diabetes Pass revealed that the data can be easily extracted from the EHR. An electronic alternative for the Diabetes pass is proposed whereby the added values for the patient are underlined. Before implementing the electronic alternative, some research should still be done, but it is thought that all the actors involved – including the patient, could easily benefit from the electronic alternative.
Keywords: Diabetes Mellitus, Electronic Health Care, Electronic Health Record, Communication, ICT

1. Introduction

Life expectancy continues to increase because of the advances in medical sciences. This results in a growing proportion of elderly people with chronic diseases and disabilities [1]. The diabetes atlas [2] mentions diabetes as being one of the most challenging health problems in the 21st century. The estimated prevalence for diabetes in Belgium was 2-5% in 2003 and is estimated to be 5-8% in 2025. Complications from diabetes result in increasing disability, reduced life expectancy and enormous health costs. Reducing the concentration of blood glucose is thought to prevent or reduce the long-term complications [2-5], whereby aspects such as self-management [6-9], empowerment, education and information [6-12] are very important. The care for the diabetic patient is mainly provided in the home environment whereby a multidisciplinary approach is favourable [5]. Communication, co-operation and information-sharing are then of utmost importance.

[1] Corresponding Author: Buysse Heidi, 5K3 – De Pintelaan 185, B-Ghent 9000, Belgium. Heidi.Buysse@UGent.be

2. The Belgian "revalidation programme for diabetic patients concerning self-regulation"

Because a multidisciplinary approach, education, information, self-management and empowerment are important, the Belgian Government provides for a "revalidation programme for diabetic patients concerning self-regulation" (Figure 1). Some information streams are hereby mandatory: 1) the General Practitioner (GP) and the home health care (HHC) professional need to communicate about the health progress of the patient; 2) the multidisciplinary diabetic team (including at least a specialist in endocrinal-diabetics, a nurse specialised in diabetes, and a dietician) has to consider teamwork, availability, and continuity of paramount importance; 3) co-operation with the GP should be thoroughly described; 4) the patient himself should have a "Global Medical Record" and a "Diabetes Pass"; 5) between the diabetic team and the GP, an individual collaboration protocol should be developed and 6) the HHC nurse should have a "nursing record".

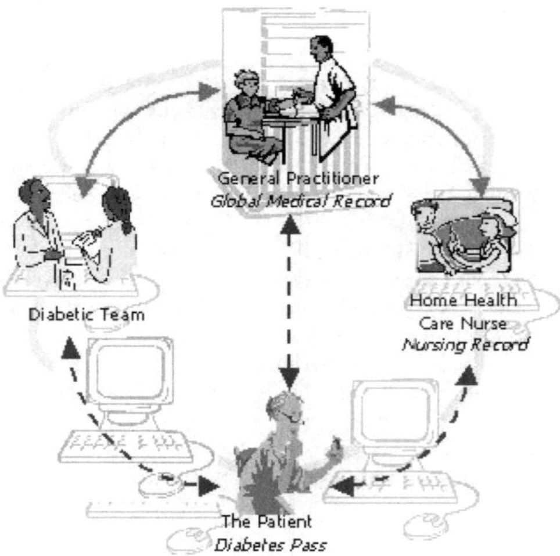

Figure 1. Communication and information-sharing as foreseen by the Belgian "Revalidation programme for diabetic patients concerning self-regulation" with the mandatory records (in *italic*)

In this revalidation programme, the Diabetes Pass introduced in March 2003 is a major component. It has several objectives and advantages:

- The patient is the owner of the pass. Every health care provider is supposed to make notes in this thin note-book pass and has at the same time insight in some data written in that pass (Table 1).
- It aims at synchronising the different treatment and care interventions.
- It includes educational information for the patient regarding the basic principles of good treatment, the prevention of complications, self-care, and first aid. It thus enhances the responsibility of diabetic patients concerning their own health.
- The pass gives diabetic patients the right for reimbursement on some consults with health professionals.

Table 1: Data extracted in the current Diabetes Pass in Belgium

Data in the paper-based Diabetes Pass	Example(s)
Personal data	Name, address...
Data concerning health care providers	Relation with patient, type of care provider
Most important health problems	Age of diagnosis, specific health problems
Drug use	Drug, start/end date, remarks
Follow-up data, e.g. treatment goals	Weight, smoking habit, physical activity, HbA1c
Information for the GP to determine the risk of a foot injury	
Pages to write down (personal) notes	
Information concerning diabetes	

3. Transition from the current Diabetes Pass to the electronic alternative

3.1. The IBBT-COPLINTHO initiative

The 2-year IBBT-COPLINTHO initiative, kicked off in January 2005, aims at introducing a platform for interactive electronic home care (eHomeCare) using innovative communication technology. It further aims at social integration and the support of people that need care in their home environment, with special attention to the support of “independent living” and the empowerment of the patient. Furthermore, there is a potential for economic return in the Flanders region (Belgium). The project takes a multidisciplinary approach to the introduction of Information and Communication Technology in HHC, in which the patient himself is the most important, central actor. After a discussion with experts in the HHC field and based on a study [13] revealing the ten prevailing disorders currently in the HHC, two target groups were selected: insulin-dependent diabetic patients and patients with multiple sclerosis. The services already available to the patients and the other actors are amongst others: voice and video communication, eAgenda and medication reminder, eScheduling, infotainment, eAlarming, eMonitoring, ePrescription, eLearning, eRecord keeping and exchange,...

In order to get feedback from the involved actors, a proof-of-concept trial is planned during Q3 and Q4 of 2006, whereby a limited set of patients and their surrounding actors will be involved.

3.2. The current situation

Till now, in Belgium no research was done concerning the use of the Diabetes Pass. However, in the light of the IBBT-COPLINTHO initiative, the possible evolution towards an electronic alternative should not be ignored.

Nowadays, the HHC nurse has to fulfil a lot of “paper work”. The “home care diary”, a small booklet, is the central tool in the daily patient information management. It

remains in the home of the patient. On a regular basis, the nurse notes down observations about the current situation and health condition of the patient. Further, it is used as a means of communication with other health care providers visiting the patient. Next to this cahier, the HHC nurse is required to record the rendered aid, which is done on a so called “working list”. This list contains personal data about all the patients visited that day. Next to these writings, there is the Diabetes Pass, a communication aid between the different care providers.
The same data is thus written at different locations. Transcription errors can be easily made and there is no guarantee that the data is up-to-date at every location whereby the patient’s health could be endangered. Furthermore, there is almost no privacy concerning the health condition of the patient. In a worst-case scenario, even the passing postman can have insight in the information; and loss of the cahier or Diabetes Pass is not unimaginable.

Analysis of the data incorporated in the different paper-based “records” showed that the data captured in the Diabetes Pass (table 1) can easily be extracted out of the electronic health record (EHR), one of the aims of the IBBT-COPLINTHO initiative.

An EHR could thus be an improvement for the actors involved in the health care process as is described in literature. Waldo [14] described the advantage of electronic capturing, storing and transmission of data whereby deciphering of the handwriting could be avoided. Cherry et al. [9] mentioned that information will be more up-to-date. Furthermore, if every authorised person has anywhere anytime access to the most recent data, they can have a more complete picture, which can also have a positive impact on the further treatment.
However, there are some drawbacks. The impact of the loss of availability is mentioned by Gritzalis et al. [15]. If patient medical data are unavailable, it may result in an unreliable service, especially for real-time emergency cases, or it may even cause suspension of the treatment if vital data signs cannot be collected. Another important issue to consider is the amount of data captured in the EHR, i.e. electronic alternative for the “Diabetes Pass”. Simmons et al. [16] comment that, if the dataset is too big, there is a greater risk for incomplete data.

3.3. The electronic alternative: possible strengths and weaknesses

In the electronic alternative, the objectives and advantages of the current paper-based Diabetes Pass will be retained and will even be improved.

3.3.1. Owner of the data

As currently outlined by the Belgian law, the patient will still remain the owner of the data stored in the EHR. At present, every health care provider should make notes and have insight in the EHR. In the electronic alternative, this will still be the case. Improvements will be that the data are always up-to-date and that the patient will not forget to take his booklet with him to every consult. Transcription errors will also be avoided as stipulated before. Because every health care provider has access to up-to-date information, the different treatments and care interventions can easily be synchronised.

3.3.2. Information-providing

In the current Diabetes Pass some rather basic and generic information and education is foreseen. Furthermore, it is possible that some "groups" are nowadays ignored.
Adolescents are of special interest. In the case of diabetes, a marked worsening of metabolic control associated with the onset and progression of complications can be detected [3;17]. As Howell et al. describe [18], the early adult transition phase often means rejecting adult control and adult authority figures. It is further important to realise that young adults who as a child faced unrealistic expectations for self-care behaviour and glucose control and had a legacy of punitive and judgmental medical encounters are especially vulnerable to "diabetes burn out".
Ethnic minorities are also a vulnerable group. A Danish study [19] suggests that limited educational background, insufficient skills regarding language, as well as a frequently marginalised social position may represent serious barriers. They are at risk for poor metabolic control.
The elderly [11] and persons with visual and/or auditive impairments could also benefit. In the electronic alternative, it should be quite easier to give "individually-adapted" information, whereby age, cultural background, educational level, etc. can be taken into account and with a possible positive influence on self-management resulting in metabolic control.

3.3.3. Reimbursement

The current pass gives the right for reimbursement on some consults with health care professionals. In the electronic alternative the Diabetes Pass will not exist anymore so the reimbursement should be provided on other grounds, i.e. the use of the EHR or more general: the eHomeCare platform.

3.3.4. Privacy, security and availability

In this whole concept, privacy and security are of utmost importance. Because of the complexity of this issue, it will not be further discussed in this paper. In the IBBT-COPLINTHO initiative however, special attention is given to this issue.
Availability is another concern. Every authorised health care provider and the patient himself, should have everywhere anytime online access to (parts of) the EHR. In Belgium there is a shift to broadband connections [20]. However, some people are still not connected to the Internet and they should not be excluded from health care. Especially in emergency cases, some information should be immediately available to everybody in one way or another. Furthermore, special attention should be drawn to the digital illiterate.

4. Conclusion and discussion

The transition from the current Diabetes Pass to an electronic alternative may represent an added value for both the patient and other actors involved. Current directives stipulated by law will still be complied with and an even more personal approach will be achieved, e.g. information and education could be adapted to the patient's current condition. Moreover, the patient will be more involved, will gain more insight, will be more independent and thus more empowered. Transcription errors will be avoided:

there will be no further problems with legibility and the data will be up-to-date. However, some drawbacks have to be mentioned. Currently, not everybody has online access nor is sufficiently motivated or educated to work with electronic records. Furthermore, in emergency cases it should be clear to everyone that the person in question is diabetic. Some health care professionals should have timely access to critical information when needed. Within the scope of the IBBT-COPLINTHO project, the patient as central actor will be more empowered. The research will focus amongst others on usability and building the bridge between the social and technological sector.

References

[1] United Nations Development Programme. Human Development Report: Cultural liberty in today's diverse world. http://hdr.undp.org/reports/global/2004/ (last accessed July 25, 2005)

[2] International Diabetes Federation. Diabetes Atlas. http://www.eatlas.idf.org/ (last accessed June 3, 2005).

[3] Hampson SE, Skinner TC, Hart J, Storey L, Gage H, Foxcroft D et al. Effects of educational and psychosocial interventions for adolescents with diabetes mellitus: a systematic review. Health Technol Assess 2001; 5(10):1-79.

[4] Fleming DM, Schellevis FG, Van C, V. The prevalence of known diabetes in eight European countries. Eur J Public Health 2004; 14(1):10-14.

[5] Capet F, Debaillie R, Van Oyen H, Tafforeau J. Diabetes - Huidige toestand in België en elementen voor gezondheidsbeleid. http://www.iph.fgov.be/epidemio/morbidat/NL/ZIE/ZIEK04r.pdf, WIV, Ministerie van de Vlaamse Gemeenschap, Admin Gezondheidszorg 1999;Episerie nr 19: 54

[6] Kaufman DR, Patel VL, Hilliman C, Morin PC, Pevzner J, Weinstock RS et al. Usability in the real world: assessing medical information technologies in patients' homes. J Biomed Inform 2003; 36(1-2):45-60.

[7] Lamminen H, Niiranen S, Kalli S. A New Media System for Supporting Asthma Self Management. The journal on Information technology in Healthcare 2003;1[4]: 269-278

[8] Buller HA. Advances in diabetes for the millennium: healthcare systems and diabetes mellitus. MedGenMed 2004; 6(3 Suppl):1.

[9] Cherry JC, Moffatt TP, Rodriguez C, Dryden K. Diabetes disease management program for an indigent population empowered by telemedicine technology. Diabetes Technol Ther 2002; 4(6):783-791.

[10] Celler BG, Lovell NH, Basilakis J. Using information technology to improve the management of chronic disease. Med J Aust 2003; 179(5):242-246.

[11] Demiris G, Finkelstein SM, Speedie SM. Considerations for the design of a Web-based clinical monitoring and educational system for elderly patients. J Am Med Inform Assoc 2001; 8(5):468-472.

[12] Montori VM, Helgemoe PK, Guyatt GH, Dean DS, Leung TW, Smith SA et al. Telecare for patients with type 1 diabetes and inadequate glycemic control: a randomized controlled trial and meta-analysis. Diabetes Care 2004; 27(5):1088-1094.

[13] Wouters, R., Paquay, L., De Vliegher, K., Debaillie, R., and Geys, L. KIT-diagnosestudie: Verpleegkundige kerninterventies in de thuiszorg volgens medische pathologieën. http://www.eyes-e-tools.com//customer/wgkfederatie/update/upload/KITdiagnosestudie.pdf (last accessed Jan. 5, 2005)

[14] Waldo B. Telehealth and the electronic medical record. Nurs Econ 2003; 21(5):245-6, 210.

[15] Gritzalis D, Lambrinoudakis C. A data protection scheme for a remote vital signs monitoring healthcare service. Med Inform Internet Med 2000; 25(3):207-224.

[16] Simmons D, Coppell K, Drury PL. The development of a national agreed minimum diabetes dataset for New Zealand. J Qual Clin Pract 2000; 20(1):44-50.

[17] Barlow JH, Ellard DR. Psycho-educational interventions for children with chronic disease, parents and siblings: an overview of the research evidence base. Child Care Health Dev 2004; 30(6):637-645.

[18] Anderson BJ, Wolpert HA. A developmental perspective on the challenges of diabetes education and care during the young adult period. Patient Educ Couns 2004; 53(3):347-352.

[19] Povlsen L, Olsen B, Ladelund S. Diabetes in children and adolescents from ethnic minorities: barriers to education, treatment and good metabolic control. J Adv Nurs 2005; 50(6):576-582.

[20] Internet Service Providers Association BELGIUM. ISPA Martkstudie - Resultaten van enquête 21 voor het tweede kwartaal van 2004. http://www.ispa.be/web/nl/?n=86&e=123&s=129 (last accessed Jan. 9, 2006).

Personalized cardiovascular risk management linking SCORE and behaviour change to web-based education

Selena Davis[a,1], Syed Sibte Raza Abidi[a], and Jafna Cox [b]
[a] Health Informatics Laboratory, Faculty of Computer Science, [b] Department of Medicine, Dalhousie University, Halifax, Nova Scotia, Canada, B3H 1W5

Abstract. The PULSE (Personalization Using Linkages of SCORE and behaviour change readiness to web-based Education) project objectives are to generate and evaluate a web-based personalized educational intervention for the management of cardiovascular risk. The program is based on a patient profile generated by combining: (a) an electronic patient data capture template (DCT); (b) the Systematic COronary Risk Evaluation (SCORE) algorithm; and (c) a Stage of Change determination model. The DCT inherently contains a set of evidence-based parameters for patient description and disease evaluation. The patient's stage of behaviour change determines messages consistent with the individual's change processes, decisional balance, and self-efficacy. The interventions are designed to address both medical and psychosocial aspects of risk management and, as such, we combine staged lifestyle modification materials and non-staged messages based on Canadian clinical guidelines to motivate personal risk management. The personalization decision logic is represented in Medical Logic Modules implemented in Java. An intelligent interactive system generates the personally relevant materials and delivers the education to the patient via the Web. An evaluation study will be conducted to determine whether web-based personalized educational strategies exert favourable influence on patient's interest, knowledge, and perceived compliance to the suggested lifestyle modifications.
Keywords: cardiovascular disease, information management, computer-tailored education, SCORE, Stages of Change, medical informatics, medical logic modules

1. Introduction

Cardiovascular diseases (CVD) place a significant burden on health professionals, patients and their care-givers, and extract significant health care costs. [1]. CVD morbidity and mortality are further contributed to by unhealthy lifestyle choices; an aspect of the condition that can potentially be addressed through risk factor modification and healthy lifestyle changes [2].

Risk factor modification is commonly addressed through patient education. This patient-directed intervention is well-evidenced and has been shown to empower patients to self-manage disease risk and improve their quality of life [3]. Despite demonstrated benefits, too few patients have disease risk reduction program opportunities or take advantage of those that are available [4]. One reason for this is

[1] Corresponding Author: Selena Davis, 6050 University Ave., Faculty of Computer Science, Dalhousie University, Halifax, NS, Canada, B3H 1W5. Selena.Davis@dal.ca; research funded by the CIHR.

that conventional health education is generally designed to include as much information as possible, often resulting in lengthy, complex and irrelevant materials for the recipient. Patients prefer to receive information that is personalized to their individual needs and situation [3] and evidence shows that personalized information is more likely to be read, remembered, experienced as personally relevant, and in turn has a greater impact in motivating patients to change to the desired behaviour [5].

Human behaviour is a key health improvement determinant. Education, when embedded in a well-designed theory-based intervention, can make a difference and is consistent with the emphasis on using evidence-based interventions in public health [6]. One model that has been successful in guiding interventions to comply with healthy behaviour changes is the Transtheoretical Model (TTM), alias Stages of Change [7].

Personalization has been an important factor in the success of patient educational programs [8]. Moreover, personalized educational interventions have been more effective than their non-tailored counterparts in changing important health behaviours [9]. Web-based patient education programs offer a viable and alternate medium for disseminating personalized education and monitoring services to patients [10].

In this paper, we present our approach and proposed system for generating personalized educational material for patients. Our approach is grounded in the observation that the efficacy of any patient educational intervention is contingent on the patient's readiness to change their behaviour. As such, we use both CVD risk assessment and behavioural change readiness tools to determine the patient's profile; based on the profile, our personalization algorithm selects relevant messages to compose the educational material for an individual patient. We use SCORE for risk assessment and Stages of Change for behaviour change readiness assessment. The educational material is derived from Pro-Change Behaviour Systems Inc. and Canadian clinical guidelines. The personalization decision logic is represented in terms of Medical Logic Modules (MLM), implemented in Java. Finally, a web-based delivery mechanism is proposed to deliver the personalized education material to the patient.

2. Background

2.1. Risk Assessment

The concept of risk assessment and reduction, introduced by the Framingham Heart Study (FHS), forms the cornerstone of preventive cardiology. A more recent predictive model of CVD events that is receiving increasing attention in Canada is SCORE [11]. This predictive equation estimates an individual's absolute 10-yr cardiovascular risk of death, and features a number of advantages over FHS including its: (i) restriction to only fatal CVD events; (ii) applicability to both coronary artery disease (MI) and stroke; (iii) ability to show changes in outcomes based on changes in risk factor values; and (iv) potential for calibration to specific populations if outcomes and epidemiological risk factor data are available for the population of interest [7].

2.2. Behaviour Change

Research and experience indicate that initiating and maintaining positive behaviour changes is challenging for most people. Prochaska's Transtheoretical Model of intentional behaviour change is a stage-based model founded on 25 years of research. The model matches the change principles and processes to each individual's current stage of change, in order to guide them through the process of modifying problem behaviours and acquiring positive behaviours [12]. The model consists of three key constructs. The first construct is the temporal dimension or Stage of Change - a characterization of a person's readiness to take and sustain action. The five Stages of Change are: precontemplation, contemplation, preparation, action, and maintenance. The second construct includes the fundamental experiential and behavioural processes of change. This dimension represents how change occurs from one stage to another. The five experiential processes include consciousness raising, dramatic relief, self-reevaluation, environmental reevaluation, self-liberation; while the five behavioural processes include helping relationships, counter-conditioning, stimulus control, social liberation, and contingency management. Each of these processes has been identified as facilitating change to the next stage when employed in messages at a stage appropriate to that process [12]. The third construct includes decisional balance and self-efficacy/temptation measures. The latter measure examines the patient's confidence to cope with a high-risk situation without relapsing to their unhealthy behaviour. The decisional balance refers to an individual's weighing of the pros and cons of changing behaviour. In general, a predictable pattern is observed between the stages of change and decisional balance suggestive of the need to place emphasis on increasing consciousness of those factors supportive of a given behaviour change.

The use of fully integrated TTM constructs to inform the design of personalized messages has been effective for intervening across a broad range of health-related behaviors [13]. More specifically, results in tobacco control studies with interventions tailored to a smoker's stage were successful more often than non-tailored interventions in promoting forward stage movement [14].

2.3. Information Personalization for Healthcare

According to DeVries et al. [15], computer-tailored education systems can conduct comprehensive assessments of health-related behaviours at an individual level. Personalizing information at this level can occur in numerous ways including: adapting the message or source of information, and/or the method of delivery. The process of how the information is personalized is based on which variables are of relevance and interest to both the provider and the consumer. For example, behavioural and clinical risk factor-specific data can be compared with clinical guidelines or with the patterns of peers. According to Jones et al., [16] the use of computer technology to personalize information requires at least five components, including: 1) a user profile, 2) a digital library containing all messages, 3) a mapping schema that generates the appropriate messages, 4) a document template for appropriate allocation and display of messages, and 5) a medium to deliver the message to the intended user.

3. Material and Methods

The following elements relate to the development of the PULSE program. The development follows the framework laid out by Jones et al. [16], with an additional component added – an evidence-based data capture template.

3.1. Patient Data Collection and Profile Generation

We use the validated, commercially available Wellsource Coronary Risk Profile as the basis for our data capture model for collecting patients' demographic, behavioural, and clinical risk factor characteristics. Cross-referencing collection parameters with the global INTERHEART Study [17] indicates that the nine risk factors accountable for over 90% of the risk of acute MI were captured in our DCT. We argue that our DCT inherently comprises a set of evidence-based parameters for patient description and disease evaluation. These parameters are used to design our objective patient profile.

The patient's profile is determined in three components: (1) CVD Risk Profile determined through the SCORE algorithm that estimates the 10-yr total cardiovascular risk of death. Patient data on age, gender, smoking, systolic BP, and total cholesterol and HDL cholesterol ratio is used to calculate the patient's risk category as percentages that are translated as follows: ≤1% (low), 1 - 5% (moderate), and >5% (high). If diabetes is indicated as a patient's personal health problem, the SCORE is adjusted as indicated by Conroy et al. [11]. The patient's risk category (e.g. high) directs the selection of risk-matched target values for all relevant risk factors; (2) Staged Risk Factor Profile depicts the patient's Stage of Change for specific modifiable risk factor behaviours. The Staged Risk Factor Profile is determined by a patient's response to questions relating to her/his readiness to change modifiable risk factor behaviours - smoking, being overweight, stress, depression, and exercise. The patient's responses infer the Stage of Change using a simple Stage Determination Model. A patient's readiness to change is placed into one of the five Stage categories; and (3) Non-staged Risk Factor Profile determined by additional risk factor values - eating practices, alcohol, LDL, Triglycerides, diastolic BP, FPG, and personal and family health history.

3.2. Message Library

In the PULSE program, we use a combination of staged lifestyle modification materials and risk-specific messages based on clinical guidelines to provide a valid use of behaviour change theory and Canadian sources of clinical and lifestyle modification education. The Staged risk management materials are commercially produced by Pro-Change Behavior Systems, Inc. The risk-matched target values for all risk factors are based on the following Canadian guidelines: cardiovascular rehabilitation and CVD prevention, diabetes, dyslipidemia, and hypertension. Also, pre-written, non-staged risk management materials were included from three sources: Heart & Stroke Foundation, Nova Scotia Cardiac Rehabilitation Centre, and Public Health Agency of Canada.

The various sourced materials are broken down into small "snippets of information", <tagged>, and stored in an SQL database. The XML <tag> for each snippet follows an indexing schematic which provides mapping ease to the patient profile for personalization purposes (e.g. <smoking>).

3.3. Decision Logic

Given a patient profile and a message library containing an assortment of education interventions, the personalization mechanism involves the selection of the most relevant set of messages based on the patient's profile. Personalization is achieved through the processing of a set of symbolic rules based on decision logic—the decision logic maps the profile elements to specific messages. We develop a rule-based inferencing engine that incorporates the decision logic. To represent our medical knowledge we use MLMs, a standard for independent units composing a series of rules in health knowledge bases. The entire decision logic, sets of MLMs, is implemented in Java and represented as a comprehensive decision tree describing each of the risk factors and risk conditions contained in the patient profile. The logic contains "if-then" rules, where the IF part of the rules contains variables for one or more patient profile elements. If the IF part of the rule is satisfied—i.e. the patient's profile matches the rule constraints then the rule fires and the THEN part of the rule becomes available for execution. Typically, the THEN part contains a list of messages that are selected as part of the patient's personalized educational material.

3.4. Display Template

An overall document structure was created to organize and present the chosen messages for each patient in a coherent manner. The *Introductory Section* provides a brief description about the personalized intervention. The *CVD Risk Profile* offers a graphical display of this patient's risk. The *Progress Page* provides a graphical display of changes in a patient's risk over time. The *Risk Factor Management* section provides information on each risk factor relevant to the patient. Each risk factor has its own section complete with an introductory brief, patient's current results, evidence-based target values, lifestyle modifications and risk management education.

3.5. Implementation & Delivery

The XML based implementation system is used to personalize the education. Input data is captured on an XSLT stylesheet and submitted by a healthcare professional in consultation with the patient at a primary care facility. The submitted stylesheet passes the patient data to the rules engine. The decision logic is implemented using Java as it facilitates the subsequent rule-based inferencing. The MLM processes the data on logic statements to determine the patient profile. Once all data is processed, the rules engine uses this information to select appropriate messages. The information is inserted into a display template and rendered in HTML, delivering the education document via a web browser. Patient's have access to their document by logging onto the website. A print copy is available for patients without access to a computer.

4. Evaluation Strategy

A pilot user study to evaluate system impact and gain insight for the outcomes evaluation design will be conducted. This user study will take place in a primary healthcare site in Halifax, NS, Canada. The subjects will include approximate forty patients between the ages 40-65 with cardiac risk factors but who do not have manifest CVD. The methodology will include a patient survey examining reaction to, learning from, usefulness of, and perceived adherence to the risk management education.

5. Concluding Remarks

In Canada, there is a realization that evidence-based standards for care are vital to effective health services. In the PULSE project we have proposed a novel computer-tailored patient education strategy that features: (a) Usage of SCORE for risk assessment; (b) Incorporation of behaviour change inputs in determining the educational content as opposed to just relying on medical data; (c) Usage of an objective patient DCT currently operational; (d) Leveraging Canadian clinical guidelines for both deriving the decision logic and the corresponding educational intervention; and (e) Personalization of educational material. The realization for personalized education information compared to generic information has led to various computer-tailored programs. We believe that quality and timely patient-directed interventions are a healthcare service that can reduce disease risks and deal with risk management by influencing changes in patients' behaviours through the provision of up-to-date and pertinent lifestyle modifications and change strategies.

References

[1] Heart and Stroke Foundation of Canada, "The Growing Burden of Heart Disease and Stroke in Canada," Ottawa, Canada, May 2003.
[2] D.G. Manuel, M. Leung, K. Nguyen, P. Tanuseputro, and H. Johansen, "Burden of cardiovascular disease in Canada," *Canadian Journal of Cardiology,* vol. 19, pp. 997-1004, 2003.
[3] T. Hoffmann, T. Russell, and K. McKenna, "Producing computer-generated tailored written information for stroke patients and their carers: system development and preliminary evaluation," *International Journal of Medical Informatics,* vol. 73, pp. 751-758, 2004.
[4] B.H. Southard, D.R. Southard and J. Nuckolls, "Clinical trial of an Internet-based case management system for secondary prevention of heart disease," *Journal of Cardiopulmonary Rehabilitation,* vol. 23, pp. 341-348, 2003.
[5] J. Brug, M. Campbell, and P. van Assema, "The application and impact of computer-generated personalized nutrition education: A review of the literature," *Patient Education and Counseling,* vol. 36, pp. 145-156, 1999.
[6] E. M. Murphy, "Promoting Healthy Behavior," Population Reference Bureau., Washington, DC, Tech. Rep. Health Bulletin 2, 2005.
[7] Canadian Association of Cardiac Rehabilitation, "Canadian Guidelines for Cardiovascular Rehabilitation and Cardiovascular Disease Prevention," Tech. Rep. 2nd edition, 2004.
[8] P. Doupi, and J. Van der Lei, "Towards personalized Internet health information: the STEPPS architecture," *Medical Informatics,* vol. 27, pp. 139-151, 2002.
[9] A. Oenema, J. Brug, and L. Lechner, " Web-based tailored nutrition education: results of a randomized controlled trial," *Health Education Research,* vol. 16, pp. 647-660, 2001.
[10] H.Q. Nguyen, and V.C. Kohlman, "Internet-based patient education and support interventions: A review of evaluation studies and directions for future research," *Computers in Biology and Medicine,* vol. 34, pp. 95-112, 2004.
[11] R.M. Conroy et al., "Estimation of ten-year risk of fatal cardiovascular disease in Europe: the SCORE project," *European Heart Journal,* vol. 24, pp. 987-1003, 2003.
[12] Cancer Prevention Research Center, "Detailed Overview of the Transtheoretical Model," Dec 15, 2004.
[13] J.A. Sarkin, S.S. Johnson, J.O. Prochaska, and J.M. Prochaska, "Applying the Transtheoretical Model to Regular Moderate Exercise in an Overweight Population: Validation of a Stages of Change Measure," *Preventive Medicine,* vol. 33, pp. 462-469, 2001.
[14] L. Spencer et al., "Applying the transtheoretical model to tobacco cessation and prevention: a review of literature," *American Journal of Health Promotion,* vol. 17, pp. 7-71, Sept-Oct. 2002.
[15] H. DeVries, and J. Brug, "Computer-tailored interventions motivating people to adopt health promoting behaviours: introduction to a new approach," *Patient Education and Counseling,* vol.36, pp.99-105, 1999.
[16] B. Jones, S.S.R. Abidi, and W. Ying, "Using Computerized Clinical Practice Guidelines to Generate Tailored Patient Education Materials," 38th Hawaii IEEE Int'l Conf on System Sciences, 2005, pp. 139b.
[17] S. Yusuf et al., "Effect of potentially modifiable risk factors associated with myocardial infarction in 52 countries (the INTERHEART study): case-control study," *The Lancet,* vol. 364, pp. 937-952, 2004.

Ubiquity: Technologies for Better Health in Aging Societies
A. Hasman et al. (Eds.)
IOS Press, 2006

Guarantying the quality of chemotherapy: from the order to the administration

S. Spahni[a,b], C. Lovis[a], M. Ackermann[c], N. Mach[d], P. Bonnabry[c], A. Geissbuhler[a]

[a] *SIM, Radiology and Medical Informatics Dpt, University Hospitals of Geneva, Geneva, Switzerland*
[b] *Nice Computing, Lausanne, Switzerland*
[c] *Pharmacy, University Hospitals of Geneva, Geneva, Switzerland*
[d] *Oncology Unit, Dpt. of Internal Medicine, University Hospitals of Geneva, Geneva, Switzerland*

Abstract. Ensuring the quality and security when prescribing drugs like chemotherapies is a complex task if one wants to cover the whole chain from the prescribing physician to the administrant nurse. At the University hospitals of Geneva, new applications covering the whole chain from the prescription up to and including the fabrication of the products have been developed in three phases and are being used in a production stage. In order to cover the "last yard" at the bed level, a fourth phase has been started with a pilot study based on labels containing RFID chips for preparations and for patients. The last phase will make use of all traceability data acquired from the prescription to the preparation to validate that the right product is administered to the right patient, and to record who is administrating it.
Keywords: Hospital Information Systems (HIS), EAN.UCC, RFID, Pharmacy, Chemotherapies

1. Introduction

Since more than three years, the production of chemotherapies for the University Hospitals of Geneva has been centralized at the pharmacy in order to guarantee the quality of the preparations. This initial step, which was already not easy to implement, was a prerequisite for the next ones: the use of computer aided systems to prescribe, produce and administer chemotherapies and the setup of a global traceability.

While several problems are addressed by this project, there is one driving force behind the whole project: quality. Indeed, this can be declined in several aspects:

- Quality of the prescription by introducing as much controls as possible already at the prescription stage and by automatically computing the doses according to in particular the demographic parameters of the patient;
- Quality of the transmission of the prescription to the pharmacy – faxes and handwritten prescriptions are not always easy to read;
- Quality of the fabrication by using standardized methods and by introducing plausibility tests during the preparation itself;

- Quality of the data acquired during the fabrication in order to be sure that a full traceability can be performed afterwards;
- Quality of the administration to the patient by validating that it is the right medication at the right time and that the medication has not to be dismissed due to a late problem detection;
- Quality of the documentation in the electronic patient record by automatically recording what was prescribed, when, by whom and when it was administered.

Taking into account all these aspects led to the design and development of a global solution for the management of chemotherapies – from the prescription to the administration. The tools that have been designed as well as the necessary data are described in the following chapters. Benefits – already visible and expected in the future are discussed in chapter 4, and future directions are briefly presented in the conclusion.

2. Methods

The problem of ensuring the quality of the process has been addressed by providing ad-hoc solutions for each category of participants: physicians, pharmacists, assistants at the pharmacy and nurses. The project has therefore been divided into four phases, each covering one specific aspect of the problem to be solved.

In the first phase, new applications enabling the prescription of single medications and producing a protocol for its fabrication by pharmacists and assistants has been developed. The main challenge of this initial phase was not the development itself but the clear understanding of the way the pharmacy works and what kind of tools were necessary to support the workflow and the quality assessment. A major task was the creation of a global database containing all substances and materials that could be used for preparations, as well as the way they can be used and the types of controls that have to be performed in order to guarantee that the preparations can be administered safely. This database is then be used by the applications developed for requesting new preparations, managing the pending requests, organizing the concrete preparations and managing the traceability before, during and after the fabrication.

Once this strong base had been established at the pharmacy, the problem of the prescription by the physicians has been addressed. The work started with the department of oncology, as they were highly interested in having an application supporting their prescription through standardized protocols and were ready to invest time in supporting the medical aspects of the project. Like for the description of the products used by the pharmacy, a lot of efforts was required to create a database of protocols that can be prescribed - including the constraints that have to be enforced when a specific protocol is instantiated for being administered to a patient. This led to some standardization of existing protocols and updating their associated documentation including required analysis, pre-medication prescription, steroids, hydration as well as chemotherapy administration information and potential side effects. This standardisation work was indeed performed in close collaboration with the pharmacy, to ensure that the desired medication can be prepared and are using products that are available.

In parallel with this work, a third phase consisting of integrating a tool aiming at reducing the risk of errors during the fabrication (cato® [1]) has been started at the pharmacy. This tool automatically gets the description of the preparations to be realized and drives the assistant during the steps of realization by validating each step and by controlling the plausibility of the result through a high precision balance connected to the tool. The final results as well as the samples truly being used are then automatically sent back to the traceability application for being stored in the global database of the pharmacy. A management of the remnants is also performed by this tool.

While these three phases, in production stage since almost one year for the first two ones, are well covering the steps from the prescription by the physician up to the checking that a specific preparation has been well produced, they do not cover the "last yard" which is the administration of the substances. The final step for the implementation of the global traceability is therefore the support of the administration by the nurses.

For this last phase, which is currently still under development, a solution based on Pocket PC is being realized. The goals of this module are the following:

- Verification that it is the right preparation for the right patient;
- Automatic control that the preparation is still administrable (used-by date);
- Verification up to the last minute that there is not a "no-go" condition that has been issued by the pharmacy following the detection of a problem during the fabrication or linked to a problem with one substance used for the preparation;
- Recording of when a specific preparation has been administered and by whom.

These functions of course require the correct identification of the three "partners": the patient, the nurse and the preparation.

While there are many ways to identify the partners, proprietary or non proprietary, we decided to start using international numbering of objects and started using GS1's EAN.UCC (formerly called EAN – European Article Numbers) coding schema [2]. All three partners have therefore their own code that can be read by the Pocket PC and validated on-line according to the information stored in the Hospital Information System (HIS): the patient and the nurse have their own permanent EAN.UCC codes and each preparation produced by the pharmacy receives its own identification (EAN.UCC) code enabling its tracking over the institution. All EAN codes are using the EAN-128 encoding scheme.

3. Results

The results of the project cover various aspects, of which the major ones are:

- Results in terms of documentation: protocols, products being used, methods for producing preparations, etc.;
- Results in terms of software modules: applications for the physicians, the pharmacists, the assistants, the nurses;
- Results in terms of organizational achievements.

It has to be noted that results presented below cover mainly phases 1 and 2, which are in production since autumn 2005. Phases 3 and 4 have only recently reached the production stage and we are too close at the time of writing of this paper to have a sound and objective view over the results of these two phases.

3.1. Documentation

It became rapidly clear that a strong separation between the software modules and the information related to the experience has to be achieved. In particular, as the way to prepare some specific substance or the precautions necessary to use some material may change, the software modules were reduced to some kind of engine being able to produce application forms and documents. The content of fabrication protocol is therefore completely controlled by the content of the database.

This of course made the description of the materials used by the software modules a critical task, but had the advantage of clearly separating the medical and pharmaceutical knowledge from the software engineering knowledge. This resulted in the creation of two linked databases, one describing the pharmacy's cupboard and one describing the protocols that can be instantiated by the clinicians. These cupboards can thus be maintained up to date directly by the people directly in charge of the information.

In parallel with the creation of these databases, a quite important work has been done for creating or updating the descriptive parts of the chemotherapy protocols (for the clinicians as well as for the nurses) in order to be able to include these parts in the documents produced by the applications.

3.2. Software modules

The HIS of the University Hospital of Geneva is a fully distributed system [3], and the software modules described here do comply with this strategy. Several modules have therefore been developed, of which we can mention:

- a module for the prescription, either through protocols (for the clinicians) or directly (for specific preparations asked by pharmacists);
- a module for generating the fabrication protocols for the assistants;
- a module for managing the traceability;
- a module for the nurses (Pocket PC).

The integration of a computer-readable identification of the preparations was a challenge: in order to promote the global traceability, we are strongly in favour of using normalized barcodes for every object – persons, vials, material, etc. We are therefore committed to use the EAN.UCC international encoding schema. The problem with the barcodes comes from the size of the printed code: 2-D formats are rather large, and data matrix is not commonly read by classical readers. As the labels stuck on the preparations are full of human readable information, we were not able to add a standard barcode on it for space reasons. We decided therefore to use labels with integrated RFID chips, for identifying the preparations as well as the patients (using a bracelet with a RFID sticker) and the nurses. The labels for preparations are printed at the pharmacy before the product is produced and joined to the raw material required for the fabrication, while the labels for patients should be produced at the admission desks (they are currently produced at the pharmacy until the trial is complete). This homogeneous solution enables therefore the use at the bedside of Pocket PC with only one RFID reader for safely getting the identities of the patient, the nurse and the preparation to be administered.

In addition to the use of RFID chips for the storage of the unique ids, wireless connected Pocket PCs are being used. Thanks to this type of connection with the HIS, nurses avoid then burden of downloading and uploading data concerning the medications they have to administer and can do the verification just at the time of the administration.

3.3. Organizational changes

The introduction of these new tools for structuring the workflow linked to the dispensing of chemotherapies has of course an impact on the organisation of several entities. Although the pharmacy already had some tools to handle the traceability of their productions, the new applications introduce additional functionality and constraints and the assistant have to follow more strictly the fabrication steps.

On the clinician side, the application does no more allow to go over certain doses. The prescription of "ad-hoc" protocols is also no more possible - or you have to do it fully manually which is clearly different from just modifying a sheet of paper!

Finally this infrastructure introduces an additional task for the nurses, validating the preparation at the dispense time through a specific application, but thanks to an effective design of the application this task should not be considered as too time consuming in relation with the benefits in terms of security that it adds.

4. Discussion

The new applications have been quite well accepted in the pharmacy as well as in the oncology department. The importance of the quality of the information is well recognized, and the tools do offer a significant improvement.

Side effects also occurred on the clinician side. The formalization and validation process requiring a significant amount of time, as every quality assessment process, there will be a homogenisation of the protocols being used in all departments prescribing chemotherapies and many variations existing mainly for historical reasons will disappear.

The introduction of the applications in the pharmacy was the start of a global process for managing all fabrications performed by the pharmacy. Several goals are progressively being achieved like sharing a single source of information about all available substances, being able to have better traceability tools, to make statistics, etc. All these aspects are contributing to the overall quality management strategy and contribute to enforce the security of the medications to patients. The fact that the pharmacy has the possibility to block the administration of a preparation even after it has been sent to the wards is also a advantage.

The validation of the administration of the drugs to the patients should also improve the global process, by drastically reducing the risk of error in the administration of a medication that should not been given – for any reason.

Finally, the fact that all these steps in the care of patients are progressively electronically documented and made available through the electronic patient record contributes to the global security that patients are waiting for.

5. Conclusion

The initial project for the support of the prescription and fabrication of chemotherapies has been successfully deployed and is well accepted by its users. The second phase, consisting at validating the administration of the medications to the patients is in its early phase and currently lacks of feed back. However with expect the have a significant improvement to the overall security of the care of the patients thanks to several achievements: better documentation in the patient record, suppression of the hand-written orders, controls at the prescription, fabrication and administration levels, etc.

Having all information in the electronic patient record also contributes to the progressive implementation of a global prescription pad for clinicians as well as nurses in order to help them in their process of taking decisions and carrying out actions **Fout! Verwijzingsbron niet gevonden. Fout! Verwijzingsbron niet gevonden.** [4].

The use of Pocket PCs by the nurses is expected to grow as new applications are being progressively installed. One can mention the selection of meals by patients, which is already deployed in some wards and will be generalized over the whole hospital.

References

[1] Computer Aided Therapy for Oncology (cato). http://www.cato.at

[2] General EAN.UCC Specifications, January 2005, Version 6.0. © 1999-2005 EAN International and Uniform Council, Inc.

[3] Scherrer JR, Lovis C, Baud R, Borst F, Spahni S. Integrated computerized patient records: the DIOGENE 2 distributed architecture paradigm with special emphasis on its middleware design. Stud Health Technol Inform. 1998; 56:15-31.

[4] [Tamblyn R, Huang A, Kawasumi Y, Bartlett G, Grad R, Jacques A, et al. The development and evaluation of an integrated electronic prescribing and drug management system for primary care. J Am Med Inform Assoc 2006;13(2):148-59.

[5] ITAC. Report to the President, Revolutionizing Health Care through Information technology. President's Information Technology advisory Committee June 2004.

[6] Bates DW, Cohen M, Leape LL, Overhage JM, Shabot MM, Sheridan T. Reducing the frequency of errors in medicine using information technology. J Am Med Inform Assoc 2001;8(4):299-308.

Address for correspondence

Stéphane Spahni
Medical Informatics Service, Radiology and Medical Informatics Department
University Hospitals of Geneva
24 rue Micheli-du-Crest
CH-1211 Geneva 14
Stephane.Spahni@hcuge.ch

3. Health Information Systems

3.1 Architectures of HIS

Ubiquity: Technologies for Better Health in Aging Societies
A. Hasman et al. (Eds.)
IOS Press, 2006

Comprehensive management of the access to a component-based healthcare information system

Christian Lovis, Stéphane Spahni,
Nicolas Cassoni-Schoellhammer and Antoine Geissbuhler
Service of Medical Informatics
Geneva University Hospitals, Switzerland

Abstract. Objective: to describe the key concepts and elements used to implement a comprehensive access management system to a distributed, component-based healthcare information system. Methods: the a priori access is based on an institution-wide policy for access rights coupled to proximity process for the granting of such rights. Access rights are explicit and externalized from the information systems components. A posteriori control is based on a centralized, exhaustive journal of accesses to all components coupled to a decentralized verification process for suspicious accesses. Results: the system has been operational for three years, initially used for the access to the computerized patient record components, and now extending to all the components of the hospital information system. The same architecture will be used for the development of the trans-institutional community health information network.
Keywords: Healthcare information systems, computerized patient record, access management

1. Introduction

Privacy protection, access rights management, and security have been discussed for many years in the field of medical informatics [1]. Meanwhile, computerized patient records (CPR) have evolved towards collaborative tools for care providers and patients in an ever-increasing networked environment, highlighting the need for complex rights managements. Providing a comprehensive, fast and simple access management system that is compatible with the daily operational work while respecting the privacy of patients is a challenging goal [2, 3], especially in a future-proof distributed CPR [4]. Moreover, the intertwining of administrative and clinical information systems within hospitals, and the creation of trans-institutional healthcare information networks require new approaches to the management of accesses to personal information.

The Geneva University Hospitals (HUG) is a 2'200-bed consortium of hospitals in four campuses and more than 30 ambulatory facilities in the state, covering the whole range of in- and outpatient care, from primary to tertiary facilities. Its hospital information system has evolved towards a component-based architecture [5] relying on key, institution-wide, transversal components, but enabling numerous, specific business-logic components and applications to respond to the needs of the users while maintaining a global coherence. The access management component is one of these transversal components. This component network architecture, based on the Web-services technology, is particularly-well suited in rapidly evolving environments, at the technological level, functional level, process level, and at the organizational level, in particular with the evolution towards trans-institutional networks.

2. Elements of the access management component

Staying close to the action while maintaining an institutional coherence is a challenge faced by many large enterprises. In order to be efficient, a priori access control must be based on an accurate picture of the role of the user at the time of access. But users move quickly within the institution, and regularly change their role. A resident in charge of patients in a ward can also be participating in a specialty consultation, or be on-call in the emergency department, situations that require different types of accesses to the hospital information system. Granting of access rights is therefore best decentralized, close to where the accesses are going to take place.

2.1. Therapeutic relationship

Meanwhile, a central authority must be in charge of defining the access rules. In the HUG, a multi-disciplinary committee sets these rules, balancing the need-to-know for providing a service with the protection of the privacy of the various stakeholders. For instance, accesses to the computerized patient record require the existence of a therapeutic relationship between the care provider and the patient.

Assessing the *a priori* existence of such therapeutic relationship is a challenge. Many elements do participate to this: there is a therapeutic relationship when a nurse or a physician is in charge of a patient, but this might be difficult to assess, when the patient calls the care provider by phone, or when providers are on call. This becomes truly difficult when a secretary has to prepare some paperwork for a care provider. Currently, the HUG accept a "proxy" to assess a very probable therapeutic relation: the care provider and the patient must, at the same time, be in the same domain, either a zone of medical responsibility (ZAM) or a physical location (ZAS). For example, the physician has an active profile in the service of Cardiology while the patient has been admitted in that service; or a nurse has an active profile in a given ward, while the patient is currently in this ward. In order to allow immediate pre- and post-encounter care, this constrained has been relaxed to allow access within 30 days before or after encounter or hospital stay.

3. Authentication

The user must log in the system and an authentication process occurs mediated through the use of a personal smartcard coupled with a personal identification number. The authentication process is validated by a LDAP server. If the contract of the collaborator with the HUG is terminated, the authentication is not possible, and all accesses are blocked.

3.1. Profiles

An access profile is defined for each possible role of a user of the hospital information system. Examples of clinical profiles include: physician chief of a medical service, physician in charge of a patient, consulting physician, charge nurse, nurse, medical student, physical therapist, and nutritionists. Accesses are only given through the intermediate of profiles. Therefore, all people having the same role in the HUG will receive the same profile. Currently, 121 access profiles have been defined, and cover all possible combination of roles and accesses in the HUG, for care and administrative or scientific purposes.

3.2. Atomic rights

A profile is made of atomic rights. These rights will be used for all components and applications used, and are, intrinsically, independent from the applications' logic. For example, the profile of "nurse" will include atomic rights to access the patient record, but also to use administrative tools for the management of the ward, to order ward material, etc. An example of atomic right is "can order radiology exam in is/her service", or "can sign a medical document in is/her service". These atomic rights are independent from the applications and describe information processing functions that can be shared by several components. Applications are prevented from embedding business logic regarding these rights. Most atomic rights will be conditional to several elements. The most common, is the medical service or the ward in which the user is working. There are currently 425 unique atomic rights.

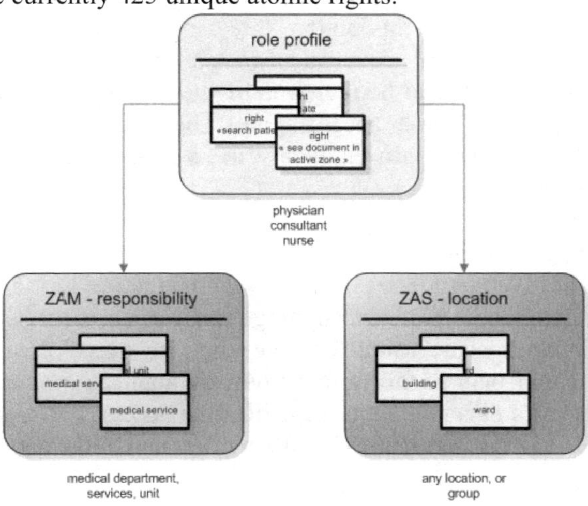

Figure 1: Medical responsibility or care location

3.3. Validity domains for profiles

Most atomic right are conditional to domains of validity. The most important and used are the medical service (ZAM= Medical Activity Zone) and the care location, such as a ward (ZAS= Care Activity Zone) (Figure 1). So, for example, the care provider working as a nurse will receive a "nurse" profile in her ward. This means that here atomic rights can only be used in her ward or for patients actually in her ward (or ambulatory consultation, etc.). The same applies to medical services. So, a physician working as a cardiologist will receive a profile of "cardiologist" for patient that is in the service of Cardiology, regardless the ward. There are other validity domains, such as a clinical study. The researcher will receive a "researcher" profile and will only be authorized on the patient list of his/her study.

3.4. Temporality

The profile is granted for a given period of time that can not exceed 12 months or the length of the contract. After 12 months, accesses are automatically removed if they are not renewed.

A user can have several active profiles at the same time. So, for instance, one can be a physician in charge of patient care, a physician chief of medical service, and a clinical researcher at the same time. However, when accessing the system, the user will have to choose his current role. There are some important constraints to this approach, and two major had to be addressed:

3.5. Escape mechanism: "breaking the glass"

Even with decentralized access granting, there are situation where the appropriateness of the access cannot be established. For example, the shared domain constraint does not allow to access to a patient that would be in another service, which could cause harm in case of emergency. Therefore, some profiles, such as medical profiles, can escape the domain constraint. We name this action "breaking the glass". In such a case, the physician is asked to justify the access and it will be granted. In order to prevent misuse, all accesses of this type have to be reviewed by the physician in charge of the service that hosted the patient. These accesses must be qualified as "appropriate", "to investigate" or "inappropriate". In the latter case, a institutionally-defined disciplinary procedure is activated.

3.6. A posteriori access control

In addition to strict access policy restricting who can access what and a mandatory review of all accesses made using the "breaking the glass" escape mechanism, all accesses to any component of the system from any application is tracked centrally. In the case of accesses to data related to the CPR, such accesses are made visible, in real-time, in the CPR, which is available to care providers and to the patient itself.

4. Results

In activity since 1999, this component handles all accesses to the clinical information system and accesses to several administrative applications. 29'000 different users have been identified, of which about 6'000 are currently active. An average of 300'000 daily authorization transactions are recorded, and about 1'000'000 events are logged in the central journal, of which xxx represent access logs.

Access profiles are granted by 80+ administrative assistants.

"Broken-glass" accesses represent about 6% of the more than 20'000 daily accesses to the CPR. This proportion is decreasing with the progressive improvement of the access granting process. These accesses are reviewed by each medical service in which the patient was taken care of at the time of access.

5. Evolving towards a trans-institutional network

The HUG's CPR can be considered as an institutional network. Moving towards a trans-institutional network, as planned and architected in the "e-toile" community health information network [6], will require several adjustments to the current model.

Firstly, the therapeutic relationship will be materialized by the physical co-existence of the professional's card and the patient's card. The access profile of the professional will describe the default behaviour of the system, but the patient will be able to refine the access rights, by choosing which information should be restricted to specific care professionals. This patient-centred approach implies that access control must be able to be applied at the document level.

Secondly, the political design of the "e-toile" project mandates a highly decentralized architecture, responding to social worries about the risk of centralized databases. Not only should the elements of the patient record remain at their source, but also the definition of the access rights and the log files documenting the accesses should not be centralized. It is nevertheless accepted that a central authority grants the access profiles to the care professionals, a mechanism already use to grant practice privileges. In this regard, the situation is similar to the one described above in the HUG.

The potential complexity of dealing with varied policies for granting and controlling accesses in a multi-institutional environment is mitigated by an ad hoc law establishing the ground rules for all stakeholders. In addition, the notion of a "physician of trust", who will serve as an info-mediator for the patient, will insure that decisions about "hiding" parts of the record to some care professionals will be taken with the full understanding of the benefits and drawbacks.

Based on the analysis of the proposed system and on pilot projects, it appears that the extension of the access management system of the HUG to the multi-institutional environment of the community health information network "e-toile" is possible, provided that the technicality of the highly distributed architecture is handled.

6. Conclusion

The HUG have developed a framework covering organisational issues, regulatory constraints and technical solutions to ensure a proper level of privacy protection while ensuring a manageable operational load for handling access rights to their hospital information system. The system, operational for 5 years, has been made mandatory for all in-house and third party application, allowing a unique concept of protection to be implemented. Managing accesses in health is a complex challenge: it implies both the control of who can see and do what with identified data, but also to provide a detailed track and trace mechanism. The most important challenges have been to develop a system that does not restrict legitimate operational activities for care providers. This has been resolved by having a centralized management system with a decentralized granting of rights and control of accesses. The difficult balance between too less and too much has been partially solved with a rather restrictive policy of access granting associated with highly controlled escape mechanisms. This model is ready to evolve towards dynamic granting of rights by the patient itself with the soon to be implemented community health information network based on a patient identification smartcard.

References

1. France FH. Security of the electronic health record. Stud Health Technol Inform 1997;43(Pt A):167-70.
2. Scherrer JR. The problems related to confidentiality and effectiveness of health care. Eff Health Care 1983;1(2):85-92.
3. Privacy, security worries could thwart health IT efforts. Med Health 2005;59(42):1, 4.
4. Blobel BG. Advanced EHR architectures--promises or reality. Methods Inf Med 2006;45(1):95-101.
5. Geissbuhler A, Lovis C, Lamb A, Sphani S. Experience with an XML/HTTP-based federative approach to develop a hospital-wide clinical information system. Medinfo 2001;10(Pt 1):735-9.
6. Geissbuhler A, Spahni S, Assimacopoulos A, Raetzo MA, Gobet G. Design of a patient-centered, multi-institutional healthcare information network using peer-to-peer communication in a highly distributed architecture. Medinfo 2004;11(Pt 2):1048-52.

Ubiquity: Technologies for Better Health in Aging Societies
A. Hasman et al. (Eds.)
IOS Press, 2006

Barriers integrating dedicated software for quality management in pain centres

A.Johannes TEN HOOPEN[a1], Pieter E. ZANSTRA[a], Egbert J. VAN DER HARING[a], Wouter S. TUIL[a] and Marten VAN WIJHE[b]
[a] *Medical Informatics, Radboud University Nijmegen Medical Centre, The Netherlands*
[b] *Pain Knowledge Centre, University Medical Centre Groningen, The Netherlands*

Abstract. Objectives: Explore the feasibility of integrating a dedicated pain centre information system as part of a quality management network with a number of different Hospital Information Systems. Material & methods: A systematic approach integrating and implementing the system in 15 selected hospital organisations (a nationwide 15% non-random sample). Results: Hospitals have widely varying policies on integration and implementation of additional clinically required 3rd party software. Financial and organisational constraints are considerable. Partial data integration could be realised in one third of the hospital organisations within the project timeframe. Linking with various types of Hospital Information Systems from the same or different vendors caused no technical difficulties. The total effort required, however, varies considerably; different versions of a HIS of the same vendor require substantial additional effort. IT departments hardly use standard rules to accept this type of systems, causing substantial increases in completion time for installation. Conclusions: Although it is feasible to integrate a local departmental system as required (download of general patient, provider and referrer data basically, and also upload of certain data) the workload of scaling to the national level is considered far beyond what is reasonable for a national quality management network for pain. Alternatives for recording and capturing data (which comply with the requirements of the national system) are currently being explored.
Keywords: HIS, Pain Clinics, Quality Control, Systems Integration.

1. Introduction

Chronic non-malignant pain is a major health problem. The lifetime prevalence of for example low back pain is 60-70%. Although most cases of low back pain recover within days or weeks, it is the up to 10% of pain problems developing into a chronic state, i.e. lasting more than 6 months, that requires substantial health care means and generates even more costs in an indirect way [1]. Estimates of at least 1 up to 2 % of GNP as cost-of-illness for low back & neck pain are considered realistic, while the cost-of-illness of low back pain alone seems to exceed that of coronary heart diseases [2]. The aging of the population is adding new dimensions to this health problem as it appears to result in higher incidence and greater susceptibility to adverse effects of analgesics [3]. The overall impact of chronic pain is however still difficult to estimate because, besides difficulties in estimating incidence and prevalence data, detailed data

[1] Corresponding Author: AJ ten Hoopen, 152 Medical Informatics, Radboud University Nijmegen Medical Centre, POBox 9101 NL-6500 HB Nijmegen The Netherlands. E-mail: AJ.tenHoopen@mi.umcn.nl

from health care organisations on chronic problems presented, procedures performed and outcome achieved are largely unavailable or have to be derived indirectly [4].

Major investment in pain treatment has occurred in the last three decades. To further the quality of pain treatment Pain Knowledge Centres (PKC) were founded in 1993 at 4 university hospitals in the Netherlands. These investments require convincing returns. Therefore PKC's started developing case documentation that allows basic quality assessment. For each chronic pain patient in every pain treatment facility, the problems, specific treatment in a given period aimed at a specific problem and outcome of that treatment should be recorded in an agreed uniform way. Based on some local experiments where automated support for case documentation was developed (example in [5]) a software tool for uniform recording in strictly controlled terminology of a minimal set of patient data was developed [6]. This is considered the only way out of the variety of existing (paper and free text) recording practices that is so characteristic of medicine in general and of pain treatment in particular, but hampering aggregation and analysis of data in quality assessment of care [7].

This paper describes and evaluates the experience implementing and integrating this software tool in 15 different hospitals (18 outpatient clinics involved). It's a pilot for a Dutch Pain Quality Management (PQM) network, and a study in scalability.

2. Material & methods

2.1. The software package

The decision to build a dedicated piece of software has been taken for several reasons. A dedicated, possibly standalone, software package that is of use in smaller and larger pain treatment facilities was considered to be of strategic importance. More or less in parallel intensivists in the ICU's discussed and analysed extensively user requirements for automated support, and evaluated implementation possibilities within the (already heavily) automated environment of the ICU and HIS [8]. Although the existing ISs for ICU's could become valuable tools in improving the quality of ICU practice, it was concluded that full implementation of specifications had still a long way to go. As a consequence additional software had to be developed for a national system to assess the quality of work in ICU's [9]. For a PQM network it is essential that not only pain diagnoses and procedures performed be recorded (which is in some global way done in all hospitals) but also *the relationship* between these concepts and with the treatment outcome. This kind of recording was not yet easily, if at all, organised within most of the available databases within the various HISs. *Extended standardised vocabulary* for the main concepts has to be used, and it was also required that at least the specialty and experience (resident, junior/senior etc.) of the professional that performed the act associated with each piece of data be recorded as well. Available systems generally did not yet go beyond simple fixed short pull down lists and time stamping at this point.

This strictly controlled data recording had to be operational in at least 4 to 6 (mostly large university) hospitals within a relatively short timeframe (about 1,5 year), to show the feasibility of a national pain quality assessment network. This timeframe did not permit extensive analysis and adaptation on the functional (application) level of different HISs. From an earlier attempt to develop a similar package with a 4GL-based tool (distributed as 'unsupported-freeware' among 18 pain treatment facilities, installation attempted by 5, only a short time and partially used in 3) the PKC's had learned a few lessons. A new tool had to be very easy and constantly available in every phase of the pain treatment process to enter data timely and 'at the source'. Instruction and support should be available & promptly delivered. Users considered it very essential to avoid irritating redundant entry of simple demographic patient data that is already available in their HIS. The recorded specific pain treatment related data should be easily made available for re-use by management and research.

The detailed description of the functional and data aspects of the developed tool can be found elsewhere [6]. The architecture is that of a client/server application, which requires just a couple of Mb on a server. If server capacity is lacking a simple PC attached to the network is in principle sufficient. The client software has to be installed, i.e. distributed with whatever mechanism a participating hospital has available.

For the exchange of data with a HIS a specific piece of middleware was to be developed for each type of HIS. The principle for a download of data, for example demographic patient data already available in the general patient database of the HIS, is depicted in the following figure. The user of the application supplies some (valid) ID of a patient, which is translated in the middleware in a hospital specific request (file access, direct database access or communication server access/standard message) which, if successful, results in data in HIS specific composition and format that is then translated into an application specific selection and format. It was expected that it would be a relatively modest task to program a couple of hospital specific pieces of middleware, which could probably be reused when the number of participating hospitals would grow. Moreover it was known that a number of Dutch hospitals had some useful communication facilities like HL7 servers.

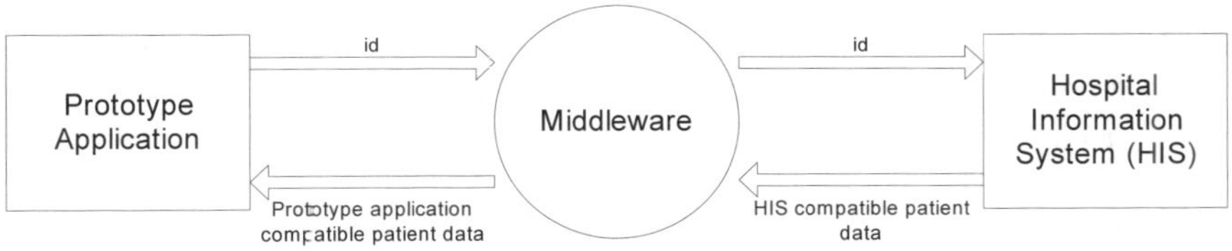

2.2. The participating hospitals

Some 15 organisations were interested in participating in the initial set up of a national quality management network, i.e. exchanging and pooling uniformly recorded data on outcome of treatment of chronic pain. These 15 organisations exploit a general or university hospital with a pain clinic at 18 locations, representing a non-random sample of just over 15% of all hospitals in The Netherlands. This 'population' was initially approached in the way described in the following section. There is some overrepresentation of larger treatment facilities (6 university and 6 large general hospitals) in this sample compared to a national enquiry [10]. An implementation of the tool is needed in at least 8-10 pain centres/clinics in different hospitals to be able to perform a planned field function study (see chapter 3, pp. 41-60 in [11]) of the tool, as a next step in developing the PQM network.

2.3. The middleware development & installation approach

To achieve a properly running (technical) implementation of the tool in de selected hospitals the following approach was used:

- initial meeting with senior staff of the pain treatment facility to
    - explain the role of the tool in the PQM-system and the PQM itself
    - demonstrate the tool and stress the importance of using uniform terminology for recording data
    - make a first inventory of facilities & functions regarding both the pain clinic and the IT-department
- contacts with the IT-department of a hospital to make an inventory of possibilities to communicate with the HIS, to distribute software, to clarify the role of the tool & middleware developers and to agree allocation of tasks like application failure handling, database management, application management and end user support
- develop the tailored piece of middleware (e.g. a Microsoft COM Object)
- apply test procedures and acceptance procedure by IT-department (if required)
- actual installation with two environments: test and production databases
- instruction to the pain centre/clinic personnel to test the ready for use installed tool with the test database

Notes of meetings and telephone or e-mail contacts were summarised in a logbook per site involved.

3. Results

In a period of two years starting early 2001 all of the 15 hospital organisations were approached as described in section 2.3.

After the initial meeting 2 organisations decided not to participate. One had a completely tailored (standalone) pain clinic system developed in the previous 15 years and would consider participation only if in some time the national PQM-network had proven to be fruitful. The other, very interested in PQM, considered the initial effort required too large. Besides that, this clinic planned a major future project to introduce an electronic patient record (EPR) that was expected to require all resources (from both clinic and hospital's IT-department) for years.

Of the remaining organisations 3 were in some explicit phase of developing and implementing their EPR. One decided to give priority to the use of the HIS vendors EPR solution in the pain centre and succeeded in participating in the PQM-network within half a year. Another was in the middle of introducing a general enterprise resource planning system throughout the organisation. This work is still ongoing. The third preferred to try the use of the proven pain clinic system (previous paragraph) and have this integrated with the local HIS and adapted to the (terminology) requirements of the PQM network. Within the timeframe of the initial setup of the PQM-network no result could be achieved for this pain clinic mostly due to financial constraints.

Of the 10 organisations that were approached for the next steps (development of middleware and installation for use in daily practice) 2 had insufficient hardware infrastructure. In both organisations an attempt to capture data on a stand alone PC was stopped because of resulting incompleteness of patient data and even the complete loss of the database and application (risk of end-user involvement in application management). For 2 other organisations out of these 10 approached, no middleware could be developed within the timeframe of the project due to financial or technical constraints. However, in one organisation the lack of data integration was largely compensated by the enthusiasm of the head of the pain clinic, which resulted in full participation in the initial PQM network.

In the end middleware has been developed for six organisations. The organisations had 4 completely different HIS vendor's: Torex-Hiscom (3 sites with different configurations), Eclypsis, McKesson and IBM. Of these organisations only two IT-departments had integrated a 3rd party-not-their-choice application before. In one case this was useful experience, leading to a specific working piece of middleware within short time. The other cases show a variety of solutions: using requests to the HIS to copy data straight forward to some file in a defined format, using different versions of HL7 (versions 2.2, 2.3) and messaging servers (e.g. Cloverleaf), and proprietary SOAP. The content of an HL7 A19 'standard' message appeared not to be the same in different hospitals that stated to use the same HL7 version.

One installed piece of middleware is not always performing at a satisfactory level. The other middleware all showed immediate responses.

Only one organisation had an extensive procedure for taking up responsibility for installation and maintenance of a 3rd party application. This included not only the properties of a piece of software and conditions needed to use it and details of application management tasks, but also installation and update procedures. Following this procedure appeared time consuming at first but speeded up the actual installation and establishing good operational & collaborating conditions. At other sites the process of working through the components of this procedure needs many contacts with at least 3 but sometimes up to 5 or 6 different functionaries.

Time between initial contact with IT-department and successful integrated properly working installation is varying from 5-6 weeks to more than one year.

Distribution of the client part of the tool is varying: manual installation on every workstation, automated procedures such as the use of an application launcher (e.g. creating a NAL object that is distributed), or scripting the application and distribution by some workplace management tool. The latter resulted in extra delay because the experience of scripting (unknown) software was still very restricted.

The availability of the tool with data integration in the pain clinic was no guarantee for continuous and proper use (as intended for the field study) by the clinicians. At the beginning of the field study two organisations did not start use of the tool because of other integration problems and requirements emerged.

The described work finally resulted in 4 pain centres using the tool in daily routine with some data integration through developed middleware, one using it without (with the risk of inconsistencies with HIS data) and one using an adapted EPR for recording an almost comparable set of data.

4. Discussion & conclusions

Our substantial efforts in a period of just over 2 years resulted in 6 out of 15 organisations that had the tool integrated at least at the data level. Of these 6 organisations 2 considered the conditions for use realised still insufficient. Of the 4 remaining 2 (large, university) pain centres had substantial additional requirements within half a year from the beginning of the use of the tool. These new requirements were in part due to the new way of accounting that has been introduced in Dutch hospitals in parallel with the work described here. This had both negative and positive consequences: it took a great deal of available time of clinics and IT-personal, but at the same time it meant a major step in extending the hardware infrastructure (LAN, PCs and servers everywhere!), making the type of data collection aimed at, i.e. 'at the source' and by the observer, feasible.

Integration of *many* applications with *one specific* HIS on the data level is sometimes reported difficult or not feasible (see for example [12]). Other attempts to integrate *one* application with *many different* HISs were not found in the literature.

The experience of this work can be summarized as "muddling through" due to a general lack of standardisation on the technical and data (and application) level. This hampers quick and easy implementation of PQM-tools needed for a national network. A shift is needed towards a proper task allocation between a central organisation for PQM and the pain centres (comparable to the organisation for ICU's [9]).

The preliminary analysis of the (still experimentally) collected actual data on pain treatment are however encouraging. They show for example a significant reduction of pain intensity with certain chronic pain problems where this is not intuitively expected by the head of a clinic. It also provides some clear defence against the accusation of being mainly 'injectionists' [1]) by showing that a lot of chronic pain conditions are treated currently also or mainly by non-invasive means.

References

[1] van Tulder MW, Koes BW, Bouter LM. A cost-of-illness study of back pain in The Netherlands. Pain 1995 Aug;62(2):233-40.
[2] Maniadakis N, Gray A. The economic burden of back pain in the UK. Pain 2000 Jan;84(1):95-103.
[3] Wilder-Smith OH. Opioid use in the elderly. Eur J Pain 2005 Apr;9(2):137-40.
[4] Latham J, Davis BD. The socioeconomic impact of chronic pain. Disabil Rehabil 1994 Jan;16(1):39-44.
[5] Vervest ACM, Stolker RJ, Schimmel GH. Computerized data-acquisition and reports in the pain clinic. The Pain Clinic 1993;6(2):83-95.
[6] Tuil WS, ten Hoopen AJ, van der Haring EJ, Zanstra PE. Quality control in management of pain. In: Roger-France FH, Hasman A, de Clercq PA, de Moor GJE, editors. E-Health in Belgium and in the Netherlands.Amsterdam: IOS Press; 2002. p. 127-35.
[7] Turner GA, Halliwell R. Data collection by acute pain services in Australia and New Zealand. Anaesth Intensive Care 1999 Dec;27(6):632-5.
[8] de Keizer NF, Stoutenbeek CP, Hanneman LA, de Jonge E. An evaluation of Patient Data Management Systems in Dutch intensive care. Intensive Care Med 1998 Feb;24(2):167-71.
[9] http://www.stichting-nice.nl/leveranciers.jsp - last accessed jan 04 2006.
[10] van Dasselaar NT, Boersma FP, de Lange JJ. Multidisciplinary pain facilities in the Netherlands: a model. The Pain Clinic 1995;8(3):211-8.
[11] Friedman CP, Wyatt JC. Evaluation methods in medical informatics. New York Berlin Heidelberg: Springer; 1997.
[12] Beuscart RJ, Souf N, Blassel N, Gourdin C, Modjeddi B. Integration Architecture: The ISAR project. In: Brender J, Christensen JP, Scherrer J-P, McNair P, editors. Medical Informatics Europe '96. Human Facets in Information Technologies.Amsterdam: IOS Press; 1996. p. 311-5.
[13] Rathmell JP. The injectionists. Reg Anesth Pain Med 2004 Jul;29(4):305-6

3.2 HIS Architectures: Examples

Ubiquity: Technologies for Better Health in Aging Societies
A. Hasman et al. (Eds.)
IOS Press, 2006

On feasibility and benefits of patient care summaries based on claims data

Thorsten BRENNECKE[a,1] Lars MICHALOWSKI[a], Joachim BERGMANN[a], Oliver J. BOTT[a], Axel ELWERT[b], Reinhold HAUX[a], Ralf SCHLICHT[b]
[a] *Institute for Medical Informatics, Technical University of Braunschweig, Germany*
[b] *intersoft AG, Hamburg, Germany*

Abstract. ***Background***: Electronic availability of health care claims data maintained by health insurance companies today is higher than the availability of clinical patient record data. ***Objective***: To explore feasibility of automatically generated patient care summaries based on claims data and its benefit for health care professionals (HCP), when no shared electronic health record is available. ***Methods***: Based on an existing claims data model for German health insurance companies, a transformation and presentation algorithm was developed. To determine the utility of the resulting summaries, a focus group session comprising HCP and insurance representatives was arranged. Properties of an information system architecture capable of providing summaries to HCP were specified. ***Results***: A set of valuable healthcare information, in particular clinical pathways, medication, and anamnesis, can be derived from claims data that fits into the ASTM specification of a Continuity of Care Record. The focus group assessed the potential benefit of the summaries as high. Major issues are partial incompleteness and a lack of timeliness due to delayed reimbursement procedures as well as privacy-preserving and practicable access methods. The specified system architecture uses web services and a web interface to provide the summaries in HL7 CDA format. An important insight was that only a timely electronic reimbursement process will lead to precise, current, and reliable claims-based summaries. ***Conclusion***: Generating patient care summaries based on claims data is feasible and produces valuable information for HCP, provided that the reimbursement process is conducted timely. Integration into a national health telematics platform will facilitate access to the summaries. Evaluation of algorithm and prototype system is underway to prove the benefit in clinical practice.
Keywords: Medical Records Systems, Computerized; Integrated Health Care Systems; Insurance Claim Reporting; Health Insurance Reimbursement.

Introduction

As shown in [1-3], using an integrated electronic health record (EHR) across several providers and regions can improve health care and reduce costs. But since heterogeneity, complexity, and autonomy of medical record systems as well as privacy concerns currently constrain practicability of a cross-institutional EHR [4,5], consumer-centric solutions like personal health records (PHR) increasingly attract attention [5], at least as an interim solution. [6] discusses benefits of a PHR supporting continuity of care under control of the patient and anticipates a future integration of both EHR and PHR. A major disadvantage of PHRs is the lack of continuously feeding

[1] Thorsten Brennecke, Institute for Medical Informatics, Technical University of Braunschweig, Muehlenpfordtstr. 23, D-38106 Braunschweig, Germany, t.brennecke@tu-bs.de

data sources. Claims databases could be a valuable feeder system [6]. Claims data comprise partly detailed information concerning the individual health care process from several providers. These data have been proved to be useful for many clinical questions, mainly by augmenting an institutional EHR with additional information [7]. Javitt et al. demonstrate that an adequate consideration of claims data for healthcare can reduce hospital admission about 19% [8]. Claims data also have been shown to be useful for tracking a patient's clinical pathways [9,10].

The aim of the research is to explore, whether compact summaries of a patient's clinical history and medication based on claims data provides a valuable and practical information alternative, especially when no integrated EHR is available. Though such a *synopsis* cannot replace an EHR, it includes cross-institutional and probably lifelong information and may additionally be used for a quick and brief overview. This approach is based on a project, which was started together with a cooperating vendor of insurance software and is currently focused on claims data of a German private insurance. The following questions are to be answered in this paper:

- Which parts of claims data are useful for health professionals and how should the data be transformed and presented?
- What are the benefits of such a synopsis for all involved parties (patient, health professional, and insurer)?
- What are the problems and restraints?
- How can a synopsis be provided to health professionals?

Material and Methods

To analyse the expectations of care providers regarding a synopsis and to specify structure and presentation of a synopsis, several health professionals were interviewed. The analysis considered the Continuity of Care Record (CCR) by the American Society for Testing and Materials (ASTM) [11] to identify relevant information categories. The identified categories have been compared with the claims data model of an established insurance application system. For information which could not directly be derived from the claims database, algorithms for automatic generation have been outlined where possible. The resulting sketch of a synopsis was presented to a focus group comprising physicians, a health care economist, a health insurance business economist, and a computer scientist. The group rated anticipated effects, costs and benefits of a synopsis on a five-stage scale separately for all involved parties (health professionals, patients and health insurers), thus an estimation of costs and benefits could be created. A formal information model based on the CCR and functional specifications for deriving the information from claims data have been developed. An information system architecture specifies software components, data flows, and work processes related to creating, configuring, providing, and using the synopsis.

Results

Results of the project are currently an information model of the synopsis based on requirements analysis and analysis of the claims data model, results from usability assessment, and an architectural concept for operating the synopsis system.

Information model of the synopsis

The resulting information model is outlined in table 1. The information categories are an outcome of an analysis of the claims data structure. The evaluation interviews consisted mostly of open questions. Beside these some closed questions were included e.g. the question to rate the clinical relevance of each information category using an ordinal scale. Thus the results became more structured and comparable. Table 1 summarizes the results of the relevance rating of each information category. Additionally it presents the coverage of each information category by the CCR and the availability in claims data. All identified information categories are covered by the CCR. Most information can be extracted by direct mapping, e.g. recent care providers, medications, and procedures. Insurers often have access to a general medical history of the patient, which has been provided by the patient initially with the conclusion of the insurance contract. Thus the data of the anamnesis might be outdated.

Table 1 – Results of the interview in comparison with the available claims data.

Information category	Rating	Available in claims data?	CCR
Basic demographic data	Required	Yes	Yes
Recent care providers	Required	Yes	Yes
Diagnoses (preferably coded)	Required	Yes (ICD-10 available)	Yes
Procedures (preferably coded)	Required	Yes (coding not available)	Yes
Encounters	Useful	Partially inferable	Yes
Relation between diagnoses, Procedures and medication	Interesting	Partially inferable	Yes
Medications	Interesting	Yes	Yes
Immunizations	Interesting	Yes	Yes
Allergies (preferably coded)	Interesting	Included in anamnesis (coding not available)	Yes
Chronicle disease (preferably coded)	Interesting	Included in anamnesis (ICD-10 available)	Yes
Anamnesis	Interesting	Yes	Yes

Usability assessment and problem identification

The focus group assessed a synopsis based on claims data to be a useful new source of information particularly for health professionals. A brief summary of the focus group results can be found in table 2 and will be discussed in the following.

Health care professionals state that a newly available care history increases efficiency of information access, but point out that mainly recent events are of interest. For data presentation, the group suggests a separated view of chronic conditions from acute diagnoses. Also an option to choose between problem-oriented and chronological view was requested.

Patients will probably benefit from an improved quality of care if the synopsis is strictly up-to-date and contains all recent encounters. The cross-institutional overview of health care history and a better control of reimbursement processes will foster the patient empowerment. In order to respect privacy, the patient must be able to fully administer access to the synopsis.

Health insurances could market the synopsis as a new innovative e-health product. An improved quality of care would also save costs, but depends substantially on the quality of the synopsis, current information, and the acceptance of the care providers.

Several problems due to the current reimbursement process have been identified. Because of the specific and mostly paper-based reimbursement procedures of private insurances in Germany, patients often cumulatively submit invoices with a delay of up to one year, a practice that is actually endorsed by certain insurers. Furthermore patients will usually wait until they have met their deductible for the year before starting to submit any documents for cost reimbursement. Thus a claims-based synopsis generally can be assumed to be out of date. Furthermore, for efficiency reasons only significant parts of paper-based invoices are captured electronically from clerks. A synopsis based on these data will obviously be outdated and imprecise and will not fully exploit its potential. Therefore, an electronic and timely reimbursement process will strongly improve usability of the synopsis, and will also reduce the insurer's costs of claims processing.

Table 2 – Rating of the synopsis by the focus group

		Health care professionals	Health insurances	Patient
Information supply	New source of information	**High effect**	**No effect**	**Small to medium effect**
	Optimized information access	**Medium effect**	**No effect**	**High effect**
Improvement of health care quality	Supporting processes in health care	**High effect**	**Great benefit**	**Small to no benefit**
	Cost savings	**No effect**	**(Possibly) high effect**	**No effect**

Architecture

The requirements analysis resulted in the following characteristics of a suitable information system architecture:

- Interoperability. Contents and presentation of the synopsis should follow the CCR, thus it is based on a commonly agreed and interoperable data format. The HL7 Clinical Document Architecture (CDA) is an already established standard [12] and was identified to be well-suited for formal representation of the synopsis.
- Patient consent and privacy. Claims data contain sensitive information about care events and diseases. Delivering the synopsis to health professionals means a new data flow added to established and legally approved data transmissions. Therefore, explicit patient consent is both initially required and also subsequently for each access or transaction regarding the synopsis. The patient should have the possibility of granting access to the synopsis in full or to parts of it to individual health professionals. He must be able to revoke these permissions at any time.
- Timeliness and data quality. In order to ensure that the synopsis is current, precise and as complete as possible, the concept requires an electronic transmission of invoices to the health insurance.

Embedding the synopsis concept into workflows requires an analysis of the reimbursement processes. In healthcare, two fundamentally different ways of clearing can be found: payment from patient to the care provider (where the patient subsequently may request reimbursement of costs from his insurance) and a direct clearing between care provider and insurance without the patient being directly involved. In the first clearing method, the patient will possibly pay some invoices by himself without passing them to his insurance, or he may delay them for a longer period. In these cases the claims data would not be available to the insurance for creating or updating the synopsis. Thus a clearing between care provider and insurance is the preferred way because it allows direct transmission of claims data. Because this is not always applicable (e.g. due to legal limitations), a concept for electronic transmission of invoices and/or claims data has to support both clearing methods in order to replace the paper based clearing completely.

In Germany, financial data flow in healthcare is fragmented and based on heterogeneous agreements between the various healthcare organizations, thus neither a nation wide standard nor a uniform concept is available. In contrast, the Swiss reimbursement model electronically supports both clearing methods: with the patient being part of the cash flow (tiers garant) or without (tiers payant). The Swiss model comprises a common data format and a concept based on a trust centre. Adopting this model, the architectural concept presented in fig. 1 was developed.

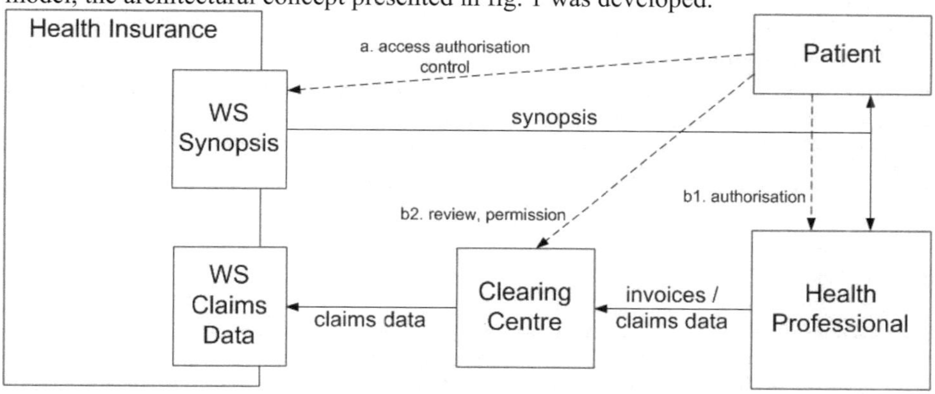

Figure 1 – Overview of the information system architecture concept (WS = Web Service)

The health insurance provides two web services (*WS Synopsis* and *WS Claims Data*). These offer a safe interface for authentication and encrypted data communication using established security protocols over the internet. A synopsis is provided in form of a CDA document, which can be accessed from a health professional authorized by the patient (a).

A health professional authorized for electronic claims transmission (b1) sends invoices and/or claims data electronically to a clearing centre, which offers the patient the option to control the process (viewing invoices and/or claims data) and to confirm further transmission (b2). The clearing centre transmits all approved claims data to the health insurance using the WS Claims Data. This way, both clearing techniques (tiers garant and tiers payant) are supported.

Discussion and Conclusion

Claims databases are assumed to be a potential data source for personal health records [6], which currently are of little use for care providers. This paper presents a system architecture to augment a personal record with claims data in form of synopsis documents. Results of the focus group analysis conducted in this project underpin the assumption that adequately processed and presented claims data can support continuity of care, if the information is current and sufficiently precise. However, such patient-centric care summaries based on claims data is intended to be an interim solution when no shared health record is available. On the other hand, Javitt et al. show that a decision support system based on claims data also can complement electronic health records with an alert function regarding guideline compliance [8]. Thus the presented architecture may be a basis for a new kind of quality management service offered from health insurances to care providers. Nevertheless, usability of the proposed concept depends on a timely transmission of claims data which can mainly be achieved with an electronic reimbursement system.

A major issue of the approach is the cumbersome handling of the synopsis by the patient needing internet access for synopsis administration, which cannot be assumed in general. An alternative paper-based consent model is suggested, which on the other hand does not allow for a smooth administration of synopsis access. Current problems regarding internet connectivity and secure smartcard-based authentication of the participants will be resolved by a regional or national health telematics platform, which is planned in Germany at present [13].

A first pilot prototype system is currently prepared and will offer access to health insurance and clearing centre via web browser and will use an authorisation method based on transaction numbers (TANs). An evaluation of the prototype system will assess the benefit of a synopsis as a new information source for care providers and will determine costs and acceptance of the proposed architecture.

References

[1] Garrido T, Jamieson L, Zhou Y, Wiesenthal A, Liang L. Effect of electronic health records in ambulatory care: retrospective, serial, cross sectional study. BMJ 2005; 330:581-5.

[2] Wang SJ, Middleton B, Prosser LA, Bardon CG, Spurr CD, Carchidi PJ, et al. A cost-benefit analysis of electronic medical records in primary care. Am J Med 2003; 114(5):397-403.

[3] Barlow S, Johnson J, Steck J. The Economic Effect of Implementing an EMR in an Outpatient Clinical Setting. Journal of Healthcare Information Management Vol 18, No. 1: 46-51.

[4] Kuhn KA, Giuse DA. From hospital information systems to health information systems. Problems, challenges, perspectives. Methods of information in medicine 2001;40(4):275-87.

[5] Waegemann CP. Closer to reality. Personal health records represent a step in the right direction for interoperability of healthcare IT systems and accessibility of patient data. Health Manag Technol. 2005;26(5):16-8.

[6] Tang PC, Ash JS, Bates DW et al. Personal Health Records: Definition, Benefits, and Strategies for Overcoming Barriers to Adoption. J Am Med Inform Assoc 2005. Pre-print.

[7] Weiner M, Stump TE, Callahan CM, Lewis JN, McDonald CJ. A practical method of linking data from Medicare claims and a comprehensive electronic medical records system. Int J Med Inform 2003;71(1):57-69.

[8] Javitt JC, Steinberg G, Locke T et al. Using a claims data-based sentinel system to improve compliance with clinical guidelines: results of a randomized prospective study. Am J Manag Care 2005;11(2):93-102.

[9] Barnato AE, Labor RE, Freeborne NE, Jayes RL, Campbell DE, Lynn J. Qualitative analysis of Medicare claims in the last 3 years of life: a pilot study. J Am Geriatr Soc 2005;53:66-73.

[10] Dendukuri N, McCusker J, Bellavance F et al. Comparing the validity of different sources of information on emergency department visits: a latent class analysis. Med Care 2005;43(3):266-75.

[11] Tessier C, Waegemann CP. The continuity of care record. Organizations unite to create a portable, electronic record for use at transfer points in patients' care. Healthc Inform. 2003;20:54-6.

[12] Dolin RH, Alschuler L, Boyer S, Beebe C, Behlen FM, Biron PV, Shabo SA. HL7 Clinical Document Architecture, Release 2. J Am Med Inform Assoc 2006;13:30-39.

[13] Blobel B, Pharow P. A Model-Driven Approach for the German Health Telematics Architectural Framework and the Related Security Infrastructure. Stud Health Technol Inform. 2005;116:391-6.

On-the-fly form generation and on-line metadata configuration – a clinical data management web infrastructure in Java

Peter BECK[a,1], Thomas TRUSKALLER[a], Ivo RAKOVAC[a], Bruno CADONNA[a] and Thomas R. PIEBER[a,b]
[a] *JOANNEUM Research Forschungsgesellschaft mbH, Institute of Medical Technologies and Health Management, Graz, Austria*
[b] *Department of Internal Medicine, Diabetes and Metabolism, Medical University Graz, Austria*

Abstract. In this paper we describe the approach to build a web-based clinical data management infrastructure on top of an entity-attribute-value (EAV) database which provides for flexible definition and extension of clinical data sets as well as efficient data handling and high performance query execution.
A "mixed" EAV implementation provides a flexible and configurable data repository and at the same time utilizes the performance advantages of conventional database tables for rarely changing data structures. A dynamically configurable data dictionary contains further information for data validation. The online user interface can also be assembled dynamically. A data transfer object which encapsulates data together with all required metadata is populated by the backend and directly used to dynamically render frontend forms and handle incoming data.
The "mixed" EAV model enables flexible definition and modification of clinical data sets while reducing performance drawbacks of pure EAV implementations to a minimum. The system currently is in use in an electronic patient record with focus on flexibility and a quality management application (www.healthgate.at) with high performance requirements.
Keywords: Medical Informatics; Medical Records Systems, Computerized; Online Systems; Databases

1. Introduction

The entity-attribute-value (EAV) approach is popular for modelling highly heterogeneous data [1]. This paper describes its adoption in a web-based, flexible clinical data management platform. This platform was developed as a basis to meet varying requirements in two different applications.

The first application is an electronic patient record system for use in hospital intranets. In most clinical applications for routine patient care the core data set rarely changes. The system therefore should be capable of handling this "static" data set for

[1] Corresponding Author: Peter Beck, JOANNEUM Research Forschungsgesellschaft mbH, Institute of Medical Technologies and Health Management, Elisabethstraße 11a, 8010 Graz, Austria, E-mail: peter.beck@joanneum.at, URL: http://www.joanneum.at/msg

routine use in an efficient way. Furthermore the system should allow extension and modification of existing as well as introduction of additional data structures for clinical study data collection. These introductions should be feasible by configuration without having to change the underlying physical database schema. The model should also be able to deal with more than one entry for a defined dataset per patient or case, which traditional EAV schemes cannot provide.

The second application built on top of the EAV system is a quality management application for chronic diseases for use over the Internet. In this application data is collected in rarely changing standardized data sets. Very high performance requirements are put on the execution of queries used for clinical quality reporting which are usually hard to fulfil with EAV databases according to [2].

In this paper we describe the approach to build a web-based clinical data management infrastructure on top of an EAV database which allows flexible definition and extension of clinical data sets as well as efficient data handling and high performance query execution.

2. Material and Methods

2.1. EAV model architecture

The EAV pattern was chosen to provide a flexible and configurable data repository. Conventional database tables still are superior in terms of performance. So a "mixed" implementation was considered which allowed dynamic dataset definitions stored in an EAV repository to coexist with static conventional tables for rarely changing core data sets.

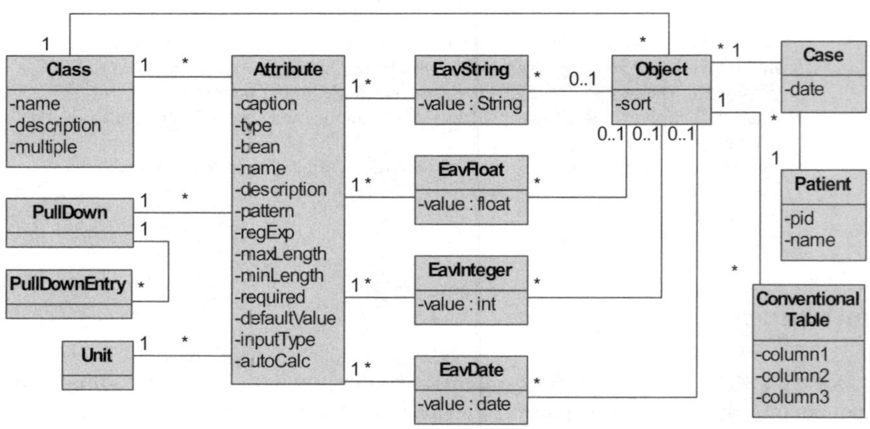

Figure 1. EAV schema in combination with conventional tables

The schema shown in Figure 1 is related to the fully object oriented model presented in [3]. It has been modified to handle also conventional tables. As a consequence a fully object oriented approach would no longer have been feasible. In contrast to the enhanced EAV model described in [4] where data tables are directly related to a patient or a case, we used an "Object" table. Hence the collection of data with 1:n relations (e.g. several foot ulcers which are treated during one visit) was possible. The field "multiple" in the "Class" table defined if multiple objects of this class were allowed for one case.

The definition of virtual classes and their attributes applied throughout the system. The definition if a data item was stored in a conventional or EAV table was made in the "Attribute" table. The "bean" field could contain the name of a conventional table to store data for this item. If this field was empty, the appropriate EAV table for the given type was chosen for data storage. A class could have attributes stored in EAV tables and different conventional tables at the same time. When a clinical data set was stored, an entry in the "Object" table was created. Then rows in the data tables were created as required, which maintained a reference to the object entry.

2.2. Data dictionary and validation

Metadata tables on the left in Figure 1 defined the data dictionary available in the system. Firstly, this information consisted of class and attribute names as well as captions and descriptions to be used in the user interface. Secondly, data types were defined. If there were several ways for data input for a data type, an "input type" could be chosen (e.g. for string values a text input field or a list to choose from). For multiple-choice items a reference to a pull-down list of entries could be assigned. Thirdly, several additional fields contained data validation information about required fields, minimal and maximal length, allowed data ranges and patterns for validation with regular expressions.

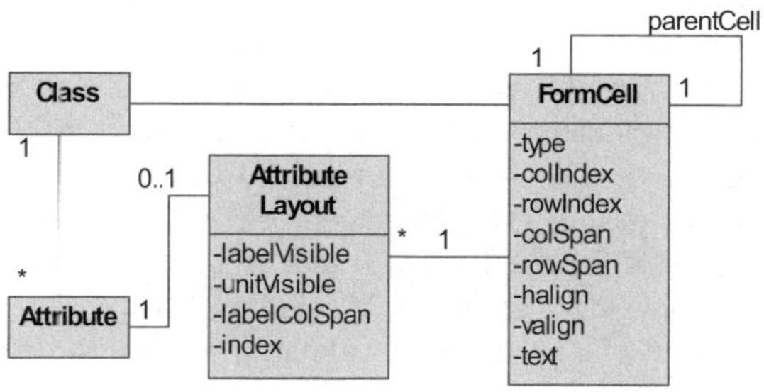

Figure 2. Dynamic frontend definition schema

2.3. Dynamic frontend definition

The dynamic user interface configuration allowed the definition of forms for the previously defined classes. Forms could be organized hierarchically. Figure 2 shows that a "FormCell" contained information about location, size and alignment of user interface elements. Additional layout information for attributes was stored in a separate table.

2.4. Implementation

The system was implemented in Java. The backend was built on top of the J2EE with Enterprise Java Beans. Session Beans were used to provide business services and Entity Beans for persistent data storage. The frontend was constructed with Struts, JSTL and several other open source libraries.

2.4.1. System architecture

According to the common practice paradigms for J2EE and Struts development three tier architecture was chosen for implementation of the system.

A core element was the data transfer object (DTO) which contained metadata information together with data to be exchanged between backend and frontend. When a data set was created or read from the database the corresponding metadata was assembled in a dedicated Session Bean and passed to the frontend together with the data values in this DTO. Hence for each item the information on how to locate it in the database, render the input widget and later validate the data was already provided.

The web application extended the Struts concept of ActionForms to dynamically deal with data submitted from a web browser form. The ActionForm was implemented based on a map and designed as a wrapper around a DTO. On form creation the DTO was assigned to the form and the map automatically set up and populated according to the information in the DTO.

The visible part of the frontend was created with Java Server Pages (JSP). In case of fully dynamic form definition as described in 2.3 the interface was rendered using JSTL tag libraries and imports of JSP fragments. However custom forms could also be provided as JSP files by developers and replace the dynamic definitions in the database. In this case additional functionality could be added to the form.

2.4.2. Data flow and validation

When the user entered data in the web browser and submitted the form, the data was received by the Struts controller and copied to the corresponding fields in the dynamic ActionForm. All ActionForm fields were defined as plain text to be able to return the original data back to the user in case of any input or transformation problems. In the ActionForm basic validation steps were taken before converting the data to the target data types and populating the DTO with the received values. Now full validation with typed data was performed in the DTO. This approach assured, that the same validation mechanisms could be used in the backend and in the frontend without having to duplicate code or data because they were present in the DTO.

In case of occurrence of one or more validation errors, the corresponding error messages were dynamically composed based on metadata and error types. The messages were presented to the user in the form together with the original user input.

If validation was successful the DTO was passed to the backend where all affected Entity Beans were updated, which resulted in SQL UPDATES automatically executed by the application server.

3. Results

The system provides fully dynamic definition of metadata. The clinical data set can be modified and extended on-the-fly. For this purpose an application client has been developed which communicates with the J2EE application server via remote service calls.

This client can be used to define classes and their attributes (Figure 3). For each attribute it allows to configure whether data should be stored in the database in an EAV or in a conventional table. Furthermore one may choose from several types and widgets for data input and define the criteria for data validation.

As an advanced data validation feature, dependencies from values in other fields can be checked. The checks can be entered as formulas in the attribute definition.

Calculated fields are another feature provided in a generalized way. For arbitrary attributes a formula may be entered. If such a formula is present the value for the attribute is calculated and updated automatically.

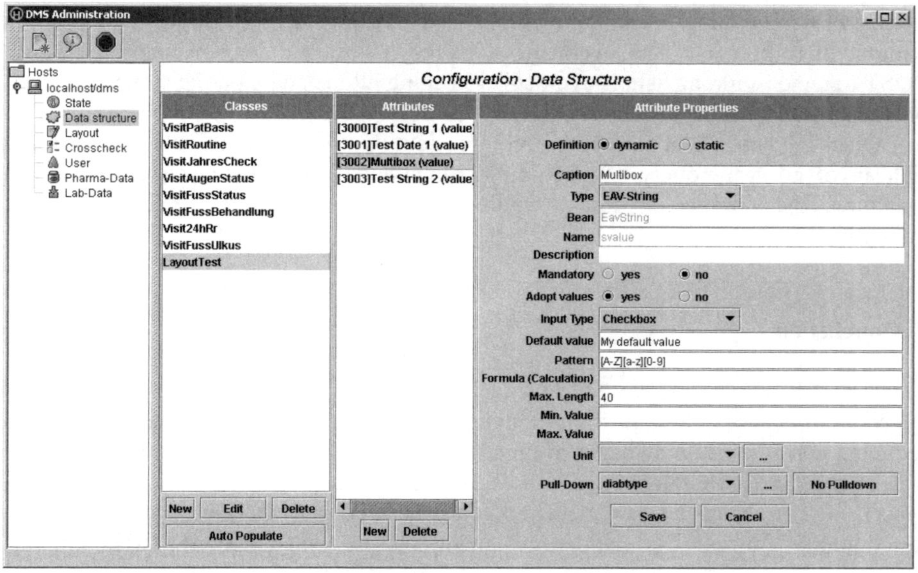

Figure 3. Remote data set configuration application client

For the dynamic configuration of user interface layout a simple editor was created, which facilitated definition of a table based layout. A simplified view is given in Figure 4. The columns contain information about position, size, label placement and visibility, orientation and static text to be displayed in the form. According to the data model also nested forms are provided, but they can not yet be defined in this editor.

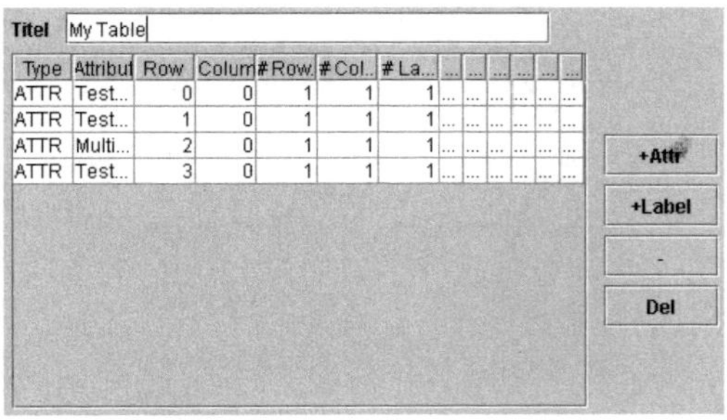

Figure 4. Simple editor for dynamic user interface layout

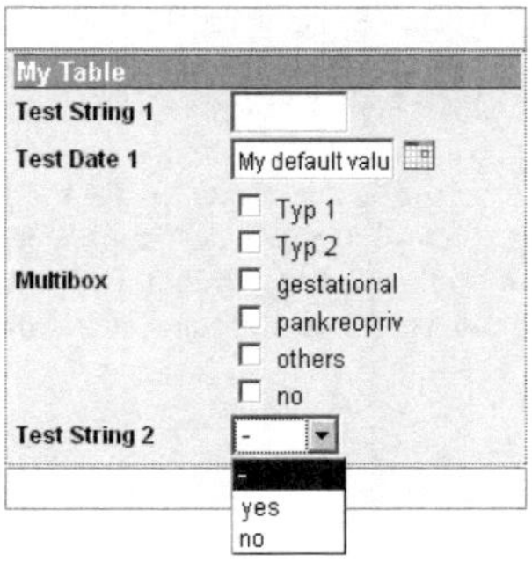

Figure 5. Online view

Figure 5 shows the resulting online dialog in the dynamically configured layout based on a dynamically defined EAV dataset.

4. Discussion

The system described in this paper uses a data transfer object and dynamic form population. This approach has several advantages compared to the web based implementation example in [5] where the metadata required for processing data in the backend is embedded in the frontend in hidden HTML fields. The first advantage is extensibility. Handling of validation and layout information would hardly be possible without this approach. The overhead of having to transmit the metadata via HTTP POST together with the data is avoided by keeping all metadata on the server and using only a single identifier per data field.

Querying pure EAV databases requires joining an additional database table per field involved in the query. By including conventional tables for rarely changing data structures the number of joins required for a typical query could be reduced dramatically.

5. Conclusion

The "mixed" EAV model with data in conventional and EAV tables is a feasible approach to enable flexible definition and modification of clinical data sets while reducing performance drawbacks of pure EAV implementations to a minimum.

The web-based implementation uses dynamically populated data transfer and form objects. These objects contain metadata in a structured and extensible way and provide functionality for data validation and manipulation. Fully dynamic user interface generation closes the loop to dataset manipulation on-the-fly.

The system currently is in use in an electronic patient record and a quality management application.

References

[1] Nadkarni PM, Brandt C. Data extraction and ad hoc query of an entity-attribute-value database. J Am Med Inform Assoc. 1998 Nov-Dec;5(6):511-27.

[2] Chen RS, Nadkarni P, Marenco L, Levin F, Erdos J, Miller PL. Exploring performance issues for a clinical database organized using an entity-attribute-value representation. J Am Med Inform Assoc. 2000 Sep-Oct;7(5):475-87.

[3] Nadkarni PM, Marenco L, Chen R, Skoufos E, Shepherd G, Miller P. Organization of heterogeneous scientific data using the EAV/CR representation. J Am Med Inform Assoc. 1999 Nov-Dec;6(6):478-93.

[4] Anhoj J. Generic design of Web-based clinical databases. J Med Internet Res. 2003 Nov 4;5(4):e27.

[5] Nadkarni PM, Brandt CM, Marenco L. WebEAV: automatic metadata-driven generation of web interfaces to entity-attribute-value databases. J Am Med Inform Assoc. 2000 Jul-Aug;7(4):343-56.

Ubiquity: Technologies for Better Health in Aging Societies
A. Hasman et al. (Eds.)
IOS Press, 2006

Log files analysis to assess the use and workload of a dynamic web server dedicated to End-Stage Renal Disease

Mohamed BEN SAID, Loic LE MIGNOT, Jean Baptiste RICHARD, Christine LE BIHAN, Laurent TOUBIANA, Jean-Philippe JAIS, Paul LANDAIS [1]

Université Paris 5 R. Descartes – UPRES EA 222- AP-HP- Service de Biostatistique et d'Informatique Médicale, Hôpital Necker-Enfants Malades- Paris- France

Abstract

A Multi-Source Information System (MSIS), has been designed for the Renal Epidemiology and Information Network (REIN) dedicated to End-Stage Renal Disease (ESRD). MSIS aims at providing reliable follow-up data for ESRD patients. It is based on an n-tier architecture, made out of a universal client, a dynamic Web server connected to a production database and to a data warehouse. MSIS is operational since 2002 and progressively deployed in 9 regions in France. It includes 16,677 patients. We show that the analysis of MSIS web log files allows evaluating the use of the system and the workload in a public-health perspective.

Key words: Web log file; workload; Dynamic Web server; n-tier architecture; system use; Multi-Source Information System; Internet; End-Stage Renal Disease.

1. Introduction

A Multi-Source Information System (MSIS) [1] was set up, dedicated to collect continuous and completed records of all patients presenting with End-Stage Renal Disease (ESRD) and their clinical follow-up. MSIS collates in a standardized representation a minimal patient record elaborated by health professionals [2].
Our project was to provide the users a tool facilitating the access to useful information concerning ESRD demand and offer of care, allowing the national sharing of the results, and supporting public health decisions at regional and national levels for adapting the offer to the demand of ESRD care.
This paper describes a method to assess the use of the system and the workload through web log analysis.

2. Methods

2.1 MSIS organizational support

MSIS organization has been described elsewhere [3]. Briefly, MSIS is part of the Renal Epidemiology and Information Network (REIN). A national committee insures

[1] Corresponding author: Pr Paul LANDAIS – Service de Biostatistique et d'Informatique Médicale – Hôpital Necker Enfants Malades – 149, rue de Sèvres – 75743 Paris – FRANCE – email : landais@necker.fr

guidance and follow-up and involves several organizations: Société de Néphrologie, Société francophone de dialyse, Paris Descartes University and Grenoble J. Fourier University, Agence de la Biomédecine, Caisse Nationale d'Assurance Maladie des Travailleurs Salariés, INSERM, Direction de l'Hospitalisation et de l'Organisation des Soins, Institut de Veille Sanitaire and representatives of patients' associations.
In each French region, regional committees involve nephrologists, decision makers, public health insurers, epidemiologists and representatives of patients' associations. A nephrologist has been elected as program coordinator. A public health and epidemiology department provides resources and expertise for methodology and epidemiological studies. Once a year, a clinical research assistant performs the quality control for every patient record.
The agreement of the French "Commission Informatique et Libertés" (data protection act 78-17) was obtained, in accordance with the European Convention 108 and the directives n°95/46/CE and 2002/58/CE.

2.2 MSIS Design and Implementation

MSIS-REIN is based on a n-tier architecture [1]. Via a web browser, the client tier connects to a middle tier that is in relation with several databases: the identification database, the production database and the data warehouse. The middle tier supports client services through Web containers and business logic services through component containers. Business logic components in the middleware support transactions toward the databases.
At the client side, MSIS relies on existing local Internet networking facilities and on a widely spread computer configuration in medical settings. Maintenance and evolutions are made centrally which reduce deployment costs and delays.
The production database was structured according to users' profiles, administrative regions, care units and patients' information:
- Users' profiles concern end users, mainly physicians, regional coordinators, clinical research assistants and MSIS administrators (users roles are summarized on table 1).
Passwords are delivered individually to the users to access to patients' data of their own unit. The regional coordinator and the clinical research assistant have access to all units within their region in order to perform data quality control. MSIS access codes (Table 2) were provided to more than 430 nephrologists and 100 codes to their collaborators. Ten clinical research assistants and 9 physicians perform the quality control at the regional level. Eight university departments of medical informatics, epidemiology or public health, insure the patient data quality control using MSIS.
- Data of each administrative region is collected into a specific protected database subset.
- Patient records comprise two parts: patient identification and patient medical record. The patient medical record is composed of three parts: medical history, aetiology of ESRD and comorbidity at start of replacement therapy; recent medical record with information about access to the care facilities and to the national kidney-graft waiting list; an annual update of the actual renal dialysis method and context of treatment.
Admission, discharge and transfer event information's are documented and updated annually on the anniversary of first ESRD treatment. A decease record file, including standard medical codification of the decease is documented when necessary.

Table 1 – MSIS users' roles.

Users profiles	**Roles**
Regional Coordinator	- Coordinates regional activities - Delivers access authorizations to the Information System - Organizes data analysis - Represents the nephrologists in the meetings with the health authorities
Regional Clinical Research Assistant	Performs : - User training and accompaniment - Data completeness checking : systematic cross matching with lists of patients treated in each unit - Data quality control : in each unit, using sampling procedures, data assessment by going back to the patient medical record - Data consolidation to allow feeding the data warehouse - Preparing feed back reports based on the region data
Nephrologists and health professionals	- Provide medical expertise and fulfills patient records - Validate patient records

Consolidated patient data are loaded periodically to a data warehouse coupled with a Geographic Information System called SIGNe [4, 5]. SIGNe provides MSIS users with dynamic analysis views in the form of tables, charts and graphics, didactic maps as well as preformatted ready-to-print reports. Interactive queries allow representing regional or inter-regional profiles such as: incidence and prevalence of ESRD; Patient treatment trajectory, transportation modes, distances and duration to reach their care unit, care units catchments areas; comorbidities, typology, and impact on ESRD health care; follow up of ESRD treatment, control of anaemia and erythropoietin treatment; distribution of dialysis treatment methods: haemodialysis, peritoneal dialysis, or transplantation according to the typology of care units: centre, medical unit, self-dialysis, home dialysis.

Table 2 – MSIS – User profiles in eight administrative regions where MSIS is deployed.

User Profile / Year	**2002**	**2003**	**2004**	**2005 semester 1**	**2005 semester 2**
Nephrologists	14	75	199	261	434
Health care professionals in Nephrology setting		13	25	30	105
Clinical Research Assistants	1	3	5	7	10
Medical informatics and/or epidemiology and/or public health professionals	3	5	7	8	10
Total	18	96	236	306	559

The log files of the JSP/Servlet web server container (Tomcat 5) keep track of every single client request and server response in a multi line text record. We studied the Web log files over a period of 17th months from June 16th 2004 to Jan 15th 2006. They consist of large text files which size 17.8 gigabytes. We first used AWK programming language [6] to process them and extract information of interest. Then we used Mysql© to re-arrange the information for analysis. Analysis was developed according to several topics: exploring the frequency of requesting of specific functionalities, of an executable program, or downloading a given document. We then rapidly moved to focus on the sessions' key identifiers and how to organize the information. It made computations easier to retrieve specific information or the MSIS user's profile [7].

3. Results

As of January 30th 2006, 16,677 patient records, 1504 transplantations, 2300 transfers from a dialysis unit to another and 4012 ESRD deaths are documented in the production database. The active file includes 11,146 patients who undergo dialysis (detailed information are presented in table 3). According to the national survey of prevalent ESRD dialyzed patients in June 2003 [8], MSIS active file includes 35 % of the nation wide ESRD dialyzed patients. As longitudinal data accumulate, annual follow-up of the ESRD cohort are progressively and systematically taking place in the regions.

We identified 17,281 user's sessions during the 17th month period of observing the log files. They correspond to 348 identified users who connected several times. This observations also means that 2 out of 3 authorized users (348/559; 62%) have been connected effectively to MSIS.

We observed the duration of the session associated with each user, the number of sessions and the total duration a user had over a period of 17 months. Cross matching with MSIS tables allowed retrieving information sorted by user's profile and by region (table 4).

Epidemiologists and public health professionals at the Biostatistics and Medical Informatics Department at Necker Hospital in Paris - France, connected to MSIS mainly to assist online users and the clinical research assistants in the regions. During the study period, they created 3271 valid sessions representing a total duration of 1641 hours 49 minutes with an average of 29:37 per session.

Consolidated data by the clinical research assistant was progressively and periodically loaded into the data warehouse for analysis and data presentation for decision making.

4. Discussion

Nephrologists direct involvement in using MSIS to update patient information varies according to the size of the region or the number of units of care. In two small regions with small number of nephrologists, the Clinical Research Assistant updates patient information using MSIS. While in regions where more nephrologists are involved in the ESRD patient care, the nephrologists and the health care professionals directly use the MSIS to update patient information.

The web log files analysis showed an important involvement among the nephrologists and their teams to directly use the MSIS and document patients' information. They

Table 3 – MSIS deployment (Jan 30th,2006). The total population in the regions where MSIS is used represents 26,211,000 inhabitants (43% of the French population).

Region	**Date of inclusion**	**Population**	**Number of cases in the active file**
Limousin	Jan 22nd 2002	711,000	374
Languedoc-Roussillon	Jun 2nd 2002	2,296,000	1,537
Champagne-Ardenne	Jan 1st 2003	1,342,000	700
Centre	Jan 1st 2004	2,440,000	1,303
Provence-Alpes-Côte-d'Azur	Jan 1st 2004	4,506,000	3,120
Ile-de-France	Nov 1st 2004	10,952,000	1,878
Midi-Pyrénées	Feb 1st 2005	2,552,000	1,425
Basse Normandie	Feb 1st 2005	1,422,000	682
Paediatrics virtual region	Jan 1st 2004	N/A	127

Table 4 – Patterns of use of MSIS in the regions during the period of 17 months between June 16th 2004 and Jan 15th 2006.

Region	**Nephrologists & Health care professionals**			**Clinical Research Assistants & Referral Nephrologists**		
	Nbr sessions	**Session duration h:mn:sec**	**Average session mn:sec**	**Nbr sessions**	**Duration h:mn:sec**	**Average session mn:sec**
Limousin	2	00:12:20	06:10	915	195:16:23	13:10
Languedoc-Roussillon	719	484:14:59	40:24	1095	984:21:24	53:56
Champagne-Ardenne	578	268:10:25	27:50	454	139:45:05	18:28
Centre	361	184:55:55	30:44	1537	1121:14:35	43:46
Provence-Alpes-Côte-d'Azur	1145	638:26:36	33:27	1111	825:38:47	44:35
Ile-de-France	1885	831:47:55	26:28	1013	706:21:48	41:50
Midi-Pyrénées	998	348:03:33	20:55	572	261:56:40	27:28
Basse Normandie	5	20:32	04:06	416	50:07:21	07:13
"Paediatrics"	200	123:25:45	37:01	75	04:22:29	03:29

created 5893 sessions and connected during 2879 hours in the study period between June 16th 2004 and January 15th 2006.
The clinical research assistants insure the data quality control and the completeness of the patient information. They heavily rely on the MSIS to document patient files, to control the data quality and to prepare statistical analyses in the region for supporting health care decision making. They connected 7188 times during 4289 hours with an average of 35 minutes during the observation period of 17 months.
The department of Biostatistics and Medical informatics at Necker Hospital in Paris accompanied the nephrologists, clinical research assistants and all the MSIS users. This technical accompaniment proved necessary not only at beginning of the loading process, but also during its growth and evolution. Professionals from the laboratory created 3271 sessions which lasted 1614 hours.

5. Conclusion

Four years of on-line use of MSIS showed that the involvement of the professionals was effective as they bring the expertise and the accuracy to the patient information record and improve epidemiology and support to decision making. The technical research assistants brought an essential contribution to obtain the completeness of the data and permitted an efficient quality control assessed by web log analysis.

Acknowledgments

The nephrologists in charge of the ESRD units in Limousin, Languedoc-Roussillon, Champagne Ardenne, Centre, Provence-Alpes-Côte d'Azur, Basse-Normandie, Midi-Pyrénées, Ile-de-France, Paediatrics are acknowledged for their fruitful cooperation and comments. JP Necker and X Ferreira are acknowledged for their skilful help. This work was supported by a grant of the University of Paris 5 and by the Agence de la Biomédecine.

[1] Landais P, Simonet A, Guillon D, Jacquelinet C, Ben Saïd M, Mugnier C, Simonet M. SIMS@REIN : Un Système d'Information Multi-Sources pour l'Insuffisance Rénale Terminale. *CR Biol* 2002;325:515-528.

[2] Ben Said M, Simonet A, Guillon D, Jacquelinet C, Gaspoz F, Dufour E, Mugnier C, Jais JP, Simonet M, Landais P. A dynamic Web application within an n-tier architecture: a Multi-Source Information System for end-stage renal disease. Stud Health Technol Inform. 2003;95 ::95-100.

[3] Ben Said M, A Multi-Source Information System via the Internet for End-Stage Renal Disease: Scalability and Data Quality. Stud Health Technol Inform 2005 ;116 :994-999.

[4] Richard JB, Toubiana L, Le Mignot L, Ben Said M, Mugnier C, Le Bihan Benjamin C, Jais JP, Landais P. SIGNe: A Geographic Information System on the Web for End Stage Renal Disease. Stud Health Technol Inform 2005 ;116 :713-718.

[5] Richard JB, Toubiana L, Le Mignot L, Ben Said M, Mugnier C, Le Bihan Benjamin C, Jais JP, Landais P. A Web GIS for Health Care Decision Support. AMIA 2005 Symposium Proceedings, Page 365-369

[6] Aho AV, Kernighan BW, Weinberger PJ. The AWK Programming Language — Eds: Addison Wesley Publishing Company – Reading Massachusetts 01867 – 1988 – ISBN 0-201-07981-X

[7] Hoffer EP, Feldman MJ, Kim RJ, Famiglietti KT, Barnett GO. "DXplain: Patterns of Use of a Mature Expert System" in AMIA 2005 Symposium Proceedings – Page 321-324 – Washington DC October 22-26, 2005 – Charles Friedman Editor.

[8] Macron-Nogues F, Vernay M, Ekong E, Thiard B, Salanave B, Fender P, Allemand H. The Prevalence of ESRD Treated With Renal Dialysis in France in 2003. Am J Kidney Dis. 2005 Aug;46(2):309-15

3.3 HIS: Integration

Ubiquity: Technologies for Better Health in Aging Societies
A. Hasman et al. (Eds.)
IOS Press, 2006

Concept Representation in Health Informatics for Enabling Intelligent Architectures

Bernd G.M.E. BLOBEL [1]
eHealth Competence Center, Regensburg University Hospital, Regensburg, Germany

Abstract. Semantically interoperable health information systems have to be based on shared knowledge and harmonised terminologies/ontologies. Therefore, knowledge representation regarding domain concepts, terms, and relationships used must be harmonised. Starting with the Generic Component Model, different approaches for representing concepts in the healthcare area are discussed, demonstrating common principles and transformation ways.
Keywords: Architecture, Models, Semantic Interoperability, Knowledge Representation, Concepts, EHR, Security Policies.

1. Introduction

Any communication and co-operation between healthcare providers must be supported by intelligently interoperable health information systems. This challenge needs to be especially met for managed care and continuity of care concepts widely introduced in most of the developed countries to improve quality and efficiency of patient's care. Interoperability might be provided at different levels. Interoperability levels are ranging from simple data exchange and meaningful data exchange with agreed vocabulary to a functional interoperability with agreed communicating applications' behaviour, or finally to a service-oriented interoperability directly invoking applications' services. Health information systems enabling such advanced co-operation mentioned above in the managed care context are characterized by openness, scalability, portability, distribution at Internet level, service-oriented interoperability as well as appropriate security and privacy services. Finally, they have to be based on standards [1].

2. Definitions and methodology: Knowledge and its representation

For understanding, communicating, and changing the reality according to our social, environmental, organisational, or technical business objectives, the reality must be observed, described and interpreted properly in a closed cycle using models. A model is a representation of something, helpful in thinking about the real world without hav-

[1] Corresponding Author: Bernd Blobel, PhD, Associate Professor; eHealth Competence Center, University of Regensburg Medical Center; Franz-Josef-Strauss-Allee 11, D-93042 Regensburg, Germany; Email: bernd.blobel@ehealth-cc.de; URL: www.ehealth-cc.de

ing to deal with every detail of reality. A purpose of models is to create knowledge. An outcome of developing mathematical models is that it helps model builders and decision makers understand the relationships between important variables in a business situation. On the other hand, description and especially the interpretation of real systems are based on knowledge. Knowledge is a combination of instincts, ideas, rules and procedures that guide actions and decisions. It is used to transform data into information that is useful in a situation. Knowledge helps users interpret and act on information. Building terms implies knowledge. Therefore, the classification of term sets deals with the ordering of knowledge. A model for classification of terms consists of items and their instances (semantic concepts, terms), relationships between terms of a terminology and the classification for explicit presentation of that relation. A concept is a formal model. It shall be uniquely identifiable, accepted by experts and users, as well as independent. A concept as a knowledge component can be specialized and generalized as components can. It provides a coherent description of domain entities, which can be identified and independently used by domain users for recording information [2]. The sum of concepts is called „ontology". An ontology provides the formalization of the domain knowledge. Knowledge representation is frequently provided in two of many possible ways: through rules (production rules, if-then rules) or frames. Production rules focus on the logic of making inferences. Frames are object-oriented approaches that focus on objects' important characteristics. A frame consists of slots identifying attributes for the particular kind of entity. The data in a frame can be used to identify which aspects of a situation are pertinent, to organise the data, and to identify exceptions. Any of the slots in a frame could have default values or references to other frames. In other words, a concept is a coherent description of domain entities (components), which are separately identified and useable for recording information. Knowledge consists of concepts that can be composed or decomposed through generalization or specialization, respectively, representing relationships between them. The definitions given are mainly based on [3, 4]. Following, analysis, conceptualization, design, implementation, and maintenance of information systems in their combination of work practices, information, people, and information technologies organised to accomplish goals in an organisation will be considered.

3. The Generic Component Model Framework

A frame is a data structure for representing a specific entity such as a concept, item or class. Frameworks are used to categorize the information that is needed in order to fully describe an enterprise, and to store that information, typically with the support of an appropriate repository tool. A framework provides a taxonomy for relating the concepts that describe the real world to the concepts that describe an information system and its implementation [5].

In the early nineties, a framework for health information systems – the Generic Component Model – has been build which has been matured meanwhile providing the basis for a series of ISO specifications and health informatics products [1, 6]. For different domains such as administrative, medical, financial, and other ones, the structure and function of a component system can be developed from driving business needs, which are expressed in informational concepts, transferred into components of an information system and properly aggregated for providing required functionalities. Such logical (platform-independent) system representation can automatically be transferred into a platform-specific system architecture to be implemented and maintained. Interrelations between views as well as composition/decomposition of components are ruled by concepts or constraint models. The Generic Component Model describes a generic ontology for the classification of application systems by their domain reference, business architecture, informational reflection (expression and functional aggregation of information), granularity and modularity of its components (concepts) and their composition/decomposition.

4. Concept Representation for Health Information Systems

Like knowledge representation in the world of objects or components, also concepts are structured (organised) in slots (see frame definition). This has been done with all existing health concept representations such as Archetypes, Arden Syntax MLMs, OCL, but also with security policy representations using formal languages such as SAML, XACML, etc. In that context, also some ontology examples such as UMLS and SNOMED will be shortly introduced first.

4.1. Ontologies

An ontology classifies concepts of a specific real world domain regarding structure, function and relationship of their components. As any conceptual framework, the purpose of the framework, the business domain, and the enterprise architecture define the classification system, which has to be comprehensive. The classes must be disjunctive, reasonable populated, based on clear decision rules for inclusion, exclusion, etc. Regarding the different purpose, different ontologies for health have been developed such as UMLS, SNOMED, Read Codes, etc. Different ontologies of a domain are interrelated. A reference ontology allows for harmonising between them as shown in Figure 1. Business models according to the Generic Component Model can directly derived from an ontology system. Using a computation-independent approach, the domain knowledge for performing a specific business has to be represented defining Business Domain, Business Process, Location, Business Organization, Event, and Business Motivation regarding meta-models, concepts and relationships.

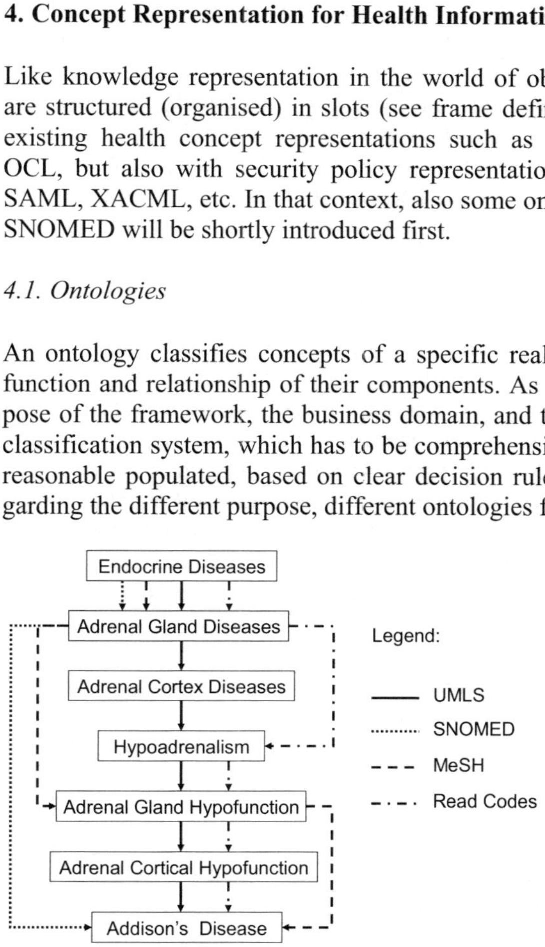

Figure 1: Concept Organisation in Different Classification Systems (after [7])

4.2. OpenEHR/GEHR Archetypes

Archetypes are another example for expressing business concepts derived from health domain related ontologies. Based on the European GEHR (Good European Healthcare

Record) project, the Australian Good Electronic Health Record project has been launched to provide internationally harmonised specifications of Archetypes as well as open source tools and implementations delivered from the global openEHR Foundation [8]. Archetypes provide information models of business concepts in health expressed by domain experts deploying their language. Archetypes can be aggregated to a higher complexity according to the composition/decomposition level given in the Generic Component Model. Therefore, existing Archetypes may represent a medical concept such as "blood pressure", the combination of concepts "blood pressure" and "blood pressure measurement", etc. The outer structure of an Archetype consists of three components called parts as presented in figure 2: header_part, body_part and terminology_ part [8]. The header part syntax requires the definition of archetype id, concept name, concept hierarchy, and finally concept description containing purpose, use, submitter, author, status and other meta-data items of that concept. The body part of an archetype contains hierarchically structured constraint statements blocks defining the domain concept. Any given valid use of the syntax is an actual Archetype. Therefore, each block corresponds to something like an "instance" of a class, rather than part of a class. Notional classes define the semantics of archetypes. Hierarchically arranged blocks of syntax correspond to composition of objects in an Archetype. The constraints are defined by references or by re-using knowledge representation explained above. The terminology part refers to the language in which that concept has been expressed first as well as to other languages the concept is available in. For classifying the concepts in different languages, the related terminologies have to be identified. Next, the terms used for knowledge representation in a certain language have to be defined. As core of domain knowledge representation, specific constraints must be declared regarding language, local constraint definitions as well as the reference terminologies deployed. Finally, for both terms and constraints bindings between the terminologies used and alternative terminologies must be given for open deployment of archetypes. The Archetype Definition Language (ADL) borrowed from CORBA's IDL [8].

header_part
    archetype identification, meta-data
body_part
    archetype definition
terminology_part
    terminology definitions and binding

Figure 2: Archetype Outer-Structure

4.3. Arden Syntax

The Arden Syntax is a knowledge representation language, developed at ASTM and now maintained by HL7 [9]. It forms a frame for knowledge representation in the field of health. The represented knowledge concepts are called Medical Logic Modules (MLMs). The Arden Syntax has been developed for enabling the sharing of computerised health knowledge bases among personnel, information systems and institutions. The Arden Syntax can be used to represent knowledge bases as a set of discrete modules, those MLMs. A MLM consists of three components called catego-

```
maintenance:
    slotname: slotbody;;
...
library:
    slotname: slotbody;;
...
knowledge:
    slotname: slotbody;;
...
end:
```

Figure 3: General MLM Layout

ries: maintenance category, library category and knowledge category. Each category is represented as a series of slots as demonstrated in figure 3.

4.4. OCL Constraint Modelling

```
package <packagePath>                                       package
context
    context <contextualInstanceName>:<modelElement>   expression
                                                            context
        <expressionType><expressonName>:                expression
            <expressionBody>
        <expressionType><expressionName>:               expression
            <expressionBody>
        ...
endpackage
```

Figure 4: OCL Constraint Modelling

OCL is a standard extension to UML that allows for querying model elements, constraining them (at modelling time) and defining query operations [10]. It enables the integration in the architectural process. Defining business rules is a challenge to be met in the General Component Model paradigm. OCL is constraining models with specific behaviour. OCL expressions consist of three parts: package context (optional), expression context (mandatory), one or more expressions. OCL constraints are organised in expressions as shown in figure 4.

4.5. Other Constraint Representations

According to the Generic Component Model, healthcare and its supporting information systems are dealing with other domains beside medicine and biology. In that context, finance, technology, legislation and security, etc. have to be mentioned. Regarding the latter one, legal and policy concepts have to be modelled. A policy covers all implications on health and health information systems such as legal, social, organisational, psychological, functional, and technical. OASIS' Security Assertion Markup Language (SAML) defines security services assigned to entities in a header-body-reference structure using XML. For formally modelling policies and ruling access control, the Extended Access Control Markup Language (XACML) has been developed at OASIS with the XML meta-language. XACML defines three top-level policy elements: <Rule>, <Policy> and <PolicySet> [11].

5. Discussion

- Concepts
  - Synonymous terms are clustered into a concept
  - Properties are attached to concepts, e.g.,
    - Unique identifier
    - Definition
- Relations
  - Concepts are related to other concepts
  - Properties are attached to relations, e.g.,
    - Type of relationship
    - Source

Figure 5: Knowledge Presentation through a Metathesaurus [7]

For expressing and sharing knowledge, the underlying concepts must be expressed properly deploying common languages, domain-specific languages, formal languages and formal models. Terms and knowledge applied can be summarised in terminologies and ontologies. A metathesaurus defines the presentation of domain knowledge as shown in figure 5. Domain-specific concepts results from business requirements. Therefore, any development framework for advanced information systems has to start with processes and methods to developing requirements and enterprise architecture, i.e. the creation of a business model. A business model collection of related architectures or blueprints of, by, and for domain experts, aimed toward capturing the business essence, but not the ICT perspective. In the next phase, a methodology has to be provided to develop, deploy, test, maintain, and integrate applications. Providing the complete development framework, the Generic Component Model describes any business model through the Enterprise View. The model's three-dimensional architecture allows for knowledge representation by concepts and their relationships including generalisation and specialisation. Alternatively, Archetypes have been established for healthcare business concept modelling. Table 1 compares different business and constraint modelling as well as representation approaches.

Table 1: Concept Representation Approaches

Approach Components	**GEHR/ openEHR**	**Arden**	**HL7 EHR/ CDA**	**SOA/MDA**	**OASIS security services**
Business modelling	Archetypes	Common language	Clinical Templates	TOGAF, CIM, MOF	--
Knowledge representation	GEHR parts	Arden Syntax Categories	HL7 CDA Structure	OCL Package	SAML/ XACML
Identification	Header (contains also ext. metadata)	Maintenance category	Header	Header	Header
Content	Body	Knowledge category	Body	Body structure Body	Body
References	Terminology	Library category	Embedded terminology Extl. refs	External reference	Extl. refs
Substructures	Blocks	Slots	Entries	UML components	Elements
Language	ADL	Semiformal language, logical ops.	XML	OCL, typed specification language	XML

6. Conclusions

For realising semantic interoperability, business concepts developed by domain experts have to be integrated into advanced architectural approaches established by informaticians. Therefore, the representation of domain concept must be harmonised by introducing unifies processes for requirements analysis, design, implementation, test, and deployment. The Generic Component Model provides a proper framework for harmonising different approaches. Nevertheless, a new and extended educational challenge has to be mastered properly. Details covering those educational requirements have been analysed. The resulting components for a corresponding curriculum are presented in [12].

Acknowledgement

The author is in debt to Thomas Beale (Deep Thought Informatics Pty Ltd, Mooloolah, Qld, Australia), Sam Heard (Flinders University, Northern Territory Clinical School, Casuarina, NT, Australia), Dipak Kalra (CHIME, University College London, London, UK) as well as the colleagues from CEN TC 251, ISO TC 215 and CORBA/OMG for kind collaborations.

References

[1] Blobel B. Analysis, design and implementation of secure and interoperable distributed health information systems. Stud Health Technol Inform, Vol. 89. Amsterdam: IOS Press; 2002.
[2] Beale T. Health Information Standards Manifesto, Rev. 2.0. Mooloolah: Deep Thought Informatics Pty Ltd.; 2001.
[3] Alter S. Information Systems – A Management Perspective. Reading: Addison-Wesley Publishing Company; 1992.
[4] Collins Compact English Dictionay. Glasgow: Harper Collins Publishers; 1994.
[5] Blevins T, Spencer J, Waskiewicz F. TOGAF ADM and MDA®.http://www.opengroup.org/cio/MDA-ADM/MDA-TOGAF-R1-070904.pdf (Last accessed January 05, 2006)
[6] Blobel B. Advanced EHR architectures – promises or reality. Methods Inf Med 2006; 45: pp 95-101.
[7] Bodenreider O. The Unified Medical Language System. Tutorial Handouts. Geneva: MIE 2005; 2005.
[8] Beale T, Heard S. Archetype Definition Language. openEHR Foundation. Chatswood: Ocean Informatics Australia; 2004.
[9] HL7 Inc.. http://www.hl7.org (last accessed January 18, 2006)
[10] Arlow J, Neustadt I: UML2 and the Unified Process, 2nd Edition. Upper Saddle River: Addison-Wesley; 2005.
[11] OASIS: http://www.oasis-open.org (last accessed January 16, 2006)
[12] Blobel B. Educational Challenge of Health Information Systems' Interoperability. Methods Inf Med 2006 (in print).

Ubiquity: Technologies for Better Health in Aging Societies
A. Hasman et al. (Eds.)
IOS Press, 2006

Integrating Radiology Information Systems with Healthcare Delivery Environments using DICOM and HL7 Standards

Bojan BLAZONA[a,1], Miroslav KONCAR[a]
[a] *Ericsson Nikola Tesla d.d., Zagreb, Croatia*

Abstract. Integration based on open standards, in order to achieve communication and information interoperability, is one of the key aspects of modern health care information systems. Interoperability presents data and communication layer interchange. In this context we identified the HL7 standard as the world's leading medical Information and communication technology (ICT) standard for the business layer in healthcare information systems and we tried to explore the ability to exchange clinical documents with minimal integrated healthcare information systems (IHCIS) change. We explored HL7 Clinical Document Architecture (CDA) abilities to achieve radiology information system integration (DICOM) to IHCIS (HL7). We introduced the use of WADO service interconnection to IHCIS and finally CDA rendering in widely used Internet explorers.

Keywords: Healthcare information Systems, DICOM, HL7 CDA, WADO, XML

1. Introduction

Healthcare delivery environments are under constant pressure to rationalize the cost of care provisioning while at the same time they have to preserve or even increase the quality of care pathways and clinical processes. In the process of evaluation how to address this stringent set of requirements, integration and integrated personalized care are recognized as the major quality component [1]. In that sense, Hospital Information Systems (HIS) and integrated healthcare information infrastructures need to address these issues by defining business processes in care delivery settings, and identify the integration mechanisms that include business scenarios and use cases, semantics and communication technology.

In the process of integration, high quality data management based on open standards represents the stepping stone in achieving the goals of integrated care. In that context, we have recognized radiology systems as one of the major medical data generators that produce a large quantity of important clinical information, which needs to be efficiently integrated in patients' medical records. Having that fact in the focus of our work, in this paper we evaluate a pilot approach to develop a generic framework for the integration of DICOM (**D**igital **I**maging and **CO**mmunications in **M**edicine) based Radiology Information Systems (RIS) to the integrated HL7v3 (**H**ealth **L**evel **7**

[1] *Contact information:* *Bojan Blazona, Krapinska 45, p.p. 93, HR-10002 Zagreb, Croatia*
email: bojan.blazona@ericsson.com

ver. **3**) enabled healthcare information systems (HIS). The developed DCM-CDA (**DICOM–CDA**) service implementation enables a reliable transformation of DICOM objects to CDA (**C**linical **D**ocument **A**rchitecture) conformant XML (E**x**tensible **M**arkup **L**anguage) documents. The access to DICOM images is based on the emerging WADO (**W**eb **A**ccess to **D**ICOM Persistent **O**bjects) definition. All of our work is fully aligned with the solution definition and services components included in the Croatian national healthcare information network that is elaborated in detail in [2].

2. DICOM and HL7 Interoperability Needs

The DICOM standard represents the evolution of the historical ACR-NEMA (The **A**merican **C**ollege of **R**adiology - **N**ational **E**lectrical **M**anufacturers **A**ssociation) sets of recommendations [3]. DICOM specifies the network communication rules and the physical data format to enable communication between modalities and humans through developed software solutions. In this context we are focusing on the DICOM supported datasets exchange, which requires specific connectors and communication enablers that use other widely accepted standards like HTTP (**H**yper**T**ext **T**ransfer **P**rotocol). This requirement has been identified as one of the key integration components, and has resulted in new additions to the DICOM standard known as WADO[4]. WADO enables the HTTP communication of DICOM objects by defining the GET and POST parameters of the HTTP request with pre-defined types of returning context e.g. image/jpeg and text/xml.

The HL7 standard today represents the foundation of many healthcare information management systems[5]. It specifies structures and mechanisms to handle administrative and clinical data without focusing on a specific healthcare domain. One of the HL7 standard packages that we refer to in our work is CDA. CDA introduces document-based specifications to exchange various clinical data in healthcare delivery environments, e.g. laboratory reports, discharge letters etc.

Based on the fact that HL7 CDA is widely used as the information layer integration solution for heterogeneous healthcare delivery environments, we have identified these specifications as the bridge of connectivity between RIS systems that are based on DICOM standard, and HL7 enabled HIS systems. However, today there are no generic data transformation mechanisms, and most of the transformations are done on a case-by-case basis. Having identified this state, we have ranged our work from HL7 and DICOM information models to communication rules and mechanisms[6], with the goal of developing the architectural framework for DICOM/HL7 CDA integration.

2.1. DICOM and HL7 Information Models

The DICOM to CDA transformation consists of semantically mapping textual and multimedia contexts. We consider the textual data mappings according to the semantic meaning, and multimedia data linking to the defined WADO services. Direct linking to WADO services enables acquiring multimedia sub context information from DICOM datasets directly from distant radiology departments.

The DICOM information model consists of IOD (**I**nformation **O**bject **D**efinition) elements. Each IOD saves data relevant to the IOD global context. For example, the patient's name, sex and ID are encapsulated in the Patient IOD. In HL7 CDA the patient relevant data is saved under the patient element, and formatted using XSD (**X**ML **S**chema **D**efinition) specifications and rules. Similarities in both information models are evident from the semantic meaning according to mapping relevant information from the DICOM dataset to the CDA Header.

The DICOM and CDA information models have slightly different base models (Figure 1). The DICOM tag and the CDA element pair define the mapping of information relevant to the CDA header. The CDA header should provide administrative information about included clinical data. In our approach we define mappings according to the pairs DICOM tag - CDA element place.

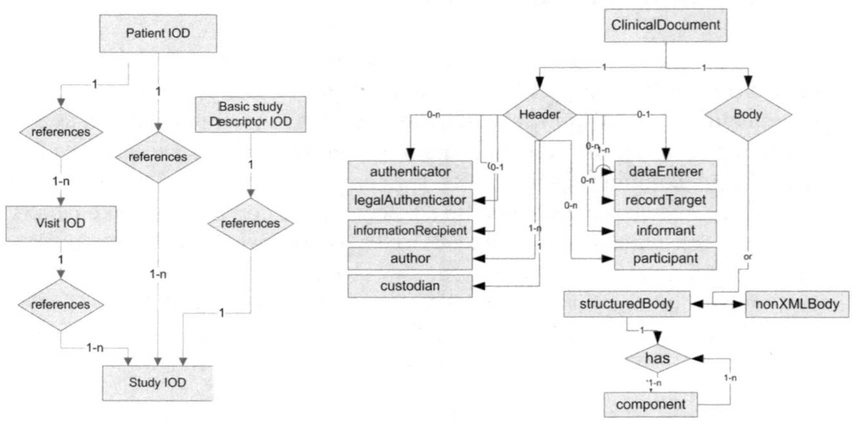

Figure 1. Information model of DICOM (left) and HL7 CDA (right). Information in Patient IOD, Visit IOD and Study IOD, according to the DICOM information model, is mapped to the corresponding place in the CDA XML element.

The CDA Body element saves clinical data which can contain textual and multimedia data. All multimedia data should be linked according to CDA external data definition. Linking presents a hyperlink to the multimedia data, and is available with simple HTTP transfer protocol.

The DICOM datasets must be divided into parts presenting separate multimedia content, providing support for multiple images management. Separate multimedia content is then available using the implementation of the DICOM WADO specification. WADO specifies HTTP GET parameters to query and retrieve modalities of underling DICOM solutions to retrieve whole data sets or separate elements (e.g. images) according to the slice number.

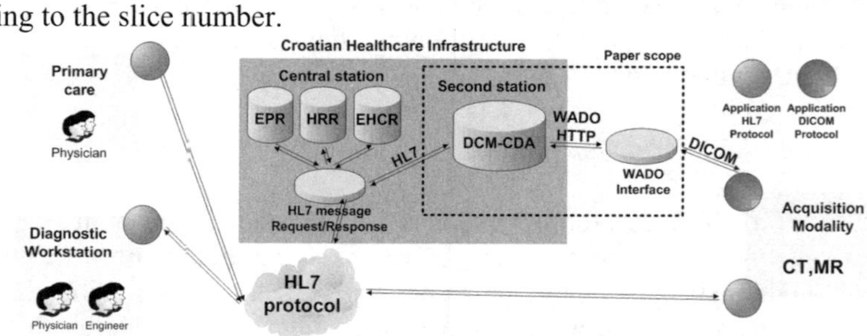

Figure 2. System architecture enabling HL7 to DICOM communication, through WADO service. Abrevations: Electronic Population Register (EPR), Health Resource Register (HRR), Electronic Health Care Record (EHCR)

2.2. DICOM Context Management for HL7 Enabled Communication

Corresponding to the HL7 protocol, a "plug-in" DCM-CDA web service upgrade is required to enable HL7 to DICOM communication. We determine this service should communicate with HL7v3 messages based on the HTTP GET request to the WADO service as illustrated on Figure 2. Note that authentication and authorization services for this framework are implemented by a third party service provider, i.e. directory services and access manager in the case of Croatian national healthcare infrastructure.

3. Results

The presented integration framework with the DCM-CDA service upgrade has been implemented in a testing environment as illustrated in Figure 3. The testing environment consisted of a DCM-CDA integration component, a web based WADO implementation and a HIS portal web application which simulated the services provided by the Croatian national healthcare infrastructure. HL7 communication with DCM-CDA was facilitated using the Web Service's implementation (Figure 4).

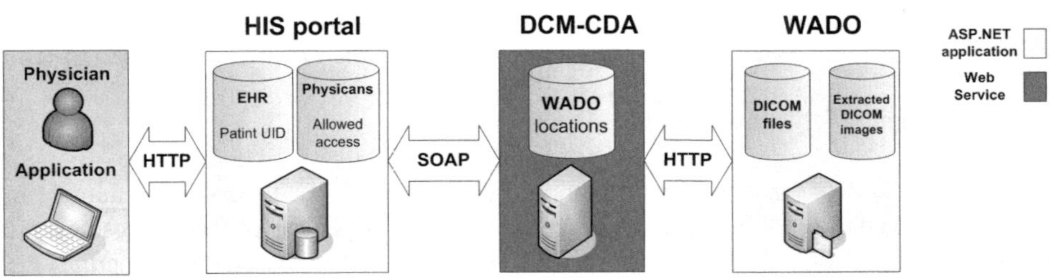

Figure 3. Testing environment consist of a HIS portal, a DCM-CDA service and a WADO application implementation.

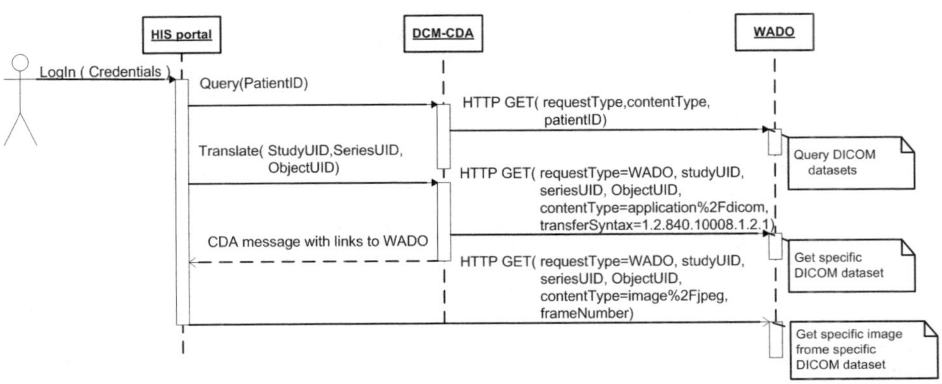

Figure 4. Sequence diagram of communication in test environment.

3.1. DCM-CDA Service Configuration and Implementation

The main responsibility of the DCM-CDA service is to reliably translate DICOM objects to a CDA conformant document. DICOM transformation to CDA and the final HTML (**H**yper**T**ext **M**arkup **L**anguage) presentation goes through the DICOM parser transforming the DICOM binary dataset to a DICOM xml presentation. MSXSL (**Mi**cro**s**oft's E**x**tensible **S**tylesheet **L**anguage) transformations processor transforms the DICOM xml presentation to a CDA document according to the DICOMtoCDA.xsl transformation definition. The final presentation is then generated in the client's user interface, which, in this case, is a simple browser. The DICOM CDA presentation is rendered according to CDA to HTML transformation as specified by the standard (Figure 5).

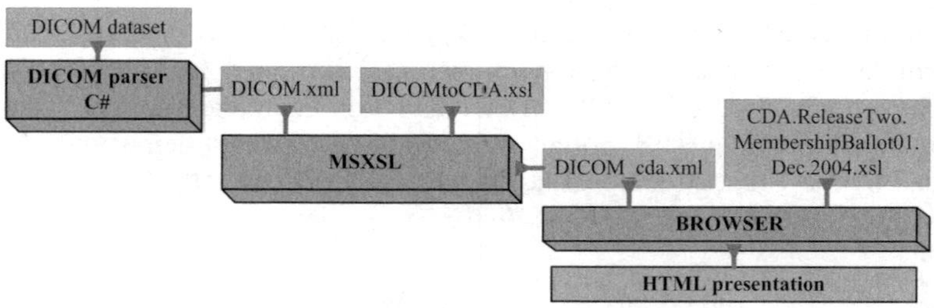

Figure 5. DCM-CDA service and dataset transformation

3.2. DICOM to CDA Transformation Flow

The DICOM datasets are first transferred to a simple XML format, as annotated in the previous chapter. The XML representation of the DICOM object represents the input for DICOMtoCDA.xsl transformation. The DCM-CDA service then generates the final DICOM CDA presentation based on this input. The DICOM to CDA transformation consists of several transformations, where each transformation element handles a different CDA sub domain (Figure 6). That makes the service modular and easy to maintain.

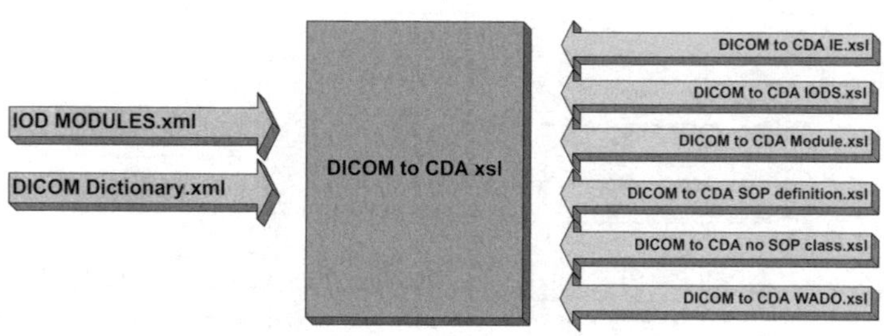

Figure 6. DICOM to CDA transformation

3.3. User-end Presentation of the CDA Document

The HTML presentation of the CDA translated DICOM dataset is generated according to the pre-defined XSLT (E**x**tensible **S**tylesheet **L**anguage **T**ransformations) transformation scheme used by the client's web browser (Figure 7). To meet with the performance and bandwidth utilization requirement, all images are retrieved directly from the WADO web application, according to the generated external resource elements in the CDA document. External resources are transformed to a HTTP request with GET method required attributes according to WADO specification.

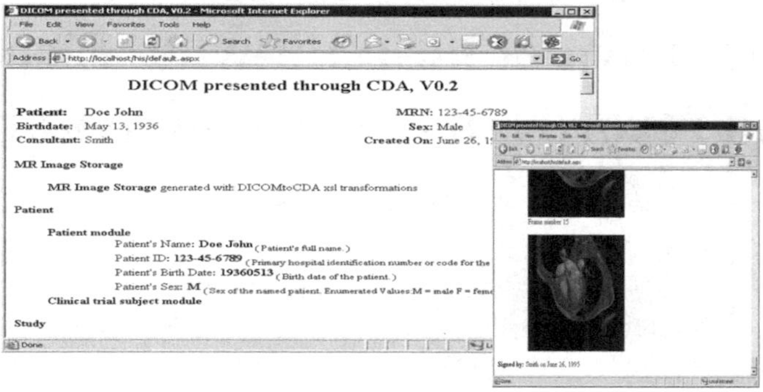

Figure 7. Sample of DICOM dataset transformed to CDA and rendered in web browser.

4. Conclusion

The developed DCM-CDA represents an easy to use, configurable and modular component which facilitates the main integration bridge from the medical imaging management solutions to the healthcare information networks. Our solution starts with the detailed analysis of information models, and facilitates WADO definitions to provide access mechanisms from patient clinical data to diagnostic images. XSLT based transformation gives the advantage of upgrades in accordance with most recent specific standard changes. Our results prove that the technical correctness of this component meets the needs of RIS/HIS integration based in the DICOM and HL7 standards. Usage of the IHCIS infrastructure to provide access to diagnostic images can be used in direct support of telemedicine applications. Readiness of the Croatian's IHCIS and further analysis of the DCM-CDA use cases, should encourage the development of the pilot project which will, in turn, give us insight to user responses.

References

[1] "Crossing the Quality Chasm: A New Health System for the 21st Century", Institute of Medicine (IOM), 2001, ISBN 0-309-07280-8

[2] M. Končar, D. Gvozdanović, "Primary healthcare information system - The cornerstone for the next generation healthcare sector in Republic of Croatia", International Journal of Medical Informatics, In Press, Corrected Proof, Available online 4 October 2005, ISSN 1386-5056

[3] NEMA. DICOM standard 2004, [Online] http://medical.nema.org/dicom/2004.html (accessed January 2006)

[4] E. Cordonnier., B. Hervault, K. Rommelse, D. Harvey, F.M. Behlen. [Online]" Early Implementations of WADO Web Access to DICOM Persistent Object", ISO TC125 / DICOM, San Diego, 2004, http://medical.nema.org/Dicom/minutes/WG-10/2004/2004-01-24/WADO-implementations-20040123.ppt, (accessed January 2006)

[5] HL7 Inc. Official Web Page, [Online] http://www.hl7.org (accessed January 2006)

[6] B. Blazona, M. Koncar, „Osnove za komunikaciju između DICOM i HL7 normi", Med. Inform. 2005, Rijeka, Croatia, pp. 47-53., ISSN 1330-1799

Ubiquity: Technologies for Better Health in Aging Societies
A. Hasman et al. (Eds.)
IOS Press, 2006

The fate of clinical department systems at the dawn of hospital-wide Electronic health records in a Norwegian university hospital

Eivind VEDVIK[a,1] and Arild FAXVAAG[a]
[a] *Norwegian Research Centre for Electronic Patient Records,*
[1] *Faculty of medicine, Norwegian University of Science and Technology (NTNU)*
Trondheim, Norway

Abstract. Objective: a) To document the presence and use of clinical department systems (CDS) in a university hospital that implemented a hospital-wide electronic health record (EHR) in 1999 and b) To compare clinical use of the CDS with that of the EHR. **Method:** Identification of CDSs in use by contacting leaders and senior physicians at clinical departments at the hospital. Identification of key properties of each CDS by interviewing users. **Results**: We identified a total of 60 CDSs, of which 53 fell in one of four categories; Journal or documentation system tailored to a department or medical specialty (19 systems), Software bundled with electronic medical equipment (control/ storage/ presentation) (14 systems), Logistics/ administration/ planning/ appointments (10 systems) and Database for medical research (10 systems). Many CDSs were described to outperform the EHR system with regard to ability to provide better patient overview and better support for registering patient data. CDSs are not integrated with the EHR and thus contain islands of data. **Conclusion**: CDSs continue to fill important roles and there is no tendency towards that the hospital-wide EHR makes CDSs obsolete.
Keywords: Electronic medical record systems. Clinical departmental systems.

1. Introduction

Hospitals are highly complex organizations established to meet the needs of patients with potential or manifest diseases in need of specialty care. Due to the nature of diseases, combined with the risks associated with the actions health personnel take to combat disease, the situation of a hospital patient is that of a person who is exposed to threats. Accordingly, the situation of hospital workers is that of experts assigned a mandate to explore, analyze, and subsequently manage diseases as risks to the health and well being of their patients.

Due to the multitude of medical conditions and the diverse repertoire of methods that has evolved and must be mastered, the work force in hospitals has grown diverse and highly specialized. This proliferation of diagnostic and therapeutic methods and the

[1]Corresponding author: Eivind Vedvik, Norwegian Research Centre for Electronic Patient Records, Faculty of medicine NTNU N-7481 Trondheim, Norway. E-mail: eivindve [at] stud.ntnu.no

corresponding work force diversity is perhaps what has contributed most to the impression that hospitals are highly complex organizations [1]. Furthermore, the increase in number of health care actors that comes with increased specialization has made the service appear less continuous. This has accentuated the demand for better coordination of care [2].

With the vision that the use of Electronic health record (EHR) systems can simplify coordination, planning, performance, documentation and evaluation of care, many countries and hospital organizations have implemented and started to use such systems. The extent to which hospitals have an EHR varies however greatly. In Europe, hospitals in the Nordic countries, Belgium and the Netherlands rank among those with the highest EHR system implementation rate [3]. Many other countries have drafted plans to develop and implement EHR systems but still lag behind when it comes to implementation. When it comes to clinical use of EHR systems that have been implemented, many investigations have found that EHR systems still lack important functionality and that much remains to be developed if the vision of the EHR system as the central hospital information handling and communication system shall be achieved [4, 5].

In parallel to efforts aiming at developing hospital-wide EHR systems, many clinical departments and specialist groups have taken initiative to development and implementation of smaller clinical department systems (CDS), typically tailored to the needs of one particular group of patients or specialists. Examples of CDSs are local quality registers and databases that are integrated with medical technology. With the assumption that CDSs are in widespread use and still occupy central roles in hospital departments we have identified and here describe the flora of CDSs in a hospital that implemented a hospital-wide EHR system back in 1999. We find that CDSs present at the time when the hospital-wide EHR was introduced still perform functions essential to patient care in the departments using them and that the CDSs outperform the EHR system with regard to ability to provide better patient overview and better support for registering patient data.

2. Materials and methods

The study was performed at St. Olavs Hospital, a 930-bed university hospital located in Trondheim, Norway. Its' EHR system (Doculive EPR, developed by Siemens Healthcare), which was implemented in 1999, now is used to document the activities of doctors, nurses and physiotherapists. Also, the hospital has implemented a PACS / radiology information system, and the hospital has had a laboratory information system since before the EHR. The core functionality of the EHR system consists of support for the production and view of health record text, and for workflow around health record documents, i.e. exchange of referral letters and medical discharge summaries. Also the EHR system supports ordering (of imaging analyses, laboratory analyses and other laboratory tests) and writing prescriptions. The hospital has not yet taken the step to remove the paper-based medical record and thus still update the paper-based medical record along with the EHR.

Data were collected between June 2004 and May 2005. The definition of a CDS used for acceptance of a system was "software made for use in performing or evaluation of health service in a medical speciality not generally available across the hospital as the EHR". Systems were identified by visiting the different wards at the hospital asking for systems fitting our definition because no complete updated database of such systems existed. Our material includes 60 CDS identified by this method as of May 2005. Nine systems come from department of ear, nose and throat & eye, eight from surgery, seven from cancer & dermatology, seven from laboratory medicine, six from medicine, six from cardiac medicine, four from anaesthesia & ER, three from paediatrics, two from radiology & imaging, two from obstetrics & gynaecology, two from neurology, two from orthopaedics and rheumatology, one from physical medicine and one from lung and occupational medicine.

Data for each CDS were collected using a structured questionnaire. The questions were developed after an initial study of a selection of CDSs and general literature of medical informatics. We attempted to identify key variables in different categories of properties. A total of 65 variables were collected for each system regarding classification, development, availability, users, population, patient data and functionality.

Registration of a particular CDS typically included a brief demonstration of both everyday use as well as more advanced functionality by either an end user or someone with administrative responsibility for the system. After getting a comprehension of the use and purpose of the system, the researcher and user filled out the questionnaire in collaboration. Definition of terms and use of clarifying examples ensured that the user felt confident about selecting the correct value to each question. If the user felt uncertain of a specific value, the question was then later addressed to someone with administrative responsibility for the system. All records were made by the same researcher.

The forms were entered into Microsoft Access 2002. SPSS 12.0.1 for Windows were used for the analysis.

3. Results

3.1. Categorisation of CDSs identified

For the purpose of classification, we described 12 categories in which we asked the respondents to group the principal function of the CDS. By this method, 53 of the 60 CDSs identified were grouped as belonging to one of four categories (table 1). Most CDSs are considered a health record tailored to a particular department / clinical specialty.

Table 1. Categorisation of CDS classification

Classification	*Frequency*	*Percentage*
Journal/ documentation system tailored to a department/ medical specialty	19	31,7
Storage/ register for biological numbers (symptoms, findings)	1	1,7
Software bundled with electronic medical equipment (control/ storage/ presentation)	14	23.3
Computer assisted real-time patient monitoring	1	1,7
Medication: Search/ dosage/ interaction/ prescription	1	1,7
Logistics/ administration/ planning/ appointments	10	16,7
Communication with public health service	0	0
Communication between patient and public health service	0	0
Database for medical research	10	16,7
Tool for medical diagnostics	2	3,3
Tool for selection/ planning of intervention	2	3,3
Tool for follow-up of treatment in progress	0	0
SUM	**60**	**100**

3.2. Use of CDSs to get patient overview

For 30 % of the 60 CDSs, users reported that their system was able to "Present patient and case history (incidents) in a more suitable or better adapted way" compared to EHR or the paper-based health record. Users of 50 % of the systems reported that their systems were able to "Perform computation/ interpretation on the collected patient data", thus enabling more efficient access to a synthesis of patient data.

3.3. Use of CDSs for registering patient data

For a medical specialty, a custom-made system can be felt more attractive compared to EHR. In 36,7 % of the CDSs, the system was able to "Ease the work of making notes by offering templates or presenting list of commonly used words and phrases". This property was seen more often in departments who performed highly specialized diagnostic or therapeutic procedures or at outpatient clinics responsible for following up on patients suffering from a particular disease or undergoing special treatment. However, more effective registration and storage of patient data did not necessarily imply that the availability of these data had improved accordingly.

3.4. CDS-EHR integration

41,7 % of the CDSs could retrieve data from other databases and thus deserving the characteristic of being an integrated system. However, only 21,7 % of the systems was able to "export their data to other patient databases". For most systems "integration" meant that data on patient demographics were obtained from the Patient administrative system. The most common means of "exporting" patient data from a CDS was to use the "copy and paste" function in the Windows™ operating system. Even more archaic, cumbersome and error prone, doctors would update the hospital-wide EHR manually by writing a summary and conclusion of the relevant data in the CDS. Hence, patient data stored in a CDS could only be reached from workstations equipped with the CDS and were beyond reach from other systems or from workstations outside the department.

3.5. Use of CDSs to generate reports

Outcome measures can be considered indicators of quality of the care given at the department. Quality improvement is the process of obtaining and analyzing data on treatment outcomes and efficiency and using these data to identify areas of possible improvement. We found that 38,3 % of the CDSs were used to create "Reports for use in quality improvement" in the ward. Furthermore, 50 % reported that their system generated "Reports being used by management and administration of the department". Reports were also used to give early alerts on possible loss of quality of care i.e. increased complication rate of given anaesthesia.

3.6. Installation of CDSs before and after the hospital-wide EHR

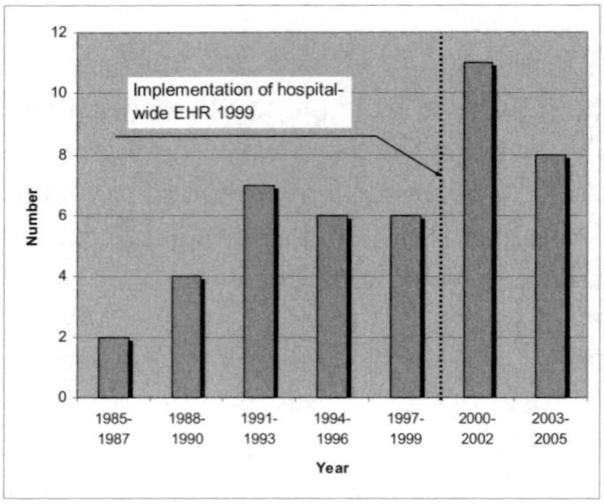

Figure 1. Year of installation. 44 of 60 CDSs

Figure 1 shows the year of installation for 44 of the 60 CDSs. The distribution ranges from 1985 until the year of registration in 2005. We found a tendency towards more systems of newer dates based on the systems in use at the time of registration in 2005. We found no tendency towards a decreased rate of new CDSs after the implementation of a hospital-wide EHR in 1999. 19 of the 44 CDSs were installed after 1999.

4. Discussion

In this report, we have described the characteristics of 60 CDSs that co-exist with the hospital-wide EHR in a 930 bed Norwegian university hospital. This flora of CDSs includes systems that have been in use and carefully upgraded for 20 years as well as CDSs recently implemented. The continued existence of CDSs that were implemented long before the EHR system and the seemingly continuous implementation of new CDSs indicate that the departments' and specialists' needs for storing and processing patient data not are met by the hospital-wide EHR system. Some of the newer CDSs have however been introduced because they accompany a digital medical imaging modality that has been purchased by the department.

For a clinical department, the care for patients with a particular disease or other patient groups depends on the ability to obtain, store, retrieve and process specific patient data. The department needs to process patient data both to document and to continuously improve the care given and for research purposes, to be able to contribute to the build up and maintenance of the knowledge base of the specialty. As of today, the EHR system meets neither of these needs. The EHR presents itself as an electronic copy of the old paper-based health record, and the EHR system seem better suited for creating documents describing care rather than a tool for support of the care provider when providing care, or evaluating care that has been provided. Facing an IT-department with an agenda to implement a hospital-wide EHR system, the particular needs of one clinical department often are ignored. In this situation, some departments have taken the initiative to development or purchase of a CDS. Compared to waiting for an update of the EHR system, this alternative gives the department better control over configuration, user interface and functionality of the final solution.

The description of the CDS as a system that can be obtained more easily than an hospital-wide EHR is in accordance with that of Ellingsen, who states that development and implementation of a complete EHR system is highly complex and time demanding and that acquiring or development of a CDS may be a less challenging alternative [6]. The department have better control on technical solution, user interface and functionality of a CDS in the pipeline. This is in great contrast to the top-down implementation of an EHR where the end-users might perceive having little or no influence. Owned and sometimes administered by the department, CDSs may be continuously developed to fit the changing needs. A CDS that has been developed locally, either by computer- savvy doctors or other staff at the department has higher chances of being followed with enthusiasm. The feeling of ownership and having a system presenting a limited but highly useful set of functionalities seems like a strong factor for the survival of many CDSs.

The existence of a diverse flora of CDSs does however also involve some undesirable effects. As the EHR becomes the legal health record, relevant patient information needs to be documented in the EHR. Parallel storage is labour intensive and the possibility for human errors is present. In a technical perspective, the diverse flora may be a great challenge for the hospitals IT department. Some smaller systems may also not meet the high standards of reliability as the hospital-wide EHR. Even though there are drawbacks concerning the co- existence of many CDSs and the EHR, the CDSs certainly continue to serve an important role in modern specialized health service. As the health services will continue to specialize and the amount of patient data will increase, the demands for more suitable systems will become stronger. If the hospital-wide EHR systems shall meet these demands, they must evolve from merely systems to support the production of documents to systems to support the production, quality control and continuous improvement of care.

5. Conclusion

We found no tendency towards that the implementation of a hospital-wide EHR system makes CDSs obsolete. CDSs in use today have true qualities not yet to be found in the general EHR. Combined with an increasing need for specialized documentation in modern hospitals, we believe that CDSs will be of great importance in many years to come. Further analysis of the factors making CDSs successful should be considered when designing and developing successful EHR systems.

Acknowledgments

This research was supported by:
Norwegian Research Centre for Electronic Patient Records (NSEP)
Faculty of medicine, Norwegian University of Science and Technology (NTNU) Trondheim, Norway

References

[1] Donini-Lenhoff FG, Hedrick HL. Growth of specialization in graduate medical education. JAMA : the journal of the American Medical Association. 2000;284(10):1284-9.

[2] Haggerty JL, Reid RJ, Freeman GK, Starfield BH, Adair CE, McKendry R. Continuity of care: a multidisciplinary review. BMJ (Clinical research ed). 2003;327(7425):1219-21.

[3] Grimsmo A. The dissemination of electronic patient records (EHR), standardisation and electronic messaging in leading countries in 2005. Proceedings Kyoto: Wonca International Classification Committee, 2005 2005 [cited 2006 January 30.]; Available from: www.nsep.no/publications/EPR_in_leading_countries.pdf

[4] Berg M. Medical work and the computer-based patient record: a sociological perspective. Methods Inf Med. 1998;37(3):294-301.

[5] Laerum H, Ellingsen G, Faxvaag A. Doctors' use of electronic medical records systems in hospitals: cross sectional survey. BMJ (Clinical research ed). 2001;323(7325):1344-8.

[6] Ellingsen G, Monteiro E. Big is beautiful: electronic patient records in large Norwegian hospitals 1980s-2001. Methods of information in medicine. 2003;42(4):366-70.

3.4 HIS and Electronic Health Records

Ubiquity: Technologies for Better Health in Aging Societies
A. Hasman et al. (Eds.)
IOS Press, 2006

Software Development for the Estimation of the Mean DRGs related Treatment Cost

Maria BOTSIVALY[1], Basile SPYROPOULOS, Michael MARINIS, Panagiotis SARANTAKIS, Georgios NIKOLOUDAKIS, Ilias BALABANIS, Nikolaos KARAGIANNIS, Irini LIMNOU, Manolis MOSCHIDIS, Aris TZAVARAS, Konstantinos KOUTSOURAKIS
Medical Instrumentation Technology Department
Technological Education Institute of Athens, Athens, Greece

Abstract. The purpose of the present study was the development and the initial implementation of a Medical Procedure Resources-Allocation and Cost-Capturing Software for the Estimation of the Mean DRGs associated Treatment Cost. The system provides means for the acquisition of health-care cost related data, based upon the actual Greek conditions, and includes Medical Equipment, Reagents, Consumables and Disposable Materials, Drugs, Man Power, Building Leasing and Infrastructure Maintenance expenditures, as the most important cost-components. The developed system was tested in the Operating Rooms (OR) and the Intensive Care Units (ICU), of a representative group of Hospitals. The obtained data have confirmed the reliability of the system for drafting DRGs in Greece; however, there is a discrepancy, between the high actual mean costs in the OR and the ICU, and the relatively low Health Insurance remuneration of the Hospitals.
Keywords: DRGs, Treatment Cost, Hospital Reimbursement, Medical Record.

1. Introduction

The Diagnosis Related Groups (DRG) system, developed initially at Yale University, has been adopted by most of the public in the EU. It seems inevitable that this will happen also in Greece, during the next few years, although this issue has not yet been publicly addressed. However, a cardinal prerequisite for the adoption of a refined DRG system is the acquisition of health-care delivery related data, based upon the Greek real-word conditions. The later seems to be both, controversial and multifaceted, due mainly to the discrepancies between the conditions in the two major cities, Athens and Thessaloniki, the rest of the mainland, and the numerous islands.

Therefore, a medical procedure cost-capturing algorithm has been developed for the major Hospital Departments and Units [1]. The algorithm allows for an appropriate correlation of the major cost components, that is, first, the mean medical equipment cost contribution, second, the average building and infra-structure maintenance cost, third, the approximate reagents and other material cost, fourth, the mean procedure related medication cost, and fifth, the weighted man-power cost, to any medical

[1] Corresponding Author : Botsivaly Maria, Botsivali_Maria@tellas.gr

activity that can be selected out of a four-level Medical Procedures Classification and/or a three-level in vitro Laboratory test Classification.

The proposed algorithm, implemented on specific department-customized software-tools, allows for an efficient and realistic approximation of single-activity cost. This in turn will constitute the input data for the formation of an appropriate medical record, that is intrinsically, however latent [2], related to the costs caused and the expected reimbursement. Thus, the updating of the patient's relevant medical data, ignites when relevant, the corresponding revision of an implicitly associated financial record, that allows for a good approximation of the individual DRG-coded case-cost.

The system was tested with data acquired on-site from a representative group of Hospitals, including a Regional University Hospital, a Provincial Hospital on an island, and a Private Hospital in Athens, and the obtained extended data will be presented.

2. Materials and Methods

In order for the estimation to become feasible the first four cost components were appropriately analyzed in a number of fundamental economical parameters such as purchase price, expected equipment life time, etc. that can be easily traced and entered into the system by anyone of the department's personnel, without being necessary for him / her to have special accounting skills, but merely good knowledge of the department's operational procedures. These parameters are appropriately transformed by the algorithm in such a way as to allow for the estimation of a daily / hourly maintenance cost and / or a daily / hourly virtual leasing rate of any item of the above mentioned cost components list that are actually used in a specific department.

The building and infra - structure maintenance cost is calculated by converting the approximate building value into a virtual annual rate. This rate, together with the maintenance personnel salaries, the overheads and the outsourcing cost can be transformed firstly into a hospital's and secondly into a department's maintenance cost per square meter

A detailed presentation of the economical parameters determined for each cost component is presented in the following figure (Figure 1).

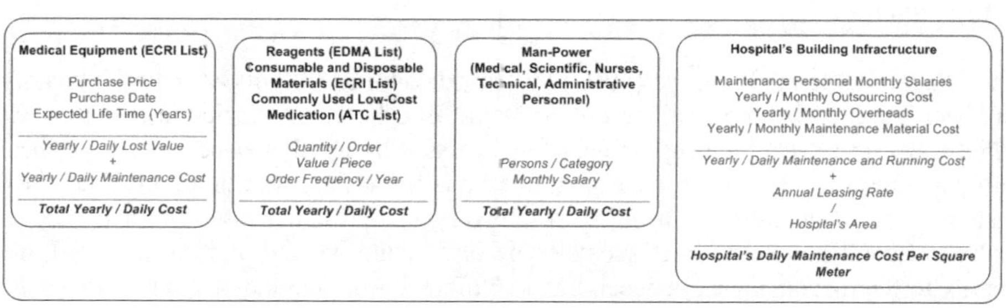

Figure 1: Cost components' elementary economical parameters

Once these parameters are filled, the mean actual department output is estimated by entering into the system the frequency with which various medical procedures take place into the specific department. Using the above features as inputs, the approximate mean cost contribution of each cost component in every medical procedure is then estimated by equally distributing their daily leasing rate and / or daily maintenance cost into the total amount of the medical procedures that daily take place into the specific department.

Although all the above mentioned cost components contribute to all medical procedures, their contribution is most of the times not equal, depending mainly on the nature of the Hospital's Department / Unit in which the procedures take place. Thus, for the final cost estimation an array of weight factors, which are presently empirically determined, is introduced into the system. The actual flow diagram of the developed procedure is demonstrated in Figure 2.

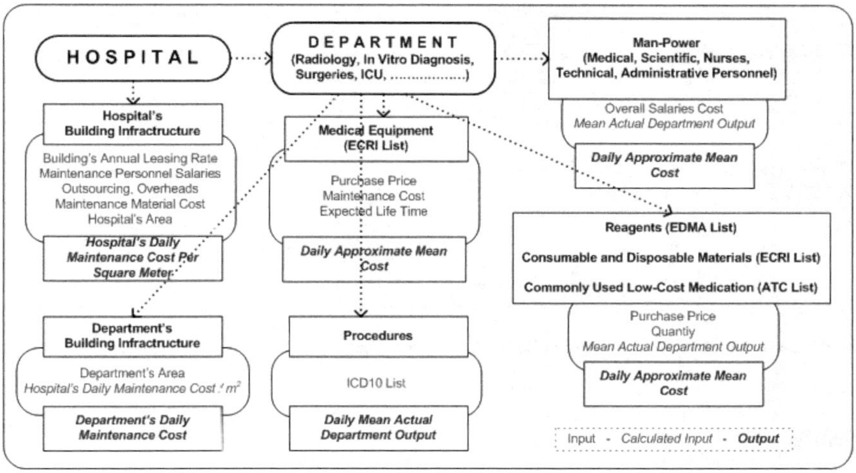

Figure 2: Flow diagram of the Developed Procedure

A number of additional special features are also included into the developed system in order to satisfy specific demands of some Departments or Units, such as, first, the sharing of resources across different departments and, second, the definition of procedure – specific resources. This is the case when, for example, some high-cost medication or material is explicitly used for a medical procedure and cannot be included in the mean calculation. Finally, the system provides for special treatment of some hospital departments, in which the medical procedures that take place present such differences in their completion time that is necessary for the corresponding cost calculations, to be made by taking into account their time duration. In the case of a "time-dependent" department the user must enter into the system the department's running hours and the mean time duration of the medical procedures that take place in it. The daily leasing rate and / or the daily maintenance cost of the cost components are the distributed into the executed procedures according to their time duration.

Microsoft Visual Basic was used for the development of the user interface and for the calculation algorithms, while a relational model was developed, using Microsoft Access. According to this model the main tables include the hospital's and department's identifications and linked tables save each department's input data, while others include medical procedure lists (ICD 9 coded), medication list (ATC coded), reagent list (EDMA coded), equipment and materials list (ECRI coded), etc. In Figure 3 a part of the entity-relationship (ER) database diagram is displayed.

Since the developed cost-capturing system utilizes standard coded vocabularies it can be easily synchronized with any EPR system, using even a simple script that would export the medical procedure code and the corresponding calculated cost estimation, while another script would import them in the EPR database, in order to create the above mentioned patient's latent financial record that allows for patient's treatment cost estimation.

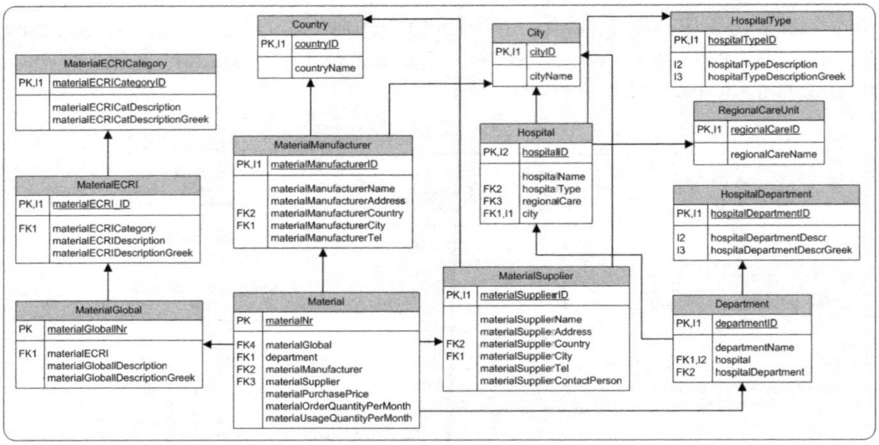

Figure 3: Part of the entity-relationship (ER) EPR Database Diagram

3. Results

Over 500 operations covering all surgical specialties have been processed and evaluated and the ICU cycle has been evaluated for nine months. The obtained data were not usually complete, and they needed further processing.

By taking into account the approximate Medical and Nursing staff structure in the Operating Department, an hourly mean cost was calculated for each Operating Room (OR). Employing a similar approach for the involved personnel outside the OR, a break–down for the mean per patient cost-basis has also been estimated. The results are presented in Table 1.

Table 1: An indicative hourly cost-basis for a mean virtual OR team and an indicative mean per patient cost-basis for the involved personnel outside the OR.

Medical/Nurse Specialty	Cost/hour €	Specialty	Cost/patient €
Surgeon A	21.90	OR Head Nurse	2.41
Surgeon B	16.40	Recovery Head Nurse	2.41
Resident	10.80	Recovery Nurses (2)	4.48
Anaesthesist A	18.33	Ward/ other Nurses (4)	8.96
Anaesthesist B	13.50	Stretcher personnel (1)	9.21
Anaesthesia Nurse	8.39	Janitors (2)	12.28
Instrumentariun Nurse	8.39	**Total**	**39.75**
Management Nurse	8.39		
Trainee Nurse	6.91		
Total	**113.01**		

A slightly modified and appropriately adapted approach was employed also for the private Hospital, however, only for the ICU, since the OR medical staff reimbursement in the private sector is based upon individually negotiated fees by case and not by salary. Therefore, the OR manpower cost statistical approach is indicative only for the Greek National Health System (ESY) Hospitals. The results are presented in Table 2.

Table 2: Example of the acquired cost data and treatment duration for a typical laparoscopic cholocystectomy and some indicative cost data for the ICU.

Mean Costs OR (Cholocystectomy)	**Euros (€)**	**Mean Costs (ICU)**	**Euros (€)**
Human resources	228.10	Mean fixed cost per day	120.00
Surgical & other material	487.87	Variable cost per patient	1,651.00
Anaesthesia & drugs	79.71	**Main sources of cost**	**%**
Medical equipment	37.01	Human resources	42.6
OR/Recovery cost	12.36	Supporting services	21.2
Procedure duration	**min**	Drugs	16.7
Pre-anaesthesia	35	Overheads	8.6
Main operation	65	Disposables	5.1
Recovery	15	Equipment employment	4.2

About 40 items are installed in an Operating Room. For all these items, according to the hospital inventory, a record was created including installation date, purchase price, expected Service duration (life) and the mean yearly service – hours required. Further, the yearly and daily lost value, the yearly and daily maintenance cost, and finally the total daily and hourly cost were calculated.

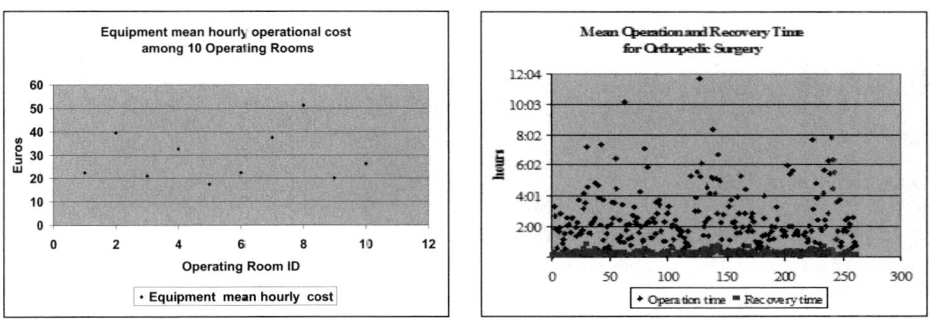

Figure 4: The variation of the equipment mean hourly operational cost among 10 different Operating Rooms serving the most common Surgical Subspecialties (left), and the variation of the duration of 261 different types of Orthopedics and Maxillar Surgery Operations and the corresponding recovery times (right).

The variation of the equipment mean hourly operational cost among 10 different Operating Rooms, all belonging to a modern regional Hospital in Greece, and, thus, serving the most common Surgical Subspecialties, is presented in Figure 4. The Mean Value is 29.05 € and the corresponding Standard Deviation (SD) is 10.87 €. The equipment mean hourly operational cost of the corresponding Recovery Room is essentially lower than the one of the OR and is approximately 8.06 €.

Further, we have calculated the mean cost for 535 different cases, classified within the most important surgical subspecialties. For example, the mean cost for 52 different Urological Operations was 812.30 € however the Standard Deviation was very high, reaching the 728.69 €, an expected outcome, due to the very wide range of variation in severity and complication, among the various cases treated in a Hospital.

4. Conclusions

Comparing the calculated data for 535 Operations covering all specialties with the official minimal and maximal nominal Reimbursement amounts for 107 frequent types of Operations, it becomes obvious that the calculation basis of the second set of data is entirely different (Figure 5).

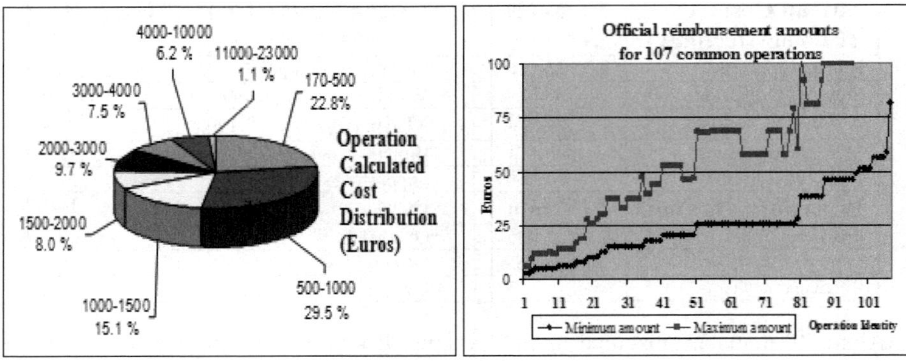

Figure 5: The Distribution in 8 cost classes of the Calculated Cost in Euros for 535 Operations covering all specialties (left), and the official minimal and maximal nominal Reimbursement amounts for 107 frequent types of Operations (right).

There is a discrepancy, between the high actual mean costs in the OR, and the relatively very low Health Insurance remuneration of the Hospitals, however, the data acquired according to our method, although not always complete, constitute a rather reliable base for drafting DRGs in Greece.

Acknowledgements

This work is part of a project co-funded by 75% from European Union and 25% from the Greek Government under the framework of the Education and Initial vocational Training Program "Archimedes".

References

[1]. Spyropoulos B. Supporting the Acquaintance of Health-Care Personnel to Operational-Cost Monitoring. CD-ROM Proceedings of TEHRE 2002 – Towards Electronic Health Record. London, United Kingdom, 2002.

[2]. Spyropoulos B., Marinis M. Design and implementation status of a system supporting hospital operational cost-monitoring. Proc. of SMDM 2002 - 24th, Ann. Meeting of the Soc. for Medical Decision Making, , Baltimore MD, USA, 2002, p. 28.

[3]. Botsivaly M., Spyropoulos B., Marinis M., Tzavaras A., Koutsourakis K., Sarantakis P. A Software-Tool allowing for Departmental Hospital Operational Cost Estimation. Proc. of EMBEC 2005 - 3rd European Conf. on Medical & Biological Engineering, Prague, Czech Republic, 2005.

[4]. Spyropoulos B, Botsivaly M, Tzavaras A, Nikoloudakis G, Balabanis I, Karagiannis N, Limnou I Acquisition of cost-data for Surgery and Intensive medicine to be employed for drafting a DRGs hospital reimbursement system in Greece. Proc. of EMBEC 2005 - 3rd European Conf. on Medical & Biological Engineering, Prague, Czech Republic, 2005.

Ubiquity: Technologies for Better Health in Aging Societies
A. Hasman et al. (Eds.)
IOS Press, 2006

Apply creative thinking of decision support in electrical nursing record

Angelica Te-Hui HAO[a], Chien-Yeh Hsu[b], Huang Li-Fang[b], Wen-Shan Jian[c], Li-Bin Wu[d], Ching-Chiu Kao[d], Mei-Show Lu[e], Her-Kung Chang[a]

[a] *Information Management Division, Department of Business Administration, Chang Gung University*
[b] *Graduate Institute of Medical Informatics, Taipei Medical University*
[c] *Institute of Public Health, National Yang Ming University*
[d] *Taipei WanFang hospital,* [e]*Nursing school of Taipei Medical University*

Abstract. The nursing process consists of five interrelated steps: assessment, diagnosis, planning, intervention, and evaluation. In the nursing process, the nurse collects a great deal of data and information. The amount of data and information may exceed the amount the nurse can process efficiently and correctly. Thus, the nurse needs assistance to become proficient in the planning of nursing care, due to the difficulty of simultaneously processing a large set of information. Computer systems are viewed as tools to expand the capabilities of the nurse's mind. Using computer technology to support clinicians' decision making may provide high-quality, patient-centered, and efficient healthcare. Although some existing nursing information systems aid in the nursing process, they only provide the most fundamental decision support – i.e., standard care plans associated with common nursing diagnoses. Such a computerized decision support system helps the nurse develop a care plan step-by-step. But it does not assist the nurse in the decision-making process. The decision process about how to generate nursing diagnoses from data and how to individualize the care plans still reminds of the nurse. The purpose of this study is to develop a pilot structure in electronic nursing record system integrated with international nursing standard for improving the proficiency and accuracy of plan of care in clinical pathway process. The proposed pilot systems not only assist both student nurses and nurses who are novice in nursing practice, but also experts who need to work in a practice area which they are not familiar with.
Keywords: Nursing Clinical Pathway, Electrical Nursing Record, Decision support

1. Introduction

The health insurance claim rule continuing changes in recently. The Diagnosis Relationship Group (DRG) and global budget has been change nursing care style in Taiwan. Also the Clinical pathway becomes a major index for nursing care. [1,2].
There are many same record with nursing record and clinical pathway record and nurses always double working for these records. Although computerize nursing record is develop for many hospitals, But defect the same style and couldn't to exchange for share

patient's data and knowledge management. Therefore, that is importance for development electronic nursing record integrated with clinical pathway.

2. Literature Review

Clinical pathway is a standard style for specify disorder that examination, treatment, nursing care, and health education since patient admission through discharge. The urology has a lot of clinical pathway in the medical center that we studied. In this kind of ward, the top three clinical pathways are Benign Prostate Hypertrophy (BPH), inguinal hernia, and Urinary tract stone. The cost analysis for these three clinical pathways in Table 1:

Table 1. Ccost analysis for three clinical pathways

Item	Patient number	Insurance cost (NT$/p't)	Nursing cost (NT$/p't)
BPH	220	49,000	2,500
Inguinal hernia	100	24,000	800
Urinary tract stone	500	40,000	800

The International nursing standard has many kinds such as North American Nursing Diagnosis Association (NANDA) [3], International Classification for Nursing Practice (ICNP) [4], and Nursing Intervention Classification (NIC).

NANDA is based on Human Response Patterns that has nine interaction models. The Nursing Interventions Classification (NIC) has 433 interventions in 27 classes for 6 domains [5]. In Taiwan, nurse will learned how to use NANDA for nursing diagnosis, and NIC to clarify nursing intervention.

An expert group verified the results. The ICNP Phenomena Classification described 87.5% of the NANDA diagnoses, 89.7% of the HHCC diagnoses and 72.7% of the Omaha System problem classification scheme. The ICNP Action Classification described 79.4% of the NIC interventions, 80.6% of the HHCC interventions and 71.4% of the Omaha System intervention scheme. The results of this study suggest that the ICNP has a sound starting structure for a unified nursing language system and can be used to describe most of the existing terminologies. Recommendations for the addition of terms to the ICNP are provided. [6]

But most nursing practitioner had been used NANDA nursing diagnosis that to describe nursing care and patient condition in Taiwan. That's why more development computer system for nursing practice.[7]

3. Innovation Electrical Nursing Record pilot structure

We using three steps to develop the pilot structure for Innovation Electrical Nursing Record (Figure 1) :

3.1. Retrospect patient record analysis

In this study, we are sampling three major clinical pathways at Urology ward in medical center such as Benign Prostate Hypertrophy (BPH), inguinal hernia, and Urinary tract stone as our major format.

We are find out 253 significant symptoms for five major nursing diagnoses since 100 patients each for three clinical pathways in past one year that sampling for retrospect record analysis. (See Table 2)

Table 2. Significant symptom analysis[8]

Nursing Diagnosis	significant symptom appear times
1. Acute Pain	64
2. Risk for Infection	27
3. Impaired Urinary Elimination	45
4. Impaired Skin integrity	50
5 Risk for Falls	67
Total	253

The decision support rule builds by retrospect record analysis and expert committee. Through expert rule, these significant symptoms will distinguish by 3 categories to major, secondary, and minor. And every significant symptom will be set a weight score for these 3 categories. The diagnosis will be produced depend on that how many major, secondary, and minor appeared.

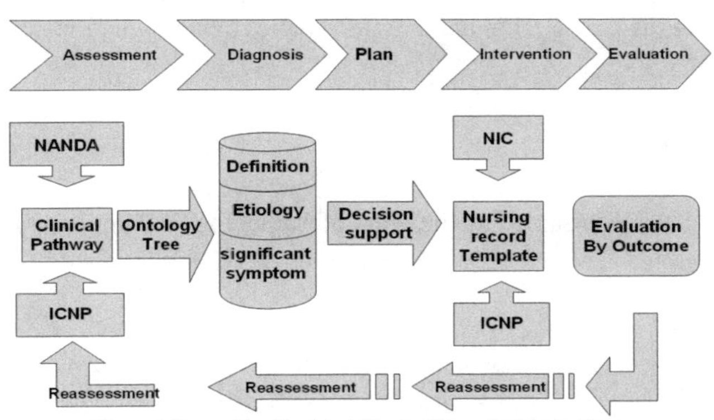

Figure 1. Innovation Electrical Nursing Record pilot structure

3.2. Integration standard for nursing taxonomy

We will integrate the taxonomy and code of NANDA, NIC, and ICNP is a relational database in these three clinical pathways for exchange different standard of ENR interface engine.

3.3. C. Building innovation Electrical Nursing Record (ENR) pilot structure

In the pilot structure, it includes nursing assessment, nursing diagnosis, nursing plan, and nursing record. We called to the Electrical Nursing Record (ENR).

The ENR pilot structure integrated clinical pathway from patient admission through discharge, There are ENR system portal (see Figure2), ENR system Main page (see Figure3), Demographic maintenance (see Figure4), Nursing assessment interface (see Figure5), Nursing care plan interface (see Figure6), Ward maintenance interface (see Figure7), Nursing diagnosis maintenance (see Figure8), Nursing intervention maintenance (see Figure9), Nursing intervention maintenance (see Figure10), Nursing Characteristics maintenance (see Figure11), Nursing assessment code interface (see Figure12), Nursing diagnosis code maintenance (see Figure13), and Decision support rule maintenance (see Figure14).

4. Future works

To develop nursing care plan is helpful for personal interaction with in environment to got health. When set a nursing diagnosis will decide what nursing intervention to use. That is potential contribution of nursing diagnosis[9]。

Under challenge in DRG, nursing care model will toward to case management style. Therefore, the nursing care planning system will base on case management. Such as Clinical pathway that clarify important exam, treatment, health education, and care plan in everyday from patient admission through discharge.

We will build up the whole ENR system based on the pilot structure in coming year and hope to bring the new goal in nursing informatics development in Taiwan.

Acknowledgments

The research project funded by the grant of Taipei medical university and WanFang hospital.

References

[1] Newell, M., Using Nursing Case Management to Improve Health Journal on Quality Improvement. 22(8):531-548. 1996.

[2] DOH. The report from nursing technical committee for Electrical Medical Record research project, 2005.

[3] Philadelphia. North American Nursing Diagnosis Association. Nursing Diagnosis Definitions and Classifications 2001-2002 (4th Ed.). 2001.

[4] International Council of Nurses (ICN). International Classification for Nursing Practice ICNP_ beta-2. Geneva, Switzerland: ICN, 2002.

[5] McCloskey, J.C., & Bulechek, G.M., Eds. Nursing interventions classification (NIC) (3rd Ed.). St. Louis: Mosby, 2000.

[6] S. Hyun & H.A. Park. Cross-mapping the ICNP with NANDA, HHCC, Omaha System and NIC for unified nursing language system development. International Council of Nurses, International Nursing Review 2002:49: 99–110.

[7] Lai Yen San. Medical diagnosis assists nursing decision support system. TMU nursing 2001:3(1):67-78.

[8] Kao Gi-Hui. New nursing diagnosis:Farseeing, Taipei, 1998

[9] Chen Yu-Zi , Computer application in clinical nursing care plan and record. VGH nursing 1994:11(1):37-43.

Figure 2. *ENR system portal*

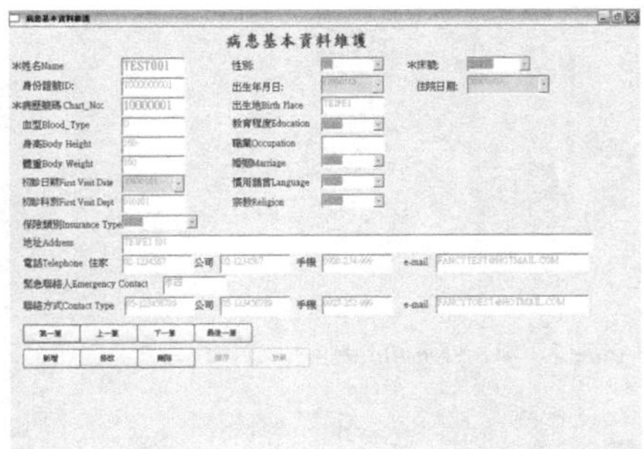

Figure 4. *Demographic maintenance*

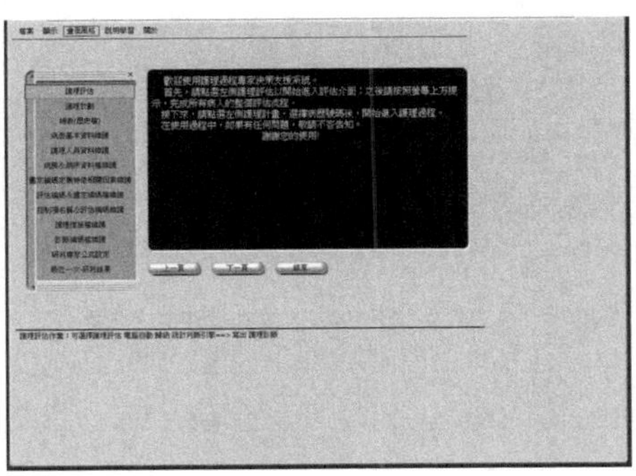

Figure 3. *ENR system Main page*

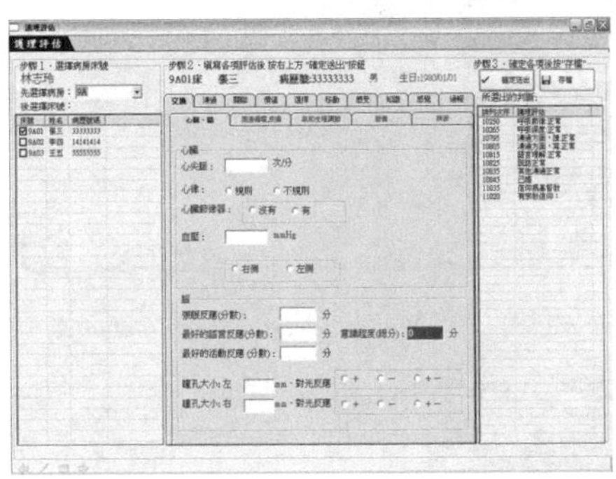

Figure 5. *Nursing assessment interface*

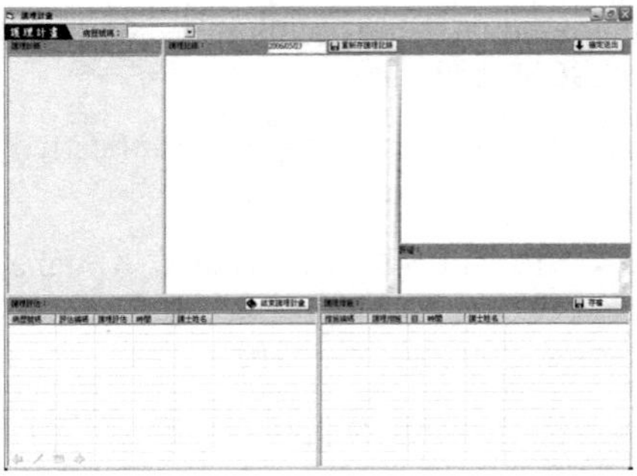

Figure 6. Nursing care plan interface

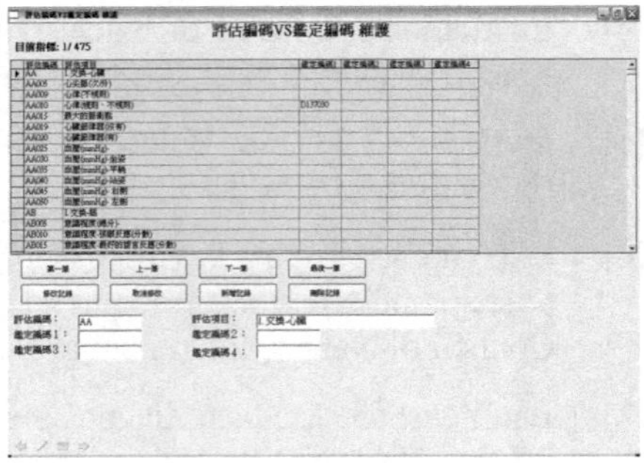

Figure 10. Nursing Characteristics maintenance

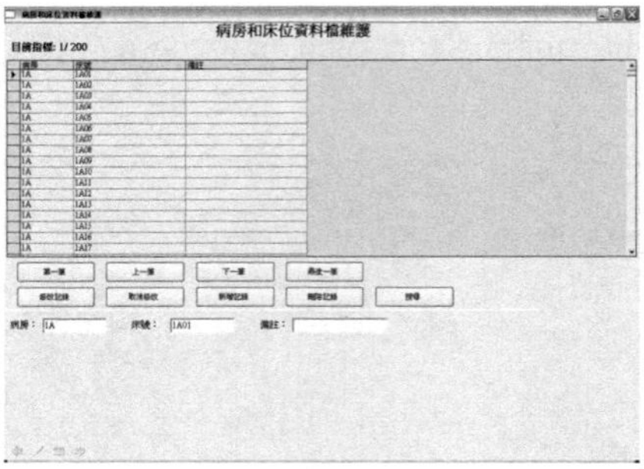

Figure 7. Ward maintenance interface

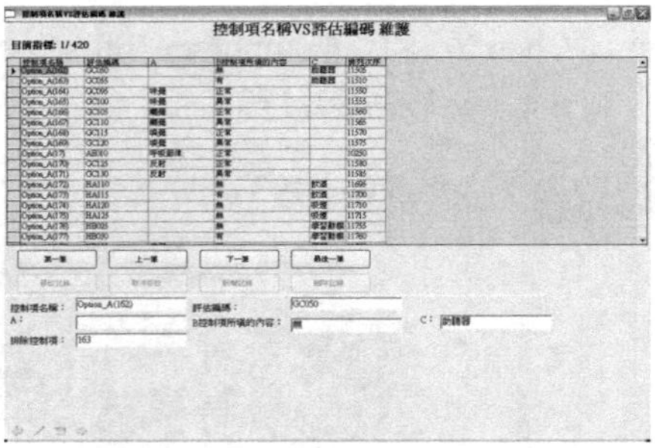

Figure 11. Nursing assessment code interface

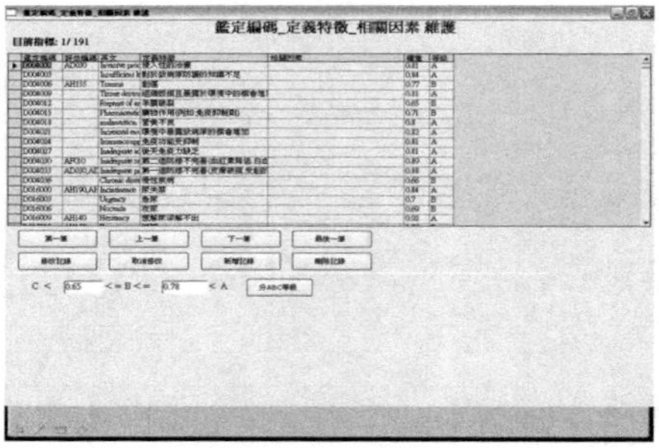

Figure 8. Nursing diagnosis maintenance

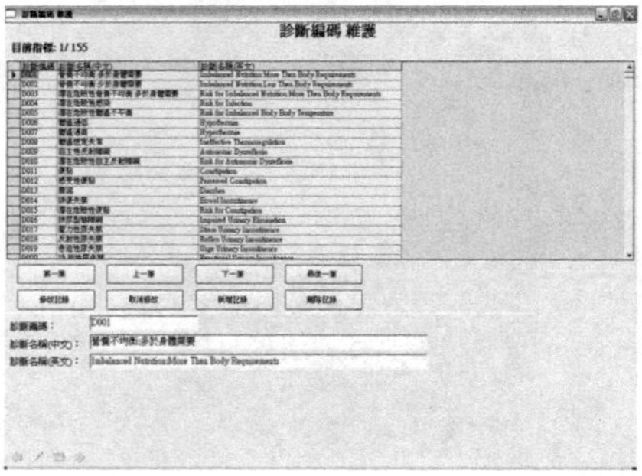

Figure 12. Nursing diagnosis code maintenance

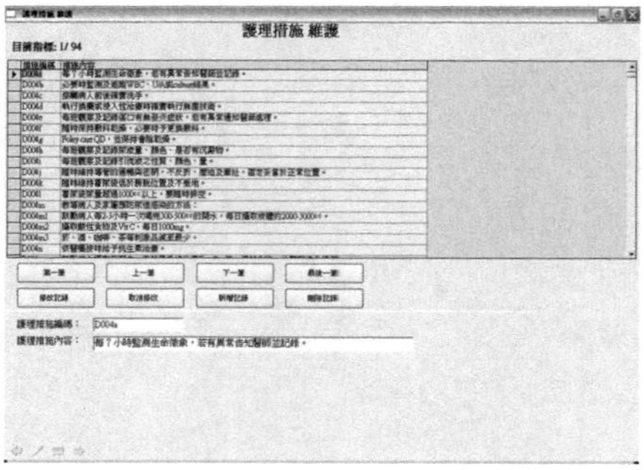

Figure 9. Nursing intervention maintenance

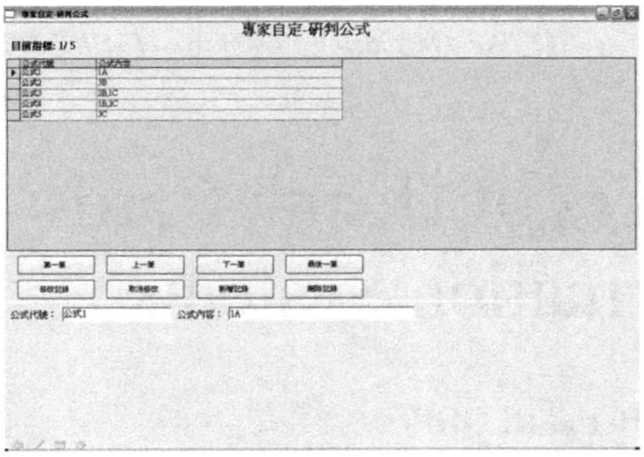

Figure 13. Decision support rule maintenanc

Ubiquity: Technologies for Better Health in Aging Societies
A. Hasman et al. (Eds.)
IOS Press, 2006

Benefits and Weaknesses of Health Cards used in Health Information Systems

Peter PHAROW[1] and Bernd BLOBEL
eHealth Competence Center, Regensburg, Germany

Abstract. The acceptance-based success of modern health information systems and health networks highly depends on the respective involvement of all relevant partners into the communication and co-operation processes characterising the medical and administrative workflow. Personal information stored in a networking environment guarantee for fast access fulfilling advanced shared care requirements whereas security token like smart cards stand for identification purposes, data protection, privacy protection, access rights, and limited person-based information storage, e.g., for emergency procedures. Linking these means of information provision allows for making use of the benefits of the different technologies without ignoring their existing weaknesses. The presented paper intends to summarise the respective categories of benefits and weaknesses allowing the reader to implement cards in health information systems as well as with the related aspects of awareness, confidence, and acceptance. Concluding this analysis, the preferred way to deal with the challenges of modern healthcare and welfare requirements shall be a well-balanced combination of cards and networks.
Keywords: Medical Informatics, Health Cards, Patient Cards, Health Networks, Electronic Health Record, Patient Involvement.

1. Introduction

For meeting the challenge of increasing quality and efficiency requirements in the healthcare and welfare domain, the respective systems of many countries are turning towards the shared care paradigm providing an integrated care. This paradigm change is bound to largely extended communication and co-operation between all healthcare establishments involved in the patient's care. Thus, new ways for designing and integrating health information systems are needed. In that context, cards in general achieve an increasing importance. The present paper deals with specific benefits and weaknesses using card technologies in healthcare and welfare [1].

2. Card Usage Benefits and Weaknesses Analysis Methods

The detailed analysis of specific benefits on the one hand and weaknesses on the other while using cards in modern integrated health information systems is performed by considering the different aspects and views on these information systems. Hereby, the

1 Corresponding Author: Peter Pharow, ehealth Competence Center, University of Regensburg Medical Center, Franz-Josef-Strauß-Allee 11, D-93053 Regensburg, Germany. Email: peter.pharow@ehealth-cc.de

scope ranges from legal and organisational aspects, interoperability features, technical constraints, and organisational requirements up to security, safety, privacy, and reliability issues. A selection of important results will be presented.

3. Smart Cards in Healthcare

In case an information system is intended to be held by a human being, an appropriate media is needed for storing data structures and applications providing the functionality required including a directed communication of respective data sets to partners. This generally applies ranging from the use of simple hardware token for restricted functions such as identity-related services up to more or less comprehensive portable information systems carried by the information subject [2].

3.1. Card Technologies and Applications Used in Healthcare

Over the last 30 years, several different technical solutions have been developed and implemented for recording, storing and using person-related administrative or medical (clinical) information in healthcare. The underlying technologies can be distinguished according to the medium deployed, according to the purpose the cards are generally used, and according to the mechanisms and functions provided.

Starting with simple paper or plastic cards (like current European Health Insurance Card) that can be written only once and read many times, memory cards have been introduced. Providing different storage capacity, magnetic stripe cards, laser-written and laser-read optical cards or even chip cards have been used to simply store the aforementioned health information. Regarding the purpose of their use, storage cards can be further distinguished; they range from tokens such as access cards, identity cards, and authentication cards up to signature cards, encoding/decoding cards, etc. Sometimes, cards are classified by their medical dedication for supporting specific treatment (e.g. DIABCARD for supporting diabetes patients [3]).

Considering structure and mechanisms applied, beside storage mechanisms, programmable processing facilities have been established forming processor cards, sometimes equipped with dedicated co-processors for supporting special complex algorithms. Processor cards are also known as smart cards. The purpose is often combined with a specific mechanism to provide the security functions required. Frequently, several purposes and functions are realised deploying the same physical cards towards a multi-functional card.

An independent electronic health information system can either be held by patients, or it is network-based. Beside the appropriate carrier of information, an environment -the infrastructure- must be provided for the authorised use of information in the sense of collecting, storing, processing, and communicating administrative and medical data. Starting in Europe, smart cards -microprocessor cards- have been used for both purposes mentioned, i.e., as Patient Data Cards (PDC) and Health Professional Cards (HPC) – meanwhile around the world (see [4], [5]).

Generally, it should be mentioned that a smart card fulfilling different purposes within a healthcare and welfare environment can be deployed in two ways. On the one hand, the card could bear all information items (in several different data structure elements) needed, in the case of PDC, e.g., all relevant administrative and medical data, as shown in this paper. On the other hand, the card can be used as an identifier for the card holder, and as a pointer providing references and linkages to the information stored in networked systems. However, a combination of those two principles is imaginable, and might be the preferred solution for future implementations of health networks and related identity token.

3.2. Card-Enabled Security Infrastructure

In Europe and beyond, smart cards are frequently used for enabling communication and application security services [2]. The basic principle reflects a certified binding of a principle (human user, organisation, device, system, application, component, or even a single object) to its electronic unique identifier or assigned properties, rights and duties, also called attributes of that principal. Communication security services concern the identification and authentication of communicating principals. In an end-to-end secure communication environment (object security), these services are used for authentication and control of access rights of principals communicating as well as integrity, confidentiality, and accountability including non-repudiation of information exchanged. For object security, security-aware principals are needed.

On the other hand, integrity and confidentiality of communicated data may also be provided at system level transparent to the application and the user following -but not requiring- the user's awareness for those security measures (channel security). Application security services deal with authorisation and access control to data and functions, but also with accountability of principals, audit track and auditing of principals and services, ensuring integrity, confidentiality of data and functions. In both concepts, notary's services have been established. Another important requirement for both communication and application security concerns the availability of information and services. For more information regarding the different security categories and underlying mechanisms, see [6], [7].

4. Chip Card-Based Systems vs. Networking Systems

In the early nineties, the German Medical Informatics Association GMDS has defined motivation and objectives as well as health-political aspects for the use of machine-readable cards in healthcare and welfare [8]. As a short-term challenge, the reliable implementation of pilots has been mentioned to demonstrate the possible use of such cards and to rationalise time-consuming administrative work in the context of patient's request for health services. Among others, the health-political aspects concern several facilities for de-centralised medical documentation, a coincidence with constitutional and data protection rights as well as with ethical principles in case the Electronic Health Care Record (EHCR) is held by the patient, a significant improvement of data quality, integrity and consistency by a unique document, and finally a significant improvement of quality and efficiency of health delivery in general.

4.1. Characteristics of Chip Card-Based Health Information Systems

Pure chip card-based health information system architectures are characterised by the chance for optimisation of medical and administrative workflow, the improvement of information flow between different healthcare providers within a shared care framework, the enhancement of information security using reliable, valid and in-time as well as in-location available data, the reduction of stress caused by repeated observations due to permanent availability of patient information, the facilitation of emergency care by directly available emergency data set, the support of prevention and intervention studies as well as the increase of autonomy and responsibility of patients and of partnership between patients and Health Professionals (HP).

However, some weaknesses can be discovered as well in a purely chip card-based information system. Among several others, the following aspects should be mentioned:

- Tough procedure of standardisation in health terminology and procedure;
- Lost of information due to the restriction of data (FIFO) caused by storage capacity limitations;
- Unsolved legal problems of information ownership in healthcare and welfare;
- Required infrastructure and appropriate interfaces to applications, and
- Impossibility of teamwork between different medical parties including pre- and post-caring activities due to the need of the physical presence of the patient with the PDC at all sites involved.

4.2. Characteristics of Network-Based Health Information Systems

In purely network-based health information systems with extended health network architecture, the required services and the existence of unique identifiers, the following advantages can be found:

- A comprehensive level of interoperability and real interactions between all parties involved into the shared care;
- Realisation of a comprehensive, complete, high-quality EHCR;
- Emergency care facilitated by directly accessible emergency data set;
- Data quality, integrity and consistency by provision of appropriate services in an extended network;
- Opportunity of pre-activities and post-activities independently or co-ordinated at the sites involved including support of prevention and intervention studies;
- Optimised medical and administrative workflow, and
- Reduced stress level caused by repeated observations.

On the other hand, several weaknesses, problems and disadvantages occur, too. Examples include, but are not limited to, the absence of a direct support of emergency care, the fact of missing ownership (possession) of data by the patient, the strong need for a trustworthy doctor-patient relationship, and the explicit relation of data security means and solutions to the information system architecture.

5. Discussion

Regarding an analysis of network-based and chip card-based health information system, series of advantages and disadvantages can be stated. The way Europe is currently pursuing with regard to the storage of administrative and medical person-related data items and structures is a combined architectural approach of both using smart cards as secure and reliable token for identity authentication and person-related security services like accountability, authorisation, access control and audit etc., as well as by introducing limited EHCR extracts located on a Patient Data Card (PDC).

Regarding the aforementioned requirements for an interoperable PDC, a series of standards has been specified at international level by ISO TC 215 “Health Informatics”. After having identified both needs and requirements for such cards, the standardisation experts have started defining a framework for card-related specification of medical and administrative content – ISO 21549 Health Informatics – Patient Health card Data ([4]. ISO TC 215 does not intend to standardise card technology but the health-related structures on such types of cards.

As said before, person-related data carried on a data card can be categorised into pure identification data (of the device itself and of the individual the data it carries relates to), administrative data related to the individual, and respective medical (clinical) data. It is important to realise that a given data card "de facto" has to contain device data and identification data (see figure 1 below); it can contain administrative data and clinical data.

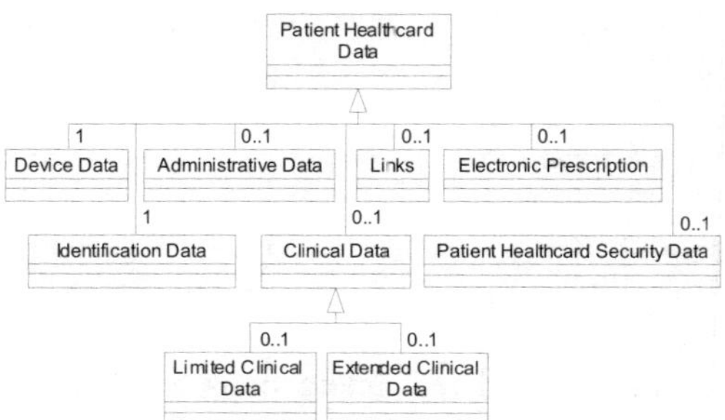

Figure 1. ISO 21549 Data Set

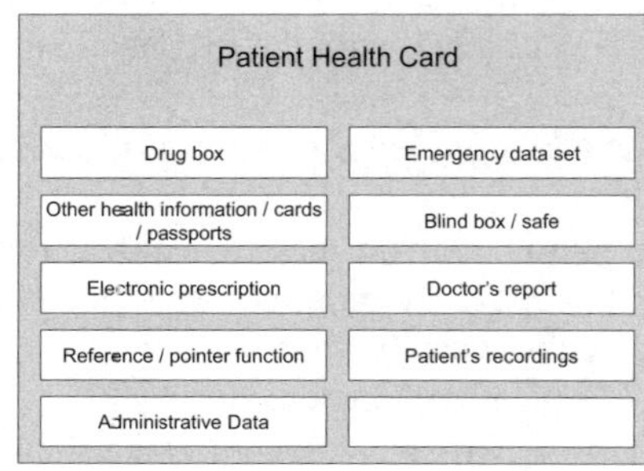

Figure 2. Functional Block of the German eGK

Furthermore, a PDC needs to significantly support the collaboration with network-based systems. For that purpose, link information has been specified. Figure 2 shows the adaptation of [4] in the context of the German Electronic Health Card which is of the aforementioned PDC type ([9]).

6. Conclusion

Patient Data Cards in the healthcare and welfare domain have several advantages that shall be considered real benefits for card holders (citizens, patients). The level of patient involvement is significantly higher as the patient is invited to take care of at least a subset of his own medical data. The patient is not any longer just a consumer; he is empowered to become a partner within the medical and administrative workflow processes. This goes along with an increased level of patient compliance. Therefore, aspects like user awareness, confidence and acceptance become increasingly important in modern healthcare and welfare.

On the other hand, the patient is responsible for a piece of technology. Certain data safety concerns exist. The lack of professionalism might make a reliable card usage complicated; it might lead to failures. Thus, information, education and training range high in future activities to make sure citizens and patients are able to use modern token technology in an appropriate manner. There are liability problems, by the way. Setting up a legal framework is very difficult because of different scopes and domains for regulations related to patients on one hand and health professionals on the other

In case Patient Data Cards are not only add-on services in a networking environment, interoperability is strictly bound to the physical presence of the patients. This will for sure cause serious problems in realising real distributed work and related interoperability by performing pre- and post-activities. In addition to what has been mentioned so far, the capacity limitations of nowadays health cards allow for storing a limited number of data structures and items.

To overcome the weaknesses listed before, there's only one choice. The use of cards and the use of networking systems have to be closely linked to each other. This allows for making extensive use of the benefits of both technologies.

Acknowledgement

The authors are in debt to their organisations, especially the colleagues representing EFMI WG SSE, EFMI WG Cards, CEN TC 251, ISO TC 215, and the German DIN for their support and their kind co-operation.

References

[1] Blobel B, Pharow P (Eds.): Advanced Health Telematics and Telemedicine. The Magdeburg Expert Summit Textbook. Series "Studies in Health Technology and Informatics" Vol. 96. IOS Press, Amsterdam, 2003

[2] Pharow P, Blobel B: Security Infrastructure Requirements for Electronic Health Cards Communication. In: Engelbrecht R., Geissbuhler A. (Eds.): Connecting Medical Informatics and Bio-Informatics. Proceedings of MIE 2005. Series "Studies in Health Technology and Informatics" Vol. 116. IOS Press, Amsterdam, 2005.

[3] Demski H., Hildebrand C., Engelbrecht R., Birkmann C.: DIABCARD, the portable patient record: pursuing routine use In: Blobel B. et al. (Eds.) Contributions of Medical Informatics to Health. Berlin, 2004.

[4] ISO 21549 "Health Informatics – Patient health card data", 2004. http:(//www.iso.org (last accessed January 16th, 2006)

[5] CEN EN 13729rev "Health informatics - Secure user identification – Strong authentication using microprocessor cards", 2006

[6] Pharow P., Blobel B., Hildebrand C.: The Role of Patient Health Cards in an Integrated eHealth Environment. Submitted to STC 2006: Integrating biomedical information - "From e-Cell to e-Patient". Timisoara, Romania, 2006

[7] Engel K, Kuziela H., Blobel B, Pharow P: Security Services for the HemaCAM Project. In: Engelbrecht R., Geissbuhler A. (Eds.): Connecting Medical Informatics and Bio-Informatics. Proceedings of MIE 2005. Series "Studies in Health Technology and Informatics" Vol. 116. IOS Press, Amsterdam, 2005.

[8] Pommerening K.: Datenschutz in Krankenhausinformationssystemen (in German). Proceedings of "Verlässliche Informationssysteme", 1995. http://www.staff.uni-mainz.de/pommeren/Artikel/vis95.pdf (last accessed January 16th, 2006)

[9] The German Specification for an Electronic Health Data Card v 1.1. PDF Document (English version). http://www.dimdi.de/de/ehealth/karte/technik/kartenspezifikation/index.htm (last accessed January 16th, 2006)

[10] Blobel B.: Analysis, Design and Implementation of Secure and Interoperable Distributed Health Information Systems. Series "Studies in Health Technology and Informatics" Vol. 89. IOS Press, Amsterdam, 2002.

Ubiquity: Technologies for Better Health in Aging Societies
A. Hasman et al. (Eds.)
IOS Press, 2006

A Socio-technical study of an Ubiquitous CPOE-system in Local Use

[a]Helle S Wenzer[1], [a]Ulrich Böttger, and Niels Boye[b]
[a] *Department of Communication and Psychology, Aalborg University, Denmark*
[b]*Randers Hospital, Randers, Denmark*

Abstract A socio-technical approach was used to study the qualitative effects of deploying a medication-CPOE (Computer-based Provider Order Entry System) at two internal medical wards on a hospital in Denmark. Our results show spatial and temporal transformations of core acts in medication work, transformation in competencies, and less time spent at the bedside for nurses and doctors, as a system - constructed for ubiquitous drug-order entries and handling - was implemented for local use. This study throws light on problems of patient continuity, patient-related and IT-system-related error-handling and time spend on core activities - when ubiquitous IT is used locally in a real physical setting with specific traditions of doing medication acts. The paper argues for the project organization to support the local renegotiation of time and place of programs of (work-)action and of collaboration with 'ubiquitous' CPOE-systems, as well as set up feed-back for maturation of software for future clinical use.
Keywords: informatics, CPOE, HCI, clinical, change management

1. Introduction

Medication is always enacted locally, in concrete physical settings involving a range of more or less globally dispersed actants, including techniques of various kinds, that all effect the outcome of patient treatment. From the view of "the cockpit crew" [1] of health care management medication and healthcare work with HIT (Healthcare Information Technology) has the general purposes of reducing errors, medical costs, give better quality and continuity of care and better cooperation between health care professionals [2]. The aim of this qualitative study was to bring in the view of the' Fire-brigade', the clinicians responsible for the medication outcome, and support their articulation of experiences as the primary users of the system. Alterations in the medication process are illustrated as new socio-technical problems for enacting medication with the CPOE-system are pointed out in this paper in order for the system and users to share responsibility, and to enhance further learning of design and change management.

Theoretically this study resides in a 'praxiographic turn' [3] that focuses on how medication is enacted in concrete spatial and temporal settings. The health care personnel as well as the CPOE-system are looked upon as 'actants' with 'programs of action', that some times interrelate, interrupt and transform each other, so that new goals, meanings, properties and competencies are exchanged [4,5,6]. This socio-technical study was done by observing the user's and the CPOE-system's role in medication work. Central situations for mediated action with the COPE system was

[1] Correspondence Helle Wentzer, Kroghstraede 1, AaU, DK-9229 Aalborg OE. wentzer@hum.aau.dk

pointed out, validated by the clinicians, and articulated by the study as socio-technical problems of organizing of health care.

2. Materials and Method

Object of study: how medication is enacted at two Danish, internal medical wards.
Method: a socio-technical study that builds on 48 hours of observation, 6 interviews with primary users, two physicians and four nurses, and an analysis of the user interface and of other documents. Based on the observations three use-scenarios for central events in the medication process were constructed (prescription, requisition, and considerations of continuation of treatment). The use-scenarios were introduced to the interviewees for four reasons: To condense, verify and generalize the observations and to trigger the memory of the interviewees.
The actants were drugs, physicians, nurses, and the CPOE-system. The CPOE-system at this stage of development is only handling "the bookkeeping" of medication. It is not offering any decision support, and is developed by a professional vendor (Systematic, Aarhus, DK) in close cooperation with the responsible Health Management Organisation (HMO) (Aarhus County, DK), who have supplied clinicians to the development project. At the time of the study the CPOE had been in use for eight months in the departments participating, replacing a paper medication scheme, called MOS. Other actants studied were the pharmacy system (electronic Danish Physicians Desk Reference (PDR), handbooks on medication, pc-tables, other staff-groups, and techniques.
Criterions of reference: Successful medication in the Danish Healthcare is commonly understood as 'the five right': the right drug, to the right patient, at the right time, in the right dose, in the right way (e.g. orally or intravenously). Errors in medication were considered as any deviation from these criterions.
The main 'program of action' for doing medication (treatment with drugs) relates to a patient's problem/diagnosis, that ideally is followed up by seven sub-programs of actions [10]: First: the indication of treatment. Second: 'the prescription' (recommendation), i.e. the choice of treatment and the patient's accept hereof. Third: 'requisition' i.e. registration/order in the CPOE-system. Fourth: 'dosage', i.e. the drug is made ready for consumption. Fifth: the 'administration', i.e. the patient is given the drug. Sixth: 'assessment' of drug effect on the patient. Seventh: considerations on how to precede, i.e. whether to continue or withdraw the drug.
The sub-programs studied: The CPOE-system is an active part of the first five medication-core-acts, but also influences the seventh core-act on how treatment is continued in the future. The present study focuses on the interaction with the CPOE-system during the core acts of: prescription (2), requisition (3), dosage (4) and continuity of treatment (7).

3. Results

The study showed transformations in the core acts (subprograms) of prescription, requisition, dosage and continuity of medication with the CPOE compared to the paper-based MOS-system. Skills and properties were exchanged and redistributed among the actants. The "medication work-model" built into the system (during the development process with co-operation between the vendor and the clinical representatives of the HMO) did not correspond to the actual work practice and lead to a transformation of both actants, as the users practice was reconfigured as well as the system was used differently than prescribed in the system design. The following show how acting with the CPOE-system participates in the transformation of three central relations, i.e. between doctor and patient treatment, between doctor and nurse collaboration, and in the coordination between hospital and future situations of treatment.

3.1 Transformation of drug prescription with the CPOE-system

Figure 1: PC-table and patient paper records of both nurses and doctors.

The use-scenario of prescription showed that the system's success or failure in use cannot be assessed isolated from the environment it participates in. The materiality of space and things challenges the use of the CPOE system. Other actants as the stationary dictaphone, the clumsy mobile PC table (figure 1), the endurance of the laptop-computers, and the walking distance between the different interactions of talking to the patient, dictating the diagnose for the patient record, and prescribing medicine for treatment in the CPOE play a decisive role in the enactment of medicine

3.1.1. Doing Patient Rounds mediated with CPOE

A more 'collective patient' appeared in the physician's way of coordinating patient rounds, with the activities of recording diagnostic information on the dictaphone for the patient record and ordering prescriptions in the CPOE. The physician would start/continue treatment of the individual patient on the background of an encounter of three and four patient at a time. The health care treatment and trust-relation between the human actants therefore to a higher degree depends on the doctor's memory and ability to keep the different patients diagnosis, prescriptions, other treatments, and plan of action apart. The CPOE may be ubiquitous in the sense that it provides global access to the system in the hospital and country. Though in its local use the space and distance between hospital beds, PC-tables and stationary dictaphones decides what is accessed, when, in what order and how. Alas the best working conditions for the users were at the stationary PC's in the ward office away from the patients and the drug storage.

3.2 Transformation of drug requisition and dosage with the CPOE-system

The CPOE software contains no facilities for computer-enhanced 'collaborative working environments". Using the paper-based system the doctor and nurse were able to collaborate, correct and negotiate patient treatments and the work-tasks of each other in achieving the 'five right'. With the CPOE this cooperation still exist, but has spread out in time, space and persons giving less foundation for negotiations and discussions.

3.2.1. Figuration of a collective clinical user

According to the healthcare model in the CPOE only doctors can perform certain tasks - among others approve prescriptions. Physicians and nurses therefore have individual passwords and different user-rights in the system. In case patients need changes in medication the nurses can register the changes, but the physician needs to log in and approve them. In practice we observed another work division and for a different reason. Login procedures were time consuming and prescription, requisition, changes in dosage, administration and assessment are done more fluently and ad hoc, according to the situation. The system was not flexible enough for supporting the mutual dependencies of physicians and nurses for making requisitions and continuing and withdrawing medication. In order for the work to flow some physicians would log in and let the nurses continue the medication work with a doctors user rights.

3.2.2. Inflexibilities and displacements in the use of the CPOE hard- and software

Login procedures were time consuming and demanded a patience of the user that fits badly with the normal pace of hospital work, and a 'team-login' in line with the mode of work was not possible as described above. 'Irrationalities' of the system, the use, and error-messages were observed. A line of technical problems on hardware also had a considerable impact on contextual use and the thrust in the system. Too few computers queued up the users and demanded a physical separation of the work tasks and the information needed to perform them adequately, relying on abilities to memorize specific combinations of drugs, doses and patients. The CPOE-system used - as the key for patient identification - the 10-digit Danish Civil Registration number. This number had to be keyed in on every access to a specific patient. On laptops with no numeric keypad, this work was time consuming and prone to mistyping.

3.2.3. Enrolling an old actant, the paper scheme

For backup reasons paper-copies of the CPOE-content were printed every 24h. Due to the instability of the system in the study period these were in use for medication procedures in average twice a week for hours. This was actually affecting the overall safety performance of drug-dispersion and gave a lot of additional work with no extra gain, thus invalidating many of the reasons for implementing a CPOE.

3.2.4. Problems of an inflexible model of medication in the CPOE.

The model in the software program contained very strict timing and inflexible control facilities designed around medication rounds four times within 24-hours (three times for intravenous medicine). This was in contrast to the dynamic medication procedures needed and the work plan of the nurses that imposed the night nurse to dispense the medicine for the next day in special trays. If the prescription of the drug was altered during the round a nurse had to identify the tray with that drug and change the dose of the drugs accordingly. When prescribing drugs in the CPOE system users can choose whether to prescribe drugs in e.g. mg, g or in number of pills. To avoid misinterpretations users are urged to register the drug doses in mg. Since the interface is not so clear in this respect this was a (novel) source of errors not experienced (to that extend) with a paper-based system, although it in principle contain the same flexibility. The basic problem was that the database underlying the CPOE system was for pharmacy-use focussed on package-variations and prices. The database contained no uniform way of declaration of strength thus allowing for the variation in the interface as well, giving foundation to errors and confusion in the work-processes.

3.3. Transformation of continuing medication with the CPOE-system

The use-scenario of 'considerations for continuation of patient care and treatment' showed considerable changes with the CPOE in two situations of use: when the patient was discharged and in need of information on how to continue mediation at home, and according to the status of the patient in the CPOE in the case the patient was hospitalized again.

3.3.1. Discharging patients with CPOE
With the CPOE system an automatic medication guide is printed for discharged patients that have to continue taking medicine in their homes. This guide includes details that are not comprehensible for the patient. It is a copy of the medication information in the CPOE and therefore the prescribed drugs are typically listed in mg and g and not in number of pills. The patients therefore have to convert the mg and g to the correct number of pills. As a solution some nurses have started to write an additional medication guide that enables the patient to dosage and take his/her own medicine, including a another actant in the nurse's work: a word-processor.

3.3.2. With-drawl or discontinuous patient care-paths with the CPOE
Traditionally paper-medication scheme put a material limit to how many drugs and how many days the consumption continuous. Because of the virtuality of electronic documents a transformation in the continuation of medication was observed with the CPOE and the medication model of the system. Medications not specifically terminated at the previous discharge were listed as active treatments at the re-admittance. This is an important issue, because it shows the backside of the new possibilities of coordinating patient trajectories virtually with information technologies: errors can live longer and reappear in different settings, also being difficult to discover. As clean ups and with-drawl are not routinely and/or automatically done with a non-paper based system, the CPOE assigned new tasks to the human actants.

3.3.3. Delegation of new tasks and redistribution of properties and competences in local use of the CPOE
The study showed a higher cognitive pressure on the physicians' and nurses' memory skills and competences of temporal and spatial coordination. The system offered no decision support, apart from a list of approved drugs to choose from. There were no CAVE-alarms (previous adverse effects) and no warnings of potential interactions. Pocketbooks (paper) were still an important actant in the medication work. Potential errors of orders not withdrawn delegate new tasks to the human actants in order to secure against future mistreatments of medication. The collective use of the CPOE in order to support workflow makes the actual collaboration in medication work more invisible. Communication between physicians, nurses and patients was not supported but demanded considerably work-around including old and new actants.

4 Discussion

The role of CPOE-systems in fostering new errors as well as reducing old ones is known and forms the debate on IT-development in health care as such [1,12,13,14,16]. The point seems to be that despite disappointments in improving the quality of health care with HIT [15], including CPOE-systems, HIT is indispensable to modern health care. The challenge is addressed from different points of action. One important issue concerns the rationality of system development as a question of politics of design [6]. Here quality of health care depends heavily on the inclusion of end-user's domain knowledge in more iterative design processes [15,18]. The argument is that transformations in socio-technical interactions, also outlined above, as well as the complexities of health care make flawless systems a priori impossible. Arguing along the line of either the system and system developers as being solemnly responsible or the user's resistance is a modernistic conception of rationality [17], that cannot grasp the transformative character of IT-implementation on communication [6] and sense-making [19]. Errors are systemic and occur at various levels of interaction. Interdisciplinary research in human-computer interaction in hospital settings can therefore inform more iterative design processes [1,15]. Errors are not an impediment to HIT-mediated health care, but can be seen as something to learn from when new understandings and solutions are symmetrically integrated into the system design and work processes.

5 Conclusion

Expectations of more quality and continuity in health care from simply implementing vendor-built COPE-systems are naive and bound to disappoint. 'Rational medication' can never be detached from local interactions between wide ranges of actants. System design and implementation has to take the pragmatic rationalities of concrete tempo-spatial interaction into consideration and negotiation. The presented STS-studies points to critical situations of enacting medication with CPOE, that demands more flexible functionalities, vendors adherence to general recommendations for human-computer interaction, and more attention to change management by the project organization.

References

[1] Koppel, R., Localio, A.R, Cohen, A, Strom, B. Neither panacea nor black box: Responding to three Journal of Biomedical Informatics papers on computerized physician order entry systems. Commentary. Journal of Biomedical Informatics 38 (2005)267-269.

[2] National IT-strategi for sundhedsvæsnet 2003-2007, Indenrigs- og Sundhedsministeriet, maj 2003.

[3] Mol, A. The body multiple : ontology in medical practice. Duke University Press. Rotterdam 2002

[4] Latour, B. On technical mediation. Common Knowledge 3, 1994:29-67.

[5] Latour, B. On Humans and Non-Humans. In: Pandora´s Hope. Harvard University Press 1999.

[6] Berg M. The Politics of Technology: On bringing Social Theory into Technological Design. In: Science, Technology and Human Values. Vol. 23 No. 4, autumn. Sage Publications Inc, 1998:456-490.

[7] Olesen F. Kommunikation og teknologisk mediering. Red. Elsass, P et al. Kommunikation og forståelse. Qualitative studies in Health care. Philosophia, Århus 1997:143-168.

[8] Svenningsen, S Electronic Patient Records and Medical Practice. Handelshøjskolen København 2001.

[9] Markussen, R, Olesen, F. Kyborger på arbejde - Elektronisk medicinordination. NIKK, 2002,2:20-22

[10] Berg, M, Langenberg C, Berg, I, Kwakkernaat, J Considerations for sociotechnical design: experiences with an EPR in a clinical context. International Journal of Medical Informatics 52 (1998) 243–251

[11]. http://www.laegemiddelstyrelsen.dk/publikationer/netpub/rapporter/brug_medicin_ (Access Sept. 20, 2005

[12] Ash, J.S., Berg, B, Coiera, E. Some Unintended Consequences of Information Technology in Health Care. J Am Med Inform Assoc. 2004;11:104–112.

[13] Shortliffe, E.H. CPOE and the facilitation of medication errors. Editorial. Journal of Biomedical Informatics 38 (2205) 257-258

[14] Horsky, J, Zhang, J, patel, V.L. To err is not entirely human: complex technology and user cognition. Journal of Biomedical Informatics 38 (2205) 264-266.

[15] Wears, RL, Berg, M. Computer Technology and Clinical Work. Still Wating for Godot. JAMA, March 9, 2005-Vol 293;10:1261-63.

[16] Cole, M, Engeström, E. A cultural-historical approach to distributed cognition. In: Distributed Cognitions. Edi. G. Salomon. Cambridge University Press 1993

[17] Latour, B. We have never been modern. Harvester, Weatsheaf 1993.

[18] Nemeth, C, Cook, R. Hiding in plain sight: What Koppel et al. tell us about healthcare IT. Journal of Biomedical Informatics 38 (2005) 262–263.
[19] Bruner, J: Actual Minds, Possible Worlds. Harvard University Press, Cambridge, Massachusetts1986

3.5 Evaluation of HIS and EHR Systems

HIS-Monitor: Quality of Information Processing in Hospitals

Elske AMMENWERTH[a,1], Frauke EHLERS[a], Bernhard HIRSCH[b], Gordon GRATL[c], Christof MACHAN[a]

[a] *Institute for Health Information Systems, UMIT – University for Health Sciences, Medical Informatics and Technology, Hall in Tyrol, Austria*

[b] *Icoserve Information Technologies, Innsbruck, Austria*

[c] *ITH - Information Technology for Healthcare, Innsbruck, Austria*

Abstract. Systematic monitoring of HIS quality is an important task; however, this task is often seen to be insufficiently supported. To support systematic HIS monitoring, we developed the HIS-Monitor questionnaire, focusing on how a hospital information system (HIS) does efficiently support clinical and administrative tasks. HIS-Monitor was applied in a feasibility study with 102 nursing participants. Results point to strengths and weaknesses of information processing in the participating departments. Based on the experiences of the feasibility study, HIS-Monitor is now further being optimized.

Keywords: Healthcare Information Systems, evaluation, questionnaire, quality, information processing

1. Introduction

The quality of information processing is an important factor for the success of health care institutions [1]. Systematic management of information systems is essential; however, HIS monitoring activities using a quantified assessment of HIS quality are missing in most healthcare institutions. Consequently, the objective of this research project was to develop and validate a comprehensive monitoring system to assess the quality of a hospital information system. This HIS monitoring system should support systematic, quantitative monitoring of HIS quality and the comparison of HIS quality between departments and institutions.

[1] Corresponding Author: Elske Ammenwerth, Eduard Wallnöfer-Zentrum 1, 6060 Hall in Tyrol, Austria, elske.ammenwerth@umit.at, http://iig.umit.at

2. Materials and Methods

Based on a review of the literature available on HIS quality and of accreditation programmes such as [2] and [3], the HIS-Monitor questionnaire was developed. Its structure comprises a combination of process steps within patient care and HIS quality criteria. HIS-Monitor addresses the following process steps: patient admission, decision-making, planning and organisation of treatment, order entry and communication of findings, execution and documentation of diagnostics and therapy, and patient discharge. Within these process steps, the following HIS quality criteria are checked: Availability of information, correctness and completeness of information, readability and clarity of information, usability of information, fulfilment of legal regulations, and time needed for information processing. For details, see e.g. [4, 5].

HIS-Monitor was tested and refined in smaller pretests. Its final version consists of 107 questions addressing the quality of information processing from the point of view of the clinical and administrative staff members involved. It uses a 4-point Likert scale for each of the three types of questions, namely:

1. "How easy it is for you to …", "How well do you feel supported …" – answer categories from "bad" to "good".
2. "How adequate is …" – answer categories from "not adequate" to "adequate".
3. "How often does it happen …" – answer categories from "seldom" to "frequently".

Figure 1 presents an extract of the HIS-Monitor questionnaire. To support the management of questions and analysis of results, the HIS-Monitor toolbox was developed.

P4.2 Documentation of diagnostic and therapeutic tasks

Imagine: You have already carried out diagnostic or therapeutic tasks for one of your patients and now want to document them. Please think especially about the information processing tools that you use for the documentation and also consider efforts for finding/accessing the tool (e.g. patient record, computer).

			tools that are predominantly used	bad seldom not adequate			good frequently adequate
1.	How easy is it for you to get an overview which tasks already have been carried out?	O this question does not apply to me	O paper-based O IT	- -	-	+	++
2.	How often does it happen to you that tasks are not completely documented?	O this question does not apply to me	O paper-based O IT	I	II	III	IIII
3.	How adequate do you consider the time needed to complete the necessary documentation?	O this question does not apply to me	O paper-based O IT	- -	-	+	++

Figure 1. Extract from the HIS-Monitor questionnaire.

To check feasibility of the HIS-Monitor questionnaire in a larger real-life context, a **feasibility study** was conducted in the Department of Internal Medicine and the Department of Surgery of a university hospital in Austria. We decided to start by evaluating the nursing subset of HIS-Monitor, as nurses are the largest professional group in a hospital. Altogether, 150 questionnaires were distributed among the 400 nurses of the Department of Surgery, and 100 questionnaires were distributed among the 300 nurses of the Department of Internal Medicine.

3. Results

After four weeks, 102 questionnaires had been returned, 93 of them being sufficiently complete for analysis, giving a return rate of around 37% with regard to the distributed 250 questionnaires.

HIS quality in the Dept. of Internal Medicine and in the Dept. of Surgery was rather comparable. Figure 2 shows the aggregated analysis of mean scores for each process step for the two departments. The mean values were calculated using the code "1" for the most negative answer (e.g. inappropriate, frequently) and "4" for the most positive answer, thus higher values indicate a "better" quality of information processing.

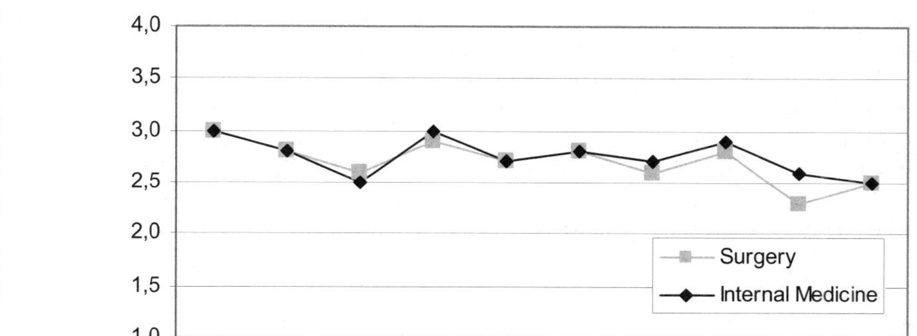

	P1.1 Appointment scheduling	P1.2 Administrative admission	P1.3 Clinical admission	P2.1 Presentation of	P2.2 Care planing	P2.3 Resource management	P3 Order entry	P4.1 Execuation of clinical	P4.2 Documentation of clinical	P5.1 Clinical discharge
Surgery	3,0	2,8	2,6	2,9	2,7	2,8	2,6	2,8	2,3	2,5
Internal Medicine	3,0	2,8	2,5	3,0	2,7	2,8	2,7	2,9	2,6	2,5

Figure 2: Analysis of mean score for the Dept. of Internal Medicine (n=42) and the Dept. of Surgery (n=51) with regard to the process steps P1.1 – P5.1. Higher values indicate "better" information processing.

A detailed analysis of the results showed that in the majority of questions, the quality of HIS was judged as positive. However, for the following questions, there were more "negative" than "positive" estimations (for this analysis, both the two negative answers and the two positive answers were combined).

The answers highlighted some problems with regard to the availability of information from other organisational units. Up to 55% (36 of 66 valid answers) of respondents indicated to have problems when they want to access information especially from other departments (to assess earlier information from his or her own department was not seen to be as problematic). Those who indicated to mostly use computer-based tools answered slightly more positive (mean=2.5, n=32) compared to those who use paper-based tools (mean=2.2, n=10) (the others did not indicate the tool used).

The readability of paper-based examination results was judged by 63% (54 of 85) as insufficient or very insufficient. The readability of paper-based drug orders was judged as problematic by 53% (45 of 85), and 66% (54 from 82) indicated that changes of drug orders were often unclearly documented. 69% (50 of 72) indicated not to be well supported in the early detection of medication errors during order entry, and 58% (38 of 66) indicated not to be well supported in the prevention of unnecessary double examinations, comparable both for paper-based and computer-based support.

Finding and booking free time slots for diagnostic or therapeutic examinations (e.g. x-ray, physiotherapy) that have to be ordered in other departments were seen as cumbersome by 56% (32 of 57) of respondents, also comparable both for paper-based and computer-based support. 63% (38 of 60) indicated to have problems to get quick information on the status of a recent examination order (e.g. x-ray-order is given, examination has begun, examination is completed, results have been transmitted). Here the judgement of those users supported by computer-based tools was better (mean=2.0, n=32) than those mostly supported by paper-based tools (mean=2.4, n=13). The others did either not answer which type of tools they use or said to use both.

71% (60 of 85) found the time needed for nursing documentation (mostly supported by paper-based tools) as partly inadequate or very inadequate, and 65% (54 of 83) complained about the frequent need to transcribe information (e.g. from one nursing plan to another). In this area, both Departments are somewhat different: The Dept. of Internal Medicine is far more advanced with regard to the quality of nursing documentation (even when using mostly the same paper-based tools), and this is reflected in HIS-Monitor (Fig. 3).

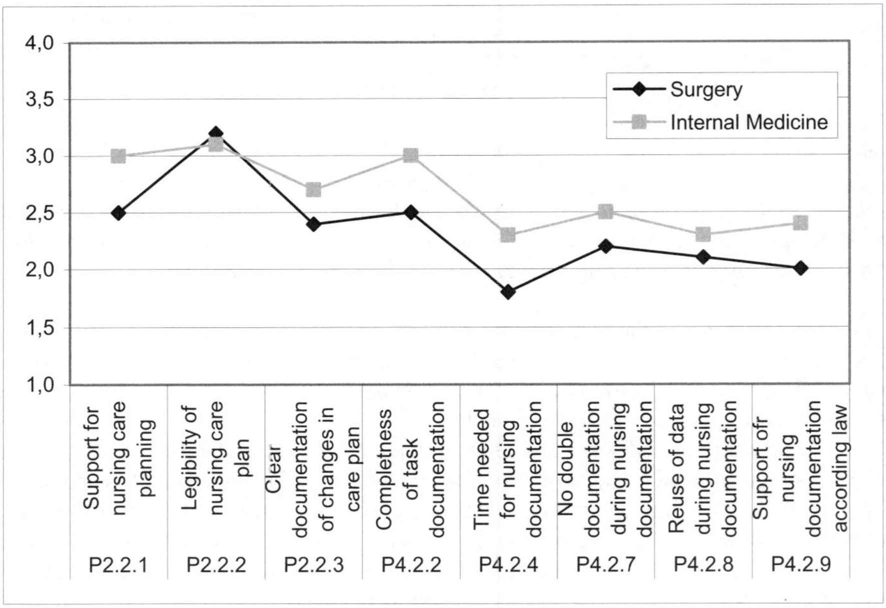

Figure 3: Analysis of mean score for the Dept. of Internal Medicine (n=42) and the Dept. of Surgery (n=51) with regard to questions on nursing care planning and nursing documentation. Higher values indicate "better" information processing.

The detailed results were discussed with representatives from nursing management. Besides the confirmation of most of the pre-defined expectations, and with information on the individual context of the departments, most results could be validated. Usefulness was also supported as nursing management now plans to conduct a time-series analysis of information processing in nursing in 15 wards based on HIS-Monitor, to measure changes after the implementation of a computer-based nursing documentation system.

4. Discussion

The results of HIS-Monitor shall help to screen HIS quality. It will not show in detail the reasons for good or bad information processing, but rather indicate where the HIS is seen as "good" or "bad" from the point of view of the stakeholder groups directly involved. We based the assessment of HIS quality on the standardized questioning of various user groups involved ("customer voices", [6]).

The mean values for all questions were between 2.0 and 3.7, thus reflecting the possible range of 1 – 4 and pointing clearly to strengths and weaknesses. The mean values for the individual process steps were between 2.5 and 3.0, reflecting the aggregation of various questions and indicating that no process step was judged as very high or very low. We found that differences in judgments (e.g. between two departments) of 0.5 or more already indicate large differences in HIS quality (cp. Figure 2).

The following application and analysis scenarios for HIS-Monitor can be imaged:

- HIS-Monitor can be used to collect information about the quality of HIS at a given point of time: HIS quality in a given area can be screened to get an impression on strengths and weaknesses of information processing.
- HIS-Monitor can be used to support a continuous monitoring of HIS quality e.g. by carrying out measurements at different points of time to assess changes or to evaluate effects of information processing tools recently implemented. We plan to do this in the near future, as we want to monitor changes over time during the implementation of a nursing documentation system.
- HIS-Monitor can be used to compare HIS quality in different organisational units, e.g. wards or even hospitals.

5. Conclusion

A first version of HIS-Monitor to assess the quality of a hospital information system was developed and tested in a feasibility study. The results of a pilot test showed that it seems feasible to assess strengths and weaknesses of a HIS by HIS-Monitor. Based on the results of the evaluation, we will now further refine and optimize HIS-Monitor before evaluating it in other settings (e.g. physicians, administrative staff). It is planned to develop a web page where hospitals can compare their HIS quality profiles in anonymous form with the quality profiles of other hospitals. In addition, time-series analysis to describe changes in

HIS quality after introduction of computer-based tools is just being prepared and will further help to assess the validity of HIS-Monitor.

Acknowledgements

We would like to thank the participants of the survey for completing the questionnaires, and Carmen Sailer for the development of the HIS-Monitor tool.

References

[1] Ward J, Griffiths P. Strategic Planning for Information Systems. 2nd ed. Chichester: John Wiley & Sons; 1996.
[2] JCAHO. Joint commission for accreditation for healthcare organizations (JCAHO). Available from: http://www.jcaho.org
[3] Deutsche Krankenhausgesellschaft. Kooperation für Transparenz und Qualität im Krankenhaus: KTQ-Katalog 4.0: Deutsche Krankenhausgesellschaft.
[4] Ehlers F, Ammenwerth E, Hirsch B. Design and Development of a Monitoring System to Assess the Quality of Hospital Information Systems: Concept and Structure. In: Engelbrecht R, Geissbuhler A, Lovis C, Mihalas G, editors. Connecting Medical Informatics and Bio-Informatics. Proceedings of Medical Informatics Europe (MIE 2005), Geneva, Aug 28 - Sep 01 2005. Amsterdam: IOS Press; 2005. p. 575-580.
[5] Ammenwerth E, Ehlers F, Hirsch B, Gratl G. HIS-Monitor: An Approach to Assess the Quality of Information processing in hospitals. Accpeted for International Journal of Medical Informatics, 2006.
[6] Ribière V, LaSalle A, Khorramshahgool R, Gousty Y. Hospital Information Systems Quality: A Customer Satisfaction Assessment Tool. In: El-Rewini H, editor. Proc 32nd Hawaii International Conference on System Sciences (HICSS-32), Maui, Hawaii, January 5-8, 1999: IEEE Computer Society Press; 1999. p. 1-9.

Ubiquity: Technologies for Better Health in Aging Societies
A. Hasman et al. (Eds.)
IOS Press, 2006

Development of Methods for Usability Evaluations of EHR Systems

Lars H. Lilholt[a,1], Signe S. Pedersen[a], Inge Madsen[b], Per H. Nielsen[b], Niels Boye[c], Stig K. Andersen[a] and Christian Nøhr[d]

[a] *Department of Health Science and Technology, Aalborg University, Denmark*
[b] *Aarhus University Hospital, Skejby Sygehus, Aarhus, Denmark*
[c] *Randers Hospital, Randers, Denmark*
[d] *Department of Development and Planning, Aalborg University, Denmark*

Abstract. Developing electronic health record (EHR) systems in Denmark is an on going, iterative process, where also a maturation process for clinical use should be considered. Convincing methodology for collecting and incorporating in the soft- and hardware knowledge and robustness for the clinical environments is not on hand. A way to involve the clinicians in the development process is conducting usability evaluations. The complexity of the clinical use of the systems is difficult to transmit to a usability laboratory, and due to ethical issues a traditional field study can be impossible to carry out. The aim of this study has been to investigate how it is possible to identify usability problems in an EHR system by combining methods from laboratory tests and field studies. The methods selected for the test design are: the think aloud method, video and screen recording, debriefing, a scenario based on an authentic patient record, and testing on the normal production system. The reliability and validity of the results is increased due to the application of method- and data-triangulation. The results of the usability evaluation include problems in the categories: system response time, GUI-design, functionality, procedures, and error messages. The problems were classified as cosmetic, severe, or critical according to a rating scale. The experience with each method is discussed. It is concluded that combining methods from laboratory test and field study makes it possible to identify usability problems. There are indications that some of the usability problems only occurred due to the establishment of an authentic scenario.
Keywords: method, usability, evaluation, participation, EHR, EMR, CPOE, data triangulation, method triangulation, health informatics.

1. Introduction

The national IT-strategy for the Danish health care sector 2003-2007 [1] sets a target for the implementation of electronic health records (EHR) systems on every hospital in Denmark based on common standards before the end of 2005. This goal has not yet been achieved and in recent years there has been an increasing criticism of the development and the implementation of EHR systems in Denmark [2], [3].

This criticism from the clinicians has also been reflected in the international literature and is pointing towards applications being unsuitable for clinical use, systems

[1] Corresponding Author: Lars Heyckendorff Lilholt, Prinsensgade 43, 2.th, DK-9000 Aalborg, lars@lilholt.biz

not supporting the clinical work practice, and lack of involvement of clinicians in the system development process [4], [5], [6]. The CPOE module is the most widely implemented part of the coming full-scale EHR systems (in Denmark), but the CPOE systems provide so far medication order entries only, thus there is still some distance to go between the actual state of EHR's and the vision of full implementation. It is therefore of importance to evaluate systems thoroughly with structured methods to provide qualified feed-back to the remaining development process.

Usability evaluation is a way of involving the end user in the system development process [7]. As the end user has to interact with the system during the test, possible usability problems can be identified, which can contribute to improvement of the system. Applying this method to the development of EHR systems can be an approach of involving clinicians in the system development.

The two archetypes of usability evaluation is laboratory test and field study. By applying the laboratory test, vital knowledge on the system-context-interaction can be missed. On the contrary it is very problematic to apply a full blown field study at a hospital due to ethical issues. Therefore the aim of this study has been to investigate how it is possible to identify usability problems in an EHR system by combining methods from laboratory tests and field studies.

The aim of the present study was to develop a structured, authentic evaluation method of clinical end user load and gains in interacting with the (medication-) CPOE-system. The method was developed on one HMO-CPOE (Aarhus County, Denmark) but will be applied in a comparative study of three to five CPOE-systems.

2. Methods

Methods from laboratory test and field study were combined utilizing experiences from Evaluation Laboratory (EVALAB) in Lille, France [8], and the EHR-center in Trondheim, Norway [9]. The focus has been on combining advantages from both types of tests to retain a large degree of control and at the same time creating an authentic situation for the clinician by including a rich clinical context.

The components in our methods were selected with the purpose that obtained experimental data could be triangulated to validate a result. Details of the specific methods described below can be found in [7], [10].

The *think aloud* method is a simple method of collecting the users' thoughts throughout a test. The user simply explains his thoughts and reflections while using the system. The think aloud method was considered as vital for the later data analysis. *Video recording* of the contextual work situation as well as *screen capture* of the CPOE system was used as a method for collecting data from the test. These recordings are essential documentation of the occurrence of usability problems during the test. Directly after the test a *debriefing* was performed, to increase the level of information concerning the test. Since the main target of this study was to evaluate the combination of methods and not to conduct a full usability evaluation itself, two CPOE experienced physicians participated as users. A scenario was scripted based on an authentic, complex patient trajectory to ensure the use of a wide range of functionalities of the CPOE system. The case concerned a 39 year old female patient suffering from cancer, admitted to a long-term medical treatment involving several stakeholders and multiple drug ordinations. In an evaluation laboratory, Skej-Lab, at Aarhus University Hospital, Skejby Sygehus, a ward was set up with a bed, a table, a chair, and a medicine chest and various other accessories. Besides the ward an office was set up with a computer from where the user should access the CPOE system, see Figure 1.

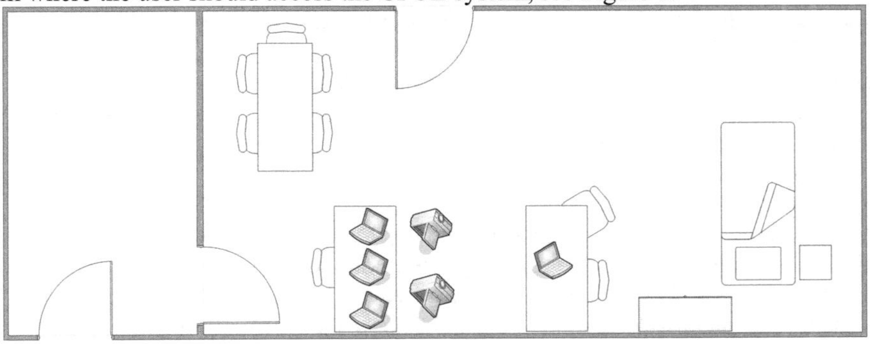

Figure 1- Illustration of the test facility setup. To the right the ward, in the middle the office, and to the left the control and debriefing area can be seen.

An actress was playing the role of the patient, partly due to ethical issues, but also to secure a similar act in each test. Since the test facility was physically located at the hospital, it was possible to use the normal production IT-system instead of a test version prototype. This should contribute to the realism of the test exposing authentic system response time etc.

After the tests the video and screen recordings were analyzed and log files were prepared. These log files were then analyzed with the purpose of identifying usability problems.

3. Results

Classification system

The data analysis of the log files identified a number of usability problems. These were classified according to Molich's rating scale for classifying severity of a usability problem into the categories: cosmetic, severe, and critical [11].

Inspired by Skov and Stage [12] a diagram to assist the quantification of usability problems according to Molich's rating scale was developed in an iterative process. The diagram shown in Figure 2 is not generic, but specifically targeted at usability problems in a CPOE system.

	Response Time	GUI Design	Functionality	Procedure	Error Message
Cosmetic	Response time of a few seconds and/or it occurs often.	The GUI design annoys or does not help the user. The user does not make any failure and/or is aware of which failure it can bring.	The functionality causes temporary problems but it is possible to perform the function, taking a few seconds extra.	The system gives rise to a complicated procedure, taking extra time from the user.	The error message does not misinform the user but it is not specific enough and causes a few seconds delay.
Severe	Response time of several seconds and/or it occurs often.	The GUI design annoys or does not help the user. The user makes a failure but identifies it.	The functionality is difficult to carry out, taking several seconds extra. The user makes a failure.	The system gives rise to an incorrect procedure.	The error message misinforms the user, resulting in several seconds delay.
Critical	Response time of more than one or several minutes and it occurs often.	The GUI design annoys or does not help the user. The user makes a failure and does not identify it.	The functionality is not possible to perform or a system crash occurs.	The system gives rise to an incorrect and critical procedure.	The error message misinforms the user, resulting in several minutes delay.

Figure 2 - The classification diagram developed in the study.

Experiences with the methods

Think aloud: Both users expressed that the think aloud method was straightforward and simple to use. During the medication process physicians are normally supported by nurses in discussing the patient situation and the specific medication, so they felt no interference in having a test monitoring person beside them during the test.

Recording: The recordings of the screen display of the CPOE system were essential for the analysis, but the video recordings of the ward and the office were less useful. The observer's notes taken during the tests were quite adequate to capture the essential information from these locations.

Debriefing: The debriefing session immediately after the test gave useful results, as the users could be more specific about and comment on some of the experienced problems during the test. The users also gave their experience of having participated in the test, both expressing that a nurse would promote the realism in the test set-up.

Scenario: The scenario play based on the authentic patient record was the key element in the test to promote the realism. The scenario play on the "ward" helped the users to acquaint themselves with the patient case and it helped them to act realistic. Both users expressed that they made considerations about the patient's record and gave serious considerations to the ordinations described in the scenario description. The name of the patient and the date of birth in the CPOE-system did not correspond to the data in the scenario. The patient name in the system was "CPOE test Woman", being twice as old as the female patient in the scenario. These system data did not promote the realism.

Scenario settings: The arrangement of the ward and the clinical staff office worked well. A pill bottle at the patient's bedside table should illustrate the patient's home medication. The label on the pill bottle did not match the drug prescription in the scenario which confused one of the users. One of the users did bring his own white coat, including reference books, stethoscope, etc., and he did use one of his reference books during the test to check some medicine.

Production system: Performing the test at the hospital using the production system definitely had an impact on the results of the tests. Halfway through one of the tests a system crash occurred, and the test could not continue. Both users experienced several times long response times.

User profile and number of users: Both users were experienced users giving the advantage that they were both able to enhance positive aspects and application-specific usability problems at the same time. Due to the system crash the second user did not complete the whole scenario.

4. Discussion

Data collection

If a system is to be used while a dialogue is going on between e.g. a clinician and a patient, an alternative to the think aloud method has to be considered. An alternative could be retrospective testing [10] where the test monitor and the user immediately after the test go through the recordings of the test together. On the basis of the recordings and the questions from the test monitor, the user should be able to recall some of his considerations when using the system. There is a risk that some information will be lost, depending on the user's ability to recall his considerations.

The video recordings of the ward and the office were of little use, and it should be considered if it is adequate to use the test observer's notes. Though, it is recommended to keep video recordings as a back-up, and as a documentation of the test issues. The triangulation of the think aloud method, the recordings, and the debriefing, has increased the reliability of the results.

The use of experienced users limited the number of false usability problems, as they have experienced the problems before and they have better knowledge of the possible functionalities in the CPOE system. This enhances the validity of the results. In the light of the system crash, resulting in only one-and-a-half scenario was tested, three or maybe more users should be considered. Increasing the number of users increase the amount of empirical data to be analyzed, so this should be a balance between resources available and the aspiration for robustness and validity of the results.

Realism parameters

The loads and gains that the CPOE system demands/give from/to the clinical end-user are evaluated through the test persons acts - as recorded by video - and through the thoughts - as reflected in the verbalisation during the *think aloud* part of the test. Loads and gains to other stakeholders in the patient trajectory and to the health care providing organisation are not evaluated by this method. The use of a script based on an authentic patient record and the use of a physical patient-artefact - utilizing an actress as patient - were essential for the clinical richness of the test-situation and hence the validity of the system evaluation, since it promoted tasks in realistic spatial and temporal relations to be acted upon by the end-user test-persons.

It is of importance that the clinical information is realistic and realistic complicated. It should not be too specialized or advanced, since it is not the medical issues in the story that are subject of test and discussions. Both test-end-users expressed that the case was realistic and that the identified usability problems during the test are so general that they could have been identified with other cases. This increases the reliability of the results.

Accessories as e.g. pill bottles have to be correct, in order to support the realism and the user's ability to play the scenario. It is recommended to ask users to bring their own white coat and other items they normally wear at their clinical work. One of the users in the test requested more office equipment which could easily be arranged with a few tables, paper and pens, and telephones. Including a nurse in the test would also promote the realism of the test, as a physician normally discusses ordinations, prescriptions, etc. with a nurse. This would at the same time open up the possibility to evaluate more applications and the support of the application to cooperative clinical work.

A prototype of the CPOE system would have given less credible results. Identified problems such as system crash and response times would not have been recognized using a test system. The use of the production system and the arrangement of the test at the hospital premises are considered to be very important parameters to promote the realism in the test.

Further development of a classification system

The diagram in Figure 2 was developed to classify the identified usability problems in this project. A more exhaustive and exclusive classification system should be developed through additional iterations. Such a classification system could be used as a support in identifying usability problems in a CPOE system, as documentation for having classified a usability problem, support a better communication and procurement to the developers of the system, and finally it could serve as a benchmarking tool for CPOE systems.

5. Conclusion

This small study indicates a path towards better methods for capturing and communicating information about the overall clinical performance and influence on the clinical process of utilizing information technology - as well as a method of quantitative estimation of the “clinical impact” of specific clinical systems. To capture both end-user acts and thoughts to the activity and performance of an IT-system, a combination of methods from laboratory test and field study should be employed in an artificial clinical milieu with sufficient complexity and richness to mimic the clinical environment. Some of the usability problems only occur due to the establishment of an authentic scenario and performing the test on the production system, instead of a test version prototype. The classification system for usability problems has proven to be supportive to the identification and communication of usability problems. The methods applied should still be fine tuned and more tests should be conducted to consolidate an optimal triangulation and to prove the usefulness of the method in comparative evaluation of clinical systems.

6. Acknowledgements

We thank Jens Kelsen and Søren Tang Knudsen from Randers Hospital for participating as clinical users in the usability tests.

References

[1] Ministry of the Interior and Health, Denmark. National Strategi for IT i sundhedsvæsenet 2003-2007, Ministry of the Interior and Health, 2003.
[2] Glerup H. EPJ-folkene skal ud til brugerne og opleve problemerne live. Dagens Medicin, Nov 18. 2005.
[3] Albinus NB. Medicinmodul tager tid fra lægernes arbejde. Dagens Medicin, Nov 4 2005.
[4] Han YY, Carcillo JA, Venkataraman ST, Clark RS, Watson RS, Nguyen TC et al. Unexpected increased mortality after implementation of a commercially sold computerized physician order entry system. Pediatrics 2005; 116(6):1506-1512.
[5] Koppel R, Metlay JP, Cohen A, Abaluck B, Localio AR, Kimmel SE et al. Role of computerized physician order entry systems in facilitating medication errors. JAMA 2005; 293(10):1197-1203.
[6] Wears RL, Berg M. Computer technology and clinical work: still waiting for Godot. JAMA 2005; 293(10):1261-1263.
[7] Rubin J. Handbook of Usability Testing, John Wiley & Sons, Inc., 1994.
[8] Evaluation Laboratory (EVALAB), University Hospital, Lille, France, http://www.univ-lille2.fr/evalab (last accessed Jan 12, 2006).
[9] Norwegian Center of Electronic Health Record, http://www.nsep.no (last accessed Jan 12, 2006).
[10] Nielsen J. Usability Engineering, AP Professional, 1993.
[11] Molich R. Brugervenligt webdesign, Teknisk Forlag, 2000.
[12] Skov MB, Stage J. Supporting problem identification in usability evaluations. Proceedings of the Australian Computer-Human Interaction Conference 2005 (OzCHI'05), ACM Pres, 2005.

Ubiquity: Technologies for Better Health in Aging Societies
A. Hasman et al. (Eds.)
IOS Press, 2006

Implementing communication systems in the community health services. The health care workers experiences

Hans Tore Mogård[a], Eli Haugen Bunch[b] and Anne Moen[c]

[a] *Research fellow, Faculty of Nursing, Sør Trøndelag University College, Norway*
[b] *Professor, Institute of nursing science, University of Oslo, Norway*
[c] *Post doctoral fellow, InterMedia, University of Oslo, Norway*

Abstract. Reengineering of the workplace through Information Technology is an important strategic issue for today's community health care. The computer-based patient record (CPR) is one technology that has the potential to profoundly modify the work routines of the care unit. This study investigates a CPR project, Gerica® aimed at allowing the health care workers in the community health care to work in a completely electronic environment. The focus of our analysis was the use of Gerica, and the health care workers interpretations of it. The rationale behind the introduction of this technology was based on its alleged capability to both enhance quality of care and control costs. This is done by better managing the flow of information within the organization. Theory of structuration is used as the conceptual vehicle to aid in widening the search to the socially constructured nature of these meaning: how people constructed their conceptions in their work setting. The present study analyzed the implementation of CPR conducted in the community health services in Trondheim, Norway. Interviews with Gerica® users demonstrate that individual interpretations vary considerably, also between users of the same application. User-resistanse was not the problem. This project was a good opportunity to understand better the intricate complexity of introducing technology in professional work where the usefulness of information is short lived and where it is difficult to predetermine the relevancy of information. Profound misconceptions in achieving a tighter fit (synchronization) between care processes and information processes were the main problems.
Keywords: Gerica, social construction, meaning, theory of structure.

1. Introduction

When work demands cooperation, one could assume that introducing an application to support cooperation would be rather straightforward. Unfortunately this seems not to be the case, see e.g. (1). Faced with collaborative software, users need to shift their interpretations from the familiar ones to novel readings to appreciate the cooperative nature of these applications, the interdependencies in work (2). Orlikowski's (3) findings suggest that when people neither understand nor appreciate the cooperative nature of collaborative software, it will be interpreted as an instance of some more familiar technology and used accordingly. This can result in counter-productive and uncooperative practice and low levels of use. An illustration by Grudin (4) of the required shift is the difference between using single-user application (such as a co-authoring tool). In the former, the interdependence must be dealt with separately by work arrangements; in the latter the co-operative nature is inherent in the application.

In this study we proceed from an assumption that a sufficiently shared understanding of the purpose and functionality of collaborative software in its particular organisational context is a prerequisite for its co-operative use. The shared understanding does not necessarily need to be articulated explicitly, it can also be conveyed by co-operative work practice. The starting points of this study were the startlingly different statements of collaborative software users, constantly encountered even within a single group. We focus on these differences within groups. How is it that health care workers who do the same kind of work in the same organisational context with the same tools, still understand (and accordingly use and tell about) them in such different ways?

Earlier studies are few. The mail focus has been on similarities within groups (5, 6) or organisations (7, 8) and differences between them. Two studies by Mackay (9, 10) focus on groups as formed by individuals. In a study of electronic mail use Mackay (9) claims that use is diverse because of the users' preferences but does not examine in dept the background of these preferences. In her study of software customisation (10) as a co-adaptive phenomenon (human behaviour affects environment and vice versa), she points out that they try to accomplish, under a complex set of influences. In this study we share Mackay's interest in individuals as active actors and also recognise the complex mutual interaction of user's perceptions and users of software in a given context.

This paper is based on a part of a larger study in the primary health services in Trondheim community, Norway, where there was chosen 2 offices of home care among 90 offices of the home care and nursing homes. Our focus were on how they had approached and appropriated a new piece of collaborative software technology, Gerica®, and what their view of technology and the applications were during the interviews. Their characterisations of Gerica® varied considerably. These variations led to exploration of what kind of conceptions of technology could be found behind these characterisations, how each conception has been formed, and whether shared meanings were being established. These socially constructed meanings (11) are reflected in how users talk about the applications and in how the applications are used for individual or co-operative purposes.

The conceptual vehicle to investigate the process of the social construction is theory of structuration (12, 13). It is used to find answers to questions such as. How are the meanings of technology constructed in action and interaction? Can an individual's interpretation of technology be better understood when viewed in the light of the resources used? What kind of a role is played by the norms which a group or individual is subject to or draws upon?

2. What is Gerica®?

The focal application used in this study, Gerica®, has been widely-used platform for developing different-time/different-place collaborative software applications. As a product, it started to evolve from the ideas of a distributed conferencing application (many to many communication) and a bulletin board (one-to-one communication) to provide sharing of information by replication databases over networks. A database in Gerica® is "a collection of related forms or semi-structured documents organized through views that sort or categorize information". The directly accessible additional functions include electronic mail, an editor, full text search capabilities, registration of ADL (activity of daily living) and daily documentations about the clients.

Gerica® differs from other client-server tools in two ways. The adjustable replication mechanism allow for distributing work while maintaining facilities for co-operation. The Gerica® documents can act as carriers of several different types of data, either included in the document or residing in separate databases linked by pointers. According to the interviews and discussions at the Trondheim community health services, most of their Gerica® applications support cooperation in better ways than other tools they tried.

Gerica® includes capabilities to support different types of applications, which would promote co-operation. DeMichelis (15) has characterized the nature of co-operation with three dimensions: coordinations, communications and collaborations. Although this characterization is not comprehensive, with Gerica® it is illustrative. Electronic mail, the broadcast applications and the discussion applications all emphasize the communication dimension. Tracking and workflow applications are typically related to co-ordination. Collaboration is the widest characterization of the three: the focus is on working together to achieve a shared goal. Almost any Gerica® applications can be used for this purpose.

3. Theory of structure and meaning construction

Theory of structuration (12, 13, 16) is used here as a sensisting device in widening the scope of the study into the social construction of technology, acknowledging the adaptation to information systems, see e.g. (17) and to collaborative software (18). The reason behind this choice is the richness and the subtlety of the theory: it encompasses the whole arena of human action and interaction, with a focus on three structural dimensions that guide action: signification, domination and legitimation. Its comprehensiveness makes it suitable to support multi-dimensional explorations into the social construction of meaning.

The main focus of this study is the structures of signification that enable our communication. From the structure point of view, our significations are conveyed by our interpretative schemes that represent institutional rules of interaction. From the point of action these structures are represented as modalities, as interpretive schemes that the actor employs in constitution of interaction. People draw on the assumptions, knowledge, or rules which may be embedded in IS (information systems), to perform or modify their tasks. Action in turn can create new structures of meaning that can alter institutionalized practices.

In this study we focus on what and how the informants told about Gerica® and its use. The purpose is to learn about the meanings attributed to Gerica®, the variation in these conceptions and possibly shared features therein. We will use the modalities of interpretative schemes, resources and norms to look into the structures, the rules and resources that facilitate or constrain meaning construction in the home care officies. Our assumption is that by looking into meaning construction as a social process from the three analytic viewpoints, the variation in the accounts can be traced and thereby shared elements found. As a process, the construction of shared conceptions is anticipated to be gradual – even though single instances might look like a flash of insight. In this process, interpretive schemes aid as knowledge is pushed into the discursive consciousness, the limits of the norms are probed and access or resources – either legitimated or through dialectic of control – enables or constrains forming and testing of different interpretations.

4. Analysis

In discussing the interpretations of the Gerica® technologies in the community health system, we consider in turn each of the domains outlined above: nature of technology, technology in use, signification, domination and interpretations.

Nature of technology. The users in this study received little official information about Gerica®, or the rationale behind its wide-scale implementation. When they were interviewed a year after the Gerica® was implemented, some users had still limited knowledge of the technology:

> "With respect to Gerica®, I know nothing about it. I don't know what it is supposed to do, and I don't know why this is better than other applications. So I have no expectations about it."

Other users had ideas about the technology, but these were either only partially correct, or somewhat incomplete:

> "I believe Gerica® is putting word processing power into statistics."
>
> "It has something to do with communications and documentation. I use it only for daily documentations for my clients"

Research in cognitive sociology (19, 20) and organisational studies (21, 22) suggests that people tend to approach the new in terms of the old. The same may be expected of people confronting new technology. In the absence of other information, they will attempt to interpret it in terms of their existing technological frames, imposing assumptions, knowledge, and expectations about a familiar technology on the unfamiliar one. This is evident in the statements above, where Gerica® is interpreted in terms of existing technologies. It is also explicit in the comments by two managers explaining why technology in general (and by implications, Gerica®) would not be useful, given their particular modes of working:

> "My workstyle is heavily interpersonal and oral. So far computers have not really saved my time. I am not interested in doing all that protocol stuff to get access. I don't want to deal with a programmer's conceptions of the world. If I wanted to be a programmer, I would have become one. I approach Gerica® with the attitude "Do I really need this?" Other folks are more trained in computers and analytic methods, and don't have the black-box mentality that I do. They tend to work quantitatively rather than textually or with narrative. And they tend to work in black and white while I work in greys. …These perceptions cloud how you see things like Gerica®. I see computers as black and white and so as not really suitable to my work."

Thus, users related Gerica® to computers in general, to particular aspects of the computer, to already familiar specific computer applications (e.g. statistics, word processing, electronic mail and files). Such reference logic led to ambiguous, incorrect, or partial images of Gerica® incongruent with those envisioned by technologists.

Technology in use. Related to inadequate training was the lack of understanding of what the technology was useful for. Where the nature of a technology is poorly understood, and in the absence of a clear directive on how to use it, and users found it difficult to know how to appropriate the technology and use it effectively.

They were also less likely to be motivated to do so. For example, consider these managers comments:

> "We are so task oriented; we are not going to think of applications and creative uses. It`s no good just putting the technology on our desks; you have to show us practical applications, something with real value to my work."

Further, in contrast to the technologists` view that existing policies and procedures around security, confidentiality, and quality were adequate, users had reservations. Principals in particular were concerned about threats to security, confidentiality, and authenticity:

> "I have concerns about what goes into the databases and who has access to them and what access they have."
>
> "The flip side of using Gerica® to share information is the risk of breaking confidentiality of clients, or undermining the value of special projects. No one would trust the security controls even if they were imposed. There are no guarantees that those who access the information will keep it confidential."

The users′ frame of Gerica® in-in-use – that it is difficult and time consuming to learn and use, and that it raised concerns about control, confidentiality, security, and liability – is reflected in their use of it. Users did not actively or significantly utilize discussion databases in their initial dealings with Gerica®. Some users found benefit in electronic documentation and file transfer, but the majority of users (at least at the time of this study) did not engage in expertise sharing and collaboration.

In these assumptions about Gerica® and its context of use, it is apparent that for Gerica® users, technology was never means to an end. Their primary focus was on client service and on consulting activities. They had to meet tight project schedules and manage a competitive and uncertain working situation. As a result, they were unwilling to invest time and energy in learning and using a technology that provided no apparent and immediate benefits. Interpretations of the Gerica® technology-in-use by these users suggest a focus and range of concerns that were not deemed problematic by the technologists.

Figure 1. Elements from the theory of structuration. Modified from Giddens (13, p. 29)

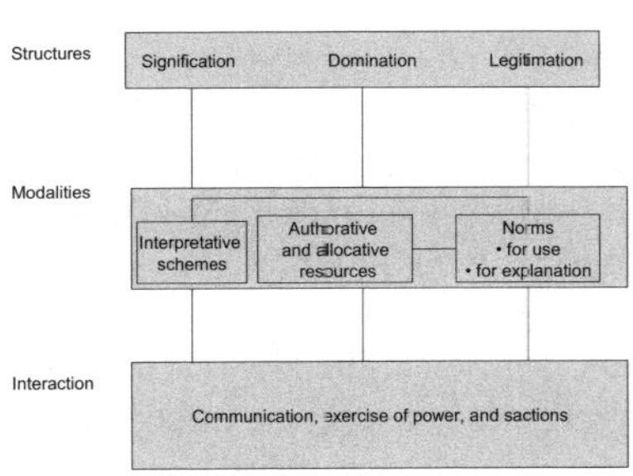

Signification. All nine interviewed were able to give a conceptual definition of Gerica® and a description of their own applications. Everybody was also aware of the other capabilities of Gerica® and gave description of possible applications. The major focus of their explanation was on the communication aspect of their applications: they saw it as an efficient way of disseminating information to the community. Their interpretations varied according to their use of Gerica® and also according to their own role in introducing Gerica® in the community, but a shared conception of daily documentation of the clients existed. The community manager emphasised the possibilities of Gerica® use for the community, displaying a somewhat wider view of Gerica® than the others.

Legitimation. The organisation of the community was efficient. Gerica® was seen to be in alignment with the new cost-cutting norm. Everybody knew her or his tasks and was reluctant to take other ones if the benefits were not tangible. The task-centredness was interpreted differently by different users: for example, the managers did not see inputting data into Gerica® as his task, especially when his colleagues did not that either. The central norms of community efficiency were clearly visible in the words and actions of both, but still resulted in different type of Gerica® use due to different interpretations of its meaning to oneself and to the community.

Dominations. The norms in the community were not against Gerica® use, but there was no official support for it either. Expansion of Gerica® depended on individuals such as the nine persons here. The resources available to these nine had not presented limits to their understanding or use of Gerica®. All claimed that they had had sufficient training and help in Gerica® use. The future promise of wider use was also mentioned. The nine were pushed for time and had different priorities for their tasks, but did not mention Gerica® as adding to their work considerably.

Interpretations. All gave clear accounts of Gerica® and how they used it. All nine had primary jobs and Gerica® was used only to the extent that it was seen to support that work. As the applications were now, they satisfied their users and were already part of the everyday life of these nine interviewees.

In summary, the power these nine had was limited only to their work. Expanding Gerica® use would have meant changes on the corporal level regarding interpretation of importance of Gerica® for the community. Through that, resources could have been re-allocated to support including new groups and the necessary norm changes could have arisen to give basis for this. The corporate management sees it now as the main communication tool for the community.

5. Discussion

This paper has explored a number of issues that change the way we think about and study the interaction of technology and organisations. Technologies are products of their time and organizational context, and will reflect the knowledge, materials, interests, and conditions at a given focus in history. There is nothing inevitable or inviolable about them (23, 24, 25).

While Giddens' (13) theory of structuration is posed at the level of society, his theory of structuration processes, describing the reciprocal interaction of social actors and institutional properties, are relevant at multiple levels of analysis. The theory of structurational model of technology allows us to conceive and examine the interaction of technology and organisations at interorganisational, organisational, group, and individual levels of analyses. This overcome the problem of levels of analysis raised by a number of commentators (26, 27, 28, 29), and underscores the value of understanding the multiple levels across which technology interacts with organisations. Only examining selected relationships, how technology influences human agents without being mindful of how users appropriate that technology – leads to a partial understanding of technology's interaction with organisations.

By moving across levels of analysis and boundaries of time and space, the theory of structurational model of technology affords an examination of technology transfer among organisations. Many of the technologies used by organisations today are not built internally. Rather they are acquired from other organisations – either customdesigned, off-the-shelf, or in some form that is part mass produced and part customized. Recognition the disjuncture in time and space between the design and use mode allows us to analyze the role of multiple organisations in developing and deploying a particular technology. A technology may be designed by one organisation, built by a second, and then transferred into a third for use. In these cases, the institutional conditions and human agents involved in technology development are different from those involved in technology use. That is, external entities – the developing organisations – play an influential role in shaping the social practices of the organisations using the technology. The theory of structurational framework affords a way of investigation not only the movement of technology through time-space, but also across organisational boundaries, potentially provided a basis for analyzing interorganisational relations of learning, influence, and dependence.

The theory of structurational model of technology does not directly deal with organisational form, which was considered an institutionalised property of organisations. Future analyses of the relationship between different organisational forms and the interaction of technology and human agency are clearly relevant. It would be useful to isolate this aspect of organisations analytically and examine how different organisational forms may engender certain kinds of technologies, and how these technologies in turn may reinforce or transform the structural configurations over time. For example, we could postulate that more or less interpretively flexible interactions with technology would be associated with the various organisational forms posited by Mintzberg (30).

6. Conclusion

This paper has proposed in alternative theoretical conceptualizations of technology which underscores its socio-historical context, and its dual nature as objective reality and as socially constructed product. This paper details and illustrates a theory of structurational model of technology that can inform our understanding and future investigations of how technology interacts with people in organisations.

The view of technology encourages investigation of the interaction between technology and organisations that seek patterns across certain contexts and certain types of technology, rather than abstract, deterministic relationships that transcend settings, technologies, and intentions. As this study shows, there are strong tendencies within institutionalized practices that constrain and facilitate certain developments and deployments of technology. In particular, understanding how different conditions influence the development, maintenance, and use of more or less interpretively flexible technologies would give insight into the limit and opportunities of human choice and organisational design.

7. Acknowledgments

This studie was supported by Trondheim community health services and Sør Trøndelag University College. The author extend a special thanks to the health care workers of Trondheim community for their participating. I am also extremely grateful to the dissertation advisor Eli Haugen Bunch, and Anne Moen for their help and support throughout the years.

References

[1] Grudin, J. Collaborative software and CSCW: Why now? In Collaborative software 1991: The potential of team and organizational computing. Software Engineering Research Centre: Utrecht, Holland, 1991; pp.3-14.
[2] Schmidt, K, Bannon, L. Taking CSCW seriously: Supporting articulation work. Computer Supported Cooperative Work, 1 (1-2), 1992; pp.7-40.
[3] Orlikowski, WJ. Learning from Notes: Organizational issues in collaborative software implementation. CSCW`92 Proceedings. ACM, New York: 1992; pp. 362-369.
[4] Grudin, J. Why collaborative software applications fail: problems in design and evaluation. In: Office: Technology and People, 4(3): 1989; pp. 245-264.
[5] DeSanctis, G, Poole, MS. Capturing the complexity in advanced technology use: adaptive theory of structuration theory. In: Organizational Science, 5 (2). 1994; pp. 121-147.
[6] Fulk, J. Social construction of communication technology. In: Journal of the Academy of Management, 36(5). 1993; pp. 921-950.
[7] Barley, SR. Technology as an occasion for structuring: Evidence from observations of CT scanners and the social order of radiology departments. Administrative Science Quarterly, 31, March 1986; pp 78-108.
[8] Bullen, CV, Bennett, JL. Learning from user experience with collaborative software. CSCW`90 Proceedings. ACM, New York. 1990; pp. 291-302.
[9] Mackay, WE. More than just a communication system: Diversity in the use of electronic mail. CSCW`88 Proceedings. ACM, New York. 1988; pp 344-353.
[10] Mackay, WE. Users and customizable software: a co-adaptive phenomenon. MIT, 1990.
[11] Berger, PL. Luckmann, T. The social construction of reality: A treatise in the sociology of knowledge. Penguin Press, London. 1967.
[12] Giddens, A. Central problems in social theory: Action, structure and contradiction in social analysis. Macmillian, London. 1979.
[13] Giddens, A. The constitution of society. Polity Press, Oxford. 1984.
[14] Patton, MQ. Qualitative ecaluation and research methods. 2nd Ed. Sage, Newbury Park, CA. 1980.
[15] De Michelis, G. Computer support for cooperative work. Butler Cox Foundation, London. 1990.
[16] Cohen, IJ. Theory of structuration theory: Anthony Giddens and the constitution of social life. St. Martin`s press, New York. 1990.
[17] Orlikowski, WJ, Robey, D. Information technology and the structuring of organizations. Information Systems Research, 2 (1): 1991; pp. 19-37.
[18] Lyytinen, K. Ngwenyama. What does computer support for cooperative work mean? A theory of structurational analysis of computer supported cooperative work. In: Accounting, Management, & Information Technology, 2 (1). 1992; pp. 19-37.
[19] Cicourel, A. Cognitive Sociology. Free Press, New York. 1974.
[20] Goffmann, I. Frame Analysis, Harper & Row, New York.
[21] Van Maanen, J. Doing new things in old ways: The chains of socializations. In: College and University Organization. New York University Press, New York. 1984; pp. 211-247.
[22] Weick, KE. The Social Psychology of Organizing. Addison-Wesley, Reading, Mass. 1979.
[23] Noble, DF. Forces of Production: A social History of Industrial Automation, New York: Oxford University Press. 1984.
[24] Perrow, C. The Organizational Context of Human Factors Engineering, Administrative Science Quarterly, 28. 1983; pp. 521-541.
[25] Zuboff, S. In the Age of the Smart Machine, New York: Basic Books. 1988.
[26] Kling, R. Defining the Boundaries of Computing across Complex Organizations, In: R. Boland: and R. Hirschheim (Eds.) Critical Issues in Information Systems Research, New York: Wiley.

[27] Leifer, R. Matching Computer-Based Information Systems with Organizational Structures, In: MIS Quarterly. 1988; pp. 63-73.

[28] Markus, ML. Robey, D. Information Technology and Organizational Change: Causal Structure in Theory and Research, In: Management Science, 34, 5. 1988; pp. 583-598.

[29] Rousseau, DM. Issues of Level in Organizational Research: Multi-level and Cross-level Perspectives, In: Research in Organizational Behavior, 7, Greenwich CT: JAI Press. 1985; pp. 1-37.

[30] Mintzberg, H. The Structuring of Organizations, Englewood Cliffs, NJ: Prentice-Hall. 1979.

Ubiquity: Technologies for Better Health in Aging Societies
A. Hasman et al. (Eds.)
IOS Press, 2006

Measurement of the Clinical Usability of a Configurable EHR

John MØLLER-JENSEN[a,1], Ivan LUND PEDERSEN[b] and Jesper SIMONSEN[c]
[a] *CSC Scandihealth A/S, Århus, Denmark*
[b] *Roskilde County Hospital, Roskilde, Denmark*
[c] *Computer Science, Roskilde University, Roskilde, Denmark*

Abstract. The objective of the project was to measure the clinical usability of an EHR configured by use of participatory design with clinicians from a neurological stroke unit in order to get input to the County's future strategy for incremental implementation of EHR. The content of the EHR was defined during a series of workshops with the clinicians after which the XML configuration files were written and deployed. In parallel with this, the participants from the University identified, prioritised and further specified a number of effects related to the clinical practice to be measured. The effects requested by the clinicians focused on improving their overview and assessment of patients as well as on more efficient coordination in three specific and highly cooperative situations, viz. nursing handover, ward round and patient conference. All three situations were measured before (using paper-based medical records) as well as during the week where the configured EHR completely replaced the paper-based medical record in order to compare a 'before' and 'after' situation. Measurements were focused on the requested effects and acquired using various techniques including questionnaires, interviews, observations, and Task Load Index (TLX) ratings. In total, 15 nursing handovers, 8 ward rounds, and 11 patient conferences involving a total of 35 patients and more than 20 clinicians were included in the measurements. Data from the project has been comparatively analysed by means of the TLX scores. Our results show several significant results, for example, during ward rounds the physicians experienced a significant improvement of TLX. The experiment has proven it possible to configure the content of an EHR that significantly improves the clinician's overview of the patient's current status in different clinical situations during the clinical process, based on the clinician's actual needs. Furthermore, the configuration process gave the County valuable experience concerning the content and management of a participatory design process as well as documentation of utility value that will be incorporated in future EHR projects.
Keywords: Participatory design, software design, strategy, evidence, computerised medical record system, usage, clinical process, effects, measure, vendor, NASA-Task Load Index, XML configuration files, case study

1. Introduction

In Denmark, the National Board of Health has published a national Conceptual Model for Communication in Electronic Health Records (G-EPJ) for highly structured problem-oriented clinical documentation in EHR systems [1, 2]. After the first tests of G-EPJ it became obvious that it was important to focus more on how such an EHR could be designed in order to make it accepted by the clinicians as a natural day-to-day tool. This insight has founded a national initiative for defining the documentation needed for standard patient workflows, but it has to be proven that such specifications can be im-

[1] Corresponding Author: John Møller-Jensen, CSC Scandihealth A/S, P.O.Pedersens Vej 2, DK 8200 Århus N, Denmark. E-mail: Jmoelle6@csc.com

plemented transparently in the clinical setting in order to specifically support the clinician's daily patient-centred routines and practices.

Roskilde County has adopted a bottom-up strategy driven by Participatory Design (PD) experiments to develop and implement their EHR systems in order to give priority to documenting the effects on clinical practice of the use of EHR systems. It has recently been proposed that an experimental strategy is required for PD engagement in large-scale public sector systems [3].

In order to measure the clinical usability of an EHR conformant with the national requirements as well as get input to the County's future strategy for incremental implementation of EHR, an experiment was conducted in the period of August to December 2005 at the Neurological Stroke Unit treating acute apoplexy at Roskilde County Hospital.

During the experiment the clinicians' need for overview and documentation was identified, followed by the configuration of the necessary templates to be implemented in the EHR that completely replaced the present paper-based medical record during a week at the neurology department.

The project group was formed by 3 partners: Clinicians from the neurological stroke unit and project managers from the EHR unit at Roskilde County Hospital, researchers from the Dept. of Computer Science at Roskilde University and business architects from the vendor, CSC Scandihealth A/S. Roskilde County Hospital's interest was to start the deployment of their new PD strategy and assess how to document clinical utility value. CSC Scandihealth's interest was to experience how to configure a clinical process EHR module in participation with clinicians and to test how their solution would work in a real clinical process. The researcher's interest was a research-in-progress project on 'evidence-based IT development' aiming at proposing a new commercial contract model where the customer's payments are dependent on measurable effects of using the vendor's system [4, 5].

The main goal of the experiment was to measure to which degree positive effects on the clinical practice could be obtained by using the configured EHR.

2. Materials and Methods

2.1. Identifying the clinicians' needs

The first part of the project (August to October) started with an initial meeting to scope the initial load of data and use of interfaces to other legacy systems during the test of the EHR. It was followed by five full-day PD workshops where clinicians from the stroke unit in cooperation with business architects from the vendor, project managers from the EPR unit and researchers from Roskilde University specified the content of the EHR.

In order to pinpoint the required effects on the clinical practice during the treatment of acute apoplexy patients, the clinicians were asked at the first workshop to identify the most critical 'information bottlenecks' during the clinical pathway of acute stroke patients. This was done in order to identify the objectives for the critical overviews that had to be configured during the experiment. Three specific and highly cooperative clinical situations were identified at the first workshop, viz. nursing handover, ward round and patient conference.

From this starting point the remaining content of the EHR was identified during the following three workshops, i.e. the structure, content and placement of clinical notes and result templates, standard plans, concept lists etc. At the final workshop the complete specification was presented and reviewed before the actual configuration of the XML-based templates and load of the templates to the EHR.

During this process the content of the EHR was elaborated in up to three iterative events. First, mock-ups were drawn on flip-over paper (figure 1). Secondly, a preliminary non-interactive prototype made with MS PowerPoint was discussed. Finally, a running prototype (figure 2) was demonstrated, discussed, and evaluated.

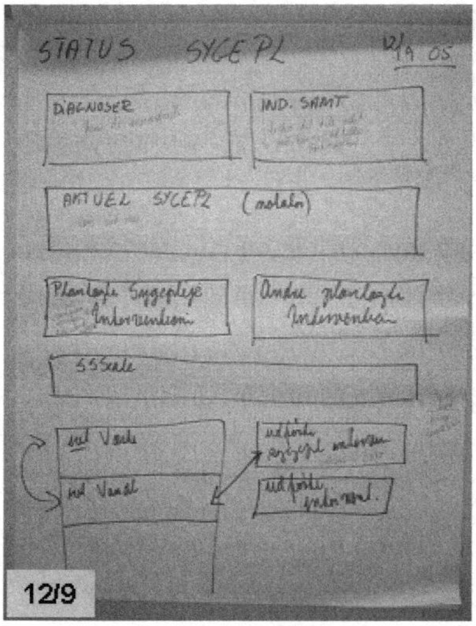

Figure 1. Photo of flip-over from the first workshop with input to overview for nursing handover

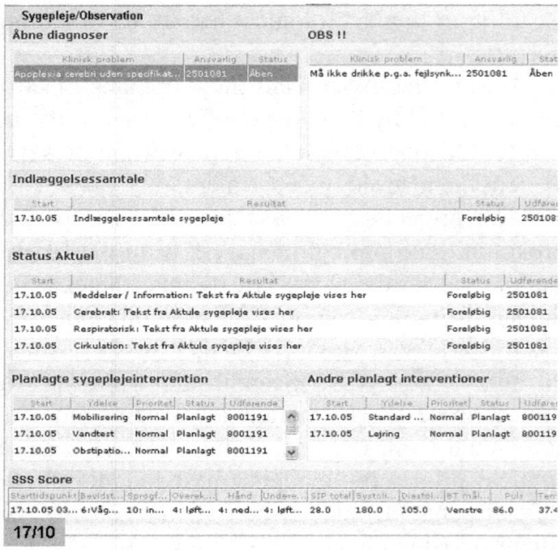

Figure 2. Screen dump from the fifth workshop showing the prototype implementation of overview for nursing handover

2.2. Identification of effect measurements

The effects requested by the clinicians focused on supporting coordinative aspects [6]. They requested an improvement of their overview and assessment of patients as well as on more efficient coordination in three specific and highly cooperative situations, viz. nursing handover, ward round and patient conference.

Nursing handover, which happens three times a day at the beginning of each nursing shift (7am, 3pm, and 11pm) and last about an hour. There is no time for the nurses that leave the ward to discuss patients with the nurses on the next shift. During the nursing handover, one nurse is designated as the team leader and provides an overview of the patients at the ward and manages the necessary coordination and exchange of information.

Ward round, which happens once every weekday and lasts for three to six hours. It includes evaluation, reviewing, and discharging of patients. The chief physician visits all patients and reviews the plans for their treatment. Usually there is no time for nurses to follow the physician during the ward round. Information exchange and coordination is obtained through the patient record and by ad hoc communication with the nurses on shift.

Patient conference, which takes place once every weekday, lasts approximately 15 minutes, and includes all clinical staff members (physicians, nurses, and therapists). An interdisciplinary assessment of each patient is carried out and plans are revised.

In parallel with defining and configuring the content of the EHR, the university researchers identified, prioritised, and further specified a number of effects related to the clinical practice that could be measured by various techniques, including questionnaires, interviews, observations, and NASA Task Load Index (TLX) ratings [7].

All three situations above were measured before the test of the EHR and during the test week.

2.3. Configuration of a clinical framework tool

In the second part of the project (November to December), the required content was configured as XML-based templates that were loaded into the clinical framework tool, CSC Clinical Suite, based on the Oracle Healthcare Transaction Base (HTB).

CSC Clinical Suite is not an EHR per se, but a clinical frame that can contain and present the clinical content as specified by the clinicians by use of XML-based templates for overviews, clinical notes, results, standard plans, work situations and structure of the patient's medical record. This makes it possible to configure a complete medical record in accordance with the clinician's requirements and is able to evolve dynamically as new requirements emerge.

The XML configuration based on the specifications from the workshops was conducted by the vendor and loaded into the EHR. After the initial load of the EHR templates, the content of the configured EHR was verified by registration of the data from actual paper-based medical records into the EHR before training the clinical staff in the use of the EHR. The need for adjustment of the content was identified during these sessions and implemented before the actual clinical test of the EHR.

2.4. Implementation and deployment of the EHR

Interfaces and initial load of data from the scoped legacy systems currently used at the hospital were established in parallel with the configuration of the EHR.

The test setup with the configured EHR included:

- All acute admitted patients in the 9-bed acute apoplexy unit within five days with a total of 15 patients
- 16 staff members used the EHR (10 received training prior to the test)
- Data included all production data, both bedside and in real work situation, 24 hours for five weekdays
- The technical setup included screens projected on the wall during nursing handover and patient conferences, stationary and portable PCs, and PDAs used for obtaining measurements at the patient's bedside
- In the weeks before the test five years of patient data from the County (a total of more than 26 million data records) were migrated to the prototype from the ADT system, laboratory system and Medication Module, and interfaces established to these legacy systems in order to receive updated data in the EHR during the test period

3. Results

Though one week of using an EHR is too short a period to establish routine use of the system, some results yield statistically significant positive effects of the EHR for each of the above clinical situations.

All participants at the nursing handover (except for the team leader) experienced a significant improved nursing plan. The number of missing information as well as the number of messages to hand over to other clinicians were significant lower during the nursing handover.

The chief physician experienced a significant reduction in mental workload on all six scales of the TLX ratings with regard to his ward round.

During the patient conference, the physicians experienced a significant reduction in their mental workload on all six scales of the TLX ratings. The nurses experienced a significant improvement on one of the TLX scales (own performance). The chief physician in charge of the conference experienced a significant better priority of tasks and division of work/responsibility for the tasks.

4. Discussion

It was of great importance that the stroke unit already had implemented detailed standard clinical guidelines for acute apoplexy patients. Everybody knew exactly what to do and when to do it, thus no data chart in the configured EHR was unknown to the clinicians.

Clinical framework-based EHRs such as CSC Clinical Suite demand heavy clinical involvement in the configuration process in order to make the system provide the expected effects, and the vendor will not be able to do this on their own.

The nurses found it easy to use the customised documentation model that was chosen and they quickly asked for a more problem-oriented model, which the EHR was able to provide to the users. CSC Clinical Suite could “follow” the user demands as the process evolved.

CSC Clinical Suite was perceived as easy to learn, to navigate and to write in and it was easy to use the standard plans of interventions defined according to the standard patient guidelines. It is also a highly flexible system that is very configurable: The configuration was changed back and forth several times during the preparations before the on-line test period. The configuration was even changed after the EHR went live, which impressed the clinicians.

Several users stated after the experiment that

- They missed the system in the days after the experiment.
- They found the juxtaposed clinical data from several systems to be very helpful in getting the desired overview of the patient.
- They had delivered better quality care but they had also spent more time recording and interpreting data.

5. Conclusion

The experiment go beyond classic IT prototyping experiments focusing on evaluations of user interfaces and interaction based on prototypes with limited functionality and small data samples. Our experiment aimed at measuring the effects of real clinical processes supported by fully functional EHR modules with complete patient records. The authors has not been able to identify description of studies similar to this study in PubMed and there are only a few references in the literature to configurable EHR's e.g. [8] based upon a future-proof EHR architecture [9].

The participatory design case study experiment has been a success in so far as it has demonstrated a realistic evaluation of a fully functional EHR solution that supports clinical work processes. The configured EHR was the result of a participatory design effort focusing on formulating and measuring desired effects of an EHR system on a selected clinical practice, the treatment of patients with acute apoplexy.

All parties in the experiment gained valuable experience from the study. The experiment has encouraged Roskilde to retain its participatory design strategy as an alternative to mainstream top-down approaches. Effects have been identified, specified, quantified and measured by the clinical staff involved that will be used as input in the coming strategy process in the county and the experiment supported the University's work on evidence-based IT development.

The clinical framework, CSC Clinical Suite, proved to be an effective tool for configuration of state-of-the-art clinical content in an EHR solution that is well received by healthcare professionals. The experiment proved that it is possible to configure an EHR dictated by the clinician's requirements for clinical overview and documentation, supporting the corporation between different healthcare professionals during the clinical pathways of critically ill patients.

Finally, the clinicians were inspired to further structure their present documentation in the paper-based medical record as a result of the experiences of the project.

References

[1] http://www.sst.dk/applikationer/epj/gepj/022_20050812/index.html (last accessed Jan. 27, 2006)

[2] Asp L, Petersen J. A conceptual model for documentation of clinical information in the EHR. Stud Health Technol Inform. 2003, 95:239-44.

[3] Shapiro D. Participatory Design: the will to succeed. In: Bertelsen OW, Bouvin NO, Krogh PG, Kyng M, editors. Proceedings of the 4th decennial conference on Critical computing: between sense and sensibility; 2005; Århus, Denmark: ACM Press, New York, NY, USA; 2005. p. 29-38.

[4] Simonsen, J. and Hertzum, M. Evidence-Based IT Development: Toward a New Contract Model for EPR Projects. In Proceedings of the 3rd Scandinavian conference on Health Informatics (SHI 2005), Virtual Centre for Health Informatics, Aalborg University, Denmark, 2005, 66-70.

[5] Simonsen J, Hertzum M. A Regional PD Strategy for EPR Systems: Evidence-Based IT Development. In: Proceedings of the 2006 Participatory Design Conference, ExpandingBoundaries in Design, Trento, Italy, August 1-5, 2006; 2006: ACM Press; 2006.

[6] Pratt W, Reddy MC, McDonald DW, Tarczy-Hornoch P, Gennari JH. Incorporating ideas from computer-supported cooperative work. J Biomed Inform. 2004 Apr;37(2):128-37.

[7] Hart SG, Staveland LE. Development of NASA-TLX (task load index): Results of empirical and theoretical research. In: Hancock PA, Meshkati N, editors. Human Mental Workload. Amsterdam: Elsevier; 1988. p. 139-183.

[8] Spidlen J, Hanzlicek P, Zvarova J. MUDRLite – health record tailored to your particular needs. Stud Health Tecnol Inform. 2004;105:202-9.

[9] Blobel B. Advanced and secure architectural EHR approaches. Int J Med Inform. 2006 Mar-Apr;75(3-4):185-90.

Ubiquity: Technologies for Better Health in Aging Societies
A. Hasman et al. (Eds.)
IOS Press, 2006

Ten years of teledermatology

Nina Eminović[a], Nicolette F de Keizer[a], Patrick J.E.Bindels[b], Arie Hasman[a]

[a]*Dept. of Medical Informatics, Academic Medical Centre, Amsterdam, the Netherlands*

[b] *Dept. of General Practice, Academic Medical Centre, Amsterdam, the Netherlands*

Abstract: Using telemedicine health professionals can communicate with each other and with their patients over a distance. Teledermatology, dermatology application of telemedicine, is one of the most often applied telemedicine applications worldwide. Various studies have been performed to evaluate the effectiveness and efficiency of and satisfaction with teledermatology. Up to now no or limited valid scientific evidence has been found that teledermatology is beneficial for any group of users. This study aimed to perceive insight into the evolution of evaluation studies of teledermatology over the past ten years in terms of the telemedicine evaluation framework by Holle and Zahlmann consisting of four continuous phases. We added the phase "post implementation studies" that evaluate teledermatology as a fully integrated service in regular care. Retrieved literature from Medline was reviewed by two reviewers independently in order to include studies and classify them into the five phases. Ninety-nine studies out of 372 found unique references were included and classified into the phases. Most represented phase was phase II with 72 (72%) studies. The number of phase II studies is continuously growing since the introduction of evaluation in teledermatology. There were eight reported RCTs found (two in phase III, six in phase IV). The number of phase III and IV studies is too low to draw conclusions about the trends in their publication and stress the need for more such studies. Phase I and post implementations studies are probably under-represented as they might often not be published separately in scientific journal papers.

Keywords: teledermatology, telemedicine, dermatology, e-health, evaluation, review

1. Introduction

Telemedicine enables health professionals to communicate with each other and with their patients even if a geographical or physical distance is present.

Dermatology is a very suitable specialty for telemedicine consultations because of its strong visual aspect and relatively low number of urgent cases. Teledermatology occurs in different forms and settings. In this study, we focus on teledermatology defined as the use of imaging and telecommunication technologies to provide skin services by a dermatologist to another health professional (GP, nurse, other specialist) or directly to a patient. Two main types of teledermatology can be distinguished: store-

and-forward and real-time teledermatology [1]. Store-and-forward teledermatology uses asynchronous data transfer technology (e.g. email) while the real-time variant is based on synchronous data transfer technologies (e.g. videoconferencing software).

Various studies have been performed to explore a variety of clinical and non-clinical outcomes related to telemedicine[2;3]. However, systematic reviews [4-7] did not succeed in showing positive effects of telemedicine which is mainly due to the lack of high quality studies in the area. Literature reviews on teledermatology are limited to general descriptions of teledermatology [8;9] or are part of telemedicine systematic reviews. Krupinski et al [10] reviewed a selection of telemedicine applications to determine the level of their maturity. The authors found that teledermatology was maturing because of the reasonably good quality and quantity of the published literature. However, although there are indications that teledermatology is one of the leading telemedicine specialties, there is lack of teledermatology specific reviews concerning evidence of its benefits.

Despite the lack of scientific evidence, the development and implementation of telemedicine and teledermatology applications continues worldwide and the number of evaluation studies on these information technology applications (IT) is increasing [11]. In order to improve and assess the quality of telemedicine evaluation studies, different evaluation frameworks have been proposed. Performing high quality studies such as randomized controlled trials in telemedicine remains difficult as telemedicine is a complex intervention depending on local variations in administration and organisation [12]. For this reason, telemedicine specific evaluation frameworks are required. The evaluation strategy of Holle and Zahlmann distinguishes itself from other frameworks as it considers telemedicine as a diagnostic or therapeutic procedure and is based on Good Clinical Practice (GCP). It describes continuous evaluation phases used for studying telemedicine procedures [13].

Aim of this study is to obtain insight in the status of teledermatology evaluation studies of the past ten years as a function of the phases defined by Holle and Zahlmann.

2. Methods

2.1. Search strategy

Published teledermatology studies were identified by searching in the Medline database (all literature from 1966 up to April 2005) using following search queries containing words and MeSH terms: dermatology, teledermatology, telemedicine, skin and electronic mail.

Literature reviews, comments, abstracts, letters and editorials were excluded. Papers in other languages than English were excluded. The next step included the manual screening of the selected papers by two independent reviewers. Papers which were not about dermatology but about another specialty (e.g. radiology, pathology) and papers where the evaluation of a specific teledermatology service was not the primary aim of the study were excluded. Papers on telemedicine application for several specialties were only included if the results were separately reported for dermatology. The selection procedure was based on the title and abstract. If the abstract of the paper was not available and the title did not contain any indications for exclusion (e.g. 'a literature review', 'teleradiology in…'), the paper was included for the further review.

Literature reviews were screened to retrieve possibly missed references. Conference proceedings were excluded only if a full paper of the same study was obtained in the selection procedure.

2.2. Classification of evaluation studies

Based on the full paper, two reviewers independently assigned the studies to one of the four phases of the Holle and Zahlmann strategy (Table 1). After communication with the authors, we added with their permission an extra phase: post implementation phase. This fifth phase should be seen as an extra category and not specifically a continuation of the four phases. Each of the five phases is determined by the study design. Other important criteria for distinguishing them are the number of participants in the study and the setting of the study. The right column of table 1 describes some of the main criteria used by the reviewers to assign each selected study to a phase.

We distinguished two main study designs: observational and intervention studies. Intervention studies were further divided into the categories: uncontrolled trial, non randomised controlled trial with the same patients as control group (e.g., a patient is first seen through teledermatology and afterwards live), non randomised controlled trial with other patients as control group, and randomised controlled trial (RCT).

A structured scoring form and scoring guidelines, which were fine-tuned on a random set of ten papers scored by both reviewers, were used to determine the phase of each evaluation study.

Table 1: Four phases of Holle and Zahlmann[13], and added fifth phase "post-implementation" with their characteristics and main criteria used in this review to classify teledermatology studies.

Phase	Study design	Usual participants	Main criteria used by the reviewers
Phase I	Exploratory, small intervention studies	Researchers and project members, simulated patients	experimental setting, i.e. teleconsultation did not result in a diagnosis or treatment plan
Phase II	Feasibility studies, can be controlled intervention studies, but rarely involve a separate control group.	Specially trained and highly motivated potential users, real or simulated patients	Field or experimental studies with potential users
Phase III	Randomised controlled trial (RCT)	Unselected sample of users with minimal training, real patients	RCT on clinical outcomes
Phase IV	(Simulation) cost studies using the benefits results from an RCT	NA	Costs are compared with any kind of benefits results from a randomized controlled trial
Post implementation phase	Telemedicine part of the regular care, observational study.	Actual users of telemedicine	Teledermatology is fully integrated in the regular care

3. Results

3.1. Teledermatology literature

The search in Medline resulted in 372 unique references. After applying the first set of exclusion criteria concerning the papers type, 181 studies were included and 191 excluded. Forty-four papers (12%) were about another specialty (radiology, pathology) or about all kinds of medical conditions including dermatology as a subset but without separate results reported). Twenty-seven studies were excluded due to the language (8 German, 5 French, 4 Norwegian, 2 Swedish, 2 Italian, 2 Spanish, 1 Russian, 1 Dutch, 1 Danish and 1 Japanese). Most of the excluded papers did not report on an evaluation of a specific teledermatology service (N=45). Five studies needed to be excluded as they were in some way duplicate publications: conference proceedings and a full version paper. Two papers that reported on the same study within the same phase were also excluded. At the end of the process, 99 studies remained included and were classified into the five phases.

3.2. Phase trends

Eleven studies (11%) studies were assigned into phase I. The majority of the studies, 72 (72%) were assigned to phase II. The number of reported RCTs in phase III was limited to two. Phase IV and the post implementations phase were represented in respectively six and eight studies.

The first teledermatology evaluation study was published in 1996. The first teledermatology studies were a phase I and phase II study (Figure 1). Since then, the number of phase II studies was dominant each year and counted up to 72 in 2005. The two phase III RCTs were published in 2004. In 2000 four cost studies were published and another two phase IV studies were published in 2001. The last published cost study dates from 2003. In these six phase IV studies, clinical outcome results deriving from an RCT were used in the cost analysis and were reported in the same paper. That brings the number of RCTs on clinical outcomes in teledermatology to 8 (8%).

Figure 1 shows that the number of phase II studies cumulates rapidly over the years, while other phases are not represented every year.

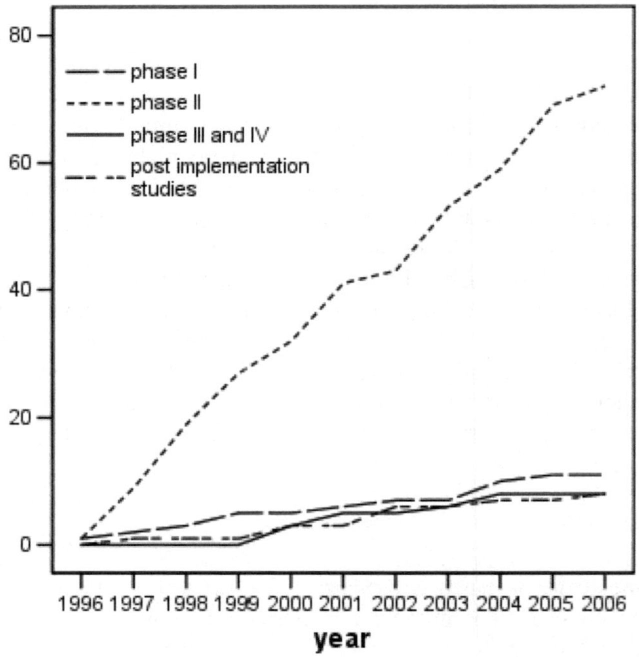

Figure 1. Cumulative number of teledermatology evaluation studies per phase, as proposed by Holle and Zahlmann [13] over past ten years.

4. Discussion and conclusions

With this study we retrieved an insight into evaluation studies on teledermatology, classified in five phases. The results show that the evaluation of teledermatology takes place in different phases and that there are no clear trends over the years for certain phases, at least not by the time the study was performed.

Our study shows that the majority of papers about evaluation in teledermatology reports on phase II (feasibility) studies. The number of phase II studies is continuously growing over the past ten years. Remaining phases are much less represented in the teledermatology research. There are different possible explanations for this. Phase I studies, usually reporting on technical issues are often not published in a scientific journal paper. It is probable that these technical evaluations are performed much more often, possibly prior to any evaluation study, but that the results of these studies are not reported separately.

More interesting from our review is the fact that the number of phase III and IV studies is very low. Evaluation of any new technology is expected to be performed more detailed and sophisticated after some basic evaluation steps have been performed such as in phase I and II. Some researchers claim RCTs in telemedicine are scarce and difficult to set up and perform as telemedicine is a complex intervention. Furthermore, the push for performing RCTs in teledermatology is possibly lower than in other technical innovations since teledermatology is already implemented in regions where the benefits are straightforward (low risks and low investment costs in comparison to medical interventions). The timing of our study is possibly too early to measure an increase of phase III and phase IV studies as these sophisticated studies are often time consuming in both performing and reporting.

Post implementation studies are, like phase I studies, not often published in scientific journals. Commercial teledermatology suppliers are probably frequently performing this kind of evaluations, but mainly for their own use.

Since no or limited valid scientific evidence about the impact of teledermatology on clinical outcome has been found [5;6], more detailed and sophisticated evaluation studies (phase III and IV) are necessary. Furthermore one should be motivated to make the results of phase I and post implementation studies publicly available.

References

[1] J.D.Whited, Teledermatology. Current status and future directions, Am.J.Clin.Dermatol. 2 (2001) 59-64.

[2] P.J.Heinzelmann, C.M.Williams, N.E.Lugn, and J.C.Kvedar, Clinical outcomes associated with telemedicine/telehealth, Telemed.J.E.Health 11 (2005) 329-347.

[3] N.Aoki, K.Dunn, K.A.Johnson-Throop, and J.P.Turley, Outcomes and methods in telemedicine evaluation, Telemed.J.E.Health 9 (2003) 393-401.

[4] F.Mair and P.Whitten, Systematic review of studies of patient satisfaction with telemedicine, BMJ 320 (2000) 1517-1520.

[5] P.S.Whitten, F.S.Mair, A.Haycox, C.R.May, T.L.Williams, and S.Hellmich, Systematic review of cost effectiveness studies of telemedicine interventions, BMJ 324 (2002) 1434-1437.

[6] D.Hailey, R.Roine, and A.Ohinmaa, Systematic review of evidence for the benefits of telemedicine, J.Telemed.Telecare. 8 Suppl 1 (2002) 1-30.

[7] W.R.Hersh, M.Helfand, J.Wallace, D.Kraemer, P.Patterson, S.Shapiro, and M.Greenlick, Clinical outcomes resulting from telemedicine interventions: a systematic review, BMC.Med.Inform.Decis.Mak. 1 (2001) 5.

[8] D.J.Eedy and R.Wootton, Teledermatology: a review, Br.J.Dermatol. 144 (2001) 696-707.

[9] G.Burg, U.Hasse, C.Cipolat, R.Kropf, V.Djamei, H.P.Soyer, and S.Chimenti, Teledermatology: just cool or a real tool?, Dermatology 210 (2005) 169-173.

[10] E.Krupinski, M.Nypaver, R.Poropatich, D.Ellis, R.Safwat, and H.Sapci, Telemedicine/telehealth: an international perspective. Clinical applications in telemedicine/telehealth, Telemed.J.E.Health 8 (2002) 13-34.

[11] N.F.de Keizer and E.Ammenwerth, Trends in Evaluation Research 1982 - 2002: A Study on how the Quality of IT Evaluation Studies Develop, Stud.Health Technol.Inform. 116:581-6. (2005) 581-586.

[12] R.Bashshur, G.Shannon, and H.Sapci, Telemedicine evaluation, Telemed.J.E.Health 11 (2005) 296-316.

[13] R.Holle and G.Zahlmann, Evaluation of telemedical services, IEEE Trans.Inf.Technol.Biomed. 3 (1999) 84-91.

3.6 HIS: Other Topics

Ubiquity: Technologies for Better Health in Aging Societies
A. Hasman et al. (Eds.)
IOS Press, 2006

Integrating anatomical pathology to the healthcare enterprise

Christel DANIEL-LE BOZEC[a,b,c1], Dominique HENIN[c], Bettina FABIANI[c], Karima BOURQUARD[d], David OUAGNE[a], Patrice DEGOULET[a,b], Marie-Christine JAULENT[a]
[a] *Inserm, U729, Paris, F-75006 France*
[b] *Univ Rene Descartes, Paris, F-75006 France ; AP-HP, Hôpital Georges Pompidou, Département d'Informatique Hospitalière, Paris, F-75015 France*
[c] *ADICAP, Paris, F-75013 France,* [d] *GMSIH, Paris, F-75013 France*

Abstract. For medical decisions, healthcare professionals need that all required information is both correct and easily available. We address the issue of integrating anatomical pathology department to the healthcare enterprise. The pathology workflow from order to report, including specimen process and image acquisition was modeled. Corresponding integration profiles were addressed by expansion of the IHE (Integrating the Healthcare Enterprise) initiative. Implementation using respectively DICOM Structured Report (SR) and DICOM Slide-Coordinate Microscopy (SM) was tested. The two main integration profiles - pathology general workflow and pathology image workflow - rely on 13 transactions based on HL7 or DICOM standard. We propose a model of the case in anatomical pathology and of other information entities (orders, image folders and reports) and real-world objects (specimen, tissue samples, slides, etc). Cases representation in XML schemas, based on DICOM specification, allows producing DICOM image files and reports to be stored into a PACS (Picture Archiving and Communication System).

Keywords: Medical Records Systems, Computerized, Information Systems, Systems Integration, DICOM, HL7, Anatomical pathology, Diagnostic Imaging, Structured Report.

1. Introduction

Information systems in anatomical pathology departments gather medical data (texts, images, etc) throughout different procedure steps from specimen processing to report editing. Since information systems are not integrated, information acquisition is time consuming with double data entry. Orders, images and reports are spread out over different systems which do not interoperate. Although standardization efforts conducted by HL7 [1] and DICOM [2] are progressing to provide integration solutions, HL7 or DICOM messages contain many optional data fields so that being DICOM or HL7 compliant does not imply direct integration. The goal of the Integrating the Healthcare Enterprise (IHE) initiative is precisely to specify how data standards should be implemented to meet specific healthcare needs and to make systems integration

[1] Corresponding Authors: Christel Daniel-Le Bozec – INSERM U729- Faculté de médecine paris V - 15, rue de l'école de médecine - 75006 Paris – France. christel.lebozec@spim.jussieu.fr
Pr Dominique HENIN – ADICAP - Service d'anatomie pathologique Hopital Bichat 75018 Paris - France

more efficient and less expensive [3]. IHE defines "Integration Profiles", that are real-word situations describing exchange of information called "Transactions", from various sources, called "Actors". IHE provides implementation guides for "Transactions", using established standards as DICOM or HL7.

In 1996, the ADICAP (Association for the Development of Informatics in Cytology and Pathology) with the collaboration of different software vendors proposed a European de facto standard for image folders [4]. Relying on this experience, the aim of ADICAP is now to promote the use of international standards (DICOM, HL7) in the development of information systems in anatomical pathology. In 2005, ADICAP launched the IHE pathology initiative in Europe.

Although specific DICOM objects are defined for pathology, some pathology-related image formats (whole-slide images, multispectral images, flow cytometry, etc) do not have applicable DICOM Information Object Definitions and a specific DICOM working group (DICOM WG26) has been recently created to address this issue. A specific working group (HL7 Pathology Special Interest Group) has been also recently created within HL7 to address the pathologists' needs, in synergy with DICOM WG26, focusing on the orders and reports aspects of the pathology workflow.

The objective of this paper is to present a methodology to integrate anatomical pathology department to the healthcare enterprise. We first describe the ADICAP efforts to model the anatomical pathology workflow in order to define new Integration Profiles. We also describe an experimentation based on the implementation of a DICOM-based XML schema to represent images and related information of cases.

2. Material and Methods

2.1. Modeling anatomical pathology workflow

ADICAP, with the collaboration of the GMSIH (Group promoting the Modernization of Hospital Information Systems in France), solicited participants to work on the Pathology Technical Framework: 7 pathologists and haematologists, 4 IT professionals, 2 professional associations and 12 vendors. Five working sessions were organized between November 2004 and January 2006.

The working group first defined the pathologists' needs and the corresponding Integration P

rofiles for anatomic pathology. They identified the Actors and Transactions involved in these profiles. Then, they reviewed the literature about order forms and reports in anatomic pathology in order to describe the main requirements for the structure and content of orders, imaging folders and reports.

Quality assurance about order forms provides a list of mandatory items [5]. Since 1993, Association of Directors of Anatomic and Surgical Pathology publishes recommendations for the reporting in many different fields [6]. A generic model of structured report can be derived from these templates. In complement, studies about quality assessment of reports provide lists of mandatory items and stress the positive role of checklists to enhance the reporting process [7,8]. According to "evidence-based pathology", only features that are reproducible and relevant – with a demonstrated diagnostic or prognostic signification – should be reported in description and corresponding evidence available [9,10]. A crucial issue is to identify a technical solution to handle templates of structured reports including findings and their evidences.

2.2. Case modeling: DICOM-based XML Schemas

DICOM Information Object Definitions dedicated to anatomical pathology are DICOM VL Photographic Image (XC) for gross imaging and VL Slide-Coordinates Microscopic Image (SM) for microscopic imaging. Moreover, in medical imaging, DICOM Structured Reporting (SR) supports the interchange of reports in which critical features shown by images can be denoted unambiguously by the observer and retrieved selectively [11]. Findings may be expressed by the observer as text, numeric measurements, or via location coordinates of specific regions of interest within images. Although DICOM is not an open technology such as XML, some studies describe XML schema representations for DICOM files [12,13]. We chose to define XML Schemas for case representation, based on DICOM Visible Light and DICOM Structured Reporting objects.

3. Results

3.1. Integration profiles in pathology

The diagnostic process in pathology (figure 1) differs from that in clinical laboratory since it relies on image interpretation. It also differs from that in radiology since it is specimen-driven and when digital imaging is performed many types of imaging equipments (gross imaging, microscopic still imaging, whole slide imaging, multispectral imaging, etc) may be involved for a single examination. Moreover, slides are always available to acquire more images, if needed. In radiology, the diagnostic process is patient-driven and an examination usually involves a single image acquisition modality.

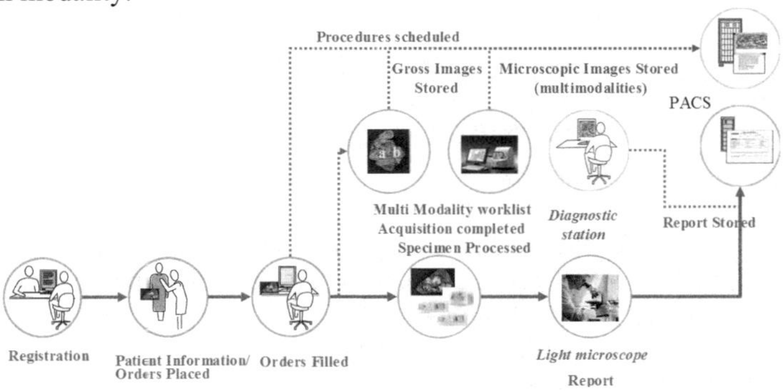

Figure 1: Anatomic pathology workflow

Two main Integration Profiles were proposed for a first IHE cycle: Pathology General Workflow (Specimen, Order and Report Management Workflow) and Pathology Image Workflow. These Integration Profiles involve 13 Transactions exchanged by 8 Actors. For each transaction, the workgroup proposes the use of the most suitable format (HL7 version 2, Clinical Document Architecture (CDA) - HL7 version 3 and DICOM).

3.2. Order, specimen and image folder

Each Order may contain one or more pathological examination(s) or case(s) possibly reported by different pathologists. Each case may contain one or more procedure step(s) (figure 2). All pertinent information to be sent within an order has been defined. The specimen identification mechanism must allow keeping track of all "real world objects": specimens, tissue samples (in blocks, in cryomolds or in cryotubes), slides, etc. Specimen, tissue samples and slides description requires a controlled vocabulary.

In pathology, the image folder (study) is defined at the level of the pathological examination or case. For each case, images acquisition may require different modalities (gross imaging, microscopic imaging, etc). A new series is created whenever an imaging procedure step is performed on a new specimen or slide.

Order: Lumpectomy + re-excision + nevus excision

Case 1: Lumpectomy + re-excision: report n°1 (study n°1)
- Specimen 1-1: Lumpectomy
  - Imaging procedure Step: Specimen radiography
  - Imaging procedure Step: Gross imaging
  - Tissue sample 1-1-1
    - Slide 1-1-1-1
    - Procedure step: Basic histology (HE)
      - Imaging procedure step: WSI
    - .../....
- Specimen 1-2: Additional tissue (re-excision)
  - Tissue sample 1-2-1
    - Slide 1-2-1-1
    - Procedure step : Basic histology (HE)
      - Imaging procedure step: Still imaging
    - .../....

Case 2: Nevus excision: report n°2 (study n°2)
- Specimen 2-1: Nevus
  - Tissue sample 2-1-1
    - Slide 2-1-1-1
    - Procedure step : Basic histology (HE)

Figure 2: Extract of the structure and content of an order: the breakdown in cases (corresponding to reports) and procedure steps and the organization of the corresponding image folders in studies and series.

3.3. Implementing DICOM files from cases

We developed a DICOM module as a part of an existing platform -the IDEM platform - dedicated to collaborative work in pathology [14]. This module generates DICOM files from cases (figure 3).

The first step consists in exporting cases (images, related information and structured report if available) into a XML file, called CaseML file. The first part of the CaseML file includes images and related technical information organized according to the specification of the DICOM object corresponding to the acquisition modality (DICOM XC or DICOM SM). The second part includes information of the structured report organized according to the specification of the DICOM SR. We chose to map each relevant morphological feature of the case report to a node of the DICOM SR structure and to use the relationship "inferred from" to express the interpretation rules applied by the pathologist during the description. We used the DICOM SR possibility to link content items to coordinates of specific regions of interest within images.

The second step consists in syntactic validation of the CaseML file against XML schema generated from the DICOM specification. For the Structured Reporting part of the CaseML, the "validator" also performs a semantic validation. A structured report is valid if it conforms to the interpretation rules expressed in the reference diagnostic classification system. An ontology of breast pathology was built using the editor Protégé and a specific plug-in was developed to extract from reference ontologies an XML Schema embedding diagnostic interpretation rules [15, 16].

The third step consists in DICOM translation. A XSLT processor generates suitable XML file for DICOM translation from the CaseML. This XML file corresponds to the type of DICOM object to produce and its format is adapted to the «xml2DICOM» library we intend to use, like «dcm4che» for example.

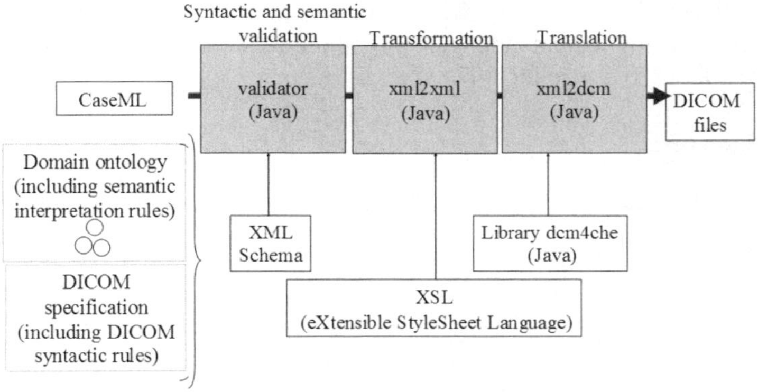

Figure 3 : DICOM module.

From XML files including all available information (images and textual data) of cases, DICOM files were produced by the DICOM module, tested with success using the DICOM Validation Toolkit [17] and also stored in a PACS (IMPAX 4.5®,AGFA).

4. Discussion and conclusion

Quality assessment studies in anatomical pathology show that each of the different steps from specimen processing to report editing may be source of errors and that information systems integration supports error reduction [18,19,20]. This work, promoted by ADICAP and done in the framework of the IHE-pathology initiative, consisted in defining the needs of systems integration in anatomical pathology.

The results show that a first significant IHE cycle in anatomic pathology could involve two new integration profiles. The main contributions of this work were to analyze the specificity of the anatomical pathology workflow with regards to laboratory and radiology workflows and to define the structure and content of cases, orders, image folders and reports. There was an issue to make explicit the links between information entities (orders, image folder, reports, etc) and real-world objects (specimen, tissue samples, slides, etc).

We successfully defined XML Schemas for case representation, based on DICOM Visible Light and DICOM Structured Reporting objects. We developed a DICOM module to export DICOM image files and reports to be stored into a PACS.

Although the main output of the anatomical pathology workflow is a timely and clear report of a diagnostic opinion, images will be more and more associated as evidence to textual reports. DICOM SM seems to be a convenient format for image archiving and communication within the anatomical pathology department. For integration to the Electronic Healthcare Record, HL7 Clinical Document Architecture seems to be the suitable format but image coding in HL7 CDA must be clarified.

With regards to the reporting process, quality assurance studies have shown that structured reports that conform to reference classification systems, allow physicians to determine treatment more appropriately, to evaluate their practice more reliably, and to compare them to other institutions more confidently [8]. An original point of this work is the attempt to embed the knowledge of reference classification systems into ontologies and to extract from these ontologies a XML schema allowing a semantic validation of the structured report.

Our perspective is that vendors implement IHE-pathology principles. Thanks to an on going collaborative work involving IHE-Pathology, DICOM WG26 and HL7 pathology SIG, an implementation guide (Pathology Technical Framework) will be available for a first IHE cycle in 2007.

Acknowledgments

This work has received a grant from the ADICAP and the Semantic Mining (NOE).

References

[1] HL7 organization : Health Level 7, http://www.hl7.org

[2] DICOM 3 Standard, NEMA and Global Engineering Group: http://www.nema.org.

[3] IHE - http://ihe.univ-rennes1.fr/whatisihe/. Accessed January, 2006

[4] Klossa J., Cordier J.C., Flandrin G., Got C., Hémet J, An European de facto standard for image folders applied to telepathology and teaching. Inter. J. Med. Information. 1998, 48:207-216.

[5] Burnett, L., D. Chesher, and Y. Mudaliar, Improving the quality of information on pathology request forms. Ann Clin Biochem, 2004. 41(Pt 1): p. 53-6.

[6] ADASP. Consultations in surgical pathology. Association of Directors of Anatomic and Surgical Pathology. Am J Surg Pathol. 1993 Jul;17(7):743-5.

[7] Campbell, F. and D.F. Griffiths, Quantitative audit of the content of histopathology reports. J Clin Pathol, 1994. 47(4): p. 360-1.

[8] Imperato, P.J., et al., Results of a cooperative educational program to improve prostate pathology reports among patients undergoing radical prostatectomy. J Community Health, 2002. 27(1): p. 1-13.

[9] Fleming KA. Evidence-based pathology, J Pathol. Jun 1996; 179(2): 127-8.

[10] Fleming KA, Evidence-based cellular pathology, Lancet, 2002; 359(9312): 1149-50.

[11] Loef C, Truyen R. Evidence and diagnostic reporting in the IHE context. Acad Radiol. 2005 May;12(5):620-5.

[12] Lee KP, Hu J. XML Schema Representation of DICOM Structured Reporting. J Am Med Inform Assoc. 2003 Mar-Apr;10(2):213-23.

[13] Tirado-Ramos A, Hu J, Lee KP. Information object definition-based unified modeling language representation of DICOM structured reporting: a case study of transcoding DICOM to XML. J Am Med Inform Assoc. 2002 Jan-Feb;9(1):63-71.

[14] LeBozec C, Zapletal E, Degoulet P, Jaulent MC. A consensus approach to maintain a knowledge based system in pathology. Proc AMIA Symp 2001; 354-8.

[15] Ptotégé. http://protege.stanford.edu/

[16] Ouagne D, le Bozec C, Zapletal E, Thieu M, Jaulent MC. Integration of multiple ontologies in breast cancer pathology. Stud Health Technol Inform. 2005;116:641-6.

[17] http://www.dvtk.org/

[18] Cowan, D.F., Quality assurance in anatomic pathology. An information system approach. Arch Pathol Lab Med, 1990. 114(2): p. 129-34.

[19] Goldstein, N.S., Diagnostic errors in surgical pathology. Clin Lab Med, 1999. 19(4): p. 743-56.

[20] Zarbo, R.J., F.A. Meier, and S.S. Raab, Error detection in anatomic pathology. Arch Pathol Lab Med, 2005. 129(10): p. 1237-45.

Ubiquity: Technologies for Better Health in Aging Societies
A. Hasman et al. (Eds.)
IOS Press, 2006

Medical Data GRIDs as Approach towards Secure Cross Enterprise Document Sharing (Based on IHE XDS)

Florian WOZAK[a,1], Elske AMMENWERTH[a], Micheal BREU[b], Robert PENZ[c], Thomas SCHABETSBERGER[a], Raimund VOGL[c] and Manfred WURZ[d]
[a] *University of Health Sciences Medical Informatics and Technology, Hall, Austria*
[b] *Leopold-Franzens-University Innsbruck, Innsbruck, Austria*
[c] *Health Information Technologies Tyrol, Innsbruck, Austria*
[d] *Icoserve Information Technologies, Innsbruck, Austria*

Abstract. Quality and efficiency of health care services is expected to be improved by the electronic processing and trans-institutional availability of medical data. A prototype architecture based on the IHE-XDS profile is currently being developed. Due to legal and organizational requirements specific adaptations to the IHE-XDS profile have been made. In this work the services of the health@net reference architecture are described in details, which have been developed with focus on compliance to both, the IHE-XDS profile and the legal situation in Austria. We expect to gain knowledge about the development of a shared electronic health record using Medical Data Grids as an Open Source reference implementation and how proprietary Hospital Information systems can be integrated in this environment.

1. Introduction

The electronic processing of medical data which is expected to improve quality and efficiency of health care services [1] will lead to an increasing amount of medical data exchanged across institutional boundaries [2].
Currently trans-institutional data exchange is only partly realized for specific fields.

Presently communication among institutions in the health care environment is based on directed data flow, implying that the intended receiver has to be known at the very beginning of the communication process. Particularly in a medical context this is only possible in rare cases, resulting in documents to be sent from the producing institution to each requester.

So in order to support patient centered care relevant medical data should be available on demand in appropriate format and after successful authorization for each institution involved in the treatment process as well as for the patient himself. In Shared Electronic Health Records (SEHRs) medical history of patients are stored in different

[1] Corresponding Author: Dipl.Ing.(FH) Florian Wozak, University for Health Sciences, Medical Informatics and Technology, Eduard Wallnöfer-Center 1, A-6060 Hall/Tyrol, Austria. E-Mail: florian.wozak@umit.at Web: http://iig.umit.at.

institutions and made accessible on demand.

Functional requirement which have been analyzed in an initial step mainly comprise security, scalability and high availability. Research revealed that distributed architectures seem to be most adequate since data and indices remain stored locally at the producing institution, which avoids single point of failure and attack.

A variety of architectures described in literature rely at least partly on centralized services, which makes them suffer from the same drawbacks as completely centralized architectures concerning security and scalability. Up to know adequate approaches for entirely distributed architectures seem to be missing.

So the aim of this work is to develop a network architecture for patient centered SEHRs exploiting the expected benefits of distributed services. The architecture is designed to comply with the cross enterprise document sharing specification (IHE-XDS) and the Austrian federal law for health telematics [3]. The main challenge is to design the architecture in the way to fulfill the requirements of those specifications.

The architecture is designed to be highly scalable and to offer standardized interfaces which simplifies the process of joining for new institutions as much as possible. The emerging field of GRID technology provides domain specific solutions for managing highly distributed architectures, particularly for the above mentioned requirements security, scalability and high availability. For this reason an architecture following the blueprint of Medical Data GRIDs seem to be most appropriate. Data GRIDs are clearly defined as

> "Data Grid is an emerging technological paradigm for the seamless access, via virtualized middleware, to heterogeneous and distributed ensembles of data storage resources"[4].

A definition, which takes into account the very specific context in which medical data is treated, is not known to us. Legal and organizational requirements in the medical domain vary vastly for every country. Data handled by a Shared Electronic Health Record are specific to a medical domain such as laboratory results or radiological images. Analysis and procession is closely bound to the originating medical domain. These specific requirements demand for adaptation of the Data GRID architecture definition. For this work we extend the original definition as follows:

> "*Medical Data GRID* is an emerging technological paradigm for the seamless access *to medical data for a patient centric Shared Electronic Health Record with special focus on legal and organizational requisitions of the environment they are operated in*, via virtualized middleware, to heterogeneous and distributed ensembles of data storage resources".

2. Methods

In an initial step functional requirements for a SEHR have been analyzed [5]. In an initial step functional requirements for a SEHR have been analyzed. Based on this analysis a technical specification has been elaborated. To ensure the highest possible level of security and data protection, functional requirements have been analyzed along with security requirements for trans institutional workflows in a SEHR.

The backbone of the methodology named SECTET [6] which was developed by the Leopold Franzens Univerity Innsbruck are UML models [7] that specify the workflows between the participating partners. A starting point is the document model that defines the basic objects that are managed and exchanged by the system. In parallel a role model is developed that reflects the permissions and capabilities of the users of the system. These models guide through the elicitation of security requirements.

2.1. *The Austrian E-Health Initiative*

In 2005 the eHealth Initiative (EHI) of the Austrian Federal Ministry for Health and Woman compiled a strategy to organize the development of the health system towards an integrated patient centered care on a long-term basis [8]. Hereby a major role is played by the Electronic Health Record (EHR; "ELGA" is the German buzzword) which is a summary of health-related data of an individual, acquired continually from either outpatient or in-patient treatment, and which is filed digitally. This electronic health record should store all these data all one's life independent of place or time, and should present them demand-oriented to all the persons involved in the medical treatment, including the patient himself.

2.2. *The IHE XDS approach*

IHE is an initiative by health care professionals and industry to improve the way computer systems in health care share information. IHE promotes the coordinated use of established standards such as DICOM and HL7 to address specific clinical needs in support of optimal patient care. Systems developed in accordance with IHE communicate with one another better, are easier to implement, and enable care providers to use information more effectively [9].

The IHE cross enterprise document sharing profile (XDS) provides an architectural approach for document sharing in heterogeneous health care environments [10]. The reference architecture proposed by XDS specification is described in Figure 1 (A).

At the Document Source medical documents are produced, which are then stored at the document Repository and made available to the end user (Document Consumer). The document Registry is for indexing and search functions, the Patient Id source provides for unique patient identification.

3. Infrastructure and Network Architecture

The proposals from the Austrian ELGA and eHealth initiative are the foundation for the implementation of the health@net [11] core architecture.
Currently we are implementing an Open Source prototype architecture based on the IHE XDS specification.
To meet legal and organizational requirements specific for Austria respective services are to be implemented to adapt the architecture according to the needs. In the following an overview of the most important services of the health@net prototype implementation is outlined. See Figure 1 (B).

To ensure greatest possible flexibility and scalability, a distributed approach following the above introduced paradigm of Medical Data GRIDs was chosen. Widespread Web service technology allows the SEHR interfaces to be designed in a generic way, which simplifies implementation for joining institutions as far as possible.

The core architecture consists of independent services responsible for storing documents and corresponding meta data, security features, unique patient identification and service discovery according to the IHE XDS specification. To guarantee a high level of security beyond the application security covered by role based access control, the entire communication is encrypted using the secure socket layer enabled hypertext transfer protocol (HTTPS), digital certificates provide mutual authentication of participating systems and each message exchanged is digitally singed using Web service security extensions [12].

Core functionality is provided by the three service groups. Document Repository, Document Registry and Patient Id Source.
Medical documents remain stored in the organization where they have been produced in the *Document Repository* (DR). Documents are exported in a proprietary format specific for the Document Source. For this reason incoming documents are converted to the Clinical Document Architecture (CDA) format internally used. The *Document Clearing* service (DC) is responsible for document conversion and mapping of locally used patient identification to the internally used unique identification.

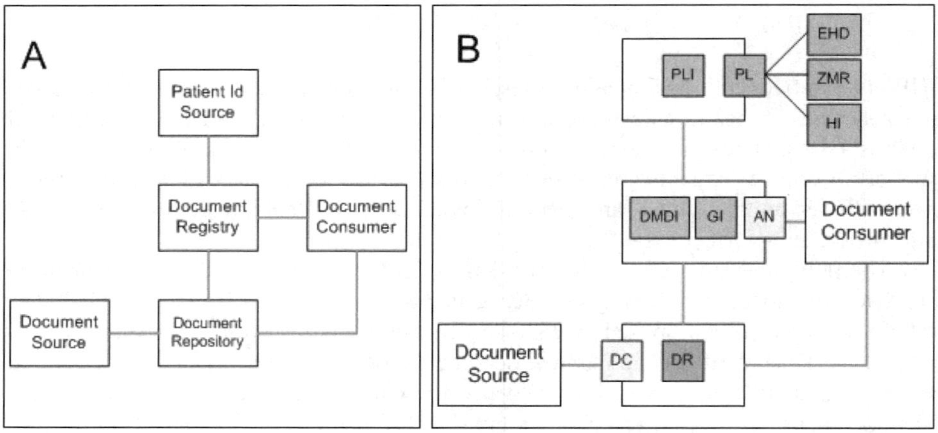

Figure 1: Health@net architecture based on the IHE-XDS reference architecture. In (A) the IHE-XDS reference architecture is shown. in (B) specific adaptations for the health@net prototype architecture are described. Document Clearing (DC), Document Repository (DR), Document Meta Data Index (DMDI), Global Index (GI), Access Node (AN), Patient Lookup Index (PLI) and Patient Lookup (PL). Governmental services for obtaining patient identification (EHD, ZMR and HI). For further information please see text.

The Document Registry service provides functionality for document search, per-document access permissions, a link to the physical location at the Document Repository as well as a service discovery unit.

Searches for documents are managed by the *Document Meta Data Index* (DMDI), which holds search relevant meta data as well as document based access permissions. This service checks each request for a document against the per-document access control lists and if permitted the request is signed. Only in case of a valid signature the Document Repository returns the requested document. The *Global Index* (GI) is responsible for finding DMDI services that hold document meta data for a specific patient identified by the unique patient ID. The Document Registry provides an Interface for Document Consumers called *Access Nodes* (AN), which generate queries to the architecture. They are defined as gateways to access the SEHR from external systems.

The Patient Id Source service group provides for unique patient identification across institutional boundaries, independent form patient's nationality. The Austrian legal situation and the fact that a unique patient identification does not yet exist requires different services, partly provided by the Austrian government to be integrated in the architecture.

The *Patient Lookup Index* (PLI) is used as interface for other services (mostly DC and AN) to obtain a unique patient identification based on demographic data. Each Austrian citizen is assigned a unique number by a governmental institution called "Zentrales Melderegister" (shown as ZMR in Figure 2). This Id is not publicly available, but a domain specific one-way derivation can be obtained for the health care sector. A second way is to take the social insurance number as unique patient identification. Social insurances (shown as HI: Health Insurer in Figure 2) provide a service to obtain the insurance number by demographic data such as first name, last name, date of birth, address.

For this reason distributed Patient Lookup services (PL) are needed to obtain this identifier transparently for the underlying architecture. Furthermore this service provides a Patient Id Source for patients without a ZMR identification, especially for non Austrian citizens.

The Austrian federal law for health telematics [3] requires a central eHealth directory service (shown as EHD in Figure 2) to be established until July 2006. At the moment medical data exchange partly makes use of a central LDAP directory of Austrian physicians known as eVGA ("Elekronisches Verzeichnis der Gesundheitsdienstanbieter"). In the SEHR architecture this directory will be used for role- and context based access control.

Currently two different approaches are being evaluated for services which allow the patient to give fine grained permissions to physicians and institutions which allows them to access specific parts of the record:

1. Distributed services, which manage and enforce access permissions assigned by the patient using roles defined by the eHealth directory.
2. A permission management system based on digital certificates issued for each permitted operation.

Due to the early stage of the evaluation this service is not implemented in the first prototype. The privacy of patient related data is temporarily solved in a way that participating institutions are bound by contract to only access data relevant for the specific treatment case.

A variety of work flows such as the adding, retrieval, version-save updates of documents and correction of misidentified patients are supported by the architecture. Their detailed description is beyond the scope of this work.

4. Discussion and outlook

The development of the network architecture follows an iterative approach which we have chosen to gradually adopt the architecture to evolving requirements of the major players. In this article the first prototype implementation is described, which of course lacks certain functionality and implements only simplified security requirements. To guarantee the highest level of data protection for patients, in the initial prototype phase physicians who test the architecture are conscientiously selected and will be bounded by contract to respect the patient's consent. The operators of the architecture commit themselves to prove the adherence to the contract by collecting random samples from logging.

The Austrian federal law for health telematics [3] requires a central eHealth directory service to be implemented until July 2006. At the moment medical data exchange partly makes use of a central LDAP directory of Austrian physicians known as eVGA (Elekronisches Verzeichnis der Gesundheitsdienstabieter). In the SEHR architecture this directory will be used for role- and context based access control.

Though the developed architecture follows the paradigm of Medical Data GRIDs, currently available GRID middleware such as the GLOBUS-Toolkit [13] is not used in this setup. The main reason is that requirements for Medical Data GRIDs, as outlined above, are currently not satisfactorily covered and only a subset of the provided GRID features would be used.

Nevertheless, the evaluation of the GLOBUS-Toolkit version 4 revealed some useful concepts. Since they mainly rely on open standards defined by the OASIS working group [14] those concepts have been integrated in the architecture.

Our implementation of the architecture follows the IHE XDS specification as closely as possible, nevertheless the legal situation in Austrian demands for specific modifications of the architecture as described above for patient identification and usage of the central eHealth register.

We expect to gain knowledge about the development of a shared electronic health record and how proprietary systems, particularly the clinical information system from the Innsbruck University Hospital can be integrated in this environment. Before the architecture can be used in productive environments technical and organizational issues have to be solved comprising extended security, financing and cooperation with other health care institutions.

References

[1] Maglaveras N, Chouvarda I, Koutkias V, Meletiadis S, Haris K, Balas EA. Information technology can enhance quality in regional health delivery. Methods Inf Med 2002;41(5):393–400.

[2] Haux R, Ammenwerth E, Herzog W, Knaup P. Health care in the information society. A prognosis for the year 2013. Int J Med Inform 2002;66(1-3):3–21.

[3] Gesundheitsreformgesetz 2005 (in German), BGBl. I Nr. 179/2004, (Dec 30 2004).

[4] Leoni L, Manca S, Giachetti A, Zanetti G. A Virtual Grid Architecture for Medical Data Using SRB. In: Inchingolo P, Pozzi-Mucelli R, editors. EuroPACS - MIR 2004 In The Enlarged Europe. vol. 1. E.U.T. Edizioni Universita' di Trieste; 2004. p. 475–478.

[5] Schabetsberger T, Ammenwerth E, Goebel G, Lechleitner G, Penz R, Vogl R, Wozak F. What are Functional Requirements of Future Shared Electronic Health Records? In: Engelbrecht R, Geissbuhler A, Lovis C, Mihalas G. European Notes in Medical Informatics (CD-Rom): Connecting Medical Informatics and Bio-Informatics; MIE2005; 2005 Aug 28-31; Geneve, Switzerland. Vol. I Nr. 1, 2005. ISSN 1861-3179.

[6] Hafner M, Breu R, Breu M, Nowak A. Modelling Inter-organizational Workflow Security in a Peer-to-Peer Environment. In: R. Bilof (Ed.): Proceedings of the 2005 IEEE International Conference on Web Services. IEEE Conference Publishing Services; 2005. p. 533–540.

[7] Unified Modelling Language [homepage in the Internet]. Needham, MA: Object ManagementGroup-UML; c1997-2006 [cited 2006 Jan 14]. Available from: http://www.uml.org/.

[8] Austrian E-Health-Initiative Strategie und Technologien (in German) [homepage on the Internet]. Vienna; Arbeitsgemeinschaft für Datenverarbeitung (ADV); c2005 [cited 2005 Dec 18]. Available from: http://ehi.adv.at.

[9] Integrating the Healthcare Enterprise [homepage in the Internet]. IHE.net; c2005 [cited 2005 Dec 19].

Available from: http://www.ihe.net.
[10] IHE Radiology Technical Framework Supplement Cross-enterprise Document Sharing for Imaging (XDS-I) [homepage in the Internet]. IHE.net; c2005 [cited 2005 Dec 19]. Available from: http://www.ihe.net/Technical_Framework/upload/IHE_RAD-TF_Suppl_XDSI_TI_2005-08-15.pdf.
[11] Schabetsberger T, Ammenwerth E, Andreatta S, Gratl G, Haux R, Lechleitner G, Schindelwig K, Stark C, Vogl R, Wilhelmy I, Wozak F. From a Paper-based Transmission of Discharge Summaries to Electronic Communication in Health Care Regions. IJMI. In press.
[12] OASIS Web Services Security (WSS) TC [homepage in the Internet]. OASIS Open; c2005 [cited 2005 Dec 11]. Available from: http://www.oasis-open.org/committees/tc_home.php?wg_abbrev=wss.
[13] Foster I, Kesselman C, Tuecke S. The Anatomy of the Grid: Enabling Scalable Virtual Organizations. International J Supercomputer Applications 2001.
[14] OASIS [homepage in the Internet]. OASIS Open; c2005 [cited 2005 Dec 11]. Available from: http://www.oasis-open.org/.

4. eHealth

4.1 eHealth: Applications and Approaches

Ubiquity: Technologies for Better Health in Aging Societies
A. Hasman et al. (Eds.)
IOS Press, 2006

A Multi-agent Approach to the Design of an E-health System

Paola Di Giacomo[a] and Fabrizio L. RICCI[b]
[a] *Center for Biomedical Research, University "La Sapienza" of Rome, Italy*
Via Antonio Nibby 18, 00161 Rome (Italy)
Paola.Digiacomo@uniroma1.it
[b] *National Council of Research, Rome, Italy*

Abstract. E-medicine covers the whole range of medical process and service. Multi-agent approach is suitable for the development of e-medicine systems. In this paper, firstly the requirements of e-medicine are analyzed and taxonomy is proposed for e-medicine systems. Secondly multi-agent approach is introduced for developing e-medicine systems, and the design of agents and the design of multi-agent structure are presented for e-medicine systems. Finally a case study is presented on a telemedicine for diabetes to illustrate the development of e-medicine systems. Then, our future work is to implement the proposed system.
Keywords: E-medicine; telemedicine; multi-agent.

1. Introduction

With the progress of information technology e-medicine has become popular in the last decade and numerous variants of e-medicine have appeared such as e-diagnosis, e-pharmaceutics, e-healthcare, telemedicine, telehealth, etc. E-medicine entered as a part of the routine clinical as early as in 1990s [1]. E-medicine delivers healthcare by integrating the information, communication and human-machine interface technologies with health and medical technologies [2]. E-medicine relies upon the technology enabler to realize the vision for health [2] and its advantage is to deliver healthcare across geographic, temporal, social, and cultural barriers [3]. E-medicine has wide applications, from diagnostics (such as teleradiology), treatment, through telesurgery or telementoring where a specialist surgeon can guide a beginner [4]. The applications of e-medicine can be classified as four areas [2], i.e. (1) lifetime health and medicine, (2) personalized health information, (3) teleconsultation and (4) continuing medical education. Moreover, e-medicine also has great impacts upon the traditional healthcare system. The national/provincial/regional health systems, health regulation, clinical programs, health institutions/organizations, and so on will be affected by e-medicine systems [3].

New technologies and methods must be explored to release the full potential of e-medicine. Multi-agent system approach has been widely used in the development of large complex systems. Agents have the autonomy and social ability, and multi-agent system is inherently multi-threaded for control as in [5],[6]. Therefore, multi-agent approach is very effective for tackling the complexity of e-medicine systems and suitable for the development of e-medicine systems.

2. Multi-agent Approach to the Design of E-medicine Systems

Multi-agent approach is effective for tackling the complexity of e-medicine systems. An agent is a computer system that is capable of independent action on behalf of user or owner [5]. An autonomous agent has the following properties [7]:

- Autonomy, an agent encapsulates some state of its environment and makes decision about what to do based on this state
- Reactivity, an agent perceives its environment and responds to changes that occur in the environment
- Pro-activeness, an agent can exhibit goal-directed behavior by taking the initiatives to satisfy the given design objectives
- Social ability, an agent interacts with other agents via an agent communication language and engages in social activities in order to achieve goals or cooperate.

A multi-agent system is a system that consists of a number of agents, which interact with each other, typically by exchanging messages through some computer network infrastructure [5]. A multi-agent system is a dynamic society made up of a great number of "intelligent agents" [8], so it is an intelligent society.

The multi-agent approach to system development consists of four steps [5], i.e.: identification of the agent' roles, identification of the responsibilities and services for the roles, determination of the goal and plan to achieve the goals, determination the belief structure of the system. These four steps can be grouped as two stages, i.e., design of agent and design of agent society (structure) [9].

2.1. Design of Agents

The task in the design of agents is to identify the roles of all agents and their responsibilities and services in e-medicine systems. This can be elaborated as follows. *Interface agent* provides guidance to an e-medicine system. Personal agent is one type of interface agents, which provides the user with a graphical interface to the multi-agent system and initiates a search or shows the results of a query to the user. *Broker agent* is an agent that knows about all capabilities of the multi-agent system. Through the broker agent the user can communicate with agents or perform a general search among all agents. *Doctor agent* maintains the schedule and appointment of a given doctor, and is aware of the doctor's times for visiting patients. *Administration agent* implements the medical administration such as the assignment of task and the cooperation between departments and agents. *Controller agent* controls the whole e-medicine systems and mediates the conflicts among the agents. *Department agent* has the knowledge of a certain medical department and manages the medical affair in the department. *Monitoring agent, diagnosis agent, therapy agent, surgery agent, consultation agent, training agent and record agent* carry out the functions of monitoring, diagnosis, therapy, surgery, consultation, training patient and medical record in a department, respectively. For example, training agent provides patients with instructions such as how to take medicine and how to take daily care. *Database wrapper agent* is an agent that controls the access to a database that contains the medical records of patients. All the communications between database wrapper agent, patient agent and doctor agent are encrypted. *Education agent* introduces the newest medical technologies and pharmaceutics, and provides e-learning for physicians or even for other agents in the e-medicine systems. *Decision support agent* integrates knowledge and provides diagnosis agent with effective decision approaches.

2.2. Design of Multi-agent Society (Structure)

Design of multi-agent society focuses on the establishment of the architecture of the multi-agent system and the interactions between the agents. Multi-agent system has three architecture, i.e., deliberative, reactive and hybrid as described in [7]. In e-medicine systems, the multi-agent system architecture tends to be hybrid. Different multi-agent systems are responsible for different specialist medical services of medical departments. A multi-agent system is a dynamic system and is similar to a society. On the one hand, the multi-agent system interacts with its external systems. On the other hand, within the multi-agent system agents interact with one another. The interactions consist of internal communication and external communication, ciphered or non-ciphered [10]. So the interaction model in multi-agent system is divided into external model and internal model [9]. The internal interactions of e-medicine systems are those not only between agents within one department, but also between different departments. The external interactions of e-medicine systems are those with other systems such as medical instrument, psychology, medical university and institute, etc. Agent in the multi-agent society of e-medicine systems are grouped control, implementation and interface, as shown in Figure 1.

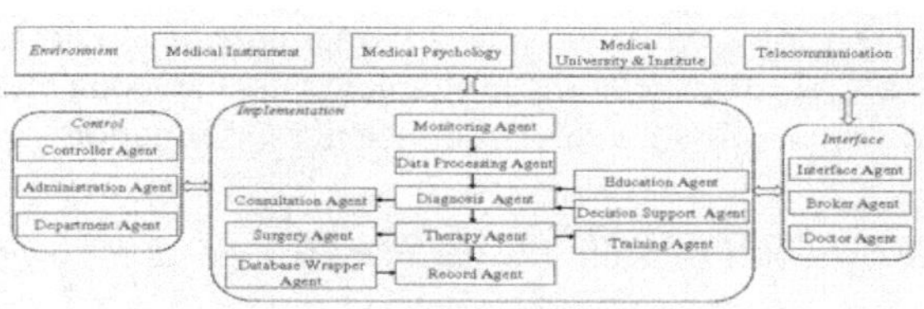

Figure 1 Multi-agent society of e-medicine systems

3. A Case Study—Telemedicine for Diabetes

To illustrate multi-agent approach to the development of e-medicine systems, a case study of telemedicine for diabetes is presented below. Diabetes is a chronic disease with a sustained elevated blood glucose level. The symptom of diabetes is that the metabolism cannot work properly because of the reduction of insulin secretion [11]. The diabetic patients need to be injected with exogenous insulin to regulate blood glucose metabolism. Moreover, patients have to perform a daily strict self-monitoring of blood glucose level, such as measuring it before every injection and recording it on diaries, together with the amount of insulin injected and the information about diet and life style [12]. Using telemedicine system to manage this healthcare process can give diabetic patients real-time monitoring and immediate therapy. Telemedicine system for diabetic patients has been studied since early 1980s [11].

3.1. Requirement Analysis of Telemedicine for Diabetes

The telemedicine system must provide following services for diabetic patients on a daily basis: **visiting** the patients and providing individual therapy, **monitoring** the patients in real time and processing the monitored data immediately, **diagnosing** the patients in term of the monitored data and making a proper therapy for the diabetic patients, **training** the diabetic patient to monitor themselves and educating the physicians to update their skills, **establishing** the patients' database the entry of which is ciphered and **providing** the diabetic patients with consultation. The system needs to interact with other systems in e-medicine systems.

3.2. Identification of the Roles and Responsibilities of Agents

In the telemedicine system for diabetes, the medical services include monitoring the patient in real time and transmitting the information to physician, then providing the patient with a corresponding therapy, and consultation to enquiries. These services are implemented by various agents and their responsibilities are detailed as follows. *Monitoring agent* monitors the diabetic patients in real time and transmits the monitored data to data processing agent. *Data processing agent* makes statistic and integrates the monitored data. *Diagnosis agent* analyses the situation and makes an accurate judgment for the patient. *Therapy agent* determines a proper therapy method. *Consultation agent* provides consultation to the enquiry of patients and contacts with diagnosis agent. *Decision support agent* provides decision support and cooperation for diagnosis agent. *Training agent* trains patients about how to take medicine and how to care himself. It implements the method of therapy agent. *Archival agent* edits and archives the patient record and therapy methods, and updates the database of the individual patients. Moreover, it integrates with the medical database and encrypts the important database. *Department agent* implements the control of the telemedicine system. *Interface agent* provides search service and information service.

3.3. Identification of the Goal and Plan to Implement

This telemedicine service is implemented by diabetes department in a hospital or can be applied at home' patients, too. The telemedicine system must not only provide the diabetic patient with immediate medical services through distant monitoring, diagnosis, therapy and consultation, but also integrate with other e-medicine systems for the functions of education, training, management, security and database.
How to achieve the goal by the multi-agent system? The multi-agent system consists of three groups, i.e. interface, implementation and control. In the implement group, there are many agents to individually and orderly carry out different responsibilities. The proposed architecture of multi-agent system is shown in Figure 2.

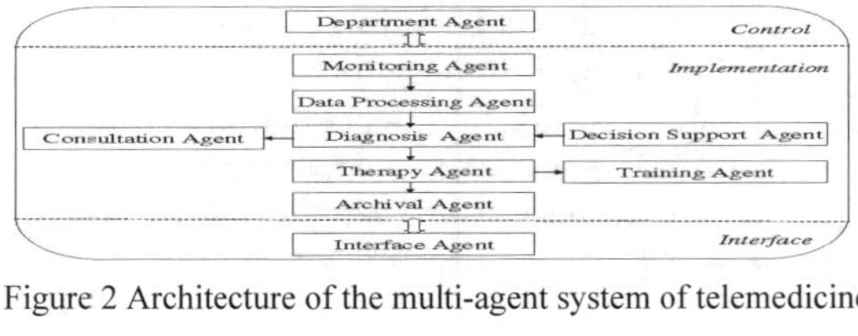

Figure 2 Architecture of the multi-agent system of telemedicine

In the multi-agent system, the control group assigns the work and mediates the conflicts between agents. The interface group keeps the link with patients and other e-medicine systems. The implementation group implements the monitoring, diagnosis, therapy, consultation and archival functions to achieve the goals.

3.4. Determination of the Belief Structure

This is to determine the information requirement for each plan and goal in the interactions between agents [5]. In the multi-agent system, the external interactions focus on the integration with other e-medicine systems and its environment, while the internal interactions focus on the cooperation between agents to realize the telemedicine process, as shown in Figure 3, where the bi-directional arrow represents the interactions. In Figure 3, the bigger dashed ellipse indicates the range of telemedicine while the smaller dashed ellipse indicates the range of the implementation group of the telemedicine system. So arrow 1 and 2 indicate that the telemedicine system interacts with its environment and other e-medicine systems, respectively. And arrow 3 and 4 indicate that the implementation group interacts with the control group and interface group in the telemedicine system, respectively.

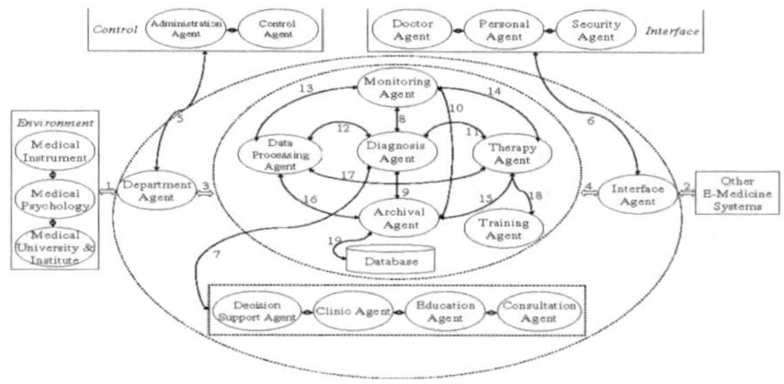

Figure 3 Interactions in the multi-agent system of telemedicine of diabetes

Arrow 5 represents the interaction between department agent and the control group in e-medicine such as administration and control agent, while arrow 6 represents the interaction between interface agent and the interface of e-medicine systems such as doctor agent, personal agent and security agent. In the implementation group, diagnosis agent plays an important role. Because diagnosis is a complex process, diagnosis agent not only interacts with other agents in the implementation group, but also integrates with decision agent, clinic agent, education agent and consultation agent, as show by arrow 7. Arrow 8-17 represents the internal interactions between agents in implementation group of the telemedicine system, respectively, which can be described by a matrix. Moreover, training agent implements the method of therapy agent, as shown by arrow 18, and archival agent directly interacts with the database, as shown by arrow 19. The interactions in the matrix are explained in a symmetrical order as follows in the Table 1. For example, **entry <1,2>** represents monitoring agent transmits the monitored data to data processing agent. And **entry <2,1>** represents data processing agent may require monitoring agent to provide other related data etc.

Table 1 Matrix of interaction of the telemedicine system for diabetes

	Monitoring Agent	**Data Processing Agent**	**Diagnosis Agent**	**Therapy Agent**	**Archival Agent**
Monitoring Agent		1, 2	1, 3	1, 4	1, 5
Data Processing Agent	2, 1		2, 3	2, 4	2, 5
Diagnosis Agent	3, 1	3, 2		3, 4	3, 5
Therapy Agent	4, 1	4, 2	4, 3		4, 5
Archival Agent	5, 1	5, 2	5, 3	5, 4	

4. Conclusion

Multi-agent system can not only integrate the medical knowledge and clinical experience and make decision support, but also adapt the system rapidly to the change in environment. In this way, the multi-agent approach can effectively tackle the complexity of e-medicine systems. Now, we will start the implementation of the system and we will adopt it in the clinical practice. Then, we are refining the problems of agents' cohabitation, service discovery and destruction of obsolete agents.

References

[1] B. H. Stamm and D. A. Perednia. Evaluating psychosocial aspects of telemedicine and telehealth systems. Professional Psychology: Research and Practice, Vol. 31, No. 2, 2000, pp. 184–189.

[2] A. B. Suleiman. The untapped potential of telehealth. International Journal of Medical Informatics, Issue 61, 2001, pp. 103–112.

[3] P. A. Jennett, and K. Andruchuk. Telehealth: 'real life' implementation issues. Computer Methods and Programs in Biomedicine, Issue 64, 2001, pp. 169–174.

[4] L. G. Kun. Telehealth and the global health network in the 21st century. From homecare to public health informatics. Computer Methods and Programms in Biomedicine, Issue 64, 2001, pp. 155–167.

[5] M. Wooldridge. An Introduction to MultiAgent Systems. John Wiley & Sons Ltd, 2002.

[6] T. Williams, C. May, F. Mair, M. Mort and L. Gask. Normative models of health technology assessment and the social production of evidence about telehealth care. Health Policy, Volume 64, Issue 1, 2003, pp.39–54.

[7] M. K. Lim and Z. Zhang. Iterative multi-agent bidding and co-ordination based on genetic algorithm. Proceeding of 3rd International Symposium on Multi-Agent Systems, Large Complex Systems, and E-Businesses, Erfurt, 7-10 October 2002, pp. 682–689.

[8] A. Garro and L. Palopoli. An XML-Agent System for e-learning and skill management. Proceeding of 3rd International Symposium on Multi-Agent Systems, Large Complex Systems, and E-Businesses, Erfurt, 7-10 October 2002, pp.636–647.

[9] M. Pankowska and H. Sroka. Business process reengineering and software agent development. Proceeding of 3rd International Symposium on Multi-Agent Systems, Large Complex Systems, and E-Businesses, Erfurt, 7-10 October 2002, pp.608–620.

[10] A. Moreno and D. Isern. A first step towards providing health-care agent-based services to mobile users. Proceedings of the 1st International Joint Conference on Autonomous Agents and Multiagent Systems, Part 2, Bologna, July 2002, pp. 589–590.

[11] O. Orlov and A. Grigoriev. Space technologies in routine telemedicine practice: commercial approach. Acta Astronautia, Vol. 51, No. 1-9, 2002, pp 295–300.

[12] M. Y. Sung, M. S. Kim, E. J. Kim, J. H. Yoo and M. W. Sung. CoMed: a real-time collaborative medicine system. International Journal of Medical Informatics, Issue 57, 2000, pp 117–126

Ubiquity: Technologies for Better Health in Aging Societies
A. Hasman et al. (Eds.)
IOS Press, 2006

A site of communication among enterprises for supporting Occupational Health and Safety Management System

E. Velonakis[a,1], J. Mantas[a,2], I. Mavrikakis[a,1]
[a]*Department of Public Health, Faculty of Nursing, University of Athens, Greece*

Abstract: The occupational health and safety management constitutes a field of increasing interest. Institutions in cooperation with enterprises make synchronized efforts to initiate quality management systems to this field. Computer networks can offer such services via TCP/IP which is a reliable protocol for workflow management between enterprises and institutions. A design of such network is based on several factors in order to achieve defined criteria and connectivity with other networks. The network will be consisted of certain nodes responsible to inform executive persons on Occupational Health and Safety. A web database has been planned for inserting and searching documents, for answering and processing questionnaires. The submission of files to a server and the answers to questionnaires through the web help the experts to make corrections and improvements on their activities. Based on the requirements of enterprises we have constructed a web file server. We submit files in purpose users could retrieve the files which need. The access is limited to authorized users and digital watermarks authenticate and protect digital objects. The Health and Safety Management System follows ISO 18001. The implementation of it, through the web site is an aim. The all application is developed and implemented on a pilot basis for the health services sector. It is all ready installed within a hospital, supporting health and safety management among different departments of the hospital and allowing communication through WEB with other hospitals.
Keywords: [Occupational Health and Safety, Workflow Management, Computer Networks, World Wide Web, File Submission, Questonnaries]

1. Introduction

The management of occupational health and safety constitutes a field which concentrates the interest of enterprises. Particularly the enterprises with a modern concept of management, not only care to comply with the legal obligations as far as concern on the occupational health and safety, but they implement a health and safety management system, in order to improve the work conditions. Quality management systems are developed and implemented to enterprises following some standards. For health and safety management, ISO 18001 is widely applied. For the development of such systems the workflow and information management between experts and enterprises is very important and interesting. The use of computer networks and the whole infrastructure is significant for this purpose. Such a system can be administered

1 Public Health Laboratory, Faculty of Nursing, University of Athens, 123 Papadiamantopoulou street, 11527 Athens, Greece, email {evelonak, imavrik }@nurs.uoa.gr

2 Health Informatics Laboratory, Faculty of Nursing, University of Athens, 123 Papadiamantopoulou street, 11527 Athens, Greece, email jmantas @cc.uoa.gr

via the WWW. An external expert can handle the information in a way to correct and to intervene for improving the quality of occupational health and safety services. We have developed such a system in order authorized users, of each enterprise, mainly the responsible for the occupational health and safety, can communicate and submit files. On the other hand, we can use comments for each submission to explain in a more executive way the files and retrieve them. For the purpose of managing and detecting some very important workflows we use questionnaires; the processing of which are very useful for management and decision making at this field.

The communication between enterprises is supported by the internet. The protocol which the internet is based is the TCP/IP[11]. The address IP is a 32-bit number fundamental to Internet addressing and routing. It is also used for addressing the servers and clients. The TCP/IP provides a realiable transport service which is used by most Internet Applications. Examples of applications are the electronic mail (e-mail), the File Transfer Protocol (FTP) and the access of Web Pages through the World Wide Web via HTTP. The evolution of the e-mail was the incorporation into the www. The TCP/IP at the network layer performs the below operations such as:

- Multiplexing of Packets
- Error Detection with a 16-bit field
- Acknowledgment Signals
- Retrasmission of Error Packets
- Flow Congestion Control

Data transmission is loss sensitive. It does not tolerate bit errors or packet loss but there are no real time requirements. It is the base of information management because we can use databases for decision support systems.

Voice over IP, video services, radio services, teleconferencing and distant learning[2] are real time applications. On those applications packet delays is important. In order to abolish this disadvantage we must accept that we may lose some packets. The protocol which supports this way of communication is the UDP. This protocol does not use acknowledgments, does not use error detection and it does not retransmit erroneous packets or contols the flow [24].

All the above applications are to some extent loss tolerant but pose additional challenges in terms of the bandwidth and timing requirements and may be used for the requirements of occupational health and safety.

In our paper we are going to present a web application [25] based on the protocol TCP/IP which was developed at the Public Health Information Lab at University of Athens. We intend with this internet web site to inform executive persons on the subject of Occupational Health and Safety. We also present some applications of this work at Hospital Information Systems[18] and Distant Learning[2].

2. Network Design Considerations

Interconnection networks play a major role in the performance of modern enterprise management. There are a lot of factors which can differentiate the design of each network.

- Performance requirements (on time and reliable decisions from the informations we collect): All operations are usually performed by explicit message passing or by accessing shared variables. To reduce the message latency, we reduce the idle time of processes and memory access time to remote memory locations [8]. The delay of information trasmission between two points is a very important aspect in case we take some useful decisions. The validity of information is also important, if we can collect information from all the enterprises we have reliable data, otherwise the latency of communication between the enterprises and the experts may cause an unreliable source of information.

- Scalability: As we add more enterprises in the network, we should proportionaly increase the network bandwidth, the I/O bandwidth and the memory bandwidth. If we are not able to use a scale factor for all the above requirements, at the network, it may become a bottleneck for the rest of the system decreasing the overall efficiency accordingly[8].
- Incremental expandability (example of ISPs): Customers are unlikely to purchase a computer network with a full set of computers and other electronic devices. More enterprises may be involved in the network until a system's maximum configuration is reached. We should find a way that the new nodes we add, do not decrease the performance. For example if we have 10 enterprises and we want to install an internet service provider we may not use a provider exactly for 10 enterprises, we should predict that after a long time we may provide our services to 15 enterprises, so we must use a provider with incremental expandability in order to reach the new requirements.
- Partition ability: It depends on the work we want to do. We devide the work in several partitions and then we dedicate certain tasks to computers of network [9]. This partition is very efficient for the performance of the network. We can use it also for the design of the network, we configure a network for the requirements of our work and we do not configure the work to the requirements of the network. For example we can use one node of the network as file server, another as database server, another as network server etc. Partition ability may also be required for security reasons [8].
- Simplicity: It is useful for the customers who understand the design and can easily exploit their performance. Otherwise the network may not be so efficient [8].
- Distance span, Locality: There are appropriate mechanisms which can reduce the noise during the transmission of the data, but there are a lot of constraints mainly on the distance between nodes on this domain. The use of optical links solve these problems equalizing the bandwidth of short and long links up to a much greater distance than when copper wire is used[8]. The locality is a very importat parameter, we must know where the enterprises are placed in a map in order[16] to design a network.
- Physical Constraints: The operating temperature control, the wiring length limitation and the space limitation are constraints for the network design. We must be careful in case we put together a lot of wires, the overheating may cause damages to the wires [8].
- Reliability and Repairability: It should be able to transmit information reliably. In addition interconnection networks should have a modular design allowing upgrades and repairs [8].
- Expected Workloads: The network design should be robust. The performance should be efficient independently of the wide range of traffic conditions [8].
- Cost Constraints: We should find an optimal solution between the cost of the implementation and the performance of the network. A solution which is probably expensive it is not necessery and efficient [8].

2.1 Interconnection Networks

At the design of a network we should predict its connectivity with other networks, such as wide area networks[17]. We may need to connect our network with a network of an enterprise and to communicate with an establishment of the enterprise, in order to find useful informations for our work. We can connect our network with leased lines of the telephone network, in order to use services such as voice over ip, teleconference, video conference etc. Interconnection networks exist between parallel and distributed systems[6], [7], [9]. In general purpose the usage of the above systems is the efficient solution of a problem. The network design help us to devide our work into small tasks and try each computer to solve these small tasks. Then we combine the solution of each node to the final[9]. For those systems the network should reduce the message latency in order to be efficient[6]. A decomposition technique based on interconnection network should care for the load balancing of the work between computers. A central unit can recursively divide the work, sending partition of work to nodes and control the overall work.

Meanwhile an application can not use general rules, the services we want the network offer to their users and the quality of services determine the parameters of a network. There are a lot of issues in network design [8], [23], [24]. First issue is how many routers we are going to use. Having a topological map of the whole network where shall we put them? The connections are going to be wireless (data links), which routing algorithm we are going to use, Dijkstra or smth else. There are also some challenging issues such as the network bandwidth and the time delay of packets. Computer networks individual for enterprises are crucial for their speeds in transfering data and for the internal communication.

3. Programming and Configuring an Internet Web Database

For the purpose of installing a web database we used the Apache Server Program and a Database Management System. The apache server enables some files of the host computer to be accessible from the Internet via the HTTP. For the communication of the database we use some script programs written in PHP language. The accessible files via HTTP are HTML forms which communicate with PHP programs[25]. For safety reasons we used some global variables (sessions) on php scripts in order only experts and executive persons from enterprises could administer the internet web database. The information on the Internet Site is accessible to everyone. On the site we can find which enterprises are interested on the domain of Occupational Health and Safety. We can find the executive persons of each enterprise which are responsible to communicate with us. We install a communication between the database and one program in java [10] in order to compute some interesting statistical values from the dataset. The internet web site which host a part of our work is located at the address http://healthandsafety.nurs.uoa.gr/sdyaergasia.php .

In order to construct an Internet Web Site, we find the system requirements and the users requirements. Then we design the web site in order every user to find easily whatever he wants on the domain of occupational health and safety. Then we configure the web site in order to achieve a communication between the site and the database. The updates for news on research in occupational health and safety are appropriate. The updates on the software are also important. The file management we present[4], uses the classic way of FTP. We insert documents, files, data and multimedia[13], [20] in a directory which is structured in such a way for displaying into the www. Using some additional attributes to the above attributes we can cluster the data into groups[19]. The development of an Internet Web Site for efficient storing and indexing documents is a pioneering research work on the domain of Occupational Health and Safety. These attributes help the users to retrieve the data easily. They are inserted by the system administrator or by the composer of digital data. The metadata are all the appropriate elements which describe the uploading documents. Some operations that a system administrator can do are the uploading of a document and the updating of metadata using a simple web interface.

We develop this internet web site in order the executive person of an enterprise be informed about the issues of Occupational Health and Safety.

We are heading the semantic web[15]. This means that the internet can support workflow models. Workflow models are abstractions revealing the most important properties of the entities participating in a workflow management system. The workflows are 3 types, Human – Oriented (a person is responsible to execute a workflow), System – Oriented (we use databases and decision support systems[5],[14],[21] in order to find answers to our queries) and Transactional workflows (consisted from a mix of tasks, some performed by humans, others by computers and support selective use of the transactional properties for individual activities or for the entire workflow). The workflow models consist from entities which interact with others in an Interconnection Network and they are participate in a management system[1], [15].

3.1 Administration by Authorized Users

The administration of internet web site could be performed via the web. The web interface is quite familiar to everyone who want to use it. We use html forms to upload documents. We insert metadata and documents in a database using php scripts. Experts on the domain of Occupationl Health and Safety are responsible to update the site with news and with the latest scientific research. The web interface which supports the administration can be seen at figure 1.

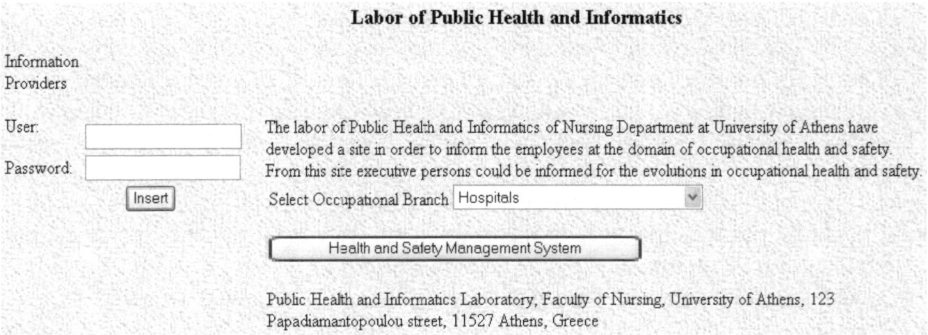

Figure 1. Web Interface for administrating and presenting information on Occupational Health and Safety.

There are some constraints at the insertion of data. We can insert only text documents, Microsoft Word Documents, and documents which can be read by the Adobe Acrobat Reader. Meanwhile we may need to insert files for educational reasons, such as power point files, video and images. On multimedia files we can insert a digital watermark in order to authenticate the constructor of the object [13]. We can insert a list of notes for each document in order to have a dialogue about certain ideas[25]. This dialogue is important for improvements in work.

The data we insert at the internet site are peer reviewed in order to be reliable. We also protect the authenticity of the digital objects inserting to them digital watermarks resistant to attacks such as compression, smoothing and geometric transforms[13] (mainly for multimedia data).

In the web site http://healthandsafety.nurs.uoa.gr/sdyaergasia.php, figure 1, developed in the Department of Public Health in the Nursing Department, University of Athens, we have configured the server to be able to find all the relevant information useful to this task. We can search the documents[19] placing keywords. There is also the possibility of browsing in all data lists.

4. An Applications on Distant Learning and at Hospital Laboratories

The development of training packages for the protection and promotion of health at the workplace and the implementation of safety procedure is another important issue of the internet web site. These packages are disseminated through the network or applied by e-learning techniques[2]. The function and the effectiveness of those packages will be evaluated at the ability of installing integrated solutions and applications to support experts on Occupational Health and Safety.

4.1 An Application at Hospital Laboratories

Although hospitals as places of employment are liable to the stipulation of law regarding Health and Safety, there is a deficit of experience for effective health and safety management system. Some hospitals attempt to develop and implement health and safety management systems according ISO 18001. We believe that the hospital information systems for health and safety management, is a very useful application [22].

Such an application has already been installed at “Agios Savvas” Hospital, to meet all the needs and requirements of the management of health and safety system [3], [4]. Our application is based on the local area network and serves the hospital needs. We have implemented a web based system in order to insert and search documents. The same application is planned to connect other hospitals, which will have a relevant interest.

Some issues of the developed software are the following:

- The policy and the objectives of the health and safety management system and any related information provided by the administration.
- A data base for all chemicals used in the hospital facilitating their properly use.
- Training packages for health and safety and environmental good practices, covering most of the activities of the hospital.

- Work health and safety procedures concerning the overall management of the system, according ISO 18001, and more specific work processes at the various departments of a hospital [18]. The health and safety procedures which are very important documents of health and safety management system, are sited in a way not only to inform employees, but also to facilitate their application and to raise comments or suggestions for their improvement.
- Date concerning controls and inspections which after elaboration induce amendments and revision of the health and safety management system.
- Information for the legal obligations, regulations, directives and recommendations on health and safety.
- Plans for the management of threatening situations and emergency plans.
- Information on the standards and the appropriate use of personal protective equipment and eventual complaint.

The software permits continuous communication among all users, according their degree of accessibility.

The evaluation of the application is permanent in order to improve the software and adapt it to the evolutionary environment.

5. Review

At this paper we have described a web application, which is administrated by authorized persons only. We intend to inform executive persons of each enterprise for the evolutions in Occupational Health and Safety. This site will inform and allow communication among enterprises. Documents and data based on ISO 18001 will be submitted, processed and presented in appropriate format in order to be useful to the interested executive persons. The corrections and the control of the whole infrastructure is significant to our work. Our target is to improve occupational health and safety, to find new policies, to design new systems, to control new procedures and finally to review the whole structure in order to be comprehensible to the users. An application has already been installed at hospital "Agios Savvas". The application will be evaluated continuously in order to achieve the standards of ISO 18001.

6. Acknowledgement:

The present study was funded through the Operational Programme for Education and Initial Vocational Training (O.P. "Education") in the framework of the project "PENED 2003 – Support of University Research Groups" with 75% from European Social Funds and 25% from National Funds.

7. References

[1] Aalst W., Hee K., Workflow Management, Models, Methods, and Systems, Cooperative Information Systems Series.

[2] Apostolakis I., Balsamos P., "Distant Learning Education of Health Officials", Health Survey November - December 2005.

[3] Velonakis E., "Health and Occupation. Occupational Health", Athens, 1990.

[4] Velonakis E., Tsalikoglou F., Occupational Health and Safety Management System in Hospital. Scientific Publications, Parisianou A.E. Athens

[5] Grinias H., Mavrikakis I., Tziritas G., "Region Growing Colour Image Segmentation Applied to Face Detection", International Workshop on Very Low Bitrate Video Coding (VLBV 01) Athens, Greece Octomber 2001.

[6] Coulouris G., Dollimore J., Kindberg T., (1994) Distributed Systems, Addison-Wesley.

[7] Date, C. J. (Translation. T. Alvas), Introduction in the Database Systems, vol A, Computer Books Publications Kleidarithmos, 6th American Publication (2003).

[8] Duato J., Yalamanchili S., Ni L., "Interconnection Networks, An Engineering Approach", Morgan Kaufmann Publishers.

[9] Elmasri R., Navathe S., Basic Principles of Database Systems, Translation – Editor Michalis Chatzopoulos, Diavlos Publications

[10] Jaworski J., JAVA 2 Platform, Unleashed The Comprehensive Solution, 1999

[11] The Internet Protocol Journal, June 2000 Volume 3, Number 2

[12] Karvounarakis G., Kapidakis S., Submission and repository management of digital libraries, using WWW, Computer Networks 34 (2000) 861-872 ELSEVIER.

[13] Mavrikakis I., Digital Watermarking applied for Digital Image Protection, Master of Science Thesis, Computer Science Department, Heraklion 2002.

[14] Mavrikakis I., Pattern Analysis with use of Support Vector Machines Applied to Image Processing. Degree Examination Paper, University of Crete, Computer Science Department, Heraklion 2000.

[15] Marinescu D., Internet - Based Workflow Management Toward a Semantic Web, Wiley Series on Parallel and Distributed Computing, Albert Y. Zomaya, Series Editor.

[16] Mason L., Vinokurov A., Zhao N., Plant D., Topological Design and Dimensioning of Agile All-Photonic Networks, Elsevier, Computer Networks 50 (2006) 268-287.

[17] Nieuwpoort R., Maassen J., Bal H., Kielmann T., Velderna R., Wide Area Parallel Computing in Java, Department of Computer Science, Vrije Universiteit Amsterdam, The Netherlands.

[18] Panagiotou E., Panagiotou G., Hospital Laboratory Information System (LIS), Epithewrhsh Hygeias November – December 2005.

[19] Park L., Ramamohanarao K., Palaniswami M., "A Novel Document Retrieval Method Using the Discrete Wavelet Transform", ACM Transactions on Information Systems, Vol. 23, No. 3 July 2005 Pages 267-298.

[20] Shirmohammadi S., Saddik A., Georganas N., Steinmetz R., Web-Based Multimedia Tools for Sharing Educational Resources. ACM Journal of Educational Resources in Computing Vol. 1, No. 1, Spring 2001.

[21] Suykens J., Gestel T., Brabanter J., Moor B. and Vandewalle J., Least Squares Support Vector Machines. World Scientific

[22] Terpos A., Research of Information Systems on Occupational Health and Safety section EL.IN.Y.A.E., Athens 2000.

[23] Walrand J., (Translation M. Anagnostou), Communication Networks, Papaswthriou Publications (1997)

[24] Walrand J., Communication Networks A First Course, Second Edition, WCB McGraw-Hill.

[25] Welling L., Thomson L., PHP and MySQL Web Development Power Programming, M. Giourdas Publishers.

Ioannis Mavrikakis is a PHD student at the faculty of Nursing, Department of Public Health, in the university of Athens. He received his master degree in 2003, at Computer

Science Department from the university of Crete. He also received, from the same university, his first degree in the field of Computer Science in 2000. His working experience during the last five years includes teaching in primary, secondary and tertiary education. His research interests include Information Systems on Occupational Health and Safety, Computer Networks, Computer Vision and Digital Watermarks.

Emmanuel Velonakis Medical Doctor. Associate professor of Prevention and Enviromental Hygiene in the Nursing Department of Athens University. Field of interest Occupational Health and Safety Management and Enviromental Epidemiology.

John Mantas Professor of Health Informatics in the Faculty of Nursing. Director of Health Informatics Laboratory.

Ubiquity: Technologies for Better Health in Aging Societies
A. Hasman et al. (Eds.)
IOS Press, 2006

A new perspective in the promotion of e-health

Eva del HOYO-BARBOLLA [a,1], Rita KUKAFKA [b], María Teresa ARREDONDO [a], Marta ORTEGA [a]

[a] *Life Supporting Technologies, Madrid, Spain*
[b] *Mailman School of Public Health. Department of Medical Informatic, Public Health Informatic. Columbia University. New York*

Abstract. This paper proposes a new model that provides assistance in understanding the reasons why individuals would use new ICT (Information and Communication Technologies) to perform a healthcare change in their lifestyle. Achieving a lifestyle change by the use of ICT is a multifold issue that can be broadly addressed by analysing, in parallel two dimensions: the individuals' attitude towards their health condition and their approach and readiness to monitor and change their attitude by the use of ICT. Our work has been to conceive and develop a model that explains the different stages the user is at both in terms of the perception of healthcare and of the use of technology to perform any desired or recommended change. In order to place the user at each dimension a set of questionnaires were designed and implemented. These questionnaires assisted us in understanding what personalised information needs to be provided according to the stage the user is at as well as to other variables (such as age, cultural background, etc). The novelty of this model is that it proposes a general framework that may be applied to the conception, design and evaluation of any e-health application. Moreover, it can be applied to different application targets (medical informatics, public health informatics, etc) and to different audiences (healthy individuals, patients, professionals, etc) as it proposes an enhanced user modelling process by taking into account both healthcare behaviour aspects as well as technological issues, which up to this moment, have not been taken into account and may be part of the explanation of e-health uptake failure in the healthcare field. The application of this model promotes the empowerment of the individuals by providing tailored information, as well as guidance, monitoring, through ICT and it will certainly make an impact on health-related behaviour. Besides, it will allow to understand some of the reasons of the success or failure of different e-health platforms. Overall, this model is likely to provide deeper insights into the process of improving e-health so it can meet ongoing individuals' needs and become an increasingly valued part of health care services.
Keywords: Healthcare behaviour models, e-health acceptance, ubiquitous information.

1. Introduction

Preventing or delaying illness and death from chronic disease is possible [1]. Many of these diseases can be prevented or ameliorated through behaviour changes [2]. At least 80% of all cardiovascular diseases and "type 2" diabetes and over 40% of cancer could be avoided through healthy diet, regular physical activity and avoidance of tobacco use, which are behaviours that can be influenced and modified through education.

[1] Eva del Hoyo Barbolla. Life Supporting Technologies, ETSI Telecomunicación, Ciudad Universitaria s/n, 28040 Madrid, Spain. E-mail: evahb@lst.tfo.upm

ICT and its applications are increasingly looked upon as a potential answer to the requirements of a modern society, with demands for better healthcare, improvements in medical outcomes, and maintenance of a relatively high quality of life, especially with the onset of chronic health conditions coming to the fore as a key issue. Furthermore, implying a view of utilising the technology as a tool for (re) addressing the prevailing state of affairs, ICT tools and applications are also seen as having a potential to support an enhanced access to health information in general and indeed, to the health system itself in particular [3]. Characteristics such as the tailoring of messages, instantaneous feedback, appeal or engagement are potential advantages that new ICT can provide (adapted from [4]) and that may be of enormous benefit to attain behaviour change.

ICT innovations have the capacity to support empowerment of individuals in managing their health concerns and acquiring the necessary resources to achieve their goals. Moreover, it can ensure ubiquitous availability of the tools and communication channels necessary to support empowerment [5]. However, although there is much emphasis on the adoption of ICT implementation in the field of healthcare (e-health) [6], [7] there is still a lack of understanding the in-depth rationale to how these applications assist individuals to effectively change lifestyle behaviours.

On the one hand, there is a scarce on studies that relate the use of ICT to change attitude towards healthcare and on the other, the studies often focus on the technical aspects of ICT failure, neglecting what has been learned from the behavioural sciences about humans and their interaction with ICT; this may be the answer to understand the added value that ICT bring over traditional channels of healthcare provision. The success of ICT implementation in order to cause a change in individual's behaviour is often grounded in behavioural science, using theories and models to identify conditions and determinants of successful use [8]. Therefore, working out why an individual would use an e-health application to change his/her lifestyle means paying attention both at the behavioural aspects both in attitude towards healthcare and in technology readiness. This paper proposes a new model to understand why individuals would use new ICT to perform changes in their lifestyle, a multifold issue that can be addressed by analysing their willingness to change health behaviours and to do so by using an e-health application.

2. Methods

Our intention is to build a system that provides personalised information according to a set of variables relevant to the user. Tailoring both the channel and the content of the message has proven essential for persuading individuals to change their health behaviour [9]. In order to personalise information, a 5-step approach was defined [10]:

- **Step 1**: Analysing the problem to be addressed and understanding its determinants
- **Step 2**: Developing an assessment tool to measure a person's status on these determinants
- **Step 3**: Creating tailored messages that address individual validation of determinants of the problem
- **Step 4**: Developing algorithms to link responses from the assessment into specific tailored messages
- **Step 5**: Creating the final health communication

A description of these steps and how they have been applied to our model follows:

Step 1:

Theoretical models are useful in predicting which patients will use e-health and in understanding what factors influence their decisions. Models also can aid in designing and evaluating the ability of specific e-health applications [11]. Currently there is no integrated model that includes a sufficiently broad set of influencing factors to understand the multidimensionality of the reasons why people use ICT to embrace a healthcare change [8]. There have been previous attempts in understanding whether a single factor may have an effect on their e-health adoption, by partially applying some of the psychological models that explain behaviour change. However, it was noticed that applying only one of the theories (i.e. the "Stages of Change") [12] was somehow limited as there are a wider variety of factors that influence a user's decision to adopt a change in his/her life in order to perform a healthcare change.

Moreover, without addressing the full range of factors, strategies to change usage behavior run the risk of being ineffective because they fail to recognize interdependencies between individual and organizational factors [8]. Our research focuses on how to promote effective change in health behaviour by means of modern ICT. Therefore, a detailed study focusing on both healthcare and ICT dimensions was carried out to understand the stage the user is at.

In the first case research developed in psychology and public health that attempts to understand the promotion of health among the populations is complex and there are a number of significant theories and models that underpin the practice of health promotion and individual's attitudes towards the change, these being mainly: the Health Belief Model [13], Theory of Reasoned Action [14] and the Theory of Planned Behaviour [15], the Trans-theoretical (stages of change) Model [16] or the Precaution Adoption process Model [17]. These theories explain health behaviour change by focusing on the individual with the principal intention of providing information either to improve knowledge or change behaviour.

On the other hand, our task is to understand why users would make use of an ICT platform (denominated from now on "e-health" platform) in order to perform a change in their attitude towards health (i.e. use the Internet or a mobile phone to help them quit smoking, to encourage them to follow a diet, etc). However, we only found adoption theories that explained why a user would make use of an ICT platform to carry out a specific ICT-work related task (i.e. adoption of spreadsheets, use of email, word processor, etc) in different environments (companies, hospitals, government agencies, etc) and situations, although not health related attitudes. Studies elaborated show the interest of individuals in using an ICT application but there is no model to explain the link between these last two (i.e. the use of ICT to perform changes in health behaviour). Our work has been to develop a model that explains the different stages the user is at both in terms of the perception of healthcare and the use of technology to perform any change.

Step 2:

In order to understand what the individual's status is regarding the variables previously explained, we designed and implemented a set of questionnaires to place the user at a stage in each of the paths so that we understand what personalised information needs to be provided according to the stage the user is at. Other factors (personality, age, etc) need to be taken into account but in this work we will focus on the stage at the healthcare and the ICT level.

Step 3:

Working together with the health professionals we modelled the structure of a medical intervention in the field for an individual placed at each of the stages in the different pathologies studied (6 month healthy-eating plan and 10000-step programme for diabetic patients), arriving to the construct of personalised messages for each case.

Step 4:

After having modelled the intervention we modelled the tool that provides the personalised messages according to the user needs. This tool is a Web-based tool in which messages are provided through different channels according to different variables. As an example, in the 10000 step program all information is delivered through mobile phone, whereas in the healthy-eating programme, information is provided or requested via the PC or the mobile phone according to different variables (stage of healthcare attitude, age, etc). The system provides dynamic tailored information that attemps to progress the individual from a given stage to the next ones.

Step 5:

The final communication strategy was designed and tested with the doctors.

3. Results

3.1. *The model proposed*

A model that takes into account in parallel both the healthcare stage of the individual as well as his/her technical readiness to perform a change using technologies was conceived and designed. The added value of this model is that it allows placing the user at a stage in both dimensions (see Figure 1).

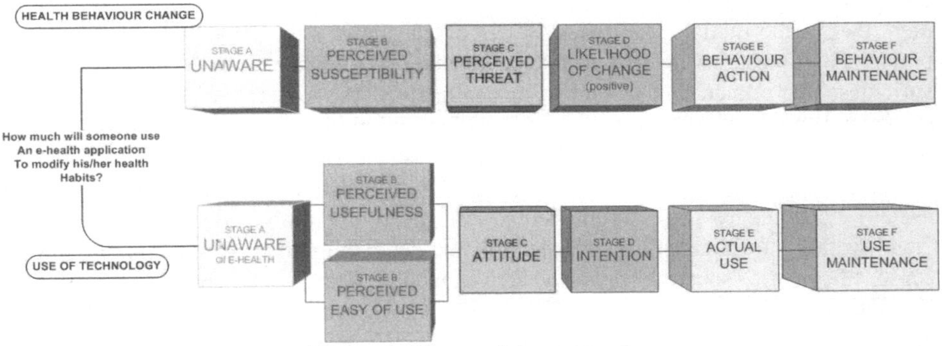

Figure.1. The model proposed

As there are many similarities across the different theories, the upper path of the model (health behaviour change) is an adaptation from an integrative model proposed by Fishbein [18]. The model proposed by this author considers and synthesizes the key variables of the models previously cited [4]. This path is appealing as it reflects the importance of intra-individual, environmental factors and self-efficacy. The main stages of the model are:

- **Unawareness:** an individual never heard about the issue and therefore may have no opinion about it.
- **Perceived susceptibility:** One's opinion of chances of getting a condition
- **Perceived threat:** the belief that one is susceptible to a specific problem
- **Self efficacy:** the belief that one has the ability to change one's behaviour
- **Attitude to change:** making a plan to change behaviour
- **Action:** implementing the plan to change behaviour
- **Maintenance:** continuation of behaviour change

This model supposes that individual's usually pass through the different stages, although the pace may be different for different individuals or different behaviours. Movement backward toward an earlier stage is also possible (although not to stage 1 obviously).

The lower path of the model is an adaptation of the Technology Acceptance Model (TAM) model proposed by Davis [19], [20] that models how users come to accept and use technology. Its stages are similar to the health behaviour model, although adapted to technology. The difference of the stages lies in these two stages [20]:

- **Perceived usefulness:** degree to which someone believes that a system would enhance the job performance.
- **Perceived ease-of-use:** degree to which a person believes that using a system would be free from effort.

3.2. *The questionnaires designed*

In order to be able to place the user in a stage a questionnaire/survey instrument was developed based on the model presented. It took into account the substantive factors influencing the adoption and readiness of individuals to perform a health behaviour change and in parallel to do it by means of an e-health application. Based on the stages identified in the model, two different sets of questions were developed, one for each branch. The questions try to find out the attitude of the user towards the two specific variables: one related to the interest in health (how the user relates to the healthcare stage) and the other related to their motivation and skills to use ICT to promote their well-being (whether the user would be ready to use an e-health application).

Although both stages are linked, two different questionnaires are presented to the user. Questionnaires are implemented in JSP, to allow conducting the user "intelligently" to the most appropriate question depending on previous answers. In this sense, there are several ways to arrive to the same stage but with different characteristics depending on the answers to key questions and these are also taken into account when tailoring the information.

The e-health questionnaire identifies whether users have knowledge, motivation or access, to make use of ICT to make a health change.

The health questionnaire (see Figure 2) assesses whether users have the awareness and the right guidelines to control their health, it they feel susceptible to suffer from that pathology and if users are following any healthcare plan. With these objectives in mind, three specific strategies to encourage self-care can be followed: offering correct and personalised information, helping users use this information and promoting ICT as a means to support healthcare by easing access to quality information and specific services, helping health problems follow-up and making the relation between patients/users and health professionals closer. Figure 2 depicts a simplified version of the health questionnaire. It shows how it relates to the attitude stages towards healthcare and the control to be exerted over it.

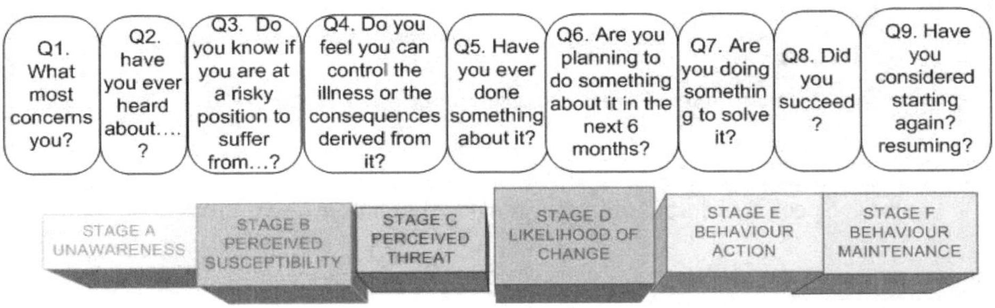

Figure.2. The healthcare attitude questionnaire

3.3. *The intervention selected according to stage*

When users have completed both questionnaires they are allocated to a stage in their attitude towards the healthcare and the use of e-health to achieve a positive change. The process that models the intervention followed by the professional is selected. Figure 3 describes what general information should be provided to a user in ACTION:

GENERAL PERSONALISED INFORMATION ACTION
-Detailed FFQ and PA questionnaires
-DIET HISTORY QUESTIONNAIRE. Environment support & barriers
-Personalised Meal plan & portions, Goal setting, Hints & tips
-Feedback on REAP provided here for awareness raising, BMI. Positive feedback if coherent.
-Focus on restructuring **cues** and **social support**
-Boost self-efficacy for dealing with obstacles
-Combat feelings of loss and reiterate long-term benefits

1. Praise the decision to change behaviour
2. GOAL SETTING. Prioritize behaviour change opportunities
3. Identify and assist in problem solving re: obstacles
4. Encourage small initial steps
5. Encourage identification of social supports

Figure 3. General protocol for an individual in ACTION

3.4. *The algorithm developed*

Two different approaches have been followed to implement the model. Both mobile phone and PC based strategies have been implemented with the algorithms designed in order to allow ubiquity for the delivery of information. The channel, the content, the tone and the format are personalised according to the user's preferences.

3.5. *Personalised information*

According to these stages, the information offered to the user is selected in line with the main features identified, and presented in the most suitable way so that a change to improve his/her health with the use of ICT can be achieved. Picture 4 shows the information for a user in Action stage (this is related to obesity as a pathology).

ACTION	
DAYLY	AGENDA, hints & tips, eaten food/mood register. Enter weight in application Physical activity register
WEEKLY	Enter weight Feedback with progress: Weight loss, improvements, barriers Appointments for face to face visits Physical activity registers. Other anthropometric tests: bioimpedance (fat %), skinfolds, etc Give feedback & identify causes for non compliance with goals. Adjustements made to the plan accordingly
Move to maintenance if behavior maintained 3-6 months	

Figure 4. Specific tailored information offered to a user in ACTION.

3.6. *Pilots to test the model*

In order to test the model developed, a system has been designed and developed, and it will be tested in several clinical pilots in the following months. A short description of the pilots follows:

- **10000 steps a day for diabetic people:** a personalised information system based on SMS (Short Message Service) and MMS (Multimedia Messaging Service) to mobile phones, has been designed and will be tested during the next months. Tailored and motivational messages are provided to the user, who is previously selected to be in either the stages of Perceived threat, Likelihood to change or Action, in order to observe changes produced in the period of monitoring (6 months).
- **Healthy dieting for individuals:** this pilot aims at creating healthy dieting habits in healthy individuals. A Web-Based platform using Flash technology has been developed and will provide customised nutritional and motivational information depending on the user characteristics (health attitude and e-health readiness).

3.7. *Evaluation of the model*

In order to assess the system in a holistic manner, both validation and evaluation need to be considered. Validation assesses the concept of the system, the need of such a system. Evaluation, on the contrary, analyses the system once implemented. It takes into account aspects such as functionality, appropriateness of scenarios, language, etc. Subjective and objective measures will ensure that the evaluation is properly carried out.

The model is currently being evaluated and validated with three different versions of questionnaires, tailored to different audiences, one for the professionals who have provide their expertise and guidance for the design of the system, one for the healthcare professionals (Doctors, Nutritionists …), and the last one for the citizens.

With this set of questionnaires the concept of the model is evaluated, finding out how e-health applications could be more effective and successful. Each question is meant to analyse a different parameter of the model.

The model is currently being tested in several clinical trials and it is being validated and evaluated at different levels and by different stakeholders (experts, professionals and users) and consistent feedback is expected from them all. With their suggestions, a refinement of the model will be proposed. Different web-based tools and applications have been built in order to provide tailored information to users according to their attitude.

4. Discussion and conclusions

Information is critical to health-related decisions and the way this information is selected and processed plays a pivotal role in the decision making of individuals. This model allows the classification of users according to their attitude towards healthcare, and their readiness to adopt the change by the use of new technologies. It provides a framework for the provision of personalised information according to these and other key variables. Besides, the application of this model promotes the empowerment of the individuals, as well as guidance, monitoring, through ICT and it will certainly make an impact on health-related behaviour.

One of its main advantages is that this model proposes a general framework that may be applied to the conception, design and evaluation of any e-health application. It allows enhancing the user modelling process by taking into account both health behaviour aspects as well as technological, which up to this moment have not been considered and could be part of the explanation of e-health underuse. It may assist both healthcare professionals and individuals to have a deeper understanding about both, the provision of healthcare and the delivery channel. However, the inclusion of other dimensions is also necessary in order to be able to effectively reach the individuals.

The model is currently being tested in several clinical trials and it is being validated and evaluated at several levels by different stakeholders (experts, professionals and users) and consistent feedback is expected from them all. With their suggestions a refinement of the model will be proposed. Different web-based tools and applications have been built in order to test the model.

Overall, this framework is likely to provide deeper insights into the process of improving e-health, so it can meet ongoing individuals' needs and become an increasingly valued part of health care services.

5. Acknowledgements

We would like to thank the PIPS project (partially funded by the EC, IST 507019) in which this model will be tested. Besides we would like to thank Ms. Joy Ngo and Ms. Gemma Peralta for all their knowledge and support towards the definition of the model.

References

[1] World Health Organisation, 2005. Facing the Facts: solving the chronic disease problem.

[2] U. S Department of Health and Human Services. The Health Benefits of Smoking Cessation: centers for Disease Control, Centers for chronic Disease Prevention and Health Promotion, Office on Smoking and Health; 1990. Report Nº (CDC) 90-8416.

[3] The Euser Project. eUser – Workpackage 1: Conceptual and Analytical Framework. D1.1: eUSER Conceptual and Analytical Framework (Part A and Part C). European Commission, 6th Framework Programme. Contract number: IST-2002-507180.

[4] Glanz, K , Rimer, B and Lewis FM, 2002. Health Behaviour and Health Education: Theory, Research and Practice. 3rd Edition. John Wiley & Sons. ISBN 0-7879-5715-1.

[5] Brennan P. F., and Safran, C. (2005). Empowered Consumers. In consumer Health informatics. Editors. Lewis D.., Eyenbach, G., Kukafka, R. et al. Springer, 2005. ISBN 0-387-23991-x.

[6] Lorenzi, NM., Riley RT., Blyth AJ., Southon G and Dixon, BG., 1977. Antecedents of the people and organizational aspects of medical informatics: review of the literature. J. Am. Med. Inform. Assoc. 4, pp. 79–93.

[7] Curtis B., Krasner, H. and Iscoe N. 1988. A field study of the software design process for large systems. Commun. ACM 31, pp. 1268–1286.

[8] Kukafka, R., Johnson S. B., Linfante, A., and Allegrante, J.P., 2003. Grounding a new information technology implementation framework in behavioral science: a systematic analysis of the literature on IT use, Journal of Biomedical Informatics, Volume 36, Issue 3, June 2003, Pages 218-227.

[9] Kukafka, R., (2005). Tailored Health Communication. In consumer Health informatics. Editors. Lewis D.., Eyenbach, G., Kukafka, R. et al. Springer, 2005. ISBN 0-387-23991-x.

[10] Kreuter M., Farrell D, Olevich L, et al. (2000). Tailoring Health Messages: Customising Communication with Computer Technology. Mahwah, NJ: Lawrence Erlbaum, 2000.

[11] Vance Wilson, E., Lankton, N.K., (2004). Modeling Patients' Acceptance of Provider-delivered E-health. Journal of the American Medical Informatics Association Volume 11 Number 4 Jul / Aug 2004.

[12] Del Hoyo-Barbolla, E., Fernandez, N., Ramirez, C. M. Tortajada, D., Ngo de la Cruz, J., Arredondo, M. T., 2005. Personalised training tool using Virtual Reality. Proceedings of the 1st workshop on personalisation for e-health. 10th International Conference on User Modelling. Edimburgh, UK.

[13] Fishbein M, Ajzen I, 1975. Beliefs, Attitudes, Intention, and Behavior: an Introduction to Theory and Research. Reading, MA: Addison-Wesley.

[14] Ajzen I, Fishbein M. Understanding Attitudes and Predicting Social Behavior. Englewood Cliffs, NJ: Prentice-Hall, 1980.

[15] Fishbein M. A theory of reasoned action: some applications and implications. In: Page MM, ed. 1979 Nebraska Symposium on Motivation. Lincoln, NE: University of Nebraska Press, 1980.

[16] Prochaska JO, DiClemente CC. Stages and processes of self-change of smoking: Toward an integrative model of change. J Consult Clin Psychol 1983;51:390 -5.

[17] Weinstein, N. D., and Sandman, P. M., 1992. A model of the Precaution Adoption Process: Evidence from home Radon testing. Health psychology, 1992, 11, 170-180.

[18] Fishbein, M., 2000. The role of Theory in HIV prevention. AIDS care, 12 (3), 273-278.

[19] Davis, F., Bagozzi, R. P., & Warshaw, P. R. (1989). User acceptance of computer technology: A comparison of two theoretical models. Management Science, 35, 982-1003.

[20] Davis, F, 1989. Perceived usefulness, perceived ease of use, and user acceptance of information technology. MIS Quarterly, 13(3), 319-340.

4.2 eHealth: Architectures and Strategies

Ubiquity: Technologies for Better Health in Aging Societies
A. Hasman et al. (Eds.)
IOS Press, 2006

E-Health Approach to Link-up Actors in the Health Care System of Austria

Thomas SCHABETSBERGER[a,1], Elske AMMENWERTH[a], Ruth BREU[b], Alexander HOERBST[b], Georg GOEBEL[c], Robert PENZ[f], Klaus SCHINDELWIG[d], Herlinde TOTH[e], Raimund VOGL[f], Florian WOZAK[a]

[a] *University for Health Sciences, Medical Informatics and Technology, Hall, Austria*
[b] *Leopold-Franzens-University Innsbruck, Innsbruck, Austria*
[c] *Innsbruck Medical University, Innsbruck, Austria*
[d] *Tiroler Landeskrankenanstalten GmbH, Innsbruck Austria*
[e] *Wiener Krankenanstaltenverbund, Wien, Austria*
[f] *Health Information Technologies Tyrol, Innsbruck, Austria*

Abstract. "Electronic health services are important" the EU commission stated in the E-Health action plan. By these means access to health care can be improved and the quality and effect of the offered medical services can be increased. By introducing the e-card in Austria, an overall link-up of nearly all health service providers of the external sector (e.g. family doctors) was achieved. In 2005 the Austrian E-Health Initiative (EHI) of the Austrian Federal Ministry for Health and Women mapped out a strategy to organise the development of the health system towards an integrated patient-centred. Hereby the electronic health record (EHR) plays a decisive role. The aim of this study is to analyse requirements for a virtual, cross-institutional and patient-centred electronic health record from the point of view of the exemplary main actors (Doctor and Patient), to define conditions, and then to evaluate the thus derived, specific concept of implementation. Aside from the two main actors regarding medical acts, namely the institution treating a patient (e.g. doctor, paramedic or nurse) and the patient receiving treatment, a row of other actors could be identified. Group assessment techniques with representatives of these actors resulted in an overview of required functions of an EHR. As a proof-of-concept an information system architecture conformable to the IHE XDS architecture for cross enterprise document sharing is currently being constructed and evaluated in the course of a pilot-project. If the core architecture fulfils the expectations, then a further extension to other hospitals and resident doctors, and subsequently also to the other actors of the health system, is planned. Since both legal and socio-technical requirements are presently not yet entirely met, and since there are also deficits from a methodical viewpoint, a complete implementation and widespread introduction will be a long term goal.

1. Introduction

"Electronic health services are important" the EU commission stated in the E-Health action plan. By these means access to health care can be improved and the quality and effect of the offered medical services can be increased [1]. By introducing the e-card, an overall link-up of nearly all health service providers of the external

[1] Corresponding Author: Dr. Thomas Schabetsberger, University for Health Sciences, Medical Informatics and Technology, Eduard Wallnoefer-Center 1, A-6060 Hall/Tyrol, Austria. E-Mail: Thomas.schabetsberger@umit.at, Web: http://iig.umit.at.

sector (e.g. family doctors) was achieved, and therefore the technical infrastructure for subsequent projects was created [2].

The processing and storage of medical data is currently however mostly geared towards the respective individual, institutional demands [3]. The transmission of data between various health service providers takes place in a provider-oriented way in the sense of a directed communication; a large part of this communication is paper-based, a smaller part is electronic via Fax, encrypted and signed E-Mail or secure Web [4-6]. From literature study it is known, that a change of this provider-oriented, directional transmission of reports and findings towards a more patient-centred provision of reports and findings would support cross-institutional information-processing, and would therefore also improve the quality and efficiency of the health system and increase the safety of medical treatment and compliance [7]. In 2005 the E-Health Initiative (EHI) of the Austrian Federal Ministry for Health and Women compiled a strategy to organise the development of the health system towards an integrated patient-centred care on a long-term basis [8]. Hereby the electronic health record (EHR) plays a decisive role. The electronic health record summarizes health-related data of an individual, acquired continuously from either outpatient or in-patient treatment, and which is saved on a digital storage medium. The most important distinguishing feature of the electronic health record in comparison to the electronic patient record is the sole control of the patient over his health record and therefore his medical data. The patient alone decides who may view and use information from his health record. Presuming that, independent of the electronic health record of an individual, there are many electronic patient records in the hands of doctors and hospitals existing parallel to each other, the EHR should constitute a higher ranking entity over all these patient records at the health care providers and should also integrate these single and distributed patient records [7, 8].

The aim of this study is to analyse requirements for a virtual, cross-institutional and patient-centred health record from the point of view of the exemplary main actors (Doctor and Patient), to define conditions, and then to evaluate the thus derived, specific concept of implementation in the course of a pilot project existing between the „Tiroler Landeskrankenanstalten“ (TILAK) and the „Wiener Krankenanstalten-verbund“ (KAV) – two major hospital holding companies from Eastern and Western Austria.

2. Methods

The realization of this study and of the general project-management is based on the 5-step-method for management of health information systems. Through systematic literature analysis and in workshops actors were identified, who are currently already exchanging data in the health service in a directed way, and who would profit as expected by a shared electronic health record in the sense of quality-improvement or financial savings. The next step was to draw up a functional requirements profile from each of the actors own viewpoint by using creativity-methods, these were then prioritized and the requirement profile was then iteratively discussed and revised with representatives of the identified actors (Delphi method). With the help of scenario techniques future scenarios were designed. Based upon this the core architecture of a distributed, virtual health record was developed by applying software-engineering methods (experimental prototyping), prototypically implemented and evaluated.

3. Actors in a cooperative health care system

Aside from the two main actors regarding medical acts, namely the institution treating a patient (e.g. doctor, paramedic or nurse) and the patient receiving treatment, a row of other actors could be identified (Fig. 1). This includes insurers, who correspondingly remunerate the medical service rendered by the person giving the treatment, and who therefore must have access to certain areas of medical documentation relevant for the account statement. Pharmacies redeem medical prescriptions (paper-based or electronic) issued by doctors and communicate these to the insurers for the account settlement. A line of communication leading from pharmacies to doctors is currently not yet established, but could play a more important role in the future (e.g. feedback to the treating institution about whether or not medications were collected). At the same time it is presumed, that institutions treating patients, insurers, national institutions or medical researchers would benefit greatly from anonymous and consolidated data in order to regulate the health system and to improve the quality of medical care.

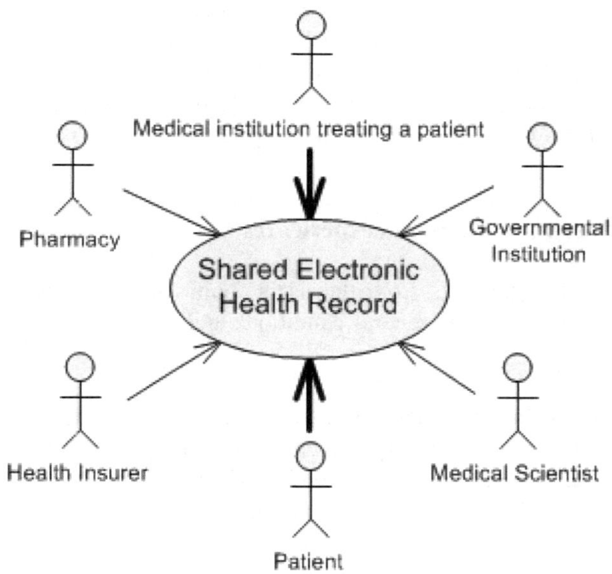

Figure 1: Schematic description of identified actors in the health system. Aside from the main actors (medical institutions treating a patient, bold type arrows) there exist a row of other actors in the health system. It is presumed, that with the help of a common database in the sense of a distributed (virtual) electronic health record, the cooperation amongst the actors could be improved and hence a lowering of costs and a quality increase could be achieved. For more information please see text.

4. Actors functional demands of an electronic health record

Several workshops and interviews with representatives of the identified actors resulted in a list of functional demands. In the following, patients and institutions treating patients are exemplary viewed as main actors and their functional demands are presented in a sequence of stated importance. With exception of the pharmacies (eprescription) all other actors are predominantly interested in non-personal, anonymous and consolidated data:

Requirements from the patient's viewpoint are:

- Guarantee of the highest possible degree of data privacy protection and security
- Access to one's own health record and control over access rights
- Medical contents should be adapted for patients
- Possibility to add personal entries and to keep medical diaries

Institutions treating patients (e.g. doctors) see the essential advantages above all, if

- Access to information relevant for treatment (medical history) can occur promptly, irrespective of the institution where this information was generated.
- The presentation of the data is adapted to the respective special field of the doctor, so that essential information can be comprehended in a short time. The possibility of retrieving an emergency data set (e.g. list of medication and diagnoses, allergies, blood group) in case of an emergency would be helpful.
- Emergency access could be carried out without explicit consent of the patient but with documentation in his medical file (the patient must later be notified of this)
- Charges can be electronically transferred to the health insurers
- Medications can be prescribed electronically (e-prescription) and if the pharmacy can then confirm whether a medication was picked up or not.

5. General conditions, requests and experiences

In addition to the demands from the users' point of view there exist technical, organizational, legal and social general conditions and requests, as well as a series of doubts voiced by users:

- Guarantee of highest organizationally and technically possible data security, high scalability/extensibility and high availability, avoidance of a „single point of failure“ or a „single point of attack“ respectively.
- Consideration/Integration of existing infrastructure and request of distributed data storage, cost-saving operation (no additional data processing centers)
- Doctors concerns regarding flood of information
- Strong request for an „Open Source“ solution
- Doubts regarding the „transparent patient“ or also the „transparent doctor“
- Non-existence of a common patient-index as well as no generally accepted common standard (neither on a communication level nor in content) are regarded as especially big challenges.
- 4-step-model for trans-institutional inquiries of medical findings as a legal issue.

The 4-step-model for trans-institutional inquiries of medical findings has been worked out and offers an area of certainty of the law and with it a legal requirement for automated, reciprocal medical inquiries amongst health service providers. The basis for retrieving information from another health service provider is the documented written consent of the patient. This is agreed upon between the communication partners in form of a contract (controlled by means of random checks). In the case of an inquiry the first step amongst the communication partners is the identification of the patient. If this allocation could be made, then a list of these patients' visits at the concerning health service provider can be requested and transmitted. In a third step the summarizing document pertaining to a certain visit, e.g. in form of a doctors or patients letter, can now be requested and transmitted. In step 4 further medical data (e.g. related reports or images) can be requested and transmitted (Fig.2.).

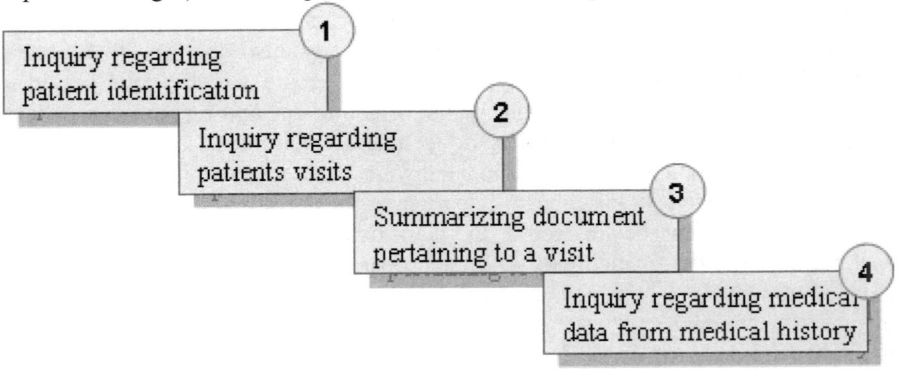

Figure 2: Graphic description of the 4-step-model for medical inquiries. Through fully- or partially automated sequential step-by-step-handling of these four levels a reciprocal medical inquiry between health service providers under preservation of data protection regulations can be realized. For closer information see text.

6. Prototype of a trans-institutional information-system core-architecture as the basis of a distributed electronic health record

As a proof-of-concept an information system architecture conformable to the IHE XDS architecture for cross enterprise document sharing is currently being constructed in a simpler way in the course of a pilot-project. Partners in this pilot-project are the TILAK and KAV (together approx. 40.000 employees and 500.000 inpatients per year). In both institutions web portals are created which provide for a patient search, their hospital stays and accompanying reports becomes possible in the respective other institution. In order to provide electronic reports and findings for the assigning resident doctors the TILAK already operates a web portal. This web portal is improved by adding appropriate input masks for entering necessary demographic data and visualizes the results of the queries. The completion of the described prototype is planned at the end of the first quarter of the year 2006. As of this time the following transactions will be possible:

- Doctors employed by the TILAK or KAV will have limited access on a trial basis to findings and reports on patients from each other.
- Resident doctors who are currently already participating in the web portal of the TILAK will have limited access on a trial basis to findings and reports available from their patients in either Tirol or Vienna.

7. Discussion and Outlook

The outlined difficulties of conventional transmission of reports and findings between health service providers are being discussed internationally by medical institutions, research institutions and lastly also by political institutions. In Austria there exists a vision of an extensive, trans-institutional electronic health record [8]. The implementation by means of the described prototype is currently being evaluated. If the core architecture fulfils the expectations, then a further improvement and extension to other hospitals and resident doctors, and subsequently also to the other actors (including the patient) of the health system, is planned and will be done within a project called health@net [9]. Although the list of conditions and requirements might not be complete. Since both legal and socio-technical requirements are presently not yet entirely met, a complete implementation and widespread introduction will be a long term goal. The arising questions concern in particular the uniform definition of relevant contents in the electronic health record, the clarification of standards for data filing, and last but not least the financing of set-up and operation. Together with its research- and cooperation-partners however, the health@net project consortium participates in initiatives on a national and international level, whose aim is the creation of those requirements needed to implement and operate a patient-centered, universally accessible and secure electronic health record.

References

[1] Council of the European Union. e-Europe Actionplan 2004. [homepage on the Internet] 2004 [cited 2006 Jan 27]; Available from: http://europa.eu.int

[2] Austrian Federal Ministry for Health and Women. The health system in Austria. [homepage on the Internet] 2006 [cited Jan 02]; Available from: http://www.bmgf.gv.at/cms/site/attachments/8/6/6/CH0083/CMS1051011595227/gesundheitswesen_in_oesterreich_2005_internet.pdf

[3] Masys D, Baker D, Butros A, Cowles KE. Giving patients access to their medical records via internet: The PCASSO experience. J Am Med Inform Assoc. 2002;9:181-91.

[4] Moorman PW, Branger PJ, van der Kam WJ, van der Lei J. Electronic messaging between primary and secondary care: a four-year case report. J Am Med Inform Assoc. 2001 Jul-Aug;8(4):372-8.

[5] Bergmann J, Bott OJ, Hoffmann I, Pretschner DP. An eConsent-based System Architecture Supporting Cooperation in Integrated Healthcare Networks. Stud Health Technol Inform. 2005;116:961-6.

[6] Simons WW, Mandl KD, Kohane IS. The PING personally controlled electronic medical record system: technical architecture. J Am Med Inform Assoc. 2005 Jan-Feb;12(1):47-54.

[7] Ball MJ, Douglas JV. Redefining and improving patient safety. Methods Inf Med. 2002;41(4):271-6.

[8] Austrian E-Health-Initiative. Austrian E-Health Strategie (in German). [homepage on the Internet] 2005 [cited Dec 05]; Available from: http://ehi.adv.at

[9] Health@Net consortium. health@net project website (in German). [homepage on the Internet] 2005 [cited Dec 05]; Available from: http://healthnet.umit.at

Ubiquity: Technologies for Better Health in Aging Societies
A. Hasman et al. (Eds.)
IOS Press, 2006

Architecture for National e-Health Infrastructure in Lithuania

Egidijus ANDRIUSKEVICIUS [a], Evaldas DOBRAVOLSKAS [b], Vytenis PUNYS [c,1], Vita SINCIENE [b], Aras VALENTINAVICIUS [a]

[a] *Hewlett-Packard Lithuania, Vilnius, Lithuania*
[b] *Ministry of Health, Vilnius, Lithuania*
[c] *Kaunas University of Technology., Kaunas, Lithuania*

Abstract. Building a patient centered National e-Health system has been a challenging task for the Health Ministry of Lithuania in 2005-2006. The first and the main task is to design and to build the infrastructure of the National e-health system. The paper presents the architecture of the system, which aims: (a) to integrate already designed and implemented information systems, (b) to communicate with other national information systems, (c) to provide services (basic ones at the initial stage) for users and (d) to support further e-health systems development in long-term perspective. The system design is based on the HISA approach. The core of the system consists of on-line data storage (including classifiers and registries), of the automation and integration layers for health care processes, of a security layer, a system control layer and a client layer. After having defined the concepts of processes and activities, of the security, of the registries and classifiers within the system and the Electronic Health Records System, then follows the implementation stage. Until the end of 2006 it will include the core of the national e-Health infrastructure and the part of the information system providing basic services for the pilot Lithuanian healthcare institutions.
Keywords: Health Information Infrastructure, National Index of Health Records, Electronic Health Record System

1. Introduction

The e-Health development in the new EU Member States has been started from a favorable position: it is possible to use the experience of similar systems developed and implemented in Europe, the US and elsewhere in the world for more than 20 years. Well-known examples are the National Program for IT in Healthcare (NPfIT) for the whole UK [1], and the INSALUD project in Spain [2]. The development of systems has been supported by activities of multitude vendors of medical equipment and IT solutions, and international organizations, e.g. European Federation for Medical Informatics (EFMI), American Medical Informatics Association (AMIA), and International Medical Informatics Association (IMIA). That fostered the active development of international standards for health informatics by relevant bodies. Most known of them are CEN/TC251 in Europe, HL7 and NEMA in the US, ISO TC215,

[1] Corresponding author: Vytenis Punys, Image Processing & Analysis Laboratory, Kaunas University of Technology, Studentu str. 56-304, LT-51424 Kaunas, Lithuania. E-mail: vytenis.punys@ktu.lt

and the initiative of users and vendors to provide methodology on "Integrating the Healthcare Enterprise" (IHE).

Having considered the experience of development and implementation of large scale IT systems [2], [3] (their scale, their division into components, their implementation strategies and corresponding experiences), which were presented to the public during the various events, like international scientific congresses Medical Informatics Europe (MIE), MedINFO, Computer Assisted Radiology and Surgery (CARS), the national-wide core of e-Health information system has been selected as the rational approach in providing cost effective and patient centered healthcare services for 3.5 million of the Lithuanian inhabitants.

2. Materials and methods

The Lithuanian Ministry of Health (MoH) assessed the IT situation in Lithuanian healthcare in 2000 and 2003. On the base of this information the e-Health implementation strategy was prepared by) as a set of political, organizational and psychological efforts to achieve better quality healthcare using information technologies [4]. The objectives of the strategy are:

- to improve population access to clear and reliable information on healthy way of living and health hazards;
- faster to deliver quality healthcare services to patients and to improve access to clear and reliable information on diseases and treatment;
- to improve professional expertise of medical personnel and instant access to reliable and comprehensive information concerning patient.
- to improve effective solutions' support measures and to develop information exchange system between healthcare institutions;
- to improve capabilities of leadership and administrators to obtain reliable information for management and planning purposes.

According to the strategy [4], a feasibility study for Lithuanian e-Health system development and the implementation roadmap was developed by the MoH [5]. The following factors were identified as critical ones for a successful implementation of the information and telecommunication technologies (ICT) in the health care system:

- In solutions being implemented, the conformity to the health informatics standards should be required to ensure openness and interoperability of the system. Starting the system development from the national level, defining standards and requirements, complicated integration tasks will be avoided in the future.
- Development of the national system should take into account the fact that healthcare institutions do have or will have their own systems, for which tools based on health informatics standards should be foreseen in order to ensure integrity of national and institutional systems and exchange of information between institutions and specialists.
- Good connection with employees of healthcare institutions should be established, and their needs should be met. Participation of physicians and solutions oriented towards their needs, special attention to analysis of business processes and active participation of all stakeholders in identifying their needs and introducing corresponding solutions should be ensured.

- The existing solutions in hospitals should be utilised, and the acquired experience should be applied efficiently.
- To enable companies to provide the services of e-Health systems development and installation in Lithuania, thus to create a competitive health information technologies market in Lithuania, as well as to provide guidance in paying special attention to experience, capacities and abilities of prospective contractors.

The roadmap divides the implementation of the e-Health system into different levels (like national, regional and local) and steps:

1. Creation of the national e-Health core with services ensuring the information exchange;
2. Set-up of the national level of e-Health system and implement **four pilot functions**, focused on Primary Healthcare level automation:
   - registration of patient visits, exchange of clinical and administrative data, which will provide a possibility to accumulate electronic health record;
   - drafting and sending referrals for consultations/treatment as well as entering and receiving the results;
   - drafting and sending of referrals for diagnostics (laboratory, imaging, etc.) as well as receiving and storing the results;
   - registration of appointments for consultation, treatment, tests, etc.
3. Implementation of **remaining functions** of the national level e-Health system:
   - receipt, archiving and exchange of medical images (e.g. radiological);
   - creating and sending e-prescriptions. This functionality will support a physician in prescribing the drugs, ensuring exchange of information with pharmacies and providing information on prescribed medicaments compensation by the Sickness fund;
   - using templates in order to create medical documents will facilitate physician's work. Methodological support during the process of treatment will improve quality of services provided;
   - drafting and sending statistical, public health monitoring reports, will facilitate activity of healthcare institutions and staff members. Analysis of accumulated information will provide the basis for more effective activity and decision making;
   - emergency medical treatment will provide possibilities to improve the quality of current emergency medical treatment services;
   - telemedicine, as provision of remote medical services, will facilitate more effective exchange of information and improve the quality of services provided.
4. Deployment of e-Health system functions in healthcare institutions (HCIs), automation of HCIs internal functions.

By designing the National e-Health core all necessary functionalities required by the following e-Health system development phases should be taken into account.

3. Results

While constructing the Lithuanian National e-Health system (LNHS) core was taken into account the exchange of information within the System and/or with the related systems interlinked to it; a basis for the operation of the functions of the e-Health system at the national level; the possibilities of expanding the functionality of the System; providing the identification of patients, medical and administrative personnel and implementing principles of granting access to the information contained in the LNHS, supporting the registers and classifiers of the main entities in the e-Health system.

Due to complex set of goals to achieve, the design of the LNHS core was performed according the Healthcare Information Systems Architecture (HISA) [6] structured approach. Both conceptual and logical specifications of individual and common services, related to the treatment of subjects of care, have been created. The services between different layers (application, middleware, bitways) have been classified. The concepts of healthcare processes, of integration of existing/future information systems, of security, of registries/classifiers and of electronic health record system (EHRS) have been set out and the system architecture has been designed.

3.1. System architecture

The architecture consists of several layers (Figure 1), each of them performs specific functions:

- The **Client Layer** acts as an interface between a user and available National e-Health System Services. For the pilot implementation, access to four main services will be available, with the full functionality, having on-line connection to the central system and minimal standalone functionality when such connection is not established.

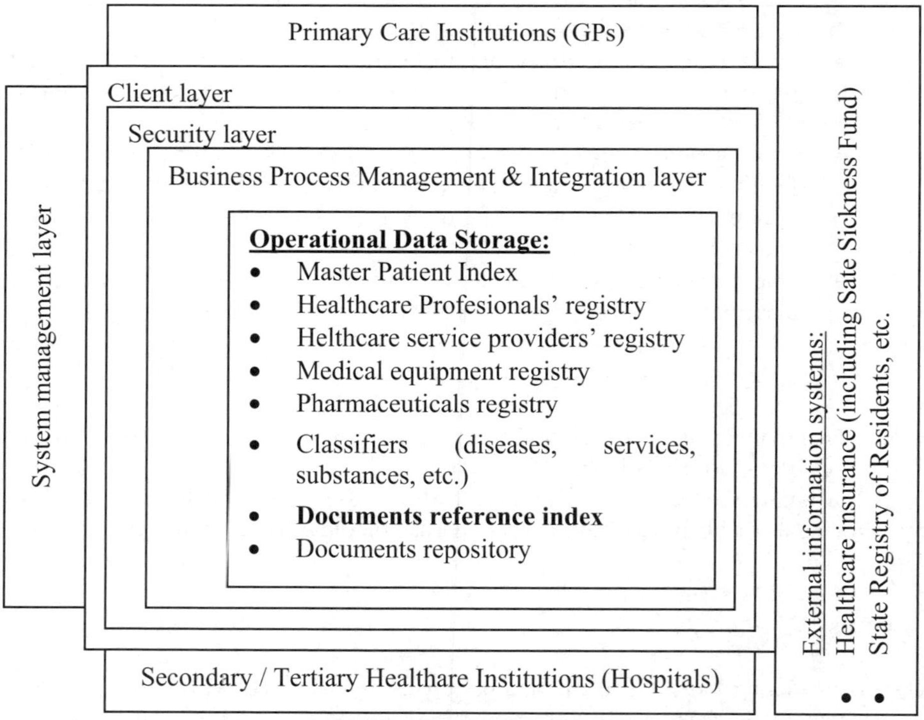

Figure 1. Architecture of LNHS core

- The **Management Layer** is responsible for overall solution network, system and service management based on the IT Service Management and IT Integration Layer framework.
- The **Security Layer** is responsible for authorization, authentication, and session control and audit services.
- The **Business Process Management & Integration Layer** is responsible for description, maintenance and execution of the e-Health business processes. It is also used for co-ordination of the technical services provided by Security Layer and Operational Data Store, and integration of the National e-Health system with the external systems (health insurance information systems, including State Sickness Fund IS "SVEIDRA", State registries of Residents, healthcare information systems, etc.) using industry standard and proprietary protocols.
- At the center of the LNHS is the **Operational Data Storage** (ODS). The ODS exposes its capabilities internally through technical services; access to these services is channeled through the combined Business Process Management & Integration Layer. The technological foundation of the data store is an Oracle database running a HL7 v3 RIM [7] compliant the electronic health record database schema and a set of the internal registries and classifiers (e.g. ICD-10, SNOMED-CT, etc.).

3.2. Operational Data Storage

The ODS includes the main registries of the system, like Master Patient Index (MPI), healthcare professionals' registry, services' providers registry, medical equipment registry, pharmaceuticals registry, etc. It provides infrastructure and services for using classifiers necessary in healthcare IT (e.g. ICD-10 for diseases, some proprietary ones for services, and SNOMED-CT in future).

Electronic Health Record System (EHRS) forms a part of the ODS. The EHRS consists of the Documents Reference Index and the Document Repository.

All the actions within the LNHS core (like patient visits, referrals to specialists, lab tests, etc.) and their results are going to be referenced in the Documents Reference Index, which stores information where the source document is stored – either in the Document Repository of the ODS, or in the external system from where it might be retrieved if certain security and data access rules are obeyed.

3.3. System security

Based on the security concept the users' roles, authorization, authentication, and session control rules have been set out. The conditions and rules for health information secure storage and secure exchange between the LNHS and information systems of healthcare institutions have been set out. The audit services covers all transactions within the LNHS. The security sub-system provides:

- Complete web-based single sign-on,
- Access and principal management for users and devices,
- Automated network and resource discovery,
- Flexible delegated administration,
- Support for SAML (Security Assertion Markup Language),

4. Conclusion

The Lithuanian National e-Health System Architecture has been developed by the extensive use of international health informatics standards and international experience of similar systems development. The core of the system provides security and integration services within Lithuanian healthcare IT domain, basic services for the users at the initial stage and enables further development of advanced services in future. It is planned to provide basic services for the pilot Lithuanian healthcare institutions at the end of 2006.

Acknowledgements

The Authors express their gratitude to dr. Kai U. Heitmann for his kind advices, consultancy and support while developing the National e-Health System Architecture.

References

[1] National Programme for IT in the NHS. (last accessed May.21, 2006) http://www.connectingforhealth.nhs.uk/
[2] Gomez-Meana A. Technology and New Healthcare Information Systems. Proceedings of the Baltic IT&T 2004 forum: eBaltic, LITTA, Riga, Latvia, 2004
[3] Bywater M . The NPfIT and England's healthcare-IT market: an assessment at year four. Br J Healthcare Comput Info Manage 2006; 23(3): 20-22.
[4] e-Health Strategy for Lithuania, 2005-2010. Vilnius. (last accessed Jan.27, 2006; in Lithuanian) http://www.sam.lt/images/Dokumentai/eSveikata/esveikata_strategija_web020.doc
[5] Feasibility Study of the e-Health system development in the healthcare sector of the Republic of Lithuania. (last accessed Jan.27, 2006; in Lithuanian) http://www.sam.lt/images/Dokumentai/eSveikata/gs_esveikata_web_version.pdf
[6] HISA. Healthcare Information System Architecture Part 1 / Healthcare Middleware Layer, CEN/TC 251, Final Draft 2, prENV 12967-1:1997
[7] HL7 version 3 Interoperability standards. Normative edition 2005 (CD). HL7 Inc., 2005

Ubiquity: Technologies for Better Health in Aging Societies
A. Hasman et al. (Eds.)
IOS Press, 2006

A National EHR Strategy Preparedness Characterisation Model and its application in the South-East European Region

Leonidas ORFANIDIS[b,c], Panagiotis BAMIDIS[a], Barry EAGLESTONE[c]

[a] *Medical Informatics Laboratory, School of Medicine, Aristotle University of Thessaloniki, Greece*
[b] *South East European Research Centre (established by the University of Sheffield and CITY Liberal Studies), Thessaloniki, Greece*
[c] *Department of Information Studies, University of Sheffield, UK*

Abstract. This paper is concerned with modelling national approaches towards electronic health record systems (NEHRS) development. A model framework is stepwise produced, that allows for the characterisation of the preparedness and the readiness of a country to develop an NEHRS. Secondary data of published reports are considered for the creation of the model. Such sources are identified to mostly originate from within a sample of five developed countries. Factors arising from these sources are identified, coded and scaled, so as to allow for a quantitative application of the model. Instantiation of the latter for the case of the five developed countries is contrasted with the set of countries from South East Europe (SEE). The likely importance and validity of this modelling approach is discussed, using the Delphi method.

Keywords National Electronic Health Records, South East Europe, characterisation Model, Preparedness

1. Introduction

Developing and establishing national electronic health record systems NEHRS is a key strategic objective in many developed countries like the UK, Australia, Canada and New Zealand as it appears for the substantial budgets allocated to major NEHRS programmes [1],[2],[4],[5]. However, initial progress of such programmes is not encouraging [3]. Programmes have encountered delays caused by both organisational and technical factors. In particular, a number of specific issues are proving to be problematic, since they seem to play a role in enabling or inhibiting such programmes. Our research has identified these problems faced by developed countries. Inevitably the above problems lead one to question the feasibility of such NEHRS initiatives for less developed countries, such as those in South Eastern Europe (SEE).

The latter question is addressed by the research presented in this paper. Specifically, it has been attempted to analyse strategies that can be adopted by SEE nations, through the development of a model is derived from the NEHRS experiences of a sample of developed countries. This model attempts to characterise individual nations in terms of those factors which determine its readiness for an NEHRS, and thus, provides an instrument with which to investigate the preparedness for such programmes of each SEE nation. Thus, the model is proposed as a method with which to identify aspects upon which a SEE nation should focus in devising a strategy towards an NEHRS which will ensure high quality healthcare data. This paper is organised as follows. Section 2 presents the research methodology that was followed. The model framework derived through this study is described and discussed in section 3. In section 4, the model is instantiated for the SEE countries and the set of developed countries showing the difference between the NEHRS readiness of the two sets of countries. Section 5 includes a description of the validation method for the Characterisation Model using the Delphi approach.

2. Material and Methods

The present investigations employ a qualitative research methodology to investigate the social phenomenon of NEHRS development in highly developed countries and the possible introduction of this technology to other countries. The research is therefore interpretive, and consequently aims to understand the context and main concerns (issues) relating to NEHRS strategy, with the aim of determining how NEHRS strategy influences or is influenced by its context and related concerns. The research is presented here as a case study focused in SEE countries as opposed to the set of developed countries. This methodological approach aims to inductively identify factors that influence the NEHRS development and their implications for healthcare data quality.

3. The Model Framework

A model framework definition is attempted in order to characterise the national preparedness for development of an NEHRS, as derived primarily from governmental reports concerning NEHRS programmes of the sample of developed countries. Thus, the main inputs are the concerns and barriers, apparent in this literature, and faced during the development and application of NEHRS strategies, by the UK, Australia, Canada, the USA and New Zealand.

3.1 Important factors as defined by five (5) developed countries

Standards (A1)

The development, adoption and enforcement of standards to cover all aspects of NEHRS emerged strongly as a necessary prerequisite to an effective NEHRS. In fact, this factor was a main theme in the NEHRS strategy documents of all of the nations in the sample. Key categories and the associated standards, which emerged from the publications reviewed, are tabulated in Table 1. In the model of this paper, it is attempted to quantify this factor as a measure of the scope of the standards adopted by a nation. By consulting Table 1, a metric is computed as the number of categories within which a nation has or plans to adopt one or more of those key standards. The minimum grade of this metric is 1, valid when no standards from none of the categories or from only one category has or is planned to be adopted. The maximum grade of this unit metric is 7 where at least one of the standards from each category has or is planned to be adopted. The rationale of the scale is discussed later in this paper, but the actual intention is that the result of this calculation will provide an indication of the infrastructure (readiness) of each country with respect to health informatics standards and conversely the extent of those areas not covered by nationally adopted standards.

Category of Standards	Name of the Standard
Medical Terminologies and Classification Standards	SNOMED, Read Clinical Terms, UMLS, GALEN, MEDCIN, CPT-4, LOINC, ICPC-2, ICD-10
Message format standards	HL7, CDA, CEN/TC 251 ENV 13606-Part 4
Digital image standards	DICOM
EHR content and structure	OpenEHR, ASTM, HL7 RIM, CEN/TC 251 ENV 13606-Part 1,2,3
Secure exchange, storage and access standards	EDIFACT, EbXML, e-GIF
Unique identifiers standards	Unique identifiers
Member of ISO/TC215 or CEN/TC215	ISO/TC215 and CEN/TC215

Table 1: Categories of the necessary health standards for the development of an NEHRS

Health network architecture (A2)

The type of established or planned network facility determines to a large extent the means by which an NEHRS must capture, store and communicate patient-related health information. Specifically, the network infrastructure facilitates or impedes interoperability between relevant information systems, and determines levels of data security that can be achieved, as well as, the ease of access. The national ICT infrastructure seems to play a key role in the selection of the appropriate technologies for the health network architecture. In particular, the availability of broadband Internet access across the country, together with the computer literacy of citizens and healthcare personnel appear to be an important factor in the selection of appropriate NEHRS solution, since these solutions must be adapted to the needs of the users and the capabilities of the national infrastructure. In particular, interconnection of all healthcare systems within a country, in both public and private health sectors, appears to be an aspiration for developed countries with a legacy of existing heterogeneous health information systems. In order to model and quantify this aspect within the model framework, five measures (subfactors) have been identified, each of which are available from statistics periodically collected by the World Bank (2003). The selected metrics are: ▪**A2.1**: Access to Broadband Internet, measured on a scale of 1 to 7, with 1 denoting no broadband access and 7 denoting wide availability. The World Bank uses a complicated function in order to provide this metric from 1 to 7. ▪**A2.2**: The number of telephone lines per 1000 citizens. In order to achieve normalisation of the metric the 1000 is uniformly divided into 7 equal ranges, in order to keep the same normalisation with other metrics of the model (i.e. metric A2.2 ranges from one (1) (1 - 142,9) to seven (7) (857.4-1000). The same grade normalisation is applied in the next three sub metrics. ▪**A2.3**: The number of personal computers owned per 1000 citizens (1 to 7 scale). ▪**A2.4**: The number of Internet users per 1000 citizens. (1 to 7 scale). ▪**A2.5**: The number of mobile phones per 1000 citizens. (1 to 7 scale). The above metrics are indicative of the level of the existing national ICT infrastructure. In particular, they determine the availability of the means to access electronically stored healthcare data.

Legislation Framework (A3)
A legislation framework is necessary for the provision of data privacy. Regulations and laws are necessary in order to define the kind of access that each kind of user will have. Moreover, legislations are necessary for the supervision of the personal health data. The laws for health data privacy are the same for the electronic commerce, digital signatures and consumer protection. The World Bank is a notable data statistics source. For legislation, they annually provide measures for the efficacy of the laws on the above areas of electronic commerce, consumer protection and digital signatures, under the specific term "Laws related to ICT use". These laws are measured on a scale of 1 (minimum) to 7 (maximum) according to the efficacy of the laws (something which was adopted in our model for the rest of the factor metrics as well). A rating of 1 means the laws are nonexistent; a rating of 7 means that the laws are well developed and enforced. These annual measures of a large number of countries on identical grading scales, enable country comparisons (World Bank, 2003) they are used as unit metrics for our model.

Percentage of GDP for Health (A4)
This aspect has not been identified in any NEHRS Plan of the five developed countries. However, it is deemed to be useful in order to present the importance and emphasis put by the various countries on the health sector. When the percentage of the GDP spending for health by a specific country is high, then this country has got the funds for the NEHRS implementation. A4 shows the amount of money that each country spends for healthcare in comparison to the whole GDP. Figures are usually presented with a percentage. For example, in 2001 Canada spent 9.5% of the Canadian GDP on health (World Health Organization, 2003). Research has shown that no country spends more than the 14% of the GDP on health. For the normalisation of this metric on a scale from 1 to 7 the very maximum found (i.e. 14%) is divided with 7 forming again 7 uniform ranges like in the case of A1-A3 (one (1) is from 0,1% to 2.0%, two (2) is from 2.1 to 4.0 etc.)

4. The model in graphical representation

In this section the model framework is instantiated on two sets of countries, namely, the set of the 5 developed nations and the set of SEE countries. Since a uniform grading scale is provided for all aspects, then by ignoring non-linearity issues, one can obtain a gross figure by adding them all up. If all sub aspects/metrics of A2 are equally taken into account (as if they were actual aspects on their own) then there are eight (8) aspects in total, each ranging from 1 to 7. Thus, the maximum grade to be expected is 56. Figure 2, shows the application of the model on SEE countries. It looks as if Greece and Slovenia are the most prepared nations of the regions as far as NEHRS development is concerned. But their preparedness is still far lower than any of the developed countries, as Figure 1 illustrates.

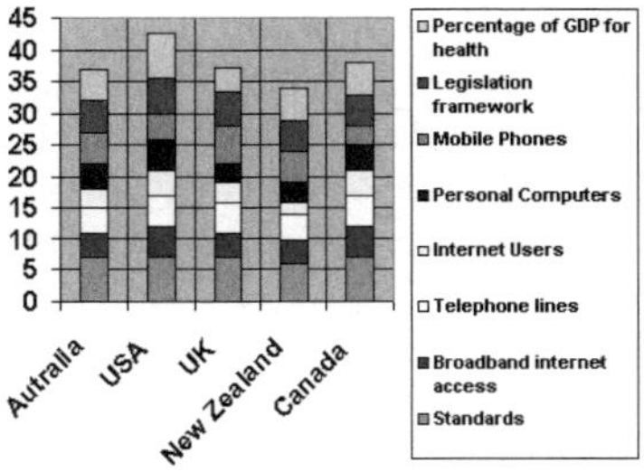

Figure1: The model for the set of developed countries.

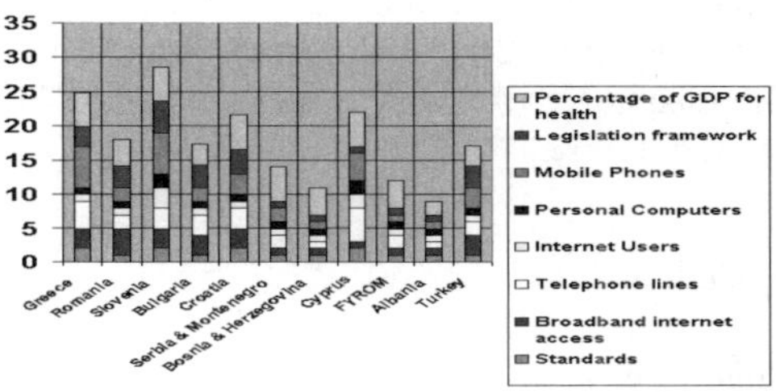

Figure2: Application of the model for SEE countries.

5. Validation of the Characterisation Model (CM)

A typical evaluation method on the CM could consist of the following steps: (i) application of the CM on a real NEHRS in preliminary stages and (ii) estimate the correctness of CM's results after the NEHRS implementation examining possible problems, weaknesses and strengths of NEHRS in combination with the CM results. However the CM cannot be applied on a real NEHRS because at the moment there is no complete NEHRS in operation, even in highly developed countries, thereby limiting the ability to validate the model. Thus, an appropriate and feasible validation method for our model is the Delphi method which is widely known on researchers trying to predict the "success" (in terms of providing good quality of results) of new models, theories and methodologies [8]. In this validation form, the model is tested and elaborated against the consensus opinions of a panel of experts within the field of NEHRS strategy. The Delphi method is based on a structured process for collecting and distilling knowledge from a group of experts using a series of questionnaires [9]. The key for the success of this validation method is the selection of the group of experts on the area of NEHRS. Although the validation phase is still in progress and until now three versions of questionnaires have been sent to the same panel of experts aiming to reach consensus on the CM aspects that have been previously presented, a number of important points have been raised. These are:

- Inclusion of a number of extra standards.
- New categories of standards, changes on the existent categories and transposition of standards from one category to another have been suggested in order to avoid standards overlapping.
- Fundamental changes on the aspect "Health Network Architecture" as some of the subaspects do not correspond to the aspect name.
- Elimination of some subaspects to avoid proper weighting problems with the other aspects.
- Inclusion of two extra aspects such as the computer literacy of the health personnel and the health network infrastructure (network's link capacity).

However, no consensus has been reached until now on whether the aspect "%GDP for health" will be considered in the model or not because for some of the experts it does not represent actual NEHRS spenditure. In addition no consensus has been reached until now on whether the unit metric "Legislation framework" represents the readiness of a country on legislation for a NEHRS or not. Finally it has been suggested to separate the NEHRS readiness of a country on categories: 1)Governmental (Laws, ICT infrastructure), 2)Health Sector (Standards, Computer Literacy of health personnel, Broadband Network Infrastructure) and 3)Citizen (computer literacy, available PCs and Internet links).

6. Conclusions

This paper aimed to investigate key concepts of the research area around the NEHRS, by developing a model framework capable of characterising NEHRS preparedness of countries. The model was lead by a set of developed countries, with the ultimate goal to allow for a south-east European perspective that could be of assistance in the regional development. Obviously, the model is not an absolute answer or solution, but rather a suggestion. The model framework has been built using the opinions, comments, criticisms, and analysis of governmental reports. Mainly, material came from the UK, Canada, New Zealand, Australia and the USA. Apart from the preparedness of a country to develop an NEHRS, the model can also locate specific points of the Infrastructure in a country that have to be improved in order to develop an effective and functional NEHRS. A critical but equally difficult part during the creation of the framework was the definition of the appropriate unit metrics. All metrics were scaled from 1-7. The reason for that was the pre-existence of the same scale in World Bank country reports for one the factors also adapted here, and the necessity to allow for uniform scaling among all aspects, so as to enable gross addition at the end. The completion of the validation phase is expected to improve the model framework substantially. However, it has to be mentioned that high accuracy of results is not a demand here because the CM is a result of a qualitative research focusing more on the description and complexity of a phenomenon (for our case the NEHRS readiness of a country) rather than an attempt at measuring it [10].To conclude, it is quite common to read in literature about (new) countries aiming at the development of an NEHRS. It is envisaged that further elaboration of this model framework could provide assistance to these countries in order to realise their NEHRS readiness, as well as, the infrastructure points that urge for improvement.

References

[1] Canada Health Infoway (2003) 'Infoway Pan-Canadian EHR Survey Phase I: Results and Analysis', www.canadahealthinfoway.ca/pdf/EHR-Survey-PhaseI.pdf, cons. Mar.2004.
[2] Health Connect (2003) 'Health Connect Project Plan 2003/05', A Health Information Network for all Australians, www.health.gov.au, con. Aug. 2004.
[3] Jones, M. (2004). 'Learning the lessons of history? Electronic records in the United Kingdom acute hospitals, 1998-2202', *Health Informatics Journal*, **Vol. 10 No. 4**, p:253-264.
[4] MidCentral District Health Board (2003) 'Information Systems Strategic Plan 2003 – 2008', www.midcentral.co.nz/pub/ISSP.pdf, Sep. 2004.
[5] National Health Services (1998), 'Information for Health: An Information Strategy for the modern NHS 1998-2005', www.doh.gov.uk/ipu/, Jul. 2002
[6] Orfanidis, L. (2005), 'Data Quality Issues in Electronic Health Records', South East European Research Centre, Internal Report, **No** 01/2005.
[7] World Bank (2003) 'ICT at a glance', www.worldbank.org, cons. Mar. 2004.
[8] Moeller, G.H. & Shafer, E.L., Ch. 39 in Ritchie and Goeldner , and Taylor, R.E. & Judd, L.L. (1994). "Delphi Forecasting" in Witt, S.F. & Moutinho, L. Tourism Marketing and Management Handbook. London: Prentice Hall.
[9] Adler, M., & Ziglio, E. (1995). Gazing into the oracle. Jessica Kingsley Publishers, Bristol.
[10] Arksey H. and Knight, P. (1999) 'Interviewing for Social Scientists', Sage Publication, London, p: 1-20.

4.3 eHealth and Economy

Ubiquity: Technologies for Better Health in Aging Societies
A. Hasman et al. (Eds.)
IOS Press, 2006

Simulation based cost-benefit analysis of a telemedical system for closed-loop insulin pump therapy of diabetes

Oliver J. Bott[a,1], Ina Hoffmann[a], Joachim Bergmann[a], Patrick Kosche[b], Christian von Ahn[b], Dirk C. Mattfeld[b], Oliver Schnell[c] and Dietrich P. Pretschner[a]

[a] *Institute for Medical Informatics, Technical University of Braunschweig, Germany*
[b] *Institute for Economic Sciences, Technical University of Braunschweig, Germany*
[c] *Diabetes Research Institute, Munich, Germany*

Abstract

Background: INCA (Intelligent Control Assistant for Diabetes) is a project funded by the EU with the objective to improve diabetes therapy by creating a personal closed loop system interacting with telemedical remote control. Cost-benefit analyses of such systems are needed to decide on the introduction of telemedical systems such as the INCA system to routine care.

Objective: To identify and apply suitable methods for a cost-benefit analysis from the perspective of the payor for health services (i.e. a health insurance company).

Methods: For the cost analysis MOSAIK-M was used, a method and tool that supports health information systems analysis and design. Two MOSAIK-M models were created during the INCA project. Both, the "As is"-model of conventional insulin pump based diabetes care, and the "To be"-model of the INCA system were parameterised with cost values. With both models a period of one year was simulated to determine the yearly costs of diabetes management and treatment for a patient who does not suffer from diabetes related complications yet. The HbA1c-value was chosen as effectiveness parameter for diabetes therapy. To determine the probability of developing complications and their probable duration the Archimedes-Model was used. It was parameterised with selected HbA1c-values anticipating the effect of INCA. The simulation results in form of years of disease within a 30-years time frame were multiplied with corresponding treatment costs from the KoDiM study.

Results: The yearly costs of conventional insulin pump treatment for a 19 year old diabetes type 1 patient with no complications are 5,907 € (German health care system). Using the INCA system would raise the yearly costs by 7,348 €. Almost all (98.53%) of the additional costs are generated by the continuous blood glucose measurement device. HbA1c-decreases from 7% (conventional treatment) to 6.5%, 6%, and 5.8 % would produce yearly savings (benefit) concerning the treatment of complications of 100.50 €, 189.20 € and 221.82 €.

Conclusions: The selected approach produces an estimation of a lower bound for cost savings. Further work is needed to improve the approximation and to include indirect and intangible costs. The INCA approach would be cost efficient from the chosen perspective, only if the costs of system operation were notably lowered.

Keywords: Medical Informatics, Information Systems; Cost-Benefit-Analysis; Model; Simulation; Telemedicine; Diabetes Mellitus; Insulin Infusion System

---

[1] Oliver J. Bott, Muehlenpfordtstr. 23, D-38106 Braunschweig (Germany). E-mail: o.bott@mi.tu-bs.de, Web: www.mi.tu-bs.de

1. Introduction

The decision for introduction of telemedical systems into clinical routine is mainly based on a positive cost-benefit relation from the perspective of payors for health services, i.e. health insurance companies. Telemedical systems are often characterized by high investments and high regular costs. Thus new telemedical approaches need to show an adequate level of benefit compared with current standards of treatment.

The telemedicine project INCA[2] (Intelligent Control Assistant for Diabetes) focuses on improvements in insulin pump therapy of diabetes through introducing a closed loop system, which combines continuous blood glucose measurement with an insulin pump by means of an intelligent personal digital assistant (the *Smart Assistant*). The idea is that the Smart Assistant controls the insulin pump using an algorithm that besides other parameters depends on continuous blood glucose measurement. This so called *Personal Loop* is embedded in a telemedical supervision of the therapy. Within this *Remote Loop* a physician monitors the glucose metabolism of each patient and gives therapy recommendations when needed. Besides analysis of clinical effects of the INCA system through clinical studies the question of suitable procedures for the health economic evaluation of such telemedicine approaches arises.

2. Materials and methods

The cost-benefit analysis is conducted from the perspective of the payor for health services and is divided into a cost and a benefit analysis (Fig. 1). The cost analysis (section 2.1) determines the yearly costs of diabetes management and treatment, but without considering diabetes related secondary diseases (complications). A better therapy of diabetes results in a decrease of the probability to develop diabetes related complications. Thus the monetary benefit of a new therapy for the chosen perspective corresponds to a reduction of the costs for treating complications. Therefore the benefit analysis (section 2.2) quantifies the reduction of costs for treating complications. For both, cost and benefit analysis, different modelling and simulation approaches were used: *MOSAIK-M* for cost analysis and *Archimedes* for benefit analysis.

2.1. Cost analysis

MOSAIK-M (MOdelling, Simulation, and Animation of Information and Communication systems in Medicine) is a method and tool for health information systems (HIS) analysis and design [1]. MOSAIK-M HIS models comprise processes, object structures, technical and human actors, and tools for information processing. MOSAIK-M models can be simulated. In INCA MOSAIK-M is used for system analysis and design. Therefore two MOSAIK-M models were created during the INCA project. The "As is"-model describes the current situation of diabetic's treatment and comprises 8 health care institutions (e.g. diabetes clinic), 25 roles (e.g. diabetologist, patient), 156 processes (e.g. opthalmological examination), and 81 object classes (e.g. insulin pump). The *"As is"-model* was the starting point for generating a model of the

---

[2] The INCA project (www.ist-inca.org; last accessed Jan. 02.2006) is funded by the EU 5th FWP (IST).

INCA telemedical system. The resulting "To be"- respectively *INCA-model* includes the system specification useful for the implementation (12 institutions, 37 roles, 341 processes, 81 object classes, 358 application models, 112 application object classes).

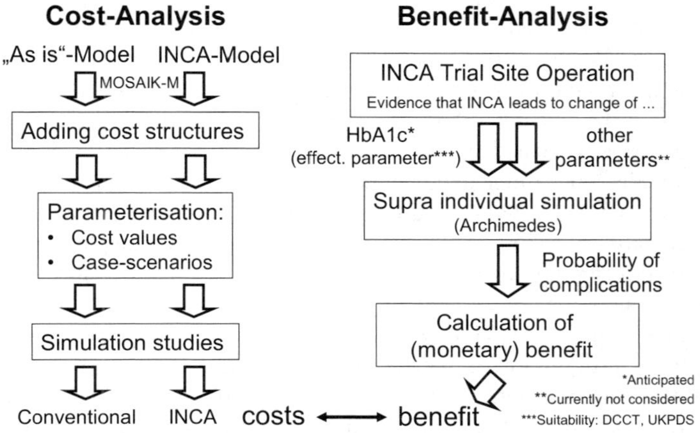

Fig. 1: Process model of INCA cost-benefit analysis

Based on corresponding enhancements of MOSAIK-M both the "As is"- and the INCA-model were extended to allow modelling costs of diabetes treatment and management. The cost parameters were acquired in September 2005 from price lists of German suppliers of medical devices, adjuvants or consumables and from official price lists for medical services in Germany [2]. To estimate costs of using INCA telemedical services cost structures of comparable commercial services were adopted.

The simulation with MOSAIK-M comprises one year for both, the "As is"- and the INCA-model. Both simulation runs focussed on insulin pump based diabetes treatment and management of a diabetic who does not suffer from diabetes related secondary diseases. A profile with no secondary diseases was chosen because the cost-benefit studies focus on the capability of the INCA system to prevent secondary diseases. The model assumes that patient and health professionals behave consistent with the evidenced based guideline of the German diabetes association [3]. The scenario was modified for the INCA-model to add usage of a Smart Assistant, and a continuous blood glucose measurement device, and online connections to the Remote Loop system (via GPRS/UMTS and the Internet). For each one-year simulation, the direct costs of diabetes treatment and management were accumulated.

2.2. Benefit analysis

The UK Prospective Diabetes Study (UKPDS) and the Diabetes Control and Complications Trial (DCCT) provide evidence that intensive glycemic control effectively slowed the onset and progression of diseases secondary to diabetes like retinopathy, nephropathy, and neuropathy [4]. Diabetes therapy should therefore result in an individual blood-glucose level that is near-normal thus reducing costs of treating complications in the long term. The HbA1c value quantifies the amount of glycosylated haemoglobin in the blood and thus gives a good estimate of how well diabetes is being managed over time [5]. HbA1c is therefore a suitable effectiveness parameter for diabetes related cost-effectiveness analyses of different treatment approaches. The clinical studies conducted during the INCA project are designed to result in statements concerning the increase or decrease of the HbA1c value in INCA intervention groups compared with control groups.

To quantify the benefit of the new telemedical therapeutic alternative, a second model and simulation based approach is utilized. *Archimedes* is a mathematical model of anatomical, physiological and pathological aspects of diabetes additionally incorporating information concerning the health care system that offers services to care for diabetes [6]. *Archimedes* allows simulating the development and progression of diabetes and diabetes related complications of a single person. To individualise the simulation, medical and other parameters of a person are entered into the model like gender, current age and weight, kind of diabetes and diabetes therapy, current HbA1c value, etc. *Archimedes* then calculates the progression of diabetes and the statistic probability of developing complications like retinopathy, nephropathy, etc. within the next 30 years. The model was positively validated against a selection of 18 randomised controlled clinical trials [7]. Thus *Archimedes* seems to be a reliable predictor for the development and progression of diabetes and its secondary diseases.

To estimate potential effects of the INCA system on the probability to develop complications a patient profile (age 19, diabetes type 1 diagnosed at age 14, normal BMI) with no secondary diseases was used to individualise the *Archimedes* model. The parameters varied between the simulations were gender of the patient and HbA1c value. Four HbA1c values were used to feed the simulation: 5.8 %, 6%, 6.5 % and 7 % (lower than 5% is normal). The results for the gender variation were averaged for each HbA1c value. The results of the eight simulation runs using the public accessible version of *Archimedes* [8] are statistical values for the probability to develop a complication (heart attack, stroke, kidney failure, eye problems, food problems) and for the probable age of developing these diseases within the next 30 years.

To estimate the potential monetary benefit of the INCA system the results of the simulation in form of years of disease were multiplied with the costs of treating these diseases. The cost factors used to quantify the costs of treating secondary diseases were taken from the German KoDiM study of 2004 that bases on data from 2001 [9].

3. Results

For the selected patient profiles (19 years old, diabetes type 1 diagnosed at the age of 14, no secondary diseases, guideline based diabetes treatment and management) a simulation was conducted for the MOSAIK-M “As is”-model and for the INCA-model:

- Simulation of conventional treatment (“As is”-model): Yearly costs of conventional treatment and diabetes management are 5,907.56 € distributed between consumables (81.2 %), devices and other adjuvants (15.4 %), and medical services (3.4 %).
- Simulation of telemedically supported treatment (INCA-model): The yearly costs of INCA related treatment and diabetes management are 13,256.02 € (consumables 80.8 %, devices and other adjuvants 17.6 %, medical services 1.6 %).

The results of the *Archimedes* simulation of the development and progression of diabetes and diabetes related complications dependent on four HbA1c values are presented in table 1. The table lists probability and the probable age of developing a diabetes related disease. All of the disease probability values are equal or lower in each case of decreasing the HbA1c value. The probable age of developing a disease remains the same or increases for each disease and each case of decreasing the HbA1c value.

Tab. 1: Results of the *Archimedes* simulation of the selected patient profile: Probability and probable age of developing diabetes related secondary diseases within the next 30 years starting at age 19

HbA1c	**Secondary disease (probability and probable age)**				
	Heart attack	Stroke	Kidney failure	Eye problems	Foot problems
7.0%	4.58%, 41.5	2.55%, 44.5	2.00%, 38.0	2.21%, 37.5	18.80%, 27.0
6.5%	4.54%, 42.0	2.55%, 44.5	1.92%, 38.0	2.01%, 38.0	14.84%, 30.0
6.0%	3.67%, 43.5	2.30%, 45.0	1.89%, 38.5	1.90%, 38.5	10.92%, 32.5
5.8%	3.21%, 44.5	2.15%, 46.0	1.89%, 38.5	1.87%, 38.5	9.40%, 33.5

The combination of the MOSAIK-M based cost factors for treating and managing diabetes in a conventional or an INCA conformant way with the cost adjusted *Archimedes* results is presented in table 2. Changes for the total costs of treating diabetes and potential diabetes related complications are presented for three potential case scenarios of decreasing the HbA1c value from 7% to 5.8%, to 6% and to 6.5%.

Tab. 2: Results of the cost-benefit analysis for three potential HbA1c decrease scenarios

HbA1c	HbA1c conventional	**7.00%**	**7.00%**	**7.00%**
	HbA1c INCA (case scenario)	**5.80%**	**6.00%**	**6.50%**
Costs	Conventional p.a. (7%)	6,287.87 €	6,287.87 €	6,287.87 €
	INCA p.a. (anticipated HbA1c decr.)	13,414.52 €	13,447.13 €	13,535.83 €
	Difference	7,126.64 €	7,159.26 €	7,247.96 €
	Additional* p.a. conventional (7 %)	380.31 €	380.31 €	380.31 €
	Additional p.a with decreased HbA1c	158.50 €	191.11 €	279.81 €
	Difference	221.82 €	189.20 €	100.50 €

* Only complications p.a. = per annum/per year

The additional costs for an INCA conformant telemedical approach would be 7,126.64 € (HbA1c of 5.8%), 7,159.26 € (6.0%), or 7,247.96 € (6.5%) higher than the yearly costs of conventional treatment, when the latter results in a HbA1c value of 7%. These costs include costs for managing and treating diabetes *and* potential complications. The effect on the yearly costs for treating complications only are 221.82 € for a HbA1c decrease of 1.2%, 189.20 € (1% decrease) or 100.50 € (0.5% decrease). These savings are the direct benefit of reducing the HbA1c value. I.e. any therapy approach that results in a corresponding HbA1c decrease and produces yearly additional costs below this threshold would produce equal or lower direct costs for treating diabetes and its complications compared with a therapy that results in a HbA1c value of 7%.

4. Discussion and conclusion

The integration of cost structures into existing MOSAIK-M HIS models requires rather low efforts. Thus the presented approach demonstrates that reusing models originally produced for the development of a HIS is an efficient way to conduct costs analyses. Other integrated modelling approaches like e.g. the ARIS approach [10] also offer functionalities to model process-cost related aspects. Essential features of the MOSAIK-M approach are its sophisticated process modelling and simulation capabilities combined with UML-based object oriented modelling and user interface prototype integration useful for user centred system design [1].

Almost all of the additional yearly costs of operating the INCA concept are resulting from the continuous blood glucose measurement device (98.53%). Especially the costs of consumables are immense when operating this device on a 24 hours 7 days per week basis. Notably reduced prices for the blood glucose monitoring device and its consumables would improve the cost/benefit relation of the INCA system dramatically.

The results of *Archimedes* are constrained on a time period of 30 years. The costs of treating diabetes complications arise in time periods beyond the 30 year time period considered here. Due to the chronic character of diabetes complications, the calculated savings are therefore a lower bound of the real savings.

The cost-benefit analysis was conducted from the perspective of the payor for health services. Potential losses of interests were not considered. Indirect costs (e.g. loss of working hours) as well as intangible costs (e.g. reduction of quality of life) were also not taken into account. Further analyses are needed to consider these aspects.

The benefit analysis presented anticipates changes of the HbA1c value when using INCA. The clinical studies are still running and final results concerning the impact of the INCA system on the HbA1c value will be available the next months. Based on their results the *Archimedes* simulation has to be repeated to finalise the cost-benefit analysis. Preliminary results of the clinical trials compare well with our assumptions.

The presented approach of using MOSAIK-M and Archimedes for cost-benefit analyses is efficient and produces information useful to discuss the introduction of telemedical systems. Its main advantage is that only few modifications of the model or its parameters are needed to simulate cost/benefit effects of new scenarios. The approach produces a conservative estimation of a lower bound for cost savings. Further work is needed to improve the approximation. To reach cost efficiency of INCA from the chosen perspective, the costs of system operation need to be notably lowered.

Acknowledgments

We wish to thank our partners E.J. Gomez and M.E. Hernando from the Bioengineering and Telemedicine Group, Universidad Politécnica de Madrid (Spain), M. Rigla, E. Brugués and A. de Leiva from the Fundacio Diabem in Barcelona (Spain), and T. Vering from Disetronic Medical Systems AG (Switzerland).

References

[1] Bott OJ, Bergmann J, Hoffmann I, Vering T, Gomez EJ, Hernando ME, Pretschner DP. The INCA-Project: Specifying a Telemedical System for Closed-Loop Therapy of Diabetes using MOSAIK-M. Stud Health Technol Inform. 2005;116:503-8.

[2] EBM2000 plus (http://www.kbv.de/rechtsquellen/4664.html, in German, last accessed Jan. 02.2006)

[3] Praxis-Leitlinien der Deutschen Diabetes-Gesellschaft (Practice Guidelines of the German Diabetes Association). Diabetes und Stoffwechsel, Suppl.2, 2002

[4] Davidson JA. Treatment of the patient with diabetes: importance of maintaining target HbA(1c) levels. Curr Med Res Opin. 2004 Dec;20(12):1919-27.

[5] American Association of Diabetes Educators. Intensive diabetes management: implications of the DCCT and UKPDS. Diabetes Educ. 2002 Sep-Oct;28(5):735-40.

[6] Schlessinger L, Eddy DM. *Archimedes*: a new model for simulating health care systems - the mathematical formulation. Journal of Biomedical Informatics 35: 37-50, 2002

[7] Eddy DM, Schlessinger L. Validation of the *Archimedes* diabetes model. Diabetes Care 26(11), 2003: 3102-10.

[8] http://www.diabetes.org/diabetesphd/default.jsp (last accessed Jan. 03.2006)

[9] von Ferber L. Die direkten Kosten der Komplikationen bei Diabetes Mellitus (The direct costs of diabetes mellitus related complications). Presentation, 39th conf. of the Deutsche Diabetes Gesellschaft (http://www.pmvforschungsgruppe.de/pdf/03_publikationen/kodim_lvf_2004.pdf in German, last accessed Jan. 03, 2006), 19.-22.05. 2004 in Hannover (Germany), 2004.

[10] Scheer A-W: ARIS-business process modeling. Berlin a.o. Springer, 2. ed., 1999.

Ubiquity: Technologies for Better Health in Aging Societies
A. Hasman et al. (Eds.)
IOS Press, 2006

E-DIMEM: an economic model to estimate the costs of e-Disease Management

Fabrizio L. RICCI[a], Luca D. SERBANATI I[b], Fabio PALAZZO[a], Daniela LUZI[a] and Giuseppe STIRPARO[a]
[a] *IRPPS-CNR, Via Nizza, 128, 00198 Rome, Italy, {f.ricci, f.palazzo, d.luzi, g.stirparo}@irpps.cnr.it*
[b] *UPB, "Politehnica" University, Spl. Independentei,313, 060032 Bucharest, Romania, luca@serbanati.com*

Abstract. In order to estimate the cost of generic clinical guidelines based on telemedicine services for chronic patients, an economic model (E-DIMEM) has been defined. This model is designed to help health suppliers and managers to plan for the future. It is essentially a workflow composed by diagnosis and therapeutic activities. The cost of each workflow activity is related to healthcare providers (activity agents), drugs, instruments, telemedicine services, and type of specialized healthcare centers used (hospital, clinics, etc.). This model performs multidimensional analysis. A web service based on the E-DIMEM has been implemented.

Keywords: telemedicine; disease management; medical protocol; healthcare costs.

1. Introduction

When organisational changes are introduced in healthcare it is necessary to evaluate their economic impact [1]. This is a complex issue especially when organisational changes are related to the introduction of ICT in the patient care process, which implies a relevant organisational "reengineering" as in the case of e-Disease Management[1] [2]. For this reason we have developed within the MobiDis project [3] some evaluating instruments which, through the use of databases used by the healthcare systems (i.e. clinical protocols, personnel databases), can help with the decisions to invest in ICT within the healthcare system. The economic model E-DIMEM (Electronic DIsease Management Economic Model) applies the activity based costing methodology [5] which considers the healthcare system as a group of processes and their subprocesses [6] on the basis of which costs and earnings are calculated. Such a model allows us to calculate the costs related to a *medical protocol*. As well known a protocol is a plan or outline of a scientific experiment, treatment or study. We intend it as a therapeutic-diagnostic path connected to a certain pathology, developed within a virtual healthcare

[1] The aim of e-Disease Management is to treat a chronic patient using telemedicine services and systems supported by healthcare local structures, avoiding and/or reducing hospitalisation [2].

system, made up of a virtual system of healthcare structures (hospitals, doctor's consulting room, etc) and the patient's home [7]. Our analysis considers the cost of a protocol for such virtual systems as the sum of the costs of various procedures, therapies, and medications contained in the protocol.
The software that implements E-DIMEM is able to find out the costs of each phase of the single medical protocol, to process the total cost and to compare it with other alternative ones pointing out its benefits in economic terms. The advantage of this system lies in the fact that every reengineering decision can be isolated and evaluated in accordance with its impact on the other single activities. For instance, an investment in ICT can determine a shorter time needed in the development of a certain activity resulting in a minor use of resources and therefore minor costs.
This paper briefly describes E-DIMEM and the software prototype that implements it.

2. The model

E-DIMEM is a linear model based on the following hypothesises: a) resources are fully used; b) every activity and action in a given clinical path is independent; c) the activity cycles are predefined, i.e. the number of repetitions is given *a priori*; d) during the protocol execution no clinical complications occur. The model is able to carry out any multidimensional analysis of the costs on the basis of three dimensions: time, place (healthcare structures) and activity [4].
The starting point is the protocol considered as an aggregation of clinical activities carried out on each patient. The protocol describes the patient healthcare process in the form of a linear flow of activities (diagnostic tests, different types of therapies, and medical visits) to be carried out according to a predefined scheduling [6]. Each activity can be subdivided into simpler actions. The order of these activities is irrelevant for the model. This analytical subdivision makes it possible to distinguish the actions that imply a cost from those that have only risks (for instance reading a medical report requires the same time independently from its quality). Therefore, besides considering the costs, our model allows us to identify also the activities, which imply a Risk Management [5].
For the actions having cost, this is calculated as a sum of partial costs of personnel, equipment and disposable material. It is also to be noted that E-DIMEM considers the cost evaluation used by the Italian National Health Service. However, E-DIMEM considers all actions performed during a medical protocol, even those that have no costs for the National Health Service. In this way the model is able to take future changes of the clinical organisation into account.
Figure 1 shows the cost of each action. Due to the importance of drugs, they have been distinguished from the other disposable material. Moreover, the percentage of general expenditures is not calculated for each action, but included in the estimation of the activities, which encompasses it.
In figure 1 the symbols indicate:

- PE_r - personnel, which carry out the clinical path, taken the professional role into account;
- EQ_i - equipment;
- EQU_t - disposable material for equipment;
- PH_h - drugs;

- MC_k - other disposable material.

The variable N indicates the number of each cost component:

- N_s - number of personnel type;
- N_i - number of equipment type;
- N_h - quantity of each type of drug;
- N_k - quantity of each type of disposable material.

Cost_ACN =		
$\sum r \in PE\ (cost\_PE_r * \sum s \in AT_r\ (N_s * time_s))$	+	Personnel cost
$\sum i \in EQ\ (cost\_AM_i * N_i) + (cost\_MR_i * N_i)$	+	Equipment cost: a) Amortisation, b) Maintenance and reparation (included assurance cost)
$\sum t \in EQU\ (cost\_AM_t * N_i)$	+	c) Disposable material for equipment
$\sum h \in PH\ (cost\_PH_h * N_h)$	+	Pharmaceutical costs
$\sum k \in MC\ (cost\_MC_k * N_k)$		Other disposable material cost

Figure 1. Cost of each action in the E-DIMEM

The cost of each activity is the sum of the costs of action components:

$$\textbf{Cost\_SRV} = \sum j \in ACN\ (cost\_ACN_j * N_j)$$

The cost of the protocol is the sum of the costs of activity components, plus the cost of standard activities provided in outsourcing:

$$\textbf{Cost\_PRCT} = \sum x \in SRV\ (cost\_SRV_x * N_x) + \sum t \in EXT_t\ (cost\_EXT_t * N_t)$$

3. The main cost components

3.1. Cost of personnel

The cost of personnel is given by the estimated time necessary to fulfil a task according to his/her role (physician, nurse, assistant, etc.). The personnel is usually an employee and his/her cost is defined by a gross rate derived by the Italian national contract, with the addition of assurance and/or reimbursement costs. It is to be noted that when the executor of the action is the patient or a volunteer the cost is nil.

3.2. Cost of equipment

The cost of equipment is calculated summing up the amortisation, maintenance and repairing costs (including insurance costs). Amortisation is given by the cost of a new equipment divided by the number of years which determine its technological obsolescence, this is different depending on medical or ICT equipments. Costs also depend on the number of people who use that equipment (patient or nurse in case of share equipments). If there are equipments of different manufacturers, the average cost is the cost of the most used equipment.

3.3.Cost of disposable material

The cost of disposable material is determined by the quantity used and by its price. In presence of material of different manufacturers, we consider the most used material. The cost of drugs is referred to DDD (Daily Defined Dose) and not to PDD (Prescribed Defined Dose), because the economic evaluation is *ex ante* and not *ex post.*

3.4. Cost of "standard" services

In a protocol there are standard therapeutic-diagnostic procedures such as hospitalisation, outpatient department visits, ECG, radiological exams. Their cost is defined by the price list of the National Health Service or by conventions made at local level.

3.5. Cost of additional services

The calculation of additional service costs, such as those of a call centre, can be obtained in forfeiting way: a) for single action b) for the entire service. In the case of services carried out in outsourcing, the second calculation is used.

4. The E-DIMEM prototype

In the MobiDis project, financed by the MIUR (Ministry of Education, University and Research), several therapeutic-diagnostic protocols were examined [3]. The chronic illnesses considered are: cardiopathy (hypertension, ischemic cardiopathy) hepatology (cirrhosis), neurology (TIA progress, ictus progress), pneumology (chronic bronchitis, chronic obstructive bronchus, emphisema, asthmatic, pulmonary-heart) and diabetes.
In order to evaluate economically the medical protocols of chronic patients (based on the use of telemedicine service and software developed in the MobiDis project), a software prototype based on the E-DIMEM model was implemented. The prototype is fully interactive and accessible through the web. The user is supported for describing his/her protocols, evaluating them and then comparing each other. In order to work with the system the users have to subscribe.

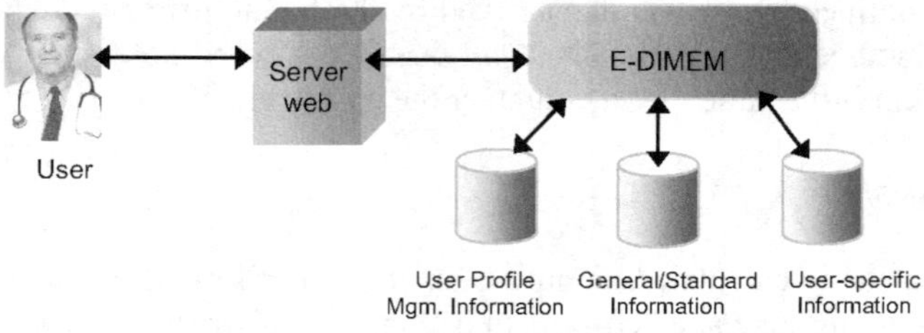

Figure 2. Prototype Architecture

Each user has his/her own information space ("User-specific Information") where he/she can store his/her protocols. Two other information spaces are also used: one ("User management") to manage subscribers system and the other ("General Information") containing valuable information for all subscribers regarding standard components of protocols, like hourly tariff of healthcare providers, drugs, costs of standard services etc. (Figure 2). Information is stored in a database and structured by some conceptual entities such as: protocol, diagnostic exam, therapy, medical visit, medical device, drug, healthcare organisation etc. The user can select from existent information or describe new instances of these concepts in economic terms and then save them in his/her database ("User-specific Information"), ready to be used in protocol descriptions. First of all he/she selects the patient type for a given pathology.

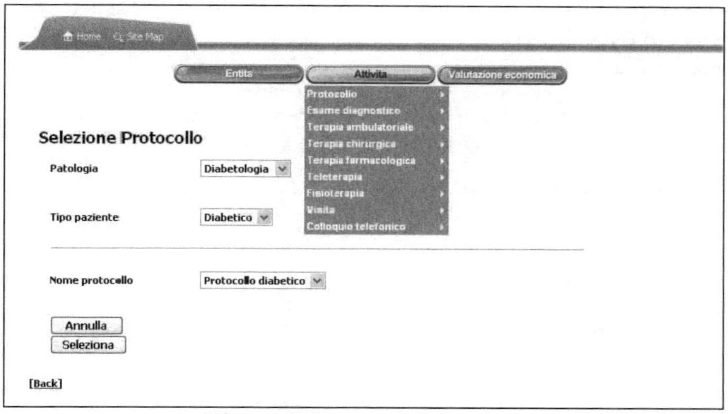

Figure 3. Protocol Selection

Figure 3 shows how the user selects an existent protocol from the "Activity" menu for the a diabetic patient. The menu items represent the activity types considered by the model.

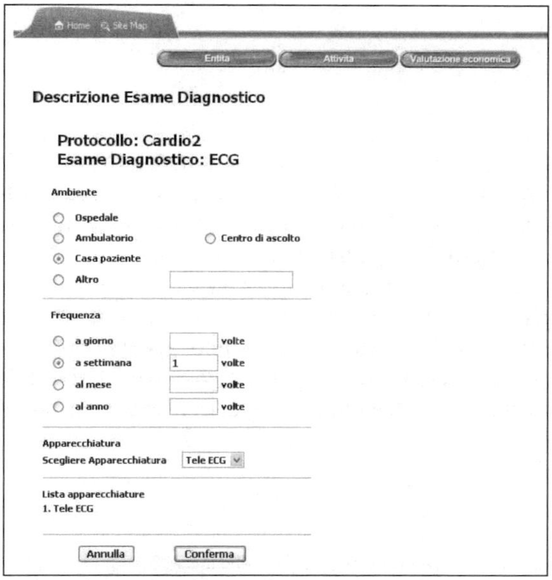

Figure 4. Diagnostic Exam Description

In order to describe a protocol, the user has to provide a list of activities, which can be diagnostic exams, therapies, and medical visits. Each of them should be further described. As an example of such an activity specification, figure 4 shows how a diagnostic exam is described providing information on the organisation where the exam is carried out, the devices used to carry it out and the frequency of its use.

When all components of a protocol are described and provided with their relative costs, the protocol evaluation can be requested, selecting "Economic evaluation" (Valutazione economica) from the menu. A final report is processed and published as shown in Figure 5. In the report the costs of personnel, equipment amortisation, maintenance and repairing, disposable material, drugs, call centres, and standard activities are given. The user may save the final report in XLS format. In this way there is the possibility to elaborate/compare this final report by using, for example a spreadsheet system (MS Excel).

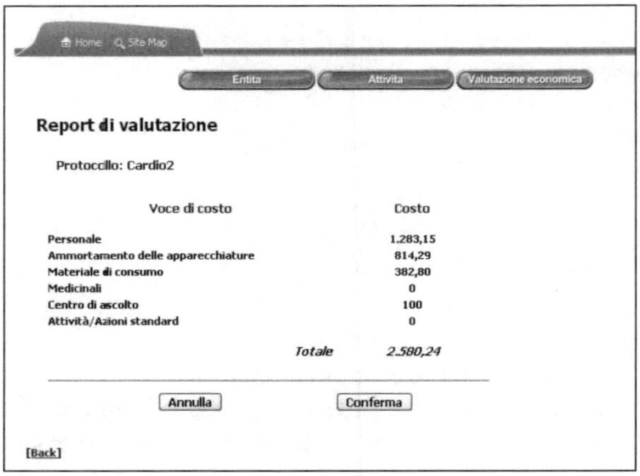

Figure 5. Final Report

5. Conclusion

E-DIMEM evaluates *ex-ante* the costs of each action, activity and related protocols in a virtual healthcare system. This makes it possible to evaluate different therapeutic-diagnostic paths considering both the clinical and the economic points of view. The comparison between alternative paths, which have the same effectiveness helps the user choosing a less expensive path and allows the evaluation in terms of costs benefits [5]. The development as well as the correct use of ICT in healthcare has to take both Disease Management and Risk Management into account in order to enhance the quality of healthcare services. In this way the advantages derive from a better control of the single actions, which can lead to cost reduction as well as quality enhancement.

References

[1] Casati.G , Programmazione e controllo di gestione delle aziende sanitarie, McGraw-Hill 2000, in Italian

[2] LeGrow G, Metzger J, E-Disease management, California HealthCare Foundation, ihealthreports, Novembre 2001

[3] Ricci FL nd ot ., Il progetto MobiDis. http://www.irpps.cnr.it/mobidis/index.php; in Italian

[4] Ricci FL and ot., Valutazione del sistema MobiDis. Deliverable MobiDis E1a.1, November 2005. Available on http://www.irpps.cnr.it/mobidis/pdf/Valutazione%20%20MobiDis.pdf; in Italian

[5] Lega F , Logiche e strumenti di gestione per processi in sanità, McGraw-Hill 2001, in Italian

[6] Casati G., Il percorso assistenziale del paziente in ospedale, McGraw-Hill 2002, in Italian,

[7] Maceratini R., Rafanelli M., Ricci F.L.: Virtual hospitalization: reality or utopia? . MEDINFO 95, Proceedings of the 8° World Congress on Medical Informatics, IMIA, 1995

4.4 eHealth and Information Sharing

Ubiquity: Technologies for Better Health in Aging Societies
A. Hasman et al. (Eds.)
IOS Press, 2006

A model for a Regional Health Information Network sharing clinical information between professionals in Britanny

Marc Cuggia[a1], Nicolas Herry[a], Delphine Rossille[a], Emmanuelle Lepage[b], Gilles Edan[b], Pierre Le Beux[a]
[a] *EA3888 – IFR140, Faculté de Médecine, Université Rennes 1*
2 rue du Pr. Léon Bernard, 35000 Rennes France
[b]*CHU Pontachaillou. Rue H. Le Guilloux 35033 Rennes Cedex*

Abstract: The purpose of this paper is to present a generic model of information system supporting healthcare networks for health professionals in Brittany. This model is aimed to develop cooperation between hospital professionals, primary care and practitioners whatever the specific pathology, by providing tools for exchanging and sharing medical-related data. The model is designed based on the heterogeneity factors revealed by a prospective survey. It includes secured exchange of nominative medical data, all other information and activities being accessible by a Web portal. Other associated tools are a synchronous collaborative platform, and a e-learning module. The first implementation, currently in use by the professionals, is presented for the existing neurology healthcare networks.

Keywords : Health information network, shared medical data, e-health, data integrity and security Introduction

Over the past ten years, France policy in the Health sector has been to encourage the development of healthcare networks and to decentralize the health management to the regional level. Healthcare networks are aimed to improve the quality of both health care delivery and health outcomes [1]. Indeed, formalizing networks and recognizing their potential for delivering better health care is nowadays recognized as a new paradigm in the health sector [2]. In this context, it was innovative in Brittany (Western region of France) to start a first experiment of health information networks focusing on the treatment of chronic neurological diseases: the NeuroBretagne Project [3]. The project's goal is to develop a network for patients suffering from chronic and handicapping neurological diseases such as Multiple sclerosis, Parkinson disease and amyotrophic lateral sclerosis. From the outset, the managers of health networks authorities have foreseen the advantages of a shared and common information system. Indeed, network efficiency depends on how information is managed. A recent study carried out in the US [4] has shown that preventable medical errors occur with alarming frequency in the US Healthcare system: over 10,000 deaths per

[1] Marc Cuggia. Faculté de Médecine. Rue L. Bernard 35000 Rennes. Marc.cuggia@univ-rennes1.fr

year are due to withheld information and bad information flow. Today, health professionals use conventional means of communication such as Phone, Fax, and mail, none of which being legally or technically secure. Let note for instance that the fax which is widely used is neither legally bound nor secured. In fact, the HAS organization (French High Health Authority Agency) considers its use for medical data transmission as strongly inadvisable. Information and Communication Technologies (ICT) can make these exchanges easier and more secure. As medical information is nowadays increasingly available in digital format, it should be accessible and shared among professionals as well as between professionals and patients.
This paper describes a generic model of a Regional Health Information Network. The model must be adaptable to any type of healthcare networks and professional systems, in accordance with model standards. We stress the need of consistency over four axes: the user's needs, computer equipment, software and medical systems interoperability. It also presents the model and its specificities. Then the first implementation of the proposed model applied to Neurology is described. Note that a second implementation is underway for perinatality, while an oncology network is being implemented in the district of Rennes.

1. Material and methods

Several existing regional healthcare networks in Brittany have been considered. Such networks include health professionals from University hospitals and private clinics, as well as private-sector specialists and general practitioners. They are aimed to bring together all Brittany health professionals for specific medical domains or pathologies. Networks in neurology, cancerology and perinatalogy contacted the Medical Informatics Laboratory of Rennes University for designing a Regional Health Information Network, aimed at facilitating their every day patient-related work. First, a prospective survey was conducted by interviewing all healthcare professionals involved in the NeuroBretagne project [3]. The survey questions were intended to understand their work habits, office work environment, computer equipment and computer skills. An additional question related to the functional needs they would like to be implemented in the future computer network. The survey data were then analyzed, leading to the identification of network design issues, the most important being heterogeneity. A general framework was defined to manage the heterogeneity factors. A generic model of the information system was set up for the health information network. The information system was modeled using UML based on the whole set of collected information. Finally the model was implemented for neurology using dynamic web technologies (LAMP: Linux, Apache, MySQL, PhP), and Open source software (SPIP based on CMS (content manager system)).

2. Results

Heterogeneity factors: The survey, conducted in 2004 [3], reveals noticeable heterogeneity. The first result shows the large diversity in the classes of health professionals (such as specialists, GP, nurses), as well as in their working places (e.g. hospitals, clinics, medical offices). Their functional needs are different although they all need to share patient-related data. Some of these needs must comply with the legal and administrative requirements on security, confidentiality and rights for patients to access their health records. Healthcare network organizations can vary depending on the medical domain (e.g. chronic pathology vs. acute pathology). Large diversity appears in computer equipments (hardware and software), in the nature and size of existing information systems (e.g. Hospitals vs. medical offices), and in the interoperability between information systems. Computer skills of health professionals as well as their software use also vary a lot (e.g. a large number of professionals use unsecured email to exchange with colleagues while very few take advantage of available medical software functionalities).

The proposed model of an information system for healthcare network:
Organization analysis: A typical patient path within the existing NeuroBretagne network is as follows: Most of the time, a general practitioner has identified the early symptoms of the disease. He refers his patient to a neurology specialist.
Once this specialist has confirmed the diagnosis, he refers the patient to the hospital, where a complete check-up is carried out. Diagnosis and therapeutic strategies are usually decided by a multidisciplinary staff. This staff identifies appropriate participants for clinical trials. Patients are also advised to join associations to meet other patients. Health professionals in NeuroBretagne use paper (postal mail or fax) to communicate with each other.
Exchanged documents are mainly appointment requests, hospital reports, and medical records. They can also be survey questionnaires, clinical trial documents. Both public and private practitioners take an active part in the clinic activity. This patients' management is in fact generalized to any pathology, non-neurological included. It is applicable to other networks. Therefore the proposed information system is strongly based on this organization model.

Information system modeling: Based on these analyses, the more important criteria to take into account in our model is security [5]. Indeed all medical nominative patient data must be transmitted in a secured way. These data will be exchanged asynchronously using secured email (SE). When professionals need to exchange information in real time (synchronous modality) to share their expert opinion, they can use for example a synchronous collaborative platform (S-CSP) in the secured mode. Sharing medical patient data can be typically resolved by adopting the Sharable Electronic Health Record (SEHR). All other information is shared and accessible through a web-based portal (WP). Obviously, conventional email (E) and S-CSP can also be used for exchanging non critical information (e.g. for e-learning activities). Finally, a final axis is considered which represents the interoperability level requirements between information systems (Table 1).

Secured email system: As e-mail was widely used, we wanted to build the information system on already acquired skills while respecting current legislation. We chose the CPSURE solution (edited by Enovacom™ [6] and approved by the French Professional Health Card - Public Interest Group. This system is a web mail service, unlike the other secured e-mail solution which uses fat clients. The web mail module uses widespread open source software (IMP-HORDE) [7]. E-mails encryption and signature are generated by a Java applet and use the Professional Health Card process. This process is based on the Asymmetric Public-key' techniques [8] [9]. Thus, this ubiquitous solution allows sending and receiving secured e-mails with the use of a web-browser only.

	Sharing data			Exchanging data		
++ ← Interoperability level requirement ← --	Free access	Access to non patient medical data	Access to medical patient data	Free access	Access to non patient medical data	Access to medical patient data
Keeping informed of professional activities	WP	WP		E	E	SE
Accessing clinical trials, guidelines and e-learning		WP			E	
Accessing members' directory		WP		E	E	SE
Accessing scientific journals		WP				
Communicating safely with other members	WP (forum)	WP (forum)		E	E	SE
Accessing or receiving hospital reports			SEHR			SE
Accessing shared medical records			SEHR			
Sharing or exchanging medical images		WP	SEHR		E	SE
Managing patient						SE

Table 1: Information system model

Web portal: The portal structure (Figure 1) contains three access levels. Level 1 is the only one accessible by the general public. It contains general information on the existing professional networks and redirects people to patients associations or support groups. Only members of the networks can access Levels 2 and 3. Level 2 provides information on professionals and activities (e.g. members' directory, schedule, pedagogical resources). Let us note that the portal can be easily managed by members and it doesn't require any specific computer skills. Level 3 connects to the professional web page of a thematic network. Each existing network has its own web publishing space while benefiting from a single CMS. At this level are found specific information of each thematic network: diagnosis or therapeutic guidelines, clinical trials, clinical training, epidemiological surveys, chat room, forum, mailing list etc. When some information is likely to be of interest to everyone, this information can be automatically published into an upper-level.

An implementation in Neurology: The proposed model has been implemented for neurology in Brittany and is currently used by neurologist professionals. In this section, selected features of the implementation are presented.

Access rights management module: Access rights to network contents can be specified. Thus, it is possible to personalize the information according to the user's profile. If a registered member is involved in a specific thematic network, he/she accesses the information of this network by default. However he/she keeps the possibility of surfing on other networks.

The structure inheritance function: The purpose of the model is to pool thematic networks into the same structure. The information system must be able to integrate new networks easily and quickly. The structure inheritance function permits a network template to be reused for another one. In addition, this function allows publishing information automatically from one level to another.

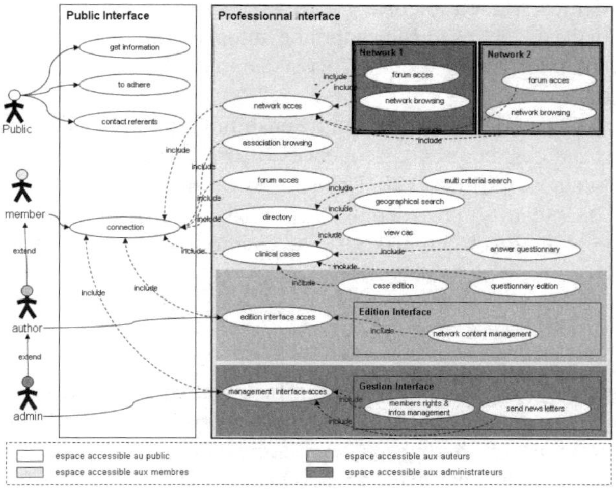

Figure 1: UML use-cases diagram for the network portal structure

E-learning module: This module answers to the continuing medical education needs. It implements the Script Concordance Test [10] well adapted to get experts opinions on clinical cases s well as to provide a means for professionals to test their competence.

Synchronous collaborative platform: This module will be used by professionals for teleconsultation. The BREEZE collaborative platform (Macromedia) was chosen. Medical information is exchanged in synchronous mode with a high security level.

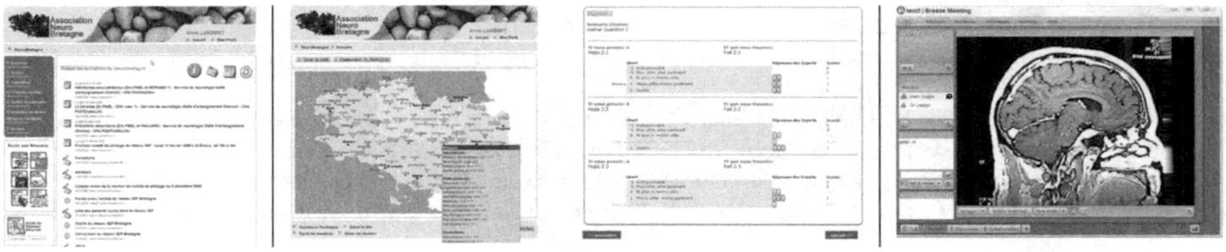

Figure 2: Screen shots of the NeuroBretagne web portal and collaborative platform

3. Discussion

The first implementation of the system has been validated by all the professionals involved (i.e. neurologists and physical therapists). The main wish of the professionals was a single and user friendly interface. Light tools were developed in order to minimize installation problems considering the large heterogeneity of computer equipments. Our system was chosen not to interact with medical software; it consists of a web system for the sharing and exchange of data. We did not want to set up a shared medical record for two main reasons: technical, organizational, legal, and ethical issues, and the current implementation of the French National Medical Shared Record. The system has some limits. When a user receives a document by e-mail, the attached file cannot be automatically integrated in the Electronic Health Record of the information system. This is due to the lack of interoperability between information systems.

Furthermore, our system does not put any constraint on the modalities of exchanges between users (procedures, controls). Sequences of the exchanges are often the same and can be modelled in a workflow [11]. Using a workflow can be very useful (it offers traceability, efficiency, clarity) provided that few exchange sequences exceptions are made. However, this kind of approach can make the information system much heavier to use.

4. Conclusion

Web technologies provide an attractive infrastructure for efficient and low cost communications in regional health information networks. The proposed information system is based on the principle of sharing and exchanging information between professionals. Patient's health information is protected and exchanged. The other kinds of information are widely shared and accessible on a web portal. We succeeded in setting up a simple system that can be instantiated in several specialties, considering the limited computer skills of the professionals. We have taken into account the heterogeneity of the computer equipment. Filling the gap with the setting up of national electronic health record, this system already supports and improves information exchanges between professionals and facilitates input in a health network

References

[1] Glasser, M., et al., *Meeting the needs of rural populations through interdisciplinary partnerships.* Fam Community Health, 2003. 26(3): p. 230-45.

[2] Conner, M., *Developing network-based services in the NHS.* Int J Health Care Qual Assur Inc Leadersh Health Serv, 2001. 14(6-7): p. 237-44.

[3] Cuggia M, Herry N, Lepage E, Edan G, Le Beux P A professional information network system for chronic neurological diseases: the Neuro-Bretagne experiment poster MIE 05 Geneva

[4] Pharow P, Blobel B. "Public key infrastructures for health."Stud Health Technol Inform. 2003;96:111-7.

[5] Sadovnick, A.D., *Genetics of multiple sclerosis in British Columbia and throughout Canada.* Ann Neurol, 1994. 36 Suppl: p. S18-21.

[6] Enovacom . Echanges électroniques professionnels. http://www.enovacom.fr/

[7] HORDE - IMP Webmail client : http://www.horde.org/imp/

[8] Pangalos, G., et al., *Developing a Public Key Infrastructure for a secure regional e-Health environment.* Methods Inf Med, 2002. 41(5): p. 414-8.

[9] Nisand, G., et al., *Improving computer security by health smart card.* Stud Health Technol Inform, 2003. 95: p. 176-81.

[10] Charlin B, Roy L, Brailovsky C, Goulet F, van der Vleuten C. The Script Concordance test: a tool to assess the reflective clinician. Teach Learn Med. 2000 Fall;12(4):189-95..

[11] Kaae, T., *A regional health network supported by organisational change. Cross sectorial electronic referral and confirmation of booking.* Stud Health Technol Inform, 2000. 77: p. 988-91.

Ubiquity: Technologies for Better Health in Aging Societies
A. Hasman et al. (Eds.)
IOS Press, 2006

Teleconsultations as a step towards hospital interoperability

Lacramioara STOICU-TIVADAR[a1], Francesco SICURELLO[b], Vasile STOICUTIVADAR[a], Victor MOGA[c], Mariana MOGA[c], Gianni PELLICANÒ[d], Giannluca RONCO[e]

[a] *University "Politehnica" Timisoara, Romania*
[b] *University of Milano-Bicocca,Italy*
[c] *County Hospital Timisoara, Romania*
[d] *Careggi Hospital, Florence, Italy*
[e]*Sirse Company, Brescia, Italy*

Abstract. The paper presents an interregional telemedicine project between hospitals from two countries that encourage good economic cooperation. It uses modern technology and diverse human resources that contribute to a better accomplishment of the goals of the project and insures continuity of the services after the ending of the project. It instantiates the vision of e-Europe regarding access to healthcare services and the increasing mobility of patients.
Keywords: teleconsultations, interoperability, regional cooperation

1. Introduction. A Bilateral Project

The e-health concept is more and more a reality, tending to become a day-to-day utility. One of the driving forces at the European level is the growing mobility of its citizens, regarding businesses, holidays, and a stronger connection with families. In this context, activities that sustain the new tendencies and realities are encouraged and find support and completion. The results are the enrichment of the physicians' role and expertise through an intense and current dialogue with a larger community of colleagues and accessing a larger palette of cases. The result is a better service for the patient, which can benefit of good practice regardless of the site where he/she is at a certain moment.

There are on the role or ended several European projects regarding teleconsultation between countries. The goals were: working to standardization of diagnosis and therapeutic methods and create a database for education and research [1], to increase the transparency in the sharing of epidemiological information among the Mediterranean countries, in the larger context of the objectives set forward by the 'Information Society' [2], analyze the value of teleconferencing for patient care and surgical education by assessing the activity of an international academic network [3].

[1] Corresponding author: Lacramioara STOICU-TIVADAR, Faculty of Automation and Computers, V. Parvan bul., no. 2, 300223 Timişoara, Romania, e-mail lacramioara.stoicu-tivadar@aut.upt.ro.

E-health is playing an important role in the EU eEurope strategy and one of the important activities to support this is the deployment of health information networks based on fix and wireless broadband and mobile infrastructures [4]. A bilateral project that sustains this vision and finances the activities has started between regions from Romania and Italy. The cooperation started between Lombardia and Toscana regions (Italy) and Timis County (Romania) with the intention of enlargement. The project had as result the development of a healthcare network between a Romanian hospital, County Hospital from Timisoara and an Italian Hospital, Careggi Hospital from Florence. The involved departments are cardiology, radiology and pathology. The University "Politehnica" from Timisoara, Romania and the University Milano-Bicocca from Milan, Italy ensured the design of the architecture of the system and the technical support.

The goal of the project is to ensure the possibility of the doctors from Romania to obtain a second opinion from Italy, about specific diseases, related to the selected departments from the County Hospital, especially for Italian patients, but not only. As a result, we expect an improvement of the quality of the medical activity.

One of the important characteristics of the project is that, based on previous good experiences [5], [6], it gathered together physicians, academics, and engineers. This insures continuity and coherence to the results of the project. Thus, another important goal is to train a group of telemedicine specialists and to develop related expertise. We will use the infrastructure as an alternative to the classical medical learning process.

2. Communication between Hospitals

2.1. The network

The communication between hospitals is based on an infrastructure that implements a virtual private network connecting the hospitals, the universities and a server farm, located in Italy, like in Figure 1.

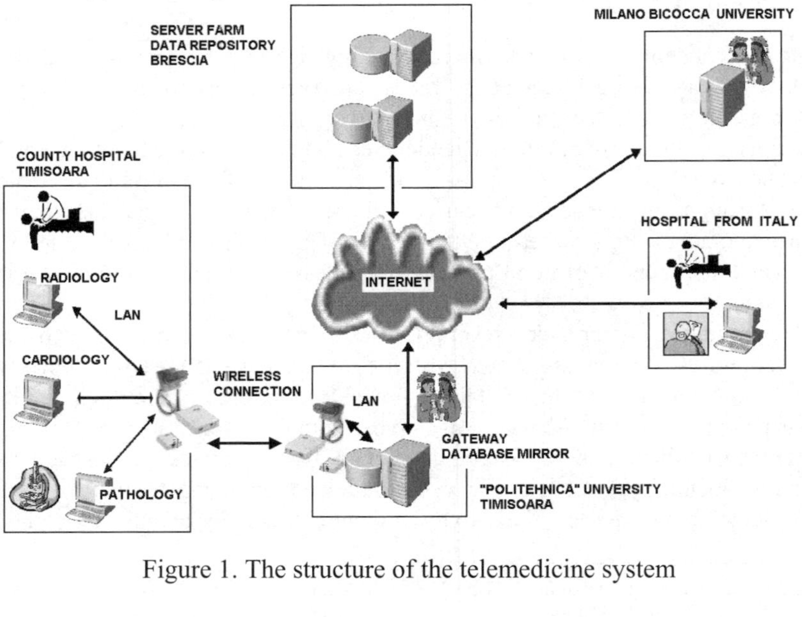

Figure 1. The structure of the telemedicine system

The support can be the Internet or an alternative using a wireless solution between the County Hospital Timişoara (CHT) and the "Politehnica" University Timisoara (PUT) because the spatial conditions for such a solution are satisfied (distance less than 2 kms and no very high buildings between them to obstruct the communication). The alternative solution was chosen to allow a quicker intervention and consequently a more efficient management of the network if the CHT has transmission problems.

2.2. Hospital Departments

The CHT, sustained by a team of enthusiastic and modern vision doctors involved three departments: cardiology, radiology, and pathology. The selection was made keeping into account the financial criteria - existing facilities regarding medical equipment, and the human factor - the cooperation, motivation and involvement of the physicians. The technical and medical teams found a common ground of intentions, objectives and language [7].

The Cardiology Department: includes 50 beds for an annually 2300 admissions/year with nine doctors that offers specific medical services. The Cardiology Department is the first in Romania in which a pacemaker implantation was performed. All these conditions determined the selection of the Department to be part of the teleconsulting projects. The Department has adequate equipment for teleconsultations, but for better results new equipment arrived: a laptop with a teleconsultation software installed (*Clinical Image Integra*) ensuring the mobility of the cardiologist to the beds and to the equipments, a web camera, headphones, a video grabber, and a high performance digital ECG.

Several examples regarding the exchanged cases are: a 56 years old woman with severe obesity (150 Kg), venous insufficiency and pulmonary embolism, for which we have presented the data as a clinical presentation, a woman with severe lymphatic diseases, for which we transmitted her images, and images of echocardiography of our database created using the *Integra* software.

The Radiology Department: is a complex radio imagistic department that offers medical services including X-rays (thoraces, bone, digestive tract and peripheral arteries and valves), echography, computer tomography and MRI. The imagistic results are stored on the server using DICOM standard. It offers services for all the departments of the hospital. It is equipped with high performance radiological devices and double displays for good visualization of images. The project supplied a laptop and the software for teleconsultations.

The Pathology Department: performs medical services for all the departments of the Hospital. The activity is focused on the analysis of the histological fragments related to various diseases (haematological diseases, liver diseases, thyroidal diseases and surgical related ones). It offers the possibility to obtain pathological and histological fragments in vivo and in vitro. The project supplied a digital camera for the microscope.

2.3. The software

The communication between partners has at its ends the software *Clinical Image Integra* ensuring functionalities specific to the clinical and administrative process: management of the electronic patient record, image processing, connection to the VPN, secure communication and authentication due to hardware keys with electronic signatures, teleconsultation sessions with videoconference facilities and electronic patient records transfer (including all associated data as images from the radiology or pathology departments or video sequences from the cardiology departments). The program and its components are developed in *Visual Basic* and *Visual C++ 6*. The interfaces are user-friendly. Two examples of interfaces are presented in Figure 2.

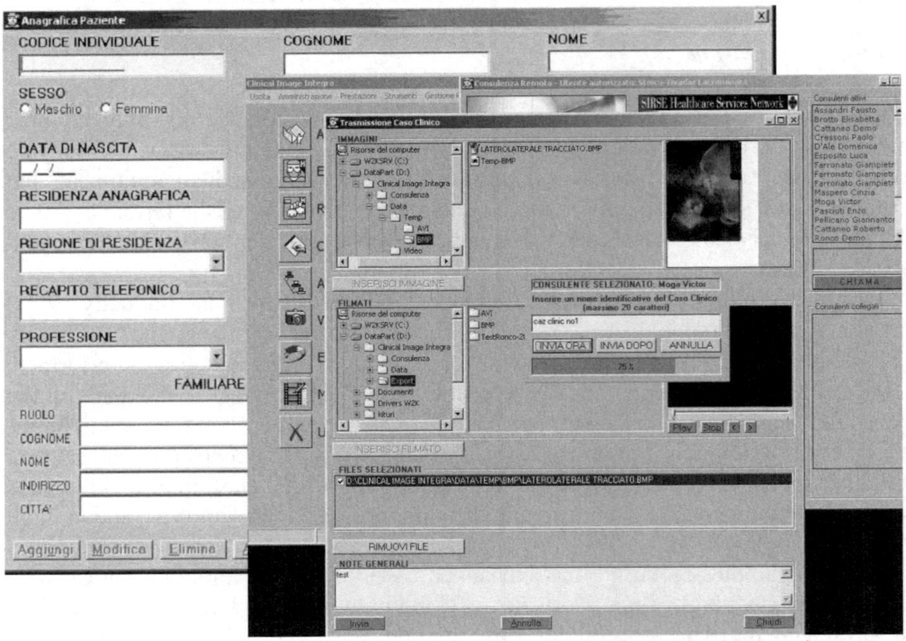

Patient file Teleconsultation session

Figure 2. Interfaces of the teleconsultation software

A teleconsultation session includes authentication (with the personalized hardware key, user and password), filling in the information for the electronic patient record, input of the external, captured information (images or video sequences), image processing - if desired, and the teleconsultation itself.

Thus, the specialist is able to communicate via videoconference, or send the required patient data and other related data. If the teleconsultation partner is not available, the content of the patient file and related data is stored on the server, and the partner will be alerted by e-mail and will download the information at a convenient moment. The images can be visualized by the partner or, with the permission given by the calling partner, can actually be transferred and stored on the partner's computer.

From current practice, this software raised several problems:

- lack of transmission large bandwidth, determined by the limited possibilities of the Internet providers of the Romanian partners; this causes accidental difficulties to information transfer;
- complexity of the software; at the beginning it seemed to be an advantage, but the users claimed a simplified version, easier to use in practice;
- transmission of the ECGs as signals is not implemented.

These aspects and previous experience [8] determined the technical staff to further develop the project in order to solve these limitations. To solve the first problem, a wireless connection between CHT and PUT was suggested. For the second issues, the staff from PUT has under development a software package, based on *Microsoft.NET* technology, in order to insure a simplified set of functionalities for the teleconsultation partners. This package contains only a simplified electronic patient record, authentication based only on user and password and a teleconsultation system without videoconference, but with associated information sending and a chat-style conversation. Other applications like *Yahoo Messenger* will be used for videoconference and classical image processing programs will be invoked for image processing, if required. The communication is based, in this new software, on invocation of a specialized *Web Service*, installed on a Server located at PUT location.

The software will include inputs from the Digital ECG and the transmission and display of the vectorial records (ECG signals). The data exchange uses the conversion of the header information in XML format. The new software structure is presented in Fig. 3.

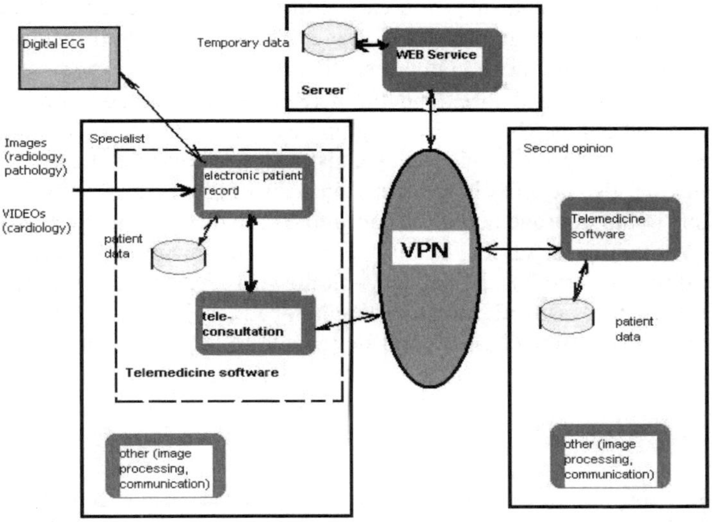

Figure 3. Structure of the new *Telemedicine* software

For the future, the intention is to add a module that will ensure the compatibility with the HL7 standard [9].

3. Results and Developments

The Italian-Romanian Telemedicine Project proves to be a useful tool in providing medical services at distance regional levels, offering the opportunities of high-level teleconsultation and videoconferences with several Italian clinics (from Lombardia, Toscana and in the future, Veneto regions).

The support offered by the activities in the project is useful for both Italians and Romanians that are performing activities in Romania, respectively in Italy, having access to their home country doctors. The idea started due to this need, when both communities increased also in Timisoara and in Florence, and now regarding the Veneto region. The activities driven by the project have just started and we estimate good research results at the end of 2006. We intend to develop researches regarding the need of such a telemedicine infrastructure, the impact of this activity in the current practice and several related statistical studies. We intend to use the infrastructure for e-learning.

The current results of the performed activities are of technical, educational, management, medical services and also social nature:

- The technical staff used modern technologies and solutions for the equipments and for the software (wireless, communication standards, and .NET technology). This experience can be used in future engineers' education.
- The hospital application can be connected to the Medicine and Pharmacy University through PUT (also using a wireless solution) and the physicians from the Hospital can directly demonstrate medical acts to the students from the amphitheatres of the UMFT.
- A good, solid team consisting of physicians, academics and technical staff was formed, that can develop further projects and can state and demonstrate by facts that the medical and technical world can cooperate for the benefits of the society (at national and European level). Not only once health information projects collapsed due to communication interdisciplinary issues.
- For the future, we intend to extend the network with neighbourhood countries. DKMT (Danube-Kris-Mureş-Tisa) region can be the umbrella for future developments of the project linking hospitals from counties in Hungary.

The project has also benefits in reducing the costs of the medical activities and the social costs, due to the improvement of the access to high qualified medical advisors, in the improvement of the scientific quality of the communication between different specialists, in direct benefits for the patients – at the cost level and in insuring mobility and availability of the medical act at different locations [10], in best practices in medical activities by removing barriers and approaching European standards.

References

[1] Fliedner, TM, Weiss, M, et al., Teleconsultation in Radiation Medicine – Results of regular telemedicine consultations between the Health Sciences Center of University of Ulm, Germany and Urals Research Centre for Radiation Medicine, Russia. Protection of the Human Environment Radiation Environmental Health, World Health Organization, Geneva, 2002.

[2] http://www.emphis.org/emphisproject, DG AIDCO Scientific collaboration with DG INFSO Euro Mediterranean Public Health Information System, 2004.

[3] Nicolas Demartines et al., Assessment of Telemedicine in Surgical Education and Patient Care, Ann Surg. 2000 February; 231(2): 282–29.

[4] Commission of the European Communities, COM 2004 (356), e-Health – making healthcare better for European citizens: An action plan for a European e-Health Area, Brussles, 2004.

[5] Sicurello, F, Simon, P, Magretti, P, Medgyesi, A, The goals and the perspectives of the common research and development work: the Italian-Hungarian bilateral programme. The Studies in Health Technology and Informatics, Health Data in the Information Society. MIE2002. IOS Press, 2002; 90: 143-147.

[6] Stoicu-Tivadar, L, Stoicu-Tivadar,V, Marginean, M, Berian, D. A Strategy for System Improvement and Data Processing in a National GP Sentinel Network. Studies in Health Technology and Informatics, The New Navigators: from Professionals to Patients. MIE2003, IOS Press, 2003; 95: 874-879.

[7] Iakovidis, I, Maglavera, S, Trakatellis, A, editors. User Acceptance of Health Telematics Applications. Looking for convincing cases. Vol. 56, IOS Press; 1998.

[8] Stoicu-Tivadar L, Stoicu-Tivadar, V, Kormos C, Oprean S. Solutions to upgrade a GP national network. Contributions of Medical Informatics to Health. Bolobel,B, Gell, G, Hildebrand, C, Engelbrecht, R. eds. AKA Berlin, 2004; 49-51.

[9] Blobel, B, Pharow, P, editors. Advanced Health Telematics and Telemedicine. IOS Press, 2003.

[10] Bushko, R, editor. Future of Health Technology. IOS Press, 2002.

A. Hasman et al. (Eds.)
IOS Press, 2006

Health information exchange on a regional level: dream or reality?

Eline R.M. Loomans[a,1], Marja Snoeijen[a], Diek Scholten[b] and Henk ten Dolle[b]
[a] Regional Commission of Healthcare Delft Westland Oostland/Nieuwe Waterweg Noord, the Netherlands
[b] Uzorg BV, Utrecht, the Netherlands

Abstract. Cooperation between healthcare institutions on the subject of Information Technology in Healthcare is an important main objective in the region Delft Westland Oostland / Nieuwe Waterweg Noord, in the mid-western part of the Netherlands. This is a densely populated area with approximately 1 million inhabitants. In 2004 a project to uniquely identify patients was initiated. The aim was to build a Regional Referral Index, containing only general patient data, which knows where patient information resides, serving as the fundament of a regional Virtual Electronic Patient Record. Initially the Regional Referral Index was filled with the patient files from two hospitals in the region. A connection was made between the Regional Referral Index and the Hospital Information Systems. Since then, the Regional Referral Index has been updated automatically. The patient files have been tested and matched in order to obtain a 'clean' patient index.

Since December 2005 the Regional Referral Index is operational and ready for further developments. Precautions are taken to guarantee patient privacy and security of data. Only authorized users will be given access to patient data. Administration and maintenance of this Regional Referral Index is arranged by establishing a Patient Information Point. Currently, patient files of the other care providers in the region are being added and pilot projects to accomplish access to medical patient information are being realised. The regional activities run parallel and in line with national initiatives, aiming to link up with these initiatives over time.

Keywords: Unique Patient Identification, Regional Referral Index, Health Information exchange, Regional Care Portal, Patient Information Point.

1. Introduction

The Regional Commission of Healthcare (RCG) in the region Delft Westland Oostland / Nieuwe Waterweg Noord (DWO/NWN) - in which all healthcare parties (care and cure institutions, patient platform, Healthcare Insurer and province) in the region are cooperating - initiates, stimulates and finances various projects. Early 2004 the Unique Patient Identification (UPI) project was initiated with the goal to create one regional database containing the general information (like name, address, residence) of uniquely identified patients, to be seen as the fundament for health information exchange. To be able to exchange information between various care providers you want to be sure you are looking at information pertaining to the correctly identified, unique patient. A so-called Regional Referral Index has been set up, comprising uniquely identified patients. Since December 2005 the Regional Referral Index, containing only general patient data, has been operational.

[1] Corresponding Author: E. Loomans, P.O. box 173, 3100 AD Schiedam, the Netherlands; upi-project@dsw.nl; www.upi-project.nl

The Regional Referral Index is now being further developed as a 'control tower' which knows where medical patient information can be found. Through a dedicated application, called Upid, authorised healthcare workers can get access to general patient information [1]. This is only the starting point for further developments towards a true Regional Care Portal: a portal which enables healthcare workers to view health information from various sources and to make use of various services - like the possibility to send referral letters to specialists or to download care protocols - at any place at any time. In the mean time pilot projects to establish access to the patients' Electronic Medication Record and GP Record have been started. Attaining access to other types of health information, like radiology and laboratory results, is also being made possible.

2. Material and Methods

2.1. Organisation and communication

A project team, with members from the various participating healthcare institutions in the region, was established. Also representatives of individual care providers, like General Practitioners and Public Pharmacists were included in the team. A project manager, appointed by and reporting to the Information and Communication Technology Steering Committee (being a part of the RCG), manages the project team. Only mandated board members of the participating care providers are members of this committee.

The project team meets on a regular basis and further communication takes place through the project website and by e-mail. The project team members are responsible for communication within their own organisation. Project management supports this when necessary.

2.2. Supplier

Early in the process it was decided that the Dutch company Uzorg, Utrecht, was the most suitable supplier for the regional plans. They had the practical experience through a similar project in the region of Utrecht.

The Upid concept boils down to the fact that on a central (regional or national) level a series of basic services should be available to simplify the digital communication between care providers and make it reliable, verifiable and scalable. These basic services are a referral index, a patient and care provider register, an authorisation structure and logging provisions. In Uzorg's point of view all these components are inextricably interconnected. A referral index is impossible without a patient register and a well-functioning authorisation structure cannot work without the adequate logging of the exchange of information. The application Upid is commonly misinterpreted as a Master Patient Index (MPI). An MPI can be defined as a system that has the responsibility to collect patient's various hospital identification numbers, perhaps from the laboratory, radiology, GP system, and so on and to keep them under a single, global identification number. The application Upid consists of a referral index, authorization mechanism, logging, care provider register and a MPI. By comparing the

definition of MPI and the functionality of Upid, one can conclude that the MPI is only a component of Upid.
However, solely the above-mentioned components are not enough to enable exchange of information in health care. To realise that, decentralised systems, such as a GP information system or a hospital information system must communicate with these basic services in a uniform way. Every information system, simply XIS, has its own internal language that the outside world may not always comprehend. Translation is then necessary. For this purpose, the so-called STUNT concept has been created, the Standard Upid iNTerface. Through special software, conversation takes place with XIS in the 'individual language' and the communication with the outside world, in this case Upid, takes place by means of generic language (HL7v3).
To enable data exchange between the different health systems, the communication is realized by standardized messages. The Medication Record, General Practitioner Deputy Record and Act Reference updates are based on the HL7v3-standard as defined in the specification of the National Infrastructure for Health Care [2,3]. To exchange other medical data such as laboratory, radiology, and discharge reports, customised messages are used. To provide a consistent data and concept reuse, the HL7v3 RIM was used as the source during the development of customised messages.
The information flow between Upid, STUNT and XIS only gets real value when applications are used that can read and write with Upid. Such an application may be an existing system: for example, the integration of the medical record of someone's own GP in an application at the GP emergency service. It will however, also become possible to develop new applications based on Upid. Those applications will be oriented specifically at supporting chain care and will need information from a large number of sources. The concept Upid shows much similarity with the national health infrastructure formulated by the National ICT Institute in the Health Care (NICTIZ). Uzorg was one of the driving forces behind the realisation of the specifications of this infrastructure and Upid is considered the first implementation of it.

All necessary components are present in Upid to provide care providers the possibility to view the medical data of a patient in correspondence with their function while taking care of the patient's privacy:

1. A referral index referring to patient-related information sources that can be found in the connected care organisations.
2. A patient register with basic information about a patient such as details about name, address and residence and a unique regional or national number. This component is also as know as the Master Patient Index of Upid.
3. A care provider register, with basic information about care providers who can retrieve data through Upid.
4. An authorisation structure that is fair to both the wishes of the patient and the professionals. The rights granted to care providers to access specific data groups and patients can be defined in advance.
5. Authentication mechanisms to determine the identity of care providers.
6. Functions for logging the exchange of data in order to be able to verify afterwards if consultations through Upid were legitimate.

2.3. Realization process

Next to the technical realization of the Regional Referral Index, activities were started to facilitate administration and maintenance of this Regional Referral Index. Therefore a Patient Information Point (PIP) was established. The PIP serves as a the main entry point for patients where they can get information and pose questions concerning their data in the Regional Referral Database and has an important role to administrate and maintain the data in the Regional Referral Index. Establishing a PIP is required by Dutch legislation concerning patient privacy.

1. A technical connection between the Hospital Information System (HIS) of the two regional hospitals (Vlietland Hospital in Schiedam and Reinier de Graaf Hospital in Delft) and Upid was realised. A STUNT was implemented to take care of communication between the systems.
2. Patient files of the last 3 years were delivered by both hospitals and read into Upid.
3. The patient files were tested and matched through a well-defined and automated process.
4. The most recent patient files were delivered, read in and cleaned.
5. The connection between the Regional Referral Index and the hospitals is operational; so when a new patient is registered or patient information is corrected in the Hospital Information System, the information in the Regional Referral Index will be automatically updated. A unique Patient Information Number is automatically assigned to newly registered patients.
6. In the short-term patient files of other institutions and care providers will be read in and matched with the existing patient data in the Regional Referral Index.

3. Results

3.1. Importing patient files

The patient file of the Vlietland Hospital contained 188,897 patients; 2.78 % was rejected after matching. The patient file of the Reinier de Graaf Hospital contained 370,098 patients; 6.23% was rejected after matching. Besides these two sets, the patient files of three other care organisations are imported, with an average of 2.5% rejected records is. Main reasons for rejection were wrong or missing address data and missing initials. Fallout results, which serve as input to correct the patient data in the Hospital Information System, were reported to both hospitals. Some errors will need to be checked personally with patients. It is expected that the patient files of the other institutions and care providers in the region will contain a lot of overlap with the patient data currently in the Regional Referral Index. At this point in time the Regional Referral Index contains approximately 500,000 patients, but is expected to grow towards around 750,000 patients soon. This number will grow with the addition of patient files from other institutions and care providers.

3.2. Establishing technical connections

The connections between Upid and the care organisations are operational, so the automatic update of the Regional Referral Index is realised. When new patient data is entered or when existing patient data is edited in the information system of the care organisation, the corresponding reference entry in the Regional Referral Index is updated.

3.3. Guaranteeing privacy

In a very early stage the project was reported to the Dutch College of Privacy Protection (CBP). All necessary precautions were taken to guarantee patients privacy:

1. Only authorised users can get access: they use a personal token, username and password. These three together ensure safe access through a protected network only.
2. Upid provides logging data: the Patient Information Point (PIP) controls and evaluates this logging data on regular basis. In addition to this, patients can ask for logging reports. Procedures are in place to follow up on this.

3.4. Complying with Dutch rules and legislation

A new Dutch law on the Use of the Citizen Service Number (BSN) in Healthcare is expected in 2006. This BSN provides every citizen in the Netherlands a unique number, to be used for various purposes, including use in healthcare. The Regional Referral Index is prepared for that; the currently used Patient Information Numbers can be seamlessly converted to the new BSN.
Recently the UZI (Unique Care Provider Identification) card has been introduced and will be gradually implemented. This card identifies Care Providers uniquely. Upid is prepared to work with this UZI card and currently used tokens can be easily replaced.

4. Discussion

Over the years, various strategies have evolved on the international and national level in order to establish health information exchange. These have evolved from using messaging principles and techniques (e.g. HL7v2, EDIFACT) to systems integration and achieving semantic interoperability (e.g. HL7v3, IHE/XDS). It is widely accepted that health information exchange should at least contain an architecture enabling identification of (1) patients (2) care providers (3) medical data and (4) the use of data. Upid was developed when propietary systems and de facto standards in The Netherlands were widespread, and relatively new (inter)national standards were about to be introduced. Therefore, many of the general principles are incorporated in Upid, such as domain models, the use of basic identification services and frameworks for dealing with security issues.

In the Netherlands, a decentralised approach has been chosen both by the National government and within the Upid product [2,3]. There has been considerable technical and legal discussion on a central versus a decentral systems approach. In the Netherlands it is recognized that a central approach would raise serious legal issues concerning patient's ownership of data and accountability of health care workers. Thus, rather than storing medical data into one regional or even national system, data remains where it is collected: the source. This approach makes migration to the new health information exchange easier. Health care workers have no need to migrate to a totally different "regional" system and software vendors only need to facilitate the connection and interoperability to the central services. Legal issues still remain, but are already addressed by national legislation and have so far not been a real problem.

On a national level a 'Referral Index', that will serve as a 'control tower' that knows where information is residing, will be established. [4] The Regional Referral Index fits that strategy perfectly well. Ultimately, the national referral index will allow a nation wide exchange of health information. It is expected to take some years before this is realised. Since 95 % of care takes place on a regional level, the region has decided to continue the developments to realise health information exchange on a regional scale. In addition to this, national developments currently only concern access to the Electronic Medication Record and the General Practitioner Deputy Record. [5]Due to experiences with Upid in the region Utrecht, it is relatively easy to add other types of health information, such as laboratory, and radiology. The regional developments correspond with the national developments so the region can 'connect' to the 'National Referral Index' once it becomes operational.

5. Conclusion

Health information exchange is a topic of great urgency because it will eventually improve care and save cost. The fact that 95% of healthcare takes place on a regional level makes it a logical choice to continue on regional level. The Regional Referral Index is an excellent starting point to continue developments concerning health information exchange. All possible precautions are being taken to be able to link up smoothly with the national developments, so the decision to continue the already started regional trajectory in line with the national developments seems a wise one. It will allow the region to become one of the frontrunners in this field in the Netherlands. Furthermore the project can serve as a learning case for other regional and national initiatives.

Acknowledgments

We would like to thank the ICT Steering Committee of the Regional Commission of Healthcare (RCG) for their support in accomplishing the Regional Referral Index, and their commitment to continue regional developments.
We would also like to thank Mr. Jaco van Duivenboden, senior staff member of Uzorg BV (Utrecht) for his contribution to this paper.

References

[1] Duivenboden, van, J.L. Upid a virtual electronic patient record. Health Information developments in the Netherlands; March 2005.
[2] Architectuurontwerp basisinfrastructuur in de zorg, november 2005, NICTIZ.
[3] Specificatie van de basisinfrastructuur in de zorg, november 2005, NICTIZ.
[4] Implementatiehandleiding, HL7v3 Zorg Informatie Makelaar, versie 2.4.1, NICTIZ, 20 april 2006.
[5] Programma van eisen voor een goed beheerd zorgsysteem (GBZ), versie 1.0.1, NICTIZ, 20 april 2006.

5. Decision Support

5.1 Decision Support: Guidelines and Protocols

Ubiquity: Technologies for Better Health in Aging Societies
A. Hasman et al. (Eds.)
IOS Press, 2006

Enabling protocol-based medical critiquing

Robert-Jan Sips [a,1], Loes Braun[a] and Nico Roos[a]
[a]*Department of Computer Science, Universiteit Maastricht, The Netherlands*

Abstract. This paper investigates the combination of *expert critiquing systems* and *formal medical protocols*. Medical protocols might serve as a suitable basis for an expert critiquing system because of the ongoing acceptance of medical protocols and the rise of both evidence-based practice and evidence-based protocols. A prerequisite for a critiquing system based on medical protocols is the ability to match the actions a physician performs in practice to actions prescribed by a protocol. Previous research has shown that this is quite difficult, due to the fact that computerized systems are unable to handle deviations from a protocol, which are common in the medical domain. Our solution to this problem is based on extracting the intention underlying a physician's action and uses the intention as the basis for matching performed actions to prescribed actions. We propose an algorithm for the intention-based matching process and we evaluate the matching algorithm on 12 cases of hyperbilirubinemia in healthy term newborns.
Keywords: Expert Critiquing Systems, Medical protocols, Intentions.

1. Introduction

In the past three decades, various authors in the medical domain have expressed their worries about the state of medical practice [4,7]. The boost of medical research in the past decades increased the difficulty of decision making for individual physicians, the "data overload has paradoxically led to a knowledge underload" [15]. To improve the quality of the medical practice and to support physicians in the process of decision making a number of tools has been proposed, including *medical protocols* and *expert critiquing systems* [4]."

Expert critiquing systems were introduced to assist physicians in decision making, without forcing them to comply to a gold standard of care [12,19]. Expert critiquing systems do this by providing critique on a physician's decisions, rather than telling him exactly what to do [12].

The first expert critiquing systems developed by Miller [12] relied heavily on user interaction, which resulted in rejection of the systems [8]. To avoid rejection, Van der Lei [19] proposed a system which gathers all its information from computer stored medical records, critiquing a physician without relying on user interaction. The current developments, such as the introduction of an *electronic patient record* (EPR) in The Netherlands, support this proposal.

A major problem with the critiquing systems proposed by Miller and Van der Lei is that these systems cannot cope with deviations from the underlying model. Moreover, the systems were not able to deal with the question *why* a physician was performing an

[1]Corresponding Author: Robert-Jan Sips, Department of Computer Science, P.O.Box 616, 6200 MD Maastricht, The Netherlands. Email rj.sips@cs.unimaas.nl.

action, which is essential to provide a grounded critique. To address both problems, Shahar *et al.* [16] and Advani *et al.* [1,2] suggest to perform critiquing by assessing the compliance of a *physician's intentions* with the *intentions behind a medical protocol*. They also stress that plan recognition is an indispensable prerequisite for the performance of critiquing.

Marcos *et al.* [10] claim that matching actions performed by a physician to those prescribed by the guideline is rather difficult. Moreover, they claim that asking the physicians to report the intentions behind the preformed actions offers no solution because of the large variance in the degree of detail, abstraction level, etc. in the reported intentions.

In the paper we will show that recent developments in languages for describing medical protocols combined with information of treatment actions and medicines provided by insurance companies in The Netherlands enables us to overcome the objections stated by Marcos *et al.*

Medical protocols were introduced to standardize medical practice. The use of medical protocols has been shown to reduce practice variations [9], improve practice quality [17], and improve the cost efficiency of medical care [11].

To improve medical protocols and to make medical protocols computer understandable, research has been conducted on the formalization of medical protocols, using formal modelling languages [14]. The most-developed examples [18] of these languages are ASBRU, EON, GLIF, GUIDE, PRODIGY and PROFORMA. We evaluated these languages to determine to what extend they can be used for critiquing systems.

For intention-based critiquing, it is important that the intentions are formulated in a computer-interpretable manner, i.e., in a formal manner. ASBRU, EON, GUIDE and PROFORMA are the only languages which specify intentions formally from which ASBRU has the most extensive intention-modelling capacity [13]. Moreover, the ASBRU protocol model is the simplest, containing only one generic plan object. In contrast to the other languages, ASBRU and PROFORMA, can model sequential and parallel execution, without introducing a new modelling concept. Therefore, we have chosen to investigate a critiquing system that uses a medical protocol described in ASBRU.

2. Intention-based treatment identification

Medical actions vary from medicine prescriptions to blood tests and from examining x-rays to surgical interventions. All of these actions can be used in a different context within different treatments. Measuring one's blood type (referred to in literature as an ABO test [3]) could be used in, for instance, a protocol containing a blood transfusion to determine the applicability of the blood for transfusion. However, in a protocol for Jaundice in newborns, the protocol can prescribe the same action to determine the likeliness of a blood-group antagonism in the newborn, i.e., the likeliness that the newborn has antiserum to its own blood [5]. Because the intention of an action can change based on the context in which the action is performed, it is hard to model the possible intentions of an action.

When we re-examine the example above, we see that in all cases the ABO test is used to determine the patients' blood group. This is a context independent intention of this action. Based on these observations, we define two types of intentions:

- **High-level intentions** Specifying an intention of an action in a specific context, e.g., "determine the possibility of blood group antagonism".

- **Low-level intentions** Specifying a context independent intention of an action, e.g., "determine blood group".

The high-level intentions relate to the intentions of medical protocols and low-level intentions relate to a physician's actions and to actions in the medical protocols.

A medical protocol formalized in a language such as ASBRU describes the clinical actions needed to realize a high-level intention, e.g., "treat hyperbilirubinemia". These protocols have the following properties.

- For each high-level intention in the protocol, there are one or more (sub)protocols which realize this high-level intention, using other high-level intentions and actions.
- The number op (sub)protocols is finite and there are no cyclic dependencies between high-level intentions.

These properties imply that each high-level intention can be transformed into one or more *execution sequences of actions*. Such an execution sequence is obtained by determining the execution through the protocol, using the execution order specified by the protocol.

To explain the trace of actions performed by a physician and reported in the electronic patient record, we have to match these actions with the actions prescribed by one of the execution sequences of a protocol. Woolf *et al.* [20] pointed out that seemingly arbitrary factors as the treating physician, the hospital or the geographical location cause physicians to divert from protocol-prescribed actions. However, when deviating from the protocol's actions, physicians can be assumed to follow the protocol's intentions [1]. Therefore, we replace both the protocol-prescribed actions and the observed action by low-level intentions.

In the Netherlands, some types of actions are thoroughly documented and restricted by the health care insurance companies. For example, the types and brand names of medicines which can be prescribed in the Netherlands are documented in the *Farmacotherapeutisch kompas* (pharmacotherapeutic compass) [6]. The pharmacotherapeutic compass groups medicines which *do the same* into *pharmaceutical groups*. For instance, Velosulin, Humalog and Novorapidwhich are in the same *pharmaceutical group*, but differ in brand names.

In order to reason about the low-level intentions of a medicine prescription, we translate medicine prescriptions to pharmaceutical groups. So the system can reason about, e.g., the prescription of "a short working blood-glucose lowering substance", instead of the prescription of Velosulin. We believe that this translation corresponds to a physician's way of thinking. In a similar way, we have used the Merck-manual [3] for translating treatment-actions to low-level intentions. Hence, using the medical literature mentioned above, it is possible to obtain a complete set of possible actions a physician might perform, linked with the low-level intentions that may be pursued by these actions.

In order to identify the medical protocols, the intentions of which are followed by the physician, we use the protocols to rewrite a high-level intention (e.g., the treatment of hyperbilirubinemia) to one or more possible execution sequences of actions and subsequently translate each action to a low-level intention using a translation table derived from the pharmacotherapeutic compass and the Merck manual. Similarly, we translate the sequence of actions reported by the physician in the EPR to low-level intentions. Next, we applied a matching algorithm to identify the execution sequence that corresponds best to the action reported by the physician.

To identify protocols followed by the physician, we define a measure for the distance between an execution sequences prescribed by the protocols and the action sequences reported in the EPR. The *distance* is defined as:

1. the number of actions in de EPR of which the corresponding low-level intentions are matched by the execution sequence of the protocol, minus
2. the number of unmatched low-level intentions in the execution sequence that occur before the last matched intention in the execution sequences.

Note that this measure does not explicitly take deviations with respect to the order prescribed by protocols into account. This increases the robustness of the matching process with respect to variations in the order performed actions are reported in the EPR. By minimizing the number of unmatched intentions prescribed by the protocol, we ensure that the correct prefix of an execution sequence is identified, which is important for medical critiquing.

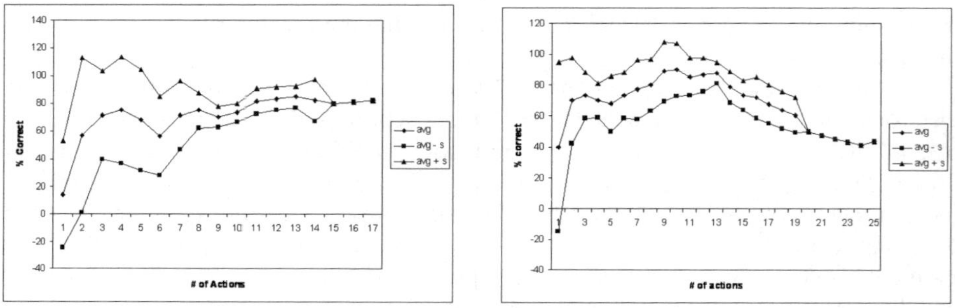

Figure 1: Average results of matching Marcos' and Mulder's actions respectively to the protocol. *s* denotes sample variance, *avg* denotes sample mean.

3. Experiments

To give a *proof of principle* of our approach described in the previous section, we implemented a prototype and evaluated the prototype using an ASBRU-modelled protocol, constructed in the Asgaard project [11]. This protocol models the diagnosis and treatment of Jaundice in healthy newborns. The protocol is based on the Jaundice protocol of the American Association for Pediatrics (AAP), which is intended for the management of Jaundice in healthy term (defined as 37 completed weeks of gestation) newborns [11]. Jaundice, or hyperbilirubinemia is a common disease in newborn babies and is caused by an elevated blood-bilirubin level [11]. For an overview diagnosis and treatment, we refer to the Merck manual [3], page 2156.

Data of treatments were obtained from Marcos *et al.*, who provided us with the test data used in their critiquing experiment [10]. Marcos *et al.* obtained their data from a pediatrician addressing patient cases. Since the pediatrician consulted by Marcos *et al.* received the protocol prior to addressing the patient cases, we asked Twan Mulder MD PhD, who works as a pediatrician / neonatologist at Maastricht University Hospital to address patient cases without prior knowledge of the protocol. In this way, we gathered solutions to 12 cases. The pediatrician consulted by Marcos *et al.* provided solutions to 7 cases and T.Mulder provided solutions to case 5 cases.

In the first experiment we measured the performance of our algorithm on all 12 cases. For each of the 12 sequences provided by the physicians, we generated new sequences by considering all prefixes of the complete sequence. The prefixes represent the available data after every step in the treatment process. Subsequently, our algorithm matched each generated sequence to the protocol. Figure 1 shows the average results of matching these sequences with the protocol for all 12 cases.

In the second experiment we measured the performance of our algorithm after changing the order of actions preformed by the physician. The rationale behind this experiment is that the physician enters the information on consultation basis, where it is not guaranteed that the actions are entered in the order in which they are performed.

4. Discussion

On inspection of the experimental results, we see that, on average, 69.4% of the actions in a test sequence are matched correctly. When we have a closer look at the misplaced action types, we see that 19 out of the 27 misplaced action types cannot be matched due to incompleteness of the protocol. The protocol is incomplete with respect to the specification of treatment for several causes of hyperbilirubinemia. When we discard the actions which cannot be matched, we observe an average performance of 80.1%.

We can also see that the average performance on short test sequences is worse than on long sequences. This is caused by an error in the protocol constructed in the Asgaard project. The first action performed by the pediatrician is often "Measure TSB value". The algorithm can only match this action to a similar action in sequences where the pediatrician has already prescribed treatment. Looking at the Merck manual [3] and the Diagnostic Compass [5], we find that the TSB value (Total Serum Bilirubinemia) is the criterium for the presence of hyperbilirubinemia and should therefore be the first action of the protocol. Thus the lack of an initial measurement of TSB indicates a gap in the protocol. When we discard the action "measure TSB value", which cannot be matched correctly due to a gap in the protocol and the actions which have no match in the protocol, we observe that, on average, our algorithm matches 95.8% of the actions in a sequence to the correct actions in the protocol.

Based on the results of the first experiment, we investigated whether there is a significant difference in the performance of our algorithm on solutions of a pediatrician familiar with the protocol and on solutions of a pediatrician unfamiliar with the protocol. The results indicate a better performance on the pediatrician consulted by [10]. To determine whether this difference is significant, we conducted a statistical test. The result of the test showed that the difference is not significant.

Similarly, we investigated whether the results of the second experiment were significantly worse that those of the first experiment. The statistical test showed the difference is not significant.

5. Conclusion

In this paper we have proposed that medical protocols can be used for medical critiquing, by translating a physicians actions, reported in the electronic patient record and actions prescribed by medical protocols to context-independent 'low-level' intentions, using the pharmacotherapeutic compass and the Merck manual. We have demonstrated through a series of experiments, that the actions performed by the physician can successfully be identified, thereby enabling medical critiquing. Future research will focus on better measures to identify the protocols used by a physician, on more extensive experiments, and on situations in which several diseases are treated at the same time.

References

[1] A. Advani, K. Lo, and Y. Shahar. Intention-based critiquing of guideline-oriented medical care. In Proceedings American Medical Informatics Annual Symposium 1998, pages 483–487, 1998.

[2] A. Advani, Y. Shahar, and M.A. Musen. Medical quality assessment by scoring adherence to guideline intentions. JAMIA, 9:S9– S97, 2002.

[3] M.H. Beers and R. Berkow. The Merck Manual of Diagnosis and Therapy, Seventeenth Edition. Merck Research Laboratories, Whitehouse Station, 1999.

[4] M. Berg. Rationalizing medical work. PhD thesis, Maastricht University, 1995.

[5] Commissie Aanvullende Diagnostiek. Diagnostisch kompas. College voor Zorgverzekeringen, Amstelveen, 2003.

[6] Commissie Farmaceutische Hulp. Farmacotherapeutisch kompas. College voor Zorgverzekeringen, Amstelveen, 2002.

[7] B.G. Druss, W.D. Bradford, R.A. Rosenheck, M.J. Radford, and H.M. Kruzholz. Quality of medical care and excess mortality in older patients with mental disorders. Arch. Gen. Psych, 58(6):565–572, 2001.

[8] A.S. Gertner. Plan recognition and evaluation for on-line critiquing. User Modeling and User-Adapted Interaction, 7(2):107–140, 1997.

[9] R.P. Giugliano, C.A. Camargo, D.M. Loyd-Jones, J.D. Zagrodsky, J.D. Alexis, K.A. Eagle, V. Fuster, and C.J. O'Donnel. Elderly patients receive less aggressive medical and invasive management of unstable angina. Archives of Internal Medicine, (158):1113–1120, 1998.

[10] M. Marcos, G. Berger, F. van Harmelen, A. ten Teije, H. Roomans, and S. Miksch. Using critiquing for improving medical protocols: Harder than it seems. In Proceedings of the Eighth European Conference on Artificial Intelligence in Medicine, pages 431–441, 2001.

[11] M. Marcos, M. Balser, A. ten Teije, and Frank van Harmelen. From informal knowledge to formal logic: a realistic case study in medical protocols. In Proceedings of the 13th International Conference on Knowledge Engineering and Knowledge Management. Springer-Verlag, 2002.

[12] P.L. Miller. Expert critiquing systems: Practice based medical consultation by computer. Springer-Verlag, New York, 1986.

[13] M. Peleg and J. Tu. Comparing models of decision and action for guideline-based decision support: a case-study approach: Stanford university, 2002. Report No.: SMI-2002-0922.

[14] M. Peleg, S. Tu, J. Bury, P. Ciccarese, J. Fox, R.A. Greenes, R. Hall, P.D. Johnson, N. Jones, A. Kumar, S. Miksch, S. Quaglini, A. Seyfang, E.H. Shortliffe, and M. Stefanelli. Comparing computer-interpretable guideline models: a case-study approach. JAMIA, (10):52–68, 2004.

[15] I. Purves. Computerised guidelines in primary healthcare: Reflections and implications. In Health Telematics for Clinical Guidelines and Protocols, pages 57–74, Amsterdam, 1995.

[16] Y. Shahar, S. Miksch, and P. Johnson. A task-specific ontology for the application and critiquing of time oriented clinical guidelines. In Proceedings of the Sixth Conference on Artificial Intelligence in Medicine in Europe, pages 51–61, 1997.

[17] S.B. Soumerai, T.J. McLaughlin, J.H. Gurwitz, E. Guadagnoli, P.J. Hauptman, C. Borbas, N. Morris, B. McLaughlin, X. Gao, D.J.Willison, R. Asingera, and F. Gobel. Effect of local medical opinion leaders on quality of care for acute myocardial infarction: a randomized controlled trial. JAMIA, (279):1392–1394, 1998.

[18] D.R. Sutton and J. Fox. The syntax and semantics of the proforma guideline modelling language. JAMIA, (10):433– 443, 2003.

[19] J. van der Lei. Critiquing based on computer stored medical records. PhD thesis, Erasmus University Rotterdam, 1991.

[20] S.H.Woolf, R. Grol, A. Hutchinson, M. Eccles, and J. Grimshaw. Clinical guidelines: Potential benefits, limitations, and harms of clinical guidelines. Bio Medical Journal, (318):527–530, 1999.

Ubiquity: Technologies for Better Health in Aging Societies
A. Hasman et al. (Eds.)
IOS Press, 2006

Management of Data Quality - Development of a Computer-Mediated Guideline

Jürgen STAUSBERG[a,1], Michael NONNEMACHER[a], Dorothea WEILAND[a], Gisela ANTONY[b], Markus NEUHÄUSER[a]

[a] *Institute for Medical Informatics, Biometry and Epidemiology, Medical Faculty, University of Duisburg-Essen, Germany*

[b] *Competence Network on Parkinson's disease, Marburg, Germany*

Abstract. Appropriate data quality is a crucial issue in the use of electronically available health data. As source data verification (SDV) and feedback are two standard procedures for measuring and improving data quality it would be worthwhile to adapt these procedures to a current level of quality in order to reduce costs in data management. This project aims to develop a guideline for the management of data quality with special emphasis on this adaptation against the backdrop of research networks in Germany, which operate registers and conduct epidemiological studies. The first step in guideline development was a thorough literature review. The literature offers many measurements as candidates for quality indicators, however, systematic assessments and concepts of SDV and feedback are missing. We assigned possible quality indicators to the levels plausibility, organization, and trueness. Each indicator must be operationally defined to allow automatical calculation. The SDV sample size calculation leads to lower numbers for sites providing data of good quality and larger numbers for sites with poor data quality. The guideline's implementation in a software tool combines two cycles, one for the adaptation of recommendations to a given study/register, the other for the improvement of data quality in a PDCA-like approach. The recommendations will address needs common to medical documentation in daily health care, clinical, epidemiological, and observational studies as well as in surveillance data bases and registers. Further work will have to supplement other aspects of data management.

Keywords: Documentation, Feedback, Quality control, Source data verification

1. Introduction

The availability of electronically stored data on patients, health care services and health technologies has increased dramatically. These data are captured within well-defined settings such as clinical trials, registers, and provider payment systems. The very fact of their availability, however, wakens the desire to use them independent of their primary function. The extended use of electronically available health data raises concerns about data quality. For example, data captured within provider payment

[1] Corresponding Author: Priv.-Doz. Dr. med. Jürgen Stausberg, Institute for Medical Informatics, Biometry and Epidemiology, Medical Faculty, University of Duisburg-Essen, Hufelandstr. 55, 45122 Essen, Germany; E-mail: stausberg@uni-essen.de.

systems are regarded as data of minor quality and often denigrated as purely administrative in opposite to data collected in clinical trials.

To assure high data quality for primary as well as for secondary usages, standard procedures for data management are needed [1]. First, data quality depends to a great extent on the motivation of the person responsible for data input. This motivation is high if the data have direct benefits for this person's daily business (e.g. for a physician who plans her/his own operation schedule), and if data input is integrated in clinical pathways. Incentives created by multiple use of the same data also increase data quality by lowering the workload for data input and by introducing second, third or even multiple awareness situations of data completeness and correctness, for example by using the operation schedule information for the operation catalogue of each surgeon. Second, any data acquisition should be accompanied by on-site quality control procedures [2]. These procedures range from automatically performed plausibility checks at data entry to working lists of missing data. Third, the central units responsible for data storage and data processing can provide further support and incentives for quality control such as feedback and source data verification (SDV). Feedback means periodic reports about data quality provided to those persons, divisions, or sites doing data input. It could be demonstrated that knowledge of somebody's own performance is an important stimulus for improvement [3]. SDV is obligatory in clinical trials for pharmaceuticals according to Good Clinical Practice. SDV means the comparison of data with their primary source, e.g. case report forms (CRFs) against a paper-based patient record. The role of SDV is threefold: It provides measurements of data quality, it enables the correction of wrong or missing data, and it also stimulates quality improvement through its controlling nature [4].

Within the Competence Network on HIV/AIDS (funded by the German Federal Ministry of Education and Research) we had been responsible for the setup and quality control of a national register for HIV-infected persons, with data transmitted from external sources like hospital information systems as well as recorded online via the World Wide Web and a data management concept including an on site monitoring with SDV. Together with our interests in low cost monitoring of clinical trials [5] the challenge of operating the HIV-register led to the idea of a computer-mediated guideline for the management of data quality.

In cooperation with other Competence Networks we initiated a project within the "Telematikplattform für Medizinische Forschungsnetze e. V." (TMF), a meta-organization working to solve logistic, technical, and administrative problems for clinical research networks. Our project aims to develop a guideline for the management of data quality in registers and epidemiological studies. This guideline is to provide recommendations about SDV and feedback adapted to the specific needs of a given register/study on the one hand and the actual quality of data on the other hand. The adaptation to the actual level of data quality is especially important because the resources available in registers/epidemiological studies typically do not allow for monitoring large data bases with a full SDV. Therefore SDV activities should be increased in cases of lower data quality and could be reduced in cases of sufficient or high data quality. The project started in July 2005. We plan to make two products available in the end, a document and a software tool, which assists a central data management in applying the guideline within their setting. In the following we present the results of a systematical literature review completed in 2005 as well as an overview of our current state in guideline development.

2. Material and methods

The project plan defines two phases: the development and implementation of the guideline and an associated software tool in the first year and their empirical evaluation in the second year. The development of the guideline follows national recommendations for medical guidelines [6]. An advisory board comprising four Competence Networks as well as the funding organization monitor and review the work. The first version of the guideline is available since spring 2006. We started by systematically analyzing the available evidence on data management through an extensive literature review and (open and semi-structured) interviews with experts. Based on this analysis our further work is focused on recommendations for

- quality indicators,
- sample size estimation of SDV adapted to the quality of data, and
- feedback (also adapted to data quality).

If necessary, amendments to the recommendations found in the literature are added by using analogies.

2.1. Literature review

The literature review comprises the topics of quality measurement in health care and industry, statistical methods of quality assurance, quality control, round-robin tests, and standards/recommendations for development and content of guidelines in health care. Sources include databases like Medline (via www.pubmed.org), search engines like www.google.de, and institutions associated with quality management. References found in the literature section of previously identified articles were added as well as articles known to the authors.

The main search was performed between July and September 2005. We used several different queries for the Medline search, for example with the headings "quality control", "data collection" and "feedback". Query results were evaluated by one of the authors using the title and abstract (if available). Potentially relevant articles were obtained in full text (if possible) and qualitatively analyzed. Medline provides about 3430 references (with duplicates). Concerning data quality, the October 2005 project report includes 132 different documents with following topics

- Data quality in health care in general - 10 documents
- Data quality in epidemiological trials - 8 documents
- Data quality in clinical trials - 35 documents
- Data quality in routine medical care - 46 documents
- Data quality in registers - 26 documents
- Secondary data analysis - 4 documents
- Data warehousing - 3 documents

The literature review did not reveal a systematic approach to data management like the one intended within this project. Moreover, for most of the techniques used in data management an empirical rationale is missing. Arts et al. propose a "framework of procedures for the assurance of data quality in medical registries" based on experiences in intensive care, which is very helpful for an overview on relevant aspects [1].

Concerning feedback, recommendations on an appropriate concept are - more or less - missing in the literature reviewed. We are therefore planning a survey about different feedback strategies with one of our partners.

2.2. Software design

We aim to design a software tool which assists in applying the guideline. On the one hand, this software tool is to enable an adaptation of the guideline to the characteristics of a given study/register. On the other hand, it is to support the adaptation of SDV and feedback to actual data quality as recommended by the guideline itself. Necessary calculations, e.g. sample size estimations for SDV, shall be made available. Therefore the application needs an interface for entering data about the study/register as well as data required for the calculation of quality indicators. The output of the application, i.e. the recommendation of the guideline for a specific setting, are to be made available in printed format and as files for automatic processing. The eXtensible Markup Language (XML) was chosen as standard format for import and export. Process modeling and software design is done with the Unified Modeling Language (UML).

3. Results

3.1. Indicators for data quality

Our literature review provides a substantial collection of possible indicators for data quality. We assigned the candidates to the three categories of plausibility, organization, and trueness, which we defined for data quality in analogy to the categories structure, process, and outcome as proposed by Donabedian for medical care [7]:

- Plausibility: agreement with previous values, concordance, consistency, distribution of values within and between study sites, known correlations, medical tests on weekend, missing values in optional variables, outliers (continuous variables), permitted values (categorical variables), preferences in last digit
- Organization: duplicates, drop-out rate, homonyms, qualification of personnel for data input, recruitment, synonyms, timeliness
- Trueness: accuracy, agreement with source data, completeness, compliance with study plan, missing values in mandatory variables, representativeness

Like Donabedian we presume that high plausibility "increases the likelihood" of good organization, and good organization "increases the likelihood" of actual trueness.

3.2. Source data verification (SDV)

SDV is often used to assess completeness or accuracy in clinical trials [8], registers [9] and electronic patient records [10]. Some authors give recommendations on the number of variables to be included in SDV, the frequency of SDV and the interpretation of disagreements [4, 11]. These recommendations, however, are not justified by the results of empirical surveys comparing different approaches. The aim of sample size calculation is the determination of the number of units (e. g. patients) and items to be assessed, and the frequency of SDV. In our concept for an adapted SDV we generate a data quality score by aggregating the quality indicators from the central study/register data base and the disagreements identified during SDV itself. The SDV should then confirm the actual category of the score, each category being associated with an expectation of a proportion of units or items with at least one missing or wrong value (p). The required precision of the measured proportion (δ) should be lower in categories representing good quality; thus we automatically generate larger sample sizes in situations with minor quality. First, we calculate an unadjusted sample size n_0 with the formula in figure 1 for $\alpha = 0.05$ and $z_{1-\alpha/2} = 1{,}96$. Second, we adapt the unadjusted sample size n_0 with the number of units that each site recorded in the previous period.

$$n_0 = \frac{p(1-p)}{\delta^2} \cdot z_{1-\alpha/2}^2$$

Figure 1. Calculation of the unadjusted sample size for SDV

3.3. Workflow

We distinguish the use cases determination of data quality in the central study/register data base, determination of the data quality as seen in SDV, SDV planning, and feedback planning. Actor is the study/register data manager. Figure 2 shows the current process model with the data manager's activities in the middle, the application's activities at the top and the local study sites' activities at the bottom. The workflow model comprises two cycles. One cycle consists of a periodic determination of input parameters by the data manager, the transfer of the results into the application, the adaptation of the guideline to the quality indicators and the study characteristics by the application, the receipt of the recommendations by the data manager, and the execution of SDV and feedback. Another nested cycle represents the stimulus of the first cycle's results for data quality, comparable to Deming's PDCA-cycle [12]. The recommendations include an evaluation of the quality of data in form of quality indicators and quality scores. That information could be used for a weak-point analysis by the central data manager as well as by each local site. The weak-point analysis should lead to the implementation of procedures to improve data quality. It is to be hoped that new data will then be collected with better quality.

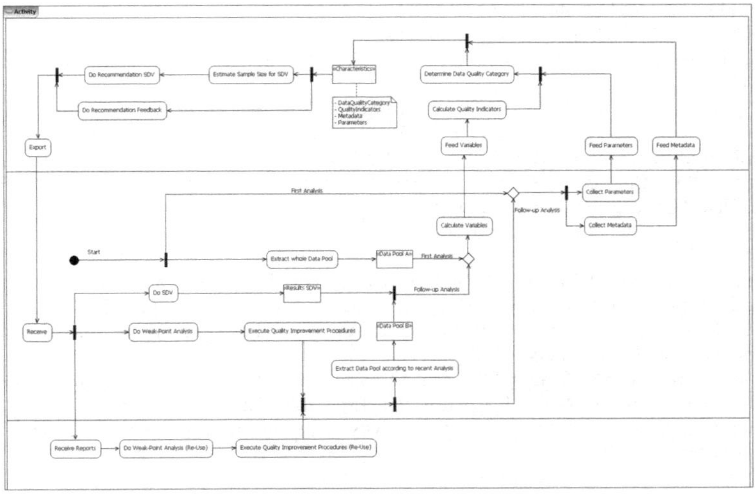

Figure 2. UML activity diagram for the application of the computer-mediated guideline

4. Conclusions and future plans

Our thorough literature review revealed the need for a systematic discussion of the underlying principles of data management and the necessity for reasonable strategies. We identified many possible indicators for data quality and operationally described them for an automatical calculation. Because of missing empirical work the aggregation of quality indicators for generating the quality score follows a pragmatical and inductive approach. Here it will be possible to consider study/register specific weights. The sample size estimation for SDV primarily focuses on the number of units and items to be verified. After setting a fixed interval between two runs of the guideline the frequency of monitor visits will depend on the sample size. For a further definition of appropriate feedback strategies we await the results of the empirical work conducted by one project partner. The computer-mediated guideline will offer a systematic approach to data management with special emphasis on SDV and feedback. One should bear in mind that other important factors, such as motivation and incentives, influence data quality. The results of the project should be one part of a more comprehensive solution surrounding the whole picture.

Acknowledgements

The project is funded by the TMF. We thank our project partners for cooperation and Katarzyna Domagala for support in literature retrieval.

References

[1] Arts DG, De Keizer NF, Scheffer GJ. Defining and improving data quality in medical registries: a literature review, case study, and generic framework. J. Am. Med. Inform. Assoc. 2002; 9: 600-11.
[2] Stausberg J. Evaluation of quality control methods for medical documentation. In: Pappas C, Maglaveras N, Scherrer JR, eds. Medical Informatics Europe '97. Amsterdam: IOS, 1997: 864-8.
[3] De Lusignan S. Using feedback to raise the quality of primary care computer data: a literature review. In: Engelbrecht R, Geissbühler A, Lovis C, Mihalas G, eds. Connecting Medical Informatics and Bio-Informatics. Proceedings of MIE2005. Amsterdam: IOS, 2005: 593-8.
[4] Khosla R, Verma DD, Kapur A, Khosla S. Efficient source data verification. Indian Journal of Phamacology 2000; 32: 180-6.
[5] Ose C, Weiland D, Jöckel K-H. How to practice low cost monitoring in clinical trials. Informatik, Biometrie und Epidemiologie in Medizin und Biologie 2004; 35: 54-62 (in German).
[6] AWMF, Ärztliches Zentrum für Qualität in der Medizin. Das Leitlinien-Manual von AWMF und ÄZQ. Z. ärztl. Fortbild. Qual.sich. 2001; 95: Supplement I.
[7] Donabedian A. The quality of care. How can it be assessed? JAMA 1988; 260: 1743-8.
[8] Weiss RB. Systems of protocol review, quality assurance, and data audit. Cancer Chemother. Pharmacol. 1998; 42: S88-S92.
[9] Maruszewski B, Lacour-Gayet F, Monro JL, Keogh BE, Tobota Z, Kansy A. An attempt at data verification in the EACTS Congenital Database. Eur J Cardiothorac. Surg 2005; 28: 400-4.
[10] Mikkelsen G, Aasly J. Concordance of information in parallel electronic and paper based patient records. Int. J. Med. Inform. 2001; 63: 123-31.
[11] Pogash RM, Boehmer SJ, Forand PE, Dyer AM, Kunselman SJ. Data management procedures in the Asthma Clinical Research Network. Control Clin. Trials 2001; 22:168S-80S.
[12] Deming WE. Out of the crisis. Quality, productivity and competitive position. Cambridge: Cambridge University Press, 1982.

Ubiquity: Technologies for Better Health in Aging Societies
A. Hasman et al. (Eds.)
IOS Press, 2006

Design of a Decision Support System for Chronic Diseases Coupling Generic Therapeutic Algorithms with Guideline-Based Specific Rules

Vahid Ebrahiminia [a,1], Christine Riou [b], Brigitte Seroussi [c], Jacques Bouaud [c], Stéphane Dubois [d], Hector Falcoff [e], Alain Venot [a]

[a] *Laboratoire d'Informatique Médicale & Bioinformatique, University of Paris 13*
[b] *Laboratoire d'Informatique Médicale, University of Rennes 1*
[c] *STIM/DPA/DSI, Assistance Publique-Hôpitaux de Paris*
[d] *RESIP, Boulogne sur Mer*
[e] *Société de Formation Thérapeutique du Généraliste (SFTG), Paris*

Abstract. Clinical Decision Support Systems (DSS) help improve health care quality. They usually incorporate an Execution Engine (EE), defined for each disease. We have designed, and present here, a generic execution engine, coupled with guideline-based disease specific rules stored in knowledge base (KB) as part of the prescription-critiquing mode of the ASTI project. This system was designed using two national guidelines for type 2 diabetes and hypertension. It takes into account the patient's clinical data, the tolerance and effectiveness of previous and current treatments and the physician's prescription made at the time. The functioning of the system has been speeded up and its maintenance made easier by indexing the KB rules according to the type of treatment they are linked to (e.g. monotherapy, etc.) and by classifying them into four categories. The EE's design formalizes generic therapeutic algorithms, leading to treatment options for cases of bad tolerance or insufficient effectiveness of the current treatment. Its applicability to other diseases was shown by applying it to dyslipidemia.
Keywords: Decision Support System, Chronic Disease, Computer Interpretable Guideline, Treatment, Therapeutic Algorithm.

1. Introduction

Electronic clinical Decision Support Systems (DSS) are efficient tools for reducing medical errors [1] and improving the quality of health care [2]. However, a recent study [3] has shown that certain features of these systems can significantly improve the clinician's practice. For system to be efficient, the decision support must be automatically provided (a) as part of the workflow, (b) at the time of decision making and (c) with actionable recommendations.

The ASTI project aims to develop a guideline-based DSS to be used by GPs for managing chronic diseases [4]. The first prototype incorporating these features was applied to the treatment of hypertension and showed promising results [5]. However, when the patient's therapeutic history was complicated, the critiquing mode of the system showed some limitations. Therefore, we developed a new model for

[1] LIM&BIO UFR SMBH 74 rue Marcel Cachin 93017 Bobigny cedex, France, vahid.nia@avc.aphp.fr

representing both the patient's treatments in electronic records and the guideline recommendations in system's KB [6]. We then tried to modify the system so that (a) it can be applied without modification to every chronic disease and (b) it requires only a limited number of rules in the KB to facilitate its development and maintenance and to increase the calculation speed. The current systems use computer-interpretable guideline models [7] but incorporate an execution engine that must be specified for each disease [8]. Our approach for satisfying these constraints has been to couple generic therapeutic algorithms with disease specific KB rules. In this paper, we present the design of this new execution engine, the principal modifications made to the KB structure, and the testing of the system's applicability to dyslipidemia.

2. Materials and Methods

2.1. Selection and analysis of two clinical guidelines

We selected two national clinical guidelines for the treatment of hypertension and type 2 diabetes [9], [10]. We attempted to obtain a generic ability for this system by using these chronic diseases in which several therapeutic options exist and a variety of decision making processes are needed to decide on the treatment. We analyzed the guidelines and extracted information describing the selection and prescription processes, and the treatment evaluation. These three steps correspond to the classic therapeutic reasoning model [11]. We identified procedures and elements playing a role in these steps and defined a relevant framework for the design of the execution engine.

2.2. Generic therapeutic strategies in execution engine

Using the framework obtained above, we identified clinical and therapeutic situations (primarily bad tolerance or the absence of treatment effectiveness) for which similar actions are recommended in the guidelines. We formalized strategies for these situations to obtain generic algorithms that interact with the KB rules that take into account the specificity of each therapeutic domain. We considered treatment characteristics, such as "type" (e.g. monotherapy), "therapeutic class", "INN", "dose" or etc., according to our model of representing medical knowledge and data [5] that formalizes the guideline-recommended actions. We also considered any changes between the patient's current treatment and the treatment suggested by the system as the intended strategy of the system. For example "substitution" of a treatment component. We checked these strategies with a domain expert and determined the exceptions for which other rules in the KB were required. Finally, we completed the design of the execution engine such that it first decides between the principal strategies and then gradually specifies them to detailed treatments.

2.3. Applicability test of the execution engine

We tested the applicability of the algorithm to other diseases by applying it to recommendations obtained from a dyslipidemia guideline [12] and checked whether this algorithm using the specified KB leads to our desired critic system.

3. Results

3.1. The architecture and main functions of ASTI critiquing mode

The architecture and main functions of this system are shown in Figure 1. Globally, the execution engine receives patient data from the electronic record, interacts with KB and builds critics and suggestions about the physician's prescription.

The first task done by the critiquing mode of ASTI is to enhance the raw clinical and therapeutic data of the patient. For example, the body mass index is calculated, some prescription lines are grouped into treatments, commercial drug names are converted to ATC codes using a national drug data-base. It also detects changes between the prescription at that time and the treatment currently being taken by patient to describe the envisioned strategies of the physician at higher abstraction levels.

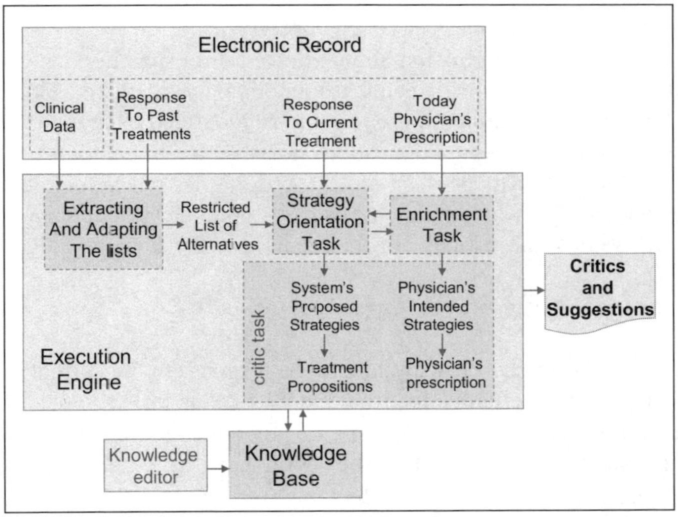

Figure 1: The main architecture of ASTI critiquing mode

The second task is to produce a temporary list of treatment alternatives. The KB rules applicable to the patient's situation are selected to give the most beneficial treatment alternatives, which are ranked according to guideline-specified priority. This temporary list is then reduced by deleting those alternatives associated with previous negative results due to intolerance, allergy or lack of therapeutic effect. This data must be available in the electronic records. Waited effectiveness also restricts the possible alternatives choices to those that are expected to result in the intended objective. This objective stated in guideline, can be accompanied by the duration of application, depends on disease severity, co-morbidities and characteristics of the patient.

The response of the patient to the currently applied treatment is taken into account in the third task to produce new strategies adapted to patient's situation. Based on these strategies and using the list obtained in Task 2, one or several possible treatment suggestions are produced. The last task consists of comparing the strategies and treatments proposed by the system with those intended by the physician allowing criticism of the system and suggestions to be made.

3.2. Organizing the rules in knowledge base

As chronic diseases may evolve, the physician may need to change the type of the treatment (e.g. for diabetes, treatments consisting of exercise and diet may be followed by oral mono, bi, tritherapy, and even insulin injections). Therefore, we group together the rules related to a given treatment type so that only those rules linked to the type of treatment taken by the patient (patient's current treatment) are used in task 1. This organization of the rules increases KB comprehensibility and maintenance. Examples of the types of treatments in diabetes are: non-pharmacological treatment alone, or with mono-, bi-, tritherapy or insulin injections.

We also classified the rules for each type of treatment according to their function:

- R1 rules are used to build the list of therapeutic alternatives and their priority ranks for various clinical and therapeutic patient situations.
- R2 rules indicate whether increasing or decreasing the dose is beneficial for some treatments; for example, increasing the dose of beta blockers is useless if no response to this class has been observed.
- R3 rules provide information about allowed drug substitutions within a given treatment type (e.g. for diabetes at monotherapy treatment, alphaglucosidase inhibitor is not prescribed if metformine is not effective enough).
- R4 rules provide information about possible treatment types if the system decides to switch to another type; for example, when monotherapy is not effective enough, what treatment type(s) should the system move to.

Figure 2 shows how these kinds of rules are used in the execution algorithm.

3.3. Defining generic strategies in therapeutic algorithm

The strategies (task 3) are chosen according to the response to the current treatment and by using the therapeutic alternative list from the R1 rules in task 2.

- If the current treatment is in the list of guideline-recommended treatments, is effective and well tolerated, it can be continued.
- If bad tolerance or insufficient effect is observed, or if the current treatment is not in the list, the treatment must be modified. Possible options are: changing the "dosage", the "INN", the "therapeutic class" or the "treatment type". In other words, the system can simply increase or decrease the dose, choose a drug from the same class or another class, or move from monotherapy to bitherapy.
- If the system decides to modify the treatment, It first checks with the R2 rules to determine whether to "increase the dose" when the therapeutic goal is not achieved. For cases of bad tolerance due to excessive effects, it may decide to "decrease the dose". The system also produces other suggestions by considering the first-ranked alternative from the list of task 2. At this stage, checking with the R3 rules may lead to some therapeutic classes in the list being added or removed, depending on the response of the patient. If at any stage of the decision-making, the obtained list becomes empty, the algorithm uses the R4 rules to add other treatment types to its suggestions and rebuilds new lists of treatment alternatives using the R1 rules linked to those types of treatment.

The system's suggestions are compared with the physician's prescriptions. No critic is made if the prescription is the same as the first ranked alternative in one of the lists. Otherwise, the strategies, the treatment type and other details of treatment are compared and the first ranked choices in the lists are suggested.

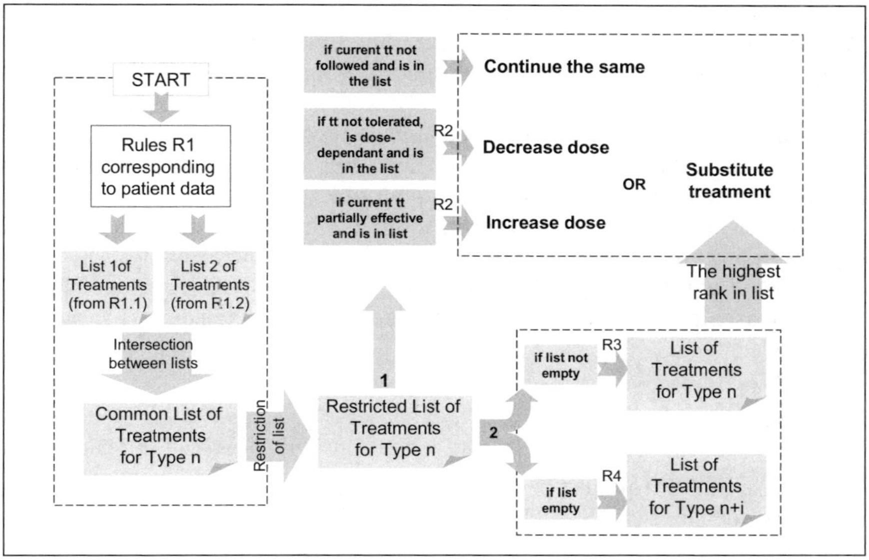

Figure 2: The main procedures for producing treatment suggestions

3.4. Application to hypertension and extension to other domains

The application of the critiquing mode of ASTI to the treatment of hypertension resulted in 27 rules, (2 initializing, 3 Life-style Changing Treatments "LCT", 17 Monotherapy+LCT, 4 Bitherapy+LCT and 1 Tritherapy+LCT). They used 20 clinical criteria to characterize the patient's clinical situations. Various treatment characteristics, such as tolerance (two values) or effectiveness (three values) are taken into account as therapeutic criteria. Example of list obtained in task 2 for a patient with heart failure due to systolic dysfunction is: rank 1; ACE inhibitor, rank 2; angiotensin-II receptor antagonists or diuretics (loop or thiazide-like) or beta blockers [9].

For dyslipidemia, the concept of treatment type was found to be satisfactory for characterizing a given treatment. We distinguished the same treatment types as those for hypertension, resulting in 14 rules. We were able to code all the therapeutic guideline recommendations using the knowledge editor developed according to the knowledge representation modes of ASTI.

4. Discussion and Conclusion

This execution engine couples a generic therapeutic algorithm with guideline-based rules in the knowledge base and satisfies to our main starting constraints. The system takes the concept of treatment type into account. This classification of treatments is classically used in guideline texts advising the physician what to prescribe at various steps of the disease evolution.

The coupling of these generic algorithms to the therapeutic knowledge base restricts the number of rules in the KB, leading to a rapid critic system and allowing easy updating and comprehension of the knowledge. This is particularly due to the rules being indexed according to their functions and treatment type.

We attempted to make the system generally applicable by using two chronic diseases, which have several therapeutic options, with a variety of decision making processes needed for choosing the treatment. However, most of the generic therapeutic strategies on which the system is based are derived from knowledge that is not explicitly written in the guidelines. Nevertheless, they follow medical common sense and seem to be sufficiently general to be applicable to other chronic diseases. Our approach for testing the generalness of the system is limited to dyslipidemia. Further evaluations in other domains are necessary to confirm this generic ability.

The method of proceeding and the reasoning of this system is close to a clinician's therapeutic reasoning, thus increasing the user's acceptance and comprehension of the system's suggestions.

Acknowledgements

The new phase of ASTI project is supported by a grant from FOPIM managed by the Haute Autorité de Santé (HAS), in France.

References

[1] Bates DW, Cohen M, Leape LL, et al. Reducing the frequency of errors in medicine using information technology. J Am Med Inform Assoc. 2001;8:299–308.

[2] Teich JM, Wrinn MM. Clinical decision support systems come of age. MD Comput. 2000;17(1):43–6.

[3] Kawamoto K, Houlihan CA, Balas EA, Lobach DF. Improving clinical practice using clinical decision support systems: a systematic review of trials to identify features critical to success. BMJ. 2005 Apr 2;330 (7494):765.

[4] Séroussi B, Bouaud J, Dreau H, Falcoff H, Riou C, Joubert M, Simon C, Simon G, Venot A. ASTI: a guideline-based drug-ordering system for primary care. Medinfo. 2001;10(Pt 1):528-32.

[5] Séroussi B, Bouaud J, Chatellier G, and Venot A. Development of computerized guidelines for management of chronic diseases allowing to position any patient within recommended therapeutic strategies. Medinfo 2004: 154-8.

[6] Ebrahiminia V, Duclos C, Toussi ME, Riou C, Cohen R, Venot A. Representing the Patient's Therapeutic History in Medical Records and in Guideline Recommendations for Chronic Diseases Using a Unique Model. Stud Health Technol Inform. 2005;116:101-6.

[7] Peleg M, Tu S, Bury J, Ciccarese P, Fox J, Greenes RA, Hall R, Johnson PD, Jones N, Kumar A, Miksch S, Quaglini S, Seyfang A, Shortliffe EH, Stefanelli M. Comparing computer-interpretable guideline models: a case-study approach. J Am Med Inform Assoc. 2003 Jan-Feb;10(1):52-68.

[8] Guideline Modelling Methods and Tools. OpenClinical, available at: http://openclinical.org/gmmintro.html (Accessed Jan 2005)

[9] Prise en charge des patients adultes atteints d'hypertension artérielle essentielle. HAS / Service des recommandations professionnelles, Juillet 2005, available at: http://www.anaes.fr (Accessed Nov 2005)

[10] Stratégie de prise en charge du patient diabétique de type 2 à l'exclusion de la prise en charge des complications. ANAES, Mars 2000, available at: http://www.anaes.fr (Accessed Nov 2004)

[11] Paolaggi JB, Coste J. Le Raisonnement médical: De la science à la pratique clinique. Paris: Editions Estem ; 2001

[12] Prise en charge thérapeutique du patient dyslipidémique. AFSSAPS, March 2005, available at : http://www.afssaps.sante.fr (Accessed Dec 2005)

5.2 Decision Support: Approaches

Ubiquity: Technologies for Better Health in Aging Societies
A. Hasman et al. (Eds.)
IOS Press, 2006

Automated Test Selection in Decision-Support Systems: a Case Study in Oncology

Danielle SENT[1] and Linda C. VAN DER GAAG
Department of Information and Computing Sciences,
Utrecht University
P.O. Box 80.089, 3508 TB Utrecht, The Netherlands

Abstract. Decision-support systems in medicine should be equipped with a facility that provides patient-tailored information about which test had best be performed in which phase of the patient's management. A decision-support system with a good test-selection facility may result in ordering fewer tests, decreasing financial costs, improving a patient's quality of life, and in an improvement of medical care in general. In close cooperation with two experts in oncology, we designed such a facility for a decision-support system for the staging of cancer of the oesophagus. The facility selects tests based upon a patient's health status and closely matches current routines. We feel that by extending our decision-support system with the facility, it provides further support for a patient's management and will be more interesting for use in daily medical practice. In this paper, we describe the test-selection facility that we designed for our decision-support system in oncology and present some initial results.
Keywords: decision-support system, test selection, patient-tailored management

1. Introduction

Since over the last decades researchers have come to understand more and more of diseases and their management, it nowadays is hard, even for medical specialists, to keep up-to-date with medical literature and with new insights of disease and new procedures. Furthermore, physicians have to be more and more aware that any mistake they make, not only will affect the patient under their care, but can also result in professional sanctions. On the other hand, the costs of treatments, tests and procedures are increasing. Hospitals have to provide care with a limited budget and yet have to keep the care at the highest possible level. To cope with the increasing complexity of medical practice and to provide for a constant level of care, medical procedures are becoming more and more standardised. Various different protocols and guidelines have been developed, for example, in which the physicians' daily

[1] Corresponding Author: Danielle Sent. Currently employed at: Information Systems Group, Faculty of Electrical Engineering, Mathematics and Computer Science, University of Twente, P.O. Box 217, 7500 AE Enschede, The Netherlands, tel: +31 53 489 5334, fax + 31 53 489 2927, e-mail: danielle.sent@utwente.nl, url: http://wwwhome.cs.utwente.nl/~sentd/

problem-solving strategies have been caught. Also, decision-support systems can assist physicians in their complex problem-solving tasks, by providing support that is tailored to individual patients.

To be accepted and integrated with other hospital information systems, a decision-support system should gain the confidence of the physicians and prove to be of added value to their daily practice. The conclusions reached by the system should be in accordance with the state of the art in medical practice. These conclusions should in addition really *support* its users, rather than only provide information they could have readily considered by themselves. To support the entire process of care of a patient, the decision-support system should not only provide information about the most probable diseases or best suitable therapy, it should also provide its user with patient-tailored information about which tests should be performed in the various phases of the patient's management. In this paper, we describe the design of such an automated test-selection facility.

Decision-support systems in medicine, where reasoning with uncertainty is quite prominent, are often based upon techniques from the field of statistics. More specifically, these systems increasingly build upon a Bayesian network for their reasoning purposes [1,2]. The test-selection algorithms currently in use with such decision-support systems are based upon the mathematical principles of decision theory. Most of these algorithms serve to select diagnostic tests in a myopic fashion. In each step, the most informative test is selected from among all possible tests, using a measure of informativeness from information theory. The user subsequently is prompted for the result of the selected test. The result is entered into the system and processed. Then the next most informative test is selected. This process continues until some stopping criterion is reached or results for all available tests have been entered.

Over the past decade, we have developed at Utrecht University a decision-support system that is aimed at patient-tailored management of oesophageal cancer. Initially, a myopic test-selection facility was added to the system. Upon working with the system, however, we felt that the resulting strategy was an oversimplification of the experts' problem-solving practice. Based upon knowledge elicited for this purpose, we designed a new test-selection facility for our system. The facility induces a strategy that is also based upon the mathematical principles of decision theory, yet more closely fits in with the test-selection strategy employed by the physicians in their daily practice. To study its practicability, we evaluated this facility in the context of our system and found highly satisfactory results. We feel that our new test-selection facility will also closely fit in with the daily strategies of physicians in other domains of medicine.

This paper is organised as follows. In Section 2, we provide some preliminaries from the field of oesophageal cancer and describe our decision-support system. In Section 3 we describe the various different components of our facility along with their background. Section 4 presents the test-selection strategy resulting from our facility in practice. We conclude the paper with our conclusions in Section 5.

2. Preliminaries

With the help of two experts in gastrointestinal oncology from the Netherlands Cancer Institute, Antoni van Leeuwenhoekhuis, a decision-support system for the staging of oesophageal cancer was developed [3]. Cancer of the oesophagus has a low incidence in the Netherlands and is often only diagnosed in a later stage of the disease. The tumour invades the oesophageal wall and may, in time, invade neighbouring organs beyond the oesophagus. When the tumour has invaded lymphatic vessels and blood vessels more specifically, it may give rise to secondary tumours, or metastases, in lymph nodes and in such organs as the liver and lungs. The depth of invasion of the primary tumour and the spread of its secondary tumours indicate the severity of the disease. In order to assess the stage of a patient's oesophageal cancer, generally a number of diagnostic tests are performed that serve to give insight in its different aspects. For patients suffering from oesophageal cancer, various different treatment alternatives are available which range from surgical removal of the primary tumour to prolong life expectancy to positioning a prosthesis for palliative care. Our decision-support system is aimed at patient-tailored management of oesophageal cancer. The system has a Bayesian network at its kernel. This network in essence provides for the staging of a patient's cancer. It models the presentation characteristics of an oesophageal tumour and describes the pathophysiological processes that influence its growth and metastasis. The network further represents the various diagnostic tests that are commonly used to gain insight in the various aspects of the tumour and the extent of its advance. The main diagnostic variable is the variable *Stage* that captures the extent to which the cancer has progressed. The network currently includes 42 stochastic variables, for which almost 1000 parameter probabilities were assessed.

3. The Test-selection Facility

A test-selection facility for a decision-support system should provide information about which tests are expected to result in the most valuable information in the present phase of a patient's management. The facility should be based upon the principles of decision theory and induce a strategy that closely fits in with the test-selection strategy employed by physicians in their daily practice. The facility should further not induce overtesting, that is, it should not select too many tests. It should also not halt the test-selection process too soon, since this may lead to misdiagnosis with possibly major consequences. Automated test selection now consists of three basic elements: a measure for establishing the usefulness of performing a test, a test-selection loop, and a criterion for deciding when to stop gathering further information. In this section, we describe these three elements for our new test-selection facility for patient-tailored management of oesophageal cancer [7].

3.1. Test-selection Measure

The usefulness of performing a diagnostic test is typically expressed numerically using a *test-selection measure*. Since the usefulness of a test may differ per patient and per phase of the diagnostic process, such a measure should be patient-tailored and dynamic in nature. It may be based solely upon the expected decrease in diagnostic uncertainty after performing a test; the test that is likely to induce the largest decrease in diagnostic uncertainty then is selected as the most useful. The measure may also include the costs associated with the test and the risks involved for the patient.

A test-selection measure that is based solely on the concept of diagnostic uncertainty in essence assigns a numerical value to the probability distribution over the main diagnostic variable. This numerical value captures the amount of information that is required to resolve any uncertainty about the diagnosis. It attains its maximum when all possible diagnoses are equally likely and its minimum when one of the diagnoses has been established with certainty. The three most commonly used measures for this purpose are the Shannon entropy, the Gini index and the misclassification error [4, 5]. We experimented with the three measures in our decision-support system for the staging of oesophageal cancer and selected the Gini index as the preferred measure for our test-selection facility.

3.2. Test-selection Loop

Most test-selection algorithms select diagnostic tests in a myopic fashion. In each step of the test-selection loop, the most informative test variable is selected from among all possible test variables to indicate the next test to perform. The user then is prompted for the value of the selected variable. The entered finding subsequently is processed. From among the yet unobserved test variables, the next most informative one is selected. We feel that the test-selection strategy that is induced by such a myopic algorithm is an oversimplification of the physicians' problem-solving strategy. A *non-myopic* test-selection algorithm, on the other hand, selects in each step a cluster of multiple test variables. Such an algorithm might easily become computationally too demanding, especially if a meaningful clustering of the test variables involves clusters of relatively large size. To reduce computation time yet retain somewhat of the idea of non-myopia, we designed an algorithm that implies a *semi-myopic* test-selection strategy by considering prespecified clusters of variables. Our test-selection facility further builds upon the arguments used by the experts for deciding whether or not to order specific clusters of tests.

For eliciting the arguments underlying the experts' test-selection strategy, we used an elicitation method that was composed of focused interviews [6]. From these interviews, we learned that the experts address a number of subgoals in a sequential manner. We felt that a more involved test-selection facility should be able to take such subgoals into account in order to result in a strategy that would appear natural to the physicians. Our experts moreover were found to order tests in clustered packages: all tests that serve to provide information for a specific subgoal are ordered simultaneously, to reduce the length in time of the diagnostic phase of a patient's management. For the latter purpose, a myopic algorithm cannot be used. Our semi-myopic algorithm now selects test variables myopically, yet prompts the user not only for the result of the most informative test variable, but for the results of the other test variables that belong to the same physical test as well. We would like to note that for our new test-selection facility, we had to provide our decision-support system with additional information about the various subgoals and physical tests involved.

3.3. Stopping Criterion

In addition to selecting appropriate tests, automated test selection involves deciding when to stop gathering further information. For this purpose, a test-selection facility is equipped with a stopping criterion. With this criterion, it decides after processing the findings entered by the user, whether or not the diagnostic uncertainty has sufficiently decreased and further tests should not be ordered for the patient. A good stopping criterion can thus lead to fewer tests being performed, which results in lower financial costs and less discomfort for the patient. A good stopping criterion should further prevent the test-selection algorithm to halt too early, which could lead to misdiagnosis.

Most test-selection facilities use a stopping criterion that is based upon a threshold value on the *uncertainty*; such a stopping criterion for example builds upon the probability of the most likely value of the main diagnostic variable. In essence, we decided to adopt this principle and to apply it to the various subgoals distinguished. Since we use the expected decrease of diagnostic uncertainty for a specific goal variable as the basis of our test selection, we chose this expected decrease for our stopping criterion. A small expected decrease then indicates that there is not much uncertainty left with regard to the most likely value. To forestall halting too early, we further designed our stopping criterion to apply a restricted look-ahead: before the test selection is actually halted, our algorithm examines whether or not a next finding could induce a shift in diagnostic uncertainty to a value above the threshold.

4. Preliminary Results

We ran some experiments with our decision-support system for oesophageal cancer using eight patient cases. We were interested mainly in whether or not our stopping criterion would indicate to stop testing at the exact step in the test-selection process at which the experts would want it to stop. We were also interested in whether or not the algorithm would result in sequences of tests that would fit the test-selection strategy employed by our experts; we were aware, however, that in our experiments we did not take the costs involved into consideration and as a consequence we did not expect a perfect match.

We presented the sequences of tests and the steps at which the test-selection process halted, as generated by our test-selection facility, to our domain experts. The experts indicated that they felt quite comfortable with the results. The sequence of tests appeared natural and the step at with the test-selection process halted was considered correct for all patient cases. The experts indicated, however, that in their daily routines they would prefer to perform 'simple' tests first, that is, tests for which the results are obtained within a short period of time. We would like to note that preferring simple tests can be readily accommodated by our algorithm by including a notion of time.

5. Conclusions

Upon working with a decision-support system in oncology, we noticed that its test-selection facility presented an oversimplification of medical practice. We designed a new facility that offers patient-tailored test selection in a way that closely fits in with the daily routines of experts in the field. Our facility is based upon mathematical principles as well as on knowledge of the arguments underlying the physician's daily test-selection strategies. Our algorithm selects the most informative variable describing a single result from a physical test, yet prompts the user for all results from the test. Our algorithm further builds upon the subgoals that our domain experts use in their test-selection strategies. Upon being presented with the test-selection sequences generated by our facility, the experts indicated that they felt quite comfortable with these sequences. Although further research on including for example the time it takes to obtain a test result should still be performed, we feel that we have taken an important step towards the acceptance of decision-support systems as valuable assistants for physicians.

Acknowledgements

This research was partly supported by the Netherlands Organisation for Scientific Research. We are most grateful to Babs Taal and Berthe Aleman from the Netherlands Cancer Institute, Antoni van Leeuwenhoekhuis, who spent much time and effort in the construction of the oesophageal cancer network and provided information necessary for the design of our test-selection facility. We would furthermore like to thank Arend-Jan de Groote, Martijn Schrage and Arjan van IJzendoorn for implementing the test-selection algorithm in our Bayesian-network toolbox.

References

[1] S. Andreassen, M. Woldbye, B. Falck, and S.K. Andersen. MUNIN. A causal probabilistic network for interpretation of electromyopgraphic findings. In: J. McDermotte, editor, *Proceedings of the Tenth International Joint Conference on Artificial Intelligence*, pp. 366 – 372. Morgan Kaufmann Publishers, Los Altos, 1987.

[2] C. Lacave and F.J. Díez. Knowledge acquisition in Prostanet, a Bayesian network for diagnosing prostate cancer. *Knowledge-Based Intelligent Information and Engineering Systems*, 2: pp. 1345 – 1350, 2003.

[3] L.C. van der Gaag, S. Renooij, C.L.M. Witteman, B.M.P. Aleman, and B.G. Taal. Probabilities for a probabilistic network: A case-study in oesophageal cancer. *Artificial Intelligence in Medicine*, 25(2), pp. 123 – 148, 2002.

[4] P. Glasziou and J. Hilden. Test selection measures. *Medical Decision Making*, 9, pp. 13 – 141, 1989.

[5] T. Hastie, R. Tibshirani, and J.H. Friedman. *The Elements of Statistical Learning.* Springer Verlag, New York, 2001.

[6] D. Sent, L.C. van der Gaag, C.L.M. Witteman, B.M.P. Aleman, and B.G. Taal. On the use of vignettes for eliciting test-selection strategies. In R. Baud, M. Fieschi, P. Le Beux, and P. Ruch, editors, *The New Navigators: from Professionals to Patients; Proceedings of the MIE2003*, pp. 510 – 515, IOS Press, 2003.

[7] D. Sent. *Test-selection Strategies for Probabilistic Networks.* Ph.D. Thesis, Utrecht University, 2005.

Ubiquity: Technologies for Better Health in Aging Societies
A. Hasman et al. (Eds.)
IOS Press, 2006

Avoiding Literature Overload in the Medical Domain*

Loes M.M. BRAUN[a,1], Floris WIESMAN[b],
H. Jaap VAN DEN HERIK[a], and Arie HASMAN[b]
[a] *MICC-IKAT, Maastricht University, The Netherlands*
[b] *AMC, University of Amsterdam, The Netherlands*

Abstract. The retrieval of patient-related literature is hampered by the large size of medical literature. Various computer systems have been developed to assist physicians during information retrieval. However, in general, physicians lack the time and skills required to employ such systems effectively. Our goal is to investigate to what extent a physician can be provided with patient-related literature without spending extra time and without acquiring additional skills. In previous research we developed a method to formulate a physician's patient-related information needs automatically, without requiring any interaction between the physician and the system. The formulated information needs can be used as a starting point for literature retrieval. As a result we found that the *number* of information needs formulated per physician was quite high and had to be reduced to avoid a literature overload. In this paper we present four types of knowledge that may be used to accomplish a reduction in the number of information needs. The usefulness of each of these knowledge types depends heavily on the specific cause underlying the multitude of information needs. To determine the nature of the cause, we performed an experimental analysis. The results of the analysis led us to conclude that the knowledge types can be ordered according to their appropriateness as follows: (1) knowledge concerning temporal aspects, (2) knowledge concerning a physician's specialism, (3) domain knowledge, and (4) a user model. Further research has to be performed, in particular on precisely assessing the performance of each type of knowledge within our domain.
Keywords: Medical Record Systems, Computerized; Information Storage and Retrieval; Quality of Health Care

1. Introduction

Medical literature is an important knowledge source for physicians. The retrieval of patient-related literature has proven to be vital to the quality of care [1]. Over the past decades the amount of medical literature has increased significantly. As a consequence, the retrieval of patient-related information has become a tedious task. Since most of the medical literature is now available in digital format, computer systems have been designed to support physicians in their information-retrieval (IR) process [2-5]. However, in general, physicians lack the time and skills required to employ such

* This research is part of the MIA project (Medical Information Agents), which is funded by NWO (grant number 634.000.021).
[1] Drs. L.M.M. Braun, MICC-IKAT, Maastricht University, P.O. Box 616, 6200 MD Maastricht, The Netherlands, L.Braun@cs.unimaas.nl.

systems effectively. Moreover, in some situations physicians are not even aware of their information needs, rendering information retrieval impossible per se. Therefore, our goal is to investigate to which extent a physician can be supported in his[2] information-retrieval process, without requiring any interaction between the physician and the system. In this way, physicians can be provided with literature without spending extra time and without acquiring specific skills.

A prerequisite for successful information retrieval is a clear information need. We define an *information need* as an expression of missing information that is needed to perform a particular task (in this case: managing a patient). A typical example of a medical information need is *Does Clarithromycin cause pancreatitis?* In previous research, we have devised a method to formulate patient-related information needs automatically (i.e., without requiring a physician to specify his information needs explicitly). To accomplish this, we have modelled general information needs into information-need templates [6]. We consider an *information-need template* to be an abstract, natural-language representation of an information need, containing data slots, which can be filled with medical concepts. For instance, the information-need template corresponding to the information need above is *Does [CHEMICAL] cause [DISEASE OR SYNDROME]?* Each data slot within an information-need template has a semantic type (i.e., a high-level description of a medical concept) indicating the type of concepts fitting the slot. The information-need template above has two data slots: one with semantic type [CHEMICAL] and one with semantic type [DISEASE OR SYNDROME]. We obtained the semantic types from the Unified Medical Language System [7].

Our method formulates information needs automatically by transforming information-need templates into patient-related information needs. This is achieved by instantiating the data slots in the templates with data from the electronic patient record (EPR) of a specific patient [6]. Each possible combination of instances represents an information need. The automatically formulated information needs can be used to initiate search actions in medical literature databases without requiring any effort of the physician. However, the number of information needs formulated per physician is too high for adequate use [6]. A high number of information needs will lead to a *literature overload*, i.e., the retrieval of an unmanageable amount of literature. Since a literature overload is undesirable, our method should be improved, i.e., the number of automatically formulated information needs should be reduced. One way to accomplish such a reduction is by employing additional knowledge.

This paper discusses four types of knowledge that may be used to reduce the number of information needs (section 2). Moreover, we describe the results of an experimental analysis that we conducted to ascertain the main cause of the multitude of information needs and to determine which type of knowledge is most useful for our purpose (section 3). Finally, we provide our conclusions (section 4).

2. Knowledge Types

Temporal aspects. For the reduction of the number of information needs, knowledge about two temporal aspects may be employed, viz. the *time* at which data are entered into the EPR and the *order* in which data are entered.

---

[2] For brevity we will use the pronoun 'he' ('his') where 'he or she' ('his or her') is meant.

Table 1a. Patient data of patient X.

Patient data	Entry time
Gentamycin	30 Jan 1976
Hypokaliemia	3 Apr 2006
Clarithromycin	12 May 2006
Pancreatitis	17 May 2006

Table 1b. Patient data of patient Y.

Patient data	Entry time
Gentamycin	8 Sep 2005
Conjunctivitis	12 Mar 2006
Budesonide	29 Mar 2006
Crohn's Disease	7 Apr 2006

Knowledge about the *entry time* may be used to determine whether a data entry is relevant. As time elapses, data might become irrelevant and might be discarded when instantiating information-need templates. For example, based on the patient data of patient X (table 1a) two information needs can be formulated: (a) *Does Gentamycin cause pancreatitis?* and (b) *Does Clarithromycin cause pancreatitis?* Since Gentamycin was prescribed 30 years ago, it seems safe to assume that no new side effects will occur at this moment. Therefore, information need (a) might be discarded. Since Clarithromycin was prescribed recently, it might be causing side effects at this moment. Therefore, information need (b) can be considered relevant and is not discarded.

The *order* in which data are entered into the EPR may be employed to detect whether information needs are nonsensical. These information needs should be discarded. For instance, consider two information needs formulated based on the patient data of patient X (table 1a): *Does Clarithromycin cause hypokaliemia?* and *Does Clarithromycin cause pancreatitis?* Since the hypokaliemia occurred before the Clarithromycin was prescribed, the hypokaliemia cannot be cause by the Clarithromycin. As a result, information need (a) may be labelled nonsensical. In contrast, the disease pancreatitis occurred after the patient was prescribed Clarithromycin. Therefore, it might be caused by the Clarithromycin, rendering information need (b) relevant.

Domain knowledge. In many hospitals, alert systems are used to forewarn physicians about potentially harmful actions. These alert systems base their alerts on medical domain knowledge. Some of the warnings fulfill certain information needs. As a result, these information needs are redundant and can be discarded. By analyzing the domain knowledge used by the alert system, it can be predicted which alerts will be issued and consequently, which information needs can be discarded. Based on the patient data of patient X (table 1a), two information needs can be formulated: (a) *Is Gentamycin contraindicated by hypokaliemia?* and (b) *Does Clarithromycin cause pancreatitis?* When the medication Gentamycin is entered in the EPR, the alert system will issue the warning: *Gentamycin is contraindicated by hypokaliemia.* As a result, information need (a) (and the corresponding literature) is redundant and should be discarded. The fact that *Clarithromycin* causes *pancreatitis* is not (yet) recorded in the standard medical knowledge sources. Therefore, the alert system will not issue a warning. Consequently, information needs (b) should not be discarded.

Specialism. Physicians engaged in multidisciplinary hospital departments (e.g., ICUs), will probably have information needs that are closely connected to their specialism. If an information need is not connected to the physician's specialism it might be discarded. Furthermore, if an information need is connected to a physician's specialism, we have to ensure that the information needs are sufficiently novel. Consider two information needs formulated based on the patient data of patient Y (table 1b): (a) *Is Gentamycin indicated for conjunctivitis?* and (b) *Is Bodesonide indicated for Crohn's Disease?* Assume that the physician in attendance is a gastroenterologist. In that case, information need (a) can probably be discarded, since it is not connected to gastroenterology. In contrast, information need (b) *is* connected to gastroenterology and should not be discarded.

Physicians engaged in hospital departments focusing on a single discipline (e.g., a cardiology department), will probably have information needs connected to other specialisms. In these departments, the knowledge about a physician's specialism should thus be used from reversed perspective.

User model. By employing a user model, the information preferences of a physician can be recorded. Consider two information needs formulated based on the patient data of patient Y (table 1b): (a) *Is Budesonide effective for conjunctivitis?* and (b) *What are the costs of Budesonide?* Suppose the physician in attendance is only interested in effectiveness of medication and not in its costs. Then information need (a) may be considered relevant, but information need (b) has to be discarded.

To prevent the physician from being provided with information he has already seen, information needs that have been presented to the physician once are not presented again in the near future. The information need *What are the side effects of Gentamycin?* is formulated for both patient X and patient Y (tables 1a and 1b). If both patients are treated by the same physician, he is provided with the same information need twice. Since this is undesirable, one of the information needs has to be discarded.

3. Experimental Analysis

3.1. Experimental Set-up

The usefulness of each of the knowledge types discussed in section 2 depends heavily on the specific cause underlying the multitude of information needs. To determine the nature of this cause, we performed an experimental analysis of the information needs formulated. The results of the analysis provide directions for determining which knowledge type is most appropriate for our purpose.

In our experiment we simulated the input of patient data into the EPR by physicians in the field of intensive care medicine. We used real patient data concerning 82 patients entered by 31 physicians. The input of one separate data element is called a *data entry*. Each data entry has a type. In our simulation, only four types of data entries were used, viz. (a) chemical, (b) disease or syndrome, (c) therapeutic of preventive procedure, and (d) diagnostic procedure. A series of data entries by a specific physician concerning one and the same patient is called an *interaction*. After each interaction information needs were formulated. For an elaborate description of our method to formulate information needs automatically, see [6].

During the simulation, two types of data were recorded for each physician separately: (a) the average number of information needs generated by each information-need template per interaction and (b) the type of each data entry.

3.2. Results and Analysis

The average number of information needs generated by each information-need template per interaction is shown in figure 1. It is clear that five information-needs templates (templates 13, 23, 26, 27, and 31) generate much more information needs than the other templates. From an analysis of these information-need templates we

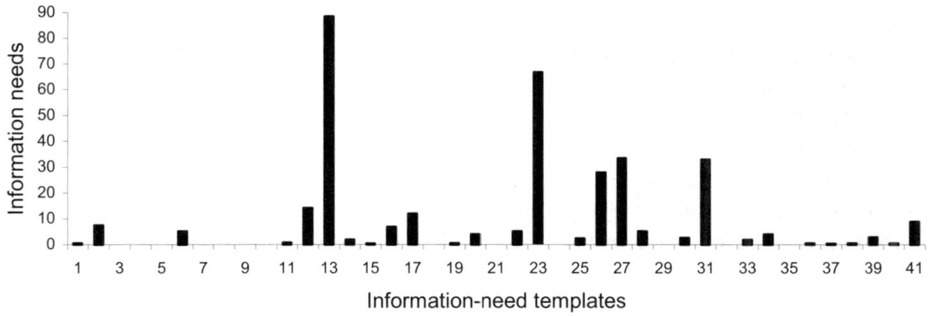

Figure 1: Number of information needs formulated per information-need template.

discovered there are two factors influencing the number of information needs formulated by a template.

The first factor is the *number* of data slots in the template. Recall from section 1 that information needs are generated by instantiating the data slots of an information-need template with appropriate patient data. Each possible combination of instances leads to an information need. The higher the number of data slots, the higher the number of combinations that can be formed by the instances of the data slots. Consequently, a high number of data slots in an information-need template leads to a high number of information needs formulated from the template.

The second factor is the *type* of the data slots in an information-need template. Recall from section 3.1 that each data entry has one of four possible data types. During our simulation we calculated the relative frequency of each data type. 62% of the data entries were of type [CHEMICAL], 25% of type [THERAPEUTIC OR PREVENTIVE PROCEDURE], 7% of type [DISEASE OR SYNDROME] and 6% of type [DIAGNOSTIC PROCEDURE]. The higher the number of data entries of a specific type, the higher the number of instances of the data slots of this type. Therefore, templates containing data slots of the most occurring data types will generate most combinations of instances, and consequently, most information needs.

3.3. Discussion

Our analysis indicated that the large number of information needs is mainly due to a combination of the number and the kind of semantic types in an information-need template. Since we cannot change the number and kind of semantic types in the information-need templates we have to change the *number of data entries* used to instantiate the data slots within an information-need template.

In this respect, the employment of knowledge about temporal aspects is probably the most appropriate knowledge type, since the use of the data entry time can be applied to all data entries. Moreover, for certain information-need templates the use of the data entry order may reduce the number of information needs even further by discarding nonsensical information needs.

Knowledge about a physician's specialism can be used to shield physicians from information needs that are not relevant according to their area of expertise. Since this knowledge applies to all data entries, it can be considered appropriate to reduce the number of information needs.

Domain knowledge is less useful, since it cannot be applied to each type of data entry. Nevertheless, it applies especially well to the data entries of the most occurring types (i.e., [CHEMICAL] and [THERAPEUTIC OR PREVENTIVE PROCEDURE]), because the topics of medication and therapy are elaborately covered in the domain knowledge.

A user model can be employed to prevent information needs from being formulated repeatedly, which may be considered a useful feature. Furthermore, the model can be used to tailor information needs to the user. However, the usefulness of this feature depends heavily on the accuracy of the user model. Since it takes time to model the user accurately, a user model requires time to reach its full usefulness. Therefore, this knowledge type is considered less appropriate for our purpose.

4. Conclusions

We presented four types of knowledge to reduce the number of information needs (section 2). The appropriateness of each of these knowledge types depends heavily on the specific cause underlying the multitude of information needs. To determine the nature of the cause, we performed an experimental analysis of the information needs formulated (section 3).

From the results of the experiment we may conclude that five information-need templates generate much more information needs than the other templates. Furthermore, we may conclude that there are significantly more data entries of the semantic types [CHEMICAL] and [THERAPEUTIC OR PREVENTIVE PROCEDURE] than of the semantic types [DISEASE OR SYNDROME] and [DIAGNOSTIC PROCEDURE].

From our analysis of the five templates generating most information needs we may conclude that the large number of formulated information needs is mainly due to a combination of the number and the kind of semantic types in the templates.

Based on the results of our experimental analysis of the information needs formulated, we have assessed the knowledge types with respect to their appropriateness in our problem domain (subsection 3.3). From this assessment, we may conclude that the knowledge types can be ordered as follows (1 is most appropriate): (1) temporal aspects, (2) specialism, (3) domain knowledge, and (4) user model. Currently, all knowledge types are still under investigation and additional research has to be performed to assess the performance of each knowledge type for our purpose.

References

[1] Gamble S. Hospital Libraries Enhance Patient Care and Save Money. Journal of the AALT 1996; 23(2): 10–12.

[2] Cimino JJ, Johnson SB, Aguirre A, Roderer N, Clayton PD. The Medline Button. Proc Annu Symp Comput Appl Med Care 1992; pp. 81-85.

[3] Miller RA, Jamnback L, Giuse NB, Masarie FE. Extending the capabilities of diagnostic decision support programs through links to bibliographic searching: Addition of 'canned MeSH logic' to the Quick Medical Reference (QMR) program for use with Grateful Med. In: Clayton P, editor. *Proc Annu* Symp Comput Appl Med Care. McGraw-Hill Inc., 1991; pp. 150–155.

[4] Rada R, Barlow J, Bijstra D, Potharst J, de Vries Robbé P, Zanstra P. OAR: Open architecture for reasoning applied to connection patient records to medical literature. In: Noothoven van Goor J, Christensen JP, editors. Advances in Medical Informatics. Amsterdam: IOS Press; 1992; pp. 287–294.

[5] Van Mulligen EM. UMLS-based access to CPR data. Int J Med Inf. 1999; 53(2–3): 125–131.

[6] Braun L, Wiesman F, Van den Herik, J, Hasman, A, Korsten, E. Towards Automatic Formulation of a Physician's Information Needs. JDIM 2005; 3(1): 40 – 47.

[7] National Center of Biotechnology Information. Unified Medical Language System. Bethesda, MD. http://www.nlm.nih.gov/research/umls/ (2005).

Ubiquity: Technologies for Better Health in Aging Societies
A. Hasman et al. (Eds.)
IOS Press, 2006

A Language Classifier that Automatically Divides Medical Documents for Experts and Health Care Consumers

Michael POPRAT [a], Kornél MARKÓ [b], Udo HAHN [a]

[a] *Jena University Language & Information Engineering (JULIE) Lab, Jena, Germany*
[b] *Medical Informatics Department, Freiburg University Hospital, Germany*

Abstract. We propose a pipelined system for the automatic classification of medical documents according to their language (English, Spanish and German) and their target user group (medical experts *vs.* health care consumers). We use a simple n-gram based categorization model and present experimental results for both classification tasks. We also demonstrate how this methodology can be integrated into a health care document retrieval system.

1. Introduction

The user population of medical document retrieval systems and their search strategies are really diverse. Not only physicians but also nurses, medical insurance companies and patients are increasingly getting access to these text-based resources. Especially health care consumers (henceforth, HCC) spend more and more time on getting information from Web resources to gather a better understanding about their personal health status, sufferings, (in)adequate treatments, medications and other interventions. Although many patients already exhibit a considerable level of expertise concerning their diseases, not every piece of relevant information from the Web is particularly suited and understandable for non-experts. To help HCCs in finding appropriate information, as well as medical experts in suppressing layman information, we realized a text analysis pipeline which consists of a simple but high-performing classification algorithm. It distinguishes the kind of natural language in which a document is written and, subsequently, divides this document space into easily understandable texts, and, into those documents that require expert-level knowledge for proper understanding. The approach we propose uses frequencies of the occurrence of typical character combinations (n-grams) only and thus refrains from heavy linguistic analysis.

2. Methods

In our experiments, we used the off-the-shelf classification software TEXTCAT which is primarily designed for determining the language of documents. The underlying method is very simple (see also [1]): the most frequent character n-grams, extracted from training documents, represent a particular category (the *category model*). In order to classify a new, unseen document a ranked n-gram list has to be extracted from the new document (the *document profile*) and has to be compared to every category model. The similarity criterion is based on a simple two-step procedure, which is applied to both, the category models and document profiles. First, the document profile is compared to a model profile pairwise. The rank-order statistics is based on the measurement of how far 'out of place' [1] a particular n-gram in the document profile is from its position in the category model. The sum of all out-of-place values for all n-grams is the document's distance from that particular category. Second, the document is assigned to that specific category to which it has the smallest overall distance.

Rank	LM_{EN}	LM_{ES}	LM_{GE}	Doc_{GE}
26	s_	es	ei	**b**
27	er	_l	**er_**	k
28	_o	de_	in	_d
29	he_	la	**te**	**er_**
30	d_	os	ie	f
31	t_	_de_	**b**	he
32	the_	_p	t_	**te**

Table 1: N-gram Based Character Language Models for English (LM_{EN}), Spanish (LM_{ES}) and German (LM_{GE}) and for a German Document (Doc_{GE}) (Rank 26-32). Bold N-grams Occur Both in the Document Profile and in the Language Models.

For the language classification task, we left all the parameters of the TEXTCAT tool 'as is', except that we always enforced a non-ambiguous classification decision. We used the language models (i.e., the category models) for English, Spanish and German that come with TEXTCAT. Table 1 depicts samples from category profiles for the language classification task. Columns 2-4 show the models for the different languages considered, column 5 represents a document profile.

3. Experimental Setting

In order to evaluate the language and target group classification results, we acquired various text corpora from the Web and from online textbooks (see Table 2). We eliminated style tags, hyper links and other tags from the Web documents and obtained plain text files. For English, we used the online version of Merck, a collection of Medline abstracts and documents from the Internet portals Netdoctor (www.netdoctor.co.uk) and Mayoclinic (www.mayoclinic.com). For Spanish, our corpora consist of the Spanish version of Merck, documents from The Scientific Electronic Library Online (www.scielo.org), the Spanish version of Netdoctor (www.netdoctor.es) and Familydoctor (www.familydoctor.es). Finally, for German, we used the German version of Merck, German abstracts indexed in Medline, and the German version of Netdoctor (www.netdoktor.de) as well as Medicine World Wide (www.m-ww.de).

	Expert	HCC
EN	Merck (16 MB)	Netdoctor (48 MB)
	Medline (3,6 MB)	Mayoclinic (65 MB)
ES	Merck (12 MB)	Netdoctor (6,5 MB)
	Scielo (182 MB)	Familydoctor (4,4 MB)
GE	Merck (48 MB)	Netdoctor (107 MB)
	Medline (3,6 MB)	MWW (21 MB)

Table 2: Survey of the Corpora Collection (Sizes in Parentheses)

Besides the language dimension (English, Spanish, German), we also distinguished between texts written for experts (articles from the Merck text books, abstracts from Medline, and documents from Scielo) and those written for HCCs (Netdoctor, Mayoclinic, Familydoctor, Medicine World Wide).

The document collection was split 10 times randomly in two sets with about the same size to obtain a training and a test set with no overlapping documents (see Figure 1, A). In the training phase, we created target group models (TGM, Figure 1, B) by extracting and ranking n-grams from typical expert and HCC documents. In contrast to language models for which a training document size between 21 KB to 132 KB is sufficient [1], the target group models had to cover a broader variety of styles for the coarse grained classification. We therefore used a larger training set (66 MB–102 MB).

In the test phase, all documents to be classified were mixed in a first step. Afterwards, TEXTCAT was used with the language models (LM_{EN}, LM_{ES}, LM_{GE},) that come with the software. Thus, we obtained a test set for the second classification task, namely the distinction between expert and HCC documents (Figure 1, C). For this purpose we also made use of TEXTCAT, though we now relied on the models that had been created in the training phase. The final result is a collection of documents that are classified according to their language, as well as to the target group they are likely to address (Figure 1, D).

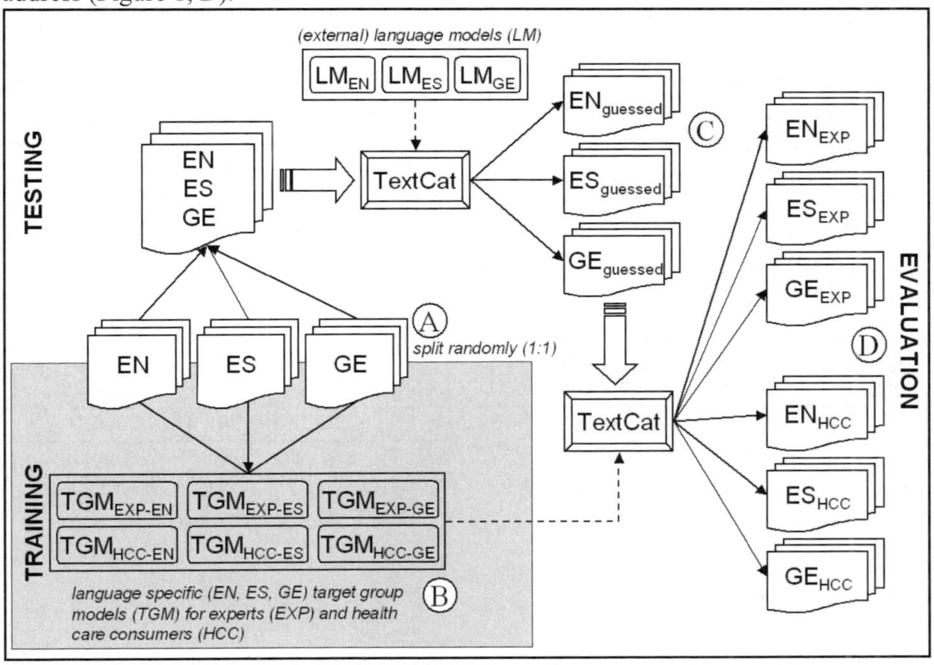

Figure 1: Experimental Setting: Document Processing Pipeline for Testing and Training

4. Experimental Results and Discussion

In our experiments, we evaluated whether the automatic procedures for the distinction between various kinds of languages and the two groups of expertise, *viz.* experts and HCCs, provided reasonable outcomes. For the language classification task, we achieved very good results (100% for English, 99,6% for Spanish and 95,4% for German). This yields a sound basis for the second test set to be used for the target group classification. It is conspicuous that the results for the German language identification task are about 5 percentage points below the results for English and Spanish. An analysis of the misclassified documents revealed that they were classified as English documents, because their predominant language was, surprisingly, English (although crawled from www.netdoktor.de). For the target group classification, we classified the documents from the second test set (i.e., after language classification) using TEXTCAT and the language specific expert and HCC models created in the training phase. Here, we attained overall classification results between 89.4% for German and 95.8% for Spanish, English (94.9%) being close to Spanish in terms of performance (see also Table 4). The accuracy results for the correct classification of HCC documents were in general higher (90.0% for German, 95.0% for English, and 98.1% for Spanish) compared to the results for expert documents (88.9% for German, 93.9% for Spanish, and 94.8% for English). Again, the overall classification results for German documents are about 5 percentage points below the results for Spanish and English. The reason for that might be the target group model for HCC documents (TGM_{HCC-GE}), that resulted from the noisy document collection we used in the training phase (see above).

The comparison of the two classification tasks shows different classification results: Firstly, language-specific classification yields better results compared to the target group classification task due to obvious reasons: guessing a language is simpler with the methods we use because substantial characteristics of a particular language are mirrored in the respective character set (umlauts in German, accents in Spanish, none of these in English). Therefore, the language models have a high discriminative power. By contrast, the distinction between expert and HCC documents is, on the one hand, a less well-defined decision task and, on the other hand, seems to require a deeper text (and content) analysis. Still, our results show that simple methods yield high accuracy figures for various document classification tasks.

English	Spanish	German
100.0%	99.6%	95.4%

Table 3: Accuracy Results for Language Classification (Coverage: 100%)

	English	Spanish	German
Expert	94.8%	93.9%	88.9%
HCC	95.0%	98.1%	90.0%
All	94.9%	95.8%	89.4%

Table 4: Accuracy Results for Expert-HCC Classification (Coverage: 100%)

5. Related Work

Comparing our experiments to others is difficult: First, apart from the simple n-gram approach, there are other classification methods with learning and classification functionality [2,3]. Second, a lot of classification experiments are based on non-medical document sets with a varying number and granularity of categories.

Karlgren and Cutting [4], e.g., use the same computational approach and run experiments on the mixed-domain English Brown corpus[1]. For the distinction between informative *vs.* imaginative texts, they come up with 96% and 95% accuracy,

[1] http://helmer.aksis.uib.no/icame/brown/bcm.html

respectively. For a four-category classification task (press, fiction, non-fiction, and miscellaneous), they achieve between 75% and 95% accuracy. Kessler *et al.* [5] employ a 13-category classification system, which achieves a classification accuracy ranging between 93-100% for the scientific category. Stamatatos *et al.* [6] compute genre categorization on a four-category system by merely comparing absolute word frequencies in general language with those from four genre-specific corpora, a task for which they achieve around 97% accuracy. The state-of-the-art performance figures for medical text categorization, however, are due to Kornai and Richards [7] who achieve 86% correct classifications of clinical verbatims on a 12-category system.
Up until now, almost no effort has been made to apply medical text classification with respect to their target group. In a study closest in spirit to our approach and objectives [8], a multi-layered classification pipeline was presented. German documents were, firstly, classified into non-medical and medical ones. Then, from the latter set, documents with clinical (e.g., pathology reports) and non-clinical (textbooks) content were determined automatically. Finally, for the non-clinical texts, their target group (experts *vs.* laymen) was identified, yielding results of 75% and 71,4% accuracy, respectively. In the present study, we could show on a larger corpus that the classification results can be improved and that the proposed method can also be applied to other languages than German.

6. Conclusion, Current and Future Work

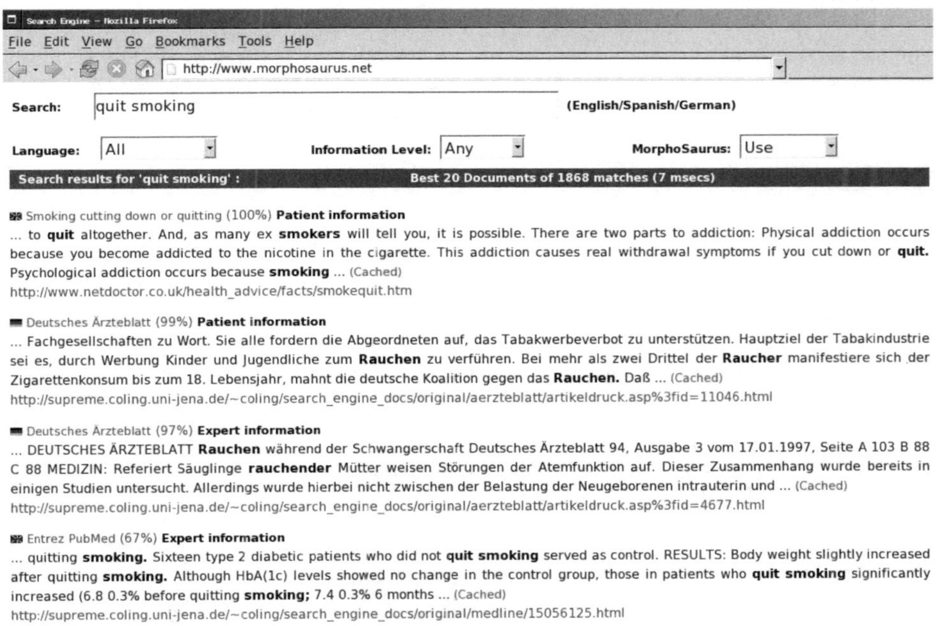

Figure 2: Screen Shot from the MORPHOSAURUS Search Engine (`http://www.morphosaurus.net`)

We here proposed a 3x2 classification system that relies on *n*-gram statistics for classifying documents according to the kind of language and their target group, *viz.* medical experts *vs.* health care consumers. Based on the good results we achieved, we applied it as an add-on for the MORPHOSAURUS indexing system [9]. We supply a search engine in order to retrieve documents that fulfill an expert's or an HCC's information need. We tested this approach using the language and target group models described in the previous sections and added the information about language and target group to the document index structure created by the MORPHOSAURUS system. The screenshot in Figure 2 shows some retrieval results for the query "quit smoking" which might be a typical request by health care consumers. Because of the multi-linguality of the MORPHOSAURUS system, not only English documents can be retrieved, but also German and Spanish ones (see [9] for details). Figure 2 shows that for both English and German documents the information about their particular target group (patients as HCCs *vs*. medical experts) is given. Even more important, with this approach, it is also possible to classify documents from text collections that are not part of the training set (here: documents from www.aerzteblatt.de) and, because of its heterogeneous readership, also divide them into their respective target groups. All in all, because of its high performance, the classification pipeline proposed in this paper indeed improves the quality in document retrieval systems that are used by a multilingual and expertise-wise heterogeneous user community. In our future work, we plan to substantiate this assumption by carrying out retrieval experiments in which experts as well as HCCs have to judge the documents returned in a result set with regard to satisfying their information needs in terms of relevancy and comprehensibility.

Acknowledgement

This work was partly supported by Deutsche Forschungsgemeinschaft (DFG), grant KL 640/5-2, and the European Network of Excellence "*Semantic Mining*" (NoE 507 505)

References

[1] Cavnar WB and Trenkle JM. N-gram based text categorization. In SDAIR'94, pp 161–175. Las Vegas, NV, USA, April 11-13,1994.

[2] Calvo RA, Lee J-M, and Li X. Managing content with automatic document classification. Journal of Digital Information, 5(2), 2004 (Article No. 282).

[3] Sebastiani F. Machine learning in automated text categorization. ACM Computing Surveys, 34(1):1-47, 2002.

[4] Karlgren J and Cutting D. Recognizing text genres with simple metrics using discriminant analysis. In COLING'94, pp 1071–1075. Kyoto, Japan, August 5-9, 1994.

[5] Kessler B, Nunberg G, and Schütze H. Automatic detection of text genre. In ACL'97/EACL'97, pp 32–38. Madrid, Spain, July 7-12, 1997.

[6] Stamatatos E, Fakotakis N, and Kokkinakis G. Text genre detection using common word frequencies. In COLING 2000, pp 808–814. Saarbrücken, Germany, 31 July – 4 August, 2000.

[7] Kornai A and Richards JM. Linear discriminant text classification in high dimension. In Hybrid Information Systems. Proceedings of the 1st International Workshop on Hybrid Intelligent Systems, pp 527– 537. Adelaide, Australia, December 11-12, 2001. Heidelberg: Physica, 2002.

[8] Hahn U and Wermter J. Pumping documents through a domain and genre classification pipeline. In LREC 2004 Vol. 3, pp 735–738. Lisbon, Portugal, May 26-28, 2004.

[9] Markó, K, Schulz, S and Hahn, U: MorphoSaurus. Design and evaluation of an interlingua-based document retrieval engine for the medical domain. Methods Inf Med, 44 (4): 537-545, 2005.

Ubiquity: Technologies for Better Health in Aging Societies
A. Hasman et al. (Eds.)
IOS Press, 2006

With good Intentions

Agnieszka Latoszek-Berendsen[a], Jan Talmon[a], Arie Hasman[b]
[a]*Medical Informatics, University Maastricht, Maastricht, The Netherlands*
[b]*KIK, University of Amsterdam, Amsterdam, The Netherlands*

Abstract. The use of intentions in computer-based guidelines may help to make them more flexible, easier to adapt to local standards, easier to evaluate and to improve. We see possibilities of using intentions in several areas: as a generator of actions which could be performed and then compared with the decision of the physician, as a method for pruning proposed actions based on current and previous patient data. In this contribution we present our approach to intention-based guidelines.
Keywords: guidelines, intentions, decision support system

1. Introduction

Clinical practice guidelines (CPGs) are "systematically developed statements to assist practitioners and patient decisions about appropriate health care for specific circumstances" [1]. Use of clinical guidelines benefits health care by improving its quality, lowering medical costs of treatments, or reducing practice variability [2]. Although it has been shown that clinical practice guidelines can improve medical care, their implementation and daily use is not always clear. This is partly because patients, physicians, insurances and managers define quality differently and evidence about the effectiveness of guidelines is incomplete [3].

Paper-based clinical guidelines are getting more complex and difficult to follow. Therefore we observe an increase of guidelines implemented in Decision Support Systems (DSSs), where the computer helps users to follow the guideline. In DSSs every step to be taken is defined in detail, the guideline is very strict and fixed and usually implemented as a flowchart. There are situations when patient characteristics or local constraints do not allow following the DSS directions. Literally, existing guidelines are not able to cope effectively with alternative actions and may generate unnecessary reminders or warnings in such situations.

There are some factors that restrict physicians' adherence to a guideline. Cabana et al. [4] reported seven general categories of barriers to physicians' guideline adherence such as the physician's knowledge (lack of awareness or lack of familiarity) and attitudes (lack of agreement, lack of self-efficacy).

Also physicians might follow a different path than the one specified in the guideline, but still work in the 'spirit' of the guideline. Winikoff et al. [5] present a simple example of a hungry cat (intelligent agent), who has a plan how to reach food on table. When circumstances change and his plan can not be accomplished, he goes for an alternative set of actions, but still with the purpose to reach his goal, getting food. He needs to change his primary plan to fulfil his intention. Similarly a physician may select an alternative treatment than specified in the guideline.

The objective of the present paper is to demonstrate a possible technique to overcome some of these problems when implementing guidelines into a DSS. We introduce intentions and show a new way of building guidelines. Development of mechanisms to reason with these intentions will be necessary.

There are other groups who implemented intention-based guidelines. Shahar, et al [6], present the use of intentions in the ASBRU language. There, intentions are temporal patterns of provider actions or patient states, to be achieved, maintained, or avoided and are inferred from care provider actions or explicitly stated by the provider. We discuss a different approach in this paper.

2. Definitions

Before specifying intentions in more detail, we define some terms commonly used in association with intentions.

Intention. An intention is a determination to act in a certain way or to do a certain thing; is a high level goal which is abstract and not measurable.

Goal. A goal is an object of one's efforts; something to reach; is situation/state dependent; is concrete and measurable.

Plan. A plan is a proposed or intended method of getting from one set of circumstances to another. They are often used to move from the present situation, towards the achievement of one or more goals. A plan contains a series of actions to be carried out.

Action. An action is a work or an activity in a specific area or field.

Purpose. A purpose is an anticipated outcome that is intended or that guides planned actions.

3. Implementation

3.1. General idea

Our system is being implemented as an addition to the GASTON framework that facilitates the development of clinical guideline application tasks [7]. It consists of three parts: a guideline editor, where intention based guidelines are created; a knowledge base, where ontologies about medical knowledge are saved; an execution engine, where the method for pruning not applicable actions is implemented.

The physician communicates with the system via an Electronic Patients Record (EPR) System. The EPR has bidirectional communication with the execution engine. Further, the execution engine exchange data with the guideline editor and the knowledge base. (Figure 1)

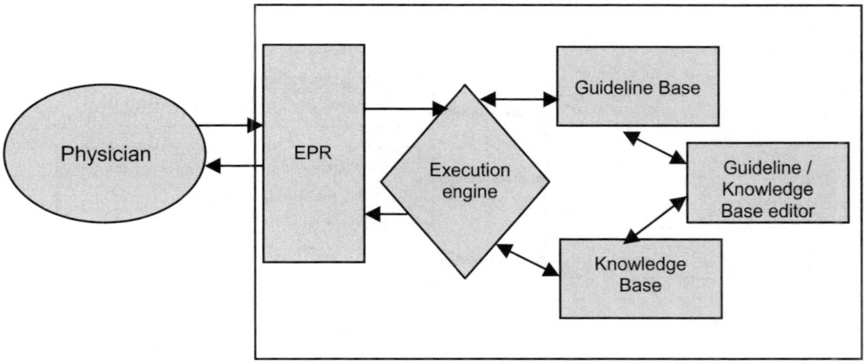

Figure 1. General architecture of the system

3.2. Guideline design

A guideline based on intentions is still represented as a flowchart. However, it will be built in a number of phases where specific parameters of the guideline and the intentions can be defined.

In the first phase, the intention of the guideline is specified and decomposed into sub-intentions. Intentions on all levels but the highest one are defined as AND/OR trees, where some intentions are carried out in parallel and others in sequence (Figure 2). The intention on a higher level is fulfilled only when the sequence of the intentions on the lower level is satisfied.

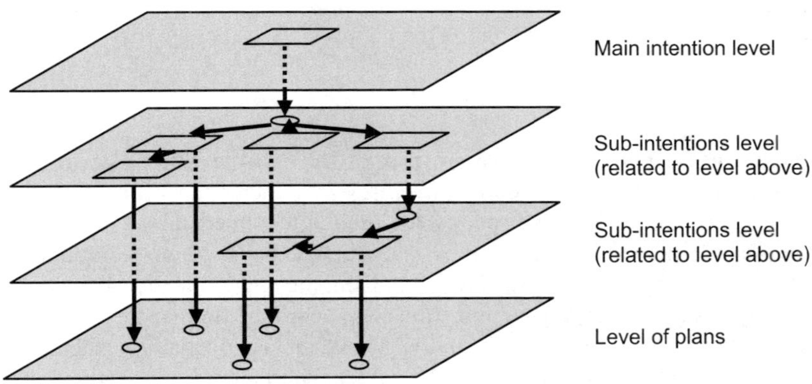

Figure 2. Multi-level approach to model guidelines with intentions.

In the second phase, for each intention an intention block is defined. An intention block contains three components: intention, goal and plans (Figure 3).

In the first component, 'intention', criteria to fulfil this intention are described. In other words, what sequence of intentions on a lower level is necessary to fulfil an intention on a higher level. The second component 'goal' describes the measurable part of an intention. The goal might be set in two ways: by the physician, based on the patient condition or by the guideline designer, based on general medical knowledge; also it might contain time constraints. The last component, 'plans', contains lists of possible combinations of actions which can be performed to fulfil this intention.

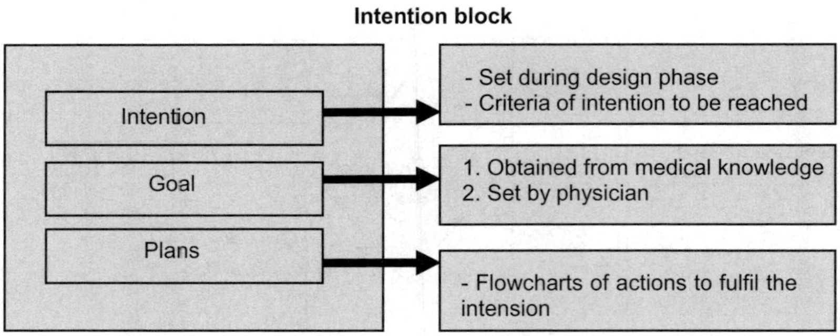

Figure 3. Schema of the Intention block.

In the last phase plans are built. By using existing guidelines and expert's knowledge, we make flowcharts of actions concerning treatments, drug dosages etc. In this stage, the designer of the guideline will be able to assign relevant plans to every intention and rank them based on general or local preferences. Because the same treatment, or kind of drug, can be used in different guidelines, storing plans in a database enables the use the same plan in different guidelines.

Both Goal and Plans use a predefined framework to represent complex temporal criteria enclosed in clinical guidelines. The framework contains a) a temporal ontology, which defines time-related concepts (e.g. time points or intervals) and their interrelationships, namely temporal operators (before, after, during, etc) that relate these concepts and are needed in order to reason about time, b) a temporal data model, c) a mechanism for executing temporal criteria (temporal query mechanism) [8].

3.3. Knowledge base

Every action has a purpose that we are using to match the actions performed by a physician with those mentioned in the guideline. In the other part of the system, which is the Knowledge Base (KB) we specify the ontology of medical acts and treatments such as drugs, laboratory tests and imaging methods, their contraindications, preconditions and purposes. The KB is to help the guideline designer implement medical knowledge within a guideline flowchart. During runtime the KB is used to retrieve the purpose of the actions of the physician as to compare them with the purpose of the actions in the guideline. So rather than compare the actions itself, we compare the purposes.

For example, if the guideline suggests MRI and the physician selects CT, as long as the purposes are the same no alarm should be generated by the system.

3.4. Guideline execution engine

As mentioned in section 3.2, during the design of the guideline, all possible plans which match criteria of the intention are placed in an intention block. It means that all possible or expected actions of the physician will be listed. However, the order on this list is not random. Based on the medical knowledge and local conditions, plans are listed in preference order, where the most common or recommended plans are at the top of the list. The number of choices can be reduced by allowing the use of only a number of the top plans.

To have possibility to give the correct advice to the user but also to generate reminders or alerts only when needed, we made a mechanism of pruning not applicable plans. Based on the plans list mentioned in the intention block and constraints, like previous experience or patient data stored in an EPR, a new list of plans is created relevant to the current patient state.

There are two potential situations where intentions and the possibility of pruning actions can make daily use of guidelines more flexible. For example, assume that in the Intention Block of the guideline plans A-F are listed as probable plans to fulfil an intention. The execution engine prunes the plans which are not relevant, based on existing constraints. Suppose only three plans result: plan A, plan C and plan F. The physician is deciding to go for plan C. In our guideline this plan is mentioned as a second best option. We might send a message to the user about plan A, which might be better, or cheaper, or more common to perform. We could also not inform physician as plan C is appropriate and within spirit of the guideline and we do not want to bother the physician unnecessary.

In another situation, the physician decides to choose a new plan W, which is not listed. The user is aware of it and wants to recommend it to the patient. Here however, we would send a note to the physician to make him aware of taking actions which are not described in a guideline.

4. Discussion

Our goal is to provide advice to the physician in a flexible way, allowing the user to take a road that may not be completely accord to the guideline, but in the spirit of the guideline. This provides the practitioner with the option to make decisions based on previous experience or the local situation. He can also take into account the patient's preferences, which is a very important issue. We implemented intentions to make guidelines more flexible in giving feedback. The use of intentions in the guideline allows the physician to deviate from the recommended action, while staying within the spirit of the guideline. By studying the use of these alternatives we can obtain insight into the preferred actions of the clinicians and determine the reasons of possible deviations from the guideline.

Designing intention based guidelines in flowcharts, where intentions are executed in parallel or in sequence, gives us control about guideline flow and makes giving relevant advice easier. Work with a heart failure guideline provided us examples of such a use of intentions. Two kinds of the medications are usually given in parallel: an ACE inhibitor and diuretics. A β-blocker, is given to the patient when the intention of the prescription of diuretics (sufficient dehydration) is fulfilled. The intentions of diuretics and β-blocker are thus considered in sequence.

To limit the number of unnecessary warnings, within each intention block a goal can be set. This can be done in two ways: by the physician based on the current patient state or by the guideline designer obtained from medical knowledge. We can show the use of this feature in a simple example.

The intention slot of our intention block may read for example: "Lowering blood pressure". The goal in this situation might be set by the physician, because the value of blood pressure that we would like to reach may vary from patient to patient.

Setting the goal by the guideline designer can be done, for example, in the heart failure guideline where the value of calcium has to be checked. The normal level of calcium for all patients should be within a fixed range and is based on medical knowledge.

Implementing intention based guidelines might help evaluating the guideline. If we see that some plan is listed as the best alternative and only some physicians recommend it to their patients, we could conclude that this plan contains actions which are not common anymore or that users might not be aware of some new possibility and need extra information or training.

We implemented the guideline for heart failure and we are evaluating its flexibility in accepting physician's choices and also in giving feedback to the user.

First, the clarity of guideline designing should be tested. Its transparency by building it in three phases; easier adaptation to local standards by modifying preference order and adding new entries in the KB with their purposes.

Second, the decision support system's feedback will be evaluated. Our focus will be to analyse how well unnecessary warnings and alarms are suppressed.

Acknowledgments

This work is financial supported by NWO, Token 2000 project MIA, grant number 634.000.021.

References

[1] Clinical Practice Guidelines. [cited; Available from: http://www.openclinical.org/guidelines.html

[2] Grimshaw JM, Russell IT. Effect of clinical guidelines on medical practice: a systematic review of rigorous evaluations. Lancet. 1993 Nov 27;342(8883):1317-22.

[3] Woolf SH, Grol R, Hutchinson A, Eccles M, Grimshaw J. Clinical guidelines: potential benefits, limitations, and harms of clinical guidelines. Bmj. 1999 Feb 20;318(7182):527-30.

[4] Cabana MD, Rand CS, Powe NR, Wu AW, Wilson MH, Abboud PA, et al. Why don't physicians follow clinical practice guidelines? A framework for improvement. Jama. 1999 Oct 20;282(15):1458-65.

[5] Winikoff M, Padgham L, Harland J, Thangarajah J. Declarative and Procedural Goals in Intelligent Agent Systems. 2002 Toulouse, France, April 22-25, 2002.

[6] Shahar Y. The Asgaard project: a task-specific framework for the application and critiquing of time-oriented clinical guidelines Artificial Intelligence in Medicine. 1998;14(1):29-51(23).

[7] de Clercq PA, Hasman A, Blom JA, Korsten HH. Design and implementation of a framework to support the development of clinical guidelines. Int J Med Inform. 2001 Dec;64(2-3):285-318.

[8] Triantafyllou E, Kokkinou E, de Clercq PA, Peek NB, Korsten HHM, Hasman A. Representation and Execution of Temporal Criteria for Guideline-Based Medical Decision Support at the Intensive Care Unit. BNAIC, Brussel 2005.

Ubiquity: Technologies for Better Health in Aging Societies
A. Hasman et al. (Eds.)
IOS Press, 2006

A method for specification of structured clinical content in electronic health records

Knut BERNSTEIN[1], Morten BRUUN-RASMUSSEN, Søren VINGTOFT
MEDIQ – Medical Informatics and Quality Management, Copenhagen, Demark

Abstract: The Copenhagen County is using clinical guidelines in the electronic health record development to provide documentation support, process support and decision support for the healthcare professionals. The electronic health record development is based on three main components: The first component is a national information model. The second component is a common classification system (SNOMED). The third key component is the so-called "clinical content". This paper describes the structured "clinical content", how it is linked to the clinical process, and how it is used to create clinical guidelines in the form of standard care plans. The Copenhagen County and MEDIQ has developed a methodology for identifying and specifying structured "clinical content" to be used in electronic health records. The method combines analyses of national clinical guidelines with local experience and practices and it heavily involves healthcare professionals. The method includes four main steps: Analyses of background material, analyses of clinical process-flow, mapping to standards (the national information model and the common classification system), and specification of the structured clinical content itself. Three secondary steps may be added to specify the clinical content in more detail: Workflow analyses, analyses of quality indicators, and decision analyses. This paper reports the experiences using the method and stresses the demand for a common exchange format and IT-tools for documenting clinical content in a formalised way.

Keywords: Documentation, clinical guidelines, clinical content, documentation support, process support, decision support, methodology

1. Introduction

One of the main goals of the current Danish IT strategy [1] for healthcare, is the dissemination of electronic health records (EHR) in hospitals. The Counties and the Central Health Authorities have agreed, that that electronic health records based on common standards is to be implemented in all Danish hospitals by 2008. The National Board of Health has developed a common standard, the so-called "Basic Structure for Electronic Health Records" [2]. This is a process-model, based on a problem-oriented way of documenting the activities, using structured data from all professional groups. Furthermore, The National Board of Health is proposing that the national classification system should be substituted with the SNOMED CT terminology [3]. SNOMED CT is currently translated into Danish.

However, experience from the past two-three years in implementing the EHR at a large scale shows that the model and the terminology are not sufficient for

[1] Knut Bernstein. MEDIQ, Østerled 8, 2100 Copenhagen Ø, Denmark. kb@mediq.dk, www.mediq.dk

dissemination and user acceptance of the EHR. A third element is required – the so-called "clinical content" [4]. The clinical content includes standard plans, standard activities, standard goals, and standard results, as well as the related decision points. The specification of clinical content results in a representation of clinical guidelines, as well as context-related information, and templates for data entry/presentation.

The clinical content can be extracted from international/national/regional/local guidelines, care plans, indicators in quality databases, textbooks, and knowledge of clinical practice. This paper presents a method for *systematic* collection and assessment of the clinical content, mapping to relevant process models and terminologies, and a specification of clinical content that can be used directly in the EHR systems. This will focus the EHR development on clinical process support, documentation support, and decision support, which is expected to be key points for the EHR acceptance by the clinical user. Furthermore, it has been shown that conforming to guidelines is a good way to improve the quality of medical care [5].

Extensive work has been made internationally regarding access to guideline resources [6, 7], models and languages for formal representing guidelines [8,9] as well as proposals for exchange formats for these guidelines [10, 11]. The concept of archetypes has been proposed for representing information similar to what we include in the clinical content specifications [12]. The archetype concept has been elaborated by the openEHR Foundation with the Archetype Object Model [13] and the Archetype Definition Language [14]. Furthermore the part of the clinical content related to presentation of data could be represented in the openEHR templates [15] which currently also being included in the CEN standard for EHR communication [16].

It should be noted that the method reported here is focused on how healthcare professionals and informaticians can cooperate specifying structured clinical content. The resulting clinical content should be expressed in a formal language in order to be processed and exchanged. Most of the clinical content probably could be represented with Archetype Definition Language or similar formalisms. However, the method does not imply the use of a specific formal language. In Denmark, national decisions regarding standardisation of clinical content representation and exchange formats have not yet been made.

2. Materials and methods

In connection with the EHR development and implementation in the Copenhagen County, a need for structured specification of the clinical content was acknowledged. This was based on the first trials using the Basic Structure for Electronic Health Records in EHRs in a clinical setting [17].

Two projects were initiated with the purpose of developing a method for specification of formalised clinical content in EHR. Both projects had heavily involvement from clinical users representing all clinical professions. The group included some experts in the selected clinical domain and some other clinicians involved in the EHR development in the county.

The first project was validating an EHR prototype focusing on converting national text-based guidelines to structured standard care plans, use of SNOMED CT as a basis for the terminology, and user acceptance of the user interface. The project was organised as nine workshops with 25 participants. The clinical topic was Acute

Coronary Syndrome, and the first version of operational standard plans was produced for this area. The first version of the methodology was also proposed.

The second project aimed at refining the proposed methodology by applying it on three areas. The work on Acute Coronary Syndrome was continued and treatment of Acute Abdominal Aorta Aneurism and Schizophrenia were selected as the other areas. The projects was organised with up to six workshops with 25 participants [18, 19]. Apart from (incomplete) specifications of the clinical content in the selected areas, a complete method was now proposed [20].

This method is reported in the following sections.

3. Results

The proposed method is divided in two parts – the overview and the refinement (see *Figure 1*). The method is iterative, in the sense that the clinical content resulting from the overview phase, can be detailed and improved by running through the refining phase one or several times. This strategy is aimed at creating useful results with a small workload, and then improving the specifications in accordance with available resources in those areas where clinical knowledge is accessible.

3.1. Step 1 – Identification of background material

The background material consist of guidelines, care plans, data descriptions of quality databases, textbooks etc. This is used as the reference for the clinical evidence and is the main knowledge basis for specification of the clinical content.

The output for the activity is a library (preferably in electronic form), with all relevant documents. This is accompanied with a directory of the documents with classification of type, evidence level, priority in the specification etc.

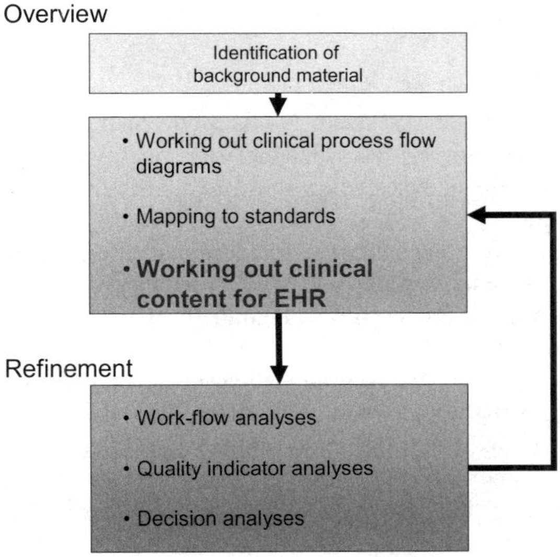

Figure 1 – The seven steps of specification of clinical content. The iteration between overview and the refinement phase is shown.

3.2. Step 2 – Working out clinical process flow diagrams

In this step the background material is analysed and combined with the participant's knowledge on clinical process flow and procedures. The result is a graphical representation of the clinical process, using symbols for start- and end-points, process, decision points, data etc.

The diagrams are worked out in several versions with increasing richness and details. The diagrams are important tools for keeping the overview and creating transparency in the development.

3.3. Step 3 – Mapping to standards

In order to re-use the data in the EHR and communicate the clinical content it is necessary to relate the information to standards. This step includes three consecutive activities: An analysis of the data representing the clinical process, a mapping to the concept model (in this case the Danish Basic Structure for Electronic Health Records), and a mapping to the terminology system (in this case SNOMED CT).

It should be noted that the two last activities should be performed by informaticians and standard experts. The clinicians should be kept in a review role.

3.4. Step 4 – Working out the clinical content

The goal with this step is to specify the clinical content in a systematic and structured way. The output from this step should be used directly to configure an EHR system. The description of the clinical content should be in accordance with clinical and management goals, and should reflect the clinical evidence identified in the previous steps.

The result of the step is specification of standard (care) plans, containing activities, containing results, as well as the related decision points. The specified clinical content will in the first place be documented in dedicated forms, which contain the information necessary to configure an EHR. Secondly the clinical content will be documented in a formal language, i.e. as archetypes using the archetype description language.

3.5. Step 5, 6, 7 – Detailing the clinical content

While it is recommended that step 1-4 is performed sequentially, the workflow analyses, quality indicator analyses and decision analyses can be performed in any sequence. Each of these analyses will refine and qualify the first version of the clinical content specifications.

These analyses are contributing with additional information from different focal points. The workflow analyses take into account which professional group performs the activities identified in the process-flow diagram. These analyses give the opportunity to change the workflow and enhance the cross-professional cooperation.

The quality indicator analyses aim at ensuring that the necessary data for the selected indicators actually is present in the EHR. This analysis can give an input to the data analyses and an update of the clinical content specifications.

The decision analyses make explicit the decision criteria and the selection possibilities in the clinical process flow. When feeding this in the EPR, the system can provide decision support and advanced reminders in relation to the standard plans. The intention is not to mimic the diagnostic decisions.

4. Discussion

The proposed method was developed and used within different medical areas, both within emergency/internal medicine, surgery, and psychiatry. Physicians, nurses and other clinical professions participated in the development.

Substantial education was needed in the first series of workshops. The participants were introduced to the complete method at the first workshops. Before the participants were performing a specific step, the information about that step was repeated. The participants were introduced to basic informatics concepts like information models, terminologies, structuring of processes and information. However, the education in these concepts was carefully adjusted to the medical area in question.

The clinicians felt comfortable using the method, and the use of flow diagrams was clearly a useful vehicle for structuring the guidelines and discussing professional differences (step 2). The diagrams typically underwent three iterations to reach the agreement on granularity and content.

The clinicians expressed that the mapping of the clinical content towards the process model was the most difficult part (step 3). However, after the mapping actually had been made, the next step (step 4) became very easy, because the specification of standard plans, their relation to the relevant diagnoses and to the standard results were laid out by the mapping activity.

During the quality indicator analyses (step 6), the clinicians teased out the specific clinical indicators, taking into account recommendation from the Danish Secretariat for Clinical Guidelines [21] and the indicators already selected by the National Indicator Project [22]. This analysis typically revealed the need for updating flow diagrams (from step 2) or the data list (from step 3)

The method also proposes forms for how the resulting structured clinical content should be documented. This documentation is to be used for configuration of the EHR system. Since the EHR-tool for entering the clinical content is not finalised, the electronic version of the clinical content has not yet been tested by the clinicians. The validation of the format of the clinical content has therefore not taken place. It is likely that the forms will need to be updated in order to get a better match with the requirements arising from the EHR-implementation.

5. Conclusion

The specification of structured clinical content is proposed as a prerequisite to enable EHR systems to present clinical guidelines dynamically at the user interface. This is expected to support the practice of evidence-based medicine.

The method presented has shown useful for transforming text-based international/national/regional/local guidelines, care plans etc. into structured specifications of clinical content. This is expressed as standard care plans or set of procedures linked to specific diagnoses. Templates for documentation of results related to the procedures are also outcomes of the specification process.

However, is not tested how the specified clinical content will work when implemented in the EHR, since the development has not yet reached this stage.

It is evident that development of the clinical content within many medical areas is a huge task. Therefore, the work has to be shared between different stakeholders in Denmark. It is also crucial to enable re-use of international guideline work. Furthermore, it is necessary that the Danish concepts are adjusted to the international standardisation work in openEHR, CEN and HL7 [15], and that we continue the search for useful IT-tools for documentation of the structured clinical content.

Consequently, it is essential that some decisions are made at a national level – based on international experiences:

- A standard for the structure of the clinical content, i.e. an information model
- A standard for an exchange format based on that model

Furthermore, there is a need for the following system development adjusted to the Danish context:

- An IT-tool for specifying the clinical content in the agreed structure
- EHR-systems that can import the agreed format
- EHR systems that can handle the formalised clinical content in the user dialog

Acknowledgments

The development of the methodology has been funded by the Copenhagen County

References

[1] Ministry of Interior and Health. National IT strategy 2003-2007 for the Danish Health Care Service.
[2] Basic Structure for Electronic Health Records (BEHR). The National Board of Health, Denmark. Copenhagen. http://www.sst.dk/Informatik_og_sundhedsdata/Elektronisk_patientjournal/GEPJ.aspx
[3] SNOMED® International. www.snomed.org
[4] Modelling and implementing Electronic Health Records in Denmark. Bernstein K, Bruun-Rasmussen M, Vingtoft S, Andersen SK, Nøhr C. International Journal of Medical Informatics, 2004.
[5] Grimshaw JM, Russell IT. Effect of clinical guidelines on medical practice: a systematic review of rigorous evaluations. Lancet 1993
[6] Guidelines International Network http://www.g-i-n.net/
[7] National Guideline Clearinghouse http://www.guidelines.gov/
[8] OpenClinical, resources on advanced knowledge management methods, technologies and applications for healthcare http://www.openclinical.org/home.html
[9] PROforma, clinical guideline representation and interchange format. http://www.acl.icnet.uk/lab/proformaspec.html
[10] GEM, the Guideline Elements Model. http://gem.med.yale.edu/default.htm
[11] GLIF, the Guideline Interchange Format. http://www.glif.org/glif_main.html
[12] Archetypes: Constraint-based Domain Models for Future-proof Information Systems. T. Beale. OOPSLA 2002 workshop on behavioural semantics.
[13] The Archetype Object Model. OpenEHR relaease 1.0. Editor:T. Beale. www.openehr.org
[14] The Archetype Definition Language (ADL), release 1.0. Editors:T Beale, S Heard. www.openehr.org
[15] A Shared Archetype Language T Beale, S Heard. www.openehr.org
[16] CEN TC251
[17] Bruun-Rasmussen M, Bernstein K, Vingtoft S, Nøhr C, Andersen SK. Quality labelling and certification of Electronic Health Record Systems. Proceedings for Medical Informatics Europe 2005, Geneva
[18] Clinical content in EHR in Copenhagen County: Acute Coronary Syndrome and Abdominal Aorta aneurism. Bruun-Rasmussen M, Bernstein K, Vingtoft S. MEDIQ and Cph. County 2005 (In Danish)

[19] Clinical content in EHR in Copenhagen County: Schizophrenia. Bruun-Rasmussen M, Bernstein K, Vingtoft S. MEDIQ and Copenhagen County 2005 (In Danish).
[20] Development of clinical content in EHR in Copenhagen County (SUFIE). Description of a methodology. Bruun-Rasmussen M, Bernstein K, Vingtoft S, Rindum JL. MEDIQ and Copenhagen County 2005 (In Danish). www.sufie.dk
[21] Danish Secr. for Clinical Guidelines http://www.sst.dk/Planlaegning_og_behandling/SfR.aspx?lang=en
[22] The National Indicator Project. http://www.nip.dk/nipUK.htm

Ubiquity: Technologies for Better Health in Aging Societies
A. Hasman et al. (Eds.)
IOS Press, 2006

Using Treemaps to represent medical data

Emmanuel CHAZARD[a,1], Philippe PUECH[a], Marc GREGOIRE[a], Régis BEUSCART[a]
[a] *Lille University Hospital, EA 2694, France*

Abstract. Confronted with the inadequacy of traditional charts, we tested the contribution of Treemaps to the representation of medical data. Treemap charts allow description of large hierarchical collections of quantitative data, on a synthetic way. Treemaps were implemented using PHP5, and were tested in the field of DRG-mining and other medical informations. From now on, this implementation is used in an interactive web-based request tool, and could be used to design interactive piloting tools.
Keywords : Medical Informatics, Computer Graphics, Treemaps, Charts

1. Introduction

Histograms, pie charts, bar diagrams and box plots are traditionally used to represent medical data. They are employed in a complex way, coupled with interminable tables, in order to render an account of the data's multidimensional structure. They show their limits with respect to the features being represented : several variables to be represented jointly, numerous classes among which the least populated cause information jamming, class overlap, very long wordings. Lastly, these frequently used tools do not exploit the interaction power that computers allow. Indeed, most end-users wish to see on the computer screen the same thing they used to read on paper sheets.

We would like to represent complex data by using areas, colors, and text labels.

2. About Treemaps

2.1. Introduction to Treemaps

Treemaps are a space-filling visualization method that allows to represent large hierarchical collections of quantitative data [1]. The principle consists in dividing the display area into a nested sequence of rectangles whose areas correspond to a quantitative variable. Treemaps have been applied to visualize files on a hard drive, as in Spacemonger [2], and to a wide variety of domains ranging from financial analysis [3], [4] to sports reporting [5]. Several distinct algorithms exist : the slice-and-dice [1], the cluster, and finally the squarrified algorithms. The squarrified algorithm [6] cuts out the available area so as to minimize the ratio of each rectangle. Thus, quantities are more easily comparable and the output looks more aesthetic. Treemaps seem to be one of the best ways to represent large hierarchical collections of quantitative data [7].

[1] Emmanuel Chazard - département d'information médicale, CHRU Lille, 2 avenue Oscar Lambret 59000 Lille France - emmanuelchazard@yahoo.fr

2.2. Problems related to scale

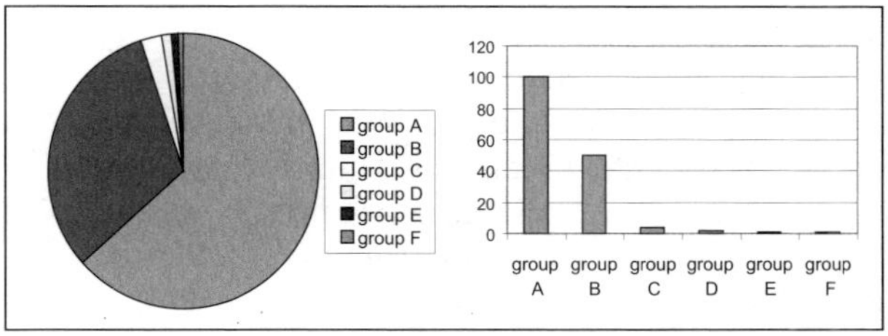

Figure 1 - ex. 1 - representing disparate quantities using usual charts

Although the usual graphs are drawn in two dimensions, they actually use only one dimension to represent quantities. As a matter of fact, in bar charts, the area is propotional to the sole height. Likewise, in pie charts, the area is proportional to the sole angle. Consequently, when a Cartesian scale is used, very disparate quantities induce reading confusion and small quantities become unreadable.

In order to illustrate this problem, we try to represent disparate quantities (going from 1 to 100), using traditional charts (Figure 1). The unidimensional charts induce difficulties of interpretation. Treemaps respond to this problem by drawing rectangles whose area is proportional to the quantity. On the Treemap chart (Figure 2), we can notice that group D is two times larger than both groups E and F.

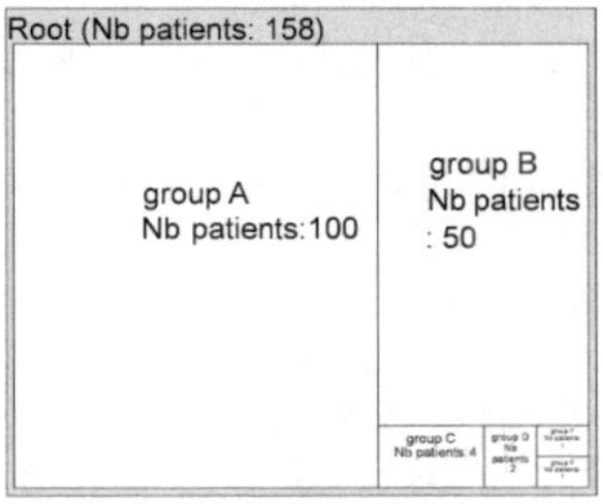

Figure 2 - ex. 1 - representing disparate quantities using Treemaps

However, the size of the rectangles is used to represent a volume (e.g. : "number of patients"), and not a density, (e.g. : "proportion of men").

2.3. Problems related to data overlap

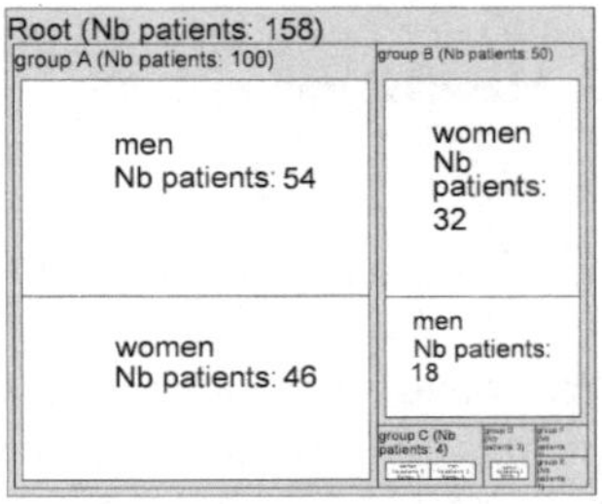

Figure 3 - ex. 2 - representing data overlap with Treemaps

The representation of data overlap is not very easy using conventional charts, and often results from a deliberate choice of the author (Figure 4). On the other hand, data overlap seems natural to Treemaps. As Treemaps consist in dividing a rectangle in several rectangles, those rectangles can be divided in their turn in a nested sequence (Figure 3).

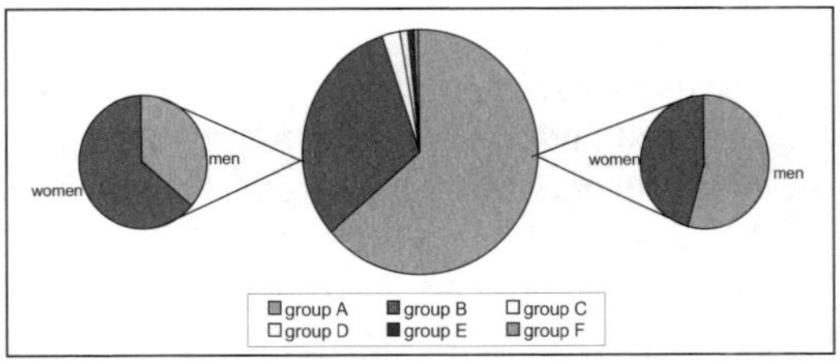

Figure 4 - ex. 2 - representing data overlap with pie charts

2.3.1. Problems related to color meaning

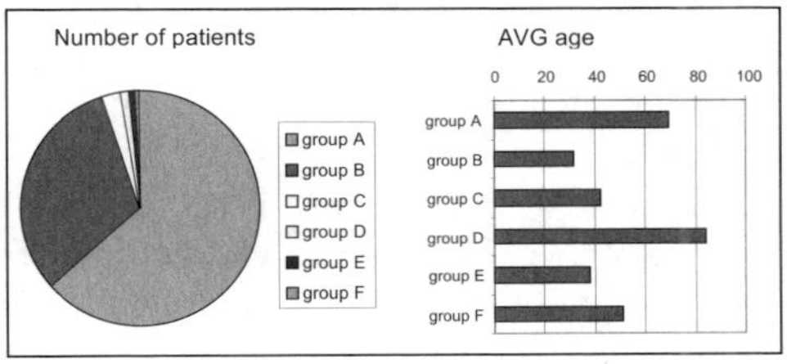

Figure 5 - ex. 3 - representing a second variable using traditional charts

Contrary to traditional charts, Treemaps can represent an additional variable using a continuous color scale. Color scale should not be used to represent a variable sensitive to volume. As continuous color scales cannot be directly interpreted, the values should appear in text labels.

In the next example, we try to represent the average age of the patient, in addition of the number of patients. Using conventional charts, we have to create a second chart (Figure 5). Using Treemaps, we can use colors (Figure 7). We can notice that the traditional charts would induce a misinterpretation of the global average age

2.4. Problems related to data labelling

Traditional charts do not allow for the use of long text labels. Moreover, the size of text labels doesn't depend on the quantity represented, which induces diffculties in reading. In this example, we try to add long text labels to pie charts (Figure 6) and Treemaps (Figure 7). One can notice that values of the variables can be easily included in text labels.

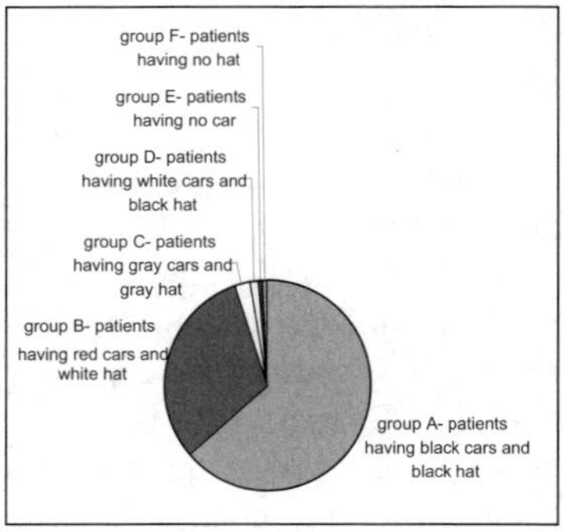

Figure 6 - ex. 4 - adding long text labels to pie charts

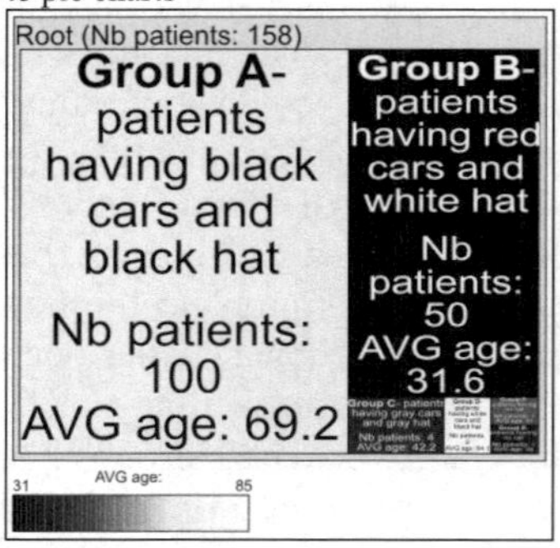

Figure 7 - ex. 3&4 - representing a variable with colors & adding long text labels to Treemaps

3. Material and methods

3.1. Program issues, output format

We wanted our program to be able to read hundred of files and to generate as many graphs on-the-fly, without any user interaction. Two output formats are available :

the JPEG format (Joint Photographic Experts Group) [8] :

- Advantages : VBA (Visual Basic for Applications) makes it possible to incorporate images of traditional formats within Microsoft Excel, such as BMP, GIF, JPEG.
- Disadvantages : no end-user interaction is possible ; the chart cannot be manually corrected ; poor quality printing

the SVG format (Scalable Vector Graphics) [9] : SVG is a XML-based format. The graph's description is written in formatted text according to XML standard. Then, a Web browser interpretes this description through a vectorial graph.

- Avantages : the end-user can zoom in and zoom out without any loss of quality ; the originator can allow end-user interaction, implementing onMouseOver effects, onClick effects, etc... [10] ; the imperfections of SVG graphs could be manually improved using some graphic softwares such as InkScape [11], or using a text editor such as Notepad, assuming SVG standards are known [9] ; high quality priting
- Disadvantages : the Web browser must be equipped with a plugin such as Adobe SVG Viewer [12], which was the case of all the hospital computers ; Unfortunately, VBA cannot load SVG graphics on the fly.

3.2. Color scale

We selected a "black-red-yellow-white" color scale (Figure 8) by respecting the following assertions :

- the scale is intuitive
- the colors can be ordered without any error
- the number of colours is rather weak, but the nuances have a strong discriminating capacity
- daltonians are not penalized
- printing in gray does not decrease legibility

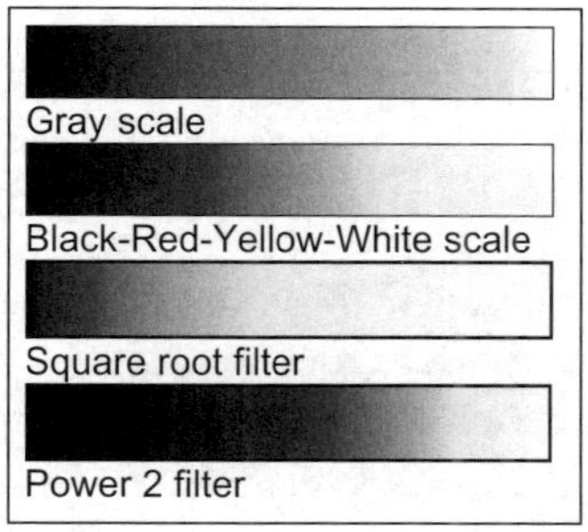

Figure 8 - color scales and filters

Canvas of increasing density could be used instead of colors, for better black & white document printing compliance. This solution cannot be adopted because of text labels. Our program also implements several nonCartesian filters.

3.3. Implementation

Several existing applications allow to generate Treemaps charts, like Microsoft Treemapper [13] (requires Microsoft Excel) or Treemap 4.1. [14] (requires Java Virtual Machine). Unfortunately, none of those applications could meet our needs.

We chose PHP5 (Php : Hypertext Preprocessor) [15] as a programming language. Its object-oriented structure [16] was particularly helpful to treat the rectangles overlap. PHP makes it possible to generate JPEG files by using the GD2 library [17], and to write text files, and thus to generate SVG, assuming SVG standards are known [9]. Our program implements the squarrified Treemap algorithm [6]. It consists in an object-oriented PHP class, which allows to reuse it in other coming programs.

4. Results

Table 1 - ex. 5 - extract from medical data : table

MDC	DRG	Nb Days	Daily Rate
MDC 40 Re-adjustment...	DRG 230 R.R. MDC - Age>=16 years – R.R. Care	201	238
MDC 40 Re-adjustment...	DRG 231 Age>=16 years – Handicap – phys dep<=12	752	176
MDC 40 Re-adjustment...	DRG 232 Age>=16 years – Handicap – phys dep>12	103	248

Our first exemple represents DRG related data. French rehabilitation care is financed by a prospective payment system, on the basis of daily rate [18]. Each day of the patient's stay is ranked into a Diagnosis-Related Group (DRG). DRGs are brought together into Major Diagnostic Categories (MDCs). A daily rate is linked to each DRG. We represent with both table (Table 1) and Treemap (Figure 10) the activity of a hole rehabilitation care hospital. Data are represented as following :

- each rectangle represents a DRG ; DRGs are brought together into MDC
- the area represents the number of invoiced days
- the color represents the daily rate of the DRG

Our second example shows geographical data. We want to represent the link between geographical origin and heavyness of the patients. Our territory is divided into 4 "life basin", themselves divided into 12 "health territory" (Figure 11). Data are represented as following (Figure 9) :

- each rectangle represents a health territory, brought together into life basins
- the area represents the number of patients comming from the health territory

- the color represents the proportion of heavy patients according to the Simplified Acute Physiologic Score [19]

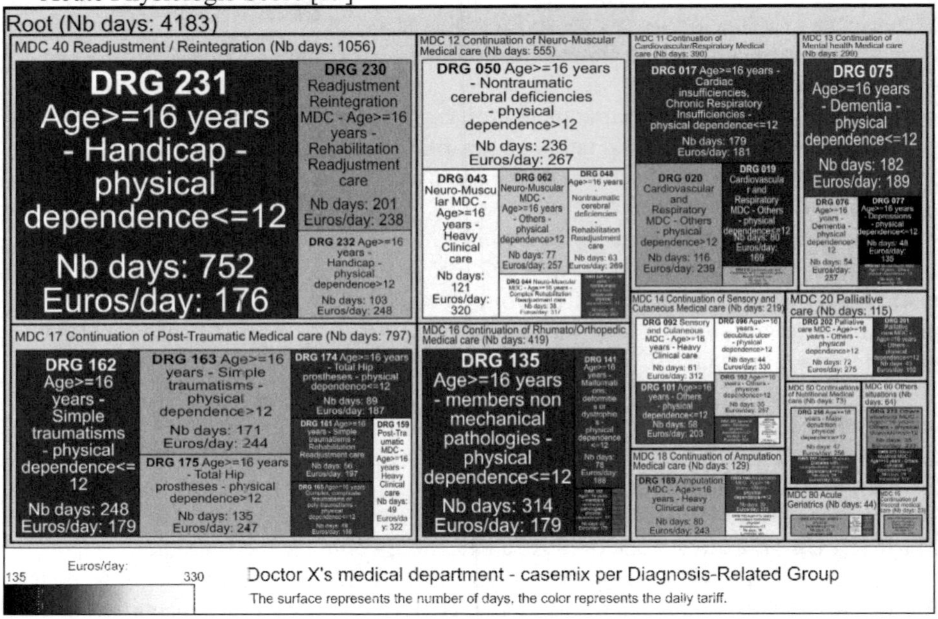

Figure 10 - ex. 5 - Complete output of the activity of a medical department, presenting simultaneously MDC, DRG, number of days, daily rate.

Figure 9 - ex. 6 - comparing number and heavyness of the patients with their geographical origin

5. Discussion

This implementation has several advantages. The use is simple and the result is immediate. Treemaps allow to represent a lot of information in the same chart ; the information is synthetic and treated on a hierarchical basis ; no additional table. They could be used to represent ratemaking data as well as geographical or epidemiological data. JPEG outputs allow to import charts into Excel generated dash-boards. SVG outputs allow interactive exploration and high quality printing.

Figure 11 - ex. 6 - geographical splitting of the territory

However, certain weak points are to be taken into account. Treemaps are a "new" way to represent data and require a certain training. Outputs should be explained to the physicians. At the moment, treemap outputs cannot be obtained with traditional softwares. Treemaps are not usable for all kind of data. As written before, the size of the rectangles should not be used to represent density variables.

They should not be used to represent chronological data. Moreover, they can hardly be used to compare successive years, the rectangles being laid according to the current decreasing order.

Treemaps should be soon integrated into quarterly dash-boards. We also designed an interactive online request tool : from now on, end-users are able to choose interactively on a webform the variables to represent, from any network-linked computer. The output is generated on the fly, the end-user can explore interactively the output on the screen by simply using the mouse.

6. Conclusion

Although Treemaps are a synthetic and at the same time a very complete mode of representation, they are seldom used in industrial applications and never used in the health sector. Nowadays in France, new ratemaking rules induce a need for synthetic piloting tools, which could be fulfilled with Treemaps. Moreover, high-level interactive tool production should be an inventive way to motivate data providers.

Acknowlegements

Nicolas Leroy (EVALAB, University of Lille), Dr François Houyengah, Dr Omolade Alao, Vincent Chouraki, Dr Laurence Lefebvre, Thomas Ponchaut, Catherine Catteau (Department of Medical Information, Regional University Hospital of Lille).

References

[1] Shneiderman, B. Tree Visualization With Treemaps: A 2-D Space-Filling Approach. ACM Transactions on Graphics, 11(1), pp. 92-99.
[2] http://www.werkema.com/software/spacemonger.html (last accessed Jan. 12, 2006).
[3] Jungmeister WA, Turo D. Adapting treemaps to stock portfolio visualization. Tech Report CS-TR-2996, Computer Science Department, University of Maryland, College Park, MD.
[4] http://www.smartmoney.com/marketmap (last accessed Jan. 12, 2006).
[5] Jin L, Banks DC. TennisViewer: A Browser for Competition Trees. IEEE Computer Graphics and Applications, 17(4), pp. 63-65.
[6] Bruls M, Huizing K, van Wijk JJ. Squarified Treemaps. In Proceedings of Joint Eurographics and IEEE, TCVG Symposium on Visualization (TCVG 2000) IEEE Press, pp. 33-42.
[7] Bederson, B. B., Shneiderman, B., Wattenberg, M., Ordered and quantum treemaps: making effective use of 2D space to display hierarchies, ACM Transactions on Graphics, 21(4), Oct. 2002, 833-854.
[8] http://www.jpeg.org/ (last accessed Jan. 12, 2006).
[9] http://www.w3.org/Graphics/SVG/ (last accessed Jan. 12, 2006).
[10] Eisenberg DJ. SVG. O'Reilly, ISBN : 2-84177-228-4
[11] http://www.inkscape.org (last accessed Jan. 12, 2006).
[12] http://www.adobe.com/svg/viewer/install/main.html (last accessed Jan. 12, 2006).
[13] http://research.microsoft.com/community/treemapper/ (last accessed Jan. 12, 2006).
[14] http://www.cs.umd.edu/hcil/treemap (last accessed Jan. 12, 2006).
[15] http://www.php.net (last accessed Jan. 12, 2006).
[16] Gutmans A, Sæther Bakken S, Rethans D. PHP 5 Power Programming. Pearson Education, ISBN 0-131-47149-X
[17] Sklar D, Trachtenberg A. PHP en action. O'Reilly, ISBN: 2-84177-231-4
[18] http://www.atih.sante.fr (last accessed Jan. 12, 2006).
[19] Le Gall JR, Lemeshow S, Saulnier E. A New Simplified Acute Physiologic Score (SAPS 11) based on a European/North American multicenter study. JAMA, 1993; 270: 2957-63.

Ubiquity: Technologies for Better Health in Aging Societies
A. Hasman et al. (Eds.)
IOS Press, 2006

Evidence in pharmacovigilance: extracting Adverse Drug Reactions articles from MEDLINE to link them to case databases

Nicolas Garcelon[a1], Fleur Mougin[a], Cédric Bousquet[b, c] and Anita Burgun[a]
[a] *EA 3888, Faculté de Médecine, IFR 140, Université de Rennes, France*
[b] *INSERM U729, Université Paris Descartes, Faculté de Médecine, Paris, France*
[c] *Hôpital Européen George Pompidou, Paris, France*

Abstract. Literature, specifically MEDLINE, is among the main sources of information used to detect whether a drug may be responsible for Adverse Drug Reactions cases. The aim of our work is to automate the search of publications that correspond to a given Adverse Drug Reactions case: (i) by defining a general pattern for the queries used to search MEDLINE and (ii) by determining the threshold number of publications capable to confirm or infirm the Adverse Drug Reaction. We applied our algorithm to a set of 620 cases from a French pharmacovigilance database. We obtained a precision of 93%, recall 70%. We determined a threshold of 3 publications to confirm an Adverse Drug Reaction case.
Keywords: Adverse Drug Reaction Reporting Systems, Algorithms, Automatic Data Processing, MEDLINE

1. Introduction

The World Health Organization (WHO) defines pharmacovigilance as "the pharmacological science relating to the detection, assessment, understanding and prevention of adverse effects, particularly long term and short term side effect, of medicines. It is gaining importance for doctors and scientists as the number of stories in the media of drug recalls increases" [1]. A high number of cases of Adverse Drug Reactions (ADR) are spontaneously reported each year by clinicians by means of a specific reporting form sent to the pharmacovigilance centers. The work of pharmacologists is then to detect which drugs induce adverse effects. Literature, specifically MEDLINE is among the main sources of information used to detect whether a drug may be responsible for ADR cases [2,3]. Nowadays, literature searches are mainly done manually. However, the exhaustive retrieval of such information may not be straightforward, for three reasons. Firstly, the query is not easy to formulate: the terms have to be translated in English (for non-English speakers), and even mapped to MeSH entry terms. Secondly, the large number of publications in MEDLINE makes the search difficult and time-consuming [4]. Thirdly, it is not easy to determine manually the level of evidence that should be the most appropriate to characterize the relation between a drug and a possible adverse effect. Therefore, as also mentioned by

[1] Nicolas Garcelon, 2 rue henri leguilloux 35033 Rennes, nicolas.garcelon@univ-rennes1.fr

other authors, (e.g. [5)], we would expect such information to be automatically retrievable from specific searches of MEDLINE. The aim of our work is to automate the search of publications that correspond to a given ADR case in the French pharmacovigilance database. The adverse effects are coded using the WHO - Adverse Reaction Terminology (WHO-ART) thesaurus. First, we defined a general pattern for building the queries used to search for MEDLINE publications. Second, publications have been filtered and we determined the threshold number of publications capable to confirm or infirm the ADR. An evaluation of our approach was performed. This work is part of the French ACI EI-Xplore project.

2. Materials and methods

The essential information in an ADR case corresponds to the association of one or several adverse effects with one or several drugs. Therefore, an ADR case can be described as a list of "active ingredient (AI) - adverse reaction (AR)" pairs, representing all the combinations between AIs and ARs. For each ADR case, every AI-AR pairs should be studied.
We want to retrieve publications for an "AI-AR" pair. We need to formulate the MEDLINE query for this pair and to filter the relevant associated publications. So, we have to:

- Translate the "AI-AR" pair in English (UMLS mapping, and mixing morphosyntaxic rules with Google® linguistic tools).
- Determine the correct MeSH terms to retrieve publications with good precision and recall in the pharmacovigilance domain.
- Analyze XML notices corresponding to the publications.

2.1. Materials

As mentioned in the introduction, the adverse effects are coded using the WHO-ART thesaurus. This thesaurus comprises 3,580 terms identified by two codes. The UMLS (Unified Medical Language System) Metathesaurus [6] is multilingual and organizes terms from multiple terminologies, including WHO-ART: synonymous terms are clustered into a unique Metathesaurus concept. Metathesaurus concepts are categorized by one or several semantic types. We used the 2004AA version of the UMLS. MetaMap [7] is a program developed by the National Library of Medicine (NLM) to map the text to the UMLS Metathesaurus concepts.

2.2. Mapping the "adverse effects" and translating "active ingredients"

Adverse effects and active ingredients were translated in English to set up the MEDLINE query.
For the French WHO-ART term used for the adverse effect:

- The equivalent English term is extracted from the UMLS, through the WHO-ART code (same concept).
- The restrict-to-MeSH [8] algorithm is applied to find the MeSH equivalent closest term(s) when no publication is found with the UMLS synonyms. (For

example: "Renal Failure Aggravated" is not a MeSH term, the restrict-to-MeSH algorithm proposes "Kidney Diseases").

For each AI terms, the first step consists of mapping the terms directly to the UMLS Metathesaurus. If the mapping is unsuccessful, the following approach is applied:

- Ignore French stop-words
- Translate terms using morphosyntaxic rules [9] (e.g.: cisplatine: cisplatin)
- Apply MetaMap
- Filter the Metathesaurus concepts so as to keep only those belonging to one of the 25 semantic types corresponding to Chemicals & Drugs.
- Apply Google® linguistic tools on terms not translated before

2.3. Query construction

The aim of this stage is to set up a query that will be sent to MEDLINE to find the publications related to an ADR case. We filter MEDLINE publications on specified AI-AR pairs, and limit the search field to pharmacovigilance.

<u>Select pharmacovigilance publications:</u> We identified two subheadings: « adverse effects » and « chemically induced » that have been used to index pharmacovigilance articles in MEDLINE and the term "drug-induced" for publications that have not been indexed yet.

<u>Select publications on specific AI-AR pairs:</u> For each UMLS concept, we add synonyms, except for molecular formulas and for some concepts that are too general (such as "SOD" which is a synonym of "acid sodium"). Then we keep only the synonyms with « [Mesh terms] » and « [Substance Name] » qualifiers by using the MEDLINE web service [10].

For each ADR case, we use the following query pattern: (« adverse effects » or « syn1 adverse effects » or …) AND («active ingredient »[Mesh terms] or « syn1 active ingredient »[Substance Name] or …) AND ((« adverse effects »[subheading] AND « chemically induced »[subheading]) OR "drug-induced"[All Fields])

2.4. Filtering publication

We used two models of co-occurrences: a simple model and a complex model. The simple model calculates the number of publications in which AI and AR are found together. The complex model calculates the number of publications in which the XML notice specifies "adverse effects" and "chemically induced" qualifiers on the same MeSH terms used to define our cases.

2.5. Evaluation

We evaluated our approach on 620 ADR cases from the pharmacovigilance database of the Hôpital Européen Georges Pompidou (HEGP). For each case in the database, all AI-AR pairs are listed. Three parameters were analyzed:

1- The mappings.

2- The performance of the information retrieval module. We calculated the precision and recall obtained with our automated extraction method by comparing our results to a manual extraction made by an expert, which is considered as the Gold Standard.

3. The supported evidence: We compared our results manually with the information given by the Vidal 2005® and the BIAM [11]. Performance was measured by calculating the area under the receiver operating characteristics (ROC) curve to determine the best model and threshold. We calculated specificity and sensitivity for both models and on 8 thresholds of co-occurrences (1, 2, 3, 4, 5, 11, 16, and 21).
The summary of the steps is shown in figure 1.

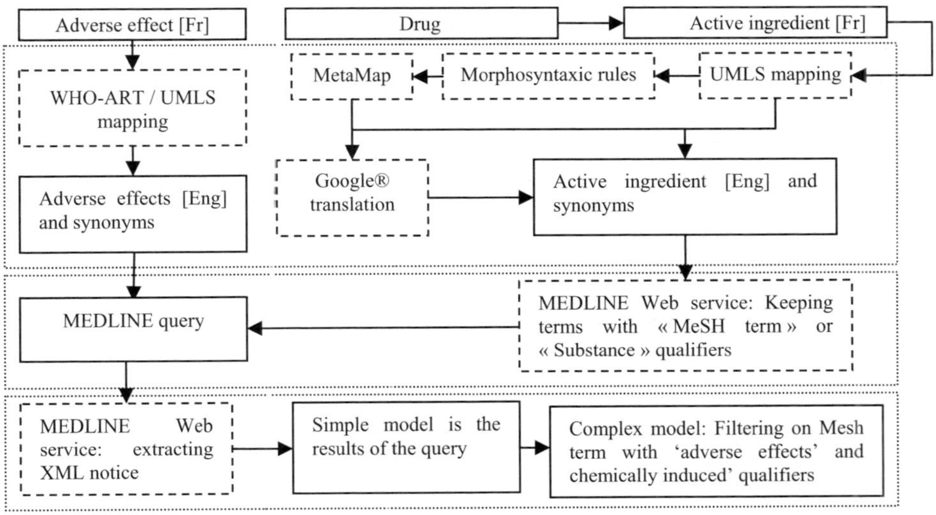

Figure 1: Schema of the different stages of the methodology

3. Results

Starting from the initial set of 620 ADR cases, 1,113 "AI-AR" pairs were obtained. All the adverse effect terms (330 different terms) were mapped successfully to the UMLS (precision=100%). We were able to translate 84% of the active ingredient (341 different terms) in English (precision=97 %).
As a result, we found 986 AI-AR pairs that were correct. We obtained 9,211 publications. We found publications for 61% of pairs with the simple model and 31% of pairs with the complex model. For pairs with publications, the simple model gives a maximum of 1,933 articles and a median equal to 7; the complex method gives a maximum of 167 articles and a median equal to 4. 50% of the publications are "Journal articles" and only 17% are "Case Reports". This can be explained by the fact that case reports represent only 25% of the publications in the pharmacovigilance domain.
An example of a result is shown in Figure 2.

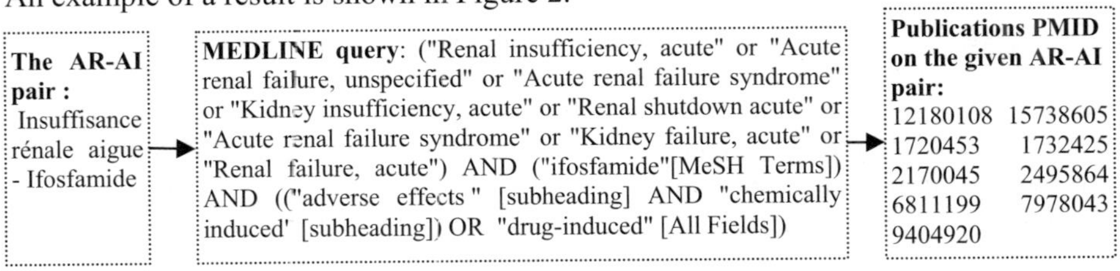

Figure 2: Example of an AR-AI pair and the corresponding result

The precision and recall measures for publication retrieval on 10 AI-AR pairs were manually evaluated for the automated search. We used the complex model to evaluate the automated search because the simple model extracts a too high number of publications to be evaluated. The automated search had a precision of 93% and a recall of 70%.
The well-known pairs have a mean value of 74 publications, a median equal to 17 (with the simple model) and a mean value of 17 publications, a median equal to 5 (with the complex model). The unknown pairs have a mean value of 6 publications (with the simple model) and 0 publication (with the complex model) and a median equal to 0 for both models.

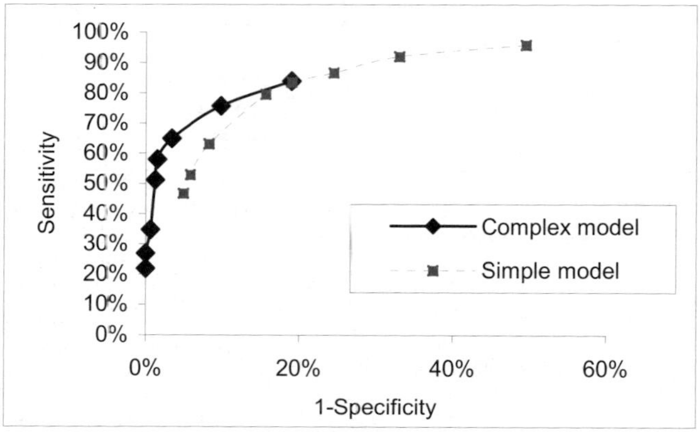

Figure 3: ROC curves for best model and threshold

The ROC performances of the two models are graphically presented in Figure 3. The specificity and sensitivity for each threshold and model are given in Table 1.

	threshold	≥1	≥2	≥3	≥4	≥5	≥11	≥16	≥21
Simple model	**Specificity**	50%	67%	76%	81%	84%	92%	94%	95%
	Sensitivity	96%	92%	87%	84%	80%	63%	53%	47%
Complex model	**Specificity**	81%	90%	97%	98%	99%	99%	100%	100%
	Sensitivity	84%	76%	65%	58%	51%	35%	27%	22%

Table 1: Sensibility and specificity of the models for each threshold

4. Discussion

The precision and recall measures of publication retrieval in the automated search show a good performance. Because specific subheadings and keywords are used in the queries that are built by our method, the automated search may be more specific than a manual query. For example, for the "cyclophosphamide - myeloid leukemia" pair, 5 publications are obtained automatically and all are relevant whereas 1,042 publications have been obtained by the manual search and most of them mention cyclophosphamide as a drug therapy for myeloid leukemia. In the complex model, filtering on the XML notices is a good means to select a sufficient number of relevant publications. But for some pairs, the manual search gives relevant articles which are not found with the automated search. This is due to the granularity of the MeSH terms used in the query. For example, a search on "Ceftazidime" may be extended with the more general term "Anti-bacterial Agents" present in the MeSH Thesaurus. We plan to exploit the hierarchical relations in MeSH to provide more citations.

We observed a correlation between numbers of co-occurrences and the evidence score given by Vidal®. The evaluation of each model to determine if a pair is true shows that the complex model is globally more specific than the simple model. To have an acceptable PPV (Predictive Positive Value), we had to use the complex model with a threshold of 3 co-occurrences given a sensitivity of 65% and a specificity of 97%. As such, the complex method is of potential interest to detect whether a drug may be responsible for ADR cases (see the case of cyclophosphamide and leukemia). The threshold must be prudently used to limit the false alarms given by anecdotal reports [12]. The simple model can be used in a search engine because of its high sensitivity. The complex model can be applied to the first model's results to rank the results of the search [13].

The work presented here is a feasibility study for automated extraction of evidence supporting drugs inducing ADR. The next step is to apply text-mining tools on all publications extracted using MEDLINE. We conjecture that we could detect ADR signals using association rules extraction on full-text literature [14] related to statistic methods on the ADR cases database [15].

References

[1] WHO Drug Information . The importance of pharmacovigilance. June 2002

[2] Stricker BH, Psaty BM. Detection, verification, and quantification of adverse drug reactions. BMJ 2004 ;329:44-7

[3] Duda S, Aliferis C, Miller R, Statnikov A, Johnson K. Extracting Drug-Drug interaction articles from MEDLINE to improve the content of drug databases. Proc AMIA Symp. 2005; 216-220.

[4] Morel P, Vandel B. Adverse drug reaction monitoring and the Internet: evaluation of the use of the Internet by French Pharmacovigilance Centres and a non-exhaustive survey of websites of interest for collecting information about adverse drug reaction. Therapie 1999 Sep-Oct;54:525-32

[5] Mendonça EA, Cimino JJ. Automated knowledge extraction from MEDLINE citations. Proc AMIA Symp 2000 ;:575-9

[6] Lindberg DA, Humphreys BL, McCray AT. The Unified Medical Language System. Methods Inf Med 1993 ;32:281-91

[7] Aronson AR, Effective mapping of biomedical text to the UMLS Metathesaurus: the MetaMap program. Proc AMIA Symp. 2001; 17-21.

[8] Bodenreider, O., Nelson, S. J., Hole, W. T., & Chang, H. F. (1998b). Beyond synonymy: exploiting the UMLS semantics in mapping vocabularies. Proc AMIA Symp, 815-9

[9] Tran TD, Garcelon N, Burgun A, Le Beux P. Experiments in cross-language medical information retrieval using a mixing translation module. Medinfo 2004 ;11:946-9

[10] http://eutils.ncbi.nlm.nih.gov/entrez/query/static/efetch_help.html .

[11] Aractingi S. Site of BIAM, a computerized drug databank. Ann Dermatol Venereol 1999 ;126:375

[12] Venning GR. Validity of anecdotal reports of suspected adverse drug reactions: the problem of false alarms. Br Med J (Clin Res Ed) 1982 ;284:249-52

[13] Greenspan R. Search engine usage ranks high. URL: http://cyberatlas.internet.com/markets/advertising/article/0,,5941_1500821,00.html [accessed 2003 Jun]

[14] Cherfi H, Napoli A, and Toussaint Y. Towards a Text Mining Methodology Using Association Rules Extraction. Soft Computing Journal. 2005

[15] Bousquet C, Henegar C, Louët AL, Degoulet P, Jaulent MC. Implementation of automated signal generation in pharmacovigilance using a knowledge-based approach. Int J Med Inform 2005 ;74:563-71

5.3 Decision Support: Other Topics

Ubiquity: Technologies for Better Health in Aging Societies
A. Hasman et al. (Eds.)
IOS Press, 2006

Towards the Use of Ontologies for Activity-Based Costing in Healthcare Organizations

Anand Kumar
IFOMIS, University of Saarland, Saarbruecken, Germany

Abstract. Activity-based costing is a methodology which provides a basis for healthcare cost optimization. A robust formal representation of such activities is needed in order to be able to draw inferences as to the costs involved in a reliable manner. An introduction to the basic ontological distinctions involved in such a representation and to the complications one faces in carrying out the task is presented in this paper.
Keywords: Activity-based costing, biomedical ontologies, process representation, clinical workflow

1. Introduction

Since Dowless's claim that activity-based costing (ABC) methods have the potential to improve resource management and thereby maximize efficiency in healthcare organizations, it has been demonstrated on a number of occasions that ABC supports better pricing practices through more accurate costing and can be used to identify underutilized resources as well as costs that can serve as targets for reduction. [Dowless 1997] Ross has pointed out that ABC can bridge the gap between the medical and financial communities and provide a foundation for performance improvement. [Ross 2004] Suthummanon *et al.* have used ABC methods for nuclear medicine; Lievens *et al.* for radiotherapy; Chambers *et al.* for dentistry; Grandlich for surgery; Crott *et al.* for endoscopy; Ridderstolpe *et al.* in cardiology; Hawkins *et al.* in breast-imaging. [Suthummanon *et al.* 2005, Lievens *et al.* 2005, Chambers *et al.* 2004, Grandlich 2004, Crott *et al.* 2002, Ridderstolpe *et al.* 2002; Hawkins *et al.* 2001]

The basic building block in order to carry out a comprehensive ABC is to identify relevant healthcare activities and to represent them digitally in such a way that their influence on costs can be measured. In previous work, we have dealt with the representation of clinical practice guidelines and ontological aspects of representation of processes [Smith *et al.* 2005, Kumar *et al.* 2004, Peleg *et al.* 2003]. This paper is a first step towards formalization of those aspects of an activity which contribute to its costs.

2. Ontological Treatment of Activities, Roles and Commodities

Continuants are entities which exist in totality at a particular instance of time, while occurrents exist at any given time only in some partial stage or phase. Thus continuants *endure* through time, while occurrents (processes) unfold themselves in time. Human being, surgeon and lung are continuants, while performance of a surgery, breathing and life are occurrents. Many of the relations represented between occurrents and continuants (formalized in [Smith *et al.* 2005]) are directly applicable to ABC. There are certain ontological characteristics of activities which we summarize as follows:

1. Activities are performed by independent continuants like human beings
2. Human beings perform activities by exhibiting specific roles (a continuant dependent on the human being who occupies the role) for example, the role of surgeon or nurse.
3. An instance of the activity across a particular time interval, and has a starting and an ending point.
4. An instance of the activity does not occur before its initiation and after its completion.
5. An instance of the activity exists when it is incomplete and ceases to exist when it is complete.

Commodities, on the other hand, are continuants with specific roles. A chemical compound has a medication role which endures until its expiry date. Activities, including healthcare services, are occurrents.

3. Standard Costing

Accountants and finance specialists have provided a variety of different ways to classify costs. One of the commonest used within healthcare organizations rests on the opposition between fixed and variable costs. Fixed costs are those which remain constant irrespective of the product manufactured or service provided, while variable costs are those which increase with increase when a product is manufactured or a service provided. While there are no costs which are absolutely fixed or variable, most of the costs are predominantly one of the other. Hospital building maintenance cost, vehicle maintenance cost, accountants' salaries, fixed component of consultants' salaries etc. are examples of largely fixed costs. Perioperative and postoperative cost, cost per bed per day, variable component of consultants' salary, salary of nurses on hire etc. are examples of largely variable costs. Ontologically, costs are dependent continuants and the occurrences of costs are occurrents. Note that cost occurrence does not always match the cost levied on the payer.

Traditionally costing has been done and reimbursement made on the basis of the consideration of continuants primarily, both independent (for example personnel, supplies) and dependent (for example roles, qualifications, diagnostic reports, images). Costs involved are, those of: Material, Labor, Depreciation and Overheads. Standardized costs are available for specific diagnoses (DRGs). For example, the average costs involved in a diagnosis of chronic cardiac failure is calculated on the basis of materials (catheters, medications, disposable goods etc.); labor costs of nursing, medical consultation, etc. which, in turn, lead to actual execution of such functions; depreciation of hospital building, beds, intensive care units, and so on, together with overheads of administration, risk management and so on. Such costing is easier to calculate and also to represent within an information system. The ontological basis of this ease of calculation is that it is easier to represent instances of continuants. Such instances, for example, an instance of *50mg. Atenolol tablet* or an instance of *hospital bed*, exist in their totality at every time instant at which they exist at all. Such instances are located in a space which can not be occupied by any entity which is not a part of the continuant instance located in it. This is not the case with occurrents, where instances of two processes can be collocated at the same location and at the same time instant.

4. Activity-based Costing

While traditional costing is a costing made predominantly on the basis of continuants, ABC is a type of costing that is carried out predominantly on the basis of occurrents. More precisely, although ABC is based ontologically on activities, it is not a costing of activities but rather of the *states* resulting from those activities. States are dependent continuants, attributes of the independent continuants which participate in the occurrent instance which gives rise to the state as its result. The main steps involved in ABC include[1]:

1. *Analyze Activities*: The activities which are involved in the provision of health-care service are identified and analyzed. Ontologically, this means that an analysis is carried out of the instances of healthcare activities created, either initiated or existing.
2. *Gather Costs*: The costs involved in the performance of activities are represented: including salaries, expenditures for machinery, etc.
3. *Trace Costs to Activities*: In this step the results of analyzing activities and the gathered organizational inputs and costs are brought together, which produces the total input cost for each activity. A simple formula for costs is provided – outputs consume activities that in turn have consumed costs associated with resources. Ontologically, outputs are states of the continuants which participated in the process instance occurrence.
4. *Establish Output Measures*: Activity unit cost is calculated by dividing the total input cost, including assigned costs from secondary activities, by the primary activity output volume; the primary output must be measurable and its volume or quantity obtained. Ontologically, consumption of activity signifies that the activity instance has been completed. Since activity instances cease to exist at their completion, this signifies that the states of the participants in an occurrence of the activity instance are the only entities which contribute to the primary activity output.
5. *Analyze Costs*: The calculated activity unit costs and bills of activity are used to identify candidates for improving the business processes.

Thus the ABC methodology would produce results such as:

[1] http://www.faa.gov/ait/bpi/handbook/chap5.htm

Unit Nursing colostomy services cost = $XYZ/ Number of nursing colostomy services
Unit Post-operative management cost for partial bowel resection = $XYZ/ Number of Post-operative management cost for partial bowel resection

The costs of all subactivities which are performed as parts of these activities are included within the ABC of a particular activity. Ontologically, the cost of a nursing activity instance in post-operative management of a patient is a part of the instance of complete post-operative management. However, ABC does not only include costs of the subactivities, but also of the substances which are consumed or used during the exercise of the activity. Thus it sees the cost of catheters used in a post-operative intensive care unit as part of its activity based costing. Ontologically this means that the cost of catheter instance is part of the cost of post-operative interventional activity. This does not mean that the continuant catheter is part of the mentioned occurrent. Continuants are never parts of occurrents or vice versa. Thus we have:

part-of (*cost of catheter*, *cost of post-operative management activity*)
not: *part-of* (*catheter*, *post-operative management activity*)

ABC applied in sectors predominantly dealing with services, especially the health sector, faces more complications than we face when dealing with manufacturing process. Some of the common complications are:

1. Activities are prepaid in toto: The activity instances are paid for, before they have been created. This is the case with advance charges placed for surgical or intensive care services. Ontologically this means that the costs are represented for a state derived from an activity instance which does not exist.

2. There are discrepancies between actual costs incurred and price paid. Activities are prepaid and these costs are not reversed if the activity is not performed. Thus the price for a partial colectomy is paid before the laparotomy occurs, before opening of abdominal activity is performed and these costs are not returned even if the procedure of partial colectomy is not performed. Ontologically, this signifies that the costs were paid for before the activity instance existed and even after the surgical procedure, when the state of the participant, the patient, does not bear the results of partial colectomy.

3. An activity is scheduled to be performed, was started, but before the completion of the activity, the activity is aborted: Ontologically, this means that the activity instance was created and was partially brought to completion, but that it ended before it being complete. The state of the participant might or might not have changed in the process. Suppose a catheterization instance with balloon angioplasty is created. There is no definitive decisive measure of when the costing should include a balloon angioplasty instance. The balloon angioplasty instance is payable whether the balloon was used for a few seconds and or for a longer duration and over a larger extent of the coronary arteries, the continuant participant of the angioplasty instance.

4. Levying costs for activities are usually not reversible, even if the results prove costs never incurred: Completion of activity instances like bilateral partial tubectomy is put into question when the patient gets pregnant. It is difficult in such cases to prove if the

activity instance was not completed or if out of natural reasons a tubal formation took place after the surgery.

5. Measurement of non-physical activities is sometimes difficult: While it is reasonably straightforward to determine the ABC of a wound redressing by a nurse, given that hourly nursing service costs, costs of material used, general cost of the outpatient department or the inpatient unit etc. are known, there are overheads related to physicians which are not easily measurable.

6. DRG based reimbursement: In many countries, the cost reimbursement is primarily calculated on the basis of diagnosis using Snomed or ICD nomenclature. The activity instances are individually not taken into account for such reimbursements.

The US Medicare program[2] uses a specific ABC variant in order to determine the cost of medical procedures, a program called Resource-Based Relative Value Scales (RBRVS). They consist of three components: professional time, practice time and malpractice expense. Costs can be assigned to procedures based on the relative value unit (RVU) assigned to the specific procedure. Ontologically, RVUs are dependent continuants, dependent on the participant who performs a given professional activity. Here again, the fact that ABC is based on the state of the patient is not always directly taken into consideration. For example, Information Systems do not regulate costing in a way that could block payments if the professional time of, say, five hours was not used to complete the specific activity instance. ABC would still provide all the payment for balloon angioplasty completion instance even when the potential activity was not actually carried out. The cost of evaluation for balloon angioplasty after catheterization and a negative decision to carry it out becomes the levied cost of the balloon angioplasty instance completion. Representation of a balloon angioplasty conglomerate procedure which contains a core balloon angioplasty procedure as one of its parts and cost allocation to the conglomerate procedure and not to the core procedure is one of the ways to manage the costing. Such representations are possible at the individual instance level, and are more complex at the class level, since at the class level certain parts of the process might not have been performed.

part-of (*coronary catheterization*, *balloon angioplasty conglomerate procedure*)
part-of (*balloon angioplasty decision*, *balloon angioplasty conglomerate procedure*)
part-of (*core balloon angioplasty*, *balloon angioplasty conglomerate procedure*)
part-of (*balloon catheter inflation*, *core balloon angioplasty procedure*)

Given that such conglomerate procedures do not have a core entity which needs to always be fulfilled in order to confirm the completion of the process, another possible representation is to depict the conglomerate procedure as a collection of subprocesses. This 'collection' involves a weaker relationship, it being stated only that an instance of the conglomerate procedure has a group of subprocesses as its parts, without representation of which procedures are exactly carried out in order to bring the activity instance to its completion. For costing purposes, this representation provides the kind of flexibility needed, though one loses the inference capability for addressing whether an activity instance is completed.

---

[2] http://www.medicare.gov/

5. Conclusion

Given all the interest raised in activity-based costing as a possible means to determine real healthcare costs and find ways and means to reduce it while optimizing service and improving outcomes, it is imperative that a formal representation of costing in general, and activity-based costing in particular, is provided. Such a formalization would make the representation of costs possible in such a way that information systems could draw inferences on their basis. With the shortage of time and other resources which current healthcare administrators face, these inferences would make is easier to identify which costs and/or revenue centers in a health care organization are performing suboptimally and what should be done in order to optimize the delivery of healthcare. This paper provides an introduction to the main issues which need to be tackled in order to provide such a formalism. Continuing work in this area would make the formalism more complete and usable for existing healthcare systems.

Acknowledgements:
Work on this paper was carried out under the auspices of the Wolfgang Paul Program of the Humboldt Foundation and also of the EU Network of Excellence in Semantic Datamining and the project "Forms of Life" sponsored by the Volkswagen Foundation.

References

[1] Dowless RM. Using activity-based costing to guide strategic decision making. Healthc Financ Manage. 1997 Jun;51(6):86, 88, 90.
[2] Ross TK. Analyzing health care operations using ABC. J Health Care Finance. 2004 Spring;30(3):1-20.
[3] Suthummanon S, Omachonu VK, Akcin M. Applying activity-based costing to the nuclear medicine unit. Health Serv Manage Res. 2005 Aug;18(3):141-50.
[4] Lievens Y, Kesteloot K, Van den Bogaert W. CHART in lung cancer: economic evaluation and incentives for implementation. Radiother Oncol. 2005 May;75(2):171-8.
[5] Chambers DW, Bergstrom R. Financial management and dental school equity, Part II: Tactics. J Dent Educ. 2004 Apr;68(4):439-45.
[6] Grandlich C. Using activity-based costing in surgery. AORN J. 2004 Jan;79(1):189-92.
[7] Crott R, Makris N, Barkun A, Fallone C. The cost of an upper gastroduodenal endoscopy: an activity-based approach. Can J Gastroenterol. 2002 Jul;16(7):473-82.
[8] Ridderstolpe L, Johansson A, Skau T, Rutberg H, Ahlfeldt H. Clinical process analysis and activity-based costing at a heart center. J Med Syst. 2002 Aug;26(4):309-22.
[9] Hawkins H, Langer J, Padua E, Reaves J. Activity-based costing via an information system: an application created for a breast imaging center. J Digit Imaging. 2001 Jun;14(2 Suppl 1):194.
[10] Smith B, Ceusters W, Klagges B, Köhler J, Kumar A, Lomax J, Mungall C, Neuhaus F, Rector AL, Rosse C. Relations in biomedical ontologies. Genome Biology, 2005; 6 (5): R46.
[11] Kumar A, Smith B, Pisanelli DM, Gangemi A, Stefanelli M. An ontological framework for the implementation of clinical guidelines in health care organizations. Stud Health Technol Inform. 2004;102:95-107.
[12] Peleg M, Tu S, Bury J, Ciccarese P, Fox J, Greenes RA, Hall R, Johnson PD, Jones N, Kumar A, Miksch S, Quaglini S, Seyfang A, Shortliffe EH, Stefanelli M. Comparing computer-interpretable guideline models: a case-study approach. J Am Med Inform Assoc. 2003 Jan-Feb;10(1):52-68.

Address for Correspondence
Anand Kumar, IFOMIS, Universität des Saarlandes, Postfach 151150, D-66041 Saarbrücken, Germany. Phone: +49-172-8984640. Email: akumar@ifomis.uni-saarland.de URL: http://www.uni-leipzig.de/~akumar/

Ubiquity: Technologies for Better Health in Aging Societies
A. Hasman et al. (Eds.)
IOS Press, 2006

From bibliometric analysis to research policy: the use of SIGAPS in Lille University Hospital

Patrick Devos[a,1], Helene Lefranc[a] , Eric Dufresne[b] and Regis Beuscart[b]
[a] *Department of Clinical Research, Lille University Hospital, France*
[b] *Department of Medical Informatics, Lille University Hospital, France*

Abstract. In French hospitals, the progressive setting up of the new rating systems has obliged the university hospitals to justify a certain amount of activities such as research, training or moreover recourse, which are specific missions of the university hospitals. In order to justify research activities, the Lille University Hospital has developed for now three years SIGAPS, a full-web bibliometric software which census and analyse, the scientific publications referenced in the Medline database. After data downloading, each article is classified on a 6 levels "quality scale derived from the impact factors. The system then performs, for a researcher or a team, a report allowing a quantitative and qualitative analysis. Started in Lille in November 2004, the inventory and analysis of data is now ending. For the period 2001-2004, 2814 articles have been published in 700 different journals. The total number of articles increased from 688 in 2001 to 757 in 2004. The mean impact factor was equal to 2.26 and 15.5 % of articles were classified as A, 20.9% as B. Those results confirm the high level of research of the University Hospital of Lille, in agreement with two other national studies which ranks our establishment at the 6th position for medical research activities among the French University Hospitals. Currently a similar evaluation has now began in the 9 other university hospital which have subscribed to the SIGAPS project. We works currently on new indicators as patents, thesis or conferences, or access to other databases as Sciencedirect or Scopus via the RIS format. The next step in the project is the implementation of a meta-base which will federate the information provided by each SIGAPS system. This meta-base will then allow us to perform comparisons between different hospitals, determine the national "sites of excellence" and create some clinical and research networks.
Keywords: Bibliometrics, Biomedical research, Databases, Pubmed

1. Introduction

In France, specific missions of a University Hospital as research, training or moreover recourse, were until then remunerated in a contractual way at 13% of their budget. The progressive setting up of the new hospital rating systems has now obliged them to justify in an indisputable way these activities. The University Hospital of Lille, from this point of view, has started a certain number of steps aiming to justify its different activities. With regard to research, two traditional approaches are generally carried

[1] Corresponding Author: Patrick Devos, Department of Clinical Research, Lille University Hospital, France, p-devos@chru-lille.fr

out : the measurement of the inputs (staff costs, technical platforms,…) and the outputs [1]. With regard to production, we were initially interested in scientific publications, a traditionally used indicator in this type of approach [2,3].

In order to list in an exhaustive and automated way the scientific articles achieved within the hospital site, the University Hospital of Lille has developed for now three years SIGAPS, a bibliometric software [4,5]. This software, entirely automated, queries normally the Pubmed server on the about 700 researchers of the site of the hospital, recovers the available data, enriches them and produces a synthetic report for a researcher, a service or a team.

Currently used in the University Hospital of Lille and in 9 other ones in France, this tool, easy to use and to administrate, allows to analyse in a very sharp way the medical research activities of a university hospital. Started in Lille in November 2004, the phase of inventory and analysis of data is now ending. The main objective of this paper is to supply the methods and the first results of the research activity of the University Hospital of Lille, considered as one of the most important in France according to recent national studies [6,7].

2. Material and Methods

In order to analyze the scientific production, SIGAPS performs several steps :

First, using the Pubmed portal, the administrator defines, for each researcher, the adequate request or PMID list. SIGAPS then interrogates the Pubmed server, retrieves the publications, verifies the existence of each article in the local database, decodes the XML and loads the data into the local database. It also interrogates the LDAP directory to find information on the authors. Requests are saved in the application in order to update automatically the database.

Secondly, "quality" of a publication is evaluated using a 6-level scale [5] derived from the Impact Factor [8,9], A corresponding to excellent journals, B very good ones, and so on. If an article is published in a journal not referenced by the ISI, the article is placed in a sixth category, the NC category (Not Classified).

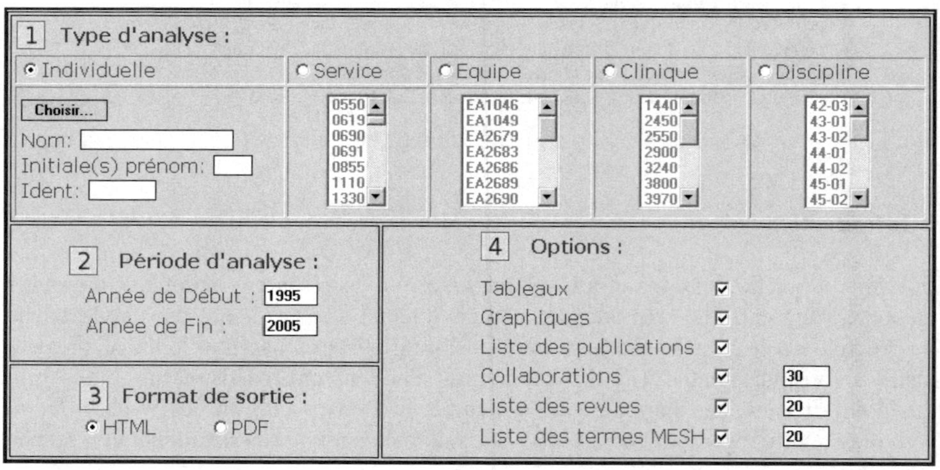

Figure 1 : SIGAPS User's Interface

Using a standard browser, we can generate a HTML report (Figure 1) for an author, a service or a research group. This report contains a series of tables and graphics as the number of publications per year, the position by category, …

Check of the data : each researcher has a login and password which allows him to connect on the server, produce his report, and validate his data. We can so add (collective names for example) or remove (homonyms) some articles. We can also correct some misspellings or names orthography.

3. Results

Researcher's inscription and data downloading started at November 2004. One year later, 700 doctors have been entered into the system, belonging to 25 research team, 104 clinical departments and 60 disciplines. All articles, since 1995 to now, containing at least one researcher from the Lille University Hospital, were downloaded from the Pubmed database. Analysis of data has been performed on the period 2001-2004.

At the establishment level, 2814 articles have been censed for the period 2001-2004, 688 in 2001, 675 in 2002, 694 in 2003 and 757 in 2004. The total number of publications seems to increase, in agreement with others studies which pointed out an increase of the publications in our hospital. Among this 2841 articles, 435 articles (15.5%) were classified A, 587 (20.9%) as B, 392 (13.9%) as C, 420 (14.9%) as D, 610 (21.7%) as E and 370 (13.1%) classified NC. Those proportions are statistically the same from the 4 years. We remark that the proportion of articles classified A or B is higher (36.3%) than we could expect (25%). The mean impact factor is equal to 2.26 (Q1=0.40, Median=1.32, Q3=3.10, P95=8.03). If we examine the distribution of the impact factor for the 4 years, we can observe that the median value seems to increase. However, the observed difference is not statistically significant.

Articles have been published in 700 different journals, of which a number of prestigious reviews. The proportion of articles published in French language is 33.7% and nearly the same for the 4 years.

The second step was to analyse and compare the production performed by individuals or teams. In this purpose, for each SIGAPS report produced, two analysis were systematically performed : a quantitative analysis and a qualitative one.

3.1. Quantitative analysis

In order to compare several researchers, research teams or disciplines on a given period of time, 7 indicators have been systematically computed and analysed :

- Nbtot : Total number of publications produced on the period of time
- NbAB : Number of publications classified A or B on the period of time
- NbNC : Number of publications classified NC on the period of time
- NbP1 : Number of publications as first author on the period of time
- NbP23 : Number of publications as 2nd or 3rd author on the period of time
- NbPL : Number of publications as last author on the period of time
- NbFRE : Number of publications in French language on the period of time

Firstly, descriptive analysis were performed on those seven parameters computed on a 4-year period (2001-2004): one quarter of the analyzed services had at most 20 publications, one quarter at least 50 publications. The proportion of articles classified A or B varied from 0 to 75% (Median=32%). The proportion of articles classified NC varied from 0 to 75% (Median=8%). The proportion of articles in French language varied from 0 to 92% (Median=32%).

Secondly, those indicators have been used to compare several researchers within a clinical department or several departments between each other. After normalization, radars representations have been used. Scientific production depending strongly on the number of doctors involved in research activities, we have taken into account the number of physicians for comparison of clinical departments or research teams.

Finally, multivariate analysis such as clustering methods were employed in order to provide typologies and define several profiles of "research activities".

3.2. Qualitative analysis

Firstly, we focused on the positions among the authors to determine whether the researcher is the investigator of the research or associated to the research. Secondly, we examined the list of collaborations provided by the application.

Here is the example of a department composed of 7 services S1, S2, … S7. The total number of articles published by this department during the 2001-2004 period of time is of 339. For 219 articles (65%) only one service has taken place, for 96 articles (28%) 2 services among 7 have taken place and for 24 articles (7%) more than 2 services. The collaborations' matrix (graph 2) allows to examine, for a certain service, the number of articles, alone or in common with another service. For example, S7 has 84 articles in total, of which 39 (46%) written alone, 30 (36%) with S1, 2 with S2, …

A schematic representation, with valued graphs representing only significant collaborations (dashed line : percentage of common articles > 15% , solid line : percentage of common articles > 30%) allows to point out strong collaborations between S2, S3, and S6 and between S1, S4 and S7 (graph 3). The examination of the publications achieved between these services allows then us to determine or confirm the common research axes of these services.

	S1	S2	S3	S4	S5	S6	S7	Total
S1	26	11	8	18	5	2	30	83
S2	11	42	17	2	5	12	2	76
S3	8	17	40	3	7	25	13	89
S4	18	2	3	26	13	1	2	54
S5	5	5	7	13	24	0	1	44
S6	2	12	25	1	0	22	9	59
S7	30	2	13	2	1	9	39	84
Total	83	76	89	54	44	59	84	339

Figure 2 : Collaboration's matrix

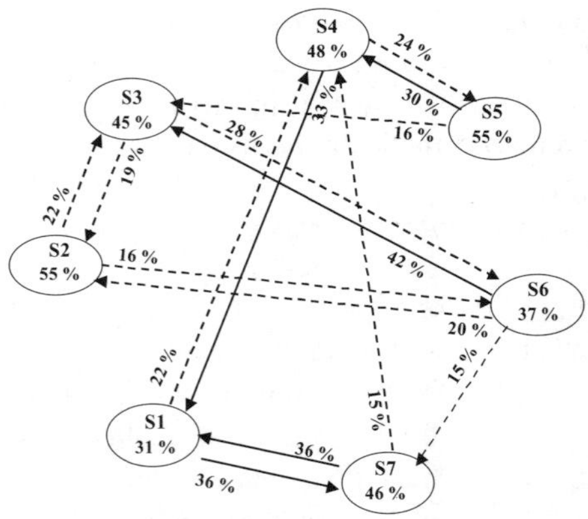

Figure 3 : Collaboration's graph

The analysis of the collaborations also allowed us to observe that some departments (radiology, molecular biology, or biostatistics for example) were very strongly implied in research activities as recourse departments for clinical ones.

Then the software provides the MESH terms the most employed in the list of articles, after deletion of non pertinent terms (as "Male" or "Human" for example). The examination of this list allows us to confirm the research's themes (or discover new ones) of the analysed structure and their type (fundamental, clinical, technical, …).

The software then provides the list of journals in which articles have been published, with their category (A, B, C, …) and the number of articles published in each journal by year. This allows us to determine the level of publication of a structure.

Finally, the whole bibliography of the analysed structure is generated by the software, according to the ICMJE recommendations (Vancouver style).

4. Discussion

The evaluation of the research activities in our establishment is now ending. The results confirm the high level of research of the University Hospital of Lille, in agreement with two other national studies which ranks our establishment at the 5th position for research activities [6,7]. The setting up of a local database and the automatic upgrade twice a month allows us to know, in real time, the number of publications produced by near 700 researchers. This analysis showed us the very wide diversity of profiles in our hospital : research team with high level of publications, clinical department with strong difficulties or emerging groups which have to be supported. From this point of view, SIGAPS will be very helpful in our knowledge of our research activities.

SIGAPS is an application developed with Internet technologies and based on a Web server. All researchers have an access to the application and their data. Each researcher, research team or clinical department can now edit its report and announce us any error, misspelling or problem. From a practical point a view, this allows us to perform a data validation, which is generally a major problem in bibliometric studies.

Journal categorization into 5 classes is presently a statistical classification based on the Impact Factor, virulently debated [10,11]. The proposed categorization, although partially correct, can with no doubt be improved.

Scientific publication is the most used indicator of research output. However, there's a lot of other indicators which could be used. Actually, SIGAPS allows researchers to declare books or chapters of books and conferences as "invited speaker". The next version of the software will permit the entry of patents or direction of thesis, indicators also regularly used in evaluation of research activities. Now the system retrieves only the publication referenced in the Pubmed database, which is the reference in the medical domain. However, some disciplines outside the clinical medical domain such as biostatistics, medical informatics or biophysics for example, have a significant proportion of the work published in more fundamental journals which are not systematically referenced in Medline. We works actually on the integration of those article referenced in other databases as Sciencedirect or Scopus via the RIS format [12] which is provided by all those portals and allows exports of data. This format is also supported by bibliometric software as Reference Manager or End Note [8]. Another way of evolution of the system concerns the integration of the number of citations per article, indicator more reliable than the impact factor of the journal in which the article has been published.

Actually there's a lot of studies concerning evaluation of research in major industrialized countries. Those macroscopic approaches, often based on the Science Citation Index [8], allows a global positioning of a city at a national level, or a country at a world level. Those studies, very useful for the Ministry of Research for example, are however useless for the management of the clinical research at a local or regional level. From this point of view, SIGAPS is a very useful and complementary approach.

In order to diffuse effectively this software, our establishment has a contract with a software company which promotes, installs and maintains the software in the different Information Systems of the hospitals who have subscribed to the SIGAPS project (currently, 9 university hospital and probably 15 at June 2006). Within each of those establishments, a similar evaluation as this described here, has now began.

The next step in the project is the implementation of a meta-base and a meta-directory which federate the information provided by each system. This meta-base will then allow us to perform comparisons between different hospitals, determine the national "sites of excellence" and create some clinical and research networks, more efficient in reply to national (PHRC, PRS, …) or European (InterReg, RTD Framework Programs) invitations to tender and, also an helpful structure for the pharmaceuticals industries for their multi-centric clinical trials.

5. Conclusion

From a "proof of concept" and a first prototype developed 3 years ago, the SIGAPS project has now matured. After successive national evaluations, reengineering and benchmarking in our establishment, the SIGAPS software is now able to produce indicators, statistics and qualitative information on clinical research realized in a university hospital. The successive subscriptions of the others university hospitals to the project will allow a large debate on biomedical research markers and then help us to enrich the software with new indicators or new functionalities asked by the news users of the software.

Since November 2005, this project is also supported for 3 years by the French Ministry of Health, which is looking for efficient tools to analyse and evaluate research activities performed in French University Hospitals.

References

[1] Godin B, The emergence of S&T indicators: why did governments supplement statistics with indicators? Research Policy 2003; 32(4): 679-691.

[2] Michalopoulos A, Falagas ME, A bibliometric analysis of global research production in respiratory medicine. Chest 2005; 128(6):3993-8.

[3] Bliziotis IA, Paraschakis K, Vergidis PI, Karavasiou AI, Falagas ME, Worldwide trends in quantity and quality of published articles in the field of infectious diseases. BMC Infect Dis 2005;5:16.

[4] Devos P, Dufresne E, Renard JM, Beuscart R, SIGAPS: a prototype of bibliographic tool for medical research evaluation. Stud Health Technol Inform. 2003;95:721-6.

[5] The SIGAPS Project : http://www.sigaps.org (last accessed May 9, 2006)

[6] http://lesrapports.ladocumentationfrancaise.fr/BRP/044000473/0000.pdf (last accessed May 9, 2006)

[7] http://www.obs-ost.fr/services/etudes_ost/virtual/10_domaine/e-docs/00/00/00/80/document_etude.phtml (last accessed May 9, 2006)

[8] Institute for Scientific Information : http://www.isinet.com. (last accessed May 9, 2006)

[9] E. Garfield, Citation analysis as a tool in journal evaluation. Science 1972, 178: 471-479.

[10] P.O Seglen, Why the impact factor of journals should not be used for evaluating research. British Medical Journal 1997, 314: 498-502.

[11] E. Garfield, Fortnightly Review: How can impact factors be improved ? British Medical Journal 1996, 313: 411-413.

[12] RIS Format Specification : http://www.refman.com/support/risformat_intro.asp (last accessed May 9, 2006)

Ubiquity: Technologies for Better Health in Aging Societies
A. Hasman et al. (Eds.)
IOS Press, 2006

Dealing with an Information Overload of Health Science Data: Structured utilisation of libraries, distributed knowledge in databases and web content

Simon Hoelzer MD[1,2], Ralf K. Schweiger PhD[1], Joerg Rieger[1], Michael Meyer[2]

[1] Institute of Medical Informatics, Justus-Liebig-University of Giessen, Germany
[2] H+ The Swiss Hospital Association, Bern, Switzerland

Abstract. The organizational structures of web contents and electronic information resources must adapt to the demands of a growing volume of information and user requirements. Otherwise the information society will be threatened by disinformation. The biomedical sciences are especially vulnerable in this regard, since they are strongly oriented toward text-based knowledge sources. Here sustainable improvement can only be achieved by using a comprehensive, integrated approach that not only includes data management but also specifically incorporates the editorial processes, including structuring information sources and publication. The technical resources needed to effectively master these tasks are already available in the form of the data standards and tools of the Semantic Web. They include Rich Site Summaries (RSS), which have become an established means of distributing and syndicating conventional news messages and blogs. They can also provide access to the contents of the previously mentioned information sources, which are conventionally classified as 'deep web' content.

1. Introduction

We live in an information age with a steadily increasing scope of products, services, specialist information and textual knowledge, but this often generates frustration when even experts can access only a small portion of the available resources. Things are even worse for the average Internet user, who for example may want to search for online information about the symptoms of a particular disease, but receives only product and price information for medicines that are supposed to be effective in treating the disease, and this as the result of an overly successful Google search that yields 1985 hits and is thus practically useless. Such scenarios are also frequently encountered in situations where research using books, specialist articles and periodicals is gradually being replaced by electronic tools, with all of their advantages, in fields such as medicine, jurisprudence and journalism. Integrating complex topical information is regarded as imperative in these fields. In many cases, the capabilities and productivity of the people involved in such activities are directly dependent on the effectiveness of their access to these electronic resources.

Actually, more information no longer necessarily results in an improved understanding in the sense of applied knowledge. Even today, only a fraction of the potential of electronic networking can actually be used, with the rest remaining hidden without consumers, users or others being aware of its existence. Increasing access to Internet resources and more complete searching reduces the probability of finding information that is actually relevant. Although new search technologies using differentiated classification criteria can yield short-term improvements in search accuracy, they always run up against their limits or even lead to distorted perception or manipulation. This has unfortunate consequences for the information society. This weakness of comprehensive Internet search services is readily apparent, even with trivial Internet searches. Google can presently access and search more than 15 billion documents (as of January 2006), but search accuracy decreases as the number of documents increases. Astonishingly enough, in most cases entering additional search terms does not improve the quality of the results.

2. Methods and Objectives

This problem can and should be approached in a variety of manners. Here the key aspects are organising or re-organising the electronic contents and the options that are available for processing the contents. The catchword 'Semantic Web' is used to promote special document markup methods used to indicate the substantive meaning (semantics) of text elements and Internet links in order to make automated knowledge processing possible. This represents a shift from simply publishing information for people (which largely represents the current situation on the Internet) to providing computer-interpretable contents. This value is created by adding specific structure to the documents and/or supplementary editorial processing, such as using meta-information and classification criteria.

Deep web

The present situation is still far removed from the ideal state of domain-specific standardisation with structured data available for every document. However, this state can be achieved in a simpler manner by taking an incremental approach and making the best possible use of presently available structures. Besides this, it is necessary to improve access to the presently poorly integrated deep web, so that it can be searched in the same manner using specialty (domain) search engines [1-4]. Access to this deep web will become increasingly important, and not only for scientific and research purposes. This is because the vast majority of textual knowledge is hidden away in database information systems (such as the editorial systems of publishing houses) and extensive libraries, most of which still cannot be accessed by contemporary search engines. Content from this deep web is primarily generated dynamically as a result of specific accesses (search requests) by users. Our current research is aimed at making these important resources accessible for intelligent searching. In this case, the structures of the underlying data sources (database models and content management system models) contribute to the previously mentioned better understanding of the meanings and interrelationships of the contents. Here as well, dependence on specific storage formats can be broken while retaining the available structural information [5]. Using such technologies can help open up the deep web and thus enable access to these essential sources of knowledge, whose extent is estimated to be more than 500 times that of the present Internet.

Rich Site Summaries: RSS

In the following discussion, we restrict ourselves to describing a new, generic technique for linking or connecting databases and other dynamic web content to current search engines and user interfaces. The RSS description standard plays a key role in this. RSS is a platform-independent format based on XML that was originally developed for exchanging messages and other types of web content [6,7]. With the currently used RSS format version (Version 2.0, as illustrated in Figure 1), this always involves storing information in structured form and making it available for automated processing by reader programs or other applications. In particular, RSS contains highly structured meta-information for these texts. RSS was primarily created to distribute messages from Internet portals, but it has developed into a standard for automated exchange of messages and human communications (weblogs and discussion forums). The format allows producers of RSS feeds, which are well-defined data streams containing references to one or more new or modified information resources, to easily and quickly inform interested parties about new contents. For readers, automated retrieval and processing of RSS feeds generates major time savings, since it eliminates the need to personally visit every website and the prepared formats can be filtered according to the specific interests of individual users. For reading and converting content from databases and CMS, users can take advantage of the fact that some of these systems already provide ready-made modules for generating RSS feeds.

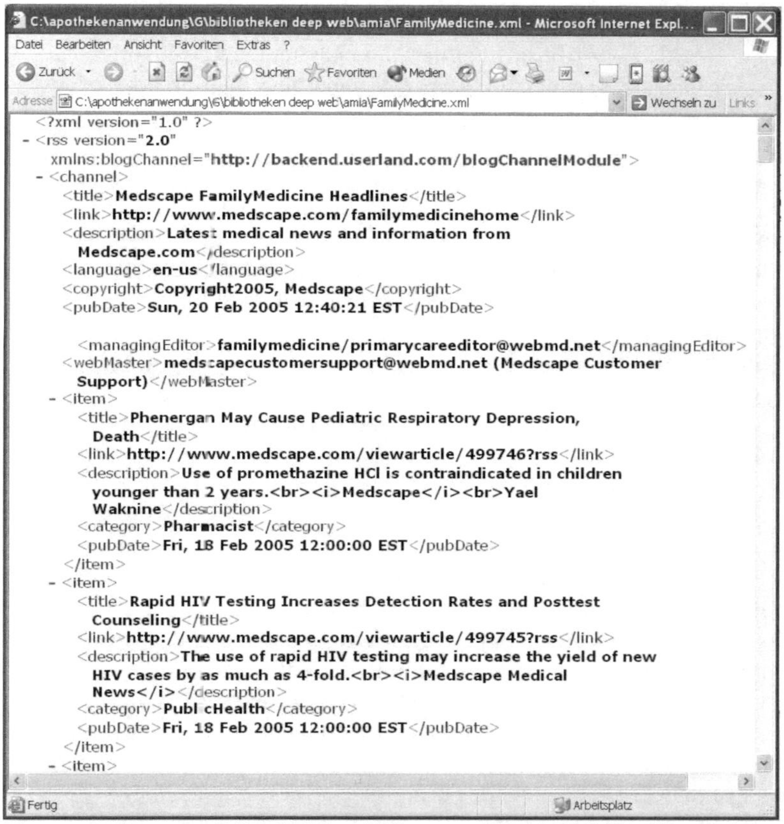

Figure 1: Sample RSS 2.0 feed (Source: WebMD / MedScape)

3. Results and Discussion

We utilise RSS as a standardized convergence point in order to make a wide variety of distributed information resources (journals, monographs, dissertations, etc.) in heterogeneous formats (XML, HTML, PDF, MDB, DB, etc.) usable and searchable in equal measure. For this purpose, a general-purpose converter has been developed in order to extract such meta-information from these sources as is deemed to be necessary. In simplified terms, a relationship is defined between the elements of the document or database and the RSS elements. In this way, we use RSS to define the essential components of a data stream that is intended to enable subsequent searches to access contents as specifically as possible. Nevertheless, differences in application parameterisation mean that it is left up to the information provider to decide whether to populate all of the RSS elements or only a subset. In the extreme case, it is even possible to transmit the entire information content of the original resource as XML or HTML. The key factor is proper assignment of contents to RSS elements.
The system has been designed such that a non-ambiguous assignment of 1 to n document database contents to a single RSS element must be defined within the specified containers. Highly differentiated case distinctions can be made in this manner according to specific input sources. In the simplest case, the structure or layout and style information (HTML or XML tagging) is assigned to the RSS tags using a direct conversion:

Example 1: <title> or <h1> or file information about the title = <title>.
Example 2: #header { width:100%; background:blue; margin:0px; padding:0px; } = <title>

Besides this, meta-information can also be employed directly.
Example 3: file information about the author, <META name="Author" content="Author Information">
plus author information and originator = <channel>, e.g. <managingEditor>
Example 4: <dc:date modified>2003-03-24</dc:date modified> = <pubDate>

For web documents, the link to the URI is made either directly or by using standardized referencing, e.g. Dublin Core (dc).
Example 5: <dc:type rdf:resource="http://dublincore.org/
usage/documents/principles/
#element" /> = <link>

A similar method is used for databases, with the field contents being read from a view or complex query and assigned to RSS tags. Alternatively, XML exports (such as mSQL DB) can be generated, with the resulting documents being processed or converted as described above (see Figure 2). There is also an update mechanism that detects changes to the original resources and initiates conversion of the resources. The net result is RSS feeds that describe and identify one or more ‘information units’. These are semantically marked in a standardized manner. The significant editorial content and meta-information (title, keywords, origin, date information, description, etc.) are then available for search queries using speciality search engines.
Besides the benefits described above, these automatically generated feeds can be searched in parallel with the innumerable subscribable news channels. A few such

valid news channels are already available, including the contents of MedScape / WebMD (see (http://www.medscape.com/pages/public/rss and Figure 1). An important factor is careful selection of the supplementary integrated feeds, which virtually expand the search space and contents in specific areas. In our example, this consists of the contents of a specialist library with an Oracle database and web documents, in this case primarily HTML and PDF. This method also allows primarily non-web (database) contents to be handled together with conventional web documents in a single application. In the medium term, the actual storage formats thus become a secondary consideration.

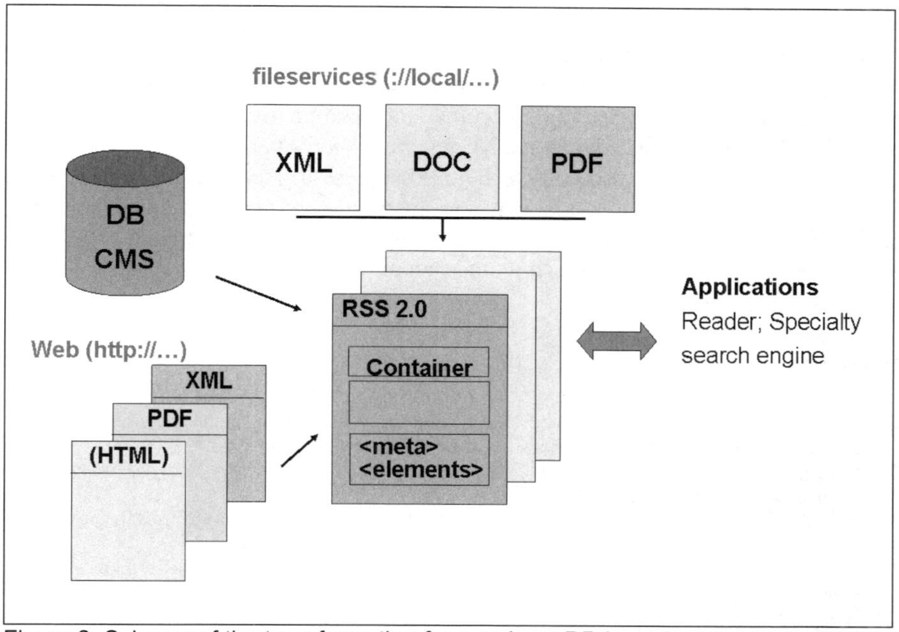

Figure 2: Schema of the transformation from web- or DB-based resources to RSS 2

4. Conclussion

Academic journals, libraries, and electronic news channels or news tickers are information media that dynamically populate themselves with new or modified contents. In the present case, the contents of these media are treated as electronic messages or news items. These messages and news items are then provided in the form of standardized, structured units (in this case RSS 2.0 feeds) for further processing (search, exploration, reading tools, etc.) This approach is based on merging the basic concept of conventional data management, such as relational databases (including content management systems), with processing partially structured (web-based) multimedia documents. Besides this, it enables even inexperienced users to access available information in a more effective and targeted manner than what was previously possible. This provides a basis for development of intelligent applications that incrementally exploit the potential added value of content markup, metadata and semantic linking. The possibility of using this sort of processing and filtering for objects such as headlines, topics, authors and associated original resources is becoming increasingly important for mastering the unrestrained flood of data and information emerging from the anarchic Internet and the deep web. This approach combines compaction of information contents into primary descriptive attributes with transformation into a standardized target format. This allows resources that are fundamentally different with regard to structure and format to be processed (handled) in a uniform manner. As a result, sources whose contents deal with identical or similar topics (e.g. from different editorial entities) can compete with each other with respect to topicality and quality, regardless of how the individual media are distributed. Various sources of information dealing with the same topic can be consulted and directly compared with each other. Such services are of considerable benefit in the medical environment, since rapid, accurate localization of information sources, in particular distributed and dynamically generated information sources, allows the quality of the individual resources to be assessed so they can be integrated into the user's routine work. In this manner, the Internet can mature into an interactive knowledge platform whose added value in many areas erases the boundaries between established information providers and the academic, non-profit sector.

References

[1] Hoelzer S, Schweiger RK, Dudeck J. Knowledge representation in digital medical documents for efficient information access and retrieval. Stud Health Technol Inform. 2003;95:451-6.

[2] Schweiger R, Brumhard M, Hoelzer S, Dudeck J. Implementing health care systems using XML standards. Int J Med Inform. 2005 Mar;74(2-4):267-77.

[3] Pschichholz H, Boeker M, Gaudes R, Schulz S, Klar R. InfoServerPlus: providing a medical library as application service. Stud Health Technol Inform. 2002;90:189-94.

[4] Hoelzer S, Schweiger RK, Boettcher H, Rieger J, Dudeck J. Indexing of Internet resources in order to improve the provision of problem-relevant medical information. Stud Health Technol Inform. 2002;90:174-7.

[5] Schweiger R, Hoelzer S, Rudolf D, Rieger J, Dudeck J. Linking clinical data using XML topic maps. Artif Intell Med. 2003 May;28(1):105-15.

[6] RSS - Introduction to XML Really Simple Syndication. http://www.xmlfiles.com/rss/

[7] Lewin J. Content feeds with RSS 2.0: Syndication goes mainstream. IBM US, http://www-106.ibm.com/developerworks/xml/library/x-rss20/

Ubiquity: Technologies for Better Health in Aging Societies
A. Hasman et al. (Eds.)
IOS Press, 2006

A Markov model to describe daily changes in organ failure for patients at the ICU

Linda Peelen[a,1], Niels Peek[a], Nicolette F de Keizer[a], Evert de Jonge[b], Robert J Bosman[c]
[a] *Dept of Medical informatics* , [b]*Dept of Intensive Care,*
Academic Medical Center, Amsterdam, The Netherlands
[c]*Dept of Intensive Care Onze Lieve Vrouwe Gasthuis, Amsterdam, The Netherlands*

Abstract. As the support and stabilization of organ function is a major goal of treatment in the Intensive Care Unit (ICU), changes in the function of organ systems are an important indicator of the progression of the disease and recovery. This paper presents how to construct a model that describes changes in organ failure of ICU patients on a day-to-day basis. The model is based on the daily Sequential Organ Failure Assessment (SOFA) scores for six organ systems and predicts, for each of these organ systems, whether failure or recovery is due on the next day, using six logistic regression equations. The joint set of equations, extended with equations for predicting ICU discharge and death, constitutes a first-order multivariate Markov model. We applied the procedure on a dataset and found that most types of organ failure are highly persistent.
Keywords: Decision-support; Intensive Care; Prognosis; SOFA score; Multivariate Markov Model

1. Introduction

A major goal of treatment in the Intensive Care Unit (ICU) is to stabilize organ function and if necessary to temporarily take over the vital functions by machinery or medication. Changes in the functioning of the organ systems are an important indicator of the progression of the disease and the probable outcome: the more organ systems with problems, the higher the probability that the patient will not survive the stay at the ICU. In contrast, if the function of organ systems is stabilized or improved, the probability of ICU survival increases. For this reason it is important to monitor the organ systems concisely and to have an estimate of future changes in organ function.

Recently, state transition (Markov) models have been used to analyze progression of chronic diseases [1, 2]. In analogy with this research, in this paper we analyze the changes in organ failure in ICU patients using a Markov model. In the ICU organ failure can be described on a day-to-day basis using the Sequential Organ Failure Assessment (SOFA) scoring system [3]. In this scoring system the degree of organ failure is expressed by six scores that indicate the function of six major organ systems: circulation, respiration, kidney function, central nervous system, liver and coagulation.

The model that is presented in this paper captures changes in organ failure in terms of changing SOFA scores. A rudimentary modeling approach is employed, where

[1] Corresponding Author: Linda Peelen. Academic Medical Center, Dept of Medical Informatics. PO Box 22700 1100 DE Amsterdam. The Netherlands. Email:l.m.peelen@amc.uva.nl

logistic regression equations are developed to predict organ failure of individual organ systems on the next day and a first-order multivariate Markov model is constructed by combining these regression equations.

The paper is structured as follows: Section 2 provides more detail on the use of the SOFA score. Section 3 describes the procedure to develop the regression equations and to combine them into a multivariate Markov model. Section 4 presents the results of applying these methods on a dataset from the ICU of the Onze Lieve Vrouwe Gasthuis (OLVG) in Amsterdam. In Section 5 we elaborate on the use of the model in practice and discuss our work.

2. The SOFA score for measuring organ failure at the ICU

The SOFA scoring system was developed in the late 1990s to describe organ failure of patients at the ICU. Although the score was developed to *describe* organ failure and not to estimate the prognosis of a patient, research has shown that the SOFA score and changes in the SOFA score are related to mortality (see e.g., [4]).

The scoring system consists of six subscores and one aggregate score, each calculated on a daily basis [3]. This proceeds as follows: for each organ system the degree of organ failure is quantified by an integer value between 0 and 4 (with 0 indicating normal organ function and 4 referring to complete failure). These scores are based on the values of one or two (mostly physiological) variables related to the particular organ system. For example, the score for the hepatic system is determined by the level of bilirubin in the blood. The total (aggregate) SOFA score is calculated as the sum of the six subscores and has a maximum value of 24. The total SOFA score gives a global impression of the patient's condition, but the subscores provide more information on the state of the organ systems and therefore may be more relevant from a clinical point of view. In our models we will therefore only use the subscores.

Table 1 Series of SOFA scores and ICU status for an example patient

Variable	Day 1	Day 2	Day 3	Day 4	Day 5
SOFA subscores					
Respiratory	4	4	3	3	-
Coagulation	0	0	0	1	-
Renal	1	1	3	4	-
Circulatory	3	4	3	3	-
Neurological	0	0	0	0	-
Hepatic	0	0	0	1	-
Total SOFA score	**8**	**9**	**9**	**12**	-
ICU status	At the ICU	At the ICU	At the ICU	At the ICU	Died at the ICU

As the scores are measured on a daily basis, we obtain a series of SOFA scores for each patient. Table 1 depicts the series of SOFA scores for an example patient. In the first two days this patient mainly has respiratory and circulatory failure. After day 2 the kidneys start failing and subsequently coagulation and hepatic dysfunction occurs. After four days the patient dies at the ICU. This example also shows the limitations of the use of the total SOFA score: the total SOFA score of 9 on both days 2 and 3 suggests that the condition of the patient remains the same, whereas the subscores reveal that renal failure develops on day 3 (as the renal subscore equals 3).

3. Modeling organ failure at the ICU

In this section we present how the SOFA scores were used to construct a Markov model that describes changes in organ failure. Section 3.1 introduces the data that we have used for the analyses. Section 3.2 describes how models were developed which predict failure for the six organ systems. In Section 3.3 we explain how these models for the individual organ systems were combined into a multivariate Markov model.

3.1 Data for the experiments

The model we present in this paper is based on data collected in the ICU of the Onze Lieve Vrouwe Gasthuis (OLVG), an 18 bed mixed medical-surgical closed-format ICU in a teaching hospital in Amsterdam. We used data from patients admitted between January 1st, 2002 and December 31st, 2004. We excluded patients admitted after cardiac surgery as in these patients the changes in organ failure are known to be very different from patients admitted for other reasons. For patients who were readmitted to the ICU during the same hospital stay we only used information on the first ICU admission.

Our dataset contained information on 1508 patients of which 248 (16.4%) died in the ICU. The median length of stay was 2 days, 587 patients (39%) were discharged or died after 1 day of ICU stay. In total 6845 records on SOFA scores were available. The total SOFA score ranged from 2 to 20 (median value 8). Almost all patients had respiratory failure during their entire ICU stay. Hepatic and neurological failure were seen less frequent (subscore > 0 in 15.8% and 18.9% of the records, respectively).

3.2 Prognostic models for failing organ systems

To create a model that describes the changes in the functioning of an organ system, we need to estimate the conditional probability that a particular organ system will fail on a specific day, given the states of that organ system, and all other organ systems, on the preceding day. Formally, such a conditional probability is written as:

$$P(system_{t+1} \mid renal_t, respiratory_t, coagulation_t, neurological_t, hepatic_t, circulatory_t), \quad (1)$$

where t refers to the day within the period of ICU admission ($1 \leq t \leq$ length of stay $(LOS) - 1$), $system_t$ gives the state of a particular organ system at time t, and $system \in \{renal, respiratory, coagulation, neurological, hepatic, circulatory\}$. As the SOFA score per organ system has five possible values, and we use the state of six organ systems at day t and one at day $t + 1$ (see Eq 1), we need to estimate 5^7 of these conditional probabilities *per day*. Even with a very large dataset we could not estimate all of these probabilities directly from the data, especially because some types of organ failure occur very rarely.

We therefore decided to reduce the number of probabilities that is to be estimated in three steps: (i) assuming the process of changes in organ functioning to be stationary, (ii) dichotomization of the SOFA scores, and (iii) use of a parametric approach to estimate the conditional probabilities. We will now briefly explain these steps:

(i) For the current analyses we assumed the probabilities in (1) to be stationary, i.e., we assumed that they are independent from the value of t. This implies that we have to estimate 5^7 probabilities in total. We restructured the dataset such that each record contained the SOFA scores on day t and the SOFA scores and the ICU status on day t+1. The number of records per patient in this restructured dataset equals *LOS* -1.

(ii) To further reduce the number of probabilities the SOFA scores were dichotomized, with $system_t = 1$ indicating failure at time t and $system_t = 0$ indicating that there is no failure. The thresholds for dichotomization were chosen based on the distribution of the data, such that the two classes were as balanced as possible. By this step the number of probabilities to be estimated was reduced to 2^7.

(iii) To estimate the conditional probabilities in (1) logistic regression equations were developed for each of the organ systems. First, a full model was constructed using the states of all organ systems on day t as covariates. Subsequently, a backward stepwise regression analysis was applied to obtain the model that minimized the Akaike Information Criterion (AIC). To prevent patients with a long length of ICU stay from largely influencing the parameters of the model, the observations in the analyses were weighted by $1/LOS$.

This procedure resulted in six logistic regression equations, one per organ system, to estimate the probabilities given in Eq. 1.

3.3 Multivariate Markov model

In the procedure described above we have considered the influence of organ failure on the functioning of each of the organ systems separately. However, it is also the combination of these findings that we are interested in. In fact, on a given day the patient can be in three possible states that are mutually exclusive: (i) the patient is at the ICU and his/her condition is characterized by the functioning of the six organ systems, (ii) the patient has died, and (iii) the patient has been discharged from the ICU.

For the first type of state we have $2^6 = 64$ possible states in our model; for both the second and third type of state we have only one possible state that patient can either be in or not. These latter two states can be described by the variables $ICUdeath_t$ and $ICUdischarge_t$, respectively. The conditional probabilities for these two variables at $t+1$, given the state of the organ systems at day t, were derived using the procedure presented in Section 3.2, resulting in two additional logistic regression equations.

Thus, the state of the patient can be described by the combination of the six dichotomized variables that represent the functioning of the organ systems, together with $ICUdeath_t$ and $ICUdischarge_t$. The changes in the condition of the patient can be seen as a Markov process: at one day the patient is in a particular state described by the eight variables mentioned above and the next day the patient is either in the same state or has moved to another state. The transition probabilities for the changes from one state to another can be derived from the eight logistic regression equations.

Once we know the state of the patient on day t, we can predict the distribution over the possible states on the next day. Note that the states in which $ICUdeath_t = 1$ or $ICUdischarge_t = 1$ are *absorbing* states: once the patient is in such a state, the probability of transition to another state equals zero.

4. Results

Figure 1 presents the results from the analyses based on the data from the OLVG. The left side of the figure provides the odds ratios obtained from the eight logistic regression equations. For all organ systems we observe that patients with failure in a particular organ system on day t have a high probability of organ failure in the same organ system on the next day. This phenomenon is called *persistence*. Furthermore, we observe differences between the extents to which organ systems are involved in the failure in other organ systems. E.g., circulatory failure on day t influences respiratory and circulatory organ failure on day $t + 1$, ICU discharge, and ICU death. In contrast, hepatic failure only influences hepatic and coagulation failure on the next day.

The right part of the figure displays a graphical representation of the multivariate Markov model obtained from combining these equations. The absorbing states "ICU death" and "ICUdischarge" are indicated by a double ellipse. Arcs indicate which organ systems influence the state at day *t*+1. It is seen that not all organ systems are equally important for prediction of the state of the patient on the next day. The maximum possible number of arcs is 48, whereas only 20 arcs are present in the current model.

Predictors system$_t$						Variable at
Renal	**Resp**	**Hepatic**	**Neuro**	**Circ**	**Coag**	**t+1**
0.8	0.6	-	0.8	0.4	-	Discharge
42.8	-	-	-	-	1.4	Renal
-	72.6	-	2.8	2.9	-	Resp
-	-	118.4	-	-	2.1	Hepatic
-	-	-	105.2	-	-	Neuro
1.8	3.8	-	-	190.4	-	Circ
1.6	-	1.9	-	-	99.6	Coag
-	-	-	-	0.7	0.8	ICU death

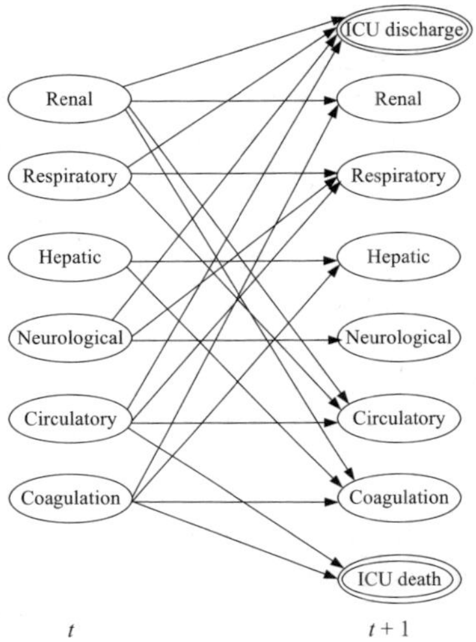

Figure 1. Results from the analyses on the OLVG data. On the left :odds ratios derived from the eight logistic regression equations. On the right: the multivariate Markov model. *Resp:* respiratory; *Neuro*: neurological; *Circ*: circulatory; *Coag*: coagulation.

5. Discussion

We have developed a Markov model to describe progression of disease in a situation where the condition is acute and state changes occur rapidly. Markov models have been applied in health care for various purposes since the 1970s [5], most prominently in cost-effectiveness analyses [6] and clinical decision analyses [7]. Recently, Markov models have been used to describe progression of disease. Salazar et al. [1] have used multi-state Markov models to analyze the progression of dementia. Saint-Pierre et al. [2] described the progression of disease in overweight asthmatic patients. The relation between organ failure and ICU survival has been modeled before using logistic regression (e.g., [4]), and dynamic Bayesian networks [8], but not using a Markov model in combination with regression equations.

The model we developed can be used for several purposes. First, the odds ratios derived from the regression equations provide us directly with information on the relationships between the organ systems, and between organ failure and ICU outcome on the next day. Second, the model can be used in daily patient care to estimate which organ systems are likely to fail in the next day, given the information from the current state of the patient. For example, in our model a patient with renal and hepatic failure on day *t* has a probability of 15% on coagulation failure on the next day. These estimates can be used to determine the treatment strategy for the patient, and also for planning of beds and technical equipment at the ICU. A third use of the model lies at the population level. For some patient groups the changes in organ failure that occur frequently are well-known, but for other patient groups this knowledge is lacking. The model can be used for simulations, to discover frequently occurring patterns and combinations of organ failure for a specific patient group.

The current model constitutes a first approach and has two types of limitations. First, we have made a set of four assumptions. We assumed that the underlying process of organ failure is a first-order Markov process, i.e., the state at day $t+1$ is completely determined by the state at day t. We also assumed stationarity, i.e., that the state transition probabilities do not change over time. For ICU care this is questionable, as the changes in the first few days of ICU stay are generally considered to be different from changes that occur later on in the ICU admission. In the future we will investigate whether it is beneficial to relax this assumption of stationarity. Further assumptions that need to be investigated are the dichotomization of the SOFA subscores and the assumption that logistic regression equations without all possible interactions between variables are appropriate to estimate the state transition probabilities.

A second limitation is the fact that we have focused on organ failure only, whereas other patient characteristics (e.g. diagnosis, admission type) may also influence the changes in organ failure. These factors can be included into the model by adding an interaction with such type of variables to the logistic regression equations.

In future work we will investigate the influence of the aforementioned assumptions and develop a model that incorporates the influence of other patient characteristics. At the conference we aim to present a model which includes the improvements mentioned above, developed and validated on a dataset from a large number of ICUs.

Acknowledgement

Niels Peek receives a grant from the Netherlands Organization for Scientific Research (NWO) under project number 634.000.020.

References

[1] Salazar JC, Schmitt FA, Yu L, Mendiondo MM, Kryscio RJ. Shared random effects analyses of multi-state Markov models: application to a longitudinal study of transitions to dementia. *Stat Med* 2005, Dec 12 [Epub ahead of print] DOI: 10.1002/sim.2437

[2] Saint-Pierre P, Bourdin A, Chanez P, Daures JP, Godard P. Are overweight asthmatics more difficult to control? *Allergy* 2006; 61(1):79-84.

[3] Vincent J-L, Moreno R, Takala J, et al. The SOFA (Sepsis-related Organ Failure Assessment) score to describe organ dysfunction/failure. On behalf of the working group on sepsis-related problems of the European Society of Intensive Care Medicine. *Intens Care Med* 1996; 22(7):707-710.

[4] Lopes Ferreira F, Peres Bota D, Bross A, Mélot C and Vincent J-L. Serial evaluation of the SOFA score to predict outcome in critically ill patients. *JAMA* 2001; 286(14)1754-1758.

[5] Beck JR, Paucker SG. The Markov process in medical prognosis. Med Decis Making 1983;3:419-458.

[6] Gaziano TA, Steyn K, Cohen DJ, Weinstein MC, Opie LH. Cost-effectiveness analysis of hypertension guidelines in South Africa; absolute risk versus blood pressure level. *Circulation*, 2005; 112:3569-76.

[7] Magni P, Quaglini S, Marchetti M, Barosi G. Deciding when to intervene: a Markov decision approach. *Int J Med Inform* 2000; 60(3):237-53.

[8] Kayaalp MM. *Learning dynamic Bayesian network structures from data.* PhD Thesis, University of Pittsburgh, 2003.

Ubiquity: Technologies for Better Health in Aging Societies
A. Hasman et al. (Eds.)
IOS Press, 2006

Online Guideline Assist in Intensive Care Medicine – Is the login-authentication a sufficient trigger for reminders?

Rainer RÖHRIG[a], Markus MEISTER[a], Achim MICHEL-BACKOFEN[b,1], Martin SEDLMAYR[c],
Dirk UPHUS[a], Christian KATZER[a] and Thomas ROSE[c]

[a] *Clinic for Anesthesiology, Intensive Care Medicine, Pain Management and Palliative Medicine,*
Universitiy of Giessen, Germany

[b] *Department of Medical and Administrative Data Processing, University of Giessen, Giessen, Germany*

[c] *Fraunhofer Institute for Applied Information Technology, Sankt Augustin, Germany*

Abstract. Introduction: Rising cost pressure due to the implementation of the DRG-System and quality assurance lead to an increased use of therapy standards and standard operating procedures (SOPs) in intensive care medicine. The intention of the German Scientific Society supported project "OLGA" (Online Guideline Assist) is to develop a prototype of a knowledge based system supporting physicians of an intensive care unit in recognizing the indication for and selecting a specific guideline or SOP. While the response of the prototype on user entries can be displayed as a signal on the used workstation itself, the location and time for a reminder of scheduled or missed procedures or reactions to imported information is a difficult issue. One possible approach to this task is the display of non acknowledged reminders or recommendations while logging on to a system. The objective of this study is to analyse user behaviour of the physicians working on the surgical intensive care unit to decide whether the login authentication is a sufficient trigger for clinical reminding. Methods: The surgical intensive care unit examined in this study comprises 14 beds. Medical care is provided by physicians working in shifts 24 hours a day, 7 days a week, with two anaesthetists at a time and an additional senior consultant during daytime. The entire documentation (examinations, medication, orders, care) is performed using the patient data management system ICUData. The authentication process of the physicians was logged and analysed. Results: Throughout the observation period from December 13th 2005 to January 11th 2006 3563 physician logins were counted in total. The mean span between logins was in 11.3 minutes (SD 14.4), the median 7 minutes. The 75% centile was 14 minutes, the 95% centile 38 min. Intervals greater than 60 minutes occurred in 75%, and greater than 90 minutes in 25% of the days. Discussion: It seems reasonable that reminders sent during authentication are able to enforce workflow compliance. It is possible to send notifications caused by external events to the physician depending on the importance of the event. Serious events with high urgency should be reliably passed using wireless pager or handheld technology. It seems that after the implementation of the prototype guideline assist further investigation is needed to monitor changes in authentication behaviour and reactions to the guideline advisory. This is also required to investigate the influence of unit's size, medical specialty and actual ward workload.
Keywords: clinical information systems, workflow management, alerts, reminder

1. Introduction

Rising cost pressure due to the implementation of the DRG-System and the obligation of quality assurance according to the legal requirements of Germany resulted in an increased use of therapy standards and standard operating procedures (SOP) for intensive care medicine [1,2].

Since its start in 1999 the documentation of the surgical intensive care unit (ICU) of the University-Hospital of Giessen and Marburg, Campus Giessen, is based on the patient data management-system (PDMS) ICUData (IMESO GmbH, Germany). The system has a modular, message based client server architecture and uses the HL7 standard [3] for information exchange.

Project “OLGA” (Online Guideline Assist) strives to develop a prototype of a knowledge-based system supporting physicians of an intensive care ward in the process of recognizing an indication for and selecting a specific guideline or SOP. Project “OLGA” is supported by the German Scientific Society (Deutsche Forschungsgemeinschaft). It will further assess the compliance of users with provided guidelines and SOPs. The prototype itself logs on to the PDMS, checks suitable indications and collects information regarding the compliance with the guidelines. Suggested or ordered procedures resulting from the use of the prototype are sent to the PDMS as a HL7 message and processed like any other regular information.

The response of the prototype to user entries can be displayed as a signal on the originating terminal itself. However, the location and point in time for a reminder of scheduled or missed procedures, or reactions to imported information (e.g. laboratory values) is a difficult issue. It’s difficult to direct the physicians’ attention to these reminders because their attention is frequently focused on critically ill patients.

The choice of now widely available mobile technology such as individual alerts on handheld computers [4] is awkward due to the restricted use of wireless LAN and cellular phones in the sensitive environment of an intensive care unit [5]. Another approach of notification is the display of non acknowledged reminders or recommendations while users are logged into the system. This technique is, however, highly dependent on the period between two logins and thus on the user’s behaviour. Hence, both factors should be examined prior to the implementation of the “OLGA”-prototype. **The objective of this study** is to analyse the period between two system authentications of physicians depending on the time of day. The aim was to investigate whether the login authentication is a sufficient trigger for clinical reminders.

2. Methods

2.1. Setting: the environment of the intensive Care Unit

The surgical ICU examined in this study contains 14 beds. In 2005 1,484 patients were treated by the unit. The SAPS II value was assessed daily for patients who stayed over 24 hours. It had an average of 30.3 points. Medical care is provided by physicians working in shifts 24 hours a day, 7 days a week with two anaesthetists at a time and an additional senior consultant during daytime. During the early shift, from 6:30 a.m. to 4:00 p.m., ward rounds are conducted by the surgeons, and transfers to other wards are organised and performed as well as scheduled diagnostic and labour intensive therapeutical procedures (e.g. percutaneous tracheostomy). The late shift from 1:30 p.m. to 10 p.m. starts with the handing over of the ward until 2:40 p.m. followed by a period with double staffed crew for handling most of the elective admissions. Conversations with patients and relatives are performed mostly during visiting hours between 4 p.m. and 6 p.m. During the night shift discharge letters and other correspondence are written as well as the treatment-schemes for the following day starting 9 p.m. till 8 a.m. At the weekend staff consists of a day shift (8. a.m. till 8 p.m.) and a night shift (8 p.m. till 8 a.m.) with approximately one hour hand over in between.

2.2. Technical-Environment: The Patient-Data-Management-System

The entire documentation (examinations, medication, orders, care) is based on the patient data management system ICUData. A PC workstation is available near each bed for documentation purposes. Due to this integration into the clinical intensive care workplace most of the examination findings or procedures are entered directly at the bedside (Figure 1). Further PC workstations are provided at the ward's central station where they are equally used by physicians and nursing staff, in the offices of the consultant and the chief nurse location. This setting allows different users with different functions to work with the real time shared patient chart [3] from different locations at the same time.

Data synchronisation is implemented online using a message based push technology [3]. Moving to a different PC or idle time exceeding three minutes (lack of mouse moving) requires a repeated user authentication with user name and password entry.

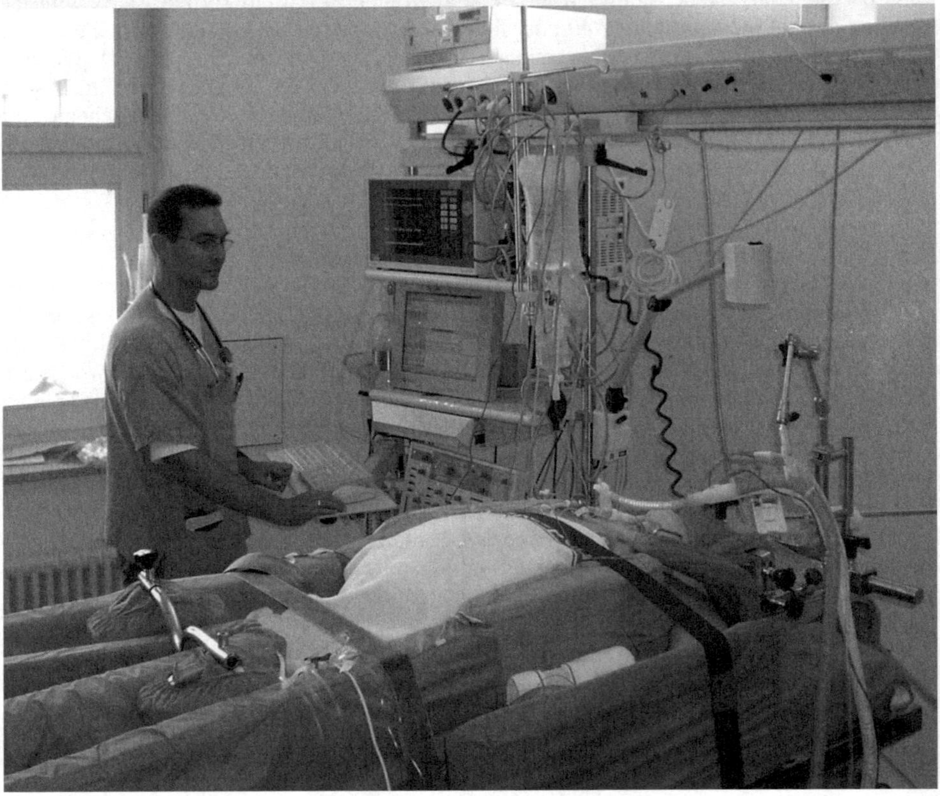

Figure 1: Integration of the PDMS bed side workstation into the ICU workplace

2.3. Logging and analysing the authentication

During the authentication procedure a request concerning the validity of the password is send to an application server of the PDMS by a HL7 defined query (MFQ^z21). The server logs all incoming queries with a timestamp into a dedicated file, which is deleted after a configurable period.

Following an agreement with the data protection officer, the employee committee, the woman's representative and the handicapped representative of the hospital to handle files containing user specific information, the application server's log files were processed with a PERL script and the queries requesting the current passwords were extracted. The encryption of the pass phrases remained untouched. From the PERL generated extract the group and the workstations of interest (physicians, bedside computers including offices and ward central) were selected. The timestamp was coded into night shift, early handover, early shift, late handover, late shift, night handover, and day shift (weekend) in order to examine the time span regarding their position in the shifting cycle.

Furthermore the data was anonymised and the auxiliary files deleted. The statistical processing was done using SPSS® (SPSS GmbH, Germany).

3. Results

The observation period described here lasted from December 13th 2005 till January 11th 2006. Between December 21st and 23rd 2005 a software problem caused a corruption of the logged data. So this period was excluded from the evaluation. During the time of investigation the beds of the intensive care unit were occupied at a rate of 89%.

Throughout the observation period there were a total of 3,563 physician logins. The interval between two logins was on average 11.3 minutes (standard deviation (SD) 14.4), the median was 7 minutes. The 75% centile was 14 minutes, the 95% centile was 38 min. The time spans, listed according to the shift in which they occur, are shown in table 1. As shown below, long periods without login occur, especially during the handover of shifts. The increased frequency of these periods in conjunction with normal mean and median values during the late handover is caused by the practice that, if there are no patient admissions, both physicians join the handover. Therefore no prescription or order is performed during this time and consequently there is no need for user authentication by any physician.

Table 1: Time span-Statistics of observation periods

Observation Period	Physician Logins	Timespan [min]			Percentile			
	n	mean	SD	Median	75%	90%	95%	99%
night shift	1,595.00	9.7	12.4	6.0	12.0	22.0	33.0	55.0
early handover	108.00	21.8	23.3	13.0	27.8	65.1	69.1	130.0
early shift	574.00	12.0	15.4	7.0	14.0	29.0	42.5	78.0
late handover	148.00	12.7	18.6	6.5	13.8	29.0	42.1	116.0
late shift	636.00	11.4	14.5	7.0	14.0	24.0	35.2	87.4
night handover	109.00	12.9	13.5	8.0	18.5	31.0	42.5	71.5
day shift (weekend)	393.00	12.9	14.3	8.0	17.0	29.6	38.6	67.0

In a detailed analysis of the login intervals according to the weekday no relevant differences were found (see figure 2). The boxes contain 50% of the data, the bars 90%. Extreme values are shown as *.

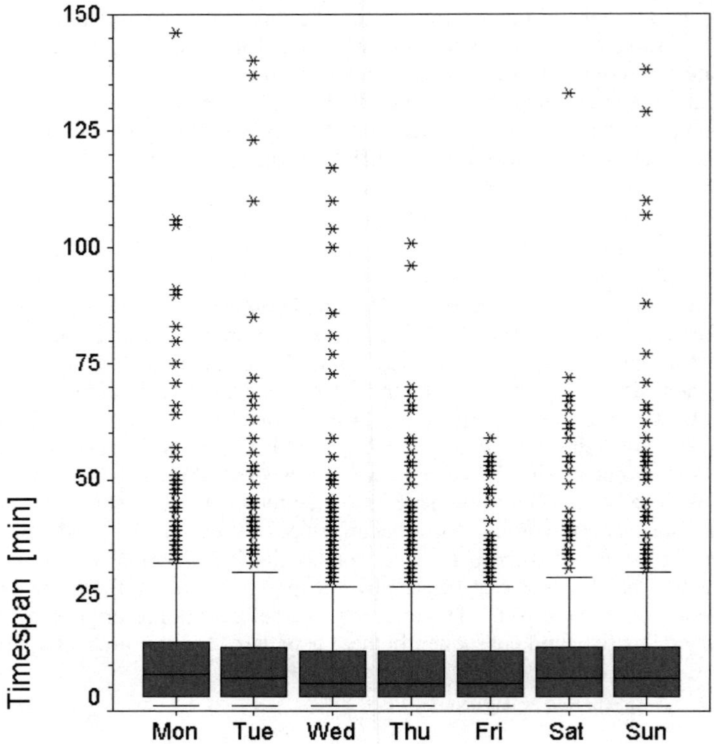

Figure 2 Time span divided by days of a week

In a further investigation, the distribution of long periods (more than 30min., 60min. and 90 min.) per day was investigated (table2). This unveils intervals greater 60 minutes in 75% and greater 90 minutes in 25% of the days.

Table 2: Counts of Long-Time span per 24h (day)

	> 30 min	> 60 min	> 90 min
n	273	55	19
Mean Value	9.4	1.9	0.7
SD	2.2	1.5	0.9
Centile			
Min	5	0	0
5%	6	0	0
25%	8	1	0
50%	10	2	0
75%	10	3	1
95%	13	4.6	2.6
Max	14	5	3

4. Discussion

The positive effects of reminders have been described in numerous works. Shea et. al. could show that even a single reminder per day could decrease the average length of stay [2] and Oniki et. al. proved that reminders at midday reduced deficiencies at the ICU. The time delays found in our own work are much shorter than the previously used times. Therefore it seems reasonable that reminders sent during authentication are able to enforce workflow compliance. The feasibility of passing reminders to the physician during authentication in a timely manner depends on the urgency of the event. Serious events could be passed to a physician mostly within a reasonable time span but with a frequency of every third day we found extended login intervals of more than 90 minutes. Therefore serious events with high urgency should be reliably passed using wireless pager or handheld technology [4,7].

Studies that used pagers and other wireless technologies found that too frequent alarms disrupted the workflow on the ICU and are annoying for the physicians [4,7,8]. Projected to our own environment it might be possible that reminders during the authentication process could distract the focus from an actual patient's problem to the reminder which might be counter productive. Another problem of wireless technology is that undeliverable messages have to be propagated using an escalation strategy [7]. The strategy described here compensates for this by replaying all events at every login that have not been acted upon by the physician.

Further work is also needed to monitor future changes in user behaviour after the implementation of the online guideline assist. This is also required to investigate the influence of ward size, medical specialty and actual ward workload.

5. Conclusions

The results of the study show that the login authentication could be a sufficient trigger for reminders to enforce user compliance to computerized guidelines. On the other hand it also shows that time-critical alerts should be send with wireless technology to a pager or personal handheld system.

Furthermore it seems that after the implementation of the prototype guideline assist further investigation is needed to monitor changes in authentication behaviour and reactions to the guideline advisory procedures.

6. Acknowledgements

This study is part of the Project "OLGA – Online Guideline Assist", supported by the German Scientific Society (DFG-Reference: RO 3053/1-1; RO 2030/2-1).
The authors thank Robert Demming from the ICUdata team for changes at the application server to enable the log functionality used by this study and Brendan Whelan for reviewing the manuscript.

7. References

[1] Martin J, Schleppers A, Kastrup M, Kobylinski C, König U, Kox WJ et. al. Entwicklung von Standard Operating Procedures in der Anästhesie und Intensivmedizin. Anästhesiologie & Intensivmedizin 2003; 44:871-876.

[2] Shea S, Sideli RV, DuMouchel W, Pulver G, Arons RR, Clayton PD: Computer-generated Informational Messages Directed to Physicians: Effect on Length of Hospital Stay. J Am Med Inform Assoc. 1995;2:58-64

[3] Michel-Backofen A, Demming R, Röhrig R, Benson M, Marquardt K, Hempelmann G: Realizing a Realtime Shared Patient Chart using a Universal Message Forwarding Architecture; Stud Health Technol Inform. 2005;116:509-514.

[4] Madhu C. Reddy, David W. McDonald, Wanda Pratt and M. Michael Shabot: Technology, work, and information flows: Lessons from the implementation of a wireless alert pager system. Journal of Biomedical Informatics, Volume 38, Issue 3, June 2005, Pages 229-238

[5] Sneiderman CA, Ackermann MJ: Cellular Radio Telecommunication for Health Care: Benefits and Risks. J Am Med Inform Assoc. 2004;11:479-481

[6] Oniki TA, Clemmer TP, Pryor TA: The Effect of Computer-generated Reminders on Charting Deficiencies in the ICU. J Am Med Inform Assoc. 2003;10:177-187

[7] Seggewies C, Bollmann H, Fleischer M, Prokosch H: Kritische Patientenereignisse IT-gestützt Erkennen und Steuern. Electronic Citation: http://www.egms.de/en/meetings/gmds2005/05gmds468.shtml

[8] Atreja A, Metha N, Jein A, Harris M: Computer Alerts for Potassium Testing: Resisting the Temptation of a Blanker Approach. J Am Med Informatics Assoc 2004;11:433-434

8. Address for correspondence

Dr. Achim Michel-Backofen
Department of Medical and Administrative Data Processing,
University Hospital of Giessen and Marburg – Campus Giessen
Klinik-Straße 23
35392 Giessen
Germany
E-Mail: achim.michel@akad.med.uni-giessen.de

Ubiquity: Technologies for Better Health in Aging Societies
A. Hasman et al. (Eds.)
IOS Press, 2006

Functional data analysis for gait curves study in Parkinson's disease

Alain Duhamel[a,1], Patrick Devos[a] , Jean Louis Bourriez[b], C Preda[a] , L Defebvre[c] and Regis Beuscart[a]

[a] *CERIM (EA2694), Faculté Médecine, Université de Lille2, France*
[b] *Neurologie et pathologie du mouvement, CHRU de Lille, France*
[c] *Neurophysiologie Clinique, CHRU de Lille, France*

Abstract. In Parkinson's disease, precise analysis of gait disorders remains essential for the diagnostic or the evaluation of treatments. During a gait analysis session, a series of successive dynamic gait trials are recorded and data involves a set of continuous curves for each patient. An important aspect of such data is the infinite dimension of the space data belong. Therefore, classical multivariate statistical analysis are inadequate. Recent methods known as functional data analysis allow to deal with this kind of data. In this paper, we present a functional data analysis approach for solving two problems encountered in clinical practice: (1) for a given patient, assessing the reliability of the gait curves corresponding to the different trials (2) performing intra individual curves comparisons for assessing the effect of a therapy. In a first step, each discretized curve was interpolated using cubic B-splines bases in order to ensure the continuous character of data. A cluster analysis was performed on the smoothed curves to assess the reliability and to identify a subset of representative curves for a given patient. Intra individual curves comparisons were carried out in the following way: (1) functional principal component analysis was performed to describe the temporal structure of data and to derive a finite number of reliable principal components. (2) These principal components were used in a linear discriminant analysis to point out the differences between the curves. This procedure was applied to compare the gait curves of 12 parkinsonian patients under 4 therapeutic conditions. This study allowed us to develop objective criteria for measuring the improvements in a subject's gait and comparing the effect of different treatments. The methods presented in this paper could be used in other medical domains when data consist in continuous curves.
Keywords: statistics, gait, Parkinson disease, curves, functional data analysis.

1. Introduction

Degenerative neurological diseases constitute a source of handicap in view of their increased prevalence with age and the aging of the population. Parkinson's disease, which principally leads to a motor deficit, is ranked second (after Alzheimer's disease) in terms of these afflictions. Its prevalence is around 2% in the over-65s and has a growing socio-economic impact [1]. Gait disorders constitute one of the major symptoms of Parkinson's disease and lead to the progressive loss of autonomy [2]. At disease onset, these disorders are limited to mild difficulty in quickening gait, which

[1] Corresponding Author: Alain Duhamel, CERIM, Faculté de Médecine, France , aduhamel@univ-lille2.fr

appears less fluid on one side. Postural instability appears later on and constitutes a true turning point in the progression of the condition. Dopatherapy and treatment with dopaminergic agonists constitutes the benchmark treatment for Parkinson's disease. The efficacy of these drugs on gait disorders has been demonstrated, with improvement in kinematic parameters (increase in speed and stride length). Deep brain stimulation also enables improvement in kinematic parameters, the intensity of which varies according to the target stimulated in the central grey nuclei. In Parkinson's disease, precise analysis of gait disorders remains essential, with several possible orientations: specifying the characteristics for diagnostic purposes, monitoring disease progression and prognosis and evaluating the effect of treatment or deep brain stimulation [3]. In gait analysis performed using a video motion analysis, data consist of continuous curves of measurements made over a gait cycle (e.g. the trace of each lower limb joint and the amplitude in degrees, expressed in different planes during a gait cycle). The current software for gait analysis automatically calculates the kinematic parameters like stride speed or cadence and allows to build a mean curve of the different trials for a subject. However, it does not include adequate statistical methods to deal with curve analysis. For example, is not possible, with the current software, to establish comparisons between different situations for a given patient (on drug versus off drug, for example) or between groups when one is studying a drug treatment or surgical therapy. Since 1998, we have collaborated with the Clinical Neurophysiology Department and the Neurology and Movement Pathology Department of the Lille University Hospital in order to develop statistical tools for application to gait analysis. In a first study, we proposed statistical methods for assessing the reliability of gait curves for a given subject and determining the range of normal values for gait curves in a given population. These methods were applied to a healthy population (60 subjects) [4]. The aim of the present study was to extend these tools for analysing gait curves in a population of patients (Parkinsonians). We focused on two problems : the reliability of gait curves and the intra individual curve comparisons. We used the functional data analysis approach [5]. The methods presented in this paper could be used in other medical domains when data consist in continuous curves.

2. Material and methods

2.1. Data acquisition

Kinematic, spatiotemporal and angular gait parameters are recorded automatically using a VICON® video motion analysis system. Fifteen spherical, retro-reflective markers are placed on the various segments of the pelvis and legs and are illuminated by stroboscopes. The trajectories in all 3 planes are recorded by 6 infrared cameras. For each subject, a series of successive dynamic gait trials were analysed during a given session. Different traces of joint position (amplitude in degrees) such as pelvic tilt, hip flexion/extension or knee flexion/extension were recorded in a discrete way for each subject. Data were expressed as a percent of the gait cycle from 0 to 100 %. To test our statistical tools, we present here the results obtained for the knee flexion/extension. Twelve parkinsonian patients were analysed. A minimum of 8 trials were recorded in four conditions, creating a minimum of 32 curves for each patient : before surgery, when the patient had not received any treatment (Off) and after acute administration of L-dopa alone (On-Dopa) - after surgery, with deep brain stimulation alone (On-Stim) and when the patient received in addition L-dopa (Best-On). In our statistical analysis, each discretized curve was interpolated using cubic B-splines bases in order to ensure the continuous character of data.

2.2. Reliability of gait curves

In clinical practice, the gait curves corresponding to different trials can be scattered and a criteria must be defined in order to decide which curves can be selected as characterising the patient. In an initial study [5] performed on healthy subjects we demonstrated how the Intraclass Correlation Coefficient (ICC) could be used as a measure of the reproducibility of determining which curves should be considered as representative of a subject's gait. ICC varies between 0 to 1 and can be interpreted as the proportion of variance due to the time-to-time variability in the total variance. In the healthy population, the gait of a subject was considered to be reproducible when a minimum of 4 curves resulted in a ICC value greater than 0.95. For parkinsonian patients, the problem is more complicated. These patients have a very jerky gait and there can be great variability between the gait curves obtained. In fact, the ICC alone did not (for certain patients) enable reliable selection of representative curves. We then proposed the use of a functional data analysis in addition to the ICC computation. For each subject, the gait curve reliability was assessed by performing the following steps : (1) let k be the number of curves for the subject, compute a k×k distance matrix on the set of the cubic B-splines interpolated curves, using integrals, since the curves are continuous; (2) perform a hierarchical classification using the WARD aggregation criteria on the distance matrix in order to identify a subset of representative curves for the patient; (3) compute the ICC to assess the reliability of the subset of curves.

2.3. Confidence bands for healthy population

The identification of the "normal" values for gait curves from a given population is essential in clinical practice. It means building a confidence band which has the following meaning: the gait curve of a subject randomly drawn from the study population has a 95% probability of falling within the confidence band. The method currently proposed by gait analysis software do not take into account the correlation that exists between the measurements and presuppose a Gaussian model for the distribution of data. Use of this method may thus result in a large distortion between the pre-specified probability and the true coverage probability. We have developed a non-parametric procedure based on the bootstrap method for building reliable confidence bands [5]. The basic idea of the bootstrap is to generate many (pseudo)samples from the original data and to use these samples to compute robust estimator. In the present study, the confidence bands for healthy population were used to assess the gait of parkinsonian patients

2.4. Intra individuals comparisons using functional analysis

The comparison of intra-individual curves is very useful for the clinician. For example, one seeks to assess the effect of L-Dopa treatment or stimulation on a given patient or to compare a patient curve to a standard curve. A first approach consists in representing the gait of a subject by a single curve, which is the mean of the reliable trials and then using the confidence bands of healthy population to visually detect the improvements in the subject's gait as a function of various treatments. However, this approach does not enable one to obtain objective criteria for measuring this improvement and comparing the effect of different treatments. We then propose the use of Functional Principal Component Analysis (FPCA) [6]. A gait curve was modelled using a continuous-time stochastic process, $\mathbf{X}=\{X_t\}_{t\in[0,\ T]}$. Let n be the number of representative curves for the patient. Denote by Y the categorical variable defined by the different therapeutic conditions (§2.1). The intra individual curves comparisons included the following steps (1) The n curves were interpolated using cubic B-splines basis functions. (2) FPCA was performed on the n B-splines to describe the temporal structure of data and to derive a finite number of reliable principal components. FPCA is a generalization of the classical principal component analysis of discrete data to continuous stochastic process. Details on this method can be found in *Preda et al.* [6]. (3) The principal components were used in a linear discriminant analysis, with the dependent variable Y, to point out the differences between the groups.

3. Results

Twelve parkinsonian patients were analyzed using the tools previously described. Because the different steps of the analysis are identical for each patient, we present here the results obtained for one patient. In figure 1a, the curves of this patient are graphically superimposed in order to visually assess reliability. We observe that two curves seem to be different from the others. Using cluster analysis (1b), we detect with no doubt that these curves (C7 and C8) can be considered as outliers. The ICC value computed on the 7 remaining curves was 0.98 and consequently, this subset of 7 curves was selected to represent the gait of the subject in the Best-On condition.

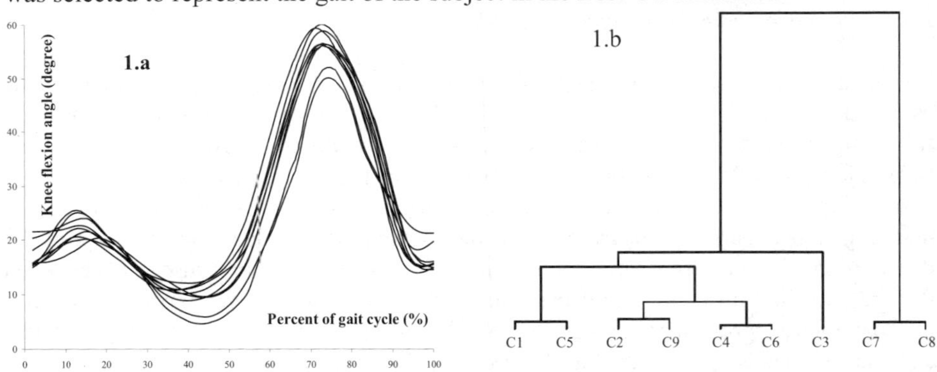

Figure 1 : Gait curves reliability of one patient in the Best-On condition

The use of confidence bands for the healthy population to compare the gait curves is illustrated in figure 2. Each patient's curve represents the mean of the reliable trials in the corresponding treatment condition (Off, On-Dopa, On-Stim and Best-On). Here, one can note that the confidence band enables us to exclude the curve in the condition Off (without treatment) from normality. However, this includes the three other curves : On-Dopa, On-Stim and Best-On. The confidence band does not, therefore, enable us to conclude as to a potential difference in efficacy between these treatments. We then used a functional data analysis approach. This procedure allows to perform a discriminant analysis of curves. In this analysis, the patient is represented by the set of the reliable curves trials in each therapeutic condition. Figure 3 shows the result obtained for the previous patient.

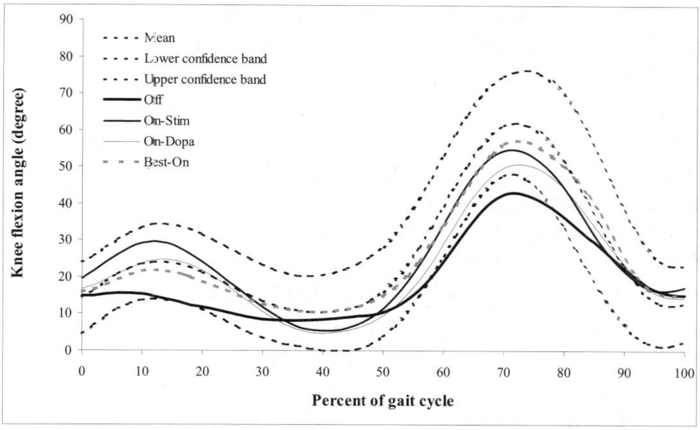

Figure 2 : intra-individual curves comparison using confidence bands

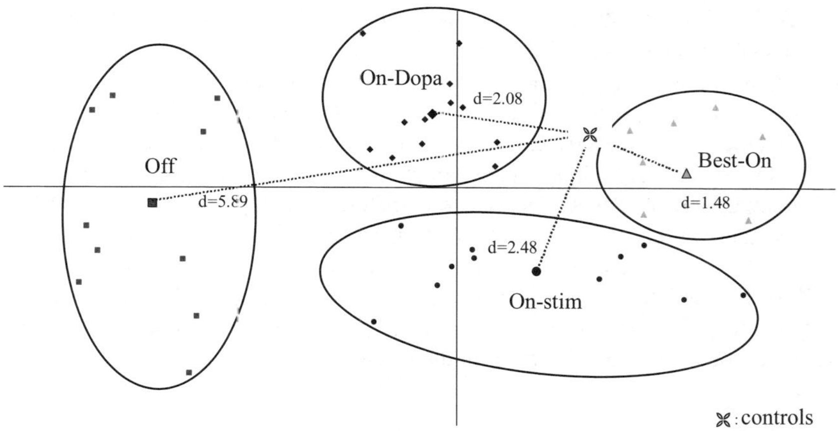

Figure 3 : intra-individual curves comparison using functional data analysis

In this plot, each point represents the projection of a curve in the first discriminant plane. The ellipses indicate the four conditions. For each condition, the mean curve of the corresponding trials is also projected (the central point of the ellipse). Here, one observes that when the parkinsonian patient is not receiving treatment, the gait curves are furthest away from the mean curve for controls (a point represents a curve, that is to say a gait cycle). The group of curves closest to the mean for controls corresponds to the trials recorded in the "Best-On" condition, that is to say with a combination of L-dopa and stimulation. A statistical test (Wilks' Lambda) enables us to show that the centres of gravity for the 4 conditions are different ($p<0.0001$). Calculation of the distances (using Mahalanobis metric) between the groups' respective centres of gravity and the mean for the controls reveals a progressive improvement in gait curves in the a parkinsonian patient - the "On-Stim" (with stimulation, no L-Dopa) and "On-Dopa" (no stimulation, with L-Dopa) conditions appear to be intermediate.

4. Discussion - conclusion

We presented statistical methods for analysing the gait curves of Parkinsonian patients. These methods make it possible to solve two problems encountered in clinical practice: the selection of reliable curves to represent the gait of a given patient and the comparison of intra-individuals curves. The gait curves are continuous processes. Classical multivariate analysis on the discrete data measurements are generally inefficient to analyse this kind of data because the number of variables (equal to the number of measurements) is greater than the number of individuals (overfit) and also because the continuous-time processes of gait curves are intrinsically smooth. Functional Principal Component Analysis (FCPA) allows to describe the temporal structure of the data and to resume the functional data by means of a finite number of principal components. These principal components do not depend on time and can be used in classical multivariate analysis such as cluster or discriminant analysis. Original contributions to the FPCA methods were developed by our team and used in this study [6]. The different algorithms used to perform FPCA were developed with R. Graphical representations of results provide the physician with support for the interpretation. Different improvements are considered :

- FFT selection for reliability : Fourier transform analysis of the curves could enable determination of those with abnormally high or low harmonics and thus their removal from the calculation.
- Measurements of influence for reliability : calculation of the influence of each curve on calculation of the mean curve (based on the principle of Cook's distance in regression). Curves with too great an influence are deleted.
- Use of the functional data analysis approach for the inter-population comparison of mean curves on different groups of patients. We will experiment the PLS (Partial Least Square) procedure.

We intend to continue this research by analysing other gait parameters such as pelvic tilt or hip flexion/extension. The methods described in this paper could be used in other medical domains when data consist in continuous curves, for example to analyse the cortical rhythm activity from the electroencephalogram.

References

[1] De Lau LM, Schipper CM, Hofman A, Koudstaal PJ, Breteler MM. Prevalence and incidence of Parkinson's disease in Europe, Eur Neuropsychopharmacol. 2005;15(4):473-90.

[2] Allan LM, Ballard CG, Burn DJ, Kenny RA. Prevalence and severity of gait disorders in Alzheimer's and non-Alzheimer's dementias, J Am Geriatr Soc. 2005; 53(10):1681-7.

[3] Defebvre L, Krystkowiak P, Blatt JL, Duhamel A, Bourriez JL, Perina M, Blond S, Guieu JD, Destée A. Influence of pallidal stimulation and L-Dopa on gait and preparatory postural adjustments in Parkinson's disease, Mov. Disord. 2002; 17 : 76-83

[4] Ramsay JO, Silverman BW. Applied functional data analysis : methods and case studies. Springer; 2002.

[5] Duhamel A, Bourriez JL, Devos P, Krystkowiak P, Destee A, Derambure P, Defebvre L. Statistical tools for clinical gait analysis, Gait Posture 2004; 20(2): 204-12.

[6] Preda C, Saporta G. PLS regression on a stochastic process. Computational Statistics and Date Analysis 2005; 48(1) : 149-158.

Ubiquity: Technologies for Better Health in Aging Societies
A. Hasman et al. (Eds.)
IOS Press, 2006

A Framework for Cohesive Healthcare Coalition Formation

Yu-N CHEAH[a,1], Yong Han CHONG[b] and Siew Lan NEOH[c]
[a, b] *School of Computer Sciences, Universiti Sains Malaysia, Penang, Malaysia*
[c] *Petronas Nasional Berhad, Petronas Twin Towers, Kuala Lumpur, Malaysia*

Abstract. The mobilisation of cohesive and effective groups of healthcare human resource is important in ensuring the success of healthcare organisations. However, forming the right team or coalition in healthcare organisations is not always straightforward due to various human factors. Traditional coalition formation approaches have been perceived as 'materialistic' or focusing too much on competency or pay-off. Therefore, to put prominence on the human aspects of working together, we present a cohesiveness-focused healthcare coalition formation methodology and framework that explores the possibilities of social networks, i.e. the relationship between various healthcare human resources, and adaptive resonance theory.
Keywords: Coalition Formation, Social Networks, Adaptive Resonance Theory.

1. Introduction

The practice of healthcare and the success of healthcare organisations are highly dependent on the expertise and experience of various healthcare human resources. Although each doctor, nurse or technician is responsible in their respective specialised task, the mobilisation of cohesive and effective groups of healthcare human resource is equally if not more important. Well-established healthcare organisations, while being aware of the need for effective knowledge and human capital management, stand to gain more when cohesive teams or coalitions are formed within the organisation for knowledge and experience-intensive tasks such as surgery and trauma management.

Efforts to define computational frameworks for coalition formation in organisations are actively being pursued, especially from an organisational behaviour or human resource perspective [1]. However, forming the right team in healthcare organisations is not always straightforward especially when human factors come into play. Traditional coalition formation approaches such as Game Theory [2][3][4] and Social-dependence Theory [5][6] have their limitations in view that they are perceived to be 'materialistic' or focusing primarily on competency, performance and pay-off.

To put prominence on the human aspects of working together, we present a cohesiveness-focused healthcare coalition formation methodology and framework. Here, we aim is to explore the potentials of social networks (that focus on the relationship between various healthcare human resources) and adaptive resonance theory (ART).

[1] Corresponding Author: Yu-N CHEAH, School of Computer Sciences, Universiti Sains Malaysia, 11800 USM Penang, Malaysia. Email: yncheah@cs.usm.my.

2. Overview of the Healthcare Coalition Formation Methodology

Our proposed methodology for healthcare coalition formation consists of two phases (see Figure 1):

1. Coalition Formation: Upon receiving a work request and relevant inputs from the user, this phase checks for the physical availability of human resources and their cohesiveness, i.e. whether or not a particular group of doctors, nurses or technicians can work together effectively and comfortably. This phase not only utilises a straightforward lookup mechanism to check the physical availability of human resources, it also ensures the cohesiveness of a candidate coalition. The latter employs a novel combination of social networks and ART. The result of this phase is a candidate coalition.
2. Scheduling: The available group of employees is checked to ensure that they are available for the duration of the task. If there is a mismatch in the schedule of the employees, an alternative candidate coalition is generated via the coalition formation phase. When the candidate coalition clears the scheduling phase, a finalised coalition is obtained.

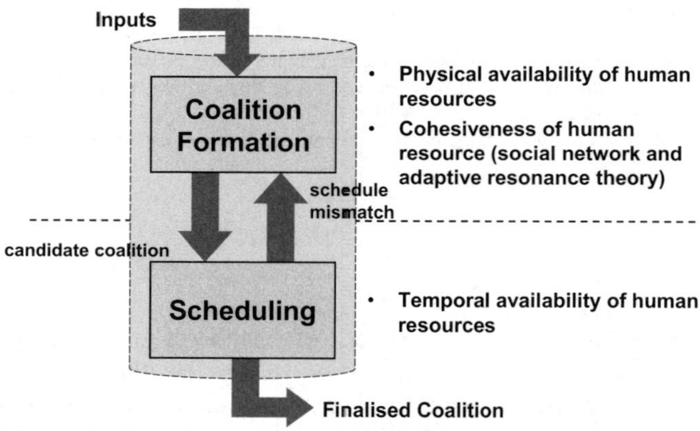

Figure 1: Overview of the Coalition Formation Methodology

Our two-phased methodology can be translated into a framework consisting of the following three layers (see Figure 2):

1. Object Layer: In general, this layer stores various resources. For our purposes, it stores human resource knowledge (e.g. of specialists, nurses, technicians, etc.). This can be viewed as a virtual community of agents represented as nodes, each having its own public and private knowledge. Public (or social) knowledge includes the agent's profile, constraints and schedule; while private knowledge refers to personality, preferences and credibility [7][8].
2. Manager Layer: Three manager components interact with the object layer agents and delivery layer interface to carry out their tasks.
   - Interface Manager: The interface manager receives requests from the user and passes on the relevant inputs to the other manager components to initiate the coalition formation activity. Upon receiving a finalised coalition as the result, the interface manager passes it on to the user.
   - Coalition Manager: The coalition manager carries out the tasks of the coalition formation phase mentioned earlier. It receives the relevant

inputs from the interface manager and identifies which human resource agents at the object layer are available to form a coalition. Following this, the coalition agent analyses the social network of the available human resources by determining their relationship values. It then applies ART on the relationship values to produce a candidate coalition.

- Scheduling Manager: The scheduling manager performs the scheduling phase of our healthcare coalition formation methodology. It checks the candidate coalition for scheduling mismatches. Mismatched human resources are sent back to the coalition manager for other coalition alternatives, while a candidate coalition that does not have mismatched schedules are returned to the interface manager as a finalised coalition.

3. Delivery Layer This layer is basically the user interface from which the user can submit requests for coalitions with accompanying inputs, and receive the finalised coalition results.

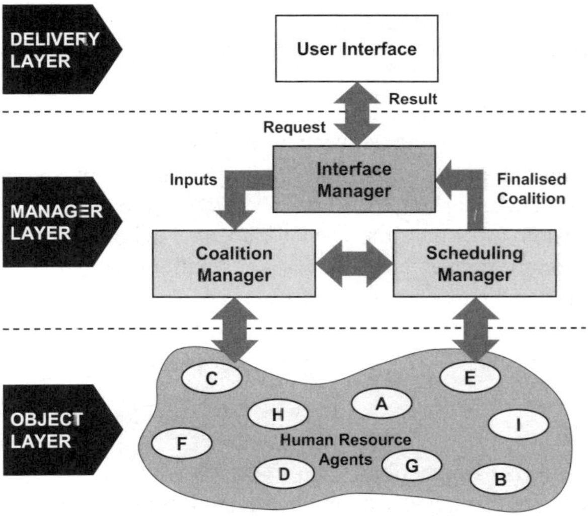

Figure 2: The Coalition Formation Framework

Presently, we aim to focus on the first phase of our coalition formation methodology, i.e. to ensure cohesiveness by employing social networks and ART. The second phase, i.e. scheduling, is beyond the scope of this paper. We now present details of the coalition manager, i.e. the main component for the coalition formation phase.

3. The Coalition Manager

The coalition manager carries out two main functions: social network management and ART.

3.1. Social Network Management

The coalition manager interacts very closely with the human resource agents at the object layer. The community of human resource agents, i.e. doctors, nurses, technicians, etc., is represented as nodes and the relationships between agents are represented as arcs (with corresponding relationship values) (see Figure 3). Their relationships are asymmetrical, i.e. the relationship doctor A has on technician B is not necessarily the same as the relationship technician B has on doctor A, as each agent has its own personality, preferences and credibility. The relationship value an agent has on another agent is a composite value derived from questionnaire-based credibility and personal trait assessments during their past interactions [8][9].

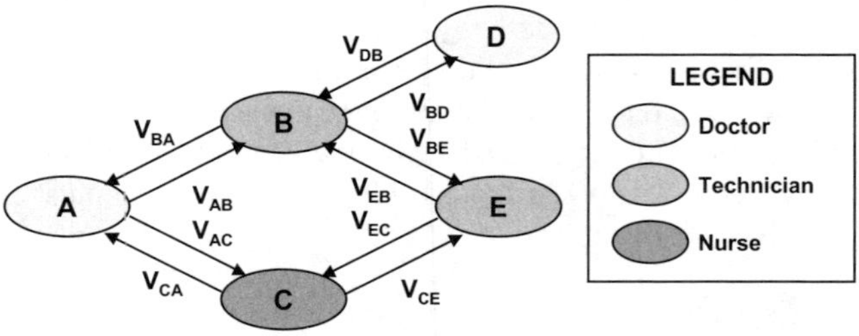

Figure 3: Nodes and arcs representing human resource agents and their relationships

Let us assume the following:

- $V_{X\text{-}|\text{-}Z}$ = relationship value between X and Z, and
- agents X and Z are coherent, i.e. having a significant amount of positive or good experience working together, if $V_{X\text{-}|\text{-}Z} > e$ (the coherency threshold).

Let us also assume three possible ways for agent Z to be connected to agent X:

1. if Z is directly connected to X ($V_{X\text{-}Z}$ = relationship value of X on Z, $V_{Z\text{-}X}$ = relationship value of Z on X), then $V_{X\text{-}|\text{-}Z} = V_{X\text{-}Z} = (V_{X\text{-}Z} + V_{Z\text{-}X})/2$,
2. if Z is indirectly connected to X via Y, then $V_{X\text{-}|\text{-}Z} = V_{X\text{-}Y\text{-}Z} = V_{X\text{-}Y} \bullet V_{Y\text{-}Z}$, and
3. if Z is connected to X via n different paths $(V_{X\text{-}|\text{-}Z})_1 \ldots (V_{X\text{-}|\text{-}Z})_n$, i.e. with multiple indirect connections, then $V_{X\text{-}|\text{-}Z} = \dfrac{\sum_{i=1}^{n} (V_{X-|-Z})_i}{n}$.

As an example, let us consider a healthcare environment with a number of human resources agents as shown in Figure 3. Doctor A has previously worked directly with technician B and nurse C; but not with doctor D or technician E. However, technician B has prior experience working with doctor D and technician E. These result in doctor A being indirectly connected to doctor D and technician E via technician B. Therefore, in order to calculate doctor A's relationship with all colleagues, various direct, indirect and multiple indirect relationships need to be considered.

Table 1 summarises the relationship values for all agents from Figure 3. By default, $V_{X\text{-}X} = 1$; and $V_{X\text{-}Y} = 0$ when $V_{X\text{-}Y} < e$.

Table 1: Relationship values for all human resource agents

Agent	A	B	C	D	E
A	1	$V_{B\text{-}|\text{-}A}$	$V_{C\text{-}|\text{-}A}$	0	$V_{E\text{-}|\text{-}A}$
B	$V_{A\text{-}|\text{-}B}$	1	$V_{C\text{-}|\text{-}B}$	0	$V_{E\text{-}|\text{-}B}$
C	$V_{A\text{-}|\text{-}C}$	$V_{B\text{-}|\text{-}C}$	1	0	$V_{E\text{-}|\text{-}C}$
D	0	0	0	1	0
E	$V_{A\text{-}|\text{-}E}$	$V_{B\text{-}|\text{-}E}$	$V_{C\text{-}|\text{-}E}$	0	1

Having obtained a set of relationship values for each human resource agent in the community, the coalition manager represents each column of Table 1 as a vector $I \equiv (I_1, \ldots, I_M)$, where M is the number of human resource agents in a community. For instance, the vector for agent A is $I_A \equiv (1, V_{A\text{-}|\text{-}B}, V_{A\text{-}|\text{-}C}, 0, V_{A\text{-}|\text{-}E})$. The M number of input vectors serve as inputs for the coalition manager to carry out adaptive resonance to generate the candidate healthcare coalitions.

3.2. Adaptive Resonance Theory

ART is basically an artificial neural network employing unsupervised learning [10]. It is characterised by comparison (input) and recognition (output) fields, a vigilance parameter and a reset module. A basic ART network is shown in Figure 4.

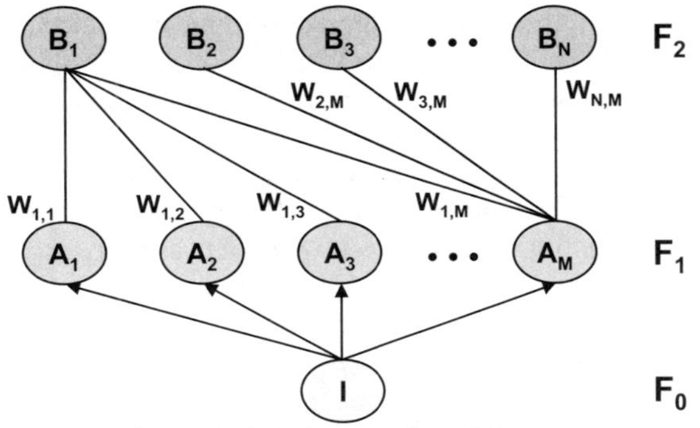

Figure 4: ART network architecture

The coalition manager maintains three levels of ART field activity vectors:

- Level F_0: This level consists of a node which represents the current input vector, I for an agent (obtained from the social network management's relationship value calculation).
- Level F_1: This level consists of as many nodes as there are human resource agents. The F_1 activity vector, i.e. the values of the nodes, is denoted as $A = (A_1, \ldots, A_M)$.
- Level F_2: Nodes at this level represents the candidate healthcare coalitions formed. The F_2 activity vector is denoted as $B = (B_1, \ldots, B_N)$.

Associated with each candidate healthcare coalition node j ($j = 1, \ldots, N$) of F_2 is a vector $w_j \equiv (w_{j1}, \ldots, w_{jM})$ of adaptive weights. The number N, which indicates possible 'slots' for candidate healthcare coalitions, may be arbitrarily large.

The ART network takes input vectors from F_1 and sends them to F_2. In the process, the weights between F_1 and F_2 are updated and the matching process of comparing the relationship value of each F_1 node with the respective weights from the previous cycle would result in certain F_2 nodes to be chosen (representing candidate healthcare coalitions). As more input vectors are put into the network, and as these inputs meet the required vigilance criteria (which determines whether the coalitions are fine-grained or general), the formation of the candidate healthcare coalitions strengthens in line with the concept of establishing resonance.

The human resource agents associated with the same candidate healthcare coalition node in F_2 are deemed coherent to each other and thus, form cohesive candidate coalitions. These candidate healthcare coalitions are then passed on to the scheduling manager that would then ensure the human resource agents, whilst being socially cohesive, are available during the period of the task.

4. Concluding Remarks

In general, our cohesiveness-focused healthcare coalition formation methodology can be viewed as a hybrid methodology that takes into account the human factor in social interactions as well as artificial intelligence techniques such as ART in order to form suitable coalitions for a particular task.

Presently, we have laid out details of the coalition manager. The development of the entire healthcare coalition formation framework is still in progress. The way forward would be to explore details of the scheduling and interface managers. We expect constraint-related techniques as well as evolutionary algorithms [11] to be relevant, especially for the scheduling manager. This would be challenging due to the complex nature of shift work in the healthcare environment. Ultimately, we hope that this framework can be integrated into existing healthcare-related groupware to capitalise on existing human resource profiles that may have already been stored. This can further enhance the quality of coalitions formed in healthcare organisations.

References

[1] Roberts-DeGennaro M. Conceptual Framework of Coalitions in an Organizational Context. Journal of Community Practice. 1997; 4(1): 91-107.

[2] Ketchpel S. Forming Coalitions in the Face of Uncertain Rewards. Proceedings of the Twelfth National Conference on Artificial Intelligence (AAAI-94), Seattle, Washington, USA. 1994; 414-19.

[3] Shehory O, Kraus S. A Kernel-oriented Model for Autonomous-agent Coalition Formation in General Environments: Implementation and Results. Proceedings of the Thirteenth National Conference on Artificial Intelligence (AAAI-96). Portland, Oregon, USA. 1996; 134-40.

[4] Klusch M, Gerber A. Issues of Dynamic Coalition Formation Among Rational Agents. Proceedings of the Second International Conference on Knowledge Systems for Coalition Operations (KSCO-2002). Toulouse, France; 2002.

[5] Conte R, Sichman JS. DEPNET: How to Benefit from Social Dependence. Journal of Mathematical Sociology. 1995; 20(2-3): 161-77.

[6] Sichman JS, Demazeau Y. On Social Reasoning in Multi-Agent Systems. Revista Iberoamericana de Inteligencia Artificial. 2001; (13): 68-84.

[7] Pěchouček M, Mařík V, Bárta J. A Knowledge-based Approach to Coalition Formation. IEEE Intelligent Systems. 2002; 17(3): 17-25.

[8] Chong YH, Cheah Y-N. Credibility-Chart and Dual-Memory Referencing for Robust and Efficient Agent Search in Coalition Formation. International Journal of Software and Information Technologies (IJSIT) Lecture Notes of International Conference on Intelligent Knowledge Systems (IKS 2004), Troy, Turkey. 2004; 1(1): 209-12.

[9] Chong YH, Cheah Y-N. Healthcare Coalition Formation Using Personal Traits and Adaptive Resonance Theory. International Conference on Intelligent Systems (ICIS 2005), Kuala Lumpur, Malaysia; 2005.

[10] Carpenter GA, Grossberg S. Adaptive Resonance Theory. In: Arbib MA, editor. The Handbook of Brain Theory and Neural Networks. 2nd ed. Cambridge, Massachusetts: MIT Press; 2003. 79-82.

[11] Neoh SL, Cheah Y-N. A Knowledge- and Agent-based Coalition Formation Framework for Surgical Team Formations and Resource Allocation. Proceedings of the Second International Symposium on Information Technology (ITSim2003), Kuala Lumpur, Malaysia; 2003.

An Approach for Generating Fuzzy Rules from Decision Trees

Amir R. Razavi[a,1], Mikael Nyström[a], Marian S. Stachowicz[b], Hans Gill[a], Hans Åhlfeldt[a], Nosrat Shahsavar[a,c]
[a] Dept. of Biomedical Engineering, Div. of Medical Informatics, Linköping University, Sweden
[b] Laboratory for Intelligent Systems, ECE Department, University of Minnesota, USA
[c] Regional Oncology Centre, University Hospital, Linköping, Sweden

Abstract. Identifying high-risk breast cancer patients is vital both for clinicians and for patients. Some variables for identifying these patients such as tumor size are good candidates for fuzzification. In this study, Decision Tree Induction (DTI) has been applied to 3949 female breast cancer patients and crisp If-Then rules has been acquired from the resulting tree. After assigning membership functions for each variable in the crisp rules, they were converted into fuzzy rules and a mathematical model was constructed. One hundred randomly selected cases were examined by this model and compared with crisp rules predictions. The outcomes were examined by the area under the ROC curve (AUC). No significant difference was noticed between these two approaches for prediction of recurrence of breast cancer. By soft discretization of variables according to resulting rules from DTI, a predictive model, which is both more robust to noise and more comprehensible for clinicians, can be built.
Keywords. Fuzzy Set Theory, Decision Tree Induction, Breast Cancer, Distant Metastasis.

1. Introduction

Breast cancer is the most common type of cancer diagnosed in women in Western countries [1]. Its prognosis is influenced by many factors such as morphological and pathological tumor specifications and biological tumor markers.

With the wide availability of medical records, they can be used as a good source for knowledge extraction. Data mining methods can be applied to them in order to predict outcomes of new patients. Extraction of predictive information from large databases is an important method with extensive capability to help physicians in their decision making [2]. However, successful data mining depends on preparing the information in data pre-processing. This includes different actions including cleaning the data, handling missing values and selecting a proper subset of data [3].

One frequently used data mining algorithm is decision tree induction (DTI). A decision tree is a classifier in the form of a tree structure and is used to classify cases in

[1] Corresponding author: Amir R. Razavi, Linköping University, Dept. of Biomedical Engineering, Div. Of Medical Informatics, University Hospital, S-58185 Linköping, Sweden. amirreza.razavi@imt.liu.se

a data set [4]. A set of training cases with their correct classifications is used to generate a decision tree that classifies each case in the test set. The resulting tree is a representation that can be verified by humans and can be used by medical staff with support of computers [5].

The decision tree can be transformed into crisp IF p Then q rules, where p and q are propositional variables whose values are either truth (T) or false (F). These rules divide the decision space into several subspaces. Partitions are non-overlapping and a slight change in the value of one of the variables may change the predicted class of that case. Additionally this is different from human reasoning. Human often use imprecise information to make decisions. In clinical medicine, it is beneficial to use formalism where the propositional variables in If-Then rules are replaced by fuzzy propositions. The binary logic based on choices "yes" and "no" is replaced with a range of possibilities. This is the basis for fuzzy set theory (FST). For DTI this can be achieved with soft discretization of variables and overlapping partitions [6]. The final goal of fuzzy logic is to provide foundations for approximate reasoning with imprecise propositions using FST as a principal tool [7, 8].

Different studies have implemented FST in the breast cancer domain. In one study researchers tried to predict the malignancy of this cancer by studying tumor specifications obtained from fine needle aspiration of the tumor using a combination of fuzzy logic and evolutionary algorithms to produce a diagnostic system [9]. In another study, the prognosis of the cancer was made by using a fuzzy k-nearest neighborhood clustering technique and the result was compared with other techniques, such as logistic regression and artificial neural networks [10]. Different fuzzy logic approaches for differentiating malignant breast tumors have been compared, too [11].

Medical diagnosis has been recognized as the most likely application domain of FST already in 1969 [12]. There are two main justifications for using FST [13, 14]:

- The real medical world is too complex for precise descriptions to be obtained, therefore approximation must be introduced in order to obtain a reasonable model.
- Human knowledge becomes increasingly important in the information era and there is a need for a theory to formulate human knowledge in a systematic way and put it into one system with information like mathematical models.

In this study, our aim is to use extracted knowledge from a breast cancer register to find the proper fuzzy sets for the predictors of recurrence of the breast cancer and transform the crisp rules to fuzzy rules. Then the fuzzy model is compared with the crisp rule model by examining their performance for one hundred randomly selected cases.

2. Materials and Methods

2.1. Dataset and Rule Extraction

Data from 3949 female patients with malignant breast cancer, mean age 62.7 years, were analyzed. In a pre-processing step, data were cleaned from outliers, missing values were handled and a subset of variables was chosen which describes the occurrence of distant metastasis or death because of breast cancer [15].

In this study, five of the most important variables were selected. Then, DTI was applied to the pre-processed data and a predictive decision tree was created. The resulting tree had nine leaf nodes with attached probability for the occurrence of the outcome.

In the tree resulting from DTI, the branches to each leaf node in the decision tree were transformed into If-Then rules. The rules are described in Table 1. In this table, instead of dichotomous 1/0 for the recurrence of the distance metastasis or death because of breast cancer, a fuzzy translation of the probability of the cancer outcome was used.

Table 1. The resulting If-Then rules from the Decision Tree Induction.

IF					THEN
PG	**T Size**	**LN**	**ERP**	**SPF**	**Outcome**
0	≤ 26	-	-	-	LoR
0	> 26	$= 0$	-	-	MedR
0	> 26	> 0	$\leq .45$	-	HiR
0	> 26	> 0	$> .45$	-	MedR
1	≤ 34	≤ 5	-	≤ 9.92	MedR
1	≤ 25	≤ 5	-	> 9.92	HiR
1	> 25 ≤ 34	≤ 5	-	> 9.92	MedR
1	> 34	≤ 5	-	-	HiR
1		> 5	-	-	HiR

Acronyms: PG: perigland growth, T Size: tumor size, LN, number of involved lymph nodes, ERP: Estrogen receptor protein, SPF: S-phase fraction, Outcome: outcome of the disease, LoR: low risk, MedR: medium risk, HiR: high risk.

Variables in the rules were predictors for the recurrence of breast cancer. Existence of perigland growth, size of tumor, involvement of adjacent lymph nodes (LNs) by malignant cells, quantity of Estrogen receptor and S-Phase fraction were present in the rules. The amount of Estrogen receptor determines the response of tumor to treatment and S-phase fraction shows the amount of cells in proliferation phase of cell cycle. The higher this quantity, the more aggressive is the tumor.

2.2. Creating Fuzzy If – Then Rules

In this step by using rules from the DTI, membership functions for the variables were defined. Representing states of variables by fuzzy sets is a way of quantifying the variables [13]. The resulting membership functions for the input variables were sigmoidal functions with their middle value on the cut off value for the corresponding If-Then rule. Membership's function of one of the variables is presented in Figure 1

The probability of recurrence were extracted from each leaf node and according to the value transformed into three groups: low risk, medium risk and high risk. The rules are presented in Table 1. Membership functions for these three groups were also constructed according to the probability of the recurrence and expert's opinion about different risk groups. The membership functions were Gaussian functions with their peaks where the risks are low, medium or high. These membership functions are used as output membership functions and are presented in Figure 2.

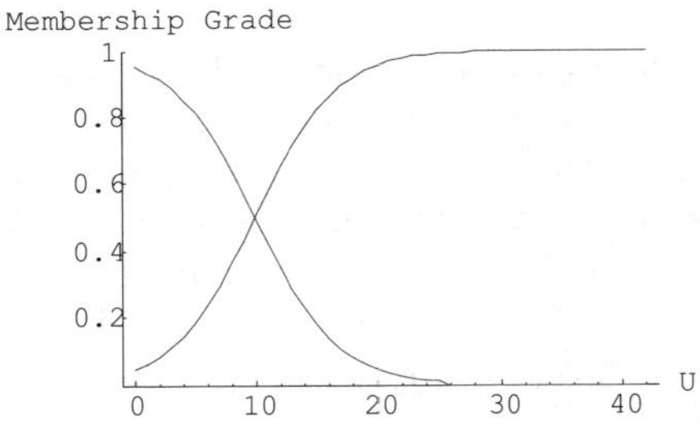

Figure 1. The membership functions for S-Phase fraction. The functions represent the If-Then rule's Boolean expressions ≤ 9.92 and > 9.92.

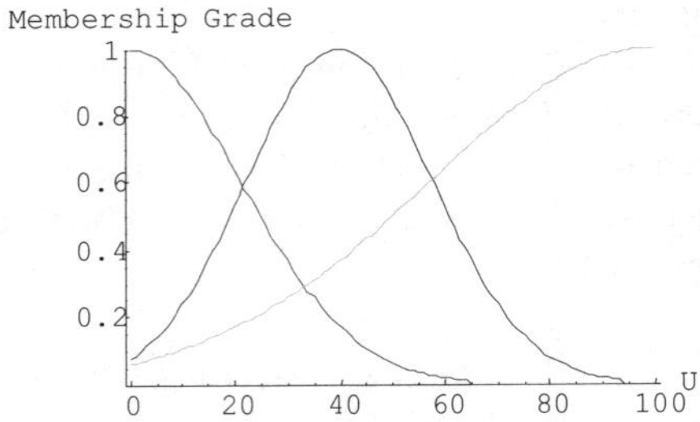

Figure 2. The membership functions for the three risk groups low, medium and high.

2.3. Model Building

A fuzzy logic inference model was constructed for all nine fuzzy If-Then rules and the corresponding input and output membership functions. Each fuzzy rule produced a fuzzy set as a partial result and later all nine partial results were combined into one global fuzzy set using max-union operation. This output fuzzy set was constructed to presents a probability for cancer recurrence for each patient between 0 and 100. To defuzzify the global fuzzy set, the smallest maximum strategy was applied (Figure 3).

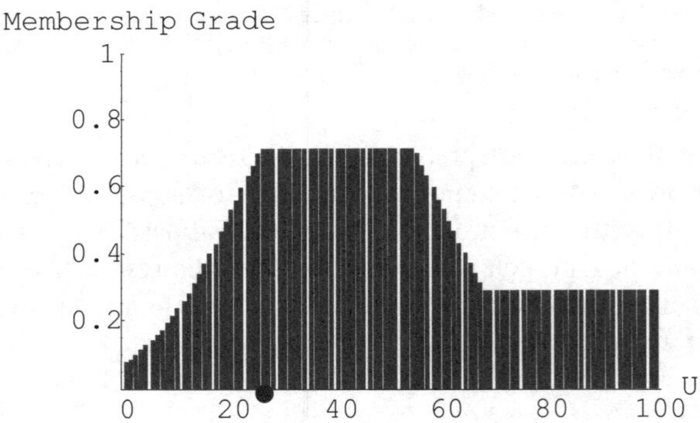

Figure 3. The fuzzy set for the occurrence of the outcome for one of the cases.

3. Result

A mathematical model built from fuzzy rules was used to classify one hundred cases, which were selected from the main database by stratified random sampling. With this method of sampling, the ratio of the occurrence of the outcome was the same in the original database and in the sample. Another mathematical model was made from crisp rules and the outcome for the same 100 cases were predicted by this model. The predictions from these two models were compared to the real outcomes using the Receiver Operating Characteristic (ROC) curves. The Fuzzy Logic Package for

Mathematica was used in this study [16]. AUC (area under the curve) was used to compare the prediction ability of the models. A ROC curve is a graphical representation of the trade off between the false negative and false positive rates for every possible cut off. The plot shows the false positive rate (100-specificity) on the X axis and the true positive rate (sensitivity) on the Y axis. The accuracy of a classification method can be measured by the AUC and an area of 1 represents a perfect test.

The model with crisp rules showed slightly higher AUC (0.67) compared to fuzzy rules (0.66) but this difference was not significant (p = 0.404) according to a nonparametric pairwise test. The ROC curves are illustrated in Figure 4.

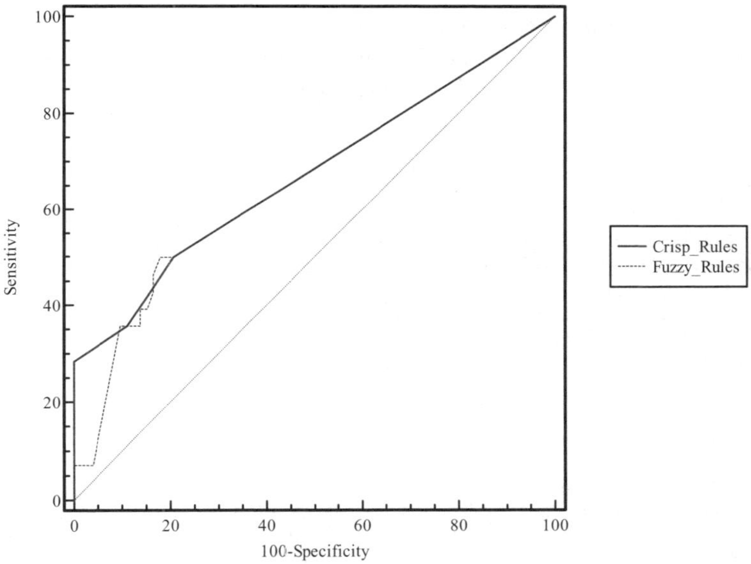

Figure 4. ROC curves for comparing different models.

4. Discussion and Conclusion

No significant difference was noticed between crisp and fuzzified rules when analysing the same cases. Fuzzy set theory has a greater capability to capture human reasoning and decision making [13]. Fuzzy expression of rules is closer to clinicians logic and this form of rules make them easier to be used by medical staff for their daily routine work. Another benefit from fuzzy modelling is that because partitions overlap and each case after prediction can be assigned to more than one class with different probabilities, it is less sensitive to noise and misclassification.

The predictive power of the two models compared is quite low, a reason being the simplified model where some available variables were omitted. In an ongoing study (not yet published) a larger model has been build, showing significantly better predictive power. However, the main contribution of this study is the approach for fuzzification of crisp rules generated from a decision tree, thereby reducing the need for domain experts for rule formulation. Knowledge elicitation from clinical experts could be a time consuming and costly process. The approach presented in this paper provides a mean to generate fuzzy rules directly from clinical data. However, the importance of clinical verification of generated rules must be stressed.

As a follow up of this study, the predictive power of the decision tree to identify high risk patients will be compared with domain experts and the same one hundred cases are going to be used. This is the main reason for choosing one hundred cases for validation, and not the otherwise often used ten-fold cross-validation technique.

Acknowledgement

This study is supported by grant No. F2003-513 from FORSS, The Health Research Council in the South-East of Sweden. Special thanks to the South-East Swedish Breast Cancer Study Group for fruitful collaboration and support in this study.

References

[1] Tejler G, Norberg B, Dufmats M, Nordenskjold B. Survival after treatment for breast cancer in a geographically defined population. The British Journal of Surgery 2004;91(10):1307-12.
[2] Antoniadis A, Lambert-Lacroix S, Leblanc F. Effective dimension reduction methods for tumor classification using gene expression data. Bioinformatics 2003;19(5):563-70.
[3] Razavi AR, Gill H, Ahlfeldt H, Shahsavar N. Canonical correlation analysis for data reduction in data mining applied to predictive models for breast cancer recurrence. Stud Health Technol Inform 2005;116:175-80.
[4] Quinlan JR. C4.5: Programs for machine learning. San Mateo, CA: Morgan Kaufmann; 1993.
[5] Podgorelec V, Kokol P, Stiglic B, Rozman I. Decision trees: an overview and their use in medicine. Journal of Medical Systems 2002;26(5):445-63.
[6] Peng YH, Flach PA. Soft Discretization to Enhance the Continuous Decision Tree Induction. In: Giraud-Carrier C, Lavrac N, Moyle S, editors. ECML/PKDD'01 workshop on Integrating Aspects of Data Mining, Decision Support and Meta-Learning; 2001; Freiburg, Germany; 2001. p. 109-118.
[7] Jang JSR, Sun CT. Neuro-fuzzy and soft computing: a computational approach to learning and machine intelligence. Upper Saddle River, NJ, USA: Prentice-Hall, Inc.; 1997.
[8] Zadeh LA. Fuzzy logic = computing with words. Fuzzy Systems, IEEE Transactions on 1996;4(2):103.
[9] Pena-Reyes CA, Sipper M. A fuzzy-genetic approach to breast cancer diagnosis. Artificial Intelligence in Medicine 1999;17(2):131-55.
[10] Seker H, Odetayo MC, Petrovic D, Naguib RN, Bartoli C, Alasio L, et al. Assessment of nodal involvement and survival analysis in breast cancer patients using image cytometric data: statistical, neural network and fuzzy approaches. Anticancer Research 2002;22(1A):433-8.
[11] Jain R, Abraham A. A comparative study of fuzzy classification methods on breast cancer data. Australasian Physical & Engineering Sciences in Medicine 2004;27(4):213-8.
[12] Zadeh LA. Biological applications of the theory of Fuzzy sets and systems. In: Int. Symp. Biocybernetics of the Central Nervous System; 1969; Boston: Little, Brown & Co.; 1969. p. 199-212.
[13] Klir GJ. Fuzzy sets and fuzzy logic: theory and applications. Upper Saddle River, NJ, USA: Prentice-Hall, Inc.; 1995.
[14] Wang LX. A course in fuzzy systems and theory. Upper Saddle River, NJ, USA: Prentice-Hall, Inc.; 1997.
[15] Razavi AR, Gill H, Ahlfeldt H, Shahsavar N. A Data Pre-processing Method to Increase Efficiency and Accuracy in Data Mining. In: Miksch S, Hunter J, Keravnou E, editors. Proceedings of the 10th conference on Artificial Intelligence in Medicine; 2005; Aberdeen, UK; 2005. p. 434-443.
[16] Stachowicz MS, Beall L. Fuzzy Logic Package for Mathematica. In. version 2.0 ed: Wolfram Research Inc.; 2003.

Ubiquity: Technologies for Better Health in Aging Societies
A. Hasman et al. (Eds.)
IOS Press, 2006

Computational Representation of Alzheimer's Disease Evolution Applied to a Cooking Activity

Audrey SERNA[a,1], Vincent RIALLE[a] and Hélène PIGOT[b]
[a] *TIMC-IMAG UMR CNRS 5525 Laboratory, Grenoble, France*
[b] *DOMUS Laboratory, University of Sherbrooke, Québec, Canada*

Abstract. This article presents a computational model and a simulation of the decrease of activities of daily living performances due to Alzheimer's disease. The disease evolution is simulated thanks to the cognitive architecture ACT-R. Activities are represented according to the retrieval of semantic units in declarative memory and the trigger of rules in procedural memory. The simulation of Alzheimer's disease decrease is simulated thanks to the variation of subsymbolic parameters. The model is applied to a cooking activity. Simulation of 100 hundred subjects shows results similar to those realised in a standardized assessment with human subjects.
Keywords: Alzheimer's Disease, Functional Performance Representation, Cognitive Modelling, Cognitive Architecture, ACT-R, Activity of Daily Living.

1. Introduction

Alzheimer's disease is characterized by the progressive deterioration of cognitive functions, causing principally loss of memory, attention deficits and confusion. Gradually, people loose their autonomy for carrying out activities of daily living (ADLs) [1]. Basic ADLs, such as bathing and eating, remain stable during the early stages of the disease. But more complex ADLs, such as preparing meals or purchasing, require higher cognitive abilities. As the disease progresses, human performances decline until family or professional caregivers are required for planning and organizing routine activities.

The evolution of Alzheimer's disease is progressive and irreversible, and follows a stepping decrease. Various scales of staging, such as the Clinical Dementia Rating (CDR) [2], provide an overview of behavioural patterns and help to make diagnosis about cognitive deficiencies. However, the person's performance assessment provided by these tests is not related to real everyday activities [3]. Several ADL observations have yet been developed to point out the cognitive abilities involved in ADLs. One of them, the Kitchen Task Assessment (KTA), evaluates the cognitive abilities involved in a cooking activity [3]. The ADL performance is correlated to the CDR rates and enables the discrimination of stages during the Alzheimer's disease evolution.

With regard to the pandemic character of Alzheimer's disease it is important to model the cognitive evolution of the disease. Cognitive modelling could help documenting the disease's evolution and predict autonomy at home. Moreover

[1] Corresponding Author: Audrey Serna. Laboratoire TIMC-IMAG, Bâtiment Jean Roger, Faculté de médecine, Domaine de la Merci. 38706 La Tronche, France. Audrey.Serna@imag.fr .

technologies for maintaining at home cognitively disabled people are called to a growing development [4][5][6]. They all require a better understanding of the people they assist. Cognitive modelling of Alzheimer's disease evolution is a primary step towards the design of a new generation of assistive devices based on cognitive technology.

This paper outlines a way to model the evolution of Alzheimer's disease through the performance of ADLs. The cognitive architecture ACT-R, chosen for the modelling, offers a framework for developing models of cognitive processes [7]. Thanks to its hybrid structure, typical errors of Alzheimer's disease can be simulated. With the example of a particular routine activity, we expose how to adjust ACT-R parameters to simulate patients.

2. Modelling Alzheimer's disease with ACT-R

2.1. Cognitive modelling and Alzheimer's disease

Cognitive architectures such as EPIC [8], SOAR [9] or ACT-R [7] are used to simulate or predict human behaviour [10]. ACT-R has been chosen due to its hybrid system. Symbolic aspects correspond to production system concepts and knowledge representation. The ACT-R theory differentiates two sorts of knowledge: procedural knowledge, for skills, and declarative knowledge, for facts. Subsymbolic aspects are represented by mathematical equations that control many of the symbolic processes, providing a way to model human errors in the performance of high-level cognitive tasks.

Patients with dementia are affected with cognitive impairments and consequently commit errors while executing daily living tasks. The cognitive deficits increase during the evolution of Alzheimer's disease. Hence, the performances on ADL decrease as the dementia develops. In other words, errors during the ADL execution occur at a higher rate with a severe dementia.

A computational representation of Alzheimer's disease can be done by reproducing the different errors that can be committed during a routine task. Dementia causes different sorts of errors, each of them can be associated to a particular representation of knowledge in ACT-R. The disease evolution is implemented in ACT-R thanks to the variation of the value of sub-symbolic parameters, which model the amount of cognitive abilities.

2.2. Memory disorders modelling

In the first stages of Alzheimer's disease, memory injuries are frequently noticed. Studies show that many patients have problems with primary memory (or recent facts memory) and with working memory [1]. In a cooking task for example, the disease induces the forgetting of some elements from the recipe (*forgetting errors*) or the confusion between various elements, like two utensils (*confusion errors*). These errors have been modelled in ACT-R using the subsymbolic system associated to declarative memory.

In ACT-R, information stored in declarative memory is organised as a network of interconnected nodes, or *chunks*. Each chunk has an activation value which controls its access in memory (Equation 1). The capacity to recall information depends on its activation. Thanks to that subsymbolic system of retrieval, ACT-R models are able to simulate errors in recall [11]. Difficulties for retrieving elements in memory are due to several factors which have been taken into account in the computation of activation.

$$A_i = B + \sum_{j=1}^{n} \frac{W}{n} S + \sum_k P_k M_{ki} + \varepsilon_2$$

Equation 1. Activation of a chunk in ACT-R

Where A_i: total activation of chunk I ; B: constant base-level activation ; W: source activation ; n: number of filled slots in the goal chunk ; S: constant strengths of association; P_k: partial matching scale ; M_{ki} : partial matching similarity and ε_2: transitory noise, generated and added every time that there is a retrieval

Forgetting errors are reproduced in ACT-R by introducing an activation threshold in the latency of elements retrieval. If a chunk cannot gather enough activation, its retrieval will exceed the latency threshold and thus will fail. Confusion errors are introduced by allowing similarity between different elements and imperfect matching in the condition part of productions (third term of Equation 1).

Forgetting errors are due to working memory disorders. In ACT-R, individual differences in working memory capacities have been modelled using the concept of capacity limitation [12]. The limited capacity of the working memory is represented by the total source activation W (second term of Equation 1) [13]. The increase of forgetting errors in Alzheimer's disease is modelled by the decrease of W value. Thus a particular value of source activation W is attributed to each stage of the disease, from 1.0 for a healthy subject to 0.6 for a subject with severe dementia.

Confusion errors are obtained by adjusting the matching scale P_k (third term of Equation 1). This scale represents the weight given to the similarity between two elements. A high value of the mismatch penalty accentuates similarities between two elements, increasing the possibility of confusion between both elements. Hence, the increase in confusion errors during the Alzheimer's disease evolution is modelled by the variation of partial matching scale P_k.

2.3. Behaviour disorders modelling

According to the medical staff, people with dementia of the Alzheimer's type suffer lack of planning, initiation, completion and judgment (*behaviour error*). For example, they will keep doing chores, without noticing that the task is over. These errors can be modelled with the subsymbolic system associated to production rules in procedural memory.

Procedural memory consists in a set of productions which is useful to perform actions. Each production is associated to a utility value (Equation 2) which is computed from past uses. If several rules can be triggered at a given time, the production rule with the highest utility is fired.

$$U_i = P_i G - C_i + \varepsilon$$

Equation 2. Utility of a production in ACT-R

Where P_i: Expected probability that production i firing will lead to a successful completion of the current objective ; C_i: Expected cost of achieving that objective, measured in time ; G: Value of the objective, measured in time and ε: Noise

This conflict resolution mechanism can be used to model phenomena such as a choice in different strategies [14]. Modelling behaviour errors in ACT-R corresponds to the creation of a conflict situation between a rule that controls a normal behaviour and a rule that controls the wrong action. Therefore to reproduce a behaviour error, the production that leads to the wrong action must be fired by the system.

To model behaviour errors, the probability that the inappropriate production achieves the current goal (P_i term of Equation 2) must be greater than the one of the expected production, in order to be triggered by the production system. As the disease evolves, the probability of the inappropriate production increases while the probability of the expected production reduces. Thus for a healthy subject, the probability to achieve the goal with the expected production is of 1.0 against 0 for the inappropriate production, whereas for a subject with severe dementia, the expected production amounts to 0.41 against 0.59 for the inappropriate production.

3. Illustration: modelling a cooking activity

3.1. Representation of the activity and the associated errors

A model of a particular ADL has been developed to illustrate the use of ACT-R subsymbolic system for modelling Alzheimer's disease and its evolution. It has been based on the observation of Alzheimer's patients completing a routine task for the KTA [3]. This standardized test provides a classification of errors committed during the performance of the ADL and measures the required level of assistance necessary to support people with Alzheimer's disease.

The KTA activity is the preparation of a pudding from a commercial package. The subjects have to measure the ingredients and stir them. Afterward, they have to cook the mix on the stove and finally, pour the hot mix into dishes.

In the model, the different stages are represented as particular goals. The cooking activity chosen in the KTA is a common one known from all the subjects. Declarative memory contains knowledge about ingredients (milk, pudding powder, etc.) and utensils (wooden spoon, pan, etc.) known by the subjects. Procedural memory is composed of rules that lead to an action in the context of the recipe. Each subtask is fragmented into basic operations coded as production rules.

The KTA puts forward six criteria (initiation, organization, performance of all steps, sequencing, judgment and safety, and finally completion) to score subjects' performance. The errors listed for each criterion can be assimilated to the three errors categories explained in section 2. The organization criterion gathers forgetting errors and confusion errors, which are due to memory disorders. The five other criteria are related to behaviour disorders. The subsymbolic system of ACT-R is used as described in section 2 to reproduce errors.

3.2. Results

The resulting system implemented in ACT-R offers the possibility to choose the level of CDR for the simulation. The simulation of a subject allows to follow the process and to localize eventual problems. Trials of 100 subjects are also possible. The repartition in percentage of subjects in each criterion and the global average score is provided as output of the simulation. For example, in Table 1 only seven people with severe dementia are independent for the organization criterion, thirty-five require verbal help, thirty-three require physical assistance and twenty-five are incapable of organizing themselves.

Criterion	Independent	Verbal help	Physical help	Not capable
Initiation	24	17	16	43
Organization	7	35	33	25
All steps	0	2	5	93
Sequencing	0	4	9	87
Judgment - safety	1	25	29	45
Completion	7	9	0	61
Average Score : 13.84				

Table 1. Simulation results for 100 people with CDR 3

Our results clearly show the bond between the cognitive assistance and the stage of the dementia. The scores in each criterion increase when the dementia worsens. Healthy subjects (CDR 0) almost never make errors whereas subjects with questionable or mild dementia (CDR 0.5 or CDR 1) make errors. In such cases verbal help is generally sufficient to have the subjects continue the task and perform it successfully. However, subjects with mild dementia who make errors are more numerous and sometimes they need a broader cognitive assistance than subjects with questionable dementia. Most subjects with moderate dementia (CDR 2) commit errors in all the criteria and they require physical assistance. Finally, subjects with severe dementia (CDR 3) almost always commit errors in each criterion, and they do not react to verbal and physical assistance. More than half are judged incapable at least in one criterion.

Finally, Table 2 shows the results presented in the KTA paper for 106 subjects and the results of the simulation of our model for 100 simulated subjects. The deviation between the two results does not exceed 0.15.

Stage of the disease		KTA results	Model results, running 100 times
CDR 0	Without dementia	?	0.01
CDR 0.5	Questionable dementia	1.75	1.69
CDR 1	Mild dementia	4.65	4.52
CDR 2	Moderate dementia	9.81	9.87
CDR 3	Severe dementia	13.88	13.84

Table 2. Global Simulation Results

4. Discussion

The results obtained with our model are quite similar to those presented in the KTA. We obtain even results more accurate: the errors repartition is displayed criterion per criterion and stage by stage during the recipe progress. However, this advantage is limited as the KTA description does not precisely provide the error repartition. It is not possible to verify in details the coherence of data relative to errors repartition. The KTA authors just present global scores for each stage of the dementia, according to subjects' CDR. Hence, in our model we supposed that the percentage of error is the same in each stage of the recipe, and that all the errors have the same chances to occur.

The hybrid architecture offered by ACT-R permits to model on one hand an ADL and, on the other hand the evolution of the dementia. Indeed, the different errors noticed in the cooking task are modelled thanks to the use of subsymbolic systems in declarative and procedural memories, and each stage of the dementia is obtained by fitting the value of ACT-R parameters.

By simulating behaviours of Alzheimer's patients, the modelling approach introduced in this article might be useful for smart homes technologies, such as intelligent assistant for tasks completion. Systems based on cognitive technologies have been designed to assist people in their routine activities [15][16]. According to the disorders of the person, the cognitive assistance system will offer an adapted support. The model could be used to simulate the occupant and to test the efficiency of cognitive assistance systems.

5. Conclusion

ACT-R is usually used to develop models for different cognitive tasks, such as problem solving and decision making, or even learning. Only a few models predict the behaviour in daily living like the use of cellular phone while driving [17]. The novelty and the interest of this work is that it presents a way to model the cognitive processes involved in the completion of an ADL.

Simulation results which have been obtained by the model are very close to results presented by the KTA. Thanks to its hybrid architecture, ACT-R allows simulating troubles due to the dementia, particularly those which are linked to memory mechanisms.

References

[1] Lezak MD. Neuropsychological Assessment. Third edition Oxford University Press; 1995.
[2] Berg L. Clinical Dementia Rating (CDR). Psychopharmacol Bull. 1988; 24:637-639.
[3] Baum C, Edwards DF. Cognitive Performance in Senile Dementia of the Alzheimer's type: The Kitchen Task Assessment. The American Journal of Occupational Therapy. 1993; 47:431-436.
[4] Rialle V, Duchêne F, Noury N, Bajolle L, Demongeot J. Health 'smart' home: Information technology for patients at home. Telemedicine Journal and E-Health. 2002; 8(4): 395-409.
[5] Tang P, Venables T. 'Smart' homes and telecare for independent living. J Telemed Telecare. 2000; 6(1): 8-14.
[6] Rialle V, Rumeau P, *et al.* Smart Homes. In Wootton R, Dimmick S, Kvedar JC, editors. Home Telehealth. London. Royal Society of Medicine Press; 2006. p. 65-75.
[7] Anderson JR, Lebiere C. The atomic components of thought. Mahwah, NJ: Erlbaum; 1998.
[8] Meyer DE, Kieras DE. A computational theory of executive cognitive processes and multiple-task performance: Part 1. Basic mechanisms. Psychological Review. 1998; 104:3–65.
[9] Newell A. Unified theories of cognition. Cambridge, MA: Harvard University Press; 1990.
[10] Cooper RP. Modeling high-level cognitive processes. Mahwah, NJ: Erlbaum; 2002.
[11] Lebiere C, Anderson, JR, Reder LM. Error modeling in the ACT-R production system. In Proceedings of the Sixteenth Annual Conference of the Cognitive Science Society. Hillsdale, NJ:Erlbaum. 1994; 555-559.
[12] Daily LZ, Lovett MC, Reder LM. Modeling individual differences in working memory performance: A source activation account in ACT-R. Cognitive Science. 2001; 25: 315-353.
[13] Lovett, M.C., Reder, L.M., & Lebiere, C. Modeling working memory in a unified architecture: An ACT-R perspective. In A. Miyake & P. Shah (Eds.), Models of working memory: Mechanisms of active maintenance and executive control. New York: Cambridge University Press; 1999.
[14] Jongman L, Taatgen NA. An ACT-R model of individual differences in changes in adaptivity due to mental fatigue. In Proceedings of the twenty-first annual conference of the cognitive science society. Mahwah, NJ: Erlbaum. 1999; 246-251.
[15] Giroux S, Pigot H, Mayers A, Lefebvre B, Rialle V, Noury N. Smart house for frail and cognitive impaired elders. UbiCog '02: First International Workshop on Ubiquitous Computing for Cognitive Aids, Göteborg (Gothenburg), Sweden. 2002; September 29.
[16] Stip E, Rialle V. Environmental cognitive remediation in schizophrenia: Ethical implications of "Smart Home" technology. Canadian Journal of Psychiatry. 2005; 50(5): 281-291.
[17] Salvucci DD. Modeling driver distraction from cognitive tasks. In Proceedings of the 24th Annual Conference of the Cognitive Science Society. Mahwah, NJ: Erlbaum. 2002: 792-797.

5.4 Decision Support: Information Retrieval

Ubiquity: Technologies for Better Health in Aging Societies
A. Hasman et al. (Eds.)
IOS Press, 2006

Strategies for Health Information Retrieval

Lina F.SOUALMIA[a,b], Badisse DAHAMNA[a,b], Benoît THIRION[a,b] , Stéfan J.DARMONI[a,b1]
[a] *CISMeF Team, Rouen University Hospital, Rouen, France*
[b] *GCSIS, PSI Laboratory, INSA & Rouen University, Mt St Aignan, France*

Abstract. Background: The amount of health data accessible on the Web is increasing and Internet has become a major source of health information. Many tools and search engines are available but medical information retrieval remains difficult for both the health professional and the patients. Objective: In this paper we describe heuristics that aim at matching as much as possible queries with the content of the documents in the context of the CISMeF catalogue (Catalogue and Index of Health Resources in French) and its Doc'CISMeF search tool. The queries are represented by terms and the content of the documents is indexed by a terminology based on the MeSH thesaurus. Results: Several operations are performed to match the terms of the terminology: natural language processing techniques on multi-words queries, phonemisation, spelling correction, plain text search with adjacency etc… Each one is tested to evaluate its contribution in matching the terminology and the indexed documents. Conclusion: The implemented heuristics contribute significantly with good results in maximising as much as possible the recall of the Doc'CISMeF search tool.
Keywords: Information Retrieval, MeSH, Internet.

1. Introduction

Internet is a major source of health information. Many people including health professionals, patients and general public, now search health care information on the Web. The access to structured medical information remains difficult when using directories such as Yahoo or search engines such as Google. Therefore many tools and applications have been developed for the healthcare professionals and until recently bibliographic databases such as Medline were available only to experts [1]. In medical information retrieval, there is a need of support. In this context, the objective of CISMeF [2] (Catalogue and Index of Health Resources in French) is to assist the health professional during the search of electronic information available on the Internet. The CISMeF health gateway describes and indexes high quality-controlled information resources written in French. We present in this paper some strategies and heuristics to match as much as possible the users' queries with the French adaptation of the MeSH and thus to reduce the silence of the system.

[1] Corresponding Author : Stéfan Darmoni, CHU de Rouen, 1 rue de Germont, 76031 Rouen Cedex France, stefan.darmoni@chu-rouen.fr, http://www.chu-rouen.fr/cismef.

2. Methods

2.1. CISMeF Metadata and Terminology

Since February 1995, the CISMeF catalogue describes and indexes a large number of health information resources (*n=15,090; Dec. 2005*). Each catalogue resource is indexed by its container using *metadata* used to improve information retrieval [3] and by its contents using the *terms* of the CISMeF terminology. CISMeF metadata are described in [2, 4]. The CISMeF terminology 'encapsulates' the French version of the MeSH thesaurus [5]. However, the MeSH was originally intended to index scientific articles for the Medline database. In order to customise it to the broader field of health Internet resources we have developed several enhancements to the MeSH since 2000. In addition to MeSH keywords and subheadings, the concepts of metaterms (*n=105*) and resource types (*n=257*) were added. As defined by the Dublin Core Metadata Initiative [4], a resource type is used to categorize the nature of the content of the resource. MeSH (term/subheading) pairs describe the topic of the resource. A metaterm (in most cases MeSH terms) is a medical specialty or a biological science, which has semantic links with one or more MeSH terms, subheadings and resource types (e.g. *cardiology, bacteriology)*. The keywords, headings and resource types are organised hierarchically. Compared to the publication types of Medline, the CISMeF resource types are more diverse, with specific resource types dedicated to electronic health resources (e.g. *association*, *clinical guidelines*). Nonetheless, the MeSH thesaurus largely inspires this list as 187 resource types (76%) are deliberately ambiguous because they are also MeSH terms (e.g. *magnetic resonance imaging*). The objective of this ambiguity is to maximise the number of search results (the Doc'CISMeF search the answers for the MeSH term and for the resource type) when the user query contains this kind of ambiguous term. Furthermore, to be as close to a standard as possible, 28 resource types (11%) are also Medline publication types (e.g. *technical report*). Each metaterm has a semantic link with one or more keywords, headings and resource types. Each term can have a set of synonyms and can belong to several trees.

Many ways of navigation and information retrieval are possible in the catalogue [6]. The most used is the *simple search* (free text interface). It is based on the subsumption relationships. If the query can be matched with an existing term of the terminology, thus the result is the union of the resources that are indexed by the term, and the resources that are indexed by the terms it subsumes, directly or indirectly, in all the hierarchies it belongs to. If the query cannot be matched, the search is done over the other fields of the metadata and in a worse case a full-text search is carried out. Contrary to Medline, the resource types and the metaterms were voluntary made ambiguous to maximize the recall (e.g. in the query *guidelines in virology*, *virology* will be recognized as a metaterm (instead of a term) and *guidelines* will be recognized as both the term and the resource type because we assume most of end users confuse content and container). We propose in the following, some enhancements for query matching.

2.2. Basic Natural Language Processing (NLP) techniques

The basic natural language processing steps developed in [7] are founded on the following operations.
Query segmentation: the query is segmented in words thanks to a list of characters and *string tokenizers*, composed by all non-alphanumerical characters (e.g.: * *$,!§;|@*).
Character normalisations: we apply two types of characters normalisation at this step. The MeSH terms are in the form of non-accented upper case characters. Nevertheless, the terms used in the CISMeF terminology are in mixed-case and accented [8]. (1) *Lowercase conversion*: all the uppercased characters are replaced by their lowercase version; "*A*" is replaced by "*a*". This step is necessary because the controlled vocabulary is in lowercase. (2) *Deaccenting*: all accented characters (*"éèêë"*) are replaced by non-accented (*"e"*) ones. Words in the French MeSH are not accented, and words in queries can be accented or not, or wrongly accented (*hèpatite"* instead *"hépatite"*).
Stop words: we eliminate all the stop words (such as *the, and, when*) in the query. Our stop words list is composed by 1,422 elements [9].
Exact expression: we use regular expressions to match the 'exact' [10] expression of each word of the query with the terminology. This step allows taking into consideration the complex terms of the vocabulary and avoiding some inherent noise generated by the truncations. The query *'**sida**'* is matched with the terms *'lymphome lié **sida**'* and *'**sida** atteinte neurologique'* but not with the terms *'gluco**sida**ses'*, *'agra**sida**e'*..
Phonemisation: the study [9] of the users' queries have shown that a great percent of no answer result from spelling mistakes. We have developed a Word phonemisation module that converts a word to its French phonemic transcription: e.g. the query *alzaymer* is replaced by the reserved term *alzheimer*, which is, has the good orthography. If the phonemic transcription of the query couldn't be matched, a spelling correction is proposed.
Spelling correction (optional): the module of spelling correction propose to the user the reserved term that has a similar phonemic (according to a score and taking into account the possible characters' inversions) with a reserved term. The query is not replaced and a correction suggestion is proposed to the user.
Bag of word: this algorithm [7] searches in the user's query the greatest set of words that corresponds to a reserved term. The reserved terms bags are formed iteratively until no possible combinations. The query *'therapy of the breast cancer'* gives two reserved words: *'therapeutics'* and *'breast cancer'* (*therapy* is a synonym of the reserved term *therapeutics).*

2.3. Heuristics to return documents from the database

The complex terms matching is more requiring than simple terms matching. The CISMeF team editorial policy concerning the queries' rewriting consists in maximising as much as possible the Doc'CISMeF recall. This approach is mainly due to the size of the CISMeF's corpus (n=15,090 vs. several million in the MEDLINE database). When all the terms of the query couldn't be recognized as reserved terms, we have implemented 5 main heuristics for information retrieval that was largely inspired by the PubMed heuristics developed to access the MEDLINE bibliographic database.

Step 1. The reserved terms: The process consists in recognizing the user query expression. If it matches a reserved term of the terminology, the process stops, and the answer of the query is the union of the resources that are indexed by the term, and the resources that are indexed by the terms it subsumes, directly or indirectly, in all the hierarchies it belongs to. If it doesn't match a reserved term, the query is segmented to seek if it contains one ore more reserved terms. The query '*enfant asthme*' is replaced by (*enfant.mr* AND *asthme.mr*), where *enfant* and *asthme* are reserved terms (*mr*). The reserved terms are matched thanks to the *bag of words* algorithm independently of the words query order.
Step 2. The documents' title: The search is performed over the other fields of the metadata. The field *title* of the documents is considered in priority. The stop words are eliminated and the search is realised over the union of the words of the query with a truncation (*) at the right in the field title (*ti*), as the following : $word_1$*.*ti* AND $word_2$*.*ti* for a 2-words query.
Step 3. Mixing the reserved terms and the titles: The system seeks if some words are reserved terms. A new Boolean query is generated with the fields reserved term (*mr*), if the word is a reserved term, and title (*ti*) if not. The query '*allergie infantile*' is replaced by the Boolean query : (*allergie.mr* AND *infantile.ti*).
Step 4. Mixing the reserved terms, all fields and adjacency in the titles : The search is processed over all the fields (*tc*) of the documents' metadata for the words that couldn't be recognized as reserved terms UNION the initial query processed over all the fields with adjacency (*at)* at *n* words with *n*=5×(nb words of the query–1). The query '*les problèmes respiratoires des enfants*' is replaced by the Boolean query [(*enfant.mr* AND *problemes.tc* AND *respiratoires.tc*) OR (*problemes respiratoires enfant.at)*]. In this query, the word *enfant* is recognized as a reserved term because it has the same sonority as the reserved term *enfants*. The words *problèmes* and *respiratoires* are searched over all the fields and the initial query *problèmes respiratoires enfants* is searched over all the fields with adjacency of 10 which means that these 3 words shouldn't be distant at more than 10 words.
Step 5. Mixing the reserved terms, all fields and adjacency in the plain texts : A plain text search over the documents with adjacency (*ap*) of *n* words with *n* = 10×(nb words of the query – 1) is realised. The query '*bronchite asthmatiforme*' is replaced by the Boolean query (*bronchite asthmatiforme.ap*) where the words *bronchite* and *asthmatiforme* shouldn't be distant at more than 10 words in the plain texts of the documents. The plain text search is possible with the Intermedia Text tool of Oracle® 9.i. which required a pre-treatment of the CISMeF corpus (~72 hours).
An intuitive scale of interpretation (from Step 1 to Step 5) is available to inform the users about their queries operations and rewritings.

2.4. Evaluation methodology

To evaluate the strategies that we have implemented, we have extracted from the Doc'CISMeF http log server of the first version of the search engine a set of 250 queries that gave no answer in the month of September 2002. The contribution of each treatment is measured. The difference with the evaluation method we have developed [9] lives in the matching of the terms with the terminology whereas here we want to match the queries with the documents of the catalogue.

3. Results

Among the 250 "difficult" queries (with no answer in 2002), the results show that a total of 176 (65%) queries give now answer(s) and 74 (35%) still not. The study of the set of the queries that give no answer shows that different reasons are possible: (a) 8 queries (10%) are matched but there is no CISMeF resources that corresponds to; (b) 27 (36%) are non corrected spelling errors returned as suggestions to the users; (c) 18 (24%) have no relationships with the medical domain and (d) 21 are unknown words. However, thanks to the heuristics (a) 27% of the queries are matched with a reserved term (Step 1); (b) 7% are matched with the title of the documents (Step 2); (c) 4% are matched with a mix of the reserved terms and the titles (Step 3); (d) 10% are matched with a mix of reserved terms, all fields and adjacency in the title of the documents (Step 4) and (e) 17% are matched with a mix of reserved terms, all fields and adjacency in the plain text of the documents. The response time is acceptable. Steps 1 and 2 response time is less than one second. Step 5 average response time ranges from 2 to 3 seconds.

Table 1. Repartition of the queries matched with documents in the CISMeF database

Operation	Number of queries with documents in return	Percentage
Step 1	57	27%
Step 2	14	7%
Step 3	8	4%
Step 4	22	10%
Step 5	36	17%
Total	176/250	65%

4. Discussion

In this paper we have presented strategies to support health information seeking using the CISMeF information gateway in the case of free queries that don't match the controlled vocabulary, i.e. that give no answer from the corpus. Simple but essential treatments such as spelling correction are processed online. McCray [11] has also presented strategies for supporting health information seeking. The major difference is in the treatment of the query and specifically in its expansion. We think that this type of query expansion (by relaxing the query) and suggestions to the users may led them too much tasks in first, choosing the expanded query and then, in navigating through the documents of the expanded queries to seek the wanted information. The second problem lives in the exponential growth of the query when it is composed by several words. Another treatment seems to us not necessary at all: the expansion of each word of the query by a set of its synonyms, derivations, and inflections. If the query contains a synonym of a reserved term, it should be replaced by the reserved term, which is more precise especially in the context of indexed resources with a controlled vocabulary. The last point concerns the search mode itself which is based on plain text search vs. indexing terms which is much more precise.

Our strategies are relatively powerful as 65% of the queries with 0 answer in 2002 give at least one answer in 2006 (see Table 1). Among the 35% of the queries with 0 answer, only 36% (n=27) are spelling errors which are not corrected. We will then focus on this problem. Furthermore, a recent Spanish study comparing 6 European health catalogues has shown that CISMeF was ranked second after OMNI in terms of precision and recall, mainly because "failure on precision may be due to exhaustive indexing" [12]. This external judgment is definitively true: we deliberately focused on maximizing recall in terms of terminology and in terms of heuristics. This approach may be explained by the relative small size of the CISMeF corpus. This Spanish study will lead to a rather serious modification of the CISMeF editorial policy. New adds-on on the CISMeF terminology or heuristics will now focus on maximizing precision (in the near future, the default query for the step 1 (reserved term) will answer CISMeF resources indexed with a MeSH major (or starred) term). The CISMeF heuristics for health information retrieval tried to improve the PubMed heuristics although the size of his respective corpus is not the same. We have introduced the step 2 (search on title) because, based on the know-how of the CISMeF Chief Librarian, this step provides very precise answers. In the step 4 (search on all metadata fields), we have introduced a search with adjacency once again to be more precise. Finally, we have also introduced the step 5 (search on plain text). This step very similar to a Google search (but more precise thanks to the adjacency) is feasible for the CISMeF Catalogue because it indexes full text resources, which is not the case for the Medline database. Nonetheless, the PubMed Website would be able to apply the step 5 to a subset of journals indexed in Medline, in particular those of PubMed Central.

References

[1] Eysenbach G, Jadad AR. Consumer Health Informatics in the Internet Age. JMIR, 2001; 3(2): e-19.

[2] Darmoni SJ, Leroy JP, Thirion B, et al. CISMeF a Structured Health Resource Guide. Methods Inf Med 2000; 39(1): 30-35

[3] Hudgins J, Agnew G, Brown E. Guetting Mileage Out of Metadata: Applications for the Library. ALA 1999.

[4] Dekkers M, Weibel S. State of the Dublin Core Metadata Initiative. D-Lib Mag. 2003; V9 n° 40.

[5] Nelson SJ, Johnson WD, Humphreys BL. Relationships in Medical Subject Headings. Kluwer Publishers, 2001; 171-84.

[6] Darmoni SJ, Thirion B, Leroy JP et al.. A Search Tool Based on Encapsulated MeSH Thesaurus to Retrieve Quality Resources on the Internet. Medl Inform Internet Med 2001; 26(3):165-178

[7] Soualmia LF. Etude et Evaluation d'Approches Multiples d'Expansion de Requêtes pour une Recherche d'Information Intelligente : Application au domaine de la Santé sur l'Internet. PhD Thesis; December 2004.

[8] Zweigenbaum P, Grabar N. Restoring Accents in Unknown Biomedical Words: application to the French MeSH thesaurus. IJMI 2002; 113-26.

[9] Douyère M, Soualmia LF, Névéol A, et al. Enhancing the MeSH thesaurus to retrieve French online health resources in a quality-controlled gateway. Health Info Libr J 2004;21(4): 253-61.

[10] Riloff E. Little Words can make Big Difference for Text Classification. 18th ACM SIGIR 1995. 130-36

[11] McCray AT, Ide NC, Loane RR, Tse T. Strategies for Supporting Consumer Health Information Seeking. Medinfo 2004; 1152-56.

[12] Abad Garcia F, Gonzalez Teruel A, Bayo Calduch P, et al.. A comparative study of six European databases of medically oriented Web resources. J Med Libr Assoc. 2005 Oct; 93(4): 467-79.

Ubiquity: Technologies for Better Health in Aging Societies
A. Hasman et al. (Eds.)
IOS Press, 2006

A method of cross-lingual consumer health information retrieval

Aurélie NÉVÉOL[1, 2], Suzanne PEREIRA[1, 2, 3], Lina F. SOUALMIA[2], Benoît THIRION[2], Stéfan J. DARMONI[1, 2]

[1] *Equipe GCSIS, Laboratoire PSI - INSA & Université de Rouen, France*
[2] *Equipe CISMeF, CHU de Rouen, France*
[3] *LERTIM, Faculté de Médecine de Marseille, France*

Abstract. Objectives: This paper presents a method of cross-language information retrieval aiming to make medical information available to patients in French and English, regardless of the query language they wish to use. Methods: We describe the two MeSH-related terminologies used in this work. We show that the French patient synonyms included in CISMeF can be automatically mapped to the English consumer-oriented health topics used in MEDLINEplus, via the MeSH thesaurus. The links between French and English patient terms thus inferred can subsequently be exploited to automatically translate patient queries. Results: 129 MEDLINEplus topics have been mapped to 142 CISMeF patient synonyms. Contextual links for cross-language retrieval have been added to the patient dedicated French information Gateway CISMeF. Conclusion: we have presented an efficient method for cross-lingual patient information retrieval in French and English, which may also be applied to other language pairs, subject to the availability of patient terminologies and of the MeSH thesaurus in these languages.
Keywords: Information Retrieval, MeSH, consumer health information, automatic translation

1. Introduction

An increasing amount of medical information is made available in electronic format, and published on the Internet. In recent years, the public has gained an easier access to these resources, and patients have also been encouraged to take advantage of this new media. The emergence of such technologies of information and communication in the medical domain calls for adequate tools to assist lay persons in their search for trustworthy health information intended for non specialist audiences, also known as Consumer Health Information (CHI).

Since 1995, CISMeF[1] (French acronym of Catalogue and Index of Medical On-Line Resources) has been selecting the most important resources of institutional health information in French [1]. It currently contains more than 15,000 resources intended for

[1] http://www.cismef.org

health professionals (e.g. evidence-based resources, practice guidelines), medical students (e.g. lecture notes), and patients (e.g. patient education handouts). As in other health gateways (such as PubMed® or OMNI) the content description of resources referenced in CISMeF uses the reference thesaurus for indexing the bio-medical literature, namely MeSH® (Medical Subject Headings). However, patients –and even health professionals- often have doubts as to the proper MeSH denomination they should use to phrase their information query. Aware that inadequate queries may result in failure to obtain the desired information, the CISMeF team has addressed this issue in previous work by refining the information retrieval algorithm used in the catalogue's dedicated search engine Doc'CISMeF [2], [3] or enriching the list of MeSH synonyms available in French [4], [4], [6].

To provide further assistance to users, cross-language MeSH information retrieval is already available in CISMeF: MeSH queries can be entered in Doc'CISMeF either in French or in English. Besides, after the result of each query, contextual links to other trustworthy health gateways are provided to help users access health information in English. Therefore, a MeSH query entered in French in Doc'CISMeF will be translated into English, and a contextual link enables users to launch the translated query in PubMed or OMNI to access corresponding resources in English. However, similar work remains to be done for patient vernacular. In fact, if many patients have a sufficient grasp of the English language to benefit from health information in English, they experience difficulties in expressing their information need in a foreign language, and particularly with very specific terms as should be used in health gateways.

2. Objectives

This work aims to make medical information available to patients in several languages, regardless of the language they wish to use to state their information need. It is mainly intended for multilingual patients who may be able to understand medical information in several languages, but find it easier to phrase their information query in one particular language. In this paper, we describe a method of cross-language medical information retrieval in order to help patients with multilingual skills accessing health information in several languages. Specifically, we intend to use terminological resources in French and English to improve cross-lingual information retrieval in the CISMeF catalogue, and provide access to patient medical information in English in response to queries formulated in French.

3. Methods

3.1. CISMeF patient synonyms and MEDLINEplus topics

In the CISMeF terminology, CHI terms are defined as MeSH synonyms that patients are more likely to use than the actual MeSH terms. Therefore, more than one patient synonym may be available for a given MeSH term. For example, "tumeur osseuse" and "cancer des os" are two patient synonyms for the MeSH term "tumeurs des os" (bone neoplasms). On the other hand, CISMeF terminology experts are reluctant to use one particular term as a synonym for more than one MeSH term. These are very rare occurrences, and such synonyms are said to be ambiguous. This choice regarding the addition of entries in the terminology results in having a larger number of synonyms (N=531) than the number of MeSH terms impacted (N=431), as can be seen in table 1.

MEDLINEplus topics are health topics specifically selected for *consumers* – this includes patients, but also any lay person looking for reliable health information. The topics are meant to cover a large range of the consumers' health interests. For this reason, some topics may relate to more than one MeSH term. For example, the topic "AIDS" relates to the MeSH terms "Acquired Immunodeficiency Syndrome" and "HIV Infections". As a result, there are less health topics (N=698) than MeSH terms impacted (N=1130), as can be seen in table 1.

3.2. Linking CHI terms

CHI equivalents to MeSH terms have been developed independently in French and English by the CISMeF and MEDLINEplus teams. Thanks to the French version of the MeSH thesaurus developed by the French Medlar centers (INSERM), links between French and English MeSH terms are already available. The effort to enrich the MeSH with CHI terms in French (CISMeF patient-oriented synonyms) and English (MEDLINEplus topics) has led to the creation of semantic links between CHI terms and MeSH terms in each language. Figure 1 shows the links existing between the different types of terms involved.

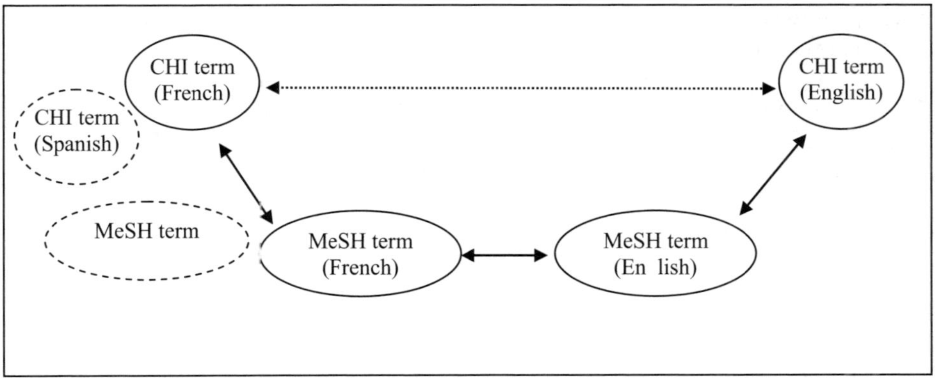

Figure 1: Method used for mapping French patient synonyms to English CHI equivalents

It is important to note that there are no explicit links between French and English CHI terms. However, the succession of three links (CHI to MeSH, MeSH to MeSH, and MeSH to CHI) makes it possible to induce semantic links between CHI in the different languages.

For example, linked to the English CHI term "second-hand smoking", we find the English MeSH term "tobacco smoke pollution", and its French MeSH equivalent, "pollution fumée tabac". There is one French CHI term linked to "pollution fume tabac", namely "tabagisme passif". Therefore, as shown on Figure 2, we can infer from these links that "tabagisme passif" is a French CHI equivalent for "second-hand smoking".

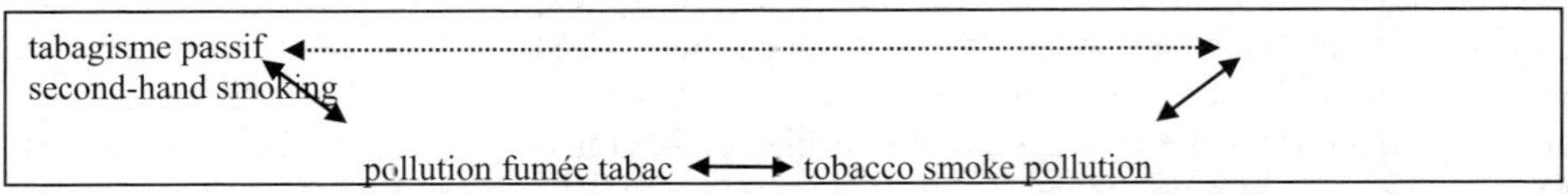

Figure 2: Sample mapping of CHI terms

4. Results

Table 1 presents the number of French (CISMeF) patient synonyms and English (MEDLINEplus) health topics used in our experiment, and the number of links that could be inferred between any pair of CHI terms. The second and third rows indicate the number and proportion of MeSH terms impacted (MeSH 2005 was the version used in this study).

	CHI (French)	**CHI (English)**	**Links created**
Rough number	531	698	280
MeSH impact (MeSH 2005)	431	1130	105
Proportion (MeSH 2005)	1.9%	4.9%	0.5%

Table 1: Number of Consumer specific terms in French and English, and links created

Table 2 provides examples of terms that could be mapped, and the links inferred. In cases where more than one CHI term was linked to a MeSH term in either French or English, more than one link could be inferred.

MeSH UID	**CHI (French)**	**CHI (English)**	**Links created**
D014028	tabagisme passif	Secondhand Smoke	tabagisme passif ↔ Secondhand

			Smoke
D001859	cancer des os tumeur osseuse	Bone Cancer	cancer des os ↔ Bone Cancer tumeur osseuse ↔ Bone Cancer
D014947	Trauma traumatisme	Injuries Wounds	trauma ↔ Injuries trauma ↔ Wounds traumatisme ↔ Injuries traumatisme ↔ Wounds

Table 2: Sample inferred links for three MeSH descriptors

Overall, 129 health topics were linked to 142 patient synonyms. As a result, 129 contextual links to MEDLINEplus topics pages will be created in French search engine Doc'CISMeF.

As an evaluation of the method, we can say that 18.5% of MEDLINEplus topics could be linked to at least one CISMeF patient synonym, and likewise, 26.7% of CISMeF patient synonyms could be linked to at least one MEDLINEplus topic.

5. Discussion

5.1. Links between CHI terms in French and English

Table 1 shows that, overall, the application of the method presented to infer links between French and English CHI terms covers only a small portion of the MeSH (0.5%). This is partly due to the limited MeSH coverage of the CHI terms (1.9% for French, 4.9% for English). In fact, the approach relies fully on the CHI information available in the terminologies it is applied to. In practice, this means that the coverage of links inferred will reflect the scope of CHI information in the terminologies used, and the overlap of CHI information between terminologies. However, one strong point of the method is that it is rather effortless as it exploits CHI information already available.

Even though the coverage of the links obtained for CHI terms in French and English is small on the MeSH scale, we consider these 129 sample links as representative of the performance of the method. All 129 links were reviewed and approved by a medical librarian.

Links such as those listed in Table 2 can be used in health search engines to answer patient queries. For example, when a patient enters the query "tabagisme passif", in addition to the list of CISMeF patient resources concerning "tabagisme passif ", French search engine Doc'CISMeF currently provides a dynamic link launching a MEDLINEplus search on "tobacco smoke pollution". Therefore, without knowing the scientific term referring to "tabagisme passif" in either French or English nor the translation of "tabagisme passif " in English, a patient can still access information on " tabagisme passif " both in French and English. In the near future, we are planning to exploit the CHI links inferred to make the dynamic links to MEDLINEplus more precise. As shown in Figure 3, a query on "tabagisme passif" entered in Doc'CISMeF will retrieve French resources on the subject and provide a link to the MEDLINEplus page specifically dedicated to "secondhand smoke", which is the exact translation in English of "tabagisme passif".

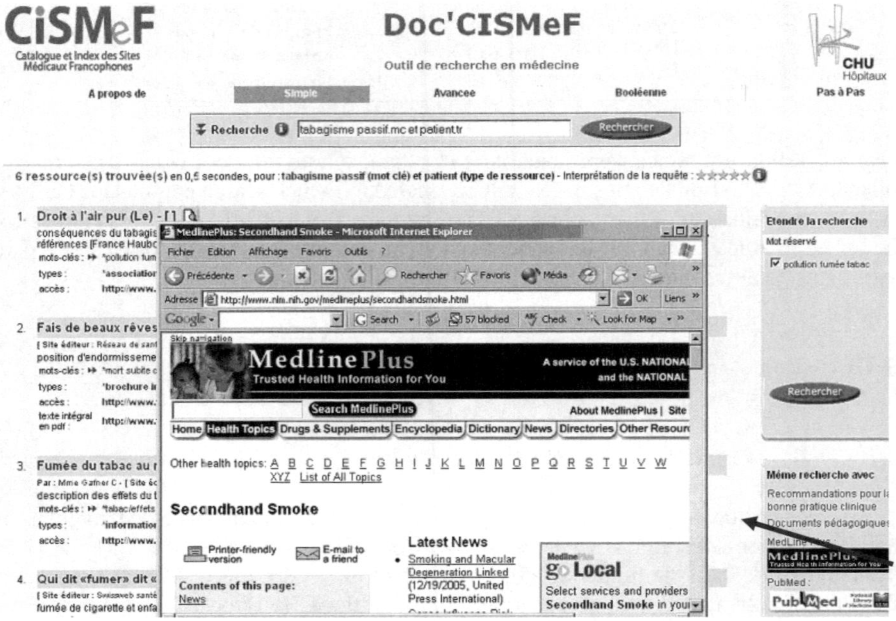

Figure 3: Prospective use of the inferred CHI links for Cross-Language Information Retrieval

5.2. Literature review

Thesauri-based Cross-Language Information Retrieval (CLIR). The problem we address, namely giving access to information in several languages for multilingual patients, answers a growing need already pointed out by Lu *et al.* [8] who developed a method of automatic translation of MeSH terms into Mandarin Chinese, and focused on the extraction of patient translations. In fact, the availability of many terminologies in the medical domain, including multi-lingual resources such as the MeSH paves the way for thesauri-based CLIR systems. A recent review of the CLIR literature [9] points out that the major drawback of thesauri-based CLIR methods is the availability of comprehensive multi-lingual thesauri. As these resources become available, thesauri-based CLIR should be preferable to query translation using either dictionaries or automatic translation software for which word-sense desambiguation in queries is difficult to overcome. A recent comparison [10] of thesaurus-based and automatic translation based CLIR systems show the superiority of the thesaurus (MeSH) based approach, but further work indicated that a combination of both methods could improve the performance.

Extending Thesauri-based CLIR. The application described above is the creation of contextual links to facilitate the interoperability between two different knowledge databases, in our example, CISMeF-patient and MEDLINEplus. However, Cimino *et al.* [11] have shown that this interoperability can be successfully extended to the creation of links between databases containing different types of information, such as a knowledge base and a patient record. In this respect, multi-lingual CHI terms can also be exploited to help patients understand their electronic files, and to search related information. In fact, while a patient is looking at his or her medical electronic record, multilingual information about the medical topics contained in the record could be made available through infobuttons. In 2005, a monolingual prototype application for health professional has been developed at the Rouen University Hospital [11], and we are planning to extend this functionality for patients and multilingual information retrieval.

Usability of contextual links for CLIR. A study of German consumer health search habits conducted in 2001 [7] showed that study participants with a good command of English did not try to search for information in this language. However, in spite of a substantial Internet experience (an average practice of 33 months), some of the participants seemed to lack understanding of search features, such as language selection in generic search tools or the use of domain dedicated engines and portals. Therefore, it seems important to inform users of the possibility for cross-language retrieval, and check whether its availability in the form of contextual links is suitable for their information needs. Previous work in the CISMeF team on contextual links to health information inserted in patient records for health professionals use received an informal positive feed-back, that will lead to business opportunities.

5.3. Perspectives

The method we used to infer links between CHI terms in French and English exploits the independent efforts of teams developing MeSH-related CHI vocabularies in various languages for which a MeSH translation is already available. The method is completely generic and can be applied to other language pairs such as English and Spanish in order to facilitate the access of bilingual patients to trustworthy consumer health information.

6. Conclusion

We have presented a method of cross-language medical information retrieval intended for patients, and we have applied it to a particular language pair, French and English. As a result, 280 links between French and English CHI terms have been inferred. The validation of the inferred links by an expert shows that the approach is relevant, and all the links will be used soon in the Doc'CISMeF search engine to launch precise cross-language patient queries.

Acknowledgements

This research was supported in part by an appointment of Prof. Darmoni to the NLM Research Participation Program sponsored by the National Library of Medicine and administered by the Oak Ridge Institute for Science and Education. The authors would also like to thank Paula Kitendaugh at the National Library of Medicine in Bethesda, MD for granting them access to MEDLINEplus topics links to MeSH terms for this work and Gaëtan Kerdelhue at the University Hospital Library of Medicine in Rouen for his help in validating the CHI links.

References

[1] Darmoni SJ, Leroy JP, Thirion B, Baudic F, Douyère M and Piot J. CISMeF: a structured Health resource guide. Meth Inf Med 2000: 39(1): 30-5.

[2] Darmoni SJ, Thirion B, Platel S, Douyère M, Mourouga P, Leroy JP. - CISMeF-patient: A French counterpart to MEDLINEplus. JMLA Journal of the Medical Library Association 2002; April;90(2):248-53

[3] Soualmia LF, Darmoni SJ. Combining different standards and different approaches for health information retrieval in a quality-controlled gateway. Int J Med Inform. 2005 Mar;74(2-4):141-50.

[4] Névéol A., Ozdowska S.: Extraction de termes médicaux à partir d'un corpus parallèle anglais/français, Actes de Extraction et Gestion des Connaissances 2005:655-64

[5] Ozdowska S., Névéol A., Thirion, B.: Traduction compositionnelle automatique de bitermes dans des corpus anglais/français alignés. Actes des 6èmes rencontres Terminologie et Intelligence Artificielle, 2005:83-94.

[6] Deléger L., Merkel M., Zweigenbaum P. Using Word Alignment to Extend Multilingual Medical Terminologies. LREC workshop, 2006: to appear

[7] Eysenbach G and Köhler C: How do consumers search for and appraise health information on the world wide web? Qualitative study using focus grups, usability tests and in-depth interviews. BMJ 2002:324;573-577

[8] Lu WH, Lin SJ, Chan YC and Chen KH: Semi-automatic construction of the Chinese-English MeSH using web-based term translation method, Proc. AMIA Symp. 2005:475-9.

[9] Chiao YC: Extraction lexicale bilingue à partir de textes médicaux comparables – application à la recherche d'information crosslingue. PhD thesis, Université Paris VI. 2004.

[10]Ruch P:Query translation by Text Categorization, Proc. COLING 2004.

[11]Cimino JJ, Elhanan G, Zeng Q: Supporting infobuttons with terminological knowledge. Proc AMIA Annu Fall Symp 1997:528-32.

[12]Pereira, S: Etude de faisabilité de différentes méthodes d'optimisation du codage médico-économique. Master Thesis, Université Paris V:2005.

Ubiquity: Technologies for Better Health in Aging Societies
A. Hasman et al. (Eds.)
IOS Press, 2006

Applying an Artificial Neural Network to Predict Osteoporosis in the Elderly

Jainn-Shiun Chiu[a], Yu-Chuan Li[b], Fu-Chiu Yu[c] and Yuh-Feng Wang[a,1]
[a] *Department of Nuclear Medicine, Dalin Tzu Chi General Hospital, Chiayi, Taiwan*
[b] *Graduate Institute of Medical Informatics, Taipei Medical University, Taipei, Taiwan*
[c] *Medical Intelligence Network, Taipei, Taiwan*

Abstract. Osteoporosis is an essential index of health and economics in every country. Recognizing asymptomatic elderly population with high risks of osteoporosis remains a difficult challenge. For this purpose, we developed and validated an artificial neural network (ANN) to identify the osteoporotic subjects in the elderly. The study population consisted of 1403 elderly adults (mean age 63.50 ± 0.24 years ranged from 50 to 91 years old, 157 male and 1246 female) randomly selected into 3 sets, 703 participants in training set, 350 participants in selection set, and the remaining 350 participants in test set. The input variables included demographic characteristics, anthropometric measurements, and clinical data. The outcome variable was dichotomous, either non-osteoporotic (T-score of greater than -2.5) or osteoporotic (T-score of -2.5 or less) groups classified by the measurement from dual energy X-ray absorptiometry. ANN was constructed with data from training and selection sets and validated in test set whose outcome variable was unknown to the network. The performance of ANN was evaluated by discrimination and calibration simultaneously. After training processes, the final best ANN was a multilayer perceptron network which determined seven input variables (gender, age, weight, height, body mass index, postmenopausal status, and coffee consumption) as significant features. The discriminatory power of ANN for test set was excellent (area under receiver operating characteristics curve = 0.82 ± 0.03). ANN also had statistically good fit represented by statistically insignificant Hosmer-Lemeshow statistic (p = 0.24). These results suggested that our final ANN concurrently had good discriminatory power and good-fit calibration. ANN can be used as a promising tool for the elderly to stratify high risk subjects into osteoporotic group.
Keywords: Neural network, Osteoporosis, Elderly.

1. Introduction

Osteoporosis is not only a healthy issue but also an economical index in every country. Recognizing asymptomatic people with high risks of osteoporosis remains a difficult challenge. The World Health Organization (WHO) has established the criteria for the diagnosis of osteoporosis based on the measurement of bone mineral density (BMD) by the dual energy X-ray absorptiometry (DEXA). Therefore, measuring BMD has been acknowledged as a useful method to predict the risk of a future osteoporotic fracture in an individual [1]. However, routine BMD measurement of all population by DEXA for screening is limited in most of the countries since the procedure is costly and the

[1] Corresponding Author: Yuh-Feng Wang, MD. Department of Nuclear Medicine, Dalin Tzu Chi General Hospital. No.2, Minsheng Rd., Dalin Township, Chiayi County 622, Taiwan. E-mail: yuhfeng@gmail.com

equipment of DEXA is not universally available. To surmount this problem, several predictive algorithms have been developed and validated for screening or monitoring.

Artificial neural network (ANN) is a computational model, being composed of parallelly nonlinear processing elements arranged in highly interconnected networks with a formation that emulates complex human thought processes such as adaptive learning, optimization, reasoning and decision making [2]. Since ANN is the most widely depicted application of artificial intelligence in clinical medicine, use of ANN to predict osteoporosis is promising. Herein, our study is the first investigation to develop an ANN model in predicting osteoporosis for the elderly and we have validated its feasibility in comparison with the measurement by using DEXA as reference method.

2. Material and Methods

2.1. Subjects

The Institutional Review Boards for Human Investigation at our hospital approved the study protocol. Subjects in this study were 2043 adults who underwent bone densitometry from 2001 to 2005. Subjects were excluded if age less than 50 years, presence of acute illness, taking medications (except steroids) known to affect bone metabolism, or any reason influencing the measurement of BMD. The final study participants consisted of 1403 elderly adults (157 male and 1246 female).

2.2. Measurements

Demographic characteristics (gender, age), anthropometric measurements (weight, height, body mass index [BMI]), and clinical data (postmenopausal status, steroid use, current smoking, alcohol drinking, coffee consumption) for all subjects before densitometry were recorded. Only one technologist was assigned to measure BMD using DEXA. All densitometrical results were interpreted by a nuclear medicine physician.

2.3. ANN construction

STATISTICA 7.0 (StatSoft, Inc., Tulsa, OK, USA) was used to generate various formulations of ANN models using a training set of subjects selected at random from the original cohort of subjects (n = 1403). We used the standard approach of cross validation to randomly separate the dataset from all participants into three parts: a training set (n = 703), a selection set (n = 350), and a test set (n = 350). Although the training set is optimized to fit ANN model, the selection set is separately used to estimate prediction error for model selection and stop training to mitigate over-learning and over-fitting. A third set known as the test set is used to validate the performance and the generation error of the final chosen model.

To choose the most excellent neural network, the automatic network designer was employed to decide an appropriate architecture. All variables were entered as continuous or nominal input variables into ANN models. The status of BMD according to the measurement of DEXA, categorized as osteoporotic (T-score of -2.5 or less) and non-osteoporotic (T-score of greater than -2.5) groups based on WHO criteria, was entered as a dichotomous output variable. During the processing procedure, the intelligent problem solver guided a large number of experiments, which were used to determine the best architecture. It could allow simultaneous comparison of different types of networks using a combination of heuristic and optimal algorithms and automatically selected the smoothing factor and the number of units for these networks. To compare the performance of networks with different input variables, the intelligent problem solver balanced error against type and diversity as criteria to select retained networks, in which case it preserved networks with a range of performance/complexity and types trade-offs. If the network file is full and the new model is inferior to the candidate for replacement, the network set will be increased in maximum size to conform a new network.

With the help of sensitivity analysis [3], predictive error ratio was calculated for each input variable according to their degree of validity. The ratio was expressed as the relation between the predictive errors for the model with a removed given input variable and the total network errors calculated based on all input variables. This index represented that the contributions made by different input variables in predicting the outcome were ranked in order of descending importance. After the network was allowed to run, the predicted outcome was correlated with the observed outcome; and if the network predicted the outcome incorrectly, by a process of back propagation, hidden weights within the network were remodified until the predicted outcome was accurate. At last, the intelligent problem solver retained the best architecture. After training was completed, the best ANN model was validated in test set whose outcome in terms of osteoporosis was concealed. Once ANN model made a prediction for each subject, the predictive performance of the network was evaluated.

2.4. Statistical analysis

Data were expressed as mean ± standard error or ratio. To compare the differences among 3 sets, Cochran Q test and Kruskal-Wallis ANOVA test were used. The statistical significance for comparisons among 3 sets was defined as p value less than 0.05. To assess the quality of classification model in clinical study, discrimination and calibration should be calculated concurrently. Discrimination measures how well a model identifies subjects correctly as two different classes; calibration can evaluate the degree of correspondence between estimated probabilities produced by a model and actual observation. The area under the receiver operating characteristics curve (AUC) was used as a measure of a model's discriminatory power. An AUC between 0.7 and 0.8 was classified as "acceptable" and between 0.8 and 0.9 as "excellent". The sensitivity and specificity at a cut-off value corresponding to the highest accuracy were also computed. Calibration was assessed using Hosmer-Lemeshow goodness-of-fit statistic (H-statistic) which divides subjects into deciles based on predicted probabilities and then computes a chi-square from observed and expected frequencies. A statistically good fit is defined as p value more than 0.05.

3. Results

The characteristics of study participants are listed in Table 1. Their mean age was 63.50 ± 0.24 years ranged from 50 to 91 years old and male to female ratio was 0.13. The final best ANN was a multilayer perceptron network with one input layer of 7 neurons, one hidden layer of 13 neurons, and one output layer with 1 neuron. Three variables (current smoking, alcohol drinking, and steroid use) were removed; and remaining 7 input variables (gender, age, weight, height, BMI, postmenopausal status, and coffee consumption) were taken as significant features. Table 2 shows their ratios and ranks of sensitivity analysis for training, selection, and test sets. In quantitative determination, we averaged the ranks of three subsets for each variable as a new score $Rank_{avg}$. In order of descending importance by $Rank_{avg}$, 7 variables were gender, postmenopausal status, weight, height, BMI, age, and coffee consumption.

The predictive performances of ANN for 3 sets are listed in Table 3. Training and test sets had excellent discrimination with good sensitivity and specificity accordantly. On the other hand, all sets had statistically good fit represented by statistically insignificant H-statistic ($p > 0.50$). These results suggested that final best ANN concurrently had excellent discriminatory power and good-fit calibration.

Table 1. Comparison of training, selection, and validation sets.

	Training	Selection	Test	p
Gender	90.45	88.29	86.00	< 0.001
Age	63.95 ± 0.34	62.40 ± 0.48	63.71 ± 0.50	0.02
Weight	57.47 ± 0.36	57.55 ± 0.53	58.32 ± 0.50	0.21
Height	153.58 ± 0.25	154.16 ± 0.39	154.33 ± 0.37	0.23
BMI	24.34 ± 0.14	24.19 ± 0.20	24.48 ± 0.20	0.51
Postmenopause	83.78	80.86	82.00	0.15
Steroid	8.96	6.00	6.00	0.048
Smoking	2.42	3.43	4.00	0.28
Alcohol	2.28	2.57	2.00	0.88
Coffee	4.98	6.57	4.00	0.27
BMD	1.01 ± 0.01	1.01 ± 0.01	1.01 ± 0.01	0.84
T score	-1.09 ± 0.06	-1.08 ± 0.09	-1.05 ± 0.08	0.88
Osteoporosis	20.63	18.86	19.71	0.13

Gender, female ratio (%). Postmenopause, steroid, smoking, alcohol, coffee, and osteoporosis are expressed as ratio (%). The unit for age, weight, height, BMI, and BMD are years, kg, cm, kg/m^2, and g/cm^2, respectively.

4. Discussion

The evolution of ANN has become more satisfactory in medical research than before. A speculative benefit of this artificial intelligence is its power to identify complex correlative interactions among clinical data and final diagnosis. It can prune excessive variables during training stage and explore latent types of clinical data in biological nature. Good results by using limited variables and ANN to assess the predictability of osteoporosis in the elderly are presented in this study. For any predictive model to be practical in making clinical decision, it should use data that are easily available to clinicians at the time of triage. In order to avoid model complexity for clinicians, it may not be necessary to add more variables even though they perhaps have some influences on predictive power. In a similar study, Rae *et al.* used multiversion system of ANN with 20 risk factors to predict osteoporosis in 274 UK women against the measurement of quantitative ultrasonography for right heel bone [4]. Their model resulted in AUC of

0.80, but our model had simpler algorithm and structure with only 7 predictors to obtain a comparable discriminatory power.

Table 2. Sensitivity analysis among training, selection, and test subsets of best ANN.

	Training		Selection		Test		$Rank_{avg}$
	Ratio	Rank	Ratio	Rank	Ratio	Rank	
Gender	1.116007	1	1.097456	1	1.222616	1	1.00
Age	1.048779	5	0.984566	5	0.993698	6	5.33
Weight	1.080328	2	0.964511	7	1.027591	3	4.00
Height	1.049166	4	1.003382	4	1.015637	4	4.00
BMI	1.047970	6	0.969516	6	1.043441	2	4.67
Postmenopause	1.052349	3	1.027129	2	1.014849	5	3.33
Coffee	1.024611	7	1.020078	3	0.959405	7	5.67

Table 3. Results of discrimination and calibration by best ANN.

	Training	Selection	Test
AUC	0.81 ± 0.02	0.75 ± 0.04	0.82 ± 0.03
p^1	< 0.001	< 0.001	< 0.001
Sensitivity (%)	76.6	72.7	78.3
Specificity (%)	71.1	70.4	73.3
H-statistic	8.95	8.73	8.00
p^2	0.35	0.27	0.24

p^1, p value derived from AUC; p^2, p value derived from H-statistic.

In many ANN investigations, data structure of input variables is seldom discussed. We can identify certain differences among three sets in Table 1. ANN was constructed from training set and internally validated successfully which represented internal generalization was well-behaved. On the other hand, ANN can extract particular input variables to assemble the model. Although variables operated in ANN model should not be explained as independent predictors as discerned by conventional statistics, they could be interpreted as part of the global function in ANN, expressing the complex nature among clinical variables. Considering 4 input variables (steroid use, current smoking, alcohol drinking, and coffee consumption) in our study might encode similar information. A model could depend completely on one, totally on the other, or on some combination of them. Then sensitivity analysis presents an associated sensibility to them. If one input variable is eliminated, the model may perform properly because other input variables still supply important information. The removed variable may be classified as of low sensitivity even though they might encrypt essential information. Therefore, our ANN selected only coffee consumption on behalf of other 3 variables. In addition, these clinical parameters are well-known having complex interrelationships that is why ANN "thought" coffee consumption could be an index variable. Moreover, a variable that encodes relatively insignificant information, but is the only variable to do so, may have higher sensitivity than any number of variables that mutually encrypt more influential information such as variable age.

The investigations of ANN to predict osteoporosis are limited in clinical medicine. In one pilot study, Ongphiphadhanakul *et al.* constructed an ANN to classify 129 Thai postmenopausal women into osteopenia or non-osteopenia subjects [5]. Their results were acceptable with reasonable sensitivity (76.2 ~ 80.0%) and low specificity (12.5 ~

33.3%). Although their goal and population was different with our study, the specificity was far less than ours. The size of participants and the predictors they selected were the possible reasons. Sadatsafavi *et al.* developed several ANN models from the dataset of 2158 Iranian postmenopausal women whose BMD values were all measured by DEXA [6]. They found that 3 to 5 input variables (age, weight, years since menopause, steroid use, estrogen use) could get good discrimination (AUC = 0.749 ± 0.017 ~ 0.818 ± 0.019) with 86.3% sensitivity and 72.1% specificity. Even though the findings and performance of their ANN model was comparable with ours, the most different emphasis was that we included the elderly whether male or female. Of note, they employed a traditional stepwise regression analysis as pre-processing method in order to select significant input variables. We believe that this superfluous procedure might lose the natural flexibility of ANN; hence, we utilized the sensitivity analysis to stress the presentation of possible results and ways to create the result in a process which involves uncertain factors by assigning different values for these factors. Through these comparisons, we can confirm that our ANN not only had appropriate design with adequate predictors but also had good performance to successfully predict osteoporosis in the elderly.

There are two limitations in this study. First, the number of female had larger ratio in our study population; this uneven distribution could lead to possible gender bias. Theoretically, ANN has its pliability to overcome this heterogeneity since no assumption of variable distribution is necessary in ANN modeling. However, osteoporosis in men is often neglected, despite the fact that one-third of hip fractures occur in men. We still need more male participants in the future study. Second, our study was carried out at a single institution. The generalization is foreseen to decline when applying any predictive tool to different populations as a result of possible confounders. Further studies in different centers for external validation should be designed to corroborate our findings and decrease potential interinstitutional variations.

5. Conclusion

Our results show that ANN could have a good performance in identifying osteoporotic subjects in the elderly and it might serve as an alternative tool to screen the individual who should arrange further work-up like DEXA. This kind of ANN approach could help clinicians to initiate other preventive actions in the future.

References

[1] Wilkins CH, Birge SJ. Prevention of osteoporotic fractures in the elderly. Am J Med 2005; 118: 1190-5.
[2] Chiu JS, Chong CF, Lin YF, Wu CC, Wang YF, Li YC. Applying an artificial neural network to predict total body water in hemodialysis patients. Am J Nephrol 2005; 25: 507-13.
[3] Hunter A, Kennedy L, Henry J, Ferguson I. Application of neural networks and sensitivity analysis to improved prediction of trauma survival. Comput Methods Programs Biomed 2000; 62: 11-9.
[4] Rae SA, Wang WJ, Partridge D. Artificial neural networks: a potential role in osteoporosis. J R Soc Med 1999; 92: 119-22.
[5] Ongphiphadhanakul B, Rajatanavin R, Chailurkit L, Piaseu N, Teerarungsikul K, Sirisriro R, Komindr S, Pauvilai G. Prediction of low bone mineral density in postmenopausal women by artificial neural network model compared to logistic regression model. J Med Assoc Thai 1997; 80: 508-15.
[6] Sadatsafavi M, Moayyeri A, Soltani A, Larijani B, Nouraie M, Akhondzadeh S. Artificial neural networks in prediction of bone density among post-menopausal women. J Endocrinol Invest 2005; 28: 425-31.

5.5 Evaluation of Decision Support Systems

Ubiquity: Technologies for Better Health in Aging Societies
A. Hasman et al. (Eds.)
IOS Press, 2006

Design and Evaluation of a Computer Reminder System to Improve Prescribing Behaviour of GPs

Jody D. MARTENS[1a], Alied van der AA [a], Bert PANIS [a]
Trudy van der WEIJDEN[a,b,c], Ron AG WINKENS[a,b], Johan L SEVERENS[d,e]

[a] *Integrated Care Unit, University Hospital Maastricht, The Netherlands*
[b] *Department of General Practice, Maastricht University, The Netherlands*
[c] *Centre for Quality of Care Research (WOK), University of Nijmegen and Maastricht, The Netherlands*
[d] *Clinical Epidemiology and Medical Technology Assessment, University Hospital Maastricht, The Netherlands*
[e] *Department of Policy and Organisation of Health Care (BEOZ), Maastricht University, The Netherlands*

Abstract
OBJECTIVE: To evaluate the implementation of a decision support system with reactive computer reminders to improve drug prescribing behaviour. METHODS: A clustered RCT with an incomplete block design was carried out in the south of the Netherlands: 25 GPs received reminders on antibiotics and asthma/COPD prescriptions, 28 GPs received reminders on cholesterol prescriptions. Prescribing guidelines were integrated into the GP information system, which was installed in the GPs practices of the intervention group. When the computer program was in use, a reminder popped up if the GP deviated from the guidelines during prescribing. Primary outcome: prescription according to the guidelines as a percentage of total prescriptions of a specific drug. Furthermore, an evaluation on the user-friendliness of the CRS in the GP's practice was carried out through questionnaires and interviews. RESULTS: Presently analyses are being carried out. Preliminary results indicate that the CRS study supported our expectations. In general, there seems to be a reduction in the numbers of prescriptions according to the advices of the computerised guidelines not to prescribe certain drugs. Final analysis will be performed shortly. In general, the Computer Reminder System was perceived as stable and user friendly. CONCLUSION: We created a stable and user friendly Computer Reminder System which was adjusted to the needs and demands of GPs. Preliminary results regarding the effectiveness of the system seem to indicate that the implementation of a Computer Reminder System with reactive reminders improves drug prescribing behaviour.
Keywords: primary health care, prescriptions, decision support, practice guidelines.

1. Introduction

Health care expenditure in the Netherlands increases each year, as it does in other

[1] Corresponding author: Jody D. Martens, Integrated Care Unit, University Hospital Maastricht, Postbox 5800, 6202 AZ Maastricht, The Netherlands. E-mail: jd.martens@caphri.unimaas.nl

European countries. Drug prescribing is an important contributor to this increase in costs, and these costs are expected to continue to increase significantly in the coming years [1,2]. Medication is not always prescribed effectively: drugs may be prescribed unnecessary, and lower-cost alternatives are not always being taken into consideration. This can be attributed to demographic factors, a trend towards new and more expensive drugs, the patients' increased awareness, and pharmaceutical advertising [3]. Furthermore, a high level of work-related stress and routines can result in an increased prescribing behaviour [4].

Clinical guidelines induce only small improvements [5]. Top-down dissemination of mono-disciplinary guidelines is hardly effective [6]. In previous research, we concluded that disseminating multidisciplinary guidelines that were developed within a certain region, only had a minimal impact on prescribing behaviour, even though GPs and specialists were involved more intensively in their development [7]. The effects of audits and feedback are small to moderate [6]. Depending on acceptance by GPs, reminders proved to be more effective in influencing the GP's behaviour in medication management [8]. It is known that computer reminders will be maximally effective in influencing prescribing behaviour when they are presented during the process of decision-making [9]. We hypothesised that this strategy could lead to a substantial change in prescribing behaviour. Furthermore we were interested in the opinion and satisfaction of the GPs with the CRS and its specific components.

2. Material and Methods

2.1 Overall design

A Computer Reminder System (CRS) with reactive reminders was developed on top of an existing GP information system in which multidisciplinary paper guidelines on prescribing were translated into computer reminders. These guidelines, which were developed by a regional committee of opinion leaders, were multidisciplinary in character and were based on the prevailing Evidence Based Medicine. The topics were selected on basis of high prevalence of the health problems and included antibiotics, asthma/COPD and cholesterol-lowering drugs [7]. A clustered Randomised Clinical Trial (RCT) with an incomplete block design was executed in the Maastricht region and the region of South Eastern Limburg in the Netherlands. GPs were allocated to one of two groups; 25 GPs received the CRS with reminders on antibiotics and asthma/COPD medication only (block A), 28 GPs received reminders on cholesterol prescriptions (block B). Both groups served as control group for one another. All GPs were aware of the fact that they were participating in a trial, but they were blind for the fact that they only received a specific subset of all available reminders.

2.2 System Design CRS

Our Computer Reminder System (CRS) consisted of three parts: a Guideline Editor in which prescription-relevant guidelines were imported using a graphical user interface; a Guideline Knowledge Base containing these guidelines and a Decision Support System (DSS) that provided reactive support (i.e. reminders), based on the developed guidelines and patient data, stored in the GP Information System's database [10].

The Guideline Editor is a graphical knowledge acquisition tool, facilitating the development of prescription guidelines. These guidelines were implemented as IF-THEN rules and flowcharts, based on patient-specific data, leading immediately to a reminder in the computerised prescription module if the corresponding guideline was not followed. The CRS was implemented on top of the existing GP Information System (GIS). When a drug was being prescribed, the DSS was activated by the GIS, which provided the DSS with the required patient-specific information (e.g. age and sex) and the prescribed drug. For all patients with asthma/COPD, cholesterol related disease, or infections for which antibiotics were prescribed, an ICPC diagnose (International Classification op Primary Care) registration and diagnosis-related information was obliged. Based on the patient's characteristics and the above-mentioned information, the DSS checked whether any of the rules fired. If the corresponding advice was not followed, a reminder was given. The content of this reminder could vary: alternative type of drugs; other doses; alternative drug administrations; specific indications; other length of prescribing; reference to specialist or not to prescribe anything. The DSS also contained a research database in which data, related to each separate prescription, was stored. This database was used to analyse behavioural changes of the GP and consisted of the patient's age and sex, prescribed drugs, diagnosis-related information and the reminders that were given by the DSS.

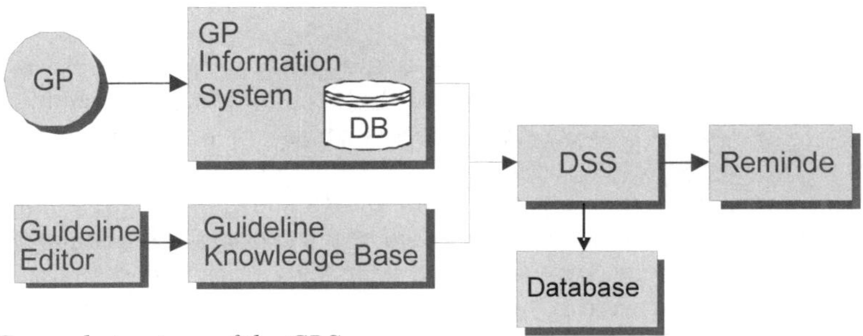

Figure 1: General structure of the CRS.

2.3 Design and validation of the computer reminders

Once the knowledge base was verified, it had to be validated. The first step in the validation process was reading the paper guidelines and making a list of possible situations in a GP's practice (e.g., 'prescribing ciprofloxacin to a 22 year old woman with cystitis').

The knowledge engineer, after having received training in how a GP will act when prescribing medication, tested the reminders together with an observer, using the following criteria:

1) Does the expected reminder appear?

2) Does only one reminder appear in each situation?
3) Is the reminder text correct with respect to content, grammar and layout?
4) Do the right screens appear?

On the basis of these test procedures, first all the non-responding reminders were listed. If these situations did not occur as a result of an erroneous knowledge base, these situations were reported to the involved programmers. Secondly, during the test procedures the guideline developer checked whether the reminders gave the same amount of relevant information compared to the information stated by the paper guidelines. Finally, two opinion leaders with expertise in computerisation were asked to exhaustively test the CRS during a month. Their comments were used to improve the functioning of the CRS and make it more user-friendly.

2.4 Outcome measures

During a 12-month period, data concerning number, type and dose of the prescribed drugs, diagnosis or complaints, age and sex were collected. The trial's central measure was: prescriptions according to the guidelines as a percentage of total prescriptions of that drug in both blocks (e.g. difference in prescribing between the blocks or the amount of reminders per GP, per year, per total of patient contacts). Furthermore, a system evaluation on the user-friendliness of the CRS in the GP's practice was carried out through questionnaires and interviews.

2.5 Data collection and analysis

The GPs were included between October 2003 and April 2004. During a 12-month period, prescription data were collected. Because of seasonal influences and the comparability of the data, we chose to select data of one year after the start of the intervention.

Around halfway the intervention year, a written questionnaire was sent to the GPs asking them about their experiences with and opinion on the feasibility of working with the CRS. Qualitative data were gathered among a purposeful selected sample of the GPs. The GPs were selected in such a way that there was a representative mix of GPs who were performing very well with the CRS and GPs who were not performing at all with the CRS (drop outs). A student in health sciences was instructed to visit the GPs and perform in-depth semi-structured interviews. Each interview was audiotaped, and the GPs' opinions were extracted from the text analyses performed by two of the authors.

3. Results

Effect evaluation

53 out of 77 GPs agreed to participate. 5 GPs dropped out of the study because of technical problems. During analysis, we received complete datasets of 48 GPs for

analysis. There were no significant differences in age and gender of the GPs between block A and block B.

First results indicate that the CRS study at least partly supported our expectations. Overall, the first results show that computerised reminders influenced the prescribing behaviour of GPs. Generally, there was a decrease in prescribing volumes according to the advices of the computerised guidelines. A more detailed description of the results regarding the effect evaluation can be found elsewhere [11].

Process evaluation

Initially, some delay in the development and implementation of the CRS occurred due to the fact that communication about several technical problems within the CRS and its technical environment was complex. Difficulties in communication and cooperation between the technical partners have impaired the developmental process of the CRS.

Because of technical problems, the disseminating of the CRS was delayed and the GPs had to wait longer for installation of the CRS. This caused dissatisfaction amongst GPs. Regarding the content and functionality of the reminder system, GPs valued a number of characteristics of the CRS positively. They valued the multidisciplinary guidelines that were being used as a basis for the reminders. The content of the advices in these guidelines were widely accepted because of the proper developmental process in which key persons had been participating. Involvement of GPs was also created during a pilot phase of the CRS. Two opinion leaders with some expertise in computerization were asked to comprehensively test the CRS during a month. Their comments were used to make adaptations in the CRS.

Apart from that, the CRS only interrupted the GPs during prescribing if they deviated from the guideline recommendations. In general, GPs appreciated the frequency in which the reminders occurred, although the reminders were ignored so now and then. Reasons for ignoring reminders were a lack of time to read the reminder, or the GP's opinion that the reminder was not suitable to the prescribing situation. Some reminders were valued as extensive and therefore caused disregarding of GPs. In general, GPs were willing to continue following the advices in the guidelines and to continue using the CRS. Using an ICPC code was not a main obstacle, since the GPs already used these codes for other purposes. The GPs liked to receive new reminders based on new guidelines, because they expected reminders to have a positive impact on their prescribing behaviour.

Serious attention was paid to supplying information about progress that was being made in improving the software and developing new guidelines and other pieces of information concerning the project. In general, GPs turned out to be favourable towards computerised advices on prescribing. Even after the completion of this study, GPs remained interested in using the CRS to support their prescribing process.

4. Near future

Depending on the effects of the trial, the CRS will be extended to all GPs in the region and, consequently, to the adhering hospitals. In these hospitals, the CRS will be implemented on top of the hospital's Computerised Drug Order Entry System. The next step will be to make the CRS nationally available by regional guidelines providers. The CRS can be used as a decision support system, but also as a management tool.

5. Discussion

Reminders might be a necessary strategy to change the GPs' behaviour [9]. A Computer Reminder System gives feedback to a GP as soon as his or her prescription behaviour differs from the behaviour recommended in the guidelines. As long as the GP does not deviate from the guideline, the system will not interfere with the GP's information system and the GP will not be interrupted by the Reminder System. We successfully developed and implemented a stable decision supporting system for prescribing of GPs. GPs valued the system positively; they valued it as user-friendly and said it adjusted to their needs and demands. Generally, it seems that GPs accepted the Computer Reminder System, which is a main condition for the effectiveness of reminder systems in general [12].

The preliminary conclusion of the effectiveness of our study, is that the implementation of a Computer Reminder System may improve drug prescribing behaviour and helps GPs to follow the guidelines. The clearest effects seem to be noticed for advices in the guidelines that discouraged prescribing certain drugs for specific diagnosis. These results are similar to other international studies about following different types of guidelines on prescription related drugs [13,14]. These results have to be confirmed by analysing all data. It is encouraging to see that computerised advices may cause effect on prescribing behaviour, since previous studies in the region about effects of disseminating paper guidelines only showed minimal effect, even though we used different ways to create involvement among GPs for this dissemination study [7].

A strong aspect of our study is that long-term effects were measured, including information about diagnosis and severity of the patient's disease. This kind of information is not often collected. Because of the disposal of these data, we were able to develop diagnosis related prescribing indicators and perform a precise and specific evaluation. Other strong aspects of the study are: a) we involved GPs in the development process of the CRS; b) the CRS interferes at the prescribing moment only. Recent literature shows the importance of connecting the CRS very close to the decision moment [12]; c) reactive reminders have been proven to be more effective because of a minimal disturbance of the prescribing process [8,9].

Based on the opinion of GPs, some general recommendations can be made concerning improvements of the CRS:

- To provide a summary of reminders at the day's end of the GP's consultation;
- To provide personal feedback on the prescription-behaviour of the GP;
- To create involvement, it is important to continue to pay serious attention to involving GPs. Involvement could be realised by involving GPs during the developmental process and by supplying general information about progress made and other piece of information concerning the project.
- To develop a clear project structure of tasks and responsibilities of the different partners in the developmental process.

6. Acknowledgments

We would like to thank both participating insurance companies, VGZ and CZ, for providing the data for this study.

7. References

[1] Stichting Farmaceutische Kengetallen; *Data en Feiten* (Foundation for Pharmaceutical Statistics; *Data and facts*). The Hague/the Netherlands; 2000-2004.

[2] Breekveldt-Postma NS, Zwart-van Rijkom JE, Egberts AC, Leufkens HG, Herings RM. Rising costs of drugs in hospitals in the period 1996-2000 and over the next few years (Article in Dutch). *Ned Tijdschr Geneeskd.* 2002 Dec 28;146(52):2547-51. Erratum in: Ned Tijdschr Geneeskd. 2003 Dec 27;247(52):2612. Ned Tijdschr Geneeskd. 2003 Jan 25;147(4):184.

[3] Dobson R. Pharmaceutical industry is main influence in GP prescribing. *BMJ* 2003; 326: 301.

[4] Evans J, Lambert T, Goldacre M. GP recruitment and retention: a qualitative analysis of doctors' comments about training for and working in general practice. *Occas Pap R Coll Gen Pract.* 2002 Feb;(83):iii-vi, 1-33.

[5] Grimshaw J, McAuley LM, Bero LA, Grilli R, Oxman AD, Ramsay C, et al. Systematic reviews of the effectiveness of quality improvement strategies and programmes. *Qual Saf Health Care.* 2003 Aug;12(4):298-303.

[6] Jamtvedt G, Young JM, Kristoffersen DT, Thomson O'Brien MA, Oxman AD. Audit and feedback: effects on professional practice and health care outcomes. *Cochrane Database Syst Rev.* 2003;(3):CD000259.

[7] Martens JD, Winkens RAG, Weijden T van der, Bruyn D de, Severens JL. Does a joint development and dissemination of multidisciplinary guidelines improve prescribing behaviour: a pre/post study with concurrent control group and a randomised subgroup. Revised Submission to BMC Health Services Research Dec 2005.

[8] Bennett JW, Glasziou PP. Computerised reminders and feedback in medication management: a systematic review of randomised controlled trials. *MJA* 2003;178:217-222.

[9] Hunt DL, Haynes RB, Hanna SE, Smith K. Effects of computer-based clinical decision support systems on physician performance and patient outcomes. *JAMA* 1998; 280: 1339-1346.

[10] de Clercq PA, Hasman A, Blom JA, Korsten HH. Design and implementation of a framework to support the development of clinical guidelines. *Int J Med Inform.* 2001 Dec;64(2-3):285-318.

[11] Martens JD, Weijden T van der, Severens JL, Clercq PA de, Bruyn D de, Kester AD, Winkens RAG. The effect of computer reminders on GPs'prescribing behaviour. *Submitted for publication.*

[12] Kawamoto K, Houlihan CA, Balas EA, Lobach DF. Improving clinical practice using clinical decision support systems: a systematic review of trials to identify features critical to success. *BMJ* 2005;330:765-8.

[13] Westert GP, Jabaaij L, Schellevis FG. Morbidity, performance and quality in primary care; Dutch general practice on stage. University of Tilburg, Bilthoven/Tilburg, The Netherlands; NIVEL (The Netherlands Institute for Health Services Research), Utrecht, The Netherlands; NIVEL, Utrecht, The Netherlands. In production, to be published in March 2006.

[14] Braspenning JCC, Schellevis FG, Grol RPTM. Kwaliteit in beeld. Medisch-technisch handelen huisartsen doorgelicht. *Medisch Contact.* 2004.

Ubiquity: Technologies for Better Health in Aging Societies
A. Hasman et al. (Eds.)
IOS Press, 2006

Clinicians' Perceived Usefulness of a Support System for Patient-centered Cancer Care

Cornelia M. RULAND, RN, PhD
Center for Shared Decision Making & Nursing Research, Rikshospitalet-Radiumhospitalet Medical Center, Oslo, Norway

Abstract. We investigated the perceived usefulness by 65 nurses and 12 physicians who had used CHOICE, a support system designed to improve patient-centred symptom management for cancer patients at the point of care. Two questionnaires addressed the following aspects: clinicians' usage patterns; ease of use; system ability to improve care planning, understanding of patients' perspectives, and patient-provider communication; attitudes towards patients' involvement in decision making about patient care; and perceived usefulness, defined as a system's ability to enhance work performance. The overall survey response rate was 78%. Clinicians reported that they had used information outputs provided by the CHOICE system on average 50% of the time, but nurses used them significantly more than physicians. The system received high ratings on all aspects of usefulness by both groups, but again, nurses provided consistently higher usefulness ratings than physicians did. There was a strong, significant correlation between patterns of use and perceived usefulness. There were no correlations between perceived usefulness and respondents' age, gender and clinical experience. Results confirm that the CHOICE system can successfully assist nurses and physicians to improve patient care for cancer patients in ongoing practice.
Keywords: User Evaluation, Computer-assisted Symptom Management, Perceived Usefulness.

1. Introduction

Support systems to assist clinicians in Shared Decision Making (SDM) and the inclusion of patients' illness experiences and preferences into symptom management are of recent date. One of these systems is CHOICE, (**C**reating better **H**ealth **O**utcomes by **I**mproving **C**ommunication about Patients' **E**xperiences), designed to assists nurses and physicians in SDM and patient-centred symptom management of cancer patients in interdisciplinary practice. Previous clinical trials on modules for cancer and rehabilitation patients have provided evidence of the systems' effectiveness to improve patient care and outcomes.[1-3] However, it is also important to evaluate the system's usefulness from the clinicians' point of view who are the actual users. To be truly helpful, Clinical Information Systems (CIS) need to be acceptable to clinicians and easily applicable in practice. Many information systems have failed because developers have neglected users' judgments about a system's feasibility, time requirements, and usefulness in clinical practice.[4;5] Clinicians work under time pressure and competing obligations, and additional tasks such as eliciting and including patients' illness experiences and preferences into patient care are not likely to be carried out if not perceived helpful. Evaluating the impact of CIS requires not only an understanding of computer technology but also of the social, behavioral and organizational processes that are affected by the introduction of such systems into existing practices. User resistance, misuse, and rejection of information systems in health care have been frequently reported. [4;5] Therefore, an evaluation of clinicians' perceived usefulness of the CHOICE system was crucial to understand its' strengths and weaknesses as a clinical support system at the point of care.

2. Materials and Methods

2.1. Theoretical Framework

Introducing support systems such as CHOICE into a clinical environment can be viewed as an innovation. How willing users are to adopt the innovation and adjust to the changes resulting from it depends on a number of factors, e.g. how well it fits into existing work processes, is compatible with professional values, able to achieve the goals for which it was developed, and ease of use. According to Roger's framework of diffusion of innovations[6] which was used to guide this evaluation study, the five most frequently studied attributes perceived by potential users to affect the rate of adoption of an innovation are: (1) relative advantage, (2) compatibility with existing values and experiences, (3) complexity, (4) ability to be tested, and (5) visibility of the results. In this evaluation study, these attributes were operationalized as follows: clinicians' patterns of use; ease of use; system ability to improve care planning, understanding of patients' perspectives, and patient-provider communication; attitudes towards patients' involvement in decision making about patient care; and perceived usefulness, defined as the system's ability to enhance work performance. These dimensions are consistent with the expected benefits of the CHOICE system that had been defined prior to its development, and that were measured in this study with the Appraisal of Usefulness Scale (AUS)[7] and Perceived Usefulness Scale (PUS)[9] described in more detail below. Figure 1 displays the theoretical framework with its key concepts, variables and instruments in this study.

Figure 1. Evaluation framework: Concepts, Variables and Instruments

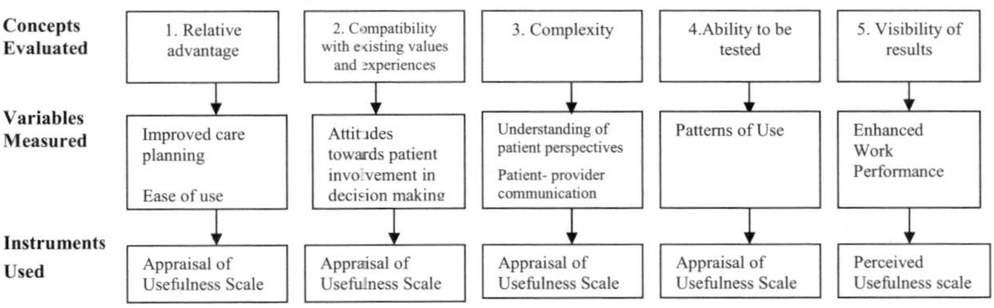

2.2. The CHOICE System

CHOICE includes (1) a comprehensive patient assessment tool for cancer-specific symptoms, functional problems and preferences along physical, functional, emotional and psychosocial dimensions; and (2) a SDM/ Care Planning component that highlights in an easy-to-use format which symptoms patients are experiencing, including their degree of bother and priority for care to patients. The evaluation study described here focused particularly on the perceived usefulness of this SDM/ Care Planning Component.

The application is contained and administered on a touch-pad, tablet computer. When using CHOICE patients are walked successively through a series of screens where they select their answers with a touch of a pen. First, patients are presented with a list of 19 problem categories from which they select those that apply. This triggers a more detailed list of symptoms under the selected categories from which the patient again selects those that apply. Patients are then asked to select the degree of bother of their selected symptoms. Finally, patients rate the importance of their problems as priorities for treatment and care (patient preferences). After the patient is done, the assessment summary can be printed and added to the patient's chart to be available to nurses and physicians for subsequent care planning. Each summary reflects the patient's individually selected problems and symptoms with their specifications of bother and priority. Figure 2 displays an example of an assessment summary.

Figure 2 : Example of Assessment summary

Assessment Summary Mar 14 2005

Patient ID

The **left column** describes the health problems experienced by the patient, the patient's, priorities for treatment /care and his or her expectations for the next week. The patient's problems are rank-ordered according to his or her priorities for treatment / care, from 0 (not important) to 10 (very important). The **middle column** displays the patient's specific symptoms associated with each selected health problem and their bothersomeness to the patient. The **right column** can be used for comments.

Please note that the assessment summary comes in addition to and does not substitute any parts of the patient's charts.

2005-03-14 11:21:16 998 TO

Problems With	Bothersomeness of Symptoms		Comments
Discomfort Priority: **9.5** Expectation: **Same**	Very	Joint ache / stiffness	
	Very	Dizziness/lightheadedness	
Mood / Feelings Priority: **8.0** Expectation: **Worse**	Very	Irritable/Angry	
	Very	Sad	
	Very	Mood Swings	
Worries / Concerns Priority: **8.0** Expectation: **Worse**	Extremely	Being away from my children	
	Extremely	About dying	
	Very	About side effects of treatment	
	Moderately	That my treatment preferences will not be respected	
	Moderately	Financial issues	
Energy Priority: **5.0** Expectation: **Same**	Very	Lack of energy	
	Very	Fatigued	

Evaluations of clinicians' perceived usefulness of assessment summaries for care planning in this study was part of a larger randomized clinical trial (RCT) to test the effect of the CHOICE intervention on patient care and outcomes. Assessment summaries were for the experimental group added to patients' charts to be available for nurses and physicians for care planning. A total of 338 assessment summaries for 73 patients were placed into patients' charts during the study period to assist nurses and physicians in patient-centered care planning. The clinician survey was distributed at the end of patient recruitment into the RCT, and after the CHOICE intervention had been in use for about one year.

2.3. Survey Sample and Procedures

This study took place in three hospital units specialized in the treatment of hematological cancer at Rikshospitalet-Radiumhospitalet in Oslo, Norway. IRB approval for the survey had been obtained as part of the RCT. Lists of names of eligible respondents were obtained from the units' head nurses and physicians. All nurses and physicians employed for at least three months prior to the survey received questionnaires in their mailbox, along with a self-addressed return envelope for in-house mail. The accompanying cover letter stated the purpose, explained the survey and asked for participation. Returning the survey was considered consent to participate. Applying the Dillman method[10] to increase response rates, two follow-up reminder letters were distributed electronically by e-mail at weeks 2 and 4 after the original mailings. The staff was also reminded about the survey by their head nurses / physicians at staff meetings.

2.4. Questionnaires

The questionnaire consisted of three sections: (1) demographic questions, (2) the Appraisal of Usefulness Scale (AUS);[7] and (3) the Perceived Usefulness Scale (PUS).[8] The 24 -item AUS is a measure of clinicians' appraisal of the usefulness of information in terms of assessment summaries produced by the CHOICE system when added to patients' charts to support care planning. The AUS has demonstrated good reliability in previous studies.[8] It was developed and tested in conjunction with two earlier clinical trials because no other evaluation instruments were available in the literature that tapped into the specific dimensions of the CHOICE intervention. To understand the usefulness of a system, it is essential to address the degree to which it achieves the specified benefits for which it was designed for. The AUS integrates attributes from the adoption of innovation literature (see Figure 1) as described above along 6 subcales: patterns of use; ease of use; system ability to improve care planning, understanding of patients' perspectives, and patient-provider communication; and attitudes towards patients' involvement in decision making about patient care. Items are scored from "strongly disagree" (1) to "strongly agree" (5), except for the "usage pattern" subscale that is scored from "never" (1) to"always" (5). Higher scores (after reversal of negatively worded items) indicate higher positive appraisal.

In addition, the 6-item PUS was used, a more generic measure of users' perceived usefulness of a CIS that has been used and found valid and reliable across a range of information systems.[8] The instrument has a strong theoretical foundation, based largely on self-efficacy theory, and perceived usefulness is defined as the degree to which a person believes that using a particular technique would enhance his or her job performance. Items are scored from strongly disagree (1) to strongly agree (5). Higher scores reflect greater perceived usefulness.

3. Results

Of the 99 distributed questionnaires 65 nurses and 12 physicians returned their answers, yielding a response rate of 79% for nurses, and 72% for physicians, or a 78% response rate overall. Reasons for not responding are unknown, but many of the non-responders were part time / night shift workers who had been less exposed to the CHOICE intervention. Respondents mean age was 36 years (SD = 10.7). Seventy-nine percent were women. Mean years of clinical experience was 8.3 (SD = 8.9), and respondents had on average worked 5.1 (SD = 6.7) years on the unit at the time of the survey. Seventy-two percent worked full time. Descriptive statistics were used to display responses to questions on the AUS. Total scores for all items were computed, and separate scores for the 6 subscales. Table 1 shows that mean scores were higher than the mid-point of possible ranges for almost all subscales, indicating that respondents appraised the CHOICE intervention positively or very positively. Also, clinicians had positive attitudes towards patients' involvement in decision making about their care. Clinicians' responses to questions about usage pattern indicated that they actively used assessment summaries for about 50% of the time. Except for the attitudes subscale in the AUS, the other subscales, total scale, as well as the PUS achieved excellent reliabilities.

Table 1: Means and Standard Deviations for AUS and Perceived Usefulness Scales (N=77)

Dimensions AUS	M	SD	Possible Range	Actual Range	Cronb. alpha
Usage patterns	12.6	4.6	5-25	5-21	0.91
Ease of Use	16.1	3.2	4-20	6-20	0.83
Improve care planning	15.7	3.2	4-20	4-20	0.92
Understanding patients' perspectives	15.9	2.6	4-20	7-20	0.78
Improving patient-provider communication	12.0	2.3	3-15	3-15	0.83
Attitudes towards patient involvement	16.2	2.7	4-20	11-20	0.62
Total appraisal score	88.7	14.3	24-120	43-110	0.94
Perceived Usefulness Scale (PUS)	22.5	5.0	6-30	7-30	0.94

Independent sample t-tests showed significant differences between nurses and physicians on several dimensions of the AUS scale and the PUS. Usage patterns indicated that nurses used assessment summaries significantly more frequently for care planning than physicians did. While the CHOICE intervention received high ratings on all other aspects of usefulness by both groups, nurses provided significantly higher usefulness ratings than physicians on system usefulness to improve care planning, understanding patients' perspectives, and total scores for both scales. There was also a strong correlation between usage pattern and PUS (r= 0.51, p<0.001), but there were no significant correlations between the AUS or PUS scores and respondents' age, gender or clinical experience.

Table 2: Differences between Nurses and Physicians

AUS	Nurses (n= 63)		Physicians (n= 12)		t
	M	SD	M	SD	
Usage patterns	13.1	4.4	9.9	4.7	2.3*
Ease of Use	16.4	3.3	14.8	2.7	1.6
Improve care planning	16.1	3.3	13.6	1.9	2.6*
Understanding patients' perspectives	16.2	2.6	14.3	2.2	2.4*
Improving patient- provider communication	12.2	2.4	11.1	2.1	1.6
Attitudes towards patient involvement	16.5	2.7	15.0	2.6	1.8
Total appraisal score	74.0	12.6	63.9	12.4	2.5*
Perceived Usefulness Scale	23.2	4.8	19.4	5.0	2.6*

* p=<0.05

4. Discussion and Conclusion

In this study we were particularly interested in getting feedback from nurses and physicians who had been exposed to a system to support them in patient-centered symptom management for cancer patients in interdisciplinary care. There is a particular need to understand the usefulness of devices to support the inclusion of patients' illness experiences and preferences and the core features that make them useful in clinical practice.

A strength of this study was that respondents based their answers on real life experiences, not just system descriptions, as has often been the case in other opinion surveys about usefulness.[10;11] There was substantial support for the usefulness of the CHOICE intervention in clinical practice. Both physicians and nurses assigned positive appraisal scores and usefulness ratings. It was however, interesting to note that nurses provided consistently higher ratings than physicians. There may be several reasons for this. Much of nursing care is directed toward symptom relief and illness management. Also, nursing frameworks have always emphasized the importance of including patients' perspectives, values, and preferences into patient care. Nurses may therefore, more easily adopt the concept of illness experiences and patient preferences. While such systems seem useful for both nurses and physicians, nurses may have particularly much to gain from systems that help them improve symptom management of chronically or seriously ill patients such as those with cancer.

In summary, results from this study confirmed that the CHOICE system can successfully assist nurses and physicians in ongoing practice to improve patient-centred symptom management of cancer patients. Information gained from this survey provides important guidance for further development and research. Given clinicians' positive evaluations, it seems promising to develop and test similar systems to support patient-centered illness management as a means to improve care not only for cancer patients, but also for a wider range of chronically or seriously ill patients.

Acknowledgements: This study was supported by the Norwegian National Research Council Grant# 154739/320. The author would like to thank members of the CHOICE research team for their assistance in conducting this survey: Glenys Hamilton, Harald Holte, Jørn Kristiansen & Heidi Sandbæk.

References

1. Ruland CM. Decision support for patient preference-based care planning: effects on nursing care and patient outcomes. *JAMIA.* 1999;6(4):304-12.

2. Ruland CM. Handheld Technology to Improve Patient Care: Evaluating a Support System for Preference-based Care Planning at the Bedside. *JAMIA.* 2002;9(2):192-201.

3.Ruland CM, White T, Stevens M, Fanciullo G, Khilani SM. Effects of a computerized system to support shared decision making in symptom management of cancer patients: preliminary results. JAMIA 2003; 10(6):573-579.

4. Anderson J, G., Aydin C, Jay SJ, editors. *Evaluating health care information systems. Methods and applications*. Thousand Oaks: SAGE publ., 1994.

5. Aronsky, D., Chan, K. J., & Haug, P. J. Evaluation of a Computerized Diagnostic Decision Support System for Patients with Pneumonia: Study Design Considerations. (2001). *JAMIA 8,* 473-485.

6. Rogers, E. M. (1983). *Diffusion of innovations.* (3rd ed.) New York, NY: The Free Press.

7. Ruland, C. (1998) Integrating patient preferences for self-care capability in nursing care: Effects on nurses' care priorities and patient outcomes. Doctoral dissertation. Case Western Reserve University, Cleveland, Ohio, USA. 198 p.

8. Davis, F. D. (1989). Perceived usefulness, perceived ease of use and user acceptance of information technology. MIS Quarterly 319-340.

9. Dillman DA. *Mail and telephone surveys: The total design method.* Toronto: John Wiley & Sons, Inc., 1978.

10. Ruland CM. A survey about the usefulness of computerized systems to support illness management in clinical practice. Int J Med Inform. 2004, Nov, 73; 797-805.
11. Dillon, T.W., McDowell, D., Saliman, F, Conklin, D. (1998). Perceived Ease of Use and Usefulness of Bedside computer systems. Computers in Nursing 16 (3), 151-56.

Evaluation of a Breast Cancer Computer Aided Diagnosis System

Antonis FRIGAS[a,b,1], Smaragdo KAPSIMALAKOU[a,c,1], George SPYROU[a,d,2], Kostantinos KOUFOPOULOS[e], Stamatios VASSILAROS[f], Aikaterini CHATZIMICHAEL[c], John MANTAS[b] and Panos LIGOMENIDES[a]

[a] *Academy of Athens, Informatics Laboratory*
[b] *National and Kapodistrian University of Athens, Nursing Faculty, Health Inf. Lab.*
[c] *National and Kapodistrian University of Athens, Medical School*
[d] *Academy of Athens Foundation for Medical and Biological Research*
[e] *Gynaikologiki Apeikonisi Diagnostic Center*
[f] *Prolipsis Diagnostic Center*

Abstract. Mammography is accepted as the most effective method to detect breast cancer. However, interpreting a mammogram is not easy for not experienced radiologists. The aim of computer aided detection techniques in breast cancer is to improve the chance that a malignant region is detected and appropriately evaluated. Breast microcalcifications have been considered as a very useful index of malignancy, which helps in the early detection of breast cancer. A system of computer aided diagnosis has been developed that is based on detailed analysis and evaluation of related features of individual microcalcifications and of formed clusters helping the doctor to make risk estimation for each microcalcification cluster as well as for isolated microcalcifications. This information is considered to be very useful to radiologists, giving them extra input before making their estimation of each case. The aforementioned system has been thoroughly tested using a number of real life cases provided from collaborating doctors. Each case, apart from the mammograms, was accompanied by a biopsy test result, the patient's demographic data and medical history. A total of 200 cases (147 benign and 53 malignant) have been examined and the results are presented as the Receiver Operating Characteristic (ROC) performance and are quantified using the ROC curve. The system is showing high levels of sensitivity identifying correctly all malignant cases.
Keywords: computer aided diagnosis, breast cancer, microcalcifications

1. Introduction

Breast cancer is the most common cancer among women [1]. In the United States 1 out of every 9 women will develop breast cancer sometime in their lifetime while in Europe this percentage is slightly lower [1,2]. Mammography is accepted as the most effective method to detect breast cancer. However, interpreting a mammogram is not easy for not experienced radiologists [3,4]. The aim of computer aided detection

---

[1] These authors contributed equally.

[2] Corresponding Author: George Spyrou, Academy of Athens Foundation for Medical and Biological Research, Soranou Efessiou 4, 115 27, Athens, Greece; e-mail: gspyrou@bioacademy.gr

techniques in breast cancer is to improve the chance that a malignant region is detected and appropriately evaluated.

Radiologists consider breast microcalcifications a very useful index of malignancy, which helps in the early detection of breast cancer [5,6]. Breast microcalcifications are viewed as bright spots with diameter varying from 0.1 to 0.3 mm. Microcalcifications usually appear in the form of clusters and can be detected on mammographic films due to their high clustering density [7-9]. However, the existence of microcalcifications in breast tissue is not always a clear evidence of malignancy. Long-standing research efforts have been made to classify breast microcalcifications as benign or malignant, based on computer-aided analysis of their structural and photometric characteristics, appearing in mammographic images [10,11].

A system, called Hippocrates-mst, has been developed and is based on detailed analysis and evaluation of related features of individual microcalcifications and of formed clusters [12-14]. This kind of information is considered to be very useful to the radiologists, giving them extra input before making their estimation of each case.

2. Method

2.1. Computer Aided Diagnosis System

Microcalcification detection and analysis is carried out with the use of an algorithm based on processing of three image data sets: a) the application of a high pass filter b) the variance normalization c) the application of an adaptive filter.

The proposed system is using a rule based decision tree in order to classify each microcalcification and categorize it according to its risk using a seven characteristics feature space for isolated microcalcifications and the LeGal microcalcification cluster categorisation [8,9] regarding the selected cluster.

The proposed methodology for microcalcification classification uses a four-way projection method in order to extract seven microcalcification characteristics including size, circularity, existence of sub dense center, brightness level, anomaly of shape, existence of apophyses and branches, existence of windings as well as those described in the Le Gal microcalcification cluster classification model. As shown in Figure 1, the algorithm examines each microcalcification cluster from four different viewing angles and is processing all possible variations of each view. The quantitative estimation of each microcalcification's characteristics is calculated in relevant normalized scales.

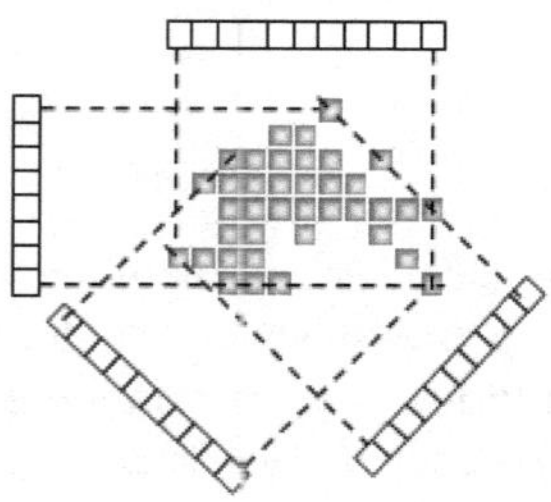

Figure 1. The four projections method

Each case is processed as follows: each mammogram has to be digitized using certain parameters (that are mentioned in detail in section 2.2), and then the user enters the patient's demographic data and medical history in the system's digital patient record. After those initial steps the user has the option to examine the mammogram(s) using the provided digital tools. These tools include facilities such as the Digital Lens, which helps users to zoom in an area of interest, or the brightness, contrast and γ-value adjustment tools that help the user reveal anomalies in a selected area of the mammogram. At the next step, the doctor can use the system's microcalcification detection algorithms and reveal the microcalcifications of a selected area (Figures 2 and 3).

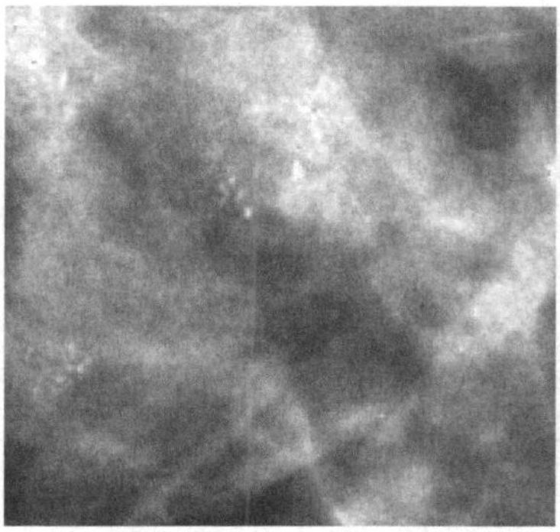

Figure 2. Unprocessed region of interest

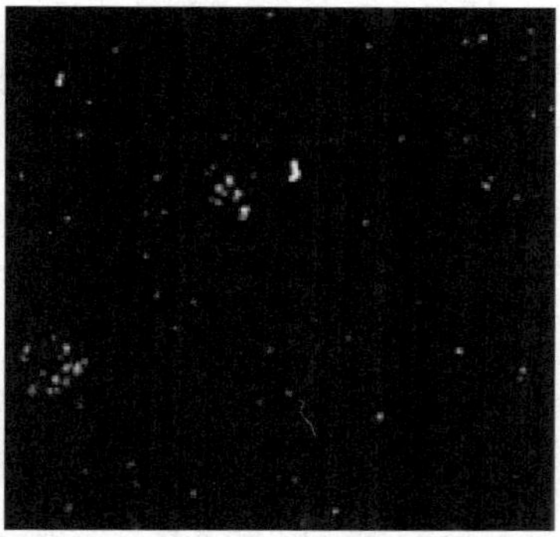

Figure 3. The region of Figure 1 after processing with microcalcification detection algorithms

The microcalcifications of this selected area are counted and categorized. Colour categorization is used for each microcalcification using a risk scale starting from 0 to 100% divided in steps of 20%. At the final step, the system provides the user with the overall risk estimation for the selected region of interest and gives a percentage that represents the risk whether this specific patient has or not breast cancer disease and whether the anomalies detected are benign or malignant. After reviewing a case, the doctor has the option to save the results, along with any notes on the case, to the system's hard disk drive or to a cd.

2.2. Evaluation Method

The sample images set contained 200 mammograms, of both craniocaudal and mediolateral views, with microcalcifications collected from April 1998 to November 2005. Each case was accompanied by a histopathological examination of resected speciment and with the patient's demographic data and medical history as well. The patients covered a variety of ages, from 37 to 73 years of age (in average 53.2 years). Indications for mammography could be routine check up, screening or follow-up.

The mammoraphic images obtained through digitization using Agfa Duoscan, with a resolution of 300 dots per inch (DPI), 12 bit greyscale colour depth and .tiff unzipped as the image format. The image size for each image was 2048X2048 while the size in megabytes was approximately 10 MB for each image. Each microcalification cluster that was relevant to the biopsy test results was annotated in the images from specialized doctors.

The results from the Hippocrates-mst system are presented in a risk percentage scale from 0 to 100. Percentage lesser than or equal to 35 means there are not enough evidence to support sending the patient for a biopsy test. Percentage from 35 to 55 declares benign state and a biopsy test is suggested, while percentage greater than 55 indicates malignancy and a biopsy test has to be made as soon as possible. In this point it must be stated that the system only makes suggestions to the doctor and it is the doctor that makes the final decision whether a case has to be sent for a biopsy test or not.

3. Results

We have separated the patients into two groups. Group A consists of 53 patients with malignant histopathological test results (biopsy: mal) and group B consists of 144 patients with benign histopathological test results (biopsy: ben). In addition, there were three cases of atypia. We have compared the two groups using independent t-test once for the age of the women as well as for the results of the CAD system's. No statistically significant differences were found between the two groups as far as the years of age concern.

The system's results in all malignant cases came above the 55%, which is the lower end of the threshold for malignancy. After calculating the mean of the CAD system's results, we found it to be 81.13 for the malignant cases of group A and 46.97 for the benign cases of group B. The range of the 95% confidence interval is calculated separately for each group as $\overline{X} \pm t_{n-1} SD/\sqrt{n}$, where $\overline{X}$ represents the mean value, t_{n-1} is the corresponding t value, SD is the standard deviation and n is the number of cases of each group. Thus, the 95% confidence interval for the malignant cases is 81.13±4.939, while the benign cases' 95% confidence interval is 46.97±3.842. Comparing those results to the biopsy test results shows that the system has a high sensitivity, classifying correctly all malignant cases; however it suffers from low specificity as it is later discussed. Nevertheless its specificity is much better compared to the doctors' decisions for biopsy referrals, as it referred to biopsy approximately 34% less patients from group B (benign cases) than the doctors did. An overview of the sample data set's descriptive statistics is given in table 1.

Table 1. CAD Results overview

	BIOPSY	N	Mean	Std. Deviation	Std. Error Mean	95% C.I
CAD Results (%)	Mal	53	81.13	17.89	2.45	76.190, 86.069
	Ben	144	46.97	23.29	1.94	43.128, 50.812

Mal: biopsy result malignant
Ben: biopsy result benign

Using the results from Table 1 for the malignant cases the Receiver Operating Characteristic (ROC) curve of figure 4 has been extracted. The area under the curve (A_z) is equal to 0.8732. The ROC curve obtained from the evaluation of benign cases has an area under the curve (A_z) equal to 0.1267. We have also created 3 categories with the CAD's results. Results with value 0-35% as category A (slope=1), 35-75% as category B/ grey zone (slope=2) and 75-100% as category C (slope=3). Patients with a result >35% were referred to biopsy.

Then we used one way ANOVA test and Bonferoni test for the group of 200 patients (malignant: 53, benign: 144 + 3 cases of atypia) with the following results: Category A and category B have no significant statistical difference (p>0.05). These results indicate that patients with a negative biopsy test were found either in category A or B.

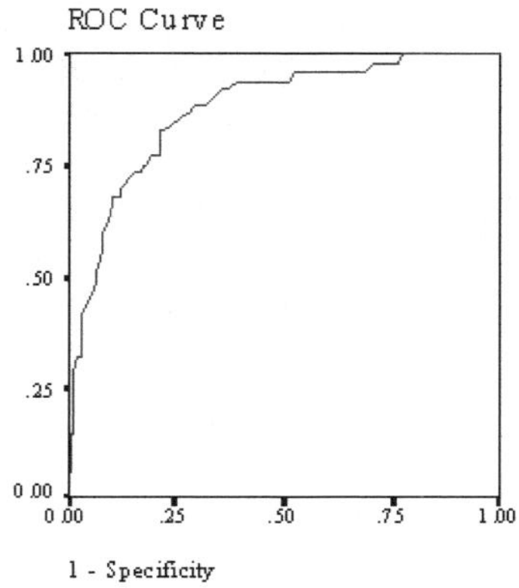

Figure 4. System's sensitivity in malignant cases

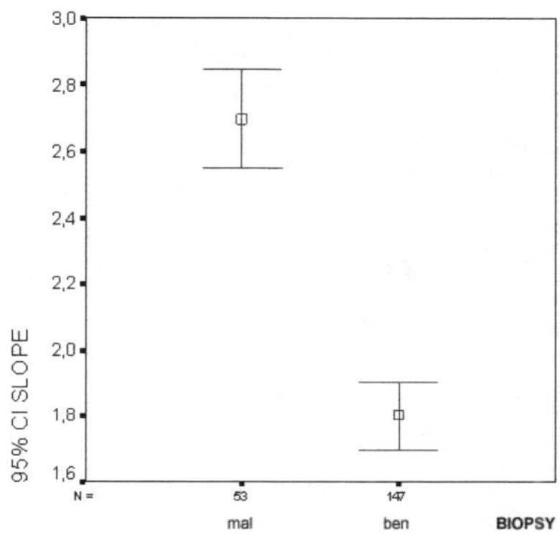

Figure 5. Graphical comparison between groups a, b and c.

Between category B and category C there is statistical difference (p<0.001), as well as between category A and category C. This means that patients with a positive biopsy test result have a result by the proposed system between 75-100%. An overview of the results is presented in table 2; and a graphical representation in figure 5.

Table 2. Comparison between groups a, b and c

CATEGORIES		mean±SE	P
A	**B**	0.09±0.063	0.45
	C	0.63±0.069	<0.001
B	**A**	-0,09±0.063	0.45
	C	0.54±0.06	<0.001
C	**A**	-0.63±0.069	<0.001
	B	-0.54±0.06	<0.001

4. Discussion

The proposed system is using a rule based decision tree, measuring seven characteristics for each microcalcification and applies the LeGal classification to the whole analyzed cluster. The laboratory evaluation results from cases supplied from collaborating doctors were encouraging enough and have shown that the system shows high sensitivity in malignant cases. The classification of all such cases was correct and no malignant cases were falsely classified as benign. It must be stated here that the correct categorisation of malignant cases was our primary goal during the proposed system's development.

In this specific set of mammograms Hippocrates-mst's results have indicated that 50 of the total number of benign cases were actually benign. If the doctors agreed to each benign indication made from the proposed system the number of unnecessary biopsies would be about 34% lower. It is reminded that all of the sample test's cases were referred to biopsy testing.

5. Conclusions

Hipppocrates-mst, a computer aided diagnosis system based on the quantification analysis of suspicious microcalcifications clusters, can help doctors to easily identify high risk microcalcifications by their specific characteristics can accurately be measured. Our future goals include the clinical use of Hippocrates-mst in radiology departments and breast clinics of Greek hospitals in Athens under a specific protocol in order to have the system tested in real-life situations with larger number of cases.

References

[1] World Health Organization, WHO Statistical Information System 2003.

[2] American Cancer Society, Cancer Facts and Figures 2004.

[3] Harvey JA, Fajardo LL, Innis CA. Previous mammograms on patients with impalpable breast carcinoma: retrospective vs. blind interpretation. AJR Am J Roentgenol 1993; 161:1167–1172.

[4] Birdwell RL, Ikeda DM, O'Shaughnessy KF, Sickles EA. Mammographic characteristics of 115 missed cancers later detected with screening mammography and the potential utility of computer-aided detection. Radiology 2001; 219:192–202.

[5] Homer MJ, Safaii H, Smith TJ, Marchant DJ. The relationship of mammographic microcalcification to histologic malignancy: radiologic-pathologic correlation. AJR Am J Roentgenol. 1989 Dec;153(6):1187-9.

[6] Stomper PC, Geradts J, Edge SB, Levine EG. Mammographic predictors of the presence and size of invasive carcinomas associated with malignant microcalcification lesions without a mass. AJR Am J Roentgenol. 2003 Dec;181(6):1679-84.

[7] Buchbinder, SS., Leichter, IS., Lederman, RB., Novak, B., Bamberger, PN., Coopersmith, H., Fields, SI. (2002). Can the size of microcalcifications predict malignancy of clusters at mammography? Acad Radiol.9(1):18-25.

[8] Le Gal, M., Chavanne, G., Pellier, D. (1984). Diagnostic value of clustered microcalcifications discovered by mammography (apropos of 227 cases with histological verification and without a palpable breast tumor). Bull Cancer;71(1):57-64.

[9] Le Gal, M., Durand, JC., Laurent, M., Pellier, D. (1976). Management following mammography revealing grouped microcalcifications without palpable umor. Nouv Presse Med.5(26):1623-7.

[10] Gavrielides, MA., Lo, JY., Floyd, CE. Jr. (2002). Parameter optimization of a computer-aided diagnosis scheme for the segmentation of microcalcification clusters in mammograms, Med Phys. 29(4):475-83.

[11] Lee S, Lo C, Wang C, Chung P, Chang C, Yang C, Hsu P. A computer-aided design mammography screening system for detection and classification of microcalcifications. Int J Med Inf. 2000 Oct;60(1):29-57.

[12] G Spyrou, M Nikolaou, M Koussaris, A Tsibanis, S Vassilaros and P Ligomenides, "A System for Computer Aided Early Diagnosis of Breast Cancer based on Microcalcifications Analysis", European Systems Science Journal (Res-Systemica) Vol. 2, Special Issue, December 2002.

[13] G Spyrou, K Koufopoulos, S Vassilaros and P Ligomenides, "Computer Aided Image Analysis and Classification Schemes for the Early Diagnosis of Breast Cancer", HERMIS International Journal of Computer mathematics and its Applications. Vol. 4. 2003, pp.175-181.

[14] Spyrou G, Nikolaou M, Koufopoulos K, Ligomenides P. A computer based model to assist in improving early diagnosis of breast cancer. 7th World Congress on Advances in Oncology and 5th International Symposium on Molecular Medicine; 2002 Oct 10-12; Crete, Greece.

5.6 Decision Support: Evaluation and Experiences

Ubiquity: Technologies for Better Health in Aging Societies
A. Hasman et al. (Eds.)
IOS Press, 2006

Comparison of the impact of cardiovascular guidelines on a working population

Pierre GILLOIS[1], Frédérique CLAUDOT[2], Gilles CHATELLIER[3], François KOHLER[1] and Marie Christine JAULENT[3]
[1] *Laboratoire SPIEAO, Faculté de Médecine Nancy I, France*
[2] *Service de médecine légale - droit de la santé, Faculté de Médecine Nancy I, France*
[3] *SPIM, Faculté de Médecine René Descarte ERM2002 Paris V, France*

Abstract The influence of guidelines is more and more important in the medicine using standardised knowledge as Clinical Practice Guideline (CPG). Objectives: 1) Determine the impact of different CPGs on a working population 2) Using the Framingham risk calculation as gold standard to check the CPG classifications. Methods This work is done in the context of two hypertension management CPGs published in 1997 and 2000, and one dyslipidemia CPG by the French agency HAS. A French regional study sample of 2817 patients was selected. The results of the classification system by CPG were compared to those given by the chosen gold standard : Framingham cardiovascular risk calculation. Results The HBP CPG concerns more of the quarter of the working population. The mean of the cardiovascular risk calculation for the patients with drug treatment is higher than for the patients with no drug treatment and for those not screened by the CPG. Some patients with a very high calculated risk, are not selected by the screening CPG based on too few or on not adequate variables. Conclusions For the two High Blood Pressure Management CPGs, the selected patients are not distributed the same way in the two CPGs. Observed agreement is poor between the two hypertension CPG versions. Screening people for specific health care should be based on risk calculation not on few variables. Feeding back our results to the CPG creator group should reduce the variability of recommendations.

1. Introduction

Clinical Practice Guidelines (CPGs) aim at: improving quality of care, reducing variation of medical practice among physicians and controlling costs. Guidelines can be implemented in Decision Support Systems to suggest treatment or diagnostic procedures to physicians through messages [1]. Moreover, it has been proved that computerized CPGs allow better accessibility to information and that output messages are more often respected [2]. A message contains the following informations: 1) the content of the message itself and 2) the grade of the recommendation defined according to the level of evidence thanks to the methodology of Evidence-based Medicine [3]. Two CPGs for the diagnosis and the management of hypertension in 1997 (V1 first version) and 2000 (V2 second version) and one for the dyslipidemia management have been published by the French Agency HAS (Haute Autorité en Santé) [4]. These CPGs have been computerized as a decision tree [5]. This approach allowed us to compare the decision obtained by two hypertension (first version and second version) and one dyslipidemia CPGs within this sample. With respect for the risk status, we compared the CPG classification to the gold standard classification provided by the Framingham cardiovascular risk calculation on working population.

2. Materials and Methods

A French sample of 2817 working people, was used to simulate the results obtained by different CPGs. This sample is a working population from a French regional agency. Three different CPGs come from the HAS, and are high blood pressure management first and second version, and the first version of dyslipidemia guideline.
First, the different contingency tables were compared as age, sex, weight, size… The CPGs were tested on our population sample. For each individual, the three different CPG classifications were stored along with its set of values. The information was "no CPG needed" if inclusion criteria were not met, and the results of the CPG in other cases.
Secondly, we compared the means of the cardiovascular risk calculation between CPG classes. There were seven ways inside the HBP CPG decision tree, which ended in two messages: "no drug therapy" (correction of lifestyle factors and monitoring) or "drug therapy needed". We analyzed by an ANOVA the seven final ways with the CV risk searching classes' effect and regrouping. We also focused on the HBP CPG and its reproducibility between first and second version calculating a Kappa coefficient. Finally, we focus on the high cardiovascular risk from the population in each CPG classes and those with more than 20 % of probability at 10 years. In this high risk selection, we also searched to explain a possible heterogeneity of CPG classifications.

3. Results

The anonymised French occupational medicine database from a governmental agency provided an one-year follow-up file with 2817 subjects. The different qualitative and quantitative values are synthesized in tables 1 and 2. This study is therefore population based.

Labels	Means	sd	Range
Weight (kg)	67	14.8	124
Size (cm)	165	8.5	50
Body Mass Index	24	4.8	40
Systolic Blood Pressure (mmHg)	129	17	127
Diastolic Blood Pressure (mmHg)	79	10	82
Serum Total Chol (g/l)	2	0.3	3
Serum HDL Chol (g/l)	0.62	0.2	2.1
Age (year)	40.5	9.9	47
Relative Cardiovascular Heart Disease Risk (%)	0.77	1.53	45.9
Blood sugar level (cg/l)	1.0	0.2	2.8

Tableau 1:Description of quantitative variables (N=2817)

All the means of these variables are near the reference means, e.g. the normal BMI between 18.5 and 24. The age ranges are consistent with the Fair Labor Standards Act.

Labels	**Number**	**Percentage**
Sex (% male)(2817)	676	24 %
Sugar > 1,27 g/l (2590)	181	7 %
Diabetic (2785)	30	1.1 %
High Chol Level (2780)	578	20.8 %
Oral Contraception (2153)	667	31 %
Drug for High Blood Pressure (2635)	134	5.1 %
Drug for High Blood Lipid (2662)	101	3.8 %
Smoker (2786)	704	25.3 %
Physic activity (2785)	1119	40.2 %

Tableau 2:Description of categorical variables

3.1. CPG outcomes

High Blood Pressure CPG Final Advice	**V1**	**V2**
Not concerned by HBP CPG	73%	
HBP + no drug therapy	19.5%	24.5%
HBP + drug therapy	7.5%	2.5%

Dyslipidemia CPG Final Advice	
Not concerned by High Lipidemia CPG	81%
High Lip + no drug	9%
High Lip + drug therapy	10%

Tableau 3:Distribution of classifications outcomes for HBP and dyslipidemia CPG (N=2817)

As the HBP screening (same inclusion criteria) didn't change between V1 and V2, the same patients were selected to be screened by both HBP CPG (table 3).

3.2. Cardiovascular risk by CPG classes

HBP final labels V1	**Risk means**	**sd**
Not concerned by HBP CPG	2.3	3.7
No drug therapy needed	6	6.5
Drug therapy needed part 1	10.5	9.7
Drug therapy needed part 2	18	8
HBP final labels V2		
Not concerned by HBP	2.3	3.7
HBP + no drug therapy	6.3	6.7
HBP + drug therapy	17.8	11.1
Dyslipidemia final labels		
Not concerned by High Lipidemia	2.2	3.5
High Lip + no drug	8.2	6.8
High Lip + drug therapy	11.9	8.9

Tableau 4:Means of CHD risk by V1 & V2 HBP and Dyslipidemia CPG classes

Using ANOVA models, we found heterogeneity between the seven ways synthesizing the final message (F ("Final seven Messages") = 125; d.f. (6-2810), p value < 0.001).

We therefore compare all of the messages paired. We expected three homogeneous classes, one for the no CPG message, another for the no drug therapy message and the final one with the drug therapy indication message, associated with a significant increase of the means of CV risk.
This special feature of two groups in one class in the HBP CPG is not found in the V2 version. The three V2 classes are intrinsically homogeneous e.g. Table 4. The dyslipidemia CPG displays the same heterogeneity between the seven ways as the V1 HBP CPG, but the merged classes are not so clear as the V2 version.

3.3. Concordance KAPPA

The same sample was used on two HBP CPGs. This table concerns only the 27% people taking care by the CPG.

Labels	V2 No Drug	V2 Drug Therapy
V1 No Drug	528	25
V1 Drug Therapy	166	45

Tableau 5: Contingency table for V1 and V2 HBP CPG (N=764), Kappa = 0.21

The table 5 displays that 75% of the HBP screened population sample are non-status changed from the V1 and V2 version. As well, the Kappa value was 0.21, which reflects poor agreement. We do not conclude that one is better than the other. To conclude that, we must consult the international Gold Standard: Cardiovascular risk calculation from Framingham.

3.4. High cardiovascular risk and CPG

We focus on the high cardiovascular risk calculation, for more than 20% of the sample. We hypothesize that these subjects are being taken care of as high risk patients by the CPG classification and will benefit of the drug therapy. In our sample, 65 subjects have more than 20% of risk (means = 28.5; sd= 9.7). We show the different distributions by the three CPGs in the table 6.

Version	**Labels**	**Num**
	Not concerned by HBP CPG	9
V1	HBP CPG No Drug Therapy	20
	HBP CPG Drug therapy	36
V2	HBP CPG No Drug Therapy	28
	HBP CPG Drug Therapy	28
V1	Not concerned by DysLipidemia CPG	8
	Dyslipidemia No Drug Therapy	15
	Dyslipidemia Drug Therapy	42

Tableau 6: Number of the high risk subject by the three CPG classes

The subjects are young (mean of age = 44 years), and predominantly female (three-quarter). There is no sex effect in these 65 subjects ($F(1;63) = 0.23$; $p = 0.63$). There is a ten-year age effect ($F(3.63) = 9.7$; $p < 0.001$). By comparing two by two the ten-year age of classes, some of them can be associated. Finally two classes persist, one

with the thirty classes and other classes of age higher than 40 years. The means of risk are 53 (sd= 25.7) and 27,5 (sd= 7) respectively for the thirty and the other classes.
For all classes from the three CPGs, we perform a search of heterogeneity inside the classes. In all the final classes, the only effect we found was the sex effect for the no screening HBP CPG.

4. Discussion

This sample of working population, mainly female, could be considered as a picture of the French population. The mean and the distribution of the measured variables show us that this is not a sample of perfectly healthy population, it is rather near of the high threshold of some variables (e.g. Mean of BMI = 24). Applying CPG shows us the real impact of the CPG message. All the variables available from our datas are not useful for each CPG. But, it shows other way to improve the global health of studied population (e.g stop tobacco, more sport…). Currently the patient selection by CPG is usually based on one, or few combined variables [6]. However the healthcare must be global for a medical field, and it must interact with all risk factors of the field. Cardiovascular risk factors are not independent of the other medical risk factors. A CPG should select a high risk patients according to a global risk calculation (using several variables), and not simply classify them, using those variables.
The screening of target patient should be based on models of prediction like Framingham equation or Laurier French one. The global risk calculation and the construction of medical field CPG (e.g. high blood pressure plus dyslipidemia… for a cardiovascular risk CPG) avoid for example to treat a patient for a specific and sharp aspect, and allows to have a "global view". The CPG classification gives a homogeneous cardiovascular risk for each of its classes. A progressive risk between the various classes is observed, between the class not concerned by the CPG (no treatment) and the others (no drug and drug treatment). The same logical progression is observed in both other classifications of other CPGs (no drug and drug treatment). The persistence of subject badly classified with high risk, indicates that the mode of subjects selection must be of primary importance, and that the strategy of selection must be improved.
This work does not show up a lack of validity. The skew of selection is avoided by using the same population for the three CPG. The construction of the set of data (working population) can be seen as a retrospective data. As our three different CPGs do not use the same set of variables, for computing, it is not a problem for characterization of the variables. The skew of information is thus avoided.
The main message of the healthcare following the CPG, is about the cost of therapy. This message concerns from 2.5 to 10% of the working population (according to the CPG). The two HBP CPGs result from two different working groups.
The knowledge used to built each CPG is also different. For screened people by the HBP CPG, three quarters of them obtain the same classification (no drug therapy vs drug therapy). The improvement of the estimation of the individual risk allows to reduce the dissimilarity.
The main message of the HBP and dyslipidemia CPG is that no drug therapy increases the quality of life by a less expensive way.
Currently, patients management is highly variable, which is due to the complexity of medical processes and the quickly changing medical knowledge.

Computerized Clinical Practice Guidelines (CPGs) improve quality of care by assisting physicians in their decision-making process [8]. So, all patients do not have appropriate care. CPGs will improve the quality of care by harmonizing patients management. At the same time, an effort to allow general practitioners access to the CPGs must be done.
Also, CPGs must be evaluated and compared to each other. For that, we must adapt them to be used on computer, and develop CPGs Knowledge Based System. The heterogeneity of the classifications indicates the points of weakness, which should be corrected in the second time. The prospective study (e.g. Framingham) provides objective criteria of analysis. Computerization of CPGs is only one of possible solutions and, as any medical decision-making systems, must be evaluated [5]. This evaluation concerns the impact on both: the processes and the outcomes of care. CPGs based system should be evaluated according to the same methods. The CPGs still need improvements. The conditions of appraisal are not real conditions, but the appraisal on other databases would show up some applicability and consequence aspect of the CPG. This demarche could be applied to other CPGs from different countries and also to different medical fields.

5. References

[1] Duff L, Casey A. Implementing Clinical Guidelines: How can informatics help? J Am Med Inform Assoc 1998;5:225-226.

[2] Gardner R, Lundscaarde H. Evaluation of user acceptance of a clinical expert system. J Am Med Inform Assoc 1994;1:428-438.

[3] Evidence-based medicine. A new approach to teaching the practice of medicine. Evidence-Based Medicine Working Group. JAMA 1992;268(17):2420-5.

[4] Diagnostic et Traitement de l'Hypertension Artérielle Essentielle de l'Adulte de 20 à 80 ans. HAS - ANAES/Service des Références Médicales/Septembre 1997; 241-277.

[5] Gillois P, Chatellier G, Jaulent MC, Colombet I, Fieschi M, Degoulet P. From paper based to electronic guideline: Application to French guidelines. Medinfo 2001

[6] Colombet I, Aguirre-Junco AR, Zunino S, Jaulent MC, Leneveut L, Chatellier G. Electronic implementation of guidelines in the EsPeR system: a knowledge specification method. Int J Med Inform. 2005 Aug;74(7-8):597-604.

[7] Marshall T. Evaluating national guidelines for prevention of cardiovascular disease in primary care. J Eval Clin Pract. 2005 Oct;11(5):452-61.

[8] Hayward RS, Wilson MC, Tunis SR, Bass EB, Guyatt G. Users' guides to the medical literature. VIII. How to use clinical practice guidelines. A. Are the recommendations valid? The Evidence- Based Medicine Working Group. JAMA 1995;274(7):570-4.

Ubiquity: Technologies for Better Health in Aging Societies
A. Hasman et al. (Eds.)
IOS Press, 2006

IT support for clinical pathways – lessons learned

R. Lenz[a,1], R. Blaser[a], M. Beyer[a], O. Heger[a], C. Biber[c], M. Bäumlein[c], M. Schnabel[c]

[a]*Institute of Medical Informatics, Philipps-University Marburg, Germany*
[c]*Department of Trauma, Reconstructive and Hand Surgery, University Hospital Giessen and Marburg, Location Marburg, Germany*

Abstract. Clinical pathways are an effective instrument to decrease undesired practice variability and improve clinician performance. IT applications embedded into clinical routine work can help to increase pathway compliance. Successfully implementing such applications requires both a responsive IT infrastructure and a participatory and iterative design process aimed at achieving user acceptance and usability. Experiences from the implementation and iterative improvement of an online surgical pathway at Marburg University Medical Centre have shown that pathway conformance actually could be improved by the use of IT. An analysis of the iterative design process has shown that future pathway projects can benefit from the lessons learned during this project. Based on these lessons a method for developing well adapted interaction mechanisms is presented, which is aimed at improving process alignment. Our goal is to build up a library of tested reusable components to reduce the number of iterations for pathway implementation.
Keywords: information systems [MeSH], clinical pathways [MeSH], process alignment, interaction design

1. Introduction

There is an increasing consensus among healthcare experts that information technology (IT) can significantly contribute to improve healthcare quality and reduce costs [1]. Clinical pathways can be used to implement guidelines in a specific setting and reduce undesired practice variability. IT applications can increase pathway compliance, as studies have shown the potential benefit of computerised decision support based on alerts and reminders [2-4]. Yet, it is also a well known fact that many IT projects fail and that multiple factors need to be considered to increase the probability for success [5]. A core reason is that an IT intervention in a healthcare setting will necessarily change a complex socio-technical system with often unpredictable results [6]. Consequently, IT projects in healthcare settings should be seen from a broad socio-technical perspective and should be accompanied with careful change management [7-10]. As the overall goal is process improvement, IT projects should be driven by demand rather than technology. Demand-driven system evolution requires a responsive IT infrastructure optimised for adaptation to changing requirements.

At Marburg University Medical Centre we tried to approximate this goal by an extensible holistic Hospital Information System (HIS) [11-13]. Responsiveness is achieved through an integrated Rapid Application Development (RAD) tool [13;14] and agile programming techniques [15] with close end user involvement [12]. An

[1] Corresponding Author: Dr. Richard Lenz, University of Marburg, Institute of Med. Informatics, Bunsenstrasse 3, 35037 Marburg, Germany, Tel.: +49 28 66205, E-Mail: lenzr@staff.uni-marburg.de

iterative and participatory software engineering process has been developed to support continuous system improvement and process alignment [7;12].

In [16] we described our approach of utilising this infrastructure to implement a guideline-based clinical pathway for patients with proximal femoral fracture at the Department of Trauma, Reconstructive and Hand Surgery. The basic idea was to support major parts of the pathway by integrating patient-specific advice according to pathway recommendations into routine documentation. In this paper we will have a closer look at the lessons learned during the iterative development of the pathway application and draw conclusions for future pathway projects.

2. Background and Objectives

The clinical pathway for patients with proximal femoral fracture was systematically developed based on the results from a prospective study of the care process in 2001/2002 and it was initially introduced without additional IT support in 2003/2004 [16]. As a result, process management could be improved significantly. However, evaluation of the pathway documentation also showed that important information, like the side of a fracture or medication details, was still often missing. In order to further increase pathway compliance and improve documentation quality, an adapted IT application was developed in 2004 which was aimed at bringing pathway conformant recommendations to the point of care by reusing online routine documentation. The pathway application is based on workflow-enabled electronic forms, in which coded data from a central database can be reused to place reminders and alerts or to automatically parameterise order sets.

In [16] we already described that pathway conformance could be improved by iteratively developing an IT application for pathway documentation. A number of iterations was required to correct "intuitive" but counterproductive first attempts. The objective of this paper is to summarise our experiences made during the iterative software development for clinical pathway documentation and to draw conclusions for future pathway projects. The goal is to derive a set of generally applicable rules or recommendations for developing online support for clinical pathways, which may help to reduce the number of iterations and improve process alignment from the beginning.

3. Methods

According to our adapted software engineering process [12] end users were intensively involved in the development of the IT application. In addition, application development was embedded in an overall change management process in order to improve integration with the clinical workflow [7]. Thereby, user participation was seen as a dialogue where both users and software engineers learned from each other. User participation covered academic detailing for effective training of clinicians during clinical routine, qualitative user surveys and usability analyses, and continuous pathway controlling.

Systematic pathway controlling included continuous monitoring of the usage rate of the clinical pathway for patients with proximal femoral fracture, deliberate pathway (variance documentation), and indicators for process quality (e.g. preoperative length of stay, time to first thromboembolism prophylaxis) and documentation quality (e.g. completeness and specificity of medication information).

In order to detect and analyse problems resulting from the IT intervention, controlling was expanded for a "methodical controlling". User feedback was evaluated and pathway documentation was compared with the actual treatment process to find out whether there were discrepancies

- between the documentation intended by the user and the actually documented data (e.g. do users unwillingly take over preselected default values?), or
- between the pathway documentation and the actual treatment process (e.g. did documented process steps actually take place?).

4. Results

The application as described in [16] is in routine use since August 2004 with several updates in 2005. Process management and documentation quality already could be improved by early versions of the application (e.g. reduced time to first thromboembolism prophylaxis, consistent documentation of the side of a fracture).

Yet, several remaining quality problems could be detected by pathway controlling, causing additional design iterations. Statements about how to continue previous outpatient medication of the patient were given in most admission reports (in 97.7% (n = 77) of the admission reports compared to 72.9% (n = 85) before IT intervention) but only a small part of these data contained enough detailed information on drug, dose, and time of intake (in 29.9% (n = 77) of the admission reports compared to 51.8% (n = 85) before IT intervention). Furthermore, there were several discrepancies between pathway documentation and the actual treatment process. Examples are:

- 5000 units of Heparin, as recommended by the pathway, were frequently documented but not applied,
- the recommended number of blood bottles was documented but not ordered,
- a bladder catheter was documented but not inserted.

The problems observed were carefully analysed and usually multiple factors contributing to a problem were detected. Some of the most influential factors concerning application design were:

- *Insufficient differentiation between different tasks*: For example documentation of previous outpatient medication and instructions for further medication on the ward were insufficiently separated.
- *Inappropriate usage of default values*: For example, taking over default values without any kind of explicit commitment led to unintended documentations.
- *Insufficient differentiation between different kinds of documentation*: E.g. prospective and retrospective documentation of activities were not distinguished.
- *Layout and wording*: For example ambiguous instructions or complex screens.

Other factors beyond application design were observed to contribute to success and failure. In particular, training and academic detailing were found to be extremely important.

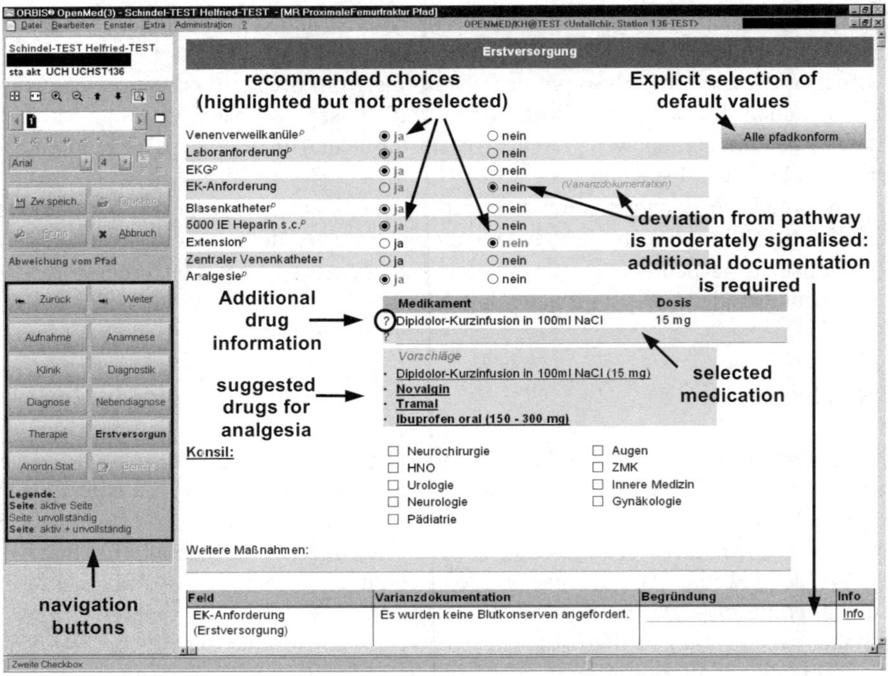

Figure 1: Form for initial treatment including medication and procedures.

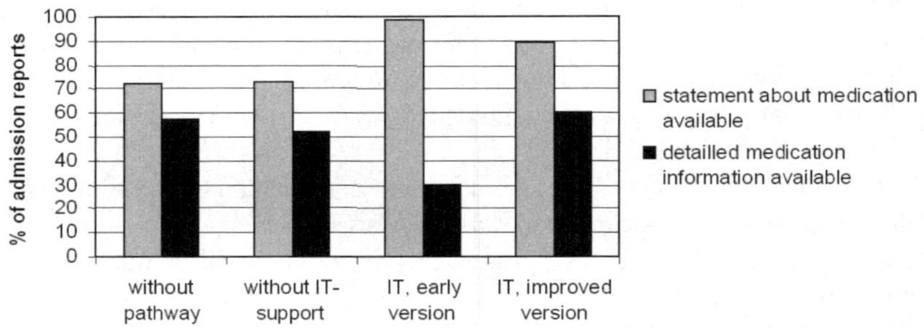

Figure 2: completeness and quality of medication information

For example, we observed that newly introduced process steps (e.g. application of Heparin during initial treatment) were in danger of being documented but not performed. To improve the pathway application, the user interface was modified:

- Data input for documentation of previous medication was explicitly separated from the medication instructions for the ward.
- Presenting medication recommendations by preselected choices was replaced by clearly highlighting the recommended choices.
- The wording of the instructions for the ward was clarified (e.g. responsibilities for ordering blood bottles were clarified by improving instructions).
- To reduce the risk of unintentionally taking over default values an additional button was introduced for explicit commitment for default values (s. figure 1).

Continuous pathway controlling showed that the modified application actually did improve pathway conformance and documentation quality. Evaluation of medication documentation shows a considerably higher rate of detailed information on how to continue previous medication (in 60.3% of the admission reports; n = 58). Figure 2 summarises the results from the different implementation phases.

5. Discussion

Many of our iterative improvements were related to interaction design. Good interaction design can greatly influence usability of an application and also speed up documentation. Lessons learned in this context were:

- *General recommendations for clearness*: Forms should not be overloaded, core messages should be highlighted, Scrolling should be avoided, dynamic build-up of forms should be used sparingly. Such general advice can also be found in the literature (for specific recommendations see e.g. [17], or for more general advice see [18]). Our participatory design process is aimed at considering user feedback as far as possible, but for interaction design professional knowledge of IT developers is required, as users might not know which options are possible and what side effects they could cause. Design decisions should be made by experienced developers according to the users' goals and led by their feedback.
- *Signalise pathway recommendations:* Documentation forms should signalise pathway-conformant decisions in an unobtrusive way. Physicians should easily recognise when pathway recommendations are not followed and when variance documentation is needed. Follow the principle *"give feedback to the user"*, so that he can see what he has already done and what still needs to be done. Application layout should clearly separate recommendations from decisions.
- *Carefully use default selections:* Some of the negative effects we initially had to deal with resulted from our intention to reduce documentation efforts as far as possible. We still believe that this is very important for pathway applications to achieve user acceptance. Yet, default selections increase the risk of unconsciously documenting process steps that did not take place. Therefore, additional effort for data entry and documentation quality have to be balanced carefully.

A core observation is that the actual patient treatment process and the online application for process documentation are necessarily closely intertwined. Process improvement is much more likely to be achieved if application development is embedded into a comprehensive change management process. Consequently, IT applications for pathway documentation should not only be viewed as accompanying add-ons. Instead, an online documentation should be viewed as a set of *speech acts* which are part of the process and actually affect the process. Speech act theory, going back to Austin [19] and Searle [20], is based on the assumption that language is not only used to formulate statements (which can be true or false), but also to perform acts, which can succeed or fail and thereby influence other acts of the communicating parties. Different speech acts, such as "declaration", "recommendation", "order", or "promise", have characteristic communication patterns that determine how different communicating partners (roles within a speech act pattern) act in order to "succeed". In our pathway implementation we observed several such typical comunication patterns which were initially not recognised as such. For example by clicking on 5000 units of Heparin, the physician actually *promises* that he will ensure that the patient receives Heparin. It is important to note that a promise is not an online order and that an explicit "uptake" is necessary to anchor this speech act within an appropriate sequence of related acts in order to be successful. Another example is the medication. In our initial implementation it was unclear whether the unspecific documentation was an order, a declaration, or a promise and which actions had to be taken as a consequence of this documentation. By introducing a clear separation of previous, current and future medication both documentation quality and process alignment could be significantly improved. An initial checking for the underlying communication pattern (e.g. *is this an order or a statement?*) could have avoided this additional iteration. Since medication is a step which typically occurs in many contexts we have also identified a candidate for a reusable module in future pathway applications.

So far we have only started to identify typical communication patterns or speech acts in the context of clinical process documentation. We plan to develop a catalogue of such patterns and a toolset of related IT interaction mechanisms, which should be helpful to increase process alignment in future pathway applications.

6. References

[1] Kohn LT, Corrigan JM, Donaldson MS. To Err Is Human. Building a Safer Health System. Washington D.C.: National Academy Press, 2000.
[2] Johnston ME, Langton KB, Haynes RB, Mathieu A. Effects of computer-based clinical decision support systems on clinician performance and patient outcome. A critical appraisal of research. Ann Intern Med 1994; 120(2): 135-142.
[3] Elson RB, Connelly DP. Computerized decision support systems in primary care. Prim Care 1995; 22(2): 365-384.
[4] Shiffman RN, Liaw Y, Brandt CA, Corb GJ. Computer-based guideline implementation systems: a systematic review of functionality and effectiveness. J Am Med Inform Assoc 1999; 6(2): 104-114.
[5] Sauer C. Deciding the future for IS failures: not the choice you might think. In: Curie W., Galliers R (Eds.). Rethinking management information systems. Oxford: Oxford University Press, 1999: 279-309.
[6] Berg M, Aarts J, van der LJ. ICT in health care: sociotechnical approaches. Methods Inf Med 2003; 42(4): 297-301.
[7] Lenz R, Büssecker F, Herlofsen H, Hinrichs F, Zeiler T, Kuhn KA. Demand-driven evolution of IT-systems in healthcare - a case study for improving interdisciplinary processes. Methods Inf Med 2005; 44(1): 4-10.
[8] Lorenzi NM, Riley RT. Organizational issues = change. Int J Med Inf 2003; 69(2-3): 197-203.
[9] Redfern S, Christian S. Achieving change in health care practice. J Eval Clin Pract 2003; 9(2): 225-238.
[10] Ash JS, Stavri PZ, Dykstra R, Fournier L. Implementing computerized physician order entry: the importance of special people. Int J Med Inf 2003; 69(2-3): 235-250.
[11] Lenz R, Kuhn KA. A strategic approach for business-IT alignment in health information systems. In: Meersman R, Tari Z, Schmidt DC (Eds.): On The Move to Meaningful Internet Systems 2003: CoopIS, DOA, and ODBASE; Springer Verlag, 2003: 178-195.
[12] Lenz R, Kuhn KA. Towards a continuous evolution and adaptation of information systems in healthcare. Int J Med Inf 2004; 73(1): 75-89.
[13] Kuhn KA, Lenz R, Elstner T, Siegele H, Moll R. Experiences with a generator tool for building clinical application modules. Methods Inf Med 2003; 42(1): 37-44.
[14] Lenz R, Elstner T, Siegele H, Kuhn KA. A practical approach to process support in health information systems. J Am Med Inform Assoc 2002; 9(6): 571-585.
[15] Beck K. Extreme Programming. München: Addison-Wesley, 2000.
[16] Blaser R, Schnabel M, Heger O, et. al. Improving Pathway Compliance and Clinician Performance by Using Information Technology. Stud Health Technol Inform 2005; 116: 199-204.
[17] Poon AD, Fagan LM. PEN-Ivory: the design and evaluation of a pen-based computer system for structured data entry. Proc Annu Symp Comput Appl Med Care 1994; :447-51.: 447-451.
[18] Preece J, Rogers Y, Sharp H. Interaction Design - Beyond Human Computer Interaction. New York: John Wiley & Sons Inc., 2002.
[19] Austin JL. How to do things with words. Oxford: Oxford University Press, 1962.
[20] Searle JR. Speech Acts. An Essay in the Philosophy of Language. Cambridge: Cambridge University Press, 1969.

Ubiquity: Technologies for Better Health in Aging Societies
A. Hasman et al. (Eds.)
IOS Press, 2006

Evidence-based practice in primary health care

Anouk DE SMEDT[a], Ronald BUYL[a] and Marc NYSSEN[a]
[a] *Vrije Universiteit Brussel, Brussels, Belgium*

Abstract. This study aims to describe primary care professionals' self-reported attitudes towards evidence-based practice (EBP), attention to information sources, perceptions of the barriers to EBP and strategies to improve insight in EBP and patient care. An e-mail invitation with link to an Internet-based survey was sent to Belgian medical doctors (MDs), nurses and paramedics. Under paramedics, we've included emergency medical technicians, firemen and medical volunteers (Red Cross). In general, respondents were supportive towards EBP and agreed that this concept improves patient care, but still, physicians claim that only 50% of their practice is evidence-based and nurses and paramedics spend respectively 59% and 54% of their time to EBP. Doctors depend mostly on clinical guidelines, the Internet and textbooks, while nurses prefer conferences and protocols and paramedics rely on courses and their own judgement. All respondents strongly rely on experimental knowledge gained through interaction with colleagues, although the majority reported that colleagues are often not supportive towards EBP. Lack of time, the overwhelming mass of literature, difficulties with implementation of evidence in to practice are the most common barriers. Nurses show lack of critical appraisal of research results and paramedics have difficulties understanding research and have limited access to computer facilities and their working environment. Communication in group and workshops are very highly valued. Nurses and paramedics are less reluctant towards the opinion of senior colleagues, audits on clinical practice and individual feedback than doctors.
EBP generally enjoys a positive attitude at every level of the health care system, but still many obstacles have to be overcome to conquer 'experience-based practice'. The most appropriate method for actual implementation of evidence-based practice at all levels of health care is to provide summaries of evidence, easily understandable protocols and web-based databases accessible from the working environment. Students should not only learn the skills related to EBP, but should be able to integrate knowledge effectively in the clinical setting and routine care. Above all, their supervisors themselves need to evolve from 'experience-based' to evidence-based practice'.
Keywords: evidence-based practice, survey, primary care, attitudes, barriers

1. Introduction

With the popularisation of the use of evidence in health care, the term evidence-based medicine (EBM) evolved into evidence-based practice (EBP).[1] The success of EBP has not only stimulated scientific research, but has also encouraged, next to medical doctors, other healthcare professionals like nurses and paramedics to aim towards the best available care. Therefore we prefer talking about evidence-based practice, as it encompasses practice in all health care domains.[2]

A number of studies, surveying the views on EBP already exist. Most of these studies focus on secondary care evidence-based practice, meaning hospital based care.

By far fewer studies deal with primary care, consisting mostly of pre-hospital care.[3] To date, evidence-based practice in primary care comprises views and attitudes of general practitioners (GPs)[4-7] and nurses. [8-10] These studies have always shown enthusiasm towards EBP, but specific problems have also surfaced. The most commonly mentioned barriers are lack of time due to a heavy workload and the limited ability to understand and implement research because of shortcomings in statistical or EBP terminology knowledge. We emphasize that the primary care team is much broader than simply general practitioners, emergency doctors and nurses.[3] A group of professionals that until now was left behind, although being very closely related to pre-hospital care are the paramedics. In this group, we've included emergency medical technicians, firemen and medical volunteers. Paramedics are frequently the first contact persons, by whom patients get in touch with health care system, but as far as our literature search showed, nothing was found on their knowledge, attitudes and barriers towards evidence-based care. Just as with physicians and nurses, paramedics are trained to aim for best practice, during their "lifelong learning". Therefore, EBP should specifically be identified to the working environment in which it occurs, especially in the case of primary care where the working environment differs from that of hospital based care.[3]

A major focus point of this paper was to review the progress that has been made in the promotion of evidence-based practice by doctors and nurses and the current status of paramedics' views towards EBP.

2. Material and Methods

2.1. Method and questionnaire

The on-line questionnaire was developed after review of previously published research utilisation questionnaires on Evidence-Based Practice (EBP).[4-10] The questionnaire tested the attitudes, barriers and usage of information sources consisting of original questions and items adapted from previous surveys. The survey was piloted with 7 nurses, 7 doctors and the Flemish Red Cross. The inclusion criterions for this survey was that respondents be involved in primary care, ranging from physicians and nurses to paramedics (emergency medical technicians, firemen and medical volunteers). An e-mail invitation with link to the Internet-based survey was sent to 22 Belgian hospitals (16 Flemish and 6 French), a medical electronic discussion group (MDF), a medical site 'Medisurf' only accessible for licensed doctors, the Erasmushogeschool of Brussels (nursing department), the Brussels fire brigade and the Flemish Red Cross. The final sample size consisted of 112 doctors, 158 nurses and 121 paramedics.

2.2. Statistical analysis

Data from completed forms were inserted into an SQL database and were then imported into a spreadsheet for analysis. Data analysis was conducted using SPSS for Windows version 13.0. Descriptive statistics were calculated to describe responses to questionnaire items. When required, the differences between 3 groups were tested with Kruskall-Wallis and between 2 groups with Mann-Whitney. The significance level (2-

sided) was set at $p \leq 0.05$.

Responses to items concerning ‘attitudes’ and ‘strategies’ were addressed using 5-point Likert scales with respectively ”strongly disagree” and “strongly agree” and “extremely useful” and “not useful at all” as anchors. 69% MDs, 60% nurses and 56% paramedics responded to the statements.

All results are summarized in table 1.

3. Results

3.1. Information sources

The overall most often consulted source by all professionals are ‘colleagues’, nevertheless, there are some remarkable differences between the groups. Physicians consult ‘Official clinical guidelines’, ‘the Internet’ and ‘textbooks’, nurses support on ‘conferences’ and ‘protocols’. Paramedics prefer ‘courses’ and rely on their ‘own judgement’. The least consulted sources are ‘workshops’ and ‘conferences’ for doctors, ‘Journals’ and ‘official guidelines’ for nurses and ‘journals’ and ‘the Internet’ for paramedics. *Pubmed* is used by 66.4% of the doctors, 31.6% of the nurses and 6.6% of the paramedics.

3.2. Attitudes towards evidence-based practice

Doctors and nurses strongly agree that EBP improves patient care (resp. 77.9%, 74.7%) and a high percentage of doctors (85.7%) claim to adapt their treatments per patient. The support that the respondents receive from their colleagues towards EBP is quoted positively by a minority of physicians (31.2%) and nurses (44.2%). A significant difference was found between doctors (59.8%) and nurses (41.1%) when comparing different information sources and nurses also tend to be less critical towards research results they read (57.9%) in comparison to MDs (41.6%).

Paramedics are very enthusiastic towards EBP, even though only a small percentage ever heard of this term (25%). They frequently follow classes to improve their skills and knowledge (82.3%) and they feel confident in emergency situations (75%). A significant difference between paramedics and nurses is ‘the initiative to look up answers for questions that rise’ (resp. 38.8%, 13.3%).

All professionals seem to agree that ‘courses on how to use research effectively’ and ‘independent institutions that provide relevant evidence-based information’ would be good initiatives (percentages in table 1).

3.3. Barriers and strategies to overcome barriers to evidence-based practice

The main barriers we noted are lack of time to search for evidence and to visit libraries. The huge mass of literature, the fact that research often reports conflicting results and the problem that research findings are not easily transferable into practice are also common barriers. Paramedics have significantly lower access (16.4%) to computer facilities and recent journals in their working area. Significant differences between nurses and paramedics are the knowledge of English and the authority to modify patients' care protocols.

Strategies to improve implementation of EBP were quoted as being 'useful' to 'very useful'. There was a striking difference between the 3 groups:
MDs depend less on support from senior colleagues (50.7%) than other professionals. Nurses find audits on the 'conformity of clinical practice' a useful strategy (73.3%). According to doctors, more research would not be useful (39.4%). Paramedics find 'individual and confidential feedback' very useful (81.6%). All agree upon the usefulness of discussion groups.

4. Discussion

This study about evidence-based practice in primary care is the first in its kind to compare 3 different professional groups. Some expected differences were confirmed. Yet, we should preserve a clear distinction between the groups regarding the academic levels when we compare the research results of this study. A majority of the respondents is still relying on traditional information sources such as colleagues, textbooks and protocols. This finding is not new[9, 11] and in the case of the study results rather disturbing, since MDs, nurses and paramedics indicated that respectively 50%, 59% and 54% of their practice is evidence-based. The information sources they use are not considered to be sufficiently updated and colleagues are still being seen as "quick and easy" sources that provide affirmation, guidance and support.[11] Moreover, most respondents also claim that their colleagues are often not supportive towards EBP. These results are very contradictory in relation to the fact that 85.7% of the doctors state that they adapt their patients' treatment to EBP. A positive outcome is the fact that doctors also rely on 'clinical guidelines'. Nurses are known for not reading journals regularly,[9, 12] which doesn't differ from our results looking for example to the percentage of *Pubmed* usage (31.6%).
They prefer protocols that contain evidence-based information in a form they can easily understand and have direct impact to their practice. We didn't find comparative research for the paramedics; our findings show that they obtain their knowledge from courses, which is a plausible result since the information is given by instructors. 20.7% of the paramedics use the Internet, which is very low in comparison with the other groups and could be related to the lack of computer facilities in their working environment that is seen as a great barrier.
More or less the same problems as revealed in other studies strike the Belgian primary care professionals when it comes to identifying attitudes and barriers towards EBP.[4, 7, 9] Physicians primarily lack time, find the mass of literature overwhelming and agree with the nurses about the fact that research reports conflicting results and that implementation of evidence into practice is a realistic barrier. Comments given by the respondents indicated that physicians still tend to continue working 'experience-based' and are too much influenced by the pharmaceutical industry and their own patients. Especially GPs, consider evidence-based practice as not financially attractive since the concept is too time consuming. This is of course not the correct attitude and not a good example for students.

Table 1 Summary of study results. Top 3 of most used information sources per profession is shown in bold. Attitudes and barriers towards EBP are indicated in numbers (percentages) of individuals who 'agreed' or 'strongly agreed' with the statements that are not the same for all groups (-). N=number of respondents, some respondents didn't answer all the questions. Strategies are indicated in numbers (percentages) of individuals who found the solutions 'useful' or 'very useful'. Significant differences are shown (*).

	Physicians	**Nurses**	**Paramedics**
Information sources (%)	**N=112**	**N=158**	**N=121**
Colleagues	**62.5**	**87.3**	**84.3**
Conferences	40.2	**75.9**	28.1
Clinical guidelines	**67**	39.9	22.3
Protocols	50	**75.3**	42.1
Workshops	31.3	43	47.1
Courses	57.1	67.1	**83.5**
Textbooks	**60.7**	49.4	44.6
Own judgement	43.8	54.4	**48.8**
Internet	**62.5**	45.6	20.7
Scientific journals	58.9	34.8	7.4
Pubmed	63.4	31.6	6.6
% of practice considered to be evidence-based	50	59	54
Barrierss towards EBP	**N=77**	**N=95**	**N=68**
- There are sufficient computer facilities and recent journals in my working environment	57 (74.1)	53 (55.8)	11 (16.4)*
- Research articles are easy understood	42 (54.6)	38 (40.0)	20 (29.4)
- It's difficult to access a library on a regular basis	62 (80.5)	56 (58.9)*	-
- The mass of literature is overwhelming	59 (76.7)	43 (45.2)*	-
- Research findings are easily transferable into my practice	27 (35.1)	22 (23.2)	-
- I have sufficient time to search for evidence	16 (20.8)	18 (18.9)	-
- It's not my task to find and suggest new treatment methods	-	15 (9.5)	19 (21.2)*
- Research literature can report conflicting results	70 (90.0)	-	-
- I have enough authority to have personal input in the care of patients	-	55 (57.9)	-
Attitudes towards EBP			
- I should follow a course to help me use research effectively	57 (74.0)	58 (61.1)	42 (61.7)
- I would feel more confident if an independent institution would provide me with relevant evidence-based information	49 (63.7)	53 (55.8)	44 (64.7)
- I believe the results of the research I read	32 (41.6)	55 (57.9)	-
- I test/compare results from different information sources I encounter	46 (59.8)	39 (41.1)*	-
- Practicing evidence-based practice improves patient care	60 (77.9)	71 (74.7)	-
- My colleagues support the concept of putting research into practice	24 (31.2)	42 (44.2)	-
- I take initiative, looking up answers for questions that raise	-	21 (13.3)	47 (38.8)*
- I adapt the treatment of patients according to evidence-based practice	66 (85.7)	-	-
- I frequently follow classes to improve my knowledge of patient care	-	-	56 (82.3)
- I feel confident about my knowledge and skills during a medical emergency situation	-	-	51 (75.0)
- During my training I heard talking about 'evidence-based practice'	-	-	17 (25.0)
Strategies to overcome EBP barriers	**N=71**	**N=90**	**N=60**
- Discussion groups with physicians, nurses and teachers	45 (63.3)	74 (82.2)	47 (78.0)*
- Support from senior colleagues	36 (50.7)	64 (71.1)	49 (71.7)*
- Audits to check conformity to clinical practice	35 (49.2)	66 (73.3)	34 (56.7)*
- More sponsored research to close the gaps	28 (39.4)	55 (61.1)	-
- Individual feedback on clinical performances	-	-	49 (81.6)

The attitude of nurses is somewhat worrying. On the one hand, a very positive finding is that nurses don't feel they lack authority to introduce changes in contrast to another study,[9] but on the other hand a minority seems to compare and critical appraise research results. Overall, paramedics are very enthusiastic to improve their knowledge and skills. Their main barriers for research are the lack of computer facilities and the difficulties they experience understanding research.

All respondents show a remarkable positive attitude towards courses that are informative on how to apply research results and would feel more confident if they were provided with reliable evidence-based information. Useful strategies perceived by the respondents put communication on a high standard with discussion groups and workshops. Unlike doctors, nurses and paramedics highly value support from senior colleagues, audits on clinical-practice and individual feedback.

5. Conclusion

Even though EBP is well accepted and seen as a tool that improves patient care, the capability to introduce EBP efficiently and effectively in daily practice is still rather poor. Nurses and paramedics clearly lack skills and understanding of EBP. Evidence-based guidelines, adapted protocols or an on-line centralised information system would be a solution for them. Work is progressing within the university to set up a learning platform to provide up to date evidence-based information for different professional levels. Discussion groups involving colleagues, doctors and medical instructors and more individual feedback and guidance to build up confidence while introducing new working methods are also considered be very helpful.

References

[1] Maggi Banning. Conceptions of evidence, evidence-based medicine, evidence-based practice and their use in nursing: independent nurse prescribers' views. J Clin Nurs 2005; 14: 411-417.

[2] A. Sophie Gosling, Johanna I. Westbrook, Rosemary Spencer. Nurses' use of online clinical evidence. J Adv Nurs 2004; 47(2): 201-211.

[3] Hugh McKenna, Sue Ashton, Sinead Keeney. Barriers to evidence based practice in primary care: a review of the literature. Int J Nurs Stud 2004; 41: 369-378.

[4] Alastair McColl, Helen Smith, Peter White, Jenny Field. General practicioners' perceptions of the route to evidence based medicine: a questionnaire survey. BMJ 1998; 316: 361-367.

[5] Catherine A. O'Donnell. Attitudes and knowledge of primary care professionals towards evidence-based practice: a postal survey. J Eval Clin Pract 2004; 10(2): 197-205.

[6] Jane M. Young, Jeanette E. Ward. Evidence-based medicine in general practice: beliefs and barriers among Australian GPs. J Eval Clin Pract 2001; 7(2): 201-210.

[7] V. Toulkidis, N.J. Donnelly, J.E. Ward. Engaging Australian physicians in evidence-based medicine: a representative national survey. Intern Med J 2005; 35: 9-17.

[8] Andrew Retsas. Barriers ro using research evidence in nursing practice. J Adv Nurs 2000; 31(3): 599-606.

[9] Kate Gerrish, Jane Clayton. Promoting evidence-based practice: an organizational approach. J Nurs Manag 2004; 12: 114-123.

[10] Hugh P. McKenna, Sue Ashton, Sinead Keeny. Barriers to evidence-based practice in primary care. J Adv Nurs 2004; 45(2): 178-189.

[11] Finlay A. McAlister, Ian Graham, Gerald W. Karr, Andreas Laupacis. Evidence-based medicine and the practicing clinician. J Gen Intern Med 1999; 14: 236-242.

[12] Carl Thompson et al. Research information in nurses' clinical decision-making: what is useful? J Adv Nurs 2001; 36(3): 376-388.

6. Information Management and Modeling of Healthcare

6.1 Health Information Management

Process Frame Instances for Integrating Strategic, Tactical and Operational Information Management in Hospitals

Lutz ISSLER[a,1], Gert FUNKAT[a], Stefan SMERS[b] and Alfred WINTER[a]
[a] *Institute for Medical Informatics, Statistics and Epidemiology, University of Leipzig, Germany*
[b] *Information Management Department, Leipzig University Hospital, Germany*

Abstract. An approach to reduce the complexity of information management is to distinguish it into strategic, tactical and operational management with each of the levels using its own software tools. In practice, the management levels are tightly connected and interact closely. The most evident information that has to be interchanged between the management levels is the information about finished, cancelled, running and planned activities or projects, respectively. We claim that the levels of information management should share information about the work done, and propose their integration by means of a network of so-called process frame instances. Process frame instances hold information about the activities of different information management levels in a structured and reusable way.
Keywords: Information Management, Hospital Information Systems, Health Information System, Organizational Models, Process Modelling

1. Introduction

There are several approaches to organize the management of hospital information systems. A helpful approach is distinguishing the tasks of information management into the levels of strategic, tactical, and operational management [1] with specific methods and software tools for each level. Nevertheless management activities at these different levels have a strong need for an extensive information exchange [2], especially across level borders.

Together with information managers from the department for information management of the Leipzig University Hospital we identified typical situations where information about activities in information management is involved. Examples for these situations are

- relating the current strategic management plan's migration plan to the projects actually initiated or completed ,

[1] Corresponding author: Lutz Issler, Institute for Medical Informatics, Statistics and Epidemiology, University of Leipzig, Haertelstr. 16-18, 04107 Leipzig, Germany. Email: lutz.issler@imise.uni-leipzig.de, World Wide Web: http://www.imise.uni-leipzig.de/

- identifying several projects with similar topics and goals consistently and using synergy effects,
- communicating the responsibilities for activities in the operational management,
- acquiring the projects assigned to a specific goal in the strategic management plan, or
- reporting all ongoing projects to the hospital's board of management.

Evidently it is crucial for information managers to be fully aware of the activities going on in the department regardless wether the activity is assigned to the strategic, tactical, or operational level of information management. A central repository containing information about the activities can efficiently support the daily work of the information managemers at the strategic, tactical and operational level. For implementing such a repository the following problem has to be solved:

- How can information management activities be described in a standardised way, regardless of the borders between the strategic, tactical and operational level?

In this paper we propose to represent an activity performed in information management by an abstraction of a process and call it a process frame instance. We describe their relationships by semantic links between them.

2. Methods

The daily work in a hospital's information management department consists of many activities like managing complex projects, running the helpdesk and preparing strategic decisions. Each of these activities can be considered as a process. Since business process models [3] focus on details inside of processes, information managers also have to have an appropriate overview on these activities or processes respectively. At least they need to know *who* performs (or performed) *which* activity with which *goal* since *when*, with which *deadline*, using which *resources* (and we expect this list not to be extensive).

We describe the activities in an information management department by a set of such properties. The set of properties can be chosen to the needs of the specific department, but should be the same for all activities in that department to guarantee a standardised description. This standardised set we call a *process frame*.

An activity is represented by an instance of the process frame. A *process frame instance* is a set of values assigned to the properties defined in the process frame. Let the tuple (PERSON IN CHARGE, GOAL, DEADLINE) be an example process frame . An example instance of this process frame might be the tuple ("Lutz Ißler", "Inventing a glossary of terms", "2005-12-15").

Process frame instances can be semantically related to each other. To express this semantic conditions on pairs of process frames can be formulated [4]. Whether two process frame instances are connected by a semantic link is determined by evaluating the semantic condition on the respective instances. Depending on the condition formulation semantic links are either directed or undirected. An example for a directed link is the "sub-project" relationship which connects the process frame instances B and A if the activity represented by B is needed to achieve the goal of the activity represented by A. Opposite to this an example for an undirected link is a link connecting two process frame instances with the same person in charge specified in their properties.

With process frame instances we abstract from the formulation of process steps usually used in business process modelling [5]. Process frame instances only describe the "outline" of a business process instance. We claim this to be sufficient to keep track of the activities in information management from a mere strategic viewpoint.

3. Results

We used process frame instances to manually represent activities in the Leipzig University Hospital's information management department together with their relationships between each other in order to prove the feasability of the concept. Here, we present a simplified cut-out of this representation. The process frame in this simplified example consists of only three properties, namely a short description of the activity, the proposed deadline for the activity to support sorting the activities by time and a reference to at least one strategic goal as stated in the strategic management plan [1].

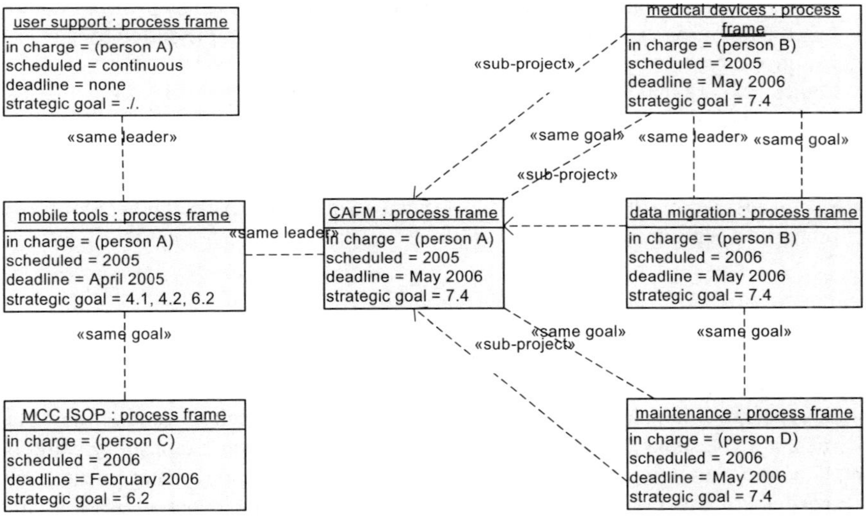

Figure 1: An example network of process frame instances with interconnections between these instances.

Figure 1 shows a sample network of several process frame instances. In the figure we used the Unified Modeling Language (UML) [6] to denote the process frame instances, but the concept is independent of a specific representation language. The figure shows several sample activities in information management based on real activities in the Leipzig University Hospital's information management department. The activities are semantically linked to each other by three different relationships: a "same leader" link if they have the same value of the "person in charge" property, a "same goal" relationship if they are assigned to fulfill the same strategic goal (denoted by the same numbers as in the corresponding strategic management plan), and a “sub-project” link if an activity contributes to another activity’s result. The sub-project link is an example for a directed relationship (from the sub-project to the main project), while the other links represent undirected relationships.

A repository of process frame instances together with the links between the instances represents information about the activities in an information management department in a standardised and structured way. A repository as shown in figure 1 can be used to

- collect the projects from the last strategic management plan, denoting which projects were finished in time, delayed, postponed, or cancelled,
- initiate projects with similar topics and goals consistent and using synergy effects if the repository contains not only finished but also planned activities,
- identify the persons in charge for activities in the operational management (in figure 1, the only activity represented from the operational management is the user support),
- easily acquire the projects assigned to a specific goal in the strategic management plan by starting at a process frame instance with a specific goal and following the “same goal” links, and
- report all ongoing projects to the hospital’s board of management by converting the repository’s contents into a list of activities, for example sorted by the deadline property value.

Figure 1 shows that activities can be represented on arbitrary levels of granularity. The “CAFM” project which introduces a Computer Aided Facility Management is split up in several sub-projects from which three are shown in the figure. The process frame instances representing the sub-projects are connected to the CAFM process frame instance by a sub-project link. Based on the desired level of detail such links can be followed or not when interpreting the repository contents.

4. Discussion

In order to support the work in a department of information management there has to be a central repository containing information about finished, cancelled, running and planned activities, and the semantic relationships of these activities. Such a repository could be the central working tool for employees on all levels of information management. Strategic managers could use the repository to get information for the preparation of a strategic management plan and to implement the results of the plan [7]. Tactical managers could plan and document their particular projects. Managers on the operational level could use the tool to report on their activities and responsibilities. In this paper we propose the representation of the activities in such a repository by process frame instances.

Process frame instances are a general concept for reflecting activities. Dealing with activities, process frame instances are closely related to the concepts of process modelling (which describes the abstract steps of a general activity) and project planning (which describes the concrete steps of a concrete activity). Process frame instances abstract from activity steps and thereby form a generalization of both process modelling and project planning. This enables them to deliver an integrated view on all activities of all management levels, regardless of the existence of process models and project plans.

Process frame instances represent all activities in the same structured and standardized way. The structured and standardized representation of activities allows for an easy access to the data and using the repository for example to formulate queries on the network of process instances, to define views on the network, or to use text-generation algorithms to create reports about the activities as natural language text. A standardised description is also helpful for knowledge-engineering and reasoning about the department's activities like project controlling and information flow analysis.

5. Outlook

In information management departments complex networks of possibly several hundred process frame instances need to be managed. An appropriate software tool would be extremely helpful. Currently, we are developing such a tool. It shall be introduced iteratively into the department for information management of the University Hospital Leipzig.

Information managers as we know them are usually overwhelmed with work. Of course no one can expect them to enter information about their activities in an additional software tool. To make the metaphor of process frame instances a useful tool we plan to extend process frame instances with the functionality to manage the documents created during an activity, for example, project plans, meeting reports, and financing plans. This way process frame instances will become the main enviroment for project file management. A similar approach was successfully taken by process-oriented knowledge management [3].

References

[1] Winter A, Ammenwerth E, Bott O, et al. Strategic information management plans: the basis for systematic information management in hospitals. International Journal of Medical Informatics 2001;64:99–109.
[2] Eppler MJ, Seifried PM, Röpnack A. Improving knowledge intensive processes through an enterprise knowledge medium. In: Proceedings of the 1999 ACM SIGCPR conference on Computer personnel research New Orleans. New York: ACM Press; 1999. 222–30.
[3] Curtis B, Kellner MI, Over J. Process modeling. CACM 1992;35:75–90.
[4] Stumme, G. Using Ontologies and Formal Concept Analysis for Organizing Business Knowledge. In: Becker J, Knackstedt R, editors. Wissensmanagement mit Referenzmodellen – Konzepte für die Anwendungssystem- und Organisationsgestaltung. Heidelberg: Physica; 2002. 163–74.
[5] White SA, editor. Business Process Modeling Notation (BPMN) – Version 1.0. BMPI.org; 2004.
[6] Unified Modeling Language Specification: Infrastructure – Version 2.0. Object Management Group; 200[illegible]
[7] Hartono E, Lederer AL, Sethi V, Zhuang Y. Key Predictors of The Implementation of Strategic Information Systems Plans. ACM SIGMIS Database, 2003;34:41–53.
[8] Rose T. Visual assessment of engineering processes in virtual enterprises. CACM 1998;41:45–52.

Ubiquity: Technologies for Better Health in Aging Societies
A. Hasman et al. (Eds.)
IOS Press, 2006

Web Services Based Syndromic Surveillance for Early Warning within French Forces

Hervé CHAUDET[a,b,1], Liliane PELLEGRIN[b], Jean-Baptiste MEYNARD[c], Gaëtan TEXIER[a], Olivier TOURNEBIZE[d], Benjamin QUEYRIAUX[a] and Jean-Paul BOUTIN[a]
[a]*DESP, IMTSSA, Marseille Armées, France*
[b]*Laboratoire d'Informatique Fondamentale, UMR CNRS 6166, Faculté de Médecine, Marseille, France*
[c]*Institut Pasteur de Guyane, Cayenne, Guyane Française*
[d]*MEDES, Centre National d'Etudes Spatiales, Toulouse, France*

Abstract. *Objectives*. Syndromic surveillance for early warning in military context needs a robust, scalable, flexible, ubiquitous, and interoperable surveillance system. *Methods*. We have designed our surveillance system as a collaborative network of web services on the basis of a skill oriented decomposition of the overall task and a formal model of epidemiological events. *Results*. The services (integration devices, epidemiologic receivers, information processing devices, GUI clients) are distributed in several locations in France and French Guiana using a secured network. *Conclusions*. This system is in operation since several months. It has already early detected two outbreaks before conventional surveillance systems.
Keywords: Public health informatics; Syndromic surveillance; Real time surveillance; Biological warfare; Web services.

1. Introduction

From a public health point of view, armed forces represent a population that is particularly exposed to natural or aggressive biological threats in unusual natural environments. In this context, the French Military Health Service, in accordance with NATO's five NBC initiatives of the Prague Summit Declaration, is developing a real time syndromic surveillance system during Forces deployments. This electronic surveillance system aims at detecting all health events with epidemic potential that concern forces in operational situation, and at contributing in the later health risk management process.

In comparison with syndromic surveillance systems met in civilian world, military syndromic surveillance is mainly characterised by mobility of the target population and a collaborative surveillance activity made by epidemiologists in several distant places. New hazardous theatres of operation may indeed be opened at any time and in any location on the earth, modifying the number and definition of the populations and

[1] Corresponding Author: Hervé Chaudet , Unité de Recherche Epidémiologique - Département d'Epidémiologie et de Santé Publique, Institut de Médecine Tropicale du Service de Santé des Armées – Le Pharo, 13998 Marseille Armées, France. Email : herve.chaudet@medecine.univ-mrs.fr

geographic areas under surveillance. Moreover, the vital continuity of service requires the scalability and redundancy of computing resources and the ability of the system for dynamic modifications. At the end, the system must be included into Command, Control, Communication, Computers, and Intelligence (C^4I) systems that need to shift from platform-centric toward network-centric systems. This shift raises issues in database integration, interoperability, and data standardisation and reuse that must be addressed [1].

A pilot project fulfilling these aims, named 2SE FAG (French acronym for Spatial Surveillance of Outbreaks within Forces in French Guiana), is being fielded incrementally. It has been conceived as a coexistence of two functionally independent networks: a declaration network for case declaration and follow-up, and an analysis network that fetch data from the previous one and which is based on a collaboration of e-services. The first network, which covers all branches of service and which is tailored for the mobility of forces in extreme conditions, has been previously described [2]. The aim of this paper is to focus on the description of the analysis network and on the way the surveillance activity has been organized and based on a formal model of epidemiological event to fit in a service-oriented architecture [3]. We present how a web service approach brings effective solutions to robustness, ubiquity, and adaptation to the surveillance target variability. We then describe the system architecture and the results of its first months of working.

2. Syndromic surveillance as collaboration of web services

Syndromic surveillance [4] for early warning is based on the continuous acquisition of health data without regard to specific diagnoses and their real-time monitoring. As we have previously shown [2], syndromic surveillance systems in a military operation context is a wide collaborative activity that involves many actors with supplementary activities and skills, and which work together in very distant places (even different continents). A natural response is a network-centric architecture gathering the main functionalities of this kind of system, which are data acquisition and organization, data integration, confidentiality, and outbreak detection [5].

In a first step, literature reviewing [6] and interviews of military experts allowed us to determine the tasks and functionalities of our surveillance system. In a second step, we have considered that these functionalities were coherently organized in different kinds of expertise characterized by the ability of performing a specific task or process (*e.g.* a quality control charts agent). Using this skill oriented decomposition strategy, we have framed the system into an agent-based distributed architecture [7]. At last, considering that typical agent architectures have many of the same features as web services [8], we have implemented each agent as a web service (henceforth WS).

On the basis of their overall specialization we have classified the WS into 4 kinds of "pluggable system modules": epidemiologic receivers (epidemiologic data servers), information processing devices (statistical or symbolic processing services), graphical user interface clients, and integration devices (knowledge integration and standards, registries). The last kind of devices ensures the services interoperability, and includes an ontology server and a service directory. The ontology server describes the concepts and relationships among concepts that represent the common language shared by the services, and covers three domains: WS description, epidemiological events and clinical features. It allows the semantic tagging of information exchanged between the services and the semantic description of service's behavior in terms of process, preconditions, postconditions, and effects.

The conceptual model we have used for representing epidemiological events is built upon a specific ontology and a logic language that we have developed previously for describing the spatio-temporal characteristics of outbreaks [9]. It considers an epidemiologic event (a case) as a tuple: event = <eventId, population, location, time>, where time is a tuple time = <apparition_date, consultation_date, notification_date, transmission_ date>, location is a georeferencing (latitude, longitude), and population defines the group at-risk of the case (unit or mission). Clinical characteristics of the case are event attributes. This conceptual model is the representational rationale for health information exchanged between WS and epidemiological information processing.

The main characteristics of the resulting architecture are that each expert is a WS hosted by a web server, a web server may offer one or more WS, a computer can host one or more servers, no restriction is made about the platform or the WS geographical location, and a WS is not exclusive to one server. It follows the usual advantages of WS: system scalability, system heterogeneity, dynamic modifications, and resources optimization.

This network modular organization constitutes a collaboration of WS for syndromic surveillance. In this architecture, epidemiological information remains on distant servers and is fetched as required by surveillance processes, with no need to store data in a main database before processing.

3. System description

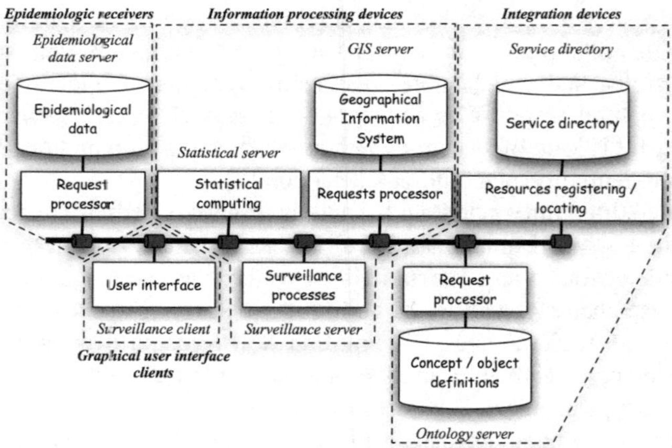

Figure 1. Schematic overview of CS^3 showing the different categories of services.

A schematic overview of our pilot system, named CS^3 ("Communauté de Services Internet pour la Surveillance Syndromique", patent pending), is shown in Figure 1. The different system components are described below. The system backbone is based on protected Intranets connected by virtual private network (VPN) over Internet. The VPN is based on SSL tunnelling with X509 authentication. This solution has the advantage of simplicity and reliability, but the drawback of X509 certificate maintenance. For this pilot project, the long distance transmissions are ensured by civil means: INMARSAT

ISDN (where required) and Internet. The services are distributed across four locations: IMTSSA, Faculty of Medicine in Marseilles, Direction Interarmée du Service de Santé in Cayenne, and Institut Pasteur de Guyane.

3.1. Standards

For the connection, communication and description we use the current standards for WS: XML, simple object access protocol (SOAP), and WS description language (WSDL). The description of WS and their orchestration/choreography rely on Web Service Modeling Ontology (WSMO) [10]. Health information is transmitted inside the SOAP envelope using ENV13606 medical data exchange standard, which in its fourth part define the message structure for the exchange of information [11] with its DTD. We have tried to enforce the CEN/TS 14796 standard [12] when defining our data structures.

3.2. Integration devices

The ontology server was developed in OWL web ontology language with Protégé 3 (http://protege.stanford.edu, last accessed 12/21/2005) and Tomcat/Axis web services solution (http://ws.apache.org/axis/). Our domain-related ontology extends the high-level DOLCE ontology (http://loa-cnr.it/DOLCE.html, last accessed 01/05/2006) with about five hundred domain-related concepts, which constitute the network metadata. The WS description is based on WSMO, and the definitions of clinical features are based on the notion of archetypes, which are expressed in accordance with the preCEN 13606 (part 2) standard. Following the philosophy of DOLCE, clinical features are qualities that take their values in quality regions ("quales"). This distinction allows the semantic description (e.g the data is a hemoglobinemia) and data typing (e.g. the data is a continuous quantitative data and is expressed in g/dl) of medical information that is exchanged between services. These services are located in the Faculty of Medicine.

3.3. Epidemiologic receivers

From an epidemiological point of view, our target population (*i.e.* Forces deployed in French Guiana) is continuously monitored within a declaration network. Physicians declare relevant cases using their desktop computer when in unit medical center or using their PDA with INMARSAT link when in mission in deep forest [2]. Relevant information is hosted into a datawarehouse, which is the epidemiological data server. This WS was developed using PHP nuSOAP, allowing data mediation. Its definition enforces the conceptual model presented above. An epidemiological event is an ENV13606 view built upon the datawarehouse, whose archetypes are defined in the ontology. Reflecting the model's components, the query model is structured to make explicit the query parts related to time, location, population, and event attributes. These services are located in French Guiana.

3.4. Information processing devices

This category of devices gathers services related to epidemiological event definition, surveillance orchestration and warning, temporal and spatial statistical analysis, geographical representation. The surveillance server is the main network component that allows the surveillance composition in terms of combination of web services (orchestration) and the definition of events under surveillance according to the conceptual model. It can be seen as a library of surveillance services that combine statistical processes on event features and a library of epidemiological events, which are defined by selection formulae on their attributes, and that may be the targets of the surveillance services. For its purpose, the surveillance server makes calls to other services as statistical programs, geographic information systems, and meteorological servers. Identification of signal abnormality (warning for outbreak) is based on quality control charts (exponentially weighted moving average and current past experience graph). Theses services were implemented with R using RSOAP (http://research.warnes.net/ projects/RStatServer/rsoap/, last accessed 12/21/2005), Mapserver (http://mapserver. gis.umn.edu/, last accessed 12/21/2005) and java. These services are located in the IMTSSA and in the Faculty of Medicine in Marseilles.

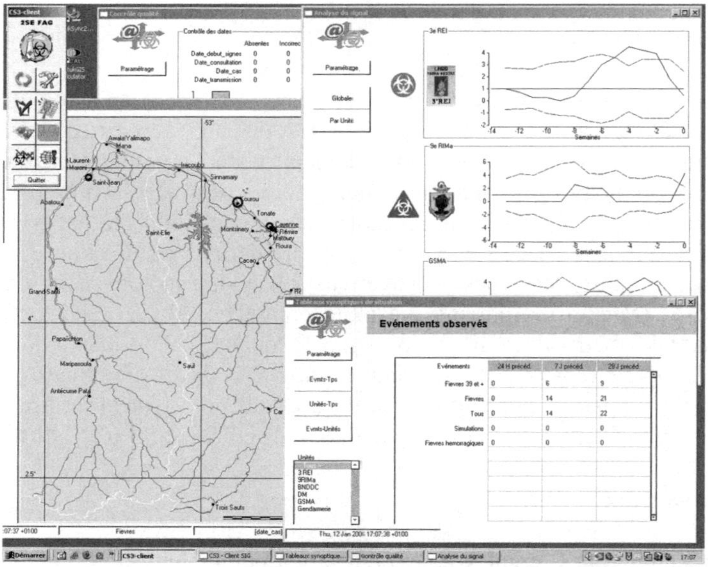

Figure 2. Example of CS^3 web service graphical user interface showing the displays related to time, space, and alerting system.

3.5. Graphical user interface clients

Currently no standard exists for making a web service accessible through web browsers, which are intended for html servers. We designed a specific graphical user interface (GUI) to interact with information processing devices and to display adequately the results of calls to WS. Figure 2 shows an example of such a GUI. Data visualization with this GUI is based on the four components of the model: event, time, place and population. It integrates a direct alerting system about the epidemiological situation based on the current past experience graph and can monitor several events and populations simultaneously. Epidemiologists in French Guiana and in Marseilles use these clients.

4. Results and discussion

This pilot system is in operation since several months. The network of web services spans two continents and allows epidemiologists in French Guiana and in France to monitors cases of fever in a population of about 3,000 persons, including forces in mission in equatorial deep forest. This system made possible a secured near real time surveillance of the population with a maximum of 10 minutes between a case declaration and its integration in the surveillance process results. It has already allowed the early detection of two outbreaks, one of dengue in last august, 15 days before its confirmation by the local Health Administration, and one of malaria, 1 week before its detection by the conventional military surveillance system.

In its current state this system fulfills our design objectives (interoperability, robustness, security, flexibility, ubiquity), but not entirely our functional objectives (integration in a C^4I) because it uses civilian transmissions and monitors, at the metropolitan level, only one theatre of operation. Partitioning the surveillance activity in several web services allows decentralizing the processes and resources involved into different structures, making easier their maintenance by specialized teams, and enforcing the collaborative aspect of epidemiological surveillance.

Restricted, in a first step, to cases of fever, the system will be extended to the data elements needed for early recognition of all biological threats. Circumscribed to French Guiana, it also will be soon extended to others forces to demonstrate its ability to monitor simultaneously several theatres of operation.

References

[1] Ceruti MG. Data management challenges and development for military information systems. IEEE T Knowl Data En 2003;15(5): 1059-68.

[2] Chaudet H, Meynard J-B, Texier G, Tournebize O, Pellegrin L, Queyriaux B, Boutin J-P. Distributed and mobile collaboration for real time epidemiological surveillance during Forces deployments. Stud Health Technol Inform. 2005;116:983-8.

[3] W3C. Web services architecture. W3C working working group note 11 February 2004. http://www.w3.org/TR/2004/NOTE-ws-arch-20040211/ (last accessed Jan 15, 2006).

[4] Reingold A. If syndromic surveillance is the answer, what is the question? Biosecurity and Bioterrorism: Biodefense Strategy, Practice, and Science 2003; 1(2):77-81.

[5] Mandl KD, Overhage JM, Wagner MM, Lober WB, Sebastiani P, Mostashari F, et al. Implementing syndromic surveillance: a practical guide informed by the early experience. JAMIA 2004;11(2):141-50.

[6] Center for Disease Control. Annotated bibliography for syndromic surveillance. http://www.cdc.gov/EPO/DPHSI/syndromic/index.htm (last accessed Jan 15, 2006).

[7] Van Aart CJ, Wielinga B, Schreiber G. Organizational building blocks for design of distributed intelligent system. Int J Hum-Comput St 2004;61(5):567-99.

[8] Huhns MN. Agents as web services. IEEE Internet Comput 2002;6(4):93-5.

[9] Chaudet H. Extending the Event Calculus for tracking epidemic spread. Artif Intell Med. In Press 2005. Epub 2005 Aug 1

[10] Roman D, Keller U, Lausen H, de Bruijn J, Lara R, Stollberg M, et al. Web Service Modeling Ontology. Applied Ontology 2005;1(1):77-106.

[11] CEN. European prestandard ENV 13606-4. Health informatics - Electronic healthcare record communication - Part 4: Messages for the exchange of information. Brussel, Belgium : CEN, 2000.

[12] CEN. CEN/TS 14796. Health informatics – Data types. Brussel, Belgium : CEN, 2000.

A. Hasman et al. (Eds.)
IOS Press, 2006

Modeling Economic Aspects of Hospital Information Systems to Give Decision Support for Strategic Information Management

Ansgar Kutscha[ab,1], Birgit Brigl[a] and Alfred Winter[a]
[a] *University of Leipzig, Institute for Medical Informatics, Statistics and Epidemiology, Germany*
[b] *Pergis Systemhaus GmbH, Ludwigshafen a.Rh., Germany*

Abstract. Information systems require strategic planning in order to adapt their functionality and quality to the needs of health care organizations. Next to effectivity, cost efficiency in supplying and operating information systems is a particular objective. Hospital information systems with their technical infrastructure, their application systems and the hereby supported business functions can be described with the help of the meta model 3LGM^2 and the 3LGM^2 tool. The meta model and the 3LGM^2 tool are extended by a generic approach to show supply and operation cost for all components of the information system and for the cost calculation between these components. This leads to the fact that all executives in hospitals are enabled to get the cost transparent which were caused by the support of the functions by the information system. The effects of planned extensions and modification of the information system can be analyzed in term of cost. In a prototypical modelling, the information system of a hospital of regular standards has been evaluated in nearly all its components and cost. An evaluation could show that information managers and executives are now delivered relevant cost information for planning, operating and control of information systems.
Keywords: Economic values, Strategic Information Management, Information Systems, Modelling, Performance Indicator

1. Introduction

Through technical progress and increasing requirements, information processing in general as well as in hospitals has become more complex. Besides other developments, standardized payment systems of hospitals require an increasing economic efficiency. This can be achieved, among other measurements, by the use of efficient information processing. In order to reach this target, strategic information management as a continuing process consisting of planning, operating and control is required.

[1] Corresponding author: Ansgar Kutscha, Rheinuferstr. 9, 67061 Ludwigshafen, ansgar.kutscha@pergis.de

Information managers in the health system are often faced with the problem to have to handle valid cost information for various questions. On the one hand, cost information can touch single components of the information system such as an application system or a hardware component. On the other hand, the information management can regard – analogously to the account of charges of hospitals – their internal customers and their company functions as cost units. This requires settling the cost of information system with the respective functions.

It is the objective of this work to deliver a performance indicator system of cost indicators to all those authorities who give directions in hospitals and to support their tasks. This way, the users shall be enabled to cut cost and to modify the information system functionally in a way it will support the business functions of the hospital efficiently and effectively. This paper presents a method and reports on first experiences made in a case study in one hospital. The method is based on the $3LGM^2$, which has been used successfully to describe HIS ([1, 2]).

2. Material and Methods

The elaboration of a performance indicator system of cost indicators has been based on three steps:

The first step develops requirements, which are to be measured by the indicators describing economic aspects of information processing. The foundation is primarily on analysis of strategic information management plans and a study of standardized queries of information managers in the health system.

As a second step usual economic indicators and indicators of documented performance indicator systems are evaluated if they meet these requirements. These results lead to the development of a generic performance indicator system.

The third step comprises the implementation on the basis of the $3LGM^2$. The $3LGM^2$ defines ontology to model information systems using three different layers. The domain layer consists of enterprise functions and entity types, the logical tool layer focuses on application systems and the physical tool layer describes physical data processing components. In contrast to other approaches the $3LGM^2$ also defines inter-layer-relationships between the layers to build integrated models of information systems. To obtain this, it was necessary to identify the indicator-relevant model section. This was followed by a specification of necessary expansion to verify the performance indicator system in the meta model. For the $3LGM^2$ tool extensions were specified to apply the performance indicator system at the $3LGM^2$ model.

The developed method was adapted and evaluated with the help of prototypical modelling.

3. Results

3.1. Requirements

Seven specific requirements of cost indicator systems were specified after this analysis: Efficient service delivery by the operator of the information system, Effective support of enterprise functions, Efficiency of in-house customer processes, Complete ascertainment of costs of information processing, Structuring in elementary types of costs, Considering of cost-cut factors, Reference of the responsibilities and operational competence of those in charge.

These specific needs are in line with general needs in terms of indicators, as e.g. formulated by KÜTZ ([3]).

3.2. Cost Indicator System for Information Processing

The cost indicator system developed for information processing consists of a hierarchical structure falling into four classes of calculation objects (see Figure 1). The cost indicators for the calculating objects of each class are in line which the generic cost types (see Table 1) which can be differentiated depending on the demand of concrete modelling. The settlement of cost figures through the hierarchy of calculation objects involves distribution keys. Through these distribution keys, the cost types (see Table 1) of each object of one layer (see Figure 1) can be added to the related objects of the upper layer.

3.3. Requirements for the 3LGM²

Proofed by an analysis the 3LGM² supports the hierarchical structure of calculation objects from the start. What is needed for an overall settlement of information processing cost is a

Project Phase	**Operation Phase**	**Life Cycle**
	Primary Performance Indicators:	
material cost (single)	material cost (periodical)	
internal personal resource (single)	internal personal resource (periodical)	
external service (single)	external service (periodical)	
planned operation time (number of periods)		
	Secondary (Computed) Performance Indicators:	
project cost in total		
project cost per period	operation cost per period	→ **total cost per period**
internal personal resource per period	internal personal resource per period	→ internal personal resource per period

Table 1: Generic plan of cost indicators

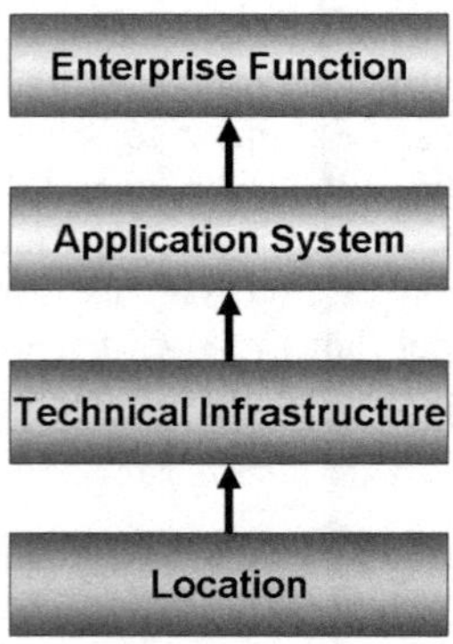

Figure 1: Hierarchical structure of the layers of object classes to calculate the settlement of information processing costs

relevant part of 3LGM² shown in Figure 2 along with the needed extensions.

In extending the 3LGM^2, the primary and secondary cost indicators as class attributes are important as well as distribution keys to settle the cost indicators between adjacent classes. Primary cost indicators are features which are directly allocated to calculation objects. Secondary cost indicators derive from primary ones.

7 Settlement functions were defined to calculate cost indicators between the objects of adjacent classes. A function for value-depending visualisation of calculation results had to be added.

To deduct efficiently the cost indicator system requirements for the 3LGM² tool were formulated:

- **Attribute editor** to define, enter and show the indicators as model specific attributes,
- **Visualisation of inter-layer relationships** to enter and show the partitions,
- Definition of modelling conditions which are able to specify the respective prerequisites to be able to use reusable cost models and to control their completion,
- An **attribute browser** to support the visualisation of indicators and to navigate through a model,
- **Import- and export functions** for data exchange with e.g. finance accounting or purchase systems.

3.4. Prototypical Modeling

The information system of a hospital of standard regular service was modeled with the help of the 3LGM² tool on all three layers as well as the relations between these layers. Consequently the primary cost data of all components of the information system were taken into account. The investment and operational costs were needed to evaluate the long term ef-

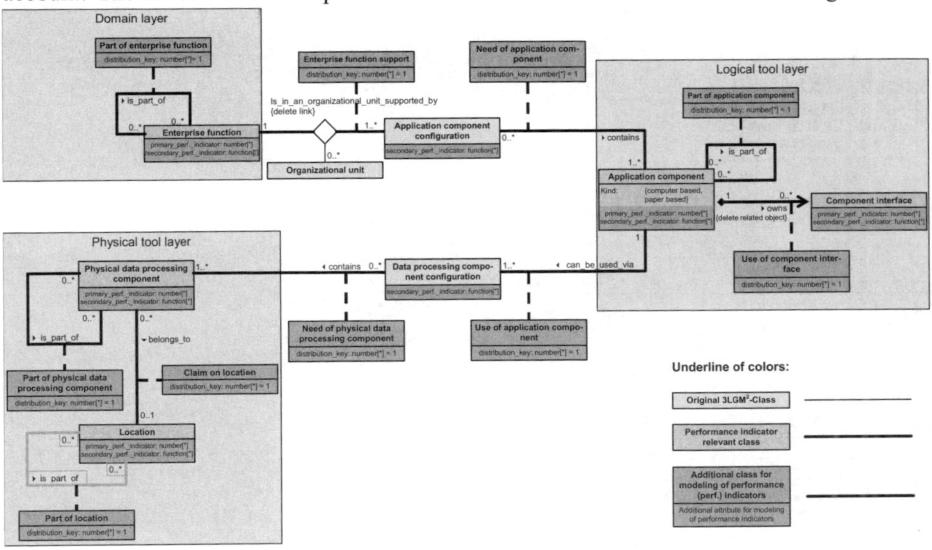

Figure 2: UML-diagram of the performance indicator relevant part of the 3LGM²-Meta model including extensions

fects of investments decisions in the model. Assuming that information systems have a limited duration of use and have to be substituted at the termination of the working life, the investment costs are based on the current replacement costs.

The replacement cost was calculated on account of present invoices, offers of comparable components or estimates of experienced consulters.

The relations between information system components describe which physical data processing components are used for the application systems and which application systems can be used to support the enterprise functions. The cost could be transferred onto the objects of superior layers according to these relations.

Technical data, single cost as value quotas as well as estimates as consensus of several executives were used as distribution keys.

3.5. Case Study

The method proposed has been applied during a case study in a 155 bed German hospital. The result is shown in

Figure 3. Five indicators are shown in this graphical model for information system in total and as an example for an enterprise function: PC: Project Cost, recalculated per year of usage; OC: Operation Cost per year, TC: Total Cost per year, PR: Personal Resources per year, WL: Planned Working Life.

These indicators had been computed for each model object. The drill down function of the 3LGM² tool enabled information managers as well as executives of the hospital to control the information system, its components and its development.

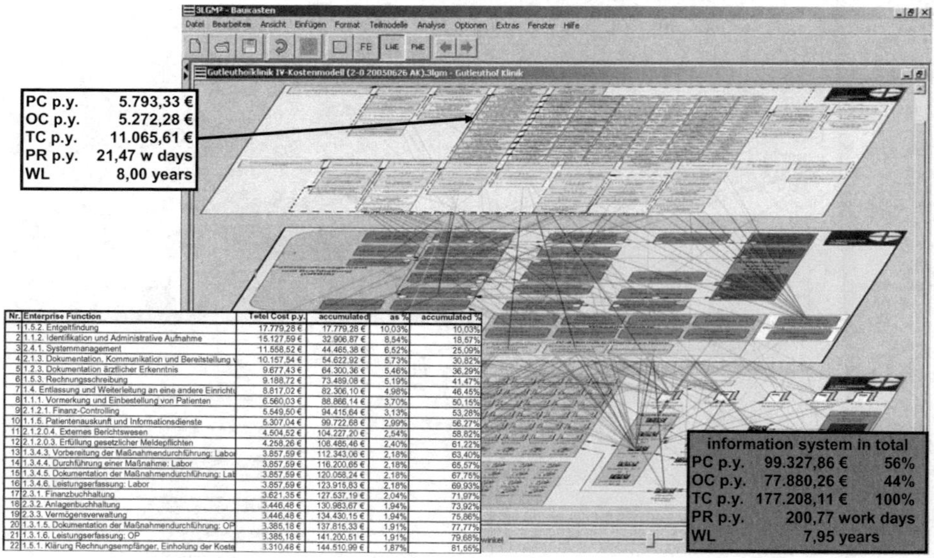

Nr.	Enterprise Function	Tetel Cost p.y.	accumulated	as %	accumulated %
1	1.5.2. Entgeltfindung	17.779,28 €	17.779,28 €	10,03%	10,03%
2	1.1.2. Identifikation und Administrative Aufnahme	15.127,59 €	32.906,87 €	8,54%	18,57%
3	2.4.1. Systemmanagement	11.558,52 €	44.465,38 €	6,52%	25,09%
4	2.1.3. Dokumentation, Kommunikation und Bereitstellung v	10.157,54 €	54.622,92 €	5,73%	30,82%
5	1.2.3. Dokumentation ärztlicher Erkenntnis	9.677,43 €	64.300,36 €	5,46%	36,29%
6	1.5.3. Rechnungsschreibung	9.188,72 €	73.489,08 €	5,19%	41,47%
7	1.4. Entlassung und Weiterleitung an eine andere Einrichtu	8.817,02 €	82.306,10 €	4,98%	46,45%
8	1.1.1. Vormerkung und Einbestellung von Patienten	6.560,03 €	88.866,14 €	3,70%	50,15%
9	2.1.2.1. Finanz-Controlling	5.549,50 €	94.415,64 €	3,13%	53,28%
10	1.1.5. Patientenauskunft und Informationsdienste	5.307,04 €	99.722,68 €	2,99%	56,27%
11	2.1.2.0.4. Externes Berichtswesen	4.504,52 €	104.227,20 €	2,54%	58,82%
12	2.1.2.0.3. Erfüllung gesetzlicher Meldepflichten	4.258,26 €	108.485,46 €	2,40%	61,22%
13	1.3.4.3. Vorbereitung der Maßnahmendurchführung: Labor	3.857,59 €	112.343,06 €	2,18%	63,40%
14	1.3.4.4. Durchführung einer Maßnahme: Labor	3.857,59 €	116.200,65 €	2,18%	65,57%
15	1.3.4.5. Dokumentation der Maßnahmendurchführung: Lab	3.857,59 €	120.058,24 €	2,18%	67,75%
16	1.3.4.6. Leistungserfassung: Labor	3.857,59 €	123.915,83 €	2,18%	69,93%
17	2.3.1. Finanzbuchhaltung	3.621,35 €	127.537,19 €	2,04%	71,97%
18	2.3.2. Anlagenbuchhaltung	3.446,48 €	130.983,67 €	1,94%	73,92%
19	2.3.3. Vermögensverwaltung	3.446,48 €	134.430,15 €	1,94%	75,86%
20	1.3.1.5. Dokumentation der Maßnahmendurchführung: OP	3.385,18 €	137.815,33 €	1,91%	77,77%
21	1.3.1.6. Leistungserfassung: OP	3.385,18 €	141.200,51 €	1,91%	79,68%
22	1.5.1. Klärung Rechnungsempfänger, Einholung der Koste	3.310,48 €	144.510,99 €	1,87%	81,55%

Figure 3: 3LGM²-model of the information system taken from a 155 bed hospital with realized inter-layer-relationships that shows calculated cost indicators and a statistic presentation

The ABC analysis over all enterprise functions is an example for feasible analysis.

4. Discussion

The specifications to model and the experiences of the prototypical use concluded in a process reference model. It is based on the $3LGM^2$ and further more practicable to model cost for information systems in facilities of health service. The process reference model describes 14 successive activities to model cost and divides them in areas information system modelling and cost modelling. For each activity, the description could be supplemented by tools to apply for implementation and work usage as well by control instruments to check the completeness, the consistency or the plausibility of the task done.

5. Conclusion

The prototypical modelling and evaluation of a hospital information system could be proved by completely materializing the configured performance indicator system

- that the performance indicator system is applicable completely, practicably and non-contradictory,
- that this leads to cost transparency which backs up decisions on the change of information system components,
- that the expenditure for the information system - and cost modelling - justifies the profit of the model and
- that - basing on the cost model - ongoing initiatives to improve efficiency of information processing (e.g. by benchmarking) were feasible.

It could also be proved that the $3LGM^2$ to model the indicator system can be extended and can answer questions concerning the raise of costs, the origin and reduction of costs. Papers concerning the balanced scorecard ([5]) also show that finance indicators themselves are not the only reason for decisions about settlements of further cost indicator types. The $3LGM^2$ extensions were thought to serve settlement of further cost indicator types such as quality indicators types or safety features of the information system components. Respecting performance indicator systems could not yet be found and practicability could not yet be proved.

Furthermore, $3LGM^2$ models according to the set-up of this paper could be suitable to underline benchmarking projects that compare facilities. Thus, the results are a basis of further going research.

References

1. Winter, A., B. Brigl, and T. Wendt, *Modelling Hospital Information Systems (Part 1)*, in *Methods of Information in Medicine*. 2003. p. 544-551.
2. Wendt, T., et al., *Modeling Hospital Information Systems (Part 2)*, in *Methods of Information in Medicine*. 2004. p. 256-267.

3. Kütz, M., *Kennzahlen in der IT*. 2003, dpunkt.verlag: Heidelberg.
4. Remenyi, D., *The effective measurement and management of IT cost and benefits*. 2000, Butterworth-Heinemann: Oxford
5. Kaplan, R., Norton, D., *The Balanced Scorecard*. 1996, Harvard Business School Press: Boston Mass.

Ubiquity: Technologies for Better Health in Aging Societies
A. Hasman et al. (Eds.)
IOS Press, 2006

Medical error management and the role of information technology – A new approach to investigating medical handover in acute care settings

Kwang Chien YEE[a], Ming Chao WONG[b] and Paul TURNER[c] (Co-authors)
[a] *Medical handover leader, Royal Hobart Hospital, Tasmania, Australia*
[b]*Researcher, and* [c]*Associate Professor, Smart Internet Technology CRC, School of Information Systems, University of Tasmania, Australia*

Abstract While identifying reasons why medical errors occur and constructing models of how to manage them has proved relatively straightforward, implementing and meaningfully evaluating solutions in 'real-world' settings has proven considerably more difficult. From an information systems (IS) perspective, although the promise of technology remains powerful, the continuing high incidence of medical errors suggest that eHealth approaches are struggling to acquire a clear understanding of the complex, dynamic and multi-layered nature of acute care settings and clinical practices, and to respond effectively to address the range of errors that actually occur. Using medical handover as a field-site, this research-in-progress paper presents an adaptation of James Reason's 'Swiss Cheese Model' to conceptualize the complex factors at play in medical errors in terms of human, system and informational elements. This research paper then examines how drawing on this model it is possible to generate and implement a methodological approach that both enhances a holistic understanding of medical error management and illuminates criteria that can be used to meaningfully identify an appropriate role for information technology in medical error mitigation. This research-in-progress paper aims to make a significant contribution to research into medical error management in 'real-world' acute care settings. This research is part of a bigger project that aims to develop, implement and evaluate an information technology artefact as part of an holistic information systems approach to improving medical error management at medical handover.

Keywords: Medical Informatics, Medical Errors, Medical Handover, IS Methodology

1. Introduction

Medical errors cost around 4500 lives in Australian hospitals every year![1] While various countries cite different statistics in regard to the frequency of medical errors in acute care settings, studies from Australia [2], USA [3] and UK [4] have all shown that medical errors remain a common occurrence resulting in significant costs. Despite considerable efforts to identify the causes and implement solutions, medical errors continue to be an ongoing problem [5], [6]. Two important factors have been thought to be important in the prevention and management of medical errors, namely, human factors and the availability of information [7]. From the human factors point of view, fatigue amongst clinicians has been identified as most important in causing medical errors [8]. It has been shown that current work practices of medical professionals; with excessively long hours, are incompatible with safe healthcare delivery [9]. The effort to reduce working hours, however, has been hindered by the concerns among health care workers on discontinuity of patient care due to inadequate clinical handovers. Recent studies have shown that inadequate medical handovers are associated with discontinuity of care and medical errors [10].

From the availability of information point of view, the information model used assumes that delivery of the "right information, to the right people, at the right place and time" will solve the problem [11]. To support this information delivery, many approaches have advocated the introduction of information and communication technologies (ICTs) as a means to implement solutions to medical error management.

At one level, these perspectives appear to suggest that there is a simple solution to medical error management - restrict medical professionals working hours and utilise technology to assist in medical handover. Indeed, this approach has merit and there are examples of where it is being recommended in guidelines [12], [13]. Unfortunately, there are always differences between theory and practice in the field and there is increasing evidence to suggest that there is a need for a more holistic understanding of the complex, dynamic and multi-layered nature of clinical practices and acute care settings in which errors occur.

This research-in-progress paper presents an adaptation of James Reason's 'Swiss Cheese Model' [7] to conceptualize the complex factors at play in medical errors in terms of human, system and informational elements. Drawing on the 'King Island Brie' Model [14] the paper then examines how this has been utilised to generate and implement a methodological approach that both enhances understanding of medical error management and illuminates criteria that can be used to meaningfully identify an appropriate role for information technology in medical error mitigation. This approach is based on research into medical handover processes on 5 general medical wards in a major public hospital in Tasmania, Australia.

The authors anticipate that this research-in-progress paper makes a significant contribution to research into medical error management in 'real-world' acute care settings. More broadly, this on-going research is part of a bigger project that aims to develop, implement and evaluate an information technology artefact as part of a holistic information systems approach to improving medical error management, especially during medical handover.

2. Models of Medical Errors Management

Traditionally, the medical field advocates personal perfectionism. This personal approach has tended to create a "blame and shame culture" in the false belief that medical errors only happen because of inadequately trained practitioners. The application of this model has, to a major extent, directed attention towards more extensive training being required for healthcare practitioners. Unfortunately, because of systemic inadequacies and human fallibilities, medical errors continue to cost lives. More recently, Professor Reason proposed the Swiss Cheese Model of medical errors, drawing our attention to latent systemic factors in error causation [7]. In this model, it is argued that there is a need to investigate and manage these latent factors in order to build a safer system [15]. This model has created a paradigm shift towards systemic factors management with the famous saying from Professor Reason: "We cannot change the human condition, but we can change the conditions under which humans work." [7]

Significantly, most parts of patient care (and, arguably most medical errors) are delivered by junior doctors and nurses who spend most of their time in hospitals. However, they have little power to reform the system that they work in. The importance of creating "medical error awareness" among healthcare workers at the "sharp end" has, therefore, recently been discussed [15]. A recent model of combining systemic and human factors simultaneously has been proposed, known as the "King Island Brie Model" [14]. If we are to change the systemic factors, then not only do we need to educate the human actors on the impact and the technicality of the new system, but also the boundaries and limitations of the new system. This concept enhances our conceptualisation of the role of informational factors as well as systemic and human factors.

While incidences of medical errors are attributed to the absence or over-abundance of information at different points in the process of delivering care, the problem of determining the distinction between what is too much or too little is fraught with difficulties. When applying the informational perspective to medical handovers, it could be argued that a good medical handover is simply the provision of right information to the right person at the right time. To date, there is little evidence of whether this is actually the case. However, it is clear from studies that discontinuity of care arising from inadequate medical handover increases the risk of preventable adverse events [10] and patient mortality [13]. While some studies [16], [17] and guidelines [13], [18] have emerged recently that investigate and/or advocate processes to improve medical handovers, the focus to date has been on ensuring the continuity of clinical information. The underlying assumption is that continuity of information equals continuity and safety of care. Information technologies designed to improve clinical information flow have often been introduced into the health care system to assist with medical handover and medical error management [11], [16], [17]. There remains, however, little or no research to establish the agreement on what is meant by 'good' or 'accurate' information and 'effective' medical handover. Consequently, the cornerstone of decision-making remains the autonomy of clinical judgement. From the experience of clinical practice we are aware that information is usually only one factor (often not the most significant) amongst many that influence medical error. The various factors that affect medical handovers, however, have not been established.

3. Developing a Methodological Approach for Medical Handover

The process of effective medical handover obviously warrants further investigation. This research project aims to obtain a clearer understanding of the complex dynamic medical handover process, to reveal the range of various factors that affect the effectiveness and efficacy of medical handover and to investigate a potential role for ICTs in medical handover. It is argued that a mixed methodology is required given the need to address the interplay among this complex set of interrelated factors, together with the need to provide "hard evidence" for clinicians and guidance for ICT design.

Qualitative research is increasingly being used in health research due to the complexity of the phenomena being studied which includes social and cultural norms and perceptions that impact on behaviour and medical practice [19]. This is particularly useful in the investigation of medical and socio-technical interactions in the field site of medical handovers. Therefore to investigate this complexity of interrelated factors a qualitative research methodology was deployed. Clinicians working within the more conventional scientific paradigm request proof of a repeatable measure of favourable outcomes in relation to any intervention in order to encourage them to participate in the change. It is therefore imperative for the research design to take into account the different audiences and their expectations. To this end, we have used a "controlled trial" approach as preferred by clinicians, and complemented this with the use of qualitative techniques to provide rich insights into the phenomena studied. As researchers in IS, this research also has to be able to provide adequate information for the designing, building, and implementing an information system through the use of a systems development life cycle [20] to guide information systems practice.

With these considerations the approach developed utilised a mixed methodology drawing on both qualitative and quantitative research techniques to investigate the underlying assumptions implicit within medical handover practices and to reveal the range of factors potentially impacting on the effectiveness and efficiency of medical handovers. This data contributes to being better able to determine the nature of, and appropriateness for any information systems to support improvements in medical handover improvement and medical error management. The approach also aims to allow the generation of research data of significance to three different stakeholders – clinicians, IS researchers and sociologists.

3.1 Details of study design

This study is being carried out over three phases.

Phase 1 involves utilising three different techniques. Firstly, 25 non-participant observations were conducted over a time frame of between 15 minutes to 45 minutes each. Some of these observations were consecutive and some of them were random in order to minimise the Hawthorn effect [21]. Secondly, 10 semi-structured interviews each consisting of 6 questions and lasting between 20 minutes to 45 minutes were conducted with the JMOs in the Department of General Medicine. These interviews were audio recorded and transcribed within 48 hours. Both the non-participant observation and the interviews were analysed using open axial semantic coding, drawing on the principles of grounded theory. A questionnaire survey was then carried out eliciting the perception of the observation and interviews. This triangulation method aims to improve dependability and credibility [22]

Phase 2 involves following three cohorts of JMOs, during the weekends and over the extended public holiday periods, when arguably, good handover is most important in ensuring patient safety and continuity of care. This phase is a repetition of what was done in Phase 1, other than the participants involved. The first cohort being the participants involved in the Phase 1 study. The second cohort is a brand new group inexperienced JMOs and the third cohort will consist of experienced JMOs towards the end of 2006, after the introduction of the new system. This comparison is valid as we will be utilizing exactly the same study protocols, researchers and analysis technique as in Phase 1. The comparison of three cohorts will provide us with "hard evidence" in regards to the effectiveness of the intervention.

In Phase 3, we intend to carry out a detailed information audit and perform a requirements analysis (conventional IS techniques) with the aim to develop and implement a technology that is designed for doctors. This trial involves not only them using the technology but critically understanding the limitations of the technology so that they can utilize their own clinical autonomy in decision-making. The participants will be educated in medical handover, information management and the use of the new information system. The effects of this custom-designed ICT, coupled with education will be addressed through the third cohort in the Phase 2 study.

3.2 Preliminary Results – Phase 1

Non-participant observation results revealed that medical handovers serve multiple functions apart from information transfer. These include discussions about difficult cases, second opinion, clarification of roles and protocols, debriefing of difficult situation and seeking supervision. These functions are important for the well-being of patients and doctors. It is imperative that any new intervention or system will allow the continuation of these various functions carried out under the broad banner of medical handovers.

Our observation study also revealed that multiple factors (cultural, environmental, organizational, information and human) affect the efficiency and effectiveness of medical handovers. Many of these factors have not been identified or clarified in previous studies [18]. More importantly, our results highlight the importance of human factors and the importance of the provision of feedback to JMOs for process improvement and change management. The analysis of the interviews confirms the observation findings of multi-factorial influences on medical handovers. The interviews suggest that JMOs think that while systemic changes are required, improvement in human performance through feedback and education is more important. The supports the idea that while JMOs often know about the procedures and limitations of the procedures, they feel disinterested and/or disempowered to be able to effect change.

Our triangulation of the observation and interview data revealed multiple factors that were not obvious or easily explicable to all participants. These “other” factors are often associated with human and cultural factors, which might be seen as cultural norms within medical practice. These have a significant impact on the effectiveness and efficacy of medical handovers. During the research it became apparent that human factors need to be taken into account when carrying out research in health informatics as participant cooperation determined the usefulness of information and analysis obtained.

Our preliminary data supports the need to include both systemic and human factors in the change process. While systemic factors are important, in order to empower and encourage JMOs to get involved in improvement in patient safety and quality of care, we need to provide them with feedback, encouragement and education. Whilst we might not be able to change human conditions, we definitely can change human practice.

We believe that our Phase 1 study has contributed significantly to the understanding of the medical handover process. The results provide a clearer understanding of the functions of handover and various factors that affect effectiveness and efficiency of medical handover. The Phase 1 study serves as an important platform for Phases 2 and 3 of the study clarify the role of ICT in medical handover and medical error management. Our project will provide designing principles for ICT in health care, which will not only be integrated into the clinical practice, but also the information system within the clinical practice framework. Our research methodology will provide adequate detailed information for clinicians, IS professionals and qualitative researchers, whilst maintaining a holistic view of the problem.

4. Conclusions

In this research-in-progress paper, we presented our views on the role of ICT in medical error management emphasising on the importance of both systemic and human factors. We then presented our mixed research methodology which will cater to the medical, IS and sociology audiences. We presented our Phase 1 research results on medical handover using this methodology. This paper aims to make a significant contribution to research into medical error management in 'real-world' acute care settings. This research is part of a bigger project that aims to develop, implement and evaluate information technology artefacts as part of an holistic information systems approach to improving medical error management at medical handover.

References

[1] Armstrong, F. How Safe are our hospitals? Australian Nursing Journal. 2004; 11: 18-21.

[2] Wilson RMcL, Runciman WB, Gibberd RW, et al. The Quality in Australian Health Care Study. Med J Aust 1995; 163: 458-471

[3] Kohn LT, Corrigan JM and Donaldson MS. To err is human: Building a safer Health system. National Academy Press. Washington, 1999

[4] Department of Health Building a safer NHS for patients. London, 2001

[5] Belloma R. Goldsmith D. Russell S. and Uchino S. Postoperative serious adverse events in a teaching hospital: a prospective study. Med J Aust. 2002; 176: 216-218.

[6] James, BC. Five Years Later – Are we any safer? Advances in Patient Safety. 2004; 1: 1-2.

[7] Reason J. Human error: models and management. BMJ. 2000; 320: 768-770.

[8] Wilson RM, Harrison BT, Gibberd RW and Hamilton JD. An analysis of the causes of adverse events form the quality of Australian Health Care Study. Med J Aust. 1999; 170: 411-415

[9] Gaba DM, Howard S. Patient Safety: Fatigue among Clinicians and the Safety of Patients N Engl J Med. 2002; 347:1249.

[10] Roughton VJ, Severs MP. The junior doctor handover: current practices and future expectations. J R Coll Physicians Lond 1996; 30: 213-214

[11] Turner P, Wong, MC and Yee KC. Medical Error Management: The Need for Information Systems Not Just Technology for Supporting Patient Safety in Hospitals. 14th Annual Health Informatics conference. Melbourne: Health Informatics Society of Australia. 2005.

[12] British Medical Association. A guide on the EWTD for junior doctors. London: BMA, 2004.

[13] Junior Doctors Committee. Safe handover: Safe patient. London: British Medical Association, 2004.

[14] Yee, KC. Impossible but doable: Substitution of Swiss Cheese with King Island Brie. 2005. Tasmania Annual Scientific Meeting, Royal Australiasian College of Physician.

[15] Reason, J. Beyond the organisitonal accident: the need for "error wisom" on the frontline. Qual. Saf. Health Care 2003; 13: 28-33.

[16] Cheah L, Amott DH, Pollard J and Watters DAK. Electronic medical handover: towards safer medical care. Med J Aust. 2005; 183: 369-372.

[17] Morris D, Baker A. Safer Clinical Handover – A further opportunity for the careconnect.sa programme to enhance clinical practice. Proceedings of the 14th Annual Health Informatics conference. Melbourne: Health Informatics Society of Australia 2005

[18] Australian Safety and Quality Council in Health Care. Clinical handover and patient safety. Canberra: Safety and Quality Council, 2005

[19] Sayre J. The patient's diagnosis: Explanatory models of mental illness. Qualitative Health Research. 2000; 10: 71-83

[20] Valacich JS, George JF and Hoffer, JA. Essentials of systems analysis and design.Prentice Hall. NJ; 2004

[21] Roethlisberger FJ and Dickenson WJ. Management and the worker. Harvard University Press. MA; 1939

[22] Denzin NK. The Research Act: A theoretical introduction to sociological methods, 2nd edn, Adldine. Chicago; 1989.

Ubiquity: Technologies for Better Health in Aging Societies
A. Hasman et al. (Eds.)
IOS Press, 2006

Human Factors Engineering for clinical applications

Marie-Catherine BEUSCART-ZEPHIR[a,1], Peter ELKIN[b] and Sylvia PELAYO[a, c]

[a] *Evalab, University Hospital and School of Medicine, Lille, France*
[b] *Mayo Clinic, Rochester, USA*
[c] *Percotec – LAMIH, CNRS, Le Mont Houy, Valenciennes, France*

Abstract. This position paper advocates the necessity for promoting a Human Factors Engineering (HFE) approach to new clinical applications in the medical informatics domain. We first describe the negative consequences of NOT using HFE methods for such sensitive applications as medication Computerized Physicians Order Entry systems. We then describe rapidly the HF technology and how it would apply to the healthcare domain. The specific and complex problem of the necessary re-engineering of existing applications is then addressed. We conclude on the mandatory cooperative characteristic of any practical HFE project for designing or re-engineering safety-sensitive clinical applications.
Keywords: Human Factors Engineering, Usability, User-Centered design, Process Engineering.

1. Introduction

In the Healthcare domain, information technology is steadily spreading through each and every working environment, and it is progressively integrated with (or substituted to) the other working devices used by healthcare professionals. Along the years, Information Technology (IT) has constantly improved the availability and reliability of administrative, logistic and medical information, thus bringing in invaluable benefits for the hospital healthcare professionals and ultimately for the patients themselves. Recently, IT tried to move one step further and gained the heart of physicians' and nurses' work, with more critical applications that directly impact the practice of medicine like Computerized Physician Order Entry systems (CPOE). The particular example of medication ordering – dispensing – administration systems is illuminating: with this new generation of systems, IT is totally mingled with physicians and nurses daily workflow through an interaction with complex expert cognitive processes such as decision making and process control. The problem in many current systems designs is that those critical clinical applications were designed utilizing the same premises used in the development of previous Hospital Information Systems (HIS) products, i.e. considering primarily the logistic process and workflow of the drugs and ignoring a user-centered design approach. The resulting applications are therefore not actually user-friendly as far as physicians and nurses are concerned.

---

[1]Marie-Catherine BEUSCART-ZEPHIR[a] Evalab, School of Medicine and University Hospital, 1, place de Verdun, 59045, Lille Cedex, France. - mcbeuscart@univ-lille2.fr - http://www.univ-lille2.fr/evalab/

This new generation of clinical applications happens to have a considerable impact on the way people work, and ultimately on the outcomes of this work, i.e. the actual patient care and its quality. Fortunately, most of the time, the consequences of the installation of these new applications are positive and meet the anticipated benefits [1, 2]. But their installation may also generate unexpected consequences, most of them bearing negative effects on the healthcare professionals work or on its outcomes [3]. These negative consequences usually express themselves through two main symptoms:

(1) Users reluctance to use the system. Unfortunately, unless this reluctance turns to overt rebellion constraining the managers to abandon the system [4], users resistance is often underestimated and rarely well analyzed and documented per se. For example, the part of this resistance attributable to actual usability problems of the application or to a global psycho-sociological negative attitude toward any technological change is rarely made clear.

(2) Negative outcomes in terms of patient care and patient's safety, because of unexpected (however sometimes temporary) increase of medical errors [5, 6].

It is worth noting that in most of the cases, these unexpected negative consequences may be attributed to a weak management of human and organizational factors. Some papers and reviews have remarkably listed and described the core organizational problems susceptible to generate such negative outcomes [7]. But most of the papers demonstrating for example an increase of medical errors after the installation of a clinical application rarely go beyond simple statement and description, and are unable to identify the true nature of the problems at hand. One can remember that after the publication of the Ross Koppel article [5], Jakob Nielsen, the "usability guru", had vigorously reacted on the Internet and clearly argued that the problems mentioned by Koppel, although not acknowledged as so, were basic usability problems that could have been fixed or prevented by proper usability engineering methods.

The Human Factors problems most commonly encountered are:

(1) paper-based records usually co-exist partly with the electronic record for a while, engendering ergonomics consistency problems, doubloons, loss of data, etc.

(2) New clinical applications may suffer from usability flaws, thus increasing the users workload [8]. They may not fit physicians and nurses thought processes, and they fail providing them the necessary overview of the patient's medical status [9].

(3) The implementation of Clinical Information systems (CIS) modifies the way co-workers communicate about the patient's case or the care process [10], constraining them to shift from a culture characterized by rich oral communications and poor written documentation to a culture of exhaustive data entry but poor oral exchanges. In this change process, they lose the efficient shared representation of the care process necessary to properly coordinate their various actions. In this way, it can be said that current clinical applications don't properly support collective and cooperative decision making and care process.

(4) CIS incorporate national and local regulations, roles and procedures, thus implementing prescribed work models that usually do not correspond to the way people and teams actually carry out their work [11].

Although there are more and more papers in the medical informatics field describing a user-centered approach to the design and installation of clinical applications, and promoting the systematic use of Human Factors Engineering methods [12], it seems that these HFE theories and methods have not actually fertilized the medical informatics field, as it has for other safety-sensitive domains such as aviation, railway transportation, or nuclear power plants. There are very few papers describing the actual use of the HFE methods in design or installation projects of clinical applications, or reporting user-centered engineering or re-engineering of existing applications [9, 13]. Elkin et al [14] reported the ability of Usability testing to improve an academic research environment. The same study team documented the importance of human factors and usability methodology in getting clinicians to document care using a problem list application server [15]. Mc Night reviewed the usability of a system used to automatically create compositional expressions from clinical statements demonstrating the impact of a user-centered design approach [16]. Brown studied the usability of an ontology for naming clinical records which led to improved access to clinical notes at the Veterans Health Administration in the United States [17].

Finally, it remains difficult to prove the efficiency of those methods and their positive benefits/costs ratio, because it is difficult to demonstrate that users' dissatisfaction led directly to errors which should not have occurred, were HFE methods having been used, leading to improved patient outcomes and decreased costs.

2. The Human Factors Engineering technology

According to PC Cacciabue [18], HFE should be defined as a "technology concerned with the analysis and optimization of the relationship between people and their activities, by the integration of human sciences and engineering in systematic applications, in consideration for cognitive aspects and socio-technical working contexts". "By this definition, Human Factors extends the concept of ergonomics, as the science of humans at work, beyond the workplace and behavioral performance to the cognitive and social aspects involved in human activity". [ibid].

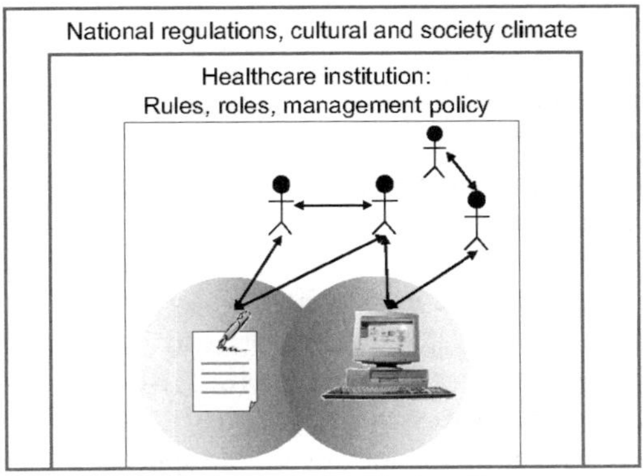

Figure 1: the work situation

Human factors engineering is fundamentally a matter of optimizing the relationship between an individual and his work situation. This work situation includes a physical environment, work devices, and other individuals or co-workers. It is constrained by a social context, represented by rules and roles edited by the institution, management policies, and at a more general level by national regulations, society and cultural climate (*see Fig1*). HFE always starts with an analysis of this work situation, aiming at diagnosing "problems" in this situation: dangerous physical environment, ineffective work devices, poor individual or collective human performances, communication or cooperation impairment, safety problems, violations of regulations, and so on. The purpose of this analysis is to propose recommendations to address the diagnosed problems. The recommendations may concern the physical environment, the work devices, or the organizational context including the way co-workers communicate and cooperate.

The core task of the HFE process is the analysis of the work situation. This analysis is multifaceted and incorporates models and methods from various human sciences. The International Ergonomics Association [19] acknowledges three domains of specialization within the discipline of Ergonomics, which it equates to HFE:

(1) physical ergonomics

(2) cognitive ergonomics, "that is concerned with mental processes, such as perception, memory, reasoning, and motor response, as they affect interactions among humans and other elements of a system. Relevant topics include mental workload, decision-making, skilled performance, human-computer interaction, human reliability, work stress and training as these may relate to human-system design".

(3) organizational ergonomics, that is concerned with "the optimization of sociotechnical systems, including their organizational structures, policies, and processes'.

The focus of HFE depends of the domain or category of work situations to which it applies. In the healthcare domain, when we deal with projects of implementing new clinical applications in the work situation, cognitive and organizational ergonomics are necessarily called for. Therefore, a Human Factors Engineering technology applied to any project of implementing or modifying a clinical application must rely on cognitive and sociotechnical models and methods. As we deal with IT-based work devices, usability methods are also mandatory.

3. The challenge of re-engineering existing applications

In the everyday life of an HFE engineer for healthcare applications, projects involving the design of a completely new clinical application are not so frequent. Most of the time, the projects HFE engineers are called for concern existing applications suffering from ergonomics problems that the users as representatives of their institution on the one hand, or the vendors and designers on the other hand perceive the need for improvement. This situation proves to be even more complex than the user-centered design of a new product. In a re-engineering project, there are numerous constraints that do not exist in a "from scratch" design project. For example new design efforts always start from a business process reengineering (BRP) perspective. Here there is a detailed look at the business processes being modeled and their goals. Benchmarks and metrics are developed to hold the feet of the design team to the fire as the design process moves from a BPR perspective to a functional specification (information models, interaction models, activity diagrams and process charts are all generated) to a detailed systems design based on sound user-centered design principles.

Usually the clinical application to be re-engineered is used in more than one medical department of a given hospital and not uncommonly in more than one hospital. Those different departments and institutions are characterized by different organizations and habits of works that interact with the product to be assessed and improved. Then the analysis of the work situation must be expanded to identify key features of the different organizations that are of interest for the re-engineering project. Similarly, organizational recommendations should be specified for each identified organizations.

When an application has been in use for some period of time in multiple sites, the usability interest of future and ancient users might be contradictory. Ancient users have overcome the sometimes painful learning process of the application. They usually want to limit the re-engineering to the problems they have identified. On the contrary, newcomers' interest could require a more radical transformation of the Human Computer Interface (HCI) to make the man-machine dialog more friendly and intuitive. Studies designed to address these issues need to run on a broad set of typical users who would experience the application.

The usability problems of the application under re-engineering are rarely simple cosmetic problems of the HCI. Usually, we deal with cognitive ergonomics problems that question the compatibility between the users expert knowledge and reasoning and the often inadequate and too poor knowledge encapsulated in the application's data model and procedures. It is an understatement to say that the Companies and their developers are reluctant both to open up their data model or to modify their interfaces.

A complete re-engineering cycle takes time (2-4 years for a CPOE like application for example) and costs money. The return on investment is qualitative as well as financial, and the return on investment is difficult to evaluate.

4. Conclusion: the HFE technology for healthcare applications: a complex and cooperative process

Speaking of Human Factors Engineering Technology emphasizes its practical nature: "engineering and human factors concentrate on the implementation in the real world and working environment" [15, p.13]. If we want to meet the challenge of a rapid adoption of the new generation of clinical applications, we need to promote an actual HFE approach to solve their current problems. The integration of the HFE Technology in a project lifecycle requires a close cooperation between three main partners:

(1) the company developing the clinical application: R&D department, designers, developers

(2) the Healthcare institutions running the application (or planning to): project managers, users representatives, HIS people

(3) the ergonomists in charge of the Human Factors Engineering.

Once the important part of the Analysis of the work situation is performed, the partners need to cooperate to design the future work situation defined by a new re-engineered application and a new re-engineered organization.

Human factors engineering as a technology and usability testing in particular have the capacity to improve the design of our systems. In this informationally intensive era as we move closer to personalized medicine in the genomic age, we will see an increasing dependency of clinical care on computerized clinical decision support. This dependency needs to be usable, understandable and must translate into benefits for our patients and the healthcare system that serves their interests.

References

[1] Ball M, Weaver C, Abbott PA. Enabling technology to revitalize the role of nursing in a era of patient's safety. Int J Med Inf. 2003;69:29-38.

[2] Bates DW, Teich JM, Lee M, Seger D, Kuperman GJ, Ma'Luf, et al. The impact of Computerized Physician Order Entry on medication error prevention. J Am Med Inform Assoc. 1999;6(4):313-21.
[3] Gorman PN, Lavelle M, Ash J. Order creation and communication in healthcare. Meth Inf Med 2003;42:376-84.
[4] Berger RG, Kichak BA. Computerized Physician order entry: helpful or harmful? J Am Med Inform Assoc. 2004;11:100-3.
[5] Koppel R, Metlay JP, Cohen A, Abaluck B, Localio R, Kimmel SE, et al. Role of Computerized Physician Order Entry Systems in facilitating medication errors. J Am Med Assoc. 2005;293(10):1197-203.
[6] Han YY, Carcillo JA, Venkataraman S, Clark RS, Watson , Nguyen T, et al. Unexpected increased mortality after implementation of a commercially sold Computerized Physician Order Entry system. Paediatrics 2005;116:1506-12.
[7] Ash J, Berg M, Coeira E. Some unintended consequences of Information Technology in Health Care: the nature of patient care information system-related errors. J Am Med Inform Assoc. 2004;11:104-12.
[8] Shu K, Boyle D., Spurr C., Horsky J., Heiman H., O'Connor P., et al. Comparison of time spent writing orders on paper with computerized physician order entry. Medinfo. 2001;10(Pt2): 1207-11.
[9] Pelayo S, Leroy N, Guerlinger S, Degoulet P, Meaux JJ, Beuscart-Zephir MC. Cognitive analysis of physicians' medication ordering activity. Stud Health Technol Inform 2005;116:929-34.
[10] Beuscart-Zéphir MC, Pelayo S, Anceaux F, Meaux JJ, Degroisse M, Degoulet P. Impact of CPOE on doctor-nurse cooperation for the medication ordering and administration process. Int J Med Inform. 2005 Aug;74(7-8):629-41.
[11] Aarts J, Doorewaard H, Berg M. Understanding implementation: the case of a Computerized Physician Order Entry system in a large dutch university medical center. J Am Med Inform Assoc. 2004;11(3):207-16.
[12] Jiajie Zhang (Guest Editor). Human-centered computing in health information systems. Part 1:Analysis and design. J Biomed Inform. 2005;38(1):1-88.- Part 2: Evaluation. J Biomed Inform. 2005;38(3):173-255.
[13] Horsky Y, Kuperman GJ, Patel VL. Comprehensive analysis of a medication dosing error related to CPOE. J Am Med Inform Assoc. 2005;12(4):377-82.
[14] Elkin PL., Sorensen B., De Palo D., Poland G., Bailey KR., Wood DL., La Russo NF. Optimisation of a research web environment for academic internal medicine faculty, J Am Med Inform Assoc. 2002; 9(5):472-8
[15] Elkin PL., Mohr DN., Tuttle MS., Cole WG., GE., Keck K., Fisk TB., Kaihoi BH., Lee KE., Higgins MC., Suermondt HJ., Olson N., Claus PL., Carpenter PC., Chute CG. Standardized problem list generation, utilizing the Mayo canonical vocabulary embedded within the Unified Medical Language System. Proc AMIA Annual Fall Symp. 1997; 500-4
[16] Mc Knight LK., Elkin PL., Ogren PV., Chute CG. Barriers to the clinical implementation of compositionality. Proc AMIA Symp. 1999; 320-4
[17] Brown SH.. Hardenbrook S., Herrick L., St Onge J., Bailey K., Elkin PL. Usability evaluation of the progress note construction set. Proc AMIA Symp. 2001; 79-80
[18] Cacciabue PC. Guide to applying Human Factors methods. London: Springer-Verlag; 2004.
[19] http://www.iea.cc/ergonomics/(last accessed Jan. 26, 2006).

Ubiquity: Technologies for Better Health in Aging Societies
A. Hasman et al. (Eds.)
IOS Press, 2006

Modeling a Health Telematics Network: Does the 3LGM² Approach assist in its Management and Operation ?

Martin STAEMMLER[a,1]
[a] *University of Applied Sciences, Stralsund, Germany*

Abstract. Health Telematics Networks (HTN) are characterized by a complex setup and interrelations. Using available tools and methods focusing on mainly one aspect, e.g. its functionality or the network infrastructure leads to a restricted view. The objective of this paper is to assess the applicability of the Revised Three-layer Graph-based Meta Model (3LGM²) - developed for modeling hospital information systems - towards health telematics networks. Having identified an approach on how to represent the hospitals effectively in the model the 3LGM² proved to support strategic management, day-to-day maintenance and documentation.
Keywords: Health Telematics, Information System Modeling and Management, Layered Architecture, 3LGM²

1. Introduction

Health Telematics Networks (HTN) are used to extend the enterprise functions of a typical hospital information system (HIS) by linking external resources and providing remote services to support diagnosis and treatment of patients locally. As such, HTN have to be integrated with each local HIS, at least to a level which guarantees seamless communication of patient data. However, the HIS, the health telematics application(s) to be used, the communication infrastructure and institutional policies appear to be different with each partner in a typical regional HTN and results in a challenging integration task.

Modeling has been used in various ways: (i) to automatically derive a network infrastructure using ICMP and SNMP, (e.g Scotty [1]) as public domain tool or commercial tools, (ii) to describe information system architectures[2, 3], (iii) to provide health specific reference information system models e.g. HISA[4], HL7-RIM[5], CORBAmed[6] and (iv) to document and analyze business processes e.g. with ARIS[7] or Bonapart[8]. However, most of the model approaches address only a part of the requirements to allow strategic management and day-to-day operation of HTN. With the 3LGM² the Institute for Medical Informatics, Statistics and Epidemiology, Leipzig [9, 10] has developed an extensive meta model together with a tool for modeling HIS. It uses three layers to represent enterprise functions on a domain layer, their mapping

[1] Martin Staemmler, University of Applied Sciences, Stralsund, Medical Informatics, Zur Schwedenschanze 15, D-18435 Stralsund, Germany, martin.staemmler@fh-stralsund.de

to application components on a logical tool layer and a physical layer to include data processing components and communication.

The objective of this paper is verify if the application scope of the 3LGM² can be extended to model a HTN and to assess its usefulness for the strategic management as well as for the day-to-day support and maintenance.

2. Materials and Methods

2.1. The Health Telematics Network (HTN)

The "Health Telematics Network for the Support of Tumor Diagnosis and Treatment in the Euroregio Pomerania" extends to more than ten hospitals in the north-eastern part of Germany and in the western part of Poland [11, 12]. In general, the hospitals involved utilize a subset of the whole set of enterprise functions available:

- teleradiological services (emergency consultation, second opinion),
- telepathological services (perioperative instantaneous section, second opinion),
- telecardiological services (ECG analysis and reporting, second opinion prior to cardiac intervention),
- teleconferencing (with two and more partners).

These interdisciplinary enterprise functions are mapped to dedicated applications, thereby introducing mutual dependencies and requirements for the network topology (point-to-point versus hub-and-spoke). Additionally, the communication is based on a variety of access technologies, like dial-up lines using up to eight ISDN channels, leased lines, Internet based communication and radio links.

2.2. The Revised Three-layer Graph-based Meta Model (3LGM²)

The 3LGM² has been primarily developed to model enterprise functions and associated applications and systems for information managers in hospitals [9, 10]. It holds three layers and defines relations to link adjacent layers.

- The domain layer models enterprise functions and entity types. Enterprise functions may use or create information about an entity of a given entity type, e.g. ADT functions use/create information of the entity patient.
- The logical tool layer holds application components, which represent dedicated tasks within an application. Communication is achieved by component to component interfaces or between a component and the user by means of an UI. As an example one could consider the ADT component sending messages to the laboratory system or the laboratory using a web-based GUI to present lab results.
- The physical tool layer contains components to support the two upper layers. This may be hardware systems (computer, network components, …) or even persons e.g. for handling a paper document based archive.
- Links between functions of the domain layer with application components of the logical tool layer reflect the relations "can support", "based on" or "triggered by" e.g. the ADT function is supported by an ADT component customized for this usage with a configuration. Similarly a link from the

logical tool layer to a processing component in the physical layer represents the mapping of an application component via a configuration to a processing component.

The 3LGM² tool does not only allow for the visualization of the above-mentioned items but provides analysis means to follow the execution of an enterprise function, as it is worked on by one or more - potentially communicating - application components and implemented using hardware and network resources. In addition, these analysis facilities motivated the use of 3LGM² for a HTN in order support questions like "What is required for providing a second opinion service for a CT-scan between hospital A and B?".

2.3. Deriving the HTN-Model

Comparing the 3LGM² target domain "hospital" with a HTN it refers not only to one hospital but to several hospitals. Three different approaches to represent this fact in the model have been investigated.

The first approach relies on using separate models for each hospital as indicated in Figure 1. As a result the models shown represent two hospitals with identical enterprise functions (F1, F2, F3) and entity types (ET1, ET2) in the domain layer, some minimal differences between the application components (AC1 … AC4) in the logical layer and a mapping on separate physical components in the physical layer (PC1 … PC8). The drawbacks of this approach are (i) a repetitive modeling for at least the domain layer and (ii) more important the inability to perform an analysis across separate models e.g. for answering the question above.

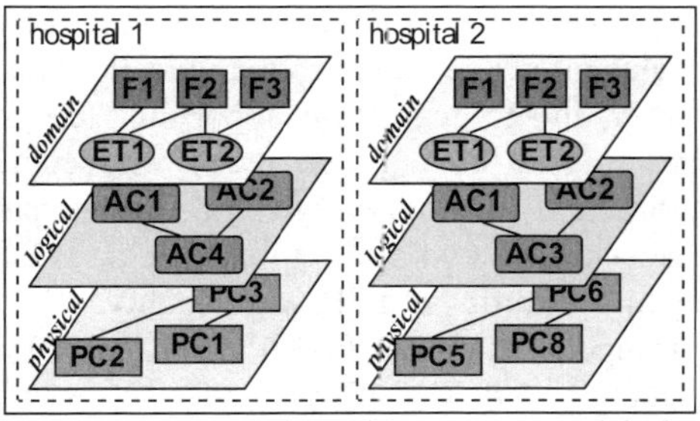

Figure 1: Approach 1 with separate models for each hospital

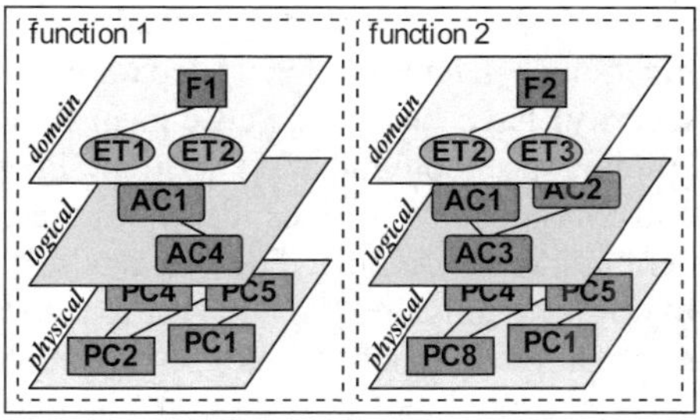

Figure 2: Approach 2 with separate models for each enterprise function

The second approach groups into separate models according to the enterprise function F1 and F2 and is shown in Figure 2. Consequently these functions may use or create information for entities types, which are quite similar. Looking at the logical and the physical layer one finds different applications and physical components reflecting the resources of each hospital provided for a particular function. This approach causes duplicated items (e.g. AC1, PC4, PC5 and PC1) and thereby conceals a clear view.

The third approach focuses on avoiding the disadvantages from the ones above, however with the implication, that the definition of a physical component had to be extended to include a hospital as a grouping element (identified by the rectangles H1 and H2 in Figure 3. From this approach it becomes obvious, which enterprise functions are supported within the HTN (as depicted in the domain layer) and their relationship to application components.

Having modeled the inter-layer relations and dependencies an answer to a particular question could result in Figure 4, which maps the function F2 to the application component AC4 and the physical component PC6 as well as the function F3 to the application component AC5 and the physical component PC1. It is worth noting, that this approach would result in two paths if an identical enterprise function is performed between two hospitals in a cooperative session.

Since this approach provides a clear view on the HTN functionality it has been chosen for modeling.

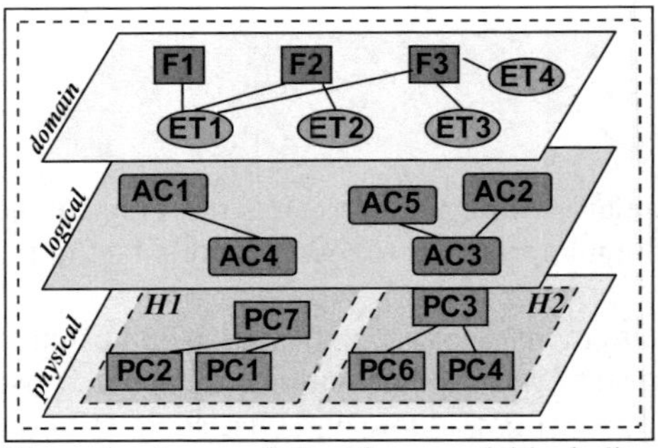

Figure 3: Approach 3 with an model including all hospitals

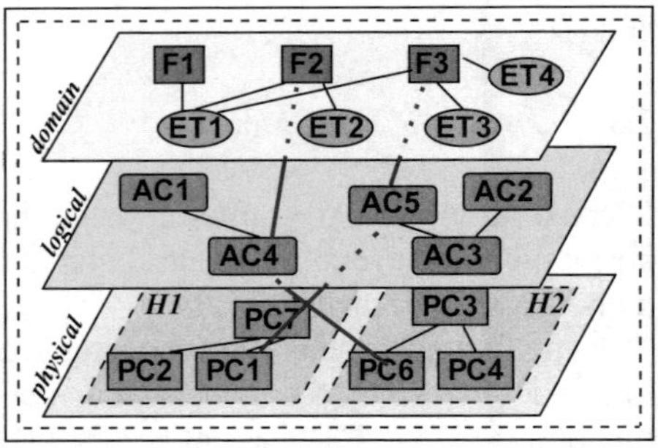

Figure 4: Approach 3 together with inter-layer relations

3. Results

Figure 5 shows the model derived as a layered view. On the top layer the enterprise functions (teleradiology, telepathology, teleconference and communication/data security) have been detailed e.g. in order entry, image generation, reporting, second opinion, pseudonymization (patient identity replaced by a pseudonym) and secure transmission. Entity types identified are e.g. order, JPEG images, DICOM images, patient and tissue sample. The two types of images reflect two different usages: (i) JPEG for viewing only and (ii) DICOM for viewing and reporting at full image quality.
Application components are for example grouped by "Web-Viewer" and "DICOM Viewer" and hold sub-components like "Query", "DICOM Download", "Viewer". Finally the physical layer depicts a view on five hospitals with a detailed representation of their individual network infrastructure. The arrows between these hospitals and communication providers exhibit different routes between partners and clearly show potential and build-in fallback communications paths in case of a fault. Additionally, Figure 5 highlights a subset of the inter-layer relations modeled according to the selected function "emergency consultation".

4. Discussion

The 3LGM² meta model and tool has been targeted for modeling HIS. One of the main purposes was to perform an assessment towards the applicability to a HTN which is characterized by several institutions forming the HTN. Investigating three different approaches resulted in an approach with a coverage of all institutions in one model, but also interpreting each hospital as a physical grouping component on the lowest layer.

This is somewhat contradictory to the definition given with the 3LGM² but simplifies the assignment of real physical components to each hospital. An alternative approach would have been to avoid the grouping and to assign real physical components by using a naming convention to a grouping for each hospital, which however would make the graphical representation less meaningful.

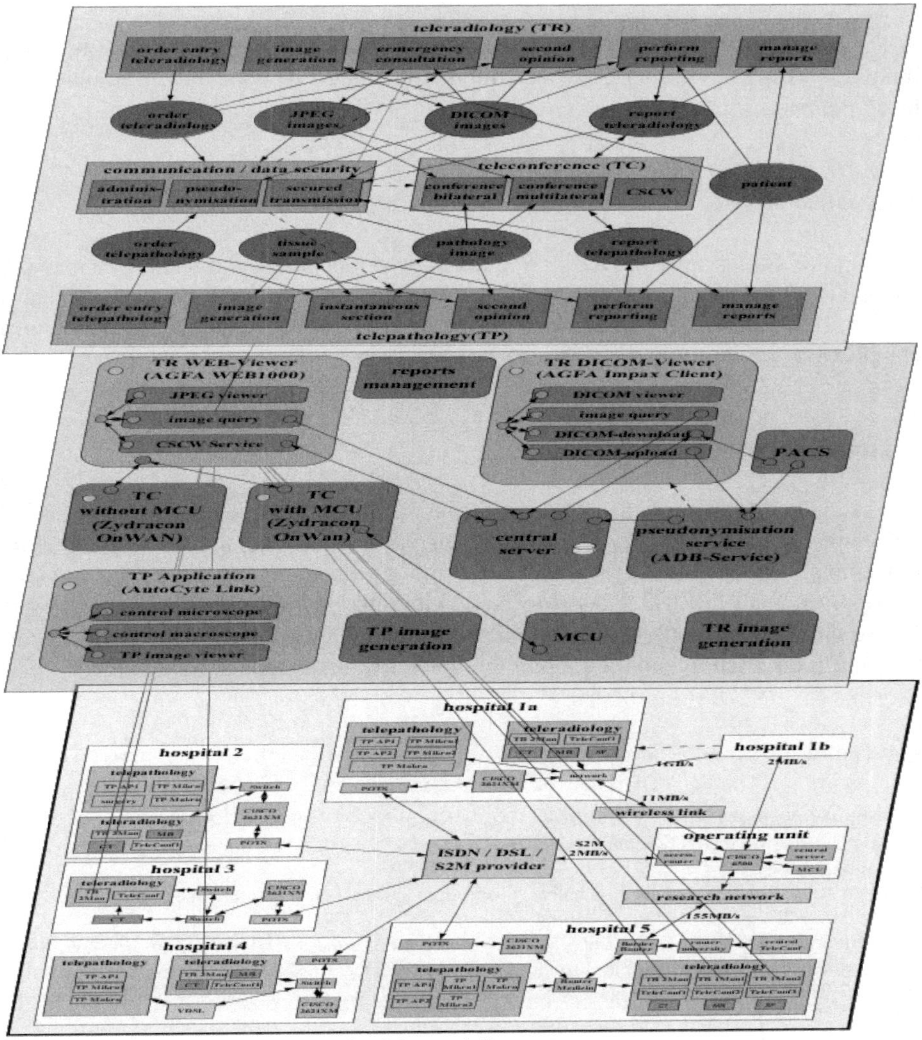

Figure 5: The three layers as a result of the modeling.

HTNs exhibit quite some diversity reflecting heterogeneous institutional policies and communication links. Using the 3LGM² tool this variety could be represented in the resulting model and additionally gave insight in potential fallback issues. Compared to typical network management tools [1], which fail to refer to applications, the 3LGM² model widens the scope from enterprise functions and their relation up to network components. With the analysis function the model serves to provide error analysis and maintenance support and can in particular be used for effective documentation. For

enhancing this usage 3LGM² would benefit from a version control mechanism and references to be kept as attributes with the items and dependencies modeled.
Compared to process oriented tools using ARIS[7] or Bonapart[8] which support simulation in time, the 3LGM² model represents only static properties of the items of each layer. Having the path identified from enterprise functions down to the physical layer, an extension to the 3LGM² tool could support simulation for enterprise functions as well.
When working with the 3LGM² tool some minor handling problems occurred, e.g. a limited workspace, missing grid lines, limited drawing functions and/or an adjustable level of detail.

5. Conclusion

The 3LGM² successfully proved its applicability to model a health telematics network. (HTN) The model obtained serves for at least three aspects: strategic management of such a HTN, day-to-day maintenance and in depth documentation of the typical complex HTN setup.

Acknowledgement

The modeling would not have been possible without access to the project "Health Telematics Network supporting Tumor Diagnosis and Therapy in the Euroregio Pomerania" which has been granted by the European Commission for regional development (INTERREG A) together with the support of the federal state Mecklenburg-Vorpommern. Furthermore, the author wants to thank Lars Kubowicz for the fruitful discussions and the support in implementing the model.

References

[1] http://www.ibr.cs.tu-bs.de/projects/scotty/ (last accessed Jan. 29, 2006).
[2] Bernus P., Mertins K., Schmidt G. Handbook on Architectures of Information Systems, Berlin, Springer, 1998
[3] Van de Velde R., Degoulet P. Clinical Information Systems, Health Informatics Series, New York, Springer 2003
[4] Ferrara FM. Healthcare Information Systems Architecture. Dudeck J., Blobel B. Lordieck W., Bürkle T. editors. New Technologies in Hospital Information Systems. Amsterdam, IOS Press 1997, pp. 1-10
[5] HL7-RIM (Reference Information Model), http://www.hl7.org -> DataModels -> RIM (last accessed Jan. 29, 2006).
[6] http://www.omg.org/technology/documents/spec_catalog.htm (last accessed Jan. 29. 2006)
[7] Seidlmeier H. Process Modelling with ARIS, Wiesbaden, Vieweg, 2004
[8] Bonapart, IntraWare AG, www.intraware.de (last accessed Jan. 29. 2006)
[9] Wendt T., Brigl B., Winter A. Modeling Hospital Information Systems (Part 1): The Revised Three-layer Graph-based Meta Model 3LGM². Methods Inf Med 2003, 42(5):pp 544-51
[10] Wendt T., Häber A., Brigl B., Winter A. Modeling Hospital Information Systems (Part2): Using the 3LGM² Tool for Modeling Patient Record Management. Methods Inf Med 2004, 43(3):pp 256-67
[11] Staemmler M., Ehricke H., Dräger J., Hosten N. Interdisziplinäres Netzwerk zur Unterstützung der Tumorversorgung in der Euroregion Pomerania. Forum der Medizin_Dokumentation und Medizin-Informatik 2004; 1(6): 14-18
[12] http://telemedizin-mv.fh-stralsund.de (last accessed Jan. 29. 2006)

6.2 Modeling Healthcare

Ubiquity: Technologies for Better Health in Aging Societies
A. Hasman et al. (Eds.)
IOS Press, 2006

Agent Based Simulations in Healthcare

Ugur BILGE[a,1] and Osman SAKA[a]
[a] *Akdeniz University Faculty of Medicine, Antalya, Turkey*

Abstract. Agent Based Simulations (ABS) is a relatively recent computer paradigm [1], [2], [3]. As opposed to "top down" conventional computer simulations, the ABS approach is a "bottom-up" modelling technique where a medium to high number of independent agents is modelled. These agents' interactions sometimes cause unexpected "emergent" system behaviour. ABS is particularly suitable in the social context such as healthcare where a large number of human agents interact and co-operate for common goals. Today ABS in the social context is often used together with the recently introduced network analysis techniques [4], [5] and network visualization tools for modelling and simulating social agents within organisations. At Akdeniz University we are starting a number of projects for applying ABS technology in healthcare. In this paper we present two of the ongoing projects in this field. Firstly we have developed a prototype simulator for the long term monitoring of Chronic Obstructive Pulmonary Disease (COPD) as a major public health problem. We present the COPD simulator, its agents, parameters and working principles. Secondly we want to apply ABS and the network analysis techniques to visualise and explore informal social networks amongst staff at the Akdeniz University Hospital to assess and evaluate properties of the organisation in terms of its ability to innovate and share knowledge. In our applications, we primarily aim to use ABS in a web-based platform to create a virtual environment for discussion, visualising and running what-if scenarios to test out various options for managing healthcare, as well as sharing information and creating a virtual community.
Keywords: Agent Based Simulations, Network Analysis, COPD, Innovation, Complex Social Systems.

1. Introduction

Agent Based Simulations are a relatively new computer modelling approach. They are suitable for real life problems where there is a medium to large number of interacting agents. Unlike conventional computer simulations, ABS puts a strong emphasis on modelling low level agents or actors of a system, and observing sometimes unexpected patterns of system behaviour. Often the development of an ABS is in itself useful and educational, and the end product is not necessarily used to predict the future, but is more to experiment with ideas, and test possible real-life scenarios in a safe environment. Swarm [1], one of the early environments built in the USA, is a platform for experimenting with ABS. Work on social systems, trading strategies and the spread of culture are simulated in the SugarScape [2] environment. Both the above approaches are abstract, displaying agents as colourful dots on graphical grids. The idea of using computer simulations as "laboratories" for many real-life problems has been gaining

[1] Corresponding Author: Dr. Ugur Bilge, Akdeniz University, Faculty of Medicine, Dept of Biostatistics and Medical Informatics, Antalya 07059, Email: ubilge@akdeniz.edu.tr

momentum [3]. TRANSIMS [6], the first of its kind, modelled Albuquerque's highway transport system with great detail at the level of each household and car. In the UK, SimStore supermarket simulator [7], [8] was developed as a business tool where supermarket customers have been modelled as autonomous agents fulfilling their shopping lists by moving in a virtual store.

In this paper we present our efforts in applying ABS technology to healthcare problems. We present two examples; firstly a COPD simulator for modelling the impact of the disease as a public health problem in the coming years, secondly we are working on the application of ABS to explore informal networks and connectivity at a university hospital.

2. Material and Methods

2.1. COPD Simulator

COPD is an umbrella term, which covers smoking-related conditions such as chronic bronchitis and emphysema and is often called ‘the hidden killer’ because of its lack of public profile. It kills approximately 30 thousand people a year in the UK [9]. Lack of awareness among healthcare staff is leaving many cases of COPD undiagnosed when early diagnosis can reduce treatment costs considerably.

The COPD simulator [10] originally motivated by National Health Service (NHS) initiatives in the UK, aimed to simulate a NHS region with a population of around 100,000 people. We are now looking for a self-contained pilot region in Turkey for applying the simulator. The simulator models a large number of people, healthcare units and random and planned events. We envisaged a simulation period of 15-20 years starting from 2000 going into 2015-2020. The simulator enables users running what-if scenarios in terms of investment decisions as well as taking a number of programmed possible actions for the management of the disease. The simulation runs in weekly simulation steps, and at every step it generates the total number of patients, their COPD stage of severity, quality of service they receive and total cost to the health service. Below are the details of the software agents that the system models.

2.1.1. Person Agents

So-called person agents (actual people in a region) are modelled using statistical distribution data available for the region. Each person agent has properties such as:

- Age, Gender, Occupation, Education
- Smoking Habits
- COPD Stage, Severity
- Diagnosis, Co-morbidities
- Willingness to come to appointments, take medicine

To make the simulation as realistic as possible, we use available data on COPD prevalence, rate of smoking, mortality rates (caused by COPD and other reasons).

2.1.2. Healthcare Units

In addition to the person agents, the model contains healthcare units such as:

- Primary Health Clinics

- Emergency Services
- Hospitals
- Pulmonary Rehabilitation Units

We are also able to model a large number of healthcare worker agents working in the above units. Healthcare units can start Smoking Cessation and Flu Immunisation schemes, open COPD Rehabilitation Centres and start Early Diagnosis initiatives including the use of Spirometry, all entailing costs to the health service, and bringing benefits to the public health, also reducing COPD mortality rates. At this stage we only need an approximate cost model for patient care at different stages of COPD, and an understanding and a simple model of benefits of early diagnosis.

2.1.3. Events

The simulator models a number of random, and some user-defined events such as:

- Weather events (dry/humid and cold spells) increasing workload for Emergency Services, Primary Health and Hospital Units
- Flu Events (light, medium or intense epidemics), similar to above
- A number of Health Service events such as staff training, restructuring, increased use of spirometry resulting in costs and potential benefits

Throughout the simulation process, patient agents are influenced by the events described above. They may develop flu, may need primary care, acute treatment, outpatient or pulmonary rehabilitation. Within the simulation, agents become satisfied or dissatisfied depending on the staffing levels and quality of service they receive, and agents die and new agents are born in line with given statistical distributions.

2.1.4. Design and Development

COPD simulator has been designed and developed by SimWorld Limited [8]. The software is written as a Java applet as it is a platform-independent and web-based technology. The simulator runs on any operating system equipped with a Java Virtual Machine. The second version is now complete and it is producing plausible estimates of number of COPD patients for the future. Users can access simulators on a website, test out their ideas, run what-if scenarios, and view outcomes.

2.2. Informal Social Networks in a Hospital

It is known that organisational forms and connectivity have a significant impact on an organisation's ability to:

- use its resources efficiently
- improve its fitness for survival
- exploit innovation and
- co-evolve with the changing environment

Theoretical work by Stuart Kauffman in connectivity in abstract networks [11] and the growing interest in social networks [4], [5] provided the background for this area. The Organisational Forms Simulator (OFS) [12] has been developed for the EPSRC funded ICoSS Project [13] at London School of Economics, UK. We used ABS to model and visualise social, informal networks in an organisation, and investigate connectivity patterns, to identify hubs and lynchpins, communications bottlenecks, primarily focusing on the "who knows who" question in a business organisation. A university research hospital such as Akdeniz University Hospital provides a suitable application domain for the simulator.

2.2.1. Organisational Forms Simulator (OFS)

Organisational Forms Simulator is a network analysis toolkit designed with agent based philosophy. It (i) displays a given network by spreading out the nodes and links, (ii) helps visualise a given network of people and their connection links in a number of "dimensions", (iii) calculates a number of network properties, such as total distance, degree of separation, distribution of skills in the network, (iv) can carry out what-if analyses by adding, deleting or moving nodes or links in the network, (v) runs what-if queries should an epidemic of good ideas start from a person or a group of individuals.

2.2.2. Methodology

We start by selecting a small department of 10 to 100 people as a pilot. The unit is preferably a relatively self-contained one with little external connections. We assume that people working in an organisation have connections with others in different "modes". So we break down each connection link and each person's competencies into a number of "dimensions". Initially four dimensions; "Team", "Business", "Technical" and "Social" aspects were used as the four dimensions of connectivity between people in the workplace. Then a web or email-based questionnaire is prepared, using the questions identified as important by the company, regarding informal networks. The OFS uses data from the questionnaire results to simulate the network. Using the OFS, we then visualise the network and examine the quantitative connectivity measures such as most utilised and most consulted people, calculated by the simulator. As the modelling approach is agent based, we can also look into the dynamic properties of organisational networks. For example we can test how long it would take for an organisation to adopt a good idea ("epidemic of innovation").

3. Results and Discussion

The prototype for the COPD simulator (Figure 1) has been developed with all known parameters and input files identified. The Graphical User Interface displays the total number of COPD patients treated by the Healthcare units.

The ICoSS Project has already used OFS (Figure 2) for exploring social networks in the work place. We are now in the process of applying the technology to hospital staff by preparing a questionnaire which reflects the aspects of social networks, information sharing and innovation in healthcare.

In developing ABS, there are sometimes difficulties, such as lack of data or a good understanding of the processes involved. When data is not available, statistical distributions of populations are used to generate agents. Where available, simulations do use real data and properties of real agents. The past data is then used to validate the system, and after that the system can be used to speculate about the future. The main strengths of ABS are:

(i) The development of ABS in itself is educational, and generates insights as a result of looking at the problem with modelling in mind. ABS enables us to look at a problem from multiple perspectives, such as from an individual's viewpoint, or the system as a whole.

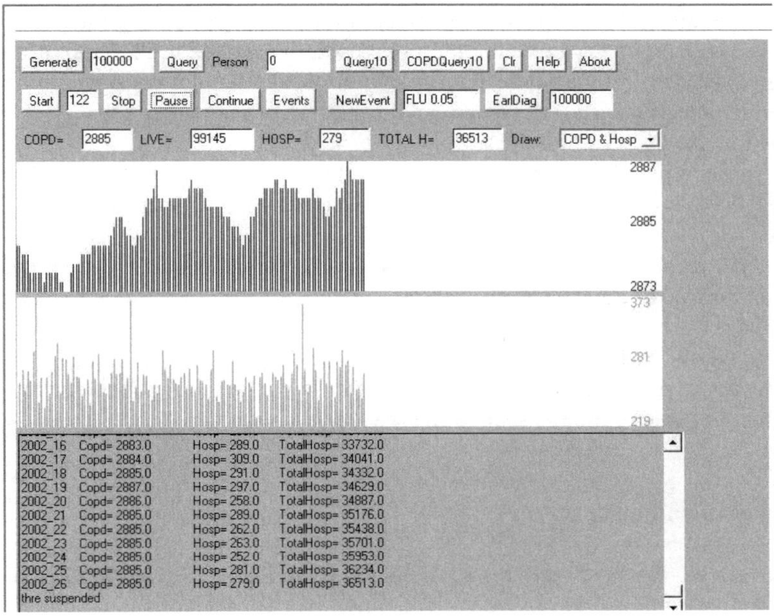

Figure 1. COPD Simulator

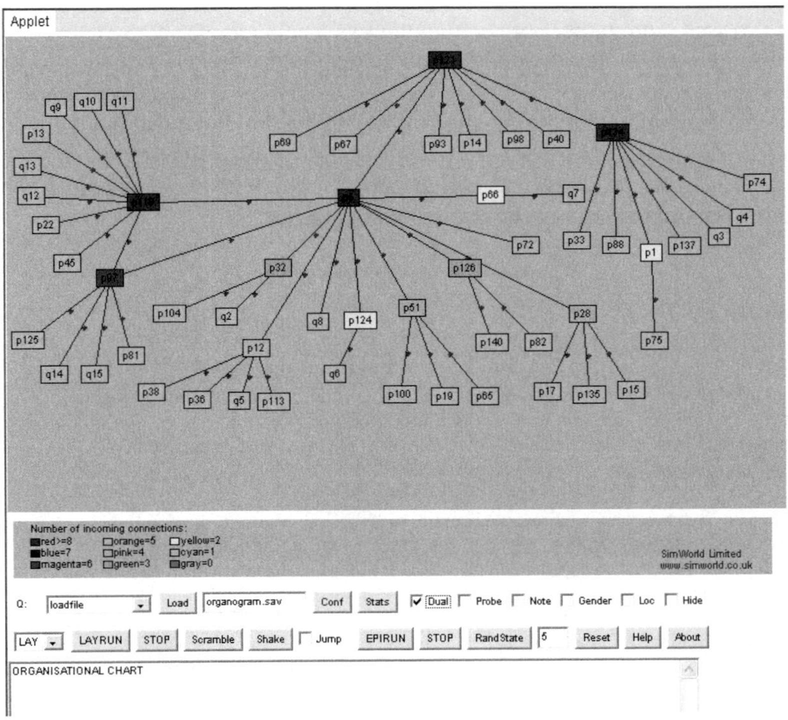

Figure 2. Organisational Forms Simulator

(ii) An additional benefit of ABS research methodology in social networks is that as we use data from interviews and questionnaires, this can be used to monitor changes in the organisation. We can then see the evolution of the pilot organisation, new and redundant connections, staff development through time, and perceptions.

The drawbacks in using ABS are:

(i) Consecutive simulation runs can provide different results. Due to the random nature of behaviour of agents and events, simulations can display radically different outcomes.

(ii) ABS technique should be used with an aim to understand a system rather than predicting how it may behave in the future.

(iii) There are no off-the-shelf software packages available. Even though a number of generic tools under development provide network visualisation and animation facilities, none provides functionality suitable for complex issues arising in the management of organisations, such as healthcare and hospitals.

4. Conclusion and Future Work

The use of Agent Based Simulations in Healthcare is new. In this paper we presented two applications of the technology. As healthcare is a complex inter-disciplinary domain it requires complex tools such as ABS. The examples we provide here are by no means limited, and in the coming years we will see more applications of ABS in healthcare. With the COPD simulator we want to set up a discussion group focusing on COPD related ideas on the same website with the simulator. We want to provide this service together with setting up a self-help patient internet community for people living with COPD. We also plan to apply the technology to other chronic disease management tasks. With the Organisational Forms Simulator, we are in the process of preparing a questionnaire for investigating the informal networks and information pathways that exist between university hospital staff. We believe the system will help us to assess the current state of the network and help the hospital management to improve connectivity in the organisation.

References

[1] SWARM (www.swarm.org last accessed Jan. 1, 2006)
[2] Epstein JM, Axtell R. Growing Artificial Societies – Social Science from the bottom up; 1996
[3] Casti JL. Would-be Worlds, John Wiley & Sons; 1997
[4] Barabasi AL. Linked – The new science of Networks, Perseus Publishing; 2002
[5] Buchanan M. Small World: Uncovering nature's hidden networks, Weidenfeld & Nicolson; 2002
[6] TRANSIMS (www.transims.net last accessed Jan. 1, 2006)
[7] Casti JL. Firm Forecast, New Scientist; 24 April 1999; 2183:42-46
[8] SimWorld Limited (www.simworld.co.uk last accessed Jan. 1, 2006)
[9] COPD International (www.copd-international.com last accessed Jan. 1, 2006)
[10] Bilge U, Honeyman A, Saka O. Agent Based Simulations for COPD, Second Middle East Conference on Healthcare Informatics, Dubai; April 2005
[11] Kauffman S. At Home in the Universe, Penguin Books; 1995
[12] Bilge U. Modelling Connectivity in the Work Place, ICoSS Project, London School of Economics (www.psych.lse.ac.uk/complexity/ICoSS/Papers/ConnectivityReportUB.pdf last accessed Jan. 1, 2006)
[13] Integration of Complex Social Systems Project (www.psych.lse.ac.uk/complexity/icoss.htm last accessed Jan. 1, 2005).

Ubiquity: Technologies for Better Health in Aging Societies
A. Hasman et al. (Eds.)
IOS Press, 2006

Change readiness research
A qualitative study of variations in participation

Anna Marie Høstgaard
Department of Development and Planning, Aalborg University, Denmark

Abstract. The Change readiness research method (CRR) has become a well-known method in Denmark to identify issues needed to be discussed on a hospital ward before implementation of a new IT-system and to start a dialogue. A precondition for a constructive dialogue, however, is a high degree of participation. The latest experiences of the CRR method were gained from its use in eight wards in the Danish Gepka project during 2003-4 (The Gepka project was established by The Danish Ministry of the Interior and Health, The National Board of Health, the County Council Society and H:S. Its purpose is to validate the "Basic Structure for The Electronic Health Record" (B-EHR) using prototypes. http://medinfo.dk/epj/proj/gepka/). In the Gepka project the participation varied from 33.3% to 78.9%. The objective of *this* study is to set out themes by which this variation can be studied. A qualitative explorative research design has been applied, where four instructions from the "Instruction for use" (Instructions for using the CRR method. Can be downloaded the Internet: (http://www.epj-observatoriet.dk/publikationer/forandringsparathed.pdf)) have been studied as themes. The methods used have been telephone interviews and direct observations. The results showed that the seven wards (one was excluded) followed the "Instructions for use" to different degrees. It was found that one instruction, in particular, seems to be especially important to follow to motivate the employees on a ward to participate in the CRR; the management of the ward must be engaged/actively involved in the project, as they are key figures when it comes to motivating the other ward employees. The aim of this study is *not* to prove a causal relationship between the degree to which the "Instructions for use" are followed and the degree of participation – it is to suggest a qualitative relationship between the two. Neither does this study try to generalize the results, as further research on more wards would be needed to do so. This study does, however, set out themes that can be a useful tool in future CRR projects in order to maximize the degree of participation. In a modified way, these themes can probably be used as a tool in other studies of human – machine interactions.
Keywords: Organizational issues, Change readiness, Electronic Health Record.

1. Introduction

The current national strategy for IT in the Danish Healthcare sector has established the goal of implementing an EHR in all Danish hospitals within the next few years (1). However, research both in Denmark and abroad has shown that not all EHR implementations are successful, and that one of the most significant reasons is that only limited resources have been applied to preparing the employees for the changes, e.g. in discussing new systems prior to implementation in order to balance clinical needs and the expectations for the new systems (2-8). During recent years we have, at Aalborg University, been developing a method that can be used to assess change readiness among hospital employees and to assist in starting a dialogue among employees and between employees and management prior to the implementation of new IT systems; the "*Change readiness research method" (CRR)* (8) the expectations for the new systems (2-8). During recent years we have, at Aalborg University, been developing a

method that can be used to assess change readiness among hospital employees and to assist in starting a dialogue among employees and between employees and management prior to the implementation of new IT systems; the “*Change readiness research method” (CRR)* (8).
The present coverage of EHR in the Danish counties and hospitals varies widely. Therefore, the national IT- strategy also contains initiatives striving to attain a coordinated development and implementation of EHR in the Danish healthcare sector (9). A precondition for this, however, is the development of a model that sets out both a common structure and a common application of concepts for the EHR, in order to make it possible to utilize information across organizations in the Healthcare sector independent of system suppliers. The National Board of Health in Denmark has developed such a model; the ”Basic Structure for Electronic Health Records” (B-EHR) (10). During the past two years, prototypes of the B-EHR have been evaluated on selected hospital wards in the “GEPKA project, where one of the areas of focus has been studied using the CRR method.

2. Conceptual framework

The Danish CRR method is based on an organisational theory developed by the American researchers Nancy M. Lorenzi and Robert T. Riley. Through intensive studies within the American healthcare sector they identified the three key components that must be addressed when implementing new IT-systems: Hardware, Software and Peopleware. Peopleware refers to the organizational or “human” aspects of implementing, e.g. EHR (3). Based on their findings Lorenzi and Riley developed a research method designed as a questionnaire for examining change readiness. At Aalborg University we have, during the past few years, adapted the American method to Danish conditions by using the method on several Danish hospital wards. A significant difference between the American and the Danish methods is that the questionnaire forms only *part* of the Danish method, as we have found that it does no more than provide indications as to the kind of problems associated with implementation of a new IT-system. Hence, in Denmark the questionnaire is regarded as a tool for providing information about factors causing resistance to new IT-systems and as a catalyst for starting a dialogue among employees and between employees and management prior to implementation. The questionnaire *must*, therefore, be followed by interviews through which it is possible to probe further into problems shown by the results of the questionnaire and to start/continue a dialogue. The main reason for the contrasting approaches is that the hospital systems in USA and Denmark have very different natures, with one of the factors characterizing the Danish health care sector being a tradition for involving staff members when assessing new tasks. The differences between the American and the Danish methods reflect the fact that they have different objectives, and these objectives have moved even further apart during the period in which the method has been used in Denmark. In the USA the method can be seen as a managerial tool to promote a change that has already been decided upon by the management, while in Denmark it can be seen as a participatory tool to identify the concerns of the employees - hereby starting a dialogue among employees and between employees and management.

3. Introduction and Objective

The most recent experience with the CRR method is from its use in the Gepka project during 2003-4, where the degree of participation of the eight participating wards turned out to be very different. It has been of great interest for us ("Us" and "we" refer to members of The EHR-Observatory that was established in 1998 of the Danish Ministry of the Interior and Health to monitor the development of EHR in Denmark: http://epj-observatoriet.dk) *to set out themes by which this variation can be studied.* Such themes can be a useful tool in future CRR projects in order to maximize the degree of participation.

4. Methods and material

A qualitative explorative research design has been applied. Four instructions from the "Instruction for use" - all directly related to the degree of participation - have been studied as themes:

Appointment of an interdisciplinary steering committee

Adaptation of the questionnaire by an interdisciplinary team from the ward

Informing the employees about the CRR

Support for the CRR by the management – all the way up

Direct observation has been the main method but, due to the fact that we have not been involved to the same extent in all the wards, it has been necessary - subsequently to completion of the Gepka studies - to hold telephone interviews with some of the wards to clarify different conditions. The wards involved in *this* study are the same wards participating in the Gepka project (table 1).

Table 1: Questionnaires handed out/in and degree of participation in the Gepka- CRR study.

Hospital	*Ward*	*Handed out**	*Handed in**	*Participation %*
1. Gentofte R	Cardio-thoracic Surgery ward R	152	84	55,3
2. Gentofte P	Cardiac ward P	208	78	37,5
3. Glostrup	Cardiac ward M	57	45	78,9
4. Herlev	Cardiac ward S	93	31	33,3
5. Amager	The Cardiac Clinic	97	52	53,6
6. Århus	Medical ward M	143	93	65,0
7. Ringkøbing	The Mother-Child Center	77	55	71,4
8. Ribe	Section P10/E3, R3, BU amb.	81	40	49,4

*Questionnaires

5. Results

Wards 1-4: Wards 1-4 all belong to the county of Copenhagen, where a *shared* steering committee, (the GEPKA steering committee), was established to take care of the CRR. The members of this committee were all employees of the county's IT-department. The steering committee worked in cooperation with the managers of the wards, who participated to differing degrees in carrying out the CRR. Because of a lack of time, the questionnaire used in the CRR was adapted for use by the steering committee with *our* help. The ward employees were not involved. On wards 1, 2 and 4 the employees were informed about the CRR at meetings with the shared steering committee. Attendance was compulsory and not many participated. On ward 3, the management of the ward took care of informing the employees and almost everybody was present. The managements of the hospitals have not been involved in the CRR in any way.

Wards 5-6: On wards 5 and 6 the managing doctor and nurse along with other employees from the ward were members of the steering committees. IT-managers were also members on ward 5. The members of both steering committees had a thorough knowledge of the Gepka project. On both wards the questionnaire was adapted for use by the steering committees with our help. The IT-members of the steering committee informed the employees about the CRR on ward 5, with only limited assistance from the members who were ward employees, while the management of the ward did this job on ward 6. The managements of the two hospitals were not visible during the project.

Ward 7: The managing doctor of the ward was member of the steering committee together with other employees from the ward, and they were responsible for carrying through the CRR. The steering committee modified the questionnaire and the committee itself informed the employees about the CRR. The hospital management was not involved.

Ward 8: No observations or interviews were made in respect of ward 8. This ward is therefore excluded from this study.

6. Discussion

Appointment of an interdisciplinary steering committee: This instruction was not followed in the case of wards 1- 4, where the shared steering committee had no members from the wards. However, the steering committee *did* cooperate with the management of each ward. The explanation for why ward 3 - despite not following the above instruction - had the highest response rate of all the participating wards might be that the manager of this ward was actively engaged in carrying out the CRR, because he was very enthusiastic about the project. The managers of the other three wards, on the other hand, found that the timing was bad, because they had too many other activities at the time etc. They felt it was an extra burden having to participate in the CRR and did not try to motivate the employees on the wards to take part. The rather low response rate for ward 5 could be explained by the "clinical" members of the steering committee not having been very visible throughout the process, as it was the IT- members of the committee who took care of conducting the CRR. This contrasts with the high response rates on wards 6 and 7 where the ward managers were *very* active, visible and engaged.

Adaptation of the questionnaire by an interdisciplinary team from the ward: This instruction was not followed by wards 1-4, as the steering committee, which included no employees from the wards, itself took care of adapting the questionnaire. As already mentioned, the employees from the wards were not involved in the adjustment process due to time pressures. The response rates were low for three of the wards, whereas ward 3 had the highest rate of all participating wards. On wards 5 to 7 the instruction was followed, but high response rates were only attained for wards 6 and 7.

Informing the employees about the CRR: On wards 1, 2 and 4, representatives from the shared steering committee held information meetings. The managers of the wards were not active at these meetings and some were not even present. Attendance at the meetings was not compulsory and not many employees participated. The manager/management of the ward explained the CRR to the ward 3 employees, most of whom were present. On ward 5, information was provided by the IT- members of the steering committee, whilst on wards 6 and 7 the management performed this function. The response rates were high for wards 3, 6 and 7, where the management of the wards provided the information. An explanation for this can be that the managers of these wards can all be described as "fiery souls", who have all been very active in the EHR debate – some for and some against. They have all found it important to encourage a debate about EHR in their respective wards and have found the CRR a useful tool for this purpose. Their personal engagement is the most likely explanation for the high participation of the employees from their wards in the CRR.

Support for the CRR by the management – all the way up: None of the hospital managements have been visible during the CRR process. This is despite the fact that the decision to participate in the Gepka project - and with this the CRR – was taken by the counties *and* the hospital managements. It appears that, after having taken this decision, the hospital managements have simply handed over the responsibility either to their IT-departments or to the ward managers. Arguments and motivation from IT departments do not, however, have the same impact on clinicians as they do if they come from hospital or ward managers. Furthermore, before delegating responsibility to the managements of the wards, hospital managements have to ensure that they both consider the project to be important and want their wards to participate. It is unlikely that the members of the ward management will be able to motivate the other employees on their ward to participate if they are not motivated themselves.

The participating wards have followed the four selected "Instructions for use" to different degrees. It seems like one instruction, in particular, seems to be especially important to follow; the management of the ward must be engaged/actively involved in the project, as they are key figures when it comes to motivating the other ward employees. In order to achieve this, the hospital management must visibly motivate and support the management of the ward throughout the process.

It is important to emphasize that it is not the aim of this study to prove a causal relationship between the degree to which the "Instructions for use" are followed and the degree of participation. This study only suggests a qualitative relationship between the two. Neither does this study try to generalize the results, as further research on more wards would be needed to do so.

7. Conclusion

Using the four instructions as themes for studying the variation in participation in the Gepka CRR study gives a good understanding of the different reasons - of a human and organisational nature - for the variation seen. It appears that it is important for the four themes to be considered in an organization before carrying out a CRR in order to motivate the employees to participate. The themes can, therefore, be a very useful tool in future CRR studies as they can lead to a higher degree of participation. They can probably also be used in a modified way in other studies of human – machine interactions.

References

(1) The Ministry of the Interior and Health. National IT Strategy for the Danish Health Care Service 2003-2007. 2003.

(2) Vingtoft S, Lippert S, Bernstein K, Bruun-Rasmussen M, Kristensen M, Nøhr C, et al. EPJ-Observatoriets Status rapport 2000. 2000.

(3) N.M.Lorenzi, Riley RT. Organizational Aspects of Health Informatics. 1995. Springer Verlag.

(4) Scott J.T., Rundall T.G., Vogt T.M., Hsu J. Kaiser Permanente's experience of implementing an electronic medical record: a qualitative study. BMJ, doi:10.1136/bmj.38638.497477.68 , Nov.2005

(5) Heeks RD, Mundy D, Salazar A. Why Health Care Information Systems Succeed or Fail: Information Systems for Public Sector Management Working Paper No.9. 1999. Institute for Development Policy and Management, University of Manchester, Manchester, UK.

(6) Southon G, Sauer C, Dampney K. Lessons from a failed information systems initiative: issues for complex organizations. 55, 33-46. 1999. International Journal for Medical Informatics.

(7) Berg M. Patient care information systems and health care work: a sociotechnical approach. International Journal of Medical Informatics 1999:87-101.

(8) Høstgaard AM, Nøhr C. Metodehåndbog i Forandringsparathed. 2004. Aalborg Universitet.

(9) Nøhr, C, Kjær Andersen S, Vingtoft S, Bernstein K, Bruun-Rasmussen M. Development, implementation and diffusion of EHR- systems in Denmark. International Journal of Medical Informatics 2005; 74:229-34.

(10) Bernstein K, Bruun-Rasmussen M, Vingtoft S, Kjær Andersen S, Nøhr C. Modeling and implementing electronic health records in Denmark. International Journal of Medical Informatics 2005; 74:213-20.

6.3 Modeling Healthcare: Communication and Patient Flows

Ubiquity: Technologies for Better Health in Aging Societies
A. Hasman et al. (Eds.)
IOS Press, 2006

A study of the communication notes for two asynchronous collaborative activities

Nathalie BRICON-SOUF[a,1], Sandra BRINGAY[b], Françoise ANCEAUX[c], Saliha HAMEK[c], Nathalie DEGARDIN[a], Catherine BARRY[b] and Jean CHARLET[d]
[a] *CERIM, Lille, France* [b] *LaRIA 2733, Amiens, France*
[c] *PERCOTEC-LAMIH-CNRS UMR 8530, Valenciennes, France*
[d] *STIM/DSI/AP-HP, Paris, France*

Abstract. *Introduction:* To build relevant tools for Health Care Professionals, we must study and understand their practices. This paper discusses the way they leave traces in the Patient Record to help asynchronous collaboration, elaborating new documents or adding annotations.
Methods: We compared the results of two studies about the various writing strategies used by the Health Care Professionals to capture knowledge in the Patient Records. The first study deals with the information written by the nurses in a textbook during homecare situations. The second one deals with the annotations leaved by all the practitioners to complete the documents of the patient record in a hospital ward.
Results: We have found some invariants in these two situations. An interpretation model based on four levels: Communication Context, Communication Object, Value of Communication and Value of Cooperation, is proposed in order to describe and to index the characteristics of the Communication Notes.
Keywords: Asynchronous and collaborative medical activities, Homecare, Annotations, Patient record

1. Introduction

Asynchronous collaboration between Health Care Professionals (HCPs) is frequent. They must produce, store, share and convey medical information. But a structured Patient Record (PR) is not sufficient enough to support their collaboration. Consequently, they use different writing strategies and different notations to keep traces of their knowledge and exchanges during asynchronous and collaborative medical activities. We had the opportunity to compare two studies of two different situations for such care communication. It seems interesting because these situations were having similarities as being asynchronous collaborations, but were very different as using two types of Communication Notes (CN): (i) the information written in a semi-structured textbook during Home Care (HC) activities; (ii) the annotations added to the PR documents in an hospital ward. Electronic settings are not yet used in these two situations. We wondered if we could find some analogies in these situations, according to the deep analysis performed, the models proposed or the users' requirements expressed. A medium term work could validate the results with other

[1] Corresponding Author: Nathalie Souf, CERIM, 1, Place de Verdun, 59045 Lille, nsouf@univ-lille2.fr

application domains. In this paper, we describe a conceptual model convenient for both types of CNs and we propose a four interpretation levels architecture to organize the concepts. We note that some of them directly deals with cooperative knowledge and are not yet taken into account to build cooperative tools in medical situations. Using this model could help us either to index the CN or to display the CN according to different points of view when using electronic settings.

2. Description of the studied situations

Information exchanges are extremely important in Health Care situations and are often supported by the PR. This PR is one of the main parts of an Hospital Information System, used by many individuals (physicians, fellow, nurses, administrative agents…), for varied tasks: medical tasks (such as diagnostics, medical decisions…) or others tasks (such as management of the activity, planning, research...). But a computerized PR is not sufficient to answer all the problems linked to the collaboration between HCPs. "*(...) there is evidence that communication problems in health-care settings are significant, collaborative work around documents is not well supported by informatics systems, and that there are insufficient in-depth, observational studies of real world communication behaviors in health care*" [1]. A short presentation of the two kind of CNs will illustrate this point.

The Home Care Chart in Homecare situations: In HC settings, we have worked with the Santelys association home care team, analyzing their HC management. HCPs work asynchronously, realizing their activities over a long period of time, cooperating between different specialities, having few meetings to share some knowledge. The HCPs in succession intervene in the patient's home, each one continuing the activities begun by their predecessors. In order to facilitate the transmission of information, the HC teams use a textbook: the "Home Care Chart" (HCC), closed to the PR used in hospitals. Lightly structured, it allows nurses, physicians, or others to describe, in natural but concise language, what is supposed to be pertinent to manage efficiently the care. The HCC captures analytic and chronological data: the visit, the activities, the deviation from care protocol, the patient health status and so on. Requests, questions, answers and decisions may also be noted on the HCC. The chart takes the place of the functional dialogues observed in synchronous situations. In this context, it is used to support cooperation at two levels of synchronization: (i) cognitive synchronization, which aims to build a mutual knowledge context, to elaborate and maintain a shared representation structure; (ii) temporal and operative synchronization which facilitates the coordination by the implementation of two functions: it ensures the task allocation between partners and it ensures the train of actions, their simultaneity, their starting or their stopping [2].

The annotations in medical activities : In the DocPatient project[2], we studied the HCPs' annotations practices in a hospital ward. HCPs are used to stick post-it, add comment, surround... These annotations support part of their daily actions and interactions and participate actively to the collaboration process. HCPs annotate when *they cannot add* without annotations their comments to the structured documents of the

[2] The DocPatient project (2002-2005) of the University of Amiens is financed by the Picardie region (France). It deals with the problematic of a document-based electronic PR. We work in collaboration with a pilot site (the hospital ward of paediatric intensive care and neonatal medicine of Amiens) and an industrial partner (UNI-MEDICINE http://www.uni-medecine.com/).

PR. It is the case of the medical forms too rigid to allow the writer to add some knowledge not anticipated by the designer (the physician fills in the very structured maternity report so he uses annotations to add a non predefined information). Therefore, an annotation is an escape clause if there is no current method to extend forms. HCPs also annotate because they *do not want* to add their comment to the document because such production is written with an intention of communication different from the initial intention of the annotated document. In so doing, annotators add information about the document rather than information that belongs to the document. It is the case when a reader annotates to keep traces of his reading (a physician underlines important results in analyses) or when several practitioners collaborate (a physician uses a post-it to ask a specialist to give his opinion). Consequently, an annotation is also an escape clause if there is no means to code the comments about the documents. To conclude, practitioners use annotations to act: either to enrich the annotated document or to build the transitory support of knowledge used to create new knowledge (recorded or not in a document). Annotations are useful, not only to keep pertinent knowledge in the PR, but also to help practitioners to collaborate when the documents show their limits. Collaboration would be reduced if we do not preserve this practice in the Electronic PR.

3. Analysis methods

In the HC situation, 10 patient records were analyzed, each one containing all the information exchanged between nurses throughout the entire care period (from 1 month to 6 months). Cooperation activities split into two major types: cooperation for action and cooperation for planning. We analyzed how such a distribution was performed [3-4]. *In the annotation situation*, our aim was to produce a component linked to a document-based electronic PR which reproduce this practice of annotations. We went in a Pediatric Unit to observe HCPs' practices of annotations. We built a first annotations model and a first tool described in [5]. We undertook a semi-realistic usability study of our tool to check with the potential users the interest of annotations. We have chosen 20 users representative of the end-users: 5 seniors physicians, 5 fellows, 5 nurses and 5 administrative agents. Each interview lasts approximately two hours. The evaluations allow us to improve our model of annotations and our tool [6].

4. Conceptual models

In both cases, we have built a conceptual model of the representation of the CN. The models detail the concepts used when a CN is produced or read and the relations linking them. To build these models, we used a pragmatic and communicational point of view: the transactions communication theory of Zacklad [7] that aims at studying the cooperative activities, from a communicational, cognitive and socio-economic point of view. We also used a collaboration point of view: the Hoc's architecture of collaboration [8] which proposes three levels of collaboration: action, plan and meta-collaboration. Communication Units and Annotations can be described within the same conceptual framework. Figure 1 illustrates the conceptual CN model obtained when merging the two models.

A CN is attached to a **target** (chart, medical record). An **anchor** is used in the case of an annotation. Its **content** corresponds to a semiotic production written by the **redactor/annotator**, in a **spatio-temporal framework of creation** (at this place, at a particular date...), according to an **intention of communication** (to transmit information, an order...) for one or more **selected readers**. A CN deals with an **object of communication**. Different **speeches acts** formulate the contents. **Readers** (sometimes different from the selected readers) read the CN within a **spatio-time framework of reading.**

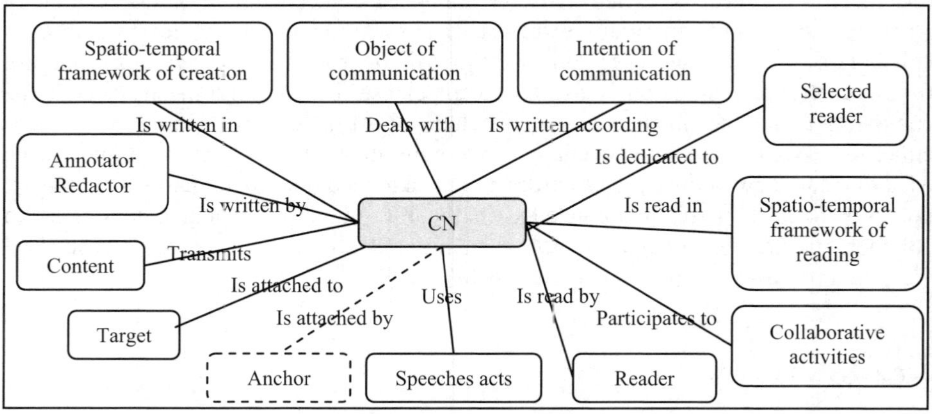

Figure 1 : Conceptual model of CN

Table 1 presents examples of some items of the model, in the two situations.

	Home care situation	Annotation situation
Note	Written Communication unit	Annotation
Target	Nurse Chart, Physician Chart	A document, part of a document, a word, an image
Anchor	None: chronology in the textbook	An arrow, an underlined passage
Spatio-temporal framework of creation	At this patient's home, at a particular date	In the box of the child, at a particular date
Object of communication	Care, Patient, Logistic	The part of the body: cardiovascular, respiratory system

Table 1 : Examples in both situations of some items of the model.

Collaborative aspects of the CN are very important but **collaborative activities** are complex. For example, when a physician uses the following CN « blood gas in two hours », if a nurse reads this message, she detects the interference with her own care whereas a fellow analyses this comment from another point of view and consequently, this note induces others collaborative activities.

5. Interpretation levels of the items of the Communication Notes

The different items of the models do not support the same kind of knowledge. We found that it was useful to classify them according to an interpretation endeavor. We determined four interpretation levels. The items classified in the first three levels can be captured as soon as a CN is written: the first level (Communication Context) contains some elements fixed by the context, the second one contains some items linked to the topic of the CN (Communication Object) and the third one contains some items anticipated by the CN's creator (Value of Communication). The last level contains the elements deduced from the use of the CN (Value of Cooperation) (*cf.* Figure 2).

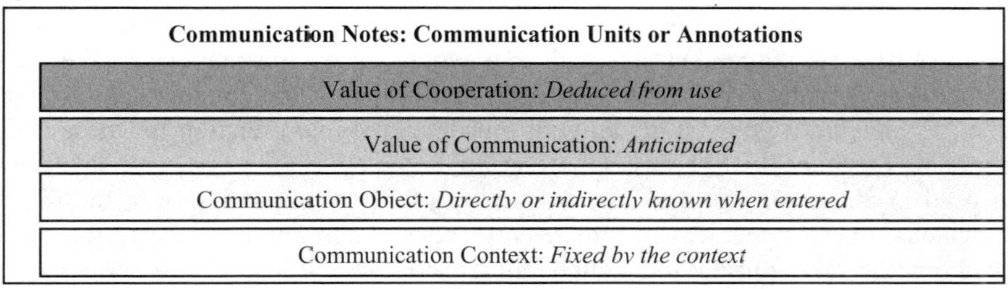

Figure 2 : Interpretation levels for Communication Notes.

Contextual level: It gives the information about the context of the production of the CN: who is speaking or writing, when does it happen, in which place, with which intention of communication, for whom… This level is used by numerous systems. Contextual information are entered in the system by the user or sometimes deduced (e.g. the date). If looking to the work done in contextual awareness in health care [9], it is this level which was mostly tried to be supported by context awareness tools.

Communication Object: The objective is to reduce the fields of possible, according to the objectives of the work: what is the CN about? Topics can be attributed to the CN through simple characteristics (*logistic*, *care management*, *patient*) according to the user's specific knowledge. As such characteristics are simple and direct, the user could give such information on this object (with key words). Even more, this information could be deduced from the communication medical context (Natural Language Processing could be applied to treat the content of the CN).

Value of Communication: The CN's creator anticipates the use of his/her production and chooses a way to express his/her need for cooperation: where and how the CN is presented, which speech acts are used... This value is sometimes expressed in hidden or implicit ways (a post-it at this place is not supposed to be seen by everyone, a nurse chooses the good textbook to report information for the physician). Speech acts are important to express some requirements for the interpretation of the readers (imperative suggests action, past perfect something done…). The presentation of information is also important to underline part of the message (bold, colorful…). Some systems propose good interfaces to manage this information but important improvements should be obtained when deeply studying this level.

Value of Cooperation: This level allows determining the interpretation of the CN for the reader: who is reading, in which context of reading, for which collaborative activity… Even if guided by the value of communication resulting from the anticipation of the author, a user of the CN integrates the communication in his/her own tasks and interprets it through the various cooperation levels mentioned by hoc's work [8]. As users have different roles, they can interpret the same CN in different prospects for cooperation. At this level, the reader uses his representation of himself in the organization and of the common ground [8] elaborated with his colleagues. This level should be a very important level to consider, so that the way of communicating the information could be adapted to the actual use of the CN. This level is often neglected, and systems are sometimes only devoted to present the information according to the anticipation of use rather than to the numerous possibilities of use.

6. Conclusion/discussion

In this paper, we have compared how HCPs communicate and cooperate with a textbook during HC situations and with annotations in an hospital ward. We have found some invariants in these two situations and proposed the communication notes concept. We have developed a model with four levels of interpretation for these CNs. This model is a conceptual one used to describe the practices: the first three levels are often used in electronic settings. The originality of our work comes from the fourth level which indicates the way the different readers interpret the cooperation tasks induced by the CN, during the collaborative activities. When a CN is described according to the different items of the four levels of interpretation, an agent (a human or a computer one) will have the necessary knowledge to use, exchange, present efficiently this CN. Next step of work would be to produce a computational model directly usable by electronic settings, to index the CNs and to present them according to different users' points of views. Ontology of cooperation is one of the perspective we hope to develop soon.

Acknowledgment

We thank the Santélys Association's team and G. Krim, in charge of the Pediatric of Amiens Hospital, for their invaluable help.

References

[1] Gennari J. H., Weng C., Benedetti J., McDonald D. W. Asynchronous communication among clinical researchers: A study for systems design. IJMI 2005; 74 : 797-807.
[2] Falzon, P. Dialogues fonctionnels et activité collective. Le Travail Humain, 1994, 57 : 299-312.
[3] Hamek, S., Anceaux, F., Pelayo, S., Beuscart-Zéphir, M. C., Rogalski, J. Cooperation in healthcare - Theoretical and methodological issues. A study of two situations: Hospital and Homecare. In N. Marmaras, T. Kontogiannis, D. Nathanael (Eds.), Proceedings of the annual conference of the European association of cognitive ergonomics - EACE'05. Athens, 2005. 225-232
[4] Hamek, S., Anceaux, F., Bricon-Souf, N., Flahou, S. La prise en charge des soins lors de l'hospitalisation à domicile: une analyse de la coopération asynchrone. JFIM Lille (France), May 2005.
[5] Bringay S., Barry C., Charlet J. Annotations for managing knowledge in the Electronic Health Record. Workshop KM&OM IJCAI-2005, Edinburgh (Scotland), 2005.
[6] Bringay S., Barry C., Charlet J. A specific tool of Annotations for the Electronic Health Record. International workshop on Annotations for collaboration, Paris (France), 24-25 November 2005.
[7] Zacklad M. Documents for Action (DofA): infrastructures for Distributed Collective Practices. Proceedings of the workshop Distributed Collective Practice: Building new Directions for Infrastructural Studies. CSCW-2004, Chicago (USA), November 6-10, 2004.
[8] Hoc J.M. Towards a cognitive approach to human-machine cooperation in dynamic situations. International Journal of Human-Computer Studies 2001, 54: 509-540.
[9] Bricon-Souf N., Newman C. Context awareness in health care : A review. IJMI 2006, Feb 17.

Ubiquity: Technologies for Better Health in Aging Societies
A. Hasman et al. (Eds.)
IOS Press, 2006

Specific classification of *e*library resources says more about users' preferences

Judas Robinson[a1], Simon de Lusignan[a], Patty Kostkova[b], Bruce Madge[c], Lesley Southgate[a]
[a] *St. George's, University of London, Cranmer Terrace, London, SW17 0RE*
[b] *City eHealth Research Ctr., Inst. of Health Science, City University, London, EC1V 0HB*
[c] *British Medical Association Library, BMA House, Tavistock Square, London WC1H 9JP*

Abstract: Background: Medical Subject Headings (MeSH) are a hierarchical taxonomy of over 42 000 descriptors designed to classify scientific literature; it is hierarchical with generic high order headings and specific low order headings. Over 1,000 resources in the Primary Care Electronic Library (PCEL – www.pcel.info) were classified with MeSH.
Methods: Each of the entries or resources in the primary care digital library was assigned up to five MeSH terms. We compared whether the most generic or specific MeSH term ascribed to each resource best predicted user preferences.
Results: over the four month period analysed statistically significant differences were found for resources according to specific key MeSH terms they were classified by. This result was not repeated for generic key MeSH terms. Conclusions: Analysis of the use of specific MeSH terms reveals user preferences that would have otherwise remained obscured. These preferences are not found if more generic MeSH terms are analysed.
Keywords: Medical Subject Headings; Libraries, Digital; Primary Health Care

1. Introduction

We chose Medical Subject Headings (MeSH) as the controlled vocabulary to index the Primary Care Electronic Library (PCEL) [1]. There are a range of alternatives we could have used: (1) A disease or clinical classification, e.g. International Classification of Diseases (ICD) [2] and the Systematized Nomenclature of Medicine (SNOMED) [3]; (2) A procedure coding system, e.g. Office of Population, Censuses and Surveys – Classification of Surgical Operations and Procedures – 4th Revision (OPCS-4); (3) Another library classification e.g. Dewey Decimal Classification hierarchy [4]; or (4) Used a metathesaurus e.g. Unified Medical Language System (UMLS); or (5) a combination of tools [5]. We selected MeSH because we wanted a system with a hierarchical structure which would enable those browsing the library to focus or broaden their search by browsing its hierarchy. There is some evidence that this orientates the user and can also reveal user choices [6]. , Using ICD-10 would provide a hierarchy [7] but has the disadvantage that it specifically classifies diseases; its clinical modification (e.g. ICD-10-CM) broadens its scope by including procedures yet does not have the breadth of MeSH or address areas like the type of publication. In addition there have been problems with discontinuation of codes between versions [8]. SNOMED CT or other clinical classification

---

[1] Judas Robinson, Primary Care Informatics, Division of Community Health Sciences, St George's - University of London, London SW17 0RE jsrobins@sgul.ac.uk

might offer advantages for linking to data within clinical records [9], but would also be difficult to link to scientific concepts or subjects. ULMS would meet our needs, however the wealth of concepts and linkages to clinical classifications within UMLS are probably too extensive for a single library resource [10].

The Primary Care Electronic Library (PCEL - www.pcel.info) contains over 1,000 abstracts of quality assured Internet resources for primary care. PCEL is the latest stage of development of digital library that has been online for 6 years [11]. Users can directly access resources or browse the MeSH tree for relevant material. Each resource has up to five MeSH subheadings or terms. These were either allocated by a qualified medical librarian or an academic supported by librarians at St. George's. We were unsure whether applying more generic (e.g. Cardiovascular disease) or specific MeSH subheadings (e.g. Myocardial infarction) was of more help to users in finding resources. We therefore decided to investigate whether generic or specific MeSH headings told us most about user preferences.

2. Method

Requests for resources were evaluated. Requests originating from identified users and search engines were separately analysed. Resource requests were compared with expected values; comparing whether the observed and expected usage was predicted by either the most generic or specific MeSH terms. Expected values were the numeric proportion of resources in each category of the MeSH tree top structure (this can be viewed and browsed from the NLM website [12]). Data was collected over a four month period. Log files were parsed and relevant data was stored in a database for analysis. We used the Chi-squared test to see whether the proportion of requested tests differed significantly from the expected proportion [13]; i.e. to indicate whether a group of resources was used disproportionately more or less than expected.

PCEL is written in ColdFusion and hosted on an Apache web server. Apache log files were used to collect data regarding site activity. In preparation for this evaluation the web site was altered so that all visits to pages relating to MeSH recorded relevant details in the log. The storage of MeSH terms was migrated to a hierarchical database architecture, using MeSH 2005 in ASCII format [14].

To test the face validity of the method results for an arbitrary taxonomy were also calculated. This was constructed by grouping resources according to their chronological entry in the database, the auto increment primary key of the table which contains resources. The results of the MeSH analysis, those of the arbitrary taxonomy, and resource requests were presented online.

3. Results

PCEL attracted 65998 requests in the four months under analysis. However, 45661 were from search engines. Approximately 31% of usage comes from identified users. As well as requests for resources, 775 requests were recorded for browsing the MeSH tree. The number of times each resource was requested over the four month period was recorded. The full list of requests for resources is presented online [15]. Although some resources were consistently popular this was the exception rather that the rule. Resources which attracted more than two requests per month each month accounted for 37.2% of the total requests.

A comparison of the distribution of key MeSH terms (both specific and generic) and all MeSH terms used is presented online [15]. It shows that broadly speaking, the distribution of key MeSH terms is representative of all MeSH terms applied to resources. The results of Chi squared analysis for the key MeSH terms (both specific and generic) of requested resources are presented online[15]. The results for specific key MeSH terms are presented in table 2.

Five of the fifteen descriptors show consistent statistical differences over the period analysed: Biological Sciences, Health Care, Physical Sciences, Information Science and Anthropology, Education, Sociology and Social Phenomena. These five descriptors cover 66.5% of PCEL resources. Only three MeSH tree top descriptors identified by the most generic key MeSH terms show consistent statistical differences over the period analysed: Anatomy, Physical Sciences and Anthropology, Education, Sociology and Social Phenomena. These three descriptors only account for 12.8% of PCEL resources. The search engine requests showed no statistically significant differences from expected results.

The results of Chi squared analysis for an arbitrary classification of requested resources is presented online [15]. Of thirty results for categories over the time period only four show statistically significant differences from those expected. Analysing all categories for the month of April did not yield a statistically significant result. None of the categories of the arbitrary taxonomy show consistent statistical differences over the period analysed.

4. Discussion

Our results show that more specific MeSH terms applied to the resources of a digital library reveal user preferences. These user preferences are not defined by the analysis of more generic MeSH terms, and can serve to guide the indexing of material for the library. The findings also emphasise that it may be more useful to apply more specific MeSH terms to resources when classifying material. Classifying into taxonomies, such as MeSH, is a long established technique. Applying this idea to digital content is a practice which has recently grown in importance; adherents arguing that taxonomies complement traditional keyword searches and help users efficiently find data [16]. Although debate exists at to whether this encourages or inhibits the creative process of finding information, controlled vocabularies are a standard resource for information retrieval [17].

Over the period analysed the observed requests for two thirds of PCEL resources were statistically different to those expected on the basis of their most specific key MeSH terms: for the majority of resources the frequency of requests could be predicted on the basis of their MeSH classification. An arbitrary classification system showed no consistent statistical differences. This strengthens the case that the results for key specific MeSH

Table 2 - Differences in requests identified by MeSH tree top category

Tree Top MeSH Descriptor	Month.	Observed(O).	Expected(E).	0-E.	P
Anatomy	Jan	12	13	-1	0.7801
	Feb	22	13	9	0.0120 *
	Apr	15	13	2	0.5767
Organisms	Jan	7	14	-7	0.0596
	Feb	10	14	-4	0.2818
	Apr	12	14	-2	0.5904
Diseases	Jan	230	190	40	0.0013 **
	Feb	228	190	38	0.0023 **
	Apr	206	190	16	0.1996
Chemicals and Drugs	Jan	17	13	4	0.2642
	Feb	4	13	-9	0.0120 *
	Apr	5	13	-8	0.0255 *
Analytical, Diagnostic and Therapeutic Techniques and Equipment	Jan	22	45	-23	0.0004 ***
	Feb	45	45	0	1
	Apr	27	45	-18	0.0060 **
Psychiatry and Psychology	Jan	33	44	-11	0.0902
	Feb	40	44	-4	0.5378
	Apr	46	44	2	0.7580
Biological Sciences	Jan	243	162	81	0 ***
	Feb	229	162	67	1.0445e-8 ***
	Apr	230	162	68	6.2534e-9 ***
Physical Sciences	Jan	7	26	-19	0.0001 ***
	Feb	8	26	-18	0.0003 ***
	Apr	11	26	-15	0.0028 **
Anthropology, Education, Sociology and Social Phenomena	Jan	18	49	-31	0.0000 ***
	Feb	29	49	-20	0.0034 **
	Apr	12	49	-37	6.1742e-8 ***
Technology and Food and Beverages	Jan	2	8	-6	0.0332 *
	Feb	3	8	-5	0.0759
	Apr	1	8	-7	0.0129 *
Humanities	Jan	3	3	0	1
	Feb	4	3	1	0.5631
	Apr	0	3	-3	0.0828
Information Science	Jan	196	170	26	0.0294 *
	Feb	225	170	55	0.0000 ***
	Apr	206	170	36	0.0025 **
Persons	Jan	2	10	-8	0.0110 *
	Feb	3	10	-7	0.0261 *
	Apr	5	10	-5	0.1121
Health Care	Jan	259	292	-33	0.0230 *
	Feb	201	292	-91	0 ***
	Apr	248	292	-44	0.0024 **
Geographic Locations	Jan	0	12	-12	0.0004 ***
	Feb	0	12	-12	0.0004 ***
	Apr	27	12	15	0.0000 ***

N = 1051 * $p<.05$, ** $p<.01$, *** $p<.001$

terms are significant. The data for individual resource requests showed that it is the exception rather than the rule for resources to be requested more than two times over the four months analysed. This excludes regular requesting of resources as a factor in trends noticed for key MeSH terms. Also no comparison was found between the sample size for the MeSH tree top descriptor and the probability that observed results differed from expected. These results confirm that the MeSH taxonomy classifies Internet resources into functional groups, functional groups associated with user preferences.

Interestingly this user preference is to a large degree absent if the most generic MeSH terms are considered. Only 12.8% compared with 66.5% of PCEL resources were shown to have statistical differences associated with generic key MeSH terms compared with specific key MeSH terms. This would indicate that more specific entries in the MeSH hierarchy are better able to indicate user preferences. This perhaps should not come as a surprise as more specific MeSH terms are a more accurate indicator of the subject under consideration.

MeSH lends itself to hierarchical searching as well as analysis. Thus the relative frequencies of requests for resources identified by key MeSH terms can be classified into these 15 groupings for analysis. Such data provides information concerning the user preferences of PCEL users and can act as a guide for indexing material in the future. The results showed that PCEL users have preferences for resources described by the MeSH tree top descriptors 'Biological Sciences' and 'Information Science' and preferences against 'Physical Sciences', 'Anthropology, Education, Sociology and Social Phenomena', and 'Health Care'. On more specific levels of the MeSH tree, in January 2005, preferences were shown for 'Computing Methodologies', 'Medical Informatics', 'Health Occupations', 'Circulatory and Respiratory Physiology', 'Nursing', 'Digestive System Diseases', 'Respiratory Tract Diseases', and 'Cardiovascular Diseases'.

The results are limited by two factors. Firstly, the size of the MeSH taxonomy exceeds the number resources indexed in PCEL: there are over 42,000 terms in the MeSH vocabulary, and of these only 942 are used for PCEL's 1000 resources. The second limitation is the number of hits PCEL receives. This was sufficient for statistical analysis in January, February and April 2005, but fell short for the month of March.

We failed to identify other studies reporting on the differences between specific and generic MeSH terms applied to a browsable classification. The National Library of Medicine permits searching and browsing of MeSH classifications [18], but this functionality is not fully integrated with the hierarchical searching of MEDLINE. Another digital resource using the MeSH taxonomy is Organising Medical Networked Information (OMNI) [19]. Although the Resource Discovery Network (RDN), to which OMNI belongs, has published evaluation reports [20], it was difficult to find evaluation reports from the RDN hubs. Further studies are needed to see if user preferences are defined by more specific levels of classification using MeSH and other taxonomies.

5. Conclusions

Assigning MeSH terms to Primary Care Internet resources and allows users to browse resources using the MeSH hierarchy provides functionality for users as well information about patterns of use for digital librarians. Although designed for periodical articles and books, the MeSH taxonomy functions with Internet resources. This analysis of MeSH terms demonstrates user preferences that would otherwise remain obscured. More specific MeSH terms are more sensitive than more generic MeSH terms in elucidating user preferences.

Acknowledgements

The authors would like to thank the library services at St. George's, University of London for help and support with MeSH indexing.

References

[1] Robinson J, de Lusignan S, Kostkova P. The Primary Care Electronic Library (PCEL) five years on: open source evaluation of usage. *Accepted for publication* Informatics in Primary Care September 2005.

[2] WHO | International Classification of Diseases (ICD) 2006, viewed 4 January 2006, <http://www.who.int/classifications/icd/en/>

[3] SNOMED International 2006, viewed 4 January 2006, <http://www.snomed.org/>

[4] Robert B. Allen. Two digital library interfaces that exploit hierarchical structure. In *Proceedings of DAGS95: Electronic Publishing and the Information Superhighway*, Boston, MA, May 1995.

[5] Jamoulle M. Euro-Med Data Survey. URL: http://www.ulb.ac.be/esp/emd/classifications.htm

[6] Mann T. Library Research Model:A guide to classifications, cataloging and computers. Oxford; OUP, 1993.

[7] *International Statistical Classification of Diseases and Related Health Problems 10th Revision*, 2003, World Health Organisation, viewed 4 January 2006, <http://www3.who.int/icd/vol1htm2003/fr-icd.htm>.

[8] Janssen F, Kunst AE. ICD coding changes and discontinuities in trends in cause-specific mortality in six European countries, 1950-99. Bull World Health Organ. 2004;82(12):904-13.

[9] *About the UMLS Resources*, 2004, National Library of Medicine, viewed 4 January 2006, <http://www.nlm.nih.gov/research/umls/about_umls.html>.

[10] Giannangelo, Kathy, and Berkowitz, Lyle, 'SNOMED CT Helps Drive EHR Success.', 2005, *JAHIMA* 76;4:66-67, viewed 4 January 2006, <http://library.ahima.org/xpedio/groups/public/documents/ahima/pub_bok1_026463.html>.

[11] Robinson J, de Lusignan S, Kostokova P, Southgate L. Developing a digital library: The evolution of the Primary Care electronic Library (PCEL). Acta Informatica Medica 2005 13;4:31-37.

[12] MeSH Tree Structures–2005,NLM,viewed 15 May 2006.<http://www.nlm.nih.gov/cgi/mesh/2005/MB_cgi>

[13] *The |STAT Handbook*, 2002, viewed 4 January 2006, <http://www.cog.brown.edu/~max/cg144/class_files/pipehandbook.htm>.

[14] *Medical Subject Headings - Files Available to Download*, 2005, National Library of Medicine, viewed 4 January 2006, <http://www.nlm.nih.gov/mesh/filelist.html>.

[15] Specific classification of resources reveals more about user preferences in a digital library, 2005, viewed 4 January 2006, <http://www.pcel.info/mie2006/>.

[16] Athey, D 2004, 'Data Classification Using a Digital Taxonomy', *ColdFusion Developer's Journal*, viewed 4 January 2006, <http://www.paperthin.com/news/upload/CFDJ-November2004-Article-taxonomy.pdf>.

[17] Adams KC 2000, 'Immersed in Structure: The Meaning and Function of Taxonomies', *Internetworking*, vol. 3, no. 2, viewed 4 January 2006, <http://www.internettg.org/newsletter/aug00/article_structure.html>.

[18] National Library of Medicine, MeSH Browser, 2006, viewed 4 January 2006 http://www.nlm.nih.gov/mesh/MBrowser.html

[19] *Organising Medical Networked Information (OMNI)*, 2006, viewed 4 January 2006, <http://omni.ac.uk/>.

[20] *The Resource Discovery Network: evaluation report*, 2002, viewed 4 January 2006, <http://www.rdn.ac.uk/publications/evaluation/evalreport02.pdf>.

Ubiquity: Technologies for Better Health in Aging Societies
A. Hasman et al. (Eds.)
IOS Press, 2006

Cancer Patient Flows Discovery in DRG Databases

Nicolas JAY[a,b1], Amedeo NAPOLI[b] and François KOHLER[a]
[a] *Laboratoire SPI-EAO, Faculty of Medecine., Nancy, France*
[b] *Orpailleur team, Lorraine Laboratory of IT Research and its Applications, Nancy, France*

Abstract. In France, cancer care is evolving to the design of regional networks, so as to coordinate expertise, services and resources allocation. Existing information systems along with data-mining tools can provide better knowledge on the distribution of patient flows. We used one year data of the French Diagnosis Related Groups (DRGs) based system to perform our analysis. Formal Concept Analysis has been used to build Iceberg Lattices of cancer patient flows in the French region of Lorraine. This unsupervised conceptual clustering method allowed us to describe patients flows with an easily understandable visual representation.
Keywords: Patient Flows, Data-Mining, Cancer, Information Systems

1. Introduction

The Cancer Plan, launched by the French government in 2003, is leading to deep changes in healthcare organization. Among measures advocated by this plan: "All establishments providing cancer care shall coordinate their expertise and services, within a set of regional Cancer Poles. Decisions regarding the deployment of major new equipment to be used on a regionwide basis shall systematically be achived through coordination among the main centers concerned... Local networks shall be developed to meet the needs for local coordination."[1].

Such goals cannot be reached without a highly cooperative information system allowing health professionals to share medical information and administrators to manage healthcare organization. But such a system is still in a development stage. The "Programme de Médicalisation des Systèmes d'Information (PMSI)" is the so-called French DRGs based information system. It could be very useful to a better understanding of the actual situation, especially to answer questions about patient flows within the healthcare system.

However it collects large amounts of data and classical statistics methods can be inappropriate to describe their complexity. In that context, Knowledge Discovery tools appear to be a good way of dealing with patient flows.

We propose in this paper a data-mining approach to explore cancer patient flows relying on Formal Concept Analysis, an unsupervised conceptual clustering method

[1] Nicolas Jay. Faculté de Médecine - 9 Avenue de la Forêt de Haye -BP 184 - 54 505 Vandoeuvre Cedex – France. Email : nicolas.jay@medecine.uhp-nancy.fr

providing interesting visualization features[2]. Concept lattices are also known as Galois lattices. Our objective is to describe cancer patient flows in the French region of Lorraine with an easily understandable visual metaphor.

2. Material and Methods

2.1. Materials

Data were extracted from the year 2003 PMSI database of the Lorraine region. We identified hospital stays related to cancer care through the use of an algorithm based on selected codes from the International Classification of Diseases (10th Revision). Since 2001, a cryptographic key can be used in the PMSI to anonymously link successive records of a same patient, whatever the location of the stay. We used this key to rebuild each cancer patient path. Data were then analyzed by subgroup according to fourteen cancer locations (i.e.: breast, lung, nervous system...).

2.2. Methods

Formal Concept Analysis (FCA) is a theory of data analysis identifying conceptual structures within data sets, introduced by Wille [3]. A strong feature of FCA is its capability of producing graphical visualizations of the inherent structures within data. This model mathematizes the philosophical understanding of a concept as a knowledge unit consisting of two parts: the extent and the intent. The extent covers all objects (or entities) that are instances of the concept, while the intent comprises all attributes (or properties) holding for all the objects under consideration.

FCA starts with a formal context defined as a triple K=(G,M,I) where G is a set of objects, M a set of attributes and I a binary relation between G and M. $(g,m) \in I$ means that the object g has the attribute m. K may be seen as a table relating objects and their attributes. The table 1 shows a formal context K_{PH} representing I, the relation between a set of four patients P={P1,P2,P3,P4}and a set of four hospitals H={H1,H2,H3,H4}. A cross indicates that a given patient had a stay in the corresponding hospital. A formal concept of K_{PH} is a pair (A,B) with $A \subseteq P$ and $B \subseteq H$ such that (A,B) is maximal with the property $A \times B \subseteq I$. A and B satisfy the following properties:

$$B=\{h \in H / (p,h) \in I \text{ for all } p \in A\} \quad (1)$$

$$A=\{p \in P / (p,h) \in I \text{ for all } h \in B\} \quad (2)$$

A is called the intent and B is called the extent of the formal concept. For example, C=({P2,P4},{H1,H2}) is a formal concept of K_{PH}. No patient can be added to the extent of C without modifying its intent. Moreover, no patient can be removed from the extent because in that case, C would not be maximal.

A subconcept - superconcept relation can be formalized as:

$$(A1,B1) \leq (A2,B2) \Leftrightarrow A1 \subseteq A2 \Leftrightarrow B2 \subseteq B1 \quad (3)$$

Equivalence in (3) comes from (1) and (2). The set of all formal concepts of a formal context K=(G,M,I) together with the order relation $\leq$ is a complete lattice and can be represented in a line diagram as shown in the left part of figure 1 for the K_{PH} formal context.

Table 1: A formal context K_{PH} *representing patients and their hospital stays*

	H1	***H2***	***H3***	***H4***
P1	X		X	X
P2	X	X		
P3		X		
P4	X	X		

Such line diagrams can be very useful in the field of knowledge discovery to understand conceptual relationships within data. A drawback lies in the number of formal concepts that increases significatively (at worst exponentially) with the size of the formal context. Stumme and al. [4] have introduced the notion of iceberg concept lattices. This approach simplifies the line diagram by keeping only the most frequent concepts. Let C=(A,B) a concept of the formal context K=(G,M,I). The support count of C is defined as *supp*(c)=|A|/|G|. Given a minimum support threshold $minsupp \in [0;1]$, C is frequent if: $supp(c) \geq minsupp$. Iceberg lattices can be viewed as filters representing the top-most part of concept lattices. Figure 1 shows the iceberg lattice K_{PH} associated with *minsupp*=0.5

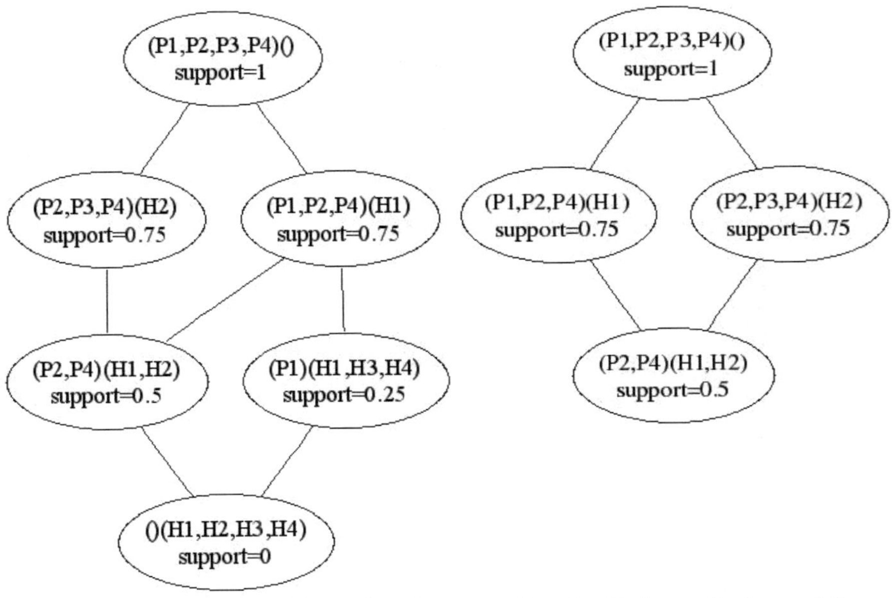

Figure 1: The lattice of the K_{PH} *formal context and its related iceberg with minsupp=0.5*

In order to discover patterns of hospital care among cancer patients in Lorraine, we built the formal context of the relation between those patients and the hospitals where they had a stay. A specific formal context was created for each cancer location. We chose the Titanic algorithm [5] to extract iceberg lattices from the formal context. Each time, several *minsupp* were tried to find the best suited representation.

3. Results

We propose in this section an example of results that we have obtained with our system. 2257 patients were selected in the original database with a lung cancer. Figure 2 shows the related iceberg lattice with a *minsupp* set at 0.005(around eleven patients). This graph can be read from left to right. The top node is a concept with an empty intent (no hospital) and whose extent contains all the patients. Second layer nodes are concepts with an intent made of one hospital. Third layer nodes illustrate cooperations between hospitals of the second layer nodes. This figure clearly shows four kinds of flows.and may be interpreted as follows. Firstly, a majority of concepts are not connected with any other. Most of the time, they represent hospitals receiving a few patients, but in some cases hospitals being geographically isolated from other major facilities. Secondly, there are three groups of connected concepts. The upper group in the figure concerns hospitals located near Metz, one of the two most populated towns of Lorraine. The middle one shows flows of patients going through two important hospitals located in Nancy, the other big town of Lorraine: The teaching hospital of Nancy (CHU) and the "Centre Alexis Vautrin", an anti-cancer center (CLCC).

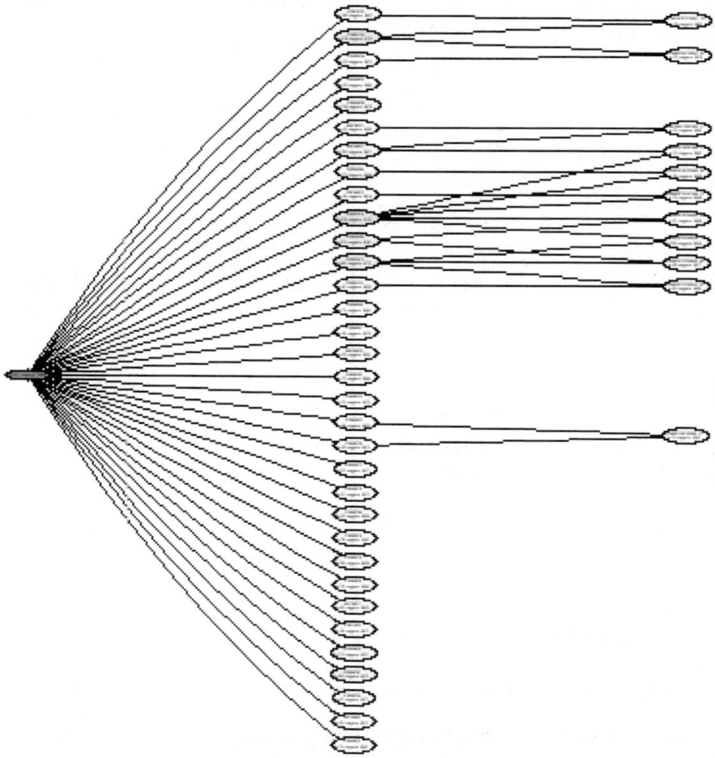

Figure 2: Iceberg Lattice showing flows of patient treated for lung cancer in Lorraine.

In the lower part of the figure, there is a group of three connected nodes. It concerns a hospital located at the Lorraine border, whose patients are also treated in the university teaching hospital of Alsace, the nearby region.

Figure 3 is a zoom of the upper part of the lattice. Each node is greyed according to its support. Labels show hospital identifiers, patient number in the concept's extent, and concept support.

The Nancy CHU (id=540002078) receives the largest number of patients (n=510) ahead of the Metz regional hospital (id=570005165) (n=370), and the Nancy CLCC (id=540003019) (n=304). Cooperation between hospitals can be considered at different levels. Of the 182 patients treated in the Epinal town hospital (id=88078051), only 10 percents (n=19) are shared with the Nancy CHU. On the contrary, half of the patients who have a stay in the Nancy CLCC (144 of 304) have been also treated in the Nancy CHU. These two institutions are the ones sharing the largest number of patients.

Since this iceberg lattice has been built with a 0.005 minimum support, all concepts involving less than 13 patients are pruned from the lattice. This means that even if they exist, smaller flows are not displayed in the lattice.

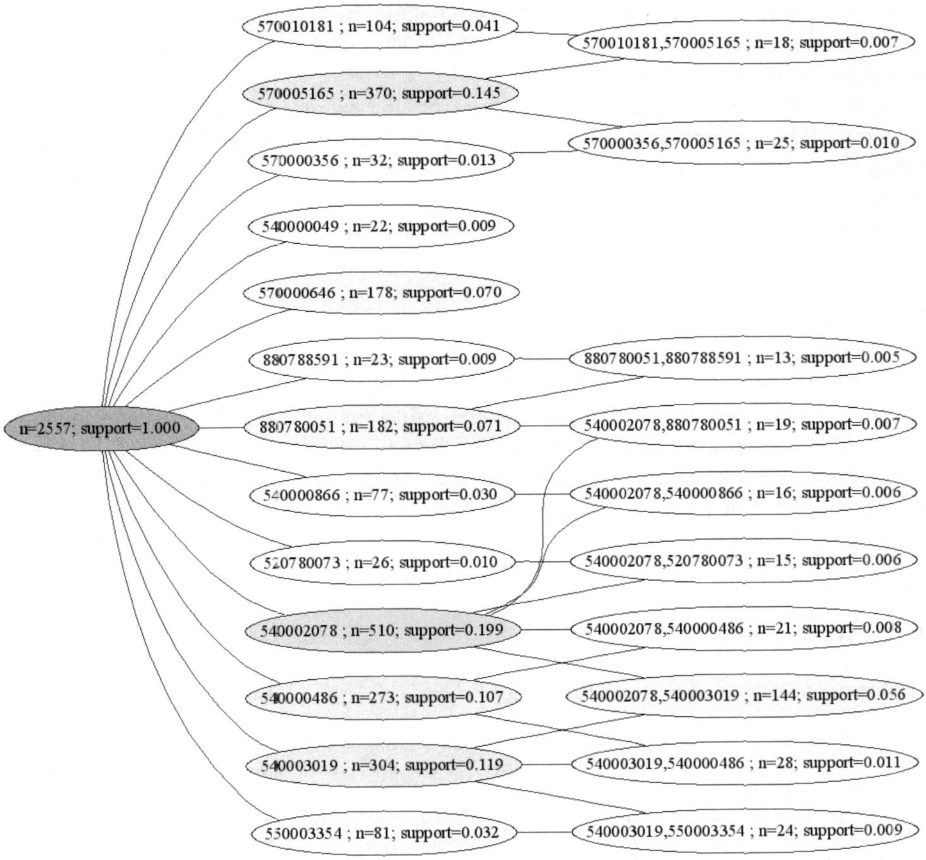

Figure 3: Zoom of the figure 2 lattice. Node labels show hospitals ID of the concept intent, number of patients of the extent, support of the concept.

4. Discussion

Our approach provides several useful features to discover and graphically represent patient flows extracted from a large database. First of all, the classification method is unsupervised and there is no use to consider the exponential number of all possible path combinations through hospitals. The analyst (an expert using the system) handles a threshold to keep the most significant information. Iceberg lattices are a hierarchical clustering method. Thus, they are an interesting alternative to geographical information systems for at least two reasons: they can symbolize flows through more than two sites, and they can highlight flows that do not necessarily rely on a geographical logic. Furthermore, iceberg lattices can be used to compute association rules.

Comparing to a close work [6] our method exploits the same type of data and relies also on the frequent itemset extraction method. While their concern is more about intra-hospital patient path, we analyzed data for a whole region. We present here a graphical approach providing end-users a global view of patient flows, whose interpretation requires only a minimum of training. We do not here take into account the sequential aspect of the problem, but we believe it is a promising way of research. Iceberg lattices could be used as a pre-processing step to refine the extraction of sequential patterns from patient paths.

Our work can be improved by exploring the nature of the discovered flows: especially when geographical proximity does not explain it, other variables should be added to the formal context. But we will then have to cope with more complex lattices. Another research direction should be to associate iceberg lattices with other classification methods, for example to deal with numeric data.

5. Conclusion

As healthcare networks are in the progress, new tools have to be assessed to deal with the growing amount and complexity of medico-economic data. Data-mining methods can be successfully used to extract new knowledge units from large medical databases. In this work, we are able to describe cancer patient flows at a regional level. This can help healthcare managers to make choices for resource allocation as well as clinicians for cooperative work.

References

[1] Mission Interministérielle pour la Lutte contre le Cancer . http://www.plancancer.fr/ (last accessed Jan. 29, 2006).

[2] Bernhard Ganter, Rudolf Wille, Formal concept analysis, Springer-Verlag, Berlin (DE), 1999.

[3] Wille R, Rival, I. editors. Restructuring Lattice Theory: an approach Based on Hierarchies of concepts Ordered Sets,.Reidel, 1982.

[4] G. Stumme: Efficient Data Mining Based on Formal Concept Analysis. In: A. Hameurlain, R. Cicchetti, R. Traunmüller (Eds.): *Database and Expert Systems Applications*. Proc. DEXA 2002. LNCS 2453, Springer, Heidelberg 2002, 534-546.

[5] G. Stumme, R. Taouil, Y. Bastide, N. Pasqier, L. Lakhal: Computing Iceberg Concept Lattices with Titanic. *J. on Knowledge and Data Engineering (*KDE) 42(2), 2002, 189-222.

[6] Dart T, Cui Y, Chatellier G, Degoulet P. Analysis of hospitalised patient flows using data-mining. Stud Health Technol Inform. 2003;95:263-8.

7. Knowledge Representation, Ontologies, Coding, Terminology

7.1 Knowledge Representation

A. Hasman et al. (Eds.)
IOS Press, 2006

Non Aristotelian categories in medicine

György SURJÁN [1]
National Institute for Strategic Health Research, Budapest, Hungary

Abstract. This paper discusses the representation of medical categories that can not be defined in Aristotelian sense. Two kinds of these categories are mentioned: the prototype and the family resemblance categories. Such categories obviously do exist in medical domain. Search on the Net was performed for free text definition for some commonly used medical categories, like 'autism', 'Burkitt lymphoma' and 'disease'. Most of the found often contradicting definitions do not follow the Aristotelian rules of definition. Many definitions describe statistical properties of the category that are often useless in individual cases. A simple way is suggested that makes possible to represent such categories in biomedical ontologies and treat them separate from better formed categories. This makes possible to revise these categories at any later stage of ontology development.
Keywords: Biomedical ontologies, prototype, family resemblance,

1. Introduction

Representations of categories of a domain that are used for computer-processing recently used to call ontologies. The principles of building ontologies go back to the fundamental work of Aristotle. In spite of the long history of theoretical work on this field, in practice, building ontologies – especially in very complex domains like medicine – is a hard work that requires co-operation of domain experts and knowledge engineers. The success of this co-operation depends of course partly on personal abilities and intelligence, but partly depends also on an objective condition: the categories that are used by domain experts must be definable. Some experts in ontology argue strongly that ontologies are not about human knowledge and human concepts [1]. This is true at least if we consider the original meaning of ontology as a branch of philosophy about existence. But if the knowledge necessary to build a domain ontology comes from domain experts they can not be independent from their own knowledge. Of course in such a co-operation it may happen that some of the concepts used by the domain experts appear to be ill-defined and have to be either corrected or rejected from the ontology. In this paper we try to investigate the following question: *Are there medical concepts that are essential for human reasoning in medicine but can not be formulated according to the requirements of a formal ontology?*

2. Formal definitions and human concepts

It is a subject of cognitive psychology to study how reality is represented in the human mind. It is obvious that language has an essential role. Words denote something that is

[1] Correspondence György Surján 1444 Budapest, Pf. 278. Hungary surjan.gyorgy@eski.hu

meaningful for us, and we use words to communicate our ideas and knowledge. (This is also the way of communication between domain experts and ontology designers when they build an ontology) The meaning of some words is apparently easy to describe in an Aristotelian way:

A bachelor *is an* adult male human *who* has not been married
A viral pneumonia *is a* pneumonia *which* is caused by viruses.

These are convincing examples that indicate the perfection of the Aristotelian way of thinking. However, there are other examples that suggest that this is not the only way, how humans – even in science – create categories or concepts. Just try to define such common things like furniture, car, or salad. In the medical domain there are also a lot of such concepts or categories that are hard to define in an Aristotelian way. Indeed, some cognitive psychologists warn us that we often use categories that are not defined – or not definable – formally. Summarising the corresponding literature Aitchison mentions two other important types of categories that are not defined in an Aristotelian sense [2]. These are the prototypes and the so-called family resemblance.

The prototypical approach asserts that membership of a class depends on the level of similarity of a given object to a prototype. E.g. an experiment performed by Elenaor Rosch [3] showed, that we have a prototypical bird, furniture, food etc. in our mind, and any particular bird, piece of furniture, dish of food is more or less similar to it. Anything that shows sufficient similarity to the prototypical bird satisfies the criterion of being a bird. The remarkable thing is that although we do not have an objective measure of similarity of these things people still usually agree on which instances are more typical members of a class and which are less.

It is a common experience among practising physicians that they hardly can diagnose a disease that they never have seen before. This is one of the reasons why personal experience has still a superior value in medicine. The prototypical way of thinking has an apparently important role in setting up medical diagnoses.

In common life, and also perhaps in medicine there are classes or categories, where membership of an instance can not be explained even by prototype theory. Wittgenstein was the first [4] who warned that there are categories like "game" that consist of so many different things, that there is neither a unique exclusive common property (i.e. necessary and sufficient attribute) nor one prototypical example for the category. The members of such classes always resemble to one or more members, but can be quite different from others. Since similarity of the different members can be related to different properties, the similarity relations form a connected graph, but not all elements are related to all others. This is the so called family resemblance. E.g. there are games, where players compete against each other; there are games that are played by cards. Patience is a card game that we play alone, so there is no competition. Bridge is a card game where players are competing. Tennis is a game where players are competing but it is not played with cards. So it is hard to find anything in common between tennis and patience, but they still a member of the class 'game' since both are similar to other games. Even if we found a common property it would not be common with other games or it will be a property of many things that are not games.

A similar situation may occur with diseases and their symptoms. It can happen that none of the cases of a certain disease depicts all relevant symptoms, so a prototypical case may never occur. E.g. the classical diagnostic criteria for rheumatoid arthritis are that 4 of the 6 possible symptoms must be present, one of them must affect 3 out of 16 joints (8 left-right pairs). Perhaps none of the patients show all symptoms and have affected all joints, so it is hard to find a prototypical case of rheumatoid arthritis. This is also something similar to family resemblance categorisation. Of course the definition of a disease and the diagnostic reasoning in clinical practice are different issues. It happens often that a clinical entity has certain formal criteria that can not be investigated in a living human (either because it is too expensive or harmful for the patient). In such cases clinicians have to use indirect symptoms and some sort of probabilistic reasoning to set up a diagnosis.

So we may expect that there are different sorts of categories in medicine:

- Some of them defined in a way that meet the Aristotelian criteria of definition
- Some of them described in an informal way although it is possible to find a formal definition (regardless to the clinical usefulness of such a formal definition)
- Some of them can not be defined in an Aristotelian sense because they are formed by prototype or family resemblance reasoning.

We have to make clear that the non-Aristotelian categories we are speaking about are different from fuzzy categories, that are also used to called non-Aristotelian [5]. While fuzzy logic obviously violates some Aristotelian rules (e.g. the law of excluded middle), fuzzy categories definitely *have* a membership criterion, just not a binary one. The categories we are speaking about have no (even a fuzzy) defined membership criterion at least we are not able to recognise it. Remarkably humans are able to use such categories and pretty well agree on what belongs to them and what not.

In the next section we investigate some categories and definitions available on the web to see the problem in more details.

3. Free text definitions in medicine

Ontology builders should consult physicians to obtain the knowledge necessary to build medical ontologies. Whether this happens via personal contacts or by reading written medical texts, physicians express themselves in natural (more precisely in technical) language. To demonstrate, how definitions expressed in free text look like, we searched the Internet (Google) for definitions of some medical categories.

The first was the definition of autism, a known psychiatric disorder. We got 22 definitions, perhaps none of them agrees with any of the others, and none of them satisfies the requirement of an Aristotelian definition. One could think that psychiatry is a fuzzy and somewhat subjective field of medicine, where it is very hard or even impossible to give precise definitions. Therefore our second search was for the definition of Burkitt lymphoma, a neoplastic disease, for which a precise histo-pathological diagnosis based on objective criteria is necessary to choose the adequate therapy. We obtained only three definitions, all of them agreed in one aspect, i.e. Burkitt lymphoma is a cancer and two of them agreed in that this is a cancer of the lymphatic system. All other attributes were totally different. (The list of definitions for both diseases can be found at http://www.eski.hu/surjan/defs.htm.) The surprising thing in these definitions is not the fact that different experts define the same thing in different ways. It comes from the nature of medicine that not only theoretical definitions may differ, but the same clinical cases can be diagnosed and treated differently. The really serious problem is that most of the obtained definitions do not help in deciding whether a certain case is an occurrence of the given disease or not.

There are many problems causing this situation, here we discuss only one what we think to be a central problem of contemporary medicine. Majority of recent medical knowledge is of statistical nature, while physicians always treat a single individual. Let we take one of the definitions of autism to show, how this fact relates to the problem of definitions:

> "Autism is disorder which *usually* appears within the first three years of life and may result in learning difficulties, speech problems and difficulty relating to people.

Does this definition help us to identify a case of autism? Let say, I am a man, who *usually* wears a tie. This statement does not define me correctly, but still may help someone to recognise me. If there is someone who *never* wears a tie this man can not be me. The above cited definition says that autism is a disorder that *usually* appears within the first three years of life. But this statistically provable statement says nothing about an individual case, since autism appears only once in the life of the patient. Either it happens within the first three years or later, this neither proves nor excludes the diagnosis of autism. The same holds for the rest of the above cited definition. Presence or absence of any of the mentioned symptoms neither excludes nor proves autism. Of course there are several well known methods in medical decision support that are able to use such statistical correlations and evaluate *all* facts known about a certain patient and infer to the *most likely* diagnosis. But these statistical facts are totally misleading if the goal is to represent facts about things that explain us what those things *are.*

In these examples it is reasonable to ask, whether the reason why these confusing definitions appear is simply the fact that physicians are not skilled enough in mathematical logic and philosophy. In all such cases it is advantageous for ontology builders to consult physicians personally and to ask further facts about the given category sufficient for a formal definition. However presumably they will be not successful in all cases. E.g. definitions of upper level categories, such as 'disease' suggest that there are important medical categories that are non-Aristotelian. Most of the definitions say that disease is a condition characterised by an abnormal body function. But it is hard to know what is abnormal, and sometimes some body function can be decreased or increased without any specific disease. We believe that disease – similarly to "game" – is a category that simply does not have any common exclusive property.

4. Representing non-Aristotelian categories in ontologies

One may conclude that such definitions can not be used in medical ontologies, and those categories that do not have an Aristotelian definition, have to be rejected from a medical ontology as vague and useless fictions. Such purism would lead to rather incomplete ontologies, essential medical entities would be missing.

As mentioned above, in many cases consultation between domain experts and ontology designers could lead to definitions that meet the Aristotelian criteria. This is very likely e.g. in case of Burkitt lymphoma. We expect this will happen very often in the coming years, and this will contribute to the development of medicine. But still it seems to be important to find ways to describe those categories in ontologies that are defined by prototype similarity and family resemblance.

Of course, existing ontology description languages do not support directly such things. But it is possible to find a work-around solution. Both prototype and family resemblance can be defined as relations in ontologies. Doing so it is possible to define categories formally based on these relations. We built a small pilot OWL ontology in Protégé environment [6] in which disease is represented as a category defined by family resemblance relations, and medical procedure as a class defined by prototypes.

Three relations were used:

Is_Prototype_Of
Protoypically_Similar_To
Resembles

None of these relations is transitive, the Is_Prototype_Of relation is asymmetric, the two other are symmetric. Using these relations we defined 'Disease' and 'Medical Procedure' formally as follows:

Anything is a disease that resembles – another – disease.
In Protégé: Disease - sufficient and necessary conditions: ∃ Resembles Disease

And

Anything is a medical procedure which is a prototype of medical procedure or which is similar to a prototype of medical procedure.
In Protégé: Medical_Procedure – sufficient and necessary conditions:
(∃ is_Prototype_Of Medical_Procedure) ⊔ (∃ Prototypically_Similar_To Medical_Procedure)

This definition allows the category to have more than one prototype.

Then we represented Infection, Injury (as something that resembles Infection) and Poisoning (as something that resembles Injury)
Either if we assert that Infection is a disease or we assert that infection resembles a disease, the RACER reasoner infers that injury, and poisoning are diseases.

We represented Auscultation and Surgical_Operation as prototypes of Medical_Procedure. Appendectomy was represented as prototypically similar to Surgical_Operation, gastrectomy, as prototypically similar to Appendectomy, blood_pressure_measurement as prototypically similar to Auscultation. The RACER classified all these categories as Medical_Procedure.

5. Discussion

The proposed work-around approach makes possible to represent non-Aristotelian categories, and entities belonging to such classes can be classified. Of course this is not a solution of the underlying philosophical problem. Stating that e.g. Auscultation is a prototype of Medical_Procedure is not less arbitrary than to leave Medical_Procedure as primitive and assert manually that Auscultation is a kind of Medical_Procedure. Smith et al argue [7], that relations – except primitive ones – should be defined in a formal ontology as well as classes. The relations we propose are by nature primitive, or dummy. Their usage is entirely left on subjective decision of the ontology builder. Still we think it is worthwhile to make a difference between formal subsumption and prototype or family resemblance classification. While the result is practically the same, it is advisable to restrict the use of the subsumption relation strictly to cases where the criteria of formal subsumption [8] are undoubtedly met. The proposed work-around solution makes possible to distinguish between Aristotelian and non-Aristotelian classes and allows revise non-Aristotelian categories anytime at a later stage of development of the ontology. This is the main practical advantage of the proposed solution.

Acknowledgements: This work was partly covered by EU FW6 IST 50750 (Semantic Mining) project. The author expresses his special thanks to Arie Hasman for commenting and careful revising the manuscript. Also thanks to the members of Hungarian Health Ontology Forum for their valuable remarks.

References

[1] Smith B, Ceusters W, Temmerman R. Wusteria. Stud Health Technol Inform. 2005;116:647-52.
[2] Aitchison J. Words in the mind 3. ed Blakwell Publishing Oxford 2003
[3] Rosch E, Mervis CB. Cognitive representations of semantic categories Journal of Experimental Psychology: General, 1975
[4] Wittgenstein L. Philosophische Untersuchungen hrsg. von Eike von Savigny. Akad. Verl., Berlin 1998.
[5] Sadegh-Zadeh K. Fuzzy Health, Illness, and Disease Journal of Medicine and Philosophy 2000, Vol. 25, No. 5, pp. 605–638
[6] http://protege.stanford.edu/
[7] Smith B, Ceusters W, Klagges B, Köhler J, Kumar A, Lomax J, Mungall C, Neuhaus F, Rector A, Rosse C. Relations in biomedical ontologies Genome Biology 2005, 6:R46
[8] Guarino N, Welty C Identity and Subsumption LADSEB-CNR Internal Report 01/2001 available at http://www.loa-cnr.it/Papers/Identity&Subsumption.pdf

Ubiquity: Technologies for Better Health in Aging Societies
A. Hasman et al. (Eds.)
IOS Press, 2006

Referent Tracking: The Problem of Negative Findings

Werner CEUSTERS[a,1], Peter ELKIN[b] and Barry SMITH[a, c]
[a] *Center of Excellence in Bioinformatics and Life Sciences, and National Center for Biomedical Ontology, University at Buffalo, NY, USA*
[b] *Department of Medicine, Mayo Foundation, Rochester, MN, USA*
[c] *Institute for Formal Ontology and Medical Information Science, Saarbrücken, Germany, and Department of Philosophy, University at Buffalo, NY, USA*

Abstract. The paradigm of referent tracking is based on a realist presupposition which rejects so-called negative entities (*congenital absent nipple*, and the like) as spurious. How, then, can a referent tracking-based Electronic Health Record deal with what are standardly called 'negative findings'? To answer this question we carried out an analysis of some 748 sentences drawn from patient charts and containing some form of negation. Our analysis shows that to deal with these sentences we need to introduce a new ontological relationship between a particular and a universal, which holds when no instance of the universal has a specific qualified ontological relation with the particular. This relation is found to be able to accommodate nearly all occurrences of negative findings in the examined sample, in ways which involve no reference to negative entities.
Keywords: referent tracking, negation, negative findings, ontology, realism, EHR

1. Introduction

Referent tracking has been introduced as a new paradigm for entry and retrieval of data in the Electronic Health Record (EHR) [1]. Its purpose is to avoid the ambiguity that arises when statements in an EHR refer to disorders, lesions and other entities on the side of the patient exclusively by means of generic terms from a terminology or ontology. Suppose that two different physicians are treating the same patient A, and that each enters into A's EHR a statement to the effect that A suffers (i) from diabetes or (ii) from a fracture of the right lower arm. Then it is in either case left unspecified whether they are referring to the same or to different entities on the side of the patient. In case (i), it is clear that only one answer is possible; yet the ambiguity as to whether each of the two physicians is referring to the same diabetes will still cause problems for software agents programmed to make inferences from the data. In case (ii) this ambiguity causes problems even for human beings, since the physicians in question might have been referring either to the same or to different fractures.

Referent tracking avoids such ambiguities by introducing unique identifiers, called IUIs or Instance Unique Identifiers, for each numerically distinct entity that exists in reality and that is referred to in statements in the record. The referent tracking paradigm

[1] Corresponding Author: Werner Ceusters, Ontology Research Group, Center of Excellence in Bioinformatics and Life Sciences, 901 Washington Street, Buffalo, NY 14203, USA. Email: ceusters@buffalo.edu. Internet: http://org.buffalo.edu.

thereby expands the entities uniquely identified for EHR purposes far beyond the current range, which is restricted to entities such as patients, care providers, buildings, machines, and so forth. A statement such as "John Doe has a fracture of the right lower arm", would be translated under the referent tracking paradigm into: '#1 has #2', the first number being the IUI for John Doe, the second the IUI for that specific fracture he is suffering from. Additional statements would then specify that '#2 is a fracture of the right lower arm' or, better, that '#2 is a #3 located in #4', together with the extra information that #3 is a fracture and #4 is John Doe's right lower arm. Expressions such as 'has', 'is a', 'located in', etc., would at the same time be replaced by the appropriate relationships from a suitable ontology [2], for which logical reasoning tools have also been defined.

In [3] we have described a framework that is able to deal with phenomena in reality that can be described by means of directly depicting statements of the sorts just described, at the same time specifying the role to be played by terminologies and ontologies in this framework. We also discussed there how information entered into an EHR system by clinicians in the usual way could be translated automatically into statements of a Referent Tracking System (RTS). One specific problem thus far left untouched is how to represent phenomena commonly called 'negative findings' or 'negative observations' within an RTS. Example statements describing such phenomena are: "no history of diabetes", "hypertension ruled out", "absence of metastases in the lung", and "abortion was prevented". Such statements seem at first sight to present a problem for the referent tracking paradigm, since there are here no entities on the side of the patient to which unique identifiers can be assigned.

2. Objectives, Materials and Methods

If referent tracking is to be accepted as a viable paradigm for the EHR, it has to be able to deal with phenomena of the mentioned sort. Our objective is thus to expand the repertoire of statements with which an RTS can deal in such a way as to allow representations of those relevant portions of reality in which something is not the case. We must do this, however, without violating the principles of Basic Formal Ontology (BFO) [4] upon which referent tracking is built. These principles counsel unqualified realism as a basis for the creation of high-quality shared ontologies in the biomedical domain. This means, most importantly, that our representations can acknowledge only those entities which exist in biological reality, and must reject all those types of putative negative entities – absences, non-existents, possibilia, and the like – which are postulated merely as artefacts of specific logical or computational frameworks.

We analysed 396 negative findings encountered in 250 sentences out of 18 patient charts from Johns Hopkins University [5]. We assumed such findings to be descriptions of real phenomena on the side of the patient and sought to classify the underlying structures and processes in terms of the various top-level categories and relations defined in BFO, taking careful account of the role of negation in the corresponding descriptions. We then explored ways to represent such phenomena by means of the types of representational units available on the referent tracking paradigm.

BFO subdivides reality into a number of basic categories. First, it distinguishes *particulars* from *universals*, the former being entities such as John Doe or the left arm fracture he suffered from last year, and the latter entities such as *person*, *fracture* and *arm*.

Table 1: Ontology-related tuple types in Referent Tracking

Tuple type	Phenomenon described
$A_i = < IUI_p, IUI_a, t_{ap}>$	Act of assignment of IUI_p to a particular at time t_{ap} by the particular referred to by IUI_a *
$R_i = <IUI_a, t_a, \mathbf{r}, o, P, t_r>$	It is asserted by the particular referred to by IUI_a at time t_a that the relationship r from ontology o obtains between the particulars referred to in the set of IUIs P at time t_r
$U_i = <IUI_a, t_a, \mathbf{inst}, o, IUI_p, u, t_r>$	It is asserted by the particular referred to by IUI_a at time t_a that the instantiation relation as defined in ontology o obtains between the particular referred to by IUI_p and the universal u at time t_r

* The subscript 'p' stands for 'particular' and 'a' for 'author'

Second, it distinguishes continuants from occurrents. Continuants are entities, such as John Doe and his left arm, that endure continuously through time. Occurrents, in contrast, unfold over a certain time span through successive temporal parts, examples being entities such as processes, actions and events. Thirdly, there is the distinction between dependent and independent entities, the former being such that they cannot exist without some instance of the latter: John Doe's height or weight, for example, cannot exist without the existence of John Doe himself.

3. Results

BFO distinguishes three major families of relations between the entities just sketched: (1) <p, p>: from particular to particular (for example: John Doe's nose being part of John Doe); (2) <p, u>: from particular to universal (for example: John Doe being an instance of the type *person*); and (3) <u, u>: from universal to universal (for example: *person* being a subkind of *organism*). [2]

Referent tracking applies BFO to the domain of EHRs, requiring: (1) that particulars are referred to by means of unique identifiers (IUIs), (2) that each particular should receive maximally one IUI, and (3) that only entities that exist are to be assigned a IUI. Real world phenomena are then represented in an RTS [3] by means of tuples of the sorts outlined in Table 1.

Table 2 lists the four headings under which negative findings can be classified when account is taken of BFO's distinction between particulars and universals and of the different types of relationships that can obtain between them. The last column of Table 2 shows the distribution of the occurrence of negative findings in the analysed sample. On the basis of our analysis we now argue that there must be included in the machinery of BFO new relations, a new family of formal <p,u>-relations which obtain whenever a given particular does not stand in some given <p,p> relation to any instance of a given universal. The relations in this family we can define more formally as follows:

p **lacks** u **at** t **with respect to identity**
=def. there is no x such that: x **identical_to** p at t and x **instance_of** u

p **lacks** u **at** t **with respect to part**
=def. there is no x such that: x **part_of** p at t and x **instance_of** u

and similarly for other <p,p> relations such as **quality_of, located_in**, **derives_from**, **has_participant**, and so on. Note that the **lacks**-relations are formal relations, analogous to instantiation or parthood. This means that they are not extra ingredients in being, but rather that in virtue of which existing entities are joined together to form larger wholes.

Table 2: categories of negative findings from the perspective of BFO

	Relation type	Type of Negative Finding	Examples	%
C1	<p, u> *	A particular is not related in a specific way to any instance of a universal at some given time	*he denies abdominal pain; no alcohol abuse; no hepatosplenomegaly; he has no children, without any cyanosis*	85.4
C2	<p, u>	A particular is not the instance of a given class at some given time	*which ruled out primary hyperaldosteronism, nontender, in no apparent distress, Romberg sign was absent , no palpable lymph nodes*	12.4
C3	<p, p>	A particular is not related to another particular in a specific way at some given time	*this record is not available to me; it is not the intense edema she had before; he has not identified any association with meals.*	2.2

* 'p' ranges over particulars, 'u' over universals

It is **lacks** that is involved in the phenomena described by means of negative findings of types C1 and C2 from Table 2. An example of type C1 arises when a patient (an independent continuant) does not exhibit a headache (a dependent continuant); on our analysis this means that the patient and the universal *headache* (both of which are from the BFO perspective full-fledged entities) stand to each other at a given time in a certain relation, namely: **lacks with respect to the relation has_quality**. C2-type phenomena receive an identical analysis, except that here the relevant relation is **lacks with respect to the relation identical_to**. If, for example, it is ruled out, for a given disorder (p) on the side of a patient, that it is a case of *primary hyperaldosteronism* (u), then it is asserted that at the given time (t) no instance of u is identical to p. Negative findings of type C3 suggest the need for a relation analogous to **lacks**, but holding not between a particular and a universal but between one particular and another. We are not yet sure, however, whether there is a need for a relation of this sort, since the corresponding cases may perhaps be dealt with in terms of the simple logical negation of straightforward statements about the corresponding particulars.

To accommodate the new **lacks** relations in referent tracking, a further tuple type is required, which we will call U^{-}:

U^{-}_{i} = <IUI_a, t_a, r, o, IUI_p, u, t_r>	The particular referred to by IUI_a asserts at time t_a that the relation r of ontology o does not obtain at time t_r between the particular referred to by IUI_p and any of the instances of the universal u at time t_r

4. Discussion

A substantial fraction of the clinical observations entered into patient records are expressed by means of negation. Elkin *et al* found SNOMED-CT to provide coverage for 14,792 concepts in 41 health records from Johns Hopkins University, of which 1,823 (12.3%) were identified as negative by human review [5]. Mutalik *et al* report the presence of 8,358 instances of UMLS concepts in 60 documents of which 571 (6.8%) were negations [6]. This is because negative findings are as important as positive ones for accurate medical decision making, and failure to document pertinent negative findings may have medico-legal consequences in case of allegations of malpractice. In 1998, an NHS Independent Review panel judged the record-keeping in a specific case to fall below the level of good practice because '*the notes make no reference to any other findings, nor of any negative ones which would be relevant when considering problems specific to diabetes. Thus no reference is made to the absence of a smell of ketones on Miss J's breath, nor any other negative indications*' [7]. In the US, Medicare and Medicaid compliance requires that the patient record should document '*specific abnormal and relevant negative findings of the examination of the affected or symptomatic body area(s) or organ system(s)*'. [8]

The sentences we studied were extracted from the patient charts by natural language parsing software sensitive to textual clues for negation [5]. Some sentences were retained erroneously because textual clues were misleading, as in: '*The patient actually answers yes, no, and sir to all questions*'. Furthermore, not all sentences containing negation are descriptions of negative findings; thus '*He has no idea why he is here*' may either refer to the positive finding of being mentally disoriented or be simply a non-clinical statement. A clear example of a sentence describing a positive phenomenon in a negative way is: '*Her workup showed that she had an MRI of the brain that was negative in 03/02*', which in fact states that the MRI was normal. Such sentences (8.3% of the sample) were not included in our analysis. Modal and similar operators were left aside in this analysis, so that for example only the italicized portion of the sentence '*He has no family history of GI malignancies* that I know of' was analyzed. This is because referent tracking has been designed to give modal aspects a second-order treatment, the discussion of which falls beyond the scope of this paper.

Some negative findings could be classified in one of the 3 categories, but describe phenomena that currently cannot be dealt with under the referent tracking paradigm. Examples are: '*no other complications of gastro-esophageal reflux disease*'.

With the introduction of the new **lacks** relations – an expanded version of the rationale for which is provided in [9] – we defend, in effect, the thesis that *negation is outside the realm of ontology* but belongs rather to the domains of logic [10], language [11] and epistemology [12]. Denial of this thesis is symptomatic of what Smith has called '*fantology'*, i.e. the false belief that the structures of logic, language and information are mirrors of the structure of reality [13]. In reality, there is only what there is. Language and logic allow us to talk and reason about what there is by using negation. But the corresponding negative expressions do not mirror anything in reality.

Thus, if a clinician describes a phenomenon on the side of a patient using the phrase 'absence of metastases in the lungs', then the corresponding assertion would be registered in an RTS using some coding along the following lines:

U^{-}_{61092} = <#23, '2005-12-27-18:40', **contains**, #678, #91, *metastasis*, 'until 2005-12-27-18:40'>,

in which #23 would be the IUI of the clinician, '2005-12-27-18:40' the time of assertion, **contains** the inverse of the <p,p>-relation **contained_in** from the OBO Relation Ontology [2], #678 the IUI of the OBO-ontology, #91 the IUI of the patient's lungs, '*metastasis*' a reference to the universal *metastasis*, and 'until 2005-12-27-18:40' a description of the time interval during which the **lacks** relation holds (in line with the provisions of EN 12338:2004 [14]). By representing this statement in some adequate logical form and by applying the corresponding inference rules further derivations can then be made, for exampe to the effect that, whatever particular there is in that patient's lung, it is not a metastasis, and that if there is a metastasis contained in some body part of the patient, then that body part is not the lung, and so forth.

Finally, it is possible to define a lacks relation that holds between universals. This would be useful for statements of the sort that all relatives of a patient are disease free or that none of his white blood cells in an examined sample exhibit a certain anomaly.

5. Conclusion

By introducing **lacks** relations of the <p,u> sort together with the new U^- tuple type, we were able to represent 99.9% of the negative findings that occur in the analysed sample (and thus, we believe, of the vast majority of negative findings that occur in EHRs in general) in such a way as to remain faithful to the principles of unqualified realism within an EHR regime based on the idea of faithfulness to clinical reality. Further research is required to assess the need for two other families of **lacks**-like relations holding, respectively, between particulars and between universals.

Acknowledgements: This paper was written under the auspices of the National Center for Biomedical Ontology (funded by the National Institutes of Health through the NIH Roadmap for Medical Research).

References

[1] Ceusters W. and Smith B. Referent tracking in Electronic Healthcare Records. In: Engelbrecht R. et al. (eds.) Medical Informatics Europe, IOS Press, Amsterdam, 2005, p. 71-76.

[2] Smith B, Ceusters W, Klagges B, Köhler J, Kumar A, Lomax J, Mungall C, Neuhaus F, Rector AL, Rosse C. Relations in biomedical ontologies. Genome Biol, 2005;6(5):R46.

[3] Ceusters W, Smith B. Strategies for referent tracking in Electronic Health Records. Journal of Biomedical Informatics. In press.

[4] Grenon P, Smith B, Goldberg L. Biodynamic ontology: applying BFO in the biomedical domain. In DM Pisanelli (ed.), Ontologies in Medicine, Amsterdam:IOS Press, 2004, p. 20-38.

[5] Elkin PL, Brown SH, Bauer BA, Husser CS, Carruth W, Bergstrom LR and Wahner-Roedler DL. A controlled trial of automated classification of negation from clinical notes. BMC Medical Informatics and Decision Making 2005;5:13.

[6] Mutalik PG, Deshpande A, Nadkarni PM: Use of general-purpose negation detection to augment concept indexing of medical documents: a quantitative study using the UMLS. J Am Med Inform Assoc 2001, 8:598-609.

[7] Health Service Ombudsman for England. Errors in the care and treatment of a young woman with diabetes. http://www.ombudsman.org.uk/improving_services/special_reports/hsc/diabetes/index.html (last accessed: December 27, 2005).

[8] Centers for Medicare and Medicaid Services. 1997 Documentation Guidelines for Evaluation and Management Services. http://www.cms.hhs.gov/MedlearnProducts/downloads/1995dg.pdf. (Last accessed: December 27, 2005).

[9] Simon J, Fielding JM and Smith B. Using philosophy to improve the coherence and interoperability of applications ontologies. In Büchel B, Klein B and Roth-Berghofer T (eds.), Proceedings of the First Workshop on Philosophy and Informatics. DFKI, Cologne: 2004, 65-72.

[10] Smith B. Logic, form and matter, Proceedings of the Aristotelian Society, Supp. Vol. 1981; 55: 47–63.

[11] Carston R. Negation, 'presupposition' and the semantic/pragmatic distinction, Journal of Linguistics 1998;34 (2):309-50.

[12] Pacitti D. The nature of the negative: Towards an understanding of negation and negativity. Giardini, Pisa, 1991.

[13] Smith B. Against fantology. In: Reicher ME, Marek JC (eds.), Experience and Analysis. Vienna 2005, p. 153-170.

[14] EN 12388:2005. Health informatics - Time standards for healthcare specific problems.

Ubiquity: Technologies for Better Health in Aging Societies
A. Hasman et al. (Eds.)
IOS Press, 2006

Enriching Medical Terminologies: an Approach Based on Aligned Corpora

Louise DELEGER[a1], Magnus MERKEL[b], Pierre ZWEIGENBAUM[ac]
[a]*INSERM U729, Paris, France*
[b]*Dept of Computer and Information Science, Linköping University, Sweden*
[c]*STIM/DSI, Assistance Publique – Hôpitaux de Paris, France*

Abstract. Medical terminologies such as those in the UMLS are never exhaustive and there is a constant need to enrich them, especially in terms of multilinguality. We present a methodology to acquire new French translations of English medical terms based on word alignment in a parallel corpus - i.e. pairing of corresponding words. We automatically collected a 27.7-million-word parallel, English-French corpus. Based on a first 1.3-million-word extract of this corpus, we detected 10,171 candidate French translations of English medical terms from MeSH and SNOMED, among which 3,807 are new translations of English MeSH terms.
Keywords: medical terminology, parallel corpora, word alignment.

1. Introduction

The UMLS Metathesaurus is an extensive vocabulary database that gathers and provides a link between different existing biomedical terminologies. But despite being multilingual, it is mostly composed of English vocabulary, and other languages such as French are under-represented in comparison to English. There is therefore a need to enrich the terminologies of the UMLS. The acquisition of new translations of English terms is required. This is the purpose of the VUMeF[2] project which aims at extending the French part in the UMLS and which provides the background for this work.

Plenty of multilingual texts can be found as regards a specific domain but exhaustive terminologies and dictionaries are far less numerous – as seen in the case of the UMLS. Hence the idea of using parallel corpora, i.e., collections of texts paired with their translations, to enlarge terminologies. So instead of employing a human translator, we can make use of existing translated texts from which translations at the term level can be extracted.

We present a methodology to acquire medical terms based on word alignment in a parallel corpus. Word alignment is a natural language processing technique and is used in several applications such as terminology development (which is the case here), automatic translation and cross-lingual information retrieval. It consists in pairing words that are translations of each other in a parallel corpus, i.e. finding for each word of a text its translation in the corresponding text in the other language.

Such a technique has only been employed in a few approaches for medical terminologies. Previous work includes [2] which derives word translations from the

---

[1] Corresponding Author: Louise Deléger, INSERM U729 15 r de l'Ecole de Médecine 75006 Paris France, louise.deleger@spim.jussieu.fr
[2] Project VUMeF (French Unified Medical Vocabulary), led by Stéfan Darmoni [1], is partially funded by the French Ministry of Research (National Network of Health Technologies).

ICD-10 parallel medical vocabulary, [3] which is based on comparable corpora, [4] which uses a statistical vector model on a corpus aligned at the document level, and [5] which is similar to ours in that it deals with word alignment in parallel medical corpora to extract term translations. However, we deal with a larger corpus and process both single-word and multiple-word alignments.

Our task involves issues such as dealing with errors in the alignment process that will spread from step to step, and detecting multi-word units. This work is outlined in the following way: based on a French-English corpus (2.1), we align sentences (2.2) and words (2.3). Medical terms are then selected (2.4) through the projection of lists of English terms from SNOMED and MeSH. We obtain a list of bilingual English-French medical terms that we review. We extract samples for evaluation purposes (2.5) and expose results (3.). We discuss (4.) and conclude (5.) on the method.

2. Material and Method

2.1. Corpus

The corpus used for this experiment is collected from the web. The web is indeed a powerful resource for building corpora, both in terms of quantity and multilinguality. The quality of such a corpus can nevertheless be questioned and this might account for a proportion of the noise detected in the results. Our corpus is gathered from a bilingual (French-English) Canadian health web site [6]. It is intended for the general public as well as for specialists and the proportion of specialized terms might therefore be lower than in resources dedicated to medical specialists. Several techniques exist for building a parallel corpus from the web [7] [8]. We generated pairs of parallel documents – i.e. documents that are translations of each other - using information contained in the document structure – HTML links to corresponding documents in the other language. This gives us 11,041 pairs of parallel documents and a total of 27.7 million words. The HTML documents acquired need to be converted to text format: this is done while keeping title, paragraph and link tags as correspondence points for sentence alignment. The resulting texts are segmented into sentences to prepare for further processing.

2.2. Sentence Alignment

The first step towards word alignment is to align the corpus at sentence-level. Sentence alignment is a mandatory task since there is not a full one-to-one correspondence between the sentences of two parallel texts. Although it is most common that one sentence in a source language corresponds to one sentence in a target language, there are instances where one sentence is translated with two – or sometimes even three or more – sentences, and this needs to be determined before working at the word level.

To do so, we use Dan Melamed's GMA (Geometric Mapping and Alignment) [9], a robust tool which performs sentence alignment of parallel texts using both statistical and linguistic techniques. It is based, among other things, on length measurements, bilingual lexicons and cognates (words sharing similar spelling and meaning).

Though sentence aligners achieve high-quality performances, any mistake at this level will be reflected at the next one - i.e. word alignment. So, in order to work on cleaner data, we attempt to automatically detect incorrect sentence alignments and bad document pairing (documents that are not parallel) using criteria such as sentence length and quality evaluation of sentence alignment.

2.3. Word alignment

Once sentences are aligned, we can proceed to word alignment. This task is far more problematic than sentence alignment. There is no true word-to-word correspondence between the words of two sentences. A word is often translated with several ones, or can be omitted in the corresponding sentence (this is typically the case for grammatical words that are specific to a language). Parallel sentences, though being translations of each other, can differ considerably in terms of structure. In that case even a human has trouble determining which words should be paired together. The results we expect are therefore on a lower level than from the previous sentence alignment task.

The issue of the type of word alignment should be raised. That is, do we want a word-to-word alignment? The objective of this work is to obtain medical terms. A term can be either a single word or a multi-word unit. A common approach is to first extract candidate terms using a separate tool - a term extractor – and then to proceed to their alignment. The originality of this work is that we do not separate the detection of candidate terms from the alignment process. In other words we use a tool that is able to detect multi-word units and to align both single words and multi-word expressions.

Word alignment systems usually derive from either statistical approaches or linguistic ones, or a combination of both. Statistical methods [10] involve co-occurrence measures and probability scores, and are especially effective on large corpora with high-frequency words but performances decrease with low-frequency occurrences. Linguistic ones [11] make use of information such as syntactic parsing. They are less robust despite being able to deal with low-frequency words. Hybrid approaches [12] [13] seem to be a good compromise.

We use the I*Tools suite to perform word alignment (developed at Linköping University, Sweden). We chose these tools partly because they are based on a hybrid approach, using both statistical and linguistic techniques. They also align multi-word units which suits our terminological purpose. They make use of resources such as co-occurrence measures, bilingual dictionaries, POS tagging and syntactical analysis. A pre-processing step is required: the corpus is tagged and lemmatized (using Treetagger [14]) and syntactically annotated (with the syntactic analyzer SYNTEX [15]). The files are transformed into XML format encoding this information. The alignment process with the I*Tools can be divided into three steps: training, automatic alignment and review of the results; each one corresponding to a specific tool of the suite. Training and review are both done with graphical, interactive tools that are fast to work with.

Training of the system is manually done using a special tool of the suite – the I*Link interactive aligner [16]. This tool proposes word pairings to the user who accepts or rejects them. The user's decisions are stored into the resources of the system and by learning from them, the performances become increasingly accurate. These resources provide training data for the automatic word aligner.

The corpus is then automatically aligned by I*Trix, the automatic aligner of the suite, using the resources created with I*Link. We obtain a list of word alignments – i.e. source words paired with target words. The system can also exploit data created during the next step (the reviewing phase). In that case, the automatic alignment is repeated after a first run and takes into account the review made by the user.

Results are reviewed with the I*View tool which enables the user to confirm, reject, or simply remove an alignment. An alignment is « removed » when it is neither an error nor a correct alignment, meaning it is a partial alignment (some parts are correct). This tool also indicates for each alignment a quality score, which enables the user to rank the alignments. The quality score used in the I*Tools is based on the frequency of the words as a pair, and the frequency of each source and target word of the pair independently in other pairs. This means that for a proposed word pair which occurs with a high frequency and where the source word and the target word only occur in this pair and not in any other suggested word pairs, we have good reasons to assume that the quality is high. On the other hand if the word pair has a low frequency and the source and target words of the pair are found in several other suggested pairs, then there is reason to be more doubtful to the suggestion. The formula is $Q=f(st)/n(s)+n(t)$, where $f(st)$ is the frequency of the word pair and $n(s)$ and $n(t)$ are the number of different word pairs in which the source and target words occur respectively.

2.4. *Term selection and review*

In practice, there is no need to review all of the results since we are only looking for medical terms. We thus retain only those likely to be of interest. We select them using English medical terminologies – namely MeSH and SNOMED CT (as extracted from the 2005AC version of the UMLS). We project these lists of terms onto the English entries of our alignment pairs and select those present in the pairs. Only then do we review the alignments. These alignments constitute French candidate translations of English MeSH and SNOMED terms. The review can be done by a linguist engineer.

2.5. *Implementation and evaluation*

The methodology described above was implemented as follows:

1. conversion of the corpus into text format;
2. sentence alignment; at this point we performed an evaluation of the quality of the alignment by checking 100 sentences randomly taken from the corpus and measuring the percentage of correct alignments (precision measure);
3. training of the automatic word aligner on a set of 600 sentences randomly taken from our corpus, by interacting manually with I*Link;
4. automatic word alignment with I*Trix. We evaluated the quality of word alignment by measuring precision on two samples: sample 1 consists of 100 word pairs randomly taken from the whole resulting pairing, and sample 2 of 100 word pairs taken from the best word alignments (alignments with a frequency higher than 1 and a good quality score – equal to or higher than 1).
5. selection of medical terms (see section 2.4);
6. review, with I*View, of the alignments for the terms selected.

3. Results

We completed steps 1 and 2 on the whole corpus, thus obtaining 1.1 million sentence pairs. As the corpus is huge, we have currently processed only part of it from step 3 to 6 – a set of 540 pairs of documents (1.3 million words) gathered in two corresponding files. From this set, we obtained 91,171 word alignments and selected 20,180 pairs of medical terms. Among these pairs, there are 4,473 different source terms (a term can have several translations), so we have a mean value of 4.5 French translations per English term. 10,171 alignments were confirmed as correct ones - by a linguist engineer (L.D.) - which gives a precision of 50.4% (see table 1) and a mean value of 2.3 correct translations per term. We count 3,752 different source terms in the confirmed alignments, meaning that 721 terms only had incorrect translations. Table 1 shows that evaluation results for the overall quality of word alignments (step 4 of 2.5) are very good for the top alignments - sample 2 (taken from the 7,366 best alignments as described in 2.5) - and average for the whole aligned corpus – sample 1. As for sentence alignment (step 2 of 2.5), we achieved a precision of 95%, which is excellent.

Table 1 Evaluation figures for word alignment

	Sample 1	Sample 2	Set of medical terms
Precision	50%	92.2%	50.4%
Errors	19.6%	4.9%	31.8%
Partial alignments	30.4%	2.9%	17.8%

A proportion of the noise in word alignment can be attributed to errors in the sentence alignment process: 17% of the incorrect alignments are due to bad sentence pairing. Other factors include errors in POS tagging, bad document pairing (in our case we observed some English-English document pairs) and low quality of the data – misspelling of words, insertion of spaces inside a word, missing spaces between words.

If we take a look at the resulting list of 10,171 confirmed medical term alignments, we notice 311 pairs that are not real translations but merely pairs of English words – i.e. the English words have not been translated. These are considered correct alignments but are of no use for our purpose, so we simply ignore them. Among the remaining 9,860, 5,235 are terms from the MeSH and 4,625 are terms from SNOMED. We concentrate on the MeSH results and look at the number of different concepts (CUI in the UMLS), terms already present in the French version of the MeSH and new translations (see table 2). New translations include morphological variants (plural, feminine gender) and synonyms. The variants being of limited interest, they could also be removed [17]. A sample of 145 MeSH terms (see figures in table 2) was also extracted for validation purposes. 79 terms had new translations that were submitted to expert validation. 64 have been validated. Examples of translations are given in table 3.

Table 2 Overview of the MeSH translations

	Terms	Concepts	Already known	New	New+valid
Complete MeSH term set	5,235	1,868	1,428	3,807	
Validation sample	145	138	66	79	64 (81%)

Table 3 Translation examples

English	French	Valid
adipose tissue	tissus adipeux	Yes
breast milk	lait maternel	Yes
reproduction rights	droits de reproduction	No

4. Discussion

Our approach presents a number of advantages as well as some drawbacks. It allows us to acquire medical terms which are actually used in French documents for certain MeSH descriptors but are not registered in the current French version of the MeSH. It does not require a human translator and makes the best of existing resources. In terms of word alignment, we are able to process noisy data quite efficiently. We do not use monolingual term extractors and we align single words and multiword expressions uniformly unlike other methods which concentrate on 1-1 and 2-2 word alignments [5].

Though being an automatic approach, it still needs human help in the process (training and validation). The success of this method is also heavily dependent on the efficiency of word alignment which is a complex task. The quality of the corpus is an important feature and its choice is a major issue. In our case, we used the Web as a resource and processed a whole website. A study of the documents would have been useful in order to best characterize the type of data acquired – which documents are intended for medical specialists, which ones are for the general public and which ones have no medical content (index pages for instance). Interesting developments of this method will include the specific search for patient-oriented translations (consumer vocabulary) which are even more lacking in medical terminologies. This can be achieved, for instance, to look for candidate translations of Medline Plus vocabulary.

5. Conclusion

We described a methodology to acquire new translations of English medical terms in order to enrich existing medical terminologies. We argued that a natural language processing technique such as word alignment is an efficient way to do so. Indeed, we were able to find a number of new translations of English MeSH terms. Moreover, it is an automatic process which only requires limited human intervention. Finally, this method raises interesting prospects such as the acquisition of patient vocabulary, and more generally its application to other parallel corpora.

Acknowledgements

We gratefully thank Stéfan Darmoni for reviewing the candidate MeSH translations, and Didier Bourigault for providing SYNTEX.

References

[1] Darmoni SJ, Jarrousse É, Zweigenbaum P, Le Beux P, Namer F, Baud R et al. Extending the French part of the UMLS. Proceedings of AMIA Annual Fall Symposium, Washington, DC, 2003.
[2] Baud RH, Lovis C, Rassinoux AM, Michel PA, Scherrer JR. Automatic Extraction of Linguistic Knowledge from an International Classification. MEDINFO98. pp.581--585
[3] Chiao YC, Zweigenbaum P. Looking for French-English translations in comparable medical corpora. J Am Med Inform Assoc 2002; 8 (suppl):150–4.
[4] Widdows D, Dorrow B, Chan CK. Using Parallel Corpora to Enrich Multilingual Lexical Resources. Proceedings of LREC 3, Spain, May 2002. pp. 240-244.
[5] Névéol A, Ozdowska S. Extraction bilingue de termes médicaux dans un corpus parallèle anglais/français. Proceedings of EGC'05, Paris, January 2005.
[6] http://www.hc-sc.gc.ca/
[7] Resnik P, Smith NA. The web as a parallel corpus. Computational Linguistics, 29, 349-380. Special Issue on the Web as a Corpus. 2003.
[8] Patry A, Langlais Ph. Paradocs: un système d'identification automatique de documents parallèles. Proceedings of TALN, Dourdan, France. 2005. pp. 223-232.
[9] Melamed D. Bitext Maps and Alignments via Pattern Recognition. In J.Véronis, editor. Parallel Text Processing: Alignment and Use of Translation Corpora. 2000. http://nlp.cs.nyu.edu/GMA/
[10] Brown PF, Pietra SAD, Pietra VJD, Mercer RL. The mathematics of statistical machine translation: parameter estimation. Computational Linguistics, 19(2), 263-311. 1993.
[11] Wu D. Bracketing and aligning words and constituents in parallel text using stochastic inversion transduction grammar. In J. Véronis, editor. Parallel Text Processing. 2000.
[12] Ahrenberg L, Andersson M, Merkel M,. A knowledge-lite approach to word alignment. In J.Véronis, editor. Parallel Text Processing : Alignment and Use of Translation Corpora. 2000.
[13] Barbu AM. Simple linguistic methods for improving a word alignment algorithm. In 7th International Conference on the Statistical Analysis of Textual Data, Louvain-la-Neuve, Belgium. 2004.
[14] http://www.ims.uni- stuttgart.de/projekte/corplex/TreeTagger/DecisionTreeTagger.html
[15] Bourigault D, Fabre C, Frérot C, Jacques M, Ozdowska S. Syntex, analyseur syntaxique de corpus. Proceedings of TALN, Dourdan, France. 2005.
[16] Merkel M, Petterstedt M, Ahrenberg L. Interactive Word Alignment for Corpus Linguistics. Proceedings of Corpus Linguistics, 2003. http://www.ida.liu.se/~nlplab/ILink/
[17] Deléger L, Merkel M. and Zweigenbaum P. Using Word Alignment to Extend Multilingual Medical Terminologies, LREC 2006 Workshop ACQSPECLEX, 9-14.

7.2 Ontologies for Medical Disciplines

Ubiquity: Technologies for Better Health in Aging Societies
A. Hasman et al. (Eds.)
IOS Press, 2006

Restructuring the Foundational Model of Anatomy

Gergely HÉJA[a1], Péter VARGA[b], Péter PALLINGER[c] and György SURJÁN[a]
[a] *National Institute for Strategic Health Research*
[b] *Eötvös Loránd University, Institute of Philosophy*
[c] *Computer and Automation Research Institute of the Hungarian Academy of Science*

Abstract. The authors present a method to convert the FMA to a description logic-based representation in OWL. The concepts denoting anatomical structures are aligned to the DOLCE formal top-level ontology, and converted to a compact core ontology in the spirit of GALEN. The paper presents the identified problems in the FMA and the main aspects of the re-modelling.
Keywords: Ontology, logic-based knowledge representation, FMA, DOLCE, GALEN

1. Introduction

Medical ontologies play an increasingly important role in medical informatics. However there are only a few of them that are both detailed enough and useful for formal reasoning. Human anatomy is often seen as a basis of all medical knowledge; therefore its formalisation is absolutely essential. The FMA [1] is perhaps the most detailed formal description of the human anatomy. However the huge number of concepts represented in it makes the knowledge management difficult, and due to its frame-based representation, formal reasoning with the FMA is practically impossible.

2. Material and methods

2.1. Foundational Model of Anatomy

The FMA is a general purpose anatomical reference ontology (not intended for end-user applications), developed by the University of Washington. It is aimed to be a reusable and extensible anatomical knowledge base. The representation of the ontology is frame based [2], and it is distributed in a mysql database dump file for Protégé [3]. The FMA is perhaps the most detailed ontology of the human anatomy, by describing the anatomical material entities ranging from molecules to the whole body, and connects them to non-material entities (e.g. geometrical shapes) required to describe structural relations. The FMA contains approximately 70 thousand anatomical concepts, associated with 110 thousand terms, which are related to one another by more than 1.5 million instantiations of over 170 kinds of relations.

[1] Corresponding Author: Gergely Héja, Szentkirályi u. 21. Budapest H-1088., heja.gergely@eski.hu

The authors of the FMA define the "disciplined" modelling in ten principles, the essence is that the ontology should model canonical anatomy of the human body, however description of anatomical variants and non-human anatomy should also be possible. *Cell* and *Organ* are the two units in terms of which other subclasses of *Anatomical structure* are defined (which constitute cells or organs, or are constituted by them). The ontology models three types of relationships between anatomical entities: class subsumption relationships, static physical relationships, and relationships describing their transformation during the ontogeny.

The high level schemata of the FMA is FMA = (AT, ASA, ATA, Mk), where AT (Anatomy Taxonomy) specifies the taxonomic relationships of anatomical entities, ASA (Anatomical Structural Abstraction) describes the partitive and spatial relationships of these concepts, ATA (Anatomical Transformation Abstraction) describes the time-dependent morphological transformations of the concepts in AT during the human life, and Mk (Meta-knowledge) comprises rules according to which the relationships in the other components of the model are represented.

At certain levels of the hierarchy meta-classes (called template classes in Protégé2000) are used to introduce new slots. For example, the class *Vertebra* is a child of the class *Irregular bone*, and an instance of the meta-class *Vertebra*. The children of a given class inherit its attributes as template slots, whereas the instances of a particular class inherit its template slots as own slots (which are not propagated), and assign specific values to them. For example, the meta-class *Vertebra* has a template slot part of which is inherited by its instance *Vertebra* as its own slot, with slot value *Vertebral column*. *Cervical vertebra* is a subclass of *Vertebra* and converts this template slot, assigning its own slot value *Cervical vertebral column* [4, p 485].

The precise and comprehensive description of the structure of the body requires frequently the use of "attributed relationships", e.g. it is necessary to specify that the esophagus is continuous with the pharynx superiorly; and its adjacency relationship with the vertebral column is anterior. Such relations are represented by reification [5]. For example, the value of the *continuous with slot* is an instance of the class *Anatomical adjacency coordinate*. This class carries template slots that describe the related structure and its relative position or coordinate that qualify its adjacency to the reference anatomical structure. Moreover, this modelling approach allows also the qualification of the relations with other additional attributes.

In our analysis we have found the following problems with the FMA:

1. Since frame-based representation is not very useful for reasoning, the FMA should be transformed to a logic-based formalism. Description logic (DL) [6] however does not allow the class and meta-class coexistence used in the FMA. Thus we have eliminated the meta-classes and assigned the template slots to the remaining classes. When the slots should not be inherited the definition of the parent concept should be modified. E.g. *Vertebra* should have the role part of restricted to (*Vertebral column* ∪ *Part of vertebral column*) which is consistent with the definition of *Cervical vertebra*.

2. The classes defining reified attributes can be converted to similar classes in the OWL [7]. Another solution would be to divide these attributed relationships into unique roles, i.e. *superior to* and *continuous with*. We decided to use this solution, because in our opinion practically every attributed relationship can be divided without losing its intended meaning and the number of roles would not increase significantly.

3. The memory requirement of the FMA is huge, Protégé requires approx. 2 GB to load the FMA into the memory after the transformation from database backend to RDF [8]. The OWL representation would be also very large, impeding automatic

reasoners. Moreover, 70 thousand anatomical concepts are hard to manage by knowledge engineers. Consequently we decided to transform the FMA to a compact core ontology, without losing too much information (see Section 2.3).

4. In some cases abstract and concrete concepts are siblings, like *Diaphysis* and *Body of rib*, children of *Body of organ*. After the restructuring it would not cause serious problems, since many concrete concepts (e.g. *Body of rib*) are deleted.

5. There are several unnecessary concepts and hierarchies, like *Subdivision of lateral meniscus*, with children *Anterior horn of lateral meniscus* and *Posterior horn of lateral meniscus*. The first concept is not necessary and it would suffice to state that *Anterior horn of lateral meniscus* is a part of *Lateral meniscus*. There are some concepts like *Subdivision of investing fascia of ankle* which have no children, thus also superfluous. The total elimination of such concepts creates long flat lists, impeding human management, consequently some organising concepts should be retained.

6. In some cases the naming (not the meaning) is reflected by the hierarchy (e.g. *Body of tongue* and *Body of rib* are siblings). Since these concepts practically have no common attributes, they should be rather classified as Part of tongue and Part of bone.

7. *Biological macromolecule* subsumes simple molecules, like *Acetylcholine*, which is an error (thus it should have the label of *Biological molecule)*. We have removed these (bio)chemical concepts from the ontology, because in our opinion they should be defined in a chemical ontology, and the anatomy should only refer to them.

2.2. Aligning the FMA to the DOLCE

We have aligned the FMA to a formal top-level ontology, to define the high-level anatomical concepts (such as *Anatomical structure*, *Body substance*), but we found that also some low-level concepts needed reclassification. The formal definition of the upper level of the FMA makes it possible to integrate it with the description of other biomedical domains (e.g. physiology, pathology). It also helps to manage the ontology by providing a foundation for formal consistency analysis. Moreover we hope that it will result in an ontology, which can better be used for formal reasoning.

The DOLCE is a descriptive upper-level ontology especially designed for ontology cleaning and interoperability [9]. Contrary to other efforts aiming at the same tasks, such as SUO [10], it is based on a sound philosophical ground, which made it well suited for our purposes. DOLCE makes a clear distinction of qualia with respect to the theory of properties and their perception.

The alignment of the FMA to DOLCE requires a clarification and decomposition. The DOLCE ontology is built on a set of highly sophisticated mereological axioms. Although the FMA contains a mereological subtheory as a part of ASA, there is no easy correspondence to observe. The insufficiencies of the FMA partonomy were recently discussed in [11]. Based on this discrepancy between the two mereological theories and driven by the rule that the clarification of the taxonomy must precede the clarification of the relations, we have decided to confine ourselves first to the alignment of the taxonomy. We only processed the subclasses of *Material Anatomical Entity* (classifying anatomical structures), which has been aligned to the endurant-sub-ontology of the DOLCE. This means the exclusion of abstractions based on physical entities (e.g. *Anatomical Space*) and higher level ideal entities like *Class Subsumptions*.

Although the FMA (as any well-designed taxonomy) contains both lower-level concepts (leaf concepts and their immediate generalisations) and upper level abstract concepts, there is a difference between the degree of the confirmation of these two levels. Even its developers noted that their sources lacked abstract, higher level generalisation [4]. Therefore while the lower level part is based on a well-established terminology, the higher level does not enjoy the same level of general acceptance in anatomical literature. This difference reappeared during the alignment process.

We have decided to classify *Anatomical cluster* and *Anatomical set* as arbitrary (purely extensional) sums (instead of collections with unity condition). Arguably they fulfill the unity condition, at least that each one of them is referred to by a single term (and sometimes even the unity of adjacency, e.g. *Back*), but this classification decision was intended to emphasise their difference from homogenous, systematic collections.

Apart from that, because there were only a few children of the *Physical Endurant* concept, the process of alignment turned out to be reduced to classify the selected concepts to the following three categories:

• *Amount of Matter*, which changes its identity when some of it parts are changed.

• *Feature*, which is a "parasitic" entity (e.g. cavity, surface). The DOLCE specification [9] explicitly mentions that although body parts could be conceived as features, we have not preferred to do it so.

• *Non-agentive Physical Object*. *Physical Object* itself is a sibling of the previous two concepts and although it could be argued that the brain has a causal role in the emergence of human agency, but certainly not the brain in anatomical sense.

Even if this discernment must be an inheritable property, it could not be performed without examining the subsumed concepts. It frequently happened that the discernment resulted in different classification of the child concepts. For example, while *Organ* is definitely a *Non-agentive Physical Object*, and *Body Substance* is *Amount of Matter*, thereby saving lots of work; the DOLCE categories several times cut across the FMA categories even at lower level. For example, the terms used for nuclear inclusions (e.g. *Interchromatin granule*) certainly behave like mass nouns, whilst most other constituents of the nucleus definitely not.

This problem was particularly present at the microscopic categories and came from the fact that the division used there (*Cell component* vs. *Cell region*) was not identical with the division *Physical object* vs. *Amount of matter*. Although *Cell region* is a "fiat subdivision of a cell", and *Cell component* has "a definable shape, bounded predominantly by bona fide boundaries and is countable" according to their descriptions in the FMA, the low-level classes of cell regions are usually not amounts of matter, in so far as they have a functional unity condition based on a given structure (e.g. *Cilium*). This fact rendered the division of cell components and regions useless for our purpose. In some cases therefore the DOLCE categories were assigned totally independently from FMA upper- and middle-level categories.

At the low-level concepts numerous examples of the phenomenon were encountered which we call 'conceptual role equivocation'. By that we mean that there are terms, each of them clearly having a unique concept associated to (e.g. *Cytoplasm*), but still playing two different conceptual roles: it can serve as a mass noun (a given amount of cytoplasm), but it has also parts, therefore serving as a name for a complex physical entity. Probably many tissue names behave similarly. This phenomenon can be seen as a problem of distinguishing names (terms) from meanings (concepts). Whilst arguably an ontology should consist of concepts, strictly speaking it is the vocabulary of the anatomy which is fixed by the tradition and which presents itself as a modelling task for the ontology engineer. Consequently there are cases, which appear frequently in the FMA, where the ontology engineer is tempted not to break the traditional term into to new ones according to their different conceptual contents.

2.3. Conversion of the FMA in the spirit of GALEN

The GALEN CRM is designed to be a reusable, application- and language-independent model of medical concepts to support EHRs, decision support systems, classification systems, etc. [12]. The key feature of its approach is that it provides a set of building blocks and constraints from which concepts can be composed (in contrast with traditional classification systems). We think that this is the most important feature of GALEN, by allowing compact and concise modelling. Our goal is not to align the FMA to GALEN, but to transform it into "a set of building blocks and constraints".

There are a lot of anatomical components (like *Articular capsule* and *Mucosa*) which appear in several anatomical entities. In the FMA these components are exhaustively listed, e.g. *Articular capsule of incudomallear joint*. In a core ontology it would be sufficient to state that every *Synovial joint* must contain an *Articular capsule*. In other cases (e.g. Mucosa) the enumeration of the corresponding entities is needed (i.e. it must be stated that *Stomach*, *Lacrimal duct*, etc. is covered by *Mucosa*).

To assert these restrictions we advise to use universal quantifier (owl:allValuesFrom), because the model must not allow e.g. that *Stomach* has a component of type *Bone*. According to the open world assumption the existential quantifier (owl:someValuesFrom) would allow such nonsense assertions. However the semantics of the universal quantifier allows that an empty role is also acceptable. Therefore it may be necessary to state also cardinality constraints. With this core model the concepts in the FMA (e.g. *Articular capsule of incudomallear join*t) can be re-created. However there are also some concepts which are siblings of composite objects (e.g. *Conus artery*, a child of *Trunk of coronary artery*). A domain expert should review such cases, because it may happen that these concepts are misplaced in the AT.

Since the human body is more or less mirror-imaged, there are a lot of anatomical entities, which have a left and a right pair. Such pairs are exhaustively listed in the FMA (e.g. *Left little finger)*. These entities have been removed, but the role laterality must have a cardinality of 1. There are other selectors (such as anterior – posterior) but it is not obligatory for the particular anatomical structures to exist in pairs (e.g. there is anterior, posterior and lateral but no medial semicircular duct). Therefore these concepts are not eliminated, only the coordinate slot should be appropriately filled in.

3. Discussion

The alignment of the FMA to the DOLCE is complete, but reviews modifying the classifications are still to be expected. We have identified two levels of problems:

1. The assignment of DOLCE categories frequently cuts across the upper and middle categories of the FMA. It is worth noting that any conceptual clarification could be of two kinds: translational or substitutional. In the first case the conceptual clarification assigns an alternative formulation to the initial problem, whilst in the latter case this alternative formulation replaces the original. We have seen that these categories were devised upon less authority than the low-level anatomical concepts, therefore the substitutional analysis of the FMA upper- and middle levels is advisable. Our approach, which results in the co-existence of two upper- and middle levels, could be in the meantime useful for ontology interoperation.

2. The hidden equivocations of the leaf concepts posed a more serious problem since any user of the ontology would presumably prefer to stick with the traditional

anatomical terminology, causing disambiguation problems. This is probably only an example of a more far-reaching problem, which deserves proper addressing. As an ad hoc solution we have always chosen one of the alternatives to avoid the ambiguity.

With the reductions explained in Section 2.3, a core model containing approximately 12000 concepts can be created. A review performed by domain experts (physicians) is necessary to guarantee that no important concept is eliminated.

After the review, the roles denoting structural relations can be filled in from the corresponding slots, or derived from the concept name (e.g. *Basal lamina of ciliary body*). Cases where the inheritance of properties is violated (e.g. *Vertebra* and *Cervical vertebra*) can be identified from the ASA (the template slots are not identical). These concepts should be redefined by a knowledge engineer.

The other concepts of the FMA which have been left out from the alignement should also be aligned to the DOLCE, however concepts like *Structural relationship* can be eliminated, since we use properties to describe such relationships.

After the core model is complete, the composite concepts could be re-created by formal methods. The resulting ontology

• can be compared to the original FMA, allowing us to find the errors made during the conversion (and to find some errors in the FMA, if they exist)

• can serve as a practical reference ontology for biomedical ontologies, i.e. the creator of the domain ontology does not need to define the composite concepts.

4. Conclusion

It has been shown that it is possible to transform the FMA to a logic-based formalism. However the alignment of the FMA with a formal top-level ontology led to some important questions. The list of concepts that should be eliminated is complete, but a review by a domain expert is necessary. After this step, the structural relations and other information can be extracted from the FMA by automatic methods. Further review by a domain expert is only needed in exceptional cases.

References

[1] Information material about FMA can be found at http://sig.biostr.washington.edu/projects/fm/
[2] Minsky ML. A framework for representing knowledge, in Winston P H (ed.) The Psychology of Computer Vision 211-277 McGraw-Hill, New York, 1975
[3] Information material about Protégé can be found at http://protege.stanford.edu/
[4] Rosse C, Mejino JLV Jr. A Reference Ontology for Biomedical Informatics: the Foundational Model of Anatomy. J Biomed Inform. 2003 Dec; 36(6): 478-500
[5] McCarthy J. First Order Theories of Individual Concepts and Propositions, Machine Intelligence 9, 1979, 129-147
[6] Bader F et al. Editors. The Description Logic Handbook (Theory, Implementation and Applications), Cambridge University Press, 2003
[7] Information materials about OWL can be found at http://www.w3.org/2004/owl/
[8] Information materials about RDF can be found at http://www.w3.org/2004/RDF/
[9] Information material about DOLCE can be found at: http://www.loa-cnr.it/DOLCE.html
[10] Information materials about SUO can be found at http://suo.ieee.org/
[11] Smith B, Rosse C. The Role of Foundational Relations in the Alignment of Biomedical Ontologies. Proc Medinfo 2004, 444-448
[12] Information materials about GALEN can be found at http://www.opengalen.org/open/crm/index.html

Ubiquity: Technologies for Better Health in Aging Societies
A. Hasman et al. (Eds.)
IOS Press, 2006

Four Ontological Models for Radiological Diagnostics

James M. FIELDING[a,b,1], Dirk MARWEDE[a,c]
[a] *Institute for Formal Ontology and Medical Information Science, Saarbrücken, Germany*
[b] *University of Paris I, France*
[c] *Department of Diagnostic Radiology, Leipzig University Hospital, Germany*

Abstract. In this paper we isolate four diagnostic models in radiology and define a set of diagnostic relations corresponding to each clinical situation. To achieve this, we describe a set of general formal ontological notions, as well as the ontological model of the imaging domain we employed in our analysis. On the basis of our results, we conclude that these diagnostic models and the relations contained therein could be applied to diagnostic situations outside of radiology as well.
Keywords: Medical Imaging, Formal Ontology, Radiology, Diagnosis.

1. Introduction

Biomedical ontologies promise to structure biomedical knowledge and thus revolutionize the electronic processing of medial data. Take, for example, the construction of ontologies in domains such as anatomy [1], surgical procedures [2], or disease classifications [3]. In each of these areas, a good deal of ontological success has been achieved. However, to date, no adequate knowledge model specifically concerned with the diagnosis of diseases – an element central to medical practice and one that involves each of the above domains – has been developed for the incorporation and mutual support of these already existing biomedical ontologies. There are, we propose, two reasons for this. In the first place, this is because, until recently, no widely accepted standardized terminology of terms used within a predominantly diagnostic domain has been available to serve as a prototype case [4]. In the second place, this is due to the particularly stubborn problems that the diagnostic situation poses for ontological analysis [5].

In this paper we take the predominantly diagnostic domain of medical imaging as our prototype. We developed an ontologically sound radiological reporting model [6] for interfacing term categories of RadLex [7], a recently published controlled vocabulary used for radiological reporting, to the Foundational Model of Anatomy (FMA) [1]. Through a set of basic relations, we established an ontological model of image domain entities, their connection to entities of FMA, as well as radiological reporting tasks in clinical practice (as set out in the DICOM standard [8]). In designing this interface, however, we discovered four different diagnostic forms employed in

[1] Corresponding Author: James M. Fielding, IFOMIS. Email: matthew.fielding@ifomis.uni-saarland.de

clinical radiology. These forms required corresponding ontological models in order to determine which relationships hold between a given diagnostic conclusion and the visual image features associated with it: whether i) there are no associated findings and thus no relations hold, ii) whether the latter serve as evidence for, iii) as conditional relevance to, or iv) as necessary consequents of the former. In this paper, we describe each of these forms in detail, and define a set of diagnostic relations that hold sway in each of the corresponding clinical situations. We conclude that, although we take radiological diagnoses as our prototype, nothing in principle should limit this analysis to radiological diagnoses in particular.

2. Materials and Methods

2.1. Entities in Biomedical Ontology

The term 'type' is used in what follows to refer to what in the knowledge representation community is occasionally expressed under the heading of 'concept' and what in the philosophical literature are called 'universals' or 'kinds.' A type is that which is general in reality. By contrast, instances are those things that are particular individuals, and thus exist in a particular time and space and relate to each other in a variety of instance-level relations [9]. The type *Abdomen* is a universal in the biomedical domain; an instance of this type is the particular andomen of a particular patient. Similarly, the type *Image of Abdomen* is a universal in the biomedical imaging domain; an instance of this type is a concrete image that can be viewed and described in its particular unique reality.

Some entities have the ability to exist without the ontological support of other entities. In the biomedical domain, these are entities such as people, organs, and bones; these entities are *Independent Entities*. On the other hand, there are entities which require the existence of the first kind of entity for their own existence, these entities are *Dependent Entities*; as a viral infection is dependent on the infected organism, so too does an image need an object imaged [9]. Within the class of *Dependent Entities*, however, we include not only the image itself, but also the individual attributes of an image (such as a particular shape, colour patch, or image quality), which depend upon a radiographic imaging materials (such as a sheet of radio-sensitive film) to bear them.

2.2. Relations in Biomedical Ontology

A major obstacle to the full interoperability of existing software ontologies has been the lack of a standardized set of top-level formal relations, applicable to the entire biomedical domain. However, recently, a set of formal ontological axioms and relations has been published within the Open Biological Ontology framework (OBO) [10]. Relations specified within this framework are distinguished according to whether they belong to the set of *genuine ontological relations* (relations such as **is_a,** holding between two entity types, and **instance_of**, which hold between a type and an instance of this type) or *domain-neutral primitive relations* (such as the instance-level **part_of** relation). In constructing our prototype ontology, we were required to an additional set of radiology-specific relations (such as **has_shape**, and **has_composition**, for radiological image features) for interfacing RadLex term categories with the categories with the FMA [6], as well as the relations defined in what follows for relating visual features and diagnoses according to the specific clinical procedure employed.

2.3. Domain of Medical Imaging

In clinical practice, the main task of the radiologist is to interpret image findings. Previous attempts at structuring radiological report content have used the term 'finding' [11,12] to refer to a startlingly wide range of image entities, from vague observations about image properties to highly interpretative diagnoses [13]. In our methodology we retain a similarly broad understanding of 'finding,' as anything visible in the image, although we designate this upper category *Imaging Finding* to highlight the fact that any finding found in a medical image refers to some entity on the image and not to the bodily entity which it resembles and from which it may take its name in controlled vocabularies like RadLex. When a radiologist arrives at the diagnostic conclusion that a liver is normal, for instance, this indicates that the *appearance* of the liver in this particular case is normal.

Not all findings are of a unified type, however, and the relations in which they stand reflect their different diagnostic roles. We consequently introduce three subcategories to *Imaging Findings*: *Visual Features*, *Anatomical Findings* and *Non-anatomical Findings*.

At the most basic level, we have the category of *Visual Feature*. As ontologists, we would like to claim that radiologists use imaging technologies in order to access distinct entities within the body directly; however, this is seldom the case. As a result of the indirect methods inherent to radiological practice, radiologists are frequently required to examine more or less distinct opacities, densities, or shadows, which may or may not represent genuine entities in the body. This includes identifiable image patterns which, though the ontological commitment they make to reality is minimal, form consistently in accordance with the nature of composite body substances. This includes what are referred to as elements of 'radiological anatomy,' which serve to orient a radiologist though they themselves do not represent *bona fide* anatomical entities.

Anatomical Findings and *Non-Anatomical Findings* are entities of the image that adequately approximate the embodied entities they represent. Anatomical entities are those entities such as body parts, organs, etc., and are typically represented in canonical anatomical knowledge bases like the FMA. Non-anatomical entities, on the other hand, are those non-canonical pathological conditions that must be diagnosed in a particular patient on a case by case basis. Descriptions of these entities can be found in controlled vocabularies like RadLex or disease classification systems like SNOMED.

3. Results

3.1. First Diagnostic Form

In the fist case, an examining radiologist may form a diagnosis without explicitly relying on any associated visual feature. While this fact applies equally to both normal and pathological evaluations in principle, in practice, it appears solely when a diagnostic test returns normal results. The appearance of a normal image is well known, and not all visual features present need be associated. In the normal case, these features are taken for granted, and the only explicit diagnostic relation that holds is between the image and the valuation itself via the **has_evaluation** relation. This relation is defined as follows:

[Anatomical Finding] **has_evaluation** *[Attribute]* **at** *t*: a relation holding between a particular anatomical finding and an assessment, at a time. For example: This liver **has_evaluation** normal **at** time of examination

3.2. Second Diagnostic Form

Should an examining radiologist discover an abnormal image, however, the abnormalities must be carefully documented. In the initial stage of this diagnostic form, each visual feature is located to a specific anatomical finding via the **has_location** relation. Note, a radiologist need not progress pass this stage of investigation; it is quite common in radiology to document undiagnosed findings and refer a patient for further testing. However, should a diagnosis be formed at this stage, it requires evidence; visual features serve as evidence for a diagnosis via the **has_interpretation** relation. These relations are defined as follows:

[Visual Feature] **has_location** *[Anatomical Finding]* **at** *t*: basic location relation holding between a particular visual feature and an anatomical finding. E.g.: This reticulonodular pattern **has_location** lung **at** time of examination

[Visual Feature] **has_interpretation** *[Non-Anatomical Finding]* **at** *t*: a relation holding between a particular visual feature and non-anatomical finding which the visual feature serves as evidence for, at a time. E.g.: This reticulonodular pattern **has_interpretation** tuberculosis **at** time of examination.

3.3. Third Diagnostic Form

An examination may also be performed in order to determine the present state of an already proven diagnosis, which no longer requires evidence, but additional information concerning the condition of the diagnosis. Thus, a direct relation between the anatomical finding and the non-anatomical is already established; here, the **has_location** relation "comes for free," as it were. In this case, however, the **has_condition_feature** relation holds between non-anatomical finding and its associated visual features. These features may serve to track the progression of a diagnosed condition by *trend* (e.g. increasing or decreasing) or *extent* (e.g. impartial or complete), or *timing* (e.g. new, residual, or chronic).

[Non-Anatomical Finding] **has_location** *[Anatomical Finding]* **at** *t*: a direct relation holding between a particular non-anatomical finding and an anatomical finding. E.g.: This tuberculosis **has_location** this lung **at** time of examination.

[Non-Anatomical Finding] **has_condition_feature** *[Visual Feature]* **at** *t*: a relation holding between a particular non-anatomical finding and a visual feature, as a property, at a time. E.g.: This tuberculosis **has_condition_feature** increasing.

3.4. Fourth Diagnostic Form

While all of the above relations function on the instance level, an image may in addition, instantiate a type-level relation between a non-anatomical finding and a visual feature via the **has_consequent_feature** relation. While this is not truly a diagnostic form, this relation needs to be modeled in a diagnostic ontology nonetheless. Because such visual features are necessarily tied to any instance of a given non-anatomical finding, the relation holds in all cases whether or not it is involved specifically in the diagnostic procedure itself, and thus may be inferred.

[Non-Anatomical Finding Type] **has_consequent_feature** *[Visual Finding Type]*: a relation holding between a non-anatomical finding type and a visual feature type, at all times. E.g.: Pulmonary Nodule **has_consequent_feature** Radiopaque.

4. Discussion

In figure 1, we can clearly see how these relations hold in different configurations depending on the diagnostic form applied. Naturally, the more we can model radiographic knowledge according to the fourth form, the more reliable radiographic diagnoses will become. Unfortunately, radiological findings can not always come with the type of analytic security afforded here, and thus, as interpretations, often rely upon variable access to evidence. In theory, with enough information and the proper methodology, it should be possible to associate a greater number of pathological entities with consequent visual features based on statistical concurrence. More research is required here. In the meantime, however, we may adopt an ontological model like the one presented here, which structures this information in a manner that allows us to accurately manage it. In this manner, we may query radiologists' diagnoses to learn which diagnostic and interpretative patterns exist between these entities in medical imaging.

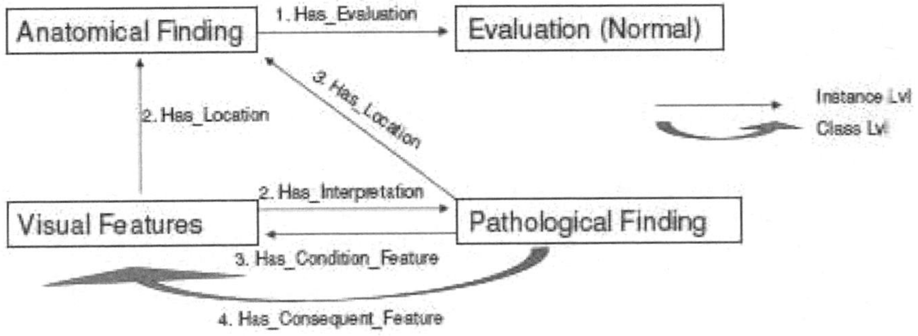

Figure 1. Diagnostic Relation Schema

5. Conclusion

Creating the *Image Finding* subcategories of *Visual Feature*, *Anatomical Finding*, and *Non-Anatomical Finding* allowed us to disambiguate the structure of radiological diagnoses according to whether i) there are no findings associated with a particular diagnosis and thus no diagnostic relations hold, ii) whether the present visual features serve as evidence for, iii) as conditional relevance to, or iv) as necessarily consequents of the given diagnosis. Although we have limited ourselves to the domain of radiological imaging, there seems no reason why, in principle, other diagnostic domains (such as the chemical analysis of body fluids, for example) should not follow a similar pattern. In all diagnostic practice, results follow a pattern of evidence, relevance, and consequence; although perhaps more emphasis is laid in some domains on the forth model than the second, as it is in the case of radiology. More research would have to be carried out here to determine whether this is the case. It is our hope that it would, and that this prototype case may serve as a solid foundation for future investigations into the ontological nature of diagnoses, which may subsequently form a shared point of reference for ontologies focused on diverse biomedical subdomains, such as anatomy [1], surgical procedures [2], or disease classifications [3].

Acknowledgements

The present paper was written under the auspices of the Wolfgang Paul Program of the Alexander von Humboldt Foundation, the Network of Excellence in Semantic Interoperability and Data Mining in Biomedicine of the European Union, and the project Forms of Life sponsored by the Volkswagen Foundation.

References

[1] Rosse C, Mejino JL, Jr. A Reference Ontology for Biomedical Informatics: The Foundational Model of Anatomy. *Journal of Biomedial Informatics* 2003;36(6):478-500.

[2] Rogers J, Roberts A, Solomon D, van der Haring E, Wroe C, Zanstra P, Rector A. GALEN Ten Years On: Tasks and Supporting Tools. *Medinfo* 2001;10(Pt 1):256-60.

[3] Spackman KA, Campbell KE, Cote RA. SNOMED-RT: A Reference Terminology for Health Care. *JAMIA* 1997: 640-644

[4] Langlotz CP. Automatic Structuring of Radiology Reports: Harbinger of a Second Information Revolution in Radiology. *Radiology* 2002; 224(1): 5-7.

[5] Fielding JM, Marwede D. The Anatomy of the Image: Diagnosis and Inference in Radio-Ontology. Forthcoming.

[6] Marwede D, Fielding JM, Smith B. Entities and Relations in Medical Imagin: An Analysis of Computed Tomography Reporting. Forthcoming

[7] RSNA. RadLex: A Lexicon for Uniform Indexing and Retrecal of Radiology Information Resources. http://mirc.rsna.org/radlex/service (last acessed Jan 29,2006).

[8] Bidgood WD, Jr., Horii SC. Introduction to the ACR-NEMA DICOM Standard. *Radiographics* 1992; 12(2): 345-55.

[9] Grenon P, Smith B. SNAP and SPAN: Towards Dynamic Spacial Ontology. *Spacial Cognition and Computation* 2004: 4(1): 69-103.

[10] Smith B, Ceusters W, Klagges B, Kohler J, Kumar A, Lomax J, Mungall C, Neuhaus F, Rector AL, Rosse C. Relations in Biomedical Ontologies. *Genome Biol* 2005; 6(5): R46.

[11] Hripcsak G, Friedman C, Alderson PO., DuMouchel W, Johnson S B, Clayton PD. Unlocking Clinical Data From Narrative Reports: A Study of Natural Language Processing. *American International Medicine* 1995; 122(9): 681-688.

[12] Friedman C, Cimino JJ. Johnson SB. A Conceptual Model for Clinical Radiology Reports. *Proceedings of Annual Symposium of Computation and Applied Medical Care* 1993: 829-33.

[13] Bell DS, Pattison-Gordon E, Greenes RA. Experiments in Concept Modeling for Radiographic Image Reports. *Journal of American Medical Informatics Association* 1994; 1(3): 249-62.

7.3 Biomedical Ontologies

Ubiquity: Technologies for Better Health in Aging Societies
A. Hasman et al. (Eds.)
IOS Press, 2006

The *Derives_From* Relation in Biomedical Ontologies

Mathias BROCHHAUSEN[a,1]
[a] *Institute for Formal Ontology and Medical Information Science, Saarbrücken, Germany*

Abstract. This paper is part of the ongoing efforts within the framework of the Open Biomedical Ontologies (OBO) ontology library to optimize the treatment of relations in biomedical informatics. When ovum and sperm fuse to form a zygote, then the latter is *derived from* the former. This *derives_from* relation is of huge importance not only in embryology but also in every other science dealing with ontogenetic processes. This study examines critically the treatment of *derives_from* in the OBO Relation Ontology in light of its conformity to the underlying processes of development and transformation in biological and medical reality.
Keywords: biomedical ontology, relations, embryology, erythropoiesis

1. Introduction

In recent years the use of ontologies for the development of medical information systems has witnessed an explosive growth [1]. However, initial attempts to use ontologies to support information management in biomedicine have focused primarily on the formulation and definition of *terms*. In ontologies and terminologies such as the Gene Ontology [2], SNOMED [3] the formal treatment of the *relations* which would link these terms has been neglected. [4] puts forward a methodology for the selection and definition of relations in biomedical ontologies, conformity with which has now been adopted as a criterion for inclusion into the OBO Foundry collaborative experiment in ontology development recently initiated by the Open Biomedical Ontologies (OBO) consortium [5]. The benefit of a relation ontology in clinical research has been argued for in [6]. I believe that the approach advanced in [4] is vitally important for the enhancement of biomedical information systems in the future; however, some aspects of its treatment of relations need to be improved if it is not to lead to mistakes when applied to biological or medical cases.

This paper deals with foundational aspects of the Relation Ontology (RO). The aim is to demonstrate that certain formal solutions presented in [4] can lead to confusing results regarding medical practice and biomedical theory. Though computational consequences are not discussed in this publication, it will become obvious that changes in the foundation will lead to changes in application and implementation.

[1] Corresponding Author: Mathias Brochhausen, IFOMIS, Universität des Saarlandes, Postfach 15 11 50 66041 Saarbrücken, Germany, mathias.brochhausen@ifomis.uni-saarland.de

The Relation Ontology distinguishes, in its initial version, ten relations between types of entities in reality, which are labeled as follows:

Foundational relations
is_a
part_of
Spatial relations (connecting one entity to another in terms of relations between the spatial regions they occupy)
located_in
contained_in
adjacent_to
Temporal relations (connecting entities existing at different times)
transformation_of
derives_from
preceded_by
Participation relations (connecting processes to their bearers)
has_participant
has_agent

It is the *derives_from* relation which is the object of the present study. This relation is of special interest given its application in the field of reproductive medicine, where its proper treatment may even bear on issues addressed in biomedical ethics [7]. From an ontological point of view, reproduction and embryological development are interesting since they provide the basis not only for the creation of new organisms but also for the coming into being of new organism parts, as well as for pathological formations such as tumors or blisters [6].

2. Ontological relations in reproduction and erythropoiesis

An informal definition of *derives_from* as conceived by the RO might read as follows:

> One universal C derives from another universal C_1, if the instances of C_1 (c_1) exist prior to the instances of C (c) and the instances C derive from instances of C_1. c derives from c_1 if c and c_1 are not identical, c inherits a biologically significant portion of matter from c_1 and the special region occupied by c in the beginning overlaps with the spatial region occupied by c_1 [4].

Obvious examples of derivation as thus defined are the relations between sperm, ovum, zygote and blastomere. The zygote *derives from* sperm and ovum. A case of derivation in which the derivate derives from two entities is called fusion. Through a mitotic cell division the zygote changes into two blastomeres.

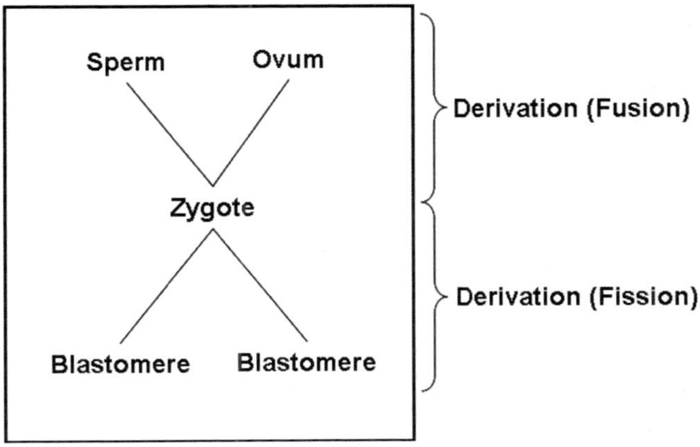

Figure 1: The relation *derives_from* in reproduction (fusion and fission), with time proceeding downwards

Cell divisions then lead progressively to the morula, and this is followed by the formation of the embryoblast, deriving from a number of cells located in the center of the morula. Other cells form the trophoblast. Thus, two fusions are taking place which structure the cells of the morula.

It is crucial to the methodology of the OBO Relation Ontology that a distinction be drawn between relations obtaining between instances, for example the **part_of** relation between John's heart and John's body, and relations obtaining between the corresponding universals or types, for example the *part_of* relation between the types *human heart* and *human body*. The goal of RO is to establish relations between universals. This reflects the basic aim of science, which is to discover general properties or general rules of nature. [4] shows that, from OBO's realist point of view, the only way to formulate relations on the level of universals is to rely on certain underlying relations between the corresponding instances. (This reflects the Aristotelian perspective, according to which knowledge of universals is derived from knowledge of instances.) It is therefore crucially important for the RO to distinguish relations on the instance level from relations on the level of universals.

Derives_from is distinguished from another temporal succession relation in RO, the relation of *transformation*, which can be defined informally as follows:

> One universal C is a transformation of universal C_1 if every instance c of C is an instance of C_1 at an earlier time, but there is not time at which c is an instance of both, C and C_1 [4].

On the level of instances, transformation is just identity: when child John is transformed into adult John, then child and adult are identical: it is <u>one and the same</u> person that has undergone a series of changes over time. When John's (healthy) lung is transformed into John's carcinomatous lung, then similarly the healthy lung and the carcinomatous lung are identical: one and the same organ has undergone a series of changes over time.

The RO distinguishes continuants, which endure identically through time, from occurrents (processes) which unfold themselves in their successive phases. We will concentrate in what follows exclusively on relations involving continuants. We will use variables as follows:

C, C_1, … to range over continuant universals
c, c_1, … to range over all continuant instances
t, t_1, … to range over instants of time

On the level of universals , transformation can now be defined as follows [4] (where *Cct* stands for "*c* **instance_of** *C* at *t*"):

> *C transformation_of* C_1 = [definition] *C* and C_1 for all *c*, *t*, if *Cct*, then there is some *t*, such that C_1ct_1, and t_1 **earlier** *t*, and there is no t_2 such that Cct_2 and C_1ct_2 only.

(Here and in what follows we use *italics* for relations between universals, and **bold face** for relations between instances.)

The treatment of derivation, in contrast, is more complex on both the instance and the type level. Provisionally, we can say that derivation occurs where continuant instances succeed each other in time but identity is *not* preserved.

The instance-level **derives_from** is a relation leading to a new entity while instance-level **transformation** is just identity. A good example of the difference between these two relations is provided by the case of erythropoiesis.

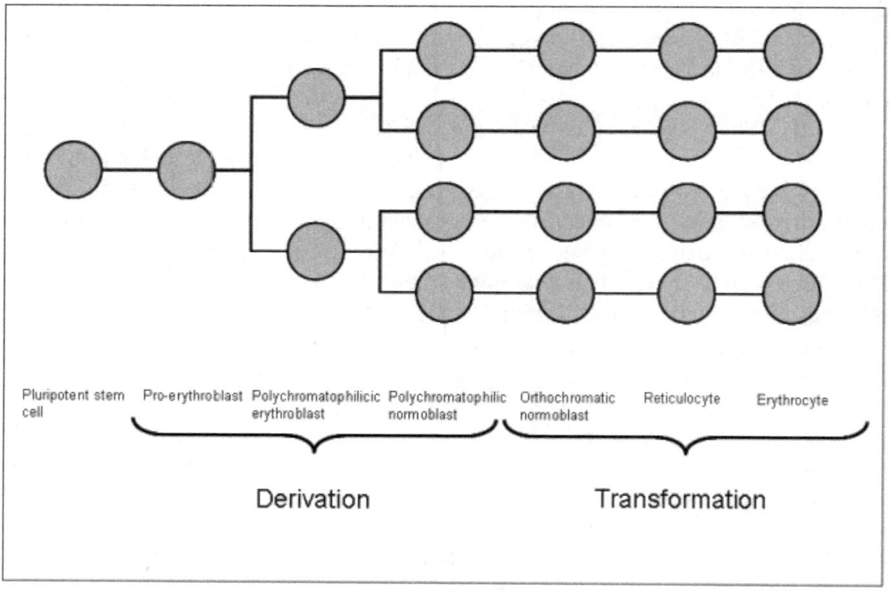

Figure 2: Erythropoiesis [9] This figure shows the development of erythrocytes through mitotic cell division (from pluripotent stem cell to polychromatic erythroblast) and maturation (from polychromatic erythroblast to erythrocyte). (Advance of time is from left to right.)

Erythropoiesis consists of both processes of mitotic cell division and processes of maturation.The first steps in this development consist of **derives_from** relations. Hemocytoblast, proerythroblast, early and late erythroblast all come into existence through mitotic cell division. Thus, fission occurs between these continuants.

The development from late erythroblast to erythrocyte is a process of maturation. Superficially one might think that only **transformation_of** (i.e.. identity) relations obtain between the continuants, involved. Yet, the ejection of the nucleus marks a **derives_from** relation. The continuants of this relation are the normoblast, the reticulocyte and the pygnotic nucleus.

3. Definition of derives_from

The definition of the derivation relation at the instance level given in [4] is:

> c **derives_from** c_1 if c is non-identical to c_1, c inherits a biologically significant portion of the matter of c_1, t such that c exists only subsequent to t and c_1 exists only prior to t.

Three modes of derivation on the instance level are then distinguished: continuation (one entity derives from another entity), fusion (one entity derives from two entities), and fission (two entities deriving from one entity).

The problems with the **derives_from** relation are connected with the inheritance of matter which is referred to in its definition ("inherits a biologically significant portion of the matter of"). For it is not clear what is meant by "a biologically significant portion". A notion which is arbitrary in this way should wherever possible be avoided in a definition.

4. Underdetermination of the *derives_from* relation

A biological example for the problems which result from the existing definition of **derives_from** is the reproduction of Ichneumon flies. Ichneumon lays its eggs in insects, insect larvae or pupae. The hatched larvae of Ichneumon feed on the host, devouring the entire interior and thereby killing the hosts.

The body mass of an ichneumon larva thus derives from the tissue of the host in a way which surely has to be classified as "biological significant". It follows that the whole ichneumon larva derives from the hosting insect larva in the sense of the definition given above. From a genetic point of view, however, the larva derives from its mother, not from the host. Thus, the definition in its present form yields the wrong result.

The larger problems slumbering beneath the superficial observation of the reproduction of ichneumon flies concern the problems which arise when we attempt to deal ontologically with cases of incorporation of biological matter which do not involve incorporation of its genetic code. The example above shows that **derives_from** does not suffice to differentiate between reproduction and ingestion, at least not in all cases known in the biological context.

But more pressing is the question whether **transformation_of** might cover all cases of continuation, which are defined as **derives_from** relations. The example [4] gives for continuation is an incorrect one from the medical point of view. [4] claims that the blastocysts **derive_from** the zygote. Yet, blastocytes do not (directly) derive from zygotes by continuation. The relations involved in the development of the zygote are in fact several fission and fusion types of the **derives_from** relation. The difference between transformation and derivation depends solely on the attribution of identity.

The question has to be raised whether continuation should simply be treated as transformation, especially considering the problems continuation produces in the biological example above. All stages of biological existence from the zygote to the dead organism should be related through **transformation_of** relations and fusion and fission types of the **derives_from** relation.

5. Conclusion

This study shows that the definition of the **derives_from** relation needs refinement. Moreover it has to be doubted whether continuation as one case of **derives_from** should be maintained at all. The curators of RO have agreed to make the necessary changes.

Basically the relations in the biomedical context are **transformation_of** and fusion or fission cases of the **derives_from** relation. The transition from a living person to a corpse is the most referred to example for continuation. But the intuition that there has to be an ontological gap between a living person and a corpse reflects a philosophical or theological attitude, not the attitude of biology and medicine. From the point of view of natural science the relations between a living organism and a dead organism are merely **transformation_of** relations.

Yet, the general outline of the RO ontology should be embraced, since many positive effects have been generated by it. In this paper we have been attempting merely to get some technicalities in a better condition. Improving the ideas developed and promoted by [4] is one way to help formal ontological reasoning to further acceptance and greater success.

References

[1] Rosse, C, Mejino Jr. JLV. A reference ontology for biomedical informatics: The Foundational Model of Anatomy. J Biomed Inform 2003; 36: 478-500.
[2] http://www.geneontology.org/
[3] http://www.snomed.org/
[4] Smith B, Ceusters W, Klagges B, Köhler J, Kumar A, Lomax J, Mungall C, Neuhaus F, Rector AL, Rosse C. Relations in biomedical ontologies. Genome Biology 2005; 6: R46.
[5] http://obo.sourceforge.net/
[6] Smith B, Kumar A, Ceuster W, Rosse C. On carcinomas and other pathological entities. Comp Funct Genom. 2005; 6: forthcoming
[7] Smith B, Brogaard B. Sixteen days. J Med Philos 2003; 28: 45-78.

I am grateful to Barry Smith, Ulf Schwarz and three anonymous reviewers for comments and suggestions for improvement.

This paper was written under the auspices of the Wolfgang Paul Program of the Alexander von Humboldt Foundation, the European Union Network of Excellence on Medical Informatics and Semantic Data Mining, and the Volkswagen Foundation under the auspices of the project "Forms of Life".

Ubiquity: Technologies for Better Health in Aging Societies
A. Hasman et al. (Eds.)
IOS Press, 2006

Tools for Czech Biomedical Ontologies Creation

Petr KOLESA[1], Petra PŘEČKOVÁ
Department of Medical Informatics, Institute of Computer Science AS CR and EuroMISE Centre, Prague, Czech Republic

Abstract. The paper describes procedures and a tool we have developed to simplify and speed up creating of Czech biomedical ontologies. Our method is based on searching for concepts in a corpus of medical texts and binding those concepts to an established international ontology. The new ontology will have two major advantages: it will be compatible with the international ontology and it will possibly cover all concepts used in the Czech healthcare.
The tool supports an author of ontology by mechanizing some routine tasks that occurs in the process of an ontology creation. It tries to learn how to identify concepts in texts and how to bind them to the ontology. The tool then displays the suggestions to a user, who can correct them and add some new ones. Based on this feedback the tool adjusts rules for concept finding and binding. To accomplish such behaviour we have employed some natural language processing methods and information extraction tools.
Keywords: Medical Informatics, Decision Support Systems, Drug Ontology.

1. Introduction

Intelligent medical systems (IMSs) help physicians to design and manage a patient-tailored treatment and provide instant suggestions or warnings about the changes in the patient's health state. In addition, they help the physicians to observe the best-known practices or institution rules during the treatment. Effects of this are both improvement of the provided health care and enormous savings. From another point of view, IMS does the routine part of the physician's work and allows the physician to work more effectively.

To achieve this goal the system is provided by corresponding medical knowledge. Such knowledge must be in the form that can be understood by a machine. Further it must both describe the given part of medicine from very basic concepts to the most advanced ideas and be reusable and versatile as creation of such knowledge representation is extremely expensive and time-consuming. For some type of medical knowledge the best representation that concerns all of the above-mentioned requirements is the *ontology*.

To the leading biomedical ontologies belong The Foundation Model of Anatomy (FMA) [1], The Systematized Nomenclature of Medicine (SNOMED) [2], and The Unified Medical Language System (UMLS) [3]. All of these ontologies has been developed in the North America and are focused on the North American conditions and procedures, and thus failing to describe the situation in the Czech public healthcare.

On the other hand, there are virtually no native Czech biomedical ontologies being developed, because such development is very expensive and time-consuming.

[1] Corresponding Author: Petr Kolesa, Pod Vodárenskou věží 2, 182 07 Prague 8.
e-mail:kolesa@euromise.cz

Regarding the environment the ontologies were designed for, they often fail when they are used in the Czech healthcare. In such a situation, no true IMS is developed and put into the commercial use and research in the field of IMS is almost impossible in the Czech Republic.

The aim of the research described in this paper is to develop methods and a tool that would be used for creating Czech biomedical ontologies. The methods and the tool focus on binding newly created Czech ontologies onto existing international biomedical ones, thus achieving compatibility and re-usability.

2. Material and Methods

The aim of the project is to answer the question whether it is possible to speed up creation of Czech biomedical ontologies. That we want to achieve by mechanizing some tasks of ontology creation, such as concept extraction and finding theirs counterparts in an international ontology. This way, we are able to reuse effort invested into development of international ontologies. Such process still requires a skilled ontology author but thanks the tool the author can work effectively. The Czech ontology created this way is not a plain translation of existing ontology but it is a creation of new ontology with the structure of original ontology, and which in addition describes well the situation in the given medical specialization in the Czech Republic. On the other hand, thanks to the tight binding to an international ontology, it is easy to connect together medical systems based on Czech ontologies to the systems based on well-established international ontologies. But it is good to remember that Czech ontology may contain some concepts that are not in the original ontology and vice versa.

In the following two sections we shortly describe two fundamental methods we use. In the third part we outline the proposed procedure of Czech ontology creation. In the last section some steps are described in greater details.

2.1. Information Extraction

The field of study called *information extraction* (IE) explores the methods used for automated automatic extraction of information from unrestricted texts. There are many tools that provide such functionality. We have chosen AMILCARE [4], [5], as it shows good performance and it is easy to integrate it with other tools. AMILCARE utilizes the learning method called supervised learning: the system is provided by the corpus of pre-annotated texts (the information we are interested in is marked in the texts). Based on the learning corpus extraction rules are generated. Next, the induced rules are applied to the texts we want to extract information from.

AMILCARE is not capable to pursue any linguistic analysis of the text by itself, which is not an issue in case the text is already well structured. However, as our preliminary experiments have shown, for the majority of the medical texts it is necessary to provide AMILCARE with additional information about the syntactic structure of the text.

2.2. Natural Language Processing

The Natural Language Processing (NLP) is a field of study that is concerned with the issues of computer processing of the natural language in the way a computer would understand (on a certain level) the language. Understanding of a natural language by computers is a very complicated task. However, some tools have been developed which make it possible to deal at least with the morphologic, syntactic and a structure-of-the-sentence level.

We use NLP tools in order to provide additional information to be exploited in the formation of the extraction rules, thus increasing the performance of the entire system. For this purpose we have used the Czech part-of-speech tagger and lemmatizer from the Prague Dependency Treebank toolkit [6], [7]. This tool is capable of extracting various linguistic information from a textual source. Currently, we use primarily the lemmatization engine and also word class identifiers.

The lemmatization process is very important for Czech texts handling, as spelling of a word changes according to the word‘s position in a nominal phrase, grammatical case and preposition used. If a text is not lemmatized, it is virtually impossible to identify similar concepts.

2.3. Ontology creation

The creation of a new ontology consists of three main phases. In the first phase a method of knowledge representation is chosen. It includes decisions about the concept representation and how to represent relations between concepts. In the second phase the concepts are identified. In the last phase discovered concepts are inserted into the ontology (relations of different types are defined among concepts).

In fact, the outlined process is very complicated and time-consuming. Thus it is reasonable to employ as much work already done in the field of biomedical ontologies as possible. Therefore we use existing the well-established international biomedical ontology as a “skeleton” for a newly created Czech ontology.

At the beginning, we choose a “skeleton” ontology and we use its structure and types of relation it defines. According to our experience, an existing international ontology fails to describe the Czech medical reality not because of a bad structure or relation types but because many important concepts are missing. Therefore we can reuse this part of the ontology.

In the second phase, a corpus of a medical text is searched for concepts. We believe that this phase can be speed up by automating it. The corpus is being processed on the fly as an author of the ontology goes through the texts. Then the tool makes suggestions about the newly identified concepts. The author is given the possibility to accept the suggestions or correct them. Based on the actions taken by the author, the tool adjusts its criteria for concept identification and continues in processing the corpus. Both NLP and IE methods we have chosen are ready for the learn-as-you-go approach.

In the third phase, we take each concept obtained in the previous phase and try to get the counterpart English concept using the Czech and English versions of MeSH [8], [9] and an online directory slovnik.seznam.cz. If this succeeds, the Czech concept is placed to the appropriate position in the ontology. If there is no English counterpart, the placement is done manually.

The proposed procedure is not a plain translation of a "skeleton" ontology. Due to the procedure the Czech ontology contains all the concepts necessary to fully cover the Czech reality in the given domain.

2.4. Concept extraction and binding

When ontology is being created from the very beginning, no computer processable resources are available but a "skeleton" ontology. A corpus contains only plain texts and is processed in the following way:

A document from the corpus is chosen and enriched with linguistic information. For this moment, such pieces of added information are part-of-speech tags and morphologic lemmas. But in general, it can be almost arbitrary information that is likely to help with concept extraction. Further, all known concepts (e.g. ones identified during processing of previous documents or from any supplemental nomenclature) are marked. Such marking is based on matching lemmatized concept names.

Next, extraction rules induced during processing of previous document are applied (for the very first document there are no extraction rules). Phrases marked by rules and concept matching are displayed to the user as suggestions. User checks the suggestions, if needed, correct them or add some new ones. Meanwhile the user is working, information extraction tools induces a new set of extraction rules to reflect corrections made by the user.

After the previous step, names of all concepts in the documents are identified. However, most concept names are in the form of compound nominal phrases and must be separated before further processing. Such separation is achieved with a heuristic approach at this moment, considering nouns, commas, conjunctions and parenthesis in the compound phrase. A parenthesis usually contains examples of a general concept that is before the parenthesis, conjunctions and commas separate concepts with containing the same noun – but in the second and consequent occurrence the noun is omitted. The heuristic algorithm tries to separate and recover the original phrases. At the end, the simple nominal phrases are displayed to the user as suggestions and the user may correct them.

Finally, the concept names are mapped to concepts and bind to the ontology. This binding can be speed up, if there are counterpart concepts in the "skeleton" ontology. This assumption is valid in our test case: we use UMLS as a "skeleton" ontology. Concept names are bound to concepts using dictionaries and multilingual nomenclatures. We use the Czech version of MeSH and the online web dictionary slovnik.seznam.cz. Translated concept names are verified against UMLSKS [10]. If the translated concept name matches any concept, it is added to the possible counterpart concepts. The list of such probable counterpart concepts is displayed to the user to choose the right one. The user has also the possibility to bind the concept manually.

3. Results

During the development, the tool has been tested on the real problem: creation of a Czech drug ontology using the corpus of drug information leaflets. First of all, we created a corpus containing approximately 200 drug information leaflets. The leaflets were chosen randomly from more than 3000 drugs distributed in the Czech Republic. The performance of the tool was measured after processing the middle of the corpus.

The reason was that concept searching is based on a machine learning technique, which provides better results as the learning set groves. Other methods employed, such as NLP and concept binding, are not affected by the user feedback in the current version of the tool. The results are reported in the same order as constituent steps occur. The result values presented in following paragraphs are in percentage and are rounded up to integers.

At first the NLP step happened. The NLP tool was able to return a lemma and a part-of-speech tag for 96 per cent of words. It is an excellent performance concerning the fact that the NLP tool was trained on a corpus of common texts (newspapers, novels, etc).

Second, it was searched for concepts in the text. The IE tool showed the recall level of 54 per cent, and precision of 93 per cent.

Third, compound phrases were disassembled to form single noun phrases. The tool was successful in 78 per cent.

Finally, it was searched for counterpart UMLS concepts. If the concept was contained in MeSH (approximately 7 percent of concepts), the counterpart UMLS concept was always found correctly. For the rest we used various on-line Czech-English web dictionaries (slovnik.seznam.cz, slovnik.centrum.cz and slovnik.atlas.cz). As these dictionaries cover the medical terminology quite well, about 64 per cent of all concepts were translated this way. Approximately 12 per cent of all identified concepts were recognised directly by the UMLS Knowledge Source Server (UMLSKS) engine, available at the UMLSKS web site [10]. This interface performs a spell checking and an approximate matching. The direct query to UMLSKS was successful when the searched word was of Greece or Latin origins and concepts names in both Czech and English differs only slightly in spelling (for instance hypokalemie -> hypokalemia). Approximately 83 per cent of concepts have been translated automatically on the whole (although disambiguation was done by the user). The tool was unable to suggest anything or all suggestions were wrong in 17 per cent of concepts. This group consisted of concepts that were contained in UMLS (9 per cent of all concepts) and of concepts that even a skilled author was unable to match to an existing UMLS concept (8 per cent).

4. Discussion

The tool is intended for creation of Czech biomedical ontologies primary. However, the developed methods are language neutral and the tool is designed in such a way that all language-dependent components (part of speech tagger, dictionaries) can be easily replaced.

The phase of IE affects the performance at most. Thus we will focus on providing the IE tool with more text structure information to improve IE.

To both information extraction and ontology creation much attention is paid worldwide. So, there are many tools that aim at such tasks. The most famous are the Protégé [11], an ontology creation and knowledge acquisition tool, and the GATE [12], a general framework for information extraction and natural language processing. However, there is a gap in between these two state-of-the-art tools. Thus we developed our tool that provides functionality of both and thus achieving user comfort an ontology user requires.

5. Conclusion

In the paper, we have discussed methods for an ontology creation and tools based on those methods. We have shown capabilities of the tool in the real example – creation of a Czech drug ontology.

Our approach to the Czech ontology creation is based on reusing a structure of an established ontology, which is usually in English, and mechanizing a concept searching in a corpus of texts. The newly created ontology is thus compatible with the original ontology. Due to the way the concepts are obtained from Czech biomedical texts, it meets needs of Czech users.

The tool has been tested on a corpus containing 200 drug information leaflets. We have shown that translation with some heuristics have shown excellent results (approximately 83 % of concepts identified in the previous stage have been translated successfully).

Our project focuses on the Czech language, however, since many problems we came across so far are beyond the language barriers, we believe that the collected knowledge will be applicable also to systems working with different languages.

The tool does not have ambitions to fully mechanize the process of text processing as a whole, nor even automate any of its part. Its goal is to support the user and reduce amount of the labour that must be done manually.

Acknowledgments

The work was partly supported by the project no. 1ET200300413 of the Academy of Sciences of the Czech Republic and by the Institutional Research Plan no. AV0Z10300504.

References

[1] The Foundational Model of Anatomy. http://sig.biostr.washington.edu/projects/fm/index.html. (last accessed Nov. 10, 2005)
[2] Systematized Nomenclature of Medicine – Clinical Terms. http://www.snomed.org/snomedct/. (last accessed Dec. 11, 2005).
[3] Unified Medical Language System. http://www.nlm.nih.gov/research/umls/.(last accessed Jan. 13, 2006)
[4] AMILCARE Homepage: http://nlp.shef.ac.uk/amilcare/lp2.html (last accessed Nov. 19, 2005)
[5] Ciravegna F. Adaptive Information Extraction from Text by Rule Induction and Generalization. Proceedings of IJCAI 2001, Seattle, 2001.
[6] The Prague Dependency Treebank 1.0. http://ufal.mff.cuni.cz/pdt/. (last accessed Nov. 12, 2005)
[7] Hajič J. Disambiguation of Rich Inflection - Computational Morphology of Czech. Prague Karolinum, 2004
[8] Medical Subject Headings http://www.nlm.nih.gov/mesh/MBrowser.html. (last accessed Jan. 13, 2006)
[9] Czech English Medical Thesaurus MeSH-CZ 2004. http://dec2.nlk.cz:4001/ALEPH/CZE/NML/CLA/MES/START (last accessed Nov. 19, 2005)
[10] UMLS Knowledge Source Server. http://umlsks.nlm.nih.gov/ (last accessed Dec. 11, 2005).
[11] The Protégé Ontology Editor and Knowledge Acquisition System. http://protege.stanford.edu/ (last accessed Jan. 19, 2006).
[12] General Architecture for Text Engineering. http://gate.ac.uk/ (last accessed Jan. 19, 2006).

Ubiquity: Technologies for Better Health in Aging Societies
A. Hasman et al. (Eds.)
IOS Press, 2006

Aligning biomedical ontologies using lexical methods and the UMLS: the case of Disease ontologies

Gwenaëlle MARQUET[a1], Jean MOSSER[b] and Anita BURGUN[a]
[a] EA 3888, IFR 140 – Faculté de Médecine, Université de Rennes1, France
[b] UMR 6061, IFR 140 – Faculté de Médecine, Université de Rennes1, France

Abstract. The process of aligning ontologies comprises two major steps: i) mapping concepts and ii) characterizing the relations between the concepts. In this paper, we present an alignment method based on a hybrid approach that reuses the UMLS knowledge base and aims at identifying patterns to characterize the relations. The proposed method consist in four steps: 1) exact matching, 2) searching for terms from one ontology that are included in terms from the other ontology, 3) identifying direct relations through the UMLS and 4) extracting syntactico-semantic patterns to infer novel alignments. This method has been applied to aligning the Human Disease ontology and the Mouse Pathology ontology resulting in 48 exact matches and 3,697 pairs of concepts for which one term is included in a term from the other ontology. 1,270 alignments are present in the UMLS. Among these, 903 are characterized by a semantic attribute. Based on these alignments, a study of the syntactic patterns has been done. Not surprisingly, the distribution of the different syntactic patterns is not sufficient to discriminate the different types of relationships found in the UMLS alignments. We have used the semantic categorization of the concepts provided by the UMLS to extract syntactico-semantic patterns. 87 novel alignments based on 6 syntactico-semantic patterns associated with isa and has_associated_morphology have been inferred.
Keywords: Ontology, alignment, lexical methods, patterns, UMLS, medical ontologies

1. Introduction

Standard ontologies have been developed in the biological and the medical domains including Gene Ontology (GO) [1], the ontologies developed in the framework of the Open Biological Ontology (OBO) project [2] and the Unified Medical Language System (UMLS) [3]. OBO ontologies include more than 50 vocabularies, covering genomics (e.g. GO, protein domain), chemistry (e.g. Physico-chemical process, Chemical entities of biological interest), anatomy (e.g. C.elegans gross anatomy, Mouse adult anatomy) and phenotype (e.g. human disease, mammalian phenotype). The UMLS is built by the merging of more than 100 vocabularies including MeSH, SNOMED and GO. It does not include the other OBO ontologies.

Different approaches exist to connect ontologies: merging [4] consists of building a single and coherent ontology, transformation [5] consists of transforming one ontology to another, and alignment [6] consists of finding correspondences between two ontologies. An alignment can be defined as a set of concepts (c and c'), coming from two ontologies (O and O'), related by a relation (R). The first step is to map concepts and the second step to characterize the relation between the concepts that have been mapped. Different tools for aligning and merging ontologies are available. For

[1] Gwenaelle Marquet, 2 rue Henri le Guilloux, 35033 Rennes gwenaelle.marquet@univ-rennes1.fr

example, PROMPT [4] and Chimaera [7] make suggestions to users based on the similarity between terms, relationships, instances and slots.

External resources such as the UMLS can be used to provide synonyms of terms as well as to provide relations between concepts. Moreover, syntactic and syntactico-semantic patterns can be used to characterize the relation between two concepts. For example, Hearst [8] identifies hyponymy relations as characterized by lexico-syntactic patterns such as 'X {including, or other} Y', where Y is a hyponym of X. Terms in the OBO ontologies are often compositional. For example, Ogren et al [9] have demonstrated that 65% of all GO terms are included in another GO term. In addition, Burgun and Bodenreider [10] have shown that 55% of the GO terms include a ChEBI term in their names. Each time a term is included in another one, it corresponds to additional information that is implicit and may not be represented by the isA and partOf relations [9].

In this paper, we present an alignment method based on a hybrid approach that (1) reuses the UMLS knowledge base and (2) aims at identifying patterns to characterize the relations. The contribution of this paper is not to propose a novel technique, but rather to combine existing techniques for aligning ontologies. We applied this method to the alignment of two OBO ontologies: the Mouse Pathology and the Human Disease ontologies.

2. Resources and methods

2.1. Resources

The Mouse pathology (MP) ontology [11] is a structured controlled vocabulary of mutant and transgenic mouse pathology phenotypes organized as "instances of" pathological processes developed by the Pathbase European Consortium. We used the revised version 1.1 (August 2005). This version contains 457 concepts.

The Human disease (HD) ontology [12] is a controlled medical vocabulary developed at the Bioinformatics Core Facility in collaboration with the NuGene Project at the Center for Genetic Medicine. We used the Disease Ontology v2.1 which contains 19,093 concepts.

An additional resource used in the alignment is the ***UMLS***. The UMLS has been developed by the US National Library of Medicine since 1986 [3]. It consists of two knowledge bases: i) the Metathesaurus®, organized by concepts (a concept is a cluster of synonymous terms), ii) the Semantic Network, a network of 135 semantic types [13]. The 135 semantic types can be aggregated into 15 semantic groups. The 2005AA edition of the Metathesaurus includes 1,179,177 concepts related by a set of 22,623,179 relations.

2.2. Methods

Our approach for aligning MP and HD consists in four steps:

- **Step 1: Exact matching**. Identical terms in the two ontologies, including the synonyms provided by the UMLS, are looked for, before and after normalization by Norm [14].
- **Step 2: Searching for terms from one ontology that are included in terms from the other ontology**. (Before and after normalization).
- **Step 3: Identifying direct relations** We use the UMLS (MRCON that contains the different String corresponding to the concept and MRREL that

contains each relationship between concepts) for identifying i) the direct relationships (e.g. CHD, RQ) and ii) the relationship attributes (e.g. isa, part_of) – which correspond in the UMLS files to RELA – between the pairs of concepts obtained in step 2.

- **Step 4: Using patterns for augmenting alignments**. This is done in three steps. First, TreeTagger [15] is used for assigning syntactic categories to words constituting the terms. Those categories include Adjective (JJ) Noun, singular or mass (NN), and Conjunction, subordinating (IN). Second, syntactic patterns corresponding to the alignments found in step 3 are extracted. For example, **NN_1 isa JJ NN_1** is a syntactic pattern. Thirdly, based on the UMLS semantic groups and the UMLS semantic types, syntactico-semantic patterns are defined.

3. Results

3.1. Exact match

Among MP concepts and HD concepts, 48 pairs are found using exact matching e.g., gangrene in MP (MPATH:10) and gangrene in HD (DOID:6712). Among those 48 pairs of concepts, 27 (57%) have been found using the UMLS synonyms. For example, Seborrheic Keratosis (MPATH:158) maps with Actinic Keratosis (DOID:15739) as both have *Senile Hyperkeratosis* as synonym in the UMLS.

Among the 457 MP concepts, 335 (73.3%) are present in the UMLS under the semantic group Disorders. Among the 19,093 HD concepts of HD, 14,206 (74.4%) are present in the UMLS concepts under the semantic group Disorders.

3.2. Inclusion

The inclusion step identifies 3,697 pairs of terms, one term belonging to an ontology and the other term to the other ontology. For example, hematoma (MPATH:121) is included in hematoma of vulva (DOID:17761). 3,100 cases correspond to MP terms included in HD terms and 597 cases to HD terms included in MP terms.

22.9% of the inclusion pairs (847 pairs) have been found using the UMLS synonyms. For example, edema (MAPTH:109) is included in the UMLS synonym *Severe oedema* which is a UMLS synonym of *Severe pre-eclampsia* (DOID:1256[illegible].

3.3. Identifying direct relations using the UMLS

Among these 3,697 pairs of concepts, 1,270 relations are also present in the UMLS Metathesaurus and 1,037 have a RELA relation in the UMLS. These 1,037 relations are "semantically" defined in the Metathesaurus. Hence, the UMLS allowed to align 1,037 pairs of concepts where one concept name is included in the other concept name. However, not all the RELA relations are semantic relations. In fact, these RELA relations can be divided in two categories: the semantic attributes (e.g. isa, associated_morphology) and the non semantic attributes (e.g. default_mapped_from, mapped_from). 903 alignments correspond to 9 semantic attributes and 134 alignments correspond to non semantic attributes. The semantic attributes are listed in table [illegible]. Four relations (*has_associated_morphology*, *isa*, *classified_as* and *associated_with*) correspond to 808 alignments. The *isa* attribute corresponds to 175 alignments while *has_associated_morphology* corresponds to 401 alignments. These two attributes have been studied in the step 4.

# alignments	Semantic attributes
401	*has_associated_morphology*
175	*isa*
117	*classified_as*
115	*associated_with*

# alignments	Semantic attributes
47	*classifies*
30	*associated_morphology*
9	*clinically_similar*
5	*clinically_associated*
4	*has_associated_finding*

Table 1: Number of alignments by semantic attributes provided by step 3

3.4. Augmenting alignment using patterns

3.4.1. isa attribute

The *isa* attribute corresponds to 89 different syntactic patterns. Among the 175 isa links, 37 alignments (21%) correspond to two syntactic patterns: **JJ NN_1 isa NN_1** (29 alignments) and **JJ JJ NN_1 isa NN_1** (8 alignments).

3.4.2. has_associated_morphology attributes

The *has_associated_morphology* attribute corresponds to 158 different syntactic patterns. Among the 401 *has_associated_morphology* links, 88 alignments (21%) correspond to four patterns: 1) **JJ NN_1 *has_associated_morphology* NN_1** (34 alignments), 2) **JJ JJ NN_1 *has_associated_morphology* NN_1** (19 alignments), 3) **NN_1 IN NN *has_associated_morphology* NN_1** (18 alignments) and 4) **JJ NN_1 IN NN *has_associated_morphology* NN_1** (17 alignments).

3.4.3. Overlap between the syntactic patterns: isa and has_associated_morphology

As shown in table 2, 47 syntactic patterns are common to *isa* and *has_associated_morphology* attributes.

	isa		*has_associated_morphology*	
	# patterns	*# alignments*	*# patterns*	*# alignments*
Identical patterns	**47**	134 (76.6%)	**47**	220 (54.9%)
Specific patterns	**42**	41 (23.4%)	**111**	181 (45.1%)
Total		175 (100%)		401 (100%)

Table 2: Repartition between identically patterns and specific syntactic pattern for the *isa* and *has_associated_morphology* attributes

We have tried to identify syntactico-semantic patterns to distinguish between patterns that are mostly associated with *isa* and patterns that are mostly associated with *has_associated_morphology*. We have studied two patterns: **JJ NN_1 *rela* NN_1** and **JJ JJ NN_1 *rela* NN_1**, which are the most frequent in our list.

3.4.4. JJ NN_1 rela NN_1

Most of the patterns corresponding to *has_associated_morphology* are associated with the combination of the semantic groups Anatomy and Disorders, more precisely the semantic types *Anatomical Abnormality* and *Pathological Function* (Table 3).

			isa		*has_associated_morphology*	
JJ(Anatomy) **NN$_1$**(Disorders) **rela** **NN$_1$**(Disorders)			4 (14%)	retinal vasculitis / vasculitis	20 (59%)	cerebral edema / edema
P1.1	**JJ** (Anatomy) **NN$_1$**(*Pathological function*)	**NN$_1$**(*Pathological function*)	2	Interstitial emphysema/ Emphysema	13	Myocardial degenerartion/ degeneration
P1.2	**JJ** (Anatomy) **NN$_1$**(*Anatomical Abnormality*)	**NN$_1$**(*Anatomical Abnormality*)	0		7	Anal fistula / fistula

Table 3: The syntactico-semantic patterns associated with *has_associated _morphology* attribute.

Most of the *isa* attributes (Table 4) are associated with the combination of the two semantic groups Concepts & Ideas and Disorders, mostly represented by the semantic type Disease or Syndrome.

			isa		*has_associated_morphology*	
JJ(Concepts & Ideas) **NN$_1$**(Disorders) **rela** **NN$_1$**(Disorders)			16 (55%)	Chronic glomerulonephritis / glomerulonephritis	7 (15%)	infantile atrophy / atrophy
P2.2	**JJ**(Concepts & Ideas) **NN$_1$**(*Disease or Syndrome*)	**NN$_1$**(*Disease or Syndrome*)	9	Other cataract / cataract	1	Coronary atherosclerosis/ atherosclerosis

Table 4: The syntactico-semantic patterns associated with isa attribute.

3.4.5. JJ JJ NN$_1$ rela NN$_1$

For the pattern **JJ$_3$ JJ$_2$ NN$_1$ *rela* NN$_1$**, 83% of the JJ$_3$ adjective corresponds to the Semantic Groups Concepts & Ideas.

3.4.6. Inferred new relations

Starting from the syntactico-semantically patterns P1.1 and P1.2 defined in step 4; we have inferred 87 new alignments. For example, mandibular hypoplasia *has_associated_morphology* hypoplasia, ureteral fistula *has_associated_morphology* fistula, acute myocardial infarction *has_associated_morphology* infarction. All the novel alignments have been judged valid.

4. Discussion and Conclusion

This study focuses on the alignment between the Human Disease ontology and the Mouse Pathology ontology. Only 21 terms are found strictly identical in HD and MP, corresponding to 11 general pathological notions such as edema and gangrene, and 10 precise disease terms such as amniotic fluid embolism and leukemia megakaryocytic. Moreover, MP and HD have different scopes. While MP terms mostly correspond to general pathological functions or lesions, HD contains disease terms and even some procedure terms. Inclusion is found in 3,697 pairs of concepts, which confirms the role played by the implicit compositional structure of biomedical terms already mentioned for GO and ChEBI [9, 10]. The contribution of external resources may be limited by their coverage. In our study, 74.4% of the 19,093 HD concepts are present in the UMLS Metathesaurus under the Semantic Group Disorder and 73.3% of the 457 Mouse Pathology concepts are present in the UMLS Metathesaurus under the Semantic Group Disorder. For example, MPATH:22: mitochondrial defect are not present in the UMLS. Although the absence of a term in the UMLS results in incomplete alignment, we have shown that 22.2% of the mappings between HD terms and MP terms have been established thanks to the UMLS synonyms. Moreover, the UMLS provided the 'RELA' characterization of 34% (1,270) of the associations that corresponded to substring relations between the terms. Most of the *isa* attributes as well as the *has_associated_morphology* attributes come from SNOMED CT.

We have explored the contribution of syntactico-semantic patterns to align HD concepts and MP concepts. We have limited our study to cases where the name of a concept from one ontology is included in the name of a concept from the other ontology. Not surprisingly, the distribution of the different syntactic patterns among the different relation attributes is not sufficient to discriminate the different types of relationships found in the UMLS alignments. In a second step, we have used the semantic categorization of the concepts provided by the UMLS in terms of Semantic Groups and in terms of Semantic Types to build syntactico-semantic patterns. Within the Disorders semantic group, the relation *has_associated_morphology* mostly corresponds to the semantic types Pathological function and Anatomical Abnormality. The same word, e.g. fibroplasia, may correspond to a pathological function (or pathological entity) and be part of a more complex term corresponding to a "disease", e.g. retrolental fibroplasia.

Although the method presented was voluntarily restricted to HD and MP, to some relation attributes, and to term inclusion, this study suggests that the compositional structure of biomedical terms as well as the UMLS knowledge base (including the semantic relations provided by ontologies such as SNOMED CT) can be used to align biomedical ontologies.

Acknowledgments

This work was supported by a grant from the Région Bretagne (20046805)

References

[1] The Gene Ontology Consortium. Gene Ontology: tool for the unification of biology. Nature Genet 2000; 25: 25-9.
[2] Open Biological Ontologies. Available at http://obo.sourceforge.net/.
[3] Bodenreider O. The Unified Medical Language System (UMLS): integrating biomedical terminology. Nucleic Acids Res 2004; 32: D267-70.
[4] Noy NF, Musen MA. PROMPT: algorithm and tool for automated ontology merging and alignment. Proceedings of AAAI 2000: 450-455.
[5] Chalupsky H. OntoMorph: a translation system for symbolic knowledge. Proceedings of KR 2000.
[6] Euzenat J. Towards composing and benchmarking ontology alignments. Proceedings of the ISWC-2003 workshop on semantic information integration 2003: 165-166.
[7] McGuinness D L, Fikes R, Rice J, Wilder S. The Chimaera ontology environment. Proceedings of AAAI 2000: 450-455.
[8] Hearst M. Automatic acquisition of hyponyms from large text corpora. Proceedings of the Fourteenth International Conference on Computational Linguistics1992; l2:539-545.
[9] Ogren PV, Cohen KB, Acquaah-Mensah GK, Eberlein J, Hunter L. The compositional structure of Gene Ontology terms. Pac Symp Biocomput 2004; 214-25.
[10] Burgun A, Bodenreider O. An ontology of chemical entities helps identify dependence relations among Gene Ontology terms. First International Symposium on Semantic Mining in Biomedicine 2005.
[11] Schofield PN, Bard JB, Booth C, Boniver J, Covelli V, Delvenne P, Ellender M, Engstrom W, Goessner W, Gruenberger M, Hoefler H, Hopewell J, Mancuso M, Mothersill C, Potten CS, Quintanilla-Fend L, Rozell B, Sariola H, Sundberg JP, Ward A. Pathbase: a database of mutant mouse pathology. Nucleic Acids Res. 2004 Jan 1; 32: D512-5.
[12] Available at http://diseaseontology.sourceforge.net/#contacts
[13] McCray AT, Burgun A, Bodenreider O. Aggregating UMLS semantic types for reducing conceptual complexity. Medinfo 2001; 10: 216-20.
[14] National Library of Medecin. UMLS Lexical Tools. Application and documentation available at http://umlsks.nlm.nih.gov
[15] Schmid H. Probabilistic Part-of-Speech Tagging Using Decision Trees. International Conference on New Methods in Language Processing Manchester, UK 1994.

7.4 Concepts and Coding: Methods

Ubiquity: Technologies for Better Health in Aging Societies
A. Hasman et al. (Eds.)
IOS Press, 2006

Towards Automated Classification of Intensive Care Nursing Narratives

Marketta HIISSA[a,c,1], Tapio PAHIKKALA[a,b], Hanna SUOMINEN[a,b], Tuija LEHTIKUNNAS[d,e], Barbro BACK[a,c], Helena KARSTEN[a,b], Sanna SALANTERÄ[d] and Tapio SALAKOSKI[a,b]

[a] *Turku Centre for Computer Science, Turku, Finland*
[b] *University of Turku, Department of Information Technology, Turku, Finland*
[c] *Åbo Akademi University, Department of Information Technologies, Turku, Finland*
[d] *University of Turku, Department of Nursing Science, Turku, Finland*
[e] *Turku University Hospital, Turku, Finland*

Abstract. Nursing narratives are an important part of patient documentation, but the possibilities to utilize them in the direct care process are limited due to the lack of proper tools. One solution to facilitate the utilization of narrative data could be to classify them according to their content. In this paper, we addressed two issues related to designing an automated classifier: domain experts' agreement on the content of the classes into which the data are to be classified, and the ability of the machine-learning algorithm to perform the classification on an acceptable level. The data we used were a set of Finnish intensive care nursing narratives. By using Cohen's κ, we assessed the agreement of three nurses on the content of the classes Breathing, Blood Circulation and Pain, and by using the area under ROC curve (AUC), we measured the ability of the Least Squares Support Vector Machine (LS-SVM) algorithm to learn the classification patterns of the nurses. On average, the values of κ were around 0.8. The agreement was highest in the class Blood Circulation, and lowest in the class Breathing. The LS-SVM algorithm was able to learn the classification patterns of the three nurses on an acceptable level; the values of AUC were generally around 0.85. Our results indicate that one way to develop electronic patient records could be tools that handle the free text in nursing documentation.
Keywords: Computerized Patient Records, Natural Language Processing, Nursing.

1. Introduction

During the past years, health-care providers have been changing paper-based patient records to electronic ones. This has, on one hand, made more data available on each patient, but on the other hand, also offered new possibilities to utilize the gathered data. However, the effects of this switch have not only been positive. It has been found that electronic charting may not provide nurses with more time for tasks unrelated to manipulating data [1,2], and that electronic systems support nurses in gathering information, but not in the active utilization of it [3].

Especially problematic is the active utilization of narrative data, in particular when the patient stays in the ward for several days, and the amount of documentation is large.

[1] Corresponding Author: Marketta Hiissa, Turku Centre for Computer Science, Lemminkäisenkatu 14 A, 20520 Turku, Finland. e-mail: marketta.hiissa@abo.fi

In intensive care units, a variety of artificial intelligence techniques can already be used to process the numerical or structured data in the patient records (e.g. [4]), but due to the lack of proper tools, the possibilities to utilize narrative data are still limited. Our approach to facilitate the utilization of narratives is to develop automatic tools that classify texts, and thus make it easier to retrieve relevant information e.g. when a nurse needs to build a general picture of narratives describing patient's breathing.

In the medical domain, classification has recently been applied e.g. to classifying texts such as chief complaint notes, diagnostic statements, and injury narratives into different kinds of syndromic, illness and cause-of-injury categories [5-8]. There is, however, little research on the automated processing of nursing narratives [9].

In this paper, we use a machine-learning approach, i.e. an algorithm that learns the classification patterns directly from pre-classified data, to classify Finnish intensive care nursing narratives. We address two issues related to designing a classifier: domain experts' agreement on the content of the classes into which the data are to be classified, and the ability of the classification algorithm to perform on an acceptable level. The classes used in this study are Breathing, Blood Circulation and Pain, and the machine-learning algorithm is the Least Squares Support Vector Machine.

2. Material and Methods

2.1. Material

The data we used were a set of Finnish intensive care nursing narratives. The documents were gathered, with proper permissions and without any identification information, in the spring of 2001 from 16 intensive care units, two or three documents per unit. In total, we had 43 copies of anonymous patient records with nursing notes written down during one day and night.

The style of the documentation varied from one nurse to another: some had written short sentences such as "*Hemodynamics ok. Very tired.*", whereas others had written long sentences in which different matters were separated with commas. In order to standardise the style of the documentation, we divided the long sentences into smaller pieces consisting of one matter or thought. This was done manually by one of the authors with nursing experience, and resulted in 1363 pieces, with the average length of 3.7 words.

2.2. Methods

We chose to classify the data according to classes Breathing, Blood Circulation and Pain. The selection of breathing and blood circulation was due to the emphasis in intensive care nursing, which is on the monitoring, assessment and maintenance of vital functions [10]. Pain was chosen because due to their critical illnesses, patients are often incapable of communicating, and nurses must assess the level of pain relying only on different kinds of behavioural and physiological indicators [11], and thus the documentation of pain was supposed to differ from the one of breathing or blood circulation. The classification process was done as three separate classification tasks, i.e. each of the classes was considered separately of the others.

In order to assess domain experts' agreement on the content of the three classes, we asked three nurses to manually classify the data, independently of each other. They were advised to label each text piece they considered to contain useful information given the specific class. All the nurses were specialists in nursing documentation; two had a long clinical experience from intensive care units (N_1 and N_2), and one was an academic nursing science researcher (N_3). We measured the agreement on the content of the classes by using *Cohen's κ* [12]. Cohen's κ is a chance-corrected measure of agreement, which considers the classifiers equally competent to make judgments, places no restriction on the distribution of judgments over classes for each classifier, and takes into account that a certain amount of agreement is to be expected by chance. It is an appropriate measure especially in situations when there are no criteria for a correct classification, which is the case in our study.

In order to test the performance of the machine-learning algorithm, we divided the data classified by the nurses into a *training set* and a *test set* so that 708 out of the total of 1363 text pieces belonged in the training set, and the remaining 655 pieces in the test set. The division was done so that text pieces from one document belonged only in one of the two sets. The machine-learning algorithm we used was a *Least Squares Support Vector Machine* (*LS-SVM*) [13], which is a technique that has been shown to yield good results on classification problems (e.g. [14]). Nine automated classifiers were trained by using the training data labelled by the three nurses, i.e. one classifier for each nurse–class pair. In order to reduce different inflection forms of the words, the data were pre-processed with the Snowball stemmer for Finnish [15]. The performance of the algorithm was measured with respect to the test data by using the *area under ROC curve* (*AUC*) [16], which corresponds to the probability that given a randomly chosen positive example and a randomly chosen negative example, the classifier will correctly determine which is which.

All the statistical analyses were performed with SPSS 11.0 for Windows.

3. Results

3.1. Agreement on the content of the classes

The amount of data the nurses included in the classes Breathing, Blood Circulation and Pain was, respectively, around 20%, 15%, and 6% of the 1363 text pieces. This reflects the extensive content of the narrative documentation, and illustrates the difficulty of finding relevant information from large amounts of text.

Table 1. The values of Cohen's κ and the respective 95% confidence intervals (CI) for the agreement between the three nurses.

	Breathing		**Blood Circulation**		**Pain**	
	κ	95% CI	κ	95% CI	κ	95% CI
$N_1 - N_2$	0.73	(0.68–0.78)	0.89	(0.85–0.92)	0.88	(0.82–0.94)
$N_1 - N_3$	0.67	(0.62–0.72)	0.81	(0.77–0.86)	0.79	(0.73–0.86)
$N_2 - N_3$	0.85	(0.82–0.89)	0.87	(0.83–0.90)	0.76	(0.69–0.83)

The comparisons between the nurses showed that the text pieces describing blood circulation were selected quite similarly ($\kappa > 0.80$ in each comparison), whereas there were more differences in selecting text pieces related to pain or breathing (Table 1). The range of the values of κ was the largest, from 0.67 to 0.85, in the class Breathing. In addition, given the classes Pain and Blood Circulation, the two clinical nurses N_1 and N_2 where the most unanimous in the classification, whereas in the class Breathing, the clinical nurse N_2 and the nursing science researcher N_3 were the most unanimous.

3.2. Learning ability of the LS-SVM algorithm with respect to the data classified by the nurses

The learning ability of the LS-SVM was tested in two ways. Firstly, we tested the LS-SVM classifiers against test data classified by the same nurse whose data was used when training the algorithm. The results showed that the algorithm was able to learn the classification patterns from the training set (Table 2). The values of AUC were in general around 0.85. The highest values, from 0.89 to 0.93, were achieved for the classification of blood circulation statements, whereas the lowest, from 0.71 to 0.81 were measured for the class Pain. Except the class Pain, the performance of the algorithm was on a similar level in the classes independently of the nurse whose classification was used to train the algorithm, also in the class Breathing, in which the disagreement between the nurses was the highest.

Table 2. The values of AUC and the respective 95% confidence intervals (CI) for the automatic classifiers.

	Breathing		**Blood Circulation**		**Pain**	
	AUC	95% CI	AUC	95% CI	AUC	95% CI
N_1	0.86	(0.82–0.90)	0.89	(0.84–0.93)	0.71	(0.61–0.80)
N_2	0.88	(0.85–0.91)	0.93	(0.90–0.97)	0.81	(0.73–0.89)
N_3	0.87	(0.84–0.91)	0.91	(0.86–0.95)	0.71	(0.61–0.80)

Secondly, we tested the classifiers against test data classified by other nurses than the one whose data was used when training the algorithm. The values of AUC were calculated for the total of six comparisons in each of the three classes, and based on these comparisons, we measured the average decrease in the values of AUC compared with the situation in which both the training and the testing data were classified by the same nurse. The average decrease in the values was 0.06 in the class Breathing, and 0.01 in the class Pain. In the class Blood Circulation, the decrease was 0.00, i.e. given a manually established reference classification, the two LS-SVM classifiers trained with other data than that of the given nurse performed as well as the classifier trained by using the classification of the given nurse.

4. Discussion

We have here assessed the agreement of three nurses on the content of the classes Breathing, Blood Circulation and Pain by using Cohen's κ, and the ability of the LS-SVM machine-learning algorithm to learn the classification patterns of the nurses by using AUC as the outcome measure. On average, the values of κ were around 0.8. According to the obtained AUC values, the LS-SVM algorithm performed on an acceptable level; the values were generally around 0.85, and the decreases in these were rather small when using test data classified by another nurse than the one who classified the training data.

The disagreement cases between the nurses appeared to be due to not only the subjective considerations and interpretations on what information was important given a specific class, but also to matters such as the different handling of non-standard abbreviations. For example, one of the nurses decided not to accept any text pieces containing non-standard and ambiguous abbreviations to belong in any class, whereas the others classified these pieces based on the meaning they thought the abbreviations could have in the context they appeared. The effects of the subjective considerations on important information were the most visible in the classes Breathing and Pain. For example, given the class Breathing, nurses disagreed on whether or not the statements related to the slime in patient's lungs should be included in the class, and given the class Pain, the disagreements were mainly due to sentences such as "*Reacts to interventions by moving extremities and furrowing eyebrows.*", which did not include any direct mention of pain.

According to the obtained values of AUC, the learning ability of the LS-SVM algorithm was on a rather high level. The results for the classes Blood Circulation and Breathing were good, and the performance in the class Pain was somewhat lower mainly due to the small amount of available data and the different nature of the documentation. The sentences in the classes Blood Circulation and Breathing contained many keywords, such as *heart rate*, *blood pressure*, and *hemodynamics* in the class Blood Circulation, and *oxygen saturation*, *respirator*, and *intubation* in the class Breathing. In contrast to these, the class Pain included non-direct statements, in which no keywords were present. Compared to the other nurses, the nurse N_2 included fewer non-direct statements in the class Pain, which explains the better performance of the automated classifier in that case. The results also showed that in the class Blood Circulation, there were no decreases in the values of AUC when using test data classified by another nurse than the one with whose classification the algorithm was trained. The decreases in the values of AUC in the classes Breathing and Pain reflect the nurses' agreement on the content of these classes; in the class Blood Circulation, the values of κ where higher than in the classes Breathing and Pain.

In order for the machine-learning based classification to be useful in the intended application area, the performance of the algorithm needs to be on a satisfactory level, but also the training data needs to be well established. In this establishment, the disagreement cases between the nurses can be handled in different ways. On one hand, they could be taken into further consideration in order to reach a consensus on them, but on the other hand, if they are considered to include valuable information about the domain, the utilization of them in establishing the training data is justified. For example, our results revealed that with the pain statements, all the nurses did not include in the class expressions such as "*Reacts against when turning him over.*", in which the nurse does not explicitly denote the presence of pain. These disagreement cases could be considered e.g. as weaker pain statements than those on which all the nurses agreed, and this information could be used to give different weights to different kinds of statements with respect to the given class.

Our results showed that the LS-SVM algorithm performed on an acceptable level. However, improvements could be gained e.g. by using more training data, and by increasing the pre-processing of the data. Here we used a stemmer to reduce different inflection forms of the words, but because Finnish is a highly inflectional language, techniques that find the real base form instead of just stemming could make the data less sparse and increase the performance of the algorithm. Another topic of further research is the automation of the pre-processing of the long sentences, which here was done manually by one researcher, and is thus a possible source of bias. Further research is also needed to assess the performance of the algorithm on other classes than the three used in this study.

5. Conclusion

Given the classes Breathing, Blood Circulation and Pain, the nurses had somewhat different opinions on the optimal content of them, and these differences could be utilized when designing an automated classifier. The LS-SVM algorithm was able to learn classification patterns from the data classified by the nurses, and its performance on unseen material was on an acceptable level. We conclude that one way to develop electronic patient records could be tools that handle the free text in nursing documentation.

Acknowledgments

We gratefully acknowledge the financial support of Tekes (grant number 40435/05) and the Academy of Finland (grant number 104639). We wish to thank Heljä Lundgrén-Laine and Filip Ginter for their contributions.

References

[1] Pierpont GL, Thilgen D. Effect of computerized charting on nursing activity in intensive care. Crit Care Med 1995; 23(6): 1067-1073.
[2] Smith K, Smith V, Krugman M, Oman K. Evaluating the impact of computerized clinical documentation. Comput Inform Nurs 2005; 23(3): 132-138.
[3] Snyder-Halpern R, Corcoran-Perry S, Narayan S. Developing clinical practice environments supporting the knowledge work of nurses. Comput Nurs 2001; 19(1): 17-23; quiz 24-6.
[4] Hanson CW,3rd, Marshall BE. Artificial intelligence applications in the intensive care unit. Crit Care Med 2001; 29(2): 427-435.
[5] Chapman WW, Dowling JN, Wagner MM. Fever detection from free-text clinical records for biosurveillance. J Biomed Inform 2004; 37(2): 120-127.
[6] Pakhomov SV, Buntrock JD, Chute CG. Using compound codes for automatic classification of clinical diagnoses. Medinfo 2004; 11(Pt 1): 411-415.
[7] Wellman HM, Lehto MR, Sorock GS, Smith GS. Computerized coding of injury narrative data from the National Health Interview Survey. Accid Anal Prev 2004; 36(2): 165-171.
[8] Chapman WW, Christensen LM, Wagner MM, Haug PJ, Ivanov O, Dowling JN, et al. Classifying free-text triage chief complaints into syndromic categories with natural language processing. Artif Intell Med 2005; 33(1): 31-40.
[9] Bakken S, Hyun S, Friedman C, Johnson SB. ISO reference terminology models for nursing: applicability for natural language processing of nursing narratives. Int J Med Inform 2005; 74(7-8): 615-622.
[10] Ward NS, Snyder JE, Ross S, Haze D, Levy MM. Comparison of a commercially available clinical information system with other methods of measuring critical care outcomes data. J Crit Care 2004; 19(1): 10-15.
[11] Payen JF, Bru O, Bosson JL, Lagrasta A, Novel E, Deschaux I, et al. Assessing pain in critically ill sedated patients by using a behavioral pain scale. Crit Care Med 2001; 29(12): 2258-2263.
[12] Cohen J. A coefficient of agreement for nominal scales. Educ Psychol Meas 1960; 20(1): 37-46.
[13] Suykens JAK, Van Gestel T, De Brabanter, J, De Moor B, Vandewalle J. Least Squares Support Vector Machines. Singapore: World Scientific Publishing, 2002.
[14] R. M. Rifkin. Everything Old is New Again: A Fresh Look at Historical Approaches in Machine Learning. PhD Thesis, Massachusetts Institute of Technology, 2002.
[15] http://snowball.tartarus.org/algorithms/finnish/stemmer.html (last accessed Jan 10, 2006).
[16] Hanley JA, McNeil BJ. The meaning and use of the area under a receiver operating characteristic (ROC) curve. Radiology 1982; 143(1): 29-36.

Ubiquity: Technologies for Better Health in Aging Societies
A. Hasman et al. (Eds.)
IOS Press, 2006

Comparison of ICHI and CCAM Basic Coding System

Susanne HANSER[1], Albrecht ZAISS, Stefan SCHULZ

Department of Medical Informatics, Freiburg University Hospital, Germany

Abstract. While diagnoses are coded by ICD across the world, there is no universally accepted coding system for procedures. In many countries there exists not even a local classification for medical procedures. As a possible solution the International Classification of Health Interventions (ICHI) has been proposed as a common denominator for an international procedure classification. We alternatively postulate a multiaxial framework for procedure classification following the French Classification Commune des Actes Médicaux (CCAM) for the generation of a procedure shortlist. We compared ICHI and CCAM Basic Coding System focusing on the appropriateness of both systems for supporting the comparability of procedure data. Considering the ongoing standardization of health terminologies and classifications, we strongly recommend to improve the ICHI structure, capitalizing on the benefits of the CCAM architecture.

1. Introduction

Information on surgical and other medical interventions constitutes an indicator for the use of health care, of hospitals in particular. But due to the insufficient standardization of the documentation in this field, structured data about medical procedures is rarely available and poorly comparable across countries and institutions. While diagnoses are coded by ICD across the world, numerous national procedure coding systems are in use. Although these traditional coding systems are considered more and more insufficient by their developers, Spain, Switzerland, and the U.S. still use ICD-9-CM [1], Germany and the Netherlands use the ICPM derivations OPS and CMSV. The implementation of the first multiaxial procedure classification, the ICD-10-PCS, intended for the U.S. for over a decade [2], it still outstanding. In the meanwhile, Canada has implemented their highly elaborated coding system CCI [3]. In 2005, France has completed the implementation of the *Classification Commune des Actes Médicaux* (CCAM [4]), a system based on the GALEN-IN-USE procedure model [5]. CCAM constitutes a viable compromise between pure multiaxial systems for post-coordinated coding (e.g. PCS, also INCP, SNOMED Int.) on the one hand, and the traditional tabular procedure lists on the other hand.

In recent years, the introduction of a uniform international medical procedure classification has been discussed again. The Australian National Centre for Classification in Health presented the *International Classification of Health Interventions*

[1]S. Hanser, Stefan-Meier-Str. 26, Freiburg University Hospital, Department of Medical Informatics, Stefan-Meier-Str. 26, D-79104 Freiburg. E-mail: hanser@mi.ukl.uni-freiburg.de

(ICHI [6]) as a proposal for an international procedure classification, primarily for countries without any procedure classification. It is also intended as a coarse-grained base classification which allows for comparing and cross-mapping between more sophisticated national procedure classifications. An alternative way could be to use the CCAM framework for the generation of a procedure shortlist.

In this paper we report on a comparison between ICHI and CCAM focusing on the appropriateness of both approaches for supporting comparability of procedure data.

2. Material and Methods

We used ICHI Alpha Version 2002, a trial version, and CCAM Système de Codification (Basic Coding System) Version V0^{bis} from 2001. ICHI is a shortlist of 1421 enumerated procedure classes, arranged in a shallow hierarchy, based on the *Australian Classification of Health Interventions* (ACHI). CCAM Basic Coding System proposes a way how to classify medical procedures by anatomical site (n=178), action ("surgical deed", n=17) and approach/method (n=14). Using this framework, the CCAM catalogue was developed, a tabular list containing about 8.000 relevant procedures, to be used for ambulatory and hospital care[2].

ICHI classes correspond to the lowest aggregation level in the Australian Classification of Health Interventions (ACHI), so-called "blocks". A block aggregates a number of more specific codes. (Confusingly, in ICHI, the term "block" denotes a single class.)

ICHI Hierarchy: ICHI presents procedures hierarchically listed according to 1.body system/specialty, 2. anatomical site, and 3. procedural type. These "axes" constitute a **hierarchical layer** above ICHI classes, which are coded by sequential numbers.

Code	Title	Hierarchical level	
X.	*Procedures on digestive system* (blocks 850-1011)	1	body system/specialty
-	*Gallbladder and biliary tract*	2	site
-	*Excision*	3	procedural type
965	*Cholecystectomy*	4	procedure

ICHI Vocabulary. ICHI titles describe procedures in varying granularity and with more or less specifity, e.g.: 555 *Transplantation of lung* versus 954 *Repair procedures on liver* which subsumes transplantation of liver for which there is no separate code. ICHI titles describe procedures with terms which are either "clinical" (according to clinical jargon) or "synthetic", e.g.: 963 *Total proctocolectomy* ("clinical") versus 953 *Excision procedures on liver* ("synthetic") - for total or partial hepatectomy.

- "Synthetic" descriptions **use** a controlled vocabulary which defines the hierarchical levels for anatomical site and procedural type. E.g.:
  X. Procedures on digestive system
  \- *Liver (anatomical site)*
  \- - *Excision (procedural type)*
  \- - - *953 Excision procedures on liver*

2 The Basic Coding system refers to CCAM's classification architecture, not to its content.

- "Clinical" descriptions **specify** descriptions with vocabulary of the hierarchical levels (anatomical site, procedural type) without controlled vocabulary. E.g.:
  X. Procedures on digestive system
  \- *Rectum, Anus (anatomical site)*
  \- - *Excision (procedural type)*
  \- - - *936 Total proctocolectomy*

CCAM is a hierarchically structured procedure catalogue used in France since 2002 for reimbursement and policy making in health care**.** Each procedure listed in this catalogue is described by using a multiaxial coding system, as well as by a standardised text, both developed according to EN 1828 [7] and the GALEN procedure model [5]. Note that in this paper we do not refer to the CCAM content itself but only to the underlying multiaxial classification architecture. This framework is referred to as "CCAM Basic Coding System" (in order to distinguish it from the "CCAM Catalogue").

CCAM Basic Coding System classifies procedures according to 1. body system/anatomical site or function, 2. action, 3. approach/method. The concatenation of the codes for these axes result in a **multiaxial code**, which gives a "synthetic" procedure description based on the code/definition tables of the CCAM Basic Coding System. E.g.:

HM	*Digestive system/ gallbladder and biliary tract*
F	*To excise (Taking out a part of the organism by cutting)*
C	*Endoscopic percutaneous approach*
HMFC => "*Excision on gallbladder OR biliary tract, endoscopic percutaneous*"	

Finer procedure descriptions, as listed in the CCAM catalogue, have a composite code: multiaxial code plus sequential number, e.g. HMFC004 *Cholecystectomy, by laparoscopy*. CCAM titles describe procedures in clinical terms, using a standardised vocabulary in standardised order. CCAM tabular list presents procedures hierarchically listed in up to five aggregation levels according to body system/anatomical site and action.

Table 1: General characteristics of ICHI and CCAM in comparison

	ICHI	**CCAM**
Publication	Pilot Version 2002	Preliminary Version 2001 Final version 2002
Organisation	NCCH	ATIH, ANAES, CNAMTS
Content based on	Australian Procedure Classification ACHI (since 1995)	CdAM, GALEN Procedure Model
Intention	Adaptation of the more detailed ACHI to international use and for countries without procedure classification	Classification of inpatient and outpatient procedures; basis for health care financing and policy making.

Method. As done earlier with classes of the OPS-301[8], a German procedure coding system [9], we assigned multiaxial CCAM codes to ICHI classes and assessed the mapping result in terms of representability and granularity. Analyzing the wording of each ICHI title, we represented ICHI codes as triples (A,P,M) according to the CCAM axes anatomical site (A), procedural type (P), and approach/method (M). This was done manually by an advanced medical student. The results were then checked and refined by a researcher (MD) with experience in medical terminologies. As an example, see our mapping of the ICHI class 875 *partial gastrectomy* to the triple (HF,F,A). HFFA is defined as (Stomach/Excision/Open approach). In this case, the resulting procedure description was slightly less specific (i.e. the granularity of the representation was slightly coarser) than the original one (" –ectomy, partial" => "excision"). An ICHI class was considered (completely) representable, if the assignment of corresponding CCAM categories for anatomy, action and approach was possible.

In addition, we compared the architecture of both classifications, regarding coding system, hierarchy, as well as the underlying vocabulary.

3. RESULTS

Mapping ICHI classes to the CCAM Coding System. A high percentage of ICHI classes were completely representable by CCAM categories, i.e. they could be mapped directly to four-digit CCAM codes (n=1114, 78.4%) due to considerable similarities between both procedure models, especially concerning anatomical site (mapping rate 99.3%). Only 307 ICHI classes (21.6%) have no or a partial CCAM code. As we performed a 1:1 mapping, various ICHI classes referring to two or more specific anatomical contents (e.g. "lung or pleura)" were mapped to a rather broad, unspecific class ("respiratory system, unspecified"), resulting in a low granular CCAM representation for these classes. Furthermore, there are varying combinations of sites, e.g. "pelvis and hip" besides "pelvis and femur". Representability of the action/procedural type axis was affected by missing CCAM codes for "*incision*" and "*other procedure*" and by imprecise ICHI procedure descriptions, e.g. "*procedure for* (e.g. *skull fracture*)". Frequently, the ICHI descriptions lack precise information which would allow an exact mapping to an approach/method such as specified in CCAM Basic Coding System. Like in traditional procedure classifications (OPS, ICD-9-CM), ICHI encodes this type of information in a largely implicit way. Thus we were unable to completely map those procedures, which in practice may be performed by using different approaches or techniques, e.g. cholecystectomy (by laparoscopic or open surgical approach).

Table 2: Mapping result ICHI to CCAM Basic Coding System. (For semantic categories A, P, M: + assignment of corresponding CCAM category possible. ∅ no CCAM match)

<table>
<tr><th>Anatomy</th><th>Action</th><th>Approach</th><th>Number of ICHI classes</th><th>% of ICHI classes (N=1421)</th><th></th></tr>
<tr><td>∅</td><td>∅</td><td>∅</td><td>3</td><td>0.2%</td><td rowspan="5">∅ anatomy:
Σ=0.7%
∅ action:
Σ=15.1%
∅ approach
Σ=15.9%</td></tr>
<tr><td>∅</td><td>+</td><td>+</td><td>7</td><td>0.5%</td></tr>
<tr><td>+</td><td>∅</td><td>∅</td><td>138</td><td>9.7%</td></tr>
<tr><td>+</td><td>∅</td><td>+</td><td>74</td><td>5.2%</td></tr>
<tr><td>+</td><td>+</td><td>∅</td><td>85</td><td>6.0%</td></tr>
<tr><td>+</td><td>+</td><td>+</td><td>1114</td><td>78.4%</td><td></td></tr>
</table>

Comparison of ICHI/CCAM architecture. Coding system: In terms of notation, it is a major drawback that the terminal ICHI classes (those which are used for coding) are identified by sequential numbers, while the chapters, subchapters and headings (which consistently correspond to body system, anatomical site and procedural type) have no codes. In contrast, CCAM provides multiaxial four-digit codes, representing body system, anatomical site, and action (procedural type). **Hierarchy**: ICHI tabulates each class according to 1) body system/ specialty, 2) anatomical site/detail, 3) procedural type. The result is a well-ordered hierarchical procedure list. In contrast, CCAM's multiaxial classification system supports different hierarchies. So it makes sense to choose a hierarchical tabulation oriented by the axis for anatomical site. **Vocabulary**: ICHI procedure descriptions (titles) are not standardised. In several cases this gives raise to ambiguities. The term "removal" for example, is used in ICHI classes for destruction or excision (e.g. of a polyp or a tooth) as well as for the removal of foreign material as devices (e.g. cerebrospinal fluid shunt), prostheses and foreign bodies. The procedural type "Application, Insertion, Removal" however subsumes only removal of foreign material. ICHI procedure descriptions are either "clinical" or "synthetic" and of varying granularity. In contrast, CCAM uses a controlled vocabulary to describe procedures unambiguously and in a consistent granularity, equivalent or slightly coarser than "synthetic" procedure descriptions of ICHI.

Table 3: Classification architecture characteristics of ICHI and CCAM in comparison

	ICHI	CCAM
Code	Sequential number	4digit multiaxial code, three axes (Catalogue: plus sequential number)
Hierarchy	1. anatomical site/specialty 2. anatomical site, detail 3. procedural type	(Catalogue:) 1. body system. 2. diagnostic/therapeutic 3. anatomical site 4. procedural type or anatomical site, detail
Title/ vocabulary	"clinical" (not controlled) or "synthetic"(controlled)	Basic Coding System: "synthetic" (controlled) Tabular list: clinical, using a controlled vocabulary in standardised order

European Standard EN 1828. ICHI does not claim to be compatible to EN 1828. "Anatomical site", on the other hand, is one of three concepts required for a procedure description, "procedural type" as a quasi-synonym for "surgical deed" another. The third obligate concept, method/approach is not required in ICHI titles.

CCAM procedure model is based on the GALEN model for surgical procedures, which specifies the model of EN 1828 Structure for Coding and Classification of Surgical Procedures: a medical procedure is described by a certain (surgical) deed on a certain anatomical site using a certain method; if necessary, a pathology concept can be used additionally. CCAM procedure descriptions cover all mandatory concepts.

4. DISCUSSION AND CONCLUSION

We found that a complete mapping is possible in a high percentage due to a high degree of similarities, especially concerning anatomical site. Incompatibilities result from undefined or residual classes in CCAM (e.g. "other"), from unspecific descriptions ("procedures for…"), and from missing information to approach / method in ICHI.

Comparing the architectures of ICHI and CCAM Basic Coding System, we found CCAM follows a well-founded architecture and is appropriate to generate classes which are consistent regarding granularity of procedure descriptions and class boundaries. This is an important desideratum for statistical purposes - provided that a total coverage is guaranteed by additional codes for residual classes ("other") [10]. With a universal modelling and an elaborated application guide [4] in combination with the simplicity of the coding system, CCAM facilitates a correct and consistent assignment of codes, which is a precondition for improved data quality. ICHI's hierarchical structure shows a rather well organized dissection of the domain, as well, suited to the needs of a statistical classification. However, these elements are not represented in a well applicable coding system. ICHI titles describe procedures in varying granularity and partially problematic ambiguity; there is also a tendency to group combinations of anatomical sites, even various ones ("of pelvis or hip" next to "of pelvis or femur"). Thus, statistical analysis using ICHI titles is poorly supported.

Considering the ongoing standardization of terminologies and classifications for health care we strongly recommend to improve the structure of ICHI by the benefits of the CCAM architecture (accordance to European Standard EN 1828, multiaxial coding system, detailed definitions/coding guidelines, controlled vocabulary, and last but not least simplicity). This means in detail 1. to adapt and further develop the Basic Coding System (e.g. by adding classes), 2. to assign multiaxial codes for "anatomy-action-approach" to ICHI classes and 3. to eliminate textual ambiguities and inconsistencies by using a controlled vocabulary for procedure description.

Even using such a framework, unresolved problems still persist. The hierarchical structure of the entities which constitute the CCAM axes is not sufficiently accounted for; this especially matters for anatomical site. E.g., the dissection of the body into anatomical regions collides with a dissection into functional systems so that multiple, part-of hierarchies are unavoidable. It may, e.g., create confusion whether to use "head" or "CNS" for a cranial CT scan. Furthermore, the anatomical site axis does not only describe body parts but also body functions (e.g. "sleep", "sensibility"). This is sensible from the pragmatic, but problematic from an ontological point of view.

References

[1] International Classification of Diseases, Ninth Revision, Clinical Modification (ICD-9-CM) http://www.cdc.gov/nchs/about/otheract/icd9/abticd9.htm Last accessed 01/2006

[2] Averill RF, Mullin RL, Steinbeck BA, Goldfield NI, Grant TM. Development of the ICD-10 Procedure Coding System (ICD-10-PCS). 3M HIS Working Paper 5-98, Wallingford. ICD-10-PCS on line: http://new.cms.hhs.gov/ICD9ProviderDiagnosticCodes/08_ICD10.asp Last accessed 01/2006.

[3] CCI Canadian Classification of Health Interventions, Canadian Institute for Health Information. http://secure.cihi.ca Last accessed 01/2006

[4] Classification Commune des Actes Médicaux, Version V0bis, 2001: Guide de Lecture - Système de Codification. CCAM Version V2, 2005, http://www.ameli.fr/77/RUB/77/omb.html. Last accessed 01/06

[5] Trombert-Paviot B Rodrigues JM, Rogers JE, Baud R, van der Haring E, Rassinoux AM, Abrial V, Clavel L, Idir H. GALEN : A third generation terminology tool to support a multipurpose national coding system for surgical procedures. Int J Med Inf 2000; 58-59:71-85

[6] International Classification of Health Interventions (ICHI), Alpha Version (Trial Version). Current (Beta) version on line: http://www3.fhs.usyd.edu.au/ncch/4.1.4.htm . Last accessed 01/2006

[7] CEN EN 1828 Health care informatics, Structure for classification and coding of surgical procedures. Brussels, December 1995

[8] OPS-301 Operationenschlüssel nach § 301 SGB V, Version 2.0 und Version 2.1, DIMDI. http://www.dimdi.de

[9] Hanser S., Zaiss A., Jakob R., Klar R. How to build a well structured procedure classification from an insufficient coding system. In: Fieschi M, Coiera E, Jack LI Y-C (Eds.). Proceedings of the IIth World Congress on Medical Informatics (Medinfo 2004); 7 - 11 September 2004, San Francisco, p.1628. ISBN 158603-4448.

[10] Ingenerf J, Giere W. Concept oriented Standardization and Statistics orientedClassification: Continuing the Classification versus Nomenclature Controversy. Meth.Inform.Med. 1998, 37 (4-5), 527-539

Ubiquity: Technologies for Better Health in Aging Societies
A. Hasman et al. (Eds.)
IOS Press, 2006

ClaML: A standard for the electronic publication of classification coding schemes

E.J. van der HARING[a,1], S. BROËNHORST[b], H. ten NAPEL[c], S. WEBER[b], M. SCHOPEN[b] and P.E. ZANSTRA[a]
[a] *Medical Informatics UMCN, Nijmegen, Netherlands*
[b] *DIMDI, Cologne, Germany*
[c] *WHO-FIC, Bilthoven, Netherlands*

Abstract. This paper proposes a number of revisions to CEN/TS 14463 (ClaML), which is a pre-standard mark-up language for the electronic publication of classification coding schemes. A CEN Taskforce in close collaboration with the WHO network carefully analysed 70 classifications from the healthcare domain. All were transformed in ClaML using a dedicated classification management tool. The proposal removes all formatting elements and adds a number of layout structuring elements. Several elements have been replaced by attributes to enforce internal consistency. A modest number of extensions are proposed to help users and authors in maintenance and version control. A pilot implementation has shown that ICD10 as one of the most complex traditional classifications can be adequately represented to produce quality printed output.
Keywords: Medical Informatics, Classification, ICD-10, Standard

1. Introduction

In 2002 the European Committee for Standardization (CEN/TC251) published a XML [1] based Technical Specification (TS) to represent the content of medical classification systems (ClaML) [2]. The goal of this TS is to simplify the electronic maintenance and publication of classification coding schemes for the use in information systems, and to create a common ground of exchange formats for classifications to enable users to compare data in a convenient and standardized way.

Rossi Mori recognizes three generations of classification schemes: (1) traditional paper-based systems (first generation); (2) compositional systems built according to a categorical structure and a cross-thesaurus (second generation) and (3) formal models (third generation) [3]. First generation classifications are still used widely and will be used far into the future, if only for statistical purposes, whilst third generation classifications have not yet reached maturity [4]. The WHO maintains and publishes a family of first generation classifications schemes (ICD, ICF, ICHI) that is widely used internationally and throughout all branches of medicine. The ICD is one of the oldest medical classification schemes and very important for statistical purposes. Although

[1] Corresponding Author: E.J. van der Haring, MI-UMCN, Nijmegen, Netherlands, e.vanderharing@mi.umcn.nl

originally arising for historical reasons, a continuing and primary characteristic of ICD is that layout of the texts of the rubrics in the classification follows certain rules [5].

WHO has expressed the need for a standard mark-up for their classifications [6]. The format should allow the generation of selections for EDP as well as printing publications from one single source. The WHO collaborating centres network has assessed [7] the suitability of the TS for these purposes. Amongst others the following problems present in the current TS were identified. It is not possible to include meta-information about a classification in the same file. The representation of information elements is not always consistent. The texts can not be uniquely identified. It completely ignores layout of texts within classifications, which makes it very hard to use the same source both for computer applications and printing hardcopy. It contains formatting elements, contradicting with the goal of separating presentation and representation. It lacks constructs that reduce redundancy present within a classification. It is not possible to record the history of changes made to a classification.

The question is if the present TS can be augmented to serve the stated WHO needs. Due to space limitations this paper focuses on the core revisions.

2. Material and Methods

To revise the current TS, CEN convened a taskforce comprising the developers of the TS and interested members of the terminology workgroup. The TS was analysed in order to improve its internal consistency and completeness.

Further inputs to the analysis included 70 classifications held by different organisations already represented to some extent in the current TS and a corpus of experimental representations involving extensions to the current TS [7]. The taskforce also undertook a detailed analysis of several classifications, with the main focus on the ICD-10, in cooperation with the national WHO collaborating centres of the Netherlands and Germany. The formatting and layout of texts within these classifications was carefully analysed and the layout constructs were categorised.

All changes to the TS were discussed within the taskforce and the WHO collaborating centres via email, web forums, and personal communication to reach consensus. Every change was illustrated by examples from ICD-10 and for the main part by experimental implementations in the Kermanog Classification Manager [8].

3. Results

3.1. Meta information

The only metadata about a classification that can be included in the current TS consists of name, title, version and date of publication. The revision extends this with the possibility to include any kind of meta information concerning the classification by the introduction of the element Meta (Figure 1). Although this can not be enforced, it is suggested to format the date of publication according to ISO 8601:2004 [9], i.e. YYYY-MM-DD. For versioning information the well-known major.minor.patch scheme is suggested. The meta information can be easily extended and converted to the format as specified by the Dublin Core Metadata Registry [10].

A number of standards or semi-standards exist defining unique identifiers that may be used to uniquely identify classification coding schemes, for example Health Coding Scheme Designator (HCD) in the CEN standard Registration of Coding Schemes [11] and ISO Object Identifiers [12] used in HL7 Version 3. The revision introduces an optional element CodingSchemeId that may be used to refer to any of these unique identifiers.

A comparison of the header information in the current TS and the proposed revision is shown in Figure 1.

```
<CodingScheme>
   <Name>icd10</Name>
   <Title>International Classification of Diseases, version 10</Title>
   <Version>10.1.0</Version>
   <Date>03-01-2006</Date>
.......................................................................
<ClaML>
   <Meta name="css" value="default.css"/>
   <Meta name="xslt" value="default.xslt"/>
   <Meta name="state" value="in production"/>
   <CodingSchemeId authority="HL7" uid="2.16.840.1.113883.6.3"/>
   <CodingSchemeId authority="CEN" uid="AA123456"/>
   <Title name="icd10" date="2006-01-03" version="10.1.0">
            International Classification of Diseases, version 10
   </Title>
   <Authors>
            <Author id="who">World Health Organisation</Author>
   </Authors>
```

Figure 1. Comparison of header information in the current TS (top) and the proposed revision (bottom).

3.2. Improving consistency and completeness

In the current TS a code in the classification may occur as the content of several elements, for example Symbol, SuperClass, etcetera. The revision introduces the attribute *code* at every place where it is possible to refer to a code in the classification (Figure 2).

```
<Class kind="category">
   <Symbol>A00</Symbol>
   <SuperClass>A00-A09</SuperClass>
   <Rubric xml:lang="en" kind="preferred">Cholera</Rubric>
</Class>
.......................................................................
<Class code="A00" kind="category">
   <SuperClass code="A00-A09"/>
   <SubClass code="A00.0"/>
   <SubClass code="A00.1"/>
   <SubClass code="A00.9"/>
   <Rubric id="_006-0103-1136-4093" kind="preferred">
            <Label xml:lang="en">Cholera</Label>
   </Rubric>
</Class>
```

Figure 2. Comparison of the representation of a class.

Figure 2 also shows the introduction of the element SubClass. In the current TS only the parent classes of a class are explicitly included, whilst subclasses of a class are only implied. The revision proposes to make the subclass also explicit.

3.3. Uniquely identified rubrics

Most classifications contain different kinds of rubrics. Usually each class has a preferred meaning, which may be further defined by inclusions and exclusions, or be explained by notes, definitions, descriptions, etcetera. In the current TS the different kinds of rubrics are introduced when they are first used. The revision adds an element RubricKinds at the beginning of the file, where each kind of rubric is defined, and only allows rubric kinds that are defined within the RubricKinds element.

In the current TS a rubric may be represented in different languages, indicated by the attribute *xml:lang*. However, those rubrics that are mutual translations can not be explicitly related to each other, as in the excerpt from ICD-10 in Figure 3. The introduction of the element Label, containing the text in a specific language for the rubric, allows the grouping of translations of the same rubric.

```
<Class kind="category">
  <Symbol>A04</Symbol>
  <SuperClass>A00-A09</SuperClass>
  <Rubric xml:lang="en" kind="preferred">Other bacterial intestinal infections</Rubric>
  <Rubric xml:lang="en" kind="excludes">foodborne intoxications, bacterial (A05.-)</Rubric>
  <Rubric xml:lang="en" kind="excludes">tuberculous enteritis (A18.3)</Rubric>
  <Rubric xml:lang="de" kind="preferred">Sonstige bakterielle Darminfektionen</Rubric>
  <Rubric xml:lang="de" kind="excludes">Bakteriell bedingte Lebensmittelvergiftun..</Rubric>
  <Rubric xml:lang="de" kind="excludes">Tuberkulöse Enteritis (A18.3)</Rubric>
</Class>
..........................................................................
<Class code="A04" kind="category">
  <SuperClass code="A00-A09"/>
  <Rubric id="_006-0103-1532-0589" kind="preferred">
          <Label xml:lang="en">Other bacterial intestinal infections</Label>
          <Label xml:lang="de">Sonstige bakterielle Darminfektionen</Label>
  </Rubric>
  <Rubric id="_006-0103-1532-3700" kind="excludes">
          <Label xml:lang="en">foodborne intoxications, bacterial (A05.-)</Label>
          <Label xml:lang="de">Bakteriell bedingte Lebensmittelvergiftungen (A05.-)</Label>
  </Rubric>
  <Rubric id="_006-0103-1533-1825" kind="excludes">
          <Label xml:lang="en">tuberculous enteritis (A18.3)</Label>
          <Label xml:lang="de">Tuberkulöse Enteritis (A18.3)</Label>
  </Rubric>
</Class>
```

Figure 3. Representation of rubrics in different languages in the current TS and the revision. In the revision the translations are explicitly related and rubrics are (optionally) uniquely identified.

3.4. History

During the development and maintenance of a classification it is important to document all changes, including when the change was made, who made the change and for what reason. For this purpose the revision introduces an element History, which records the author and date of a change. Again for the date it is suggested to use the format as defined in IS0 8601:2004 [9].

3.5. Layout

The revision introduces a number of elements allowing layout information within rubrics to be specified, including paragraphs, tables, lists and fragments. The first three are self-explanatory and are not further described. The element Fragment may be used to represent 'repetition'-layout that is especially seen in ICD-10 (Figure 4).

A06.8 Amoebic infection of other sites
Amoebic:
· appendicitis
· balanitis+ (N51.2*)

....

```
<Class code="A06.8" kind="category">
   <SuperClass code="A06"/>
   <Rubric id="_006-0105-1030-1314" kind="preferred">
               <Label xml:lang="en">Amoebic infection of other sites</Label>
   </Rubric>
   <Rubric id="_006-0105-1030-3423" kind="includes">
               <Label xml:lang="en">
                           <Fragment type="lhead1">Amoebic</Fragment>
                           <Fragment type="litem">appendicitis</Fragment>
               </Label>
   </Rubric>
   <Rubric id="_006-0105-1031-3611" kind="includes">
               <Label xml:lang="en">
                           <Fragment type="lhead1">Amoebic</Fragment>
                           <Fragment type="litem" usage="etiology">balanitis</Fragment>
                           <Reference usage="manifestation">N51.2</Reference>
               </Label>
   </Rubric>
</Class>
```

Figure 4. Example from ICD-10 illustrating layout in ICD-10. (The fragment attribute *usage* is used here to describe dagger/asterisk. This attribute is outside the scope of this paper)

4. Discussion

Hoelzer et al [15] describe an extension to the current TS. Unfortunately their effort has not been contributed to CEN and seems to have been used for their own experiments and applications only. Although they published a paper about their approach, the full specification of their extension is not publicly available. A disadvantage of their approach is that it divides the contents of the ICD-10 between more than 1600 XML documents, greatly increasing the risk of consistency problems and undermining the exchange of classifications, which is the main purpose of the standard.

In this proposal several elements from the TS have been replaced by attributes in order to make the specification more internally consistent. The revision also removes the two formatting elements, <i> and <b> and instead introduces a number of layout structuring elements. Subsequent formatting of these elements for display should be achieved by using separately defined style sheets, for example Cascading Style Sheets [13] or XSLT [14]. The layout structure of texts has been added in such a way that each text is meaningful even when the layout elements are ignored.

Although the subclasses of a class may be derived by examination of other classes, the order in which those subclasses should appear can not be. This is especially a problem with multi-hierarchical classifications in which a class has multiple superclasses. In the revision the element Class contains the complete environment of the class in the classification, with both immediate super- and subclasses explicitly stated, including ordering. Further, making the complete environment of a class explicit means the information contained within the element Class is sufficient to completely define a class.

5. Conclusion

Relatively modest revisions in the TS have proven to drastically improve expressive power and internal consistency of the representation. First experimental implementations of the revised TS show that it can represent the layout of texts within the ICD-10 and format these comparable to bookformat.

References

[1] Extensible Markup Language (XML) 1.0 (Third Edition). W3C 2004 February, http://www.w3.org/TR/2004/REC-xml-20040204

[2] CEN/TS 14463 Health Informatics - A syntax to represent the content of medical classification systems. Brussels: CEN; 2002.

[3] Rossi Mori A, Consorti F, Galeazzi E. Standards to support development of terminological systems for healthcare telematics . Methods Inf Med 1998 Nov;37(4-5):551-63.

[4] Zanstra PE, van der Haring EJ, Cornet R. Introduction of a Clinical Terminology in The Netherlands - Needs, Constraints, Opportunities. nictiz 2003, http://www.nictiz.nl/kr_nictiz/uploaddb/downl_object.asp?atoom=2128&VolgNr=1

[5] History of the development of the ICD. WHO 2000, http://www.who.int/entity/classifications/icd/en/HistoryOfICD.pdf

[6] WHO business plan for medical classifications 2005-2010. WHO 2005, http://www.who.int/classifications/BuisinessPlan.pdf

[7] ten Napel H, van der Haring EJ. Dutch ICD-10 and ICF in a CEN Technical Standard Format for version control and maintenance. WHO 2003 October, http://www.rivm.nl/who-fic/Colognepapers/cologne35.rtf

[8] Classification Manager. Kermanog 2005, http://www.kermanog.com/clam/index.html

[9] ISO TC154. Data elements and interchange formats -- Information interchange -- Representation of dates and times. Geneva: ISO; 2004

[10] The Dublin Core Metadata Registry. Dublin Core Metadata Initiative 2006, http://dublincore.org

[11] CEN EN 1068:2005 Health Informatics - Health care Information Interchange - Registration of Coding Schemes. Brussels: CEN; 2005.

[12] Mealling M. A URN Namespace of Object Identifiers, RFC 3061. NN 2001 February, http://rfc.sunsite.dk/rfc/rfc3061.html

[13] Cascading Style Sheets, level 1, W3C Recommendation. W3C 1999 January 11, http://www.w3.org/TR/1999/REC-CSS1-19990111

[14] Extensible Stylesheet Language (XSL) Version 1.0. W3C 2001 October, http://www.w3.org/TR/2001/REC-xsl-20011015/

[15] Hoelzer S, Schweiger R, Dudeck J. Transparent ICD and DRG coding using information technology: linking and associating information sources with the eXtensible Markup Language. J Am Med Inform Assoc 2003;10(5):463-9.

7.5 Concepts and Coding: Systems

Ubiquity: Technologies for Better Health in Aging Societies
A. Hasman et al. (Eds.)
IOS Press, 2006

The International Classification of Primary Care (ICPC-2): an essential tool in the EPR of the GP

Marc Verbeke M.D.[1], Diëgo Schrans M.D., Sven Deroose M.D., Jan De Maeseneer M.D. Ph.D
Department of General Practice and Primary Health Care, Ghent University, Belgium

Abstract. The International Classification of Primary Care (ICPC) has become a standard all over the world. It became a standard tool to classify the important elements in the Electronic Patient Record (EPR) of the GP: reasons for encounter (RFE) reflecting the patient's view, process of care (decision, action, intervention or plans) reflecting the care process, and the assessment (diagnosis or health issue) reflecting the doctor's view. ICPC-2 is fully compatible with structuring data in the episode of care model and it's reflecting the essential elements of each patient/provider encounter. To implement ICPC-2 in the EPR a Thesaurus has been developed in Belgium with double encoded clinical labels. The implementation is now mandatory for labeled EPR systems in Belgium. The use of ICPC 2 ay improve the accessibility and use of on-line Expert systems and Guidelines.
Keywords: Medical Informatics, ICPC, primary care, thesaurus, medical record.

1. Introduction

The International Classification of Primary Care (ICPC), developed by the ICPC Working Party, broke new ground in the world of classification when it was published for the first time in 1987 by WONCA [6], the World Organization of National Colleges, Academies, and Academic Associations of General Practitioners/Family Physicians, now known more briefly as the World Organization of Family Doctors (Wonca). For the first time health care providers could classify, using a single classification, three important elements of the health care encounter: reasons for encounter (RFE), diagnoses or problems, and process of care. Problem orientation of the medical record and linkage of encounters over time enables classification of the episode from the beginning with an RFE to its conclusion with a more defined problem, diagnosis, or disease.

[1] Corresponding Author: M.Verbeke UZ.1k3, De Pintelaan 185, B9000 Gent, Belgium. Marc.verbeke@Ugent.be.

2. The choice of the International Classification of Primary Care (ICPC)

A classification relevant to primary care is necessary. It should be logically organized to:

- reflect the concerns of patients (not the immediate assumptions or beliefs of their doctors)
- characterize episodes of care
- be easy to routinely use
- be inexpensive
- be relevant to any primary care setting
- be expandable
- be connected to existing coding and other classification approaches
- be readily deployed in the evolving electronic health record.

ICPC with its linkage to ICD-10 which is sanctioned by World Health Organization and included in the US National Library of Medicine's Unified Medical Language System (UMLS) is the only existing classification scheme that meets these standards [9], [10].Using ICPC with it's ordering principle in a defined domain and based on the high prevalence of common diagnoses in family practice, gives much more power and possibilities in care management, than could ever be reached by using only terminologies or vocabularies as SNOMED or Read codes. Another reason to prefer ICPC is the international recognition and the inclusion in the WHO-FIC (WHO Family of International Classifications). ICPC acts as the link between other classifications (ICD, ICF, ICHI, ATC) in the FIC.

2.1. History and future

The first version of ICPC published in 1987 is referred to as ICPC-1 [6]. In 1998 Wonca published a revised version: ICPC-2 [7] with inclusion and exclusion criteria attached to the classification rubrics, and a mapping to ICD-10. ICPC-2e is an electronic version from year 2000 of the revised and corrected chapters 10 and 11 of the ICPC-2 book [13]. In 2003 WHO recognized ICPC-2 as a WHO related classification for the recording of data in primary care.

A new version numbering system was introduced by the WICC Update group in November 2005. The latest version of ICPC-2e will always be available at the web site: http://www.kith.no/templates/kith_WebPage____1062.aspx [2]. Information about ICPC history, background, member countries etc. can be found on the WICC web pages [1].

ICPC-3 will be developed in the coming years: nothing has been decided yet but there are some accepted points of discussions:

- ICPC structure will probably be maintained due to relationships with ICD within WHO-FIC
- including patient preferences
- mapping to ICF and/or including new rubrics for this?
- process codes to be revised and maybe including objective findings and investigation
- revising prevention rubrics

- mapping to other classifications?

2.2. Structure of ICPC

ICPC is based on a simple bi-axial structure: 17 chapters based on body systems on one axis, each with an alpha code, and seven identical components with rubrics bearing a two-digit numeric code as the second axis.

Component 1 provides rubrics for symptoms and complaints. Rubrics in this component can be used to describe presenting symptoms, and are valuable for describing the problem under management when the condition is as yet ill defined. Component 7 is the diagnosis/disease component in each chapter. This component will be the one most often used when you have sufficient information to arrive at a diagnosis in the medical record. Within this component 7 are five subgroups, which are not numerically uniform across chapters: infectious diseases, neoplasms, injuries, congenital anomalies, and other diseases. Components 1 and 7 in ICPC function independently in each chapter and either can be used to code patient RFEs, presenting symptoms, and diagnoses or problems managed. Components 2-6 (process-codes) are common throughout all chapters, each rubric being equally applied to any body system.

The structure of ICPC represents a move away from the combined anatomical and etiology based structure of ICD. For example, where ICD includes a separate chapter for neoplasm's, one for infections and infestations, and another for injuries, such problems are distributed among chapters in ICPC, depending on the body system to which they belong.

3. Belgian Bilingual Thesaurus and clinical labels

Primary care physicians have an immediate need for a simple and honest way to routinely record and retrieve data reflecting their perspective. The primary care perspective must be incorporated into the nation's data standards and electronic health records. Clinical research and a fully integrated health information system cannot be sustained without practical, easily used primary care data standards.

Because of the mandatory implementation of ICPC/ICD classification systems within the labeling procedure in 2006 for Belgian software systems in primary care, a user-friendly instrument to encode data had to be created in order to make classification systems acceptable and useful. Implementing only a classification without search instruments would never be accepted by family physicians who don't know anything about coding systems. Coding data in the 'background' of the EPR and showing only the clinical label as the doctor wants it, is the best way to make it acceptable. For this reason, under the authority of the federal government, a bilingual Belgian Thesaurus has been developed in use of GP's reporting and encoding the data in the EPR. A fruitful collaboration of two university departments of Primary Care (Ghent University Ugent for the Dutch part and Brussels University ULB for the French part of the Thesaurus) has been installed for this purpose. Implementation of the Thesaurus with double encoding system is mandatory for the labeling of the EPR system by the government in 2006 [5].

The Belgian labeling system is a way to guarantee quality, inter-operability, communication possibilities and standards in the EPR systems in combination with a free market of software systems in Health Care. The doctor gets financial incentives for using a software system that has been labeled by the government.

3.1. Position of a Thesaurus in reporting and organizing data in the EPR

A Thesaurus creates a bridge between the words (or in the EPR: parts of words) in the GP's thoughts in describing symptoms, complaints, assessment or other data to be reported in the patient record and the scientific correct way of describing clinical concepts, used in organizing the record-system with all it's possibilities of communication, research and other management issues.

Translating steps:

- doctor's vocabulary linked to
- terminology of clinical concepts used in reporting data in EPR and linked to
- clusters of concepts linked to
- classification systems:
  - primary care ICPC
  - ICD if more granularity is needed
- Organization of EPR (prevention, risk management, expert systems etc.) based on these classifications in use.
- Classifications linked to guidelines, research, administration, insurance systems and communication with other health care providers

3.2. Development of a Thesaurus

The starting point of this Belgian Thesaurus was a Thesaurus created in Amsterdam by the department of Prof. H. Lamberts (UvA), in collaboration with Ghent University, Department of General Practice and Primary Health Care [11]. But this Thesaurus has been refused as such by developers and users especially because of the lack of clinical labels useful as text in the EPR. Search terms and ICD/ICPC labels (the content of the Amsterdam Thesaurus) are not always the preferred text in a medical record. The Belgian GP needs also a perfect bilingual system by which all information (codes and labels) became completely interexchangeable from one language to another (French/Dutch) and from one software system to another.

For this reason each collection of search terms (doctor's vocabulary) has been linked to a specific 'clinical label' (terminology of clinical concepts) useful as terminology in the EPR and to the appropriate pair of classification codes (clusters of concepts): ICPC-2-ICD10. Each clinical label is perfectly bilingual French/Dutch. The search terms are specific but different for each language. The classification mapping is according the rules of the Wonca International Classification Committee (WICC) [8].

In order to avoid as much as possible "double or identical labels" that could be linked to different pairs of codes, we have rewritten the labels in a way each description, each label has been formulated according the content of the mapped ICD /ICPC codes. This system improves the accuracy of the chosen label and mapped codes.

In ICPC chapter X and Y (female and male genital systems) a lot of labels are identical, but the EPR can know which one has to be taken according patients gender.

According to ICPC philosophy we prefer to encode diagnosis as 'manifestation' instead of 'etiology'. Example: 'tuberculous prostatitis' has been mapped to ICPC/ICD Y99/N51.0 (Genital disease male other/Disorders of prostate in diseases classified elsewhere) but not to A70/A18.1 (Tuberculosis/Tuberculosis of genitourinary system). This point of the Thesaurus has to be evaluated by the users to know if both viewpoints (etiology and manifestation) have to be implemented or not. The disadvantage to keep both would be two identical 'clinical labels' expressing the same concept but with different codes. Within the context of clinical research, this could be an important issue.

3.3. Results [4]

About 49.000 different clinical labels have been created. Most of them are unique and compatible with the content of the appropriate pair of ICPC/ICD codes. By using this system the GP has a vocabulary and terminology system at his disposition, which encodes the reported data automatically without the GP necessarily knowing the codes. This encoded information is useful to organize patient record systems for prevention (inclusion and exclusion criteria to a specific preventive program), risk calculation, bi-directional communication with other health care providers and even expert systems as access to guidelines, working flow charts, assistance in prescriptions etc.

Besides this creation of a bilingual and double encoded Thesaurus with added clinical labels, expert systems have been indexed with ICPC codes to make it 'on-line' accessible by using the assessment (the 'A' of SOAP) codes from the EPR. This makes guidelines accessible with a language independent system, avoiding typewriting errors and incorrect search terms. The Belgian GP can use the French or Dutch search terms from the Thesaurus to search in English guidelines, in this case: the 'EBM Guidelines' from Duodecim in Finland [3]. Even process codes (in 'O' or 'P' of SOAP) can be the trigger of access to the guidelines. The key for this link is the used ICPC code, but maybe in the future also the mapped ICD code could be used. A correct granularity offered by ICPC and the mapping to ICD is an essential tool to organize this, and can hardly be done, if not impossible by using only terminologies, vocabularies or ICD.

Also procedures are encoded with a list of about 1500 different most common procedures done or ordered by the GP. In ICPC component 2-6 all codes are chapter-independent. For this reason the same rule has been respected and all of the 1500 procedures has been added chapter independent as extensions of the ICPC process codes *30 to *69. The encoded procedures are very helpful in organizing the process of taking care by the GP and in Belgium the GP has certainty that all procedures reimbursed to the patient and reported mandatory in his EPR, are within that list. This makes the list of procedures quite adapted to the Belgian situation, but could very easily be changed to fit for other countries where GP's are doing different things.

3.4. Future developments.

A thesaurus useful in daily practice will never be finished: medical concepts are changing, new concepts are developing, other information can become more important, language is dynamic and GP's are dynamic, search terms will be missing and errors will be detected, classifications will be updated, other classifications could be added (i.e. ICF: International classification of functioning, disability and health) [12] and so on, so the Thesaurus will always be changing.

Other classification systems will be added in communication, administration and organization of health care. The Thesaurus is only a part of the tool the GP needs within the care process, but it's an important tool in the organization of the EPR and it improves the possibility to a better quality of care. The better quality of encoded data means also an improvement of epidemiologic data and scientific studies in primary care in Belgium.

3.5. Conclusion

A Thesaurus is a good instrument to give GP's the freedom of reporting in the EPR the data in their own language and words without loosing all the opportunities and help in the management of taking care offered by language independent classification and coding systems. Even GP's who are 'allergic' to codes and classifications are pleased by the extra value offered by using a Thesaurus.

Acknowledgment

The development of the Belgian Thesaurus has been made possible by the grant of the Belgian Federal Administration of public health.

References

[1] http://www.globalfamilydoctor.com/wicc/icpcstory.html
[2] http://www.kith.no/templates/kith_WebPage____1062.aspx
[3] http://www.ebm-guidelines.com (last accessed Sept. 30, 2005).
[4] www.health-telematics.be/thesaurus.
[5] https://portal.health.fgov.be/portal/page?_pageid=56,4280434&_dad=portal&_schema=PORTAL&_menu=menu_5_6_1
[6] Lamberts H, Wood M, eds. ICPC. International Classification of Primary Care. Oxford: Oxford University Press, 1987.
[7] ICPC-2. International Classification of Primary Care. Second edition. Oxford: Oxford University Press, 1998.
[8] International Classification of Primary Care ICPC-2-R, Revised second edition, WONCA International Classification Committee, Oxford University Press 2005; ISBN 978-019-856857-5
[9] Banff Declaration calls for ICPC in the United States. Wonca News, April 2004. Okkes IM.
[10] The Banff declaration and the information needs of Primary Care and Family Medicine in the USA. http://annalsfm.highwire.org/cgi/qa-display/short/annalsfm_el;48#175, 16 Dec 2003
[11] Becker HW, Van Boven C, Oskam SK, Hirs WM, Lamberts H, Verbeke M, et al, eds. ICPC2-ICD10 Thesaurus. Version May 2003. Amsterdam, Den Haag etc: Academic Medical Center/University of Amsterdam, Ministry of Health, Dutch College of General Practitioners, Dutch Institute for Public Health and Environment RIVM, Free University of Brussels, University of Ghent, 2003.
[12] ICF publication by WHO: International classification of functioning, disability and health, ICF. Geneva 2002
[13] Okkes IM, Jamoulle M, Lamberts H, Bentzen N. ICPC-2-E. The electronic version of ICPC-2. Differences with the printed version and the consequences. Fam Pract 2000; 17: 101-6.

Ubiquity: Technologies for Better Health in Aging Societies
A. Hasman et al. (Eds.)
IOS Press, 2006

Using SNOMED CT Codes for Coding Information in Electronic Health Records for Stroke Patients

Judith van der Kooij[a,1], William T.F. Goossen[a], Anneke T.M. Goossen-Baremans[a], Marinka de Jong-Fintelman[a] and Lisanne van Beek[a]
[a] *Acquest, Koudekerk aan den Rijn, the Netherlands*

Abstract. For a project on development of an Electronic Health Record (EHR) for stroke patients, medical information was organised in care information models (templates). All (medical) concepts in these templates need a unique code to make electronic information exchange between different EHR systems possible. When no unique code could be found in an existing coding system, a code was made up. In the study presented in this article we describe our search for unique codes in SNOMED CT to replace the self made codes. This to enhance interoperability by using standardized codes. We wanted to know for how many of the (self made) codes we could find a SNOMED CT code. Next to that we were interested in a possible difference between templates with individual concepts and concepts being part of (scientific) scales. Results of this study were that we could find a SNOMED CT code for 58% of the concepts. When we look at the concepts with a self made code, 54,9% of these codes could be replaced with a SNOMED CT code. A difference could be detected between templates with individual concepts and templates that represent a scientific scale or measurement instrument. For 68% of the individual concepts a SNOMED CT could be found. However, for the scientific scales only 26% of the concepts could get a SNOMED CT code. Although the percentage of SNOMED CT codes found is lower than expected, we still think SNOMED CT could be a useful coding system for the concepts necessary for the continuity of care for stroke patients, and the inclusion in Electronic Health Records. Partly this is due to the fact that SNOMED CT has the option to request unique codes for new concepts, and is currently working on scale representation.
Keywords: Controlled Vocabulary, Coding System, Medical Records Systems, Electronic Health record, Medical Informatics, SNOMED CT, Electronic Messages, HL7 v3

1. Introduction

Health care is a domain for which information plays an important role. This goes not only for registration and declarations, but even more for the direct care for patients. Especially for the documentation of patient records there is no univocal universal language. At this moment classifications that are meant for classifying diseases or treatments are used for

[1] Corresponding Author: Judith van der Kooij, Dorpsstraat 50, 2396 HC Koudekerk aan den Rijn, the Netherlands, acquest@acquest.nl

various purposes, both administrative and clinical. Most of the time, these classifications are not developed for the registration of direct care for patients, but more for statistics and billing. For that reason, those classifications are not always suitable for the registration of daily care observations and activities.

Under the authority of NICTIZ, the abbreviation of the Dutch name for the National IT Institute for Healthcare [1], a study has been conducted with the main goal to determine the usefulness of a broad introduction of a clinical terminology in The Netherlands. In the report of this study a clinical terminology has been described as: the collection of standard terms with their synonyms, which can be used in direct patient care to record all symptoms, circumstances, interventions, diagnosis, results and the decision making [1]. To make information technology work in health care one needs to accept standardisation of the data, as well as the vocabulary and the electronic messages, next to security of medical information. In the Netherlands, NICTIZ takes responsibility for this [2].

According to NICTIZ standardisation is necessary to realize clinical data exchange between different Electronic Health Record (EHR) systems, all with their own characteristics. NICTIZ chose Health Level Seven version 3 (HL7 v3) [3] to be the standardisation methodology for messaging. In several projects in which HL7 v3 was used, we identified the need for an additional coding system, next to existing coding systems, because unique codes were needed. At first we decided to invent new codes as a temporary solution to this problem; however, this is an undesirable situation in the long run, due to maintenance and interoperability issues. For this reason an explorative study was carried out for using SNOMED CT codes as standardized clinical terminology

SNOMED Clinical Terms (SNOMED CT) is a dynamic, scientifically validated clinical health care terminology and infrastructure [4]. By applying SNOMED CT coding, data can be captured, shared and aggregated in a consistent way across specialities and domains of care. Due to the SNOMED CT Core clinical terminology, SNOMED CT can be used for electronic medical records, ICU monitoring, clinical decision support, medical research studies, clinical trials, computerized physician order entry, disease surveillance, image indexing and consumer health information services [4].

In this study we tried to replace the present codes, self made or from existing coding systems, with SNOMED CT codes in the care information models we created for an EHR for the continuity of care of stroke patients. A care information model is often also referred to as a template [3, 5]. The care information models, or templates, that were developed for the EHR for stroke patients, integrate knowledge, terminology and coding, and information models, with HL7 v3 as the underlying standardisation method [5, 6]. These models were derived from (paper) records given to us by the care professionals that were involved in the project for the EHR for stroke patients. After creating these templates they were shown to the care professionals for validation and feedback. If necessary they were corrected so they would correspond to the domain of stroke.

The goal of this study was to explore the usefulness of SNOMED CT for uniquely coding the medical information in the templates for the stroke domain. This is part of the ongoing development of standards, messages and the EHR for stroke patients [7]

The following research questions were formulated:

1. For how many of the codes of the clinical concepts in the care information models can we find unique SNOMED CT codes?
2. Is there a difference for individual concepts and concepts representing scientific scales or measurement instruments with clinimetric characteristics?

The results of the study were not only quantitative results but also qualitative results and recommendations. An important recommendation is to perform 'real life' experiments with SNOMED CT in an EHR and in HL7 v3 messages. As a result of this, one can underpin decisions about the introduction of SNOMED CT as the national standard terminology for coding in EHR and HL7 v3 messages.

2. Method

2.1. Background

We were involved in a project on creating an Electronic Health Record (EHR) for the complete chain of care for stroke patients. We gathered all the information that is recorded for the stroke patients in the present situation. Then this information was organised. This organisation resulted in 84 templates. In such templates (validated) scales or instruments, observations or actions are described in detail [5]. These represent best practice, are Health Level 7 compliant, support the uptake of standardized terminologies and facilitate technical implementation in both electronic messages and clinical information systems.

One of the paragraphs of the templates describes the mapping table from the domain to the HL7 Reference Information Model and message models. In this mapping table all items that are recorded in an EHR receive a unique code. A unique code is needed to exchange the information with other systems; this is called semantic interoperability [3]. Preferably these unique codes would be adopted from existing coding systems, so first we tried to look for an appropriate code in IDC10, ICF and ICNP, in order of preference. Next to that, LOINC was also searched for unique codes. However, most items could not get a unique code from these existing coding systems. This resulted in self made codes, which are unique but do not correspond to an existing coding system, are difficult to maintain, and, due to lack of standardization, will only support partial interoperability.

White and Hauan [8] discuss the way the LOINC coding system can adequately represent instruments and scales. They argue that a particular important aspect of coding is to maintain the psychometric or clinimetric properties of instruments and scales Although this work was carried out with another coding system, their criterion refers to the domain content and therefore, we believe, is relevant for any coding system applied: the meaning of the concept in the scale should precisely be represented in the wording and in the coding.

After creating the 84 templates, NICTIZ wanted to test if SNOMED CT codes could replace the self made codes with standardized codes. There have been other studies carried out to test the breadth of SNOMED CT [9, 10, 11].

In the study of Campbell et al. [9] three potential sources of controlled clinical terminology were compared (READ codes version 3.1, SNOMED International, and Unified Medical Language System (UMLS) version 1.6) relative to attributes of completeness, clinical taxonomy, administrative mapping, term definitions and clarity (duplicate coding rate). The authors assembled 1929 source concept records from a variety of clinical information taken from four medical centres across the United States. The source data included medical as well as sample nursing terminology. The study showed that SNOMED was more complete in coding the source material than the other schemes, because SNOMED covered 70% compared to READ covering 57% and UMLS 50%. From this study it could be concluded that SNOMED was more complete, had a compositional nature and a richer taxonomy.

Chute et al. [10] reported a similar result when evaluating major classifications for their content coverage. For their study the clinical text from four medical centres was sampled from inpatient and outpatient settings. This resulted in 3,061 distinct concepts. These concepts were grouped into Diagnoses, Modifiers, Findings, Treatments and Procedures, and Other. Each concept was coded into ICD-9-CM, ICD-10, CPT, SNOMED III, Read V2, UMLS 1.3, and NANDA. When coding the concept, the reviewers also scored the concepts: 0 = no match, 1 = fair match, 2 = complete match. Result of this study was that SNOMED had a broader coverage than any of the other coding systems used in this study. SNOMED received the highest score in every category, including Diagnoses (1.90), and had an overall score of 1.74.

Wasserman and Wang [11] found a concept coverage of 88,4% when evaluating the breadth of SNOMED CT terms and concepts for the coding of diagnosis and problem lists within a computerized physician order entry (CPOE) system. When they took the relevance of the 145 terms that could not be coded with SNOMED CT into account, they could even conclude that the concept coverage of SNOMED CT was 98,5%.

Although these three studies [9, 10, 11] showed that SNOMED CT has a rather high concept coverage, we expected to find SNOMED CT codes for about half (50%) of our self made codes. This 50% assumption was based on our experiences with looking at other existing coding systems used for these templates and the vast amount of self made codes that where necessary for all clinical details for stroke patients during their full episode of care [7].

2.2. Description of the study

From the American College of Pathologists we received a licence for the project, material about the structure, the contents of SNOMED CT and instructions on how to search for concepts, terms and their corresponding codes. After studying this material, all items from the mapping tables of the templates have been systematically searched for in SNOMED CT. We decided not only to search for codes for items that had a self made code, but also for items that had a code from another coding system, like ICD 10. We did this to find out how complete SNOMED CT is for our purposes.

To make sure we used SNOMED CT in the right way, we used the knowledge of experts. Like, for example, the document on how to search in SNOMED CT written by Casey [12]. Based on this document we developed a search strategy; this strategy was as follows:

1. translate the existing Dutch concepts in the mapping table in the templates into English (this was done, as a requirement from NICTIZ, during the construction of the templates);
2. start searching with the translated concept, as mentioned in the mapping table of the templates;
3. when there is no perfect match, search for a concept on the SNOMED CT hierarchical levels above;
4. when there is no perfect concept, search with synonyms;
5. when still no perfect match can be found, search using the SNOMED hierarchy from top down.

When we look at the care information model for body temperature, for example, we reported the following search strategy. The model for body temperature consists of two concepts: body temperature and method of measuring. For the first concept the term 'body temperature' was entered in SNOMED. This resulted in a hit: 386725007: body temperature. This resulting term fully represented the concept, so the SNOMED CT code was accepted. Then the second concept, method of measuring, was entered in SNOMED. This generated the following result: 371911009: measurement of blood pressure using cuff method. This was not the right concept. The hierarchical levels that lied above were also about blood pressure, so no result could be found via this strategy. Then we entered a kind of synonym: measurement of body temperature. This only generated a concept related to ovulation, which was not what we were looking for. The next step in our strategy was to search top down. For this we entered the following terms respectively: body temperature, measurement and method. All three remained without a result. So, no SNOMED CT code could be found for 'method of measuring' of body temperature.

For every template a short report has been made. In this report, for every item, it was reported what the search results were and what result was chosen as the right one for the concept put in the search, together with a motivation for choosing this SNOMED CT concept. Next to that, other remarks on the search or the search results were reported. When in a search a synonym was needed, this synonym and its accompanying search results were reported. An expert on medical terminology reviewed the complete search report and indicated which search results should be accepted and which should be rejected. Next to this, the expert also corrected codes by advising to use another SNOMED CT code. Sometimes neurologists were approached to come up with synonyms for a certain item, which apparently was also not clearly formulated in the original Dutch wording.

The final SNOMED CT codes were added to the mapping table in the templates, so the original codes, self made or from another coding system and the SNOMED CT codes can be used next to each other. The adjusted templates have been sent to Portavita, a company that works on the SNOMED CT implementation of the EHR for stroke patients for DWO in Delft, The Netherlands. This information system is an EHR for all health professionals involved in the care of a stroke patient. This means that all (medical) information that comes from the GP, the hospital, the rehabilitation centre, the nursing home, and home health care is put in one record. When this system is implemented, the messages, containing patient related information, will be sent from system to system, while using SNOMED CT codes next to the original codes. This way the applicability of SNOMED CT in the EHR and in HL7 v3 messages will be tested. This will also lead to another report in the near future.

3. Results

For this research 84 templates need to be coded. From these 84 models, up to today, 32 have been coded. These 32 care information models that have been coded contained 207 concepts or items. From these 207 items 87 could not be coded with SNOMED CT (42%). For 120 (58%) a SNOMED concept ID code has been found. For 178 (88,1%) items there was agreement either for the code that was found or for the fact that no code could be found.

For this result a remark must be made. The items in the care information models are grouped in a HL7 v3 way: by using Organizers and Batteries. For example: the nursing assessment contains 2 Organizers, nursing record and decubitus. The nursing record contains 7 templates, each with its own items and Batteries to group these items. SNOMED CT does not support this grouping principle, so none of the Organizers and Batteries can be coded with SNOMED CT. And although these Organizers and Batteries all need a unique code we can not expect SNOMED CT to add these codes because they do not represent medical information, just grouping of medical information. For this study we decided not to code Organizers and Batteries and to leave them out of the calculation for the results.

When we make a distinction between self made codes and codes from an existing coding system we see the following result. For the 164 self made codes 90 could be coded with SNOMED CT (54,9%). For the 43 codes from an existing coding system, like ICD 10, 32 could be coded with SNOMED CT (74,4%).

To answer our second research question we also looked at the difference between SNOMED CT codes found for templates with individual items and templates representing scientific scales or measurement instruments. The results are presented in Table 1.

Table 1: Difference between SNOMED CT codes found for care information models with single items and care information models representing scientific scales.

Amount of care information models	**Amount of concepts**	**Amount of SNOMED CT codes**	**Percentage of SNOMED CT codes**
6 scientific scales	50	13	26%
26 models with individual concepts	157	107	68%
32 in total	207	120	58%

Next to these quantitative results we would also like to report some relevant qualitative findings. First, items of the care information models which exist of two (or more) combined concepts do not always correspond with one concept ID within SNOMED CT. Our solution was to report all the separate codes. For example, when examining the family history of a stroke patient one wants to know if stroke at an early age runs in the family. For this concept we needed three SNOMED CT codes: one for stroke, one for age and one for young. So sometimes concepts correspond to two (or more) SNOMED CT concept ID's; the question is whether it is allowed to combine two (or more) separate ID's.

Second, the terminology used in the clinical area can not always be found in SNOMED CT. In this case we tried to find a concept ID that represents the concept that lies behind the terminology used by the care professional. Third, the English translation of the Dutch items did not always result in a SNOMED CT code. Then we tried synonyms. Fourth, items with a left or right indication, like 'fingers extensors left', can not always be found in SNOMED CT. We solved this problem by using the attribute in an HL7 v3 class in which, for instance, location can be entered. However, this does not solve the terminological issue. On the other hand, this makes the addendum left and right superfluous in the terminology. For example, 'finger extensors left' and 'finger extensors right' can be replaced by just one item, namely 'finger extensors', and the left and right are covered in the information model. Fifth, as mentioned earlier, the Organizer and Battery concepts are not supported by SNOMED. We decided not to search for SNOMED codes for the Organizer and Battery concepts anymore. Sixth, the degree of detail within SNOMED CT is very different. Some concepts are coded within detail and other concepts just have one code for the concept itself and no codes for the underlying details of the concept. For example, regularity and constancy of breathing are common observations for stroke patients, however no concept ID's and codes can be found in SNOMED CT. Yet, frequency and profound breathing can be found. Seventh, the items in the care information models are mainly observations, in SNOMED these need to be observable entities. Though, most items can only be found as clinical findings instead of observable entities. Eighth, in SNOMED CT the use of stimulants, like alcohol, marihuana, and cigarettes is defined as abuse of these stimulants although the use of stimulants is not always considered as abuse, if used with care. Using SNOMED CT here would imply a wrong meaning: use is intended to document, and abuse is implied by the terminology. For now we decided not to use the SNOMED codes for these items.

4. Discussion

In our study we could find 58% of the items in SNOMED CT. Until now we coded items for only 32 templates; 52 still need to be finished. The percentage of items found might be higher when all models are coded, although we expect the result of these 32 care information models to be representative for all models. It might even be that the percentage of SNOMED CT codes found will be lower after coding the items of all templates because 17 of the 52 models that still need to be coded are scientific scales or measurement instruments. For these scales or instruments, the concepts need to be exactly the same as the scoring items of the tests or scale [8], including the answering possibilities. Hardly any tests are included in SNOMED CT As was shown above only 26% of the concepts coming from scientific scales or instruments could be found in SNOMED CT against 68% of the individual concepts.

The current 58% is slightly disappointing compared to similar studies [9, 10, 11], however, we believe that we have dealt with difficult concepts in our study. These concepts are difficult in two ways: first is the very large granularity of the clinical concepts necessary for the care of stroke patients. These are usually very fine grained details of muscle functions, body position, thought processes etcetera, and in some instances expressed in quite awkward wording. Second, the templates are often especially made to have an accurate representation in HL7 v3 messages of scales, which have specific clinimetric characteristics [5, 8]. These clinimetric characteristics require an accurate equivalence between the concepts as used in practice and in the clinical terminology used for the unique coding [5, 8].

In addition, translation errors and or cultural differences might be a reason for this lower percentage. SNOMED CT has predominantly been built for the English language realm. We worked with clinical concepts in Dutch, given to us by clinicians, often in their local wording. They might use slightly different terms that could not be translated one to one from Dutch to English.

For a more reliable result we could ask the care professionals, who gave us the (medical) information on which we based the concepts in the templates, to review the results of our search in SNOMED CT. They are the best to check if the concepts they use in daily practice are well represented by the concepts we found in SNOMED CT. Next to that they might also be able to formulate synonyms for concepts without a SNOMED CT code, although the translation from Dutch to English might still cause a bias.

5. Conclusion

We could answer the question as to how many of the concepts from the care information models, that where initially developed for the EHR and HL7 v3 messages for stroke patients' continuity of care, could receive a SNOMED CT code. Currently, with a quite difficult set of concepts in the area of stroke care, we could find an overall coverage of 58%. It is important to have standardized clinical terminology such as SNOMED CT applied in the EHR and HL7 v3 messages instead of reinventing the wheel by using self made codes. The motivation for this is to truly achieve semantic interoperability in the exchange of patient information [2, 3]. Based on this study we could replace these self made codes for a bit over half of the clinical concepts.

Despite the result of this study we can conclude that SNOMED CT has several characteristics that make it useful to continue its application in stroke care, and after further testing, in the Dutch national infrastructure. First, SNOMED CT has the possibility to request inclusion and coding of concepts that could not be found in SNOMED CT. This means that expansion is an option. Second, SNOMED CT has an ongoing project for scale representation that takes the clinimetric aspects of scales and concepts into account. Third, SNOMED CT is working on further internationalisation in order to meet European requirements.

The clinical use in the Electronic Health Record system, which is built at the moment, and in HL7 v3 messages will reveal additional information about the usability of SNOMED CT in the clinical care for stroke patients and is therefore recommended. Of course it is also recommended to have more clinical areas than just the area of stroke researched in such a way. A final recommendation would be to tackle the translation issue.

Acknowledgments

We thank the care professionals for providing the information for the care information models; a special thanks to the neurologists Dr. Weinstein (Amsterdam), Dr. Swen (Delft) and Dr. Dippel (Rotterdam). We also thank Portavita, developer of the EHR for stroke patients and implementer of the HL7 v3 messages, for working together on this project. The American College of Pathologists are acknowledged for offering a SNOMED CT licence to carry out different tests for the stroke patients EHR and message development. Anne Casey, Davera Gabriel, Deb Konicek and David Markwell for teaching us how to search SNOMED, for answering our questions and for considering adding concepts to SNOMED CT. This study could be carried out through funding from the Dutch National IT Institute for Healthcare (NICTIZ).

References

[1] Nationaal ICT Instituut in de Zorg, NICTIZ (2003). Introduction of a clinical terminology in the Netherlands, Needs, Constraints, Opportunities. Leidschendam, NICTIZ.

[2] Nationaal ICT Instituut in de Zorg. Web documents [Online]. [cited January 10, 2006]; Available from: URL: www.nictiz.nl.

[3] Health Level 7. Web documents [Online].[cited January 10, 2006]; Available from: URL: www.hl7.org.

[4] SNOMED Clinical Terms. Web documents [Online]. [cited December 28, 2005]. Available from: URL: www.snomed.org.

[5] Kooij van der J, Goossen WTF, Goossen-Baremans ATM, Plaisier N. Evaluation of documents that integrate knowledge, terminology and information models. Article submitted for publication.

[6] Goossen WTF. Templates: an organizing framework to link evidence, terminology and information models in the nursing profession. In: de Fatima Marin H, Pereira Margues E, Hovenga E, Goossen W, editors. E-Health for all: designing a nursing agenda for the future. Proceedings 8th International Congress in Nursing Informatics NI 2003. Rio de Janeiro, Brazil, E-papers Serviços Editoriais Ltd, pp. 461 - 465.

[7] Reuser L, Goossen WTF, van der Heijden H. Call for ICT within stroke service care. In: Runnenberg J, et al, editors. Health Information Developments in the Netherlands by the year 2003. 6th ed, pp. 52-56.

[8] White TM, Hauan MJ. Extending the LOINC conceptual schema to support standardized assessment instruments . Journal of the American Medical Informatics Association 2002; 9 (6):586-599.

[9] Campbell JR, Carpenter P, Sneiderman C, Cohn S, Chute CG, Warren J. Phase II evaluation of clinical coding schemes: completeness, taxonomy, mapping, definitions, and clarify. Journal of the American Medical Informatics Association 1997; 4:238-51.

[10] Chute C, Cohn S, Campbell K, Oliver D, Campbell JR. The content coverage of clinical classifications. Journal of the American Medical Informatics Association 1996 May/June; 3:224-233.

[11] Wasserman H, Wang J. An applied evaluation of SNOMED CT as a clinical vocabulary for the computerized diagnosis and problem list. AMIA 2003: Annual Symposium Procedure; 2003. p.699-703.

[12] Casey A . Finding terms in SNOMED CT® - supplement to browser guidance 2005 June. Internal report SNOMED CT.

Ubiquity: Technologies for Better Health in Aging Societies
A. Hasman et al. (Eds.)
IOS Press, 2006

SNOMED CT in multidisciplinary clinical practice – evaluation of usefulness for classification and coding of care-planning procedures

Anna VIKSTRÖM[1] and Gunnar NILSSON[2]

Karolinska Institutet, the Department of Neurobiology, Caring Sciences and Society, Center for Family Medicine, Stockholm, Sweden

Abstract: Studies of the usefulness of SNOMED CT in everyday clinical practice have received little attention. The purpose of this study was to evaluate the usefulness of SNOMED CT in a multidisciplinary clinical setting. Another aim was to examine how well the terminology of care-planning procedures in a local development project could be mapped to SNOMED CT, and to explore the terminological surroundings in terms of superordinate concepts, subordinate concepts and coordinate concepts. A term list of 66 non-analysed types of procedures, considered necessary in the documentation of care planning, had been compiled in a local development project. The Clue-browser, version 5.5, was used to browse for matching concepts in SNOMED CT, and the degree of correspondence was assessed for each item. Of the 66 procedures, 72% corresponded completely and 22% corresponded partly to concepts in SNOMED CT. No match was found for 6% of the procedures. More subordinate concepts) were lacking in 30% of the concepts. The study showed that SNOMED CT could be mapped fairly well to care-planning procedures in a multidisciplinary setting such as advanced home health care. The terminological surroundings were extensive and stable for superordinate concepts (stem concepts), but unstable regarding subordinate concepts. SNOMED CT and the Clue-browser were feasible for choosing terminology with machine-readable codes for a specific area where there is no terminology in Swedish.
Keywords: medical informatics, SNOMED CT, advanced care-planning

1. Introduction

1.1. SNOMED CT

The Systematised Nomenclature of Medicine Clinical Terms (SNOMED CT) is a multidimensional, concept-based terminology. It is expected that SNOMED CT will play an important role in clinical information systems [1]. In 2005, SNOMED CT contained about 366,170 concepts and 993,420 terms. The procedure hierarchy in version 3.4 of SNOMED CT consists of 30,796 concepts. The definition of procedure

1 *Anna Vikström anna.vikstrom@sll.se*
2 *Gunnar Nilsson gunnar.nilsson@ki.se*

concepts is "concepts that represent the purposeful activities performed in the provision of health care" [2].

1.2. Previous studies

Several evaluations of SNOMED CT terminology coverage in different areas have been performed [1,3,5,]. One study investigated subsumption of Description Logic-Based Terminologies in a case study, and described the total presence of superordinate and subordinate concepts in SNOMED CT. It was found that 73% of the concepts had no subordinate concepts (i.e. leaf-concepts), and out of 73,627 concepts with one or more subordinate concepts, 31.5% had a single subordinate concept. There was no information in that study on the number of superordinate and subordinate concepts in the procedure category in SNOMED CT. Another study described matching narrative problem list entries that had been compiled by clinicians and that were mapped to SNOMED CT [3]. A total of 108 (11.9%) of a sample of 5024 entries remained unresolved even with expert manual review and coding, largely due to an inability to determine the original meaning of the terms. A further study showed that SNOMED CT covered 86.3% of single concepts present in Food and Drug Administration-approved oncology drug indications, and 11.3% of the indications were covered completely [4]. Coverage was best for concepts describing diseases, anatomy, and patient characteristics. However, implementation of SNOMED CT in clinical information systems and in everyday clinical use has received little attention.

Concept based terminology sources are useful in projects in which clinical information systems are developed. If such sources are lacking, development of a terminology for different domains that relate to the information system is a time-consuming activity. However, implementation of SNOMED CT demands decisions regarding where and how the terminology should be linked to the information model of a system [6]. This process also includes selecting which parts of SNOMED CT to use in a certain function.

No established concept-based terminology for health care procedures is available in Swedish. For that reason, SNOMED CT can be evaluated in this area as a reference terminology despite the fact that it is not translated into Swedish.

1.3. Study aims

The aim of this study was to evaluate the usefulness of SNOMED CT in a multidisciplinary clinical context regarding how well the procedures in a project could be mapped to SNOMED CT. The aim was also to explore the terminological surroundings in terms of superordinate concepts, subordinate concepts and coordinate concepts.

2. Method

2.1. The clinical part of the study

A pilot study within a local development project in Stockholm County Council aimed to map the terms and concepts used locally to an established domain terminology in order to avoid work with terminology as much as possible while still having a concept structure connected to the locally used terms. This ongoing project (Sams), with the general aim of developing care and computer-support for caring processes in advanced home health care in a multiprofessional setting, was used for the evaluation of SNOMED CT.

In the project, there was a need for procedure-concepts in a two-level hierarchy to be used in a care-planning component in the IT-system. For that purpose, a term list of 66 non-analysed types of procedures, considered necessary in the documentation of care-planning and originating from three participating health care units, had been put together in an initial part of this development project. These terms were selected, as they were often used by more than one category of health care professional and were thus needed for communication.

2.2. Mapping procedure

The 66 terms used in the evaluation (describing procedures/measures) were:

Activation
Acupuncture/Transcutaneous electrical nerve stimulation
Alert
Antibiotic therapy - infusion or injection
Application of dressing, splint, sutures
Assistive devices
Bilicheck
Biopsy on bladder
Blood pressure
Care planning
Catheter procedure
Chemotherapy regimen
Communication relating to death and dying
Congestive heart failure monitoring
Continuous infusion in/out
Dilation of oesophagus
Discharge visit
Dispensing medication (in a dosett)
Drainage of intra-abdominal collection
Elimination
Enema
Enterostomy management and care
Evaluation of pain
Evaluation visit
Follow-up visit
Group exercise
Head circumference
Housing assessment
Inhalation
Initial visit
Injection
Intake of food/fluid
Length
Massage
Measurement of oxygen saturation
Medical status
Nasopharyngeal suction
Nursing procedure
Nutrition – supplement, eating, training oral stimulation, massage of abdomen, finger feeding
Oxygen supply
Oxygen, ordering
Patient education
Postoperative care
Preparation for treatment
Providing material
Pulse
Putting on/taking off anti-embolic stockings
Respiratory care
Social planning
Specimen collection
Stoma care
Supervision
Support
Teaching activities
Telephone encounter
Temperature
Total parenteral nutrition on/off
Transfusion
Tube feeding of patient
Vaccination
Verbal communication health care professional
Verbal communication patient interventions
Verbal communication relative
Weight
Weight without clothes

The concepts behind each of the 66 terms were analysed briefly, and the terms were translated into English and mapped manually to SNOMED CT. Mapping is defined as "assigning an element in one set to an element in another set through semantic correspondence" [7]. The degree of correspondence was assessed for each item. The assessment scale used was "yes" if the code completely captured the meaning of the concept, "partly" for matches that approximated the concept meaning, and "no" if no reasonable match existed in SNOMED CT, which is in concordance with a similar study [3]. Only single concepts in SNOMED CT were used and accepted as a complete match.

All subordinate, superordinate and coordinate concepts were listed and counted for each of the 66 terms that were assessed as 'completely' and 'partly' corresponding to SNOMED CT. A subordinate concept was defined as a narrower concept that is either a specific concept or a partitive concept and a superordinate concept was was defined as 'a broader concept that is either a generic concept or a comprehensive concept [8]. The definition used for a coordinate concept was 'a concept that differs from other concepts at the same level in a relation regarding at least one distinguishing characteristic' [9]

Clue version 5.5, including both the free-text search and the graphical presentation of concept relations in the browser, was used to browse and search for matching concepts in SNOMED CT.

3. Results

Seventy-two percent of the 66 procedure concepts corresponded 'completely', and 22% 'partly', to concepts in SNOMED CT, while no match was found for 6% of the concepts. All concepts had a minimum of one and a maximum of four superordinate concepts that corresponded to the item. Subordinate concepts (i.e. leaf concepts) were lacking in 30% of the procedure concepts. The number of subordinate concepts (when present) varied between one and 64. Twelve concepts had one subordinate concept, six had two, while one to five concepts had between three and 64 subordinate concepts (Fig. 1).

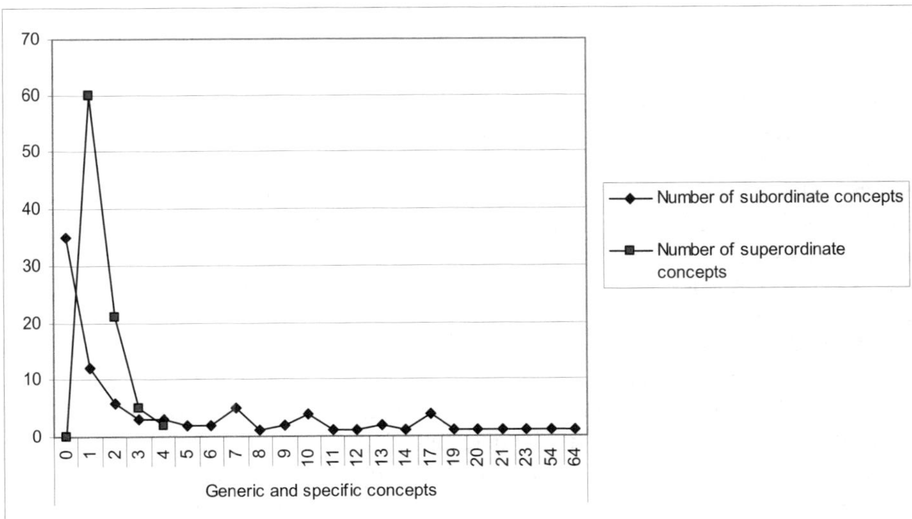

Figure 1 The number of superordinate and subordinate concepts related to a concept found in SNOMED CT

Coordinate concepts were found in 20 (30%) of the 66 procedure concepts. There were mainly one or two coordinate concepts, but in three cases there were three coordinate concepts. An example of a coordinate concept is 'infusion in/out that was mapped to both 'attention to intravenous infusion' and 'continuous infusion of therapeutic substance', where 'attention', 'therapeutic substance' and 'intravenous' were the distinguishing characteristics that differentiated the concepts.

SNOMED CT in the Clue-browser was feasible for choosing terminology with machine-readable codes for this area.

4. Discussion

A vast majority of the 66 procedures corresponded to SNOMED CT, 72 % completely and 22 % partly. However, no match was found for 6% of the concepts, and subordinate concepts were lacking in 30% of the concepts.

The study investigating Subsumption in Description Logic-Based Terminologies showed similar results concerning subordinate concepts: 72.3 % of all concepts in SNOMED CT had a single superordinate concept and 19.8 % had two superordinate concepts, as compared with our study where 68 % had a single superordinate concept and 21% had two superordinate concepts [1]. Another study of 80 concepts concerning 'Guidelines for Evaluation and Management of Chronic Heart Failure' that were mapped to SNOMED CT also showed similar results, with 71.2 % coverage in SNOMED CT, and 100% coverage for concepts representing 'plans' [5]. The aim in the Sams-project was for every care-planning concept to have a subordinate concept. This was not completely fulfilled, and whether total coverage of everyday clinical terms such as in our study can be achieved can be questioned. However, in our study most of the concepts without a match were of a very comprehensive nature.

There is still an urgent need for evaluation of SNOMED CT in everyday clinical use, i.e. end-user usability. However, our study shows that it is possible and feasible to find and select concepts and terms from SNOMED CT for a specific care-planning function in an information system. The 'terminological surroundings' found in our study provide an indication of what selections must be made in association with implementation. For example, it must be decided which and how many subordinate and coordinate concepts will be used in the information system. A weakness in our study is that the selection of concepts was rather small, and that only one person made the judgements in the mapping process. Other than that, however, we believe that the method used was appropriate for the purpose of our study; it has been used in previous studies, and it is easy to reproduce in similar studies measuring correspondence between concepts.

Previous studies have shown that SNOMED CT has a rich content [4,5,10]. Some SNOMED CT concepts have a very large amount of subordinate concepts: 8,034 concepts (11%) have 10 subordinate concepts or more, and 150 have more than 99 subordinate concepts. There is still the question of how to refine the content to be used by clinicians in an information system. Although subsets are available, the vast majority of concepts and terms remain. Few studies have dealt with the usefulness of SNOMED CT in a clinical setting and the selection of content connected to the information system, or with the reuse of such clinical data. These areas require further research.

5. Conclusions

The study showed that SNOMED CT could be mapped fairly well to care planning procedures in home health care. The terminological surroundings were extensive and stable for superordinate concepts, but unstable regarding subordinate concepts.

SNOMED CT and the Clue-browser were feasible for choosing terminology with machine-readable codes for a specific area where there is no terminology in Swedish.

Acknowledgement

This study was by part funded by the European Community under the FP6 programme (Semantic Mining NoE 507505).

References

[1] Bodeneider O, Smith B, Kumar A, Burgun A. Investigating Subsumption in DL-Based Terminologies: A Case Study in SNOMED CT. Proceedings of the First International Workshop on Formal Biomedical Knowledge Representation, 2004 June 2-5; Whistler, BC Canada. California: AAAI 2004.

[2] http://www.snomed.org/snomedct/what_is.html What is SNOMED CT? (last accessed Dec. 30, 2005).

[3] Penz FEJ, Brown SH, Carter JS, Elkind P L, Nguyen VN, Sims SA, Michael J Lincoln. Evaluation of SNOMED Coverage of Veterans Health Administration Terms. Medinfo 2004 Sept 7-11; San Francisco, USA. Amsterdam: IOS Press 2004 IMIA

[4] Brown SH, Bauer BA, Wahner-Roedler DL, Elkin PL. Coverage of Oncology Drug Indication Concepts and Compositional Semantics by SNOMED-CT(R). Proceedings of AMIA Symp. 2003 Nov 8-12; Washington DC, USA.

[5] Dykes C.P, Currie M.L, Cimino JJ. Adequacy of evolving national standardized terminologies for interdisciplinary coded concepts in an automated clinical pathway. Journal of Biomedical Informatics 2003; 36(4-5):313-25.

[6] Markwell D. SNOMED CT Advanced Tutorial SNOMED CT conference September 9, 2005.

[7] Draft international standard ISO/DIS 17115 ISO/TC 215 Health informatics Vocabulary for terminological systems 20041207.

[8] ISO 1087-1:2000, Terminology work Vocabulary Part 1: Theory and application

[9] http://app.socialstyrelsen.se/termbank The Swedish national board of health and welfare Terminology database (last accessed Nov.5, 2005)

[10] Chute CG, Cohn S P, Campbell K E, Oliver D E, Campbell J R. The content coverage of clinical classifications. For The Computer-Based Patient Record Institute's Work Group on Codes & Structures. JAMIA 1996; 3(3): 224–233.

7.6 Terminologies

Ubiquity: Technologies for Better Health in Aging Societies
A. Hasman et al. (Eds.)
IOS Press, 2006

Mapping of the WHO-ART terminology on Snomed CT to improve grouping of related adverse drug reactions

Iulian Alecu[a,1], Cedric Bousquet[a], Fleur Mougin[b] and Marie-Christine Jaulent[a]
[a] INSERM U729, Université Paris Descartes, Faculté de Médecine, Paris, France
[b] EA 3888, IFR 140, Faculté de Médecine, Université de Rennes I, France

Abstract. The WHO-ART and MedDRA terminologies used for coding adverse drug reactions (ADR) do not provide formal definitions of terms. In order to improve groupings, we propose to map ADR terms to equivalent Snomed CT concepts through UMLS Metathesaurus. We performed such mappings on WHO-ART terms and can automatically classify them using a description logic definition expressing their synonymies. Our gold standard was a set of 13 MedDRA special search categories restricted to ADR terms available in WHO-ART. The overlapping of the groupings within the new structure of WHO-ART on the manually built MedDRA search categories showed a 71% success rate. We plan to improve our method in order to retrieve associative relations between WHO-ART terms.
Keywords: WHO-ART, SNOMED CT, UMLS, Pharmacovigilance

1. Introduction

The Pharmacovigilance is "the science and activities relating to the detection, assessment, understanding and prevention of adverse effects or any other drug-related problems". The World Health Organization (WHO) defines a signal in Pharmacovigilance as "any reported information on a possible causal relationship between a drug and an adverse drug reaction (ADR), the relationship being unknown or incompletely documented previously" [1].

Statistical data analyses are run on ADR databases to detect such signals and are based on reliable and reproducible coding of ADR. WHO-ART (World Health Organization – Adverse Reaction Terminology) and MedDRA (Medical Dictionary for Drug Regulatory Activities) are the terminologies used in pharmacovigilance for data coding and data statistical analysis. The medical relevant grouping of ADR terms is necessary in order to detect a threshold frequency or disproportionate incidence [2], to identify important drug-related events and to support the recognition of new syndromes [3]. Some authors have argued about the relevance in data analysis of groupings stated by the structure of the terminology [4, 5].

WHO-ART has a hierarchical structure with restricted multiple inheritance. It has three levels but a large part of it is actually organized in only two levels and the terms with different levels of generalization may be siblings. Consequently, grouping similar terms based on WHO-ART hierarchy will provide either very large clusters or very small ones. For specific medical topics, that do not have corresponding groupings in its

[1] Corresponding author: Iulian ALECU, 15, rue de l'Ecole de Médecine, 75006 PARIS, FRANCE. Iulian.Alecu@SPIM.jussieu.fr

hierarchy, MedDRA proposes 13 manually built Special Search Categories (SSC).

This paper focuses on issues related to WHO-ART. We present a new method that aims to automatically group related medical conditions. Groups further referred in this paper as WHO-ART Search Categories (WSC) play the same role as the SSCs in MedDRA.

Our objective is to enhance the WHO-ART structure in order to facilitate the precision and flexibility of the semantic clustering, keeping in mind that an important amount of work has already been realized by the creation of medical terminologies. A polyhierarchical structure of WHO-ART will allow multiple views of the data [6] and will consequently integrate the SSC notion. This kind of structure will allow the evolution of WHO-ART towards a more versatile terminological system: an ontology.

Grouping related terms can be performed within a classification task (concepts sharing similar features are grouped under the same higher level concept). In knowledge engineering, formal definitions of terms[2] are required for automated classification [5,7]. Our main assumption is that automated classification of WHO-ART may be achieved by adding new properties to WHO-ART terms, using the information embedded in the hierarchy of a more sophisticated terminological system, such as Snomed CT. The Unified Medical Language System® (UMLS®) offers the mapping capabilities necessary for this purpose.

Section 2 and section 3 describe respectively the material and our approach for structuring WHO-ART. Section 4 presents an appraisal of WHO-ART groupings compared to MedDRA SSCs. We discuss the automated grouping in the fifth section.

2. Material

Our material is composed of WHO-ART (2004 third quarter) and Snomed CT terminologies. Snomed CT is available through UMLS (2005AA).

WHO-ART[3] is organized on three hierarchical levels[4]: 1) the preferred terms (PT) level is recommended for the coding of ADRs, 2) 31.3% of PTs are grouped in high level terms classes (HLT), 3) at the most general level, PTs are grouped according to 32 system organ classes (SOC). The rest of PTs (68.7%) are linked directly to a SOC.

SOCs group terms according to anatomy (e.g. “Gastrointestinal disorders”) and problems (e.g. “Neoplasm”). Within each SOC, the hierarchy is strict so that none of the ADR case reports under a SOC will appear twice and interfere with statistical data analysis. Analysis groupings based on the native WHO-ART leads either towards too large semantic clusters (an average of 58 terms for SOCs) or too specific, granular ones (an average of 4 for HLTs).

Snomed CT has a polyhierarchical structure and covers a large part of the medical field [8] but it has a broader scope than ADR coding.

MedDRA is a terminology dedicated to the coding of ADRs that includes all WHO-ART terms. The resulting WHO-ART structure is evaluated against the SSCs built by MedDRA (7.1) editors[5] and available through UMLS. The SSCs are delivered separately and they are static. For instance, the use of the “PAIN” SSC is not adapted for the retrieval of case reports related to an “abdominal pain”, and a more specific list

---

[2] Formal definitions aim to represent medical concepts in a formal logical language.

[3] Developed by the WHO collaborating center for international drug monitoring.

[4] HLTs and PTs are linked to included terms (IT) that are synonymous or more specific.

[5] Maintenance and Support Services Organisation (MSSO).

has to be created manually.

At the implementation level, the classification task is based on the Description Logics (DL) principles. The Web Ontology Language (OWL) is a standard of knowledge representation that facilitates greater machine interpretability and is an implementation of DL. Protégé is one of the most actively developed OWL enabled ontology editor [9]. Racer is an inference engine performing automatic classification [10].

3. Methods

We propose a generic three steps methodology to structure WHO-ART by taking into account the Snomed CT hierarchy.

Firstly, we create a mapping list between WHO-ART preferred terms and their Snomed CT synonyms. This step relies on the fact that an UMLS concept groups synonymous terms. Fung [11] argues that UMLS view of synonymy is not a totally independent assertion of synonymy, but a representation of views of all concept-based source vocabularies. In UMLS, terms with very close meanings (eg. the Snomed CT terms "thrombotic disorders" and "thrombosis") might be included in the same concept (eg. the UMLS concept "thrombosis"). We observed that the semantic differences in UMLS for WHO-ART terms are quite large so that two terms never appear in the same UMLS concept so that, the non strict synonymy is not considered as an obstacle here.

Secondly, starting from the list of Snomed CT terms obtained during the previous step, we rebuild only the part of Snomed CT structure strictly including all the terms in the mapping list (Figure 1).

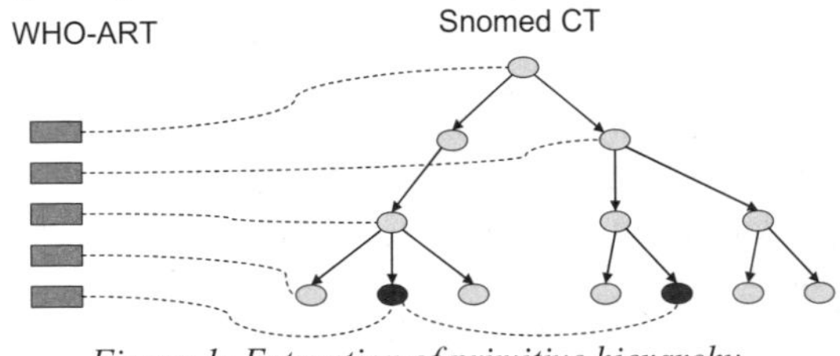

Figure 1: Extraction of primitive hierarchy.

This structure gives us the relative positions of corresponding synonymous Snomed CT terms and the primitive concepts required for the construction of the ontology, i.e. concepts which are not defined, but assumed to satisfy some basic properties, expressed as axioms. These axioms are based on the medical domain common sense and, in our case, they correspond to the generalization/specialization relations represented as a hierarchy.

Thirdly, WHO-ART terms definitions are written in OWL and automatically classified. At the beginning of this step, we have all the elements to build the OWL ontology: the granular and polyhierarchical Snomed CT as primitive hierarchy and the Snomed CT concepts as primitive concepts.

We propose a "shallow" way to write the description logic definitions (Figure 2) considering a single property: the synonymy. A WHO-ART concept ($Concept_{WHO}$) definition will be the union of the existential restrictions of this property on all synonymous Snomed CT concepts. The choice of the union operator instead of the intersection is due to the fact that it brings more results in term of grouping. The necessary condition (partial) asserts the fact that the defined concept belongs to the WHO-ART hierarchy. This way, the primitive concepts hierarchy (Snomed CT) is clearly separated from the defined concepts (WHO-ART).

$Concept_{WHO}$ =
(*is_synonym_with* some *Snomed_CT_term1*)
or (*is_synonym_with* some *Snomed_CT_term2*) …

Thrombosis $arterial_{WHO}$ is defined by[6]:

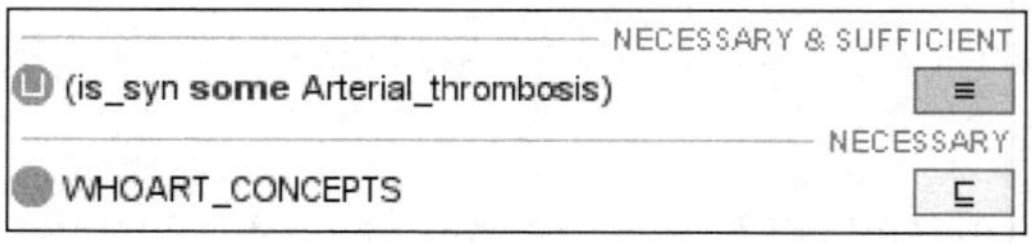

Figure 2: Formal definitions

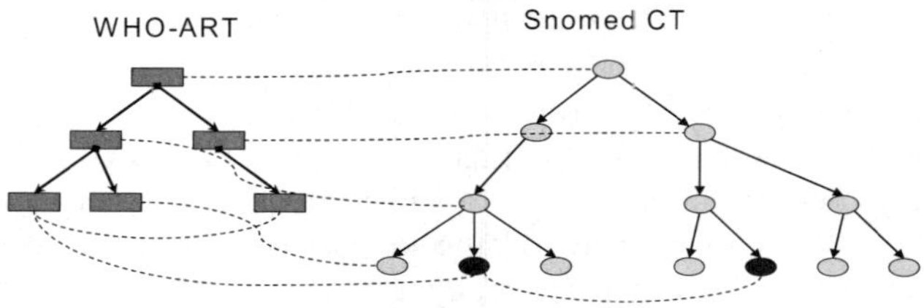

Figure 3: Structure of WHO-ART after automated

The resulting knowledge representation model will be further referred as an ontology. The definitions are further processed by an inference engine (RacerPro 1.8) which compares the hierarchical neighborhood of each concept. The result of this process is a new WHO-ART structure (Figure 3).

The main idea of the validation of the WHO-ART structure is to compare the already existing SSCs with homologous WSCs built automatically from the ontology. We first restricted the MedDRA SSCs to terms that are included in WHO-ART and that have a Snomed CT synonym. For instance, the terms included in the restricted SSC "thrombosis" and in the sub-tree "thrombosis" in the WHO-ART structure are compared. Some SSCs do not correspond to single WHO-ART terms. In that case, we consider Snomed CT concepts whose meanings would cover this SSC and we create a WSC defined as the union of synonymous Snomed CT high level concepts. For example, the definition of "UPPER GI BLEEDING/PERFORATION" is:
"(is_synonym_with some *upper gastrointestinal hemorrhage) or (is_synonym_with* some *gastrointestinal perforation)".*

4. Results

In the first step of the methodology, 85.9 % (1,597) of WHO-ART terms were successfully mapped to one or more Snomed CT synonyms.

The ontology contains 7,357 classes including 1,596 defined classes involving 2,977 existential assertions. The primitive concepts hierarchy represents 1.6 % of

[6] This view is taken from the Protégé interface.

Snomed CT (366,170 concepts). It takes an average time of 4 minutes for RacerPro to classify this ontology. The major achievements are the following:

- The hierarchy is refined. For instance, "Thrombosis arterial" is a sibling of "thrombosis carotid" in WHO-ART; after classification, the second one becomes a child of the first one.
- The polyhierarchy objective is reached. For example, after processing our ontology, the WHO-ART term "Thrombosis arterial" is classified under "Thrombosis" and under "Vascular disorder".
- A new WSC can be built on the fly by composition when it does not correspond directly to a WHO-ART term. For instance, the WSC "Disorders of pregnancy" groups several terms among which "abortion", "eclampsia", "hydramnios", "uterine atony", and "uterine spasm".

A corresponding WHO-ART concept was found for 6 SSCs. For two of them, a WSC was defined as a composition of high level Snomed CT concepts. There are 5 SSC for which we have not been able to find a corresponding concept in WHO-ART or Snomed CT, nor an obvious composition of Snomed CT high level concepts. There are 220 concepts in the restricted SSCs. Only 121 PT were found in the 8 matched/recomposed WSC. The matching SSCs contained 79 well classified concepts out of 121 PT (71%).

When the SSC name is identical to a WHO-ART concept, we obtained good groupings (89% for "OEDEMA") as well as bad groupings (25% for "CARDIAC ARREST"). The main reason for bad groupings is that our method relies on "is_a" relationships, whereas some relationships within SSC are not "is_a". For instance, "sudden death" (SSC term) is not a cardiac arrest, but it may cause one. Therefore, using this method, "sudden death" is not classified under "cardiac arrest".

In the case where the SSC name does not have a WHO-ART homologue, the composition approach does not always provide a solution. For the "HEMORRHAGE" SSC, we have not yet been able to find an equivalent concept or a conveniently small union of concepts to give a definition expressing the same notion ("disorders characterized by hemorrhage") in Snomed CT.

5. Discussion and conclusion

Cimino argued that terminologies should comply with desiderata such as formal definitions and polyhierarchy [6]. This paper presents a method to transform the WHO-ART terminology from a classification system with no formal inference to a knowledge based nomenclature enabled to inference capabilities. This is achieved by reusing an already existing resource [12] rather than building the ontology from scratch. The method benefits from the fact that Snomed CT embeds in its hierarchy different semantic views of a same term. Our main achievement is to obtain meaningful groups of terms by classification; the relevance of these groups being compared to the SSC manually built.

The method is a three-step process. The 85.9 percentage of the mapping step is still insufficient but could be improved naturally since UMLS is reviewed regularly. To improve this percentage independently of the review frequency of the UMLS, we started a work on approximate matching by browsing others terminologies in UMLS. The method still needs to be evaluated regarding its effects on classification. The mappings provided by UMLS present some limits. Terms can be considered synonymous although they are on different axes in Snomed CT (e.g. "Thrombosis", a "Qualifier value" and "Thrombotic disorders", a "Clinical finding" are synonymous in UMLS). A further improvement of our ontology will be to assign axis information to each WHO-ART term.

The method is successful for the SSC that have corresponding classes in Snomed CT ("PAIN" WSC mapped to "Disorder characterized by pain"). The method should be improved to take into account other SSC such as "HAEMORRHAGE", where the groupings relies more on associative relations than on "is_a" generalization/specialization relations.

The method allows also to easily building WSC when a higher class exists in Snomed CT corresponding to the semantic clustering needs. Practically, a Snomed CT high level concept can be used for data aggregation on pharmacovigilance case reports coded using WHO-ART.

Even though OWL expressivity is not entirely used, this implementation supports a better evolution of the ontology. Further steps are to add information specific to pharmacovigilance within the formal definitions. Once the structure is refined, we can create associative relations between diseases and clinically associated signs, symptoms and abnormal results of laboratory exams. For example, this can be done by adding to the definition of "hepatitis" the fact that it is associated with "jaundice" and "elevated transaminases".

Acknowledgements: We acknowledge the WHO Collaborating Centre for International Drug Safety Monitoring for providing the WHO-ART sources.

References

[1] Edwards IR, Biriell C. Harmonisation in pharmacovigilance. *Drug Saf* 1994;10(2):93-102.
[2] Bate A, Lindquist M, Edwards IR, Orre R. A data mining approach for signal detection and analysis. *Drug Saf.* 2002;25(6):393-7.
[3] Brown EG. Effects of coding dictionary on signal generation a consideration of use of MedDRA compared with WHO-ART. *Drug Saf* 2002;25(6):445-52.
[4] Yokotsuka M, Aoyama M, Kubota K. The use of a medical dictionary for regulatory activities terminology (MedDRA) in prescription-event monitoring in Japan (J-PEM). *Int J Med Inform.* 2000 ;57(2-3):139-53.
[5] Bousquet C, Lagier G, Lillo-Le Louet A, Le Beller C, Venot A, Jaulent MC. Appraisal of the MedDRA conceptual structure for describing and grouping adverse drug reactions. *Drug Saf.* 2005;28(1):19-34.
[6] Cimino JJ. Desiderata for controlled medical vocabularies in the twenty-first century. *Methods Inform Med* 1998; 37:394-403.
[7] Rossi Mori A, Consorti F, Galeazzi E. Standards to support development of terminological systems for healthcare telematics. *Methods Inform Med* 1998; 37(4-5):551-63.
[8] Spackman KA, Campbell KE. Compositional concept representation using SNOMED: towards further convergence of clinical terminologies. *Proc AMIA Symp.* 1998;:740-4
[9] Knublauch H, Musen MA, Rector AL. Editing Description Logic Ontologies with the Protégé OWL Plugin. *International Workshop on Description Logics (DL2004)*, Whistler, Canada, 2004.
[10] Haarslev V, Möller R. Description of the RACER System and its Applications. *Proceedings International Workshop on Description Logics* (DL-2001).
[11] Fung KW, Hole WT, Nelson SJ, S. Srinivasan, T. Powell, L. Roth. Integrating SNOMED CT into the UMLS: An exploration of different views of synonymy and quality of editing. *J Am Med Inform Assoc* 2005; 12(4):486-494.
[12] Musen MA. Dimensions of knowledge sharing and reuse. *Comput Biomed Res* 1992;25(5):435-67.

Ubiquity: Technologies for Better Health in Aging Societies
A. Hasman et al. (Eds.)
IOS Press, 2006

Knowledge acquisition for computation of semantic distance between WHO-ART terms

Jimison IAVINDRASANA[a], Cédric BOUSQUET [a,b,1] and Marie-Christine JAULENT [a]
[a] *INSERM U729, Université Paris Descartes, Faculté de Médecine, Paris, France*
[b] *Hôpital Européen Georges Pompidou, AP-HP, Paris, France*

Abstract. Computation of semantic distance between adverse drug reactions terms may be an efficient way to group related medical conditions in pharmacovigilance case reports. Previous experience with ICD-10 on a semantic distance tool highlighted a bottleneck related to manual description of formal definitions in large terminologies. We propose a method based on acquisition of formal definitions by knowledge extraction from UMLS and morphosemantic analysis. These formal definitions are expressed with SNOMED International terms. We provide formal definitions for 758 WHO-ART terms: 321 terms defined from UMLS, 320 terms defined using morphosemantic analysis and 117 terms defined after expert evaluation. Computation of semantic distance (e.g. k-nearest neighbours) was implemented in J2EE terminology services. Similar WHO-ART terms defined by automated knowledge acquisition and ICD terms defined manually show similar behaviour in the semantic distance tool. Our knowledge acquisition method can help us to generate new formal definitions of medical terms for our semantic distance terminology services.
Keywords: terminology service; natural language processing, UMLS, semantic distance; pharmacovigilance

1 Introduction

Data mining large pharmacovigilance databases for drug safety signal generation has been described as "finding a needle in a haystack" [1]. This is partly due to the fact that similar medical conditions can be described in different ways [2]. In order to standardize the description of medical conditions, the WHO Collaborating Centre for drug safety monitoring has developed the WHO-ART (World Health Organization – Adverse Reaction Terminology) medical terminology to classify ADR (Adverse Drug Reaction) terms. It is organized in 3 hierarchical levels organized around 32 system organ classes (SOC) [3]. We expect that performances of different signal detection algorithms used in Pharmacovigilance can be improved if the algorithms would extract similar cases according to the knowledge carried by each WHO-ART term. Although, classification allows to group terms with close meanings under a same high level concept, a problem remains persistent: the WHO-ART classification is static and provides a single point of view. If two WHO-ART terms are related to the same

[1] Corresponding author: Cédric BOUSQUET 15, rue de l'Ecole de Médecine, 75006 PARIS, FRANCE. cedric.bousquet@spim.jussieu.fr

medical condition but belong to a different SOC, they cannot be grouped using only the hierarchical structure. Our main contention is to assume that the grouping of terms is more efficient when performed by a semantic distance approach rather than a classification one.

Semantic distance is an appropriate method to find medical terms expressing similar conditions [4]. The application of semantic distance computation approach depends on the representation of meaning and requires formal definitions (FD) to be available in order to express WHO-ART terms meanings. We showed in a previous work that such FDs can be expressed in a multi-axial terminological system such as SNOMED [5] to facilitate further computation of the semantic distance. To measure a distance between two FDs, they are projected on each SNOMED axis. We used then a Lp norm based on a weighted combination of the distances on each axis to compute a semantic distance between the two FDs [4]. The weight associated to each axis is predefined. The acquisition process of FD is usually a manual process. The aim of the present work is to propose a methodology for semi-automatic FD acquisition of ADR terms. The methodology is based on existing terminological systems and relies on knowledge extraction from UMLS and natural language processing (NLP) techniques. Medical terms have the particularity of being composed of one or more words which can take 3 morphological variations: inflexion {*leukocyte* + *s*}, derivation {*leucocyt* + *ic*} and compositional structure {*leuk* + *em* + *ia*} [6]. Natural language processing (NLP) technique like morphosemantic analysis allows decomposing complex terms into smaller meaningful units or morphemes [7] and is useful to acquire FDs.

The semantic distance tool described in a previous work [4] provides a relevant framework to evaluate the obtained FDs. Indeed, the tool allows retrieving the *k nearest neighbours* of a term according to this distance. The semantic distance of the neighbours has to be evaluated by an expert.

This paper is organized as follows: section 2 details material and methods used in this work as well as the evaluation framework. Section 3 presents results of our formal definition method acquisition for the WHO-ART terminology as well as results obtained by applying the semantic distance on these formal definitions. Section 4 discusses limits of this work and future directions.

2 Material and Methods

2.1 Material

Nine hundred and twenty one French WHO-ART terms have been extracted from the French pharmacovigilance database. Their equivalents in English were obtained from the UMLS (version UMLS 2005AA) and stored in a MySQL database. This database is composed of a first set of 539 terms (SNOMED International set) for which a corresponding term has been found in the 1998 version of SNOMED called SNOMED International (SNOMED Int). The second set of 382 terms (NO MAPPING set) is composed of terms with no correspondence in SNOMED Int.

SNOMED Int contains 11 axes organized in a hierarchical structure: disease and diagnoses (D), topography (T), morphology (M), function (F), living organisms (L), chemical (C), physical agents (A), occupation (J), social context (S), relational procedures (P) and general terms (G) [5]. With SNOMED Int, we can combine elements from different axes to formally define a complex clinical situation. The MRCONSO and MRSAT tables of the UMLS were respectively used to obtain the equivalent of a WHO-ART term in SNOMED Int and the projection of this term along the SNOMED Int axes.

We used the *Morfessor* tool for the NLP morphosemantic analysis, a parser suited for English terms [8]. *Morfessor* works on morphemes lists; it doesn't need any corpus annotation as it is the case for other tools such as *Dérif* [9] for French morphosemantic analysis.

A set of 392 ICD-10 (International classification of diseases) terms from the cardiovascular chapter were manually defined according to the SNOMED axes in our previous study on semantic distance [4]. This ICD set was used to compare measures of semantic distance between similar ICD and WHO-ART concepts.

2.2 Method: WHO-ART terms formal definition acquisition

FDs are WHO-ART terms *translated* into SNOMED Int. An FD is a list of elementary concepts belonging to some SNOMED Int axes and characterized by modifier terms in some cases. For example The WHO-ART concept *Acute Pancreatitis* was formally defined as a combination of the following SNOMED Int concepts: *Pancreas*, NOS (T-65000) + *Acute* (G-A231) + *inflammation* (M-41000) where *Acute* is a modifier term to precise the inflammation type. The methodology for formal definition acquisition followed several steps:

- WHO-ART concepts were mapped to SNOMED Int concepts in UMLS. When UMLS proposes several synonyms we kept in the database the synonym providing the maximum of information for morphosemantic analysis. For example, *Incoordination* is projected in UMLS on *Abnormal coordination* (F-A4202) and *Incoordination* (F-A4202). We have chosen *abnormal coordination* for a better morphosemantic analysis when this term has no FD provided by the UMLS.
- The SNOMED Int set was used to extract FDs from UMLS. 1) When the WHO-ART concept was mapped on a SNOMED Int concept from axis D, projections on other axes were retrieved thanks to relationships expressed in the MRSAT table. If no projections were available, the definition was declared non informative and rejected. For example projection of the WHO-ART concept *retinitis* on the SNOMED Int concept *retinitis* with no projections on other axes is not useful. 2) When the WHO-ART concept was mapped on a SNOMED Int concept from other axes, the corresponding projections were also retrieved to get additional knowledge about this concept. However SNOMED Int concepts without projections were sent to the expert for evaluation because the FD may already be complete. For example the WHO-ART concept *vomiting* is mapped to the SNOMED Int concept *vomiting* on the Function axis and does not need more decomposition on the other SNOMED Int axes.
- We used the morphosemantic analysis tool *Morfessor* for term segmentation into morphemes for the NO MAPPING set. In the SNOMED Int set, terms from the disease axis with no projections on other SNOMED Int axes were also parsed with *Morfessor* to get additional knowledge. Either simple terms composed of SNOMED Int concepts (example 1 below: *Bladder incontinence*) or complex terms composed of many morphemes (example 2 below: *Arteriospasm*) were processed. The morpheme list could be created manually or obtained from a corpus [8] or a thesaurus [7]. These morphemes were associated with concept from

one of the SNOMED Int axes. For example the suffix *–itis* was associated with the *inflammation* concept on the morphological axis [10].

- The first decomposition is a term decomposition in separate words (example 1): *Bladder incontinence* (F-72055) = *Bladder, NOS* (T-74000) + *Incontinence, NOS* (F-01400).
- The second decomposition is a compound word decomposition in morphemes (example 2): *Arteriospasm* (F-39730) = *arteri* (T-41000) + *o* + *spasm* (M-02590). One major problem is to avoid unnecessary decomposition. For example, the *lymphocyte* concept could be decomposed into *lymph* + *o* + *cyte* where *lymph* is a radical of lymph and *cyte* is a morpheme indicating the cells. To avoid this Hahn *et al* introduced the concept of "subword" to prevent decomposition errors [11]. A subword is a part of word which should not be decomposed into smaller units. A similar method was implemented by Lovis *et al* who added as a new morpheme any new word that could not be correctly segmented while keeping its meaning [12]. The decomposition step is a cyclic process. We updated the morpheme list when necessary and reused *Morfessor* for a new analysis.
- Abbreviations and eponyms were replaced manually by the corresponding concepts. For example, *DLE* was replaced by *discoid lupus erythematous* and *Lyell's syndrom* was replaced by *bullous eruption*.
- We excluded some components of FD like *drug* or *disorder* as in *Drug induced fever* or *Deglutition disorder* knowing that all WHO-ART terms are actually drug related disorders.

• The evaluation of FD consisted of the application of the semantic distance through a web based system. The semantic distance has been described in a previous work [4]. It is provided by the LP-norm which is the combination of the weight associated to each axis and the semantic distance of every component of the FD on different axes. In this work a web terminology server using J2EE technology has been developed and permits to navigate through formal definitions, compute semantic distance between two WHO-ART terms, obtain the *k nearest neighbours* of a WHO-ART term and calculate a semantic distance matrix between terms of some WHO-ART SOCs. For a given WHO-ART term, the semantic distance of the found neighbours was evaluated by an expert.

3 Results

Table 1: Formal definitions repartition

WHO-ART terms	***Total***	***FD example***
WHO-ART terms having FD provided by UMLS	321 34.8%	DUODENAL ULCER HEMORRHAGE = T-58200 Duodenum, NOS M-38000 Ulcer, NOS M-37000 Haemorrhage, NOS
WHO-ART terms formally defined using *Morfessor*	320 34.7%	RETINITIS = retin ⊕ itis = T-AA610; Retina, NOS; M-40000; Inflammation, NOS;

WHO-ART terms	*Total*	*FD example*
WHO-ART terms formally defined manually	117 12.7%	HEMOTHORAX = T-29050 Pleural cavity, NOS M-37000 Haemorrhage, NOS
Formally defined WHO-ART terms	758 82.2%	

Our ICD FDs from the cardiovascular chapter were loaded with the WHO-ART FD. Figure 1 shows the *k nearest neighbours* of the *PHLEBITIS* WHO-ART concept in a web browser. The HTML page was generated by the server. This concept is close to the *Phlebitis and thrombophlebitis* ICD concept. Every concept is associated to its semantic distance to the PHLEBITIS concept. For example semantic distance between *PHLEBITIS* and the *Cerebral arteritis, NEC* ICD concept is equal to 0.260. When the WHO-ART concept to retrieve is not related to a cardiovascular disorder, ICD concepts are not part of the k-nearest neighbours.

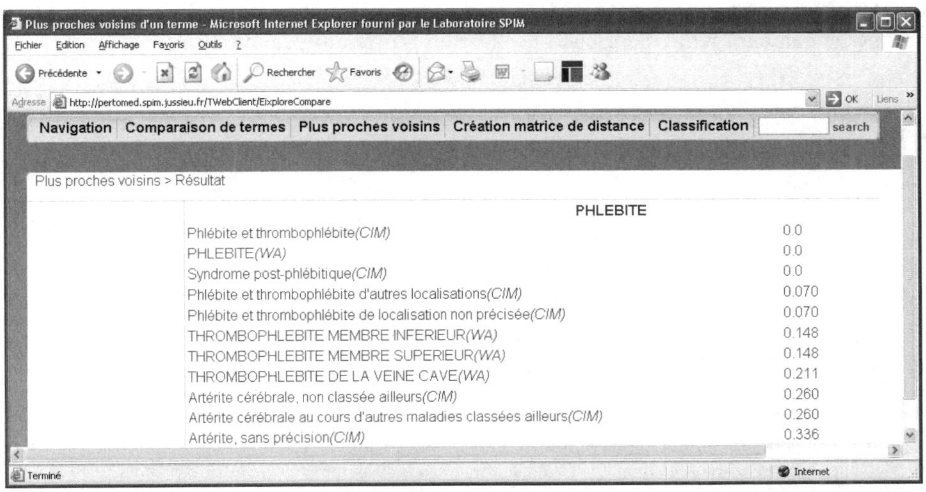

Figure 1: k-nearest neighbours of the PHLEBITIS WHO-ART concept. WA means WHO-ART and CIM is the French abbreviation for ICD.

4 Discussion

We described in this paper a methodology for acquiring WHO-ART terms formal definitions using the SNOMED Int. Some WHO-ART terms could not be automatically defined. Morphosemantic analysis can not capture appropriate medical knowledge for all WHO-ART terms. Not all FDs extracted from UMLS were complete or relevant. The data model contains only elementary concepts sometimes refined by modifiers (axis G). Some relationships between elementary concepts are not expressed given the limited number of modifiers from the general terms axis of SNOMED Int we used. A new version of SNOMED terminological system (SNOMED CT) is available in UMLS and contains additional terms, a more extended conceptual model and formal definitions based on description logic. This kind of terminological system should improve formal definitions of ADR terms.

However the knowledge representation of WHO-ART terms was appropriate to the semantic distance objective. WHO-ART acquired FDs showed a similar behaviour compared to the manually described ICD definitions. The method can ease the process of building large dictionaries of FDs for our semantic tool. About one third (34.8%) was provided by UMLS and about one third (34.7%) was defined by morphosemantic analysis. Indeed when vocabularies are very large such as WHO-ART or SNOMED this process can speed up the definition process.

5 Conclusion

FD associated with semantic distance computation is a first step for terminological clustering. A first limitation related to the quality of formal definitions model was highlighted and additional experiments with other terminological systems are necessary in order to improve the overall performances. In the continuation of this work, the whole WHO-ART terminology will be formally defined. We intend to implement SNOMED CT instead of SNOMED Int in order to improve formal definitions. The morpheme database will be enriched for a better morphological analysis.

References

[1] Edwards IR. Adverse drug reactions: finding the needle in the haystack. BMJ 1997;315:500

[2] Brown EG. Effects of coding dictionary on signal generation: a consideration of use of MedDRA compared with WHO-ART. Drug Saf 2002;25(6):445-52.

[3] Uppsala Monitoring Center. The WHO-ART Adverse Reaction Terminology. 2003 http://www.umc-products.com/graphics/3149.pdf (last accessed Jan. 5, 2005).

[4] Bousquet C, Jaulent MC, Chatellier G, Degoulet P. Using Semantic distance for efficient coding of medical concept. Proc AMIA Symp 2000;:96-100.

[5] Lussier YA, Rothwell DJ, Cote RA. The SNOMED model: a knowledge source for the controlled terminology of the computerized patient record. Methods Inf Med 1998;37(2):161-4.

[6] Zweigenbaum P, Darmoni SJ, Grabar N. The contribution of morphological knowledge to French MeSH Mapping for informational Retrieval. Proc AMIA Symp. 2001;:796-800.

[7] Grabar N and Zweigenbaum P. Language-independent automatic acquisition of morphological knowledge from synonym pairs. Proc AMIA Symp. 1999;:77-81.

[8] Creutz M, Lagus K, Linden K, Virpioja S. Unsupervised morpheme segmentation and morphology induction from text corpora using Morfessor 1.0. Rapport technique A81:Publications in Computer and Information Science, Helsinky University of Technology. 2005

[9] Namer F, Zweigenbaum P. Acquiring meaning for French Medical Terminology: contribution of Morphosemantics. Medinfo. 2004;11(Pt 1):535-9.

[10] Pacak MG, Norton LM, Dunham GS. Morphosemantic analysis of -ITIS forms in medical language. Methods Inf Med 1980;19(2):99-105.

[11] Hahn U, Honeck M, Piotrowski M, Shulz S. Subword segmentation-Leveling out morphological variations for medical document retrieval. Proc AMIA Symp 2001;: 229-33

[12] Lovis C, Baud R, Rassinoux AM, Michel PA, Scherrer JR. Medical dictionaries for patient encoding systems: a methodology. Artif Intell Med 1998;14(1-2):201-14.

Ubiquity: Technologies for Better Health in Aging Societies
A. Hasman et al. (Eds.)
IOS Press, 2006

Construction of a semi-automated ICD-10 coding help system to optimize medical and economic coding

Suzanne PEREIRA[a,c], Aurélie NÉVÉOL[a], Philippe MASSARI[b], Michel JOUBERT[c] and Stefan DARMONI[b]
[a] CISMeF Team., LITIS, Rouen, France
[b] Medical Informatics Service, CHU de Rouen, France
[c] LERTIM., Marseille, France

Abstract. Introduction : In order to measure the medical activity in hospitals, physicians are required to code manually information concerning a patient's stay using ICD-10. This requires trained staff and a lot of time. We propose to help speed up and facilitate the tedious task of coding patient information. Methods: we show two methods. First, we propose an automated ICD-10-based coding help system using an automated MeSH-based indexing system and a mapping between MeSH and ICD-10 extracted from the UMLS metathesaurus. Secondly, we propose the use of drug prescriptions to complete the previous coding with the use of a mapping between a given prescription drug and the relevant ICD-10 codes (in compliance with the drug approval). Results : the results of a preliminary experiment indicate that the precision of the indexing system is 40% and the recall is 30% when we compare to an economic rules-based coding and to a descriptive coding. Discussion: moreover, we show that the use of prescription coding is relevant as the recall reaches 68% when the Vidal tool is used. Conclusion : Then, it is very interesting to complete the coding obtained automatically by the indexing/mapping system by the coding obtained from the prescriptions.
Keywords: Abstracting and indexing; ICD-10; MeSH; computerized Medical records systems.

1. Introduction

The PMSI (French equivalent to The Medicare Prospective Payment System (PPS)) was introduced by the French government in 1996, as a way to change hospital practice through financial incentives that encourage more cost-efficient management of medical care [1]. Each patient's stay is classified into a GHS (French equivalent to a Diagnosis Related Group (DRG)) [2] according to information documented by the physician in the Medical Record : diagnoses, Complications and Comorbidities, healthcare procedures etc…The hospital is paid a relative cost for the GHS.

Moreover information from a patient's stay are coded to allow an automatic processing. The nomenclatures are the International Classification of Diseases (ICD-10) [3] to code diagnosis and the Healthcare Common Procedure Coding System (HCPCS) to code healthcare procedures. These codes constitute the medical and economic coding of patients medical records. The coding process is extremely important, coding an incorrect principal diagnosis or failing to code a significant secondary diagnosis can have a substantial impact on reimbursement for the hospital

Physicians are required to perform the coding manually or with the help of a navigation tool within a nomenclature or of a lexical research tool [4]. This requires trained staff with a good knowledge of the economic coding rules and of the classification used to code. As a result, coding is a time consuming activity and must be performed together with patient care that is and should remain a priority for medical staff.

In the framework of a French project, VUMeF [5], we built a semi-automated ICD-10 coding help system to code medical records in ICD-10. Physicians writing their patient record can use this tool to obtain a preliminary coding of the record. Then they can validate or specify some of the ICD-10 codes recommended. It will help reducing coding time and training time.

2. Material and Methods

2.1. An automated MeSH-based indexing system

The objective of this work is to extract the relevant diagnosis codes from an unstructured free text document namely a hospital summary report. Because an automated ICD-10-based indexing tool doesn't exist, less direct solutions have been developed using an indexing system based on a different terminology and a mapping to ICD-10 [6]. Then, we choose to use an automated MeSH-based indexing system developed in our laboratory [7].

MeSH (Medical Subject Headings) is the controlled vocabulary used by NLM to index articles from biomedical journals for the MEDLINE/PubMED® database. The French version of the thesaurus includes 22,995 terms and 61,000 synonyms (among which 4,000 were added by CISMeF), in its 2005 version. MeSH descriptors are organized in 15 thematic trees (e.g.: anatomic terms).

The indexing system can find textual elements referring to MeSH terms in a medical record. Then it allocates a score depending on the text length and the frequency of each term to each MeSH terms extracted. This score is called indexing score.

2.2. A mapping between MeSH and ICD-10

The mapping between MeSH and ICD-10 provides a list of ICD-10 terms from a list of MeSH terms supposed to be equivalent. This mapping was extracted from the UMLS metathesaurus (Unified Medical Language System) [3] which clusterizes more than 100 medical controlled vocabularies and classifications (MeSH, SNOMED, ICD-10…). It provides links between several vocabularies : some codes have the same concept unic identifier (CUI) in the two classifications (e.g : the MeSH term “Abdomen, acute” is mapped to the ICD-10 term “Acute abdomen” as they have the same CUI C0000727 in UMLS).

The goal of the ICD-10 classification is to make up the list of all the diagnoses. It's a five level hierarchy. On the first level ICD-10 contains 21 sections representing all the morbidities classified by functional apparatus and associated to a letter (e.g : E : Endocrine, nutritional, and metabolic disease). Each section is divided into groups, themselves divided into subgroups containing all the ICD codes in 3-character

categories and 4-character categories. The classification contains more than 18,000 alphanumerical codes and about 50,000 terms.
Our mapping contains 4,801 mapped terms among which 1598 distinct pairs MeSH CODE - ICD-10 CODE (a MeSH term and its synonyms are assigned the same MeSH code).

2.3. Taking drug prescriptions into consideration

A tool developed by Le Vidal company provides a mapping between a given drug prescription and the relevant ICD-10 codes (according to the drug approval). We used this mapping to determine the ICD-10 coding from the drug prescription.
Among all the relevant ICD-10 codes for each prescription drug, only a few codes correspond to the diagnosis contained in the summary report. We also proposed an ordering by relevancy score based on two ideas.
- the first idea is that if a large number of prescription drugs have for indication a same ICD-10 code, this code should probably be extracted by our system.
- the second idea is if this code is often coded by physicians, this code is more likely to be extracted. Therefore, we use the ICD-10's prevalence in the Rouen University Hospital. Formula [1] takes these parameters into account in the score calculation :

$$\text{Prescription Score } (C_{\text{ICD-10}}) = \left(\frac{x-1}{N} + \frac{p}{N}\right) * 100 \qquad [1]$$

$C_{ICD\text{-}10}$:a given ICD-10 code; x : number of prescription drugs that have $C_{ICD\text{-}10}$ as an indication; p : Prevalence (=frequency) of $C_{ICD\text{-}10}$ at the Rouen University Hospital; N : Number of drugs in the prescription.

The higher the score the more likely a diagnosis is a diagnosis actually treated with the prescription and its code have to be coded for this hospital summary report.

2.4. Evaluation

2.4.1. Two evaluations

ICD-10 code	Indexing Score
J45.9	66,76
J46	33,33
O60	0,05
L50.5	0,00

Table 1. ***Sample prescription coding of a summary report***

ICD-10 code	Prescription+indexing Score
R06.0	100
J44.9	73,85
J45.9	66,77
B44.9	25,42
O60	0,05
L50.5	0,00

Table 2. ***Sample added coding of the same summary report***

Two evaluations have been carried out. First, we evaluated the coding provided by our system indexing/mapping (see table1). The automated coding was compared to two manual codings : a priori and a posteriori. First, we compare the coding which is a descriptive coding to an economic and medical coding which is the effective coding made by the physicians after having written the patient record (MC1). Secondly, we compared the automated coding to another descriptive coding which is a re-evaluation of the automatic coding made a posteriori by a medical coding expert (MC2).

Secondly, we evaluated the coding of our system indexing/mapping taking drug prescriptions into consideration with the same references. Table 2 shows that we completed the list of ICD-10 codes extracted by our system with the ICD-10 codes found with the drug prescription. For each ICD-10 score we added the indexing score and the prescription score. We have considered that these scores have the same weight (each with a maximum of 100).

2.4.2. The corpus of Hospital Summary reports

A sample of 100 hospital summary reports written by physicians in the precedent year were used for this evaluation. 50 reports come from the cardiology service of the Hospital and 50 from the pneumology service. This choice was driven by the areas of expertise of the medical coding expert. They have been extracted through the electronic health record system of the Rouen University Hospital (1.080.384 patients and 182.808 summary reports in 2005). A hospital summary report details the disease history, the healthcare procedures and the drug prescription.

2.4.3. Evaluation methodology

We computed the precision and the recall for each coding comparison (automatic vs effective and automatic vs. re-evaluated) [8]. Precision and recall are the usual measures used in information science. Precision is the ratio of the number of relevant records retrieved (codes extracted by our system and by the reference) over the total number of irrelevant and relevant records retrieved (codes extracted by our system). Recall is the ratio of the number of relevant records retrieved over the total number of relevant records in the database (codes extracted by the reference). We have calculated precision and recall using the related codes (codes sharing the same parent in the ICD-10 hierarchy ex: A10.0 and A10.1 ; A10-A19 and A11). Taking these codes into consideration allows to determine the coding close proximity to the reference coding.

3. Results

We show in the two tables the results of the comparison between the automated coding to MC1 and MC2 before and after taking drug prescriptions into consideration.

The first table (Table 3) shows that our automated coding system as described in section 2.1 and 2.2 provides a precision of 43% and a recall of 30% if compared to MC1 and a precision of 38% and a recall of 30% compared to MC2. Our system is able to extract only a third of the reference codes which constitute themselves only a third of the automatic coding when we only take into consideration the coding performed by our system indexing/mapping.

Table 4 shows that the contribution of the ICD-10 codes extracted from the drug prescriptions as described in section 2.3 results in a precision of 5% and a recall of 61% if compared to MC1 and a precision of 4% and a recall of 68% compared to MC2. Thus we determine 61% of the economic and medical codes and 68% of the descriptive codes that should be coded for these summary reports.

Measures	Economic and medical coding (MC1)	Descriptive coding (MC2)
Precision	0.43	0.38
Recall	0.30	0.30

Table 3. ***Results of the evaluation provided by our system indexing/mapping in comparison with the 2 references (MC1 and MC2)***

Measures	Economic and medical coding (MC1)	Descriptive coding (MC2)
Precision	0.05	0.04
Recall	0.61	0.68

Table 4. ***Results of the evaluation provided by our system indexing/mapping taking drug prescriptions into consideration in comparison with the 2 references (MC1 and MC2)***

4. Discussion

4.1. System performance

The evaluations show in terms of recall that it is very interesting to take the prescriptions into consideration to code in ICD-10 the patient's stay as table 2 indicates that the contribution of the ICD10-codes extracted from the drug prescriptions results in significant increase of the recall of the summary reports coding. Unfortunately, only a less than 5% of the coding obtained in this case corresponds to the reference coding. Precision is very low as too many codes are presented to the physician (an average of 375 per report) compared to the number of ICD-10 codes coded by the expert (4). A possible explanation is that 29% of the prescription drugs are prescribed for an indication that is not indexed in the drug approval [9]. We have tested a method introducing a threshold (0, 5, 10 and 20) : if a ICD-10 code score is lower than the threshold, the code is not considered in the evaluation (not shown in this article). But we have not demonstrated that the noise decreases as the threshold increases. The results show that this automatic coding system needs several improvements to reduce the number of codes extracted and to allow an utilization in Hospitals.

4.2. Comparison to other systems

The results are encouraging as a similar work using SNOMED instead of the MeSH with the SNOCODE tool (MedSight Informatique Inc, 99) [6]. The SNOCODE tool provides a precision of 25% compared to MC1 and MC2 and a recall of 46% compared to MC1 (vs.61% for our system) and 46% compared to MC2 (vs. 68% for our system).

4.3. Indexing and mapping

Many factors have an influence on our system's performances. First of all, the indexing system provides a precision of 50% [7]. The MeSH/ICD-10 mapping is limited to 8% of the ICD-10 classification codes in contrast with physicians that can code with the whole classification. This limitation is due to the differences of structuration and goal of the classifications that provide a loss of information in the mapping. Nonetheless,

among the 1,000 ICD-10 codes most frequently coded in the Rouen Hospital, 53.5% are mapped into MeSH codes and belong to our MeSH/ICD-10 mapping.
Moreover, the wording of the reports can affect the results. Indeed, medical records are made to describe a patient's stay and enable the care continuity of the patient in contrast with the budgetary objective to measure the medical activity in hospitals. The education of the physicians (knowledge of the economic coding rules and of the classification used to code) can also affect the results.

4.4. Perspectives

We will proceed with the development of an automated SNOMED indexing system. With a mapping between SNOMED and ICD-10, it will provide an ICD-10 coding of summary reports. SNOMED International, the Systematized NOmenclature of MEDicine [9] is a terminology elaborated to index precisely medical records as SNOMED can describe diagnosis and healthcare procedures. To extend our work, we contemplate adding HCPCS coding to our tool, as a mapping between MeSH and HCPCS is being developed.

5. Conclusion

This evaluation study shows that our two approaches are complementary. Then, it is very interesting to complete the coding obtained automatically by the indexing/mapping system by the coding obtained from the prescriptions.

Acknowledgements

This work is part S.Pereira's PhD Work. It is partly granted by Le Vidal Company. We acknowledge the CISMeF team for their help in medical terminology expertise and P.Massari for his help as a coding expert.

References

[1] Dubois -Lefrère J., Coca E., Maîtriser l'évolution des dépenses hospitalières : le PMSI. éd Berger-Levrault; 1992.
[2] Ministère de la santé, de la famille et des personnes handicapées. Manuel des groupes homogènes de malades V9 de la classification V7.9 de la fonction groupage, 2004.
[3] Zweigenbaum P., Encoder l'information médicale : des terminologies aux systèmes de représentation des connaissances. ISIS; 1999.
[4] TOTHEM, un outil d'aide au codage selon la CIM10. J.Chevallier[1] V.Griesser[2], L.Brunel
[5] Darmoni SJ., Jarrousse E., Zweigenbaum P., Le Beux P., Namer F., Baud R., et al. VUMeF: extending the French involvement in the UMLS Metathesaurus. AMIA Annu Symp Proc, 2003;:824.
[6] A. Buemi, M. A. Boisvert, R.A. Côté. Transcodage SNOMED - CIM-10 : proposition pour une meilleure indexation des dossiers médicaux. Actes JFIM2002.
[7] Névéol A, Rogozan A, Darmoni SJ., Automatic indexing of online health resources for a French quality controlled gateway. Information Management & Processing; 2005.
[8] Pereira S., Evaluation of several medical terminologies to optimize medical and economic coding using medical records. Master dissertation; July 2005.
[9] Chalumeau M., Treluyer JM., Salanave B., Assathiany R., Cheron G., Crocheton, et al. Off label and unlicensed drug use among French office based paediatricians, Arch Dis Child. 2000 Dec;83(6):502-5.
[10] Lussier YA., Rothwell DJ., Cote RA., The SNOMED model: a knowledge source for the controlled terminology of the computerized patient record. Methods Inf Med., 1998 Jun;37(2):161-4.

Ubiquity: Technologies for Better Health in Aging Societies
A. Hasman et al. (Eds.)
IOS Press, 2006

Interactive Visualization and Navigation of Complex Terminology Systems, Exemplified by SNOMED CT

Erik Sundvall[a,1], Mikael Nyström[a], Håkan Petersson[a], Hans Åhlfeldt[a]
[a] *Department of Biomedical Engineering, Linköpings universitet, Sweden*

Abstract. Free-text queries are natural entries into the exploration of complex terminology systems. The way search results are presented has impact on the user's ability to grasp the overall structure of the system. Complex hierarchies like the one used in SNOMED CT, where nodes have multiple parents (IS-A) and several other relationship types, makes visualization challenging. This paper presents a prototype, *TermViz*, applying well known methods like "focus+context" and self-organizing layouts from the fields of Information Visualization and Graph Drawing to terminologies like SNOMED CT and ICD-10. The user can simultaneously focus on several nodes in the terminologies and then use interactive animated graph navigation and semantic zooming to further explore the terminology systems without loosing context. The prototype, based on Open Source Java components, demonstrates how a number of Information Visualisation methods can aid the exploration of medical terminologies with millions of elements and can serve as a base for further development.
Keywords: Terminology, Information Visualization, SNOMED CT, Medical Informatics

1. Introduction

Large terminology systems with complex intertwined structure can be hard to navigate and get acquainted with. Expandable trees like in Figure 1, from the *Clue* browser bundled with the SNOMED CT[2] distribution, work fine for some tasks, but have limitations when showing structures allowing multiple parents. It is also easy to run out of screen space and lose context if trying to expand and focus on several nodes at once. The goal of the *TermViz* prototype is to provide an alternative terminology exploration tool. The prototype development described here is a project that intersects the research fields of *Information Visualisation, Graph Drawing* and *Medical Terminology Systems.*

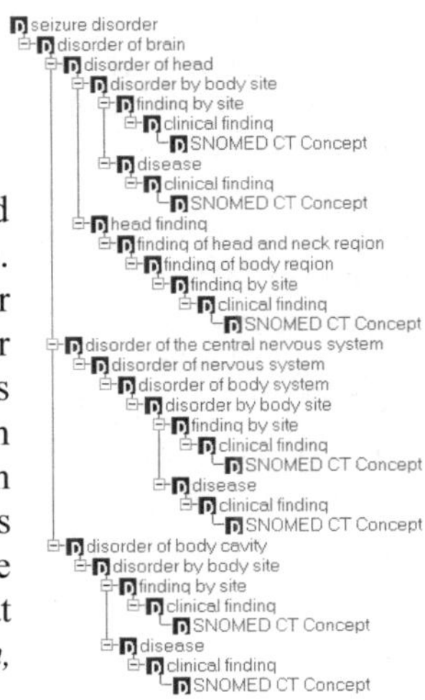

Figure 1: *Expandable tree.*

Information Visualisation (IV) aims to reduce the cognitive effort required to understand abstract information by engaging the human visual perception system, which is often under-utilized in more traditional text-oriented systems. Humans can rapidly scan, recognise, and recall

[1] Corresponding author: Erik Sundvall, Linköping University, Dept. of Biomedical Engineering, SE-581 85 Linköping, Sweden. erisu@imt.liu.se, http://www.imt.liu.se/~erisu/
[2] http://www.snomed.org/

images, and can easily detect changes in size, colour, shape, movement, or texture [1]. If structural changes are smoothly animated at a rate that the perceptual system can track, rather than instantly changed, then the cognitive effort of getting oriented in a new modified scene is reduced [2]. An enjoyable user interaction is considered a key element in IV [1]. IV's importance in medical applications has been highlighted by Chittaro [3].

Graph Drawing (GD) is a subfield of mathematics and computer science concerning for example graph theory and layout algorithms. GD also presents algorithms designed to fulfil "aesthetics criteria" like minimizing the number of edge crossings, edge bends and total graph area [4, 5]. The *Graphviz*[3] suite is a common GD tool used for example when visualizing parts of the *Gene Ontology*[4] and *Semantic Web*[5]. Most of the Graphviz-based systems do however mainly present static non-interactive graphs. A static graph generator without edge crossing minimization is also available in *Clue*.

The data sets (terminology systems) in TermViz are directed graphs or networks where the data elements have inherit relations and thus fit into the subfield ***Graph Visualisation*** [6] which is the intersection of *Information Visualisation* and *Graph Drawing.*

The objective of this paper is to present and discuss the features of the TermViz prototype.

2. Material and methods

The terminology systems to be visualized were converted to and stored in *RDF*[6] format. Components from the visualization toolkit, *prefuse*[7], were extended, modified and stringed together to create the visualizations. A surrounding graphical user interface (GUI) with menus, buttons etc. and a caching graph loader bridging the terminology storage and prefuse was created for the application. The storage is accessed remotely over a network connection. This enables us to have several visualizing front-ends (including Java Applets) accessing a single terminology server. The architecture also allows accessing and aggregating information from multiple networked information storages of different kinds (for example RDF storages, Relational Databases and UMLS[8]) by creating suitable graph loaders and defining appropriate queries.

The *SNOMED CT* version used was delivered as tables in three "flat files" containing concepts, descriptions and relations. *TermColl* is a collection of English and Swedish versions of five[9] terminology systems, prepared and imported into a tree-structured database by one of the authors (MN). *TermColl* has so far been used for generating a medical English-Swedish dictionary [7]. Mappings[10] from SNOMED CT to ICD-10 were also converted to RDF and stored.

---

[3] http://www.graphviz.org/
[4] http://www.geneontology.org/GO.tools.shtml
[5] http://www.w3.org/2001/11/IsaViz/
[6] Resource Description Framework, http://www.w3.org/RDF/ We used Kowari as storage,http://kowari.org
[7] http://prefuse.org/ - prefuse is intentionally spelled in lower case by its creators.
[8] http://www.nlm.nih.gov/research/umls/
[9] **ICD-10** and **ICF** (by WHO) **MeSH** (by NLM). **NCSP** (NOMESCO Classification of Surgical Procedures, by the Nordic Medico-Statistical Committee). **KSH97-P** (Swedish Primary Health Care Version of ICD-10, by the Swedish National Board of Health and Welfare).
[10] Crossmaps from SNOMED CT to ICD-10 delivered with the SNOMED CT distribution

prefuse[7] is a toolkit for interactive visualisation aimed at Java programmers. It divides the visualization task into a sequence of logical steps that can all be modified by the programmer. Usability studies have been conducted to ensure the toolkits' effectiveness and usability [8].

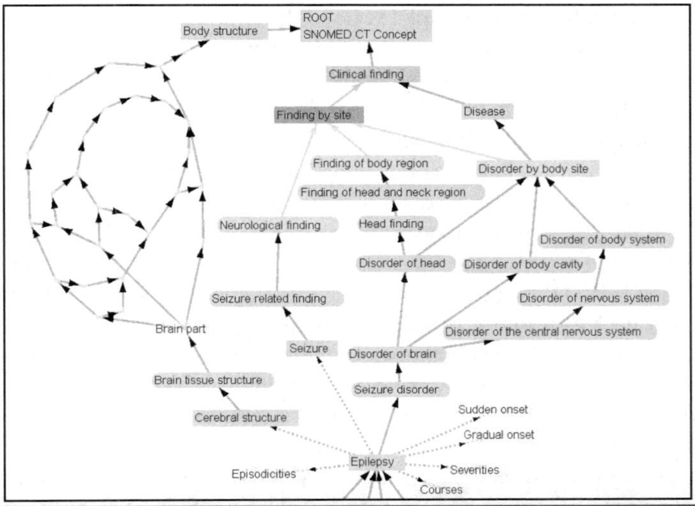

2a

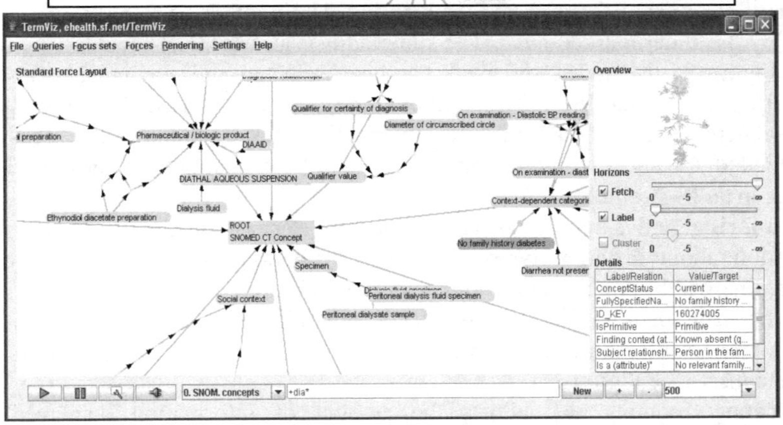

2b

Figure 2a: *A subset of concepts related to "Epilepsy" in SNOMED CT. Solid lines represent IS-A relations. Purple*[11] *nodes have been focused by user selections. When the cursor enters a node it becomes highlighted in orange, incoming links/edges turn yellow and outgoing turn red.* ***Figure 2b:*** *Parts of the surrounding GUI.*

3. Prototype Description[11]

In TermViz the user can start with an empty screen or by showing a set of predefined root nodes. A node can be expanded by manually selecting to show in- and/or outbound links. Some queries can return thousands of nodes and some SNOMED CT concepts have thousands of incoming links, so the number of hits returned by a query or expansion is bounded by a user-selectable upper limit.

[11] If you are reading a copy without colours, refer to corresponding author's homepage for color images.

Queries: By entering query text in the field at the bottom of the screen (Figure 2b) the user can find nodes matching the query. The query syntax[12] of the Lucene search engine used resembles that of Google. When issuing a query the results are added to the already visible nodes. Upon query submission the contents of the text and limit fields are inserted into one of the predefined query templates and then submitted for execution. The available query templates have descriptive short labels that are shown in the drop-down list in front of the query field. Templates can for example perform the task "Return all SNOMED CT concept nodes that have any description containing the word supplied in the search field".

Advanced users can modify or add new query templates and preselect a suitable graph loader for the query. This was designed with the purpose of making TermViz an easily evolvable system that can be extended by end users [9].

Query templates can take a node ID as input instead of free-text terms. Such queries can be used for tasks like "show all ICD-10 nodes that are crossmapped from this selected SNOMED CT node".

Focus sets and automatic graph traversal: A query returns a set of nodes and puts them in a *focus set.* Nodes can also be manually added or removed from a focus set. In Figure 2a the outgoing IS-A links, from all focused (purple[11]) nodes, have been automatically climbed as far as possible so that a natural root node has been reached. During the climb newly found nodes are continuously being loaded and visualized, and every node is assigned a *degree-of-interest* (DOI) value that decreases when the number of steps from focused nodes increases. The number of steps to climb can be limited by the slider marked *fetch.* The slider adjusts at what DOI-level to stop fetching.

In Figure 2a the node "epilepsy" has been selected and expanded to also show other relationships than IS-A. To avoid cluttering the view, the nodes added this way are not expanded further unless they are manually focused (added to the focus set) by being clicked. *Cerebral structure* and *Seizure* have been manually focused in Figure 2a, resulting in an IS-A climb.

Rendering and Layout: Different renderers (templates) can be used to convey information about different types or states of relations and nodes. Different line patterns, line widths and fill and colours can be configured.

By hovering over an edge or node, details about it are shown in the *Details* view at the bottom right of Figure 2b. A highlighting of the item and its neighbours occurs simultaneously (see Figure 2a).

Semantic zooming [6] adjusts the amount of information displayed, while geometric zooming only adjusts size. A simple form of semantic zooming is also illustrated in Figure 2a where only the focused nodes (purple) and their closest neighbours have text labels. The *Label* slider can be used to decide to what DOI-level labels should be rendered. This way we compress the structure between *Brain Part* and *Body Structure.*

The only layout algorithm currently used in the TermViz prototype is a force based physics simulation. Nodes exert "anti-gravity" repelling each other and links act as springs pulling connected nodes towards each other. A "drag force" (friction) is also active to stabilize the system. The simulation usually results in a fairly balanced self organizing graph structure where the forces balance each other. The force simulation can be stopped at any time using the pause button and nodes can be rearranged manually. In Figure 2a such manual rearrangements has been done to reduce size and

---

[12] http://lucene.apache.org/java/docs/queryparsersyntax.html

improve readability. Individual nodes can also be pinned down at specific positions during force simulation. This is currently illustrated by using sharp instead of rounded node corners. Force simulation parameters can be adjusted during runtime. Increasing the anti-gravity may for example increase readability if the graph is too dense.

4. Discussion

TermViz is useful as a terminology search and browsing tool in its current state, but there is ample room for improvements.

Schneiderman's task list: Schneiderman [1] lists seven desired tasks that an IV system should perform (written in *italics* below). Many of these tasks are accomplished by TermViz.

Smooth *zooming* and panning is available by mouse operations. *Filtering* is available in both queries and the semantic zooming functions. *Details-on-demand* are shown by hovering over the node or edge of interest. Expansion of individual node relations can also be preformed.

The *overview* task is partly accomplished by the available parallel zoomed out view, but no general overview of the entire data collection is available. A limited structural overview can be created by executing a query fetching the root nodes of the terminologies and their descendants a number of steps down. The possibility to *relate* items is partly inherit in the node-link structure of the application area and it is also possible to issue queries to find relations. Schneidermans *history* and *extraction* tasks are not yet available in the prototype.

Problems with force based layouts: Layouts based on force simulations have inherit problems by being non-deterministic. Different runs of a layout algorithm should ideally not produce radically different results since that violates the desired objective preserving the mental map of the user. [6] Algorithms describing how to create more predictable force based layouts are available [10] but have not yet been implemented and tested in TermViz.

The *Graph Drawing* aesthetic rule of "minimizing the number of edge crossings" has been shown to be a prioritized rule for attaining readable graphs [12]. The force based layout algorithms are not optimized reduce crossings but perform fairly well in many situations due to the laws of physics.

Most of the TermColl terminologies are actually simple trees where nodes have a single parent. There are several more efficient layouts than force layout for such graphs. DOI-trees [13] would probably work well in this application. The possibility for, and consequences of, combining graphs with different layouts on the same screen (including mapping links between the graphs) would be interesting to investigate. Ideally there should be many layouts available for the user to choose from.

Semantic zooming and rendering improvements: By grouping sets of nodes and edges with low DOI and replacing the whole group with a special cluster node, the number of visible items can be reduced. This is a semantic zooming method, referred to as *clustering*, that can improve readability and performance. The structure between *Brain Part* and *Body Structure* in Figure 2a could for example be turned into one or a couple of cluster nodes. Clustering would improve TermViz and its implementation is planned.

Adding symbols, patterns, more colours and shapes to the rendering would make it possible to convey more detailed information in the main graph.

Future and Applications: Obvious future development possibilities of TermViz would be to implement the above discussed improvements. Other possibilities would be tailoring TermViz to enable visualization of UMLS[8], FMA [13] and Gene Ontology/OBO[14] projects. It would also be possible to enable editing functions to support maintenance and authoring of terminologies and mappings between them. Methods from the prototype are currently being adapted and applied to other tools; Users of an archetype editor [14] will be supported in finding and selecting appropriate terminology bindings when creating archetypes[15] intended for structured clinical data entry. The ontology alignment tool *SAMBO*[16] has also started incorporating methods from the TermViz project.

Acknowledgements

This work was performed in the framework of the EU-funded Network of Excellence entitled Semantic Interoperability and Data Mining in Medicine. We would also like to thank J. Heer et al, the authors of *prefuse*.

References

[1] Shneiderman B. The eyes have it: A task by data type taxonomy for information visualizations. Proceedings of the 1996 IEEE Symposium on Visual Languages; 1996 Sep 3-6; Boulder, Colorado, USA. Washington DC: IEEE Computer Society; 1996. p. 336-343

[2] Robertson GG, Mackinlay JD, Card SK. Cone trees: Animated 3D visualizations of hierarchical information. CHI '91: Proceedings of the SIGCHI conference on Human factors in computing systems; 1991 Apr 27 - May 2; New Orleans, Louisiana, USA. New York: ACM Press; 1991. p. 189-194

[3] Chittaro L. Information visualization and its Application to Medicine. Artificial Intelligence in Medicine. 2001;22(2):81-88.

[4] Battista GD, Eades P, Tamassia R, Tollis IG. Graph drawing: Algorithms for the visualization of graphs. Prentice Hall; 1998

[5] Kaufmann M, Wagner D, editors. Drawing Graphs: Methods and Models (Lecture Notes in Computer Science, 2025) Springer-Verlag; 2001

[6] Herman I, Melançon G, Marshall MS. Graph Visualization and Navigation in Information Visualization: a Survey. IEEE Transactions on Visualization and Computer Graphics. 2000 Jan-Mar;6(1):24-43

[7] Nyström M, Merkel M, Ahrenberg L, Petersson H, Åhlfeldt H. Creating a Medical English-Swedish Dictionary Using Interactive Word Alignment. Submitted to BMC Medical Informatics and Decision Making

[8] Heer J, Card CK, Landay JA. prefuse: a toolkit for interactive information visualization. In Proc. CHI 2005, Human Factors in Computing Systems; 2005 Apr 2-7, Portland, Oregon, USA. Available from: http://jheer.org/publications/2005-prefuse-CHI.pdf

[9] Fischer G. Beyond "Couch Potatoes": From Consumers to Designers and Active Contributors. First Monday. 2002 Dec; 7(12). Available from: http://www.firstmonday.org/issues/issue7_12/fischer/

[10] Bertault F. A force-directed algorithm that preserves edge-crossing properties. Information Processing Letters archive, Volume 74 , Issue 1-2 , April 2000

[12] Purchase HC. Which Aesthetic has the Greatest Effect on Human Understanding? Lecture Notes In Computer Science; Vol. 1353. In Proc. 5th International Symposium on Graph Drawing Pages: 248 – 261, 1997

[13] Heer J, Card SK. DOITrees revisited: Scalable, Space-Constrained Visualization of Hierarchical Data. In Proc. Advanced visual interfaces 2004, Gallipoli, Italy. P. 421-424 Available from: http://jheer.org/publications/2004DOITree-AVI.pdf

[14] Forss M, Hjalmarsson J. Development of an archetype editor: A tool for modelling structure in electronic health records. (Master's thesis, 2006) http://urn.kb.se/resolve?urn=urn:nbn:se:liu:diva-6205

[13] http://sig.biostr.washington.edu/projects/fm/

[14] http://www.geneontology.org/ and http://obo.sourceforge.net/

[15] Archetypes as defined by openEHR, http://www.openehr.org/

[16] http://www.ida.liu.se/~iislab/projects/SAMBO/

Ubiquity: Technologies for Better Health in Aging Societies
A. Hasman et al. (Eds.)
IOS Press, 2006

Cross-Lingual Alignment of Biomedical Acronyms and their Expansions

Kornél MARKÓ [a], Philipp DAUMKE [a], Udo HAHN [b]

[a] *Medical Informatics Department, Freiburg University Hospital, Germany*
[b] *Jena University Language & Information Engineering (JULIE) Lab, Jena, Germany*

Abstract. We propose a method that aligns biomedical acronyms and their definitions across different languages. The approach is based upon a freely available tool for the extraction of abbreviations together with their expansions, and the subsequent normalization of language-specific variants, synonyms, and translations of the extracted acronym definitions. In this step, acronym expansions are mapped onto a language-independent concept-layer on which intra- as well as interlingual comparisons are drawn.

1. Introduction

The understanding of abbreviations in biomedical texts is very important for natural language processing applications, such as information extraction or information retrieval systems. This is witnessed, in particular, for protein and gene expressions from biomedical texts (as well as the relations between them). Those expressions frequently consist of acronyms, but their definitions in the text might differ from the ones found, e.g., in an external database, such as ARGH, AcroMed, or SaRAD (cf. [1] for an overview).

Multiple expansions for a unique acronym, or multiple acronyms for a unique definition, tend to create difficulties when we try to match natural language expressions to a standardized vocabulary such as the UMLS or MeSH. In an information retrieval scenario, unresolved acronyms will possibly lead to a loss of precision: Does "*AD*" refer to "*Alzheimer's Disease*" or to "*allergic dermatitis*"? The problem of ambiguity becomes even harder, when multilingual documents are made available to a search interface, which is the case for most Web search engines. As a consequence, the acronym "*AD*" can have the German expansion "*atopische Dermatitis*", Spanish "*aurícula derecha*", Portuguese "*agua destilada*", and many more. On the other Hand, the German acronym equivalent to "*Alzheimer's Disease*" is "*AK*" ("*Alzheimer Krankheit*") or "*MA*" ("*Morbus Alzheimer*") and for Spanish "*EA*" ("*enfermedad de Alzheimer*").

There has been extensive research on the automatic extraction of short-form/long-form pairs (abbreviations and acronyms mapped to their expansions/definitions) within one language [2,3,4,5,6]. Different ways how abbreviations are actually used in written (medical) language have been studied [7], however, little attention has been paid on how acronyms behave across languages, which is the focus of this paper.

2. Analysis of Terms into Subwords

We propose a method that automatically aligns acronyms and their definitions across different languages. It is based upon a dictionary the entries of which are *equivalence classes* of subwords, i.e., semantically minimal units [8]. These equivalence classes capture intralingual as well as interlingual synonymy. As equivalence classes abstract away from subtle particularities within and between languages and reference to them is realized via a language-independent description system they form an *interlingua*.
Subwords are assembled in a multilingual lexicon and thesaurus, with the following considerations in mind:

- Subwords are listed with their attributes such as language (English, German, Portuguese, Spanish) or subword type (stem, prefix, suffix, invariant). Each lexicon entry is assigned one or more morpho-semantic identifier(s) representing its equivalence class, the MID.
- Semantic links between synonymy classes are added. We subscribe to a shallow approach in which semantic relations are restricted to a paradigmatic relation *has-meaning*, which relates one ambiguous class to its specific readings, [1] and a syntagmatic relation *expands-to*, which consists of predefined segmentations in case of utterly short subwords.[2]

Figure 1 depicts how source documents (top-left) are converted into an interlingual representation by a three-step procedure. First, each input word is orthographically normalized in terms of lower case characters and according to language-specific rules for the transcription of diacritics (top-right). Next, words are segmented into sequences of subwords from the lexicon (bottom-right). Finally, each meaning-bearing subword is replaced by a language-independent semantic identifier, a MID, which unifies intralingual and interlingual (quasi-)synonyms, thus producing the interlingual output representation of the system (bottom-left). In Figure 1, bold-faced MIDs co-occur in both document fragments.

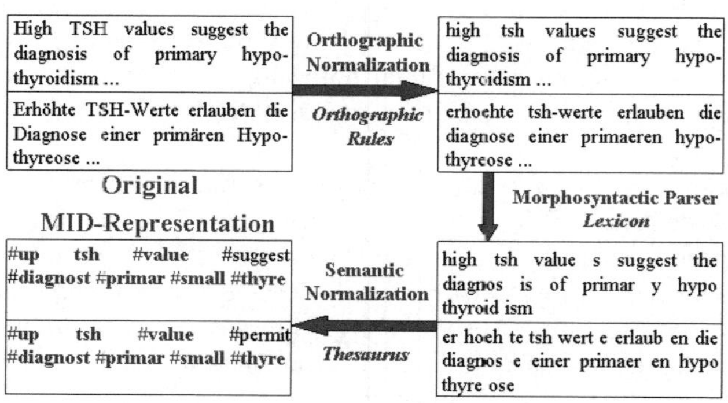

Figure 1: Morpho-Semantic Indexing (MSI)

The combined subword lexicon currently contains 72,513 entries, with 22,067 for English, 22,497 for German, 14,888 for Portuguese and 13,061 for Spanish. All of these entries are related in the thesaurus by 20,990 equivalence classes.

[1] For instance, {*head*} → {*zephal,kopf,caput,cephal,cabec*} OR {*leader,boss,lider,chefe*}
[2] For instance, {*myalg*} → {*muscle,muskel,muscul*} AND {*schmerz,pain,dor*}

In earlier experiments on cross-language information retrieval [8], we have shown the usefulness of representing medical documents on an interlingual layer. To be able to properly account for acronyms, we here adapt previous work on automatic acronym detection to the needs of our interlingual lexicon representation approach.

3. Extracting Biomedical Acronyms

Schwartz and Hearst [5] offer a simple and fast algorithm for the extraction of abbreviations and their definitions. The algorithm achieves 96% precision and 82% recall on a standardized test collection, thus, performs at least as good as other existing approaches [2,3,4,6]. Generally, the process of identifying abbreviations and their full forms can be seen as a two-step method: the extraction of possible short-form/long-form (SF-LF) pairs and the validation of SF-LF terms among the list of possible candidates in a sentence.

1. Extraction of possible SF-LF terms: SF-LF pairs are identified by the adjacency to parentheses. The two basic patterns *LF (SF)* and *SF (LF)* are thereby distinguished. A short form has following characteristics: it contains between 2 and 10 characters, it has a maximum of two words, at least one character is a letter and its first character is alphanumeric. The long form must immediately appear before or after the corresponding short form and the maximum number of words is constrained by min(|A|+5, |A|*2), with |A| being the number of characters in the corresponding SF. In practice, the first pattern *LF (SF)* proved to occur more frequently. Only if a criterion for an *LF (SF)* pattern is not fulfilled (e.g., more than two words inside the parentheses), the second pattern *SF (LF)* is tried.

2. Identifying the correct SF-LF term: A set of rules is used to identify the correct SF-LF pair out of a set of possible candidates. Most importantly, each character in the short form must match a character in the long form. Characters of the short form must appear in the same linear order as in the long form. Furthermore, the first character of the SF has to be the same in the LF. Finally, all LFs are removed which are shorter than the corresponding SF, or which include the corresponding SF within one of their single words.

For our experiments, we used corpora taken from heterogeneous WWW sources, including MEDLINE abstracts. With over 250m words the derived English corpus was much larger than those for the other languages involved (37m for German, 14m for Portuguese, and 11m for Spanish, cf. Table 1). Using the algorithm described above, we collected over 1.2m abbreviations together with their long forms for English, about 30,000 for German, and about 8,000 for Portuguese and Spanish. In contradistinction to the other languages, the English corpus included a large number of expert-level MEDLINE abstracts. As a consequence, every 200th token in the collection was classified as an acronym. For the other languages (for which the corpora included a higher amount of health care consumer information), this ratio is much smaller (0.06 to 0.08 percent of the corpora).

After the acquisition of SF-LF pairs, the long forms were processed with the morpho-semantic indexing (MSI) procedure as depicted in Figure 1. Upon prior manual inspection of document samples we observed that English long forms also frequently occur in German, Portuguese, and Spanish texts. Therefore, we had to decide which lexicon we wanted to use for MSI. Finally, we segmented the long forms using every lexicon involved. Those language hypotheses were kept for which the underlying lexicon yielded complete lexical coverage with regard to the specific long form. If there were more than one remaining language hypothesis, the document language (if not English) was preferred over English.

This procedure led to over one million SF-LF pairs completely covered by the MSI procedure for English (83%), 26,770 (89%) for German, 7,065 (83%) for Portuguese, and 4,723 (61%) for Spanish (cf. Table 1). In the following, we will only focus on this (sub-)set of extracted abbreviations. Figure 2 gives an impression of how frequent unique SF-LF pairs occur in the corpora considered, for each language condition. 61% to 76% of all acronyms extracted occur only one time, 12% to 23% appear two times, whilst five or more occurrences are found for 6% to 12% of all SF-LF pairs. As depicted in Table 2 (Column 2), 212,470 unique SF-LF pairs were generated for English, 4,276 for German, 3,934 for Portuguese, and 2,037 for Spanish. Column 3 of the table shows the average number of corpus occurrence for each unique SF-LP pair. After the morpho-semantic normalization of long forms (MSI, as shown in Figure 1), the number of unique SF-LF pairs decreased to about 190,000 for English (more than 3,600 for German and Portuguese and about 1,900 for Spanish). Accordingly, the number of tokens per type increases, as depicted in the fifth column of Table 2. As an example, morpho-syntactical variants in long forms such as in *"CTC"-"computed tomographic colonography"* and *"CTC"-"computed tomography colonography"* are unified.

Language	**Corpus Tokens**	**Acronyms**
English	250,258,039	1,253,311 (0.5%)
MSI-Covered		1,033,929 (82.5%)
German	37,105,363	29,967 (0.08%)
MSI-Covered		26,770 (89.3%)
Portuguese	13,904,790	8,532 (0.06%)
MSI-Covered		7,065 (82.8%)
Spanish	11,103,066	7,714 (0.07%)
MSI-Covered		4,723 (61,2%)

Table 1: Corpus and Acronym Extraction Statistics

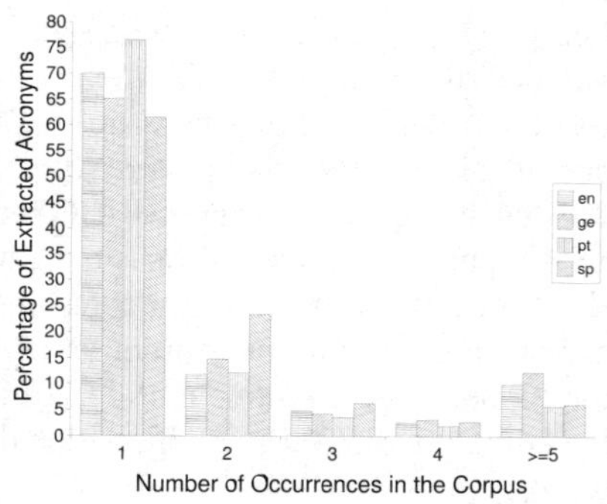

Figure 2: Distribution of SF-LF Occurrences in each Corpus

Language	**Surface**		**MSI**	
	Unique	**Ratio**	**Unique**	**Ratio**
English	212,470	4.87	189,639	5.45
German	4,276	6.26	3,653	7.33
Portuguese	3,934	1.20	3,633	1.95
Spanish	2,037	2.32	1,911	2.47

Table 2: Effects of Morpho-semantic Normalization in Terms of Unique SF-LF Pairs and Tokens per Type

Intra-Lingual Phenomena

Two basic phenomena have to be considered when we inspect the results for one given language: First, one short form can have multiple long forms, and, second, one long from can have multiple short forms. An example for a SF ambiguity is given with "*ABM*" mapped to "*acute bacterial meningitis*" or to "*adult bone marrow*". Table 3 (Columns 2-4) shows the numbers of different long forms for each short form, both for the baseline condition (lower-case surface form) and the MSI condition. For English, 82,501 unique short forms were extracted. The average number of long forms associated to unique SFs decreases from 2.56 to 2.30 for MSI, as expected. The same relationship can also be observed for the other languages we considered. The second phenomenon is also observable in all languages involved in our experiments. For example, the noun phrase "*acid phosphatase*" has nine different abbreviations in the English corpus processed (case insensitive): "*AcP*", "*acPAse*" "*ACP-ase*", "*Acph*", "*ACPT*", "*AP*", "*APase*", "*AphA*", and "*APs*". Table 3 (Columns 5-8) depicts the numbers describing this phenomenon. For English, a total of 184,639 different long forms were extracted, embedded in 212,470 different SF-LF pairs (cf. Table 2). Thus, each LF is associated to 1.15 SFs, on the average. For the MSI condition, there are less different long forms, hence, the ratio slightly increases, for all languages.

	LF/SF			**SF/LF**			
	Surface		**MSI**	**Surface**		**MSI**	
Language	**# SF**	**Average**		**# LF**	**Average**	**# LF**	**Average**
English	82,501	2.56	2.30	184,639	1.15	154,693	1.23
German	2,954	1.45	1.24	4,187	1.02	3,515	1.04
Portuguese	2,517	1.56	1.44	3,798	1.04	3,395	1.07
Spanish	1,450	1.41	1.32	1,979	1.03	1,825	1.05

Table 3: Number of Long Forms for each Short Form (*LF/SF*, Columns 2-4) and Number of Short Forms for each Long Form (*SF/LF*, Columns 5-8) in Absolute (#) and Relative (*Average*) Numbers

Inter-Lingual Phenomena

Identical SF-LF Pairs. The first observation we made is that quite often SF-LF pairs are identical across languages on the surface level. Especially common or technical English terms also appear in other languages, such as "*WHO*" and its expansion "*World Health Organization*", "*PCR*" and its definition "*polymerase chain reaction*", or "*IL*" associated to "*interleukin*". In numbers (cf. Table 5, Row 2), we found 584 identical SF-LF for English-German, 181 for English-Portuguese, 192 for English-Spanish, 35 for German-Portuguese, 40 for German-Spanish, and 106 for Portuguese-Spanish (the latter sets also may contain some English SF-LF pairs).

Identical SF, Different LF. One way of identifying possible translations of long forms is to collect those long forms, which are connected to a unique short form at the surface level. For example, if an English document contains "*WHO*"-"*World Health Organization*" and a German document contains "*WHO*"-"*Weltgesundheits-organisation*", the long forms can be regarded as possible translations of each other. For English-German, 100,915 of these pairs can be extracted, for English-Portuguese 151,037, for English-Spanish 109,568, for German-Portuguese 2,468, for German-Spanish 1,709, and for Portuguese-Spanish there are 3,454 of these hypothesized translations (Table 4, Row 3). Of course, these sets also contain syntactical variants and a large number of false positives, since short forms are used differently across languages. Therefore, we switched our perspective to the interlingual layer of long form representations.

Surface	**EN-GE**	**EN-PT**	**EN-SP**	**GE-PT**	**GE-SP**	**PT-SP**	**Total**
I(SF),I(LF)	584	181	192	35	40	106	1,138
I(SF),D(LF)	100,915	151,037	109,568	2,468	1,709	3,454	369,151
MSI	**EN-GE**	**EN-PT**	**EN-SP**	**GE-PT**	**GE-SP**	**PT-SP**	**Total**
I(SF),T(LF)	2,479	665	573	81	110	250	4,158
D(SF),T(LF)	3,212	3,982	2,136	328	290	207	10,155

Table 4: Statistics on Cross-Lingual Acronym Extraction: Results for Identical (I), Different (D) and Translations (T) of Short Forms (SF) and Long Forms (LF)

Identical SF, Translation of LF. In this condition we examined those cases, in which short forms were identical and long forms were different at the surface level, but identical at the interlingual layer, comparing SF-LF pairs extracted from the different source corpora. As a result, we obtained lists of bilingually aligned terms, such as English "*acute lymphatic leukemia*" linked to the German "*akute lymphatische Leukämie*" via the shared short term "*ALL*". As an example, 2,479 translations were generated for English-German using this heuristics (cf. Table 5, Row 5).
Different SF, Translation of LF. In this scenario, we examined those cases, for which the long forms were identical or translations of each other (identical at the interlingua layer), but with different short forms. This captures interesting constellations such as English "*AIDS*" ("*acquired immune deficiency syndrome*") aligned to Spanish or Portuguese "*SIDA*" ("*síndrome de inmunodeficiencia adquirida*"). We collected 207 of these translations for Portuguese-Spanish, up to 3,982 for English-Portuguese (Row 6).

4. Related Work and Conclusion

Several different techniques for the automatic extraction of abbreviations and their definitions from biomedical text have been developed up until now [2,3,4]. Our approach for the multilingual alignment of acronyms is tied up to these precursors. By translating long forms into an interlingual layer, acronyms and their definitions are made comparable across different languages with a high coverage.
Since our work focuses on cross-language information retrieval [8], we are interested in the cross-lingual mapping of lexical entries. Hence, we incorporated those extracted SF-LF pairs in our subword lexicons, for which the long form is a translation of another, at least one, long form in a different language (after mapping on the interlingua layer). For the monolingual mapping of short forms to long forms, we decided to additionally collect those language-specific SF-LF pairs, which occur at least 2 times on the layer of the interlingua. As a result, the lexicon size for all languages considered increased from initially 72,513 entries to 138,343 lexical items.
Acknowledgements. This work was supported by Deutsche Forschungsgemeinschaft, (KL 640/5-2), and the European Network of Excellence "Semantic Mining" (507505).

References

[1] Wren JD, Chang JT, Pustejovsky J, Adar E, Garner HR, Altman RB. Biomedical term mapping databases. Nucleic Acids Research, 33:D289– 293, 2005
[2] Adar E. SaRAD: a simple and robust abbreviation dictionary. Bioinformatics, 20(4), 2004
[3] Chang JT, Schütze H, Altman RB. Creating an online dictionary of abbreviations from MEDLINE. JAMIA, 9(6):612–620, 2002
[4] Pustejovsky J, Castaño J, Cochran B, Kotecki M, Morrell M. Automatic extraction of acronym-meaning pairs from MEDLINE databases. In Proc MedInfo, 10(Pt.1): 371–375, 2001
[5] Schwartz A, Hearst M. A simple algorithm for identifying abbreviation definitions in biomedical text. In Pacific Symposium on Biocomputing, pages 451–462, 2003
[6] Yu H, Hripcsak G, Friedman C. Mapping abbreviations to full forms in biomedical articles. JAMIA, 9(3):262–272, 2002
[7] Liu H, Aronson AR, Friedman C. A study of abbreviations in MEDLINE abstracts. In Proc AMIA Symp: 464–468, 2002
[8] Markó K, Schulz S, Hahn U: MorphoSaurus - Design and Evaluation of an Interlingua-based, Cross-language Document Retrieval Engine for the Medical Domain. Meth Inf Med. 4/2005(44): 537-545

Ubiquity: Technologies for Better Health in Aging Societies
A. Hasman et al. (Eds.)
IOS Press, 2006

An Ontology driven collaborative development for biomedical terminologies: from the French CCAM to the Australian ICHI coding system

Jean M. RODRIGUES[a,1], Alan RECTOR[b], Pieter ZANSTRA[c], Robert BAUD[d], Kerry INNES[e], Jeremy ROGERS[b], Anne M. RASSINOUX[d], Stefan SCHULZ[f], Beatrice TROMBERT PAVIOT[a], Huib ten NAPEL[c], Lucienne CLAVEL[a], Egbert van der HARING[c], Ceu MATEUS[g]

[a] *University of Saint Etienne France,*
[b] *University of Manchester, United Kingdom*
[c] *Radboud University of Nijmegen Medical Centre, The Netherlands*
[d] *University Hospitals of Geneva, Switzerland*
[e] *National Centre for Classification in Health, Australia*
[f] *Freiburg University Hospital, Germany*
[g] *Nova University of Lisbon, Portugal*

Abstract. The CCAM French coding system of clinical procedures was developed between 1994 and 2004 using, in parallel, a traditional domain expert's consensus method on one hand, and advanced methodologies of ontology driven semantic representation and multilingual generation on the other hand. These advanced methodologies were applied under the framework of an European Union collaborative research project named GALEN and produced a new generation of biomedical terminology. Following the interest in several countries and in WHO, the GALEN network has tested the application of the ontology driven tools to the existing reduced Australian ICHI coding system for interventions presently under investigation by WHO to check its ability and appropriateness to become the reference international coding system for procedures. The initial results are presented and discussed in terms of feasibility and quality assurance for sharing and maintaining consistent medical knowledge and allowing diversity in linguistic expressiveness of end users.

Keywords: Ontclogy, Biomedical, erminology, Coding system, Healthcare, Surgical procedures.

1. Introduction

Since the beginning, clinical terminological systems, classifications and coding systems have been developed by independent, divergent and uncoordinated approaches which produced non reusable systems on overlapping fields for different needs. For some decades, several broad pre-coordinated or compositional systems have been proposed

[1] Corresponding Author: 650 St Jean Bonnefonds, France.

to users targeting different goals. The most well known are the UMLS (Unified Medical Language System) [1], SNOMED international [2], Read Clinical Classification Version 3 [3], LOINC [4] for clinical laboratories, DICOM SDM [5] for imaging, SNOMED CT [6], Convergent Medical Terminology (CMT) [7].

On the other hand most of developed countries have lasted to maintain, update and modify their own coding systems for procedures, as well as national adaptations of ICD, in order to manage and to fund their healthcare delivery. The most significant efforts were done in Australia with ACHI (Australian Classification of Health Interventions) or ICD10 AM [8], and in France with CCAM (Classification Commune des Actes Médicaux) [9].

In the last 15 years a new paradigm has been established thanks to advances in computer science and artificial intelligence, giving raise to parallel, mostly uncoordinated activities. They are based on terminology server architectures making available knowledge bases built upon multi-hierarchies of concepts associated with terminological subsumption These knowledge representations are often named ontology [10] which is a term originating from metaphysics. In fact they are implementations of formal logic describing the meaning of terms. These representations have been supported by specific software tools and natural language processing techniques, including the linkage of lexicons from different national languages [11] to the knowledge representation The most important achievements are GALEN (Generalised Architecture for Languages, Encyclopaedias and Nomenclatures in Medicine) [12], SNOMED CT [13] and FMA (Foundational Model of Anatomy) [14]. Finally WHO has initiated the revision of ICD which will take advantage of such achievements to enhance the interoperability of products of family of international classifications (ICD, ICF and classification of procedures) by the year 2011.

2. Galen and CCAM

The new French coding system for surgical procedures, CCAM, has been developed using a dual methodology: the traditional domain expert consensus and the formal representation GALEN [15]. The GALEN formal representation consists of around 52000 entities, with 800 included links. In order to represent the 7,478 CCAM surgical procedures in the GRAIL (Galen Representation and Integration Language) formal language, 2,400 concepts descriptions and 59 semantic links were used. Amongst these 2,400 descriptions, 1297 are from the semantic category Anatomy, 271 from Pathology, 231 from Device, 186 from Deed. Amongst the 59 semantic links 27 are for the 4 previous categories and 7 only for Pathology. The development of CCAM required an effort of 1.5 person century and a GALEN representation [16] with a NLP interface [17] under the constraint of conformity with the European standards EN 1828 [18] and EN 12264 [19]. This representation allowed to perform consistence checking with multiple inheritances of the rubrics proposed by the domain experts consensus and their modification for around 20 %. It also increased the existing model with around 700 new entities. Among results presented elsewhere [17], the most stable conclusion is the inability of terminology alone to produce consistent knowledge as exemplified by Table 1.

Table 1. Cervical myelography (English): Myélographie cervicale (French ,CCAM)

Intermediate representation

```
MAIN imaging
    ACTS_ON spinal cord
        IS_PART_OF cervical spine
    ACTS_ON_1 Nerve: root of spinal nerve
        IS_PART_OF cervical spine
    ACTS_ON_2 subarachnoid space
        IS_PART_OF cervical spine
    BY_MEANS_OF Xray Device
    BY_TECHNIQUE injecting
        ACTS_ON contrast medium
            HAS_DESTINATION subarachnoid space
                IS_PART_OF cervical spine
        BY_APPROACH TECHNIQUE inspecting
            ACTS_ON peritoneal cavity
            BY_MEANS_OF laparoscope
        WITH_GUIDANCE_BY laparoscope
```

Formal grail representation

```
ImagingProcess which
hasSpecificSubprocess Injecting which
    involves Material which
        isActedOnSpecificallyBy Displacement which
            isBetaPartitiveConnectionOf BodyCavity which
                isSolidRegionOf CervicalSpine
                isSpaceBoundedBy Arachnoid
                isSpaceBoundedBy PiaMater
hasContextOfUse Radiology
            hasRadioluscence radioopaque
    playsClinicalRole SurgicalRole
actsMultiplyOn SpinalCord which
    isSolidRegionOf CervicalSpine
actsMultiplyOn SPET_root of spinal nerve which
            isSolidRegionOf CervicalSpine
actsMultiplyOn BodyCavity which[2]
            isSolidRegionOf CervicalSpine
            isSpaceBoundedBy Arachnoid
            isSpaceBoundedBy PiaMater
hasPhysicalMeans XrayMachine
```

Generations in natural language

English: Imaging the spinal cord of the spinal spine, the roots of the spinal nerves of the cervical spine and the subarachnoid space of the spinal spine by injecting contrast medium in the subarachnoid space of the cervical spine with a Xray machine.

French: Image de la moelle épinière de la colonne vertébrale cervicale, des racines des nerfs spinaux de la colonne vertébrale cervicale et de l'espace sous arachnoïdien de la colonne vertébrale cervicale par injection de produit de contraste dans l'espace sous arachnoïdien de la colonne vertébrale cervicale avec un appareillage radiologique.

3. ICHI Galen

ICHI, the International Classification of Health Intervention, has been designed by the Australian National Centre for Classification in Health. Its primary purpose is to meet the needs of countries which use ICD-10, but do not presently code intervention/procedure data, and who need access to a simple classification system. ICHI contains 1,420 procedure codes and is based on the *Australian Classification of*

Health Interventions (ACHI) or ICD10 Australian Modification. ACHI contains about 6,000 procedure codes [8]. Following the interest for the Galen CCAM developments in several countries and in agreement with the WHO Classifications and terminology unit, Department of measurements and health information systems (MHI), the Galen network extended to Portuguese speaking domain specialists decided to test the method on the available ICHI beta version with the following objectives:
- To demonstrate that a compositional approach to developing WHO terminologies is feasible and cost-effective ;
- To demonstrate a more collaborative approach to building international terminologies, including some open developments ;
- To demonstrate that automatic translations based on an ontology make the translation process more transparent and of better quality assured ;
- To provide a report on technical feasibility and costs to inform future decision making.

4. Material and Method

We started by the ICHI initial rubrics' linguistic expressions to construct a semi-formal representation in the two modelling centres of the University of Saint Etienne Medical Informatics centre (France) and the Radboud University of Nijmegen Medical Centre (The Netherlands), using a web based software made freely available by Kermanog: Claw Inc. This step allows to check the conformity to the 2 European standards EN 1828 and EN 12264.
The second step was the translation into the GRAIL formal representation language, done by the representation manager centre of the University of Manchester Medical Informatics group (UK), where the knowledge encoded was checked against formal logic rules.
The following step was natural language generation (NLG) done by the University Hospitals of Geneva Medical Information Service (Switzerland), producing English, Portuguese and French controlled linguistic expressions. An example of the input and of the output is summarised in table 2.

Table 2 ICHI Galen analysis

Initial rubric in English ICHI-Chapter XV, 1381 Immobilisation of fracture or dislocation of spine
Natural language generations
English : Immobilizing of the spine, for treatment of a dislocation lesion of the spine, without closed reduction...
Portuguese : Imobilização da coluna vertebral, para tratamento de uma luxação da coluna vertebral, sem redução fechada ...
French : Immobilisation de la colonne vertébrale, pour traitement d'une luxation de la colonne vertébrale, sans réduction fermée ...

The last step consisted of a cross-validation study between the controlled and the initial linguistic expressions done by the University of Saint Etienne Medical Informatics Centre (France), the Freiburg University Hospital (Germany), and the Nova University of Lisbon (Portugal).

5. Discussion

We restrict the discussion of the added value of ontology driven tools for biomedical terminologies to the following example stressing the inconsistent usage of the term arthroplasty across the different rubrics of the same biomedical classification (Table 3)

Table 3 QA Inconsistent usage of the same term: Arthroplasty

Initial rubric in English ICHI-Chapter XV	Natural language generations
1518 Arthroplasty of knee	**English**: Repairing of the knee joint **Portuguese**: Reparação da articulação do joelho **French**: Réparation de l'articulation du genou
1519: Arthroplasty of knee with bone graft to femur or tibia	**English**: Replacement of the knee joint with a prosthesis of the knee joint, with installation of bone graft tissue in the femur or the tibia by means of harvesting bone **Portuguese**: Substituicão da articulação do joelho com una protese da articulação do joelho, com instalação de transplante de tecido de osso no fémur ou na tíbia por meio de colhida /colheita do osso'. **French**: Remplacement de l'articulation du genou par une prothèse de l'articulation du genou avec mise en place de tissu osseux greffé sur le fémur ou le tibia au moyen de prélèvement osseux
1489 Arthroplasty of hip	**English** : Repairing of the hip joint . **Portuguese**: Reparação da articulação do quadril/anca . Reparação plástica ou substituição da articulação da anca com uma prótese ? **French**: Réparation de l'articulation de la hanche. Réparation plastique ou Prothèse de hanche?

Within the rubrics 1518 and 1519 the same term arthroplasty is used once to define a limited plastic repairing of the joint and once to define a complete replacement of the knee with a prosthesis. But the 2 rubrics do exist and may allow the separation of these two different intervention classes. Unfortunately, for the hip joint there is only one rubric with the term arthroplasty, and it is not possible to separate these two types of interventions. This example shows that it is not possible to develop a consistent biomedical terminology without ontology driven tools when the semantic interoperability becomes the top priority of the healthcare systems.

6. Conclusion

This work has demonstrated the feasibility to apply ontology driven tools and a collaborative approach to the development of an international terminology. Such an organisation allows better quality assurance by sharing and maintaining consistent medical knowledge and allowing diversity in linguistic expressiveness at the level of end users. It is still necessary to convince the decision makers that it can be faster and cheaper by multiplying the number of modelling centres working in their local language and investing in few web based tools platforms than relying on the traditional bottomless domain expertise.

The healthcare systems are becoming more and more knowledge-based and knowledge communication between healthcare professionals from different linguistic groups and their training are mainly provided through computers. The terminology they use must be able to allow their expressiveness and in a sense their ambiguity but must be constrained by a formal logic representation base within the computer to insure the safety of healthcare delivery.

The process we have developed for classifications and coding systems shall be tested for any clinical terminology.

Acknowledgements

We wish to thank the WHO classifications and terminology unit Department of measurements and health information systems in Geneva for granting us the opportunity to apply the GALEN tools to one of the major FIC tools presently under investigation.

References

[1] McCray AT, Nelson SJ. The representation of meaning in the UMLS. *Methods Inf Med* 1995; 34(1-2):193-201.

[2] Iggulden P,Price C. SNOMED Clinical Terms – Looking forward to the next phase *Br J Healthcare Comput Info Manage* 2001; 18(10):20-22.

[3] Read Clinical Classification (RCC, Read Codes). Updated by NHS Centre for Coding and Classification. Crown copyright, 1990.

[4] Logical Observation Identifiers Names and Codes(LOINC). See :http://www.loinc.org/]

[5] DICOM see http://www.xray.hmc.psu.edu/dicom/dicom_home.html

[6] SNOMED Clinical Terms. College of American Pathologists.see http://www.snomed.org/

[7] Dolin RH. Kaiser Permanente's Convergent Medical Terminology. Testimony to the National Committee on Vital and Health Statistics, Subcommittee on Standards and Security. May 22, 2003. http://ncvhs.hhs.gov/030522sstr.htm]

[8] National Centre for Classification in Health see http://www3.fhs.usyd.edu.au

[9] Agence Technique de l'Information Hospitalière see http://www.sante.atih.gov.fr

[10] Gruber T. A translation approach to potable ontology specifications. Knowledge Acquisition 1993; 5(2):199-220.

[11] Smith B, Ceusters W. Towards industrial strength philosophy: how analytical ontology can help medical informatics. Interdisciplinary Science Reviews 2003; 28:106-11.

[12] Rector AL, Johnson PD, Tu S, Wroe C, Rogers J. Interface of inference models with concept and medical record models. In: Quaglini S, Barahona P, Andreassen S, editors. Artificial Intelligence in Medicine Europe (AIME); 2001; Cascais, Portugal: Springer Verlag; 2001. p. 314-323.

[13] Dolin RH, Huff SM, Rocha RA, Spackman KA, Campbell KE. Evaluation of a « lexically assign logically refine »strategy for semi-automated integration of overlapping terminologies.J Am Med Informatics Assoc 1998 ;5 :203-213

[14] UWDA; University of Washington Digital Anatomy A reference ontology for biomedical informatics:the Foundational Model of Anatomy (FMA).see http://fma.biostr.Washington.edu

[15] http://www.opengalen.org/

[16] Rector AL, Zanstra PE, Solomon WD, Rogers JE, Baud R, Ceusters W, et al. Reconciling Users' Needs and FormalRequirements: Issues in developing a Re-Usable Ontology for Medicine. IEEE Transactions on Information Technology in BioMedicine 1999;2(4):229-242.

[17] Trombert-Paviot B, Rodrigues J-M, Rogers J, Baud R, van der Haring E, Rassinoux A-M, Abrial V, Clavel L, Idir H. GALEN: a third-generation terminology tool to support a multipurpose national coding system for surgical procedures. International Journal of Medical Informatics 2000; 58–9: 71–85.[18] CEN AFNOR EN NF 1828:2002. Health informatics – Categorial Structure for classifications and coding systems of surgical procedures .[19] CEN AFNOR EN NF 12264:2005. Health informatics – Categorial Structure of concept systems.

8. Education

8.1 Health and Biomedical Informatics Education

Ubiquity: Technologies for Better Health in Aging Societies
A. Hasman et al. (Eds.)
IOS Press, 2006

The Redesign of the Medical Informatics Master of Science Course at the University of Amsterdam

Monique W. M. Jaspers[a] and Arie Hasman[a]
[a] *Department of Medical Informatics, Academic Medical Center- University of Amsterdam*
Amsterdam, The Netherlands

Abstract. Objectives: To describe our new two years Master of Science (MSc) program starting in September 2006 at the University of Amsterdam- Academic Medical Center, The Netherlands. **Methods:** We elaborate shortly on the mission, organizational structure and new contents of this new MSc course in medical informatics. **Results:** Through the years, our medical informatics university program underwent some major revisions of which the transition from a four years course into a three years BSc program and a two years MSc program has been the most fundamental. The new MSc program is aimed at (international) baccalaureates in medical informatics, computer science, medicine, health sciences, and biology. Besides, health care professionals or professionals with a background in computer science may enter the program. The program length is two years, comprising four study semesters of 30 European Credits each (*EC*, 1 EC corresponding to 27 hours study load), equalizing 120 EC in total of which 48 EC are reserved for the master's thesis. **Conclusions:** With the new set up of the MSc program, that will be offered in English, we hope to both accommodate the learning needs of our own baccalaureates and to attract international baccalaureates and other professionals to this course. Our ultimate aim is to bring forth medical informatics specialists who are well equipped to make significant contributions to the field.
Key words: Medical informatics; Health informatics; Curriculum; Education; Training

1. Introduction

The increased awareness of the need for professionals who can contribute to the development, evaluation and adaptation of ICT applications supporting health care research and delivery has encouraged universities to install specialized educational programs in medical/health informatics. In 1990, at the University of Amsterdam- Academic Medical Center a four years full-fledged medical informatics university program was likewise introduced as a discipline next to the medical science university program. During these 15 years, about 170 students graduated from this program [1].

In 1999 in Bologna, a joint declaration was signed by the European Ministers of Education of 29 countries binding them to standardize the structures of higher education systems by 2010 in a system of academic grades that are easy to understand and to compare [2]. The main goal is to enhance the employability and mobility of European students, teachers and researchers and to increase the international competitiveness of European higher education. The standardization process is meant to let the higher education systems in Europe converge towards a more transparent system whereby the different national systems use a common framework for BSc, MSc and PhD degree programs.

To accomplish these aims, besides this common framework, the European Ministers adopted the European Credit system (EC system), developed under the 1988-1995 Erasmus program [3]. This EC system enhances flexibility for student-exchange programs between institutes for higher education, curriculum transparency, and is helpful in reflecting about curriculum structures, student workload and learning outcomes [4].

To harmonize with the Bologna agreement, we had to transform our four years medical informatics curriculum into individual BSc and MSc programs of which the new MSc program is described in this contribution.

2. The mission of the MSc program

In restructuring our 4 years medical informatics program into a 3 years BSc and a 2 years MSc course, we followed the IMIA recommendations for dedicated programs medical informatics at the BSc and MSc level [5]. IMIA recommends a (minimum) study load of a medical informatics MSc program of 60 EC. For several reasons, we decided to offer an MSc program of 120 EC. First, in contrast to a BSc program that should focus on a practice-oriented application of knowledge, for programs leading to an MSc degree, the objective is to provide an education of a scientific character, focussing on methodologies. The aim is to bring forth graduates who are able to contribute to the scientific development of the medical and health informatics domain [6]. We therefore felt that our program should heavily emphasize research methodologies and advanced statistics and that our medical informatics masters should have profound knowledge of the latest theories, methods and techniques of the health/medical informatics domain. Besides, medical informatics graduating at the MSc level should be skilled in the *application* of chief methods and principles of scientific research and should have a deep understanding of the research outcomes for the development of advanced applications to be used in medical or health care practice. In our new MSc program, students will therefore from the very beginning be involved in ongoing medical informatics research. Besides to that they are to conduct a large MSc research project on their own. Second, the growing competition in the field of dedicated MSc programs in medical informatics let us decide to set up a MSc program equal in length of other medical/health informatics MSc programs in Europe and the US of which most take two years to complete. Third, an earlier survey among our master graduates inquiring them on their working areas [in 1] had revealed that at least 35 percent of them are involved in managing health information system teams responsible for the deployment of information resources, in projects related to a portion of the information system function, and/ or in the development of a health organization's vision of how (part) of the health information system of a particular institution is to be organized. The survey results also revealed that managing teams and decision making as regards a health care organization's information management was insufficiently covered in the last version of our integrated medical informatics course. The job profiles of our former students let us decide to reserve a quite prominent and distinct place in the MSc program for education focused on health care organizational settings, project and team management, logistics in the organization of health care, steering of and change management in health care organizations, and pros and cons of various ICT infrastructures and architectures in this context.

Finally, we anticipated applicants for our MSc program to come from a wide variety of backgrounds with insufficiently deep skills in particular areas of medical informatics. The broadness of the medical informatics field, reflected in the many subjects to be covered in MSc curricula according to the IMIA recommendations required an MSc program accommodating the learning needs of all these candidates. We felt that all types of students would benefit from advancing their medical informatics knowledge in a broad sense as this would enhance their employability. Besides a broad education would help them in deciding on the final specialization they envision achieving.

3. Premaster program

Besides our own baccalaureates that may prolong their training by our MSc program, other types of baccalaureates and professionals may enter our new medical informatics MSc course. Candidates for our MSc program may come from a variety of backgrounds, among which health care or computer-science. These candidates may have profound skills and practice in their area of expertise but may lack sufficient knowledge and skills in specific areas of the broad field of medical informatics. We therefore constructed a premaster program for these baccalaureates and professionals. The aim of the premaster program is to remedy their knowledge deficiencies and practical skills in medical informatics so as to qualify them for our two years MSc program. The goal is to impart required knowledge of medicine, health and biosciences, health system organization and biometry to baccalaureates or professionals with an informatics-oriented degree, whereas the necessary mathematics-, computer science-/ informatics- and biometry knowledge and skills are imparted to baccalaureates or professionals with a health care-oriented degree. The premaster program starts in January and September of each year and takes at the most 30 EC to complete. To enable candidates to follow the main part of the premaster program by distance-learning, courseware has been developed for the Blackboard virtual learning environment This learning environment has been introduced at the University of Amsterdam in 2001, and proved very successful as communication platform between students and faculty and students mutually. Blackboard in particular is an economic means for distributing the course materials and exercises of the remedy programs to international students. Successful completion of these dedicated programs gives access to the MSc program.

4. The structure of the MSc Program

The MSc program in medical informatics lasts two years and is in English. The study load of the program is 120 EC, 1 EC equalizing 27 hours of lecturing, practical work and independent hours of study. Figure 1 outlines the structure of the MSc program.

Each year is divided into two 20 weeks semesters according to an 8-8-4 week's model (see Figure 1).

1st semester					2nd semester				
Current Issues in Medical Informatics 6 EC	Knowledge Representation & Reasoning in Medicine 6 EC	Advanced Data Analysis in Medicine 6 EC	Biomedical Information Systems Engineering 6 EC	Internship I 6 EC	Biomedical Research & Evaluation Methodology 6 EC	Organizational Settings of Health Care 6 EC	Health Care Logistics & Information Systems 6 EC	Information & Process Modeling in Health Care 6 EC	Internship I 6 EC
Elective Course I 6 EC	Elective Course II 6 EC			Scientific Master	Research & 48 EC	Project Thesis			

Figure 1: Overview of courses in the MSc program with respective European Credits (EC).

The program covers both core modules in medical informatics (a total of 48 EC), as well as elective modules, constituting a specialization in the medical informatics field (a total of 12 EC). Each semester in the first year ends with an internship (of 6 EC each) in the AMC, a teaching hospital that is fully integrated with the Faculty of Medicine. The main idea of these internships is that students are confronted with real-life information issues arising at clinical and outpatient departments in this university hospital center and advance their medical informatics skills in a real health care setting. Students are thus involved in ongoing clinical and medical informatics research projects from the very beginning of their training.

Students spend another 48 EC on their MSc thesis work including a research traineeship. During the research traineeship students practice various aspects of scientific research by taking part in a research project, typically a larger ongoing project within departments of the AMC-UvA or other institutes and companies associated with medical or health care practice, both in the Netherlands and abroad, among which our IPHIE partner universities [7]. Whereas the focus of the internships mainly is on the application of medical informatics methods and techniques, an additional goal of the research traineeship is that students learn to process and critically appraise both the scientific literature as well as their own research outcomes in the context of the latest developments in the medical informatics field. After graduation, a selected group of students may start a PhD research project. The average length of a PhD project is 4 years.

4.1 Course contents

The first year starts with a module *Current issues in medical informatics I.* This module serves two purposes. For baccalaureates in medical informatics, this module has the intention to increase their knowledge of the broad spectrum of subjects that are currently studied in medical informatics such as bio-informatics, signal and image processing, decision support techniques, health information systems and to acquaint students with the scientific approach to research. Yet, for master students who completed a premaster program, the first module is used to further remedy these students' gaps in knowledge and practical skills in these similar medical informatics domains. The second module *Knowledge representation and reasoning in medicine* deals with types and anatomy of medical knowledge, terminology systems, theory and practice of decision support tools, formalization of medical guidelines and probabilistic reasoning. The module *Advanced data analysis in medicine* elaborates on the biometry knowledge of the students. This module covers maximum likelihood theory, empirical Bayesian methods, multilevel analysis and advanced regression analyses methods such as non-linear, mixed-effects regression and regression trees. The module *Biomedical systems engineering* elaborates on the BSc module in software engineering. The focus is on advanced concepts, methods and techniques of systems engineering to be used in various phases of system design, software architectures for complex software systems and on quality control.

The second semester of the first year starts with *Biomedical research and evaluation methodology* presenting a broad range of sophisticated research designs to evaluate the effects of medical or ICT interventions. A variety of methods from health technology assessment are covered and their similarities and differences discussed: formative/summative, subjectivist/objectivist, quantitative/qualitative and usability evaluation methods, outcome measures to evaluate effectiveness, efficiency and usability of interventions, and health technology assessment and cost study designs. The module *Organizational setting of health* focuses on health care organization structures, designs and strategies, organizational dynamics and IT organization. Logistical concepts for (re)designing health care processes and clinical paths, patient logistics, work flow management, and IT support in this context are covered in the module *Health care logistics and information systems*. The last module of the first year *Information and process modeling in health care*'s focus is on health information management, health care IT strategies and architectures, and on security aspects in health care.

The second year of the MSc course starts with two elective modules. Students choose two out of a variety of modules to specialize themselves in a specific medical informatics domain, preferably the domain of the student's master research thesis. Students thus obtain in-depth knowledge in the research area of their interest. Bio-informatics, signal and image processing, decision support techniques, patient-centered information systems including telemedicine applications, and public health informatics are at the focus of these electives.

5. Status of the MSc program

The accreditation of this 120 EC MSc program by the Accreditation Organization of The Netherlands and Flanders for university programs was achieved in December 2005, after 20 months of preparation. Dutch candidates entering our MSc program will (partly) be funded by the Dutch Government for a maximum period of 60 months. Applicants from the European Union may apply for a student performance-related scholarship (maximum of 48 months) of the Dutch government. At present, 9 Dutch applicants follow an individualized premaster program. Since we currently have 21 baccalaureates in the 3rd year of our BSc program, we expect about 30 students to enter our MSc program in September 2006. The implementation of the MSc program requires an equivalent of six full-time faculty positions. Approximately 30 faculty members with backgrounds in medical informatics, computer science, medicine, public health, epidemiology, biostatistics, health care management, will contribute to the program on a part-time basis. The department of Medical Informatics that performs applied and theoretical research in this field on a national and international level is responsible for a major part of the MSc course. Strong links to medical informatics research projects thus exist within the course. Besides, our medical informatics master course will be offered in a collaborative effort by the AMC- an academic hospital and the University of Amsterdam. This means that the MSc course will be embedded in an environment offering great opportunities to link research projects with the practice of health care so that new insights can be shared among both faculty and students from the outset.

Our MSc program is still unique in The Netherlands. We yet combine our own expertise and efforts with our partner universities within IPHIE [7] and hope to continuously improve our course. The challenge will continue to be bringing forth medical informatics specialists who will make significant contributions to the field.

References

[1] Jaspers MWM, Fockens P, Ravesloot JH, Limburg M, Abu-Hanna A. Fifteen years of medical information sciences: The Amsterdam curriculum. Int J Med Inf. 2004; 73: 465-77.

[2] The Bologna Declaration in Higher Education, Joint declaration of the European ministers of education convened in Bologna, June 1999. See also http://europa.eu.int/comm/education/index_en.html (last accessed May 9, 2006).

[3] ECTS- European credit transfer system. http://europa.eu.int/comm/education/copenhagen/index_en.html (last accessed May 9, 2006).

[4] Hovenga EJS. Globalization of health and medical informatics education- what are the issues. Int J Med Inf. 2004; 73: 101-109.

[5] Recommendations of the International Medical Informatics Association (IMIA) on education in medical informatics. Methods Inf Med. 2000; 39: 267-77. See also http://www.imia.org (last accessed May 9, 2006).

[6] Hasman A, Haux R. Curricula in medical informatics. In Hovenga ES, Mantas J, editors. Global Health Informatics Education. Amsterdam: IOS 2004: 63-74.

[7] Jaspers MWM, Gardner R, Gatewood L, Haux R, Schmidt D, Wetter T. The International Partnership in Health Informatics Education: IPHIE, lessons learned from 6 years of experience, Methods Inf Med. 2005; 1: 25-31.

Ubiquity: Technologies for Better Health in Aging Societies
A. Hasman et al. (Eds.)
IOS Press, 2006

Medical Education and Role of Medical Informatics

Izet Masic[a], Ahmed Novo[a], Selim Toromanovic[b], Almir Dzananovic[a], Zlatan Masic[c], Alisa Piralic[a], Sejla Deljkovic[a]

[a] *Cathedra for Medical Informatics, Medical Faculty, University of Sarajevo, Bosnia and Herzegovina, imasic@lol.ba, www.imasic.com*
[b] *Nursing College, University of Bihac, Bosnia and Herzegovina*
[c] *Technical University Vienna, Austria*

Abstract. Bosnia and Herzegovina (BiH), as developing country in transition, has to shift from traditional ways of learning to the transformation of the university education in accordance with Bologna process and educational standards in European Union. In the light of these changes authors conducted research at bio-medical faculties in Sarajevo in order to address issues of the education of future physicians and especially role of medical informatics in their under and post graduate studies and continuous medical education. As per given results in this study, current quality of medical education at biomedical faculties, University of Sarajevo, is unsatisfactory due to several reasons and some among others are those traditionally seen as "computer literacy". Problems are determined and recommendations are given for decision makers to support transformation of BiH medical educational system to have physicians, dentists, pharmacists and nurses who posses the knowledge, skills and attitudes required to be competent in medical informatics if they wish to incorporate into their practices systematic approaches for promoting and maintaining the health of defined populations.
Keywords: medical education, medical informatics education, quality

1. Introduction

The emergence of medical informatics as a discipline is due in large part of advances in computing and communication technology, to an increasing awareness that biomedical knowledge and clinical information about patients are essentially unmanageable by traditional paper-based methods, and to a growing conviction that the process of knowledge retrieval and expert decision making are as important to modern biomedicine as the fact base on which clinical decisions or research plans are made [1].
Medical Informatics is rapidly developing scientific field that deals with resources, devices and formalized methods for optimizing the storage, retrieval and management of bio-medical information for problem solving and decision making [2,3]
In the future physicians will be expected to be more effective than is now the case in acquiring, managing and utilizing information for clinical decision-making and to be committed to using systematic approaches for promoting and maintaining the health of both individuals and the populations of which those individuals are members. Of curse, on the state level consensus must beset on the knowledge, skills an attitude that students should posses prior to graduation from medical school. Before graduation student should

demonstrate ability to retrieve, manage and utilize biomedical information for solving problems and decision making that are relevant to the care of individuals and populations. Also, they should understand roles of other health professionals and need to collaborate with others in placing patients in focus of care.

1.1. Problem formulation

The argument that medical informatics should be central feature of the medical curriculum rests on the following premise: "To support health care, life-long learning, education, research and management, medical students should be able, at the time of graduation, to utilize biomedical information for: formulating problems; arriving at strategies for solutions; collecting, critiquing and analyzing information; taking action based on findings; and communicating and documenting these processes and the results [1]".
The methods, tools and resources developed through medical informatics often helps physician accomplish tasks that they were already doing, enabling them do so more efficiently or in entirely new ways. Other applications of information technologies allow physician to accomplish tasks that were not previously possible. Medical education is a life-long or career-long process beginning with medical school. Support of life-long learning with information technology requires more then computer literacy. In short successful medical school graduate should be able to do the following:

- Demonstrate knowledge of information resources and tools available to support life-long learning;
- Retrieve information;
- Filter, evaluate and reconcile information;
- Exhibit good information habits – attitudes that support the effective use of information technologies.

Using medical informatics specific roles have clinicians, educators, researchers and health managers. The clinician must acquire information about patient, make clinical decisions based on available information and document and relay findings. Educator must be able to select and utilize information resources whether for health professionals or patients demonstrating practical knowledge of instructional technologies and resources available via Internet, CD-ROM, video teleconferencing and other media. Physician –researchers must understand sources for data and employ methods of decision theory to help formulate testable hypothesis and they must collect, organize, analyze and interpret the data. Physician must understand and manage costs, manage and work effectively in groups and effectively manage themselves. They also must understand their roles within the context of the overall health care system [4,5,6,7,8,9].

1.2. Activities at Medical Faculty in Sarajevo

In year 2002 at Cathedra of Medical informatics of Medical faculty in Sarajevo conducted first tele-exam in the history of education in Bosnia and Herzegovina between Podgorica (Montenegro) and Sarajevo (Bosnia and Herzegovina). In October 2003, University of Sarajevo began with Distance learning education, opening University Distance Learning

Centre at University of Sarajevo (UTIC – University Tele-informatics Centre). Opening the University Distance Learning Centre, as coordination body and leader in all activities in connection to Distance learning, has provided opportunity for development and growth of this kind of lifelong education. Pilot project was realized during three past school years, theoretical and practical education of subject Medical informatics are adapted to the new concepts of education using world trends of education from the distance. One group of students was included in the project finalized by electronic exam registration and electronic exam on 20th June 2005, publicly, in the Physiology amphitheatre of the Medical faculty in Sarajevo. Students' satisfaction was very high [10,11,12].

2. Materials and methods

In the period March – April 2006, 491 students from Medical faculty, Dentist faculty, Pharmacology faculty and Nursing College, University of Sarajevo, were interviewed. For the survey structured questionnaire was use, and piloted two moths before commence date of the research. Surveyors were specially educated students from final year of Medical faculty as per identical methodology of collecting data.

3. Results

In the survey 491 students were interviewed: 98 – first year, 81 – second year, 79 – third year, 84 – fourth year, 78 – fifth year and 71 – final year of study. Sex structure of interviewed student was: 64% of female and 36% of male (Figure 1.).

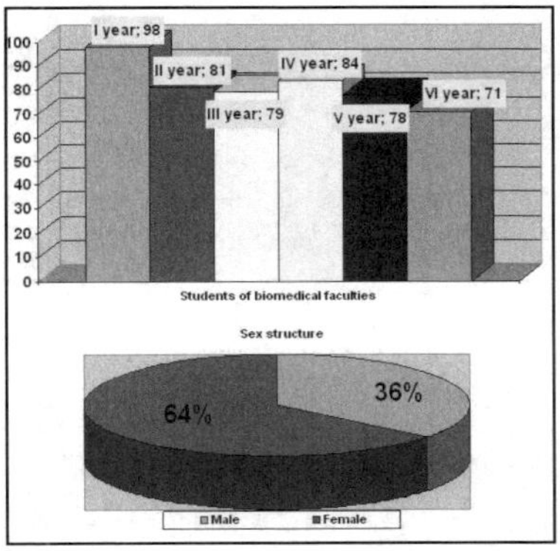

Figure 1. Structure of interviewed students

In this study authors were keen to know computer literacy of the students and its influence on quality of medical education and further work of young physicians.

Next questions were in regards do the students have opportunity to use computer at the faculty and how they use it. Number of students who do not have opportunity to use computer is 235, 186 do not know, 49 students have opportunity but they do not use it, 19 use it rare and 2 students use faculty computers almost every day. 18% of student use computer for fun, 26 for education, 33% for retrieval information and 23% for communication (Figures 2. and 3.)

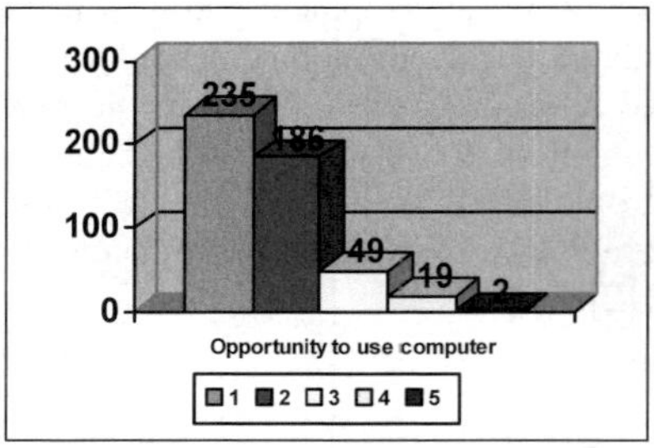

Figure 2. Opportunity to use computers

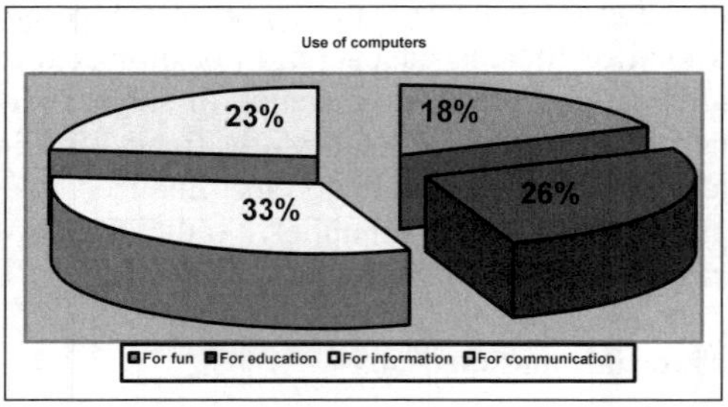

Figure 3. How students use computers

Teaching process at the Cathedra for medical informatics 53 % of students assess as appropriate, good or excellent but 47 % are not happy with it. Students also proposed what physicians should know more before graduate: MS Office 33%, Internet applications 28%, Distance learning and telemedicine 20% and 19% feel comfortable with the level of knowledge they posses (Figures 4. and 5.).

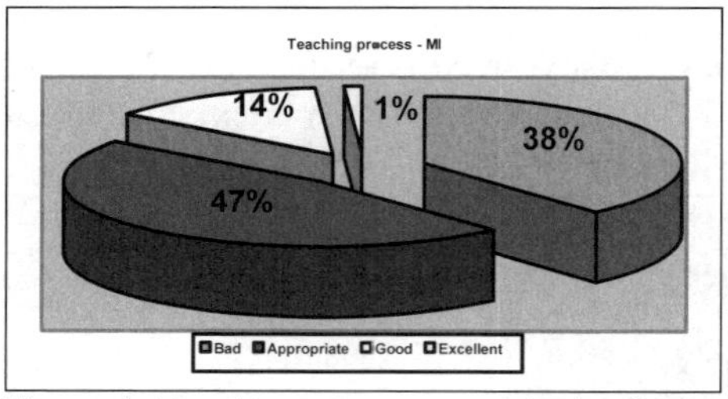

Figure 4. Teaching process at cathedra for Medical Informatics

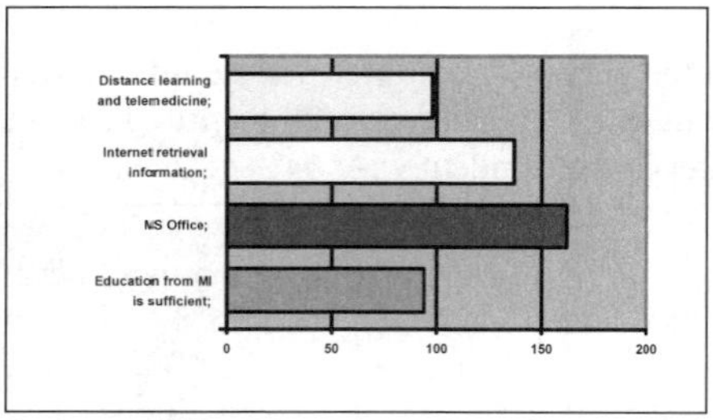

Figure 5. What physicians should know more before graduate?

Key question was: Do you feel that after you graduate you can work without help of experience physicians in primary health care? Results were shown in figure 6 per year of study.

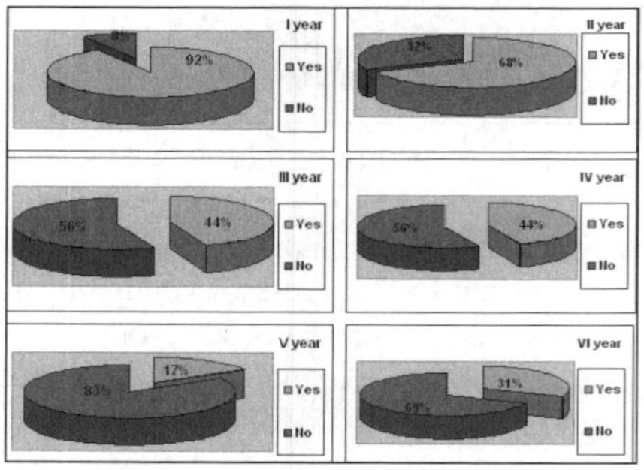

Figure 6. Do you feel comfortable to work alone?

Students also assessed quality of the education at bio-medical faculties at University of Sarajevo as per years of study. On the scale from 1 to 5, average grade is 2.88, first year – 3.72, second year – 2.88, third year – 2.79, fourth year – 2.87, fifth year – 2.53 and final year – 2.46. Students also assessed quality of the libraries, equipment and teaching process giving scores from one to five (Figure 7.).

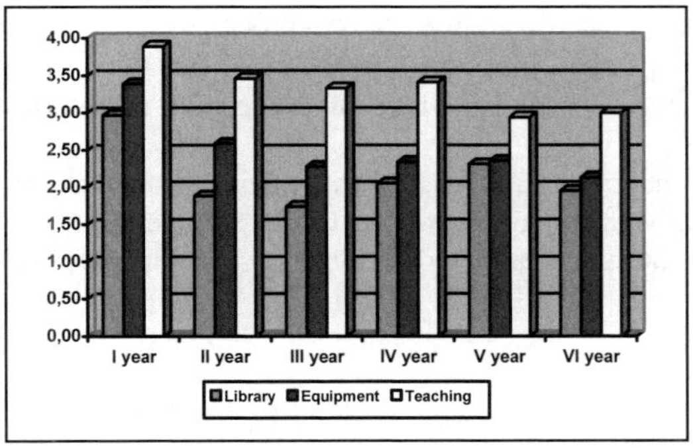

Figure 7. Quality of the libraries, equipment and teaching process

4. Discussion

Having in mind number of enrolled student at bio-medical faculties at University of Sarajevo divided per years of study we have representative sample covering also sexual structure of the students and physicians in BiH. It is very sad that students do not have opportunity to use computers commonly but it is promising that almost 60% of student use IT to develop knowledge and improve education. Teaching process at appropriate cathedras for medical informatics is not sufficient as per students' point of view and it is obvious that decision makers have to change current process. Also student gave comment how they see libraries, equipment and teaching ability of university staff.
Very discouraging is that almost 70% of student about to graduate do not feel comfortable to work without help in family medicine units.

5. Conclusion

Medial informatics is the application of computers, communications and information technology and systems to all fields of medicine – medical care, medical education and medical research. Content and design of educational programs must be with evolving societal needs, practice patterns and scientific patterns. Medical informatics is asked by decision makers, users and patients to be proven safe and beneficial. Rigorous evaluation of ICT applications in health care and education are necessary and of great importance for users and all stakeholders in health system.
Based on the results of the survey main conclusions are:

- Reform of education system must be performed as soon as possible in accordance with needs and possibilities;

- Assure continuous quality of education (internal and external evaluation);
- Determine golden standards in education, what is minimum what graduate should have in terms of knowledge and skills;
- Adjust medical curriculums with countries in the region and in EU;
- Teachers and educators must be evaluated regularly;
- Student must be involved in all reform processes;
- Enlarge volume and content of the practical education;
- In accordance with available funds improve work of libraries and better equip classrooms.

Concept of quality incorporates at least three dimensions and has three different meanings. Those are: comparative meaning in terms of the level of perfectionist, quantitative meaning in terms of the level accessed and appropriateness for certain purpose.

6. References

[1] Contemporary Issues in Medicine: Medical Informatics and Population Health. Association of American Medical Colleges. 1998; 3-14

[2] Greenes RA, Shortliffe EH. Medical Informatics: An emerging discipline and institutional priority. JAMA 1990; 263:1114-20.

[3] Shortliffe EH, Perreault LE, Wiederhold G, Fagan LM, eds. Medical Informatics: Computer applications in health care. Reading, MA: Addison-Wesley. 1990 (Chapter 1, page 20).

[4] Medical Informatics as a Discipline. Introduction to Medical Informatics. Available on URL address www.cpmc.columbia.edu Information accessed on 15 December 2005.

[5] Engelbrecht R, Ingenerf J, Reiner J. Educational standards - terminologies used. Stud Health Technol Inform. 2004;109:95-113.

[6] Masic I, Bilalovic N, Karcic S, Kudumovic M, Pasic E. Telemedicine and telematics in B&H in the war and Post war Times. European Journal of Medical Research, 2002; 7(supl.1): 47.

[7] Hovenga J. Bricknell L. Current and Future Trends in Teaching and Learning. Stud Health Technol Inform. 2004;109:131-142.

[8] Mantas J. Future Trends in Health Informatics – Theoretical and practical. Stud Health Technol Inform. 2004;109:114-127.

[9] Mantas J. Comparative educational systems. Stud Health Technol Inform. 2004;109:8-17.

[10] Masic I, Medical Informatics Education in Bosnia and Herzegovina. IMIA Yearbook 2004:204-8.

[11] Masic Z, Novo A, Masic I, Kudumovic M, Toromanovic S, Rama A, Dyananovic A, Bande I, Basic M, Guso E, Balta E. Distance learning at bimedical faculties in Bosnia and Herzegovina. Stud Health Technol Inform. 2005;116:267-72

[12] Masic I, Novo A, Kudumovic M, Rama A, Dzananovic A, Guso E, Basic M. Web based distance learning at Faculty of Medicine of Sarajevo University. Bosnian Journal of Basic Medical Sciences 2006; 6(2):71-5.

Ubiquity: Technologies for Better Health in Aging Societies
A. Hasman et al. (Eds.)
IOS Press, 2006

Developing an Interactive Approach in Teaching Medical Informatics

Diana LUNGEANU[a 1], Rochelle E. TRACTENBERG[b] and George I. MIHALAS[a]
[a] *Department of Medical Informatics, University of Medicine and Pharmacy, Timisoara, Romania*
[b] *Department of Biostatistics, Bioinformatics and Biomathematics, Georgetown University, Washington DC, US*

Abstract. A new masters program in medical informatics is proposed for development at the University of Medicine and Pharmacy in Timisoara. Given the rapidly changing technology itself and its deployment in biomedical science, the master's program curriculum has to be multidisciplinary, comprehensive and coherent in conveying the concepts, as well as the interdisciplinary character, of medical informatics (MI).
We describe the rationale and methods for a pilot study to develop a new, interactive approach in teaching MI. The study is being conducted within the existing MI course offered for the medical students in order to evaluate its impact on instruction and determine if a larger scale design is feasible. Two teaching teams of four instructors have been assigned to one of two tracks in our pilot study: traditional instruction or interactive instruction. After one term we have gained important information about how the structural and instructional aspects of the pilot design may influence confidence and attitudes.
Keywords: Medical Informatics, Education, Teaching Methods

1. Introduction

A two-year masters program in medical informatics (MI) is proposed for development at the University of Medicine and Pharmacy in Timisoara. This decision came in response to the current explosion in complexity in, and attention to, issues in health care quality, cost-effectiveness, access to health information, and medical research, especially in the context of our aging society. There is a need for professionals educated at the intersection of computer science, medicine, statistics, cognitive science, health economics, and medical ethics — *the MI professionals*. In order to meet these interdisciplinary demands in the educational approach we intend to partner with the Faculty of Computer Science from the Polytechnic University in Timisoara in the establishment of the new program.

The curriculum has to be inter- and multi-disciplinary, comprehensive, and coherent at the same time. It has to employ the best ways of effectively conveying the concepts and the interdisciplinary character of medical informatics. The field is dynamic, so the *overall objectives of the educational process* will be to: (1) provide an understanding of how computer technology can contribute to improving the quality in

[1] Corresponding Author: Diana LUNGEANU, Department of Medical Informatics, University of Medicine and Pharmacy Timisoara, P-ta E. Murgu 2, 300041-Timisoara, Romania, *dlungeanu@umft.ro*, Tel/Fax +40-256-490288

healthcare; (2) support the development of critical thinking skills; and (3) enable these professionals to continue to learn and to assume responsibility for their further development in the future. Such a curriculum cannot be developed in a single step. In fact, both the content and the delivery must be carefully considered. The content will be based on other MI programs [1-3], classical books of MI [4, 5], the Recommendations of the International Medical Informatics Association [6], and international experience [7]. Moreover, in 1997-1998 the Department of MI at the University of Medicine and Pharmacy in Timisoara coordinated a European Tempus Programme which focused on identifying the needs of different specialists in the Romanian health informatics market and the ways the universities curricula should be adjusted to meet these needs [8]. This is a large amount of content, and the traditional instructional stance would be to cover it all within the two years of the program. Instead, with appropriate attention to the structure of the curriculum, we may be able to accommodate it in ways that, rather than presenting a huge pile of material for students to simply consume, can be integrated across courses and other instructional opportunities to promote deeper learning and engagement with the materials. We wish to promote learning, not coverage.

The teaching in our universities is generally one-way, based mainly on traditional lectures with examinations typically at the end of the term. We believe a new approach should be employed for this interdisciplinary program with interactivity being a characteristic of the whole program, rather than of one or some particular course(s).

Taking all these into consideration, we started a pilot program in the context of the existing one-semester course in MI for medical students, with two main aims: (1) to test a new approach in teaching MI, mainly based on a high degree of teacher-student interaction in the teaching process itself and on a formative evaluation procedure for students' learning; (2) to evaluate changes in student (subjective) ratings of self-confidence with MI and attitudes towards the discipline and the course as indicators of the effectiveness of our changes to the instructional character of the course.

2. Material and Methods

2.1. Developing a New Approach in Teaching

Figure 1 presents the pilot program in the context of the masters MI program, emphasizing its aims and outputs for the long-term objective of the masters program. The aim of the pilot program is to test an instructional approach based on a high degree of interaction and on a formative evaluation procedure. The expected outcomes consist of: a higher level of self-confidence in using computers; a better attitude towards MI in general and the course in particular; and a better understanding of the course material.

Three important issues were addressed for designing the new approach in teaching MI: (a) restructured course material, but following the same main chapters as before; (b) more interactive teaching methods; (c) students' evaluation procedure aimed at assessing the deep learning and conducted during the entire semester, so providing both formative and summative assessment.

The long-term objective is to transfer the experience gained during the pilot study to the larger task of designing a successful MI masters program with a holistic, modern teaching approach, rather than a collection of heterogeneous courses.

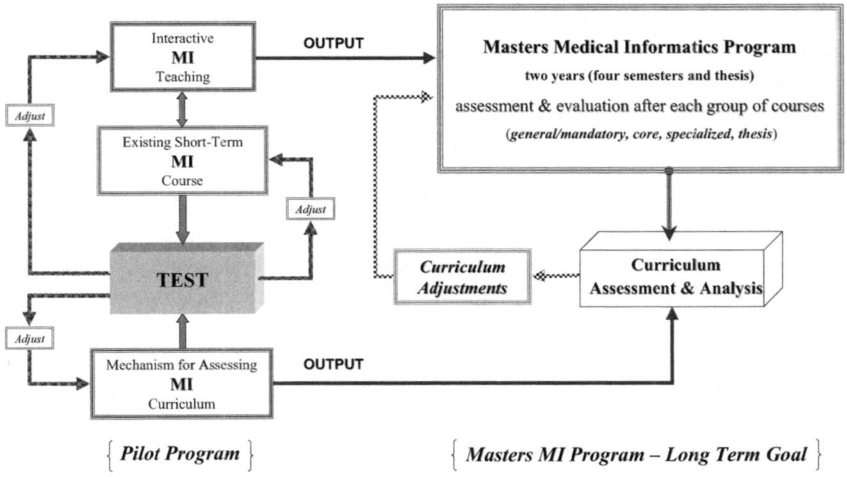

Figure 1. The pilot study in the context of the MI Masters Program's global structure. The output of the pilot study consists of both a new teaching approach for the masters MI program and a preliminary study for evaluation and analysis of the curriculum during the first generation of masters' program students.

For *good quality teaching*, there should be a good compatibility and aligning between: the curriculum (what we *want*), the teaching methods (how we *teach*), the assessment procedures and methods of reporting (how we *assess*), and the climate we create. A problem largely discussed in the literature is the teaching approach we should employ for making the transition between surface learning to deep learning [9-11].

At our university the courses are based on large-class teaching – especially the lectures have a high student-staff ratio (about 60-80 students). As it is difficult to really engage in dialogue in these conditions, we propose to take advantage of the setting to provide the necessary information in a "classical" fashion (basic concepts, general ideas, theory, etc.), while encouraging discussions at the seminar & computer room work in groups of 10-12 students. Shortliffe, too, suggests the lecture format for the basic MI concepts [5].

Table 1 presents the proposal for a more interactive course of MI in contrast to the present approach. In the pilot course, the instructors will not gauge the progress of the class in terms of the amount of material that is 'covered' [12]; instead they will allot time for discussing major topics in a more detailed manner, using the minor topics as opportunities to extend thinking patterns that are a part of the presentation of the major topical material. Additionally, the instructors in the pilot course will focus more on the depth of approach (based on the feed-back from seminar), so that the faculty should increasingly be perceived "less as disseminators of knowledge and more as facilitators of learning" [13]. Another important aspect we propose to change is the assessment of student work: in the pilot course the assessments are aimed at deep learning: weekly home-work will be assigned, and then problem-solving strategies and difficulties will be discussed at seminars. Further, instead of a single end-of-term exam the pilot course will involve three examinations during the semester, with the last one assessing their capacity of literature research and synthesizing (T1, T2, T3 respectively in Table 1 and Figure 2). One final attribute of the pilot course that is very different from the traditional version is the level and type of communication that is encouraged between students and between students and the instructors throughout the term. We use *Moodle*, an open-source educational application [15], as a standard communication tool for encouraging/pursuing questions beyond the classroom and throughout the whole semester. Moreover, incorporating this application into the course compels the students to use IT in their daily work. In Romania, most medical students do not have personal computers or Internet access at home and have to come to the University for computer work. Thus, incorporating *Moodle* into the instructional environment promotes familiarity with this technology.

In Romania medical students are selected based on their results in admittance and baccalaureate examinations and then quasi-randomly assigned to parallel "study series" of about 60-80 students. One of the two existing courses remained in traditional instruction environment and assessment and the other became the pilot course; thus the grouping was naturalistic, and not specifically randomized.

2.2. Designing and Testing a Mechanism for Assessing Teaching Effectiveness

The general scheme for teaching effectiveness' assessment is presented in Figure 2. Our mechanisms for assessing effectiveness combine quantitative measures (grades) and qualitative ones (attitude, change in attitude, comfort level) [16]. We also explore the possible inclusion of students' background (technical/non-technical) and gender as covariates in future models.

Table 1. *The present traditional vs. the pilot approach in teaching MI – 14 weeks/semester. In order to keep the programs independent, the two teaching teams are completely separated, while keeping them multi-disciplinary*

Current traditional approach Group 1 — 84 students enrolled *Background of the teaching team: 1-mathematics/physics, 2-informatics, 1-medicine*	**Pilot approach** Group 2 — 84 students enrolled *Background of the teaching team: 2-computer science, 1-mathematics, 1-medicine*
Large classes — lectures Classical one-way directed lectures – 2 hours/week, 1 chapter (main topic) / lecture, trying to cover as much as possible	**Large classes — combined lectures & discussions** 1 hour detailed presentation + discussion of 1 or 2 topics of interest, addressing any practical concerns that came up at the seminars 1 hour "moving forward" lecture – handouts, outline of a chapter, suggested readings, topic-related homework assignment
Practical work Practical computer room work under supervision — groups of 10-12 students, 2 hours/week (no required homework)	**Seminar & computer work** Seminar & computer room work – encourage discussions to put forward questions, suggestions for topics to be detailed, weekly homework — groups of 10-12 students, 2 hours/week
Examination is mainly summative, at the end of the semester: (a) practical skills - ECDL-based[14] (b) in-class 90 min written examination: a 10-question quiz; 3-4 simple problems; short essays on 2 important MI topics	**Formative & summative evaluation** Students' evaluation is conducted during the semester – formative feedback — open for discussions after every assignment + give the graded papers back. T1 – computer-based examination (15 min) – to assess basic knowledge of IT concepts T2 – in-class open-books problem solving – 45 min – to assess the capacity of applying theory to practical situations T3 – take-home paper – the subject posted 1 week before the due date – to assess capacity to synthesize – a learning experience in itself

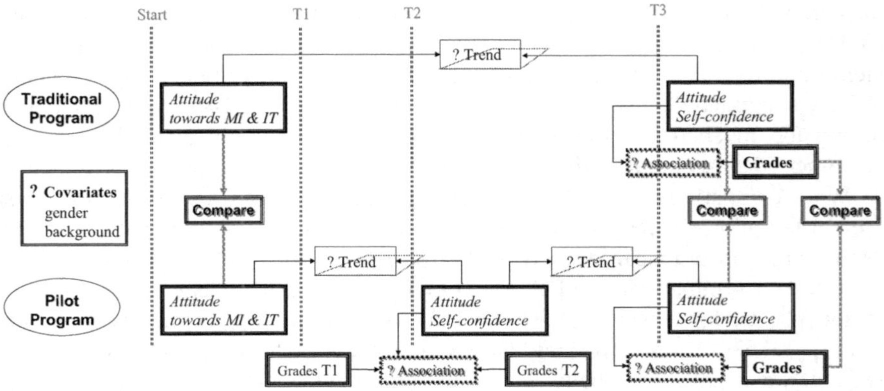

Figure 2. The design we use for data collection and analysis. T1, T2, and T3 are the examinations mentioned in Table 1. Attitude towards MI and IT was investigated at beginning of the semester, before any material had been presented. Other anonymous questionnaires are given after T2 (for the pilot program) and T3 (for both programs).

For assessing the program effectiveness we developed questionnaires that reflect: a) the level of self-confidence with respect to MI technology; b) students' attitudes towards MI and the course; we also use the final grades. The final questionnaires (comprising 40 questions) include questions like: *Your knowledge about the use of information technology and computers in medicine have improved (Not at all – Very little – Partly – Much – Very much); Your practical skills in using computers and data synthesizing or looking for information on Internet have improved (Not at all – Very little – Partly – Much – Very much); Practical applications/homework helped course understanding (Not at all – Very little – Little – Quite a lot – Completely); The pace for lectures was adequate (No – No opinion – Yes); The pace for seminars/practical activities was adequate (No – No opinion – Yes).*

3. Results and Conclusions

This paper is mainly focused on the rationale and methods of the pilot study, but after one term we have already gained important information. Attitudes and confidence were assessed in both courses at the start of the term, and there was no significant difference between the two groups of students: they all were a little confused (e.g. not sure why such a course had been included in their curriculum) but generally positive towards using information technology tools.

More than halfway through the term, the instructors' informal and subjective perceptions of the two courses reflected a sense that the students in the pilot group seem to have a greater level of self-confidence towards the student-teacher relationship. At the T2 assessments in the pilot course (evaluations as well as examination) students expressed some displeasure about their work-load (weekly homework, examinations during the semester, etc.) because their other courses were not so demanding during the semester and moreover, many students reported they knew that other students (in other study series or in previous years) do/did not have to work so much every week. When comparing the attitude of pilot course students towards the course at T2 and T3 (Mann-Whitney U test), we found statistically significant improvement at T3 for almost every item: e.g. volume of course material and work ($p=0.031$), structure of the course ($p<0.001$), usefulness of homework ($p=0.001$), course attractiveness ($p=0.014$), audience-appropriate level of the course ($p=0.001$). At T2, responses to open-ended questions suggested that students perceived the course to involve too much work, requiring too much time. However, when approaching the subject of their personal development they seem pleased and, sometimes, surprised at their own achievements.

When comparing attitudes at T3 across the two programs (Mann-Whitney U test), we found the pilot was perceived as being more structured ($p=0.016$) and leading to a better progress in practical skills ($p=0.025$); however, the students in pilot program perceived their seminar pace and clarity as being worse than students in the traditional program ($p<0.01$ for both items). The instructors' subjective perception was that both programs' students were working harder and with dedication, but self-reported attendance was significantly better for the pilot group (Mann-Whitney U, $p=0.001$) and final grades for the pilot group were a little higher (mean=7.24 compared to 6.55; ind-samples t-test, $p=0.037$) on a scale from 0 to 10 (10 = best).

We conclude that this "reshaping" of the teaching process is difficult, implying a completely different teaching philosophy; however there are indications that students will benefit from this shift. The dynamic nature of the MI curriculum surely warrants greater consideration of a more dynamic instructional orientation.

References

[1] Medical Informatics at University of Heidelberg and University of Applied Sciences Heilbronn. http://www.mi.fh-heilbronn.de/en/ (last accessed Jan. 15, 2006).

[2] Stanford Medical Informatics. http://www.smi.stanford.edu (last accessed Jan. 15, 2006).

[3] Columbia University. Medical Informatics. http://www.dbmi.columbia.edu (last accessed Jan. 15, 2006).

[4] van Bemmel JH, Musen M, editors. Handbook of Medical Informatics. Heidelberg: Springer; 1997.

[5] Shortliffe EH, Perreault GW, editors. Medical informatics: computer applications in health care and biomedicine (2nd Ed). New York: Springer, 2001.

[6] International Medical Informatics Association, Working Group 1: Education in Health and Medical Informatics. Recommendations on education in health and medical informatics. http://www.imia.org/wg1/rec.pdf (last accessed Jan. 15, 2006).

[7] Jaspers MWM, Gardner RM, Gatewood LC, Haux R, Schmidt D, Wetter T. The international partnership for health informatics education. Methods Inf Med 2005; 44: 25-31.

[8] Lungeanu D, Dimitriu R, Bazavan M, Sima R, Richards B, Neame R, Murphy G, Mihalas G. Medical informatics educational tasks seen from practical perspective. In: Kokol P, editor. Medical Informatics Europe '99. Proceedings of the European Conference on Medical Informatics. Amsterdam: IOS Press; 1999: 497-499.

[9] Biggs J. Teaching for quality learning at university: what the student does (2nd Ed). Philadelphia: SRHE and Open University Press, 2003.

[10] Fink LD. Creating significant learning experiences: an integrated approach to designing college courses. San Francisco: Jossey-Bass, 2003.

[11] Mezirow J, editor. Learning as transformation: critical perspectives on a theory in progress. San Francisco: Jossey-Bass, 2000.

[12] Brookfield SD, Preskill S. Discussion as a way of teaching: tools and techniques for democratic classrooms. San Francisco: Jossey-Bass, 1999.

[13] Diamond RM. Designing and assessing courses and curricula: a practical guide (2nd Ed). San Francisco: Jossey-Bass, 1998.

[14] ECDL - European Computer Driving Licence Foundation. http://www.ecdl.com/ (is being rebuilt, last accessed Jan. 15, 2006).

[15] Moodle (Course Management System). http://moodle.org/ (last accessed Jan. 15, 2006).

[16] Tractenberg RE, Chaterji R, Haramati A. Assessing and analyzing change in attitudes in the classroom (in press), Assessment and Evaluation in Higher Education.

8.2 Education and Networking

Ubiquity: Technologies for Better Health in Aging Societies
A. Hasman et al. (Eds.)
IOS Press, 2006

The INFOBIOMED Network of Excellence: Facilitating Training and Mobility in Biomedical Informatics in Europe

Guillermo de la Calle, Victor Maojo, Mario Benito

Biomedical Informatics Group, Artificial Intelligence Lab., Facultad de Informática, Universidad Politécnica de Madrid, Campus de Montegancedo s/n, 28660 Boadilla del Monte, Madrid, Spain
{gcalle, vmaojo, mariobr}@infomed.dia.fi.upm.es

Abstract. In this paper we present the work carried out within INFOBIOMED, the European Network of Excellence (NoE) in Biomedical Informatics, to facilitate training and mobility in the area. An analysis of past experiences in both Medical Informatics and Bioinformatics has led to various actions. In this scenario, we have elaborated three actions: (1) a survey of the training and mobility situation, needs and actions to be taken, (2) a Biomedical Informatics course database (ICD), and (3) a Mobility Brokerage Service (MBS). We describe the use of Web Services to build the MBS, designed to facilitate the exchange of professionals within the consortium, belonging to ten European countries. The goal of INFOBIOMED is to expand these initiatives to other NoEs and institutions within the European Union.
Keywords: Biomedical Informatics, Medical Informatics, Bioinformatics, Training, Mobility, Web Services

1. Introduction

Over the last few years, many academic, industrial and institutional initiatives have launched a new area: Biomedical Informatics (BMI), at the intersection between Medical Informatics (MI) and Bioinformatics (BI). Traditionally these disciplines have evolved without interacting. As regards, it has been proposed [1] that MI could provide the methods, tools and lessons learned in more than forty years of experience in developing clinical applications. Meanwhile, BI could provide a scientific basis that could be quite important to advance scientific knowledge in areas such as physiology and the underlying pathological processes of diseases. In particular, the new concept of "Genomic Medicine" could benefit from a strong support and interaction with BMI.

One of the most important constraints to advance towards molecular medicine and BMI is the shift in education that will be required for physicians to learn how to interpret and manage genomic and genetic information in their clinical routines. The knowledge and reasoning methods that will be needed in this new concept of medicine may be partially provided by informatics tools. Physicians should learn how to use those tools but it does not sound reasonable to make them fully understand, for instance, how to interpret microarray results or SNPs (single nucleotide polymorphisms) information. A different issue will be how to create the methods and build the informatics tools that will be needed to facilitate the work of practitioners. For that purpose, different backgrounds and expertises will be needed.

INFOBIOMED [2] is a European Network of Excellence (NoE) –funded by the European Commission (EC)– composed by organizations and universities from ten different countries. Its main objective is the establishment and strengthening of the BMI community at European level in order to support individualized healthcare. Four pilots focused on different problems have been established to demonstrate the benefits of this new approach. The training of new BMI experts is one of the most important challenges. For that reason, the INFOBIOMED mobility program is focused in promoting exchanges between organizations.

In this paper we present some of the work that has been carried out over the last few years to enhance training and mobility in BMI. We present the Mobility Brokerage Service, developed at the UPM to enhance the publication of training and job opportunities (both offers and demands) over the Web. While these initiatives were originally internal to the INFOBIOMED NoE, they are being expanded to make them available to the whole BMI community.

2. Background

Proposals for a stronger interaction between MI and BI professionals have been recently made. Collaborative efforts may be beneficial to exchange research results and methods as well as computing tools [3]. Some of the underlying issues below both disciplines have common points that could be shared and exchanged. This was the central focus of the BIOINFOMED study [4, 5], supported by the European Commission: to study synergies between BI and MI and elaborate a R&D agenda for the forthcoming programmes at the European Commission. Associations such as the American College of Medical Informatics are promoting efforts to analyze this issue.

Different programs and initiatives have been developed over the last decades to support training in both areas. The rise of BMI introduces new challenges that have been addressed by US universities such as Stanford or Columbia, with BMI programs that integrate both MI and BI and an increasing number of institutions in Europe. Related approaches have been taken in Europe, such as at the Karolinska Institute in Sweden, where a PhD Program in medical bioinformatics has been launched. The purpose is "to build up competence in bioinformatics with special emphasis on biomedical and clinical applications". In this sense, the European Commission has strongly supported the field and academic initiatives in BMI, since an early conference that took place in Brussels, December 2001, to support BMI. Within these initiatives, the various Network of Excellence funded by the e-health division, DG IST, have created their own strategies to support training in BMI.

Another EC-funded initiative is the SYMBIOmatics project [6]. This project is funded by the Sixth European Framework Programme for Research and Technological Development. It is coordinated by the EBI (UK) and is composed by nine partners from six different European countries. The main objective is to summarize in a white paper the state-of-the-art in BMI in Europe. This information will provide the appropriate input to future European scientific and funding policy. Some results of this study and the work done by the different e-health NoEs is expected, at the time of writing this paper, to be presented in a wide Open meeting in July 2006 in Brussels,

3. Methods

Various initiatives to study and support training and mobility have been carried out within the members of the INFOBIOMED Consortium. These initiatives are three: a) a survey of training courses, infrastructures and actions needed in the field, at the European level, b) elaboration of a BMI course database, and c) a Web Service-based Mobility Brokerage Service (MBS), to facilitate exchange of students and professionals. This paper is focused on the MBS, which is presented below:

3.1. The Mobility Brokerage Service (MBS)

A Web Service-based brokerage service was set up and implemented to integrate distributed information related to mobility. Previously, a survey was carried out to identify the main barriers to solve and the features that the MBS should include. Results from this survey were adopted in the system. The MBS includes Web Services technology and facilitates different kinds of queries and information retrieval. More details about the implementation and visual aspects are given in the next sections.

After analyzing the information collected in the survey, some interesting results appeared. For almost 75% of participants, the main barrier to participate in mobility activities was the lack of funding. The lack of information about host offers and lack of time appeared as other important drawbacks. Based on these results, some practical solutions have been deployed to address these two principal drawbacks. To supply the lack of funding, a Mobility Funding System (MFM) has been approved by the Training and Mobility Committee (TMC) of INFOBIOMED. Basically MFM consists on offering economic incentives to foster exchanges between partner staffs. On the other hand, a Mobility Brokerage Service (MBS) has been implemented to deal with the lack of information about mobility offers.

As pointed out above, another important barrier for the consolidation of BMI as a novel synergical discipline is the current lack of information about education possibilities. It might not be easy for researchers to find specific opportunities to learn and consolidate their expertise on this discipline. The mobility action plan intends to promote exchanges of experts among European institutions. Both host organizations and participants would be benefited with the exchanges. To facilitate mobility opportunities searches and communicate host organizations with candidates, a Mobility Brokerage Service (MBS) has been developed within the INFOBIOMED Consortium. The MBS has been envisaged as a secure web on-line marketplace to promote mobility opportunities. The MBS has been designed following a novel and incipient technology. Its system architecture is service-oriented [7] and based on XML and web services.

3.2. Web Services concepts

Web Services (WS) is an emerging technology arisen to address the problem of exchanging information among existing and new applications over Internet. They are based on consolidated technologies such as HTTP and XML for transporting and representing data, respectively. Numerous attempts for integrating distributed programs appeared during the last 30 years, such as Common Object Request Broker Architecture (CORBA), Distributed Component Object Model (DCOM), Remote Procedure Call (RPC), and Java Remote Method Invocation (RMI). Three new protocols have arisen related to Web Services: WSDL (Web Services Definition Language) [8] is a W3C specification, SOAP (Simple Object Access Protocol) [9], and UDDI (Universal Description, Discovery, and Integration) [10].

There exist several implementations of these specifications and protocols, as well as Web services toolkits developed by different companies and organizations. For inexperienced users, WS are more complicated to set up and deploy than classical systems, but they offer interesting advantages. For instance, WS are platform-independent, allowing interconnecting heterogeneous applications and systems without problems. WS can also be combined with other WS.

3.3. Implementation

The MBS has being implemented as a web application. JavaServer Pages (JSPs) and Servlets are used to generate dynamically web pages with the information requested by users. The web server and the servlet container use *Apache Jakarta Tomcat 5.0*. This is an open and collaboratory environment. For storing data *MySQL 4.0.20* has been selected. *MySQL* is the most popular open source DataBase Management System (DBMS), particularly suitable for developing web applications.

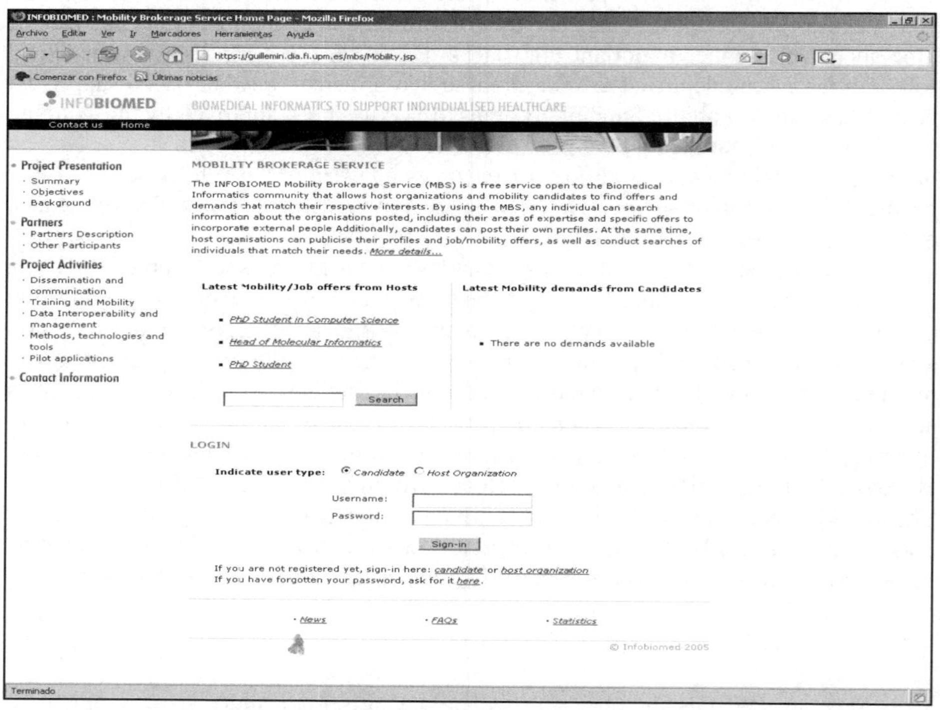

Figure 1. The INFOBIOMED Mobility Brokerage Service main page.

The internal functionalities are solved by using Web Services. We have defined several WS for different heterogeneous tasks (e.g., identification and validation of the users, managing candidates and hosts organizations information). We address traditional security issues such as identification, authentication, authorization, integrity and confidentiality of the data by using WS-Security. Security communications are addressed by using the *Secure Socket Layer* (SSL) protocol.

The MBS is completely integrated and accessible through the INFOBIOMED official web site. Figure 1 shows a screenshot of the application. WS technology allows to store the databases in the same computer or access them over the Internet. For instance, a web application might be running at the official site server and each WS might be placed on different computers, remotely. Similarly, databases with the information about candidates and host organizations can be distributed. Access is transparent to users by using a common and uniform interface.

The MBS allows users (candidates and host organizations) to register into the system, manage their own data, make suggestions and, finally, leave the application. Users can also look for offers or demands depending on their role. The system presents two significant functionalities for registered users:

- Mobility opportunities and job offers can be searched using textual searching approaches or by graphically navigating the MBS by means of visual maps.
- Each time a position offer or demand is registered on the system, an automatic matching process is performed. When a match is found, an email is sent to the interested actors, inviting them to visit the site to review the information.

Since Web Services are based on HTTP and XML standards, no problems arise with proxies and firewalls. Communications are completely transparent in this sense.

3.4. The Matching System

As exposed above, the *matching system* is a facility offered by the MBS to automatically find the opportunity or candidate wished. The keywords defined by the training topic thesaurus are used to improve the effectiveness and accuracy of the automatic searches. Keywords are classified in various main categories and subcategories. When a new offer or demand is introduced in the system, users are asked to indicate a set of keywords which define the area that fits better to their desires.

Each time an offer is stored in the database, the *matching process* is automatically performed. The system tries to find matches between the new offer and the rest of existing demands. Additionally, searches are filtered using other criteria such as the kind of collaboration sought, the starting and ending dates or duration. Analogous situation happens when a demand is introduced. The system looks for the offers which satisfy the demand. Anytime a match is found, the user is notified by email. An email will invite each user to visit the MBS and check the news. The new matches are displayed using an animated icon.

4. Conclusions

INFOBIOMED has addressed BMI education in the context of the European Union. Following a survey, a course database and a MBS were implemented to address these issues. This paper has shown some of its main features.

WS technology provides different possibilities for integrating data sources. Within the NoE, four pilot applications are being developed, in the areas of pharmainformatics, genomic and microbiology, genomic and chronic inflammation, and genomic and colon cancer. Issues such as integration from different and heterogeneous data sources, common to all pilots, can be solved using similar approaches, such as service-oriented architectures. In the framework of INFOBIOMED, the Mobility Brokerage Service is the main approach used to promote mobility in the BMI. Also, it is expected that the others European e-health NoEs use the system developed for their own mobility programs.

Acknowledgments

The present work has been funded by the European Commission (FP6, IST thematic area) through the INFOBIOMED NoE (IST-507585).

References

[1] Maojo V, Kulikowski CA. Bioinformatics and Medical Informatics: Collaboration on the Road to Genomic Medicine? Journal of the American Medical Informatics Association 2003 Nov–Dec; 10 (6): 515-522.
[2] http://www.infobiomed.org / http://www.infobiomed.net (last accessed Jan. 10, 2006)
[3] European Commission, DG Information Society, Applications relating to Health, http://www.cordis.lu/ist/ka1/health/home.html (last accessed Jan. 10, 2006).
[4] Martin-Sanchez F, Iakovidis I, Nørager S, Maojo V, de Groen P, Van der Lei J, et al. Synergy between Medical Informatics and Bioinformatics: Facilitating Genomic Medicine for Future Healthcare. Journal of Biomedical Informatics 2004. Feb; 37(1):30-42.
[5] http://bioinfomed.isciii.es (last accessed Jan. 11, 2006)
[6] http://www.symbiomatics.org (last accessed Jan. 11, 2006)
[7] Erl, T.: Service-Oriented Architecture. Prentice Hall, 2004.
[8] http://www.w3.org/TR/wsdl (last accessed Jan. 10, 2006)
[9] http://www.w3.org/TR/soap/ (last accessed Jan. 10, 2006)
[10] http://www.uddi.org/ (last accessed Jan. 10, 2006)

Ubiquity: Technologies for Better Health in Aging Societies
A. Hasman et al. (Eds.)
IOS Press, 2006

Multimedia and Physiology: a new way to ensure the quality of medical education and medical knowledge

Yvon Lessard[a,1], Pridi Siregar[b], Nathalie Julen[b], Jean-Paul Sinteff[b], Pierre Le Beux[c]

[a] *Department of Medical Physiology, IFR 140 GFAS, University of Rennes 1, Rennes, France*

[b] *Integrative Bio Computing (IBC), Rennes, France*

[c] *Department of Medical Informatics, IFR 140 GFAS, University of Rennes 1, Rennes, France*

Abstract. Background: since the eighties and the existence of virtual campuses, the value of computers in distance education has been acknowledged. The development of information and communication technologies is driving at discriminating distance education and on-line education. Purpose: the aim of the" Campus Numérique de Physiologie" is not to reproduce an on-line copy of classical textbooks but to put at students' and physicians' disposal the huge possibilities of multimedia resources for an active and easier understanding of complex physiopathological phenomena. Methods: the on-line course materials were created using both original IBC-made and registered trade-mark software tools. Multiscale modelling and corresponding knowledge bases were implemented by mathematicians, biologists and software engineers from Rennes. The website, which is accessible through a server of the French Virtual Medical University, was developed in the language HTML/PHP connected to a MySQL database. Results: the content managing system is consistent with classical home page facilities and multicriteria browser. Interactive resources are freely available for the site's users. Two- and three-dimensional simulations born out of mathematical qualitative and quantitative models at the molecular, cellular or organic level keep students active with regards to fundamental mechanisms by interactively manipulating the simulation environment. Conclusion: authors comment the already available course materials which should stimulate the creation of new documents following a validation by a qualified commission of the "Société de Physiologie". Providing evaluation tests, teachers anticipate that the increasing content of this virtual campus will allow users to gain a complete understanding and an integrative view of many physiopathological mechanisms.
Keywords: E-training/ Physiology/ Multimedia/ Medical Education.

1. Introduction

Medical school teachers have for many years criticized the lecture method commonly used for teaching basic science. Students are frequently frustrated by the attempt to acquire huge bodies of information in very limited time periods. Time consuming laboratory exercises have been eliminated from the curriculum of almost all schools and the number of lectures hours has increased. Dedicated teachers have tried a number of ways to help students learn and control their own learning. Educational change has been especially noticeable in medicine. There is an acknowledged need for student-centered learning which has shifted the focus from the teacher to the learner. There are various alternatives to the traditional didactic lectures. Modern approaches to teaching

[1]Corresponding author: Yvon LESSARD Service de Physiologie Médicale, Faculté de Médecine, 2 Avenue du Professeur Léon Bernard 35043 Rennes Cedex France. e-mail : yvon.lessard@univ-rennes1.fr.

and learning basic science include collaborative learning, problem-based learning and the use of computers. Computers and information and communication technologies (TICs), the Internet and Internet-related resources are of special interest in physiology[1]. Human physiology and anatomy constitute the foundation of medicine and there is a universal agreement that physicians must have knowledge of these basic subjects. Many difficult concepts in physiology are truly learned only when the student's brain converts heard or read words, static pictures and diagrams into moving models[2]. Available computer technology allows the use of dynamic models, making learning more efficient. Computers are now sufficiently powerful and the Internet sufficiently fast to allow fast distribution of multimedia materials which are especially useful for teaching physiology[3]. Furthermore, the interactive capability of computer-based instruction keeps the student involved so that learning is more interesting and not purely passive. This concept represents a step forward after the initial north-american models of "virtual campus" and "virtual universities" which, also using internet-based technology, originate from the sole objective of distance education, exchange and home delivery of information[4].

This is the global conceptual context in which we have decided to create in Rennes a national "Campus Numérique de Physiologie" (CNP), freely accessible for every one and particularly elaborated for medical students. The latter are willing to complete more deeply their knowledge in physiopathology. Practitioners also feel the need for reactivating their long-term memory about the details of biological phenomena with complex spatio-temporal development. We wanted to give every teacher in physiology the ability to submit didactic materials using the wide possibilities of computers and Internet technology. Since the CNP has been created recently, multimedia materials are, at the present, limited to the domain of the cardiovascular physiopathology. After describing the methods and technics for producing pedagogical resources, this article contains examples of some materials made available on-line as well as comments regarding our experience as teachers and authors.

2. Materials and methods

2.1.Organization of the virtual campus

For the first time in June 2000, the french ministries of national education and research have launched the concept of "Campus numérique"[5] which is somewhat similar to the north-american internet-connected courses made available for distance education through various "virtual campuses" and "virtual universities". The joint thinking of university teachers has led to the setting up of the french-speaking virtual medical university (UMVF) which provides various medical disciplines with the possibility of creating their own "Campus numérique". The CNP was born in December 2004 under the responsibility of two teachers of the Faculty of Medicine of Rennes and validated in January 2005. Learning materials are the result of the collaborative work of teacher members of the “Société de Physiologie” (SP), association which gathers researchers and teachers in Physiology of the faculties of medicine, pharmacy and biological sciences of the french, swiss, belgian and canadian universities. Teachers can provide raw data and also elaborate scenarii. After being worked-out, pedagogical contents must be validated by an independant committee of teachers of the SP. The conception of the campus was thought out by a physiologist (YL) of the medical school of Rennes and its technical realization was carried out with the support of computer specialists of the company Integrative Bio Computing (IBC S.A.R.L.) and of the Laboratory of Medical Informatics (LIM). Pedagogical contents include texts, static images (radiography, magnetic resonance imaging (MRI), etc…), sounds (voice, cardiac and pulmonary sounds), dynamic recordings (echogram, videogram), dynamic two- or three-dimensional simulations. As we will see in the discussion, learning materials of the CNP are not intended to become a surrogate for classical forms of teaching. TICs and multimedia are uniquely used as tools for learning, providing learners with complementary knowledge in fundamental sciences necessary to scientific and medical education.

2.2.Structure of the campus

The website is accessible through the UMVF site at the following adress : http://www.umvf.org, virtual campus heading. There is also a direct Internet access : http://www.campus-physiologie.org. On the homepage one can find the classical headings of a website : contacts, news, forum, links, access to pedagogical items, etc… Authors have to get a free registration before submitting an announcement or a new teaching resource.

2.3.Tools for the construction of the website

A LIM PC computer was dedicated to function as the web server. On-line documents are transmitted over the Internet using the open source server Apache operated on the Mandrake Linux operating system. Asymmetrical digital subscriber line (ADSL) or any connection with high data transmission rate is needed between server and learners' computers to view the different multimedia teaching resources which sometimes represent heavy files. Course materials can be added or manipulated with Mambo, an open source content management system (CMS). Teaching materials were converted to HTML/PHP, the standard for web clients using open source softwares. Mambo own HTML and PHP internal editors allow the dynamic management of the data. Mambo is linked to an open source MySQL knowledge base where all the generated data files are saved. Open source tools were used as frequently as possible for financial reasons.

2.4. Multimedia tools

Animations were created using Director (Macromedia development tool). In order to view Director animations, the ShockWave (Macromedia) last version plug-in (directly available on the CNP site) can be downloaded by learners using Netscape, Internet Explorer as well as Mozilla Firefox browsers. Indeed, the CNP site is accessible to Macintosh or PC computers using various operating systems : Macintosh OS (Apple) , Windows 98 to XP (Microsoft) or Linux (Unix family). Different movie players as QuickTime or Windows Media Player can be used to view .mov or .avi movies.

Figure 1 is a schematic recall of the technical functioning of the CNP website.

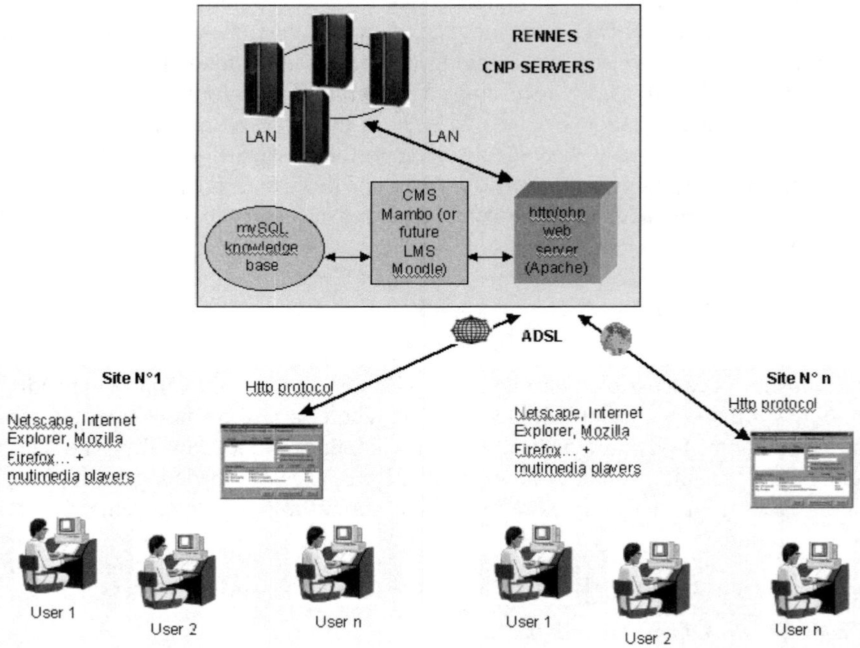

Figure 1 : technological structure of the CNP website

3. Results

At the present time, about 20 pedagogical items are available on the CNP in the sole domain of the cardiovascular physiopathology. We have developed 2D or 3D animations in sections which are known to be of special difficulty for the students. For example, learners can start the electric cardiac activation process : using an original IBC-made 3D and chromatic simulation program[6], they are able to follow the step by step depolarization and repolarization pathways along the fast conducting system, the endocardium and epicardium. 3D visualization of the cardiac electrical signal moving forward is essential to understand the corresponding genesis of the normal and pathologic human heart EKG[7] which can be studied in any standard lead. In the same way, student can watch in details hard to understand 3D phenomena such as a myocardic repetitive reentry process associated with the genesis of a ventricular tachycardia, or such as the successive or slightly asynchronous mechanical events of the cardiac cycle and the corresponding cardiac hemodynamic recordings. Another pedagogical interest of the site lies in the possibility of watching explanatory videos of some technical process, such as the EKG recording material connecting up. An attempt was made to make the teaching resources outlines uniform in structure and appearance among different sections of the course. Each pedagogical content includes at least two frames : one for the explanatory text and one for the multimedia material. Different buttons allow the learner to display more detailed explanations about the phenomena, to add sound, music, to hear the reading of the text, to navigate along the different pages of the session.

4. Discussion

Computer-assisted learning is generally assumed to be worth and to have a positive influence on student learning[3, 8-9]. Nevertheless, in agreement with the views of the teachers of the SP and in opposition to classical virtual universities, it was decided before its development that the CNP would not be a web-based course aimed at distance learning. Choosing to provide learners only with complementary teaching resources, we wanted to keep the possibility of face-to-face Physiology courses for on-campus students : the will of a physical classroom in preference to a virtual one is expressed by many when a direct interactive aid by dedicated teachers is possible[8, 10].

Though the CNP is for the moment moderately interactive, medical first-cycle students of Rennes and several already graduated practitioners are highly enthusiastic about the complementary use of multimedia resources to improve the understanding of physiopathological mechanisms and the possibility to re-active their memory before or after clinical reasoning learning or practical workshop sessions. This statement is in agreement with a recent survey[11] showing that medical students mostly use Internet in order to find additional learning information. Nevertheless, the CNP is in permanent evolution in order to improve its technical effectiveness and to come up to the users' expectation[12-13]. Both aspects will be explored by questionaries. To reach this goal, the actual CMS Mambo will be replaced as soon as possible by a learning management system (LMS). LMS are aimed to offer functionalities allowing easy construction of a questionnaire. A part of the teaching quality of the CNP will be warranted by the teachers validating committee and by the administrator's control of the hardware and software efficiency and updating. Since it has been shown that didactic fashion is efficacious in basic sciences and Physiology [9] and that understanding is the most useful in clinical reasoning[14-15], original IBC-made software tools will be used to develop movies from 3D colour images and photographs in the jpeg file format to give the learner the possibility of a real interactive navigation in a three-dimensional cardiovascular environment. Well designed computer-aided learning is useful for conceptually difficult topics[16, 3]. Furthermore, an original IBC-made cardiac simulator is in preparation, soon at teacher's disposal to create different scenarii of pathological affects. The CNP and the teaching resources development will also probably entail several servers and a local-area network (LAN) between them.

5. Conclusion

Developing computer assisted learning applications is a lengthy process. Innovators within traditional courses who have often produced creative and high quality material to supplement existing courses are in a minority[13]. We hope that the rising success of the CNP will enhance the creativity of the teachers but the development of teaching scenarii using multimedia and the multiplication of the teaching contents of the CNP for the next years depends on both the vitality of the members of the SP and the acceptation of the different teaching projects to be granted by the UMVF.

Acknowledgements

We acknowledge the members of the GIP of the UMVF, the members of the SP, the president of the University of Rennes I and the dean of the Faculty of Medicine of Rennes for supporting the development of the Campus Numérique de Physiologie. The study was funded in part by grants from the GIP of the UMVF, from the University of Rennes I and from the Faculty of Medicine of Rennes.

References

[1] Davis MJ, Wythe J, Rozum JF, Gore RW. Use of world wide Web server and browser software to support a first-year medical physiology course. *Am J Physiol* 1997 ; **272** : S1-14.

[2] Lilienfield LS, Broering NC. Computers as teachers : learning from animations. *Am J Physiol* 1194 ; **11** : S47-54.

[3] Ryan M, Mulholland CW, Gilmore WS. Applications of computer-aided learning in biomedical sciences : considerations in design and evaluation. *Br Biomed J Sci* 2000 ; **57** : 28-34.

[4] Childers P, Delany P. Wired world, virtual campus: universities and the political economy of cyberspace. *Works and Days* Spring/Fall 1994 ; **12** : 61-73.

[5] Hotte P, Leroux R. Technologies et formation à distance. *Revue STICEF* 2003, **10** : 1-14.

[6] Siregar P, Sinteff JP, Julen N, Le Beux P. An interactive 3D anisotropic cellular automata model of the heart. *Comput Biomed Res* 1998 ; **31** : 323-47.

[7] Lessard Y. A CD-Rom for teaching "Vectorcardiography and genesis of EKG". 70th annual meeting Société de Physiologie, Quebec, Canada, 18-20 septembre 2002. *Eur J Physiol* 2002, **444** : R49.

[8] Reding R, Denef J-F, Parmentier P, Lebrun M. Accès, compétences et opinions des étudiants en médecine vis-à-vis des technologies de l'information et de la communication. *Pédagogie Médicale* 2001 ; **2** : 242-49.

[9] Devitt P, Palmer E. Computer-aided learning : an overvalued educational resource ? *Med Educ* 1999 ; **33** : 136-39.

[10] Motiwalla L, Tello S. Distance learning on the internet : an exploratory study. *The Internet and Higher Education* 2000 ; **2** ; 253-64.

[11] Ricard J-D, Lejoyeux M, El-Ghoneimi A, Matheron S, Maillard D, Crickx B, Dreyfuss D. Utilisation des nouvelles technologies de l'information et de la communication par les étudiants en médecine. Enquête de pratique et mise en situation. *Pédagogie Médicale* 2005 ; **6** : 112-22.

[12] Valle R, Petra L, Martinez-Gonzalez A, Rojas-Ramirez JA, Morales-Lopez S, Pina-Garcia B. Assessment of student performance in problem-based learning tutorial sessions. *Med Educ* 1999 ; **33** : 818-22.

[13] Greenhalgh T. Computer-assisted learning in undergraduate medical education. *BMJ* 2001 ; **322** : 40-44.

[14] Eva KW. What every teacher needs to know about clinical reasoning. *Med Educ* 2004 ; **39** : 98-106.

[15] Woods NN, Brooks LR, Norman GR. The value of basic science in clinical diagnosis : creating coherence among signs and symptoms. *Med Educ* 2005 ; **39** : 107-12.

[16] Ward JPT, Gordon J, Field MJ, Lehman HP. Communication and information technology in medical education. *Lancet* 2001 ; **357** : 792-96.

Ubiquity: Technologies for Better Health in Aging Societies
A. Hasman et al. (Eds.)
IOS Press, 2006

ENN-ICS – Implementation and Evaluation of a Multilingual Learning Management System for Sleep Medicine in Europe

Brigitte KNOBL[1,*], Teresa PAIVA[2] Dieter JUNGMANN[3], Rico BÖHME[4] and Thomas PENZEL[1]

[1] *Sleep Laboratory, Hospital of Philipps-University, Baldingerstr. 1, D-35033 Marburg, Germany, (knobl@mailer.uni-marburg.de)*

[2] *Centro de EEG e Neurofisiologia Clínica, Lda, Lisbon, Portugal*

[3] *Robotron Datenbank-Software GmbH, Dresden, Germany*

[4] *Dept. of Business Management and Economics, University of Technology, Dresden, Germany*

Abstract. A new web based network aims at the improvement of health care in Europe by integrating advanced e-learning and e-publishing technologies for the training of medical doctors, nurses, and students. The field of application is sleep physiology and sleep medicine. Based on a multilingual, multimedia communication system, ENN-ICS Centre offers direct access to medical information for users, i.e healthcare professionals and citizens, in Europe and worldwide. The use of XML supports the development of media independent contents for multiple target groups. Editorial and distributive processes are supported by customized central editorial, content management and learning management systems (CMS, LMS). ENN-ICS e-health services are evaluated by selected user groups in North, Middle and Southern Europe using reliable and scientifically accepted validation instruments. The compliance with essential quality requirements and criteria is tested and verified by using online questionnaires based on the DISCERN questionnaire for evaluating patient information, the HON principles for health-related websites and the GMDS catalogue of quality criteria for electronic publications in medicine. The system architecture and its exemplary applications can be used as a model for future e-health services dealing with neurological and other medical topics.

Keywords: sleep disorders, sleep medicine, e-health service, learning management system, e-learning

1. Introduction

Sleep disorders produce high medical and societal costs with significant repercussions on health care budgets and society well being. Although affecting 25% of the general population [1], sleep complaints are often disregarded by healthcare systems, physicians, patients and the general public. Their negative impact upon mental, psychological and body functions as well as short and long term consequences on quality of life and well-being of affected individuals point out the need for adequate health education measures in sleep medicine. Taking into account the economic impact in terms of direct and indirect costs in view of high injury and accident risks associated with untreated sleep disorders, the e-health project ENN-ICS (European Neurological

* Corresponding Author: Brigitte Knobl, Sleep Laboratory, Hospital of Philipps-University, Baldingerstr. 1, D-35033 Marburg, Germany, knobl@mailer.uni-marburg.de

Network – Interactive Communication System) considers the relevance and importance of improving sleep in the European population by generating and providing innovative methods for educating and training different target groups [2–4].

As knowledgeable societies can handle their problems more effectively, ENN-ICS intends to build an interface between the scientific community and the clinical practice mediating the scientific state-of-the-art gathered by current research results in the field of sleep medicine. Its strategical concept implies disease prevention by providing essential information to citizens and patients. Nevertheless, its main focus is on educational aspects placing emphasis on e-training and e-learning systems.

ENN-ICS products – planned partly as paid service, partly free of charge - will be integrated in services at national, European and, in the future, on worldwide level. Multilinguality is realized in the project phase covering the world languages English, German and Portuguese. The principles "production in one-time charge - multi-functional, multinational, multilingual usability" and "single source – multiple media – multiple usage" are inherent project contributions to primary objectives of EU policy favouring financial discharge in health care systems.

2. Learning platform and web based e-learning infrastructure

Based on open XML (extensible markup language) standards, the electronic management of medical contents is realized by a central editorial system [5]. This system combines and integrates an Authoring Tool, a Content Management System and a Learning Management System supporting efficient, cost saving multilingual and multimedia production and distribution. ENN-ICS tools, services and products comprise interactive web sites, didactically well-shaped, multimedia-based, self-directed learning courses, and collaborative group learning in virtual classrooms using synchronous and asynchronous tele-tutoring techniques. CDs/DVDs and printed media can be orderd at the e-commerce centre that will be integrated after successful validation in the phase of market entry.

The system architecture allows flexible archival storage, re-use, re-configuration, realtime updating and distribution of learning contents. Being edited according to predefined didactical XML-schema-structures, all contents are delivered to a central content repository and then distributed by the XML data based Content Management System (CMS) to offline and online interfaces giving access to learners and interested users. Modifications referring to editions, revisions and translations are performed at the central editorial system. This central editing and content configuration process is the pre-condition for the easy use of multiple publishing functionalities such as generating HTML files for web pages and CD/DVD distribution or PDF files for electronic or traditional print media (fig. 1).

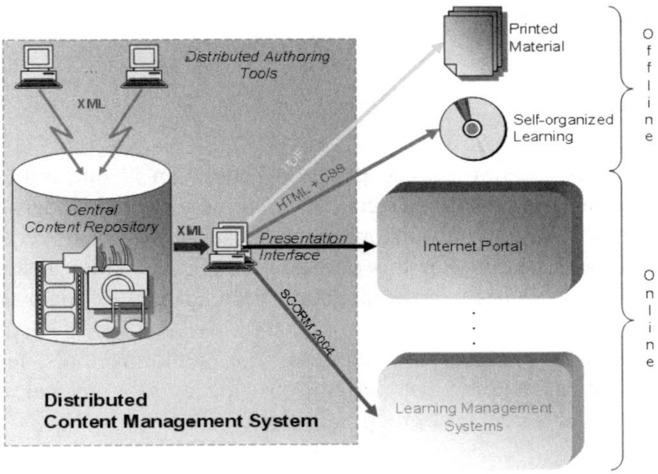

Figure 1: Architecture of the electronic management and distribution of contents[6]

Web based training (WBT) is supported by the learning management system (LMS) offering essential functionalities such as learners enrolment in courses, learners progress data management, synchronous and asynchronous communication tools for involved learners and peers, evaluation by tutors as well as tutor access to course statistics.

The ENN-ICS web portal (www.ennics.org) is configurated as universal interface offering access to all services and tools. Project partners have access to a password restricted working area during the project phase. Authors and editors will find access to the CMS, users and learners to the LMS and to online evaluation questionnaires, clients to products offered by the e-commerce centre, and citizens and patients to general information on sleep disorders (fig. 2).

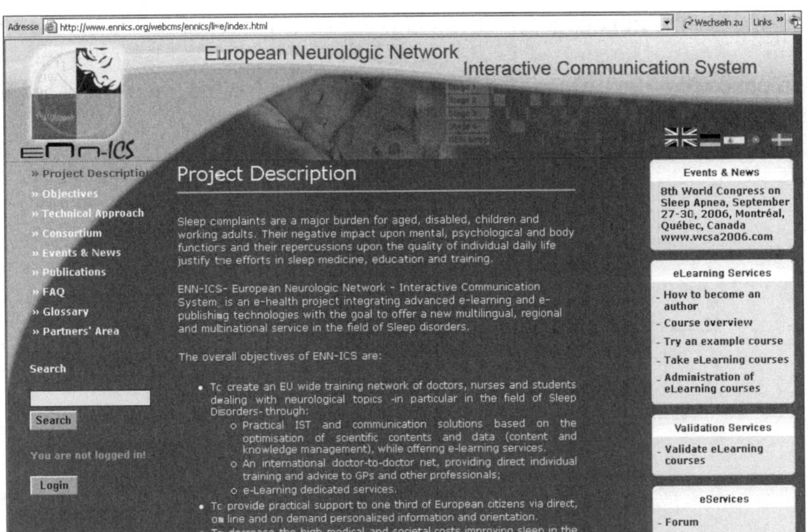

Figure 2: ENN-ICS web portal offering interfaces to the password restricted working area, the CMS, the LMS, evaluation questionnaires and e-commerce services

3. Evaluation and quality assurance

3.1. Validation strategy

The creation of a still non existing service that will benefit up to one third of the European citizens suffering from disturbed sleep and/or chronic sleep disorders as well as the planned service for teaching and supporting healthcare professionals require the assurance of high quality standards. For this purpose, reliable validation instruments regarded as gold standards have been chosen and have been adapted for the evaluation process.

Several ENN-ICS tools, products, and services - i.e. authoring tools, technical usability of services, contents, and the web portal in its capacity as a medical e-health web site - are evaluated by different user groups (fig. 3).

What will be evaluated?		Who will evaluate? 1st Step Evaluation	2nd Step Evaluation
ENN-ICS web portal	⇨	Project partners	HON Foundation
LMS	⇨	Project partners	Users
CMS	⇨	Project partners	Editor(s)
Authoring Tool	⇨	Project partners	Authors
Contents			
- Patient Information	⇨		DISCERN, Users
- eLections	⇨		Users
- CD-ROM	⇨		Users
- Printed Products	⇨	ENN-ICS Consortium	Users

Figure 3: Selected user groups evaluating ENN-ICS Sleep Tutorial

The ENN-ICS Sleep Tutorial comprising e-lections on sleep disorders listed by the International Classification of Sleep Disorders (ICSD) [7] for training medical doctors, students and nurses is evaluated by selected user groups in Sweden, Germany and Portugal representing the different health care systems in North, Middle and Southern Europe (fig. 4).

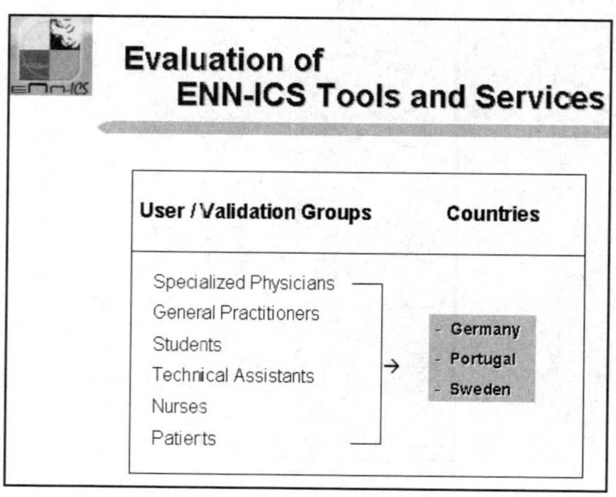

Figure 4: *Selected user groups evaluating ENN-ICS Sleep Tutorial*

3.2. Identification and adaption of adequate validation instruments

The following validation instruments have been adapted to be used in the evaluation process:

– DISCERN Questionnaire	for evaluating patient information
– HON Principles	for evaluating health-related web portals / websites
– GMDS Criteria Catalogue	for evaluating electronic publications / training courses / ENN-ICS services

The DISCERN questionnaire is used by the ÄZQ (Ärztliches Zentrum für Qualität in der Medizin = Medical Centre for Quality in Health Care) for validating online patient information in Germany, e.g. the "Patienten-Ratgeber" (Online Patient Information) of the German Sleep Society (DGSM) [8-9]. It offers elaborated quality criteria for online consumer health information on treatment choices. The HON Principles were developed by the Hon of the Net Foundation (HON) to guide both lay users and medical professionals to reliable sources of healthcare information in the internet. The HON Foundation participated actively in the development of the EU document "eEurope 2002: Quality Criteria for Health-Related Websites" and integrated all 8 HONcode Principles into its result [10-11]. The GMDS (Deutsche Gesellschaft für Medizinische Informatik, Biometrie und Epidemiologie / German Society for Medical Informatics, Biometry and Epidemiology) is actively involved in the definition of the medical curriculum and of the content of CME (continuous medical education) for physicians and for certified further professional qualification of medical informaticians, biometricians, epidemiologists and medical documentalists. The working group "Computer Based Training (CBT)" developed a catalogue of quality criteria for electronic publications in medicine [12].

These validation instruments are used to test, assess and verify the compliance with essential quality requirements and criteria. The DISCERN Questionnaire is used to assess the reliabilty of patient information whereas HON Principles are selected to test the compliance with the following requirements:

- Principle 1. Information must be authoritative
- Principle 2. Purpose of the websitePrinciple 3. Privacy – ConfidentialityPrinciple 4. Information must be documented: Referenced and datedPrinciple 5. Justification of claimsPrinciple 6. Website contact detailsPrinciple 7. Disclosure of funding sourcesPrinciple 8. Advertising policy

The Criteria Catalogue for Electronic Publications in Medicine developed by the GMDS are used for assessing the compliance with domain competence, competence in software engineering, media competence, design competence, and didactic competence. In addition, the HON Principles and the GMDS Criteria Catalogue serve as guidelines for the further development of ENN-ICS tools and platforms.

Adapted online questionnaires for evaluation have been implemented at the ENN-ICS website.

4. Conclusions

ENN-ICS Centre is validated by selected user groups in Germany, Portugal and Sweden. These three countries represent three very different examples of health care systems in Europe. Testing a system in these countries gathers feedback which will be applicable in a larger number of countries or regions. The scope of the validation as carried out in ENN-ICS is on the medical contents and on the tools. The results of the validation of ENN-ICS services allow to continuously revise contents, editorial and technical aspects, ergonomics, design, and didactics. All ENN-ICS partners contribute to the stepwise refinement of the system development using their specific competence area in accordance with the quality requirements defined by DISCERN, HON, and GMDS.

5. Acknowledgement

The ENN-ICS project presented is funded by the European Union (Project ENN-ICS, European Neurological Network – Interactive Communication System, No. 517418).

6. References

[1] Robert-Koch-Institut (Ed.): Schlafstörungen. Gesundheitsberichterstattung des Bundes. Heft 27, 2005.
[2] Knobl B, Paiva T, Jungmann D, Böhme R, Penzel T: ENN-ICS – Multilinguales Lernmanagement-System für die Schlafmedizin. In: Jäckel (Ed.): Telemedizinführer Deutschland, Bad Nauheim, Ausgabe 2006, 229-232.
[3] Paiva T, Penzel T (Eds.): European Neurological Network. IOS Press, 2000
[4] ENN-ICS – European Neurological Network – Interactive Communication System. Online: http://www.ennics.org/ (last accessed Jan. 18, 2006).
[5] Böhme R, Kalb H, Petzoldt P, Schoop E. Der E-Learning Redaktionsleitstand: Zentrale Koordination verteilter Produktions- und Einsatzprozesse im E-Learning. In: Engelin M, Meissner K (Eds.): Virtuelle Organisationen und Neue Medien GeNeMe 2005, Dresden, 2005, 385-397.
[6] Gersdorf R, Schoop E. Content Management für die Unterstützung verteilter Redaktionsprozesse im E-Learning. In: Bogaschewsky R, Hoppe U, Klauser F, Schoop E, Weinhardt Ch: Impuls Research Report 4, Osnabrück 2003.
[7] American Academy of Sleep Medicine (Ed.). The International Classification of Sleep Disorders. Diagnostic and Coding Manual. Second Edition. Westchester, IL, 2005.
[8] DISCERN Online - Quality Criteria for Consumer Health Information. Online: http://www.discern.org.uk/index.php (last accessed Jan. 13, 2006).
[9] Deutsche Gesellschaft für Schlafforschung und Schlafmedizin (Ed.): Patienten-Ratgeber – Schlafstörungen und ihre Behandlungsmöglichkeiten. Online: http://www.dgsm.de/ (last accessed Jan. 18, 2006).
[10] HON Code of Conduct (HONcode) for medical and health Web sites: Online: http://www.hon.ch/HONcode/Conduct.html (last accessed Jan. 13, 2006).
[11] Communication from the Commission to the Council, the European Parliament, the Economic and Social Committee and the Committee of the Regions - eEurope 2002: Quality Criteria for Health Related Websites, COM(2002) 667 final: Online: http://www.eu.int/information_society/eeurope/ehealth/doc/communication_acte_en_fin.pdf (last accessed Jan. 13, 2006).
[12] Schulz S, Klar R, Auhuber T, Schrader U, Koop A, Kreutz R, Oppermann R, Simm H. Qualitätskriterien für Elektronische Publikationen in der Medizin. Online: http://www.imbi.uni-freiburg.de/medinf/gmdsqc/ (last accessed Jan. 13, 2006).

8.3 Patient Education and Consumer Informatics

A. Hasman et al. (Eds.)
IOS Press, 2006

What makes an "informed patient"? The impact of contextualization on the search for health information on the Internet

Laurence ALPAY[a,1], John VERHOEF[a, b], Pieter TOUSSAINT[a], and Bertie ZWETSLOOT-SCHONK[a]
[a]*Clinical Informatics, LUMC, Leiden, The Netherlands*
[b]*Department of Physical Therapy, LUMC, Leiden, The Netherlands*

Abstract. The Internet is nowadays a powerful medium and can help patients to become better-informed citizens. Increasingly, citizens are searching for health information on the Internet. The Internet-based resource often acts as a virtual healthcare professional. The effectiveness of the communication between the patient and the virtual healthcare professional depends partly upon the presence of contextual information. The issue here is to what extent the contextualization of information is needed for effective information seeking and for the person's understanding of the retrieved/received information. The impact of contextualization on information search also closely relates to the person's cognitive resources. Using a theoretical communication framework (Te'eni 2001), we explore contextualization in a health website, and discuss the above issues and the possible relevance of contextualized information on the informed patient.
Keywords: informed patient, health information, contextualization, Internet

1. Introduction

The much-heralded term "informed patient" conveys the idea of self decision-making and improved communication (e.g. knowing which questions to ask) [1]. The informed patient is expected to know how to communicate with his/her healthcare professional and understand the information provided. The increasing use of the Internet makes this situation the more complex, especially when the patient is communicating with a "virtual" healthcare professional (mediated by a website for example). Indeed, there is a limited control and insight on how users comprehend the health information they have found in a website. The presence of contextual information can help improve the user's understanding of the health message, and support an effective communication between the patient and the virtual healthcare professional.

This paper explores the issue of the contextualization of information on the patient's understanding of the retrieved information from the Internet. This introduction presents the notion of the informed patient, and highlights the use of the Internet by the informed patient in order to acquire health related information, and its influence on the

[1] Corresponding Author: Dr. L.L. Alpay, Clinical Informatics, LUMC, Building 2, Room S-05-048, Postzone S-05-P, Postbus 9600, 2300 RC Leiden, The Netherlands L.L.Alpay@lumc.nl.

healthcare communication. The paper then presents a theoretical model of communication whose central feature is that of the contextualization of information. Further, the possible applicability of the model to the healthcare domain is examined, using as an illustration a website which we implemented to incorporate initial contextualized information. The last part of the paper discusses relevant aspects related to the contextualization of information, of importance to the informed patient, and outlines directions for further work.

1.1. The Informed Patient

In order to become informed, patients need to access the appropriate information. Various driving forces can explain why patients and citizens seek to be informed about their health. The first driving force comes from the lay persons. Being informed about one's own health suggests the emergence of a "health self" ideal, and thus driving the endless pursuit of information in order to achieve that ideal [2]. A second driving force comes from the patients themselves. Patients are becoming more informed and knowledgeable about their health conditions. There is a shift from passive information access to active citizen participation. People with long term illness such as diabetic or asthma patients will tend to have specific information needs with regards, for example, to the prevention, management and treatments of their health problems which can change overtime. In addition, healthcare professionals have to manage an increasing workload and an aging population with chronic illnesses. It is to the benefit of both parties that patients become better informed about their health conditions. Finally, insurance companies and governmental parties have an interest in encouraging citizens and patients to take their health in their own hand and to live a "healthy life style". Such an attitude to one's own health has relevant economic implications and can reduce the costs of services.

1.2. The Internet as a Tool for Information Seeking

Patients' information needs often change and evolve in response to the situation they find themselves in and to how they decide to respond to that situation. Indeed, information seeking does not occur in a vacuum but rather arises from and is conditioned by the circumstances in which the information seeker is [3]. While strong personal ties are still people's primary information sources, the Internet is a popular tool, which offers unlimited possibilities for finding health information. The access to the Internet brings a shift in the access to knowledge and also to empowerment.

1.3. Communication in Healthcare

Introducing Information and Communication technology (ICT) in healthcare practices inevitably affects communication between citizens/patients and healthcare professionals. Indeed, a new health consumer identity is emerging, sometimes referred to as the "on-line self-helpers" [4]. Healthcare professionals are also experiencing an identity shift, moving from the authoritative figure to that of facilitator. Their knowledge is often mediated on the Internet via health related websites. The virtual healthcare professional often has limited insight about the user's understanding of the information given. The contextualization of the information can help improving the user's understanding and sustain an effective man-machine communication. The next section describes a theoretical model of communication whose central theme is that of contextualization.

2. Model of Communication and its Applicability in Healthcare

2.1. Theoretical Model of Communication

The use of ICT in healthcare communication has impact on the effectiveness of the communication. In his model of communication, Te'eni [5] introduces the notion of "communication complexity" as the communicator' sense of the cognitive effort needed to ensure effective communication. Effective communication[2] results from the use of cognitive resources to overcome the difficulties in understanding and uncertainty about the message. Another construct of this model is that of "mutual understanding" which requires that the communication be comprehensible according to the sender's meaning. Central to the model is the idea of context. Contextualization requires cognitive resources and is seen as a communication activity aimed at increasing mutual understanding by reducing complexity. In other words, contextualization is about providing contextual information to explain a core message.

Communication complexity is affected by the heterogeneity of the communicators involved, that is, in the terminology they used to communicate. It appears that heterogeneity triggers communicators to incorporate more contextual information in order to improve mutual understanding. Contextualization, if effective, reduces communication complexity and thereby increases mutual understanding (see Figure 1).

Figure 1 Contextualization within the communication process

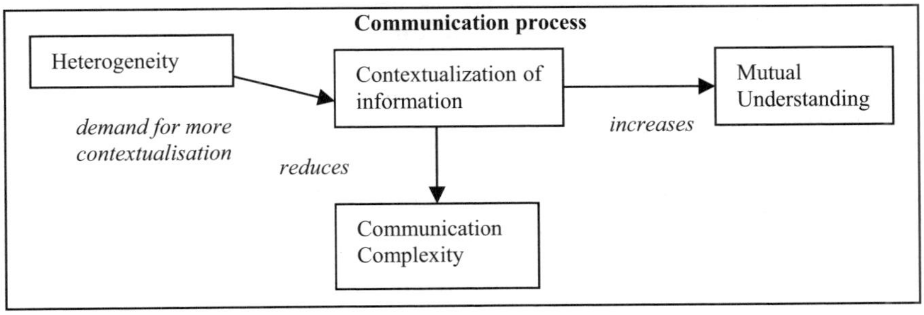

2.2. Possible Applicability in Healthcare

The Te'eni model is of relevance to our study of ICT-based healthcare communication and informed patients. The communication processes we are interested in centre around the task of information seeking within a health related website. This is a communication setting where the patient is the requester of information and the healthcare professional (via ICT) acts as the information provider. We are engaged in investigating which cognitive requirements and constraints need to be taken into

[2] Either face-to-face or man-machine communication.

account in ICT enabled communication in healthcare, which aims to enhance the emerging, informed patient.

Within this man-machine communication, contextualization is about providing information, which is explanatory and supportive to convey the health message, and the user's understanding of the message. We view contextualization of information as a construction to the cognitive requirements and constraints put upon the information seeker. The contextual information will increase the likelihood that the user (as receiver of information) will understand the message he/she has received via the Internet. In practice, information seekers might often start searching using a general search engine. According to Te'eni, contextualization should be used only when needed. However, this is not always the case. For example, a search with Google™ can easily retrieve thousands of results, some of which irrelevant to the user's information needs. Problems are that 1) the contextual information is not always in tune with the user's own initial query (e.g. health problem) and 2) the contextual information might not point out to appropriate solutions for the user's initial problem. This is a kind of "randomized" contextualization of information. In this situation, the contextualization of information can be counter-productive since it overloads the receiver unnecessarily and diverts from the important information.

Within the boundaries of a health website, contextualization is more constrained but not necessarily present as needed. As Te'eni stresses contextualization is necessary when a message is liable to be misunderstood. This is especially true when dealing with health related information, which may impact on someone's well-being. Within a given website, contextualization may be more easily "tailored" to accommodate the user's information needs. Information seekers of health information on the Internet tend to shift back and forth between "randomized" and "tailored" contextualization.

In practice, visitors of health websites will have different levels of knowledge and understanding about the healthcare topic they are searching for. The heterogeneity of the pair user-machine will depend on how much the user knows about and understand the information presented to him/her by the system. Of interest here is the notion of cognitive distance i.e. the difference in knowledge between the user and the website which will affect the complexity of the communication. A user, for whom the cognitive distance with the website is high, will need contextualization to a greater extent and will request more information. Such a user will tend to have less knowledge and understanding of the topic he/she is searching for. In contrast a user with a low cognitive distance will have less need for contextualization of information. Typically, an informed patient or a healthcare professional will be the latter kind of user.

2.3. Contextualization in the SeniorGezond website

SeniorGezond[3] (SeniorHealth in English) contains information about fall incidences and possible related interventions, and incorporates some *initial* contextualization of information [6]. It is structured around problems and contains four levels of information. The top level 'Causes of falling' includes frequently occurring problems in the domain of fall incidences e.g. dizziness and lack of balance. The second level below 'Solutions' focuses on possible interventions and advices associated with the causes of falling. Solutions are supported by the third level 'Products and Services' e.g. walking aids, fitness programmes. From the 'Products & Services' level users can

[3] http://www.seniorgezond.nl (in Dutch)

select addresses and insurance information (Supportive facts) if needed. These levels provide a structure that starts with general information and ends with specific suggestions. Within the website, the *links* between the four different levels form the backbone of the information structure and are considered an intrinsic factor of the initial contextualization.

According to some research [2], patient's knowledge is often grounded in everyday health and life experiences. The importance of this 'experiential knowledge' over medical expertise has relevant implications with regards to how to contextualize information. Indeed, someone with little experiential knowledge may have a higher demand for contextual information in order to reduce complexity. In SeniorGezond, for example, "practical" solutions have been implemented for contextualization. The level 'Solutions', which focuses on possible interventions and advices, incorporates practical information, often taken from everyday experiences. Furthermore, within the website, the user can directly choose from a list of 'Practical problems', that is, daily life situations where falls can happen such as 'Fear of falling while taking a shower'.

3. Discussion and Further Work

More research work on the contextualization of information and its impact on the informed patient remain to be carried out. We have reported in this paper how some aspects of contextualization have been implemented in SeniorGezond. Our work in progress [7] is now to assess *empirically* whether the impact of contextualization (or lack of it) on mutual understanding depends on the user's cognitive distance. Based upon the Te'eni model, the hypothesis here is that a user with a low cognitive distance (i.e. a person knowledgeable in the field) will have a high level of mutual understanding whether or not they use contextualization. Conversely, a user with a high cognitive distance (e.g. a non-knowledgeable person) is expected to have a low mutual understanding with no contextualization (e.g. a missing level of information, a missing link between the levels of information) and a high mutual understanding with contextualization. In other words, we explore the homogeneity and heterogeneity of the duo user-website. In our forthcoming study, participants are assigned to groups (knowledgeable vs. non-knowledgeable) depending on their prior knowledge of the domain. Within both groups, participants are given exposure or no exposure to contextualization, which results in four experimental groups. As outcome parameter, we use the participants' understanding of the domain after visiting the website.

There are other aspects related to contextualization of information that are of importance to the informed patient, such as the person's information literacy and health literacy [8]. A lack of both skills can inhibit the emergence of the informed patient identity [1]. Closely related to these is the issue of lifelong learning. Becoming an informed patient is a lifelong learning process. This is especially true for patients with long-term illness such as diabetic patients. We foresee that the contextualization of information has a role to play in enhancing the website visitor's health literacy and eventually his/her lifelong learning development. A better comprehension of the terminology employed through contextualization can reduce the cognitive distance between user and website and increase mutual understanding. In SeniorGezond for instance, attention has been given to provide clear definition and explanation to the terms used in the domain of fall incidences.

The notion of "information journey" [3] is also relevant to the contextualization of information. While Attfield *et al* [3] studied it in a broad scope; we focus on the information journey within an electronic health resource such as a health website. We take the contextualization of information as an accompanier and a "tuner" to a person's information journey. In SeniorGezond for instance, the contextualization of information through the information structure can be viewed as supporting the user's information journey from general (the causes of falling) to specific information (products to purchase). Gathering good quality and contextualized information can ensure that informed patients be fully equipped to pursue his/her information journey.

4. Conclusions

The informed patient will not emerge naturally or easily within the existing structures and relationships between patients and healthcare professionals [1]. It is our view that the contextualization of information can act as a facilitator to sustain communication between the patient and the virtual health professional, and thus to enhance the informed patient. The effectiveness of this communication is in part determined by the complexity of communication processes including contextualization of information and cognitive distance. Given the cognitive resources needed for information seeking, contextualization of information can be viewed as a construction to the cognitive requirements and constraints put upon the information seeker. More research is needed to investigate the impact of contextualization on information seeking on the Internet and on the person's understanding of the information retrieved. The model of communication we are using whereby contextualization is central provides a sound framework in order to pursue this line of research.

Acknowledgements

Thanks for Dov Te'eni (Tel Aviv University, Israel) and Regina Overberg (LUMC) for their comments, and for Andries de Man (LUMC) for current development of the website SeniorGezond.

References

[1] Henwood F, Wyatt S, Hart A, Smith J. 'Ignorance is bliss sometimes': constraint on the emergence of the 'informed patient' in the changing landscapes of health information. Sociology of Health Illness 2003; 25(6): 589-607.

[2] Kivits J. Researching the 'Informed Patient': the case of online information seekers. Information, Communication & Society 2004; 4: 510-530.

[3] Attfield S, Adams A, Blandford A. Patient information needs: pre and post consultations. Health Informatics Journal 2006; 12: 165-177.

[4] Ferguson T. Health online and the empowered medical consumer. Jt Comm J Qual Improv 1997; 23(5): 251-7.

[5] Te'eni D. Review: A cognitive -affective model of organizational communication for designing IT. MIS Quarterly 2001, 25(2): 251-312.

[6] Alpay LL, Toussaint PJ, Ezendam NPM, Rövekamp T, Graafmans W, Westendorp R. Easing internet access of health information for the elderly users: design considerations and usability. Health Informatics Journal 2004; 10(3):185-194.

[7] Alpay LL, Verhoef J, de Man AJM, Zwetsloot-Schonk JHM. Communication between patients and virtual healthcare professionals: does the contextualization of information matters? Abstract Proceedings of the 3rd International Conference of the European Association for Communication in Healthcare (EACH); 2006 September, 5-8, Basel, Switzerland. (http://www.each-conference.com).
[8] Kerka S. Health Literacy beyond basic skills. Eric Digest 2003. http://www.ericdigests.org/2004-1/health.htm (last accessed 21st December 2005).

Ubiquity: Technologies for Better Health in Aging Societies
A. Hasman et al. (Eds.)
IOS Press, 2006

The Effect of a Multimedia Health Educational Program on the Postoperative Recovery of Patients undergoing Laparoscopic Cholecystectomy

Antonia STERGIOPOULOU[a,1], Konstantinos BIRBAS[b], Theophanis KATOSTARAS[c]
Mariana DIOMIDOUS[c], John MANTAS[a]

[a] *Laboratory of Health Informatics, Faculty of Nursing, National and Kapodistrian University of Athens, Greece*
[b] *Department of Surgery, Faculty of Nursing, National and Kapodistrian University of Athens, Greece*
[c] *Department of Public Health, Faculty of Nursing, National and Kapodistrian University of Athens, Greece*

Abstract. Aim of this study is the evaluation of the impact of preoperative informative session using a Multimedia Health Educational Program (MHEP) on patients undergoing elective Laparoscopic Cholecystectomy (LC) for cholelithiasis, preoperative anxiety and postoperative pain and nausea. Sixty consecutive patients scheduled for elective LC were considered for enrolment in the trial. Patients were assigned randomly to four groups: Group A included 15 patients, preoperatively informed regarding LC through the MHEP presented by a Registered Nurse (RN). Group B included 15 patients preoperatively informed through a leaflet (designed and developed using the exact contents of the MHEP). In Group C, there were 15 patients who were being informed verbally from the RN. Finally, the control Group D included 15 patients, who had the conventional preoperative information about the operation and postoperative course by the attending surgeon and anesthesiologist, as every other patient included in groups A, B, C. Preoperative assessment of patient's knowledge about cholelithiasis and LC was performed after informative session, and was based on a specifically developed "closed, true-false" questionnaire. Preliminary results suggest that conventional information provided by the attending surgeon (Group D) is inadequate. Specifically developed informative sessions with the contribution of MHEP seems to be effective on reducing preoperative anxiety and postoperative pain, in patients undergoing elective LC.

Keywords: Patient education, Preoperative Information, Multimedia Health Educational Program, Laparoscopic Cholecystectomy

[1] Antonia Stergiopoulou, Fokionos 2 str. Nikaia 18454, Greece E mail: toniastr123@hotmail.com

1. Introduction

Patient education is defined as the process of influencing patient behavior and producing the changes in knowledge, attitudes and skills necessary to maintain or improve health. Implementation of structured informative session could also change the preoperative anxiety status of candidates to surgical procedures and improve their postoperative recovery. Many different approaches to pre-operative education have been used, including (a) group lectures, (b) individual teaching, (c) printed information, (d) learning packages and (e) audio-visual presentations [1]. Each of the above methods has its own limitations. There is a need to evaluate and improve traditional methods and consider new approaches. Computerized information programs for patients are beginning to gain popularity. Suitably designed and implemented information systems can promote effective learning, offer consistent information time after time, give a more stimulating and interesting presentation, making maximum use of available media and human senses, allow people to learn at their own pace and offer limitless repeat options [2], [3].

2. Materials and Methods

2.1. Purpose

Aim of this study is the evaluation of a structured informative session using a Multimedia Health Educational Program (MHEP) in patients undergoing elective LC. The study focuses on the impact of preoperative informing on (a) patient's preoperative anxiety and (b) patient's postoperative pain and nausea [4].

2.2. The multimedia health educational program

MHEP is a specifically developed multimedia health educational product based on Toolbook Asymetrix, version 8.5, Macromedia Company. The tool was designed in order to be flexible, dynamic, portable and cheap, permitting easy navigation even to someone unfamiliar to a computer machine.
Content of MHEP was selected in close collaboration with expert surgeons in this surgical field and was written in simple Greek of senior high school grade avoiding confusing medical terms. Structure and hierarchy of knowledge contained in MHEP is presented in order to compare with the flow of knowledge during conventional informative sessions provided to candidates to LC, by the attending surgeon. Evaluation of MHEP was performed by a group of independent laparoscopic surgeons. Concurrently, we developed a leaflet and a personalized presentation using the exact (visual and other) contents of MHEP.

2.3. Participants and data collection

From July to December 2005, 60 consecutive patients, candidates to elective LC for cholelithiasis, were considered for enrolment in the trial. Exclusion criteria were: (a) patients older than 75 years and younger than 18, (b) patients with an American Society of Anaesthesiologists (ASA)[5] physical status score greater than 2, (c) patients unable to understand Greek, (d) patients with serious sight and deaf impairment, and (e) patients undergoing LC combined with another laparoscopic or open procedure, simultaneously. Informed consent for participation in the trial was obtained and the trial was approved by the Administrative and Scientific Council of the Patras University Hospital, Patras, Greece and the "Attiko" University Hospital, Athens, Greece.

At the day of the admission, demographic data and medical history were acquired by all patients. Patients were assigned randomly to four groups: **Group A** included 15 patients, preoperatively informed about the scheduled operation through the MCD, presented by the Registered Nurse (RN). **Group B** included 15 patients preoperatively informed through the leaflet. The information leaflet was designed and developed using the exact contents of the MHEP and was delivered to the patients without the presence of the RN. However, she confirmed the compliance of patients. In **Group C**, there were 15 patients who were informed verbally from the RN and finally, the control **Group D** included 15 patients, who had the conventional preoperative information by the attending surgeon and anesthesiologist, as every other patient included in groups A, B, C. The information leaflet and the MHEP was available to patients for as how long as they wished (usually, 20-30 minutes). Four hours after the completion of each informative session of each group, RN collected the completed patients' questionnaires which had given to them, earlier. Patients in group D fulfilled the same questionnaire having received only the conventional verbal information by the surgeon and/or the anesthesiologist as every other patient included in groups A, B, C. All the data were collected in a randomized way.

Assessment of preoperative knowledge about cholelithiasis and LC, was performed, using "closed, true-false" questionnaire, specifically developed. Each question was scored equally with the others yielding a maximum score of eleven. Evaluation of preoperative anxiety was conducted using the six items of the translated Amsterdam Preoperative Anxiety Scale and Information Scale (APAIS) on which scores could range from 6 to 30 – subdivided by Anxiety Scale with a scoring range from to 4 to 20 and the Need-for-Information Scale scored from 2 to 10.[6] Postoperative pain and nausea scores were measured using a Numerical Rating Scale (NRS) scale, 8 and 16 hours after the patient had returned to the wards. The NRS[7] scale consists of 11-points (where 0 indicates no pain or nausea at all, and 10 the most severe pain and nausea imaginable).

3. Results

There were 34 (56, 7%) women and 26 (43,3%) men enrolled in the trial. Thirty nine patients (65%) had a previous operation in their history. Thirty six (60%) patients with ASA I (where ASA I denotes a patient presenting for surgery who is otherwise healthy) and 24 (40%) with ASA II (refers to a patient with mild comorbidity) were included in the study, respectively. The mean age of patients was 51.5 years (range: 18-75). Twenty patients (33.3 %) were familiar with the use of computers. All patients were informed preoperatively about the procedure, by their surgeon and signed an informed consent for the procedure.

The collected data were inserted firstly in a backward model of linear regression analysis (Table 1.1 to 1.5). Then, the variables that followed statistical significance of p value<0.05 in the single linear regression were inserted in the backward model of multiple regression analysis (Table 2). The statistical System SPSS 13 for Windows was used for statistical analysis.

4. Discussion

Effective informative sessions provided by the health professional, require specifically developed educational tools. In this study, we investigated the effect of MHEP on patient's postoperative recovery. In the linear regression analysis, structurally informed patients (groups A, B, C) achieved a better knowledge score regarding LC, answering correctly to more questions, comparing to group D[2], ($p<0.01$, $r^2=0.3$) as it is shown in Table 1.1 This finding suggests that, at least in Greece, quality of personal informative communication between surgeons and patients is relatively poor. Younger patients answered correctly in more questions than the older ones. Also, more educated patients achieved a better score. Patients with a previous exposure to computers had also more correct answers than those who did never used one, independently of the educational tool used. Finally patients who were being informed by the MHEP achieved a higher score of correct answers than oral information acquired by the RN or the doctor. Ng SKS et al [8] found that provision of preoperative information regarding the recovery process leads to significant anxiety reduction. It is also known that preoperative information would reduce preoperative anxiety as Miller et al [9] suggests. Our results, in the linear regression analysis, proved that conventional information[3] increases patient anxiety. Specifically patients informed only by their doctor had a higher APAIS score than the patients of groups A,B,C. ($p<0.05$, $r^2=0.06$).(see Table 1.3) Higher APAIS score means that the patient felt more preoperative anxiety. Choi-Kwon et al [10] underlines in his study that health professionals do not know what exactly patients want to know about stroke. Specifically, the Need-for Information-Scale (APAIS II) according to Table 1.2 is higher in patients informed conventionally ($p=0.004$, $r^2=0.119$). The patients with a high APAIS II score feel that they have not been informed adequately. In our study, patients who were informed through a structured informative session and especially patients in group C expressed smaller need for preoperative information than the patients of control group D. We can assume that at least in Greece, patient's preoperative need for information is not full filled during conventional informative sessions.

It is widely accepted that preoperative information induces rapid postoperative recovery. Giraudet et al reports a strong positive effect of patient education to the reduction of postoperative pain [11]. In our study, patients who were informed through a structured informative session with or without educational tools[4] reported less postoperative pain and nausea during the first 16 hours (p=0,021 and p=0,039 respectively).(seeTable1.4-1.5)

In multiple regression analysis, as shown in Table 2, previous computer familiarization resulted to better performance of the patients regarding their knowledge score,

[2] patients conventionally informed by the attending surgeon/anesthesiologist

[3] Group D

[4] Who belonged in groups A, B, C

independently of the educational tools used. Not surprisingly, patients older in age, achieved lower knowledge score than younger patients. Both the linear and the multiple regression analysis did confirm the statistical knowledge significance for the MHEP. In multiple regression analysis, patients informed by the MHEP and the presence of the RN had a higher knowledge score comparing to the four groups ($p<0.001$ $r^2=0.41$). It is hoped that the MHEP can help by improving the quality of health care, increasing patient knowledge and provide multiple therapeutic possibilities by reducing preoperative anxiety and postoperative pain and nausea in increasingly aging societies. As this study has not finished yet, we feel that this significant relationship will become more evident in the near future.

5. Conclusion

Use of MHEP in structured preoperative informative sessions, in patients undergoing LC has been proven effective as far as the learning transfer in concerned. However, the impact of MHEP on preoperative anxiety and postoperative pain and nausea is less obvious. Further double blind control studies with broader sample is necessary to establish definitive conclusions. On the other hand, specifically developed MCDs for different populations (regarding gender, age or educational level) have to be tested in clinical practice to provide therapeutic approach to personalised needs.

6. References

[1] Enzenhofer M., Bludau H., Komm N. et al. Improvement of the Educational Process by Computer-based Visualization of Procedures: Randomized Controlled Trial. J Med Internet Res [serial online] 2004; 6(2): e16. Available from. URL: http://www.jmir.org/2004/2/e16/.

[2] Campbell F.A., Goldman B.D., Boccia M.L., & Skinner M. The effect of format modifications and reading comprehension on recall of informed consent information by low-income parents: a comparison of print, video and computer-based presentations. Patient Education and Counseling, 53, 205-216.

[3] Knee K., Jacobs M. More than a wise patient care decision, computer-based patient education makes good business sense. Patient Education and Counseling. Available from. URL: www.oncologyinteractive.com/site/pdf/business_sense.pdf. Accessed August 9, 2004.

[4] Van den Bosch JE, Moons KG, Bonsel GJ, Kalkman CJ. Does Measurement of Preoperative Anxiety Have Added Value for Predicting Postoperative Nausea and Vomiting. Anest Analg 2005;100:1525-1532.

[5] Walker R. Asa and Cepod Scoring. World Anesthesia Issue 14(2002) Article 5: Page 1 of 1.

[6] Moerman N, Frits S.A., M11uller M., Oosting H. The Amsterdam Preoperative Anxiety and Information Scale (APAIS). Anesthesia & Analgesia. Vol 82(3), March 1996, pp 445-451.

[7] Douglas G.S. Dawn M.E. et al. Efficacy of gabapentin in treating chronic phantom limb and residual limb pain. JRRD.Vol 42, Num 5, Pages 645-654.

[8] Ng SKS, Chau AWL, Leung WK. The effect of pre-operative information in relieving anxiety in verbal surgery patients. Community Dent Verbal Epidemiology [serial online] 2004; 32:227-35. Available from. URL: www.blackwell-synergy.com/links/doi/10.1111/j.1600-0528.2004.00161.x/pdf.

[9] Miller K., Wysocki T., Cassady J., Cancel D., Izenberg N. Validation of Measures of Parents' Preoperative Anxiety and Anesthesia Knowledge. Anesthesia & Analgesia 1999; 88:251-7.

[10] Choi-Kwon S., Lee S. K., Park H. A., Kwon S.U., Ahn J.S. and Kim J.S. What stroke patients want to know and what medical professionals think they should know about stroke: Korean perspectives (2005) Patient Education and Counseling, 56 (1), pp.85-92.

[11] Giraudet -Le Quintrec JS., Coste J., Vastel L., Pacault V., Jeanne L., Lamas JP., Kerboull L. Positive Effect of Patient Education for hip surgery: a randomized trial. Clin Orthop Relat Res. 2003 Sep;(414):112-20.

Table 1.1 Single linear Regression for the variable "Knowledge Score".

Variables	Knowledge Score P value (Adj.R^2) A[5], B(CI of B)[6]
Informed	*0.000 (0.3)* *6 , 2.3 (1.4_3.2)*
Group A	*0.046(0.068)* *8.1, 0.9 (0.016_1.784)*
Group C	*0.014(0.113)* *8.7, -1.1 (-1.9_-0.23)*
Group D	*0.000(0.3)* *8.4 , -2.3 (-3.2_-1.4)*
Computer users	*0.000(0.18)* *7.25, 1,7 (0.8_2.6)*
Educational level	*0.001(0.169)* *6.3, 0.337 (0.150_0.525)*
Age	*0.002(0.134)* *10.2 , -0.04 (-0.07_-.01)*

Table 1.2 Single linear Regression for the variable "APAIS II Score[7]".

Variables	APAIS II Score P value (Adj.R^2) A, B (CI of B)
Informed	*0.004 (0.119)* *6.4, -2.6* *(-4.34_-0.860)*
Group C	*0.003 (0.166)* *4.7, -2.5* *(-4.11_-0.88)*
Group D	*0.004 (0.119)* *3.86, 2.6* *(0.860_4.34)*

Table 1.3 Single linear Regression for the variable "APAIS".

Variables	APAIS Score P value (Adj.R^2) A, B (CI of B)
Informed	*0.035 (0.059)* *17.4, -4.1* *(-7.99_0.312)*
Group D	*0.035(0.059)* *13.3, 4.1* *(0.312_7.99)*

Table 1.4 Single linear Regression for the variable "Pain Score".

Variables	**Pain Score** in the first 16h P value (Adj.R^2) A, B (CI of B)
Informed	*0.021 (0.073)* *4.55, -0.88* *(-1.61_-0.138)*
Group D	*0.021 (0.073)* *3.65, 0.87* *(0.138_1.618)*

Table 1.5 Single linear Regression for the variable "Nausea Score".

Variables	**Nausea Score** in the first 16h P value (Adj.R^2) A, B (CI of B)
Informed	*0.039 (0.056)* *2.2, -0.989* *(-1.925_-0.053)*
Group D	*0.039(0.056)* *1.2, 0.989* *(0.053_1.925*

Table 2. Multiple Regression Analysis for the variable "Knowledge score".

Variables	Knowledge Score P value (Adj.R^2) A, B(CI of B)
Computer users	*0.004 (0.41)* *9.61, 1.13(0.38_1.88)*
Group A	*0.007(0.41)* *9.61, 1.01(0.28_1.73)*
Total Age	*0.005 (0.41)* *9.61,-0.038(-0.06_-0.012)*

---

[5] constant
[6] Confidence Intervals of B
[7] Need-for-Information Scale

Ubiquity: Technologies for Better Health in Aging Societies
A. Hasman et al. (Eds.)
IOS Press, 2006

A New Health Strategy to Prevent Pressure Ulcer Formation in Paraplegics using Computer and Sensory Substitution via the Tongue

Alexandre MOREAU-GAUDRY[a,c,1], Anne PRINCE[b], Jacques DEMONGEOT[a,c] and Yohan PAYAN[c]

[a] *University Hospital of Grenoble, France*
[b] *University and Medical Center of St Hilaire du Touvet, France*
[c] *TIMC-IMAG-CNRS Laboratory, Grenoble, France*

Abstract. Pressure ulcers are recognized as a major health issue in individuals with spinal cord injuries and new approaches to prevent this pathology are necessary. An innovative health strategy is being developed through the use of computer and sensory substitution via the tongue in order to compensate for the sensory loss in the buttock area for individuals with paraplegia. This sensory compensation will enable individuals with spinal cord injuries to be aware of a localized excess of pressure at the skin/seat interface and, consequently, will enable them to prevent the formation of pressure ulcers by relieving the cutaneous area of suffering. This work reports an initial evaluation of this approach and the feasibility of creating an adapted behavior, with a change in pressure as a response to electro-stimulated information on the tongue. Obtained during a clinical study in 10 healthy seated subjects, the first results are encouraging, with 92% success in 100 performed tests. These results, which have to be completed and validated in the paraplegic population, may lead to a new approach to education in health to prevent the formation of pressure ulcers within this population.
Keywords: Spinal Cord Injuries, Pressure Ulcer, Sensory Substitution, Health Education, Biomedical Informatics.

1. Introduction

A pressure ulcer is defined as an area of localized damage to the skin and underlying tissue caused by pressure, shearing, friction or a combination of these factors [1]. Its prevalence, which ranges from 23% to 39% in adults with spinal cord injuries (as reported in two recent retrospective and one prospective studies), remains high in this population [2-4]. Located near bony prominences such as the ischium, sacrum and trochanter, pressure ulcers are recognized as the main cause of rehospitalization for patients with paraplegia [5]. Their treatment, which can be medical or surgical, is always long, difficult and expensive. Thus, this pathology appears to be a major health issue for this population.

---

[1] Corresponding Author: Alexandre Moreau-Gaudry. TIMC-IMAG-CNRS Laboratory. GMCAO Team. Faculty of medicine, Alexandre.Moreau-Gaudry@imag.fr

Because of the need to identify effective intervention strategies that provide health education for skin care management and the prevention of pressure ulcers for individuals with paraplegia, a new approach has been developed using computer-aided instruction as an educational tool for promoting independent skin care [6]. As with traditional approaches, the main purpose of this strategy is to control the pressure applied at the seat/skin interface in order to prevent the formation of pressure ulcers. Nevertheless, to our knowledge, no approach has been developed that compensates for the sensory loss for paraplegics in the buttock area which would allow them to lead a similar lifestyle to that of healthy subjects in term of pressure ulcer prevention. Healthy subjects prevent this pathology by moving the body in a conscious or subconscious manner according to a “signal” arising from the buttock area. The main purpose of this research is to define a new health strategy that prevents the formation of pressure ulcers in individuals with paraplegia by simulating the natural performance of a healthy subject via use of sensory substitution. The first results related to the feasibility of this approach in healthy seated subjects are reported in this work.

2. Material and Methods

2.1. Which organ is best suited to sensory substitution?

The concept of sensory substitution has its origins in the works of P. Bach-y-Rita, who has studied this area for 30 years and, in particular, sensory substitution for blind people. Indeed, he states: “we do not SEE with the eyes”. The visual image does not go beyond the retina, but is turned into patterns of pulses along nerves and is carried to the brain [7]. To demonstrate this theory, a human-machine interface was developed: the Tactile Vision Substitution System (TVSS), which allows visual information to be emitted from a TV camera to an array of stimulators in contact with the skin on one or several parts of the body (abdomen, back, and forehead). After training with this device, blind people are able to perform complex perceptual and “eye”-hand coordination tasks. Furthermore, when the human-machine interface is moved from one area of skin to another, there is no loss of correct spatial localization. Thus, the trained blind subject does not perceive the image on the skin, but in fact locates it correctly in space.

More recently, P. Bach-y-Rita has focused on the tongue as the most ideally suited organ for the human-machine interface because of outstanding tactile and physical properties. He also reported the possibility of developing a cosmetically acceptable interface into an orthodontic retainer thanks to recent technological improvements [8]. For these reasons, and because the sensory and motor nervous pathways for the tongue are preserved in paraplegia, this organ was chosen for sensory substitution in spinal cord injury individuals.

2.2. The implementation of sensory substitution

A new device has been developed to compensate for sensory loss in paraplegics in the buttock area. It consists of three components (Figure 1). The first is a pressure mapping system which allows real-time acquisition of the pressure applied on the

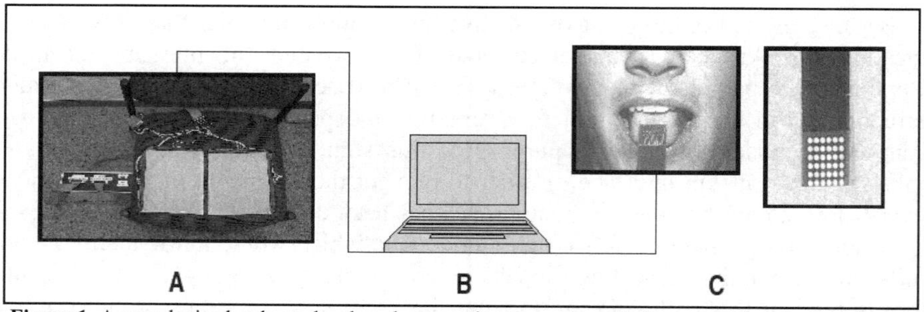

Figure 1. A new device has been developed, attempting to compensate for sensory loss in paraplegics. The pressure mapping system (A) allows real-time acquisition of the pressure applied on the seat/skin interface. The computer (B) enables communication between the pressure mapping system and the Tongue Display Unit (TDU). From the laptop, electrotactile stimuli can be delivered to the tongue surface via a flexible electrode array (C) placed in the mouth. This new device allows paraplegics to receive informative signals on their tongue that originate from the buttock area.

seat/skin interface. The second is the human-machine interface. In a preliminary experiment, this device was the Tongue Display Unit (TDU) developed by P. Bach-y-Rita and colleagues. The third device is a laptop, which has been programmed to enable communication between the pressure mapping system and the TDU. From the laptop, electrotactile stimuli can be delivered to the dorsum of the tongue via a flexible electrode array (6x6), which is placed in the mouth and lightly held between the lips. Each electrode can be activated independently.

This whole device thus allows electro-stimulation of parts of the tongue according to a "signal" received from the pressure applied on the buttock area of the paraplegic. This information, that is lost after spinal cord injuries, is therefore felt again via the tongue. We hope and claim that the paraplegic will be able to develop an adapted behavior according to the electro-stimulated information in order to prevent the formation of pressure ulcers.

2.3. An evaluation of the feasibility of this approach

In this preliminary work, the feasibility of the approach was evaluated by the subject's capacity to adopt an adapted behavior, with changes in pressure, according to electro-stimulated information. Our objective was therefore to demonstrate this capacity as without it, sensory substitution is of no use for this purpose. The electro-stimulated signal is not efficiently interpreted, since there is no change in pressure at the seat/skin interface, and therefore there is no possibility of relieving the cutaneous area of suffering and of preventing the formation of pressure ulcers.

After obtaining informed written consent from each subject, this evaluation was performed with 10 healthy seated subjects, at rest. For each subject, the TDU was first calibrated. This step was necessary as the perception of electrotactile stimuli on the tongue is subjective and decreases from the anterior to the posterior part of the tongue. To perform this first step, a C++ program was developed which enabled this calibration to be both user-friendly and interactive.

Following this, healthy subjects were asked to move their chest according to the felt electro-stimulated direction. More precisely, when the six electrodes placed in front of the electrode array were activated, subjects were asked to move their chest forward

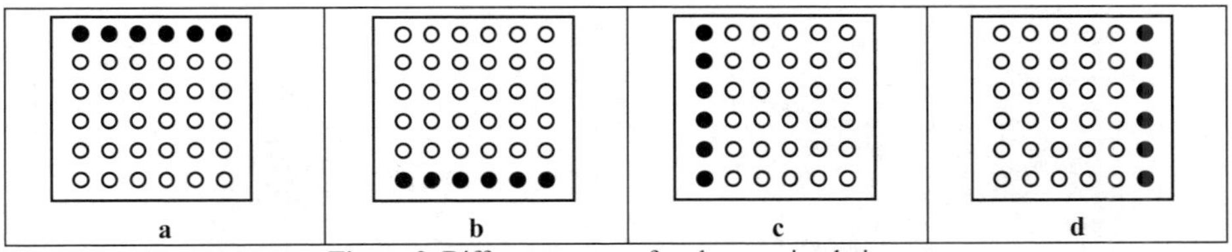

Figure 2. Different patterns for electro-stimulation.

When the six electrodes placed in front of the electrode array were activated, subjects were asked to move their chest forward (a). In the same way, they were asked to move their chest backward, to the left or to the right if electrodes situated behind (b), on the left (c) and on the right (d) were activated.

(Figure 2a). The instructions for movements were similar for stimulation towards the back (Figure 2b), and the left (Figure 2c) or the right sides (Figure 2d) of the electrode array.

The pressure map applied at the seat/skin interface was then recorded in real-time. After each movement following the electro-stimulated direction, a new record of the pressure map was improved. By computing the differences between the two pressure maps (before versus after electro-stimulation), we determined whether or not the movement was adapted to the electro-stimulated information. In the first case, the result was marked as “one”, and otherwise, as “zero”. For each subject, this experiment was carried out 10 times, thus obtaining a total score out of 10. The entire procedure was performed in an automated manner with specifically developed software.

3. Results

No difficulty was reported during the calibration stage. The electro-stimulation of the tongue was very well accepted. This procedure was completed by each seated and healthy subject. The mean score was 9.2, with a standard deviation at 0.79.

This encouraging result demonstrates three main points: first, the subject has a strong perception of the electro-stimulated information; second, this information is both meaningful and correctly interpreted; and third, the action resulting from the interpreted information is adapted, with changes in pressure, to the electro-stimulatory information.

4. Discussion

To our knowledge, there has been no study to date that uses sensory substitution via the tongue in order to prevent the formation of pressure ulcers in paraplegics. One of the main difficulties of using sensory substitution reported by P. Bach-y-Rita was the development of a practicable user-friendly human-machine interface. Indeed, it is evident that the TDU used in this first experiment is not practicable for daily use by individuals with paraplegia. As this device has to be wholly accepted by paraplegics, it is therefore essential to solve this problem. We developed, in collaboration with the CORONIS SYSTEMS® and the GUGLIELMI TECHNOLOGIES DENTAIRES® companies, a new cosmetic interface incorporated into an orthodontic retainer, with wireless transmission from a laptop to the tongue device unit (Figure 3). A clinical evaluation of this prototype will start shortly. Collaboration has also been established with the VISTA MEDICAL® Company in order to improve the pressure mapping system (Figure 4).

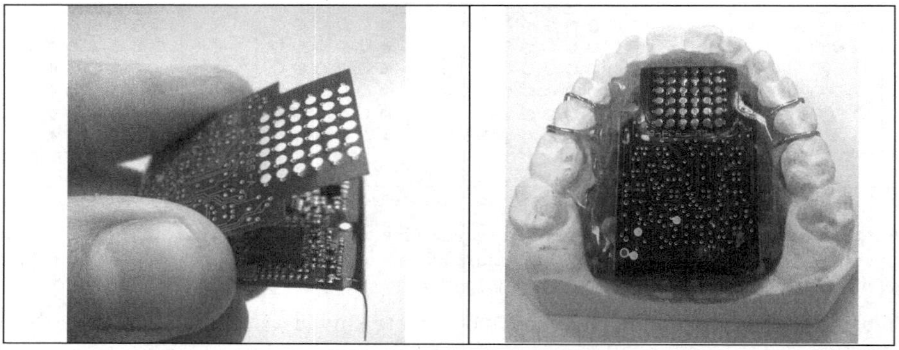

Figure 3. A new cosmetic interface.

On the left, the electronic interface with wireless transmission from/to the laptop is shown. On the right, this system is integrated into an orthodontic retainer, which is placed into the upper part of the mouth (bottom view).

The major finding of this study is that a healthy subject is able to acquire an adapted behavior, with changes in pressure, according to electro-stimulatory information on the tongue. In other words, communication from the organ of sensory substitution towards the region of sensory loss is achieved. The reverse communication from the buttock area to the tongue has not yet been evaluated. If this reverse communication is validated, it would then enable the simulation of the whole conscious or subconscious loop defined in the healthy subject by perception of a stimulus coming from the buttock area, analysis of this signal and action adapted to the signal in order to correct the cause of this alert. The evaluation of this complete cycle will be the purpose of a further clinical study.

Finally, it will be necessary to specify the type of information that will be applied through tongue electro-stimulation. The following questions need to be examined. Should low-level information be used as pressure that is applied to the seat/skin interface, with the subject having to interpret this information and moving in accordance with the stimuli in order to prevent the formation of pressure ulcers? Or should high-level information be used, such as an optimal direction of movement computed in an automated way from pressure maps (by minimizing a cost function of time and pressure, for example)? To obtain the best "therapeutical" acceptance, the most appropriate type of information would be the one that which would lead to a reflex-adapted behavior in the paraplegic. Further clinical evaluations should enable us to determine the most suitable information for electro-stimulation.

5. Conclusion

Because pressure ulcers are still a major health issue for individuals with spinal cord injuries (as shown in the high prevalence of this pathology in this population), it is necessary to develop new health strategies to address this problem. This paper reports the principles of a new approach using computer and sensory substitution via the tongue and the first performed evaluations of this technique. Initial encouraging results with healthy seated subjects, which have yet to be confirmed with paraplegics, provide the possibility of developing new strategies in health education for persons with spinal cord injuries in order to prevent the formation of pressure ulcers.

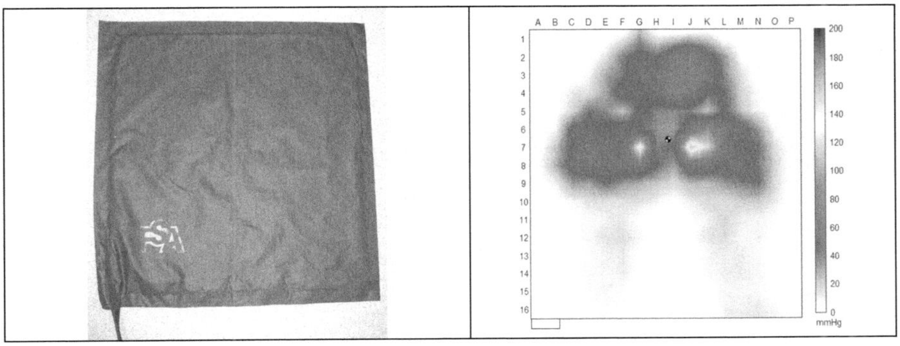

Figure 4. The pressure mapping system of the VISTA MEDICAL® Company.

On the left, the pressure mapping system is shown. On the right, a colorized example of a pressure map at the seat/skin interface in a healthy subject is shown. In this example, the light gray colors circled by the dark gray colors correspond to the areas of maximum pressure.

References

[1] http://www.epuap.org/gltreatment.html (last accessed Jan. 30, 2006).

[2] Garber SL, Rintala DH. Pressure ulcers in veterans with spinal cord injury: a retrospective study. J Rehabil Res Dev 2003; 40(5): 433-41.

[3] Raghavan P, Raza WA, Ahmed YS, Chamberlain MA. Prevalence of pressure sores in a community sample of spinal injury patients. Clin Rehabil 2003; 17(8): 879-84.

[4] Nair KP, Taly AB, Maheshwarappa BM, Kumar J, Murali T, Rao S. Nontraumatic spinal cord lesions: a prospective study of medical complications during in-patient rehabilitation. Spinal Cord 2005; 43(9): 558-64.

[5] Cardenas DD, Hoffman JM, Kirshblum S, McKinley W. Etiology and incidence of rehospitalization after traumatic spinal cord injury: a multicenter analysis. Arch Phys Med Rehabil 2004; 85(11): 1757-63.

[6] Pellerito, Jr, JM. The effects of traditional and computer-aided instruction on promoting independent skin care in adults with paraplegia. Occup Ther Int 2003; 10(1): 1-19.

[7] Bach-y-Rita P. Brain Mechanisms in Sensory Substitution. New York, Academic Press, 1972: 182.

[8] Sampaio E, Maris S, Bach-y-Rita P. Brain plasticity: 'visual' acuity of blind persons via the tongue. Brain Res 2001; 908(2): 204-7.

8.4 Health and Clinical Management

Ubiquity: Technologies for Better Health in Aging Societies
A. Hasman et al. (Eds.)
IOS Press, 2006

Reducing dropouts in outpatient care through an SMS-based system

Andrea BATTISTOTTI[a], Silvana QUAGLINI[b] and Enrico CUOCO[a]
[a] *Azienda Ospedaliera Pavia - Sistema Informatico Aziendale, Pavia, Italy*
[b] *Dept. Computer Science and Systems, University of Pavia, Italy*

Abstract. The objective of this work is to reduce the number of outpatients dropouts, i.e. appointments that, unexpectedly, are not attended by the scheduled patients, thus causing disturbances to the regular ambulatory workflow, waste of resources, and, eventually, longer waiting lists for ambulatory care. A collateral, but not less important result is the improvement of the relationship between citizens and the public Hospital Company. The method proposed consists in sending short instant messages to remind the appointment. After the promising estimates obtained by a preliminary cost-benefit analysis, a commercial product was purchased for the communication infrastructure, while a software module has been developed for interacting with the legacy system, in order to retrieve the information necessary to compose the text of every sms and send it to the citizen. The system has been implemented since seven months, and its benefits have been measured. Dropouts trend, stable at 7-8% during last years, is clearly decreasing, while the number of citizens providing their mobile number to the healthcare booking facilities is constantly increasing. Our conclusion is that the cost of the system will be amortised in a very short time, leading to a significant cost saving and, hopefully, to shorter waiting lists.
Keywords: Medical Informatics, Dropout, ICT, Sms, Cost-Benefit analysis

1. Introduction

Lack of communication among healthcare professionals is now well-recognised by the medical community as one of the main problems impairing healthcare delivery quality, causing medical errors, and thus leading to both health and economical damages. But also lack of communication between citizens and healthcare delivery organisations deserves attention. More than influencing individual health outcome, it may jeopardise the outpatients' service flow, creating annoyances to citizen and waste of resources to the healthcare service. Very common problems in this context are the so-called "dropouts", i.e. visits that have been booked but not attended by patients, without any notice, i.e. without cancellation or shift. As a consequence, healthcare professionals must cope with unexpected "holes" in their scheduled workflow. Since overbooking is not a common practice in healthcare routine, these holes represent waste of time and resources, and, importantly, concur to lengthen waiting lists. Dropouts may also be hospital company's fault, e.g. sudden personnel unavailability (e.g. illness), out-of-work instrumentation, unavailable room, etc., but they represent a negligible percentage. The most common cause is that a citizen/patient simply forgets to attend the visit or decides to not attend the visit but does not care for noticing the hospital company. Interesting studies about the dropout phenomenon can be found in [1], where related costs for the UK National Healthcare System are estimated to pound 266 million annually; [2], where dropouts causes are analysed; [3] and [4], showing the effectiveness of telephone or mailed reminders. We propose a solution based on Information and Communication Technology (ICT) on top of the so-called sociotechnical approach [5]. This term indicates a methodology that potentially integrates technical design, policy, and organizational performance into a whole. In this framework, considering the current problems affecting the "image" of public health for outpatients in our country, waiting lists emerge as a leading issue. Reasons for long waiting lists are manifold and world-wide spread. Some of them are not modifiable by ICT: first of all the scarcity of economical resources, then political and deontological problems such as relationships between public and private organisations and individual behaviour of professionals. Eventually, bottlenecks in workflow organisation contribute to worsen the situation. Dropouts are one of these bottlenecks and the aim of the paper is to show how ICT can help to reduce them.

2. The Social Context

2.1. The Italian Healthcare System

In order to understand the economical burden caused by dropouts, we briefly illustrate the Italian Healthcare System. It is a national, public system financed by general taxation, where universal healthcare coverage to the whole population is a key characteristic. Essential health services are provided free of charge, or at a minimum charge. In the last years there has been a growing decentralisation of power to the 20 Italian Regions, while the Italian State retains limited supervisory control and continues to have overall responsibility for the National Healthcare Service (NHS) to assure uniform and essential levels of health services across the country. Thus there are three different levels of public authority: the Central Government, the Regions, and the Local Healthcare Agencies (LHAs). The central government annually allocates national health fund to Regions which, in turn, organize services that are designed to meet the needs of their specific populations, define ways to allocate financial resources to all the LHAs within their territories, monitor LHAs' activities, and assess their performance. Importantly, costs sustained for delivered services are refunded "post-hoc", i.e. after specific documentation of the effective delivery has been produced. Based on criteria of efficiency and cost–quality, LHAs might provide care either directly, through their own facilities (directly managed hospitals and territorial services), or by paying for the services delivered by providers accredited by the Regions, such as independent public (hospital companies and university-managed hospitals) or private structures. In particular, in our province, located in Northern Italy, the LHA strongly collaborates with the hospital company "Azienda Ospedaliera della Provincia di Pavia" (AOP), that is the organisation where our system has been implemented.

2.2. The AOP HospitalCompany

The following numbers gives an idea of the AOP extent. In the same time, they allow appreciating the quotes of outpatients and inpatients services:

- 11 multiple purpose outpatients services; 8 hospitals;
- about 3,000 employees; about 1,000 beds;

- about 50,000 hospital admissions/year;
- 900,000 outpatients visits/year; 100,000 accesses to ER /year.

Concerning financial dimension, data from 2004 show 106,679,000 € for in-hospital activities (75%) and 143,049,000 € for outpatients activities (25%).

3. The Dropouts Problem at AOP

Outpatients visits are managed almost entirely by a centralised service for visit booking, the CUP (Centro Unico Prenotazione). Two authors of this paper (AB and EC) have full access to the CUP database and performed all the statistics. CUP manages 900 appointment books, for a total of about 410.000 bookings per year. Out of them, 380,000 are delivered and then refunded according to the above mentioned regional regulation. On the contrary, the remaining 30,000, i.e. 7.3%, are not delivered and consequently not refunded. Thus, they represent a waste for the company. These figures are quite similar to those computed in other Italian Healthcare Organisations, for example at the HLA of Bologna (8%), which also launched a promotional campaign for decreasing dropouts [6], and to those reported in other countries: a range from 5.5% to 15% is reported in [4].

3.1 The Financial Burden

Table 1 shows, as an example, some of the services provided by the AOP, with the associated monetary value, the number of dropouts in a sample day, and the consequent monetary loss (we used the regional pricing list as a proxy of the real service value).

It is world-wide agreed that the *physiological* percentage of dropouts in a healthcare system is about 2% - 3%, and it is very difficult to decrease this value in practice: thus, remaining below this threshold indicates an efficient system. Thus, the AOP, having an average rate of 7%, should recover about 5% of dropouts to be defined "efficient" from this point of view.

Table 1- Counts derived from data of a specific day (july 20th, 2005)

service code	description	booked	delivered	dropouts (%)	value (€)	loss (€)
87.37.1	bilateral mammog.	21	19	2 (9.5)	43.9	87.8
87.44.1	chest RX	37	32	5 (13.5)	17.04	85.2
89.13	neurol. exam.	37	35	2 (5.4)	16.53	33.06
...	...	...	...	...	...	...

From the AOP legacy system, we made an estimate of the losses from 2001 to 2004, amounting to 780000, 684000, 748000, 764000€ respectively. From the same database, the average number of dropouts has been calculated as 150 per day. The average value of a service is 20€, leading to 3000€ loss per day. Thus, recovering 5% of dropouts corresponds to saving about 500.000€ per year.

4. The proposed solution

To decrease dropouts, it is essential to improve the communication between citizens and AOP, sensitizing citizens to notice AOP in case of impossibility to attend the visit and offering recall services and cancellation facilities free of charge. Possible means to reach citizens are synchronous (telephone) and asynchronous (e-mail and sms). Calling a patient by telephone is effective [3,4], but very expensive and arduous. Sms and e-mails have similar characteristics, but the latter is more likely to reach more people. National data (year 2003) show 17 millions of e-mail addresses vs 60 millions of mobiles. Our statistics on dropouts show that the higher proportion is due to people in the age range 20-40 years, the same range where mobiles are most diffuse. Moreover, we don't know if e-mail is checked every day, while we are quite sure that mobile is. This is crucial: the reminder must be seen not too early, because the citizen could forget the appointment again, and not too late, because he could have taken another commitment coincident with the appointment time. Thus, we decided that sms is the right means and that *two days before* is the right moment to send the reminder. Before implementing the system, a cost-benefit analysis has been performed. Then, to set up the system, two types of issues, organisational and technical, were faced.

4.1 Cost-Benefit Analysis

After a careful investigation of the market, we took into consideration the commercial product SMSjob (by Kernelsoft). Its cost is 8000€. Considering that not only AOP but also HLA could use it for other purposes, we accounted for half its cost, i.e. 4000€, and we attributed this cost to the first year only. Considering 500 sms/day at a unitary cost

l'informatore

Fondatore: Carlo Natale

bisettimanale del territorio

ISSN 1594-2856

informatore@ievve.com

ANNO 81 * GIOVEDÌ 25 AGOSTO 2005 * N° 59

€ 1,00

Intanto è stato concluso l'acquisto della seconda Tac per Vigevano: costerà 1,2 milioni di euro, la fornirà la Philips

Un sms per ricordare la data dell'esame

Progetto dell'Azienda Ospedaliera: con meno prenotazioni a vuoto si riducono le liste d'attesa

VIGEVANO – Capita che prenoti un esame o una visita e poi te ne dimentichi perché passa molto tempo oppure magari hai deciso di rivolgerti ad un'altra struttura. Raramente ci si ricorda di avvisare l'ospedale della disdetta, così nell'erogazione delle prestazioni restano "buchi" che fanno allungare le liste d'attesa. Ora dovrebbe capitare molto meno, grazie ad un sms.

L'Azienda Ospedaliera infatti ha deciso di inviare ai pazienti un messaggino sul cellulare per ricordare la prenotazione qualche giorno prima della data fissata. Si tratta di un progetto realizzato a costo zero. L'ha ideato un dipendente, che ne ha fatto oggetto della sua tesi di laurea in ingegneria informatica. Intanto è stato finalmente concluso l'acquisto della seconda Tac per l'ospedale di Vigevano: la fornirà la Philips, costerà 1,2 milioni di euro.

A pagina 4

Il messaggio inviato sul cellulare per ricordare la prenotazione dell'esame

Figure 1 – Promotional campaign: title is "an sms to recall the visit date"

of 0,065€, the first year cost will be 11,500€, while the next years cost will be about 8200€, accounting for sms and maintenance. Reasoning on a daily basis, this amounts to 50€ during the first year and 35€ in the future. Since the average monetary loss of a dropout is 20€, it's sufficient to recover 2-3 dropouts/day to amortise the system. It's easy to calculate that with 3 less dropouts/day, corresponding to 2 percentage points less, the system is completely amortised in 10 months, while with 9 less dropouts/day, corresponding to 6 percentage points less, the system is amortised in 3 months.

4.2 Organisational issues

The first problem to be solved has been the retrieval of the mobile phone numbers of the citizens. Since it is not mandatory for a citizen to provide his mobile number (if any) to AOP, we found that only 12% of the registered citizens did. Thus, the initial sample of people eligible for reminders was 66,000 out of 554,000. Contemporary, an advertising campaign was initiated, addressed both to the general population and the

AOP operators, and the proportion of citizens providing their number has constantly grown. While in august (the second month of the system implementation) we sent 840 sms, reaching 11.2% of the citizens that would have benefit from the reminder, in november we sent 6134 sms, reaching 18.2% of the citizens. The campaign exploited also some of the main newspapers of the province and Figure 1 shows one of them.

4.2.1 Legal and privacy issues

Since we are not sure that the sms will be read by the target person, the message must be effective, but not too explicit. For example, the name of the person must not appear, as well as the motivation of the medical appointment. Taking also into account the maximum length of a regular sms (160 characters) the solution found has been the following one: *"The Azienda Ospedaliera of Pavia recalls you the appointment of 05/09/2005. If you want to cancel, please call free 800448800. Thank you for your collaboration"*

4.3 Technical issues

From the technical point of view, the system architecture is shown in Figure 2. SMSjob has been purchased as the core of the system. It may provide several functionalities [7], but to our purposes it is just an sms gateway, managing an SQL database for storing personalized setups and checking the collected data. The queue of messages to be sent may be written in a mailbox with a special configuration for SMSJob, or in a text file, to be saved in a pre-defined folder, or through JDBC/ODBC connection, to write

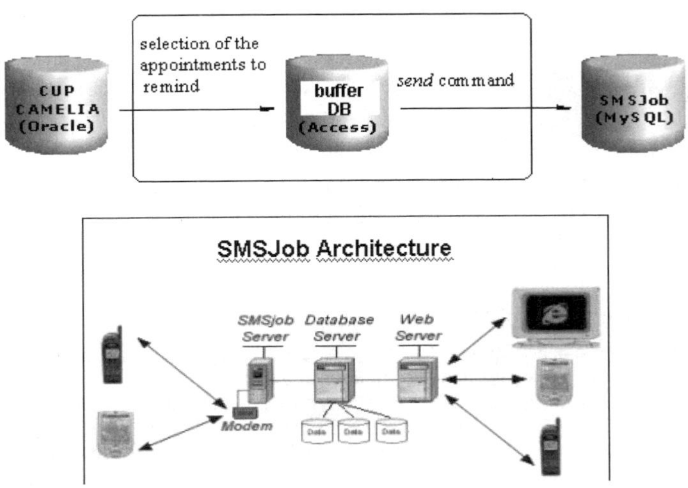

Figure 2 – The system architecture

directly in the *sending* tables of SMSJob. A software package, based on MS ACCESS and Visual Basic, has then been developed, to query the legacy system CAMELIA (the CUP database) in order to find, every day, the eligible citizens to be reminded, the related phone numbers and to compose the individualised message. Once the information has been retrieved, it is packaged and sent to SMSJob, which, every day at eleven 'o clock, sends the messages. The entire procedure is authomatic.

5. Results

Results obtained in the first 6 months of the system implementation are very encouraging. Figure 3 shows the percentage of dropouts in the various months of 2005 (system started running in june) with respect to the average in the past 4 years. While the latter never goes below 7%, the former approaches 5.5%. Even though these percentages seem close, the importance of this result may be appreciated considering

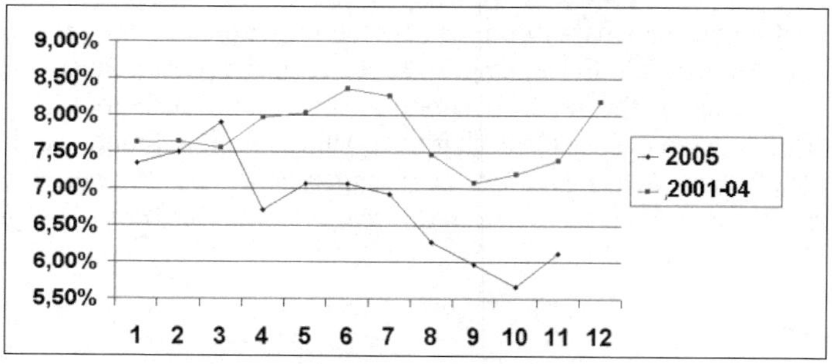

Figure 3 - Dropout percentage in 2005 vs the average in the past 4 years

the analysis in the paragraph 4.1. If the number of citizens providing their mobile number increases at the current rate, we plan, in the next years, to amortise the annual cost of the system in one month. It's too early to see a positive effect on waiting list: it will be our next outcome to check for.

6. Conclusions

We are collecting positive feedback from citizens, some of them even send a message back to thank for the reminder. Few negative feedbacks are due to wrong phone number such that the sms reaches someone other, that may be annoyed by the message itself. However, benefits are largely overcoming costs and drawbacks, thus we think that the proposed system can be a model also for other healthcare organisations.

References

[1] Dockery F; Rajkumar C; Chapman C; Bulpitt C; Nicholl C. The effect of reminder calls in reducing non-attendance rates at care of the elderly clinics. Postgrad Med J 2001; 77: 37-39

[2] Pal B, Taberner DA, Readman LP, Jones P. Why do outpatients fail to keep their clinic appointments? Results from a survey and recommended remedial actions. Int J Clin Pract. 1998 Sep;52(6):436-7

[3] Grover S, Gagnon G, Flegel KM, Hoey JR. Improving appointment-keeping by patients new to a hospital medical clinic with telephone or mailed reminders. Can Med Assoc J. 1983 Nov 15;129(10):1101-3.

[4] Reti S. Improving outpatient department efficiency: a randomized controlled trial comparing hospital and general-practice telephone reminders.N Z Med J. 2003 Jun 6;116(1175)

[5] Berg M, Aarts J, Van der Lei J. ICT in health care: sociotechnical approaches. Methods Inf Med 2003; 42(4): 297–301

[6] http://www.ausl.bologna.it/labcitsal/disdette.html (last accessed Jan. 30, 2006).

[7] http://www.kernelsoft.it/kernelsoft/smsjob.jsp?LANGUAGE=2

Ubiquity: Technologies for Better Health in Aging Societies
A. Hasman et al. (Eds.)
IOS Press, 2006

Evaluation of a Discussion Forum for Knowledge Sharing Among Emergency Practitioners: A Social Network Approach

Janet Curran[a,b,1] , Syed Sibte Raza Abidi[b]
[a]*CIHR Child Health Clinician Scientist Training Program, IWK Health Centre, Halifax, NS, B3K 6R8, Canada*
[b]*Health Informatics Lab, Faculty of Computer Science, Dalhousie University, Halifax, NS, B3H 1W5, Canada*

Abstract. Peer to peer knowledge sharing is recognized as a key contributor to the development of expert practice for health care professionals. Emergency departments with access to extensive expertise, such as in urban hospital settings, present greater potential for rich collaborative learning opportunities as compared with rural settings where expertise is at times scarce. Collaborative technologies such as electronic discussion boards may assist in leveling the "knowledge" playing field and increase opportunities for the growth of a strong social network for emergency clinicians. A social network perspective is used to explore the effectiveness of a discussion forum to support knowledge sharing among emergency practitioners in rural and urban emergency departments in Nova Scotia..
Keywords: social networks, emergency practice, knowledge transfer

1. Introduction

Healthcare knowledge is vast; it exists in a variety of modalities and is dispersed across the healthcare organization. Research supports that people rely heavily on their network of relationships to find information and solve problems. Practitioners generally cite colleagues or peers as a key source of practice knowledge [1]. Clinical practice in an emergency department (ED) demands knowledge from different sources in terms of both published best evidence and case-based experiences of peers. In practice, ED practitioners work in teams and collaborate to acquire and share their knowledge in order to address the knowledge gaps inherent within the healthcare system. This is similar to Wenger's [2] concept of community of practice (CoP) where members interact and learn from each. Interactions and relationships with colleagues facing similar problems and clinical scenarios enhance learning and the development of clinical expertise. As such, social relationships are important in acquiring expert practice knowledge.

The ED work environment is subject to a number of unique challenges in relation to sustained knowledge sharing relationships. The unpredictable nature of patient flow

[1] Corresponding Author: Janet Curran, 5980 University Ave, IWK Health Centre, Halifax, NS, Canada, B3K 6R8 email: janet.curran@iwk.nshealth.ca

and the wide variation in patient presentation leads to multiple interruptions and may impact knowledge sharing among peers. A number of disparities also exist between rural and urban settings. EDs with access to extensive expertise, such as urban hospital settings, possess greater potential for rich collaborative learning opportunities as compared with rural settings where expertise is at times scarce. Comparisons of information seeking behaviors among rural and urban primary care clinicians indicate that both have the same needs for information and rely on colleagues and personal libraries as their main sources of information [3]. However, rural clinicians' access to information sources is limited by isolation, inadequate library sources, and limited access to onsite specialist practitioners.

This myriad of challenges led to the development of a proof of concept project in which a web based environment was established to facilitate intra and inter-organizational knowledge exchange among ED clinicians from 9 rural and 2 urban health centres. The site was populated with 12 self-directed content modules focused on pediatric emergency care issues. An online discussion board was added to facilitate communication among practitioners participating in the project. Discussion was guided by a group of clinicians who were recognized for their clinical expertise and were responsible for development of the 12 content modules. Interaction on the discussion board was organized at two levels: **Topics** corresponded to the themes of the content modules (pediatric trauma, diabetic emergencies, poison management etc) and **Subjects** within topics allowed for more focused discussion related to the parent topic. Participants moved through the content modules at a pace that suited their individual learning needs and workload. Participation on the discussion board was optional. However, content experts (CE's) encouraged participants to post questions and share their clinical experiences or challenges on the discussion board under the relevant **Topic** heading. A threading feature organized the series of replies to a principle subject message. Threading of messages allowed information seeking and associated information sharing episodes to stay connected. All interactions were tagged with the authors name and time of posting.

In this paper we present the results of our investigations into the efficacy of an electronic discussion board as a medium for knowledge exchange among rural and urban ED practitioners. The representation of social relationships in a CoP can be useful in depicting the flow of information in the community [4]. As such, social network analysis methods will be used to measure the effectiveness of the discussion forum in terms of the information seeking and sharing patterns of urban and rural practitioners. The primary research question under investigation is: How do emergency clinicians in rural and urban ED's seek or share information in an online discussion forum? *Information seeking* is characterized as any person looking for information from a clinical (patient specific) or health systems perspective. *Information sharing* is characterized as any person sharing information from a clinical or health systems perspective. The following hypothesis will be tested to address the primary question:

- The extent to which practitioners participate in an online discussion forum as a means for seeking information is a function of geographic location
- The extent to which practitioners participate in an online discussion forum as a means for sharing information is a function of geographic location

2. Methods

The online discussion forum created an opportunity for emergency practitioners to engage in dialogue around topics that were relevant to their practice. To this end, interaction data for the 18 month project period was extracted from the discussion forum and we used UCINET 6.19 [5], a social network tool, to calculate social network measures for the relationship data. Social network analysis (SNA) provides a means for mapping and analyzing relationships among people and/or organizations. Network data are generally defined by two elements; actors (nodes) and relationships (ties) that connect a pair of actors [6].

3. Results

3.1 Network Sample

One hundred eighty eight (188) multidisciplinary practitioners from 9 rural and 2 urban EDs participated in the project. Forty seven percent (N=89) of participants accessed the discussion board at least once and 72% of those (N=64) posted at least one message. However, for the purposes of this analysis only the postings that included a seeking or sharing reference to patient related issues were included in the analysis. Therefore the number of participant actors in the network was 42. Content modules that were supported by more than one content expert (CE) were merged into one actor profile (CE1, CE2, CE3 etc) resulting in a total of 12 actor profiles for the CEs. These changes brought the total number of actors in the network to 54 (k) and the total number of possible connections for each actor to 53 or (k – 1). Distribution of actors in the seeking and sharing networks included 12 CE's, 24 rural participants (dispersed across 8 different sites) and 18 urban participants (dispersed across 2 sites). Data in the sharing and seeking networks is asymmetric (directed) therefore there were a total of k * (k-1) or 2,862 possible relationships or ties in each network.

3.2 Knowledge seeking and knowledge sharing networks

We begin our analysis on relationships by developing a seeking and sharing matrix based on the interaction data from the discussion board. We used NetDraw to visualize the knowledge seeking and sharing activities within the virtual community (Fig 1 & Fig 2, respectively). A visual inspection of the social networks suggest that the majority of nodes in both networks are connected by at least one relation but there is a small subset of nodes in the core that have multiple ties, which may indicate that they have central roles in the flow of information in the network.

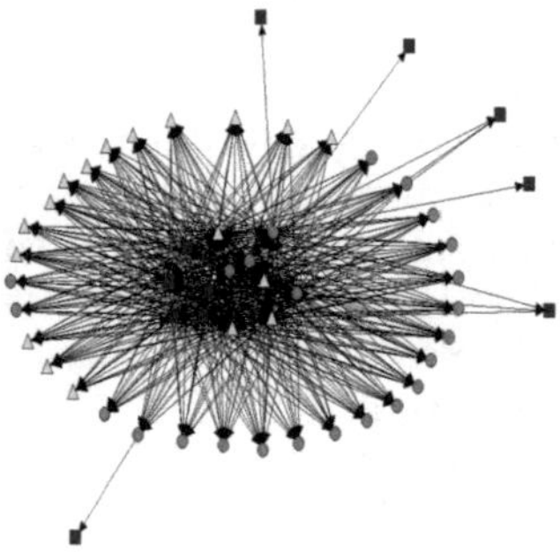

Figure 1. Information Seeking

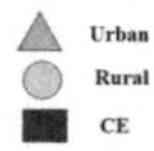

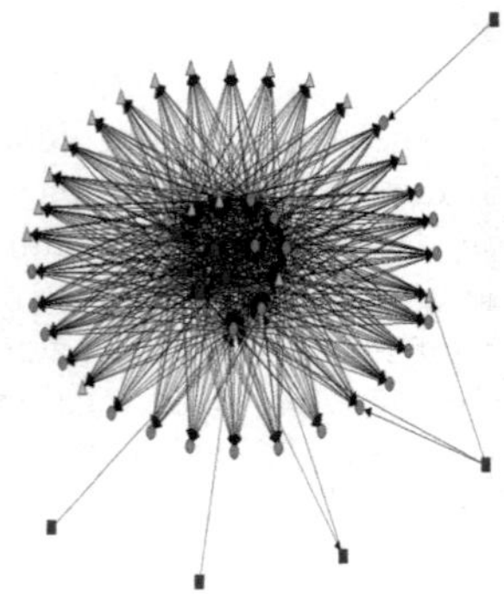

Figure 2. Information Sharing

3.2.1 Density

The density of a binary network reflects the percentage of all possible ties that are actually present in the network. Density can provide an indication of the rate at which information diffuses through the network. We partitioned both matrices into blocks to explore the interaction pattern for each group (rural, urban and CE). Table 1 and Table 2 show density measures for each relationship block in the seeking and sharing network. A density ANOVA using the Structural Blockmodel [5] was run to look for differences in interaction patterns. Although density measures tended to be higher in the CE-Rural and the CE-Urban relationships for both networks these were not found to be significant.

Table 1. Density Blocking for Seeking Network

	Rural	**Urban**	**CE**
Rural	.125	.146	.069
Urban	.222	.222	.037
CE	.417	.417	.000

Table 2. Density Blocking for Sharing Network

	Rural	**Urban**	**CE**
Rural	.250	.250	.010
Urban	.1667	.1667	.000
CE	.441	.421	.000

3.2.2 Centrality

Measures of centrality depict the distribution of useful/active actors in a network; identifying actors who are in favorable or prestigious positions [7]. Simple measures of centrality such as degree centrality have been found to be the most stable centrality measures [8]. Therefore we used Freeman's degree centrality to identify out-degree (influence) and in-degree (prestige) measures to provide a summary of the contribution of each actor. This measure reflects the number of interaction partners. For example, if practitioner X connected with 10 different practitioners when seeking information the out-degree centrality for this actor was "10". In contrast, if 5 practitioners connected with practitioner X in the seeking network, then the in-degree centrality for him/her was “5”. The range of out-degree and in-degree for our seeking network was 0 to 48 and 0 to 13, respectively. The range for the out-degree of our sharing network was 0 to 42 and in-degree, 0 to 14. We see the range and variability (mean and standard deviation) of out-degree is much larger than the range for in-degree. This is reflective of a population that is heterogeneous with out-degree (influence) in structural positions.

Table 3. Degree Centrality Measures for Seeking and Sharing Network

	CE	**Rural**	**Urban**	**Sig. p-value**
Total - N (%)	12 (22.2)	24 (44.4)	18 (33.3)	
Seeking Outdegree – mean (SD)	17.5 (21.63)	6.33 (14.72)	9.55 (17.85)	.201
Indegree – mean (SD)	2.33 (3.53)	11.87 (0.33)	12.27 (0.574)	.002
Sharing Outdegree – mean (SD)	18.16 (21.05)	10.37 (18.16)	6.83 (15.72)	.220
Indegree – mean (SD)	0.25 (0.621)	14.04 (0.86)	13.88 (0.471)	.000

To test our project hypothesis we calculate the out-degree and in-degree mean and standard deviation for the 3 groups of participants in both networks (Table 3). We use UCINET to perform an ANOVA with a significance based upon a permutation test and we note that there was no significant difference in the seeking and sharing behavior (out-degree) of clinicians who participated in the discussion forum. A significant difference was found for the in-degree measure of the seeking and sharing network. ($p=.002$, $p=.000$). However, it would appear that the difference does not lie with the rural and urban participants. We note that these results should be viewed with caution, as it is unclear if all assumptions for ANOVA are met.

4. Discussion

It was hypothesized that geographic location would influence information seeking and sharing behavior. It was anticipated that rural practitioners would express a greater number of information seeking ties than their urban counterparts, as the higher number of expert clinicians in the urban centres would make seeking in an online environment redundant. Out-degree measures for the seeking network did not support this hypothesis. It is possible that while the number of clinical experts present in the urban centers is higher, the chaotic nature of the ED work environment may not support sustained meaningful discussions. An asynchronous discussion forum may provide clinicians from both rural and urban centres with an opportunity to reflect on their practice and seek feedback from their peers. It was also thought that urban practitioners would express a greater number of knowledge sharing ties as practitioners from urban EDs experience a higher volume of pediatric patients and thus would have a larger repertoire of experiential knowledge to contribute to the network. However, the out-degree measures for the sharing networks did not support our second hypothesis. Both rural and urban practitioners used the online discussion board to share their practice knowledge with their peers.

It is interesting to note that mean in-degree measures in both networks for rural and urban participants are much higher than the out-degree measures. Actors with high in-degree measures receive information from many sources and are said to have high prestige. These results would indicate that although a small number of actors actively participated in reaching out, the information that was sent out was widely distributed to all actors. This reflects the potential for information to flow through the network.

This project presented a novel solution for enhancing knowledge exchange among emergency clinicians. Block densities and centrality measures would suggest that CE's took a leadership role in establishing ties with rural and urban practitioners participating in the project. The presence of "experts" in the online community may have been a limiting factor for some participants concerned with exposing knowledge gaps. It is possible that over time as comfort with the technology grows and trust develops the number of ties will increase.

An inherent weakness in using SNA to measure interaction in a discussion forum is the inability to measure passive participation. In this project the criterion used to capture a relationship was active posting on the discussion board. This issue may be addressed with a revision to the discussion board design to include logs to track the number and title of posted messages opened.

5. Conclusion

The busy pace of an ED practice setting is not always favorable for sustained real-time information seeking interactions. This project demonstrates that an electronic discussion forum can provide an effective means for bringing rural and urban practitioners together to exchange practice knowledge. Further research is required to better understand the contextual and individual factors that influence participation. Supporting more effective knowledge-sharing practices through enhancement of a social network will have a great impact on the quality and safety of health service delivery in EDs in Nova Scotia.

References

[1] Dawes M, Sampson U. Knowledge management in clinical practice: a systematic review of information seeking behavior in physicians. International Journal of Medical Informatics 2003; 7: 9-15

[2] Wenger E. Communities of Practice: Learning, Meaning and Identity. Cambridge University Press, 1998.

[3] Dorsch J. Information needs of rural health professionals: a review of the literature. Bull Med Libr Assoc 2000; 88(4):346-354

[4] Garton L, Haythornthwaite C, Wellman B. Studying Online Social Networks. JCMC 1997; 3(1):1-45.

[5] Borgatti, S.P., Everett, M.G. and Freeman, L.C. 2002. Ucinet for Windows: Software for Social Network Analysis. Harvard, MA: Analytic Technologies

[6] Huisman M, Van Duijn M. Software for statistical analysis of social networks. 6th International Conference on Logic and Methodology: 2004 August 16-20; Amsterdam

[7] Hanneman RA, Riddle M. Introduction to social network methods. University of California, Riverside (published in digital form at http://faculty.ucr.edu/~hanneman/) 2005

[8] Zemljic B, Hlebec V. Reliability of measures of centrality and prominence. Social Network Analysis 2005;27:73-88

Ubiquity: Technologies for Better Health in Aging Societies
A. Hasman et al. (Eds.)
IOS Press, 2006

On Neural Network Classification of Otoneurological Cases on the Basis of Recognition Results of Vestibulo-ocular Reflex Eye Movements Signal

Martti JUHOLA[a,1], Heikki AALTO[b] and Timo HIRVONEN[b]
[a] *Department of Computer Sciences., University of Tampere, Finland*
[b] Department of Otorhinolaryngology, Helsinki University Central Hospital, Finland

Abstract. We constructed a signal analysis and recognition system to compute revealing results from vestibulo-ocular reflex eye movements and their stimulation head movements of impulsive type that were generated as passive head movements with a special device designed by us for this purpose. Further, we implemented perceptron neural networks to separate healthy subjects from patients suffering from dysfunction of vestibulo-ocular reflex in either ear. We gained high classification accuracies with this method ready for routine use.
Keywords: Otoneurology, vestibulo-ocular reflex, eye movements, signal analysis, pattern recognition, neural networks

1. Introduction

Vertigo and balance problems originating from otoneurological reasons, e.g. inner ear diseases, are common. Eye movement tests give a means to investigate vertiginous patients provided that efficient software is available. In the past decades, we have constructed software [1,2] for the eye movements types used at balance laboratories. The latest one [3] concerns the signal analysis of eye movements generated by the vestibulo-ocular reflex (VOR), in which a patient's eye movements are compared to the corresponding stimulating movements of the head. For patients, a difference between the stimulations and their responsive eye movements is often possible to find with our program. We also classify the preceding results with neural networks.

Early analysis techniques of eye movement signals were designed on basic signal analysis theory and use of threshold values to determine beginnings and ends of eye and stimulation movements from eye movement and stimulation signals [4]. They were prone to get stuck in electromyographic noise caused by facial muscles or artifacts like eye blinks, since the adaptive use of thresholds was difficult for greatly varying eye movement signals. We introduced syntactic pattern recognition methods for the eye movement studies set out in the 1980s. The principal types in otoneurology are saccades, nystagmus and smooth pursuit applied since the 1970s and newer VOR eye

[1] Corresponding Author: Martti Juhola, Department of Computer Sciences, 33014 University of Tampere, Finland. Martti.Juhola@cs.uta.fi

movements. Saccades, also used to the calibration of VOR signals, are fast eye movements performed when one abruptly moves the gaze from a target to another.

Inasmuch as the signal analysis programs constructed for eye movements produce medically important results from otoneurological patients, there was a chance to directly classify patient cases and to separate them from healthy persons by applying classification methods. We formed a perceptron neural network program to separate healthy subjects from a group of mainly patients that were operated because of acoustic neuroma known to affect vestibulo-ocular reflex. Our objective was to show that the whole system starting from the signal analysis program and ending to the neural network classification of such patient cases can aid diagnostics at a balance laboratory.

2. Signal analysis and pattern recognition

2.1. Material: head movement stimulation and eye movement recording

To test the vestibulo-ocular reflex, head shaking is exploited. It can be produced either actively by guiding the subject to shake the head or passively by mechanically shaking the head with a special device prepared for this purpose. We noticed that the latter was better since it was possible to yield head movements with the exactly predefined angles and the pace desired that could be repeated for all subjects similarly. The head was horizontally moved along with the vertical axis from the right to the left or vice versa. There was a short (0.7-1.2 s) randomly varying interval between such impulsive stimulation movements. Since the gaze was fixed to the target, all stimulated head movements produced quite symmetric eye movements in the opposite direction. When the distance of the head from the mark on the wall was constant (1.4 m), we were able to calculate amplitudes of the head and eye movements in angular degrees. First the calibration measurement was performed by supporting the head as steady, but asking to move the gaze between two small fixed marks that were located in front of the subject on the opposite wall. When the interval between them was suitably measured, we obtained calibration saccades of 20°. We calculated the corresponding amplitude value from the recorded calibration signal of performed saccades of the subject. The ratio of 20° and the average saccade amplitudes measured was the calibration coefficient with which the actual VOR signal was multiplied to calibrate it into angular degrees.

To achieve impulsive head movements a relaxed but alert subject was seated in a clamped chair. The head was moved with two push rods attached to the boxer's helmet tightly set on the subject's head. An electronic DC motor used the push rods via an appropriately installed gear [5]. The whole mechanism constructed was computer-controlled, where, naturally, stimulation movements were carefully accomplished and their angles limited to cause no harm for the subject. The set-up is depicted in Fig. 1.

We recorded an eye movement signal and its stimulation (head movement) at 400 Hz for 80 s with the amplitude resolution of 13 bits (one for the sign) corresponding to ±10 V of the AD converter. Eye movements were electro-oculographically measured. Nevertheless, the calibration signal was first recorded as mentioned above. Head movement signals were recorded with a potentiometer, which directly gave amplitudes of the head movements as values from the AD converter. Signals were lowpass filtered with a nonlinear median-hybrid filter to dampen noise above circa 70 Hz. We recorded signals from 22 healthy persons randomly chosen as well as from 22 patients (Fig. 2).

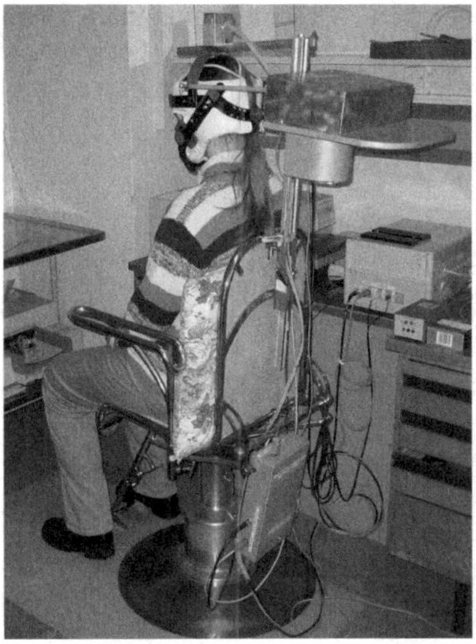

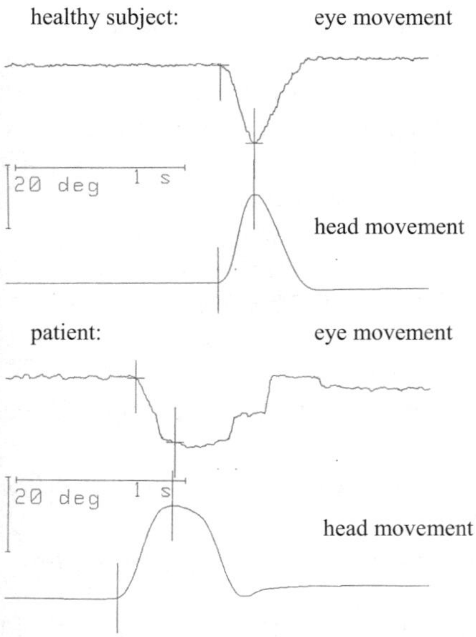

Fig. 1. VOR test: the subject sits in a chair looking at a small target on the wall and her head is rotated slightly.

Fig. 2. The curves down are responses to the left and the curves up stimulation impulses to the right.

2.2. Syntactic recognition of signals

The aim of the syntactic analysis was to approximately localize eye movement waveforms apart from the rest of the signal. Eye movement signals were first divided into consecutive segments of 20 ms (8 samples) from which angular velocity values were computed as slopes of linear regression. These velocity values were transformed into a string of symbols of set $\{a, b, c\}$ to be input to a syntactic parser program. The segments were transformed according to their velocities. If a velocity was greater than 10°/s, it was determined to be symbol a, if it was less than -10°/s, it was written as symbol b, but otherwise as symbol c. The bound of 10°/s applied in the literature [6] aids the definition of a valid eye movement beginning or conversely its end.

The string was input to the parser program, whose task was to recognize such substrings that represented actual eye movement waveforms, but to discard artifacts and noise. Since the head movements were quite noiseless signals, we were able to compute head movement waveforms from them by using a simple velocity profile procedure, in which the beginnings and end of the head movements were extracted from the signal with the bound of 10°/s and in accordance with the condition that a stimulation impulse was always shortly (less than 40 ms) ensued by its response.

The following formal grammar defines all possible "sentences" that can be generated to represent valid substrings as VOR eye movements. The rules of (1) map eye movements from the left to the right and those of (2) in the opposite direction. The grammar commences from start symbol S that is one of the nonterminal set $\{S, A, B\}$. By repeatedly using the rules the grammar can generate single exceptional symbols b

or *c* as in-between symbols "inside an eye movement" in (1). Likewise in (2), single exceptional symbols *b* or *a* may occur. In this way the exceptional single symbols represent occasional uneven segments within eye movements and are possibly caused by low frequency noise. In a way, this is a syntactic model of VOR eye movements.

	$S \to baaA$			$S \to bccB$	
	$S \to caaA$	$A \to bb$		$S \to accB$	$B \to bb$
(1)	$A \to aA$	$A \to cb$	(2)	$B \to cB$	$B \to ab$
	$A \to caA$	$A \to bc$		$B \to bcB$	$B \to ba$
	$A \to baA$	$A \to cc$		$B \to acB$	$B \to aa$

The grammar derives an example sentence of a VOR eye movement as follows.

$$(3)\; S \Rightarrow baaA \Rightarrow baaaA \Rightarrow baaaaA \Rightarrow ba^4caA \Rightarrow ba^4caaA \Rightarrow ba^4caaaA \Rightarrow ba^4ca^3bb$$

2.3. Computation of eye movement results

The exact locations of the VOR beginnings from the two first segments (*ba* in (3)) and those of the tops (*abb* in (3)) (or bottoms like in Fig. 2) were computed as the point where the velocity exceeded 10°/s or reduced below -10°/s. Gain and latency parameters are essential for VOR. First, gain g_1 was computed from a ratio of slopes given by linear regression for the two velocity curves. Second, we calculated gain by amplitude as follows. The maximum of the angular velocity curve (the first derivative of the positional eye movement or head movement) was defined. Such a maximum was located within the 100 ms after the beginning of a VOR eye movement or its corresponding impulsive head movement. The location of the maximum velocity was computed for both the eye movement and head movement signal pertaining to each impulsive pair. The ratio of these two maximum velocity values is gain by amplitude g_2. Third, gain g_3 was computed as the ratio of the means of the velocity values of an eye movement and its stimulation. Both g_1 and g_3 were computed during the 100 ms from the beginnings of the eye movement and head movement signals [3].

$$(4)\quad g_1 = \frac{m\sum_{i=1}^{m} v_s^i v_r^i - \sum_{i=1}^{m} v_s^i \sum_{i=1}^{m} v_r^i}{m\sum_{i=1}^{m} v_s^i - (\sum_{i=1}^{m} v_s^i)^2} \qquad g_2 = \frac{\max\limits_{i=1,..,m}\{v_r^i\}}{\max\limits_{i=1,...,m}\{v_s^i\}} = \frac{v_r^{\max}}{v_s^{\max}} \qquad g_3 = \frac{\frac{\sum_{i=1}^{m} v_r^i}{m}}{\frac{\sum_{i=1}^{m} v_s^i}{m}} = \frac{\sum_{i=1}^{m} v_r^i}{\sum_{i=1}^{m} v_s^i} \qquad l = \frac{p_r - p_s}{f}$$

Here the subscript r denotes m=40 (for f=400 Hz and 100 ms duration) velocity values v_r of the response and the subscript s denotes those v_s of the stimulation. We also computed latency l between the beginnings of each VOR eye movement and its impulsive stimulation with the formula, where p_r is the beginning sample of the response, p_s is that of the stimulation and f is the sampling frequency.

To prepare our data, we first rejected such cases as possible outliers that at least one of the three gain values was farther away than two times the standard deviation from the corresponding mean along with each gain variable; for the latency it was not necessary, since in the signal analysis step no latency over 40 ms was accepted as an unreasonably long. This procedure discarded approximately 10 % of the cases.

Means and standard deviations (Table 1) were computed from the four result values of the 22 healthy and 22 patients. All 22 patients suffered from a unilateral disorder, i.e. there was a dysfunction caused, e.g., by an operated tumour in the area of the vestibular nerve or other verified deficiency: 13 patients out of 22 had a lesion on the right side and 9 had it on the left side. We computed the results of the groups of these 13 and 9 patients. Since these numbers were small, we united the disordered sides so that there were two sides from the same 22 patients: the disordered and the intact.

Table 1. Average test results (mean and standard deviation) of the 22 healthy subjects (the directions separated) and 22 patients (their worse and better side or ear separated). The numbers n of all response-stimulation pairs, three gains and latency are given.

Eye movements	n	**Gain** g_1	**Gain** g_2	**Gain** g_3	**Latency** l [s]
Healthy: right - left	422	1.02±0.15	1.08±0.16	1.09±0.22	0.021±0.012
Healthy: left - right	420	0.96±0.14	1.03±0.14	1.03±0.22	0.021±0.012
Patients: intact side	341	0.85±0.19	0.93±0.18	0.89±0.29	0.023±0.012
Patients: disordered	332	0.59±0.21	0.62±0.21	0.58±0.27	0.024±0.012

The healthy were capable to respond with gains close to 1, but the disordered sides of the patients were much worse and their intact sides were somewhat weaker than those of the healthy. The differences were significant between the healthy and patients according to t test at the significance level 0.001. The numbers of response-stimulation pairs were 19 per signal pair on average for the healthy and 15 for the patients.

3. Neural network classification on the basis of signal analysis results

Our aim was to separate the patients from the healthy by applying percpetron neural networks with the backpropagation training algorithm (implemented in Matlab) as the final step in Fig. 3. We prepared a set of several alternatives concerning the only free parameter of the structure, viz. the number of hidden nodes. The numbers of the input and output nodes were four and two in conjunction with the result variables in Table 1, i.e. the three gains and latency, and the two classes, either healthy subject or patient.

Referring to the results of Table 1, it might have been possible to separate the patients from the healthy by establishing the computation, on one hand, on both disordered and intact sides of the patients and, on the other hand, on both directions of the healthy. Nevertheless, the classification was unsuccessful if the intact sides were included. Two first principal components computed from the data of the four input variables explained 79 % and 13 % of the total variability, respectively. Though the cases of intact sides were mostly different from those of the disordered sides, they much resembled those of the healthy. Thus, we took the intact cases and randomly chose halves from the two subsets of the healthy. All the healthy cases were not taken in order to balance both classes roughly of the equal size. If a class distribution is very bias, this probably impedes the perceptron network to learn minority classes.

In order to treat the data for neural networks, the values of all four input variables were normalized to contain the same scale. We executed the learning and testing process by using 10-fold crossvalidation, where each of ten test sets consisted of circa

10 % of both classes (the healthy and patients). Every such learning situation with the rest 90 % of the cases and its test were run ten times to average possible slight differences in results arisen from random initializations of the networks. Thus, one hundred runs were accomplished for each network altogether. We addressed perceptron networks of one hidden level with 2 to 20 nodes, each posing a sigmoidal threshold function. No more than 500 learning epochs were run, and the adaptive learning rate and momentum coefficient were utilized. Results obtained from a few networks are given in Table 2, in which no specificity or negative prediction values are given, because these are symmetric to the sensitivity and positive prediction values, respectively, in the opposite class on account of the two-class division. The results showed the classification to be highly successful. Their differences were small, but the best alternatives appeared to the hidden node numbers of 4 to 20.

4. Discussion

We elaborated a syntactic pattern recognition method with neural network classification for the diagnostic aid of otoneurological patients. The results produced were substantially good. Our next objective will be to enlarge the dataset in order to cover more patient cases. Perhaps it is also possible to separate different otoneurological diseases from the healthy and each other.

Table 2. Percentages of sensitivities, positive prediction values and total accuracies.

Hidden nodes	**Sensitivity: healthy**	**Sensitivity: disordered**	**Pos. pred. of healthy**	**Pos. pred. of disordered**	**Total acc.**
2	92.4±14.8	81.3±22.3	86.0±15.8	85.7±24.0	87.5±11.8
4	95.0±6.1	86.3±8.1	90.0±5.5	93.5±7.3	91.2±5.5
6	94.8±6.5	87.0±8.4	90.4±5.7	93.4±7.6	91.3±5.8
8	94.7±6.5	87.5±7.8	90.7±5.4	93.4±7.5	91.5±5.6
12	95.0±6.3	87.8±7.9	91.0±5.5	93.6±7.1	91.8±5.5
16	94.9±6.2	88.3±8.1	91.3±5.6	93.5±7.2	92.0±5.7
20	94.7±6.1	88.7±7.8	91.6±5.5	93.3±7.1	92.1±5.6

References

[1] Juhola M. A syntactic method for analysis of saccadic eye movements. Pattern Recogn 1986; 19: 353-9.

[2] Juhola M. A syntactic analysis method for sinusoidal tracking eye movements. Comput Biomed Res 1991; 24: 222-33.

[3] Juhola M, Aalto H, Hirvonen T. A signal analysis technique of vestibulo-ocular reflex stimulated with impulsive head movements. Submitted to Annals Biomed Eng 2005.

[4] Baloh RW, Sills AW, Kumley WE, Honrubia V. Quantitative measurement of saccade amplitude, duration and velocity. Neurology 1975; 25:1065-70.

[5] Aalto H, Hirvonen T, Juhola M. Motorized head impulse stimulator to determine angular horizontal vestibulo-ocular reflex. J Med Eng Techn 2002; 26: 217-22.

[6] Collewijn H, Smeets JB. Early components of the human vestibulo-ocular response to head rotation: latency and gain. J Neurophysiol 2000; 84: 376-89.

9. Miscellaneous

9.1 Publication and Presentation

Ubiquity: Technologies for Better Health in Aging Societies
A. Hasman et al. (Eds.)
IOS Press, 2006

Publication Bias in Medical Informatics evaluation research: Is it an issue or not?

Christof MACHAN[a], Elske AMMENWERTH[a], Thomas BODNER[b]
[a] *Institute for Health Information Systems, UMIT – University for Health Sciences, Medical Informatics and Technology, Hall in Tyrol, Austria*
[b] *Department of Biological Psychiatry, Innsbruck Medical University, Innsbruck, Austria*

Abstract. The phenomenon of publication bias has probably existed since results of scientific research are being published. Positive and / or statistically significant results seem more likely to be published than negative and / or insignificant results. However, it is unclear if there is a remarkable impact of publication bias in medical informatics evaluation literature and how aware researchers are of its effect. We conducted a small-scale study in order to find out what the ratio of papers describing positive results vs. negative results is, tried to find enough studies to a certain subject to carry out a meta-analysis and assess publication bias by statistical methods, and finally examined reviews and meta-analyses for their results and their quality. A random sample of 86 studies showed a remarkably high percentage of descriptions of positive results (69,8%). 19 (36,6%) of the analyzed 54 reviews and meta-analyses came to a positive conclusion with regard to the overall effect of the analyzed system, 32 (62,5%) were inconclusive, and only one review came to a negative conclusion. Quantitative assessment of publication bias for health informatics studies was found difficult due to the low number of comparable studies. Although there is no clear evidence for a great impact of publication bias in medical informatics evaluation literature, further research should be carried out.
Keywords: Medical informatics, evaluation, assessment, publication bias

1. Introduction

Publication bias is a phenomenon that has probably existed as long as results of scientific research are being published. In brief, it is defined as the publication or non-publication of results depending on their direction and statistical significance. Positive and / or statistically significant results seem more likely to be published than negative and / or insignificant results. In addition other kinds of biases can have an influence on scientific work: the inclusion of results depending on language, availability, cost, familiarity (publications only of one's own field), outcome (selective inclusion of results in a primary study). Primary (qualitative or quantitative) studies can as well be affected as reviews that either statistically combine quantitative results of primary studies (meta-analyses), quantify (systematic reviews) or summarize (narrative review) results of other studies. [1]

A number of studies investigating the effects of publication bias have been carried out – also in the field of biomedical informatics – giving an overview of its existence and risk factors [2] or assessing to what degree the direction and significance of results influence publication. [3, 4]

As regards narrative reviews, standardized objective methods for assessing publication bias do hardly exist. For meta-analyses as the quantitative combination of results by statistical methods, a number of approaches such as the funnel plot [5] or the "Trim and Fill method" [6] have been described.
Influential journals have implemented guidelines developed by specialized working groups to make sure that work published is of high quality. Examples for that are the QUOROM statement concerning the quality of meta-analyses of randomized controlled trials [7] and the MOOSE checklist for meta-analyses of observational studies [8]. Both of them demand that the issue of publication bias be dealt with.
In order to assess the existence of publication bias, its influence on meta-analyses and the awareness of authors in medical informatics evaluation research we conducted a small-scale study to answer three study questions:.

1. How does the percentage of evaluation studies describing positive results compare to those describing mixed or negative results?
2. To what extent is a statistical assessment of publication bias for evaluation studies in health informatics possible?
3. How can the quality of reviews and meta-analysis in health informatics with regard to the issue publication bias be described?

2. Methods

According to our three study questions we applied three different methods to get a comprehensive overview on publication bias in medical informatics evaluation research. Any of them is based on a search in a database specialized in medical informatics evaluation literature. The database (http://evaldb.umit.at) – hosted by the Institute for Health Information Systems at the UMIT – comprises 1.035 references to papers published between 1982 and 2002 that describe evaluation studies [9].

2.1. Study question 1: Ratio of positive vs. negative results: How does the percentage of evaluation studies describing positive results compare to those describing mixed or negative results?

If there is a great influence of publication bias in the population one should assume that a significantly higher number of studies describing positive results than negative results has been published. Comparable statistics were presented by [2].
In order to answer that question a random sample of 86 references was drawn out of the database (http://evaldb.umit.at) on the basis of the following criteria:

- **Research type**: explanative
- **Evaluation approach**: more quantitative
- The full version of the paper had to be available in case the abstract did not allow for a clear interpretation of findings

For every abstract it was investigated if the authors found unambiguous results and if so, if they were positive or negative as regards the hypothesis the publication was based on. This categorization formed the foundation for a quantitative analysis. For abstracts that showed no clear results, qualitative categorization was done in order to find out more about the reasons.

2.2. Study question 2: Statistical assessment of publication bias: To what extent is a statistical assessment of publication bias for evaluation studies in health informatics possible?

A funnel plot is a simple way to depict the phenomenon of publication bias in a sample of studies. It can be briefly described as a scatter plot classifying every single study by effect size and study quality. The assumption is that fewer studies with low quality and low effect size or describing a negative effect are published. The plot can visualize this sign of publication bias. [5]

We decided to try to draw a funnel plot for evaluation studies on CPOE (computerized physician order entry), because a lot of studies evaluating such systems have been published during the last years. Their potential to significantly improve patient care or to even save lives has been described frequently.

In consequence we had to find a certain number of studies evaluating CPOE systems, measuring the same effect and providing quantitative data for effect size and study quality. We first searched the database already described (http://evaldb.umit.at) database to find the evaluation studies using the following criteria:

- **Information System**: CPOE: physician order entry system, drug prescription system
- **Evaluation approach**: more quantitative

The found studies were categorized based on the effect(s) they were measuring (e.g. time intervals, length of stay, costs), and the effect measured in most studies was selected for further investigation and for drawing a funnel plot. Five different papers are given as a limit to be able to assess publication bias by using a funnel plot by [5]. After deciding on the effect, we extended our search to PubMed for further papers on CPOE evaluation and measuring the selected effect to be sure not to overlook important studies.

2.3. Study question 3: Quality of reviews and meta-analyses: Are there more systematic and narrative reviews or more meta-analyses referenced in the database? Do most of them draw positive, neutral or negative conclusions? Is the issue of publication bias being dealt with by the authors or not?

The sample for this investigation comprised all 52 reviews in the database (http://evaldb.umit.at). Thus, the following search statement was used:

- **Research type**: review (systematic overview on studies)

By going through the abstract, each reference was identified as either review (systematic or narrative) or meta-analysis combining results quantitatively and either reporting positive, neutral or negative results.

In a second step we looked for hints for the handling of publication bias by the authors of meta-analyses:

- Do the abstracts of meta-analyses of randomized controlled trials conform to the QUOROM statement?
    - Is there a list of databases searched?
    - Have selection criteria for publications to be included or excluded been described?

- Are there any hints in full versions available (the QUOROM statement demands a clear description "of potential biases in the review process" [7]).

3. Results

3.1. Results of approach 1: Ratio of positive vs. negative results

In our sample of 86 abstracts, 60 (69,8%) showed positive results, 12 (14%) negative and 14 (16,3%) could not be attributed to either of these categories. Typical phrases found in abstracts indicating positive results were e.g. (depending on the outcome variable measured): "is a useful system for improving", "time showed significant reductions"…; a typical phrase found to describe negative results was e.g. "the system was less accurate than"; A typical indication that an abstract could not be attributed to either of the categories was that no comparison could be made, that no conclusions could be drawn, that positive and negative results were balanced or that no difference could be found.

3.2. Results of approach 2: Feasibility of a meta-analysis

In the first part of the search, the evaluation database was searched, and 46 CPOE evaluation studies were found in total. Table 1 shows the categorization for these references in terms of their outcome variable.

Table 1: Frequencies of effects measured by 46 CPOE evaluation studies. More than one effect can be measured by one study

category	number
time intervals	6
transcription errors	2
length of stay	3
costs	8
medication errors	6
drug interactions	2
redundant tests	4
acceptance	7
other	17

Except for the category "other" (which contains any reference with a unique outcome variable) the category comprising most references is "costs" with 8 references. Taking into account that there are many different kinds of costs further categorization was done that revealed that 5 times prescription costs were measured, twice total costs, once costs per employee and once total costs for the hospital (it was possible of course that one publication measured more than one factor). In a second step, a query for prescription costs and CPOE in PubMed (compiled of several keywords) retrieved 132 results. 11 of them (minus 2 that had already been found in evalDB) also seemed to measure prescription costs in a general or special context. Further results on the attempt to assess publication bias in this sample will be available for the conference.

3.3. Approach 3: Quality of reviews and meta-analyses

The search as described in 2.3 delivered 52 references, 46 (88,5%) of them narrative or systematic reviews and 6 (11,5%) meta-analyses. Only 1 review came to an overall negative conclusion with regard to the effect of the analyzed system (1,9%), 4 meta-analysis and 15 reviews (together 36.6%) came to a positive conclusion, but the major part of studies could neither draw positive nor negative conclusions (32 studies, 61,5%. For a graphical depiction see figure 1.
1 of 6 meta-analyses was based on observational studies, 5 on randomized controlled trials. The abstracts of 2 of these 5 meta-analyses did not contain a description of data sources and review methods, not even mentioning how many trials they were based on, thus not conforming to the QUOROM statement. One of the other 3 was based on 98, one on 16 and one on 7 RCTs.

Figure 1: Frequencies of reviews / meta-analyses by study type (review or meta-analysis) and outcome (positive, neutral or negative conclusion with regard to the effect measured by analysed studies). (n=52)

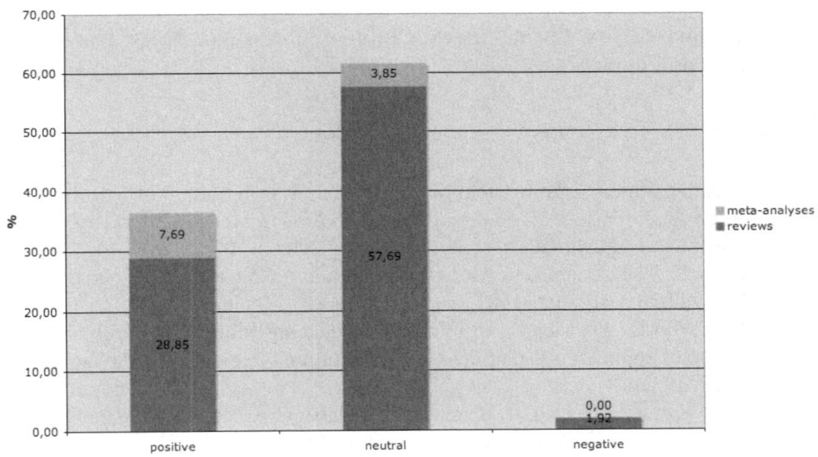

4. Discussion

In our random sample of 86 evaluation studies a high number of publications describe positive results (69,8 %). The chance that this is coincidental is remarkably scarce. Assuming that the probability for a publication to describe positive results was 50% (which is of course a hypothetical value) randomly drawing 60 positive studies of 72 (total of positive and negative publications found) would show a p < 0,00001%, thus be very unlikely. This can be seen as an indicator for publication bias.
For a quantitative analysis of possible publication bias, there needs to be a sufficient number of comparable studies (studies evaluating the same type of system using the same evaluation criteria). In our analysis, we found that even for a frequently evaluated system (CPOE), it is difficult to get such a number in order to carry out a serious meta-analysis. This indicates that in medical informatics evaluation literature, it seems hard to find a number of studies that precisely assess the same effect by employing the same measurements. This and a sometimes low quality of studies was also described by [10].

Results of the attempt to combine results of our queries in a funnel plot to assess possible publication bias will be presented at the conference.
In our sample of 52 reviews, it was obvious that narrative and systematic reviews (88,5%) are the preferred methods for combining results of primary research. Possible explanations are that they require less statistical knowledge and are less time consuming compared to meta-analyses. In contrast to the effect seen in the sample of 86 primary studies, effects of publication bias in reviews and meta-analyses could hardly be found. 61.5% neither drew positive nor negative conclusions. From the 5 meta-analyses that were based on an analysis of RCTs, only 3 conformed to the QUORUM statement as regards their abstract.

5. Conclusions

This small-scale study should contribute to the discussion whether publication bias in medical informatics evaluation studies is an issue worth reflecting on. Although striking evidence for the existence of publication bias was not found we conclude there are indications that rectify further investigation of the topic. More results may be presented at the conference.

References

[1] Rothstein H, Sutton A, Borenstein M. Publication Bias in Meta-Analysis. In: Rothstein H, Sutton A, Borenstein M, editors. Publication Bias in Meta-Analysis, prevention, assessment and adjustments. Chichester: Wiley; 2005. p. 1-7.

[2] Dickersin K. The existence of publication bias and risk factors for its occurrence. JAMA 1990;263(10):1385-1389.

[3] Dickersin K, Min YI, Meinert CL. Factors influencing publication of research results. Follow-up of applications submitted to two institutional review boards. Jama 1992;267(3):374-8.

[4] Olson CM, Rennie D, Cook D, Dickersin K, Flanagin A, Hogan JW, Zhu Q, Reiling J, Pace B. Publication bias in editorial decision making. Jama 2002;287(21):2825-8.

[5] Sterne J, Becker B, Egger M. The Funnel Plot. In: Rothstein H, Sutton A, Borenstein M, editors. Publication Bias in Meta-Analysis, prevention, assessment and adjustments. Chichester: Wiley; 2005. p. 75-99.

[6] Duval S. The Trim and Fill Method. In: Rothstein H, Sutton A, Borenstein M, editors. Publication Bias in Meta-Analysis, prevention, assessment and adjustments. Chichester: Wiley; 2005. p. 127-145.

[7] Moher D, Cook DJ, Eastwood S, Olkin I, Rennie D, Stroup DF. Improving the Quality of Reports of Meta-Analyses of Randomised Controlled Trials: The QUOROM Statement. Onkologie 2000;23(6):597-602.

[8] Stroup DF, Berlin JA, Morton SC, Olkin I, Williamson GD, Rennie D, Moher D, Becker BJ, Sipe TA, Thacker SB. Meta-analysis of observational studies in epidemiology: a proposal for reporting. Meta-analysis Of Observational Studies in Epidemiology (MOOSE) group. Jama 2000;283(15):2008-12.

[9] Ammenwerth E, de Keizer N. An inventory of evaluation studies of information technology in health care: Trends in evaluation research 1982 - 2002. Methods Inform Med 2004.

[10] Whitten P, Kingsley C, Grigsby J. Results of a meta-analysis of cost-benefit research: is this a question worth asking? J Telemed Telecare 2000;6 Suppl 1:S4-6.

Ubiquity: Technologies for Better Health in Aging Societies
A. Hasman et al. (Eds.)
IOS Press, 2006

The multiple faces of the e-patient, if not disabled

M. Cristina MAZZOLENI[a,1], Franco CORBELLA[b]
[a] *S,Maugeri Foundation IRCCS, Pavia, Italy*
[b] *Policlinico San Matteo IRCCS, Pavia Italy*

Abstract. The term e-health entered common speech a few years ago , while the term "e-patient" has recently appeared on the healthcare scenario. The aim of the present paper is to describe the nature of the so called "e-patient" from different points of view, through a review, not systematic, of the literature . A profile, though not totally exhaustive, of the current e-patient has been drawn, in an attempt to report uncertainties and worries that should not be underestimated. Comments are provided on the asymmetry between the evaluation effort around the internet world and corresponding face-to-face world. Disabled patients are almost excluded from the e-patient family due to the inaccessibility of most of the health related web sites.
Keywords: e-patient, e-health, internet

1. Introduction

The term e-health entered the common speech a few years ago: politicians speak about the e-health plan due to the European Commission initiatives, journalists write about e-health projects and researchers publish using "e-health" in their papers, vendors offer "e-health" solutions or systems.

The term e-patient has recently appeared in the healthcare scenario, pointing out the no longer marginal role the patient has gained in the e-health realm.

2. Aim

Aim of the present paper is to describe the nature of the so called "e-patient" from different points of view, through a review, not systematic, of the literature, in order to answer the following questions: Who are the e-patients? Which are the critical issues for this relatively new role the patient is playing?

3. People who seek for health related information on the web

Health related information is one of the most sought topics on the web. In [1] the authors estimated that, in 2003, globally a minimum of 6.75 million health-related

[1] Corresponding Author: M.Cristina Mazzoleni, Fondazione Maugeri IRCCS, Via Ferrata 8- 27100 Pavia Italy. cmazzoleni@fsm.it

searches were being conducted on the web every day. Cyberdialogue [2] estimated that, by 2005, 88.5 millions of adults would have surfed the web seeking information about diseases and therapies, shopping for health products or communicating with health providers through online channels.

Great interest has grown during the last ten years about the quality of the information a lay person can find on the web. The interest has been expressed mainly in form concern: The World Heath Organization published a leaflet – available on the web as well – in which it is clearly stated that "The Internet is a valuable source of health information .... when used properly ..". The European Union developed the "Quality Criteria for Health Related Web Sites", and many other initiatives have been brought up - Internet Healthcare Coalition, Health On the Net Foundation - in order to spread the culture of an ethical approach to the distribution of information on the web. In the near future, the result of the WRAPIN - Worldwide online Reliable Advice to Patients and Individuals - European project [3] will help the e-patient determine the reliability of documents by checking the ideas contained against established benchmarks

Echoing the above mentioned worries, in the late nineties, probably beginning with [4] a certain not small number of studies were carried out and published to explore and divulge the status of the quality of the information available on the internet for the general public. Small and large projects were published whose aim was to inspect information related to different specialities and topics: oncology [5] , osteoporosis [6] , eating disorders[7], quit smoking programs [8],drugs [9], etc. The common result was that the quality was, and is, extremely variable.

A few years later, a new research trend began more related to modalities of quality evaluaation [10]. Not only correctness and completeness, often evaluated versus some golden standards such as guidelines, together with transparency and honesty must be evaluated, but also the language, the content intensity and presentation according to the different purposes for which a site is accessible and accessed [11] .

Papers about this problematic aspect which the e-patient must face are published in prestigious international scientific journals in the medical domain and in more specific medical informatics journals. A new term has even been coined by G.Eysenbach: "infodemiology". This term is used only to describe the information available from the web, a fact that leads to the idea, by contrast, that misinformation does not exist or is rare case elsewhere. Few studies have been published on the comparison web versus printed information [12] and the results don't condemn the web.

If, on one hand, the "offer" of information has been quite largely evaluated, on the other hand the demand and the behaviour of this kind of e-patient is not widely known. Despite the general concern about reliability of health information on the web, the users don't seem to be very interested in this aspect: in an observational study [13] about consumer search and appraisal of information on the www, no participants checked any "about us" section of websites, disclaimers, or disclosure statements, pillars of any recommendation for the users. Slightly different results were found in a study based on interviews [14] in which only one quarter of the respondents say they always check the source, date, and privacy policy of a site. As pointed out in [15] it is clear that there is a variety of skills among health consumers about the proper way to search for information on the web and moreover there have been few studies in the literature, and hence probably educational sessions in reality, that have sought to educate the consumers .

The fact that there is such a low knowledge and awareness of the problem among both the content providers and the e-patients community can mean that the proper target audience for so many papers has been missed.

Outcome reports for the e-patient after having searched and eventually found information on the web are difficult to find. Most of the times the results describe only the subjective perception of the e-patients.

4. People who interact with others via ICT and the web in particular

4.1 People who attend virtual communities and support groups

Virtual communities are social networks formed or facilitated through electronic media, and the internet in these years, where people with common interest "meet" to share experiences, ask and give information, provide and find emotional support. These communities can be moderated by some healthcare professional or not moderated at all and they are very numerous. These support groups have certain benefits for users who may not be able to or do not want to attend face-to-face sessions. In [16] positive findings for the e-patients are reported as to "more knowledge about medication", "more ability to discuss subjects they felt unable to discuss elsewhere", "less isolation", and others issues. These kind of results refer to the population using the virtual community, nothing is said about the people (around 50% of the initial sample) who have got in touch with it, but haven't exploited the supplied services. It is curious that a clear statement in the paper points out that " ..our findings do not support a view of the internet as harmful" as if this were the common opinion to disprove. On the other hand, challenges are surely present when comparing virtual communities to traditional communication [17] such as : the exclusion of certain groups through the digital divide, the misinterpretation of online messages lacking in visual or aural cues, and the dissemination of inaccurate information. A different interesting perspective is described in [18] where the authors invite the healthcare personnel to learn from observing and communicating with the e-patients in virtual communities, " it may prove invaluable in your future work".

4.2 People who are "monitored" in some structured way

The medical informatics community seems to share the opinion that "telemedicine is expected to make it possible to link medical expertise with patients in the most distant locations-providing clinicians with valuable new tools for remote monitoring, diagnosis, and intervention [19]. It is symptomatic that researcher have been writing sentences like this for twenty years. Now the interest is more focused on the patient. Quite a large number of projects has been carried out in the last ten years to reduce the distance between the patient and the medical expertise, and, more recently, to take advantage of the most popular technologies such as mobile phones and SMS; Tools for the management of patients with diabetes [20] , with asthma[21], with heart failure [22]; Tools for enhancing the compliance to therapy[23] and for administering virtual speech therapy and many other. Most of the published papers conclude using "may" or "might" or similar conditional form when summarizing the benefits for this kind of e-patient and of this e-health practice. In the meantime, other papers warn against another new source of medical errors, the “telemedicine error” [24] ,while others [25] point out that “not always the inter-physician agreement on in-person assessments is significantly better than the agreement of the telemedicine system, even if not high “.

4.3 People who spontaneously start a “treatment” via the web

The web can be a way to reach the “hard–to-reach“audiences, thanks to its 24-hours-a-day availability and its “privacy” compared with the one of a waiting room. Some public web-based services have been implemented to try to influence the health behaviour of people with specific problems such as, for example, heavy drinkers [26], people needing/looking for cognitive behaviour therapy program to control depression and anxiety [27], etc. The findings show a high attraction rate and a positive effect on the evaluated outcomes (for the examples quoted here, alcohol consumption and depression and anxiety score) for those who have completed the programs. But in the meantime two problems are outlined: the attrition of the programs and the high drop out; no data are supplied in order to compare the attrition and drop out in equivalent face to face interventions.

5. People who could use ICT and the web to overcome their limitations

There is no reason to think that people with sensory or motor impairments should have less need for multiple features offered by the web, in the terms described in the previous paragraphs. Or, on the contrary, they could be the ones who benefit more from web services, having limitations due to their impairments. Alternative input output devices, even very sophisticated, have been studied and some are on the market, but they are not sufficient, by themselves, to surf the web.

Guidelines for accessible and usable web publishing have been produced [28] and disseminated. Nevertheless, this issue doesn't seem to have particularly interested the medical informatics community. Searching PubMed and looking for “disabled AND accessibility AND (internet OR web OR www)” in the title or abstract only five papers appear.

In the retrieved papers the situation described is not at all good in terms of accessibility of health-related web sites. Moreover, checking for structural accessibility through the requirements of WAI guidelines (priority 1 and 2) is not enough to ensure usability for differently impaired persons. Subjective-oriented measures on the field are necessary [29] for an effective evaluation.

Moreover, using ' "Disabled Persons"[MeSH] AND "internet"[MeSH Terms] ' , a sample of 86 papers is retrieved from PubMed, and among these only few (21) focus on the direct usage of the web by impaired people and , more or less directly, address the problem of accessibility. The design of web based services that takes into account the accessibility issues could make the difference in the future for many people, when both the internet pervasiveness for health related purposes and the mean age of the internet users will be even higher: The e-patient of today could be excluded tomorrow .

6. Conclusions

Who are the e-patients today? They are people, who need health related information, social support, support to change life-styles, systems that help them to monitor and maintain their health status when discharged from the hospital. These people, if not disabled, can find, taking advantage of the web, something that may meet their expectations. A profile, may be not totally exhaustive, of the current e-patient has been drawn, in an attempt to report uncertainties and worries that shouldn't be underestimated, and some comments are provided on the asymmetry between the evaluation effort around the internet world and the one around the face-to-face world. What happens on the web is often, more or less expressly, compared with a golden standard represented by the ideal healthcare personnel, always expert, up-to-date, available at walking distance when needed, able to act at best, able and willing to communicate in the most proper and effective way. For this last issue for example - that so much worries as to the internet world - it must be noted that, at least in Italy, physicians receive during their regular curriculum no education on topics such as communication with patients. Far from saying that the healthcare personnel is not expert and face-to-face facilities are useless, the point, in our opinion, is that we have to consider the variability, in terms of quality and availability, of the "real world". The strength of the "e-world" is its capacity to reach the hard-to-reach people and to shorten the distances between providers and e-patients with benefits for the single person and the entire society, and this must have a proper weight when looking for a balance.

A doubt arises: is the opulence of papers published acting as a double-edge sword? While looking for the excellence of the services offered to the e-patients and discussing their weacknesses, is there the risk to monopolize the attention almost only on problems and to atrophy the good opportunities that already are available to the e-patient on the web, transforming the web in an unusable good tool for e-patients?

From Jacques Carelman 1969
Unusable objects

What is really today hard to accept is the low attention paid to the needs of disabled people. Poor accessibility constitutes a new barrier for those who could greatly benefit from being e-patients. The Medical Informatics Associations should promote the accessibility to reduce this digital divide that really prevents a large number of people from being less emarginated and probably better cared for thanks to the ubiquitous nature of the web.

References

[1]Eysenbach G, Kohler Ch. What is the prevalence of health-related searches on the World Wide Web? Qualitative and quantitative analysis of search engine queries on the internet. AMIA Annu Symp Proc. 2003;:225-9

[2] www.cyberdialogue.com

[3] http://www.wrapin.org/

[4] Impicciatore P, Pandolfini C, Casella N, Bonati M. Reliability of health information for the public on the World Wide Web: systematic survey of advice on managing fever in children at home.BMJ. 1997 Jun 28;314(7098):1875-9.
[5]F. Meric, E.V. Bernstam, N.Q. Mirza, K.K. Hunt, F.C. Ames, M.I. Ross, H.M. Kuerer, R.E. Pollock, M.A. Musen and S.E. Singletary, Breast cancer on the world wide web: cross sectional survey of quality of information and quality of information and popularity of websites, *BMJ* **324** (2002), pp. 577–581]
[6] Wallace LS, Turner LW, Ballard JE, Keenum AJ, Weiss BD. Evaluation of web-based osteoporosis educational materials. J Womens Health (Larchmt). 2005 Dec;14(10):936-45.
[7] Murphy R, Frost S, Webster P, Schmidt U. An evaluation of web-based information. Int J Eat Disord. 2004 Mar;35(2):145-54
[8] Bock B, Graham A, Sciamanna C, Krishnamoorthy J, Whiteley J, Carmona-Barros R, Niaura R, Abrams D. Smoking cessation treatment on the Internet: content, quality, and usability. Nicotine Tob Res. 2004 Apr;6(2):207-19.
[9]Graber MA, Weckmann M. Pharmaceutical company internet sites as sources of information about antidepressant medications. CNS Drugs 2002;16(6):419-423]
[10]Eysenbach G, Powell J, Kuss O, Sa ER. Empirical studies assessing the quality of health information for consumers on the world wide web: a systematic review.JAMA. 2002 May 22-29;287(20):2691-700.
[11]Roberts JM, Copeland KL. Clinical websites are currently dangerous to health. Int J Med Inform. 2001 Jul;62(2-3):181-7
[12] Mazzoleni MC, Marello A, Giorgi I, Marchi A.Web versus press with a high circulation: an Italian case study.Stud Health Technol Inform. 2003;95:684-9.
[13] [Eysenbach G..How do consumers search for and appraise health information on the world wide web? BMJ 2002; 324:573-7
[14]S. Fox and L. Rainie, Vital Decisions, Pew Internet and American Life Project, Washington, DC (2002). http://www.pewinternet.org.
[15]Peterson G, Aslani P, Williams KA How do Consumers Search for and Appraise Information on Medicines on the Internet? A Qualitative Study Using Focus Groups. J Med Internet Res 2003;5(4):e33 URL: http://www.jmir.org/2003/4/e33/
[16]Powell J, McCarthy N, Eysenbach G. Cross-sectional survey of users of Internet depression communities.BMC Psychiatry. 2003 Dec 10;3:19
[17]White M, Dorman S. Receiving social support online: implication for health education. Health Education Research; 2001 vol 16, N°6; 693-707
[18]Lester J, Stephanie Prady S, Finegan Y, Hoch D. Learning from e-patients at Massachusetts General Hospital *BMJ* 2004;328;1188-1190
[19] Spivack RN. Innovation in telehealth and a role for the government. Stud Health Technol Inform. 2005;118:32-42.
[20] Bellazzi R, Arcelloni M, Ferrari P, Decata P, Hernando ME, Garcia A, Gazzaruso C, Gomez EJ, Larizza C, Fratino P, Stefanelli M. Management of patients with diabetes through information technology: tools for monitoring and control of the patients' metabolic behavior. Diabetes Technol Ther. 2004 Oct;6(5):567-78.
[21] Anhoj J, Moldrup C.Feasibility of collecting diary data from asthma patients through mobile phones and SMS (short message service): response rate analysis and focus group evaluation from a pilot study. J Med Internet Res. 2004 Dec 2;6(4):e42
[22]Louis AA, Turner T, Gretton M, Baksh A, Cleland JG. A systematic review of telemonitoring for the management of heart failure. Eur J Heart Fail. 2003 Oct;5(5):583-90. Review
[23] Smart N, Haluska B, Jeffriess L, Marwick TH. Predictors of a sustained response to exercise training in patients with chronic heart failure: a telemonitoring study.Am Heart J. 2005 Dec;150(6):1240-7
[24]Demiris G, Patrick TB, Mitchell JA, Waldren SE. To telemedically err is human. Jt Comm J Qual Saf. 2004 Sep;30(9):521-7
[25] Kim HM, Lowery JC, Hamill JB, Wilkins EG. Accuracy of a web-based system for monitoring chronic wounds.Telemed J E Health. 2003 Summer;9(2):129-40
[26]Linke S, Harrison R, Wallace P. A web-based intervention used in general practice for people with excessive alcohol consumption. Journal of telemedicine and telecare 2005;11(Suppl 1): S1:39-41
[27] Christensen H, Griffiths K, Groves C, Korten A.Free range users and one hit wonders: community users of an Internet-based cognitive behaviour therapy program. Aust N Z J Psychiatry. 2006 Jan;40(1):59-62. [28]Web Accessibility Initiative. Available at: http://www.w3c.org/WAI/
[29]Federici S, Micangeli A, Ruspantini I, Borgianni S, Corradi F, Pasqualotto E, Olivetti Belardinelli M. Checking an integrated model of web accessibility and usability evaluation for disabled people.Disabil Rehabil. 2005 Jul 8;27(13):781-90.

Ubiquity: Technologies for Better Health in Aging Societies
A. Hasman et al. (Eds.)
IOS Press, 2006

Computing Latent Taxonomies from Patients' Spontaneous Self-Disclosure to Form Compatible Support Groups

Eduard HOENKAMP[a,1] and Regina OVERBERG[b]
[a] *Nijmegen Institute for Cognition and Information, Nijmegen, the Netherlands*
[b] *Leiden University Medical Centre, Clinical Informatics, Leiden, The Netherlands*

Abstract. A growing number of Internet sites provide patients with an opportunity to share experiences about their illness. Unfortunately, it depends on the luck of the draw whether patients find compatible others for mutual support. To improve this situation, this paper demonstrates how tools from Information Retrieval can be deployed to discover patients with compatible stories more systematically. We show how finding compatible support is like finding relevant documents on the Internet, and also in what way it is different. We designed an algorithm to produce a 'latent' taxonomy over patients' self-disclosures to aid them in forming support groups. At the end we briefly consider the result of this taxonomy based on only the patients' stories, with one based on explicit questionnaires.
Keywords: Information Retrieval, support groups, illness stories, clustering, latent taxonomy.

1. Introduction

Often the most valuable encounters happen by chance. This is as true for encounters in real life, as it is for encounters on the Internet. The latter are often metaphorical encounters, as when happening on an interesting web page that thus far defied a directed search. Or, especially, when finding a story that relates to one's own life.

In our research we are interested in stories of particular interest to a very special group of people: women diagnosed with breast cancer, and who reach out to others to share their stories with. We would like to improve these women's chances of finding compatible support in their predicament. Depression, anxiety, and fear are the heavy psychological toll that follows a diagnosis of breast cancer and its treatment. But several studies have shown that sharing experiences may help patients better cope with these ramifications as depression [1], anxiety, and fear [2]. The present research is part of a larger effort to study the sharing of experiences on the Internet [3].

To this end, we designed a way to compute a taxonomy that can partition stories on the Internet in groups of similar stories. Such a partitioning could help patients to find compatible support groups by method rather than by chance. Our approach is to view the problem from the standpoint of Information Retrieval (IR) and the way search engines find documents based on a query. The key is to view a patient's self-disclosure as a very elaborate query to search for similar stories. The present paper explains how

[1] Corresponding Author: Montessorilaan 3, 6525 HR Nijmegen, the Netherlands. hoenkamp@acm.org.

such a search can be accomplished using techniques that have proven their value over the last four decades of IR research.

2. An Information Retrieval Approach to Match Stories Related to Self-Disclosure

When users search the Internet, they usually type a few words into a browser, and wait for documents that presumably pertain to the words they just typed. Often they add to or change the words (the query) a bit to get more precise results. Conceptually, the user defines a very small document (containing only the keywords) and asks for documents that are closest in meaning to this mini-document. The documents returned depend on the search engine's metric to measure distance in meaning based on the words that the documents contain. In the present paper, a patient's diary or other written self-disclosure is seen as a very elaborate and precise query. From an IR standpoint, then, finding similar stories amounts to returning stories that are close to this very elaborate query. In the next paragraphs we will have to cut a few corners for ease of exposition as this is not a tutorial on IR, but we would like to keep the paper nonetheless self-contained. We will explain three techniques used in the computation: *term weighting, dimension reduction,* and *document clustering*.

2.1. The document space and the 'bag-of-words' approach in IR

The search engines that most people are familiar with take little heed of the structure of documents, and mostly consider a document as just a collection, or bag, of words. Much of the relative success of search engines comes down to a clever balancing of word frequencies within the document, and the distribution of these words over all documents under consideration. This has lead to the definition of the *document space.* This space is a vector space that has the words (more generally the 'terms') as coordinate axes, and the documents as points in that space. For every document the coordinates are computed by sometimes complex formulas, called *term weighting*, which attempt to balance the importance of a word in the document and its importance to distinguish a document from other documents. One could e.g. take just the frequency of nouns in a document, because the more a noun appears in a document (such as 'patient'), the more it is about the concept expressed by this noun. But if the same noun appears in every document of the collection, it does not discriminate well between documents. So usually a value is taken that is proportional to its frequency in the document, but inversely proportional to the number of documents it appears in. This so-called *tf/idf* (term frequency / inverse document frequency) metric is the one we used. For each document the *tf/idf* value is computed for each word it contains, and these will be the coordinates of the document. The whole document collection, the *corpus*, can thus be represented by a matrix, with columns for documents, and rows for words, and with the *tf/idf* values as elements of the matrix (the word-by-document matrix). This way, IR computations become linear algebra manipulations on the *document space*.

2.2. The patient's story as a very elaborate query

There are several computations that have become standard besides the *tf/idf* measure. First, the distance between documents is usually taken as the cosine between the document vectors. Hence, the documents that a search engine returns are those that make a small angle with the query. Other computations are used to avoid the lexicon problem (the problem of synonymy and polysemy). Most notably, a technique of *dimension reduction* is applied: Instead of taking the very high dimensional space where each word represents a separate dimension, the space is reduced to lower dimension of *latent semantic factors*. The technique is comparable to a factor analysis on high dimensional data (see our overview of such techniques in [4]). This way, for example synonyms that originally are different dimensions can reduce to one dimension representing the underlying meaning. The end result is a word-by-document representation that is the state of the art in search engine technology.

Recall that a search engine takes a query and returns documents relevant to the query by measuring their distances to the query. Likewise, for a given story, the distances to other stories is measured, and the ones with the shortest distance are presumed to tell a similar story. So, conceptually, the only difference with a routine web search is that the query is much more elaborate and precise, namely a whole story instead of two or three keywords.

As an illustration of query elaboration, figure 1 depicts part of the document space with several stories projected on the unit sphere in 3D, so that it can be visualized [5]. The dimensions chosen are the largest latent semantic factors of the original space [6]. Suppose a new story were incrementally added to the document space. The first paragraph could be close to Story 1, but as more of the story is added, it travels through the document space, to end near Story 5 when it is complete.

If similar stories were all that is required, a straight forward clustering technique, such as *k-means* with cosine distance would be appropriate. And finding compatible others would be analog to finding relevant documents given a query. So in the example of figure 1, stories 4, 5 and 7 would be selected as compatible with the new story. This is not necessarily the most compatible from the patient's viewpoint, as will be explained in a moment.

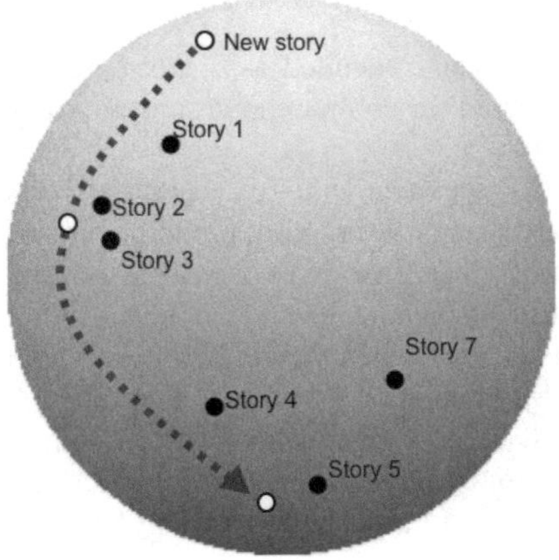

Figure 1. Part of the document space containing several stories, and after dimension reduction to the three largest latent semantic factors. The arrow shows how a new story might travel through the document space, were it added in increments, from the first paragraph to the complete story at the arrowhead.

3. Computing a Latent Taxonomy over Illness-Stories

It might seem that the quest for finding compatible support was solved in the previous section: just look up the people with similar stories. But not quite. The operative word here is 'compatible'. Maybe that someone compatible should have some part of her story in common, but should differ in others. For example, perhaps patients should be in the same age group, but a woman just diagnosed with cancer may want to talk to someone who is already under treatment. To find groups that are the same in some respect, but different in others, requires a taxonomy that partitions the stories according to certain properties. For this to work the properties do not have to be explicit, they just have to be identifiable so that combinations can be used to define the partition of interest. In other words these properties may be latent as long as they can be used in the computation. This section will show how such a *latent taxonomy* can be found by deriving a *hierarchical clustering* of illness stories. We will first describe the data we used, and then the algorithm to cluster the data.

3.1. The Corpus

For the experiment we used a carefully chosen [3] corpus of forty-nine illness stories published on a website (cf. 'De Amazones' [7]). The authors were all female breast cancer patients, who voluntarily submitted their stories for publication on the Internet. The website offers no author guidelines, so the women could freely choose the structure and length of the story about their illness. The women shared their experiences with breast cancer, often chronologically, starting with cancer symptoms, seeing a physician, hearing the cancer diagnosis, choosing the best combination of treatments, and undergoing those treatments. Most women also describe their feelings about being a breast cancer patient, and the changes in their daily lives pertaining to work, family and friends.

3.2. Hierarchical Clustering for a Latent Taxonomy

The corpus was downloaded and encoded as a word by document matrix. For the presumptive taxonomy to stand for compatible groups we suggest the following steps:

1. *Derive the word by document matrix using the tf/idf weighting.* Search engines usually apply other steps to preprocess the documents such as stemming and stop-word removal. They also may use sophisticated weighting schemes. All we use here is just *tf/idf*.
2. *Apply dimension reduction.* In the IR literature, this is usually as simple as a singular value decomposition of the word by document matrix, retaining the eigenvectors with highest eigenvalues as new dimensions[8]. Here we use a computationally much cheaper method, which gives also better results, namely noise reduction based on wavelets. (That is the method we proposed in [4]). However, the precise nature of the reduction is not important for the present paper.
3. *Compute a hierarchical clustering* based on the cosine distance.

Any cluster algorithm depends on three choices: First, the measure of similarity (e.g. we already chose the cosine distance). Second, the number of clusters required. For *n* objects this number must be somewhere between one (all objects in the same cluster) and *n* (each object is its own cluster). Third, a distance measure defined over groups of objects. Often *nearest neighbor* is used; the minimum distance between objects in each group. For our data we explored [6] several group distances, and Figure 2 shows the results for four of them. As can be seen, the furthest-neighbor technique gives a reasonable clustering. Ward looks good as well, but it is not an appropriate method in case of the cosine distance. In brief, for the kind of documents under consideration, we suggest the cosine distance with *furthest-neighbor* hierarchical clustering as the most promising.

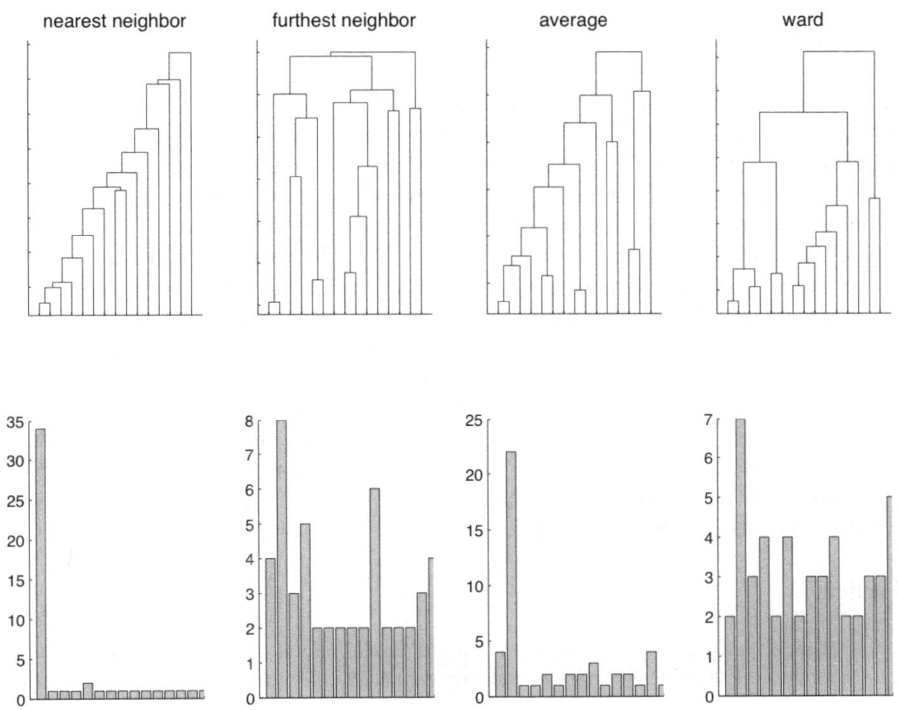

Figure 2. Hierarchical clustering of the documents. The horizontal axes represent document clusters. For legibility only the top15 clusters are shown. In the top row the vertical axis indicates the dissimilarity among clusters; in the bottom row the vertical axis gives the number of documents in each cluster.

Of course we don't know what properties distinguish these clusters, hence we called it a latent taxonomy (analogous to the latent factors in factor analysis). Yet, any processed story is subsumed in this taxonomy. Moreover, any new story can be placed in this taxonomy by finding the nearest story already in the taxonomy.

A manifest taxonomy was constructed by carefully analyzing the content of the stories. (We used the term 'latent taxonomy' before because the technique does not make the underlying content of the diaries explicit.) The analysis comprised 59 items including 51 yes/no questions, ranging over medical history such as examinations, treatments, and prognosis, to items such as coping with feelings, interactions with family, and social consequences of the illness. In that way, the participants can be compared more uniformly on items that may be present in some stories and absent in others. The details of that study will be published elsewhere, but we can qualitatively compare those results with the study presented here. From both the latent taxonomy and the manifest taxonomy we selected closest pairs and most distant pairs. We found that for these extreme cases, the similarity of the stories coincided with the more elaborate comparison of their authors.

This result is promising: The advantage of an algorithm is that it can inspect vast amounts of stories that would be otherwise impossible to collect, thus raising the odds to find compatible support groups substantially above chance level.

4. Conclusion

More and more people turn to the world-wide-web to let others know what happens in their lives. This paper reports on our venture to build taxonomies for people who want to share their illness-stories. It is intended at forming compatible support groups using proven tools from the area of Information Retrieval. It does so by taking a story as query to find similar stories; using in effect the story as the author's proxy. This lead to our algorithm for constructing the intended taxonomy. The variables in this algorithm: term weighting, dimension reduction, and cluster technique, are under-constrained however. This is good news and bad. The bad news is that it is still too early to know which part of the taxonomy defines compatibility. The good news is that we will have the chance to communicate with women participating in our study to narrow down the choices for these variables. So for the very special group we studied, women diagnosed with breast cancer, we will pursue this potential for increasing the odds of finding compatible support. For these women, further research could mean building virtual support groups to complement friends and family. And in the absence of those, even the difference between a network of friends, and potential isolation.

References

[1] Winzelberg AJ, Classen C, Alpers GW, Roberts H, Koopman C, Adams RE, Ernst H, Dev P, Taylor CB. Evaluation of an internet support group for women with primary breast cancer. Cancer 2003;97:1164-1173.

[2] Lieberman MA, Golant M, Giese-Davis J, Winzlenberg A, Benjamin H, Humphreys K, Kronenwetter C, Russo S, Spiegel D. Electronic support groups for breast carcinoma. Cancer 2003;97:920-925.

[3] Overberg R, Toussaint P, Zwetsloot-Schonk B. Illness stories on the Internet: Features of websites disclosing breast cancer patients' illness stories in the Dutch language. Patient Education and Counseling In Press.

[4] Hoenkamp E. Unitary operators on the document space. Journal of the American Society for Information Science and Technology, 2003: 314--320.

[5] Hoenkamp E, van Dinther G. Live Visual Relevance Feedback for Query Formulation. 28th Annual International ACM SIGIR Conference on Research and Development in Information Retrieval. New York: ACM Press: Sheffield, 2005: 611-612.

[6] Deerwester S, Dumais S, Landauer T, Furnas G, Harshman R. Indexing by Latent Semantic Analysis. Journal of the American Society of Information Science, 1990: 391--407.

[7] De Amazones. Jonge vrouwen met borstkanker. [The Amazons. Young women with breast cancer.] http://www.de-amazones.nl downloaded 17 March 2005.

[8] Eckart G, Young G. The approximation of one matrix by another of lower rank. Psychometrika 1936;1:211-218.

9.2 Biomedical Imaging

Ubiquity: Technologies for Better Health in Aging Societies
A. Hasman et al. (Eds.)
IOS Press, 2006

Generation of 4D CT Image Data and Analysis of Lung Tumour Mobility during the Breathing Cycle

Heinz HANDELS[a1], Rene WERNER[a2], Thorsten FRENZEL[b1], Dennis SÄRING[a3], Wei LU[c1], Daniel LOW[c2], Jan EHRHARDT[a4]

[a]*Department of Medical Informatics, University Medical Center Hamburg-Eppendorf, Germany*
[b]*Hermann Holthusen Institute of Radiotherapy, AK St. Georg Hamburg, Germany*
[c]*Mallinckrodt Institute of Radiology, Washington University School of Medicine, St. Louis, USA.*

Abstract. The mobility of lung tumours during the breathing cycle is a source of error in radiotherapy treatment planning. Spatio-temporal CT data sets can be used to measure the movement of lung tumours caused by breathing. Because modern CT scanners can only scan a limited region of the body simultaneously at different times, patients have to be scanned in segments consisting of multiple slices. For studying free breathing motion multislice CT scans can be collected simultaneously with digital spirometry over several breathing cycles. The image data set is assembled by sorting the free breathing multislice CT scans according to the couch position and the tidal volume. But artefacts can occur because there are no data segments for exactly the same tidal volume and all couch positions. In this paper, a non-linear registration method is used to interpolate and reconstruct 4D CT data sets from multislice CT scans in high quality. The non-linear registration estimates a velocity field between successive scans, which is used to reconstruct a 4D CT data set by interpolating data at user-defined tidal volumes. By this technique, artefacts can be reduced significantly. Furthermore, the reconstructed 4D CT data sets are used for studying the motion of lung tumours during the respiratory cycle. The reconstructed 4D data sets of 4 patients were used to quantify the individual lung tumour motion as well as to estimate the tumour's appearance probability during a breathing cycle.

1. Introduction

The mobility of lung tumours caused by breathing is a source of error in radiotherapy treatment planning in the thorax region and the upper abdomen. For conventional conformal radiation therapy accounting for lung motion requires to enlarge the safety margins. As a consequence the volume of irradiated healthy tissue is increased substantially. We generate 4D CT image to analyse the spatial-temporal behaviour of anatomical structures and lung tumour mobility during the respiratory cycle.

There are several approaches to handle the problem of tumour mobility in radiotherapy planning. One approach is to use breath hold devices in order to

[1]Corresponding Author: Heinz Handels, Department of Medical Informatics, University Medical Center Hamburg-Eppendorf, Martinistraße 52, 20246 Hamburg, Germany. Email: h.handels@uke.uni-hamburg.de, WWW: www.uke.uni-hamburg.de/imi

immobilize the patient [1]. Via these techniques a significant reduction of target motion could be shown. This can also be achieved by respiratory gating [2]. Gating techniques do not directly compensate for breathing motion, so the radiation beam is switched off whenever the target is outside a predefined window. Often a combination of breath-hold techniques and respiratory gating is used. Such systems are currently in clinical use, for example deep inspiration breath hold [3] or active breath control [4]. Breath-holding techniques have the potential to reduce the effects of breathing motion [5,6], however, in practice they are limited by the fact that many patients cannot tolerate holding their breath. Furthermore, gating techniques are significantly increasing the expense of time for the patient and physician. Some attempts to explicitly account for free breathing motion are underway but suffer from little existing knowledge regarding the spatial-temporal behaviour of anatomical and pathological structures involved. In [6] the tracking of the tumour motion is done by a combination of external infrared emitters on the patient's surface and implanted gold fiducials. The position of the gold fiducials is computed repeatedly by x-ray imaging. Although first approaches exist to solve the technical problems arising for motion adaptive radiation therapy, an accurate non-invasive tracking method for following the tumour motion is still needed.

In this paper we reconstruct artefact reduced 4D CT image data using a non-linear registration method to analyse the mobility of lung tumours and to identify biometrical parameters that could be applied to radiotherapy planning.

2. Material and Methods

Modern CT scanners cannot scan a big region of interest simultaneously. With one rotation of a multislice CT scanner it is only possible to acquire segments of multiple neighbouring images [1]. A multislice CT scanner is operated in cine mode to acquire repeated scans per couch position over several free breathing cycles. Simultaneously the patient undergoes digital spirometry measurements. 15 scans were acquired at each couch position before the couch was moved to the adjacent position. For each couch position a data segment consisting of 12 slices was generated. This process was repeated until the entire thorax was scanned (16-19 couch positions) resulting in more than 2500 images per patient. To associate the CT scans with tidal volumes, simultaneous digital spirometry measurements were acquired. The patients were instructed to breath normally during the entire scanning sequence [7].

In a first approach a 4D image set was generated by sorting free breathing multislice CT scans according to user defined tidal volume bins. For a given tidal volume a 3D CT data set was reconstructed by examining the spirometry record to determine which CT scan was acquired at a tidal volume closest to the desired volume at each couch position. Hence, by this technique a nearest neighbour interpolation is performed. To prepare a 4D data set the 3D data sets for a scale of tidal volumes were arranged in series (fig. 1).

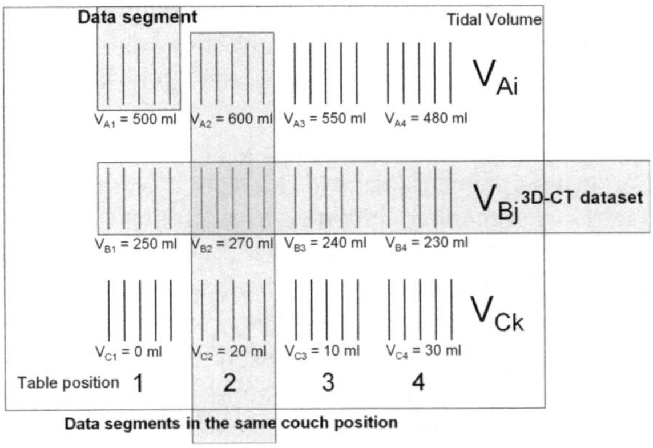

Fig. 1: CT data set measured at 4 table or couch positions each with 3 data segments containing 5 slices.

Free breathing causes the problem that there are no acquired data for exactly the user defined tidal volume. This induces artefacts similar to motion artefacts in 3D CT [8]. This is especially noticeable when viewing sagittal reconstructions of the boundary between diaphragm and lung: the reconstructed diaphragm boundary shows an apparent and striking break-up (fig. 2, left).

An optical flow based reconstruction method was developed to avoid such artefacts caused by missing CT scans for the desired volume. In a first step, for each couch position the optical flow is determined between the scans measured with tidal volume closest to the selected one. Here, a non-linear registration algorithm computes a velocity field describing the motion of corresponding pixels.

The initial hypothesis of the used optical flow based registration method is that the pixel intensities of time varying image regions remain constant [9]. Using this approach, the velocity field $\boldsymbol{v}$ can be described by

$$\boldsymbol{v} = -grad(I)\frac{\partial_V I}{\|grad(I)\|^2}, \quad (1)$$

where $grad(I)$ is the spatial image gradient. However, equation 1 is ill-posed and only the motion component in direction of the local brightness gradient of the image intensity function may be estimated [9]. As a consequence, the flow velocity cannot be computed locally without introducing additional constraints. In our implementation the necessary regularization is done by Gaussian smoothing of the velocity field. Hence, the velocity field is computed by an iterative algorithm similar to the demons-based registration [10].

Furthermore, problems occur near segment borders. This is caused by the existence of voxels that do not have corresponding voxels in the considered couch position, because some structures may change from one data segment into another segment during the respiratory cycle. To overcome this problem CT scans at adjacent couch positions are also taken into account for the registration process.

Based on the estimated velocity field $\boldsymbol{v}$ corresponding voxels are defined in image segments measured at two successive tidal volumes nearby the user-defined tidal volume of the interpolated data segment. The interpolated image values are computed as weighted average of the grey values of corresponding voxels. This technique is applied to overcome the problem that the intensity conversation assumption might not be fulfilled and structures contained only in one of the interpolated images.

The resulting 4D CT data set is used to analyse the respiratory motion. In a first step, the lung, the skin and the bronchial tree are segmented for any reconstructed tidal volume using region growing techniques and interactive correction. Surface models of the segmented organs are generated in order to enable the visualisation of the 4D breathing motion. Anatomical landmarks, e.g. the branches of the bronchial tree, are determined interactively and the trajectories of the selected points are analysed and visualised.

Furthermore, the segmented 4D data sets are used to calculate the optical flow of the organ surfaces and landmarks. Therefore, a non-linear registration similar to the algorithm described in [11] is performed. The resulting velocity fields are used to approximate the trajectories of points on the organ surface. So, the maximum displacement of any surface point can be calculated and regions with large respiratory motion are identified.

3. Results

Four tumour patients were examined with a 12 slice spiral CT scanner. 16-19 segments, each consisting of those 12 slices, were scanned at 15 different times of the breathing cycle. The position within the breathing cycle was measured using a digital spirometer. (3D +t) CT data sets were reconstructed by the nearest neighbour and the optical flow based interpolation.

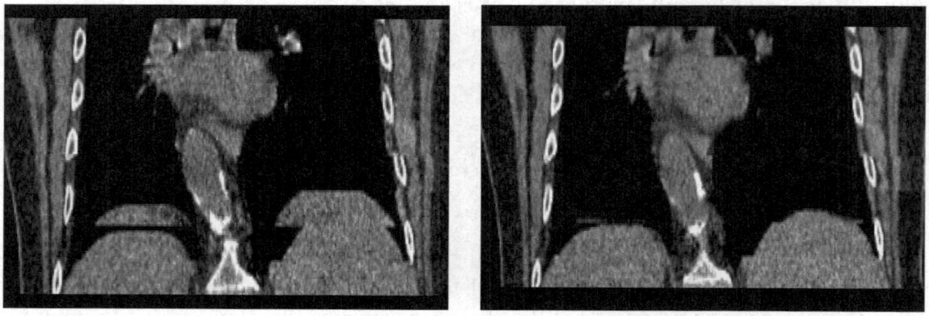

Fig. 2: Artefacts at the diaphragm with the standard nearest neighbour method for the reconstruction of 4D CT data (left) and improved reconstruction after optical flow based interpolation (right)

The first reconstruction method shows "steps" at the edges of neighbouring segments that were not scanned exactly at the same period of the breathing cycle (fig. 2, left). With the optical flow based interpolation method described above the artefacts are reduced significantly on visual inspection (fig. 2, right). The comparison between slice changes inside a data segment and slice changes at segment borders enables a quantitative evaluation. For the four patient data sets the artefacts could be reduced by 31.8 %, 29.9 %, 30.7% and 41.6% (mean over all data segments).

The reconstructed 4D data sets are used to quantify organ displacements and to visualise the thoracic organ motion. Therefore, 3D images are computed at volumes corresponding to equidistant times, and the lung tumour and inner organs like the lungs and the liver are segmented at each time step. Based on the determined velocity field, distances between organ surfaces at different respiratory states are computed and trajectories of landmarks and surface points are analysed. Especially, the mass centre of the lung tumour is computed and traced at different time steps to analyze the tumour's mobility. The 3D trajectory of the tumour's mass centre is projected in craniocaudal, (CC), anteroposterior (AP) and lateral (LA) direction to compare the tumour motions of different patients (fig. 3).

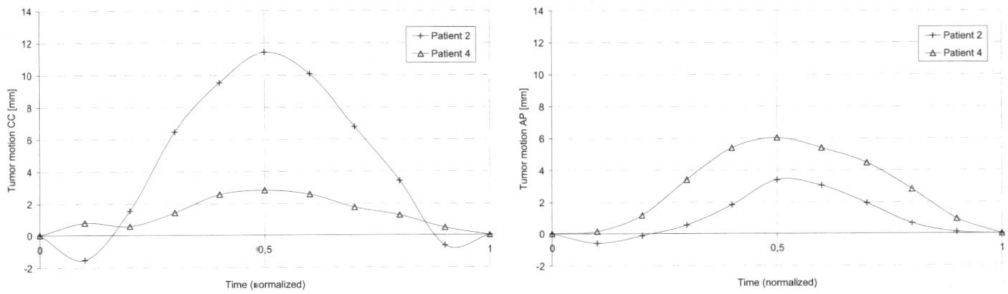

Fig. 3: Motion of the mass centre of the two tumours shown in figure 4 in craniocaudal (CC) and anteroposterior (AP) direction. The tumour of patient 2 shows a much stronger motion in CC direction than the tumour of patient 4 (left), while its motion in AP direction is smaller than the tumour of patient 4 (right).

Furthermore, probabilities of appearance are computed in 3D for the lung tumour as well as for inner organs. The estimated appearance probabilities visualise the movement of the tumour during the respiratory cycle in one static image (fig. 4). Furthermore, the knowledge about 3D appearance probabilities of the tumour can be used to define specific individual and motion oriented safety margins in radiation therapy.

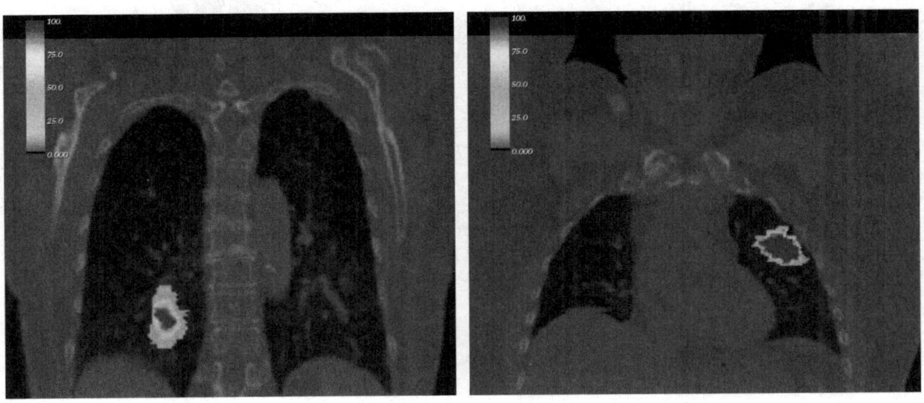

Fig. 4: Colour-coded estimated appearance probabilities of lung tumours of patient 2 (left) and patient 4 (right) displayed in a 2D slice.

4. Conclusion

The optical flow based registration method presented significantly improves the image quality of reconstructed 4D CT data sets at user-defined tidal volumes. The reconstructed 4D image data sets are used to model and analyse the influence of respiratory motion on the motion of lung tumour and inner organs. The results presented are based on the analysis of four patient data sets. The appearance probability distribution changes from patient to patient depending on the localisation of the tumours in the lung and the individual differences in their breathing motion. In further research the number of patient data sets with lung tumours will be increased and correlations between the respiratory state and the motion of skin markers will be investigated in order to predict the abdominal organ motion from external, non-invasive tracking data. Furthermore, it will be of interest to analyse the breathing behaviour of different patients in more detail in order to identify typical breathing patterns.

References

[1] Lax I, Blomgren H, Naslund I, Svanstrom R. Stereotactic radiotherapy of malignancies in the abdomen. Methodological aspects. Acta Oncol 1994; 33:677–83.

[2] Kubo H, Hill B. Respiration gated radiotherapy treatment: a technical study. Phys Med Biol 1996;41.

[3] Rosenzweig K, Hanley J, Mah D, et al. The deep inspiration breath-hold technique in the treatment of inoperable non-small-cell lung cancer. Int J Radiat Oncol Biol Phys 2000; 48: 81–87.

[4] Wong J, Sharpe M, Jaffray D, et al. The use of active breathing control (abc) to reduce margin for breathing motion. Int J Radiat Oncol Biol Phys 1999; 44:911–19.

[5] Schweikard A, Glosser G, Bodduluri M, et al. Robotic motion compensation for respiratory movement during radiosurgery. Comput Aided Surg 2000; 5:263–77.

[6] Shirato H, Seppenwoolde Y, Kitamura K, et al. Intrafractional tumour motion: lung and liver. Semin Radiat Oncol 2004; 14:10–18.

[7] Low D-A, Nystrom M, Kalinin E, Parikh P, Dempsey J-F, Bradley J-D, Mutic S, Wahab S-H, Islam T, Christensen G, Politte D-G, Whiting B-R. A method for the reconstruction of four-dimensional synchronized CT scans acquired during free breathing. Med Phys 2003; 6:1254–63.

[8] Frenzel T, Werner R, Ehrhardt J, Säring D, Lu W, Low D, Handels H, Reconstruction of 4D-CT datasets from 3D data segments acquired during free breathing. Proceedings of the 14th International Conference on Medical Physics, 909 – 910, 2005.

[9] Horn BKP, Schunck BG. Determining optical flow. Artificial Intelligence Aug. 1981; 17:185–203.

[10] Thirion J-P. Image matching as a diffusion process: an analogy with Maxwell's demons. Medical Image Analysis 1998; 2(3): 243–60.

[11] Ehrhardt J, Säring D, Handels H., Interpolation of temporal image sequences by optical flow based registration. In: Reinhardt, J.M., Pluim, P.W. (eds.), Image Processing, SPIE Medical Imaging 2006, San Diego, Vol. 6144, 61442K1-61442K8, 2006.

Ubiquity: Technologies for Better Health in Aging Societies
A. Hasman et al. (Eds.)
IOS Press, 2006

Kernel Methods for Melanoma Recognition

Elisabetta LA TORRE[a,1], Tatiana TOMMASI[a], Barbara CAPUTO[b], Giovanni E. GIGANTE[a]
[a] *University of Rome La Sapienza, Physics Department, Italy*
[b] *NADA/CVAP, KTH, Stockholm, Sweden*

Abstract. Skin cancer is a spreading disease in the western world. Early detection and treatment are crucial for improving the patient survival rate. In this paper we present two algorithms for computer assisted diagnosis of melanomas. The first is the support vector machines algorithm, a state-of-the-art large margin classifier, which has shown remarkable performances on object recognition and categorization problems. The second method, spin glass-Markov random fields, combines results of statistical physics of spin glasses with Markov random fields. We compared the two approaches using color histograms as features. We benchmarked our methods with another algorithm presented in the literature, which uses a sophisticated segmentation technique and a set of features especially designed for melanoma recognition. To our knowledge, this algorithm represents the state of the art on skin lesions classification. We show with extensive experiments that the support vector machines approach outperforms the existing method and, on two classes out of three, it achieves performances comparable to those obtained by expert clinicians.
Keywords: Melanoma Recognition, Computer Assisted Diagnosis, Support Vector Machines, Kernel Methods.

1. Introduction

Malignant melanoma is a significant public health problem. Its incidence is rising faster than that of any other cancer in the US and in Europe [1, 2]. Early detection and treatment are critical and result in improved patient survival rates. The most used diagnostic technique is called Epiluminescence Microscopy (ELM). It is a non-invasive technique that allows for a detailed surface analysis of a suspicious skin lesion by using hand-held device emitting incident light from a light source penetrating the epidermal skin layer. The diagnosis of early melanoma is based on simple observation of images by dermatologists, who commonly use the ABCD (Asymmetry, Border, Color and Dimension) method as clinical guide. So, the diagnosis depends heavily on the physician's level of expertise.

An automatic system for melanoma recognition would constitute a valuable support for physicians in every day clinical practice and should reproduce the perceptual and cognitive strategy followed by doctors. The last years have witnessed numerous research on this topic; a key factor for the development and evaluation of these systems is the availability of a statistically significant database. To our knowledge the state of the art in melanoma recognition was presented by H. Ganster et

[1] Corresponding Author: Elisabetta La Torre, Università La Sapienza, p.le A. Moro 5, 00185, Rome, Italy (elisabetta.latorre@uniroma1.it).

al. [3]. That paper presented a large database, accompanied by: (a) a segmentation algorithm for isolating the potential melanoma from the surrounding skin [3]; (b) a set of features containing shape and radiometric features as well as local and global parameters, calculated to describe the malignancy of a lesion, from which significant features are selected by application of statistical feature subset selection methods [3]; (c) a nearest neighbor classification algorithm [3]. In that work the authors concentrated particularly on the segmentation techinque and the features selection process. Here we focus instead on the classification algorithm, and we propose the use of kernel methods for recognition of skin lesion images. Specifically, we focus our attention on two algorithms: Support Vector Machines (SVM) [4] and Spin Glass-Markov Random Fields (SG-MRF) [5]. SVM is a state-of-the-art large margin classifier, where the optimal separating surface is defined by a linear combination of scalar products between the view to be classified and some support vectors [4, 5]. SG-MRF is a fully connected MRF which integrates results of statistical mechanics with Gibbs probability distributions via non linear kernel mapping [6]. The database on which we will run our experiments is the same introduced by Ganster et al. [3]; our classification algorithms use binary masks determined by the segmentation algorithm developed in [3], and color histogram features. The choice of color histograms as feature types reproduces one of the criteria followed by dermatologists for diagnosis. We performed several series of experiments for selecting an optimal feature descriptor and we replicated the experimental setup used in [3] for a benchmark evaluation. Our results show that SVM obtains remarkably better performances than SG-MRF and Ganster's classification method. Furthermore, on two classes out of three, SVM achieves recognition results comparable to those obtained by skilled clinicians.

In summary, the contributions of this paper are: (1) The introduction of kernel methods for melanoma recognition, via two approaches: a probabilistic method, and a well known state-of-the-art classifier. For this second algorithm particularly, we studied in depth the classification performances with different kernel types. (2) The benchmark with a method presented in the literature [3], which to the best of our knowledge represents the state of the art in this field: on the same database and using the same segmentation masks, we had an improvement of more than 20% of the experimental results. Moreover our results are very stable and reliable because are obtained as mean value on five different partitions.

The rest of the paper is organized as follows: Section 2 describes some basic knowledge on SG-MRF theory and section 3 briefly explains the SVM algorithm. Section 4 reports on the experiments performed. The paper concludes with a summary discussion and some possible directions for future research.

2. Spin Glass - Markov Random Fields

In this section we shortly describe the probabilistic method which constitutes one of the kernel methods proposed here for classification. This technique was introduced first for 3D object recognition [6], and was then applied to microcalcification detection with promising results [7]. The interested reader will find a comprehensive discussion in [6].

Consider n visual classes Ω_j, $j = \{1, \dots n\}$, and a set of k observations $\{x_j^1 \dots x_j^k\}$, $x \in \Re^m$, random samples from the underlying, unknown, probability distribution $P(x)$ defined on $\Re^m$. Given an observation $\hat{x}$, our goal is to classify $\hat{x}$ as a sample from Ω_{j^*}

one of the Ω_j visual classes. Using a Maximum A Posteriori (MAP) criterion and Bayes rule we have j = argmax $_j$ $P(\Omega_j / \hat{x})$ = argmax $_j$ $P(\hat{x} / \Omega_j)$, where $P(\hat{x} / \Omega_j)$ are the Likelihood Functions (LFs); the prior probabilities $P(\Omega_j)$ of the classes are assumed to be constant.

Spin Glass-Markov Random Fields (SG-MRFs) [6] are a new class of MRFs which connect SG-like energy functions (mainly the Hopfield one [8]) with Gibbs distributions via a non linear kernel mapping. The resulting model overcomes many difficulties related to the design of fully connected MRFs, and enables to use kernels in a probabilistic framework. The SG-MRF probability distribution is given by

$P_{SG\text{-}MRF}(x / \Omega_j) = (1/Z) \exp[-E_{SG\text{-}MRF}(x / \Omega_j)], \quad Z = \Sigma_x \exp[-E_{SG\text{-}MRF}(x / \Omega_j)]$

with $E_{SG\text{-}MRF} = -\sum_{\mu=1}^{p_j} [K(x, \tilde{x}^{(\mu)})]^2$, where the function $K(x, \tilde{x}^{(\mu)})$ is a generalized Gaussian kernel [5] $K(x,y) = \exp\{-\rho\, d_{a,b}(x,y)\}$, $d_{a,b}(x,y) = \Sigma_i \,|x_i^a - y_i^a|^b$ and $\{\tilde{x}^{(\mu)}\}_{\mu=1\ldots p_j}, j \in [1, n]$ are a set of vectors selected (according to a chosen ansatz [6]) from the training data that we call *prototypes*. The number of prototypes per class must be finite, and they must satisfy the condition $K(\tilde{x}^i, \tilde{x}^k) = 0$, for all i, k = 1,... p_j, $i \neq j$ and j = 0,... n (the interested reader can find a detailed discussion regarding SG-MRF in [6]). Thus using SG-MRF modelling, the Bayes classifier (1) will become $j^* = \text{argmin}_j\, E_{SG\text{-}MRF}(\hat{x} / \Omega_j)$.

3. Support Vector Machines

In this section we briefly describe SVM in the two class case. For further details and the extension to multiclass settings we refer the reader to [4].

Consider the feature vector $\boldsymbol{x} \in \Re^N$ and its class label $y \in \{-1, +1\}$. Let $(\boldsymbol{x}_1, y_1)$, $(\boldsymbol{x}_2, y_2)$, ... , $(\boldsymbol{x}_m, y_m)$ denote a given set of m training examples. If we assume that the two classes are linearly separable, there exists a linear function $f(x) = \boldsymbol{w} \cdot \boldsymbol{x} + b$ such that for each training example $\boldsymbol{x}_i$, it yields $f(\boldsymbol{x}_i) \geq 0$ for $y_i = +1$ and $f(\boldsymbol{x}_i) \leq 0$ for $y_i = -1$. The optimal separating hyperplane is the one which has maximum distance to the closest points in the training set. Mathematically this hyperplane can be found by solving a constrained minimization problem using Lagrange multipliers α_i (i = 1, ... , m). It results in a classification function $f(x) = \text{sgn}\,(\sum_{i=1}^{i=m} \alpha_i y_i\, \boldsymbol{w} \cdot \boldsymbol{x} + b)$, where α_i and b are found by using an SVC learning algorithm [4]. It turns out that a small number of the α_is are different from zero; their corresponding data $\boldsymbol{x}_i$ are called support vectors.

SVM can be extended to nonlinear problems by using a nonlinear operator $\Phi(\cdot)$ to map the input feature vectors $\boldsymbol{x}_i$ from the original $\Re^N$ into a higher dimensional feature space $\mathcal{H}$ by $\boldsymbol{x} \rightarrow \Phi(\boldsymbol{x}) \in \mathcal{H}$. Here the mapped data points of the two classes become linearly separable. Assuming there exists a kernel function K associated with the inner product of the desired nonlinear mapping such that $K(\boldsymbol{x},\boldsymbol{y}) = \Phi(\boldsymbol{x}) \cdot \Phi(\boldsymbol{y})$, then a non linear SVM can be obtained by replacing $\boldsymbol{x} \cdot \boldsymbol{y}$ by the kernel $K(\boldsymbol{x},\boldsymbol{y})$ in the decision function, obtaining then $f(x) = \text{sgn}\,(\sum_{i=1}^{i=m} \alpha_i y_i\, K(\boldsymbol{x}_i, \boldsymbol{x}) + b)$. This corresponds to constructing an optimal separating hyperplane in the feature space.

In this paper we consider four kernel types :

1. Polynomial kernel ("poly") $K(\boldsymbol{x}, \boldsymbol{y}) = (\gamma * \boldsymbol{x} \cdot \boldsymbol{y})^d$
2. Generalized Gaussian kernel ("gengauss") $K(\boldsymbol{x}, \boldsymbol{y}) = \exp\{-\gamma * |\boldsymbol{x}^a - \boldsymbol{y}^a|^b\}$

3. Gaussian kernel ("gauss") $K(x, y) = \exp\{-\gamma * | x - y |^2\}$
4. Chi-squared kernel ("chi") $K(x, y) = \exp\{-\gamma * \chi^2(x, y)\}$.

4. Experiments

In this section we present experiments that show the effectiveness of kernel methods for melanoma recognition. To this purpose, in a preliminary step, we ran a first series of experiments for feature selection. Then we used the selected features for an extensive set of classification experiments. In the rest of the section we describe the database used, the experimental setup and our experimental findings.

Database: We performed our experiments on the database created by the Department of Dermatology of the Vienna General Hospital [3]. The whole database consists of 5380 skin lesion images, divided into three classes (these numbers are not perfectly coincident with those reported in [3], where the database is said to be of 5363 images, but this difference should not affect the comparison between the two algorithms): 4277 of these lesions are classified as clearly benign lesions (Class 1), 1002 are classified as dysplastic lesions (Class 2) and 101 lesions are classified as malignant melanomas (Class 3). The lesions of the classes 2 and 3 were all surgically excised and the ground truth was generated by means of histological diagnosis [3]. In order to have statistically significant results, we ran experiments with five different partitions, then we calculated the mean and the standard deviation of the obtained recognition rates. This procedure has been adopted for all the experiments which are reported in this paper.

Experimental Setup: The three key components for an automated melanoma recognition algorithm are: segmentation, features extraction and classification. We describe below the general approach followed in this paper for each of these steps:
Segmentation: We used the segmentation method developed by Ganster et al. [3] on this database. It consists of a binary mask determined by several segmentation algorithms combined together with a fusion strategy. This choice allows for a fair comparison between Ganster's technique and ours.
Feature Extraction: In the ABCD-rule of dermoscopy, the color distribution in the skin lesion is one of the discriminant features for clinical melanoma recognition, thus we used color histograms as features. The color histogram was computed by discretizing the colors within the image and counting the number of pixels for each color. We performed several experiments for selecting the best features, namely using hue, rg, RG, RB and GB color histograms. The resolution of the bin axes was varied for each representation, consisting of 8, 16, 32, 64 (for bidimensional histograms we chose the resolution of each axis with the same bin value). We found that the GB representation obtained the best results for all the bin values, thus we used it in all the following experiments.
Classification: We used SG-MRF and SVM algorithms (see section 2 and 3 respectively). For SG-MRF we learned the kernel parameters during the training stage using a leave-one-out strategy [6]. For SVM we used the four kernel types described in section 3. The kernel parameters were chosen via cross validation.

Classification Experiments: All the experiments were performed respecting the procedure reported by Ganster et al. [3]. The training set consisted of 270 images (90 for each class); the test set consisted of the whole database [3]. Note that training and test set are not disjoint; once again we underline that this follows the procedure proposed in [3] which allows for benchmarking. We used the GB features and we ran experiments for 8, 16, 32 and 64 resolution of bins per axes, with five different partitions for training and test set, using SG-MRF and SVM with four different kernel types. Table 1 reports, for SG-MRF and SVM, the recognition rates for each class averaged on five partitions. We also report the average of the recognition rate obtained class by class ("Mean Class"), and the overall recognition rate ("Overall"). For sake of clarity we also report the results obtained in [3]; note that these results were obtained on a single run.

Table 1. Recognition results for the classification experiments on three classes of lesions obtained from Ganster et al. [3] and with SG-MRF and SVM methods with different kernels. We report the recognition rates for the three classes, the overall and the mean recognition rates. Results obtained with SG-MRF and SVM are mean values from five different runs with their standard deviations. Class 1, Class 2, Class 3 identify the benign, dysplastic and malignant lesions respectively.

	Ganster et al. [3] (%)	SG-MRF (%)	SVM (%)			
			poly	**gauss**	**gengauss**	**chi**
Class 1	59	48.6 ± 4.2	80.1 ± 13.0	71.9 ± 11.1	96.2 ± 4.0	68.6 ± 17.7
Class 2	53	38.8 ± 3.4	15.7 ±13.7	24.8 ± 12.7	11.0 ± 1.8	22.4 ± 7.5
Class 3	73	94.1 ± 3.4	29.5 ± 20.4	45.0 ± 28.5	89.5 ± 0.9	62.6 ± 19.7
Mean Class	61	60.5 ± 17.0	41.8 ± 19.6	47.2 ± 13.6	65.6 ± 27.4	51.2 ± 14.5
Overall	58	47.7 ± 2.9	67.1 ± 7.8	62.6 ± 6.2	**80.2 ± 2.8**	59.9 ± 12.9

A first comment is that SVM, with the generalized Gaussian kernel, obtains the best result with respect to Ganster's method and SG-MRF. The overall recognition rate is of 80.2% to be compared with a 58% obtained by Ganster and 47.7% obtained by SG-MRF. This proves the effectiveness of this technique for melanoma recognition. A second comment is that SVM performance varies considerably as the overall recognition rate goes from a minimum of 59.9% for the chi-squared kernel to a maximum of 80.2% for generalized Gaussian kernel. It is also interesting to note that, for the overall recognition rate, the kernels which obtains the worst performances tend to have the highest standard deviations, while the kernel with the best performance has the smallest one. This illustrates the importance of doing kernel selection in the training phase; the low standard deviation of the SVM's best result also shows the stability of our findings. A final remark should be made on the poor performance of SG-MRF. This might be due to the dimension of the training set for each class; it might be possible that the probabilistic method needs a higher statistic in order to estimate properly the energy function.

Table 2 reports the confusion matrix for SVM with generalized Gaussian kernel and the confusion matrices obtained by Ganster and the one obtained by dermatologists, both reported in [3]. We see that for class 1 and class 3 SVM outperforms Ganster's method and is comparable with the dermatologists' performances. It is very interesting to note that, in contrast, SVM performs poorly on class 2, which corresponds to dysplastic lesions. This might be explained considering that here we are using only color information, while Ganster used a selection of different features and dermatologists used the ABCD rule. It is thus possible that color information only is not discriminant enough in order to recognize correctly dysplastic lesions, while it seems to be effective for separating benign and malignant lesions. In the future we will explore this issue by testing different types of informations.

Table 2. Confusion matrices for different classification methods: (a) Confusion matrix for the SVM results with the "gengauss" kernel. The number of images reported are mean value of the number obtained from 5 different partitions; (b) Confusion matrix obtained with the Ganster's method [3]; (c) Confusion matrix obtained from clinical diagnosis, performed from expert dermatologists of the Department of Dermatology at the Vienna General Hospital [3]. In the tables 1, 2, 3 identify the three classes corresponding to benign, dysplastic and malignant lesions respectively.

(a)	**Assigned**		
True	**1**	**2**	**3**
1	4112.6	112.6	50.8
2	874.8	110.0	17.2
3	10.4	0.2	90.4

(b)	**Assigned**		
True	**1**	**2**	**3**
1	2500	1347	410
2	324	531	155
3	14	12	70

(c)	**Assigned**		
True	**1**	**2**	**3**
1	4161	94	9
2	42	960	8
3	6	19	78

5. Conclusions

In this paper we proposed the use of kernel methods for melanoma recognition, with two approaches: SG-MRF and SVM. For this second algorithm particularly, we studied in depth the classification performances with different kernel types. The experiments showed that SVM, with the generalized Gaussian kernel, obtains an improvement of more than 20% with respect to the results presented in [3], which to the best of our knowledge represents the state of the art of the field. Moreover, on two classes out of three, SVM achieves recognition results comparable to those obtained by skilled clinicians. In the future we will conduct similar experiments with different descriptors, such as gray-level textural features and shape descriptors, in order to test the effectiveness of different types of information and to eventually reproduce the ABCD method followed by the dermatologists in every day clinical practice.

References

[1] D. S. Rigel and J. A. Carucci, Malignant Melanoma: Prevention, Early Detection, and Treatment in the 21st Century, CA Cancer J Clin 2000; 50:215-236.
[2] E. De Vries, F. I. Bray, J. W. W. Coebergh and D. M. Parkin, Changing Epidemiology of Malignant Cutaneous Melanoma in Europe 1953-1997: Rising Trends in Incidence and Mortality but Recent Stabilizations in Western Europe and Decreases in Scandinavia, Int. J. Cancer 107: 119-126; 2003 Wiley-Liss, Inc..
[3] H. Ganster, A. Pinz, R. Rhrer, E. Wildling, M. Binder and H. Kittler, Automated Melanoma Recognition. IEEE Transactions on Medical Imaging, vol. 20, n. 3, march 2001.
[4] V. Vapnik. Statistical learning theory. Wiley and Son, 1998.
[5] B. Scholkopf, A. J. Smola, Learning with kernels, 2001, the MIT Press.
[6] B. Caputo,A new kernel method for object recognition: spin glass Markov random fields, PhD thesis, Stockholm, November 2004. Available at http://www.nada.kth.se/~caputo
[7] B. Caputo, E. La Torre, S. Bouattour, G.E. Gigante, A New Kernel Method for Microcalcification Detection: Spin Glass- Markov Random Fields, Proc. of MIE02, Budapest, August 2002.
[8] D. J. Amit, Modeling Brain Function, Cambridge University Press, Cambridge, USA, 1989.

Ubiquity: Technologies for Better Health in Aging Societies
A. Hasman et al. (Eds.)
IOS Press, 2006

Pre- and Intraoparative Processing and Integration of Various Anatomical and Functional Data in Neurosurgery

Roland METZNER[a,1], Urs EISENMANN[a], Christian R. WIRTZ[b] and Hartmut DICKHAUS[a]
[a] *Department of Medical Informatics, University of Heidelberg, Germany*
[b] *Neurosurgical Department, University of Heidelberg, Germany*

Abstract. A software system is presented, capable of integrating various information sources for neurosurgical procedures. These include anatomical data such as a standard 3D DICOM image stacks, atlas data, as well as functional information (e.g. fmri, MEG, EEG).
The system is programmed in C++ using Open GL for visualisation, and was developed in a close cooperation with a neurosurgical department to match existing needs.
Preoperatively the data may be combined, registered by a rigid or elastic matching process, enriched by user specified planning information such as annotations or trajectories, and visualised in a standard fashion using different segmentation schemes, interactive rotation, zooming etc. Selected portions of the gathered and generated information may then be exported for neuronavigation input in DICOM or a vendor specific format.
Intraoperatively, this information may, on the one hand, be simply used as an integrational part of the routinely used navigational data. On the other hand, the system is also capable of interacting with the navigational system to integrate the actual spatial information of the ongoing procedure into the preoperative data, thus allowing further planning and visualisation beyond the scope of the navigational unit. Furthermore, intraoperatively updated information such as intraoperative MR images or electrophysiological data may be integrated and correlated to the existing information.
Ongoing developments comprise redistribution of the relevant data for injection onto the screen of the navigational system or into the optical pathway of the (3D capable) microscope display/ocular. This also includes information about the actual automatically optimized accuracy of the navigation process in relation to the head markers in use.
Keywords: Medical Informatics, Image Processing, Neuronavigation, Multimodal Oparation Planning, Neurosurgery.

1. Introduction

Neurosurgical procedures in the brain require thorough planning and a discreet course of action. Therefore, comprehensive information about the patients anatomical and - if available - functional conditions in the area of interest of the brain are indispensable

[1] Corresponding Author: Dr. Roland Metzner, Department of Medical Informatics, University of Heidelberg, Im Neuenheimer Feld 400, D-69120 Heidelberg, Germany, roland_metzner@med.uni-heidelberg.de

(see e.g. [1] and [2]). Effective processing and a favourable and consistent presentation of data throughout the different phases of planning and performing the intervention can considerably support the neurosurgeon. This includes intraoperative information for updating the preoperative data to yield valuable support for the decision process. This may comprise morphological data such as intraoperative MRI [3, 4] or electro-physiological stimulation to identify eloquent brain tissue in tumour resection or chronic pain treatment [e.g. 5].
We therefore present a software system called Multimodal Operation Planning System (MOPS 3D, [6]) capable not only of standard processing and viewing features but also of integrating into the intraoperational workflow of neuronavigation and intraoperative imaging or electrophysiological testing.

The main application domain of the system are procedures such as tumour or epilepsy surgery [e.g. 7] where the lesions are located in the vicinity of eloquent brain structures such as speech or motor areas.

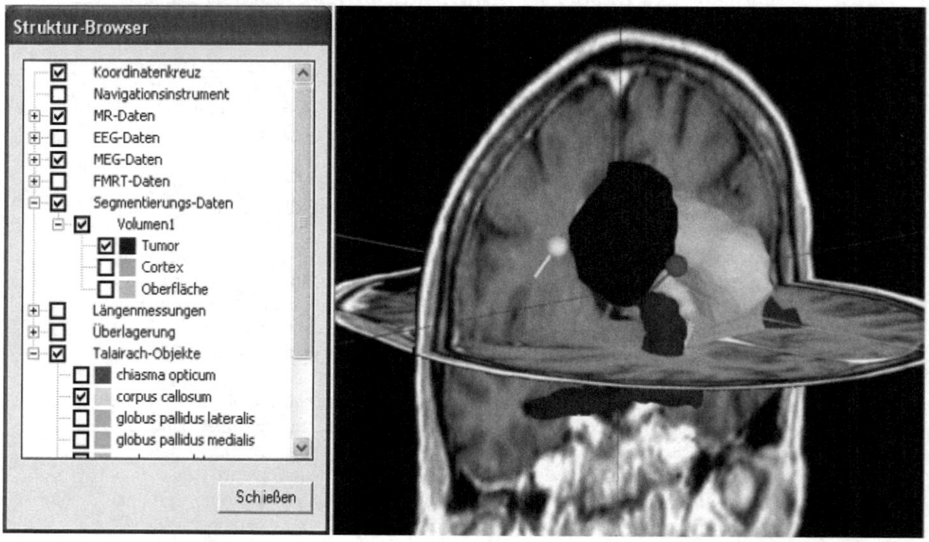

Figure 1: viewing of different modalities: morphological and functional MRI, atlas objects, MEG-dipoles, tumour; on the left hand side checkboxes for selecting/deselecting modalities and assigning colours.

2. Material and Methods

MOPS 3D was programmed in C++ with Open GL as a visualisation interface to achieve good performance on standard PCs or notebooks with 512 MByte of working memory and a middle to upper class graphics adapter.

The whole development process took place in close cooperation with the Neurosurgical Department of the University Hospital of Heidelberg. In this way, the needs of the neurosurgeons could be taken into account in all of the phases (such as planning, implementing and testing of the system). This contributed to broad acceptance among the clinicians.

The plug-in type architecture of the system allows easy integration and testing of new algorithms and visualisations because new modalities may simply be superimposed onto the existing data.

Currently, input to the system may be provided in the form of DICOM or ANALYZE image datasets, atlas data from a digitized Talairach-Tournoux atlas [8], functional information maps from the BrainVoyager (functional Magnetic Resonance Tomography: fMRI) or BESA (Magnetoencephalography: MEG, Electroencephalography: EEG) systems or the Stryker-Leibinger navigation system (image and contour data from segmented objects).

This information may be viewed in an integrated fashion with the possibility of assigning different colour codes, choosing appropriate thresholds for the functional information, and selecting or deselecting part of the data (see Figure 1).

Additional processing features include rigid and elastic matching of different imaging modalities, segmenting algorithms allowing an automatic (e.g. head surface) or interactive (e.g. cortex) modus operandi involving region growing, histogram based methods or watershed approaches. The user may also choose between different visualisation schemes such as triangulation or voxel based representation.

Furthermore the surgeon may add annotations such as measured distances between anatomical atlas objects and a tumour or other critical structures, or he might draw trajectories indicating the planned advancement throughout the operation.

The integrated information may then be exported (in part or as a whole) to a suitable format, namely DICOM, ANALYZE or a vendor specific format (Stryker-Leibinger so far). Furthermore, VTK objects (Visualization Toolkit) are supported to provide input for an augmented reality visualisation in the display of the operation microscope.

Due to the possibility of acquiring online information about the navigational coordinates of the used tools intraoperatively, the system may also serve as a monitoring device to visualise and document differences between actual and preoperative data. This also includes matching of intraoperative MRI imaging data and the digitization of electrode grids used for intraoperative lesion delimitation and electrophysiological testing and monitoring. The fusion functionality is integrated into a flexible and expandable framework (within MOPS 3D), which makes use of the broad variety of algorithms of the Insight Segmentation and Registration Toolkit (ITK).

Figure 2: left: combination of tumour, fMRI and atlas objects (colour coded in the original images) after elastic matching; right: 8x4 electrode grid used for intraoperative electrocorticography & depiction on a brain phantom after digitization.

3. Results

The system offers a broad variety of processing and viewing possibilities to the planning neurosurgeon. Much of this is accomplished in a fashion similar to the interface modern commercial neuronavigation systems provide, and surgeons are used to elements such as 3D and split views, zooming, panning, segmenting, and matching of different anatomical data. Interfaces to functional and atlas data and to the internal data format of the mostly used navigation system in the neurosurgical department of Heidelberg (Stryker-Leibinger) enable MOPS 3D to serve as a powerful tool for viewing and integrating multimodal data. Options to selectively view these modalities, and to assign desired colors, help with keeping track of the sometimes vast information. The option of saving all of the generated data and reviewing it via an integrated database access (ODBC) adds to the clinical utility.

In a rather simple context, the system may be used as a mere planning tool by importing the available data (such as morphological, atlas and functional data (MRI, CT, fMRI, MEG, EEG, Talairach-Tournoux)), performing rigid and elastic matching of all of these modalities, and exporting the preferred data after adding planning information such as trajectories or annotations. As output formats comprise standard options like DICOM and ANALYZE; these may simply be imported to any navigation system. The support of the internal format of the system commonly used for complex procedures (Stryker-Leibinger), on the other hand, allows better separation and visualisation of the different categories of objects on the target system.

Furthermore, MOPS 3D may also offer its enhanced viewing and processing capabilities throughout the whole operation as information about the position of the actual navigation instrument (pointer or microscope) can be transferred to provide a parallel real-time view. The registration process may be carried out simultaneously with that of the navigation system, thus minimizing the additional effort.

Moreover, digitization of electrode grids allows for a visual comparison of preoperatively acquired functional information - such as motor or speech areas adjacent to a tumour - with intraoperative electrophysiological data. Tracking of these electrodes significantly eases the workflow as the electrodes may be removed for further intervention on the brain, but still remain visible within the system in combination with actual navigation information.

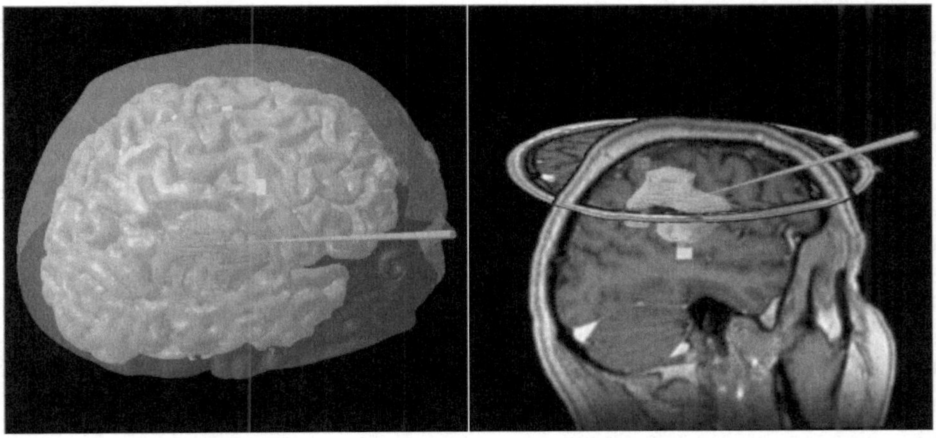

Figure 3: intraoperative views in conjunction with the navigational instrument pointing at the tumour; fMRI activation superimposed; transparent skin with navigational markers on the left hand side.

As the neurosurgical department in Heidelberg is equipped with an MRI scanner inside the operating room (Siemens Magnetom Open), intraoperative imaging data may be used to match all of the previously collected information onto the actual anatomical situation to compensate for brain shift effects. Whereas an optimised combination of ITK-based algorithms for the actual image data led to very good rigid matching results within about 3 minutes, further evaluation is being carried out at the moment to minimise the calculation time for an elastic matching process.

On the whole the features of preoperatively combining multimodal data to export the desired synopsis to the navigation system is being used quite regularly within complex tumour resection surgery (about 30 patients so far, mostly with fMRI as funtional data, some with MEG). Due to the restrictions of the elastic matching and the tracking of electrode grids being a new feature, these intraoperative features are being evaluated at the moment. Nevertheless the intraoperative handling of the system (e.g. viewing capabilities, registration and navigation process parallel to the commercial navigation system) has been tested and approved by the neurosurgeons.

4. Discussion

The flexible terms of use and the close cooperation with the neurosurgical department throughout the development process has yielded high acceptance of the presented system MOPS 3D (Multimodal Operation Planning System) on the part of the surgeons.

When being used as a planning tool preoperatively, the handling corresponds to that of modern neuronavigation systems but includes additional functionality such as the ability to elastically match atlas objects onto the patient's brain, or adjust thresholds for functional data. On the other hand, the system may also accompany the procedure itself, exchanging navigational data and allowing for visual monitoring of the advancement in relation to the collected data. As this may also include intraoperatively acquired anatomical (MRI) or functional (electrocorticographical) data, visual monitoring of the planned procedure is supported.

5. Conclusion

The system is well accepted among neurosurgeons in the University of Heidelberg. It is frequently used to preoperatively integrate morphological and functional information, and export these in DICOM format for neuronavigation in cases of tumours close to eloquent brain areas (some 30 patients so far).

The intraoperative use of the system connected closely to the navigational unit is currently being evaluated. Ongoing developments, such as offering an automated updated display of the actual navigation accuracy or of selected 3D objects superimposed in the microscope display, [9] may contribute to increasing acceptance. Further development has to be done to make elastical matching of intraoperative image data available within an acceptable time interval and to evaluate the new possibilities of supporting electrophysiological testing intraoperatively.

References

[1] Jannin P, Fleig OJ, Seigneuret E, Grova C, Morandi X, Scarabin JM. A data fusion environment for multimodal and multi-informational neuronavigation. Comput Aided Surg. 2000;5(1): 1-10.
[2] Braun V, Dempf S, Tomczak R, Wunderlich A, Weller R, Richter HP. Multimodal cranial neuronavigation: direct integration of functional magnetic resonance imaging and positron emission tomography data: technical note. Neurosurgery. 2001 May;48(5): 1178-81.
[3] Wirtz CR, Knauth M, Staubert A, Bonsanto MM, Sartor K, Kunze S, Tronnier VM. Clinical evaluation and follow-up results for intraoperative magnetic resonance imaging in neurosurgery. Neurosurgery. 2000 May;46(5): 1112-20.
[4] Black PM, Moriarty T, Alexander E 3rd, Stieg P, Woodard EJ, Gleason PL, Martin CH, Kikinis R, Schwartz RB, Jolesz FA. Development and implementation of intraoperative magnetic resonance imaging and its neurosurgical applications. Neurosurgery. 1997 Oct;41(4):831-42; discussion 842-5.
[5] Brown JA, Barbaro NM. Motor cortex stimulation for central and neuropathic pain: current status. Pain. 2003 Aug;104(3):431-5.
[6] Eisenmann U, Dickhaus H, Metzner R, Wirtz CR. [A computer system for planning and carrying out neurosurgical interventions]. Biomed Tech. 2002;47 Suppl 1 Pt 1: 49-52.
[7] So EL. Integration of EEG, MRI, and SPECT in localizing the seizure focus for epilepsy surgery. Epilepsia. 2000;41 Suppl 3:S48-54.
[8] Ganser KA, Dickhaus H, Metzner R, Wirtz CR. A deformable digital brain atlas system according to Talairach and Tournoux. Med Image Anal. 2004 Mar;8(1): 3-22.
[9] Aschke M, Wirtz CR, Raczkowsky J, Wörn H, Kunze S. Stereoscopic augmented reality for operating microscopes. CARS 2003: 408-413.

9.3 Professionalism

Ubiquity: Technologies for Better Health in Aging Societies
A. Hasman et al. (Eds.)
IOS Press, 2006

Progressing professional maturity in health informatics

Jean ROBERTS[1]
University of Central Lancashire, UK

Abstract : This paper looks at ways in which the profession of health informatics is gaining definition, through the actions of both the BCS Health Informatics Forum and the UK Council for Health Informatics Pprofessions. It explores the some of the ways the current situation was achieved, how the lessons can be applied internationally and the challenges yet to be met. Opportunities for further progress within the existing landscape of technology and health care delivery and management are outlined. The paper documents actions taken and planned and the rationale of why, with supporting evidence of similar contributory work to date. It concludes by stating that the profession must remain flexible and dynamic, and nurture its professional members if it is to continue to develop within an ever-changing scenario.
Keywords: :health informatics, professionalism, regulatory body, brand image, inclusivity

1. Introduction

In the UK, six working communities have been defined within informatics to support the health domain; typically those in Information and Communications Technologies, health records management, knowledge management, information management, clinical informatics and senior management. These were described by the late NHS Information Authority [1] and adopted by the registration body UK Council for Health Informatics Professions (UKCHIP) and the British Computer Society Health Informatics Forum (BCSHIF). Typical of the eclectic mix of individuals involved, the domain identity is not yet cohesive, but input to consultations and initiatives relating to informatics in support of health shows a consistency and over-arching focus for a learned society (BCS) and one registration body (UKCHIP), and has implications internationally for professional mobility.

This paper will explore initiatives designed to capture the domain view about current and future developments and to consider steps to be taken to create a mature profession of health informatics (HI).

BCSHIF holds and annual scientific congress, well-attended quarterly meetings, has vibrant geographic and specialist group programmes, seeks and communicates coordinated views on topics of particular interest and is deemed by the parent BCS to be a very active

[1] Corresponding Author: Jean Roberts, Health Informatics Unit, Lancashire School of Health and Post-Graduate Medicine, University of Central Lancashire, PRESTON, PR1 2HE. JRoberts1@uclan.ac.uk

Forum within its current structure of six. UKCHIP has recently held its first elections and has (as at April 2006) around 1000 fully processed registrants.

2. Existing criteria

2.1. Presentation

Well-documented by Brennan [2] and Burns [3], information is an integral and necessary component of effective health care delivery, management and research. However the challenges of matching existing staff to a collective hierarchy under the NHS Agenda for Change programme [4] and the definition of occupational standards for all the UK home countries by UKCHIP [5] is showing how difficult it is to determine skills, competencies, roles and responsibilities. These processes are necessary to recognize existing staffs appropriately and also to plan the way forward. Whilst the process takes cognizance of the international Skills for the Information Age (SFIA+) definitions for IT professionals [6], they do not apply in enough detail for all the communities within the domain.

2.2. Sharing

There is a visible willingness amongst HI practitioners in the field to share their experiences, evidenced by specialist/members group meetings, collaborative projects and the nationwide spread of informatics developments at both specialty levels and as ‘early adopters’ within the English National Programme for IT (NPfIT) led by Connecting for Health (CfH)[7]. However, the sharing of experiences is frequently amongst ‘trusted friends’, as reported by Szczepura [8] rather more than as a result of corporate instruction as evidenced in the RADICAL STEPS [9] consultation series. CfH is over two years into exciting major informatics developments across England and there are welcome signs of effective communication of progress, experiences and pitfalls emerging.

2.3. Strengths

A generic technological trend in functionality and facilities is putting more pressure on HI specialists to distinguish themselves. For example, BCS investigation of out-sourcing led by Sparrow [10] resulted in some concerns within HI. Specialist criteria proved to be relevant to the health domain, such as the need for fast responses to operational activities for which distance/proximity is crucially important, and activities requiring ‘employee agility’. Thus the health domain could not be complacent but did have some unique strengths to counter pressures to out-source off-shore.

2.4. Opportunities

Informatics for health supports many functional areas, notably:

- Direct patient care

- Order Communication and Test Results reporting, to support care
- Facilities management at an operational level; ensuring hospitals beds, diagnostic facilities, primary care capacity and the like are available as required
- Strategic planning of the development of resources; for clinical specialties and to support the hierarchy of care demanded by a population
- Monitoring/measuring (of activity and finance) to identify value for money, efficient/effective service delivery, and efficacious evidence-based practice
- Emerging cross-professional and cross-sectoral working together; involving increasingly voluntary agencies, social services, lifestyle managers and the 'patient' themselves

2.5. Data

We use the Korner principle that 'data should be only be collected without which it is deemed impossible to manage properly'. Information is collected as near to source as possible and then aggregated and anonymised as necessary to satisfy functions that are more patient-distant. Collective reuse and re-versioning of data is possible as a result of partnership working, and must be underpinned by information governance, using information in a fit for purpose manner. Anecdotal history shows that inappropriate data can skew the decisions made relating to the areas above.

UKCHIP was established, in part, to put an onus for 'good' data on informatics staff handling it, in addition to responsibilities held by the data originators. UKCHIP stresses the obligation to work to a Code of Conduct, even when it runs counter to corporate plans, giving strength to the individual when challenges to employers are (occasionally) necessary.

2.6. Legacy

Previous developments have resulted in legacy solutions that still have a useful lifecycle. They can be interconnected in many ways, including that where standardized interfaces are developed within a national strategy (see Scottish NHS strategy [11]); or developed selectively (as with pre-existing European technology standards and some Local Service Provider offerings in England); or by a radical replacement programme such as NPfIT CfH. Throughout such a turbulent transition, however it is carried out, staff need to be supported, their skills and competencies recognized (for example by the UKCHIP fast-track grandparenting entry scheme). Communication channels are opened for the expression of views (for example through think-tank/consultations like RADICAL STEPS [9]) and kept informed (through UK Health Informatics Community [12] and increasingly CfH [7]).

2.7. Benchmarking

In any major programme of investment it is useful, retrospectively, to quantify the changes made as a result of HI activities. For this, a baseline/benchmark is required against which progress can be measured and feasible local targets set. Evidence of this is sadly lacking. Searches for operational case studies to showcase success have only highlighted a limited number to date. This may be because of the staggered UK programmes and more instances will surface over time. Awareness and sharing of the methods used to implement them and benefits realized may go a long way to stimulating similar positive informatics implications in other areas and countries.

2.8. Landscape

A changing technology landscape is evident across the UK. With external drivers such as the e-government initiatives [14] nationally and across Europe, and citizens becoming more able and demanding about technologies to communicate with public bodies and enquire about personal circumstances; the health domain must review professional information and refine their work patterns and roles to keep up with the wider 'game'.

Clinicians will be faced with informed patients and must be able to consider their researches in respect of feasibility/availability within the NHS and clinical appropriateness to the condition. This challenges those in direct patient care, particularly the family physician (GP) gatekeeper, to harness technologies at least equally as effectively as the public. As the professionalism of HI increases then all who use any informatics in their daily activities must be seen to work to the same standards of good practice – knowledge workers, technologists, information managers and those with specialist clinical informatics responsibilities. In addition, the new generation of professionals must be taught and developed to the same standards. Emerging systems developments must adopt the same rigor. The academic community and commercial providers must work to and pass on the same codes of conduct and practical principles to students and in their clients' systems. UKCHIP registration and continuing professional development requirements establish the baseline and scrutinize steps taken to keep individual knowledge and skills contemporaneous and fit for practice.

Cross-cutting re-stratification of health and social care into multi-disciplinary teams to address standards/targets relating the child health, older people and those with mental health challenges will change the landscape of HI immeasurably. The profession must continue to share its experiences if it is to be in a position to respond to the changing environment.

2.9. Models for moving forward

Using a co-operative inquiry model Oates [14] that recognises experiential knowledge to explore the current concerns and outcomes of HI development, the RADICAL STEPS series of think-tank/consultations [9] has provided a vehicle for free speech and communication that has not necessarily been comfortable for those receiving its outputs. It has demonstrated that constructive criticism and opportunities to share experiences can contribute to forward movement. The RADICAL STEPS series, from which 'STRIDES' emerged as positive case studies became more evident, started as a Delphic exercise (where ultimate goals were not stated in advance) , under 'Chatham House' rules of non-attribution, collating individual observations, with rapportage confirmed by the participants before a position paper was released. The most recent instantiation has an additional triangulation of views through interim wider consultation, required because of logistical difficulties in getting all who wished to comment physically to one place. Added value from the additional stage is under review against the formal position statement produced.

2.10. Drivers for Change

For example, the Wanless 2 report [15] focused on a number of population concerns that will require collective action to solve and will necessitate progress monitoring over time using informatics trackers, looking at epidemiology and management performance criteria. A contemporary issue, such as the potential Avian Flu pandemic coming towards the UK requires measures in place to track specific cases, project likely incidence and recheck as further data emerges from current outbreaks, monitor availability of vaccines, at risk populations and the available trained and protected professional workforce. This will not be possible unless effective systems and competent informatics professionals are in place to provide support to those taking clinical and public health action.

2.11. Roadmap

A clear road map of the intended way in which informatics will support care delivery and management in each home country will help to confirm the likely needs for informatics specialists. Unfortunately, the political imperatives for organizational change (such as Primary Care Trust mergers and the designation of Foundation Trusts and outsourcing of hospital management) may impact on the direction as currently stated. In addition, the significant proportion of informatics staffs who have personal ad hoc protection rather than role placement under Agenda for Change [4] makes for a volatile workforce at risk of leeching to the private sector and out of health.

2.12. Belief

One feature that is very strong in UKCHIP registrants and many potential registrants is that informatics can make a significant positive difference within the health domain. This is evidenced by the case studies in the RADICAL STEPS position papers, the annual national Health Computing (HC) and Specialist Group conference proceedings, the specialist trade press, external reports such as the Kennedy Report [16] and as award winners in schemes such as Health Information Technology Excellence Awards (to 2005) and the e-Health-Insider Awards.

3. Conclusion

In order to maximize the realizable effects of informatics in support of the health domain, the HI profession must ensure it is flexible enough to move with the volatility of the domain, including the impending convergence with social care and the constant public sector reorientations. This is not an easy feat! The profession must nurture its expert members by improving their collective identity, facilitate their learning experiences, evicting those who do not practice HI in a proper manner, ensuring that the means to develop professionally is made available to all who wish to take it up and informing those externally, who may have an impact, about the specialist capability and sensitivity necessitated by Informatics in support of Health. BCSHIF and UKCHIP are continuing to contribute to development and recognition of quality professionals in the field.

4. References

[1] Department of Health, Making Information Count, 2002
[2] Brennan S, The NHS IT Project: The biggest computer programme in the world, Radcliffe Publishing, 2005
[3] Burns LR et al, The Business of Healthcare Innovation, Cambridge University Press, 2005
[4] www.dh.gov.uk/PolicyAndGuidance/HumanResourcesAndTraining/ModernisingPay/AgendaForChange/fs/en
[5] www.ukchip.org , Internal Working Document MG129
[6] Skills for the Information Age www.sfia.org.uk
[7] NHS Connecting for Health, www.connectingforhealth.nhs.uk
[8] Szepura A, Biggerstaffe, D, Roberts J, Fagan R et al in Department of Health, Optimising the use of existing ICTS to inform health & social welfare professionals and carers in the community, 121/7182, , 2003
[9] RADICAL STEPS, www.bcshif.org
[10] Sparrow E et al Offshoring : A challenge or opportunity for British IT Professionals, www.bcs.org/BCS/News/PressReleases/2004/november
[11] Scottish NHS eHealth Strategy, www.show.scot.nhs.uk/imt
[12] Health Informatics Community website, www.informatics.nhs.uk
[13] e-government portal , www.localegov.gov/uk
[14] Oates, B Co-operative Inquiry : Reflections on Practice, Jnl Business Research Methods, Vol2 Issue 2, October 2004
[15] Wanless D for HM Treasury, Securing Good Health for the Whole Population, 2004
[16] Learning from Bristol: the report of the public inquiry into children's heart surgery at the Bristol Royal Infirmary, 1984-1995. (Cm.5207-I), London: Stationery Office, 2001
* all sites accessed October 2005 unless otherwise specified

Ubiquity: Technologies for Better Health in Aging Societies
A. Hasman et al. (Eds.)
IOS Press, 2006

Bridging information gaps between primary and secondary healthcare

Habibollah PIRNEJAD[a,1], Arjen P. STOOP[b], Marc BERG[c]
[a, b, c] *Institute of Health Policy and Management, Erasmus University Medical Centre, Rotterdam, The Netherlands*

Abstract. Medication errors are harmful and costly for healthcare systems. Recent studies show that a major part of these errors are due to the problems in transferring the patient current medication-data between primary and secondary healthcare. Recent ICT development promises to improve the communication between primary and secondary healthcare. In order to find out the constraints that may hamper a communication project's productivity, an IT configuration for building a medication-data communication network between primary and secondary healthcare in the Netherlands was followed applying qualitative methods. We analysed some important problems that project faced and conclude that problems with the data integration and saving the data integrity are important challenges for the project to maintain its objectives.

Keywords: Medication Error, Computer Communication Network, ICT, the Netherlands

1. Introduction

Many of medical errors are "Medication Errors" [1]. In the US, it has been estimated that 1-2% of patients admitted to hospitals are harmed as a result of medication errors and 7000 patients die a year because of them [2, 3]. The Royal Dutch pharmacists society (KNMP) has suggested that as many as 131,000 hospital admissions, 8.2 per cent, per year occur in the Netherlands due to adverse drug reactions. These admission would cost the Dutch society approximately 186-430 million Euro per year [4].

Amongst the sources for medication errors, limited or impaired access to "patient current medication history", specially when patient is moving from one level of healthcare to another, is of great importance [5]. A recent systematic review by Canadian researchers of 22 studies including 3755 patients showed that errors in current medication histories at hospital admission were "disturbingly common and potentially harmful to patients". Such errors occurred in up to 67% in the studies, which were published between 1966 and April 2005 [6].

One of the promises that recent ICT development has raised is improving the communication of patient data especially across the healthcare boundaries. Despite all IT potentials, however, fulfilling such a promise has not usually been straightforward. IT projects are complicated at different levels, thereby their productivity is challenged [7]. For an IT communication project it is of paramount importance to safeguard the integrity of the interchanged data as well as the ability to integrate different pieces of

[1] Corresponding Author: H. Pirnejad, P.O. BOX: 1738, 3000 DR, Rotterdam, The Netherlands. E-mail: h.pirnejad@erasmusmc.nl

data together. The productivity of such a communication project, then, can be challenged by complications at these levels.

This study focuses on several important issues that complicated data integrity and data integration in an IT project aiming at medication-data communication, and thereby challenge its productivity. Our ambition in this study is to raise points that can benefit similar projects. Evaluating the medication-data transaction process, this paper indicates *information gaps* between the Dutch primary and secondary healthcare that may induce medication errors. The early development of an ongoing IT project, which aims at crossing these information gaps and improve medication errors, is sketched out. Through qualitative methods, then, some important challenges that may curb this kind of ICT configuration in improving medication errors are explored.

2. Methods

We approach the process of medication-data exchange between primary and secondary care providers as building a *loop[2]-process*, within which patient medication-data is being circulated, updated, but may also suffer from attrition or distorted. IT accordingly is expected to accelerate circulating and updating of the data whilst prevent its attrition and distortion. We interviewed care providers from this loop and observed their work. Our interviews were semi-structured, in depth, one-by-one and face-to-face. We held 10 interviews, with each one lasting 1.5 to 2 hours. Community and hospital pharmacists' works were observed for about 4 hours as well. Moreover, we analysed the relevant documents including project initiation documents and updated about the changes in the project through emails, phone calls and direct talks to project managers.

3. The study environment

The study environment is Almere, a city near Amsterdam in the northwest of the Netherlands, where a project, TUMA[3], has been launched in order to exchange medication-data between primary care providers, GPs and community pharmacists, and secondary care providers, hospital pharmacists and medical specialists. In this region, from 115 GPs nearly all of them use the same GP Information System (GPIS). All of the 17 community pharmacies use the same pharmacist information system from the same vendor. Since the GPISs and pharmacist information systems share the same server, they build an "application specific network" through which primary care providers can easily share some part of patient data including patient medication-data. Though GPs and community pharmacists already communicate to some extent in the Netherlands, Almere is quite unique in the sense that in one region there exists this level communication between all GPs and community pharmacists. "We in primary care always check each other's work [on patient medication]. This is normally done both by our information system and also through direct observation of the prescriptions. If we see there is something wrong, we just pick up the phone and call to the GP for more clarification. Every time a prescription is filled, the information system generates an automatic message that informs and updates the GP." [CP1-04][4]

---

[2] With the word "loop" we would like to emphasize the end-to-end closeness of the process.

[3] TUMA means Trans-Mural Exchange of Medication data in Almere.

[4] We quote our interviewees with abbreviations plus the code number of their interview: PL= Project Leader, PM= Project Manager, CP= Community pharmacist, HP= Hospital Pharmacist,

Another remarkable feature of Almere is existence of only one general hospital in this region; patients who need specialist attention are referred to this hospital. TUMA, then, is building a communication network on such an ideal condition.

4. Medication-data communication between primary to secondary healthcare

In the Netherlands every patient has a GP as a family doctor who acts as a gatekeeper between primary and secondary care. Every patient also has his own community pharmacist who fills his prescriptions. At the primary care level pharmacists have the responsibility for taking care of their patients' medication safety. Registration of the patient data in the Dutch GPISs is based on the episodes, i.e. based on every time a patient consults his/her GP for a new medical problem. Except for diagnosis, coding with ICPC[5], and medication-data, coding with ATC[6]-classification, most data entries are made in free text format [8]. To write a referral letter, a GP may have to draw related information from several episodes in his information system. Due to the time and effort this takes for GPs the letters usually do not contain enough information for medical specialists [9].

At the hospital level, secondary care providers need a patient medication profile from primary care in order to avoid medication errors and offer high quality clinical care. Nevertheless, they do not have direct access to this data. At this point the medication-data flow comes to a halt. It does not cross the secondary care border; the first *information gap* between primary and secondary patient care is thus created. To fill this information gap, patients most frequently are the source for their medication related information at the Dutch hospitals. However, it is always possible that patients – or their relatives – do not remember all the currently using drugs or confuse between look-alike drug names. In addition, it is possible that hospital care providers fail to take an accurate history of patient medication or some part of information is missed while history taking or when it is handed over amongst care providers. A hospital pharmacy told us a story in this regard. "A nurse failed to register a drug name (Methoteroxate) while she was taking the drug history from a patient, only because the drug had been used in intervals. The patient got cystitis during his hospitalisation and a physician ordered him Cotrimoxazole. As soon as the Cotrimoxazole started for the patient, his condition suddenly got worse and turned to dangerous one with leucopoenia and other signs of Methoteroxate toxicity. Such a dangerous condition happened to the patient, only because the nurse failed to take an appropriate drug history from the patient. Our information system, on the other hand, failed to react to this drug interaction because Methoteroxate had not been entered to it." [HP2-02]

When a patient is discharged from a Dutch hospital, he receives prescriptions that should be filled by a community pharmacist. Beside this early contact, additional information, including the diagnosis, procedures, and changes on patient medication are sent to the primary care providers through a discharge letter. Previous studies, however, show that this process usually takes a long time and in most cases when patients contact their GPs after discharge, the GPs usually are unaware of the last changes in the patients' medication [10]. This delay, which is also the case for the Almere region, creates the second *information gap* between primary and the secondary Dutch healthcare. This information gap, in its turn, may hamper the quality of patient care and induce medication errors. Moreover, without access to the information from secondary care, both primary care providers and patients will have been left in limbo

[5] International Classification for Primary Care.

[6] Anatomical Therapeutic Chemical Classification System.

[11]. "After discharge [from the hospital], patients most frequently do not know what to do with the drugs they were using before hospitalisation. They don't know whether they have to use them together with their discharge medications or they have to stop their using. They usually ask us, because we are supposed to take care of their medication safety. But we cannot help them because we do not know the reasons for the changes." [PM/CP1-01]

5. The role of TUMA

TUMA is building a communication network between primary and secondary healthcare. The communication network is a "Virtual Private Network" (VPN) that connects the server of the primary information systems to the server of the hospital pharmacist information system. At the centre of this communication network named "eHealthNet", there is a "Central-Patient-Index" system to secure one-by-one match of patients' files between the primary and the secondary healthcare. Through this communication network, patients' medication profiles, including patient's current medications list and a summary of his medical record, will be exchanged by *OZIS,* an EDIFACT[7]-based communication protocol. TUMA has set a mechanism that every time a patient is admitted to the hospital an enquiry message is sent to primary care by the hospital pharmacist information system. The reply message, the patient medication profile, will be integrated to the hospital pharmacist information system. In near future, the information from primary healthcare care will also be available to care providers at two wards of the hospital. A link is designed in the Electronic Patient Record (EPR) interface for this purpose. Whenever a care provider clicks on this link, a message will go to the pharmacy information system, then an XML message containing patient medication-date will come back automatically to the EMR. The project finally aims at bridging over the second information gap. Discharge medications and a summary of the patient medical file will be sent through the hospital pharmacist information system –in a similar way – to the local health network as a patient is discharged from the hospital.

6. Implications

We touched upon the idea that the medication-data should ideally be circulated within a closed-loop. Medication-data is changed, updated, and handed over in this closed-loop. In this multi-stage process, it is important for care providers to receive patient medication-data in a *timely* fashion and *accurately,* with safeguarded integrity. In the Dutch healthcare system, this loop is far from closed especially when the medication-data is supposed to cross the borders between the primary and secondary healthcare. Currently, patients are considered as a link between the primary and the secondary care parts of the "medication-data loop", filling the information gaps in these points, while we know that patients get it wrong 28-38 % of the time [12].

Moreover, in the discussion of data communication among healthcare providers, the relevance of *interoperability* of care providers' information systems comes into the front stage. In TUMA, this can be translated into integration of the medication-data from the community pharmacist information system into the hospital pharmacist information system and vice versa in a way that the data can be processed by the both sides information systems. Despite the high aspiration of an integrated solution,

[7] The Electronic Data Interchange For Administration, Communication and Transport.

however, a fully integrated information system is hard to find [13]. One reason is that many software products have been built and acquired from heterogeneous sources during a long period of time, and the systems have differences in implementation and architectures [14]. This has been the case for a couple of ICT projects that have recently been launched on the base of OZIS protocols in the Netherlands. TUMA, as explained, is quite unique among them. Tracing the progress of TUMA, nonetheless, we observed that the project faced problems in integrating different parts of the exchanged data. "One main reason that the project fell behind its timetable is the problems we had in integrating the medication-data to the hospital pharmacist information system, a software functionality problem. To solve the problem, we consulted other projects' people that have already worked with the same way of data transaction. To our surprise, we found out that they only use the system for inquiring the data from primary care and then transfer the data manually to their hospital information system." [PL-03]

There are at least two reasons why TUMA has faced problems regarding integration of different parts of patients' medication-data. Firstly, there are drugs that are in use solely in one level of healthcare. For example, 'Nexium', a drug that is prescribed for a peptic ulcer disease in the hospital, is not being used in the primary care. Therefore, if a patient on this drug is discharged from the hospital, his medication has to be changed into another proton pump inhibitor, such as Omeprazole, at the primary care. Secondly, the hospital pharmacist information system works with more comprehensive "Medication Dispensing and Administration Coding system" than community pharmacist information system. A drug that is ordered once a day orally in primary care might be changed into intravenous form and in a distributed dosing schedule according to a hospital wards' routine. Automatic integration of the medication-data in such conditions can be problematic and information systems thus produce errors. As a result, there have to be manual steps in the process of integrating different parts of patient data. Yet, there is no exact idea what portion of the medication-data has to be integrated manually. However, existing integration problem, even if it is limited to a small part of medication-data, will not let the medication-data process loop be closed and error production in some cases may facilitate [15].

In the discussion of circulating medication-data in a closed-loop, the integrity of medication-data not only depends to the integration of different parts of patient data produced in different healthcare levels, but also to safeguarding its integrity while it circulates within each healthcare level solely. This means that the medication-data should be protected from being damaged or lost while it is registered or communicated among care providers inside primary or secondary healthcare levels. In Almere region, we have seen that within primary healthcare, there is a good integrity for medication-data. However, such integrity does not exist for the medication-data at the hospital. We have presented a story of a nurse negligence that shows how the integrity of the medication-data can be hampered within the hospital. Despite being common, these kinds of errors do not usually gain attention and are often ignored if they do not lead to immediate or serious disadvantages [16]. Therefore, TUMA's success in preventing medication errors will to some extent bound to extent to which the medication-data integrity is protected at the hospital.

For TUMA stakeholders, the weakest link in the medication loop is the patient, which has to be substitute by ICT. Yet, the patient should play a core role in verifying the accuracy of his current medication-data. There are at least three reasons for this claim. First, some patients have to fill their prescriptions in a pharmacy other than their designated community pharmacy. Second, Over-The-Counter Drugs (OTC) are not registered in any information system. Last but not least, some patients make changes in their medication administration plan by themselves. This information and changes on patient medication are important and only can be obtained through asking the patient. In a study by Van der Kam et al [12] on medication-data exchange between GPs and pharmacists, for the drugs reported by the patient-only there was no difference between electronic communication and paper-based communication. Ignoring the patient's role in completion and updating the current medication-data, therefore, will lead to missing some part of patient medication-data and thereby damage the data integrity. TUMA has not planned to prevent this data attrition and in that sense it will not succeed to close the medication loop.

Taken together, reducing medication error is depended not only on being timely, but also on saving the integrity and integrating different pieces of data to gather. TUMA doubtlessly improves the timeliness of medication-data transaction among care providers. However, its success in reducing medication errors is challenged by the problems with integrating different pieces of the data and safeguarding its integrity.

Acknowledgement

The authors would like to thank the TUMA managers for their cooperation and the RITHM group of the iBMG of Erasmus University MC for their useful comments on an earlier draft of this paper.

References

[1] Leape LL, Brennan TA, Laird N, Lawthers AG, Localio AR, Barnes BA, et al. The nature of adverse events in hospitalized patients. Results of the Harvard Medical Practice Study II. N Engl J Med 1991;324(6):377-84.

[2] Barber N, Rawlins M, Franklin BD. Reducing prescribing error: competence, control, and culture. Qual Saf Health Care 2003;12 Suppl 1:i29-32.

[3] Dean B, Schachter M, Vincent C, Barber N. Prescribing errors in hospital inpatients: their incidence and clinical significance. Qual Saf Health Care 2002;11(4):340-4.

[4] Boysen M. Delivering pharmaceutical care in the Netherlands: Practice and challenges. The Pharmaceutical Journal 2004;273:757-9.

[5] Leape LL, Bates DW, Cullen DJ, Cooper J, Demonaco HJ, Gallivan T, et al. Systems analysis of adverse drug events. ADE Prevention Study Group. Jama 1995;274(1):35-43.

[6] Tam V, Knowles R, Cornish P, Fine N, Marchesano R, Etchells E. Frequency, type and clinical importance of medication history errors at admission to hospital: a systematic review. CMAJ 2005;173:510-15.

[7] Wears RL, Berg M. Computer technology and clinical work: still waiting for Godot. Jama 2005;293(10):1261-3.

[8] Lagasse R, Desmet M, Jamoulle M, Correa G, Roland M, Hoyois P, et al. EUROPEAN SITUATION OF THE ROUTINE MEDICAL DATA COLLECTION AND THEIR UTILISATION FOR HEALTH MONITORING. Research. Bruxelles: Université Libre de Bruxelles; 2001. Report No.: 1998/IND/2011.

[9] Bal R, Mastboom F, Spiers H, Rutten H. The product and process of referral: Optimizing GP-specialist communication through ICT. Int J Med Inform;In press (2006).

[10] Schiphorst P, Bernaad H, Muller N. Medicatie bij opname in en ontslag uit ziekenhuis. Vergelijking van vier bronnenvan medicatiegegevens (Admission and discharge medications. Comparison of four sources of medication data.). Pharm Weekbl 1992;127:128-132.

[11] Preston C, Cheater F. Baker R, Hearnshaw H. Left in limbo: patients' views on care across the primary/secondary interface. Qual Health Care 1999;8(1):16-21.

[12] Van der Kam W, Meyboom de Jong B, Tromp T, Moorman P, van der Lei J. Effects of electronic communication between the GP and the pharmacist. The quality of medication data on admission and after discharge. Family Practice 2001;18(6):605-9.

[13] Berg M. Medical Work and the Computer Based Patient Record: A Sociotechnical Perspective. Methods Inf Med 1998;38:294-301.

[14] Mykkanen J, Porrasmaa J, Rannanheimo J, Korpela M. A process for specifying integration for multi-tier applications in healthcare. Int J Med Inform 2003;70(2-3):173-82.

[15] Koppel R, Metlay JP, Cohen A, Abaluck B, Localio AR, Kimmel SE, et al. Role of computerized physician order entry systems in facilitating medication errors. JAMA 2005;293(10):1197-203.

[16] Leape LL. Error in medicine. JAMA 1994;272(23):1851-7.

Ubiquity: Technologies for Better Health in Aging Societies
A. Hasman et al. (Eds.)
IOS Press, 2006

The Demographic Bias of Email as a Survey Method in a Pediatric Emergency Population

Brett W. Taylor

Dalhousie University / IWK Health Centre, Halifax, Nova Scotia, Canada

Abstract. Email has been considered as a communication medium between patients and clinicians in pediatric emergency departments, but the demographic bias involved in using email has not been fully explored. We developed a paper based survey to explore access, willingness to participate and the demographic bias of email within our parent population. METHODS: To 1733 possible subjects, 1200 surveys were distributed with a return of 1018, a survey response rate of 85%, and a population response rate of 59%. RESULTS: Subjects from families with incomes less than $60,000 per year had lower access rates (OR = 0.40, 95% CI [0.25,0.62]), as did those with lower education (OR = 0.37, [0.17, 0.81]). Employment outside of the home was associated with increased email access rates (OR = 1.79, 95% CI [1.19, 2.70]). Visible minority status was associated with an increased willingness to participate (OR = 1.84, 95% CI [1.10, 3.06]) as was low education (OR = 2.12, 95% CI [1.04, 4.32]). The population of theoretical responders to an email based quality assurance process would have been significantly different from the base population of adults accompanying children to our emergency department as a result of these biases. CONCLUSIONS: We have demonstrated a degree of demographic bias in email access rates, negatively affecting those individuals with lower income, less employment, and lower education. Email based surveys directed at parents in pediatric emergency departments should include questions on income, employment and education in order to permit those who analyze the data to correct for these variables. More research is needed to confirm these findings.
Keywords: Electronic mail; Demography; Emergency service, hospital; Medical Informatics

1. Introduction

Health care delivery in a pediatric emergency department (PED) is complex and involves many partners, with some tasks performed by the emergency team and some delegated to others. Care of most conditions involves a period within the PED followed by a course of ambulatory management; care, in other words, that may be initiated by the emergency physician, but is often facilitated by a visit to a pharmacist, delivered by the parent and followed up by a completely separate health care team, often the family physician. Substantial time lag may occur between the time when the emergency physician / patient interaction is over, and those times when ambulatory therapy starts, benefits or adverse effects are encountered, and compliance or educational issues are uncovered. This time lag represents a temporal disconnect or asynchrony between the parent's experience and the physician's expectations regarding the efficacy of care.

Communication between parents and clinicians across this disconnect is often very difficult, and this leads to various challenges towards the delivery of quality care to patients. For instance, it has been noted that it is extremely difficult to evaluate practice patterns, parent education, compliance, parent satisfaction and many other aspects of patient care. We therefore argue that a reliable communication medium between patients and clinicians is needed in order to establish constructive feedback about the quality and efficacy of healthcare services provided to them. Patient feedback, in other words, can serve as a practice auditing tool in an attempt to examine and streamline the clinical practices of clinicians.

Attempts to provide this feedback have met with limited success. Paper based surveys are inhibited by the costs of data entry and printing. Discharge summaries and similar reports are generally paper based, precluding collation of information by the receiving physician. Further, the delay between the delivery of information and the index visit is such that the context is diluted, if not lost entirely. Too often, the paper trail simply fails to reach the emergency physician at all. Chart audits do not interrogate that portion of the PED care pathway that is "downstream" from the department. Virtually the only mechanism that provides effective feedback between parent and physician is the error management processes that are put in place following a parent complaint. This feedback mechanism is (fortunately) rarely used and, of course, very affected by selection bias.

Given these problems, we propose the use of electronic mediums, such as emails and web-based portals, as a means to communicate with patients and to get their feedback through electronic surveys.

Email has been considered as a mechanism for communication with physician groups (1), and with patients following a visit both to a Canadian pediatric emergency department (2) and to an American general emergency department (3). These publications have not been particularly positive; traditional mail seemed to be as good as email (1), and telephone seemed better (2,3). However, these studies were directed at obtaining medical follow-up, not the general survey data required for feedback purposes. The failure rate of medical follow-up communications must necessarily be very small, much smaller indeed than is required for the acquisition of good quality population survey data.

We believe that the evaluation of email as a part of a feedback loop involving parents must begin with an assessment of the implementation barriers. The most obvious of these are the issues of access, demographic bias, and willingness to participate. Data regarding internet access shows marked regional and temporal variation, and in Canada range from as high as 79% (3) to as low as 21% (5). Demographic parameters that may affect email access include age (6,7), income (8,10), socioeconomic status (5,9), and education (11), although variables such as gender, race, family make-up and employment have not been well explored. Willingness to participate in email communication for medical purposes ranged from 63% to 72% of participants (2,4). Data regarding the impact of various demographic parameters on willingness to participate is sparse.

Given the wide variation in the responses to some of these variables, a survey was designed to determine the level of access, level of interest, and demographic bias of email as a communication strategy for quality assurance purposes in the population served by our department. In this paper, then, we report the results of a paper based survey of adults who accompany children to the PED at the IWK Health Centre in Halifax, Nova Scotia, Canada. Specifically, our research questions were:

- How do demographic variables affect email access rates within that population of adults who accompany children to a pediatric emergency department?
- How is willingness to share email address for the purpose of quality assurance affected by the demographics within the same population?
- How would theoretical responders, defined by ability and willingness to respond to an email survey, differ from the base population from which they are drawn?

2. Method

The Isaac Walton Killam (IWK) Health Centre in Halifax, Nova Scotia, Canada is the only tertiary care pediatric facility for maritime Canada. The IWK Emergency Department is a dedicated tertiary care PED, serving an immediately local urban and semi-urban community of 400,000 people incorporating the Halifax Regional Municipality. The department is also a referral PED, serving a maritime population of roughly 2 million. The department sees roughly 30,000 visits per year.

This was an observational study, using a paper based survey tool. We surveyed the adults who accompanied patients to the IWK Emergency Department for 4 weeks in October and November 2005. Only one adult from each emergency department visit was recruited; in cases where there were more than one adult (eg. both parents attending with a child), we asked that only one fill out the survey, and that where possible that the subject was a current caregiver. Consent was implied by the return of a completed survey document. The survey was structured to be anonymous. Data was entered into a Microsoft Access® database. This study received approval from the IWK Research Ethics Board.

Social and economic data included visible minority status, income group, employment, single parent status and education. Personal information included age and gender. We also collected data on access to email and willingness to share email for quality assurance purposes. Except for age, all data was stored as categorical variables. Interest in sharing email address for study purposes was recorded as a 5 point Likert scale. Data was analyzed using logistic regression with Minitab® software.

3. Results

1733 potential subjects were seen at the IWK during the study period; to these 1200 surveys were distributed with a return of 1018, a survey response rate of 85% and a population response rate of 59%. Of the 1018 surveys, 1011 of the respondents identified themselves as parents or guardians, 4 as other relatives, 2 as having some other relationship with the patient, while in 1 returned survey the relationship was not specified. As we expected, therefore, this was principally a survey of parents in our department. Because our survey was anonymous, we have no mechanism for identifying and statistically handling our non-responders.

In returned surveys, questions were well answered. 98 demographic questions were unanswered, out of a possible 6108 (1.6%). Questions about income were significantly more likely to be unanswered compared to the rest of the group (2.9% unanswered compared to 1.0% , $p < 0.01$) as were questions about visible minority status (2.7% unanswered compared to 1.0%, $p < 0.01$). Two questions about email access were not answered (0.2%). Among those with email access, 100% of questions about willingness to share email address were answered.

Overall 810 subjects had email access (79.7%). Of these, 79.5% were willing to participate ("likely" or "very likely" to be interested in sharing their email addresses for the purposes of quality assurance), 14.7% were "unlikely" or "very unlikely" to share, and 5.8% were unsure. The proportion of the population that demonstrated both access to email and a willingness to participate in an email survey, therefore, was 63.3%.

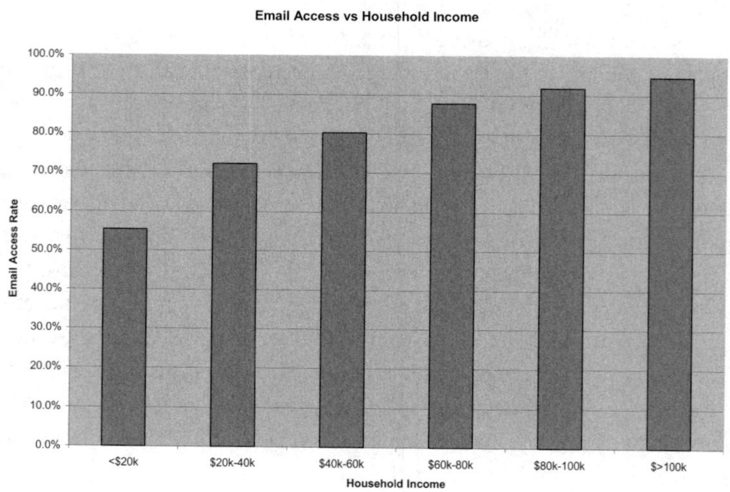

Figure 1: *Relationship between Household Income and Email Access Rates.*

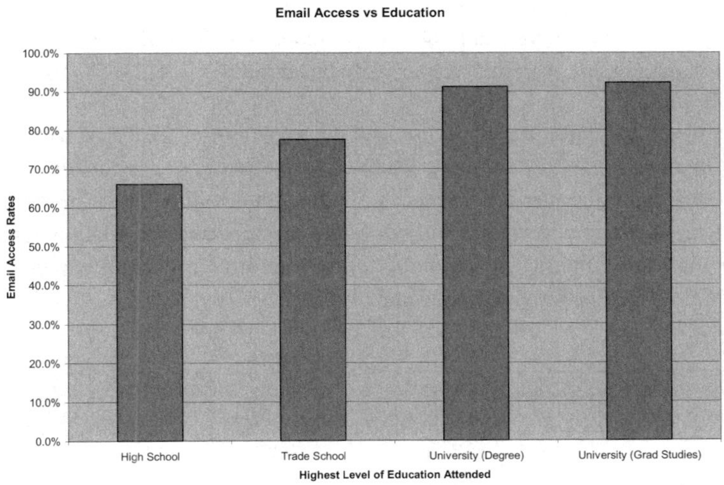

Figure 2: *The relationship between Education and Email Access Rates.*

Initial analysis revealed that lower email access rates were associated with single parent status (70.0% vs 82.3%, p<0.001), failure to be employed outside of the home (65.6% vs 83.7% p<0.001), age < 30 years (73.2% vs 81.7%, p<0.01) and visible minority status (73.5% vs 82.5%, p<0.01). Income and education were both highly linked to email access as well (see Figures 1 and 2, p<0.001, p<0.001).

However, logistic regression narrowed the number of parameters which independently predicted email access to three factors: income, employment outside of the home and education. Incomes less than $60,000 per year were clearly linked to lower email access rates (Table 1). Employment outside of the home was linked to higher access rates (OR= 2.08, 95% CI [1.03, 2.41], p < 0.05). When income and age were restructured as binary categorical variables ("less than $60,000" or not and "less than 30 years" or not) a relationship was demonstrated between access and education (OR "high school only" = 0.37, 95% CI [0.17, 0.81]; OR "trade school" = 0.47, 95% CI [0.22, 0.99]). A relationship between access and age, gender, single parent or visible minority status was not found.

Income	Odds Ratio	P value
<$20,000	OR = 0.11 95% CI [0.04,0.30]	< 0.001
$20,000 to $40,000	OR = 0.23 95% CI [0.10, 0.56]	< 0.005
$40,000 to $60,000	OR = 0.29, 95% CI [0.13, 0.67]	< 0.005

Table 1: Income categories less than $60,000 are negatively related to access rates.

Analysis by logistic regression showed a positive relationship between interest in participating in an email survey and both visual minority status (OR = 1.84, 95% CI [1.10, 3.06] and having no more than high school for education (OR = 2.12, 95% CI [1.04, 4.32].

Individuals who had both email access and were willing ("likely or very likely") to participate in an email survey can be considered hypothetical responders to such a survey, and it is in this group that the consequences of demographic bias can be most clearly seen. The responder subgroup was significantly different from the base population in terms of the proportion that was female, young, held visible minority or single parent status, was unemployed, low income or poorly educated (Table 2).

Proportion of Population that is/has:	Base Population	Hypothetical Responders	% Change	P
Female	77.10%	75.90%	-2%	<0.01
<30 years old	23.30%	21.40%	-8%	<0.01
Visible Minority	16.90%	14.10%	-17%	<0.01
Single parent status	20.50%	17.20%	-16%	<0.01
Unemployed	21.80%	17.70%	-19%	< 0.01
household income <$20k	12.20%	8.30%	-32%	< 0.01
household income <$40k	32.20%	25.50%	-21%	< 0.01
education high school or less	27.30%	21.60%	-21%	< 0.01

Table 2: A comparison of the demographics of the base population vs (hypothetical responders.

4. Discussion

Our population had an email access rate close to the maximum found in the literature, and an impressive willingness to participate. Despite this, slightly less than two thirds of the population are classified as potential responders to an internet survey. Further, it is clear that demographic factors would influence the data obtained using an email delivered survey.

We did not find that gender, single parent or visible minority status were independent predictors of access, nor, as others have, that age per se had an impact. These demographic labels seemed in our study to be proxies for other parameters that did independently predict access: income, employment and education. The lowest income families in our study had email access rates of only 55%, compared with 95% of the highest income group. When email access rates and interest are combined, responses to a hypothetical email survey would reach 75% from high income families, but only 44% from the lowest. The population that would have responded to an email survey would have been significantly older, richer, better educated, more male, more white, and less likely to be unemployed than the general emergency department population (Table 2). The needs of the most vulnerable in our population would clearly have been under represented.

It is easy to hypothesize mechanisms for these findings; income is clearly required to obtain technology and service, and both employment and advanced education increases the opportunity for contact with email outside of the home. In the case of gender and age, it may be that email has become a more mature product than was the case at the times of previous studies, that it is no longer preferentially a product for use by young males. We are also dealing with a relatively young group of adults; our oldest subject was 58, and the median age was 35 years, and this may have influenced our findings.

In our study population, lower education and visual minority status was unexpectedly associated with an increased willingness to participate, and this fact offset to some extent the differences in email access in these groups. Thus, in our population, the differences between the hypothetical response population and the base population are significant and meaningful, but not so large that appropriate management cannot to some extent ameliorate them. This balancing bias is probably not present in all emergency populations. We would argue that any population survey based on email must be structured to permit researchers to correct for income, employment and education, probably by including questions that specifically elicit this information from the subjects. Ultimately, the bias of email surveys will be determined by examining data on responders and non-responders of a real, not hypothetical, electronic survey. This research needs to be done.

5. Conclusions

We have demonstrated a degree of demographic bias inherent in an email survey process for a pediatric emergency department parent population. This bias negatively affects those individuals with lower income, less employment, and lower education. In order to ensure that future surveys do not result in an under representation of the needs of our most vulnerable populations, questions on income, employment and education should be included. More research is needed to confirm these findings using email surveys, rather than paper based tools.

References:

1. Akl EA, Maroun N, Klocke RA, Montori V, Schunemann HJ. Electronic mail was not better than postal mail for surveying residents and faculty. J Clin Epidemiol. 2005 Apr;58(4):425-9.

2. Goldman RD, Mehrotra S, Pinto TR, Mounstephen W. Follow-up after a pediatric emergency department visit: telephone versus e-mail? Pediatrics. 2004 Oct;114(4):988-91

3. Ezenkwele UA, Sites FD, Shofer FS, Pritchett EN, Hollander JE. A randomized study of electronic mail versus telephone follow-up after emergency department visit. J Emerg Med. 2003 Feb;24(2):125-30.

4. Wong BM, Yung BM, Wong A, Chow CM, Abramson BL. Increasing Internet use among cardiovascular patients: new opportunities for heart health promotion. Can J Cardiol. 2005 Mar 15;21(4):349-54.

5. Partridge MR. An assessment of the feasibility of telephone and email consultation in a chest clinic. Patient Educ Couns. 2004 Jul;54(1):11-3.

6. Keil, JM The digital divide: Internet and email use by the elderly, Med Inform Internet Med. 2005 Mar;30(1):19-23

7. Wolf JM, Ritter M, Weiss AP, Akelman E, Access and use of the internet in a hand surgery population, Hand Surg. 2004 Jul;9(1):29-33.

8. Parekh SG, Sodha S, McGuire KJ, Bozentka DJ, Rozental TD, Beredjiklian PK. The digital divide phenomenon in a hand surgery outpatient clinic. Clin Orthop Relat Res. 2004 Apr;(421):54-9.

9. Mead N, Varnam R, Rogers A, Roland M, What predicts patients' interest in the internet as a health resource in primary care England? J Health Serv Res Policy. 2003 Jan;8(1):33-9.

10. Bowen D, Predictors of women's Internet access and Internet health seeking, Health Care Women Int. 2003 Dec;24(10):940-51.

11. Cotton SR, Gupta SS, Characteristics of online and offline health information seekers and factors that discriminate between them. Soc Sci Med. 2004 Nov;59(9):1795-806.

Address for correspondence:
Dr. Brett Taylor
c/o Emergency Department, IWK Health Centre
5850/5980 University Avenue
P.O. Box 9700 , Halifax, Nova Scotia, Canada
B3K 6R8

9.4 Evaluation and Lessons Learned

Ubiquity: Technologies for Better Health in Aging Societies
A. Hasman et al. (Eds.)
IOS Press, 2006

Artificial Neural Network Versus Subjective Scoring in Predicting Mortality in Trauma Patients

Adrian PEARL[a,b], Reuben CASPI[b] and David BAR-OR[a]
[a] *Swedish Medical Center, Trauma Research Dept. Englewood,Colorado. USA*
[b] *University of Derby In Israel., Ramat Efal Campus.Israel*

Abstract. Objective: Current methods of trauma outcome prediction rely on clinical knowledge and experience. This makes the system a subjective score, because of intra-rater variability. This project aims to develop a neural network for predicting survival of trauma patients using standard, measured, physiological variables, and compare its predictive power with that obtained from current trauma scores. Methods: The project uses 7688 patients admitted to the Swedish Medical Center, Colorado, U.S.A. between the years 2000-2003 inclusive. Neural Network software was used for data analysis to determine the best network design on which to base the model to be tested. The model is created using a minimum number of variables to produce an effective outcome predicting score. Initial variables were based on the current variables used in calculating the Revised Trauma Score, replacing the Glasgow Coma Scale (GCS) with a modified motor component of the GCS. Additional variables are added to the model until a suitable model is achieved. Results: The best model used Multi-Layer Perceptrons, with 8 input variables, 5 hidden neurons and 1 output. It was trained on 5881 cases and tested independently on 1807 cases. The model was able to accurately predict 91% patient mortality. Conclusions: An ANN developed using pre-hospital physiological variables without using subjective scores resulted in good mortality prediction when applied to a test set. Its performance was too sensitive and requires refinement.
Keywords: Artificial Neural Networks, Mortality Prediction

1. Introduction

Outcome measurements are a core tool in improvement in the quality and standard of care provided by medical facilities. As part of these models, an effective trauma registry provides a method by which critical data can be managed, and the standard outcome and severity scores are calculated.[1] The definitive outcome is the probability of mortality (death) or its inverse, the probability of survival. These probabilities are calculated from a number of clinical factors and scoring systems[2].

Major measurement scores have been based on at least one factor derived by application of clinical experience. The most common factor is the Glasgow Coma Scale (GCS), a score based upon three elements of motor, verbal and eye movement.[3] Correct application of GCS can be affected by drugs or other medical conditions, so it becomes a question of experience as to the value applied in any given situation. The system is effectively subjective, dependent on interpretation, in which clinicians with widely varying experiences assess patient conditions. This dependency therefore restricts the ability to obtain extremely high correlations between injury and outcome predictions.[4]

The application of artificial neural networks (ANNs) in outcome prediction has become increasingly prevalent in physiological modeling, and has started to appear as a methodology in predicting trauma outcome. ANNs are mathematical models constructed on the basis of organic neural systems. These networks are flexible systems which are increasingly used in predictive modeling due to the ability of the ANN to learn and improve. This occurs by a methodology known as feed-forward/back propagation, in which the artificial neuron adds weights according to positive or negative deviation from a training set of data from which it 'learns'.

The study hoped to remove the bottle-neck caused by using a subjective scoring system in outcome prediction by applying artificial neural network analysis to trauma registry data. The goal was to determine if an alternative scoring system could be found that was simple and accurate in predicting mortality. This was performed as part requirement for a Master of Science degree in computing.

2. Materials & Methods

2.1. Development of the Artificial Neural Network

There are a number of software packages available to perform a neural network analysis. This project used Tiberius (ver. 2.7.2) by Dr Philip Brierley (www. philbrierley.com). This software was selected due to the ease of building the neural network by specification of input and output variables. Tiberius uses multilayer perceptron (MLPs) methodology. Multilayer perceptrons are feed-forward neural networks trained with the standard back-propagation algorithm. The algorithm consists of two steps.

In the *forward pass*, the predicted outputs corresponding to the given inputs are evaluated. In the *backward pass*, partial derivatives are propagated back through the network. The chain rule of differentiation gives very similar computational rules for the backward pass as the ones in the forward pass. The network weights can then be adapted using any gradient-based optimization algorithm. The whole process is iterated until the weights have converged[5].

The initial neural network was designed using five input variables and one output variable. It consists of three layers, one being a hidden layer of neurons. The numbers of neurons within the hidden layer affect the number of degrees of freedom in the optimization process, and therefore the performance of the model. The extra neuron increases the non-linearity of the model. Therefore a higher value can be used to extract a more complicated feature.

2.1.1. Patient Population

Swedish Medical Center (SMC) became recognized, by the American Surgeons Committee on Trauma, as a Level I Trauma Center in June 2002. Previously it was classified as a Level II trauma center. The analysis was conducted on a subset of the SMC Trauma Registry. Study data was taken from the years 2000-2003. Total number of cases during this period entered into the Registry was 7689 cases, approximately

2000 injuries per year. Variables extracted from the Registry included patient demographics, physiological variables, diagnostic and procedural treatment codes, and the RTS & RPS scores for comparison against the ANN derived model. All patients registered during that time period were included in the study with no exclusion criteria.

2.1.2. Creating New Variables for analysis

As Neural Networks are good at classification problems, the best format for data is to be in a binary format. New variables that are thought to affect survival rates were created to allow coding into the ANN model. These new variables are created, and coded based upon the categories of the Revised Trauma Score, the age predictors used in TRISS and the ability to obey simple commands. These variables will then be used to replace the scalar variables initially used in the neural network to test for improved performance. The following variables were created as defined below:

LowSBP – Systolic Blood Pressure less than 40
LowRR - Respiratory rate shallow (less than 10)
OCmd – Ability to obey simple commands: 0 = mGCS 1-5: 1 = mGCS=6
PedAge – Age of patient class for under 16 years: 0 = >16; 1= ≤ 16
ThirdAge – Age of Patient greater than 55 years: 0 = ≤ 54; 1= >54

2.1.3. Statistical Analysis of Performance by Discrimination (ROC curve analysis)

Discrimination is the ability of the model to separate the population into two groups. In this instance we discriminate between those who live or die. Receiver Operator Curves (ROC) are independent of outcome prevalence, and are a useful tool in the performance evaluation for separating two populations. A ROC plot is the graph of all observed (1-specificity, sensitivity) pairs. Each point on this empirical plot can be represented by a 2x2 contingency table. Two different tests on the same patient can then be compared.[6]

3. Results

Initial splitting of the data gave us 5881 cases in the training set (Trauma Registry numbers from 14460 – 20699), and 1807 cases for the test set (Registry numbers 20700-22690). The training set can be assumed to cover mostly 2003, while the test set data will be the years 2000-2002 inclusive.

3.1. Demographic Analysis

Analysis of the training set showed two peaks in age distribution, with the first peak of injuries occurring among those aged 20 years, and a second peak for those aged 80 years. Of the sampled population, 44% were female, 56% male. In this sample 213 deaths were recorded, 91 of whom were aged over 55 years old, 54 of these were accounted for by those aged 76 years or more. There were 13 recorded deaths for those aged 16 years and under. Of the 1807 cases included in the test set, 45% were female, 55% male. They also exhibited the bimodal peaks in age distribution at 20 years and 75 years. There were 61 cases of death recorded in the test set, with 21 being recorded in the over 75 years age range, and 6 in the under 16 years.

The best performing models were tested replacing the SBP, and Respiratory Rate with binary variables for low SBP and low Respiratory Rate. These were coded using the Revised Trauma Score (RTS) scoring table as a basis. The training set was then used with a second model taking into account the new variables of lowSBP, lowRR, as replacements for SBP and Respiratory Rate, and used age markers. A third model of 8 variables was created that combined the scale variables, SBP, RR and age, and in addition used the markers for lowSBP, lowRR and the over 55 years.

Using the Kolmogorov-Smirnov test, an alternative to chi-squared, comparison is made between the cumulative distribution functions, *F(x)* and *G(x)*, of the two populations. In addition Lift, a measure of the effectiveness of a predictive model, calculated as the ratio between the results obtained with and without the predictive model, is included. A summary is shown in table 1 for this model.

Table 1: Summary of ANN model input variables.

Input	Info.Value	No. Cat	Purity	Gini	K-S	K-S Split	10% Lift1	10% Lift2	Lin.Correlation
SBP	1.587	181	0.885	0.309	0.338	1.000	4.176	1.017	0.007
PULSE	0.894	134	0.903	0.320	0.327	60.000	4.426	1.465	0.031
RR	0.690	80	0.884	0.249	0.291	12.000	1.808	1.584	0.000
OCMD	0.201	2	0.861	0.222	0.222	0.000	1.617	1.630	0.007
AGE	0.166	102	0.863	0.126	0.120	61.000	1.978	1.138	0.002
LOWSBP	0.137	2	0.861	0.182	0.182	0.000	1.531	1.517	0.005
LOWRR	0.099	2	0.861	0.153	0.153	0.000	1.484	1.469	0.004
3rdAge	0.040	2	0.861	0.097	0.097	0.000	1.293	1.279	0.001

Using this model, trained with the training data-set, the system was able to correctly predict 83.1% of those with traumatic injuries who died, and 80.9% of those who lived. The best Root Mean Square Error (RMSE) obtained was 17.99 The model was then applied to the test data set. Results showed that 90.8% of those who died from traumatic injuries were predicted by this model. 73.3% who lived were also correctly classified, with a best RMSE of 17.96. Summary data is reproduced in table 2.

Table 2: Summary of ANN model input variables – Test Set.

Input	Info.Value	No. Cat	Purity	Gini	K-S	K-S Split	10% Lift1	10% Lift2	Lin. Correlation
SBP	1.375	157	0.904	0.249	0.382	1.000	4.495	1.787	0.007
PULSE	1.087	111	0.907	0.310	0.349	51.000	4.690	1.276	0.033
RR	0.702	43	0.879	0.330	0.386	1.000	3.429	1.041	0.009
AGE	0.659	100	0.871	0.173	0.181	72.000	1.194	1.044	0.003
LOWRR	0.329	2	0.863	0.282	0.282	0.000	1.929	1.878	0.013
OCMD	0.138	2	0.862	0.181	0.181	0.000	1.582	1.632	0.005
LOWSBP	0.137	2	0.862	0.176	0.176	0.000	1.650	1.593	0.005
3RDAGE	0.106	2	0.862	0.161	0.161	0.000	1.444	1.402	0.004

The coefficients calculated by the ANN were then used to calculate the probability of survival according to the following equation.

$$P_{survival} = 1 - 1/(1 + e^{b})^{7}$$

Where b is calculated from:

b =b1(pulse)+b2(lowRR)+b3(RR)+b4(SBP)+b5(lowsbp)+b6(OCmd)+b7(3rdAge)+b8(Age)

The probabilities were then plotted, along with models derived from logistic regression, RTS & Revised Probability of Survival (RPS) in an ROC curve analysis- see figure 1.

4. Discussion

Demographic analysis of the training set shows that the population being assessed shows no bias in gender or age distribution. However with only 213 recorded deaths it is hard to create a well-trained model to predict accurately survival through traumatic injury. Frequency distributions of physiological variables considered important showed that both Systolic Blood Pressure and Respiratory Rates showed bimodal distributions, while the Pulse Rate showed a normal distribution. This justified the creation of two binary markers, lowSBP and lowRR, to use instead of scaler variables.

Figure 1: ROC Curve for Survival Predictors.

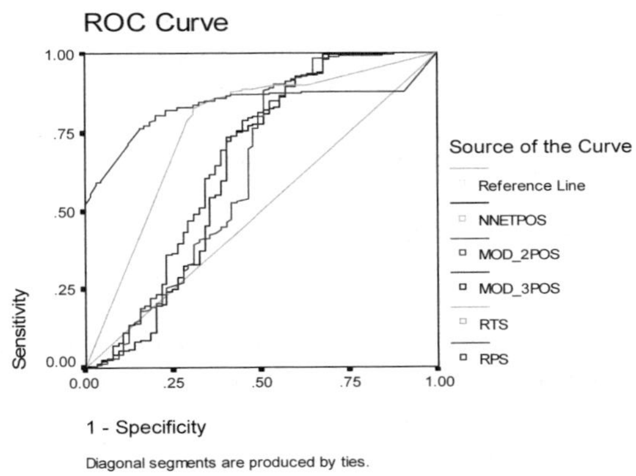

The two new models – mod_2 (using lowSBP & lowRR markers) and mod_3 (using variables and markers for SBP and RR) were compared by logistic regression with the standard measures of RTS and RPS. As the ROC curve shows (figure 9), neither model is as effective as current models are in prediction at the low end of the graph of low sensitivity and specificity. Since the model containing the variables SBP, RR and the markers for lowSBP and low RR showed slightly better performance under logistic regression this model was chosen for the final test, and the NNet software was allowed to train for 150,000 epochs on the training data using the variables of OCmd, Pulse, SBP, RR, Age, 3rdAge, lowSBP & lowRR. The model did show improvement in selection performance, able to correctly predict 81%-83% of the cases. When applied to the test set, the model predicted 91% of the cases that died, 52/65 cases. On those it failed, the majority were elderly patients. Other factors may influence death.

With calculation of the probability of survival, the ANN model was added to the ROC plot – figure 9. It shows to be marginally better then model_2 but slightly worse than the logistic regression for its basic design. It still suffers the same poor performance at low levels of sensitivity and specificity.

5. Conclusion

In this study, we have used ANNs to assist in identifying predictors of survival in a population-based database. While a number of previous ANN-based studies were conducted, these used population-based registries, with large numbers of inputs[8,9]. This study concentrated on a number of factors previously considered to give good correlation in survival prediction, and introduced a new component based on one factor in the Glasgow Coma Scale easily identifiable – the ability to obey simple commands. By using clear, measurable variables, and keeping the number of inputs to single figures, it is possible to produce a system that can be used "in the field" for triage.

Previous studies have used large numbers of input variables, often including ISS or GCS scores. Since it is known that there is a good relationship between ISS and probability of survival, using these variables may well prompt ANN models into finding the same relationship and thus predictive scope. Likewise, by using GCS, systolic blood pressure (SBP) and respiratory rate (RR), the components of the RTS score, we also promote a similar relationship. An additional complication to using many input variables is that the model can not be easily tailored for triage, and it is difficult to assess which variables are essential to the model. While using the weighting values used in the model can give an indication, it would seem more prudent to operate with fewer input values, and if necessary, add more.

Knowing that the GCS components were shown to be critical yet problematic, the total GCS was replaced by the motor component alone, mGCS, which has been shown to be an effective predictor of survival. A new mGCS was created that simplified the component into two groups, 1-5, problems understanding simple commands, and 6 able to obey simple commands. By converting the motor component of the Glasgow Coma Scale to a binary (yes/no) variable, we make it easier for the ANN to use this variable in assessing its usefulness in detecting the trends in our data. However we can lose the resolution of the mGCS component by making it a binary variable, even though we achieve the objective of removing subjective (clinically trained) scoring.

While the model may not have performed optimally, observed trends suggest that using smaller numbers of variables can create Artificial Neural Networks capable of accurately predicting survival in traumatized patients. Although increasing the number of inputs improved the performance, there were still only five physiological factors required to operate the model, and the extra variables used were identifiers of classes within these factors. Using a different population size has the potential to improve predictive ability by this model and enable creation of a true outcome scoring system.

References

[1] Boyd DR, Lowe RJ, Baker RJ, Nyhus LM.Trauma Registry. New Computer method for multi-factorial evaluation of a major health problem. JAMA 1973 Jan 22;223(4):422-8

[2] Gunning & Rowan. Outcome Data and scoring systems. BMJ 1999;319:241-244

[3] Teasdale & Jennett. Assessment of Coma and impaired consciousness. Lancet 1974;ii:81-4

[4] Champion HR, Sacco WJ, Copes WS, Gann DS, Gennarelli TA, Flanagan ME. A revision of the Trauma Score. J Trauma 1989;29(5):623-9

[5] Haykin, S. (1994), Neural Networks: A Comprehensive Foundation, NY: Macmillan

[6] Lett RR, Hanley JA and Smith JS. The Comparison of Injury Severity Instrument Performance Using Likelihood Ratio and ROC Curve Analysis. J Trauma, 38(1) 1995:142-148

[7] Boyd CR, Tolson MA, Copes WS. Evaluating Trauma Care: TRISS method. J.Trauma 1987;27:370-378

[8] DiRusso SM, Sullivan T, Holly C, Cuff SN, Savino J. An artificial neural network as a model for prediction of survival in trauma patients. J.Trauma 2000 Aug;49(2):212-220

[9] Nimgaonkar A, Karnad DR, Sudarshan S, Ohno-Machado L, Kohane I., Prediction of mortality in an Indian Intensive Care unit. Intensive Care Med. 2004 Feb;30(2):248-53

Ubiquity: Technologies for Better Health in Aging Societies
A. Hasman et al. (Eds.)
IOS Press, 2006

Clinical Pathways Development and Computer Support in the EPR: Lessons learned

Thomas BÜRKLE[a], Thomas BAUR[b] and Norbert HÖSS[c]
[a] *Department of Medical Informatics, University Erlangen, Germany*
[b] *Department of Medical Informatics, University Münster, Germany*
[c] *Department of Neurosurgery, St. Barbara Hospital Hamm-Heessen, Germany*

Abstract. This paper refers to a project for development and optimization of a clinical pathway for the Lumbar Nerve Root Compression Syndrome. A special focus is taken on computer support for pathway development and implementation. An innovative combination of "rapid prototyping" of a workflow model and a 2 level approach using round robin methods in a large group and individual semi-structured interviews is presented. The method focuses on process optimization instead of process modeling and concentrates on areas of the workflow which may be optimized. Critical parts of the optimized clinical workflow have been implemented inside a commercial electronic patient record system.

1. Introduction

Germany has converted its inpatient financial reimbursement system to DRG based reimbursement. Although this process is not yet complete (budget cuts for the hospitals are limited to a certain percentage until 2009), major effects such as reduced length of stay and bed count as well as rapid change in hospital ownership have been observed already [1]. A further drop of the total number of hospitals is expected and recently the first university hospital has been completely privatized and sold to a commercial company [2]. Thus German hospitals strive intensively for optimized and cost conscious patient treatment and improved cost effectiveness. Clinical process management and clinical pathways have become central issues [3]. Workflow Management Systems have been discussed before already [4,5]. Integration into the Electronic Patient Record has been discussed [6] but rarely been realized [7]. Clinical pathways have been advocated and established in countries with DRG based reimbursement system before [8,9,10]. Pathway implementation however is a time consuming and difficult task and despite many publications in Europe the amount of patients treated along a clinical pathway is still limited [9].

In this paper we focus on rapid development of an optimized clinical pathway. Our question was

How can we rapidly derive an optimized clinical pathway for patient treatment and which computer support within an electronic patient record (EPR) system is desirable?

The environment for this research was the Neurosurgical Department of Münster University Hospital MUH. This department is concerned with the surgical treatment of injuries, tumors and malformations of skull, brain, spine and nerves. Clinical pathways are advised for those diseases which affect a large part of the respective patient population. The Neurosurgical Department selected the Lumbar Nerve Root Compression Syndrome LNRCS which affects approximately 180 of their patients per year. Pathway optimization has many facets. The Neurosurgical Department was interested in improved quality of patient treatment by means of an improved clinical workflow and treatment standardization.

2. Methods

2.1. Rapid Prototyping of a workflow model

Multiple methods and tools have been advised for the analysis and documentation of workflows in hospitals (see e.g. [3]). It was obvious that there would be some variability in current treatment of LNRCS patients in the Neurosurgical Department. However, our primary research question was workflow optimization and not so much an exact model of the current LNRCS workflow. We would however soon need a first workflow model to discuss improvements with the clinicians. Therefore we decided for "rapid prototyping" of the current workflow model.

Instead of a lengthy analysis of current workflow activities we simply started rapid prototyping with a guideline of the German Association of Neurosurgeons [11]. This guideline comprises an elementary diagnostic and treatment algorithm which is textually enhanced to explain differential diagnosis and treatment recommendations. Another prerequisite for rapid prototyping is some sort of workflow management tool. Here we decided for the Adonis® software from BOC company which allows visualization of hierarchically organized workflows and supports workflow simulation. The LNRCS guideline workflow was implemented using Adonis® and the results were discussed with an experienced senior clinician of the Neurosurgical Department (author N.H.). The guideline workflow was then adapted to the local reality of the department according to his advise.

2.2. Optimization of the path

A clinical pathway must be adapted to local requirements and consented by all members of the treatment team. The clinical pathway should describe an optimized treatment path which ensures maximum effectivity and optimized treatment quality. We used the workflow model derived from 2.1 to discuss potential improvements of patient treatment with all clinicians. Multiple methods to reach a consented opinion have been described e.g. for guideline development [12]. We decided to use a dual layered Delphi like process.

On the first level, discussion about optimization has been promoted in a large joint meeting between physicians, nurses, physiotherapists and other involved persons. There "round robin" methods have been used to collect proposals for optimization from group members. Proposals have then been grouped together according to the part of the workflow which should be optimized, and the group was asked to prioritize those areas of potential improvement.

On the second level, individual semi-structured interviews have been performed. A total of 15 interviews has been conducted with different representatives from the same group. Structuring of interviews was based upon the prioritized optimization proposals from the first level. Open question technique was used. If for example in the first level someone had put up a proposal to clarify discharge criteria in order to assure that patients are not discharged too early or too late after surgical treatment, in the second level the interviewer would ask "Which criteria in your opinion could be used to determine if the patient can be discharged from hospital?" and "Which value for each of these criteria you gave us would be appropriate to guarantee a sufficient health status for discharge?".

2.3. EDP support of an optimized clinical pathway within an EPR

We considered appropriate feedback as a main priority in the optimization process. Feedback was not only given in repeated meetings of the first level group, but also via intranet. MUH provides an intranet site for their employees. The workflow visualization tool Adonis® permits the export of static HTML pages for any workflow. Starting with the first rapid prototype workflow (see 2.1) all workflow models were timely published in the intranet. During pathway development both the continuously elaborated current LNRCS workflow and the developing optimized workflow model were simultaneously on display. All participants were invited to join the discussion and to supply comments by email or during personal interviews. Cooperation in pathway development was actively encouraged in talks and meetings. After project finalization only the consented optimized workflow was on display.

MUH runs a hospital wide network with more than 2000 clinical workstations and operates the commercial Orbis® EPR system from GWI, an Agfa subsidiary. The Orbis® EPR enables the development of own applications [13,14]. This system formed the basis for an implementation of critical parts of the LNRCS clinical pathway. We mentioned that the emphasis of our work has been workflow improvement. Therefore we decided not to implement the complete LNRCS clinical pathway in our EPR system. Optimization activities concentrated rapidly on two parts of the pathway where a potential for process improvement was seen by all participants. Those were the pre-admission phase of the patient (who usually had one or more outpatient appointments before admission for surgery) and the treatment after surgery which should result in discharge of a patient who would be able to care for himself in his home environment with no need for readmission.

Two applications have been implemented in house inside the Orbis® EPR system to support diagnosis and treatment of LNRCS according to the optimized clinical pathway [15].

The first application had the goal to guide the physician and the patient through various outpatient appointments up to the day of admission. An electronic form which acted like a checklist supported patient documentation in this time period. The computer would prompt for documentation of the appropriate findings to make sure that this patient was suffering from LNRCS. When information was missing or further diagnostic tests were required, the application would fork to pre-filled electronic requests for those procedures (e.g. additional radiology tests or preoperative medical second opinion in patients where anesthesia fitness was in question). In any stage of the pre-admission phase the physician in charge would see which steps of the pathway had been completed and which were still missing. There are cases of patients being seen in outpatient clinic and then being referred back to the general practitioner to complete the required diagnostic tests. There the EPR application was able to print a checklist for the patient to make sure that all those tests had been completed and to let him know what he would need to bring with him when he came back as an inpatient.

The second application concerned appropriate and standardized assessment of patients after surgery. It comprised an electronic progress note to be completed for each patient on each postoperative day and an electronic report form. The computer would prompt for the assessment of mobility and remaining back pain symptoms using structured documentation and recognized clinical bedside tests such as Lasegue's test. Afterwards an electronic report could be invoked which showed an overview of patient progress during the past days at a glance so that the clinician would immediately see how fast mobility and pain symptoms had been improved. Thus a standardized criterion for the appropriate discharge timing would be available.

3. Results

We said that our goal was the development of an optimized LNRCS clinical pathway. A somewhat unconventional approach using "rapid prototyping" instead of comprehensive workflow analysis for the development of the first workflow models resulted in a first prototype pathway model within 6 weeks from project start. The two level approach of pathway optimization led in 3 months to an improved and optimized clinical pathway and demonstrated two important areas for workflow improvement. Those were the pre-admission phase and assessment of the optimal timing for patient discharge. The optimized clinical pathway has been modeled in a commercial workflow modeling tool and published in the hospital intranet.

Furthermore, development resulted in two applications within the MUH commercial EPR system to support clinicians in diagnosis and treatment of LNRCS patients according to the optimized clinical pathway. These applications have been used for about 10 patients with LNRCS, but use has ceased due to several problems.

One problem was of technical nature and concerned the availability of new patients demographic data inside the hospital EPR system. The clinical pathway would start with the first outpatient appointment of the patient. The EPR system, however, had been mainly used for inpatient care until then. Although outpatient data was imported via interface from admissions into the EPR system, we observed in a series of cases that this data import came too late for the physician who did the outpatient clinic. Being not able to start documentation immediately using the electronic form it was frustrating to do some paper documentation first and the advantages of pre-filled request forms for further diagnostic tests could not be realized. Within a trial period of several weeks this problem could not be sufficiently resolved because it would have meant complex reorganization of administrative outpatient workflow.

The second problem concerned the clinical pathway itself, although only in some patients. The optimized pathway intended that the patient who was admitted for surgery would have completed all necessary tests and diagnostics in outpatient clinic or at his GP. As a consequence, however, this meant that during the neurosurgery outpatient clinic an anesthesist would be needed in standby to assess those patients in whom there was concern for anesthesia e.g. due to cardiac problems. This problem was discussed with the anesthesia department during pathway development and an intermediary solution was found for a trial period. A final solution for this problem however cannot be easily advocated because improvement of the treatment pathway for LNRCS patients can have negative effects on other departments of the hospital.

4. Discussion

As shown in the previous parts of this paper, there have been positive and negative experiences during clinical pathway optimization and computer support.

We can definitely recommend the method of "rapid prototyping" for clinical pathways, e.g. based on a structured guideline and some input of one or few senior clinicians. The resulting first prototype pathway model may not reflect all variations of the existing workflow but it offers a tremendous source for discussion and feedback to all participants during pathway development. Rapid prototyping requires something to start with. We used a German LNRCS treatment guideline on level S1 (expert round only) which dates back to 1998. In our experience there is no need for highest guideline quality or actuality. As an alternative, structured textbook knowledge could probably be used as well. The general idea is to start the discussion process soon and to enhance the current workflow model during this discussion process. In our experience an electronic visualization tool is urgently recommended for "rapid prototyping". The visualization tool should support electronic pathway publication e.g. in the intranet to enhance feedback. Paper printouts will be very unhandy in comparison.

We had positive experiences with a two level approach for pathway development which has some similarities to the Delphi method. Compared to the classical Delphi method we interviewed only a small number of participants in the second round, thus reducing the effort for the study. On the first level, as many persons as possible concerned with the treatment of the respective patients should be invited to ensure acceptance of the clinical pathway. When working with the large group we would recommend appropriate methods to prevent everyone from just consenting the opinion of a leading senior physician. In our case, a card-based round robin method did the job and enabled a fast approach to essential pathway problems. Later, some sort of feedback to this large group, e.g. by publishing the actual development status in the intranet, is helpful. The second level, consisting of individual interviews, is helpful to work out details in those areas where improvement seems feasible. Those individual interviews promote the impact of individual opinions and help to avoid influence of others. They may also help to clarify the actual workflow which due to "rapid prototyping" may still comprise faults or simplifications.

We have not been able to establish the permanent use of EPR based tools to document patient treatment along the optimized LNRCS clinical pathway. The reason for failure was mainly the non-availability of patient data at the time of the first outpatient visit. But this as well as the observation that any clinical pathway development will rapidly involve more persons and departments than previously anticipated led also to some generic issues regarding clinical pathways. There have been success reports about "localized" clinical pathways in departments. But we do think that institutions who are willing to invest heavily into clinical pathways are probably better off when department structures are softened and the patient is guided throughout treatment by means of a care manager who is responsible for the adherence to the diagnostic and treatment protocols given in the pathway. Obviously such changes in hospital structure are anything but easy to implement inside large institutions such as university hospitals.

5. Conclusion

This paper describes exemplarily the complete process from current workflow, workflow optimization, development of an optimized clinical pathway and computer support for the clinician to treat patients along the clinical pathway. It demonstrates that many facets need to be tied together to achieve success.

We have presented the use of two innovative methods for pathway development and optimization, namely "rapid prototyping" of a clinical workflow and a 2 level approach using round robin and individual semi-structured interviews to focus upon areas of the workflow which may be optimized.

We have failed to establish the clinical pathway in our EPR system due to technical and organizational reasons. But this failure has shed light on the implications which rapidly evolve from clinical pathway implication: Priorities in organizing hospitals have to be put under consideration and the current departmental organization may have to be replaced slowly by a patient or pathway driven organization where pathway managers will be responsible for optimum treatment of the individual patient.

References

[1] Clade H. Krankenhäuser/Fallpauschalen: Umsetzung läuft planmäßig. Deutsches Ärzteblatt 2006:1-2: C13-C15.
[2] Flintrop J. Die Vollversorgung in der Fläche ist das Ziel. Deutsches Ärzteblatt 2006:1-2: C9-C10.
[3] Zapp W (ed): Prozessgestaltung im Krankenhaus. Economica Verlag Heidelberg 2002.
[4] Dazzi L, Fassino C, Saracco R, Quaglini S, Stefanelli M. A patient workflow management system built on guidelines. Proc AMIA Annu Fall Symp. 1997;:146-50.
[5] Dadam P, Reichert M. Towards a new dimension in clinical information processing. Stud Health Technol Inform. 2000;77:295-301.
[6] Schadow G, Russler DC, McDonald CJ. Conceptual integration of guidelines and workflow into the electronic health record. Stud Health Technol Inform. 2000;77:807-11.
[7] Blaser R, Heger O, Beyer M, Schnabel M, Biber C, Bäumlein M, Lenz R. Erfolgsfaktoren zur Umsetzung klinischer Pfade. http://www.egms.de/en/meetings/gmds2005/05gmds470.shtml
[8] Hindle D. Clinical pathways: a serious business. Health Management Bulletin 1997 Vol. 1
[9] Hindle D, Yazbeck AM. Clinical pathways in 17 European Union countries: a purposive survey. Aust Health Rev. 2005 Feb;29(1):94-104.
[10] Roeder N, Hensen P, Hindle D, Loskamp N, Lakomek HJ. Clinical pathways: effective and efficient inpatient treatment. Chirurg. 2003 Dec;74(12):1149-55.
[11] Leitlinien der Dt. Ges. f. Neurochirurgie: Lumbale Spinalkanalstenose. AWMF 2005. http://www.uni-duesseldorf.de/WWW/AWMF/ll/008-007.htm
[12] Arbeitsgemeinschaft der Wissenschaftlichen Medizinischen Fachgesellschaften (AWMF), Ärztliche Zentralstelle für Qualitätssicherung "ÄZQ" (eds). Das Leitlinienmanual. Zeitschrift für ärztliche Fortbildung und Qualitätssicherung Suppl 1, 2001, Urban&Fischer Verlag München 2001.
[13] Kuhn KA, Lenz R, Elstner T, Siegele H, Moll R. Experiences with a generator tool for building clinical application modules. Methods Inf Med 2003;42(1):37-44.
[14] Bürkle T, Ganslandt T, Tübergen D, Menzel J, Kucharzik T, Neumann K, Schlüter S, Müller M, Veltmann U, Prokosch HU. Guideline based structured documentation: the final goal? Stud Health Technol Inform. 2002;90:256-61.
[15] Baur T, Höß N, Wassmann H, Bürkle T. Critical pathway: workflow optimization of the lumbar nerve root compression syndrome. http://www.egms.de/en/meetings/dgnc2004/04dgnc0153.shtml

Ubiquity: Technologies for Better Health in Aging Societies
A. Hasman et al. (Eds.)
IOS Press, 2006

Removal of paper-based health records from Norwegian hospitals: Effects on clinical workflow

Jan Tore LIUM[a,1] and Arild FAXVAAG[b]
[a] *Norwegian university of science and technology, Trondheim, Norway*
[b] *Norwegian Research Centre for Electronic Patient Records, Faculty of medicine, Norwegian university of science and technology, Trondheim, Norway*

Abstract. Several Norwegian hospitals have, plan, or are in the process of removing the paper-based health record from clinical workflow. To assess the impact on usage and satisfaction of electronic health record (EHR) systems, we conducted a survey among physicians, nurses and medical secretaries at selected departments from six Norwegian hospitals. The main feature of the questionnaire is the description of a set of tasks commonly performed at hospitals, and respondents were asked to rate their usage and change of ease compared to previous routines for each tasks. There were 24 tasks for physicians, 19 for nurses and 23 for medical secretaries. In total, 64 physicians, 128 nurses and 57 medical secretaries responded, corresponding to a response rate of 68%, 58% and 84% respectively. Results showed a large degree of use among medical secretaries, while physicians and nurses displayed a more modest degree of use. Possibly suggesting that the EHR systems among clinicians still is considered more of an administrative system. Among the two latter groups, tasks regarding information retrieval were used more extensively than tasks regarding generating and storing information. Also, we observed large differences between hospitals and higher satisfaction with the part of the system handling regular electronic data than scanned document images. Even though the increase in use among clinicians after removing the paper based record were mainly in tasks where respondents had no choice other than use the electronic health record, the attitude towards EHR-systems were mainly positive. Thus, while removing the paper based record has yet to promote new ways of working, we see it as an important step towards the EHR system of tomorrow. Several Norwegian hospitals have shown that it is possible.

Keywords: Medical Records Systems. Computerized/*utilization

1. Introduction

Whatever the cause might be, the health care sector in a few, small European countries have achieved a remarkably high degree of penetration of electronic health record (EHR) systems. In Norway more than 90% of primary care physicians and 90% of the

[1] Corresponding Author: Jan Tore Lium. Norwegian Research Centre for Electronic Patient Records. Medisinsk teknisk forskningssenter, 7489 Trondheim. Norway. Jan.tore.lium@iot.ntnu.no

hospitals have implemented an EHR [1, 2]. Furthermore, an increasing proportion of lab reports, referral and medical discharge letters are exchanged via the National health care network [3, 4]. The propulsion towards increased electronic storage and communication of health data and further integration of EHR systems in clinical workflow is nurtured and closely watched by the Norwegian directorate for health and social affairs in concert with national standardization bodies, the national IT-healthcare industry, health informatics communities in Norwegian universities and the health care sector itself [4].

In a hospital, the health record should be considered both a tool for health personnel and a legal document which use is strictly regulated by law. Implementing an EHR system is a necessary, but not sufficient step towards replacing the legal, paper based health record with an electronic version. To avoid loss of clinically important documents which only exist on paper, these must be reproduced and stored in the EHR. The process of scanning paper health record documents and making these available to clinicians via the EHR is now being enacted at numerous Norwegian hospitals. In a study from the first Norwegian hospital to take this step, physicians reported that removal of the paper-based health record and subsequent total dependence on the EHR system alone had made a few clinical tasks more cumbersome but others more effective. In this study, most physicians were satisfied with the use of the system as a whole, but some physicians reported a negative impact on the performance and the quality of the department's work. Despite some unwanted, negative effects Lærum concluded that the process of removing the paper-based health record was possible without a major negative impact on clinical practice [5].

The effects of introducing an EHR system and removing the paper-based electronic health record might depend on the size of the hospital, nature of work at the department, functionality in the EHR system introduced and preparedness, ability and willingness of the hospital organization to adapt to the changes introduced [6]. Based on an assumption that it is more cumbersome to use an EHR system to introduce organizational changes at larger compared to smaller hospitals we have followed "paperless hospital" projects throughout the Norwegian hospital landscape. We here bring the preliminary results from a survey conducted to explore the use of EHR-systems at selected hospital departments deprived of the paper-based record.

2. Material and methods

2.1. The survey

An adapted version of a questionnaire developed by Lærum et al was used in the study [7]. The main feature of the survey is the description of a set of tasks commonly performed at hospitals (24 clinical tasks for physicians, 19 for nurses, and 23 tasks for medical secretaries). For each task, the respondent is asked to rate the degree of use and performance compared to previous routines. Examples of tasks for physicians are: "Review the patient's problem", "seek out specific information from patient record", "write prescriptions" and "complete sick leave form". Also included in the survey are questions about demographical data, self rated computer experience, availability and problems with computers, detailed user satisfaction, and an assessment of the system as a whole.

The respondents included physicians, nurses and medical secretaries from three medical, one surgical and three dermatology departments from six different hospitals in Norway. At all departments the paper based medical record was removed from clinical workflow, and all three different hospital EHR-systems in Norway was represented. The time since the paper-based medical record had been removed differed among the hospitals. One having eliminated its paper based record in 2001, while others were in the process or just had started working paperless. In total, 64 physicians, 128 nurses and 57 medical secretaries responded. The response rate was 68%, 58% and 84% respectively.

2.2. Analysis

We used SPSS 12.0 for windows for statistical analysis of the survey. The analyses of the questionnaire were performed separately for each question, using the nonparametric analysis Kruskal-Wallis or Mann-Whitney U. Correlations were calculated using Spearman's Rank Order Correlation.

3. Results

3.1. Some physicians reported diminished efficiency compared to the situation before the paper-based health record was removed.

In general, respondents from the 3 dermatology departments reported lower degree of use than the other departments. The dermatology departments belonged to three large university hospitals that also had implemented a different EHR-system than the other hospitals participating in the survey. When inquired about change of ease compared to previous routines the dermatology departments scored considerable lower at least for certain tasks. For instance, while the EHR-system among physicians in all departments where used routinely to both review the patient's problems and seek out specific information from the patient records, more than 50% of the respondents from the dermatology departments reported a negative impact on the performance of their work compared to previous routines (figure 1). In contrast, only a small proportion of the non-dermatology respondents from other hospitals reported a decrease for either task. This difference was, as far as we can tell from the survey, not due to differences in computer skills or access to and problems with computers. Also, the various EHR systems had to a large degree the same functionality supported.

3.2. Large differences between physicians, nurses and medical secretaries with regard to EHR-use

Even though a detailed comparison can not be made due to different tasks and the nature of the work, the overall impression was that medical secretaries used the EHR system far more than both physicians and nurses. When asked about use, the median response by medical secretaries was always or most of the occasions for 19 of 23 tasks. Also, when asked to rate the performance of completing the tasks compared to previous routines, medical secretaries overall responded highest.

As for physicians, the results indicated a difference between tasks regarding *generating* information and tasks regarding *retrieving* information. While the EHR system was used extensively to retrieve information, they were generally utilized to a limited degree when it came to generate and store information. The main exception being entering daily notes, where 85% reported to use the EHR system always or most of the occasions. Still, despite varying degree of use, for most tasks the majority of physicians were positive to the change of ease of performing tasks compared to previous routines. However, as we have seen, exceptions exist.

Nurses were the group that reported the lowest degree of use. Still, we noticed the same tendency as for physicians. Tasks regarding information retrieval were used more than tasks regarding generating and storing information. For example, tasks regarding obtaining various tests results were used routinely. In contrast, only 23% reported to use the EHR-system more than half of the occasions to keep a list of short notes about each patient. Still, despite reporting a modest degree of use, nurses generally were positive to the changes imposed by the EHR-system.

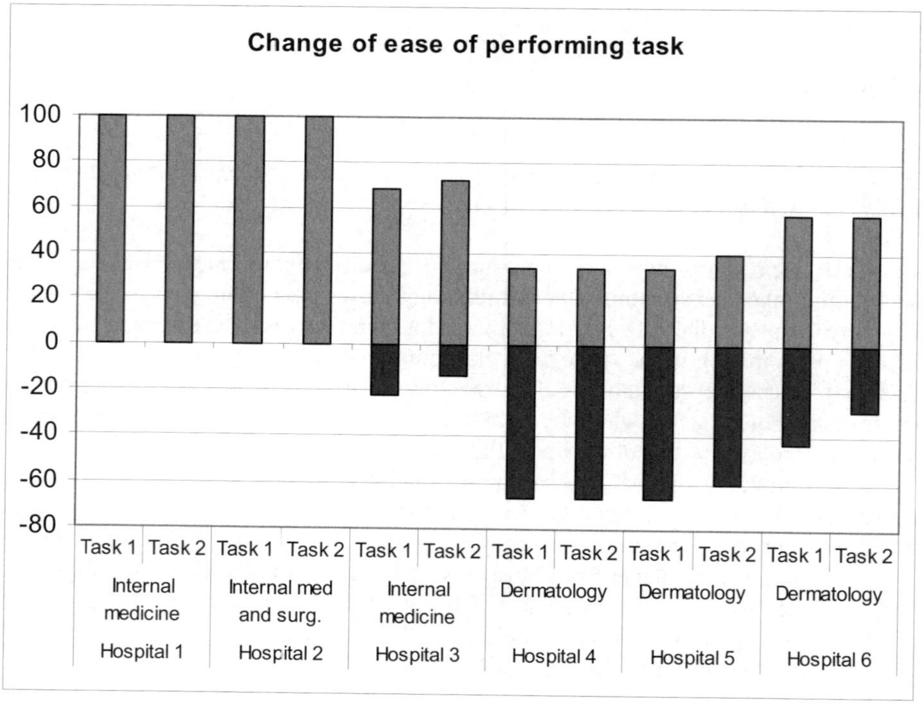

Figure1: Change of ease of performing the tasks compared to previous routines. The positive part of the bar represents percentage of respondents that reported an increase in performance, the negative part percentage that reported a decrease. Task 1: Review the patients' problems. Task 2: Seek out specific information from the patient records

3.3. Generally positive attitude towards EHR-systems

Despite considerable differences in both use and perceived change of ease at the different hospitals, most of the respondents where positive when asked about the overall impact of the system at the various departments. However, when asked about system-specific user satisfaction, respondents where much more satisfied with the part handling regular electronic data than scanned document images (Wilcoxon Signed Rank Test; p<0,001).

4. Discussion

In this report we have presented preliminary results from a survey among Norwegian hospital departments working without the paper based medical record. The results show considerable differences both between professions and hospitals. While the medical secretaries display extensive usage for most tasks, physicians and nurses generally report high usage in tasks regarding retrieving information. Tasks they have to perform using the EHR system since the paper based medical records no longer is available. The reasons for this are not clear. However, a possible reason might be that the functionality offered by the EHR systems in Norway to date is more directly relevant for medical secretaries than physicians and nurses. So, while working with the EHR-system is directly relevant to the job of medical secretaries, it is mere a support for the main task of physicians and nurses. That is, curing and caring. Thus, despite the ambitions, our results suggest that to date the EHR-systems is still more of an administrative system than a system supporting the main jobs of clinicians. Hence, in line with Lærum et.al [8], our results do not indicate any major change of routines compared to the days of the paper based medical record. The lack of an electronic medical chart in the EHR systems, a function much wanted by clinicians, is a possible explanation.

As for the different hospitals, we observed large differences in use and perceived change of ease of performing the different tasks. Still, even though the three departments that displayed lowest degree of both use and satisfaction use a different EHR system than the others, we argue that pointing the blame to the system is too simplified. Instead, we argue in line with Berg [7] that the introduction of EHR-systems, and likewise removing the paper based medical record, should be seen as a mutual transformation process where the technology and the organization influence each other. Thus, we do not rule of the technology as an influencing factor, but argue that focusing solely on the system will lead to inferior explanations. A possible reason for the differences between hospitals might be the time that has elapsed since going paperless and the pre-paperless situation. For one of the hospitals in this study, a similar investigation was carried out in 2002 [3, 6]. Comparison of these data with data obtained in this study clearly shows that both physicians' and nurses' use of EHR have increased. The majority of the remaining hospitals in our survey had just embarked on the process of becoming paperless.

The three hospital departments that displayed the lowest degree of use were dermatology departments from large university hospitals. There may be several reasons for this, one being that they use a different EHR system than the other hospitals. The observed difference might also be due to differences in the nature of the medical work, and of the speed at which the EHR-systems were implemented. For more than five years the university hospitals had been using EHR in parallel with a paper based record, suggesting the department had adapted their routines to take advantages from the strengths of the paper record as well as those of the computerized system. When the paper-based medical record subsequently was withdrawn, this was felt as a loss amongst the clinicians. A contrasting situation was found in the hospital with the greatest degree of EHR use, where the transition went straight from a paper-based record to a paper-deprived EHR-system. In this hospital, users of the EHR-system perceived that they had gained a novel important tool. Thus, the situation before removing the paper-based health record may be more important than the time since implementing an EMR, and may also to some degree explain why one of the hospitals that most recently implemented an EHR and went paperless displayed the highest overall degree of use. Supporting this view, recently gathered qualitative data from one of the hospitals point to word of mouth as an important influencing factor for use of mandatory functionality amongst clinicians. Still, another factor might be that the dermatology departments typically have more patients with complicated case-histories, and thereby have greater need of historical data. Thus, having to work more with scanned images that is regarded as more cumbersome than regular electronic data.

5. Conclusion

Even though the intention of achieving higher efficiency, quality and new ways of delivering health care remains to be fulfilled, our results lend support to the conclusion that removal of the paper-based health record is feasible. The results obtained from the university hospital departments are however worrying and warrants more thorough analyses.

References

[1] National programmes and plans for implementing health information technologies: Europe. [cited 2006 Jan. 30]; Available from: http://www.openclinical.org/nationalHITprogrammesEurope.html

[2] Grimsmo A. The dissemination of electronic patient records (EHR), standardisation and electronic messaging in leading countries in 2005. Wonca International Classification Committee 2005 [cited 2006 Jan 30]; Available from: www.nsep.no/publications/EPR_in_leading_countries.pdf

[3] Norsk Helsenett AS. [cited 2006 Jan. 30]; Available from: http://www.norskhelsenett.no/tiki-view_articles.php

[4] SH dir. [cited 2006 Jan. 30]; Available from: http://www.shdir.no/samspill/indikatorer/4_elektronisk_epikrise_og_henvisning_20188

[5] Laerum H, Karlsen TH, Faxvaag A. Effects of scanning and eliminating paper-based medical records on hospital physicians' clinical work practice. Journal of the American Medical Informatics Association: JAMIA. 2003;10(6):588.

[6] Berg M. Implementing information systems in health care organizations: myths and challenges. International Journal of Medical Informatics 2001;64(2-3):143.

[7] Laerum H, Faxvaag A. Task-oriented evaluation of electronic medical records systems: development and validation of a questionnaire for physicians. BMC medical informatics and decision making [electronic resource]. 2004;4:1.

[8] Laerum H, Karlsen TH, Faxvaag A. Use of and attitudes to a hospital information system by medical secretaries, nurses and physicians deprived of the paper-based medical record: a case report. BMC medical informatics and decision making [electronic resource]. 2004;4:18.

[9] Berg M. Patient care information systems and health care work: a sociotechnical approach. International Journal of Medical Informatics 1999;55(2):87.

Author Index

833, 839
Aalto, H. 947
Abidi, S.S.R 235, 941
Ackermann, M. 241
Åhlfeldt, H. 581, 851
Alecu, I. 833
Allaert, F.-A. 201
Alpay, L. 913
Altmann, U. 139
Ammenwerth, E. 335, 377, 415, 957
Anceaux, F. 713
Andersen, S.K. 341
Andriuskevicius, E. 421
Antony, G. 477
Arredondo, M.T. 404
Artoisenet, C. 181
Aud, M.A. 45
Back, B. 789
Balabanis, I. 307
Bamidis, P. 427
Bar-Or, D. 1019
Barry, C. 713
Bartz, C. 157
Battistotti, A. 935
Baud, R. 863
Bäumlein, M. 645
Baur, T. 1025
Beale, T. 221
Beck, P. 271
Bellazzi, R. 119
Ben Said, M. 277
Benito, M. 893
Berg, M. 1003
Bergmann, J. 265, 435
Bernstein, K. 515
Beuscart, R. 522, 543, 569
Beuscart-Zephir, M.-C. 685
Beyer, M. 645
Biberc, C. 645
Bilge, U. 699
Bindels, P.J.E. 362
Birbas, K. 920
Blaser, R. 645
Blazona, B. 292
Blobel, B.G.M.E. 145, 285, 320
Bodner, T. 957
Böhme, R. 905
Bonnabry, P. 241
Bosman, R.J. 555
Botsivaly, M. 91, 307
Bott, O.J. 265, 435
Böttger, U. 326
Bouaud, J. 483
Bourquard, K. 371
Bourriez, J.L. 569
Bousquet, C. 528, 833, 839
Boutin, J.-P. 666
Boye, N. 326, 341
Braun, L.M.M. 471, 497
Brennecke, T. 265
Breu, M. 377
Breu, R. 415
Bricon-Souf, N. 713
Brigl, B. 672
Bringay, S. 713
Brochhausen, M. 769
Broënhorst, S. 801
Broeren, J. 51
Bruun-Rasmussen, M. 515
Buck, J. 215
Burggraeve, P. 181
Burgun, A. 528, 781
Bürkle, T. 1025
Buyl, R. 651
Buysse, H. 229
Cadonna, B. 271
Cano, D.M. 173
Caputo, B. 983
Caspi, R. 1019
Cassoni-Schoellhammer, N. 251
Ceusters, W. 133, 741
Chambaz, A. 111
Chang, H.-K. 313
Charlet, J. 713
Chatellier, G. 639
Chatzimichael, A. 631

Chaudet, H. 666
Chazard, E. 522
Cheah, Y.-N. 575
Chen, C.-I. 77
Chiu, J.-S. 609
Chong, Y.H. 575
Choy, C.-S. 77
Chu, T.-B. 77
Claudot, F. 639
Clavel, L. 863
Coenen, A. 157
Cohen, O. 201
Collyda, C. 99
Corbella, F. 963
Courtney, K.L. 45
Cox, J. 235
Cuggia, M. 449
Cuoco, E. 935
Curran, J. 941
Dahamna, B. 595
Daniel-Le Bozec, C. 371
Darmoni, S.J. 595, 601, 845
Daumke, P. 857
Davis, S. 235
De Clercq, E. v, 181
de Jonge, E. 555
de Jong-Fintelman, M. 815
de Keizer, N.F. 362, 555
de la Calle, G. 893
de Lusignan, S. 719
De Maeseneer, J. 809
De Moor, G. 229
De Smedt, A. 651
Defebvre, L. 569
Degardin, N. 713
Degoulet, P. 371
Degroote, K. 181
del Hoyo-Barbolla, E. 404
Deleger, L. 747
Deljkovic, S. 879
Demiris, G. 45
Demongeot, J. 926
Deroose, S. 809
Devos, P. 543, 569
Di Giacomoa, P. 389
Dias, G.S. 125
Dickhaus, H. 105, 989
Diomidous, M. 920
Diplaris, S. 99
Dixon, M. 51
Dobravolskas, E. 421
Dostal, C. 187
Dubois, S. 483
Dudeck, J. 139
Dufresne, E. 543
Duhamel, A. 569
Dzananovic, A. 879
Eaglestone, B. 427
Ebrahiminia, V. 483
Edan, G. 449
Ehlers, F. 335
Ehrhardt, J. 977
Eisenmann, U. 989
Elkin, P. 685, 741
Elwert, A. 265
Eminović, N. 362
Engel, K. 145
Fabiani, B. 371
Falcoff, H. 483
Faxvaag, A. 298, 1031
Fensli, R. 195
Fernández, C.A. 173
Ferrazzi, F. 119
Fielding, J.M. 761
Frenzel, T. 977
Frigas, A. 631
Funkat, G. 661
Garcelon, N. 528
Garde, S. 215, 221
Geissbuhler, A. 241, 251
Gigante, G.E. 983
Gill, H. 581
Gillois, P. 639
Goebel, G. 415
Goossen, W.T.F. 815
Goossen-Baremans, A.T.M. 815
Grabar, N. 111
Grabe, N. 105
Gratl, G. 335
Gregoire, M. 522
Hägglund, M. 71, 167
Hahn, U. 503, 857
Hamek, S. 713
Handels, H. 977
Hansen, F.Ø. 195
Hanser, S. 795
Hanzlicek, P. 187
Hao, A.T.-H. 313

Hasman, A v, 362, 497, 509, 873
Hassinen, M. 85
Haugen Bunch, E. 347
Haux, R. v, 265
He, Z. 45
Heard, S. 221
Heger, O. 645
Heinonen, M. 85
Héja, G. 755
Henin, D. 371
Herry, N. 449
Hiissa, M. 789
Hirsch, B. 335
Hirvonen, T. 947
Hoelzer, S. 549
Hoenkamp, E. 969
Hoerbst, A. 415
Hoffmann, I. 435
Hong, W.-H. 157
Höss, N. 1025
Høstgaard, A.M. 705
Hovenga, E. 215
Hsu, C.-Y. 313
Huang, W.-K. 77
Huber, S. 105
Iavindrasana, J. 839
Innes, K. 863
Issler, L. 661
Jais, J.-P. 277
Jaspers, M.W.M. 873
Jaulent, M.-C. 111, 371, 639,
Jay, N. 725
Jian, W.-S. 313
Jonckheer, P. 181
Joubert, M. 845
Juhola, M. 947
Julen, N. 899
Jungmann, D. 905
Kao, C.-C. 313
Kapsimalakou, S. 631
Karagiannis, N. 307
Karsten, H. 789
Katostaras, T. 920
Katz, F.R. 139
Katzer, C. 561
Knaup, P. 207, 215
Knobl, B. 905
Koch, S. 25, 71, 167
Kohler, F. 639, 725
Kolesa, P. 775
Koncar, M. 292
Kosche, P. 435
Kostkova, P. 719
Koufopoulos, K. 631
Koulouris, E. 91
Koutsourakis, K. 91, 307
Kuan, C.-F. 77
Kukafka, R. 404
Kulikowski, C.A. 19
Kumar, A. 537
Kutscha, A. 672
La Torre, E. 983
Lafontaine, M.-F. 181
Landais, P. 277
Latoszek-Berendsen, A. 509
Le Beux, P. 449, 899
Le Bihan, C. 277
Le Mignot, L. 277
Lee, J. 45
Lefebvre, C. 111
Lefranc, H. 543
Lehtikunnas, T. 789
Lenz, R. 645
Lepage, E. 449
Lessard, Y. 899
Li, Y.-C. 77, 609
Liargovas, D. 91
Li-Fang, H. 313
Ligomenides, P. 631
Lilholt, L.H. 341
Limnou, I. 307
Lium, J.T. 1031
Lo, Y.-S. 77
Loomans, E.R.M. 461
Lorant, V. 181
Lovis, C. 241, 251
Low, D. 977
Lu, M.-S. 313
Lu, W. 977
Lund Pedersen, I. 356
Lungeanu, D. 885
Luzi, D. 441
Mach, N. 241
Machan, C. 335, 957
Madge, B. 719
Madsen, I. 341
Maglaveras, N. 99
Magni, P. 119

Mantas, J. 395, 631, 920
Maojo, V. 19, 893
Marinis, M. 307
Markó, K. 503, 857
Marquet, G. 781
Martens, J.D. 617
Martin-Sanchez, F. 125
Marttila-Kontio, M. 85
Marwede, D. 761
Masic, I. 879
Masic, Z. 879
Massari, P. 845
Mateus, C. 863
Mattfeld, D.C. 435
Mavrikakis, I. 395
Mazzoleni, M.C. 963
Mead, C.N. 151
Meister, M. 561
Merkel, M. 747
Mertika, K. 91
Metzner, R. 989
Meyer, M. 549
Meynard, J.-B. 666
Michalowski, L. 265
Michel-Backofen, A. 561
Mihalas, G.I. 885
Mitkas, P.A. 99
Moen, A. 347
Moga, M. 455
Moga, V. 455
Mogård, H.T. 347
Molina, P.C. 173
Møller-Jensen, J. 356
Moreau-Gaudry, A. 926
Moschidis, M. 307
Mosser, J. 781
Mougin, F. 528, 833
Müller, D. 105
Napoli, A. 725
Neoh, S.L. 575
Néri, C. 111
Neuber, K. 105
Neuhäuser, M. 477
Névéol, A. 601, 845
Nielsen, P.H. 341
Nikoloudakis, G. 307
Nilsson, G. 824
Nøhr, C. 341
Nonnemacher, M. 477
Novo, A. 879
Nyssen, M. 651
Nyström, M. 581, 851
Oliveira, J.L. 125
Orfanidis, L. 427
Ortega, M. 404
Ouagne, D. 371
Overberg, R. 969
Pahikkala, T. 789
Paiva, T. 905
Palazzo, F. 441
Pallinger, P. 755
Panis, B. 617
Pappas, C. 99
Payan, Y. 926
Pearl, A. 1019
Pedersen, S.S. 341
Peek, N. 555
Peelen, L. 555
Pelayo, S. 685
Pellegrin, L. 666
Pellicanò, G. 455
Penz, R. 377, 415
Penzel, T. 905
Pereira, S. 601, 845
Petersson, H. 851
Pharow, P. 145, 320
Pieber, T.R. 271
Pigot, H. 587
Pilz, J. 207
Piralic, A. 879
Pirnejad, H. 1003
Pommerencke, T. 105
Poprat, M. 503
Přečková, P. 775
Preda, C. 569
Pretschner, D.P. 435
Prince, A. 926
Puech, P. 522
Punys, V. 421
Quaglini, S. 935
Quantin, C. 201
Queyriaux, B. 666
Rakovac, I. 271
Rantz, M.J. 45
Rassinoux, A.M. 863
Razavi, A.R. 581
Rector, A. 863
Rialle, V. 587

Riandey, B. 201
Ricci, F.L. 389, 441
Richard, J.B. 277
Rieger, J. 549
Rigby, M. 65
Riou, C. 483
Roberts, J. 997
Robinson, J. 719
Rodrigues, J.M. 863
Roger France, F.H. v, 32
Rogers, J. 863
Röhrig, R. 561
Ronco, G. 455
Roos, N. 471
Rose, T. 561
Rossille, D. 449
Ruland, C.M. 624
Rydmark, M. 51
Sacchi, L. 119
Saka, O. 699
Salakoski, T. 789
Salanterä, S. 789
Sarantakis, P. 307
Säring, D. 977
Scandurra, I. 71, 167
Schabetsberger, T. 377, 415
Schadow, G. 151
Schindelwig, K. 415
Schlicht, R. 265
Schnabel, M. 645
Schnell, O. 435
Scholten, D. 461
Schopen, M. 801
Schrans, D. 809
Schuler, T. 221
Schulz, S. 795, 863
Schweiger, R.K. 549
Sedlmayr, M. 561
Segura, J.A.M. 173
Sent, D. 491
Serbanati, L.D. 441
Serna, A. 587
Seroussi, B. 483
Severens, J.L. 617
Shahsavar, N. 581
Sicurello, F. 455
Simonsen, J. 356
Sinciene, V. 421
Sinteff, J.-P. 899
Sips, R.-J. 471
Siregar, P. 899
Skubic, M. 45
Smers, S. 661
Smith, B. 133, 741
Snoeijen, M. 461
Soualmia, L.F. 595, 601
Southgate, L. 719
Spahni, S. 241, 251
Spyropoulos, B. 91, 307
Spyrou, G. 631
Stachowicz, M.S. 581
Staemmler, M. 691
Stausberg, J. 477
Stergiopoulou, A. 920
Stibrant Sunnerhagen, K. 51
Stirparo, G. 441
Stoicu-Tivadar, L. 455
Stoicutivadar, V. 455
Stoop, A.P. 1003
Stut, W. 57
Sundvall, E. 851
Suominen, H. 789
Surján, G. 735, 755
Talmon, J. 509
Taylor, B.W. 1009
ten Dolle, H. 461
Ten Hoopen, A.J. 257
ten Napel, H. 801, 863
Tervo, H. 85
Texier, G. 666
Thalheimer, M. 207
Thirion, B. 595, 601
Tommasi, T. 983
Toromanovic, S. 879
Toth, H. 415
Toubiana, L. 277
Tournebize, O. 666
Toussaint, P. 913
Tractenberg, R.E. 885
Trombert Paviot, B. 863
Truskaller, T. 271
Tuil, W.S. 257
Turner, P. 679
Tyrer, H.W. 45
Tzavaras, A. 91, 307
Uphus, D. 561
Valentinavicius, A. 421
van Beek, L. 815

van Bemmel, J.H. 3
Van Casteren, V. 181
Van den Herik, H.J. 497
van der Aa, A. 617
Van der Gaag, L.C. 491
van der Haring, E. 863
van der Haring, E.J. 257, 801
van der Kooij, J. 815
van der Lei, J. v
van der Weijden, T. 617
Van Maele, G. 229
van Steen, M. 57
Van Wijhe, M. 257
Vandenberghe, H. 181
Varga, P. 755
Vassilaros, S. 631
Vedvik, E. 298
Velonakis, E. 395
Venot, A. 483
Verbeke, M. 809
Verhoef, J. 913
Vicente, J. 125
Viejo, M.R. 173
Vikström, A. 824
Vingtoft, S. 515
Vogl, R. 377, 415
von Ahn, C. 435
Walker, D.M. 151
Wang, Y.-F. 609
Wartena, F. 57
Weber, S. 801
Weiland, D. 477
Wenzer, H.S. 326
Werner, R. 977
Wiesman, F. 497
Winkens, R.A.G. 617
Winter, A. 661, 672
Wirtz, C.R. 989
Wong, M.C. 679
Wozak, F. 377, 415
Wu, L.-B. 313
Wurz, M. 377
Yee, K.C. 679
Yu, F.-C. 609
Zaiss, A. 795
Zanstra, P. 863
Zanstra, P.E. 257, 801
Zvarova, J. 187
Zweigenbaum, P. 747
Zwetsloot-Schonk, B. 913